W9-ALV-169

Diesel Engine Reference Book

Diesel Engine Reference Book

Second Edition

Edited by
Bernard Challen
Rodica Baranescu

Published on behalf of:
Society of Automotive Engineers, Inc.
400 Commonwealth Drive
Warrendale, PA 15096-0001

Butterworth-Heinemann
Linacre House, Jordan Hill, Oxford OX2 8DP
225 Wildwood Avenue, Woburn, MA 01801-2041
A division of Reed Educational and Professional Publishing Ltd

A member of the Reed Elsevier plc group

First published 1984
Second edition 1999

Library of Congress Cataloguing in Publication Data
Diesel engine reference book/edited by Bernard Challen, Rodica Baranescu.
p. cm.
Includes bibliographical references
ISBN 0 7506 2176 1
1. Diesel motor I. Challen. Bernard. II. Baranescu. Rodica.
TJ795.D437
621.43'6-dc21 98-55395
CIP

400 Commonwealth Drive
Warrendale, PA 15096-0001 USA
Phone: (724) 776-4841
Fax: (724) 776-5760
E-mail: publications@sae.org
http://www.sae.org

ISBN 0 7680 0403 9
Library of Congress Number 98-89287
SAE Order No. R-183

Printed and bound in Great Britain by The Bath Press, Bath

Contents

Foreword

It is now 15 years since the *Diesel Engine Reference Book* first appeared and in the meantime the diesel engine industry has undergone more product development than probably at any time in its 100 years history. It is very timely, therefore, that this classic reference book should be updated to reflect the significant advances that have taken place in diesel engine design, improving performance and economy, environmental attributes, reliability and durability.

The demand for the diesel engine continues to increase worldwide as it consolidates its position as the prime mover for tractors, medium and heavy trucks, buses, locomotives, ship propulsion, auxiliary power generation and many industrial applications. Additionally, the diesel engine continues to challenge the gasoline engine as the power unit for the passenger car market particularly in Europe although exhaust emissions legislation, and specifically limits on particulate emission, threatens the survival of the diesel in this sector in the USA and Japan.

Much of the advancement has been driven by exhaust and noise emission legislation which is both increasing in severity and broadening from automotive engines to cover virtually all applications. The impact of such legislation is extending from North America, Europe and Japan to cover most of the world, as other developing nations encounter serious environmental problems.

Fuel economy remains the prime factor favouring the application of the diesel in many markets. In the smaller engine sizes where the gasoline engine has made dramatic improvements in economy through improved combustion, lean-burn operation, and advancements both in fuel injection equipment and electronic control systems, the diesel engine has been able to maintain its advantage. This has been achieved principally by the development of the direct-injection combustion system for smaller and smaller cylinder sizes, where traditionally the indirect-injection engine, and particularly the Ricardo Comet system, dominated the market. Other trends in this sector are the move to 4-valve operation, greater use of light alloy materials for the engine structure, charge cooling, cooled EGR (exhaust gas re-circulation) and variable geometry components such as the turbocharger, inlet manifold and valve-train.

The specific output of larger medium-speed diesel engines continues to rise steadily. Despite increases in mechanical loading, overhaul lives of 20 000 hours, or 1 million miles in locomotive applications, are being achieved with improved reliability and lower manufacturing costs.

These advancements in the diesel engine have only been possible by the development of advanced fuel injection systems, generally with electronic control systems, offering higher injection pressures with much more accurate control of fuel metering and injection timing. The unit injector is already well established, and common rail fuel injection systems are expected to find widespread application especially in automotive engines, providing additional benefits in noise reduction and design flexibility—allowing greater freedom in selecting the number of cylinders and cylinder configuration.

The engineer involved in diesel development is increasingly under pressure to achieve a higher quality product in a shorter time frame. This has resulted in the rapid development of sophisticated analytical tools to enable greater optimization at the design stage.

This new edition has been comprehensively revised and extended to cover these latest advances in diesel engine design and development.

I extend my congratulations to the joint editors Bernard Challen and Rodica Baranescu who together with the impressive team of world-renowned experts have succeeded in producing an outstanding volume.

As a young apprentice engine designer at Ruston & Hornsby I found Pounder's original book of 1955 an invaluable aid; later the first edition of this book, which my old colleague Les Lilly was responsible for, embodied 30 years additional knowledge. I am sure this new edition, bringing us to the state-of-the-art at the end of the century, will continue to be the essential reference book for all engineers involved in diesel engine design, development and application.

JOHN MCCLELLAND
Managing Director
AVL Powertrain Engineering Ltd

Foreword

With the publication of this second edition the name of Leslie Lilly will pass from the world of technical books, into obscurity. And so, while the opportunity remains, please consider for a moment the achievement of a modest man whom I came greatly to respect. In many ways Leslie Lilly was the archetypal researcher, the 'back-room boy', tall, unassuming, but driven by an inner conviction to achieve what was to be achieved. He was well into retirement from Ricardos when I met him in the late 1970s, singing in his church choir and active in the charitable work that meant so much to him. I was commissioning for the publisher Butterworths, following a trail that I hoped would lead to the rebirth of a classic that had been too long out-of-print, C.C. Pounder's *Diesel Engine Principles and Practice.* I knew when the path eventually led to Lilly that there would be no snap decision, and that every consideration would be weighed like every decision taken in the engine laboratories in which he had worked since 1928 when, as he recalled, '. . . parts not infrequently flew out of the doors . . .' while one stood '. . . beside the engine, or with more discretion in front of it.'

When he agreed to accept the challenge, I found myself working with a man of wisdom and a quite unmalicious humour, of determined energy, and above all an integrity that enabled him to assemble and earn the trust of an impressive cast of contributors from the UK, Europe and America. From the diffidence of my first meeting with this long-retired engineer, there emerged a man who was discovering a second lease of life. Through firmness and diplomacy, the massive manuscript of the first edition took shape, the work of more than thirty experts—and in the days before E-mail and personal word-processors. It was handed over on the agreed date, and the expertise of lifetimes became once more available to the diesel fraternity worldwide.

One of the rewards of publishing is its encounters with the authors and editors who emerge from the backrooms of technology to become the unforgettable *personae* of a working life. I am grateful to the publishers, my one-time employers, for this opportunity to pay tribute to Leslie Lilly.

DON GOODSELL

Preface to the second edition

This edition represents the many changes that have appeared in the diesel engine field. Not since Diesel's conception of the engine have there been so many substantial changes in diesel design and applications. Some of these are captured in this edition, so that readers will find most of the coverage from the first edition but major new sections covering the impact of electronics and modern control theory as well as the developments in fuel injection systems. Some of the original chapters have been revised, either by their original authors, or by others.

The editors have attempted to include information relevant to interested engineers, both those who are new to the field and also those who are long-term practitioners. Such is the diversity of knowledge required today, that the various specialisms require detailed study and the editors would like to express their sincere gratitude to the authors and their companies who have contributed to this revised edition.

The editors are both past Chairs of the SAE Diesel Engine Committee and have a wide range of friends, colleagues and acquaintances who are active in the design, development and application of diesel engines. We would like to recognize the special efforts that many of these people have made to assist in the production of this new edition. Not all have been chapter authors, but many have played supporting roles of various types, providing advice and guidance.

The linkage of this book with the original *Pounder's Diesel Engine Principles and Practice* continues, with the chapter on marine engine applications reproduced here from the current edition of Pounders.

One of the editors (Bernard Challen) was privileged to work with Leslie Lilly at Ricardo for many years and was pleased to follow in his footsteps as an editor of this volume. Tribute is paid elsewhere to Leslie but the present editors would like to record their homage to him in providing a firm foundation on which to extend and develop this book. They hope to have lived up to the original objectives to be useful and informative.

The editors hope that this revised edition will appeal—and be useful—to a wide range of readers and so they have attempted to organize the material in logical sections, in addition to providing some of the most current material available. Some of this will perhaps change rapidly, especially that affected by political decisions, such as exhaust emissions and especially particulate legislation. Apart from this area, the book contains much factual and descriptive diesel material that will endure and provide help to many of us as we practise our profession. We trust that you find the book both useful and enjoyable reading.

Bernard Challen
Rodica Baranescu

Preface to the first edition

This book was originally commissioned in order to provide a modern version of *Pounder's Diesel Engine Principles and Practice* and as an addition to the publisher's series of reference books. In the event, it has ended up as a new book, since all the chapters have been rewritten, although some include still valid material from the old book. The engines covered include all types manufactured in Europe, the USA and Japan. The authors of the various chapters have been drawn not only from the UK but from Europe and the USA and include many widely-known authorities. Several new chapters have been added to cover the developments of the twenty-two years since the last edition of Pounder's book.

Thermal and mechanical loads in cylinder heads, pistons and liners have increased dramatically because power from the same cylinder size has about doubled. The development of new cooling methods and new design techniques has been necessary to enable such higher loadings to be sustained. Therefore new chapters on 'Thermal Loading' and on 'Pistons, Rings and Liners' have been included.

The increasing worldwide concern with the control of unpleasant smoke and harmful emissions from the exhaust outlets of vehicle engines and also with the control of the noise level, has resulted in legislation to limit smoke, emissions and noise and in the setting up of standard test methods of measurement in several countries, of which the USA was the first. These developments are explained fully in the chapters on 'Exhaust Smoke Measurement and Regulation', 'Exhaust Emissions' and 'Automotive Engine Noise'. The need to improve performance whether as regards power output or as regards lower fuel consumption or preferably both, though they are not always found together, has led to research into other engine systems to find out their potential. This research is covered in the chapter on 'Compound and Other Engine Systems'.

The chapter on 'Modern Health Monitoring Methods' deals with the subject of the application of continuously recording health monitoring equipment to marine two-cycle low-speed engines in large merchant ships in order to reduce manning levels and maintenance costs. In fact, a more general interest in the application of health monitoring to smaller engines now exists.

The SI system of units is used throughout the book but with some relaxations, as suggested by Ricardo Consulting Engineers, to bring them more into line with general worldwide usage. A list of those units in use in this book which differ from SI units, is given at the end of the book. For those units most commonly employed, this means that power is expressed in kilowatts (kW), though horsepower (bhp) may also be quoted; rotational speed is in revolutions per second (rev/s), though revolutions per minute (rev/min) may be given also; fuel consumption is in grammes per kilowatt hour (g/kW.h); and pressure is in bars (bar).

Since it is not possible to describe in the chapters concerned more than a few of the engine makes available, a list of manufacturers and types of engines, arranged according to nationality, is given at the end of the book. A list of the various main institutions, associations and specialist firms particularly concerned with diesel engines is also included.

It is my hope that this new reference book will be useful to students and lecturers in universities and technical colleges, to the personnel of manufacturing firms and research and development organizations, and to all diesel engine users and operators.

I would like to acknowledge and thank the directors of Ricardo Consulting Engineers and my many friends in this firm for their generous support and help without which this book would not have been possible. I worked there for forty-eight years, and count myself fortunate to have arrived at a time when the diesel engine was only just starting to be applied to vehicles, and at a time when Sir Harry Ricardo was experimenting with many single cylinder engines both single sleeve valve and poppet valve. The fact that parts not infrequently flew out of the doors lent spice to the testing of these engines, especially since one stood beside the engine or with more discretion in front of it, but not sitting in a sound-proofed control room as is today's practice. A great variety of forms of combustion chamber were being patented at this time, among them the Ricardo 'Comet' chamber which in an advanced form is now used in so many passenger car engines. I think we were ahead of most other people for we reached 21 bar b.m.e.p. before the commencement of the second world war on two supercharged single cylinder engines, one a single sleeve valve and the other a poppet valve. I am grateful that I have spent my working life with such a pioneering firm.

I would like to make a special mention here of the late John Hempson who died unexpectedly after a short illness only three years after retiring. It was he who suggested I might like to take on the editing of this book. I am grateful for his help and for his chapter on 'Modern Instrumentation'. I would also like to thank all the authors for the very worthwhile chapters they have contributed and especially those who have undertaken the writing of several chapters. Many companies and establishments have supplied information and in some cases enabling chapters towards this book and I am very grateful for their contributions. I am grateful also for receiving help and encouragement from the publishers and from Don Goodsell in particular. Finally I must thank Miss M. C. Lines and Mrs. M. E. Holmes for the large volume of secretarial work they have cheerfully carried out.

L R C LILLY

Contributors

Chapter 1 The theory of compression ignition engines

Frank J. Wallace DSc PhD FIMechE FSAE

Retired Professor of Engineering from Queen's University Belfast and the University of Bath. His research interests have been in the diesel engine field throughout his career and he has published widely. He has updated the chapter for this edition.

Chapter 2 The theory of turbocharging

Neil Watson PhD CEng MIMechE

Deceased-Formally of Imperial College, London, he was responsible for research in IC engines. He published numerous papers on turbocharging and was the author of a reference book on this subject.

Chapter 3 Compound and other engine systems

Professor Frank J. Wallace (see Chapter 1)

Chapter 4 Diesel combustion and fuels

Dr Jon Van Gerpen PhD PE
An Associate Professor of Mechanical Engineering at Iowa State University, he has been active in diesel combustion and emissions research, simulation and modelling engine processes.

Dr Rolf Reitz PhD FSAE

He is currently Wisconsin Distinguished Professor in the Mechanical Engineering department at the University of Wisconsin-Madison. Previously he was a research scientist at the Courant Institute of Mathematical Sciences in New York, Princeton University and a staff research engineer at the General Motors Research Laboratories.

Chapter 5 Thermal loading

Dr Cecil C.J. French DSc(Eng) FEng FIMechE FIMarE

A past President of IMechE and former director of Ricardo Consulting Engineers, now working as an independent consultant. He has presented and published numerous papers on a wide range of engine-related subjects. He has updated the chapter for this edition.

Chapter 6 Thermodynamic mathematical modelling

Anthony Joyce BSc DCAe CEng CPhys MRAeS MInstP

Anthony Joyce is the MD of Systems Studies Ltd. He has been responsible for developing models of automotive systems, including drivelines and cooling systems.

Chapter 7 Computational fluid dynamics

Dr Rolf D Reitz (see Chapter 4)

Chapter 8 Modern control in diesel engine management

Richard K. Stobart MA CEng MIMechE MSAE

A graduate of Cambridge University, Richard Stobart has spent time with British Rail and Ricardo. He now works for Cambridge Consultants Ltd/Arthur Little where he is the Automotive Business Manager, responsible for Electronic Control Development and Design. He has presented and published numerous papers on Control-related subjects in Europe and the United States.

Chapter 9 Inlet and exhaust systems

Charles A. Beard CEng FIMechE MRAeS

Charles A. Beard was a senior consultant with Ricardo Consulting Engineers and worked in design and development. Bernard Challen has updated the chapter.

Chapter 10 Design layout options

Original chapters by Charles A. Beard. Bernard Challen has updated this chapter.

Chapter 11 Fuel injection systems

Professor Michael Russell MSc CEng FIMechE

Chief Research Engineer, Lucas Diesel Systems. Formerly Professor of Automobile Engineering at the Institute of Sound and Vibration Research, Southampton, he has published widely on diesel injection and combustion characteristics, especially in relation to noise.

Caterpillar Inc.

Caterpillar Inc. is the world's largest manufacturer of construction and mining equipment, diesel and natural gas engines and industrial gas turbines.

Max R. Lanz

Max Lanz is the Marketing Manager at Robert Bosch Corporation. He was educated at the Eidgenoessische Technische Hochschule in Zürich, Switzerland. The material is consolidated from various authors within Bosch.

Chapter 12 Lubrication and lubricating oils

A.R. Lansdown MSc PhD Ccem CEng FRSC FIMechE FInstPet

Dr Lansdown has worked in the Canadian petroleum industry

and in the Ministry of Aviation and the Ministry of Technology. He was Manager and subsequently Director of the Swansea Tribology Centre for 20 years. He is the author of four books and over thirty other publications on lubrication.

Chapter 13 Bearings and bearing metals

C. Evans PhD and **J.F. Warriner** CEng MIMechE MIMarE
C. Evans has worked in all aspects of bearing development and manufacture. Co-author J.F. Warriner has worked on large engine development and is the author of several papers on engine bearings. Reviewed by Mike Neale of Neale Consulting Engineers Ltd.

Chapter 14 Pistons, rings and liners

Robert Munro PhD ARCST CEng MIMechE
Formerly Director of Engineering at Wellworthy Ltd, now a consultant. He has presented many papers in the field of diesel pistons, rings and liners. He has updated the chapter for this editon.

Chapter 15 Auxiliaries

Governors and Governor Gear—**M. Davies,** formerly with Woodward Governor.

Starting Gear and Starting Aids—**D. Hodge** CEng FIMechE MIEAust, formerly with GEC.

Heat Exchangers—**M.K. Forbes** MA CEng FIMechE, formerly with Serk Heat transfer.

Chapter 16 Aircooled engines

Dr I.G. Killmann Dr techn Dipl-Ing MSAE MVDI and **P. Tholen** Dipl-Ing

I.G. Killmann joined Klöckner-Humboldt-Deutz (KHD) in 1970 where he became Director of Research. He was made Vice President of AVLK Prof List Ges. mbH in Graz in 1980 where he was responsible for engine research.

Paul Tholen started his professional career in 1950 at Klöckner-Humboldt-Deutz (KHD). His work was mainly connected with air cooled diesel engines. Before retiring in 1982 he was mainly engaged with advanced engine systems. This chapter was updated by Rodica Baranescu.

Chapter 17 Crankcase explosions

S.N. Clayton CEng FIMechE FIMarE

S.N. Clayton was formerly with Lloyd's Register. This chapter was edited by Rodica Baranescu.

Chapter 18 Exhaust smoke, measurement and regulation

Contributed by **Ricardo Consulting Engineers.** Ricardo is an independent consultancy, based at several locations in England and the US. Established in 1919, the Group's core business is reciprocating engines and their related technologies.

Chapter 19 Exhaust emissions

Contributed by **Ricardo Consulting Engineers** (see Chapter 18).

Chapter 20 Engine noise

Raymond Farnell and **Duncan Riding**
Raymond Farnell is Senior Noise Engineer at Perkins Technology. He has been responsible for the noise engine reduction of many engines, combining practical testing and development techniques.

Duncan Riding is Section Manager at Perkins Technology. He has been involved with the development and refinement of many engines, working mainly from the finite element and analytical area.

Chapter 21 Larger engine noise and vibration control

Stanley Walker

Stanley Walker is currently Principal Engineer, Truck Products, at Barry Controls. He is responsible for the design and development of diesel engine mounts, cab mounts, and isolation components for Class 5 to Class 8 trucks worldwide.

Chapter 22 Passenger car engines

H.S.-H. Schulte Dr-Ing PhD Dipl-Ing (MSME)

Dr Schulte has worked at the Aachen Technical University, FEV Motorentechnik and at Mercedes-Benz. He is now the Executive Vice President and Chief Technical Officer of FEV Engine Technology in Auburn Hills, Michigan.

Chapter 23 Trucks and buses

David Merrion and **K.E. Weber**

David Merrion has spent his entire career at Detroit Diesel Corporation. He is past president of the Engine Manufacturers' Association, a fellow of the SAE, a member of ASME, a Buckendale Lecturer and a member of the US Environmental Protection Agency Technical Advisory Committee.

K.E. Weber has over 20 years experience in advanced heavy-duty diesel engine design and development at General Motors and Detroit Diesel Corporation. She is currently the manager of Applications Documentation at DDC.

Chapter 24 Locomotives

Humphrey Niven BSc (Eng) CEng FIMechE

A specialist in research and development in the field of large engines, he formerly worked with Paxman Ltd. He is currently with Ricardo Consulting Engineers where he is responsible for all aspects of large engine work.

Chapter 25 Dual fuel engines

Roger Richards

Roger Richards is Principal Engineer with Ricardo Consulting Engineers. He has worked on various engine areas, from Stirling engines through alternative fuels and artificial intelligence applied to condition monitoring. He is responsible for the development of novel techniques for improving efficiencies and emissions of large engines.

Chapter 26 Marine engine applications

Doug Woodyard

After experience as a seagoing engineer with the British India Steam Navigation Company, Doug Woodyard held editorial positions with the Institution of Mechanical Engineers and the Institute of Marine Engineers. He subsequently edited *The Motor Ship* journal for eight years before becoming a freelance editor specializing in shipping and marine engineering.

Chapter 27 Condition monitoring

Michael Plint PhD CEng FIMechE and **Tony Martyr**

Michael Plint founded the company Plint & Partners and for the following 30 years was deeply involved with engine testing and the developing of test equipment. He held positions of visiting

Professor at Austrian and American universities. Michael Plint died in November 1998.

Tony Martyr has worked extensively with marine and automotive engines. He is currently Operations Manager with Ricardo Test Automation. He is co-author with Michael Plint on a standard work on engine testing.

Editors

Bernard Challen MSc CEng FIMechE FIEE MIOA FSAE

Former Technical Director, Ricardo Consulting Engineers, now an independent consultant to the automotive industry worldwide. He has presented and published numerous papers covering various aspects of engineering. He writes for several engineering publications, organizes and participates at engineering conferences and workshops in Europe and the United States.

Dr Rodica Baranescu PhD MSc FSAE MASME

Dr Baranescu was previously Professor at the Technical University 'Polytehnica' Bucharest Romania. She moved to the USA in 1980 and joined Navistar International Transportation Corp. where she is chief engineer in the Engine Division. She is Chair of the Alternative Fuels Committee of the Engine Manufacturers' Association.

Part 1

Theory

1 The theory of compression ignition engines

Contents

1.1 Introduction

1.1.1 Historical

Although the history of the diesel engine extends back into the closing years of the 19th century when Dr Rudolf Diesel began his pioneering work on air blast injected stationary engines, and in spite of the dominant position it now holds in many applications, e.g. marine propulsion and land transport, both road and rail, it is today the subject of intensive development and capable of improvements. These will guarantee the diesel engine an assured place as the most efficient liquid fuel burning prime mover yet derived.

Before 1914, building on the work of Dr Rudolf Diesel in Germany and Hubert Akroyd Stuart in the UK, the diesel engine was used primarily in stationary and ship propulsion applications in the form of relatively low speed four-stroke normally aspirated engines.

The 1914–18 war gave considerable impetus to the development of the high speed diesel engine with its much higher specific output, with a view to extending its application to vehicles. Although the first generation of road transport engines were undoubtedly of the spark ignition variety, the somewhat later development of diesel engines operating on the self or compression ignition principle followed soon after so that by the mid 1930s the high speed normally aspirated diesel engine was firmly established as the most efficient prime mover for trucks and buses. At the same time with the increasing use of turbocharging it began to displace the highly inefficient steam engine in railway locomotives while the impending 1939–45 war gave a major impetus to the development of the highly supercharged diesel engine as a new aero engine, particularly in Germany.

Since the 1939–45 war every major industrial country has developed its own range of diesel engines. Its greatest market penetration has undoubtedly occurred in the field of heavy road transport where, at any rate in Europe, it is now dominant. It is particularly in this field where development, in the direction of turbocharging in its various forms, has been rapid during the last twenty years, and where much of the current research and development effort is concentrated. However, a continuous process of uprating and refinement has been applied in all its fields of application, from the very largest low speed marine two-stroke engines, through medium speed stationary engines to small single cylinder engines for operation in remote areas with minimum attendance. There is little doubt that it will continue to occupy a leading position in the spectrum of reciprocating prime movers, so long as fossil fuels continue to be available and, provided it can be made less sensitive to fuel quality, well into the era of synthetic or coal derived fuels.

1.1.2 Classifications

The major distinguishing characteristic of the diesel engine is, of course, the *compression-ignition* principle, i.e. the adoption of a special method of fuel preparation. Instead of relying on the passage of a spark at a predetermined point towards the end of the compression process to ignite a pre-mixed and wholly *gaseous* fuel–air mixture in approximately *stoichiometric* proportions as in the appropriately named category of *spark-ignition* (SI) engines, the *compression ignition* (CI) engine operates with a *heterogeneous* charge of previously compressed air and a finely divided spray of *liquid* fuel. The latter is injected into the engine cylinder towards the end of compression when, after a suitably intensive mixing process with the air already in the cylinder, the self ignition properties of the fuel cause combustion to be initiated from small nuclei. These spread rapidly so that complete combustion of all injected fuel, usually with air-fuel ratios well in excess of stoichiometric, is ensured. The mixing process is crucial to the operation of the Diesel engine and as such has received a great deal of attention which is reflected in a wide variety of *combustion systems* which may conveniently be grouped in two broad categories, viz.

(a) *Direct Injection (DI) Systems as used in DI engines,* in which the fuel is injected directly into a combustion chamber formed in the clylinder itself, i.e. between a suitably shaped non-stationary piston crown and a fixed cylinder head in which is mounted the fuel injector with its single or multiple spray orifices or nozzles. (See *Figures 1.1* and *1.2*.)

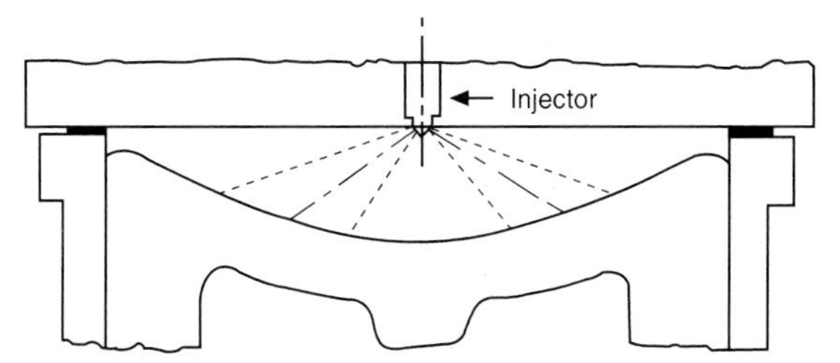

Figure 1.1 Quiescent combustion system. Application–Four-stroke and two-stroke engines mostly above 150 mm bore (*Benson and Whitehouse*)

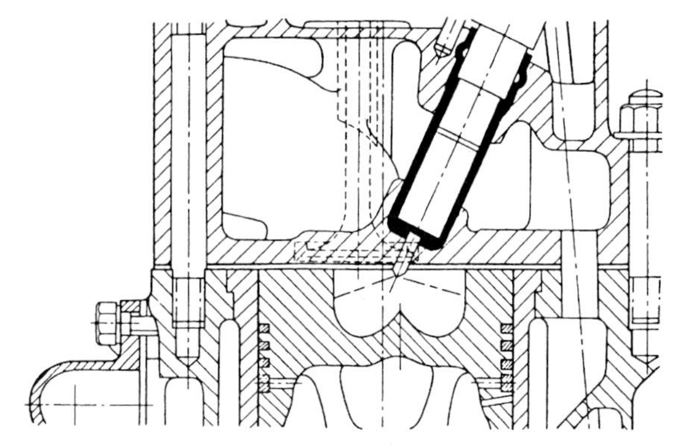

Figure 1.2 High swirl system. Application to virtually all truck and bus sized engines, but increasingly also to the high speed passenger car engine

(b) *Indirect Injection (IDI) Systems as used in IDI engines* in which fuel is injected into a prechamber which communicates with the cylinder through a narrow passage. The rapid transfer of air from the main cylinder into the prechamber towards top dead centre (TDC) of the firing stroke promotes a very high degree of air motion in the prechamber which is particularly conducive to rapid fuel–air mixing. (See *Figure 1.3*.)

Combustion systems are described in more detail in Chapter 4 and generally in Chapters 22 to 29 describing engine types. A further major subdivision of diesel engines is into *two-stroke* and *four-stroke* engines, according to the manner in which the gas exchange process is performed.

1.2 Two-stroke and four-stroke engines

An even more fundamental classification of diesel engines than that according to combustion system is into two-stroke or four-stroke engines, although this latter classification applies equally to spark ignition engines and characterizes the gas exchange process common to all air breathing reciprocating engines. The function of the gas exchage process, in both cases, is to effect

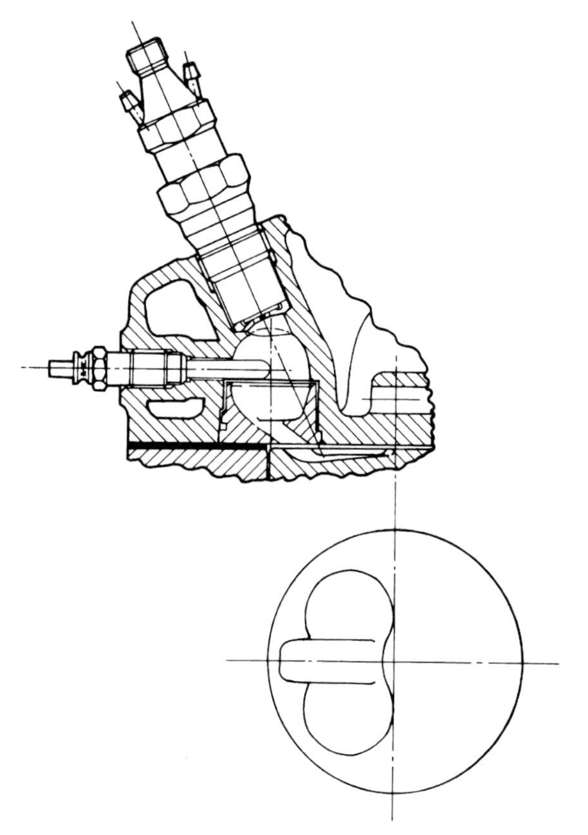

Figure 1.3 Prechamber system–compression swirl. Application–traditionally to high speed passenger car engines but now increasingly replaced by direct injection engine

expulsion of the products of combustion from the engine cylinder and their replacement by a fresh air charge in readiness for the next working cycle.

1.2.1 Two-stroke engines *(Figures 1.4a, b, c)*

In two-stroke engines combustion occurs in the region of top dead centre (TDC) of every *revolution*. Consequently gas exchange also has to be effected once per revolution in the region of bottom dead centre (BCD) and with minimum loss of expansion work of the cylinder gases following combustion.

This implies that escape of gas from the cylinder to exhaust and charging with fresh air from the inlet manifold must occur under the most favourable possible flow conditions over the shortest possible period. In practice the gas exhange or SCAVENGING process in two-stroke engines occupies between 100° and 150° of crank angle (CA) disposed approximately symmetrically about BDC.

Two-stroke engines may be subdivided according to the particular scavenging system used into the following sub-groups.

1.2.1.1 Loop scavenged engines (Figure 1.4a)

This is the simplest type of two-stroke engine in which both inlet and exhaust are controlled by ports in conjunction with a single piston. Inevitably this arrangement results in symmetrical timing which from the standpoint of scavenging is not ideal. In the first instance the 'loop' air motion in the cylinder is apt to produce a high degree of mixing of the incoming air with the products of combustion, instead of physical displacement through

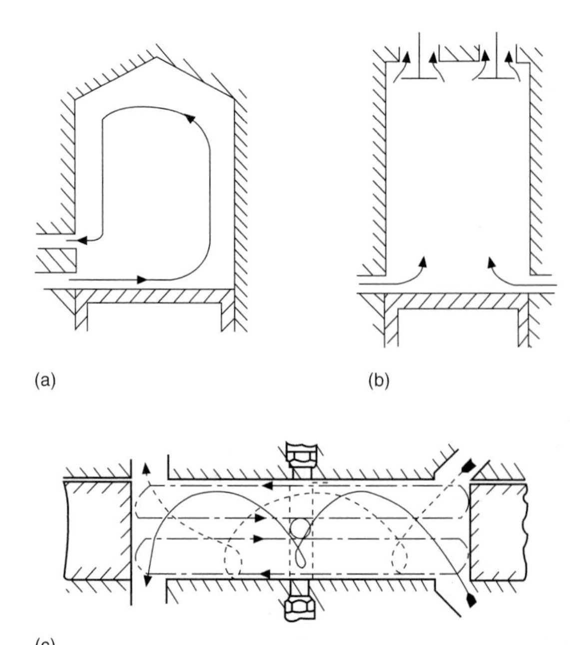

Figure 1.4 Two-stroke engines: (a) Loop scavenged engine; (b) Exhaust valve-in-head engine; (c) Opposed piston engine (*Benson and Whitehouse*)

the exhaust ports. As a result the degree of charge purity (i.e. the proportion of trapped air) at the end of the scavenging process tends to be low.

A second adverse feature resulting from symmetrical timing is loss of trapped charge between inlet and exhaust port closure and susceptibility to further pollution of the trapped charge with exhaust gas returned to the cylinder by exhaust manifold pressure wave effects. The great advantage of the system is its outstanding simplicity.

1.2.1.2 Uniflow scavenge single piston engines (Figure 1.4b)

In engines of this type admission of air to the cylinder is usually effected by piston controlled ports while the products of combustion are exhausted through a camshaft operated exhaust valve. Such systems are preferable from the standpoint of scavenging in that the 'uniflow' motion of the air from the inlet ports upwards through the cylinder tends to lead to physical displacement of, rather than mixing with, the products of combustion thus giving improved charge purity at the end of the scavenging process. At the same time it is now possible to adopt asymmetrical timing of the exhaust and inlet processes relative to bottom dead centre (BDC) so that, with exhaust closure preceding inlet closure the danger of escape of fresh charge into the exhaust manifold present in the loop scavenge system is completely eliminated. This system has been adopted in a number of stationary and marine two-stroke engines.

1.2.1.3 Uniflow scavenge opposed piston engines (Figure 1.4c)

In engines of this type admission of air is effected by 'air piston' controlled inlet ports, and rejection of products of combustion by 'exhaust piston' controlled exhaust ports. The motion of the two sets of pistons is controlled by either two crankshafts connected through gearing, or by a signle crankshaft with the 'top' bank of pistons transmitting their motion to the single crankshaft through a crosshead-siderod mechanism. By suitable offsetting of the cranks controlling the air and exhaust pistons asymmetrical timing can be achieved.

It is evident that this system displays the same favourable characteristics as the exhaust valve in head system, but at the expense of even greater mechanical complications. Its outstanding advantage is the high specific output per cylinder associated with two pistons. However, the system is now retained only in large low speed marine, and smaller medium speed stationary and marine engines. In high speed form it is still employed for naval purposes such as in some fast patrol vessels and mine searchers, although its use in road vehicles and locomotives is discontinued.

1.2.2 Four-stroke engines *(Figure 1.5)*

The vast majority of current diesel engines operate on the four-stroke principle in which combustion occurs only every other revolution, again in the region of top dead centre (TDC), and with the intermediate revolution and its associated piston strokes given over to the gas exchange process. In practice the exhaust valve(s) open well before bottom dead centre (BDC) following the expansion stroke and only close well after the following top dead centre (TDC) position is reached. The inlet valve(s) open before this latter TDC, giving a period of overlap between inlet valve opening (IVO) and exhaust valve closing (EVC) during which the comparatively small clearance volume is scavenged of most of the remaining products of combustion. Following completion of the inlet stroke, the inlet valve(s) close well after the following bottom dead centre (BDC), after which the 'closed' portion of the cycle, i.e. the sequence compression, combustion, expansion, leads to the next cycle, commencing again with exhaust valve opening (EVO).

The main advantages of the four-stroke cycle over its two-stroke counterpart are:

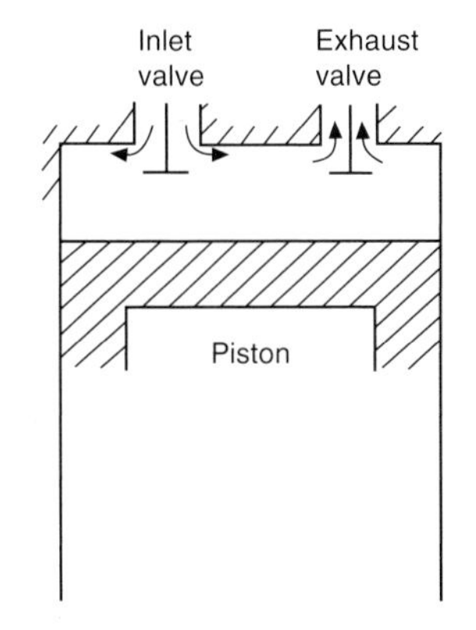

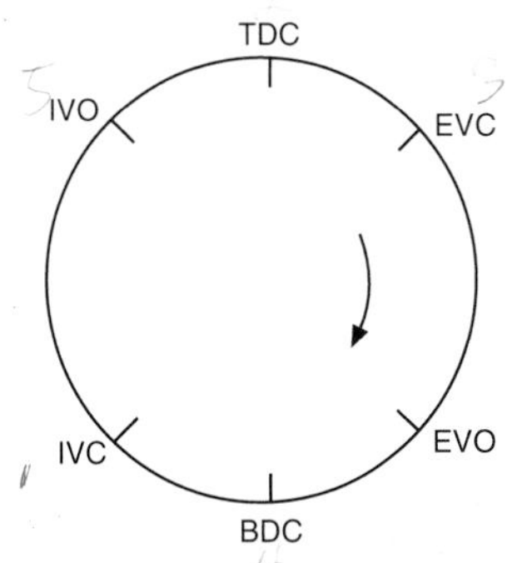

Figure 1.5 Four-stroke engine (turbocharged)

(a) the longer period available for the gas exhange process and the separation of the exhaust and inlet periods—apart from the comparatively short overlap—resulting in a purer trapped charge.
(b) the lower thermal loading associated with engines in which pistons, cylinder heads and liners are exposed to the most severe pressures and temperatures associated with combustion only every other revolution.
(c) Easier lubrication conditions for pistons, rings and liners due to the absence of ports, and the idle stroke renewing liner lubrication and giving inertia lift off to rings and small and large end bearings.

These factors make it possible for the four-stroke engine to achieve output levels of the order of 75% of equivalent two-stroke engines. In recent years attention has focused particularly on three-cylinder high speed passenger car two-stroke engines as a possible replacement for conventional four-cylinder, four-stroke engines with considerable potential savings in space and weight.

1.2.3 Evaluation of power output of two-stroke and four-stroke engines *(Figures 1.6a and b)*

In order to determine the power developed within the engine cylinder as a result of gas forces acting on the piston as opposed to shaft power from the output shaft, it is necessary to have a record of the variation of gas pressure (p) with stroke or cylinder volume (V) referred to as an Indicator Diagram (or *p-V Diagram*). This used to be obtained by mechanical means, but such crude instrumentation has now been completely replaced by electronic instrúments known as pressure transducers. It is also generally more convenient to combine the pressure measurement with a crank angle (CA) measurement, using a position transducer in conjunction with a suitable crank angle marker disc, and subsequently convert crank angle to stroke values by a simple geometric transformation.

The sequence of events for the two cycles may be summarized as follows:

(a) *Two-stroke cycle* (asymmetrical timing)

1–2	compression	Closed Period	360°CA
2–3	heat release associated with combustion	Closed Period	
3–4	expansion		
4–5	blowdown	Open Period	
5–6	scavenging	Open Period	
6–1	supercharge	Open Period	

(b) *Four-stroke cycle*

1–2	compression	Closed Period	720°CA
2–3	heat release associated with combustion	Closed Period	
3–4	expansion		
4–5	blowdown	Open Period	
5–6	exhaust	Open Period	
6–7	overlap	Open Period	
7–8	induction	Open Period	
8–1	recompression	Open Period	

In both cases the cycle divides itself into the *closed period* during which power is being produced, and the *open* or *gas exchange* period which may make a small positive contribution to power production or, in the case of the four-stroke engine, under conditions of adverse pressure differences between inlet and exhaust manifold, a negative contribution. In the case of the four-stroke engine the area enclosed by the p-V diagram for the gas exchange process, i.e. 5–6–7–8, is known as the *pumping*

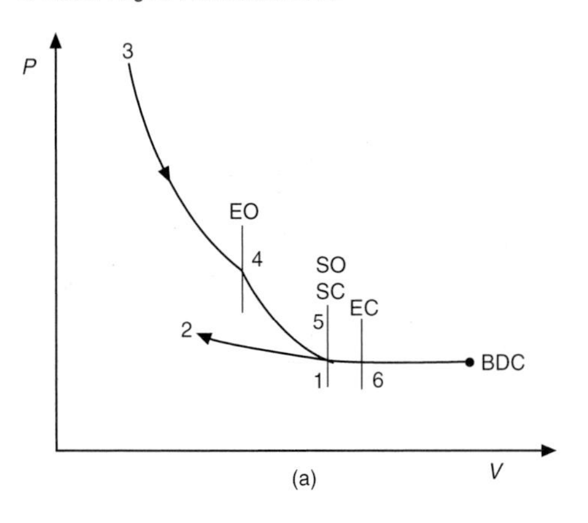

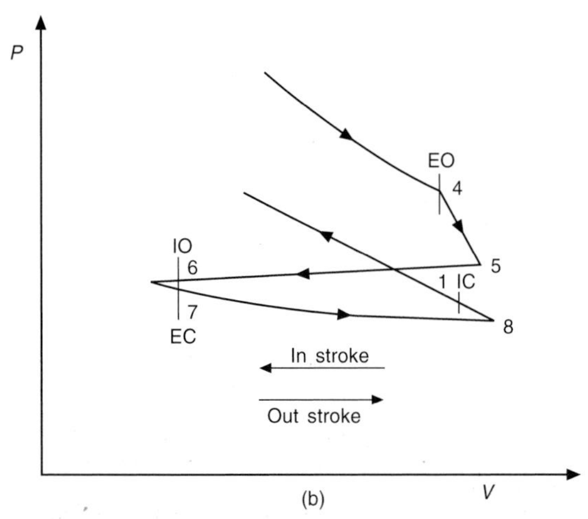

Figure 1.6 Gas exchange period. *p-V* diagrams: (a) Two-stroke (asymmetrical timing); (b) Four-stroke. (*Benson and Whitehouse*)

loop which may contribute positive or negative loop work to the work associated with the power loop. *Figures 1.6a and b* are typical p-V diagrams of the open or gas exchange period for two-stroke and four-stroke engines.

In both types of engine the cyclic integral expression leads to the so-called indicated *mean effective pressure*

$$p_{\text{ind}} = \frac{\int p dV}{V_{\text{swept}}} = \frac{\int dW}{V_{\text{swept}}} \tag{1.1}$$

where $\int dW$ represents the cyclic work with the distinction that the cycle occupies 360° for two-stroke and 720° for four-stroke engines.

The power may then be evaluated from the following expressions:

$$\dot{W}_{\text{two-stroke}}\ (\text{kW}) = \frac{p_{\text{ind}}\ (\text{bar})\ V_{\text{swept}}\ (\text{m}^3) N_{\text{e}}\ (\text{rev/s}) n}{10^{-2}} \tag{1.2a}$$

or

$$\dot{W}_{\text{four-stroke}}\ (\text{kW}) = \frac{p_{\text{ind}}\ (\text{bar})\ V_{\text{swept}}\ (\text{m}^3) N_{\text{e}}\ (\text{rev/s}) n}{2 \times 10^{-2}} \tag{1.2b}$$

For purposes of comparison with air standard cycles (see section 1.3), it is appropriate to use the *effective* swept volume $(V_{\text{swept}})_{\text{eff}}$, i.e. that associated with the closed period only rather than the geometric swept volume V_{swept}. In the case of two-stroke engine, with the gas exchange period occupying up to 150°CA, $(V_{\text{swept}})_{\text{eff}}/V_{\text{swept}}$ may be considerably less than unity while for four-stroke engines it varies between close to unity and 0.8 (approx.).

Similarly the volumetric compression ratio (CR), which again is crucial in air standard cycle calculations, is usually based on the effective swept volume

$$\text{i.e.}\quad (CR)_{\text{eff}} = \frac{(V_{\text{swept}})_{\text{eff}} + V_{\text{clearance}}}{V_{\text{clearance}}} \tag{1.3a}$$

rather than the geometric value

$$(CR)_{\text{geom}} = \frac{V_{\text{swept}} + V_{\text{clearance}}}{V_{\text{clearance}}} \tag{1.3b}$$

Finally, indicated thermal efficiency η_i or indicated specific fuel consumption i.s.f.c. are evaluated from the expression

$$\eta_i = \frac{\dot{W}(\text{eqn (1.2a) or (1.2b)})}{\dot{m}_f\ (\text{kg/sec})\ CV(\text{kJ/kg})} \tag{1.4a}$$

where $\dot{m}_f$ is the rate of fuel flow to the engine and CV is the lower calorific value of the fuel and

$$\text{i.s.f.c.} = \frac{\dot{m}_f \times 3600}{\dot{W}}\ \text{kg/kW hr} \tag{1.4b}$$

1.2.4 Other operating parameters

(a) *Air-fuel ratio*

The combustion process is governed in large measure by the air fuel ratio in the cylinder, expressed either in actual terms

$$\text{i.e.}\quad (A/F) = \frac{\dot{m}_{a_t}\ (\text{kg/sec})}{(\dot{m}_f)\ \text{kg/sec}} \tag{1.5a}$$

where $\dot{m}_{a_t}$ is the rate of trapped airflow to the engine *or* relative to the chemically correct or stoichiometric air fuel ratio for the particular fuel, i.e. excess air factor

$$\varepsilon = \frac{(A/F)_{\text{actual}}}{(A/F)_{\text{stoichiometric}}} \tag{1.5b}$$

In practice, for most hydrocarbon fuels

$$(A/F)_{\text{stoichiometric}} \doteqdot 14.9 \tag{1.5c}$$

and, depending on the combustion system used, the limiting relative air fuel ratio for smokefree combustion at full load is in the range

$1.2 < \varepsilon < 1.6$

being lower for IDI than for DI engines.

(b) *Gas exchange parameters*

For two-stroke engines, in particular, it is vitally important to make a distinction between the trapped rate of airflow $\dot{m}_{at}$ and the total rate of airflow supplied to the engine $\dot{m}_a$. This arises from the fact that the scavenging process in two-stroke engines is accompanied by substantial loss of air to exhaust, partly through mixing with products of combustion and partly through short-circuiting (see section 1.3.1) and leads to the definition of *trapping efficiency* as

$$\eta_{tr} = \frac{(\dot{m}_a)_t}{(\dot{m}_a)} \tag{1.6a}$$

or its reciprocal, the scavenge ratio

$$R_{SC} = \frac{\dot{m}_a}{(\dot{m}_a)_t} \tag{1.6b}$$

In practice $1.1 < R_{SC} < 1.6$.

For four-stroke engines, particularly those with small valve overlap, e.g. in road traction, it is safe to assume that all the air delivered to the engine is trapped in the cylinder, i.e. $\eta_{tr} \doteqdot 1$. However, due to charge heating during the gas exchange process and adverse pressure conditions in the cylinder, it is likely that the volumetric efficiency η_{vol} defined as

$$\eta_{vol} = \frac{\text{volume of air trapped under inlet manifold conditions}}{\text{swept cylinder volume}}$$

$$= \dot{m}_a \frac{RT}{10^2\, p} \Big/ V_{swept} \tag{1.7}$$

(where T and p are respectively the inlet manifold temperature (°K) and pressure (bar)) is considerably less than unity. Clearly for the highest specific output, both the relative air fuel ratio ε and the volumetric efficiency should be as close to unity as possible.

1.3 Air standard cycles

It will be clear from the foregoing sections that the real processes in the diesel engine cylinder, particularly those of fuel preparation, combustion and gas exchange are extremely complex and require sophisticated computational techniques which are discussed in a number of specialist texts.[1,2,3]

Air standard cycles which are discussed in most elementary textbooks, provide a useful basis for comparing actual engine performance expressed in terms of indicated mean effective pressure (p_{ind}, eqn (1.1) and indicated thermal efficiency (η_i, eqn (1.4a) with corresponding values for highly idealized cycles, based on certain drastic simplifying assumptions as follows:

(a) the mass of working fluid remains constant throughout the cycle, i.e. gas exchange and fuel addition are ignored;
(b) the working fluid throughout the cycle is pure air treated as a perfect gas;
(c) the combustion and gas exchange processes are replaced by external heat transfer to or from the working fluid under idealized, e.g. constant volume or constant pressure conditions;
(d) compression and expansion processes are treated as adiabatic and reversible, i.e. heat transfer and friction effects are completely neglected;
(e) at any point of the working cycle, cylinder charge pressure and temperature are completely uniform, i.e. spatial variations in their values as for instance during combustion or scavenging, are completely neglected.

The most commonly used air-standard cycles are as follows (*Figures 1.7a, b and c*):

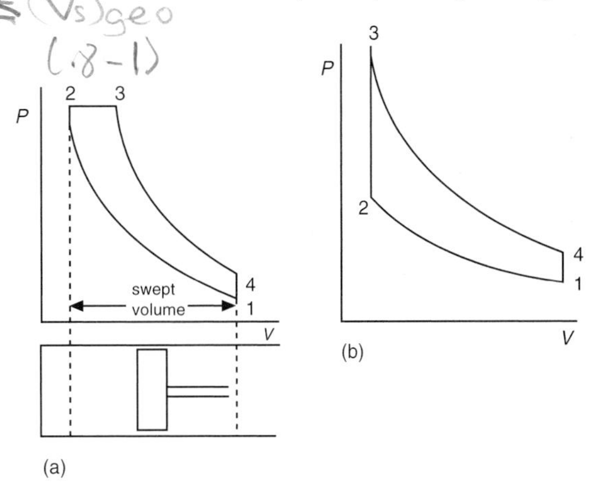

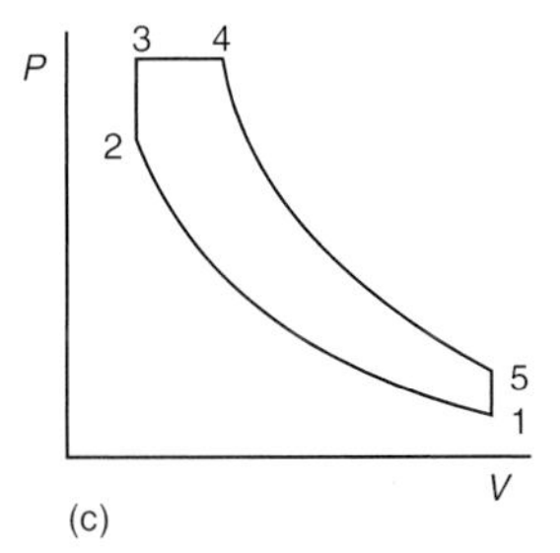

Figure 1.7 Air standard cycles: (a) Constant pressure cycle; (b) Constant volume cycle; (c) Dual combustion or composite cycle

(a) *The constant pressure or diesel cycle (Figure 1.7a)*
Here combustion is simulated by constant pressure heat addition (2–3), and blowdown, followed by scavenge, by constant volume heat rejection 4–1. Compression 1–2 and expansion 3–4 follow the isentropic state relationships for a perfect gas. This particular cycle has, in the past, been used as a reference cycle for the 'classical' Diesel engine with air blast injection giving a rather long injection and hence heat release period, corresponding to 2–3. It has, however, little relevance to the modern diesel cycle.

(b) *The constant volume or Otto cycle (Figure 1.7b)*
Here combustion is simulated by constant volume heat release 2–3, and the blowdown-gas exchange sequence once again by constant volume heat rejection 4–1. Again compression 1–2 and expansion 3–4 are isentropic.

Traditionally this is the reference cycle for spark ignition (SI) engines, but it has distinct validity as a reference cycle for diesel engines, particularly under light load conditions when the heat release period is short so that the assumptions of zero heat release duration implied by the constant volume process 2–3 does not introduce excessive errors.

(c) *The 'dual combustion' or composite cycle (Figure 1.7c)*
This represents a combination of the constant pressure and constant volume cycles and is intended to provide a closer approximation to actual diesel cycles than either of the above ideal cycles. It is particularly appropriate where comparisons are to be made with actual diesel cycles on the basis of the maximum cylinder pressure p_{max} obtained during the heat release period, i.e. for engines operating in the mid-to full load range.

1.3.1 Theoretical expressions for air standard cycles

In the following derivations it will be assumed that the compression ratio CR corresponds to the effective compression ratio $(CR)_{eff}$ of the engine, eqn (1.3a), and that the isentropic index γ, i.e. the specifc heat ratio for air as a perfect gas, has the constant value $\gamma = 1.4$.

1.3.1.1 The constant pressure or diesel cycle (Figure 1.7a)

From basic engineering thermodynamics:

$$\text{compression work } W_{12} = \frac{+p_1V_1 - p_2V_2}{\gamma - 1} \tag{i}$$

(note this is negative)

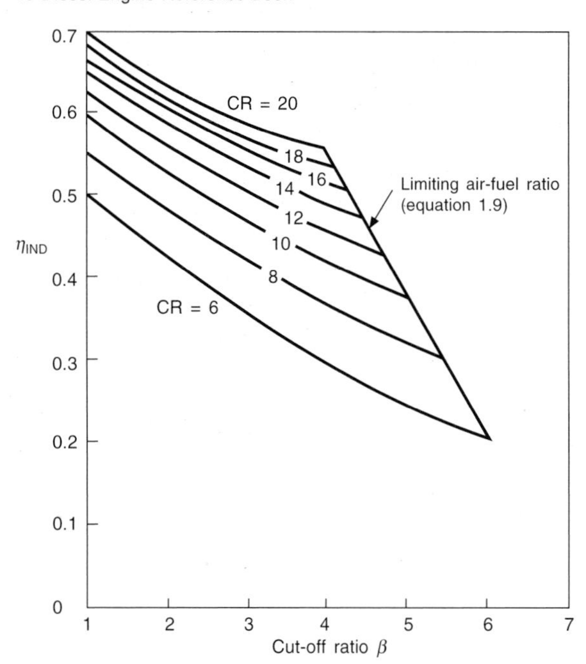

Figure 1.8 Constant pressure cycle. Indicated efficiency *vs* cut-off ratio (eqn 1.8)

constant pressure work

$$W_{23} = p_2 (V_3 - V_2) = p_2V_2(\beta - 1) \quad \text{(ii)}$$

where β = volume ratio $\dfrac{V_3}{V_2}$

constant pressure heat transfer

$$Q_{23} = mC_p(T_3 - T_2)$$

$$= \frac{p_2 V_2}{RT_2}\left(\frac{\gamma}{\gamma - 1} R\right)\left(\frac{V_3}{V_2} - 1\right)T_2$$

$$= p_2 V_2 \left(\frac{\gamma}{\gamma - 1}\right)(\beta - 1) \quad \text{(iii)}$$

$$\text{expansion work } W_{34} = \frac{p_3V_3 - p_4V_4}{\gamma - 1} = \frac{p_2V_2\beta - p_4V_1}{\gamma - 1} \quad \text{(iv)}$$

$$\text{nett work } W_{\text{nett}} = W_{12} + W_{23} + W_{34}$$

$$= \frac{p_1V_1 - p_2V_2}{\gamma - 1} + \frac{p_2V_2\beta - p_4V_1}{\gamma - 1} + p_2V_2(\beta - 1) \quad \text{(v)}$$

but

$$p_2 = p_3 = p_1\left(\frac{V_1}{V_2}\right)^{\gamma} = p_1(CR)^{\gamma},\ V_2 = \left(\frac{V_1}{CR}\right)$$

$$p_4 = p_3\left(\frac{V_3}{V_4}\right)^{\gamma} = p_2\left(\frac{\beta V_2}{V_1}\right)^{\gamma} = p_2\left(\frac{\beta}{CR}\right)^{\gamma} \quad \text{(vi)}$$

Substituting from (vi) for p_2, V_2 and p_4 in (v) and (iii) and writing for the ideal efficiency of the constant pressure (CP) cycle:

$$(\eta_i)_{CP} = \frac{W_{\text{nett}}}{Q_{23}} \quad \text{(vii)}$$

(vii) eventually reduces to

$$(\eta_i)_{CP} = 1 - \left(\frac{1}{(CR)^{\gamma-1}}\right)\frac{\beta^{\gamma} - 1}{\gamma(\beta - 1)} \quad (1.8)$$

The volume ratio β is an indication of the air–fuel ratio A/F at which the engine is operating, since to a first approximation

$$Q_{23} = mC_p(T_3 - T_2) = p_2V_2\frac{\gamma}{\gamma - 1}(\beta - 1)$$

$$= m_f(CV) \quad \text{(viii)}$$

where m_f is the mass of fuel burnt

$$\text{But } m_{\text{air}} = m_1 = \frac{p_1V_1}{RT_1} \quad \text{(ix)}$$

$$\text{whence } A/F = \frac{p_1V_1}{RT_1}\Bigg/\left[\frac{p_2V_2\frac{\gamma}{\gamma - 1}(\beta - 1)}{(CV)}\right]$$

$$= \left(\frac{1}{CR}\right)^{\gamma-1}\frac{CV}{RT_1}\,\frac{1}{\frac{\gamma}{\gamma-1}(\beta - 1)} \quad (1.9)$$

Assuming that the limiting air–fuel ratio is the stoichiometric ratio $(A/F)_{stoich}$ eqn (1.5c), it is possible to find a limiting value of the volume ratio β for any given compression ratio CR from eqn (1.9). This is shown in *Figure 1.8* indicating the behaviour of eqn (1.8) with different values of compression ratio CR and 'cut off' ratio β, including the position of the 'limiting line' for stoichiometric combustion.

Indicated efficiency $(\eta_i)_{CP}$ is seen to increase rapidly with volumetric compression ratio CR and to decrease with increasing values of the cut off ratio β, i.e. with decreasing air-fuel ratio, being a minimum, for any value of CR on the limit line, and a maximum for a cut off ratio $\beta = 1$.

Efficiency is not the only consideration appertaining to cycles. Specific output also has to be taken into account so that the relationship between indicated efficiency, specific output and compression ratio is equally important. The specific output is best measured in terms of the mean effective pressure defined by eqn (1.1) relative to the trapped pressure p_1. The calculation is as follows:

For any assumed value of the cut-off ratio β the equivalent air–fuel ratio A/F may be calculated from eqn (1.9), giving the heat input to the cycle as

$$Q_{in} = m\frac{CV}{A/F} = \frac{p_1V_1}{RT_1}\,\frac{CV}{A/F} \quad \text{(i)}$$

With indicated efficiency $(\eta_i)_{CP}$ from eqn (1.8), the indicated work output of the cycle is given by

$$\int dW = Q_{in}(\eta_i)_{CP} \quad \text{(ii)}$$

and the mean effective pressure, from eqn (1.1) becomes

$$p_{ind} = \frac{\int dW}{V_{swept}} = \frac{\int dW}{V_1\left(1 - \frac{1}{CR}\right)} = \frac{\frac{p_1V_1}{RT_1}\,\frac{CV}{A/F}(\eta_i)_{CP}}{V_1\left(1 - \frac{1}{CR}\right)}$$

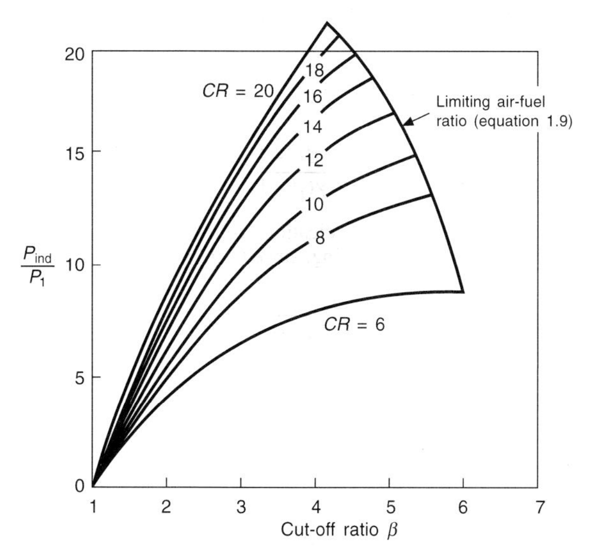

Figure 1.9 Constant pressure cycle. Indicated mean effective pressure *vs* cut-off ratio (eqn 1.10)

$$\text{or} \quad \left(\frac{p_{\text{ind}}}{p_1}\right)_{\text{CP}} = \frac{\dfrac{CV}{RT_1}\,\dfrac{1}{A/F}\,(\eta_{\text{i}})_{\text{CP}}}{\left(1-\dfrac{1}{CR}\right)} \qquad (1.10)$$

Equation (1.10) is represented by *Figure 1.9* and shows that, for any given compression ratio CR, efficiency decreases with increasing specific output, with a minimum value again on the limit line.

Figure 1.9 may be used both for naturally aspirated engines for which the trapped pressure p_1 is approximately equal to atmospheric pressure p_a as well as for supercharged engines with a supercharge (or boost) pressure ratio given approximately by $R_B = (p_1/p_a)$ (> 1).

1.3.1.2 The constant volume or Otto cycle (Figure 1.7b)

As already stated this cycle has only limited applicability to diesel engines, mainly under part load conditions. Heat transfer now occurs under constant volume conditions, both for the 'combustion' process 2–3 and the 'gas exchange' process 4–1. Nett cycle work

$$W_{\text{nett}} = W_{12} + W_{34}$$

$$= \frac{p_1V_1 - p_2V_2}{\gamma - 1} + \frac{p_3V_3 - p_4V_4}{\gamma - 1} \qquad \text{(i)}$$

constant volume heat transfer

$$Q_{23} = mC_v(T_3 - T_2) = \frac{p_2V_2}{RT_2}\,\frac{R}{\gamma - 1}\left(\frac{p_3}{P_2} - 1\right)T_2$$

$$= \frac{p_2V_2}{\gamma - 1}(\alpha - 1) \qquad \text{(ii)}$$

where $\alpha = \dfrac{p_3}{p_2}$

But $p_2 = p_1(CR)^\gamma,\ V_2 = \dfrac{V_1}{CR} = V_3$

$$p_4 = p_3\left(\frac{1}{CR}\right)^\gamma = p_2\,\alpha\left(\frac{1}{CR}\right)^\gamma = p_1\,\alpha \qquad \text{(iii)}$$

Substituting from (iii) in (i) and (ii) and writing for the ideal efficiency of the constant volume (CV) cycle

$$(\eta_{\text{i}})_{\text{CV}} = \frac{W_{\text{nett}}}{Q_{23}} \qquad \text{(iv)}$$

(iv) eventually reduces to

$$(\eta_{\text{i}})_{\text{CV}} = 1 - \left(\frac{1}{CR}\right)^{\gamma - 1} \qquad (1.11)$$

Equation (1.11) demonstrates that the efficiency of the constant volume cycle is a function of compression ratio CR only, and unlike the constant pressure cycle, independent of the level of heat addition, as expressed by the pressure ratio $p_3/p_2 = T_3/T_2 = \alpha$ (see *Figure 1.10*).

It is generally quoted in support of arguments to raise compression ratio in spark ignition (SI) engines.

1.3.1.3 The 'dual combustion' or composite cycle (Figure 1.7c)

As already stated, this cycle tends to approximate more closely to actual diesel cycles than either the pure constant pressure or constant volume cycles as described above. It lends itself particularly well to the representation of limited maximum cylinder pressure, as expressed by the pressure ratio $\alpha = p_3/p_2$ often specified in real diesel cycles, and to assessment of the effect of increased or retarded heat release, as expressed mainly by the volume ratio $\beta = V_4/V_3$.

The evaluation of cycle efficiency follows a similar pattern to that adopted above:

$$\text{nett cycle work} = W_{12} + W_{34} + W_{45}$$

$$= \frac{p_1V_1 - p_2V_2}{\gamma - 1} + p_3V_3(\beta - 1) + \frac{p_3V_4 - p_5V_5}{\gamma - 1} \qquad \text{(i)}$$

Constant volume heat transfer

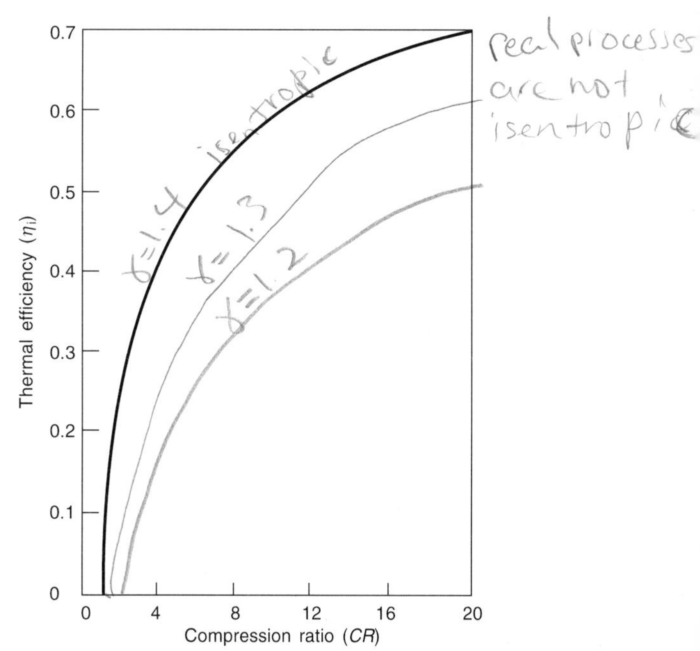

Figure 1.10 Constant volume cycle. Indicated efficiency *vs* compression ratio

$$Q_{23} = mC_v(T_3 - T_2) = \frac{p_2V_2}{\gamma - 1}(\alpha - 1) \quad \text{(ii)}$$

(see constant volume cycle, above).
Constant pressure heat transfer

$$Q_{34} = mC_p(T_4 - T_3) = p_3V_2\frac{\gamma}{\gamma - 1}(\beta - 1) \quad \text{(iii)}$$

(see constant pressure cycle, above).

But $p_2 = p_1(CR)^\gamma$, $V_2 = \dfrac{V_1}{CR} = V_3$, $p_3 = p_4 = p_2\ \alpha = p_1(CR)^\gamma\alpha$

$$V_4 = V_3\beta = \frac{V_1}{CR}\beta, \quad p_5 = p_4\left(\frac{\beta}{CR}\right)^\gamma = p_3\left(\frac{\beta}{CR}\right)^\gamma \quad \text{(iv)}$$

Substituting from (iv) in (i), (ii) and (iii) and writing for the ideal efficiency of the dual combustion cycle

$$(\eta_i)_{DC} = \frac{W_{nett}}{Q_{23} + Q_{34}} \quad \text{(v)}$$

eqn (v) eventually reduces to

$$(\eta_i)_{DC} = 1 - \left(\frac{1}{CR}\right)^{\gamma-1} \frac{\alpha\beta^\gamma - 1}{(\alpha - 1) + \gamma\alpha(\beta - 1)} \quad (1.12)$$

Inspection of eqn (1.12) shows that with $\beta = 1$, characteristic of the constant volume cycle, the efficiency reduces to that given by eqn (1.11) for that cycle, becoming a function of compression ratio CR only, while with $\alpha = 1$ characteristic of the constant pressure cycle, the efficiency reduces to that given by eqn (1.8) for that particular cycle. The dual combustion cycle may therefore be regarded as embracing, at the same time, those other ideal cycles.

As in the case of the diesel cycle, it is of interest to establish the relationship between specific output, represented by the mean effective pressure ratio

$$\frac{p_{ind}}{p_1}$$

the efficiency $(\eta_i)_{DC}$ and compression ratio CR. A maximum pressure ratio

$$\frac{p_{max}}{p_1} = \frac{p_3}{p_1} \quad (\textit{Figure 1.7c})$$

has to be fixed, in the first instance. This in conjunction with the compression ratio CR will yield the pressure ratio

$$\alpha = \frac{p_3}{p_2}$$

from

$$\frac{p_{max}}{p_1} = \frac{\alpha p_2}{p_1}\alpha(CR)^\gamma = PR \quad \text{(v)}$$

Substituting for CR in terms of PR from eqn (v) in the efficiency expression eqn (1.12) leads to the alternative expression, in terms of the pressure ratio PR

$$(\eta_i)_{DC} = 1 - \left(\frac{\alpha}{PR}\right)^{(\gamma-1)/\gamma} \frac{\alpha\beta^\gamma - 1}{(\alpha - 1) + \gamma\alpha(\beta - 1)} \quad (1.12a)$$

The equivalent air–fuel ratio A/F may also be determined by a procedure equivalent to that adopted for the diesel cycles from (ii) and (iii) above

$$Q_{23} + Q_{34} = \frac{p_2V_2(\alpha - 1)}{\gamma - 1} + \frac{p_3V_2\gamma(\beta - 1)}{\gamma - 1} = m_f(CV) \quad \text{(vi)}$$

where m_f is the mass of fuel burnt, i.e.

$$m_f = \frac{m_{air}}{A/F} = \frac{m_1}{A/F} = \frac{p_1V_1}{RT_1}\Big/ A/F \quad \text{(vii)}$$

Substituting for m_f from (vii) in (vi) provides an explicit solution for β, α already being known from (v) in terms of the stipulated max. pressure $p_{max} = p_3$ resulting eventually in

$$A/F = \left(\frac{1}{CR}\right)^{\gamma-1} \frac{CV}{RT_1} \frac{\gamma - 1}{(\alpha - 1) + \gamma\alpha(\beta - 1)} \quad (1.13a)$$

$$= \left(\frac{\alpha}{PR}\right)^{(\gamma-1)/\gamma} \frac{CV}{RT_1} \frac{\gamma - 1}{(\alpha - 1) + \gamma\alpha(\beta - 1)} \quad (1.13b)$$

As before, the concept of the limiting air–fuel ratio may be applied, with A/F = $(A/F)_{stoichiometric}$, to give a limiting value of β for any given value of CR.

Finally, with mean effective pressure given by $(p_{ind})_{DC} = \int dW/V_{swept}$ and applying similar arguments to those for the diesel cycle

$$\left(\frac{p_{ind}}{P_1}\right)_{DC} = \frac{\dfrac{CV}{RT_1}\dfrac{1}{A/F}(\eta_i)_{DC}}{\left(1 - \dfrac{1}{CR}\right)} \quad (1.14)$$

Equations (1.12), (1.13) and (1.14) have been combined in *Figure 1.11* to give a single representation of indicated efficiency and mean indicated pressure as a function of compression ratio (CR) the upper limit of mean effective pressure being set by the limiting air–fuel ratio. The assumed value p_{max}/p_1 has been set at 68 corresponding to a max. cylinder pressure of 68 bar for a naturally aspirated engine or 136 bar for a supercharged engine operating with a boost ratio of 2:1. The relationship between compression ratio CR, the pressure ratio

$$\frac{p_3}{p_2} = \frac{p_{max}}{p_2} = \alpha \text{ for } \frac{p_{max}}{p_1} = 68 \text{ and the cut-off ratio } \beta,$$

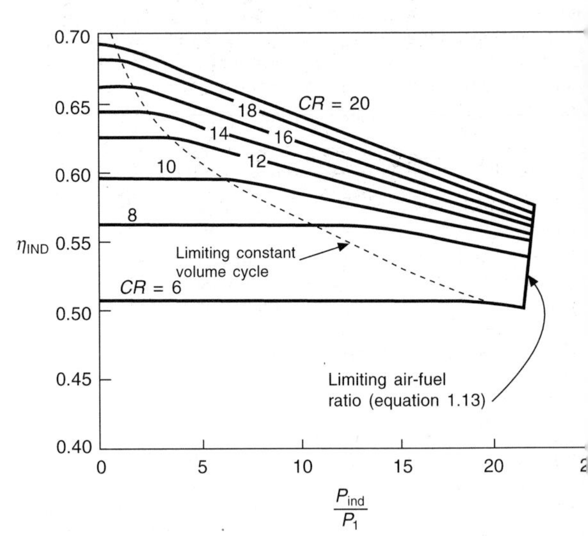

Figure 1.11 Dual combustion cycle. Indicated efficiency (eqn 1.12) *vs* indicated mean. Effective pressure (eqn 1.14) $P_{max}/P_1 = 68$

the latter for limiting air–fuel ratio, is summarized in *Table 1.1*.

Table 1.1

CR	6	8	10	12	14	16	18	20
a	5.53	3.71	2.71	1.11	1.69	1.40	1.19	1.02
β_{lim}	1.35	1.73	2.12	2.50	2.91	3.31	3.73	4.16

Figure 1.11 dispels the widespread misconception that the constant volume cycle for which $\beta = 1$ is the most efficient possible air standard cycle. For a given maximum pressure ratio PR and a given indicated mean effective pressure ratio

$$\frac{p_{\text{ind}}}{p^{1}}$$

Figure 1.11 shows clearly that the constant volume cycle is the least efficient, and that cycle efficiency increases progressively as the constant pressure cycle is approached.

1.3.2 Further comments on air standard cycles

1.3.2.1 Interpretation of results so far

The three reference cycles described above, i.e.
(a) the 'classical' constant pressure diesel cycle;
(b) the constant volume Otto cycle;
(c) the 'dual combustion' cycle.
have been, and are still, widely used as useful reference cycles for the assessment of the actual performance of diesel engines. Their limitations, resulting mainly from the drastic idealizations involved in their formulation have already been stated. Actual performance will therefore inevitably fall far short of that predicted by air standard cycle theory, the quantitative effects of the many departures from ideal conditions being the subject of a later section of this chapter. Nevertheless, *Figure 1.11* in particular, for the dual combustion cycle, provides some extremely valuable insight into the limits of performance of high output diesel engines.

It is clear that limiting indicated mean effective pressures, irrespective of compression ratio, are of the order of 22 bar for a naturally aspirated engine, with proportionally higher values for supercharged engines, assuming that the trapped cylinder temperature can be maintained constant by aftercooling. Typical high output engines now operate at boost pressure ratios approaching 3:1 and with very generous aftercooling, suggesting a theoretical limiting IMEP of the order of 66 bar. This figure has to be drastically reduced to account for

(a) the need to operate with actual air–fuel ratios of the order of 22:1 instead of the assumed limiting air–fuel ratio $(A/F)_{\text{stoic}} = 14.9$

(b) the real effects of
(i) pumping and heat losses during the gas exchange process;
(ii) heat losses during the closed period, particularly during the period of highest cylinder temperatures, i.e. combustion and the early phases of expansion;
(iii) inevitable irreversibilities during combustion;
(v) variable specific heats a function of temperature and composition.

Hence one would expect to achieve a limiting IMEP of rather less than half the theoretical value, i.e. 30 bar in a highly developed four-stroke diesel engine operating at a boost ratio of 3:1, and even less in a two-stroke engine due to its generally shorter effective stroke and the need for even more generous air–fuel ratios to ensure satisfactory scavenging and freedom from thermal overloading.

Likewise the values of ideal efficiency based on *Table 1.1* and eqns (1.12) and (1.14), showing that with the limiting air–fuel ratio a value of 0.570 (57%) is reached for the highest compression ratio of 20:1 when operating on the dual combustion cycle, while for the pure constant volume cycle and for the same compression ratio, but with zero fuelling, the corresponding efficiency is 0.696 (69.5%), are clearly far above practically realizable values. Nevertheless indicated efficiencies approaching 46% have been achieved in practice and may well be exceeded in future, with continuing improvements in suppression of heat loss to coolant[4], fuel injection equipment and turbocharging arrangements.

1.3.2.2 Other theoretical cycles and representations

(a) *The modified Atkinson cycle (Figure 1.12)*
It has already been indicated that maximum cylinder pressure is an important limiting factor in improving diesel engine performance. In practice values of p_{max} in excess of 150 bar are rarely permissible for mechanical reasons. Inspection of *Figure 1.11* shows that the more the 'trapped' pressure p_1 (approximately equal to the boost pressure) is raised to raise output, the more the volumetric compression ratio CR will have to be reduced if the limiting value of p_{max} is not to be exceeded. Equation (1.12) and *Figure 1.11* indicate that this inevitably has an adverse effect on efficiency. The Atkinson cycle seeks to redress this effect partially by operating with an expansion ratio ER well in excess of the compression ratio CR. In practice this is achieved not by lengthening the stroke of the piston, but by early inlet closing, and late exhaust opening (*Figure 1.12*) the latter being delayed until the cylinder gases have fully expanded to the trapped pressure p_1.

It may be shown, returning to the notation adopted for the dual combustion cycle, with CR denoting compression ratio as

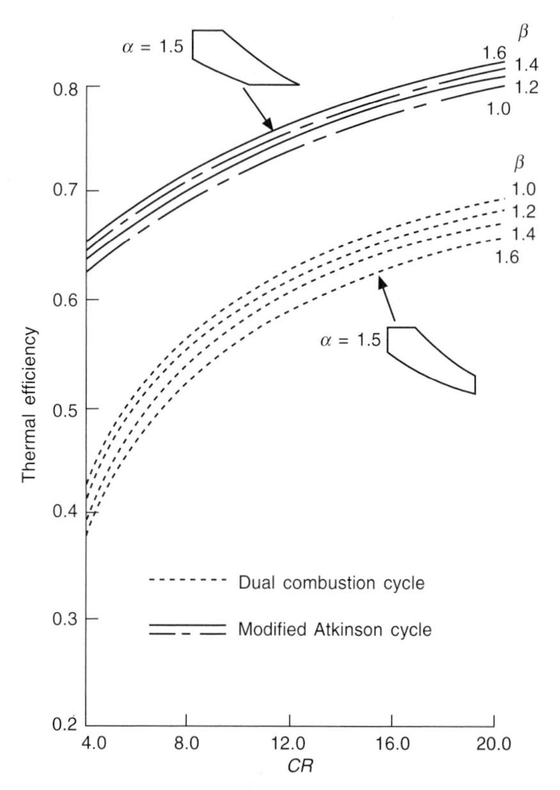

Figure 1.12 Atkinson cycle (*Benson and Whitehouse*)

before, but with expansion ratio ER leading to expansion terminating at p_1 not, as before p_5 where $p_5 > p_1$ (*see Figure 1.12*) that the indicated efficiency of the idealized Atkinson cycle becomes:

$$(\eta_i)_{AT} = 1 - \left(\frac{1}{CR}\right)^{\gamma-1} \frac{\gamma(\alpha^{1/\gamma}\beta - 1)}{(\alpha - 1) + \alpha\gamma(\beta - 1)} \tag{1.15}$$

This expression differs from eqn (1.12) for the dual combustion cycle only in that the numerator of the second term on the RHS, with $\alpha > 1$ and $\beta > 1$, is clearly smaller than that in eqn (1.12), i.e. $\gamma(\alpha^{1/\gamma}\beta - 1) < \alpha\beta^{\gamma} - 1$, leading inevitably to $(\eta_i)_{AT} > (\eta_i)_{DC}$ for equal values of CR, α and β. The effect is illustrated in *Figure 1.12*.

Attempts have been made to take advantage of the improved efficiency of the Atkinson cycle in certain highly boosted diesel engines[5].

(b) *The Carnot cycle (Figures 1.13a and b)*

It is well known that Carnot formulated the theoretical conditions for the most efficient conversion of heat energy to useful work. The Carnot cycle consists of two isothermal processes, viz. heat addition to the cycle Q_{23} at the maximum cycle temperature $T_2 = T_3 = T_{max}$ and heat rejection from the cycle Q_{41} at the minimum cycle temperature $T_4 = T_1 = T_{min}$, these two processes being joined by a reversible adiabatic compression 1–2 and a reversible adiabatic expansion 3–4.

The cycle is represented in the p-V diagram, *Figure 1.13a*, and for completeness also in the temperature-entropy, or *TS* diagram, *Figure 1.13b*.

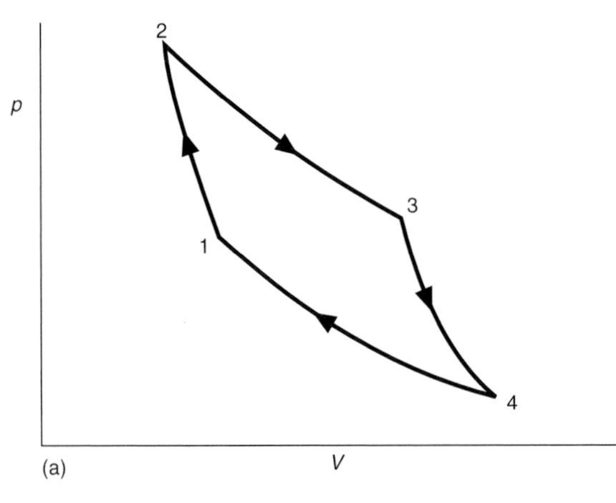

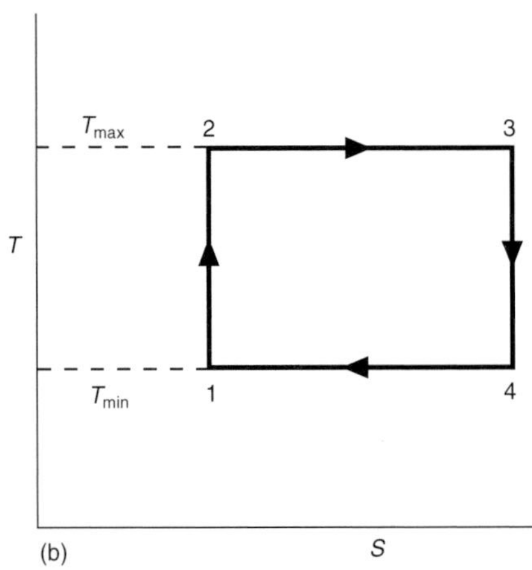

Figure 1.13 Carnot cycle: (a) *p-V* diagram; (b) *TS* diagram

The theoretical arguments relating to the Carnot cycle are fully presented in any standard text on engineering thermodynamics[6,7]. The Carnot efficiency, which may be regarded as the ultimate theoretical criterion for all work conversion processes, is given by

$$\eta_{CARNOT} = 1 - \frac{T_{min}}{T_{max}} \tag{1.16}$$

Typically, for the dual combustion cycle of *Figure 1.7c* the minimum and maximum temperatures with

$$CR = 20, \quad \frac{p_{max}}{p} = 68$$

and on the limiting line are given by $T_{min} = 290$ K and $T_{max} = 5450$ K. *Figure 1.11* gives an efficiency, for this condition, of $(\eta_i)_{DC} = 0.57$. The corresponding Carnot efficiency would be

$$\eta_{CARNOT} = 1 - \frac{390}{5450} = 0.928$$

Actual maximum temperatures in diesel cycles are of the order of 2500 K giving Carnot efficiencies of the order of 0.84 or 84%.

It is obvious that the Carnot criterion has very little direct relevance to diesel conversion efficiencies. Furthermore, the cycle is wholly unsuitable for adoption in reciprocating prime movers as it would lead to absurdly large compression and expansion ratios, as well as wholly artificial maximum cylinder pressures, for realistic output levels. Under limiting heat release conditions (approximately 2800 kJ/kg air) and for a maximum temperature of 4820 K, the Carnot cycle requires a peak pressure

$p_{max} \doteqdot 146\,000$ bar.

1.4 Basic thermodynamics of real gases

The air standard cycle treatment as outlined in the previous section was based on sweeping simplifying assumptions, first and foremost that of the working fluid being air treated as a perfect gas. In order to deal with cycle calculations on the more realistic basis developed in section 1.6, it is first necessary to deal with certain fundamental aspects of the behaviour of the working fluid in real cycles, both under non-reacting and reacting conditions (i.e. without or with combustion).

1.4.1 Gas properties

In full cycle simulation programs the cylinder contents before, during and after combustion are treated as a mixture of chemically pure gases. In simpler calculations not specifically concerned with emission formation, and where air–fuel ratios well above stoichiometric are used, it is sufficient to limit the species under consideration to oxygen (O_2), nitrogen (N_2), carbon dioxide (CO_2) and water vapour (H_2O); where fuel rich combustion has to be considered, carbon monoxide (CO) has to be included, while for emission calculations additional species such as oxides of nitrogen and unburnt hydrocarbons would also have to be considered. In general, for a mixture of gases characterized by mol fractions $x_1, x_2, \ldots, x_i$ of species $1, 2, \ldots, i$ where

$$x_1 = \frac{n_1}{n}, \quad x_2 = \frac{n_2}{n}, \ldots, \quad x_i = \frac{n_i}{n}$$

in which n refers to the total number of mols in the maxture and $n_1, n_2, \ldots, n_i$ to the individual number of mols of each species, the equation of state may be written:

$$p_i V = n_i R_0 T \tag{1.17}$$

where p_i = partial pressure of constituent i;
V = volume of mixture;
R_0 = universal gas constant (kJ/kg mol K);
T = absolute temperature (°K).

while the partial pressure p_i for species i is given by

$$p_i = x_i p \tag{1.18}$$

and the resultant total pressure of the mixture by

$$p = p_1 + p_2 + \ldots p_i = \sum_{i=1}^{i} p_i \tag{1.19}$$

The equivalent molecular mass M of the mixture is given by

$$M = x_1 M_1 + x_2 M_2 + \ldots + x_i M_i = \sum_{i=1}^{i} x_i M_i \tag{1.20}$$

where $M_1, M_2, \ldots, M_i$ are the molecular masses of the individual constituents.

The specific internal energy of the mixture on a molar (or kg mol) basis is given by

$$u = x_1 u_1 + x_2 u_2 + \ldots + x_i u_i$$

$$= \sum_{i=1}^{i} x_i u_i \text{ (kJ/kg mol)} \tag{1.21}$$

where $u_1, u_2, \ldots, u_i$ are the specific internal energies (kJ/kg mol) of the individual constituents while similarly the specific enthalpy may be expressed as

$$h = x_1 h_1 + x_2 h_2 + \ldots + x_i h_i$$

$$= \sum_{i=1}^{i} x_i h_i \text{ (kJ/kg mol)} \tag{1.22}$$

In connection with eqns (1.21) and (1.22) it is convenient to introduce the concept of specific internal energy at absolute zero (u_0) and specific enthalpy at absolute zero (h_0) so that

$$u = u_0 + u(T) \tag{1.21a}$$

and

$$h = h_0 + h(T) \tag{1.22a}$$

where $u(T)$ and $h(T)$ are functions of temperature only both of which reduce to zero at absolute zero giving

$$u_0 = h_0 \tag{1.23}$$

While, from the semi-perfect gas relationship, again for the mixture as a whole, and again on a molar basis

$$h - u = p\text{v}_{mol} = R_0 T \tag{1.24}$$

This latter relationship implies that only one of the two functions $u(T)$ and $h(T)$ has to be defind as a power series in T to determine both $u(T)$ and $h(T)$.

Defining the term $h_i(T)$ for constituent i as a power series of the form

$$h_i(T) = R_0[a_{i,1}T + a_{i2}T^2 + \ldots a_{ij}T^j]$$

$$= R_0 \sum_{j=1}^{j} a_{ij} T^j \tag{1.25}$$

it is generally sufficient to end the power series at $j = 5$. Equation (1.24) again applied to constituent i then leads to

$$u_i(T) = R_0 \left[\sum_{j=1}^{j} a_{ij} T^j - T \right] \tag{1.26}$$

Finally, eqns (1.25) and (1.26) together with eqns (1.22a), (1.21a) and (1.23) lead to the following expressions for the specific internal energy and enthalpy of constituent i:

$$u_i = u_{i,0} + u_i(T)$$

$$= u_{i,0} + R_0 \left[\sum_{j=1}^{j} a_{ij} T^j - T \right] \tag{1.27}$$

and since $u_{i,0} = h_{i,0}$ (eqn 1.23)

$$h_i = u_{i,0} + R_0 \sum_{j=1}^{j} a_{i,j} T^j \tag{1.28}$$

Combining eqns (1.27) and (1.28) for the specific internal energy and enthalpy of constituent i with the mixture eqns (1.21) and (1.22) it is possible to evaluate the specific internal energy and enthalpy of the mixture, given the mol fractions $x_1, x_2, \ldots, x_i$. Since in actual cycle calculations we are dealing with a cylinder charge consisting of n mols of mixture where n itself is a function of the instantaneous chemical composition of the charge as determined from separate combustion equations (see section 1.4.2), we may finally write, for the actual internal energy U and enthalpy H of the mixture, using eqns (1.21), (1.22) with (1.27) and (1.28)

$$U = nu = n \sum_{i=1}^{i} x_i \left\{ u_{i,0} + R_0 \left[\sum_{j=1}^{j} a_{ij} T^j - T \right] \right\} \tag{1.29}$$

and

$$H = nh = n \sum_{i=1}^{i} x_i \left\{ u_{i,0} + R_0 \sum_{j=1}^{j} a_{ij} T^j \right\} \tag{1.30}$$

The terms $u_{i,0}$ and a_{ij} have been determined for all the common species associated with hydrocarbon combustion CO, CO_2, H_2, H_2O, O_2, N_2 and are available in tabulated form[1]. Appropriate computer subroutines have been written to make possible the direct evaluation of eqns (1.27) and (1.28) for any species i at any given temperature T.

1.4.2 Combustion

The details of the combustion process are extremely complex and can be dealt with adequately only in specialist texts. Here we shall confine ourselves mainly to the question of energy release during the combustion process (*Figure 1.14*).

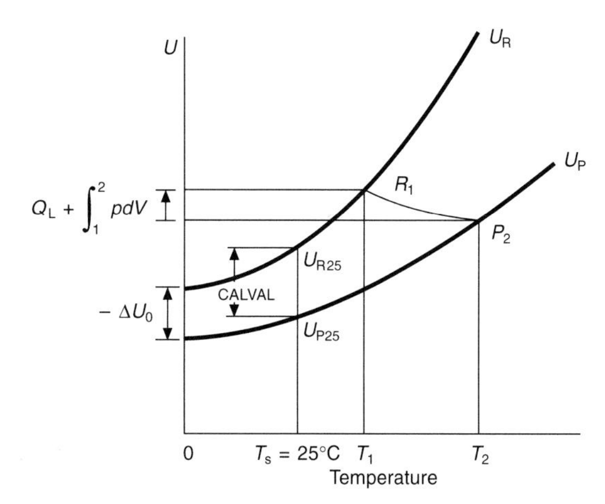

Figure 1.14 Combustion energy diagram

Consider the combustion process in an engine cylinder starting from reactants at state 1 and leading to products at state 2. During the process there is a heat loss to the combustion chamber walls given by Q_L while the work done by the gases on the piston is given by $\int_1^2 pdV$.

Simple application of the First Law of Thermodynamics to this non flow system leads to

$$U_{2p} - U_{1R} + Q_L + \int_1^2 pdV = 0 \qquad (1.31)$$

where U_{2p} and U_{1R} are the internal energies of products and reactants as obtained by the methods outlined in section 1.5.1, as functions both of composition and temperature. U_R and U_P are shown in *Figure 1.14* as functions of temperature T, with the initial state 1 and final state 2 of the combustion process described by eqn (1.31) clearly indicated. Note that the temperature rise $T_2 - T_1$ due to combustion is due not to an external heat input but due to the change in internal energy between reactants and products, the latter being primarily carbon dioxide (CO_2) and water vapour (H_2O) both of which have large negative internal energies of reaction. This internal energy of reaction for the products involved in the process 1–2, expressed as an effect at absolute zero temperature, is shown as $-\Delta U_0$, in *Figure 1.14*, i.e. the distance separating U_p and U_R at absolute zero.

Instead of the heat of reaction $-\Delta U_0$, it is more usual to employ the calorific value, CALVAL, derived from constant volume bomb experiments, with fixed initial and final temperature of products and reactants (usually 25°C).

Equation (1.31) may be adapted to this case by omitting the work term and equating the heat loss term to m_f. CALVAL where m_f is the mass of fuel burnt:

$$U_{25p} - U_{25R} + m_f.\,\text{CALVAL} = 0 \qquad (1.31a)$$

Subtracting (1.31a) from (1.31) we obtain

$$(U_2 - U_{25})_P - (U_1 - U_{25})_R + Q_L + \int_1^2 pdV - m_f.\text{CALVAL} = 0 \qquad (1.32)$$

By changing the datum of the internal energies from absolute zero to 25°C (an operation simply performed using eqn (1.29)), eqn (1.32) becomes, in differential form for use in small steps, and with $u_{i,25}$ denoting internal energy of species i relative to 25°C

$$\sum_{i=1}^{i} d(n_i u_{i,25}) + dQ_L + pdV - dm_f\text{CALVAL} = 0 \qquad (1.32a)$$

where

$$\sum_{i=1}^{i} d(n_i u_{i,25}) = \sum_{i=1}^{i} (n_i du_{i,25}) + \sum_{i=1}^{i} u_i dn_i$$

$$= \sum_{i=1}^{i} n_i \overline{C}_{v_i} dT + \sum_{i=1}^{i} \overline{C}_{v_i} T dn_i \qquad (1.32b)$$

in which the mean specific heat $\overline{C}_{v_i}$ derives from the basic formulation, eqn (1.27), i.e. $u_i = \overline{C}_{v_i} T$.

The calorific value CALVAL, for complete combustion, is independent of the air–fuel ratio, and is therefore assigned a constant value. The change in mol number $\sum_{i=1}^{i} dn_i$ is determined by the combustion equations and may be found as follows:

For unit mass of a hydrocarbon fuel C_nH_m and for complete combustion under stoichiometric conditions, the combustion equation may be written:

$$\underbrace{\left[\frac{(1-CF)H}{1.008} + \frac{CF \cdot C}{12.01}\right]}_{\text{fuel}} + \underbrace{\left[\frac{1-CF}{4.032} + \frac{CF}{12.01}\right]\left[O_2 + \frac{79}{21}N_2\right]}_{\text{air}}$$

$$= \underbrace{\frac{(1-CF)H_2O}{2.016}}_{\text{water vapour}} + \underbrace{\frac{CF}{12.01} \cdot CO_2}_{\text{carbon dioxide}} + \underbrace{\left[\frac{1-CF}{4.032} + \frac{CF}{12.01}\right]\frac{79}{21}N_2}_{\text{nitrogen}} \qquad (1.33a)$$

where CF = carbon fraction of fuel C_nH_m

$$= \frac{12.01\,n}{12.01n + 1.008m}$$

Putting $x = \dfrac{1 - CF}{4.032}$ and $y = \dfrac{CF}{12.01}$

eqn (1.33a) becomes

$$(4xH + yC) + (x+y)\left(O_2 + \frac{79}{21}N_2\right) = 2xH_2O + yCO_2$$

$$+ (x+y)\frac{79}{21}N_2 \qquad (1.33b)$$

If, as is more usual, combustion takes place with excess air, expressed by the excess air factor $\varepsilon > 1$, eqn (1.33b) becomes

$$(4xH + yC) + (x+y)\varepsilon\left(O_2 + \frac{79}{21}N_2\right)$$

$$= 2xH_2O + yCO_2 + (x+y)(\varepsilon - 1)O_2 + (x+y)\varepsilon\frac{79}{21}N_2 \qquad (1.33c)$$

For both eqns (1.33b) and (1.33c) the change in the number of gaseous mols between reactants and products is given by

$(n_p - n_R) = x$ per unit mass of fuel,

or, for increment dm_f of fuel burnt

$$\sum_{i=1}^{i} dn_i = x\, dm_f \qquad (1.33d)$$

The case of fuel rich combustion ($\varepsilon < 1$) leads to the formation of carbon monoxide (CO) and results in lower heat release, since the fuel calorific value CALVAL no longer applies. However, in most cases, combustion takes place with adequate excess air.

Given the rate at which fuel burns, $\dot{m}_f$ (see section 1.1) the incremental fuel mass $dm_f = \dot{m}_f\, dt$, may be evaluated step-by-step, and eqns (1.32a), (1.32b), (1.33b) or (1.33c) and (1.33d) solved to determine the changes in the state and composition of the cylinder charge, i.e. temperature change dT, mol number change

$$dn = x\, dm_f$$

and composition change expressed as a decrease in the number of mols of $O_2\,[dn_{O_2} = -1(x+y)\,dm_f]$, increase in number of mols $CO_2\,[dn_{CO_2} = y\,dm_f]$ increase in number of mols of $H_2O\,[dn_{H_2O} = 2x\,dm_f]$. The corresponding pressure change, dp, is finally obtained by application of the semi-perfect gas equation of state

$$pV = \sum_{i=1}^{i} n_i GT$$

or

$$p\,dV + Vdp = \sum_{i=1}^{i} dn_i \cdot R_0 T + \sum_{i=1}^{i} n_i R_0 \cdot dT \qquad (1.33e)$$

The determination of the state changes during combustion is thus vitally dependent on a knowledge of the burning rate $\dot{m}_f$. This question will be discussed briefly in section 1.6.1a.

1.4.3 Dissociation and reaction kinetics

The equations of section 1.4.2 assume that the product composition is uniquely determined by the chemical composition of the fuel and the air–fuel ratio. In practice two effects occur which render these assumptions inaccurate, particularly at very high temperatures.

Dissociation is the term used to describe the tendency of gases such as water vapour, H_2O, and carbon dioxide, CO_2 to undergo a partial decomposition process according to expressions such as

$$2H_2O \leftrightharpoons 2H_2 + O_2 \quad (1.34a)$$

$$2CO_2 \leftrightharpoons 2CO + O_2 \quad (1.34b)$$

The degree of dissociation is itself a function of the absolute temperature T and the excess air factor ε and tends to be most severe at or near stoichiometric combustion conditions ($\varepsilon = 1$). A dissociation constant is used to define the degree of decomposition when equilibrium has been reached.

Reaction kinetics is the term used to describe the rate at which chemical reactions proceed, both in a forward and reverse direction under non-equilibrium conditions.

Reaction kinetics is a highly complex topic which has been the subject of intensive research particularly in relation to noxious gases, and arising from the introduction of increasingly severe legislation in all advanced industrial countries[8].

The first of the noxious emissions to be investigated in detail was a mixture of nitirc oxide (NO) and nitrogen dioxide (NO_2) generally described as NO_x. Its formation involves a complex series of reactions, usually described as the Zeldovich mechanism in which nitrogen becomes an active rather than a passive participant in the combustion process as previously described by eqns (1.33a) to (1.33d).

The Zeldovich mechanism is expressed in the following form:

$$O + N_2 \leftrightharpoons NO + N \quad (1.35a)$$

$$N + O_2 \leftrightharpoons NO + O \quad (1.35b)$$

$$N + OH \leftrightharpoons NO + H \quad (1.35c)$$

where the symbol $\leftrightharpoons$ indicates that the process may proceed in either direction, as in the dissociation eqns (1.34a) and (1.34b). The process may also be expressed in the form of a rate equation

$$\frac{d(NO)}{dt} = k_1^+[O][N_2] + k_2^+[N][O_2] + k_3^+[N][OH] - k_1^-[NO][N] - k_2^-[NO][O] - k_3^-[NO][H] \quad (1.36)$$

where k_1^+, k_2^+ and k_3^+ are rate constants for the forward reactions expressed by eqns (1.35a), (1.35b) and (1.35c) respectively, while k_1^-, k_2^- and k_3^- are the rate constants for the corresponding reverse reactions.

In practice the conditions favouring the formation of oxides of nitrogen in diesel engines are a combination of high temperature and the presence of high concentrations of excess oxygen during the combustion process. The most effective countermeasures devised so far are retarded injection timing and exhaust gas recirculation (EGR) from the exhaust to the inlet manifold via an electronically controlled EGR valve. EGR drastically reduces the rate of formation of oxides of nitrogen as described by eqn (1.36).

Other important emissions are

(a) *hydrocarbons,* usually the result of either incomplete combustion or traces of lubricating oil carried over into the combustion chamber, or a combination of both;
(b) *particulates,* essentially unburnt carbon combined with sulphur compounds. These are considered to be highly carcinogenic and are, like oxides of nitrogen and hydrocarbons, the subject of detailed legislation.

For a full discussion of emission formation in diesel engines, the reader is referred to Heywood, *Internal Combustion Engine Fundamentals*, Chapter 11, (McGraw Hill)[8] which also contains a comprehensive list of references.

1.5 Real diesel engine cyclic processes

This section is confined to a largely qualitative description of the differences between idealized and actual diesel cycles. The final section (section 1.6) of this chapter will deal with an outline of current methods of cycle analysis.

1.5.1 Closed period

It is evident that the sequence of events constituting the idealized cycles described in section 1.3 on air standard cycles, does not represent the processes which actually occur in the diesel engine.

The compression process, instead of being reversible and adiabatic, with fixed index of compression $\gamma = 1.4$, will be irreversible due to frictional effects, non-adiabatic due to heat transfer from cylinder walls to gas (not pure air since the trapped charge will inevitably be slightly polluted by residual products of combustion) in the early stages of compression, followed by heat transfer in the reverse direction in the later stages of compression, while finally the specific heat ratio γ, used previously as the index of compression, will vary continuously during compression due to the changes in gas properties with temperature, as described in section 1.4.1 above. Similarly the expansion process will differ even more markedly from the stipulated adiabatic reversible process due to the intense heat transfer from gas to cylinder walls and the even more marked gas property changes associated with higher gas temperatures and the much more heterogeneous composition of the cylinder charge following combustion.

Finally, the combustion process itself is a far more complex sequence of events than the simple constant volume and/or constant pressure processes used in air standard cycles. The full sequence of events in the engine cylinder, consisting of

(i) generation of air motion during the open period; particularly in the form of organized swirl and/or controlled turbulence levels in high speed engines for transport purposes, and establishment of the air flow pattern near TDC;
(ii) injection of fuel with control of timing and duration and possible modulation of injection rate near TDC;
(iii) mixing of fuel and air under conditions determined by (i) and (ii) above;
(iv) initiation of combustion following the delay period, and subsequent combustion first under pre-mixed and later under diffusion conditions;
(v) separation of products and reactants under the combined action of density changes in a rapidly changing velocity field—buoyancy effects—with further fuel–air mixing and combustion occurring simultaneously;

is clearly extremely complex and not amenable to full analysis.

Conventionally the process is treated as consisting of at least three distinct periods (*Figure 1.15*) viz.

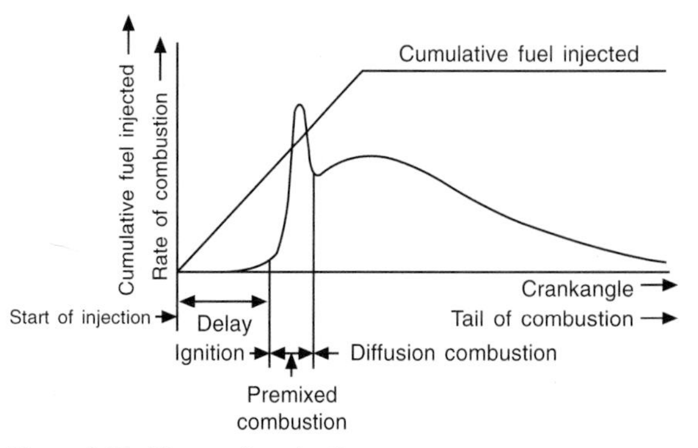

Figure 1.15 Phases of combustion process

(i) the delay period
(ii) the premixed burning phase (chemically controlled)
(iii) the diffusion burning phase (controlled by mixing rate).

Many workers have constructed combustion models of,[9,10,11] representing these processes with various degrees of refinement ranging from purely empirical to quasi-fully analytical representations. However, in so far as such models are intended for use in cycle simulation programs, the permissible degree of complexity is limited by considerations of computing time and cost. References 8 and 10 also provide examples of available programs.

When combustion rates are known during the premixed or diffusion phases, the combustion equations of section 1.4.2 may be applied on a step-by-step basis to solve for changes in state.

Figure 1.16a illustrates burning rates for turbocharged engines, as derived by careful analysis of experimental cylinder pressure—crank angle records by a procedure which is essentially the inverse of that described in section 1.4.2, throughout combustion under high load and low load conditions. The delay period is considerably longer for low load conditions, resulting in a higher proportion of the injected fuel remaining unburnt. When premixed burning commences the chemically controlled rate of combustion is extremely high in all cases, but in the low load case the large accumulation of unburnt fuel during the delay period leads to a characteristic sharp peak, as opposed to the high load case where the premised phase merges almost immediately—due to the relatively much smaller accumulation of fuel and the rapid exhaustion of this limited supply—with the mixing controlled diffusion phase. Clearly for the low load case rates of burning during the prolonged latter phase are much lower than for the high load case.

Figure 1.16b shows burning rates based on a simple mathematical model due to Austen and Lyn[12], based on successive elements of fuel (as shown in the top part of the figure) being subject to a progressively shortening delay period during which unburnt fuel accumulates up to the start of combustion. This unburnt fuel, indicated by the shaded areas of fuel elements 1, 2, 3 in the bottom part of the figure, then burns extremely rapidly under the chemically controlled conditions of the premixed phase, resulting in the initial sharp heat release peak. Thereafter combustion of the still unburnt portions of the early fuel elements 1, 2, 3 and the entire contents of the later elements occurs at the much slower rates characteristic of the mixing controlled diffusion phase, resulting in the long heat release tail of *Figure 1.16b.*

Figure 1.16c shows the same succession of processes, but now represented by simple mathematical expressions for the premixed and diffusion phases due to Whitehouse and Way[9], derived from a single zone combustion model with instantaneous mixing of products and reactants at every stage of the process.

For the purposes of experimental heat release analysis, which has a vital bearing on the interpretation of the combustion process, the following procedure based on records of cylinder pressure versus crank angle, is used. The state changes occurring in the cylinder during the combustion process, again using a single zone model, may be described by the differential forms of the energy equation (First Law of Thermodynamics) and of the semi-perfect gas equation with due allowance for the effect of changes in gas properties due to variations in composition and temperature as described in sections 1.4.1 and 1.4.2.

The energy equation is expressed in terms of the apparent or nett heat release rate dQ_n/dt, i.e. the difference between the true

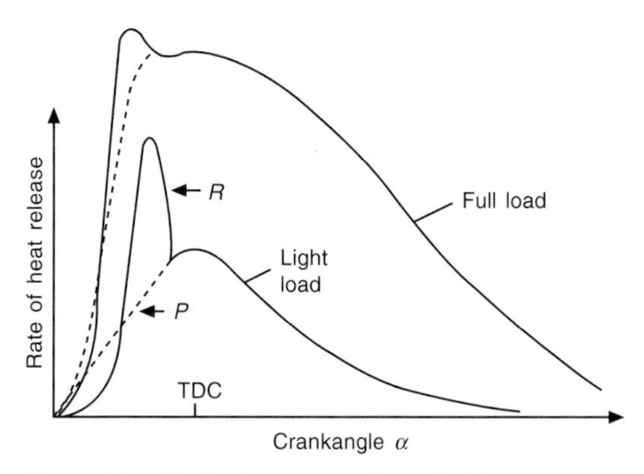

Figure 1.16 (a) Burning rates—light and full load

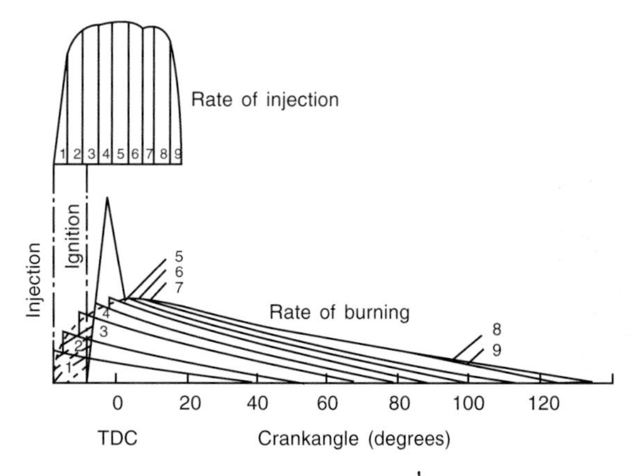

Figure 1.16 (b) Diagrammatic heat release rates (*Lyn*)

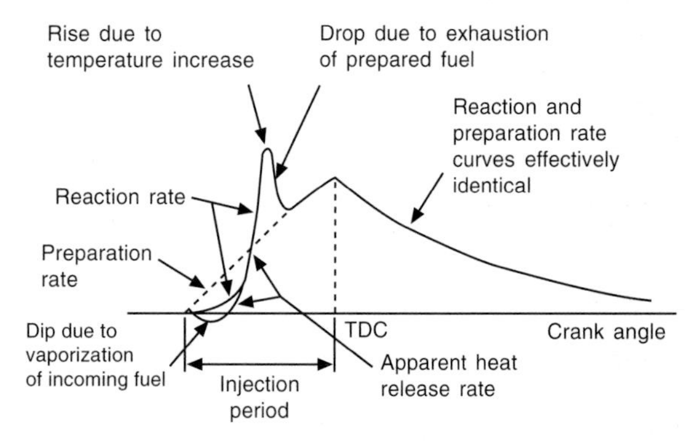

Figure 1.16 (c) Heat release rates (*Whitehouse and Way*)

heat release rate dQ_{hr}/dt and the heat transfer rate dQ_{ht}/dt based on the bulk gas temperature T and the wall temperature T_w (see section 1.6). It assumes the following form for a small time or crank angle increment:

$$dQ_n = dQ_{hr} - dQ_{ht} = p\ dV + dU = p\ dV + mC_v dT \qquad (1.37)$$

where U = internal energy, C_v = specific heat at constant volume and the other symbols have their usual meaning.

The equation of state, expressed in differential form, and neglecting the small changes in mass due to combustion of successive fuel elements, dm_f, becomes:

$$\frac{dp}{p} + \frac{dV}{V} = \frac{dT}{T} \qquad (1.38)$$

Equations (1.37) and (1.38) may be combined to form the following expression for the nett heat release dQ_n:

$$dQ_n = \left[\frac{\gamma}{\gamma - 1} p\ dV + \frac{1}{\gamma - 1} V\ dp\right] \qquad (1.39)$$

The specific heat ratio, γ, should be adjusted, step-by-step, in line with gas property changes.

1.5.2 Open period

The open period during which the products of combustion formed during the closed period, are largely expelled and replaced by fresh air, is not represented at all in the various air standard cycles of section 1.3.3. Instead a single constant volume heat rejection process returns the cylinder charge from its end of expansion state to that at the beginning of compression. It is clear that the actual events during the open period, as briefly described in section 1.2.3, will lead to very significant departures from the idealized air standard cycles of section 1.3. Furthermore, they will differ greatly as between four-stroke and two-stroke engines.

Figures 1.17a and b represent typical gas exchange processes on a pressure-crank angle ($p - \theta$) basis for four- and two-stroke engines respectively.

1.5.2.1 Four-stroke engines

The four-stroke engine of *Figure 1.17a* is assumed to be of the naturally aspirated type, with the inlet and exhaust manifold pressures approximately constant and equal to atmospheric pressure.

1–2: *Blowdown* (EVO to BDC_1)
During the interval between EVO and (BDC), products of combustion escape rapidly from the cylinder due to the large available pressure difference between cylinder and exhaust manifold and the rapid opening of the exhaust valve.

2–3: *Exhaust* (BDC to IVO)
The cylinder pressure at (BDC), is likely to be still in excess of atmospheric. During the subsequent return stroke of the piston, with only the exhaust valve open, the products of combustion continue to be expelled from the cylinder by the action of the piston initially accelerating rapidly towards $(TDC)_2$. This leads to a substantial and initially increasing positive pressure difference between the cylinder gases and the exhaust manifold, which generally diminishes again in the region of mid stroke.

3–4: *Overlap* (IVO to EVC)
During this period, on either side of $(TDC)_2$ the inlet valve rapidly opens, while the exhaust valve is rapidly closing. The function of the overlap period in turbocharged engines, is to ensure that the clearance volume, i.e. the small cylinder volume

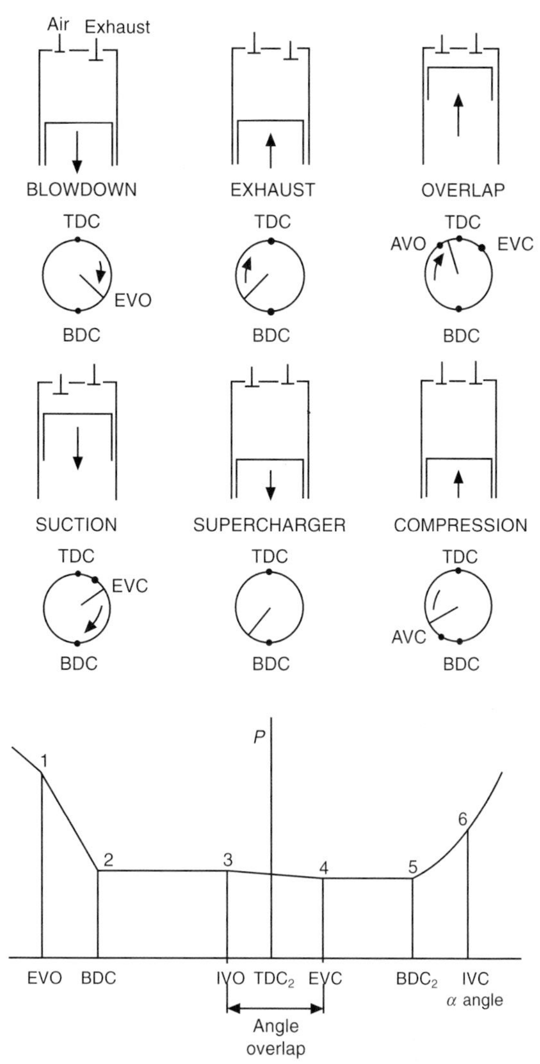

Figure 1.17 (a) Gas exchange four-stroke (*Benson and Whitehouse*)

when the piston is at or near the TDC position, is cleared, as far as possible, of the remaining products of combustion, by displacement of these 'residuals' by the incoming air. This period varies from very small values (< 40° CA) for road transport engines to substantial values (≑ 120° CA) in medium speed highly turbocharged engines.

4–5: *Induction* (EVC to BDC_2)
In naturally aspirated engines most of this period during which the fresh air charge is drawn into the cylinder, takes place under conditions of subatmospheric pressure, due to the pressure drop across the inlet valve. This pressure drop tends to reach a maximum value near mid stroke, with pressure returning to near atmospheric towards $(BDC)_2$.

5–6: *'Pre-Compression'* (BDC_2 to IVC)
During this final period which is usually quite short, a small degree of pre-compression of the air charge takes place with some resultant increase in cylinder pressure before the inlet valve finally closes.

The sequence of events described above constitutes the 'pumping loop', as is obvious with reference to the pressure volume diagram, *Figure 1.6b*, which was briefly discussed in section 1.2.3 in connection with the evaluation of indicated work. In naturally aspirated engines, due to the excess of cylinder pressure during the exhaust stroke over that during the induction stroke, this pumping loop is invariably associated with negative work, while under favourable conditions in turbocharged engines it may make a positive contribution to cycle work.

1.5.2.2 Two-stroke engines (Figure 1.17b)

The two-stroke open period process has been briefly discussed already in section 1.2.3. Only two-stroke engines with asymmetrical timing will be considered; to avoid confusion an exhaust valve-in-head, inlet ported engine (section 1.2.1.2) is chosen. Unlike the four-stroke engine which depends largely on piston displacement action to effect the gas exchange process, the two-stroke engine receives no such assistance from the motion of the piston and instead has to rely on a positive pressure difference between inlet and exhaust manifold to force fresh air into the cylinder through the inlet valve or ports while simultaneously products of combustion escape through the exhaust valve or ports. The time available for the gas exchange process is greatly reduced, compared with four-stroke engines, varying between approximately 110° CA for large low speed engines to nearly 200° for very small, high speed engines.

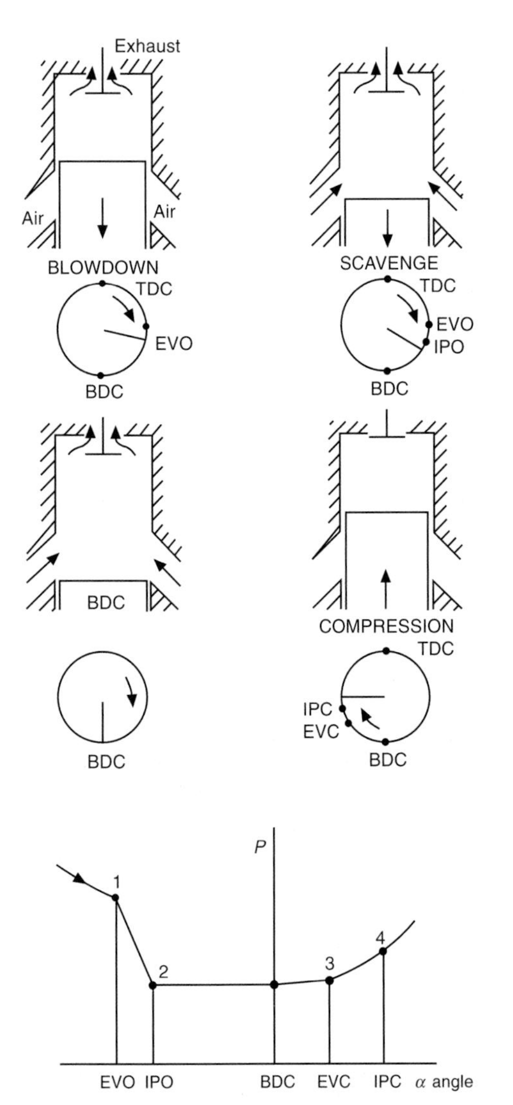

Figure 1.17 (b) Gas exchange two-stroke (*Benson and Whitehouse*)

The sequence of events is as follows:

1–2: *Blowdown* (EVO to IPO)
During this period, generally of approximately 30° CA duration, a large part of the products of combustion is discharged rapidly under the action of the large pressure difference between cylinder and exhaust manifold and the rapidly opening exhaust valve.

2–3: *Scavenge process* (IPO to EVC)
This corresponds to the overlop period in the four-stroke engine, the remaining products of combustion in the large cylinder volume as a whole—as opposed to the clearance volume—leaving through the exhaust valve having to be replaced as completely as possible by the air entering through the inlet ports. The degree of 'scavenging' achieved is very critically dependent on the characteristics of the inlet and exhaust arrangements, and on the internal flow characteristics within the cylinder, as well as the existence of a positive pressure gradient across the engine. A useful indicator of the degree of scavenging achieved is provided by the two limiting theoretical criteria of '*perfect mixing*' and '*perfect scavenge*' (*Figure 1.18*).

In both processes it is stipulated that charge pressure and volume remain constant during the gas exchange process. In the case of perfect mixing it is assumed that successive increments of incoming air mix instantaneously with the mass of gas in the cylinder, m_{cyl}. As a result the composition of the latter changes continuously during gas exchange, but complete charge purity can only be approached asymptotically. It may be shown that the expression for charging efficiency η_{ch} expressed as the mol fraction of pure air in the cylinder charge, in terms of the scavenge factor λ expressed as the ratio of volume of air supplied to trapped volume becomes

$$\eta_{ch} = 1 - e^{-\lambda} \tag{1.40a}$$

while the corresponding expression for perfect scavenge in which the incoming air charge physically displaces the products of combustion becomes

$$\eta_{ch} = \lambda \quad \text{for } \lambda < 1 \tag{1.40b}$$

and $$\eta_{ch} = 1 \quad \text{for } \lambda > 1 \tag{1.40c}$$

Figure 1.18 clearly indicates these limiting cases with perfect scavenge representing the most favourable case, allowing a pure charge with a scavenge ratio $\lambda = 1$. Practical charging efficiencies are intermediate between perfect mixing and perfect scavenge, tending towards the latter for uniflow scavenged engines, and the former for simple loop scavenged engines. The literature on scavenging is substantial[13–19].

3–4: *Recompression* (EVC to ICP)
As in the case of the four-stroke engine this period is generally quite short, and merely provides a small degree of precompression of the air charge in the cylinder before inlet port closure.

The question of a pumping loop, as in the case of four-stroke engines, does not generally arise with two-stroke engines. *Figure 1.6a*, section 1.2.3, illustrated diagrammatically the gas exchange process in the p-V diagram. If cylinder pressure during part of the scavenge process before BDC is at a lower level than during part of that process after BDC, a small amount of negative loop work may result. However, because of the much shorter duration of the scavenge process compared with the gas exchange process in the four-stroke engine, and the generally smaller cylinder pressure fluctuations during the gas exchange process, the pumping loop work, negative or positive, remains very small.

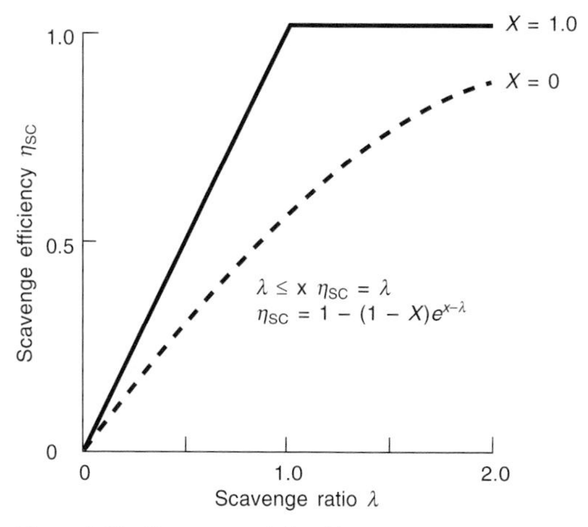

Figure 1.18 Scavenge relationships

1.6 Detailed cycle analysis methods

In view of the complexity of full cycle calculations, as embodied in the majority of current computer programs, only a very condensed account can be given here.

In general, the technique is to solve the governing equations of state and of conservation of energy and mass on a step-by-step basis, using small crank angle increments $d\theta$ (usually $0.5° < d\theta < 2°$ CA) depending on the particular part of the cycle to be evaluated, with particularly short steps for those processes where very rapid changes of state occur, e.g. combustion and gas exchange during the overlap period.

The equations of state for mixtures of gases have already been discussed (section 1.4.1) as have also the combustion equations and their application to the determination of changes of state of the cylinder gases (section 1.4.2).

The calculation sequence invariably divides itself into two major portions, for the closed and open period, starting with assumed trapped conditions of the cylinder charge [(trapped pressure p_t, trapped temperature T_t, charge purity $(\eta_p)_t$)] at the beginning of compression. Stable values of these trapped conditions are obtained only after several successive cycles have been evaluated.

A common feature of these step-by-step calculations not previously referred to is the inclusion of a heat transfer term dQ_L to allow for heat transfer from cylinder gas to wall or vice versa (see eqn 1.32a). The calculations depend on a knowledge of

(i) local heat transfer coefficients;
(ii) gas temperature T_g (usually assumed uniform throughout cylinder);
(iii) cylinder wall temperature T_w (usually assumed constant over the surfaces of cylinder head, piston, but varying spatially with distance from top of cylinder liner along the latter.

The local heat transfer coefficient is determined wih the aid of one of the standard correlations, the best known being those of Eichelberg, Annand and Woschni[20,21,22]. The Annand correlation, probably the most widely used, is of the form

$$\dot{q} = a\frac{k}{D}(R_e)^b(T_g - T_w) + C(T_g^4 - T_w^4) \qquad (1.41a)$$

where a, b and c are constants to be determined for particular engines, R_e is a quasi-Reynolds number related to piston speed, cylinder bore and mean gas density and viscosity. Although wall temperatures T_w fluctuate cyclically, they are generally assumed to be constant. Strictly speaking the wall temperatures themselves should be the subject of a further iteration procedure, as for trapped temperature, in which heat flow from gas to cylinder walls, through the cylinder walls and finally from the latter to coolant and radiation are balanced over several cycles.

The total incremental heat loss dQ_L for any crank angle step $d\theta$ is obtained by writing

$$dQ_L = \sum_{s=1}^{s}(\dot{q}_s A_s)\frac{d\theta}{6N_E} \qquad (1.41b)$$

in which s = total number of separate surfaces to be considered; A_s = instantaneous value of exposed surface area A_s (fixed for cylinder head and piston, but varying for the liner); $d\theta/6N_E$ = time increment for crank angle increment $d\theta(°)$ with engine speed N_E in rev/min, and where the appropriate wall temperature T_{w_s} has to be inserted in the rate eqn (1.41a).

The calculation sequence for the closed and open period will now be briefly described.

1.6.1 Closed period

1.6.1.1 Generalized calculation procedure

The governing equations are the non-flow energy eqn (1.32a) together with the semi-perfect gas state eqn (1.33e), i.e.

Non-flow energy equation:

$$\sum_{i=1}^{i} n_i \bar{C}_{v_i}\, dT + \sum_{i=1}^{i} \bar{C}_{v_i}\, T\, dn_i + dQ_L + p dV - dm_f\, \text{CALVAL} = 0 \qquad (1.42)$$

in which the first two terms represent the change in internal energy of the cylinder charge, dQ_L the incremental heat loss from the cylinder, $p\, dV$ the piston work, and dm_f. CALVAL the fuel heat release.

Equation of state in differential form:

$$p\,dV + Vdp = \sum_{i=1}^{i} R_0(dn_i) + \sum_{i=1}^{i} n_i R_0 \cdot dT \qquad (1.43)$$

Equations (1.42) and (1.43) may be solved, using appropriate numerical procedures, for the temperature and pressure increments dT and dp, for any crank angle step $d\theta$, i.e.

$$\text{time step } dt = \frac{d\theta}{6N_E}$$

1.6.1.2 Application of generalized equations to separate phases of closed period

(a) *Compression*

Starting from assumed trapped conditions at inlet closure, p_t, T_t, together with an assumed degree of charge purity determining the composition of the trapped charge, eqns (1.42) and (1.43) are solved step-by-step, in suitably simplified form. In eqn (1.42) the heat release term dm_f. CALVAL is omitted while in both eqns (1.42) and (1.43) the terms representing change of composition, i.e.

$$\sum_{i=1}^{i} \bar{C}_{v_i}\, T\, dn_i \quad \text{and} \quad \sum_{i=1}^{i} R_0\, T\, dn_i$$

may be omitted.

(b) *Combustion*

For the purposes of analysis as used in cycle simulation, eqns (1.42) and (1.43) have to be used in their full form. Before proceeding to solve them for the pressure and temperature increments dp and dT, it is necessary to specify the fuel burning rate (FBR) from which can be derived the burnt fuel increment dm_{fb} for any calculation step involving a crank angle increment $d\theta$ during the combustion period. It is convenient to express the fraction of fuel burnt, FB, up to any crank angle θ from start of combustion (SOC) in terms of a dimensionless crank angle defined as

$$\tau = \frac{\text{crank angle } \theta \text{ from start of combustion}}{\text{total crank angle occupied by combustion } \theta_{comb}}$$

Marzouk and Watson[23] have formulated the following widely used shape function for the fraction of fuel burnt, FB, in which FB is represented by two sub-functions $(FB)_{pm}$ and $(FB)_{diff}$ for the premixed and diffusion phases of the combustion process, respectively:

$$(FB)_{pm} = 1 - (1 - \tau^{K_1})^{K_2} \tag{1.44}$$

and

$$(FB)_{diff} = 1 - \exp(-K_3 \tau^{K_4}) \tag{1.45}$$

where K_1, K_2, K_3 and K_4 are empirically derived constants having the following suggested form for turbocharged truck diesel engines:

$K_1 = 2.0 + 1.25 \times 10^{-8} (ID \times N)^{2.4}$

where ID = ignition delay in ms

$K_2 = 5000$

$K_3 = 14.2/F^{0.644}$

where F = equivalence ratio ($F > 1$ for fuel-rich combination.)

$K_4 = 0.79 \ K_3^{0.25}$

Equations (1.44) and (1.45) may be combined for the process as a whole by a phase proportionality factor β such that

$$FB = \beta(FB)_{pm} + (1 - \beta)(FB)_{diff} \tag{1.46}$$

where β is a function of the ignition delay and equivalence ratio F of the form

$$\beta = 1 - aF^b/ID^c \tag{1.47}$$

in which

$0.8 < a < 0.95$

$0.25 < b < 0.45$

$0.25 < c < 0.50$

Figure 1.19 shows typical fuel burning rate curves for a typical turbocharged truck diesel engine operating over a range of speeds and loads[23].

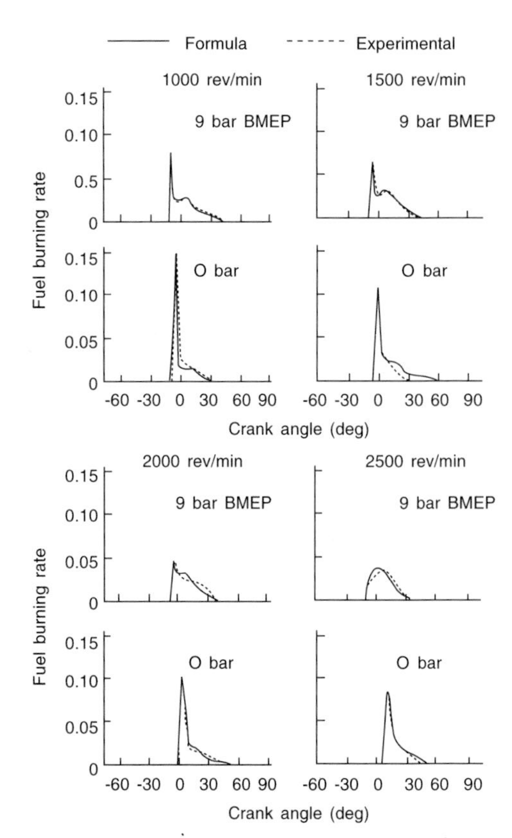

Figure 1.19 Fuel burning rates in a DI truck diesel engine under variable load/speed conditions (Watson and Marzouk)

1.6.2 Open period (gas exchange process)

This has already been discussed qualitatively in section 1.5.2. The calculation technique again involves the solution of appropriate differential equations on a step-by-step basis. The main difference between the calculations for the closed and open period is that the non-flow energy eqn (1.42) is now replaced by the unsteady flow energy equation which allows for inflows and outflows of gas to and from the cylinder.

The gas exchange process, like the combustion process, is associated with major changes in the composition of the cylinder charge, but these are now connected with the incoming mass increment dm_{in} (usually pure air) and outgoing mass increment dm_{ex} (usually products of combustion) each of known composition, for the case of simultaneous inflow and outflow. The magnitude of these mass flow increments dm_{in} and dm_{ex} is a function of the instantaneous flow rates $\dot{m}_{in}$ and $\dot{m}_{ex}$ which in turn are functions of the geometric characteristics of the inlet and exhaust valves (or ports, when appropriate), of the state of the gas entering or leaving and the available pressure difference across the valves (or ports). In most high speed engines and many medium and slow speed engines, the inlet and exhaust manifold pressure, p_{im} and p_{em} are subject to substantial cyclic pressure fluctuations which leads to further complications in that the change of state of the gases in the inlet and exhaust manifolds has to be calculated by further step-by-step procedures. For this purpose two techniques are available:

(a) the so-called filling and emptying technique which treats the manifolds as having a uniform state at any one instant[24].
(b) wave action calculations, usually by the method of characteristics[25] in which the manifold state is evaluated on the basis of complex pressure waves propagating from the inlet or exhaust valves and subject to repeated partial reflection and transmission at branches, junctions, etc.

For the purposes of this chapter the inlet and exhaust manifold pressure p_{im} and p_{em} will be assumed to be known.

1.6.2.1 Generalized calculation procedure (Figure 1.20)

The equations governing the most general case of simultaneous inflow, dm_{in}, and outflow, dm_{ex}, are as follows:

(a) Unsteady flow energy equation

(Replacing eqn (1.42) for the closed period and assuming perfect

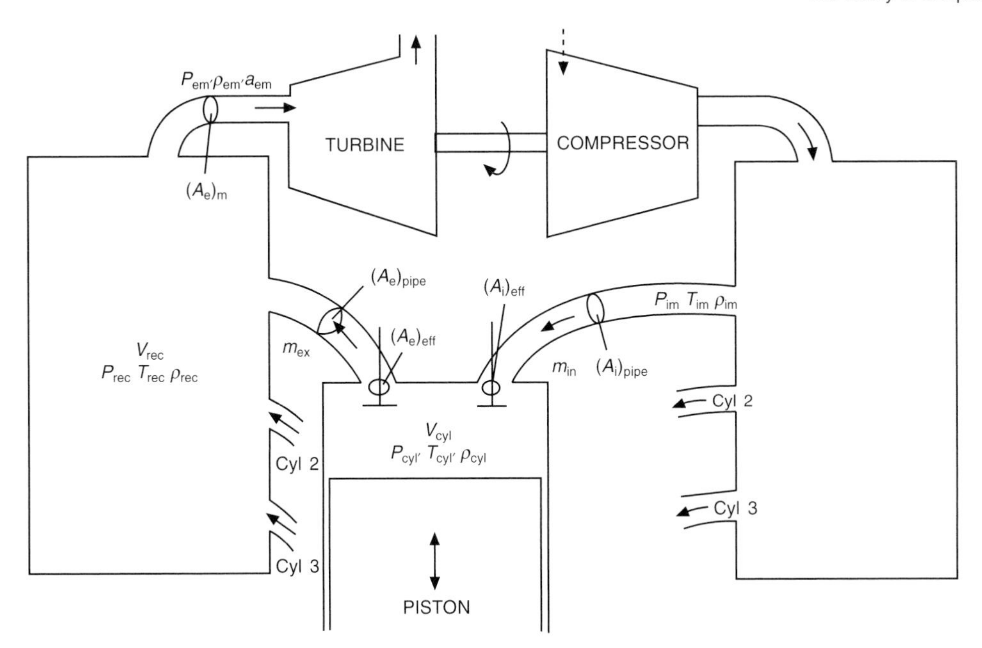

Figure 1.20 Simplified system for filling and emptying

mixing of the incoming mass increment dm_{in} and the cylinder charge m_{cyl})

$$\underbrace{d(m_{cyl}\overline{C}_{v_{cyl}}T_{cyl})}_{\text{change of internal energy of cylinder charge}} + \underbrace{dQ_L}_{\text{heat loss}} + \underbrace{p\,dV_{cyl}}_{\text{pison work}}$$

$$-\underbrace{dm_{in}\overline{C}_{pim}T_{im}}_{\text{incoming enthalpy}} + \underbrace{dm_{ex}\overline{C}_{p_{cyl}}\overline{T}_{cyl}}_{\text{leaving enthalpy}} = 0 \qquad (1.48)$$

where $\overline{C}_{v_{cyl}}$ is the mean specific heat at constant volume of the cylinder gases (eqn (1.32b)) expressed on a unit *mass* and not a *molar* basis. The conversion is easily effected via eqn (1.20), the mass basis being more appropriate since inflows and outflows are invariably calculated on a mass basis.

$\overline{C}_{pim}$ is the mean specific heat at constant pressure of the incoming gases (usually air), again on a mass basis.

$\overline{C}_{p_{cyl}}$ is the mean specific heat at constant pressure of the cylinder ($= R_{cyl} + \overline{C}_{v_{cyl}}$) characterizing the enthalpy of the leaving gas.

(b) Equation of state for semi-perfect gases

This is again expressed on a mass rather than a molar basis (see eqn (1.33e)) in the form

$$pdV + Vdp = d(m_{cyl}R_{cyl}T_{cyl})$$
$$= R_{cyl}T_{cyl}dm_{cyl} + m_{cyl}R_{cyl}dT_{cyl} + m_{cyl}T_{cyl}\,dR_{cyl} \qquad (1.49)$$

The gas constant R_{cyl} is expressed on a unit mass basis and replaces the universal molar gas constant G, where

$$R_{cyl} = \frac{G\sum_{i=1}^{i} n_i}{\sum_{i=1}^{i} n_i M_i}$$

which enables dR_{cyl} to be evaluated in terms of changes in cylinder gas composition, expressed by $\sum_{i=1}^{i} dn_i$.

(c) Equation of mass continuity

$$dm_{cyl} = dm_{in} - dm_{ex}$$

Equation (1.50) in turn enables the change in composition of the cylinder charge to be calculated since, for the incoming air

$$(dn_{O_2})_{in} = \frac{21}{100}dn_{in} = \frac{dm_{in}}{28.97}\frac{21}{100} \qquad (i)$$

where 28.97 = molecular mass of air; 21/100 = mol fraction of oxygen in air

$$(dn_{N_2})_{in} = \frac{79}{100}dn_{in} = \frac{dm_{in}}{28.97}\frac{79}{100} \qquad (ii)$$

The corresponding mol number changes associated with the outgoing charge are

$$dn_{ex} = \frac{dm_{ex}}{\sum_{i=1}^{i} x_i M_1} \qquad (iii)$$

where $x_1, x_2, \ldots x_i$ are the instantaneous mol fractions associated with the different constituents of the cylinder gases, having molecular masses $M_1, M_2, \ldots, M_i$.

From continuity of mol numbers we may then write, for each species

$$(dn_i)_{cyl} = (dn_i)_{in} - (dn_i)_{ex} \qquad (1.51a)$$

or, for the total mol number change

$$\sum_{i=1}^{i}(dn_i)_{cyl} = dn_{cyl} = dn_{in} - dn_{ex} \qquad (1.51b)$$

The system of eqns (1.48), (1.49), (1.50) and (1.51) may then be solved, step by step, for the changes in cylinder temperature dT_{cyl}, pressure dp_{cyl}, mass dm_{cyl}, as well as overall mol mumber change du_{cyl}.

It should be remembered that the above formulation applies to the case of perfect mixing (section 1.5.2, *Figure 1.18*) as opposed to the alternative limiting case of perfect scavenge. Appropriate alternative formulations for this or other scavenging models may be found in the literature.

1.6.2.2 Application of gas exchange equations to different flow regimes during open period

In principle the gas exchange process may now be evaluated by proceeding step-by-step through the various phases of the open period, as described in section 1.5.2.

These phases are characterized by the following flow regimes:

(a) *outflow only*
four-stroke engines—blowdown and exhaust *(Figure 1.17a)*
two-stroke engines—blowdown only *(Figure 1.17b)*

(b) *simultaneous inflow and outflow*
four-stroke engines—overlap period *(Figure 1.17a)*
two-stroke engines—scavenge period *(Figure 1.17b)*

(c) *inflow only*
four-stroke engines—induction and supercharge period *(Figure 1.17a)*
two-stroke engines—supercharge period *(Figure 1.17b)*

(During the supercharge period both types of engine may suffer outflow through the inlet valve or port, giving reversal of pressure gradient).

Cases, (a), (b) and (c) will now be briefly described:

(a) *Outflow only*: The inflow terms in eqns (1.48), (1.50) and (1.51) of the general formulation are now omitted. In every other respect the calculations proceed as described in section 1.6.2.1.

(b) *Simultaneous inflow and outflow.* Here the equations of section 1.6.2.1 apply in full.

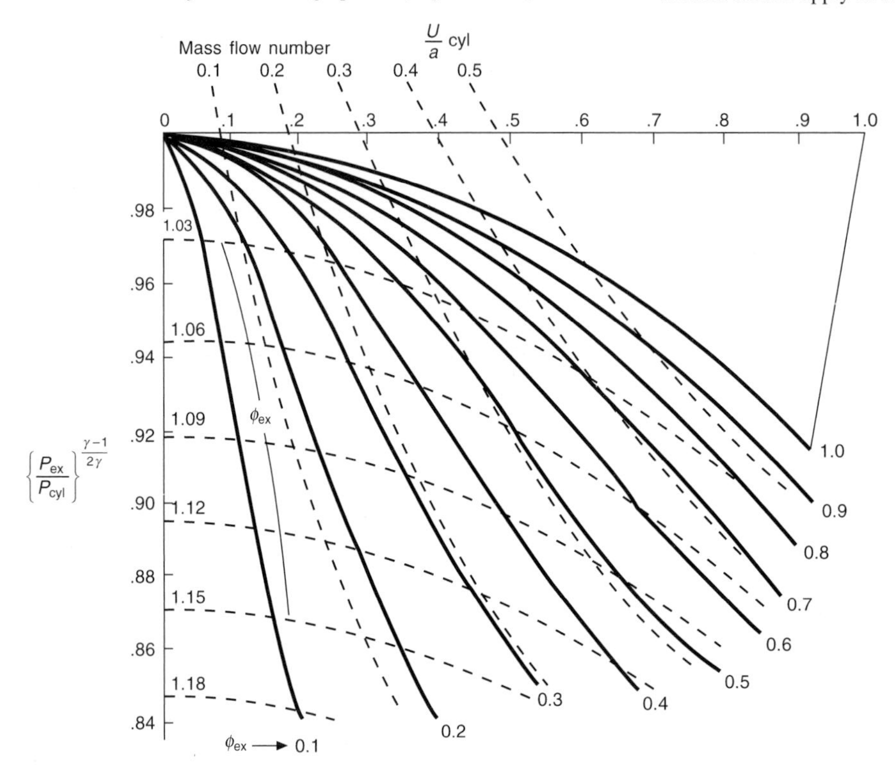

Figure 1.21 (a) Outflow chart

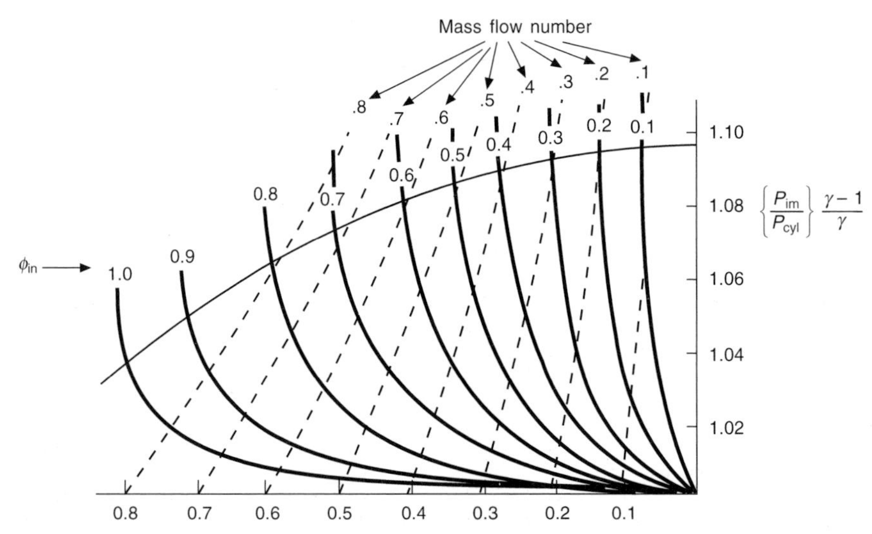

Figure 1.21 (b) Inflow chart

(c) *Inflow only*: The outflow terms in eqns (1.48), (1.50) and (1.51) are now omitted. Otherwise the procedure of section 1.6.2.1 is followed.

1.6.2.3 Determination of inflow and outflow rates $\dot{m}_{in}$ and $\dot{m}_{ex}$

The mass increments dm_{in} and dm_{ex} in the governing equations for the open period depend on a knowledge of the instantaneous flow rates $\dot{m}_{in}$ and $\dot{m}_{ex}$ such that, for a crank angle increment $d\theta$

$$dm_{in} = \dot{m}_{in} \frac{d\theta}{6N_E} \quad (1.52a)$$

$$dm_{ex} = \dot{m}_{ex} \frac{d\theta}{6N_E} \quad (1.52b)$$

These flow rates in turn are functions of the instantaneous effective flow areas of the appropriate valves or ports, usually expressed in dimensionless form as

$$\phi_{in} = \frac{(A_{in})_{eff}}{A_{cyl}} \text{ and } \phi_{ex} = \frac{(A_{ex})_{eff}}{A_{cyl}}$$

and the state of the gas upstream of the valve or port, i.e. T_{im}, p_{in} for inflow and T_{cyl}, p_{cyl} for outflow. These relations are described in detail in the literature[25] and may be expressed in generalized form as

$$\frac{\dot{m}_{in}\sqrt{T_{im}}}{p_{im}} = f\left[\phi_{in}, \frac{p_{im}}{p_{cyl}}\right] \quad (1.53a)$$

and

$$\frac{\dot{m}_{ex}\sqrt{T_{cyl}}}{p_{cyl}} = f\left[\phi_{ex}, \frac{p_{cyl}}{p_{ex}}\right] \quad (1.53b)$$

A graphical representation of these inflow and outflow conditions is given in *Figures 1.21a and b*. In cycle analysis programs they are represented by the numerical equivalent of these so-called boundary charts.

1.6.3 Completion of calculation sequence

Completion of the open period calculations for a cycle leads to the trapped conditions, p_t, T_t and charge composition which form the initial conditions for the next cycle (see section 1.6.1.2a). Since these conditions were an estimate for the first cycle, the computation has to be continued for several cycles (usually three) until the trapped conditions, and hence the cycle conditions as a whole, have stabilized.

Cycle simulation is now universally adopted throughout industry as an invaluable development tool and is subject to continuous refinement in modelling techniques. Nevertheless, the vast majority of programs in use still rely on simple, single zone models for in-cylinder processes, including combustion and gas exchange, although scavenging models for two-stroke engines tend to be of the two or three zone type[17].

However, the need for more detailed spatial modelling with respect to certain in-cylinder flow processes has become more and more evident over the years. Computational fluid dynamics (CFD) which solves the three-dimensional fluid flow equations for continuity, momentum and energy (the so-called Navier Stokes equations) on the basis of finite difference formulations applied to a large assembly of cells making up the relevant control volume, is increasingly employed to elucidate the processes of valve and port flows, in-cylinder air motion and even the complete sequence of events during the combustion process of fuel injection, droplet break-up and evaporation, initiation of combustion and propagation of the flame front within the fuel spray. However, such programs make very large demands on computer memory and computing time, so that up to now it has not proved possible to incorporate them in conventional cycle simulation programs.

References

1 BENSON, R. S. and WHITEHOUSE, N. D., *Internal Combustion Engines*, Vols. 1 and 2, Pergamon Press, Oxford (1979)

2 RICARDO, H. R. and HEMPSON, J. G. G., *The High Speed Internal Combustion Engine* (5th edn), Blackie, Oxford (1968)

3 'Symposium on Diesel Engine Combustion', *Proc. I. Mech. E.*, London (1970)

4 KAMO, R. 'Cycle and Performance Studies for Advanced Diesel Engines', *ERDA Conference on Ceramics for High Performance Applications*. Newport, Rhode Island (1977)

5 ZAPPA, G. and FRANCA, T., *A Four-stroke High Speed Diesel Engine with Two Stage Supercharging and Variable Compression Ratio*, CIMAC, Vienna 1979, Paper D19 (1979)

6 WALLACE, F. J. and LINNING, W.A., *Basic Engineering Thermodynamics*. Pitman Publishing Co. Bath (1970)

7 ROGERS, G. F. R. and MAYHEW, Y. M., *Engineering Thermodynamics, Work and Heat Transfer*, Longmans (1967)

8 HEYWOOD, J. B., *Internal Combustion Engine Fundamentals*, McGraw-Hill (1988)

9 WHITEHOUSE, N. D. and WAY, R. J. B., 'A Simple Method for the Calculation of Heat Release Rates in Diesel Engines based on the Fuel Injection Rate', *SAE paper 71034*, Detroit (Feb. 1971)

10 SHAHED, S. M., CHIU, W. S. and LYN, W. T., 'A Mathematical Model of Diesel Combustion', *Proc. I. Mech. E. Conference on Combustion in Engines*, Cranfield (1976)

11 MEGUERDICHIAN, M. and WATSON, N., 'Prediction of Mixture Formation and Heat Release in Diesel Engines', *SAE paper 780225*, Detroit (Feb. 1978)

12 AUSTEN, A. E. W. and LYN, W. T., 'Relation between Fuel Injection and Heat Release in a Direct-Injection Engine and the Nature of the Combustion Processes', *Proc. I. Mech. E.*, 1960–61, **175** (1961)

13 BENSON, R. S. and GALLOWAY, K., 'An Experimental and Analytical Investigation of the Gas-Exchange Process in a Multi-cylinder Pressure-charged 2-stroke Engine', *Proc. I. Mech. E.*, **183**, Pt. 1 (1968–69)

14 RYTI, M., 'Computing the Gas Exchange Process of Pressure Charged Internal Combustion Engines', *Proc, I. Mech. E.*, **182**, Pt. 3L (1968)

15 WALLACE, F. J. and CAVE, P. R., 'Experimental and Analytical Scavenging Studies on a Two-Cycle opposed Piston Engine', *SAE 710175*, Automotive Engineering Congress, Detroit (Jan. 1971)

16 RIZK, W., 'Experimental Studies of the Mixing Processes and Flow Configurations in Two-Cycle Engine Scavenging', *Proc. I. Mech. E.*, **172**, No. 10 (1958)

17 BENSON, R. S. and BRANDHAM, P. T., 'A Method for Obtaining a Quantitative Assessment of the Influence of Charging Efficiency on Two-stroke Engine Performance', *Int. J. Mech. Sci.*, **11**, pp. 303–312 (1969)

18 NAGAO, F. and SIMAMOTO, Y., 'The Effect of Crankcase Volume and the Inlet System on the Delivery Ratio of Two-stroke Cycle Engines', *SAE 670030*, Automotive Engineering Congress, Detriot (Jan. 1967)

19 SCHWEITZER. P. H., *Scavenging of Two-stroke Cycle Diesel Engines*, The Macmillan Co., New York (1949)

20 EICHELBERG, G., 'Some new Investigations on old Combustion Engine Problems', *Engineering*, **148**, 463 and 547 (1939)

21 ANNAND, W. J. D., 'Heat Transfer in the Cylinders of Reciprocating Internal Combustion Engines', *Porc. I. Mech. E.*. **177**, No. 36, 973 (1963)

22 WOSCHNI, G., 'A universally applicable Equation for instantaneous Heat Transfer in the Internal Combustion Engine', *SAE Paper 670931*.

23 WATSON, N. and MARZOUK, M., 'A non-linear digital simulation of turbocharged diesel, engines under transient conditions', SAE 770123 (1977)
24 WATSON, N. and JANOTA, M., *Turbocharging the Internal Combustion Engine* MacMillan (1982)
25 BENSON, R. S., GARG, R. D. and WOOLLATT, D., 'A numerical Solution of unsteady flow Problems', *Int. J. Mech. Sci.*, **6** (1964)
26 *The Thermodynamics and Gas Dynamics of I.C. Engines, Chapter 18.* Memorial Volume to the late Prof. R. S. Benson, Pergamon Press, Oxford

2 The theory of turbocharging

Tip velocity $V_t = 300-350\ m/s$

$= 72{,}000-83{,}000\ rpm$

2.1 Introduction

The purpose of supercharging is to increase the mass of air trapped in the cylinders of the engine, by raising air density. This allows more fuel to be burnt, increasing the power output of the engine, for a given swept volume of the cylinders. Thus the power to weight and volume ratios of the engine increase. Since more fuel is burnt to achieve the power increase, the efficiency of the engine cycle remains unchanged.

A compressor is used to achieve the increase in air density. Two methods of supercharging can be distinguished by the method used to drive the compressor. If the compressor is driven from the crankshaft of the engine, the system is called 'mechanically driven supercharging' or often just 'supercharging'. If the compressor is driven by a turbine, which itself is driven by the exhaust gas from the cylinders, the system is called 'turbocharging'. The shaft of the turbocharger links the compressor and turbine, but is not connected to the crankshaft of the engine (except on some experimental 'compound' engines, see Chapter 3). Thus the power developed by the turbine dictates the compressor operating point, since it must equal that absorbed by the compressor.

The essential components of the 'turbocharger' are the turbine, compressor, connecting shaft, bearings and housings. The advantage of the turbocharger, over a mechanically driven supercharger, is that the power required to drive the compressor is exctracted from exhaust gas energy rather than the crankshaft. Thus turbocharging is more efficient than mechanical supercharging. However the turbine imposes a flow restriction in the exhaust system, and therefore the exhaust manifold pressure will be greater than atmospheric pressure. If sufficient energy can be extracted from the exhaust gas, and converted into compressor work, then the system can be designed such that the compressor delivery pressure exceeds that at turbine inlet, and the inlet and exhaust processes are not adversely affected.

The process of compression raises temperature as well as pressure. Since the objective is to increase inlet air density, charger air coolers (heat exchangers) are often used to cool the air between compressor delivery and the cylinders, so that the pressure increase is achieved with the maximum rise in density.

Figure 2.1 shows the ideal dual combustion cycle of a diesel engine in naturally aspirated and turbocharged form. Since the inlet and exhaust pressures are above ambient, and more fuel is burnt in the engine, the cylinder pressure throughout the cycle, and particularly during combustion, is substantially higher for the turbocharged cycle. The compression ratio of the engine must be reduced to prevent an excessive maximum cylinder pressure being reached. *Figure 2.2* compares naturally aspirated and turbocharged ideal dual combustion cycles, when compression ratio is adjusted for the same maximum cylinder pressure. Since reducing compression ratio lowers cycle efficiency, and may make the engine difficult to start, there is a limit to how low a compression ratio can be used in practice.

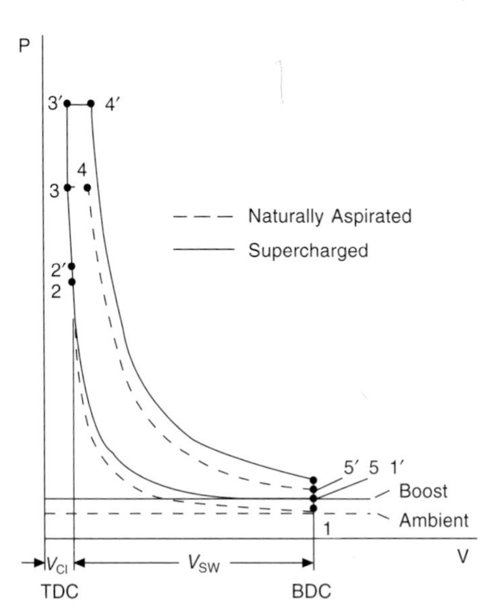

Figure 2.1 Comparison of supercharged and naturally aspirated air standard dual combustion cycles having the same compression ratio

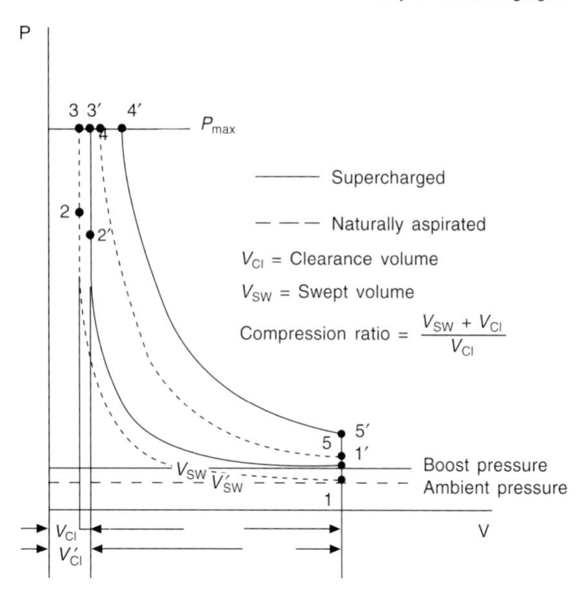

Figure 2.2 Comparison of supercharged and naturally aspirated air standard dual combustion cycle having same maximum pressure but different compression ratio

Turbocharging increases power by increasing the work done per engine cycle. Thus brake mean effective pressure (b.m.e.p.) increases. *Figure 2.3* shows trends in b.m.e.p. for four-stroke and two-stroke engines, and the compressor pressure ratios used. The increase in b.m.e.p. in the 1960s occurred due to the widespread adoption of turbochargers on industrial, marine and rail traction engines. This trend continued in the 1960s as charge air cooling became more popular and engines were redesigned to accept higher compressor pressure ratios. Although these trends have slowed in recent years, b.m.e.p.'s are still increasing and many experimental engines have run at much higher values than those shown in *Figure 2.3.* It is inevitable that ratings will increase further in the search for lower manufacturing cost per horse-power. The alternative approach of increasing power output by increasing speed is unattractive, due to the rapid rise of mechanical and aerodynamic losses, and the corresponding fall in brake thermal efficiency.

Turbocharging of large two-stroke engines, four-stroke medium speed engines, and high speed truck and passenger car diesel engines all make different requirements on the turbocharger. In the following section, different types of turbochargers will be described, followed by a description of the different types of turbocharging systems used to deliver exhaust gas to the turbine. Later sections deal with charge air cooling and the application of turbochargers to the different classes of engine described above. Chapter 2 is restricted to analysis of conventional turbocharging systems. Non-conventional systems are presented in Chapter 3.

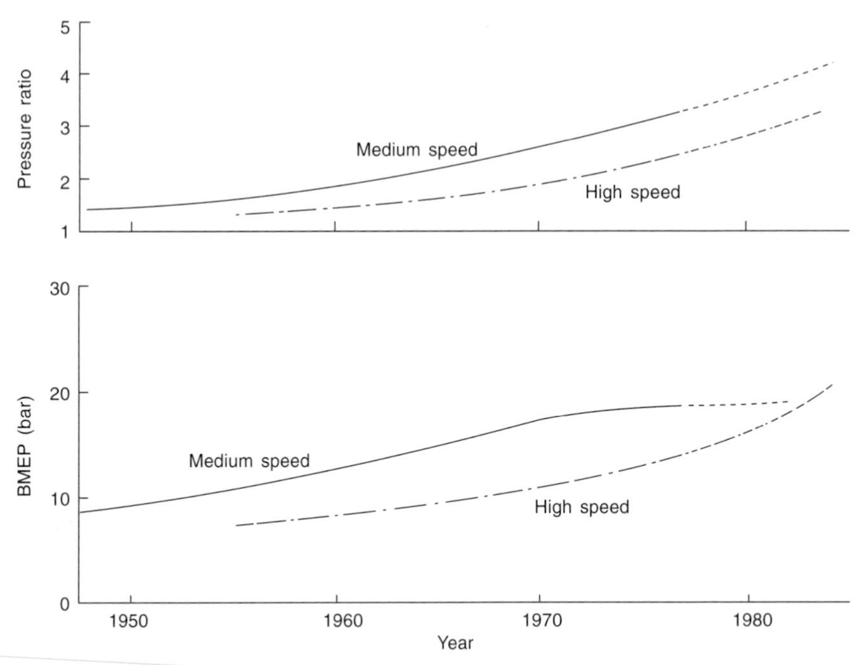

Figure 2.3 **(a)** Four-stroke diesel engine–increase in b.m.e.p. and compressor ratio

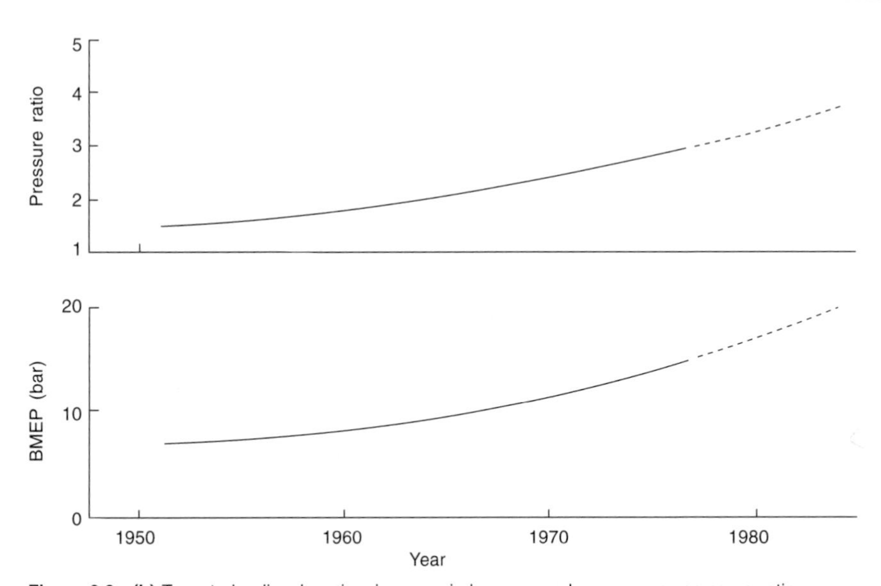

Figure 2.3 **(b)** Two-stroke diesel engine–increase in b.m.e.p. and compressor pressure ratio

2.2 Turbocharging

2.2.1 Turbochargers for automotive diesel engines

Turbochargers in this class are used for passenger car diesel engines rated at 45 kW upwards to larger special heavy truck and construction vehicles rated at up to 600 kW. The most important design factors are cost, reliability and performance. To keep cost low, the design must be simple, hence a single stage radial flow compressor, and a radial flow turbine are mounted on a common shaft with an inboard bearing system (*Figure 2.4*). This arrangement simplifies the design of inlet and exhaust casing and reduces the total weight of the turbocharger.

2.2.1.1 Compressor

The compressor impeller is an aluminium alloy (LM-16-WP or C-355T61) investment casting, with a gravity die-cast aluminium housing (LM-27-M). The design of the impeller is a compromise between aerodynamic requirements, mechanical strength and foundry capabilities. To achieve high efficiency, and minimum flow blockage, very thin and sharp impeller vanes are required, thickening at the root (impeller hub) for stress reasons. It is

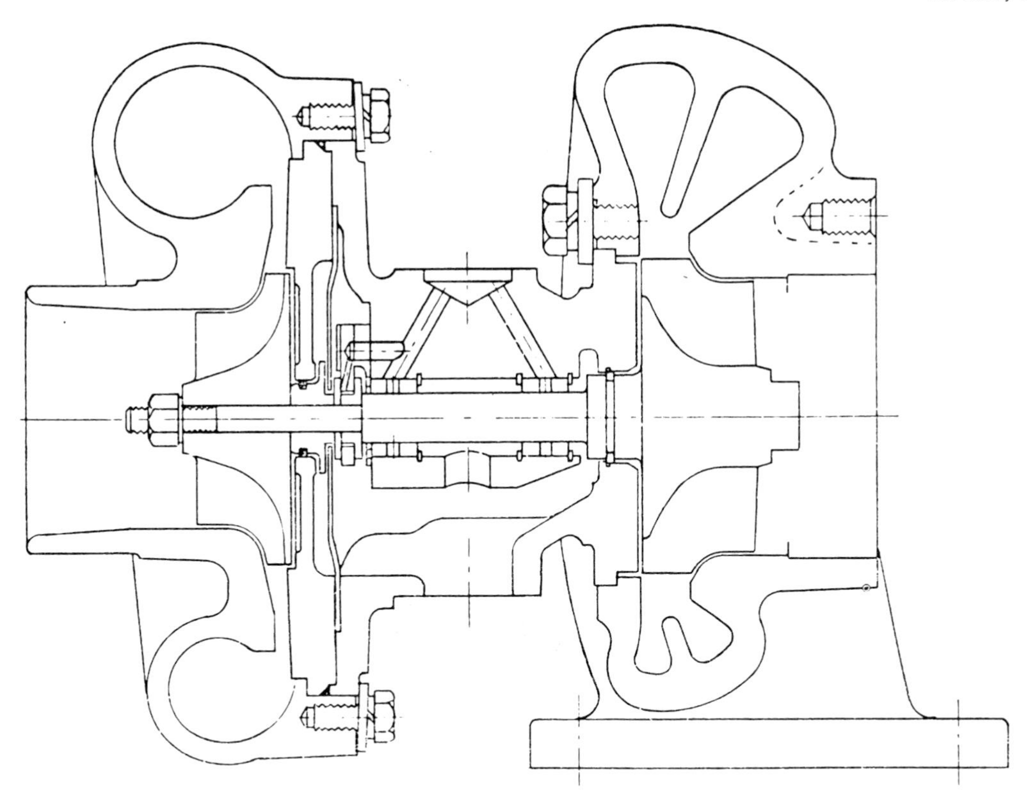

Figure 2.4 An automotive diesel engine turbocharger

Figure 2.5 Automotive turbocharger compressor impeller, with splitter blades.

common practice to use splitter blades (*Figure 2.5*) that start part way through the inducer, in order to maintain good flow guidance near the impeller tip without excessive flow blockage at the eye. Until recently the impeller vanes have been purely radial so that blades were not subjected to bending stress. However most recent designs incorporate backswept blades at the impeller tip since this has been shown to give better flow control and reduces flow distortion transmitted through from impeller to diffuser.

Typical design point pressure ratios fall in the range of 2 to 2.5:1, requiring impeller tip speeds of 300 to 350 m/s, hence small units of typically 0.08 m tip diameter rotate at 72 000 to 83 000 rev/min. In order to match wide differences in air flow requirements from one engine to another, a range of compressor impellers is available to fit the same turbocharger. These will be produced from one or two impeller castings, but with different tip widths and eye diameters generated by machining as shown in *Figure 2.6,* and matched with appropriate compressor housings. Usually up to ten or more alternative 'trims' are available but since the impeller tip diameter is unchanged and the hub diameter at the impeller eye is fixed by the shaft diameter, the flow passage variations alter the efficiency as well as flow characteristics of the impeller.

The compressor can be a loose or slight interference fit on the shaft, clamped by the compressor end nut. Impellers of most turbochargers are balanced before assembly onto the shaft, so that components can be interchanged without rebalancing.

Vaneless diffusers (*Figure 2.4*) are used on all except very high pressure ratio compressors. Relative to the alternative vaned designs, the vaneless diffuser is slightly less efficient due to a longer gas flow path and poorer flow guidance, but has a substantially wider range of high operating efficiency. This is important in truck and passenger car applications where engine speed, and therefore mass flow range, is large. The volute acts not merely as a collector of air leaving the diffuser, but is usually designed to achieve a small amount of additional diffusion in its delivery duct. Generally the volute slightly overhangs the diffuser (*Figure 2.4*) in order to reduce the overall diameter of the turbocharger. The volute and impeller casing are invariably formed as a single component.

2.2.1.2 Turbine

Radial inflow turbines are universally used, usually friction or electron beam welded to the shaft. The turbine wheel must sustain the same high rotational speed of the compressor and operate at gas temperatures up to 900 K. The turbines are investment cast in high temperature creep resistant steels, such as 713 C Inconel. Its properties exceed the requirements but it

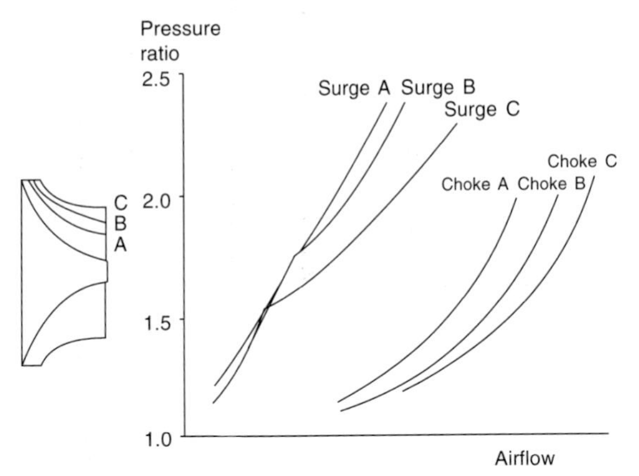

Figure 2.6 A range of compressor 'trims', machined from the same impeller casting

is a readily available material, from the gas turbine industry. The turbine housing must stand the same high gas temperature as the impeller but is not subject to rotating stress. However it should be strong enough to sustain a rotor burst. Sand or shell cast SG iron (spheroidal graphite nodular) is used in most applications, and is free of scaling at temperatures up to 900 K. Ni-resist is used for higher temperatures but is more expensive, and prone to cracking.

Vaneless stators are used except for a very few high pressure ratio applications. These are cheaper, and enable the gas angle at stator exit to vary to some extent with the mass flow, giving high efficiency over a very wide flow range. On the rare occasions when stator nozzles are incorporated in turbocharger turbines of this size, a nozzle ring is cast in Ni-resist or the nozzle vanes are fabricated from nickel-chromium alloys. The gas angle at rotor inlet is controlled by the nozzel vane angle or the geometry of the vaneless housing. In the latter case (*Figure 2.7*) the inlet scroll and vaneless stator are combined. The cross-sectional area of a section A defines the tangential gas velocity for a given mass flow rate and inlet density. The scroll is designed to spill the mass evenly around the circumference of the rotor. Again, for a given mass flow rate and density the radial component of velocity entering the impeller is fixed by the cross-sectional area at the rotor tip. This, and area A control the flow angle at rotor entry. Altering this flow angle alters the effective flow capacity of the turbine. Thus a range of casings with varying area A, or nozzle stator rings with varying blade angle, are available to match each rotor design to the requirements of a particular engine.

Gas flow through a turbine is predominantly accelerating, whereas in a compressor the gas diffuses. Fluid dynamic flow control is much easier to achieve in an accelerating flow, hence turbine design is less critical than compressor design. The turbine is more tolerant of mass flow variation, hence a single rotor may be used with a number of different area vaneless casings. Only a few turbine rotor trim variations are needed to cover the requirements of a large range of engine sizes.

Most turbine casings have twin entries as shown in *Figure 2.4*. The twin entries are used to separate the exhaust gas pulses coming from cylinders whose exhaust valves are open at the same time. This for example, the exhaust manifold of a typical six-cylinder engine will have two pipes, each connecting three cylinders to a twin entry turbine. Single entry casings are used on passenger car engines in order that a single waste-gate may be used to bypass some gas around the turbine at high engine

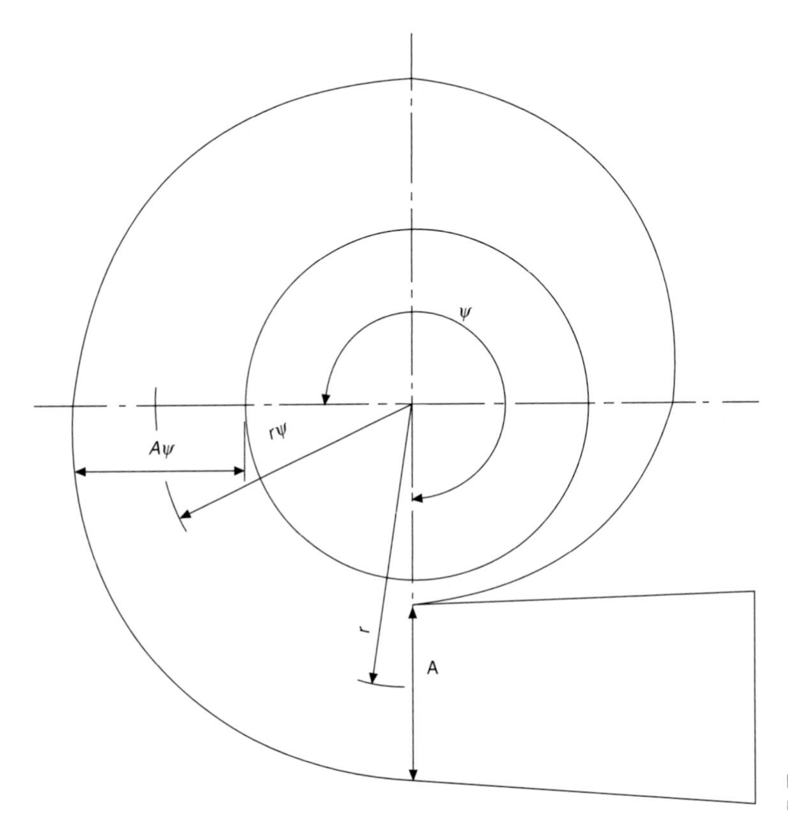

Figure 2.7 Geometric specification of a vaneless radial turbine housing (volute)

speeds. This reduces turbine and compressor power and hence boost pressure, avoiding the tendency for the turbocharger to overboost the engine at full speed and load (see section 2.6.7).

Since flow guidance is less critical than in the compressor, the turbine impeller has fewer blades. For stress and manufacturing reasons, the vanes are thicker than compressor vanes. They must withstand the pulsating gas pressure developed in the exhaust system of the engine. To reduce rotating inertia the back of the impeller is cut away between the blades (*Figure 2.4*). Although this slightly reduces efficiency, the benefit in terms of reduced stress and faster acceleration is substantial. Considerable rotor to housing clearance is essential, due to thermal effects, large thrust bearing clearance and the build-up of tolerances between components. Thus efficiencies are lower than those normal in gas turbine practice, due to leakage and clearance losses.

The turbine inlet flange mounts the turbine to the engine via the exhaust manifold. This avoids the need for expansion joints in the exhaust manifold.

2.2.1.3 Bearings and centre casing

For low cost, simplicity and ease of maintenance, the bearing system must be designed to use the lubricating oil of the engine. All automotive turbochargers use simple journal bearings, since ball bearings are more expensive, have a short life at very high speeds and are difficult to replace. The inboard (between compressor and turbine) bearing location imposes a short distance between bearings which, when combined with very light radial loads and a heavy overhung mass at one end only (the turbine), leads to complex bearing behaviour. In addition the very high rotational speed dictates that the rotor runs through both the first and second critical vibration frequencies of the rotating assembly. Thus bearing design is primarily concerned with the stability of the system.

Two fully floating sleeve bearings are used, free to rotate at a speed less than the rotor speed, and determined by bearing clearances. The outer oil film imparts an additional degree of damping into the bearing system, but both the shaft to sleeve to housing clearances are large for improved stability. Typical clearances are 0.02 to 0.05 mm between shaft and sleeve and 0.07 to 0.1 mm between sleeve and housing. These large clearances mean that oil filtration down to about 20 μm only is required; easily met by full flow paper engine oil filters. A large oil flow rate is required, due to large clearances and the need to cool the turbine end bearing. Bearing material is typically leaded bronze with added tin flashing.

Axial thrust loads are relatively light, except when an exhaust brake is used, upsetting the compressor-turbine pressure balance. A simple thrust assembly is located at the cooler, compressor end, with flat or tapered lands, made from sintered leaded bronze.

The power required to overcome bearing losses is quite significant, typically being 5 to 10% of turbine power at full speed. More important is the fact that this percentage increases at lower engine speeds, which is a disadvantage in vehicle applications where high boost pressure is difficult to achieve at low engine speed. Lower viscosity oil would help but is incompatible with the requirements of the engine.

The bearing sleeves run directly in a bore in the high grade grey iron centre casting of the turbocharger. This casting also acts as an oil drain and holds the compressor and turbine housings. For good oil drainage the casing must be installed with the oil inlet and drain at top and bottom, in the vertical position, with the rotor shaft horizontal. The oil seals at compressor and turbine ends are a difficult design problem, due to the need to keep frictional losses low, and to large movements due to large bearing clearance and adverse pressure gradients under some conditions. The piston ring seals shown in *Figure 2.8* are typical, the rings being a press fit in the housing. They act as a labyrinth seal due to small clearance between the stationary ring and the sides and base of the groove in the rotor shaft. In practice they leak unless care is taken to keep as much oil as possible away from them. Rotating flingers are designed into the shaft for this reason and an oil shield is frequently fitted at the compressor since a depression can be generated in the compressor when the engine is idling.

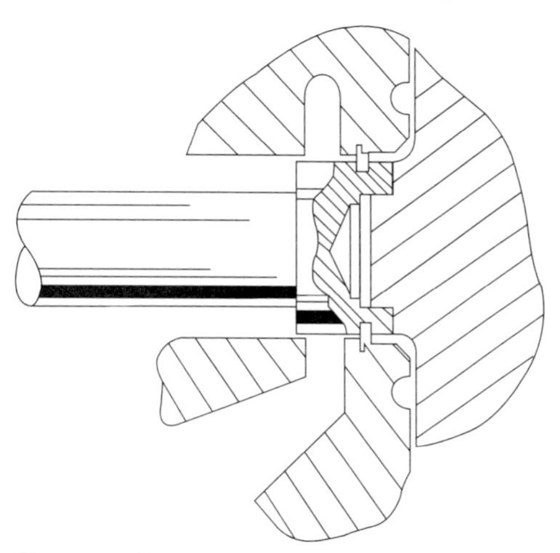

Figure 2.8 Piston ring oil seal, located in a stepped bore

The compressor and turbine housings can usually be rotated so that the compressor delivery and turbine inlet can be located in convenient positions. Set screws or stainless steel V-clamps are used to hold the components together.

The performance characteristics of typical small turbochargers are presented in section 2.3.3.

2.2.2 Small industrial and marine engine turbochargers

Turbochargers designed for small industrial and marine engines, though larger than those of large truck engines, are similar in concept to the automotive turbochargers described above. Radial flow compressor and turbines are used, with an inboard bearing arrangement (*Figure 2.9*). Apart from the larger size, they are required to have greater durability and higher efficiency. Thus the designs are usually more complex and expensive.

Engines designed for these applications operate over a smaller speed range than truck engines, and at greater b.m.e.p., hence higher compressor pressure ratio. It follows that the flow range required from the compressor is smaller, hence vaned diffusers are used. Vaned turbine stator nozzles are also used. This results in higher design point compressor and turbine efficiency. A range of diffuser nozzle angles and turbine stator blade angles are available for matching a basic turbocharger to a particular engine.

The maximum size is governed by precision casting limitations for the radial flow turbine rotor, currently about 300 mm, although most units in this class are smaller. Turbine housings are simple volutes designed to deliver the flow evenly around the circumference of the stator nozzle ring, the latter generating the design gas flow angle at rotor inlet. The turbine housings are supplied in uncooled or water cooled form. Although cooling is undesirable thermodynamically, it is sometimes required for safety reasons due to the potential danger of hot exposed surfaces in small engine rooms.

Bearings are of similar design to those of automotive units, except that clearances, relative to turbocharger size, are smaller. Sometimes cooling air is bled from the compressor to the rear

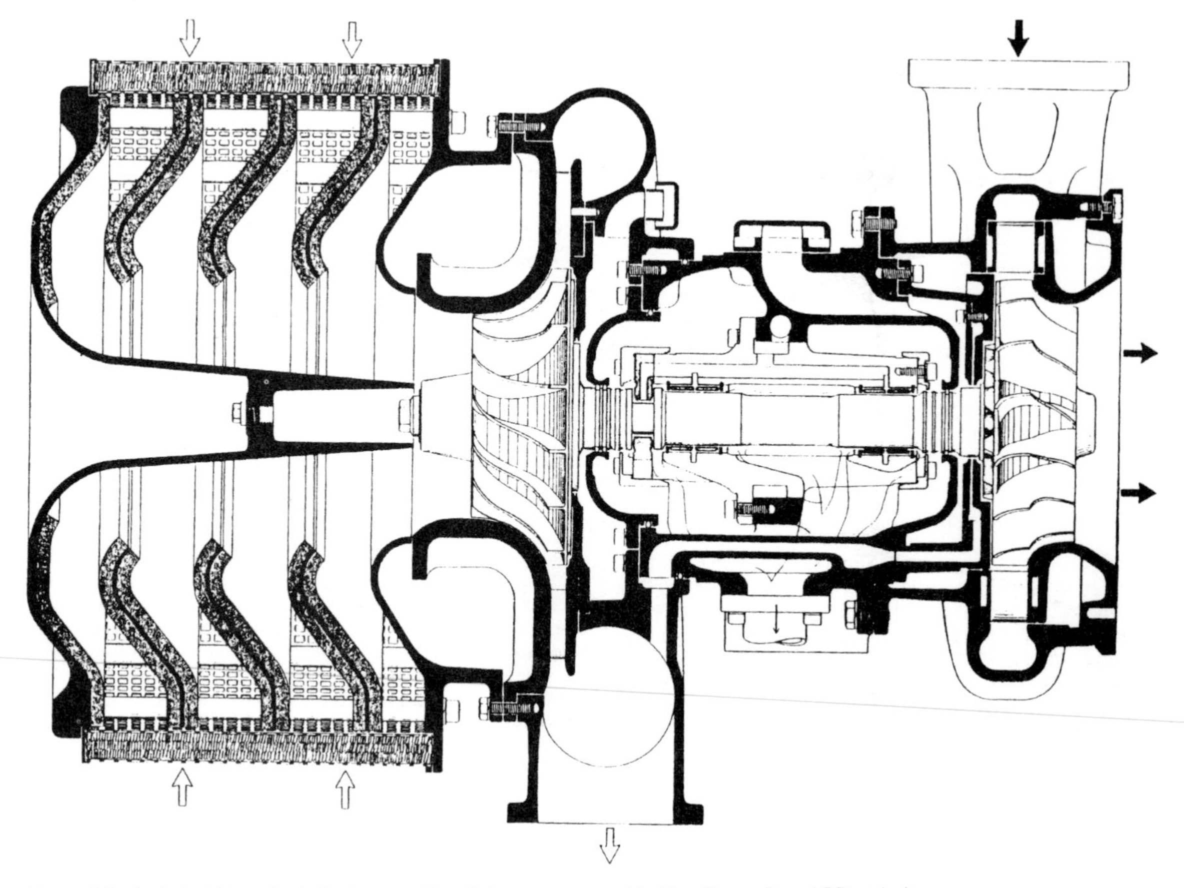

Figure 2.9 An industrial engine turbocharger with radial compressor and turbine (Brown Boveri RR series)

of the turbine hub and bearing area. This also helps prevent exhaust gas leaking down the back of the turbine wheel and reaching the bearings. These techniques help keep the hot end bearing cool, preventing serious oil oxidation deposits. Like the smaller units previously described, the lubricating oil system of the engine is also used for the turbocharger. Since bearing clearances are smaller, rotor movements are small and conventional labyrinth oil seals can be used at the compressor and turbine ends of the rotor shaft.

Turbochargers of this type are made in relatively small numbers, by batch production, hence their cost is high relative to automotive units.

2.2.3 Large industrial and marine engine turbochargers

These turbochargers are characterized by having axial flow, single stage, turbines and are fitted to the majority of large industrial and marine engines, both four- and two-stroke. The duty cycles of these engines are more arduous than that of automotive engines and they tend to spend much more of their operating time at high load. Furthermore the consequences of failure are more serious, particularly on a marine engine. As a result, although every attempt is made to keep the designs simple, the primary objectives are a very high level of reliability, high efficiency and versatility to cover a great range of engine types and sizes at reasonable cost. However, design variations from one manufacturer to another are greater than is the case with smaller turbochargers.

Figure 2.10 is a cross-section of a typical large turbocharger, with a radial flow compressor and axial flow turbine. The compressor impeller is made in two separate parts, the inducer and main part of the impeller. The inducer is usually machined from a steel casting or an aluminium forging, and is splined or keyed to the shaft. The impeller is machined from an aluminium forging except for very high pressure ratio requirements when titanium is used due to its superior high temperature properties. The advantage of the two piece compressor is ease of machining, but an additional benefit is some impeller vane damping provided by friction at the inducer-impeller contact surfaces. Compressor diffusers are vaned for high efficiency.

The turbine disc is machined either as an integral part of the shaft or is shrunk on to the shaft. The rotor blades may be cast, forged or machined from a high temperature creep-resistant steel such as Nimonic 80A or 90. Welded joints or 'fir-tree' roots are used to fix them to the disc, the latter design being more common on high pressure units since they provide a degree of vibration damping and allow a wider selection of blade and disc materials to be considered. Additional vibration damping can be provided by wire lacing the blades. The turbocharger manufacturer will offer a range of 'trims' or flow capacities with each basic design of turbocharger by varying blade (stator and rotor) height and stator blade angle.

A disadvantage of the axial flow turbine is that it complicates the design of the gas inlet and outlet. The gas inflow section is particularly important hence this is usually located on the end, allowing generous curvature in the inlet ducts to the stator blades for minimum flow distortion and loss. The turbine exit duct acts merely as a collector, hence a compact design can be used,

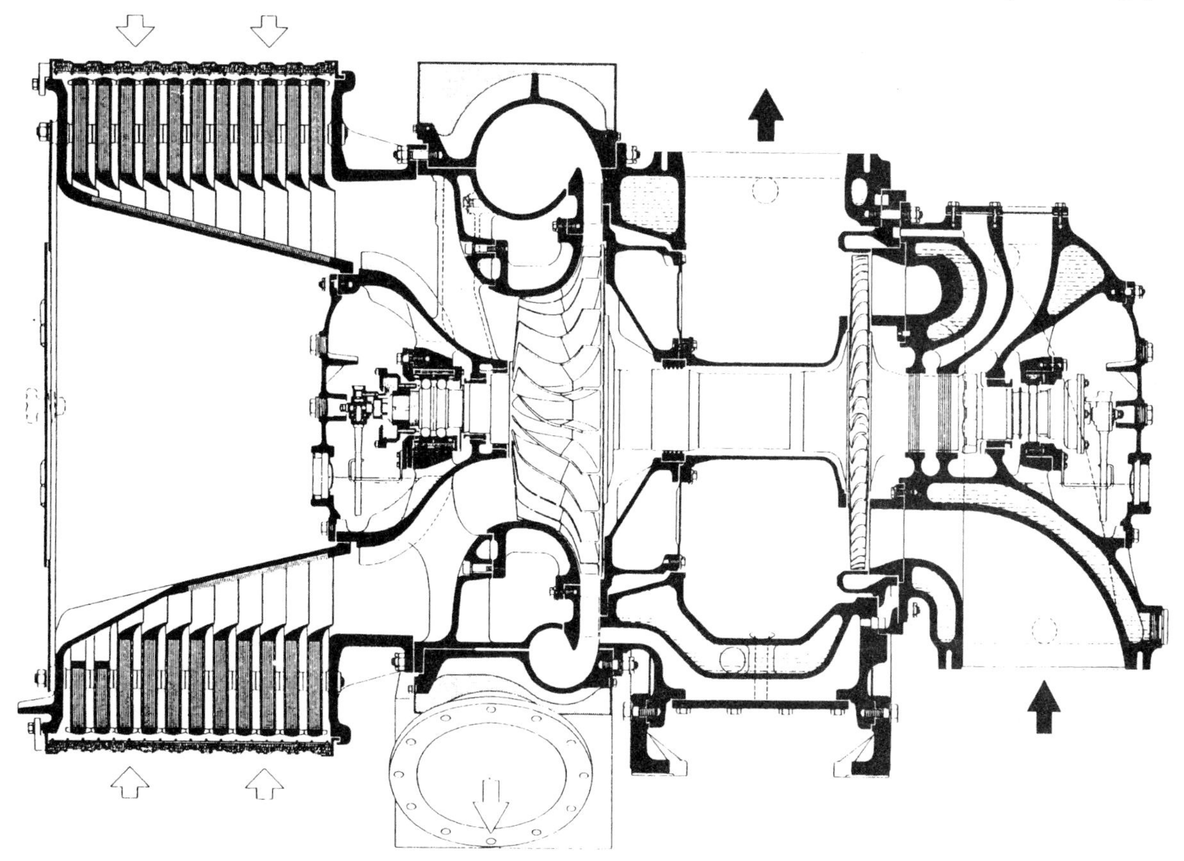

Figure 2.10 A large turbocharger with radial compressor and axial turbine (Brown Boveri)

minimizing turbocharger length. However, a recent trend is to utilize some exhaust diffusion to increase turbine expansion ratio and power output.

Most of the larger turbochargers in this class have outboard rolling element bearings (i.e. outside the compressor and turbine, *Figure 2.10*), with their own oil supply, and resilient mountings to prevent brinelling. The advantages of this are stable shaft mounting and low dynamic loads due to the wide bearing spacing, small bearing diameter, low rolling resistance and good access for bearing maintenance. The use of separate oil supplies for the turbocharger and engine enables a lower viscosity oil to be used, further reducing bearing friction. Low pressure ratio turbochargers use simple rotating steel discs, partially immersed in the oil, to pick up and deliver the oil to the bearings, but with higher bearing loads and speeds, gear pumps are used to spray oil on the bearings. Plain or sleeve bearings are sometimes available as an option and are preferred for durability although their frictional losses are greater.

Turbocharger design is simpler with inboard bearings since this gives greater freedom to design low loss intake ducts. Fewer components are required and the turbocharger is shorter, lighter and cheaper as a result (*Figure 2.11*). The disadvantage is a less stable bearing system and higher bearing loads. Fully floating sleeve and multi-lobe plain bearings are used, with well damped mountings for stability; the rotors must still be carefully balanced. Relative to rolling element bearings, higher oil pressure and greater oil flow rates are required and the combination of large diameter and width means that frictional losses are greater.

With either bearing system, the turbine outlet casing is the main structure to which the other components are bolted, and incorporates mountings to the engine. The casing is usually water cooled. Bolted to it is the water cooled turbine inlet casing, incorporating the bearing housing (for outboard bearings) and its oil reservoir. Single, two, three and four entry turbine inlets are available, manufactured from high grade cast iron. Between turbine inlet and outlet casings, provision is made for mounting the turbine stator nozzle ring. The compressor inlet and outlet casings are aluminium alloy castings.

The compressor inlet casing incorporates webs to support the bearing housing if outboard bearings are used. These webs must be carefully designed to be far enough away from the impeller to avoid impeller vane excitation. The casing also houses a combined air filter and silencer on most larger turbochargers (*Figure 2.10*). Sound waves originating at the compressor intake are reflected and reduced in intensity by baffles lined with sound absorbing material.

2.3 Turbocharger performance

The performance of turbochargers can be defined by the pressure ratio, mass flow rate and efficiency characteristics of the compressor and turbine, plus the mechanical efficiency of the bearing unit. In this section we will look at the efficiency of compressors and turbines leading into a description of typical turbocharger performance maps.

2.3.1 Compressor and turbine efficiency

The work output from (or input to) a turbomachine can be found from the first law of thermodynamics. From this law the steady

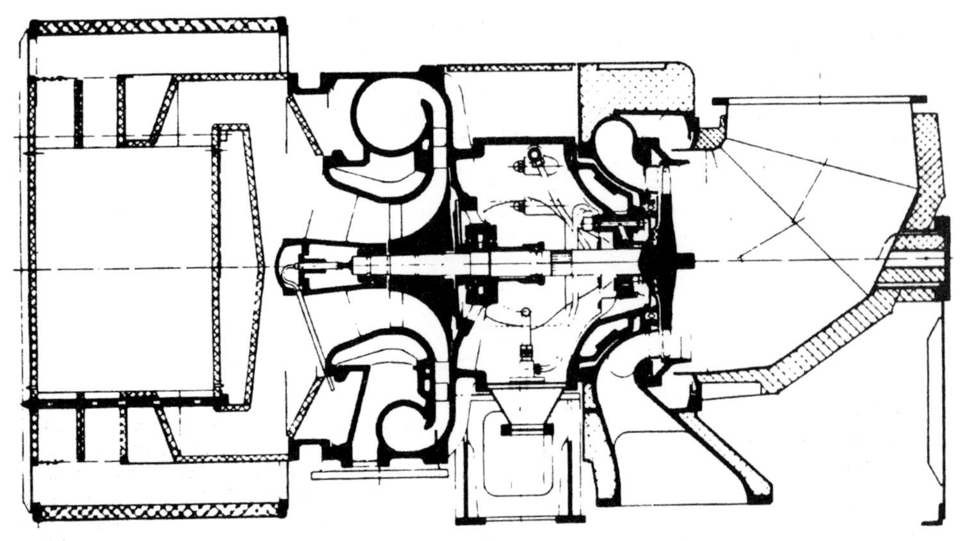

Figure 2.11 A large turbocharger with axial turbine but inboard bearings (M.A.N.)

flow energy equation may be derived. A turbomachine has one inlet and one outlet port. The steady flow energy equation becomes:

$$\dot{Q} - \dot{W} = \dot{m}[(h_2 + KE_2 + PE_2) - (h_1 + KE_1 + PE_1)] \quad (2.1)$$

where $\dot{Q}$ = heat transfer rate (+ ve to the system);
$\dot{W}$ = work transfer rate (+ ve by the system);
$\dot{m}$ = mass flow rate;
h = specific enthalpy;
KE = specific kinetic energy;
PE = specific potential energy;
suffixes 1, 2 = inlet and outlet ports respectively.

Denoting the stagnation enthalpy (h_0) as

$$h_0 = h + KE \quad (2.2)$$

and neglecting changes in potential energy and heat transfer, since these terms are small, this becomes

$$-\dot{W} = \dot{m}[(h_{02} - h_{01}) \quad (2.3)$$

Both air and exhaust gas are considered as perfect gases. Hence they obey the equation of state.

$$Pv = RT \quad (2.4)$$

where P, v, R and T denote pressure (absolute), specific volume, gas constant and temperature respectively. The specific heat capacity at constant pressure (Cp) for a perfect gas is given by:

$$Cp = dh/dT \quad (2.5)$$

Thus eqn (2.3) becomes

$$-\dot{W} = \dot{m}Cp[(To_2 - To_1) \quad (2.6)$$

where To denotes stagnation (or 'total') temperature, the temperature of a gas if brought to rest. Relative to the free stream temperature (T) of a gas moving at velocity V,

$$To = T + V^2/2Cp \quad (2.7)$$

The second law of thermodynamics tells us that specific entropy is related to the specific heat transfer

$$ds > dQ/T \quad (2.8)$$

The second law can also be used to show that ideal adiabatic compression or expansion takes place at constant entropy.

One definition of the efficiency of a compressor is the power required for ideal, adiabatic compression divided by the actual power required in a non-ideal, non-adiabatic compressor, working with the same inlet pressure and temperature and outlet pressure.

$$\eta_{is} = \frac{\text{isentropic power}}{\text{actual power}}$$

Hence η_{is} is termed the isentropic efficiency of the compressor. From eqns (2.3), (2.6) and (2.8),

$$\eta_{isTT} = \frac{ho_{2_{is}} - ho_1}{ho_2 - ho_1} \quad (2.9)$$

and

$$\eta_{isTT} = \frac{To_{2_{is}} - To_1}{To_2 - To_1} \quad (2.10)$$

where suffices 'is' and 'TT' denote 'isentropic' and 'total to total', meaning an efficiency based on total temperature values. Note that the work required by a non-ideal compressor exceeds that of an isentropic compressor, hence the exit air temperature To_2 is higher than $To_{2_{is}}$.

For isentropic compression, pressure and temperature are related by the expression

$$\frac{Po_2}{Po_1} = \left(\frac{To_{2_{is}}}{To_1}\right)^{\gamma/(\gamma-1)} \quad \text{where } \gamma = Cp/Cv \quad (2.11)$$

Hence eqn (2.10) may be rearranged as

$$\eta_{isTT} = \frac{(Po_2/Po_1)^{(\gamma-1)/\gamma} - 1}{\dfrac{To_2}{To_1} - 1} \quad (2.12)$$

An evaluation of compressor efficiency based on total-to-total temperature rise, implicity assumes that the kinetic energy leaving the compressor can be made use of in the following machine components. This is true in a gas turbine since the gas velocity is maintained through the combustion chamber to the turbine, where it does useful work. However, air delivered from the turbocharger compressor to the inlet manifold of an engine is brought almost to rest, without doing useful work. This loss of kinetic energy should be considered as a loss of compressor efficiency relative to the ideal of a negligible exit gas velocity. At inlet to the compressor, air is accelerated from rest, into the compressor eye without introducing inefficiencies, hence the inlet ambient temperature can be used (total and static temperatures are equal). Thus a more appropriate definition of compressor efficiency is

$$\eta_{isTS} = \frac{(Po_2/Po_1)^{(\gamma-1)/\gamma} - 1}{To_2/To_1 - 1} \tag{2.13}$$

This is the total-to-static isentropic efficiency and will usually be a few percentage points lower than the total-to-total isentropic efficiency. Unfortunately some turbocharger manufacturers quote total-to-total values, some without declaring the basis of their measurement.

Manipulation of eqns (2.6), (2.8) and (2.11) gives the following relationships for compressor power $(\dot{W})$

$$-\dot{W}_c = \dot{m}CpTo_1\left[\left(\frac{Po_2}{Po_1}\right)^{(\gamma-1)/\gamma} - 1\right]/\eta_{isTS} \tag{2.14}$$

The negative sign results purely from the thermodynamic sign convention of work being done by the system being considered positive, and work done on the system as negative. Thus the power required to drive the compressor is a function of the mass flow rate $(\dot{m})$, inlet air temperature, (To_1), pressure ratio (Po_2/Po_1), compressor efficiency (η_{is}) and specific heat at constant pressure. Equations (2.13) and (2.14) show that low compressor efficiency not only increases the power requirement for a given pressure ratio, but also increases the delivery temperature (T_2) and therefore reduces the air density leaving the compressor. It is important to achieve high compressor efficiency for both reasons.

The isentropic efficiency of the turbine may be expressed as the actual power output divided by that obtained from an ideal adiabatic (isentropic) turbine operating with the same inlet pressure and temperature.

$$\eta_{is} = \frac{\text{actual power}}{\text{isentropic power}} \text{ (turbine)} \tag{2.15}$$

This expression may be developed in a similar manner to the compressor to give

$$\eta_{isTT} = \frac{1 - To_4/To_3}{1 - (Po_4/Po_3)^{(\gamma-1)/\gamma}} \tag{2.16}$$

and

$$\eta_{isTS} = \frac{1 - To_4/To_3}{1 - (Po_4/Po_3)^{(\gamma-1)/\gamma}} \tag{2.17}$$

Kinetic energy leaving the turbine is wasted through the exhaust pipe, hence the total-to-static efficiency is again most appropriate, although not always quoted by the turbocharger manufacturer.

The power output of the turbine is given by:

$$\dot{W}_t = \dot{m}CpTo_3\left[1 - \left(\frac{P_4}{Po_3}\right)^{(\gamma-1)/\gamma}\right]\eta_{isTS} \tag{2.18}$$

Thus the power developed by the turbine is a function of its inlet temperature (To_3) mass flow rate $(\dot{m})$, expansion ratio (P_4/Po_3), efficiency (η_{isTC}) and specific heat capacity of the exhaust gas (Cp).

2.3.2 Non-dimensional representation of compressor and turbine characteristics

The mass flow rate, efficiency and temperature rise $(\Delta T = T_{02} - T_{01})$ of a compressor or turbine can be expressed as a function of all possible influencing parameters, as follows:

$$\dot{m} = \eta_{is}, \Delta T = f(Po_1, Po_2, To_1, N, D, R, \gamma, \mu) \tag{2.19}$$

where N, D and μ are rotational speed, diameter and the kinematic viscosity of the gas respectively, and $\gamma = Cp/Cv$. These can be reduced, using dimensional analysis, to the following non-dimensional groups:

$$\frac{\dot{m}\sqrt{To_1 R}}{Po_1 D^2}, \eta_{is}, \frac{\Delta T}{To_1} = f\left(\frac{ND}{\sqrt{RTo_1}}, \frac{Po_2}{Po_1}, \frac{\dot{m}}{\mu D}, \gamma\right) \tag{2.20}$$

For the compressor, the value of γ remains constant, except for a very small variation with temperature, hence the last term may be ignored. γ does vary with air-fuel ratio, but its influence on turbine performance is small and is therefore also ignored. Fortunately the Reynolds number $(\dot{m}/\mu D)$ also has only a small effect on performance and can be ignored. A relationship between η_{is}, $\Delta T/To_1$ and Po_2/Po_1 has already been given (eqns 2.12 and 2.16), hence eqn (2.20) may be reduced to

$$\frac{\dot{m}\sqrt{To_1 R}}{Po_1 D^2}, \eta_{is} = f\left(\frac{ND}{\sqrt{RTo_1}}, \frac{Po_2}{Po_1}\right) \tag{2.21}$$

(For a turbine, suffixes 3 and 4 replace 1 and 2).

For a particular turbocharger, the diameter remains constant and, for turbocharger applications the gas constant remains fixed, hence the variation in performance with running conditions is given by

$$\frac{\dot{m}\sqrt{To_1}}{Po_1}, \eta_{is} = f\left(\frac{N}{\sqrt{To_1}}\right), \frac{Po_2}{Po_1} \tag{2.22}$$

The complete performance map of the turbomachine (*Figure 2.12*) can be shown by plotting a graph of $m\sqrt{To_1}/Po_1$ against

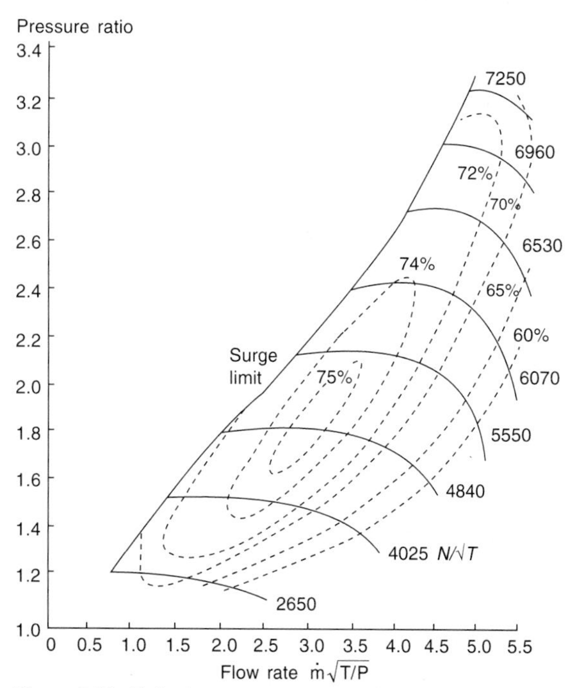

Figure 2.12 Turbocharger compressor performance map

Po_2/Po_1 showing lines of constant $N/\sqrt{To_1}$ and efficiency (η_{is}). The advantage of this presentation is that it uniquely describes the performance of the turbomachine, regardless of inlet conditions (pressure and temperature). However, the terms in eqn (2.22) are no longer truly dimensionless.

2.3.3 Compressor performance

A typical performance map, from a turbocharger compressor designed for a medium speed engine, is shown in *Figure 2.12*. The central area is the stable operating zone, bounded by the surge line on the left (low mass flow rates), and a regime of high rotational speed and low efficiency on the right (high mass flow rate).

A detailed explanation of the causes of surge has yet to be fully accepted, but it is clear that when the mass flow rate through the compressor is reduced whilst maintaining a constant pressure ratio, a point arises at which local flow reversal occurs in the boundary layers. Complete flow reversal can develop in a 'stall cell' somewhere in the compressor, with normal flow elsewhere. Once several stall cells have developed the complete flow pattern can break down hence mass flow and pressure fall. A stable flow pattern becomes re-established at a lower pressure ratio, allowing the mass flow to build up again to the initial value. The flow instability repeats in a surge cycle. The compressor must not be asked to work in this region of operation but it must be realized that surge is influenced by the complete intake, compressor and inlet manifold system. Thus the surge line drawn by the turbocharger manufacturer is only a guide, and in practice it varies from one engine installation to another.

The area of high rotational speed and low efficiency is a result of choking of the limiting flow area in the compressor. Extra mass flow can only be achieved by increasing rotational speed, which must be limited by stress constraints. If the diffuser chokes, rather than the rotor, then compressor speed eventually rises substantially with little increase in airflow. This is likely to be the case with a vaned diffuser, as fitted to the compressor whose characteristics are shown in *Figure 2.12*.

Constant efficiency loops are also shown in *Figure 2.12*. Note that these tend to be parallel to the surge line.

2.3.4 Turbine performance

An axial flow turbine characteristic based on the same pressure ratio versus mass flow parameter is shown in *Figure 2.13*. The most evident feature is the way that the lines of constant speed parameter ($N/\sqrt{To}$) converge to a single line of almost constant mass flow parameter. This flow limit is caused by the gas reaching sonic velocity and choking the inlet casing or stator nozzle blades. This choked flow will remain constant (for constant inlet conditions), regardless of rotor speed. At pressure ratios below the choking condition, the effective flow area of the turbine, and hence mass flow rate, will be influenced by the rotor. Thus rotational speed influences mass flow rate unless the stator chokes.

In a radial flow turbine, rotational speed of the rotor influences the pressure at stator exit, due to centrifugal effects. Thus the overall pressure ratio (stator inlet to rotor exit) at which the stator chokes is dependent on rotor speed. *Figure 2.14* shows pressure ratio versus mass flow rate curves for a radial flow turbine, with stator nozzles, illustrating the variation of choked pressure ratio and mass flow rate with rotor speed.

Nozzleless radial turbines exhibit the widest variation of mass flow with rotor speed. However, it is rare for the data obtained by the turbocharger manufacturer to cover the full pressure ratio versus mass flow range along each constant speed line, even for these turbines. It reduces test time and is rarely required

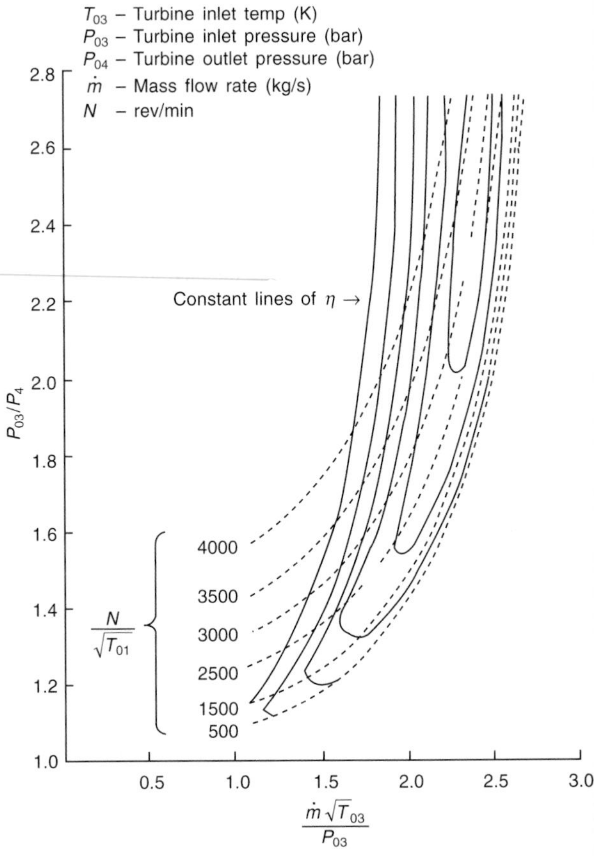

Figure 2.14 Radial flow turbine performance map

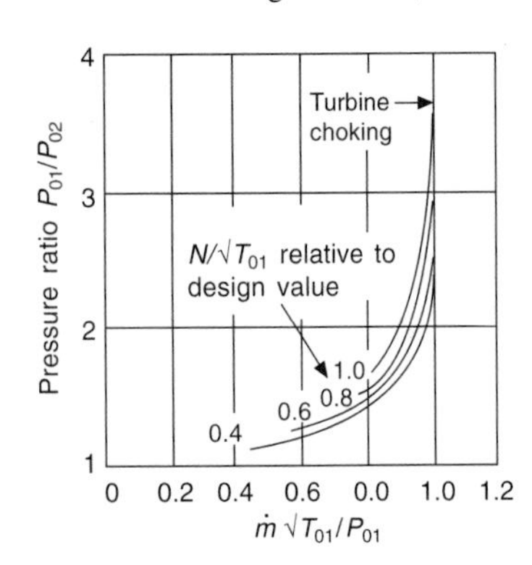

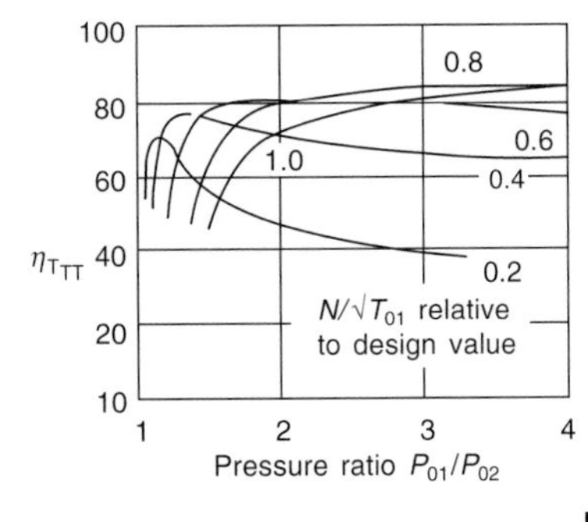

Figure 2.13 Axial flow turbine performance map

by the user, for the following reason. For a given turbocharger, it is possible to link the compressor and turbine characteristics and plot the equilibrium running line on compressor and turbine maps. This comes from balancing the compressor and turbine power eqns (2.14 and 2.18), and imposing a common rotational speed, factors that are implicit in any experimental turbocharger test. This equilibrium line is shown in *Figure 2.15* and runs across from a low speed line at low pressure ratio to a high speed line at high pressure ratio. The equilibrium line is based on steady-state testing and theory, and is not necessarily the operating regime if the turbine inlet conditions are pulsating.

All turbocharger manufacturers present compressor data in the form described in section 2.3.3, but there is less uniformity of turbine data. For example, some manufacturers follow gas turbine practice and transpose the ordinate and abscissa of *Figures 2.13* and *2.14* to those of *Figure 2.15*. Further variations occur in presentation of turbine efficiency data. The problem arises from the fact that the operational area of the turbine occupies such a restricted area on the pressure ratio versus mass flow parameter map. It is possible, but inconvenient, to superimpose lines of constant efficiency with a radial turbine (*Figure 2.14*), but quite impossible for an axial flow turbine. It is simpler to present efficiency on a separate diagram, but different manufacturers tend to use different diagrams.

Most common is a plot of turbine efficiency against blade speed ratio (u/c). u is the tip speed of a radial turbine rotor or the blade velocity at the mean blade height of an axial rotor. The rotor blade velocity (u) is non-dimensionalized by dividing by a theoretical velocity (c) that would be achieved by the gas if it expanded isentropically from the turbine inlet condition to the turbine exit pressure. Lines of constant pressure ratio or speed parameter ($N\sqrt{To}$) are superimposed on the map (*Figure 2.16*). This presentation happens to be useful for the turbocharger manufacturer when matching compressor and turbine rotor diameters but is inconvenient for an engine manufacturer. He desires a graph whose axes represent parameters relevant to his engine. Thus some turbocharger manufacturers plot turbine efficiency against pressure ratio, showing lines of constant turbocharger speed (*Figure 2.15*).

In practice diagrams such as *Figures 2.15* and *2.16* show the product of turbine and mechanical efficiency (bearing losses predominate) since it is difficult to separate them. Indeed it is surprisingly difficult to achieve accurate turbine efficiency measurements, due to heat transfer, non-uniform and unsteady flow effects, etc. As a result most turbocharger manufacturers have developed their own standard test techniques whose results can be reliably compared across their range of turbochargers. However these results cannot be directly compared with those of another manufacturer who, for example, tests with a different standard turbine inlet temperature and therefore measures a

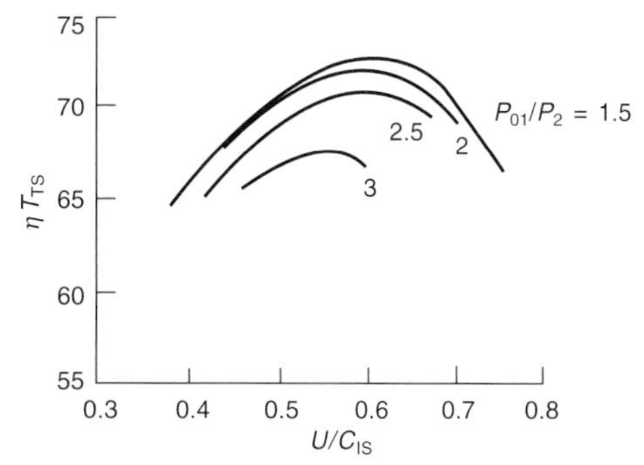

Figure 2.16 Turbine efficiency map

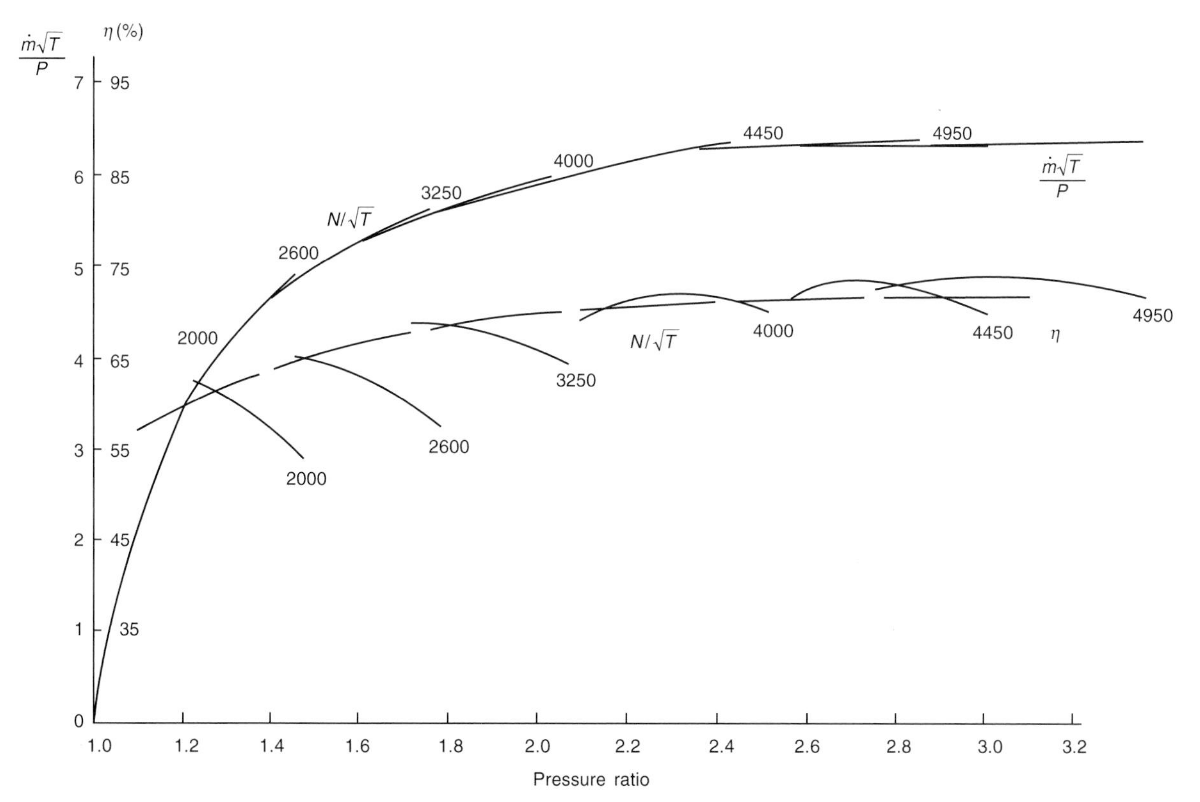

Figure 2.15 Turbine performance map, alternative presentation

different isentropic efficiency due to different heat transfer. As a result several turbocharger manufacturers do not reveal their turbine efficiency data since unrepresentative comparisons with those of another manufacturer could easily be made by a user.

2.4 Turbocharging systems–principles

Successful design of a turbocharged diesel engine is highly dependent on the choice of system for delivering exhaust gas energy from the exhaust valves or ports, to the turbine, and its utilization in the turbine.

Virtually all the energy of the gas leaving the cylinders arrives at the turbine. Some is lost on the way, due to heat transfer to the surroundings, but this is unlikely to exceed 5% unless water cooled exhaust manifolds are used, and will usually be much less. However, the design of the exhaust manifolds between the exhaust valve and turbine influence the proportion of exhaust gas energy that is available to do useful work in the turbine. An important parameter is the pressure in the exhaust system. Equation (2.18) shows that turbine power increases with pressure ratio (Po_3/P_4), hence the exhaust manifold pressure should be high. However, this implies that the pistion has to push the combustion products out of the cylinder against a high 'back-pressure', reducing the potential power output of the engine. Various turbocharging systems have been proposed to rationalize these apparently conflicting requirements. The most commonly used will be described herein. More complex systems that have been developed for special purpose applications are described in Chapter 3.

2.4.1 The energy in the exhaust system

The ideal thermodynamic cycles of engine operation were presented in Chapter 1. *Figure 2.17* shows the energy potentially available in the exhaust system, with an ideal cycle. The exhaust valve opens at BDC, point 5, where the cylinder pressure is much greater than the ambient pressure at the end of the exhaust pipe. If the contents of cylinder at EVO were somehow allowed to expand isentropically and reversibly down to the ambient pressure (to point 6), then the work that *could* be done is represented by the cross-hatched area 5-6-1. This work could be recovered by allowing the piston to move further than normal as shown in *Figure 2.17*. However, this requires an engine with an exceptionally long stroke and in practice it is found that the additional piston friction offsets the work gained by an ultra-long expansion stroke.

The work represented by area 5-6-1 is therefore potentially available to a turbocharger turbine placed in the exhaust manifold. It is called the 'blow-down' energy, since it involves the combustion products being 'blown-down' from cylinder pressure at point 5 to atmospheric pressure at point 6, when the exhaust valve opens. *Figure 2.17* represents a naturally aspirated engine. Consider now an ideal turbocharged four-stroke engine, as shown in *Figure 2.18*. Turbocharging raises the inlet manifold pressure, hence the inlet process (12–1) is at pressure P_1, where P_1, is above ambient pressure P_a. The 'blow-down' energy is represented by area 5-8-9. The exhaust manifold pressure (P_7) is also above the ambient pressure P_a. The exhaust process from the cylinder is represented by line 5, 13, 11, where 5, 13 is the 'blow-down' period when the exhaust valve opens and high pressure gas expands out into the exhaust manifold. Process 13, 11 represents the remainder of the exhaust process, when the piston moves from BDC to TDC displacing most of the gas from the cylinder to exhaust manifold. This gas is above ambient pressure and therefore also has the potential to expand down to ambient pressure whilst doing useful work. The potential work

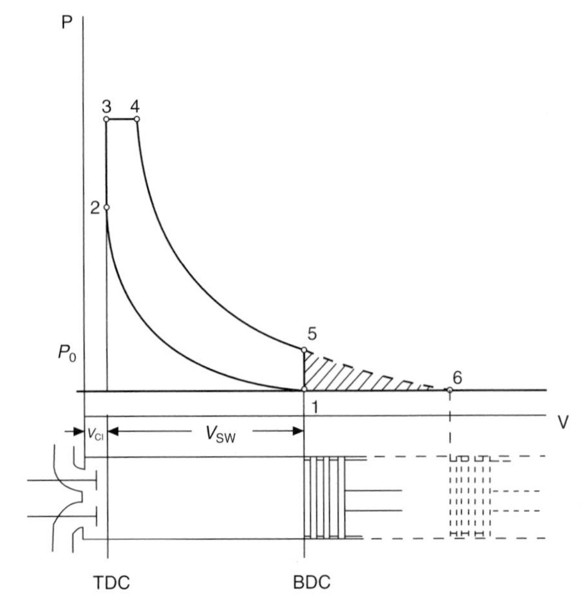

Figure 2.17 Ideal limited pressure cycle–naturally aspirated

that could be done is represented by the cross-hatched area 13-9-10-11. This work is done by the piston but could be recovered by a turbine in the exhaust. It will be called the piston pumping component of exhaust energy.

The maximum possible energy available to drive a turbine will be sum of areas 5-8-9 and 13-9-10-11, but it is impossible to devise a practical system that will harness all this energy. To achieve this, the turbine inlet pressure must instantaneously rise to P_5 when the exhaust valve opens, followed by isentropic expansion of the exhaust gas through P_7 to the ambient pressure ($P_8 = P_a$). During the displacement part of the exhaust process, the turbine inlet pressure would have to be held at P_7. Such a series of processes is impractical.

Consider a simpler process that would occur if a larger chamber were fitted between the engine and turbine inlet in order to damp down the pulsations in exhaust gas flow. The turbine acts as a flow restrictor creating a constant pressure (P_7) in the exhaust manifold chamber. The available energy at the turbine is given by area 7-8-10-11. This is the ideal 'constant pressure turbo-charging system'. Next consider an alternative system, in which a turbine wheel is placed directly downstream of the engine, very close to the exhaust valve. The gas would expand directly through the turbine along line 5-6-7-8, assuming isentropic expansion and no losses in the exhaust port. If the turbine were sufficiently large, both cylinder and turbine inlet pressure would drop to equal ambient pressure before the piston has moved significantly from BDC. Thus the piston pumping work would be zero during the ideal exhaust stroke and area 5-8-9 represents the available energy at the turbine. This is the ideal 'pulse turbocharging system'.

In practice the systems commonly used and referred to as constant pressure and pulse systems are based on these principles but are far from ideal. They will be described below.

Although the diagrams illustrating the energy available at the turbine are based on the four-stroke engine cycle, similar diagrams can be constructed for two-stroke engines. Apart from the change in valve or port timing, the work done by the piston is replaced by energy transferred from the compressor to scavenge air. For clarity, scavenge in the overlap period in the case of a four-stroke engine, has been ignored.

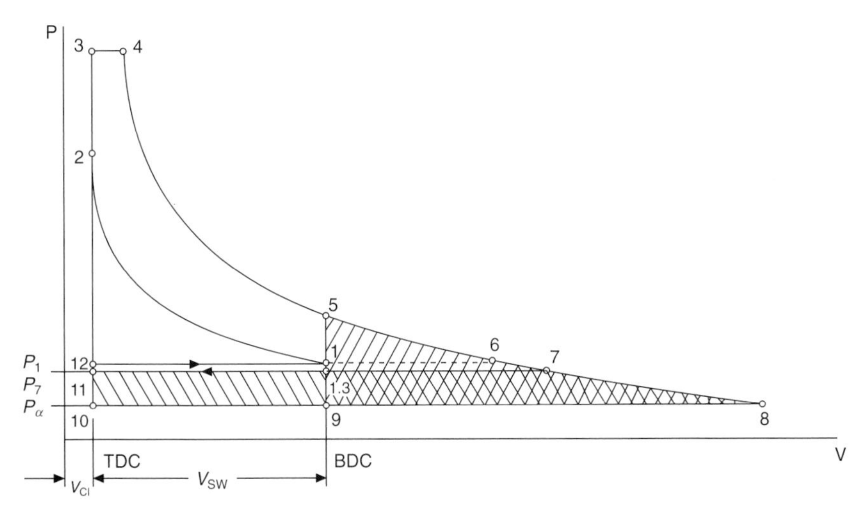

Figure 2.18 Ideal turbocharged limited pressure cycle

2.4.2 Principles of constant pressure turbocharging

With constant pressure turbocharging, the exhaust ports from all cylinders are connected to a single exhaust manifold whose volume is sufficiently large to ensure that its pressure is virtually constant. The unsteady exhaust flow processes at the cylinders are damped into a steady flow at the turbine. Only one turbocharger need be used, with a single entry from the exhaust manifold, but frequently several smaller units are fitted so that a reasonable boost pressure can be obtained in the event of a turbocharger failure. A major advantage of the constant pressure system is that turbine inlet conditions are steady and known, hence the turbine can be matched to operate at optimum efficiency at specified engine conditions. The main disadvantage is that the available energy entering the turbine is low, since full advantage has not been taken of the pulse energy. In *Figure 2.18* area 7-8-10-11 denotes energy available to the turbine, hence the energy represented by area 5-7-13 cannot be used. This energy is not lost, since energy loss only occurs by heat transfer, but since no work is done during the pressure reducing process 5-7 it represents a loss of potential turbine work.

Typically, a constant pressure exhaust manifold will consist of a large diameter pipe running along the exhaust side of an engine, with each exhaust port connected to it via a short stub pipe. On a 'vee' engine, the large bore manifold will usually lie between the banks with the inlet valves and manifolds arranged to be on the outside.

The volume of the exhaust manifold should be sufficient to damp pressure pulsations down to a low level. Thus the volume required will depend on cylinder release pressure (point 5) and frequency of the exhaust gas pulsations coming from each cylinder in turn. Pulse amplitude will be a function of engine loading (b.m.e.p. or boost pressure), the timing at which the exhaust valve or port opens, turbine area and exhaust manifold volume. Frequency will be dependent on the number of cylinders.

The effect of engine speed will be less significant since the duration of the exhaust process from each cylinder will be relatively constant in terms of crank angle, rather than time, and the turbine area will be chosen as a match at the operating speed and load. Thus it is inappropriate to give an exact rule that the manifold volume should be x times the total swept volume of the engine. Clearly x will be larger on an engine with few cylinders than on an engine with many cylinders, and its value will be some compromise between an acceptable total volume for installation and the volume required to damp out the pulses. For guidance however, it can be stated that the volume would normally be in the range of 1.4 to 6 times the total swept volume of the engine.

If the exhaust manifold volume is not sufficiently large, the 'blow-down' or first part of the exhaust pulse from a cylinder will raise the general pressure in the manifold. Since all cylinders are connected to the same manifold, it is inevitable (if the engine has more than three cylinders) that at the moment when the blow-down pulse from one cylinder arrives in the manifold, another cylinder is nearing the end of its exhaust process. The pressure in the latter cylinder will be low, hence any increase in exhaust manifold pressure will impede its exhaust process. This will be particularly important where the cylinder has both inlet and exhaust valves or ports partially open and is relying on a through-flow of air for scavenging the burnt combustion products. A rise in exhaust manifold pressure at this time is virtually inevitable in an engine with more than three cylinders, unless the volume is large. This will be particularly important on a two-stroke engine, since if the exhaust pressure exceeds inlet pressure during 'scavenging', the engine cannot run at all.

Any heat lost from the exhaust manifold will result in reduced energy available at the turbocharger turbine, therefore it is sensible to insulate the manifold. When the large surface area of the manifold is considered, it is not surprising that such insulation can significantly increase the boost delivered by the turbocharger compressor.

From a purely practical point of view, the exhaust manifold is simple to construct although it may be rather bulky, particularly relative to small engines with few cylinders. However, for large engines with many cylinders, the convenience of being able to join all cylinders to a common exhaust manifold with a single turbocharger on top, or at either end is useful. A major disadvantage of the constant pressure system arises from the use of an exhaust manifold having a large volume. When the engine load is suddenly increased or a rapid engine speed increase is required, the pressure in the large volume is slow to rise. Hence the energy available at the turbine increases only gradually. Turbocharger, and therefore engine response, will be poor. The poor response of the constant pressure turbocharging system restricts it from consideration for applications where frequent load (or speed) changes are required.

The turbocharging system will affect the engine through three parameters only, the boost pressure and temperature in the inlet manifold and the pressure in the exhaust manifold. Hence it is these factors that must be examined when considering a turbocharging system. The effect of the first two are obvious. The importance of the exhaust manifold pressure depends on whether the turbocharged engine is a four-stroke or a two-stroke. The subject will be discussed in detail later, but it is useful to consider, when comparing turbocharging systems, that several arrangements might enable a certain boost pressure to be developed. The merits of the different systems can then be compared by considering the exhaust manifold pressure developed and its effect on engine performance.

By considering the energy balance for the turbocharger when running with a constant exhaust pressure it is simple to derive a relationship between the exhaust manifold pressure (P_3) and the boost pressure (P_2). Compressor power must equal the product of turbine power and turbocharger mechanical efficiency, hence

$$\dot{W}_c = \dot{W}_t \times \eta_{\text{mech}} \quad (2.23)$$

The mass flow rate through the turbine ($\dot{m}_t$) must equal the mass flow rate through the compressor ($\dot{m}_c$) plus fuel flow rate ($\dot{m}_f$), if piston blowby is neglected, hence

$$\dot{m}_t = \dot{m}_c + \dot{m}_f \quad (2.24)$$

or

$$\frac{\dot{m}_t}{\dot{m}_c} = 1 + \frac{1}{AFR}$$

where AFR = air-fuel ratio.

Combining eqns (2.14), (2.18), (2.23) and (2.24) gives

$$\left[\left(\frac{P_{02}}{P_{01}}\right)^{(\gamma a-1)/\gamma a} - 1\right] = \left[1 - \left(\frac{P_4}{P_{03}}\right)^{(\gamma e-q)/\gamma e}\right] \times \left[1 + \frac{1}{AFR}\right] Cp_e/Cp_a\,[T_{03}/T_{01}]\,\eta \quad (2.25)$$

where η = overall turbocharger efficiency.

Thus the relationship between the inlet manifold pressure (P_{02}) and exhaust manifold pressure (P_{03}) is a function of the overall turbocharger efficiency (η), the turbine inlet temperature (T_{03}) and, to a lesser extent, the air-fuel ratio (AFR).

The air–fuel ratio at full load, will be governed by thermal loading or the onset of black smoke in the exhaust. The turbine inlet temperature will also be dependent on air–fuel ratio, the amount of cool scavenge air passing through the cylinders and heat loss from the exhaust manifold.

Equation (2.25) is plotted, for a compressor pressure ratio of 2, in *Figure 2.19*, showing the dominant effects of turbocharger efficiency and turbine inlet temperature on the pressure ratio between inlet and exhaust manifolds. It is important to realize that if the turbine inlet temperature is held constant (for metallurgical reasons), then higher turbocharger efficiencies are required to maintain a favourable scavenge pressure ratio at higher boost pressures.

Although the turbocharger efficiency (and air–fuel ratio, via the turbine inlet temperature) governs the relationship between exhaust and inlet pressures, it is the turbine area that controls

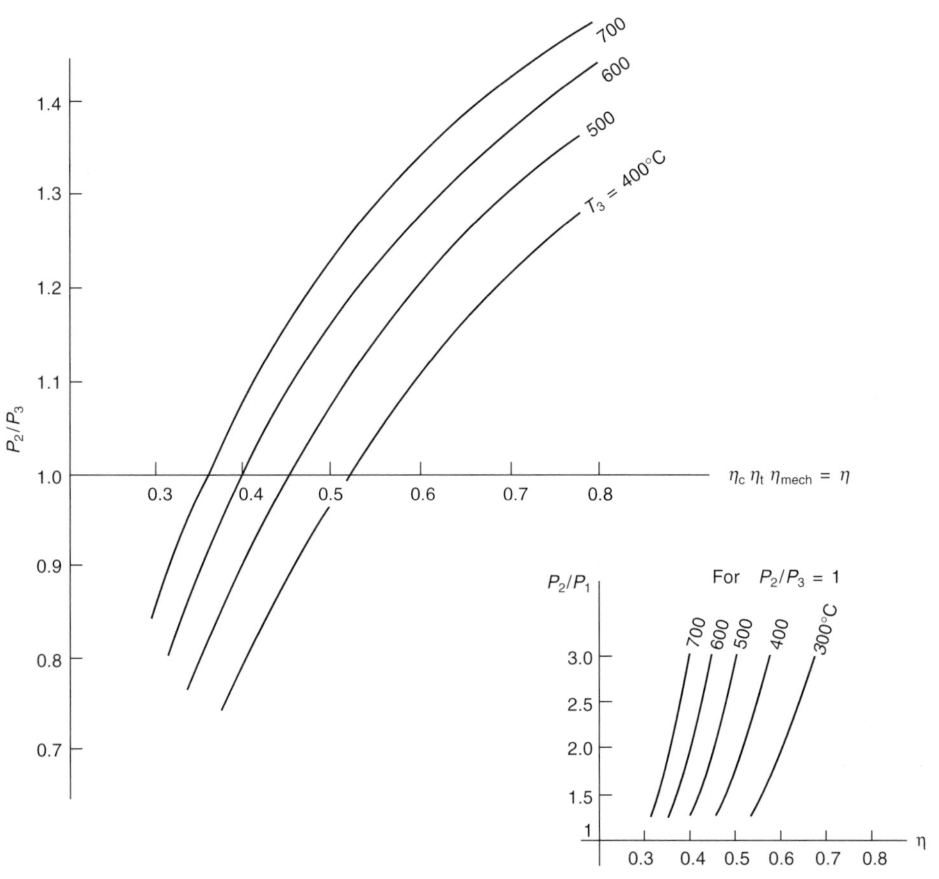

Figure 2.19 Inlet/exhaust manifold pressure ratio as a function of turbocharger efficiency and turbine inlet temperature (compressor inlet temperature = 300°K, pressure ratio = 2:1)

the exhaust pressure itself, since it acts as a restricting orifice. By fitting smaller and smaller turbines to a four-stroke engine it is theoretically possible to develop very high exhaust pressures and correspondingly high levels of energy available at the turbine. This should allow very high boost pressure to be obtained (subject to the relationship described above). However the practical effects of turbine speed and inlet temperature limitations combined with decreasing turbocharger efficiency at high pressure ratios and poor scavenging, prevent very high pressure ratio being developed in practice. Thus it is the turbine area that principally decides the exhaust manifold pressure and the air–fuel ratio and turbocharger efficiency that fix the relationship between inlet and exhaust pressure at the cylinders.

The choice of boost pressure for a specific engine, although developed by the turbocharger, will be limited by specific constraints imposed by engine design. For example the engine will be designed to sustain a certain level of cylinder pressure, thermal loading, etc. Thus the engine designer will have some clear idea of the boost pressure required, and the turbine area will be matched to provide it.

With constant pressure turbocharging, the amount of scavenge air that passes through the cylinder (expelling exhaust residuals out of the clearance space) is directly governed by the pressure drop between inlet and exhaust manifold and the valve overlap period, if the engine is a four-stroke. A two-stroke engine is not self-aspirating and hence this pressure drop is the only factor that expels the exhaust gas and charges the cylinder with fresh air. This difference has major implications concerning the turbocharging system. Therefore the application of the constant pressure system to two- and four-stroke engines will be discussed separately.

2.4.2.1 Constant pressure turbocharging of four-stroke engines

Four-stroke engines are self-aspirating. They have a discrete intake and exhaust stroke. Virtually regardless of the pressure in the exhaust manifold, piston motion during the exhaust stroke will displace most of the gas, thus a four-stroke engine will run with a high 'back-pressure'. However, the situation is undesirable for three reasons. Firstly, work is done by the piston in expelling exhaust gas resulting in less useful engine power output and lower efficiency. Secondly, if the exhaust pressure exceeds the inlet pressure, a considerable quantity of residual gas will be left in the cylinder, reducing the volume of fresh air drawn in during the next intake stroke. Thirdly, some blow-back of combustion products (residual gas) into the intake manifold may occur during valve overlap, resulting in an undesirable build-up of carbon deposits. Clearly it is desirable to avoid developing an exhaust manifold pressure greater than that in the inlet manifold (compressor delivery).

Since naturally aspirated engines run with virtually equal inlet and exhaust pressures, no significant scavenging of the residual gas takes place. When turbocharging, advantage can be taken of the potential difference in manifold pressures to generate a scavenge air throughput to clear the cylinder of residual combustion products. Thus a pressure drop between intake and exhaust is desirable, especially during the period of valve overlap. The magnitude of the pressure drop required to achieve good scavenging without an excessive and wasteful throughflow of air, will be dependent on the amount of valve overlap used on the engine. Valve timing will itself depend on the primary application of the engine, since an engine with large valve overlap will not run well over a wide speed range.

Consider the idealized intake and exhaust process shown in *Figure 2.18*. During the intake process (12-1) the pressure on the piston crown (the boost pressure P_1) exceeds the crankcase pressure on the underside (ambient pressure P_a). Hence useful work will be done by the compressed fresh air on the pistion (denoted by the area 12-1-9-10). During the exhaust stroke (the displacement 13-11), gas pressure on the piston crown (P_7) again exceeds crankcase pressure (P_a) but the piston motion opposes the resultant force, hence the piston is doing work on the exhaust gas.

The net gain or loss (gain in this case) of useful work during this 'gas exchange' process will be given by the area 12-1-13-11. This work benefit is only gained at the expense of compressor work, although that is desirable since it in turn is derived from exhaust gas energy normally wasted. This is one reason why a turbocharged engine may be more efficient than a naturally aspirated engine. High turbocharger efficiency will raise the inlet manifold to exhaust manifold pressure ratio (P_2/P_3; *Figure 2.19*) and increase this work gain, *Figure 2.20* shows the direct effect of this gain on engine specific fuel consumption[1].

The present discussion of constant pressure turbocharging has related only to full engine load and speed. However, engines for most applications are required to operate at part load for much of the time. The engine is controlled via its fuel injection system, hence under part load conditions the *volume* of fresh air drawn in to the cylinder will not change significantly but the quantity of fuel injected will. Thus the principle change will be to the air-fuel ratio and hence, as far as the turbocharger is concerned, the exhaust temperature. The lower exhaust temperature is equivalent to reduced energy arriving at the turbine, hence turbine and therefore compressor work drops. The boost pressure is therefore lower resulting in a reduced *mass* of fresh charge being drawn in to the cylinder. Hence the boost pressure, mass flow rate and turbine inlet temperature all fall. The drop in boost pressure is to be expected and will not create any problems unless the exhaust pressure has not dropped by a comparable amount.

Consider again the energy balance for the turbocharger (eqn 2.25). At low loads, the turbine inlet temperature will drop and, since the mass flow rate and pressure ratios will also have fallen, the turbocharger is probably operating off the design point conditions on the turbine and compressor characteristics. Clearly at part-load engine operation, the pressure drop between inlet and exhaust will deteriorate and will eventually become negative. Scavenging will be impaired, the gas-exchange work will become a loss and hence power output and efficiency will fall. Thus the constant pressure system is not ideal for part load operation.

Similar conditions apply at high load and low speed operation of the engine (i.e. automotive applications). The turbine area is too large at these conditions, leading to a small compressor pressure rise and low efficiency performance. However, the situation is not as serious as that at part load, since the air-fuel ratio and hence turbine inlet temperature remains high.

Examples of the application of constant pressure turbocharging to four-stroke engines are given in later chapters.

2.4.2.2 Constant pressure turbocharging of two-stroke engines

The two-stroke engine is neither self-aspirating nor self-exhausting. It relies on a positive pressure drop between the inlet and exhaust manifold in order to run at all. The scavenging process, in which fresh charge is forced in and residual gas out, is the key to a successful two-stroke engine. It follows that the two-stroke engine is far more dependent on a reasonable pressure drop being developed across the cylinder than is the case with the four-stroke engine. The four-stroke engine can work with an adverse pressure gradient, the two-stroke will not.

Several scavenging systems have been developed for two-stroke engines. Those in common use are the cross-loop and uniflow scavenge systems and are described in later chapters.

To obtain reasonable scavenging, it is essential to pass some

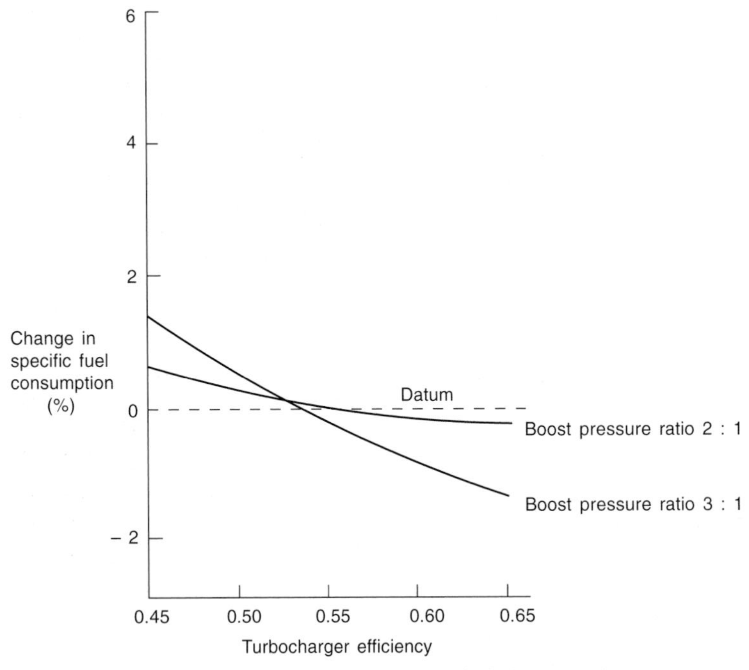

Figure 2.20 Effect of turbocharger efficiency on specific fuel consumption, constant pressure system (Ryti and Meier)

air right through the system and to waste, since considerable mixing of air and combustion products takes place. The amount of 'excess air' required will depend on the scavenging efficiency of the system being used and many other factors. It is therefore impossible to quote general rules, but the amount will vary from 10% to about 40% (from uniflow to cross scavenging).

The consequence to the turbocharger is twofold. Firstly, the gas in the exhaust manifold will be diluted with cool scavenge air lowering the turbine inlet temperature. Secondly, a penalty must be paid for compressing the excess air since, although it will be expended through the turbine in due course, only part of the energy expanded will be regained, due to compressor and turbine inefficiencies. Thus not only does the turbocharger have a more difficult job than on a four-stroke engine, since it *must* provide a positive pressure drop across the cylinder, but it is required to work under the adverse conditions mentioned above. It was for these reasons that turbocharged two-stroke diesel engines first appeared with some form of compressor assistance, and many now operating on the constant pressure system, still retain it.

The general equation for the energy balance of a turbocharger under constant pressure operation (eqn 2.25) and the curves that result (*Figure 2.19*) are just as valid for the two-stroke engine as the four-stroke, but the exhaust temperature will be lower. Typical values of mean turbine inlet temperature on highly rated two-stroke and four-stroke engines at full load might be 400°C and 500°C respectively. *Figure 2.19* (inset) shows that for a pressure ratio of 2.5:1, a four-stroke engine will perform adequately if the turbocharger efficiency is 50% or more, whilst the two-stroke engine will require a turbocharger efficiency of over 55%.

A large, well-designed turbocharger can develop 60–65% overall efficiency, but some allowance of about 5% must be made for a reduction in efficiency in service (due to fouling—build up of dirt and carbon). In practice this means that a well-designed engine having a good scavenging system, with little pressure loss, will run satisfactorily at full load provided that it has a well matched, efficient turbocharger. Large, slow speed engines are in this category at the present time but no small, higher speed two-stroke engine can run with a constant pressure turbocharging system without some other aid to scavenging. This is due to the lower efficiency of very small turbochargers and the high pressure losses in the scavenge ports and valves of a high speed engine.

At part load, the turbine inlet temperature will fall due to the lower air–fuel ratio. Turbocharger efficiency will probably also fall since the turbocharger would usually be matched for optimum working conditions near full load (although this depends on engine duty). It is obvious from *Figure 2.19* (inset) that if the turbine inlet temperature drops below 300°C and the turbine efficiency below 55%, then the engine will stop. Thus all two-stroke engines using the constant pressure turbocharging system require some additional aid for scavenging for starting and part load operation. In the past, the underside of the piston was used in cross-head type engines, as a compressor placed in series with the turbocharger compressor. As engine and turbocharger design have improved the contribution of the scavenge pump to overall compression has reduced, so that simple electrically driven fans are used today. These are switched off once the engine load and speed are such that the turbocharger can provide the necessary positive pressure differential across the cylinders.

The use of the constant pressure turbocharging system on two-stroke engines is usually restricted to large engines, since most other engines are required to operater over a wide load and speed range, and also because of the requirement for high turbocharger efficiency (which is easier to meet with large turbochargers). However, a few medium sized and smaller two-stroke engines have used the system. Most of these engines use some form of Roots or screw (Lysholm) blower as a scavenge pump, driven direct from the crankshaft. Typically however, most two-stroke engines with Roots or screw scavenge blowers are either pulse turbocharged or are not turbocharged at all.

2.4.3 Principles of pulse turbocharging

The majority of turbocharged engines use the pulse, not the constant pressure turbocharging system. However, the pulse system that has been developed is not the pure impulse system described in section 2.4.1 which is impractical, but a system that tries to make some use of the available energy of pure pulse and constant pressure systems. The objective is to make the maximum use of the high pressure and temperature which exists in the cylinder when the exhaust valve opens, even at the expense of creating highly unsteady flow through the turbine. In most cases the benefit from increasing the available energy will more than offset the loss in turbine efficiency due to unsteady flow.

The key to the pulse system is to try to use the additional (relative to a constant pressure system) energy represented by area 5-7-13 in *Figure 2.18*. This requires the turbine inlet pressure to suddenly rise to P_5 when the exhaust valve first starts to open, then fall along line 5, 6, 7. Consider an exhaust port connected to a turbine by a very small exhaust manifold, the turbine also having a small effective flow area. Initially, before the exhaust valve opens, the manifold may be at atmospheric pressure. As the exhaust valve begins to open, exhaust gas flows from cylinder to manifold under the influence of the large pressure drop. Mass flow rate rapidly increases as the valve opens. Since the turbine acts as a flow restriction, the pressure in the manifold builds up, governed by the difference in flow into the manifold and out through the turbine. Thus exhaust manifold (turbine inlet) pressure tends to rise towards cylinder pressure.

Once the peak mass flow rate into the manifold has passed, exhaust gas may be flowing out through the turbine faster than the flow rate from cylinder to manifold, hence manifold pressure gradually drops, reaching the atmospheric value shortly after the exhaust valve closes. This unsteady flow process is shown in *Figure 2.21*, for a single cylinder, four-stroke engine[2]. The exhaust manifold pressure rises from atmospheric pressure to a peak just after BDC, akin to movement from P_9 to P_5 in the ideal cycle diagram, and then falls back to atmospheric pressure after EVC, akin to movement from P_5 to P_8. Note that during the exhaust stroke of the piston, from BDC to TDC, exhaust gas is pushed out from cylinder to manifold. It passes through the turbine doing useful work. The actual exhaust process (*Figure 2.21*) is similar but not identical to the ideal process of *Figure 2.18* in which the exhaust manifold pressure would instantaneously rise to P_5, gradually fall to P_7 then remain constant until the exhaust valve closed. Thus a considerable protion of the blow-down (area 5-8-9) and piston pumping (area 13-9-10-11) energies are made available to the turbine. Relative to the constant pressure system, the energy available to the turbine is greater.

The pulse turbocharging system requires the exhaust manifold to be as small as reasonably possible in order that the turbine inlet pressure should rapidly rise to almost equal cylinder pressure when the exhaust valve opens. It is also desirable to open the exhaust valve rapidly for the same reason. In practice valve train inertia and cam stresses limit valve acceleration and turbocharger location influences exhaust manifold volume. The turbocharger must be mounted as close to the cylinders as possible to reduce the length of the exhaust manifold. The cross-sectional area of the manifold cannot reasonably be significantly less than that at the exhaust valve when the latter is at full lift. These together define the minimum exhaust manifold volume. Virtually all turbocharged engines are multi-cylinder, hence several exhaust ports must be connected to the turbocharger turbine. To keep exhaust manifold volume small, the exhaust ports are connected to the turbine by short, narrow diameter pipes. Rules for joining pipes from various cylinders together will be given later.

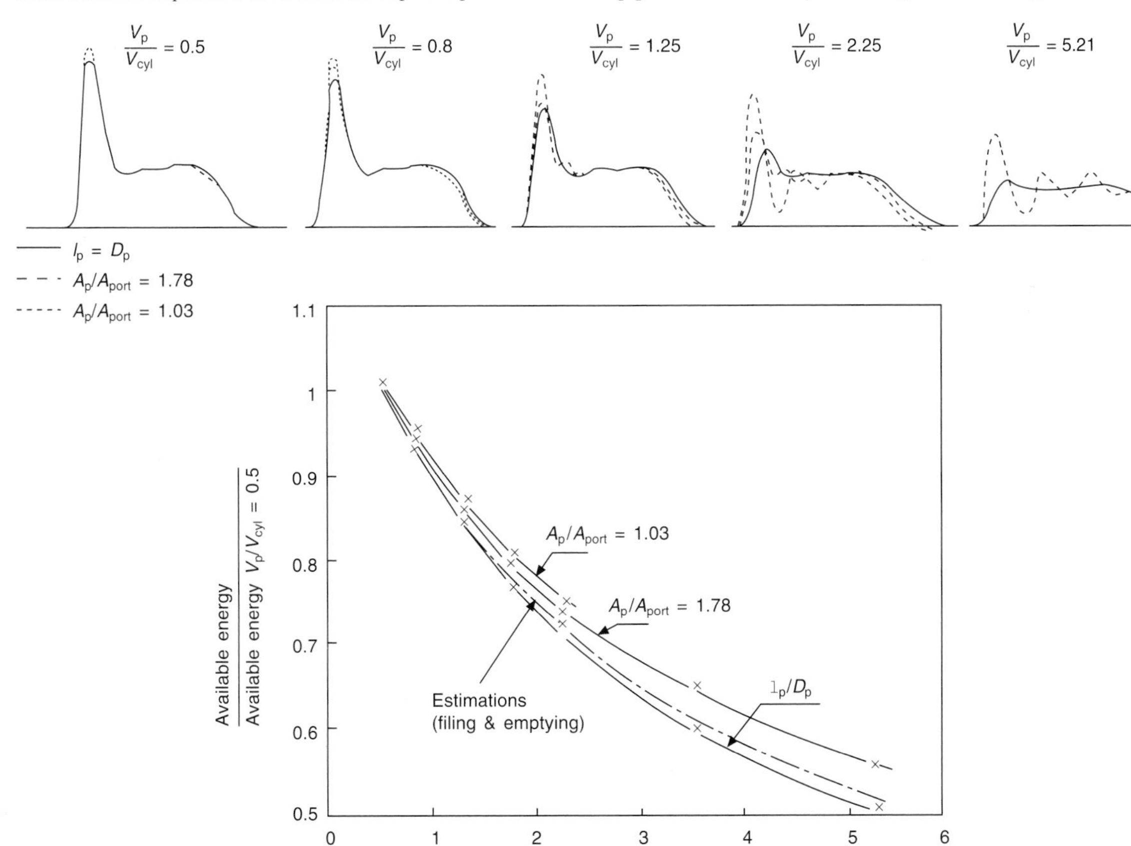

Figure 2.21 Effect of exhaust manifold shape on available energy (Janota)

A typical example illustrating the gain in available exhaust gas energy when employing narrow pipes is shown in *Figure 2.21*. This test data was obtained on a single cylinder, loop-scavenge, high speed two-stroke engine with three types of exhaust manifolds, a compact volume (pipe diameter = pipe length) and two narrow pipes (pipe area/cross-section port area ratio of 1.78 and 1.03). All the tests refer to the same engine speed, b.m.e.p., boost pressure, air flow rate and turbine area. The main diagram shows how turbine available energy varies with exhaust pipe volume (non-dimensionalized over datums of available energy if manifold volume equals half of cylinder volume), for all three manifolds. Note that increasing exhaust manifold volume by a factor of 5 halves turbine available energy. It is also interesting to note that there is little difference between the three types of manifold (volume or long narrow pipe) provided that their volumes are equal. The top set of diagrams shows the pulsating exhaust manifold pressure, for the three types of manifold, with five gradually increasing total exhaust manifold volumes. In the first two cases (exhaust manifold volume = 0.5 or 0.8 times cylinder volume), the type of manifold, whether a long narrow pipe or a plenum, has little effect. However, in the last two diagrams there are substantial differences. These differences are caused by pressure wave reflections in the very long narrow pipes.

If the exhaust pipe is narrow, then the cross-sectional area of the pipe itself is sufficient for pressure to build up at the exhaust valve end as the exhaust valve opens. A pressure pulse is built up (hence the term pulse turbocharging). This pressure pulse or wave, travels at sonic velocity along the pipe to reach the turbine. Thus energy is being transmitted along the pipe at sonic velocity to the trubine. At the turbine the pressure wave is reflected with reduced amplitude since the turbine is a partial flow restriction. Thus a pressure wave is generated by the gas initially released at high pressure at EVO by the cylinder, which travels forwards and backwards along the pipe with gradually diminishing amplitude. This differentiates the performance of a long narrow pipe from a plenum of the same total volume, although the available energy at the turbine is not greatly different (*Figure 2.21*).

The important difference is shown in *Figure 2.22*. Here the effect of varying pipe length on the timing of the reflected pulse is shown. In case 1, the reflection occurs after the exhaust valve has closed causing no problem, but this is a rare case since it can only occur with an exceptionally long manifold. More common is case 4, in which the reflection time is very short relative to the valve opening period. Case 2 is the serious one, that can occur with long pipes, of the reflected pulse raising exhaust pressure at the valve or port, during the scavenge period. The turbocharger position and exhaust pipe length must be chosen to avoid this situation, or scavenging will be seriously impaired.

In *Figure 2.22* the pressure waves are shown on a time basis, but the figures can be considered on a crank-angle basis. It will be then evident that as engine speed changes (horizontal scale magnifies or reduces), the effective time of arrival at a reflected pulse, in crank-angle terms, will vary. Hence the exhaust pipe length is critical and must be optimized to suit the speed range of the engine. The interference of reflected pressure waves with the scavenging process is the most critical aspect of a pulse turbocharging system, particularly on engines with a large number of cylinders. It is on these engines that the wave travel time, when expressed in terms of crank-angle, will be longest since:

reflection time $t = \dfrac{2l}{a}$ (from valve to turbine and back)

where a = speed of sound and l = pipe length.

If the engine speed is N (rev/min), the reflection time or lag in terms of crank angle ($\Delta\theta$) is given by:

$$\Delta\theta = N \times 360 \times 2l(a \times 60) = 12lN/a$$

The speed of sound (a) may be expressed as:

$$a = \sqrt{\gamma RT}$$

where T is the mean gas temperature. Hence:

$$\Delta\theta = \frac{12lN}{\sqrt{\gamma RT}} \tag{2.26}$$

Evidently if the engine speed is high and the pipe length long, $\Delta\theta$ will be large. An automotive diesel running at 2000 rev/min with a 450°C exhaust temperature and a pipe length of 0.8 m would experience a lag of 35 degrees. However at low speed the lag would be less (18 degrees at 1000 rev/min). Only if a very long exhaust manifold were used would direct reflection be as bad as case 2. *Figure 2.22*, on such an engine. Reflections can also occur from closed exhaust valves at other cylinders.

The basic method of increasing (or reducing) the available energy at the turbine is common to both pulse and constant pressure systems. The available energy is governed by, amongst other factors, the pressure at the turbine inlet or the exhaust manifold pressure. This in turn is controlled by the flow area of the turbine. By reducing the turbine area, pressure will be increased and vice versa (*Figure 2.23*). This will be illustrated, by example, in section 2.6.7.2.

So far, factors influencing the available energy at the turbine have been considered. Pulse turbocharging increases available energy but creates a highly unsteady flow through the turbine. How does the turbine behave under unsteady flow, and what is its efficiency? The energy is contained within the exhaust gas flow and can be found by applying the turbine power eqn (2.18) instantaneously and integrating over an engine cycle (*Figure 2.24*). By this means the variation in turbine inlet temperature and pressure plus turbine efficiency are accounted for.

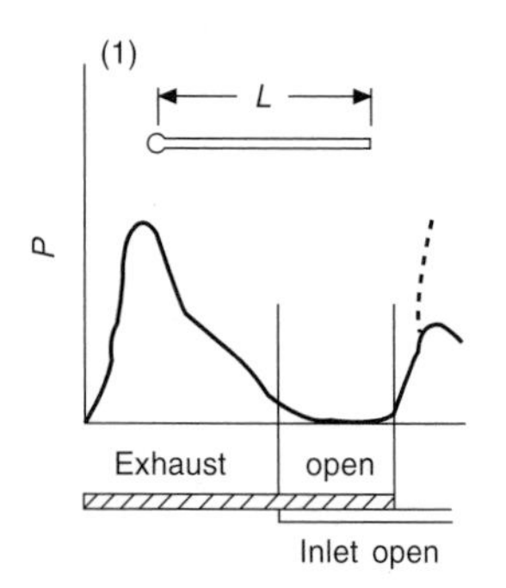

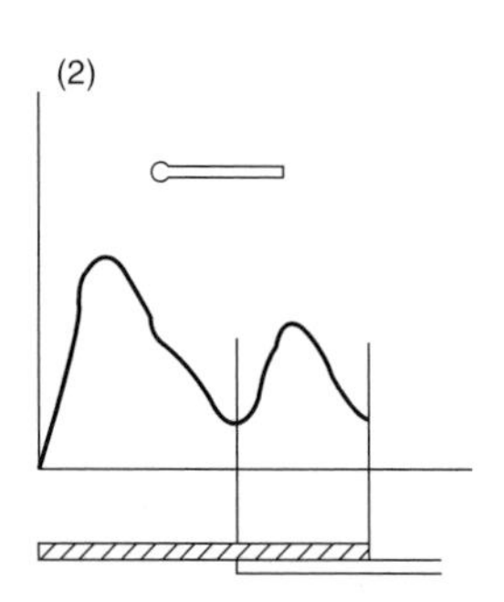

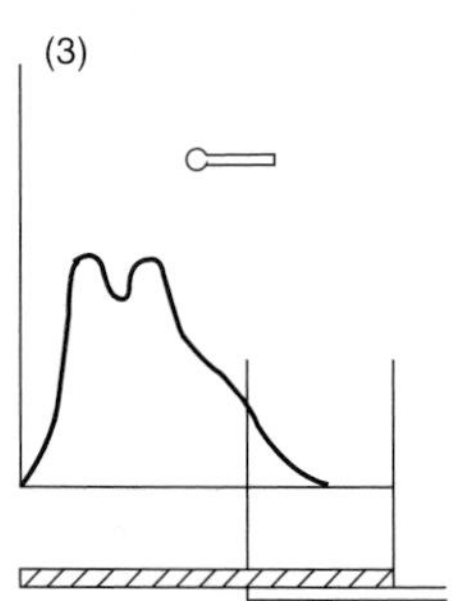

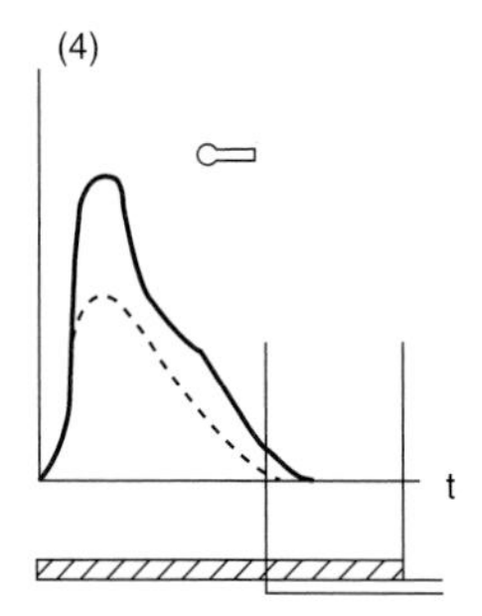

Figure 2.22 Effect of exhaust pipe length on pressure wave reflections relative to the valve overlap period

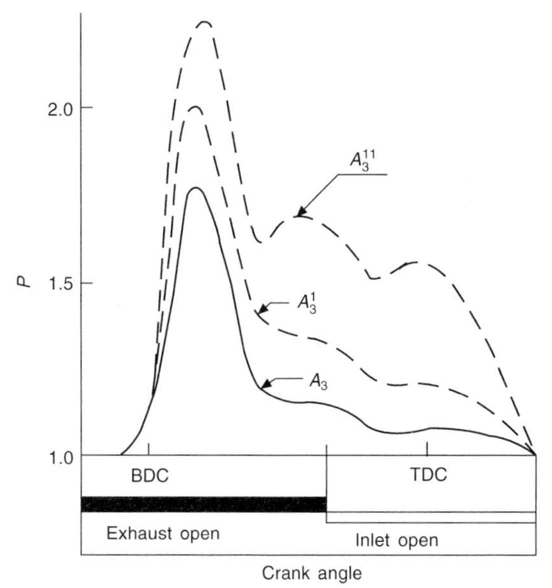

Figure 2.23 Effect of reducing turbine area on instantaneous pressure at turbine entry ($A_3 > A_3' > A_3''$)

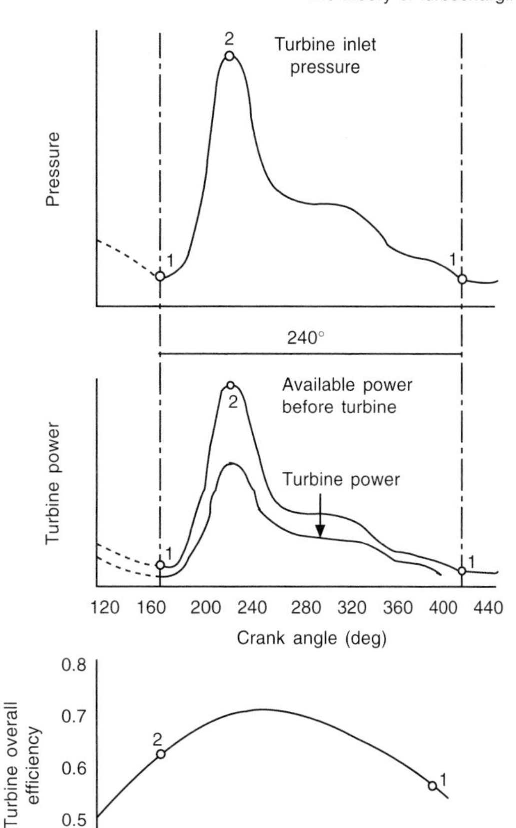

Figure 2.24 Turbine inlet pressure, power and efficiency curves

In *Figure 2.24* (*lower*) a typical turbine efficiency curve is shown. If quasi-steady flow is assumed, then the instantaneous turbine efficiency will be that corresponding to the appropriate instantaneous value of blade-speed ratio (U/C_s). The values of efficiency coinciding with the peak and a low point of the pressure diagram are illustrated. For much of the remainder of the engine cycle the turbine will be windmilling and decelerating. When the next exhaust pulse arrives, part of its energy will be used to accelerate the turbocharger again. Thus energy is wasted due to a low average turbine efficiency resulting from the very unsteady flow and windmilling no-flow conditions.

On a multi-cylinder engine, narrow pipes from several cylinders can be connected via a single branch manifold to one turbine. Consider the three-cylinder four-stroke, automotive engine shown in *Figure 2.25*. Due to the phase angle between cylinders the opening periods of the exhaust valves follow successively every 240° with very little overlap between them. Thus a steady 'train' of pressure pulses arrive at the turbine, virtually eliminating the long periods of pure windage (*Figure 2.26*), although the average turbine efficiency will remain lower than that obtained with a correctly matched constant pressure system (operating near the peak of the efficiency curve). The remaining important point to consider is the exhaust pressure close to the valves, during the valve overlap (scavenging) period. As with the constant pressure system, a good pressure drop between inlet and exhaust manifold during the period when both valves are open is important in the case of a four-stroke engine with significant valve overlap and vital for a two-stroke engine. In *Figure 2.26* the pressure history in the inlet and exhaust manifolds (at a valve) is shown, the pressure drop during the period of valve overlap being cross-hatched. Clearly, at the running condition shown, the pressure drop is satisfactory.

The diagrams in *Figure 2.25 and 2.26* have been deliberately drawn for a three-cylinder four-stroke engine with short, narrow exhaust pipes and automotive type valve timing. Such an engine is rarely made, but is a convenient unit on which to present some important aims of pulse turbocharging, namely, the way to increase average turbine efficiency by reducing windage periods, whilst avoiding interference with scavenging of one cylinder due to the effect of the blow-down pulse from another. The pressure pulse exhausting from a cylinder travels along the manifold until it reaches a junction. At the junction it divides into two pulses (each of smaller magnitude due to the effective area increase) one travelling down each adjacent pipe. One pulse will travel towards the turbine, the other will arrive at the exhaust valve of another cylinder. It is the latter pulse, from cylinder number 3 (*Figure 2.25*), that has arrived near cylinder 1 just at the end of the scavenge period of cylinder 1, that could be a problem. If it had arrived earlier (perhaps due to shorter exhaust pipes) it would have interfered with scavenging. This type of interference due to the *direct* action of a pressure wave from another cylinder, is quite separate from the action of a pressure pulse *reflected* from the turbine, whether the latter started from cylinder 1, 2 or 3.

The pipe lengths in *Figure 2.25* have deliberately been kept short so that the reflected pulses are almost superimposed on the initial pulse (condition 4 in *Figure 2.22*). This will not always be the case so the effect of both types of possible pulse interference must be considered. It should be mentioned that if the direct pulse meets a closed exhaust valve at cylinder 1 (as it has in the figure) it will be reflected and will eventually arrive at the turbine some time after the first component of the pulse from cylinder 3.

The overall pressure wave system that occurs in a manifold such as that in *Figure 2.25* could be very complex, with pulses propagating from each cylinder, pulse division at each junction, total or partial reflection at an exhaust valve (dependent on whether it is closed or not) and reflection from the turbine. Since the turbine is a restriction, not a complete blockage, the

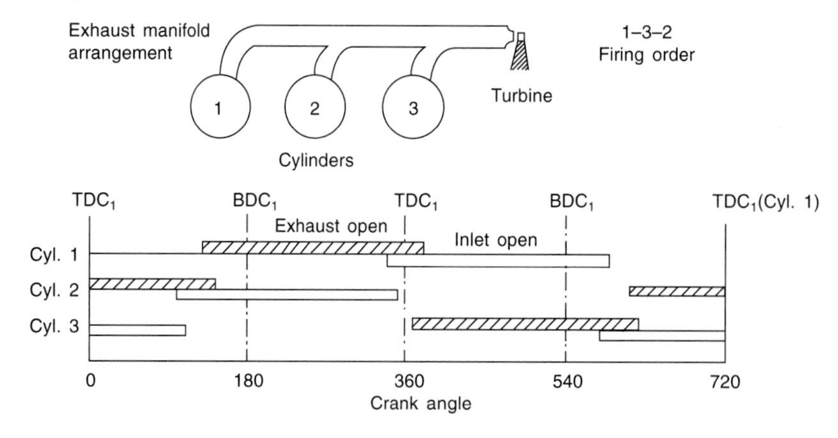

Figure 2.25 Pulse system–three-cylinder four-stroke engine with automotive type valve timing

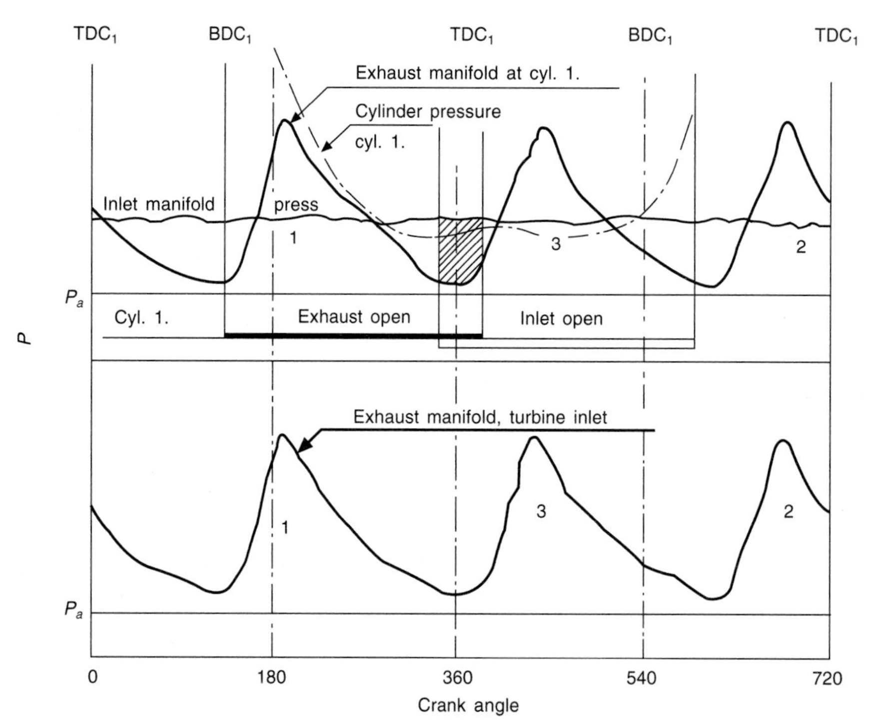

Figure 22.6 Exhaust pressure diagrams–three-cylinder four-stroke engine with automotive type valve timing

reflection from it is only partial. Fortunately, pulse division at junctions weakens the pulses (due to the area increase) and hence the system of pressure waves is weakened at each point, and successively with each reflection. This it is usually the direct pulses and those from the first reflections that are important. The complexity of the pressure wave system does, however, make theoretical calculations a difficult process.

Most engines have four or more cylinders, but it is convenient to consider a six-cylinder engine next. *Figure 2.27* shows valve timing of a typical automotive six-cylinder diesel engine and its firing order. It is obvious that if all six cylinders were connected to a single entry turbine via narrow pipes, the pressure waves from each cylinder would significantly interfere with the exhaust processes of each other during valve overlap and the exhaust stroke, thus increasing piston pumping work. The effect would be poor engine efficiency. A two-stroke engine might not operate under these conditions. The difficulty can be avoided by simply connecting the cylinders in two groups of three, either to two different turbines, or separate entries of a single turbine. If the correct cylinders are grouped together, then the pressure pulse system in each group will be the same as that shown in *Figure 2.26*. From *Figure 2.27* it is clear that cylinders 1, 2 and 3 may form one group and cylinders 4, 5 and 6 the other, but the arrangement would differ if the firing order were changed. It may be concluded that the six-cylinder engine is similar to the three-cylinder, from the turbocharging point of view, but turbine performance may be slightly worse due to the losses associated with the join of two sectors of a divided entry turbine.

It is disadvantageous to connect more than three cylinders to a single turbine entry. Thus, for the four-cylinder engine shown in *Figure 2.28* pairs of cylinders (1-2 and 3-4) would be connected to a double entry turbine. On engines with other numbers of cylinders, the general rule will be to connect cylinders whose firing sequences are separated by 240° crank angle (in the case of four-stroke) and 120° (two-stroke) to a turbine inlet, and select those cylinders whose exhaust processes are evenly spaced out. However, this is not always possible. For example, on a vee-form engine, the vee angle will introduce an additional phase difference to the firing intervals between cylinders. In such cases, the more basic rule of avoiding direct pressure wave

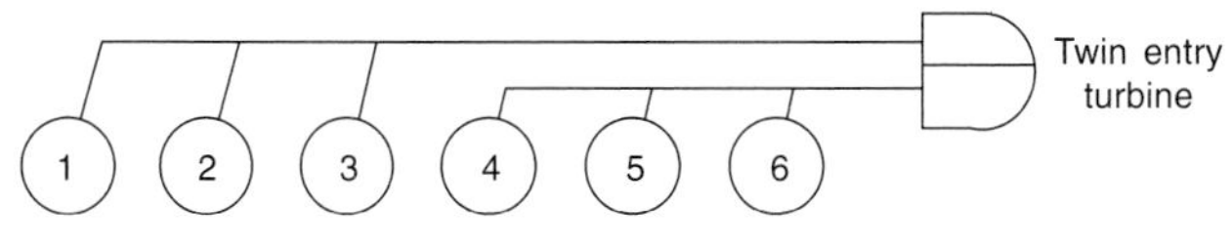

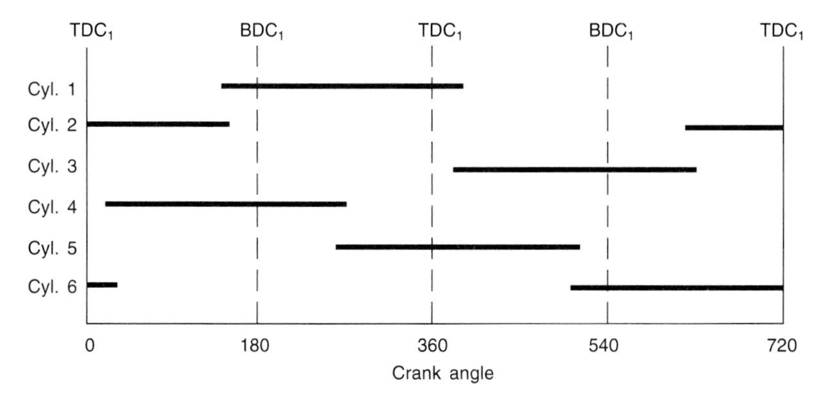

Figure 2.27 Exhaust valve timing for an automotive six-cylinder four-stroke engine

interference must be observed. Some exhaust manifold arrangements for common firing orders on several engine forms are given in *Figure 2.29*. In several cases (7, 8 and 12 cylinder engines) an additional option will be a choice between two twin-entry turbochargers or one four-entry unit. The former option is more commonly used, since end of sector losses in the turbine are reduced and two smaller turbochargers are often cheaper and easier to instal.

Before discussing the application of the pulse system to four-stroke and two-stroke engines in detail, some general points relating to the pulse system are appropriate. The principal advantage of the pulse over the constant pressure system is that the energy available for conversion to useful work in the turbine is greater. However, this is of little value if the energy conversion process is inefficient. The operation of radial and axial flow turbines under partial admission (i.e. multiple entries), unsteady flow conditions is complex. The single pulse developed in *Figure 2.24* clearly will result in low average turbine efficiency due to a long windage period, and quite significant mass flow when the equivalent steady-state turbine efficiency is known to be low. Thus the benefits of pulse energy will be lost by low turbine efficiency. Of course, it is difficult to use the alternative 'constant pressure' system with a single cylinder engine, the example has merely been given to illustrate the penalty when one cylinder of a multi-cylinder engine is connected to a single turbine entry (unfortunately this is necessary on some engines, for example, the five-cylinder engine in *Figure 2.29*).

With three cylinders to a turbine entry (*Figure 2.26*) the average turbine efficiency will be much higher since windage is almost eliminated. The efficiency is better still if the valve timing permits a larger overlap by having longer exhaust periods (290°) as is the practice in medium speed diesel engines. However, turbine efficiency, averaged over the unsteady flow cycle, will be lower than that obtained in a well-matched steady-flow system. If two cylinders are connected to a turbine entry the average turbine efficiency will be lower than would be the case with three cylinders, since (short) windage periods would exist (see *Figure 2.28*). Thus the pulse turbocharging system is most suitable for those engines whose exhaust manifolds may connect groups of three cylinders to a turbine entry. However, even if this is not

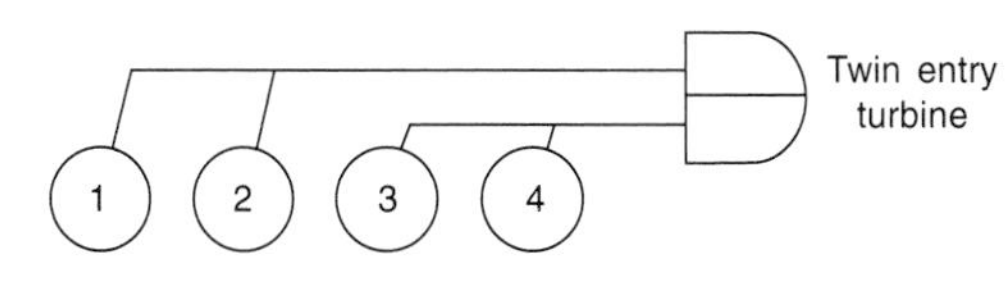

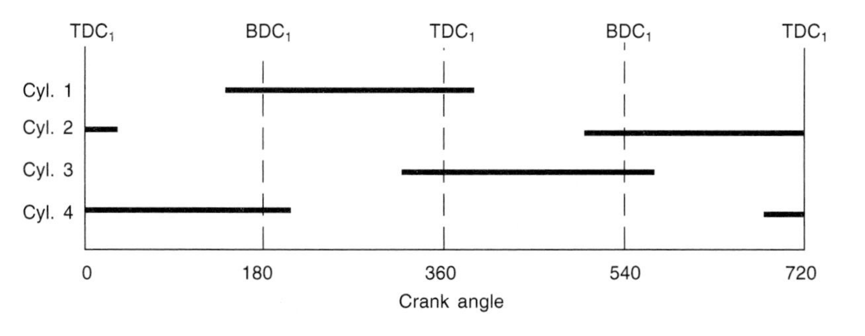

Figure 2.28 Exhaust valve timing for an automotive four-cylinder four-stroke engine

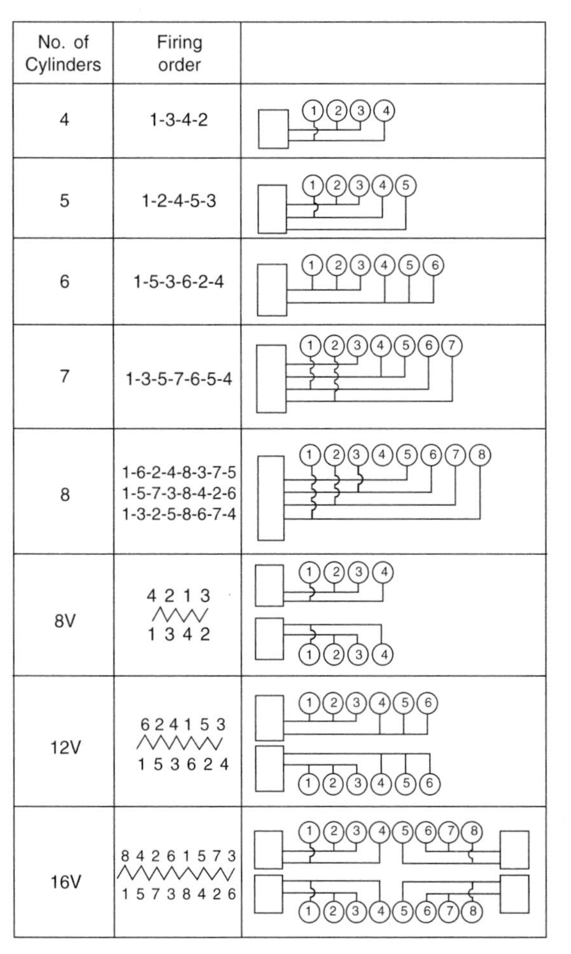

No. of Cylinders	Firing order	
4	1-3-4-2	
5	1-2-4-5-3	
6	1-5-3-6-2-4	
7	1-3-5-7-6-5-4	
8	1-6-2-4-8-3-7-5 1-5-7-3-8-4-2-6 1-3-2-5-8-6-7-4	
8V	4 2 1 3 1 3 4 2	
12V	6 2 4 1 5 3 1 5 3 6 2 4	
16V	8 4 2 6 1 5 7 3 1 5 7 3 8 4 2 6	

Figure 2.29 Exhaust pipe arrangements for four-stroke engines

possible, the loss in turbine efficiency due to partial admission and unsteady flow, is usually more than offset by the additional energy available at the turbine, hence the pulse system is by far the more widely used.

The pulse system has several other attractive features over the constant pressure system. If the system is properly designed, it will usually be possible to arrange for the pressure just downstream of the exhaust valve to fall substantially below the cylinder and inlet manifold pressure during the valve overlap period (*Figure 2.26*). Hence reasonable scavenging can be obtained even at low engine load when the exhaust gas temperature, turbine efficiency and hence boost pressure are low. For the same reason if the overall turbocharger efficiency falls slightly (due to fouling in service), scavenging is not seriously impaired. From the practical point of view, the pulse system is attractive on engines with small numbers of cylinders, since the exhaust system is simple and compact. On larger engine, with many cylinders, the manifolds become very complex and expansion joints become a problem, hence the constant pressure system becomes more convenient. For many applications a further advantage of the pulse system is superior acceleration. The small volume of the exhaust manifold results in rapid transfer of energy by pressure waves to the turbine.

One of the important disadvantages of the pulse system has already been touched upon, namely poor performance when one or two cylinders only are connected to a turbine inlet, particularly if the pressure ratio is high. In these arrangements, the pressure downstream of the exhaust valves may be close to atmospheric before the valve opens. Hence, little of the first part of the pressure pulse will be used effectively at the turbine. The higher the supercharging pressure ratio, the higher the cylinder pressure will be at the moment when the exhaust valve opens, hence the more significant this loss will be. At the same time turbine efficiency will be poor due to partial admission and the highly unsteady flow. Another important disadvantage of the pulse system has also been mentioned, namely a problem of poor scavenging if pressure waves arrive at an exhaust valve at the wrong time. Some engines, having long pipes, may even be speed limited to avoid this happening. Undesirable pulse interference may occur on only one cylinder, in which case that cylinder may have low air flow and will run at a very rich fuel-air ratio unless the fuel pump is adjusted to compensate. In the latter case, overall power output is naturally reduced.

Usually, a turbocharger correctly matched to an engine operating with the pulse system will use a larger turbine than would be fitted for constant pressure operation. With the pulse system, the mass flow through the turbine is intermittent, taking place over shorter time intervals, hence the turbine must be sized to accept a high instantaneous flow rate, especially for two-cylinder/pipe and one-cylinder/pipe groups. Thus the ideal size ratio between turbine and compressor is different in pulse and constant pressure operation and often a larger frame size turbocharger will have to be used with the pulse system. This can lead to less efficient operation of the compressor in the low mass flow range.

The choice between pulse or constant pressure turbocharging system is governed by engine duty, performance plus economic and maintenance considerations. In practice, these criteria mean that the constant pressure system is used exclusively on very large, highly rated two-stroke engines and some industrial medium speed engines. On these engines the ratings are such that very large pressure pulses would be generated with the pulse system. Since most of the exhaust pulse energy coincides with the peak of the pulse, matching this point with high instantaneous turbine efficiency is important. In practice, it is difficult to maintain high turbine efficiency when the pressure ratio exceeds 3:1, hence turbine efficiency will be low if exhaust pressure pulse amplitude substantially exceeds this value. This is what happens on very highly rated engines, hence constant pressure systems operate with higher turbine efficiency, more than offsetting their lower available energy. The principle advantages and disadvantages of pulse and constant pressure turbocharging are summarized in *Table 2.1*.

Various designs of multiple entry turbines have been used. Today almost all turbocharger manufacturers use similar designs, dependent on whether the turbine has axial or radial flow. With axial flow turbines the toroidal inlet area to the stator is equally divided into 2, 3 or 4 separate sectors, each connected to separate pipes from groups of engine cylinders. Radial turbines with stator blades are available with single or twin entries, in the latter case (*Figure 2.30*) with the periphery of the rotor divided into two 180° sectors. Radial turbines with nozzleless stators are also available in single or twin entry form. In this case the twin entry systems are divided meridionally (*Figure 2.30*). This has the advantage of eliminating pure windage periods in any of the rotor passages and hence improving turbine efficiency when the flow in one entry is momentarily zero and the flow in the other is larger.

2.4.3.1 Pulse turbocharging of four-stroke engines

The important factors that must be considered when using the pulse turbocharging system on a four-stroke engine have been mentioned in the previous section. Most significant of these are the effects of pressure waves on the exhaust and scavenging process and the boost developed by the turbocharger under pulsating turbine inlet flow.

Table 2.1 Summary of turbocharging systems

Pulse turbocharging

Advantages
High available energy at turbine
Good performance at low speed and load
Good turbocharger acceleration

Disadvantages
Poor turbine efficiency with one or two cylinders per turbine entry
Poor turbine efficiency at very high ratings
Complex exhaust manifold with large numbers of cylinders
Possible pressure wave reflection problems (on some engines)

Applications
Automotive, truck, marine and industrial engines; two-and four-stroke; low and medium rating (e.g. up to 17–18 bar b.m.e.p. on four-stroke engines)

Constant pressure turbocharging

Advantages
High turbine efficiency, due to steady flow
Good performance at high load
Simple exhaust manifold

Disadvantages
Low available energy at turbine
Poor performance at low speed and load
Poor turbocharger acceleration

Applications
Large industrial and marine engines operating at steady speed and load, highly rated; two- and four-stroke

Pulse converter turbocharging (simple and multi-entry types)

Advantages
Good performance on engines normally pulse turbocharged with two (or one) cylinders per turbine entry

Disadvantages
Poor performance at very low speed and load
Only suitable for engines with certain numbers of cylinders (e.g. four, eight, sixteen)

Pulse converter turbocharging (SEMT-modular system)

Advantages
Simple exhaust manifold

Disadvantages
Poor turbocharger acceleration (between pulse and constant pressure systems)

Figure 2.26 showed exhaust, cylinder and inlet manifold pressures measured on a four-stroke engine in which the cylinders are connected in groups of three. The valve timing is typical of automotive engines (medium speed engines have longer valve overlap). The diagram represents the almost ideal case of pulse operation for several reasons. First, the exhaust pressure wave has built up rapidly, increasing energy available at the turbine. The pressure drop between cylinder and valve is low for much of the exhaust process, minimizing the effect of kinetic energy being generated and then wasted. The pressure in the exhaust manifold (which was measured near the exhaust valve) has fallen to below the intake manifold pressure before the inlet valve begins to open, and continues to fall. The favourable pressure drop between inlet and exhaust is maintained for the full period of valve overlap, creating scavenge air flow, with the maximum pressure drop coinciding with the period when the combined (inlet and exhaust) valve area is a maximum. At the turbine inlet, the flow will be unsteady but windage has been eliminated since the pressure is above ambient at all times. It follows that turbine efficiency is good, being little worse than that of the constant pressure system.

The only danger evident from *Figure 2.26* is the fact that pressure pulse 3 (directly from cylinder 3) arrives just as the exhaust closes. This does no harm in this case, but would if it had arrived earlier. Thus if the period of exhaust valve opening were made significantly longer, undesirable pulse interference might occur. Scavenging the exhaust products of a four-stroke engine is not difficult, but the effect of exhaust pressures during the blow-down period, valve timing and valve overlap also have a significant effect in terms of the work done by the piston during the gas exchange process. The ideal conditions are to have the peak of the blow-down pulse occurring close to the bottom dead centre position of the piston, followed by a very rapid pressure drop to below the boost pressure level during the expulsion stroke.

If two cylinders are connected to a turbine entry the direct pressure pulse from one cylinder will usually arrive substantially after the exhaust valve of the other has closed. Problems are more likely to arise from the reflected pulse (from the turbine) affecting an individual cylinder of a large engine with long valve overlap, running at high speed. For example, cylinder No. 8 of the eight-cylinder engine shown in *Figure 2.31* might receive a reflected pulse after a considerable time lag due to the long pipe length (eqn 2.26). In *Figure 2.31* the blow-down pulse from cylinder 8 is reflected from the close-end of cylinder 1, and is strengthened by a joining reflection from the turbine. It then arrives at the valve during the critical period of valve overlap, raising the cylinder pressure slightly above the inlet manifold pressure for almost the whole 'scavenge' period. This can only be avoided by keeping the exhaust pipes as short as possible (for example, by using a centrally mounted turbocharger on the engine of *Figure 2.31*). If it is impossible to shorten the pipes, poor performance from that particular cylinder must be accepted or a change made to a pulse converter (see next section) or constant pressure system. Obviously the same pulse reflection problem can also occur on engines with three or one cylinder connected to a turbine entry.

With two cylinders connected to a turbine entry the pressure diagram at the turbine will be less favourable than would be the case if three equally phased cylinders were joined. The exhaust pressure drops to ambient (for a short period of time, between pulses) introducing windage losses, etc., hence the average turbine efficiency will be lower than that obtained with groups of three. However, unless a pulse reflection problem occurs, the exhaust pressure diagram will encourage good scavenging and a favourable gas-exchange pumping loop, hence the arrangement is widely used. The increased energy available at the turbine over the constant pressure system will usually more than offset the loss in turbine efficiency unless the engine is running under very high boost pressures.

Occasionally it may be necessary to connect one cylinder alone to a turbine entry, for example, on five-and seven-cylinder engines as shown in *Figure 2.29*. Fortunately, these engines are quite rare since the turbine inlet conditions are highly unfavourable making energy conversion very inefficient. However, it will be simple to obtain a pressure drop during scavenging and a favourable pumping loop, if the overall turbocharging system is efficient enough to develop a reasonable level of boost pressure. The performance of engines using groups of one and two cylinders to a turbine entry will always be poor, but this may have to be accepted if other reasons prevent the adoption of constant pressure turbocharging. The only alternative is the pulse converter system (section 2.4.4).

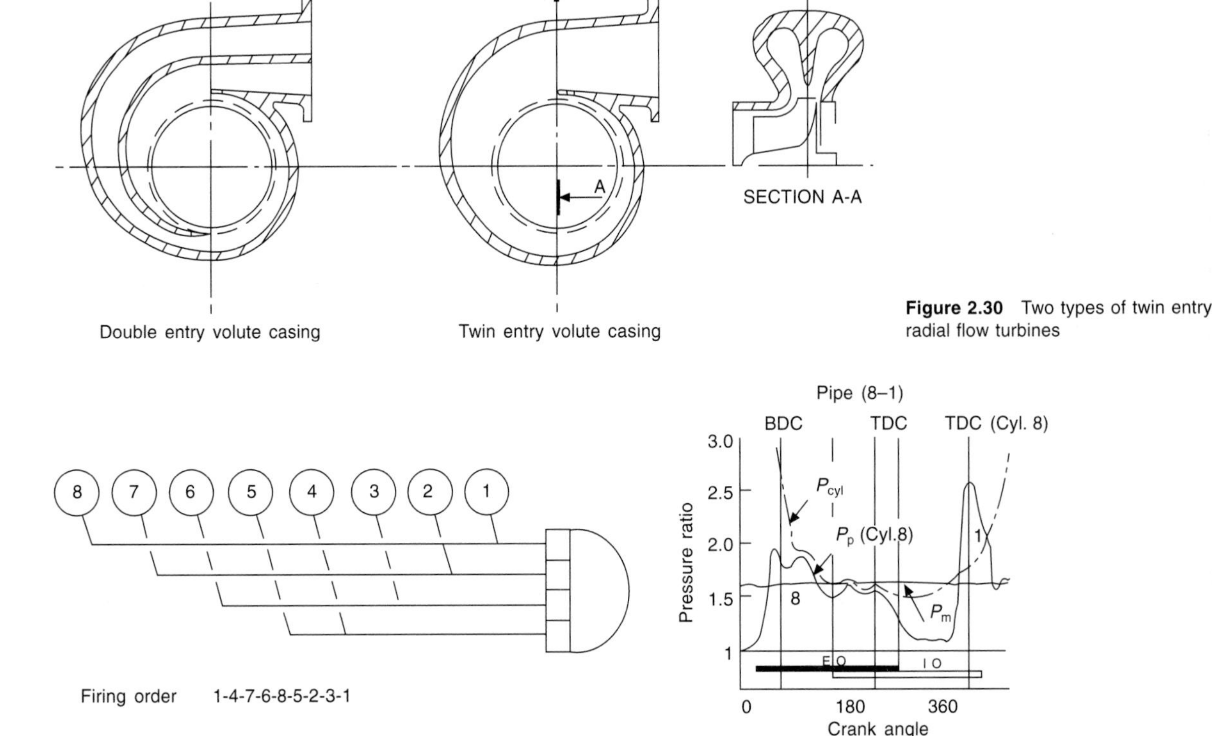

Figure 2.30 Two types of twin entry radial flow turbines

Figure 2.31 Pulse turbocharging arrangement on an eight-cylinder engine

Increases in turbocharger efficiency will improve the pressure drop from inlet to exhaust manifold with pulse turbocharging as with the constant pressure system. Fuel consumption improves due to better scavenging and a more favourable piston pumping loop. *Figure 2.32* illustrates the relative change in SFC with turbocharger efficiency (as measured under steady conditions on a turbocharger test stand) for a medium speed four-stroke engine, with groups of two or three cylinders connected to each turbine entry[1]. This also shows the better performance obtained with the three-cylinder group arrangement.

2.4.3.2 Pulse turbocharging of two-stroke engines

The effectiveness of scavenging is the key to successful and efficient operation of the two-stroke engine. It has also been shown that, with the constant pressure system, the turbocharger is not self-supporting at low engine load. During the scavenge period the piston of the two-stroke engine moves only slightly and hence no work is done by the piston that could be utilized to supply energy lacking in the exhaust system.

Turbocharging with the pulse system provides a large proportion of pressure energy during the blow-down period to the turbine and very effective scavenging at part load operation. This is achieved by keeping the exhaust manifold volume as small as possible and avoiding long pipes which could cause pulse reflections that interfere with scavenging. Thus two-stroke engines operating with the pulse system have turbochargers fitted very close to the cylinders, one turbocharger serving only two or three cylinders.

Comments made in previous sections, regarding the efficiency at which the turbine converts the available exhaust pulse energy when one, two or three cylinders are connected to a turbine entry, related to both two- and four-stroke engines. The only major difference in this respect is that the shape of the exhaust pressure diagram is somewhat different, due to the long scavenge period of the two-stroke engine. The diagram (*Figure 2.33*) tends to consist of quite distinct 'blow-down' and 'scavenge period'.

If a two-stroke engine has a good scavenging system and the exhaust system is designed to enable the pressure drop between inlet and exhaust manifolds to be large during the scavenge period, then the pulse system enables the engine to run in a self-sustained condition without any auxiliary scavenging aid. The combination of good scavenging, high energy available at the turbine and efficient turbocharger operation enables the two-stroke engine to run, self-sustained, over the whole operating range.

On some (rare) occasions it is possible to use the reflected pressure pulse to aid the gas-exchange process. If the combination of engine speed, exhaust pipe length and valve timing is such that the lag of the reflected pulse is a little less than the full opening period of the exhaust valve, then the reflected pulse can raise the exhaust pressure, just as the valve is closing. This will force the cylinder pressure to rise and equal the inlet manifold pressure when inlet and exhaust valves close, increasing the mass of fresh charge trapped in the cylinder by a significant amount.

Due to the momentum of the inflowing gas, it is sometimes, possible for the cylinder pressure to exceed both inlet and exhaust pressures in a carefully designed system. However, if the reflected pressure wave arrives somewhat early, when the exhaust valve area is large and scavenging is not yet complete, then the mass trapped in the cylinder will have a high residual gas content. Furthermore combustion products may be blown right through into the inlet manifold where carbon deposits may build up. This situation should not be accepted, but may occur at low

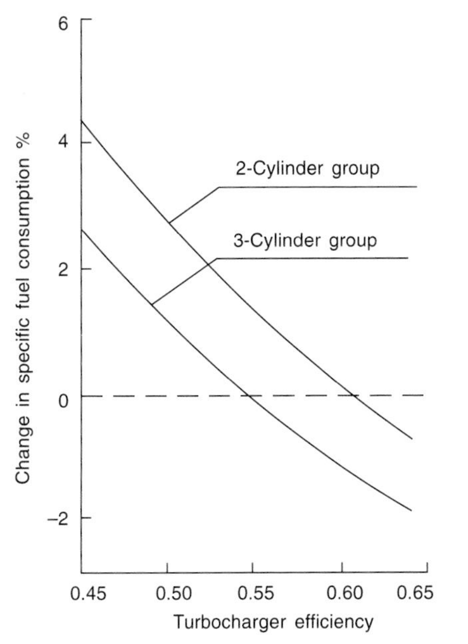

Figure 2.32 Effect of turbocharger efficiency on specific fuel consumption, pulse system. 3:1 boost pressure ratio (Ryti and Meier)

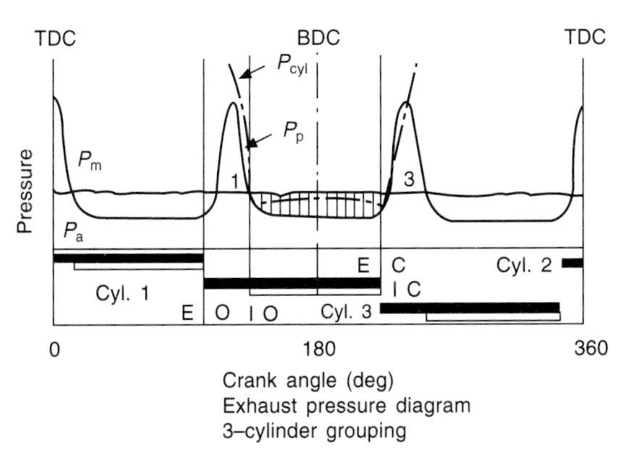

Figure 2.33 Exhaust pressure diagram for a two-stroke engine with three cylinders per turbine entry

engine speeds, when the pulse lag (from eqn 2.26) becomes shorter. Clearly when attempting to use this reflected pulse, the effect of pressure wave dynamics must be checked over the complete speed range of the engine.

The major disadvantage of the scheme outlined above can be avoided if use is made of the direct pressure pulse from an adjacent cylinder, rather than a reflected pulse originating from the cylinder in question. For example, with a three-cylinder group (*Figure 2.33*), the phase interval between cylinders (120°) is close to the opening period of the exhaust valve. If the exhaust period is rather long, and the direct pressure wave travel time is very short, then pressure pulse 3 might arrive slightly earlier, at the end of the exhaust period, raising the mass of fresh air trapped in the cylinder as described above. The lag between exhaust valve opening (cyl. 1) and pulse 3 arriving has two components, the phase angle between cylinders and the pressure wave travel time. The first will dominate and is constant in terms of crank angle, hence the effect of engine speed is small.

2.4.4 Principles of pulse converter and other turbocharging systems

Pulse converter turbocharging systems have been developed to improve the performance of those engines that suffer from low turbine efficiency with pulse turbocharging due to long windage periods and partial admission (multiple turbine entries) losses. They attempt to preserve the advantages of the pulse system, with its inherent high available energy and unsteady flow at the exhaust port, with steadier and more efficient flow at the turbine.

A pulse converter system, in its simplest form, is shown in *Figure 2.34*, applied to a four-cylinder engine. A conventional pulse manifold is used, but a carefully designed junction connects the two branches of the manifold to a single entry turbine. *Figure 2.35* shows pressure diagrams recorded from an automotive two-stroke engine with the pulse converter of *Figure 2.34* and a conventional pulse system. By connecting all four cylinders to a single turbine inlet, windage periods between exhaust pulses are totally avoided. Turbine entry conditions are not steady, as per the constant pressure system, but the very low efficiency operating points of the pulse system are avoided. The junction is designed to minimize pressure pulse transmission from one branch of the exhaust manifold to the other, thereby avoiding a blow-down pulse from one cylinder destroying the scavenge process of another. This is achieved by accelerating the gas as it enters the junction, reducing its pressure at the junction, and minimizing its effect on the other branch. In *Figure 2.35*, an exhaust pressure pulse from cylinder 3 arrives at cylinder 1 at the end of its scavenge period, but the junction has reduced its amplitude.

Varying the pipe cross-sectional areas at the inlets to the junction can control the influence of pressure pulses in one branch on the pressure in the other, but substantial area reductions must be avoided or turbulent mixing at the junction will reduce available energy at the turbine.

Pulse converters of this type are fitted to many medium speed diesel engines, particularly those with 4, 8, 16 and other awkward (for pulse turbocharging) numbers of cylinders. Invariably the normal pulse-type manifold is used with the pulse converter joining the branches to a turbine. In the case of an eight-cylinder engine, two pulse converters are used, each connected to one of the two entries of the turbine. A sixteen-cylinder engine must use four-pulse converters with two twin entry turbochargers or one four entry unit. See *Table 2.1* for a summary of the advantages and disadvantages relative to pulse and constant pressure turbocharging.

Alternatively a multi-entry pulse converter (Multi-stoss, Brown Boveri patent) can be used, with three or four manifold branches joining in the pulse converter and to a single entry turbine. Provided that the firing intervals of the cylinders are equally spaced, then in principle the more cylinders connected to the turbine, the steadier the junction and turbine inlet pressure will be. *Figure 2.36* compares exhaust pressure diagrams with a four-entry multi-pulse system and a conventional pulse system, on a medium speed V8 engine. The pressure fluctuation at turbine inlet is very small, hence the turbine is operating under conditions just as favourable as a constant pressure system. Although the turbine inlet pressure is not following the 'ideal' path as discussed with reference to *Figure 2.18*, the system does fully utilize the available energy associated with area 5-7-13 in *Figure 2.18*. This energy is transmitted to the pulse converter in the pressure pulse shown in *Figure 2.36*, where it is largely converted to kinetic energy in the converting inlet section of the pulse converter. Most of this kinetic energy is transmitted directly to the turbine, although some is made unavailable to the turbine due to inefficient mixing at the junction.

By creating an almost steady pressure in the junction, the influence of a blow-down pulse from one cylinder on the

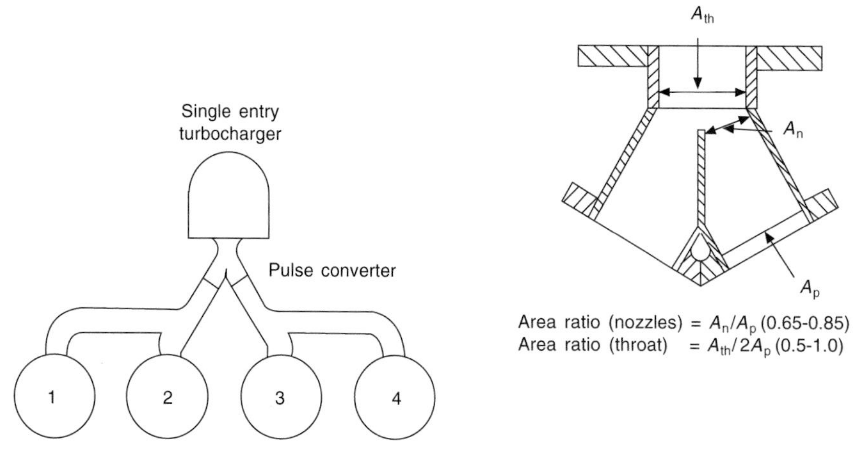

Figure 2.34 A simple pulse converter and its application to a four-cylinder engine

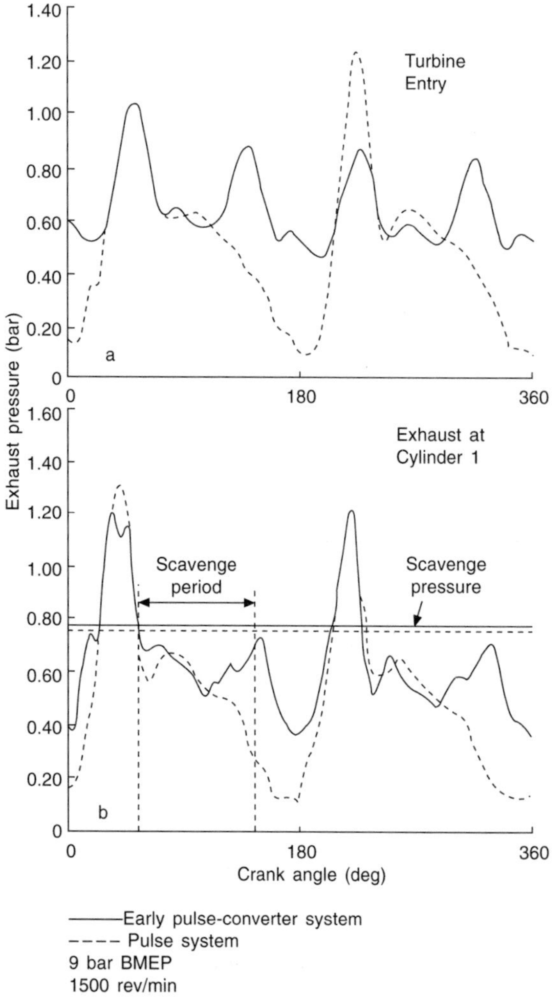

Figure 2.35 Comparison of exhaust pressure diagrams on a four-cylinder, two-stroke engine, pulse and pulse converter turbocharging

scavenging process of another, is reduced. For example, *Figure 2.36* shows a substantial inlet to exhaust manifold pressure difference during the valve overlap period of this medium speed engine.

The modular pulse converter concept developed by SEMT's even closer in concept to a constant pressure system, but avoids the use of a large-volume manifold. The exhaust port from each cylinder is connected to a single exhaust manifold of relatively small diameter running past all the cylinders (system d, *Figure 2.37*). Each exhaust port is connected to the main pipe by a pulse converter (*Figure 2.37, top*). The presssure in the main pipe is kept low by maintaining a high gas velocity in the pipe. Thus the pressure pulse energy at the exhaust port is transformed to kinetic energy in the pulse converter which is retained and transmitted to the turbine by the main pipe. The large number of cylinders connected to the main pipe help reduce flow unsteadyness within it and hence at turbine entry. Single pipe systems similar to this have been used on medium speed engines in the USA for many years, but these use larger diameter main pipes with no significant area reduction at the junctions from each exhaust port. In concept, they are closer to conventional constant pressure systems. The advantage of the SEMT system is not specifically better performance, although this can be achieved, but simplicity relative to a pulse system, and less bulk than a normal constant pressure system.

2.4.4.1 Four-stroke engines with pulse converters

The advantage of higher average turbine efficiency due to pulse converter junctions with some cylinder arrangements improves the performance of four-stroke engines. However, the junction areas have to be carefully selected if the engine has a long valve overlap period, to avoid harming the scavenging process.

Figure 2.38 shows pressure diagrams measured at an exhaust port of a highly rated medium speed four-stroke engine. The three diagrams were obtained with pulse converters having different pipe cross-sectional areas at the entry to the junction. In the top diagram no area reduction from a normal pulse type manifold was used at the junction. In the lower diagrams converging sections were used, reducing area to 70% and 50% respectively With no area reduction, junction pressure is high and the blow-down pulse from cylinder 2 causes the pressure near cylinder 1 to rise above the boost pressure towards the end of the overlap period of cylinder 1. Reducing pipe area to 70% at the junction reduces pressure at the junction and improves

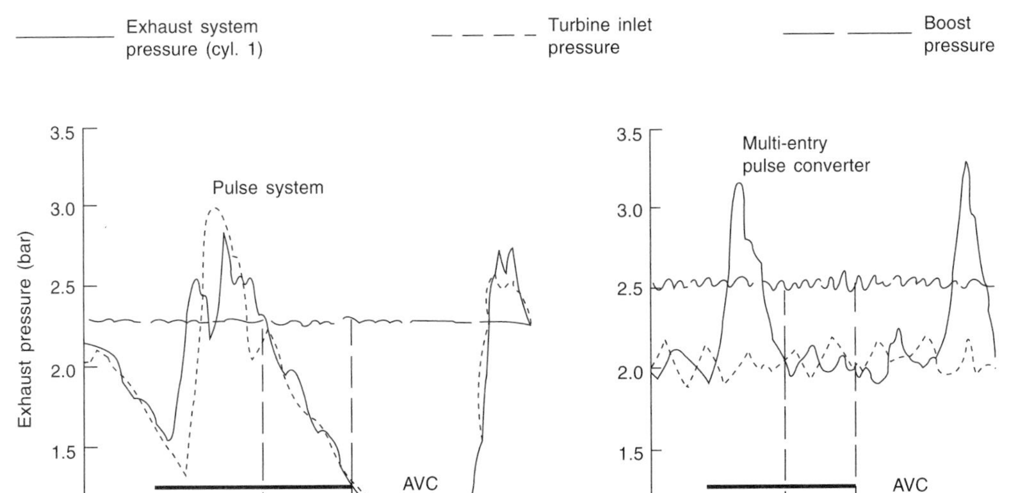

Figure 2.36 Comparison of exhaust pressure diagrams with pulse and multi-entry pulse converter turbocharging systems. V8 engine (Meier)

the pressure drop across cylinder 1 during valve overlap (centre diagram). However, reducing pipe further is counterproductive because this forms a significant flow restriction for cylinder 1, raising exhaust manifold pressure in that branch of the exhaust manifold. This reduces the pressure drop across the cylinder during most of the scavenge period. Thus there is an optimum, in this case with the 70% area junction.

Automotive four-stroke engines have little valve overlap and negligible scavenge flow, and are therefore less susceptible to pressure pulse interference. The optimum area at the junction, relative to normal pipe diameter, will be greater.

2.4.4.2 Two-stroke engines with pulse converters

Two-stroke engines are much more sensitive to exhaust pressure during the scavenge period, but in certain cases, pulse converters can be beneficial. In the case of the four-cylinder automotive (truck) engine of *Figure 2.35* the exhaust pressure diagram measured near the valves, shows pressure rising to almost equal the boost pressure at the end of the scavenge period.

In the case of the two-stroke engine this can be beneficial, since it raises cylinder pressure and the mass of fresh air trapped at port closure, provided that the pressure drop across the cylinder is satisfactory earlier in the scavenge period. This situation can be achieved with four, eight and sixteen-cylinder engines due to their convenient combination of firing interval and inlet port opening period. Note (*Figure 2.35*), that the pulse converter design has prevented the pulse that arrives at valve closure, from exceeding boost pressure at its peak.

2.5 Charge air cooling

2.5.1 Charge cooling principles

The principle reason for turbocharging is to increase the power output of an engine without increasing its size. This is achieved by raising the inlet manifold pressure, hence increasing the mass of fresh air drawn into the cylinders during the intake stroke and allowing more fuel to be burnt. However, from the basic laws of thermodynamics we know that it is impossible to compress air without raising its temperature (unless the compressor is cooled).

Since we are trying to raise the density of the air, this temperature rise partly offsets the benefit of increasing the pressure. The objective must therefore be to obtain a pressure rise with a minimum temperature rise. This implies isentropic compression. Unfortunately due to inefficiencies in practical compressors, the actual temperature rise will be greater than that of an isentropic machine. The more efficient the compressor, the nearer the temperature rise approaches the isentropic temperature rise.

Denoting states 1 and 2 as the inlet and outlet to the compressor, the air density ratio (compressor exit over inlet) is given by

$$\frac{\rho_2}{\rho_1} = \frac{p_2}{p_1}\left[1 + \frac{1}{\eta_c}\left(\left(\frac{p_2}{p_1}\right)^{(\gamma-1)/\gamma} - 1\right)\right] \qquad (2.27)$$

Equation (2.27) is plotted in *Figure 2.39* for a range of pressure ratios and compressor efficiencies. Several interesting points emerge. Firstly, the benefit obtained by raising inlet manifold pressure is almost halved due to the accompanying temperature rise in the compressor (dependent on compressor efficiency). Secondly, the advantage of high compressor efficiency in helping to hold the boost temperature down is relatively small, but worthtwhile. Thirdly, in absolute terms, the benefit that could be obtained by cooling the compressed air back to near ambient conditions is substantial, and increases with pressure ratio. Clearly it is attractive to try and cool the air between compressor delivery and the intake to the cylinders.

A further advantage of charge of cooling is that lower inlet temperature at the cylinders will result in lower temperatures throughout the working process of the engine (for a specified b.m.e.p.) and hence reduced thermal loading.

By using a charge air cooler after the compressor and before the intake to the cylinders, the density of the air entering the cylinders will be increased, enabling more fuel to be burnt. The reduction in temperature achieved in the cooler will be a function of the temperature of the cooling medium available and the effectiveness of the cooler. Both these subjects will be treated in detail in a later chapter, but some elementary facts can be presented here. The effectiveness of an intercooler may be expressed as

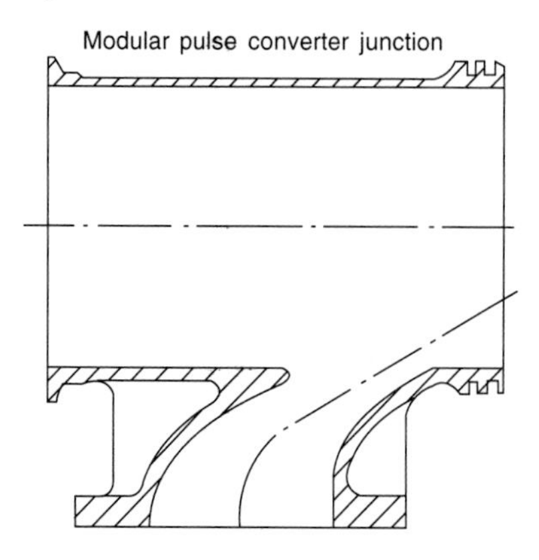

Firing order 1. 3. 5. 7. 9. 8. 6. 4. 2. 1.

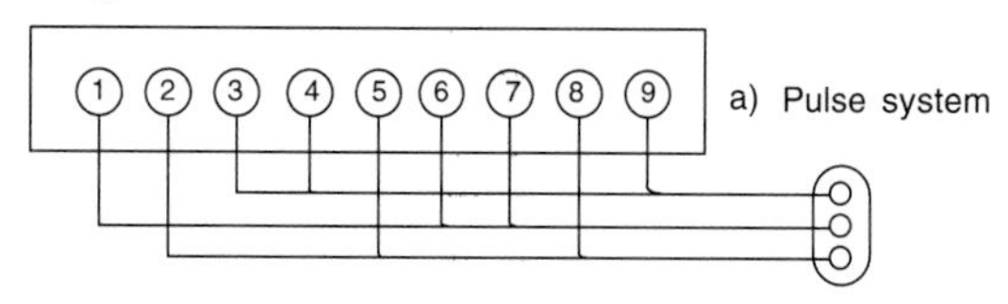

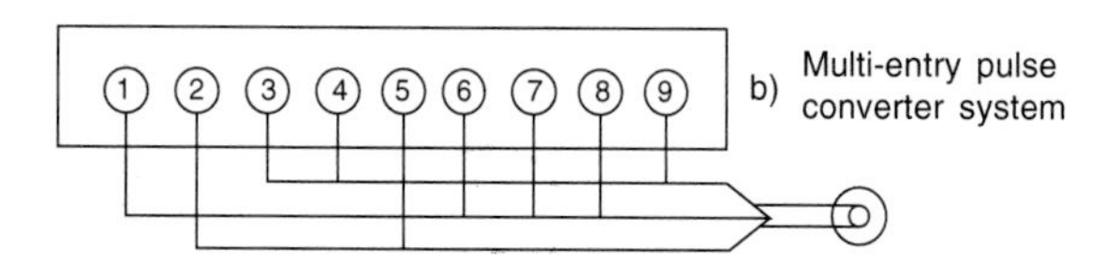

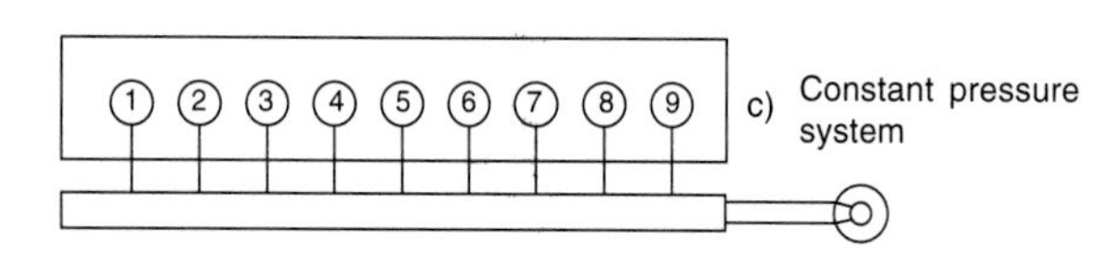

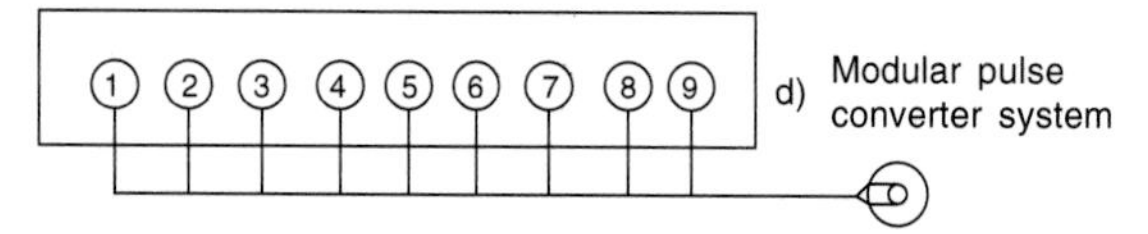

Figure 2.37 Modular pulse converter (SEMT Pielstick) and pipe arrangements with various turbocharging systems on a nine-cylinder engine

$$\varepsilon = \frac{T_2 - T_3}{T_2 - T_w} \left(\text{i.e. } \frac{\text{actual heat transfer}}{\text{maximum possible heat transfer}} \right) \qquad (2.28)$$

where T_2 = hot air inlet temperature (from compressor);
T_3 = 'hot' air outlet temperature (from cooler);
T_w = temperature of cooling medium (at inlet).

The effectiveness (ε) is sometimes called the thermal ratio. From eqn (2.28) it is evident that the effectiveness and the temperature drop between the 'hot' air inlet and the 'cold' cooling medium govern the extent of cooling achievable. Clearly the cooling medium should be as cold as possible, hence a supply of cooling water at ambient temperature will be more useful than the engine's own cooling water system. Secondly, the higher the inlet gas temperature the more useful the cooler will be, hence a charge cooler will become more attractive at higher boost pressure (*Figure 2.39*).

From eqn (2.28) inlet temperature to the cylinders is given by:

$$T_3 = T_1 \left[1 + \frac{1}{\eta_c} \left\{ \left(\frac{p_2}{p_1} \right)^{(\gamma-1)/\gamma} - 1 \right\} \right] (1-\varepsilon) + \varepsilon T_w \qquad (2.29)$$

In *Figure 2.40* the effect on density ratio is shown. Compared to the uncooled case, in which the temperature rise in the compressor offsets half of the benefit due to increased pressure, the cooler enables the density ratio to be increased up to 80% of the pressure ratio.

The advantages of intercooling are clear, but although common (particularly on highly rated engines), it is not universally adopted. Aftercooling does have some disadvantages. From the thermodynamic point of view, the only problem is that air flow through the cooler results in a pressure loss, since narrow flow passages are required for effective cooling. This will result in some offset against the density increase from the cooling. The pressure drop through an intercooler will be a function of its size, detail design and mass flow rate. A very large intercooler will be awkward to install on an engine, may cost more than a compact one but will probably have a higher effectiveness. The subject is discussed in detail in Chapter 15.

The second disadvantage is a more practical problem. A source of cold air or preferably water (due to its higher heat transfer coefficient) must be available. This may be easy to arrange, for example in marine applications, but is not always possible. If the boost pressure is low and the available coolant relatively warm (such as the engine's own cooling water system on an automotive unit), charge cooling will produce only marginal benefit and at full power only. A third disadvantage will be cost, but the benefits of charge cooling will outweigh the additional cost on all but low rated engines.

Air to air charge cooling may be adopted in locations where a cooling water supply is not available, and high ambient temperatures make installation of a local closed system unattractive. In these cases, radiator cooling of the engine oil and water cooling system is usually arranged, directly behind a remote air-to-air charge cooler. Very large coolers may be used since the spacial limitations imposed when mounting a cooler directly on the engine are removed. Thus a very high effectiveness (up to 0.95) may be achieved, with consequent benefits in engine performance.

An interesting alternative air-to-air charge cooling system for vehicle engine uses bleed air from the turbocharging system to drive the cooling air supply fan (*Figure 2.41b*)[3]. Around 5–10% of the airflow through the compressor is used to drive an impulse type air turbine, built around the circumference of a fan. The fan provides cooling ambient air for the charge cooler. The advantage of the system is that cooling air flow tends to increase as it is required by the engine.

Air-to-water cooling systems can use the normal water cooling system of the engine (*Figure 2.41c*) or a separate closed water cooling system with its own water-to-air radiator (*Figure 2.41d*). The advantage of the former system is simplicity of installation, but cooling is limited by the high water temperature (typically around 90°C). Indeed charge air heating is likely to occur at low speed and low load. The indirect system shown in *Figure 2.41d* has a greater cooling potential, since the water temperature can be set lower than that of the normal engine cooling system, is more compact on the air side than system (*Figure 2.41a*), but involves the expense of two heat exchangers.

A comparison of engine performance with the two most compact systems (system b, air-to-air tip fan and system c, air-to-water engine coolant), is given in *Figure 2.42*[3]. The air-to-air systems shows a clear benefit in achieving a lower cylinder intake air temperature due to the lower temperature of the coolant, and the consequent benefit in engine performance.

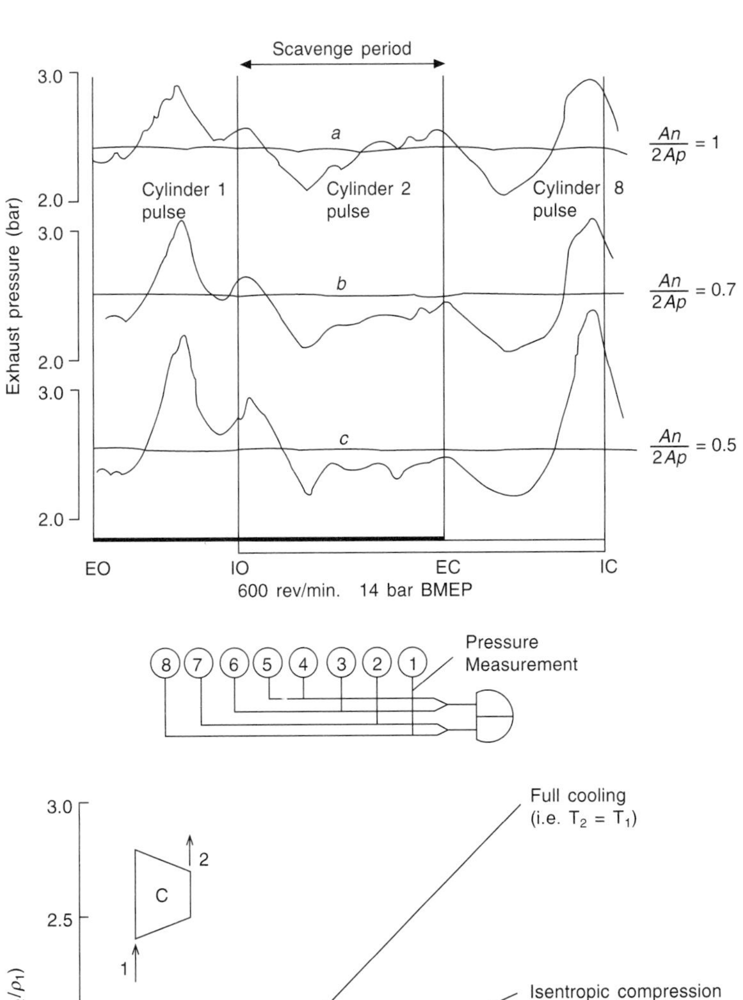

Figure 2.38 Exhaust pressure diagram from a four-stroke medium speed engine with pulse converters (Ruston)

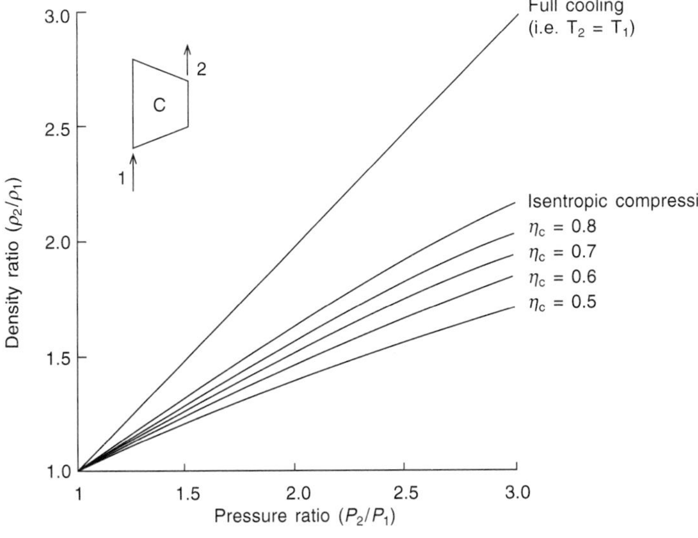

Figure 2.39 Compressor density ratio as a function of pressure ratio and compressor isentropic efficiency

Charge cooler design and construction is covered in Chapter 15.

2.5.2 Charge air cooling and engine performance

For a fixed power output, intercooling will reduce charge temperature and hence thermal loading. Efficiency will also improve slightly since to achieve the same mass of air trapped in the cylinder, the boost pressure need not be as high and air-fuel, ratio can increase. Naturally if the thermal loading is reduced, reduction of the heat lost to coolant will also aid thermal efficiency, although marginally. However, the additional cost of the intercooling equipment must be offset by a higher power output. Since the charge density increases with intercooling, more fuel can be burnt raising the power output in proportion to the density (all other factors being assumed equal). Thus, from *Figure 2.40*, the use of an intercooler of 0.7 effectiveness permits an increase of power output from 1.5 to 1.8 times the naturally aspirated power output at 2:1 pressure ratio. At a 2.5:1 pressure ratio the increase is from 1.75 to 2.20 times the naturally aspirated power output. However, in practice thermal loading or other factors might limit the increase.

Consider first engine performance with and without a charge air cooler, with no change in maximum fuelling. Results from an engine operating over a wide speed range are shown in *Figure 2.43*. Intercooling obviously reduces inlet manifold temperature substantially and this effect is followed right through the cycle, lowering heat transferred to the cylinders and the turbine inlet

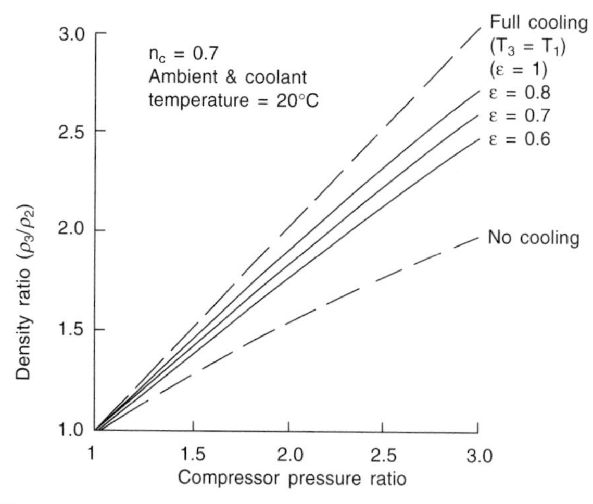

Figure 2.40 Intercooler effectiveness and density ratio, as a function of pressure ratio of the compressor

temperature. The latter results in a drop in specific energy at the turbine, hence boost pressure will also reduce, but not enough to offset the density gain due to the lower charge temperature. Thus air flow increases and overall, the total turbine power does increse. The air-fuel ratio will be weakened, leading to a reduction of specific fuel consumption of near 6%. The gain in b.m.e.p. will be similar. However, both improvements will occur largely at low engine speeds, where the charge air cooler is most effective and higher compressor efficiencies are obtained.

Consider next the more realistic situation in which advantage is taken of charge cooling to increase fuelling, although not to the extent that minimum air-fuel returns to its previous value (*Figure 2.44*). Again with no change in turbocharger match, b.m.e.p. increases from 16 to 19.5 bar (22%) and specific fuel consumption is reduced by 6%. These results are obtained with fuelling adjusted to achieve no increase in thermal loading of the combustion chamber. However, the maximum cylinder pressure has risen from 105 bar to 118 bar (12.5%). In this particular test, the movement to increased air flow through the compressor has moved the characteristic point at full power to an area of lower compressor efficiency. Thus air-fuel ratio at full power is much the same with and without the charge cooler.

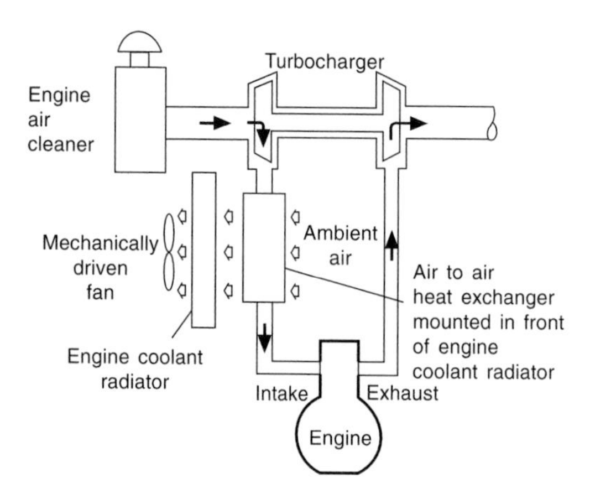

a) Mechanical fan drive air-to-air intercooling system

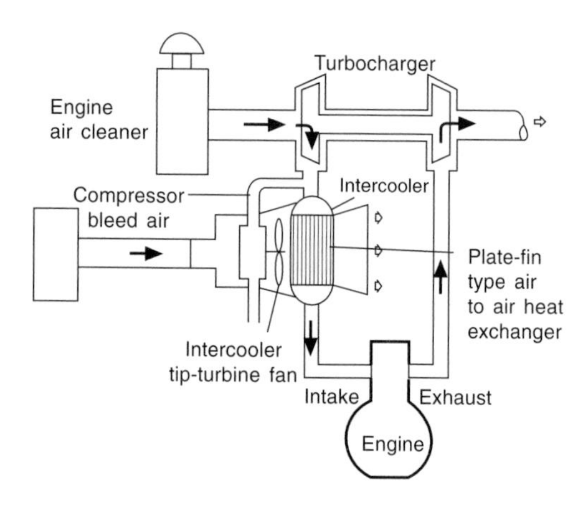

b) Air-to-air tip turbine fan intercooling system

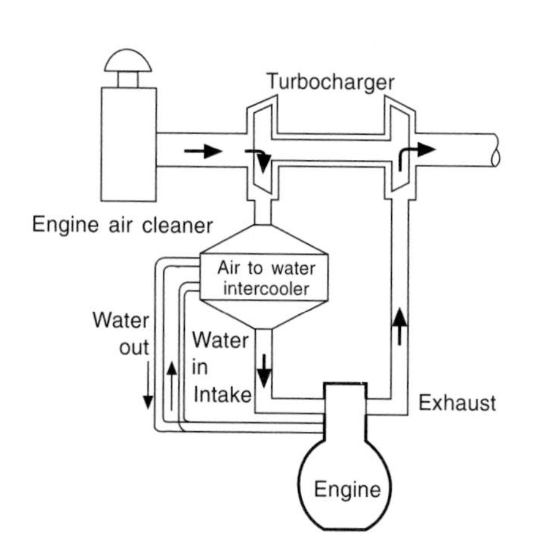

c) Air-to-water intercooling system (using engine cooling system)

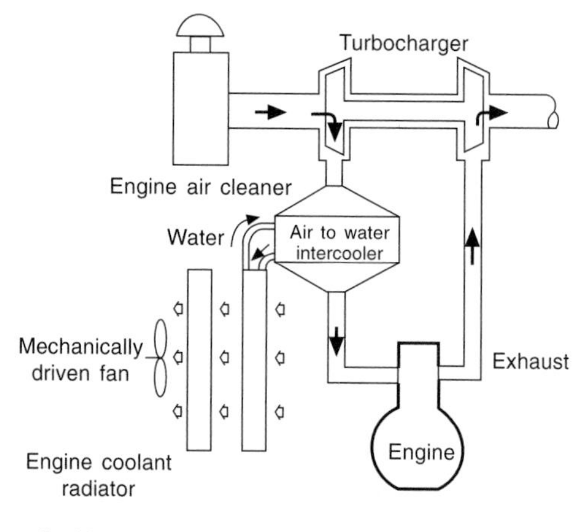

d) Air-to-water intercooling system (using closed water cooling system)

Figure 2.41 Alternative charge air cooling systems

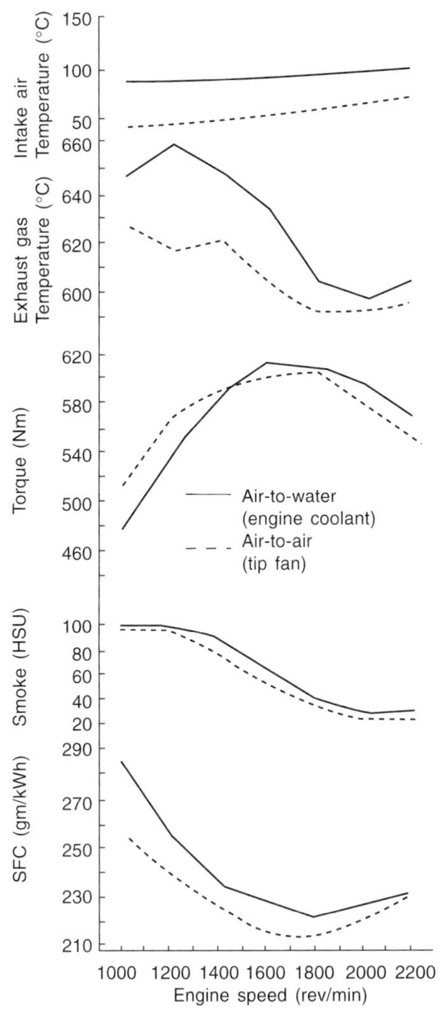

Figure 2.42 Comparison of engine performance with ambient air and water from the engine cooling system as charge air coolants. (Adapted from McLean and Ihnen)

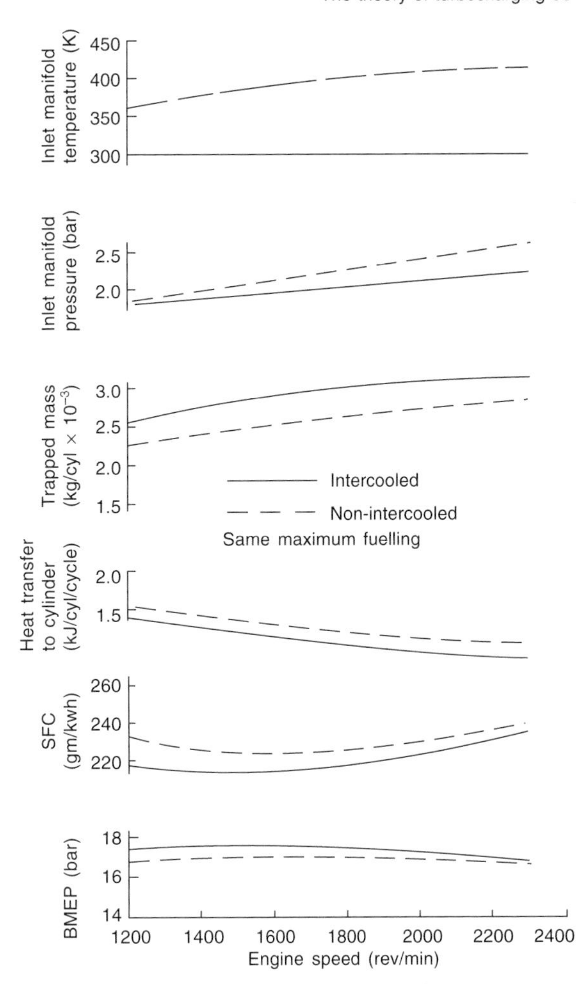

Figure 2.43 Effect of intercooling on engine performance

Analysis of these results shows that charge cooling enables substantial power increases to be achieved without increasing thermal loading. Specific fuel consumption may also benefit due to the indicated brake mean effective pressure rising without a corresponding increase in engine frictional losses, and to the weaker air-fuel ratio. The clear problem is that the maximum cylinder pressure rise with power output, though fortunately not by the same ratio.

2.6 Turbocharger matching

2.6.1 Introduction

Naturally aspirated diesel engines are capable of operating over wide speed ranges, of the order of 3:1 in truck engine and 5:1 in passenger car engines. The maximum useful speed will usually be limited by poor volumetric efficiency, the inertia of reciprocating parts or, in the case of some small, high speed engines, high frictional losses and sometimes poor combustion. An engine that is designed for variable speed operation will usually exhibit some deterioration in performance both at extreme low and high speed. This is due to high gas friction losses in the inlet valves, the use of valve timing optimized in the mid-speed range and a gradual mismatch between fuel injector characteristics and swirl. However, the useful speed range can be wide, since reciprocating machinery is well suited to cater for a wide range of mass flowrate.

The performance of turbomachines is very dependent on the gas angles at entry to the impeller, diffuser and turbine rotor. The blade angles are set to match these gas angles, but a correct match will only be obtained when the mass flow rate is correct for a specified rotor speed. Away from this 'design point' the gas angle will not match the blade angle and an incidence loss occurs due to separation and subsequent mixing of high and low velocity fluids.

These losses will increase with increasing incidence angle, hence turbomachines are not well suited for operation over a wide flow range. Their use as superchargers is due to their high design point efficiency and their ability to pass high mass flow rates through small machines.

It is clear that a turbomachine is not ideally suited to operate in conjunction with a reciprocating machine, hence the combination of diesel engine and turbocharger must be planned with care. 'Matching' of the correct turbocharger to a diesel engine is of great importance and is vital for successful operation of a turbocharged diesel engine. The overall objective of turbocharger matching is to fit a turbocharger with the most suitable charac-

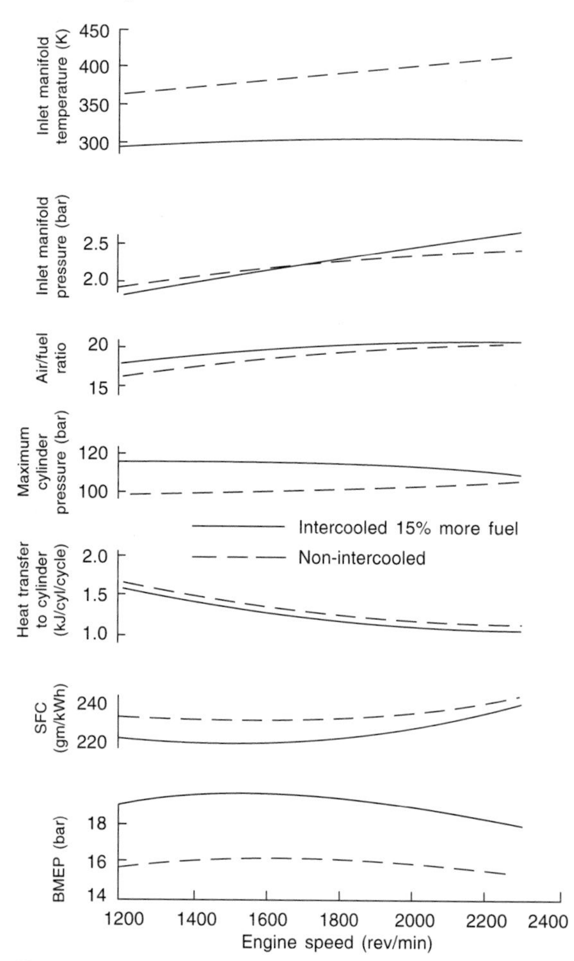

Figure 2.44 Effect of intercooling on engine performance with fuel delivery increased

teristics to an engine in order to obtain the best overall performance from that engine. The turbocharger will not be operating at its high efficiency flow condition over the complete working range of engine speed and load. It follows that it is possbile to 'match' the turbocharger correctly only at a particular point in the operating range of the engine. For example, if an engine is required to run for most of its life at constant speed and full load, the turbocharger will be chosen such that its high efficiency operating area coincides with the pressure ratio and mass flow requirements of the engine at that condition. If the engine is required to operate over a broad speed and load range, then a compromise must be made when matching the turbocharger. This compromise will principally be governed by the duty for which the engine is required.

The basic size of the turbocharger will be determined by the quantity of air required by the engine. This will be a function of swept volume, speed, rating (or boost pressure), density of air in the inlet manifold, volumetric efficiency and scavenge flow. If these parameters are known an initial estimate of the air mass flow rate may be made:

$$\text{for a four-stroke engine } \dot{m} = \simeq \frac{N}{2} \times V_{sw} \times \rho_m \times n_{vol} \quad (2.30)$$

With little valve overlap, the volumetric efficiency will be less than unity and may be estimated from values obtained under naturally aspirated operating conditions or from previous experience. With large valve overlap, the clearance volume will be scavenged and some excess air will pass into the exhaust.

The boost pressure (P_m) will have been estimated for the engine to produce its target power output, subject to expected thermal and mechanical stresses. By assuming a value for the isentropic efficiency of the compressor (total to static) the boost temperature T_m may then be estimated.

The isentropic efficiency of the compressor may be taken from compressor maps. If the compressor efficiency is quoted on the total-to-total basis a correction of a few percentage points down on the total to static value is required.

For an engine employing aftercooling the boost temperature T_m has to be reduced according to eqn (2.29). From known values of ambient temperature and pressure, and required compressor pressure ratio, the mass flow rate at maximum power can be estimated.

By looking at the basic guidelines presented in the turbocharger manufacturer's literature, or the complete compressor characteristic curves (*Figure 2.45*), a basic 'frame-size' of turbocharger may be selected. The final choice of compressor will be made bearing in mind the complete operating lines of the engine over its whole speed and load range, superimposed on the compressor characteristic. The compressor 'trim' or diffuser will be chosen to allow a sufficient margin from surge whilst ensuring that the operating lines pass through the high efficiency area. This will be discussed in detail later (section 2.6.2).

Once the basic frame size and compressor have been established the turbine must be matched by altering its nozzle ring, or volute (if it is a radial flow machine). The effective turbine area will change, raising or lowering the energy available at the turbine and hence adjusting the boost pressure from the compressor.

Turbocharger matching, particularly on a new engine, can be a lengthy process since many dependent parameters are involved. Although the basis is to match the turbocharger to the engine, it may well be necessary to improve the performance of the combination by engine design changes at the same time. Some such changes, adjustment of the fuel injection system, for example, are obviously essential. Others, that may be desirable, include

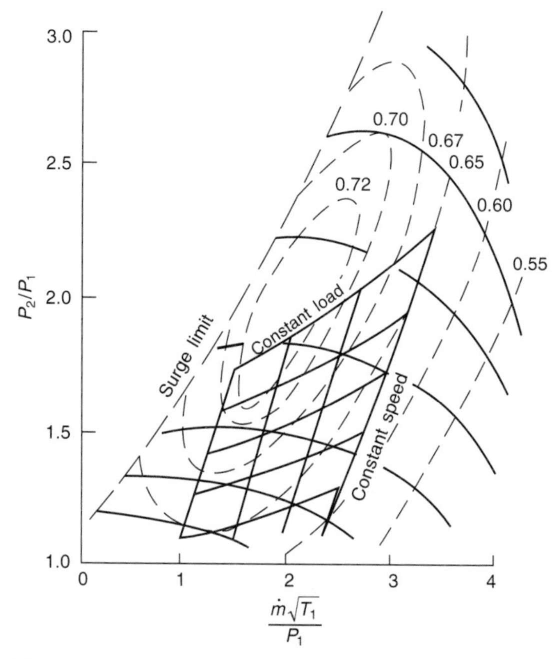

Figure 2.45 Compressor characteristic, with air flow requirements of a four-stroke truck engine superimposed with constant compressor speed and efficiency lines shown also

alteration to the valve area and timing. It will not be possible to discuss all these factors in detail and the broad principles of matching fuel injection systems, swirl, optimization of valve timing, etc. are not substantially different from those pertaining to naturally aspirated engines.

Most large industrial highly rated turbocharged engines are designed as such from the outset today, and it is less common for a naturally aspirated design to be uprated by turbocharging in this class of engine. The designers will know from experience approximately what valve timing, injection pressures, etc. will be suitable, although factors such as these will be modified in the light of subsequent testing. Manufacturers of smaller engines, however, sometimes turbocharge engines that were originally designed for naturally aspirated use. Fortunately, if modest power gains are sought (up to 50%), major redesign is seldom necessary since the maximum cylinder pressure may be held down by an engine compression ratio reduction, and thermal stresses are not usually a major problem. However, if greater power gains are required the engine will probably require major redesign.

2.6.2 Air flow characteristics of engine and turbocharger

2.6.2.1 Four-stroke engines

The air flow rate through a turbocharged (non-aftercooled) diesel engine will be a function of the engine speed, compressor delivery air density and the pressure differential between intake and exhaust manifolds during the period of valve overlap. If the engine is run at constant speed, but steadily increasing load, then the mass flow rate will increase approximately with the increasing charge density. The air flow through the engine may be superimposed on a turbocharger compressor characteristic, as shown in *Figure 2.45*, the slope being governed by the density ratio. When matching a turbocharger to an engine with this operational requirement, the objective will be to choose a compressor such that the constant engine speed line falls through the middle of the high efficiency area of the compressor map. If an aftercooler is fitted then, as load increases the cooling effect will increase charge density more rapidly for a corresponding boost pressure, hence the slope of the constant engine speed air flow line on the compressor characteristic will be less steep than those shown in *Figure 2.45*.

Consider next an engine running at constant load but increasing speed. As the engine speed increases so will the volumetric flow rate of air. The effective flow area of the turbocharger turbine remains almost constant hence turbine inlet pressure will rise. As discussed in section 2.4, the result is an increase in energy available for expansion through the turbine and hence increased boost pressure at the compressor. Thus the constant load line of the engine will not lie horizontally on the compressor characteristic, but will rise engine speed (*Figure 2.45*), the slope depending on whether the engine is aftercooled or not. If the engine is required to operate over a range of speeds and loads (e.g. an automotive unit), then a set of constant speed and constant load lines may be drawn on the compressor characteristic to represent the operating range (*Figure 2.45*). The complete engine characteristic must lie between the compressor's surge line and the limit imposed by low efficiency or possibly turbocharger overspeed at high mass flow rates.

The margin between surge and the nearest point of engine operation must be sufficient to allow for three factors. Firstly, pulsations in the intake system may well induce surge when the mean flow lies clear of the nominal surge line. Secondly, if the air filter becomes excessively blocked in service, the air flow rate through the engine will reduce, but the turbine work will be maintained by a hotter exhaust as the air-fuel ratio gets richer. Thus boost pressure may not fall and some movement of the engine operating line towards surge may occur. A larger movement towards surge will result if the engine is operated at altitude. The effect of altitude operation on turbocharged engine performance is discussed in section 2.6. The combined effect of the three factors affecting surge margin will vary from one engine for one application, to another. In general, however, a margin of at least 10% (of mass flow rate) between surge and the nearest engine operating lines should be allowed. On engines with a small number of cylinders a 20% margin will be sometimes required.

Generally, the turbocharger turbine can operate efficiently over a wider mass flow range than its compressor. It follows that it is more important to examine the engine air flow plotted on the compressor map than on the turbine map. This is indeed fortunate since if the turbine is operating under the pulse system with highly unsteady flow, then it is not realistic to plot a 'mean' value on the turbine map. The result can be quite misleading. To accurately assess the operating area on a turbine map would require a plot of instantaneous gas flow and pressure ratio over a full range of engine operating conditions. The information is very difficult indeed to measure accurately.

The compressor may be matched initially by choosing the best combination of impeller and diffuser such that the engine operating characteristics lie within the guidelines given above. Final matching will depend on the type of power or torque curve required from the engine for a certain application, and will be discussed later. However, it has been pre-supposed that the turbine is able to provide sufficient power to drive the compressor and produce the air flow conditions discussed. Since the useful flow range of the turbine is wider than that of the compressor, the turbine supplied by the turbocharger manufacturer will inevitably be able to cope with the necessary mass flow. Whether it produces sufficient power depends on its efficiency and the turbine area (since this dominates the energy available for useful expansion). The turbine nozzle ring or volute controls its effective area. Thus the effective area at the turbine will be adjusted (by changing the components mentioned above) to achieve the desired boost level from the compressor. If turbine area is reduced, then the compressor boost pressure and the mass flow rate will go up, the former by a larger amount than the latter since charge temperature will also rise. The effect, in terms of the engine operating lines superimposed on the compressor characteristic, is shown in *Figure 2.46*.

2.6.2.2 Two-stroke engines

The air flow characteristics of a two-stroke engine will depend on whether the engine is fitted with a turbocharger alone or has an auxiliary scavenge pump or blower. Consider first the situation when a turbocharger is used on its own. During the period when the inlet ports are open, the exhaust ports or valves will also be open and the air flow rate will depend on the pressure drop between intake and exhaust manifolds. The physical arrangement is analogous to flow through two orifices placed in series. The mass flow versus pressure ratio characteristic for steady flow through two orifices in series is a unique curve and it follows that the two-stroke engine will exhibit a similar characteristic. Thus the engine operating line, when superimposed on the compressor characteristic will be a unique curve, almost regardless of engine load or speed (*Figure 2.47*). Fortunately the compressor characteristics are well suited to this type of demand, and it is relatively easy to match the turbocharger. If an aftercooler is fitted (which it is on most engines of this type) the mass flow versus pressure ratio curve will have a different slope (smaller gradient) which suits the compressor characteristic even more.

If scavenge pumps or compressors are placed in series with the turbocharger compressor on a two-stroke engine, then the air flow characteristic will be dominated by the scavenging device. For example, a reciprocating scavenge pump will exhibit much the same flow characteristics as a four-stroke engine,

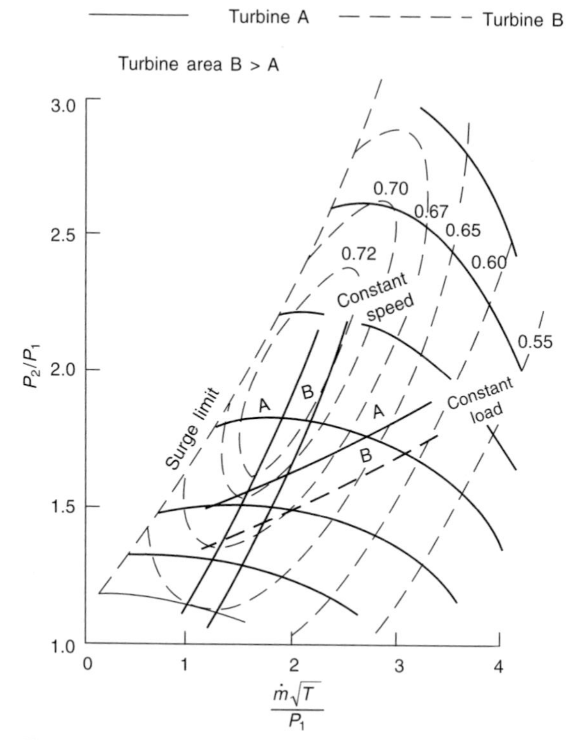

Figure 2.46 The effect of turbine matching on compressor match

hence when constant load and speed operation lines are plotted on the compressor characteristic, the result is little different from that shown in *Figure 2.45*. Somewhat similar characteristics are obtained if a rotary compressor (e.g. Roots blower) is used, although small differences will result from different compressor

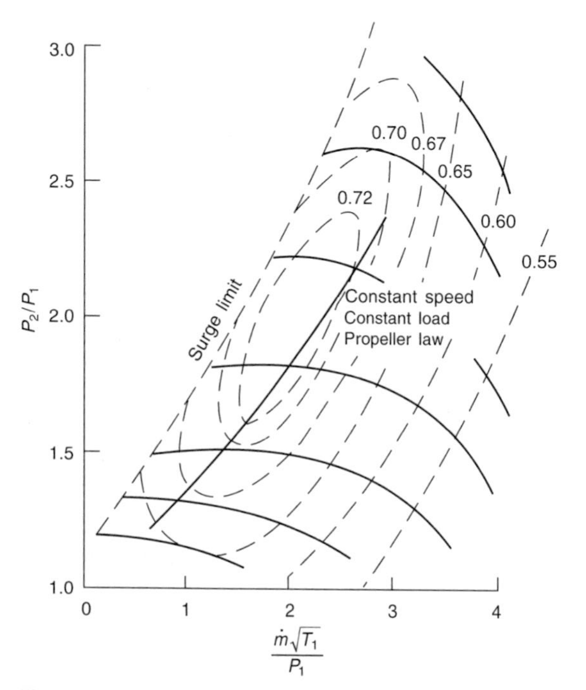

Figure 2.47 Compressor characteristic with air flow requirements of a two-stroke engine superimposed (with no additional compressor)

characteristics. In general, however, the matching problem will not be significantly different from that of a four-stroke engine and the comments previously made apply. The additional difficulty that does arise is that the scavenge blower must be matched to the engine at the same time, as a compromise must be reached between scavenge blower and turbocharger work.

Generally the turbocharger will be expected to do as much of the compression work as possible. Depending on the quality of the scavenging system, turbocharger efficiency, use of pulse or constant pressure turbocharging, etc., the turbocharger may well be able to provide sufficient boost pressure at full power but not at low load. The subject has been discussed in detail in section 2.4. It follows that the work division between turbocharger and auxiliary compressor may well be governed by the requirement for adequate scavenging (and hence engine performance) for part load operation. The final balance will vary from engine to engine since many different factors are involved.

2.6.3 Matching for constant speed operation

The most common application requiring a constant speed diesel power source is electricity generation. It is a relatively simple requirement from the point of view of turbocharger matching.

The basic compressor characteristics resulting from this varying load at constant speed application is shown by a constant speed line on *Figure 2.45*. Small adjustments in turbine area will affect both the boost and exhaust pressures, hence pumping work and the resultant fuel consumption of the engine. Within the limits of acceptable mechanical and thermal loading of the engine, turbine matching will be used to achieve optimum performance and fuel consumption. *Figure 2.48* shows the effect to turbine area changes on the specific fuel consumption with varying load at constant speed. Matching will be a compromise between performance and low and high load. If the diesel generator is required for the base-load operation, then the turbocharger will be matched at the rated load (full line in *Figure 2.48*). Otherwise a larger nozzle ring or volute will be fitted to improve part load efficiency. The small nozzle increases piston pumping work and hence increases fuel consumption, except at high load, when the engine benefits from the extra air which results from more turbine and compressor work.

Fine tuning of the turbine match produces the smaller changes in engine performance shown in *Figure 2.49*. This is a more highly rated engine (17 bar b.m.e.p.) with large valve overlap. The flow characteristics of the three turbine trims used are given in *Figure 2.50*. In this case, the smaller turbine, although benefiting from increased available energy, has a lower efficiency than the larger trims (being cut down from a larger design). In addition the high exhaust pressure level developed by the small effective flow area of the turbine, has an adverse effect on the scavenge

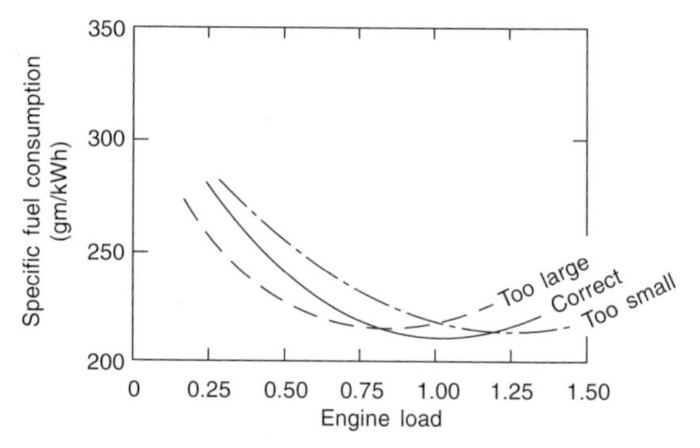

Figure 2.48 Optimum turbine matching for fixed load, constant speed

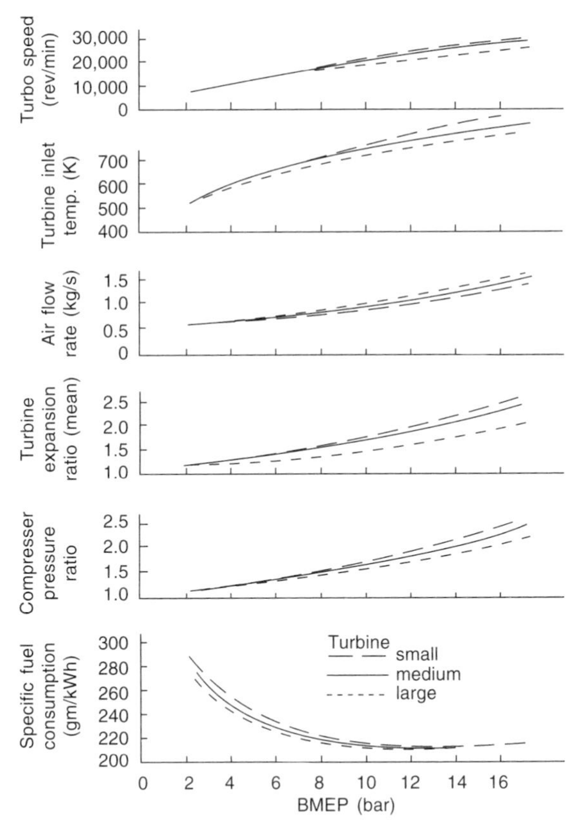

Figure 2.49 The effect of small turbine area change on engine performance at constant speed

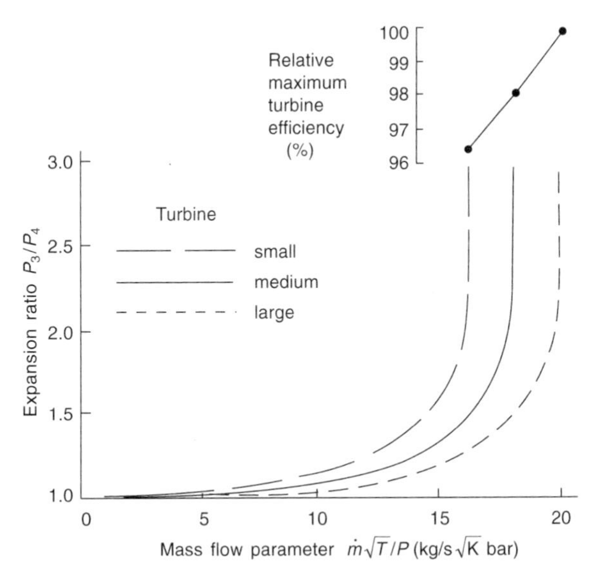

Figure 2.50 Flow characteristics of turbine trims from *Figure 2.49*

air flow during valve overlap. These effects combined to impair engine performance with the smaller turbine trim.

In *Figure 2.51* the engine air flow rates are superimposed on the compressor map, with each of the three turbine trims. Clearly the choice of compressor is correct with the medium turbine,

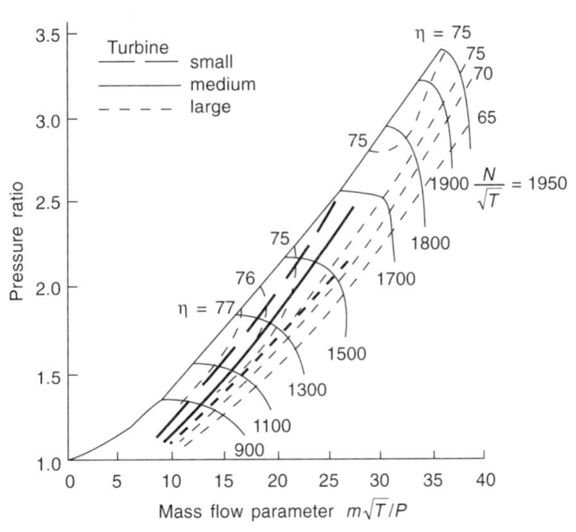

Figure 2.51 The effect of turbine matching on compressor requirement for constant engine speed operation

but a smaller compressor trim would be needed to avoid surge with the smallest turbine, and a slightly larger variant to achieve optimum compressor efficiency with the largest turbine.

2.6.4 Matching the marine engine

The required power versus speed characteristics of the marine engine is governed by the performance of the propeller, and will therefore depend on whether a fixed or variable pitch propeller is used.

The characteristics of the fixed pitch propeller are such that the power requirement increases with the cube of the speed (the well-known 'propeller law').

$$\dot{W} \alpha N^3 \tag{2.31}$$

Thus b.m.e.p. increases with speed squared. It happens that the output characteristics of the turbocharged engine are ideal for this application, hence matching is a case of optimization rather than compromise, since the compressor pressure ratio rises with engine speed as well as load.

2.6.4.1 The four-stroke engine with fixed pitch propeller

Figure 2.52 (line 1-2-3) illustrates a typical operating line on the compressor characteristic for fixed pitch propeller operation of a four-stroke engine. If the turbocharger is correctly matched, the compressor is working in its area of reasonably high efficiency at all engine speeds and loads, but if highly rated, the surge margin may be governed by mid-speed performance, due to the 'waist' shown in the surge line.

When matching at full speed the turbine will be matched to produce maximum engine power output, subject to thermal and mechanical limits. Generally the result will also be minimum specific fuel consumption, exhaust temperature, etc. will be density). If different turbine areas are tried, various boost pressures will be developed and the 'propeller law' working line will move across the compressor map. The compressor diffuser (or the complete assembly) will be changed to ensure that the operating line falls through the optimum efficiency area with a sufficient surge margin. When matching, the power output, specific fuel consumption, exhaust temperature, etc. will be monitored but generally there will be no major conflict between variables.

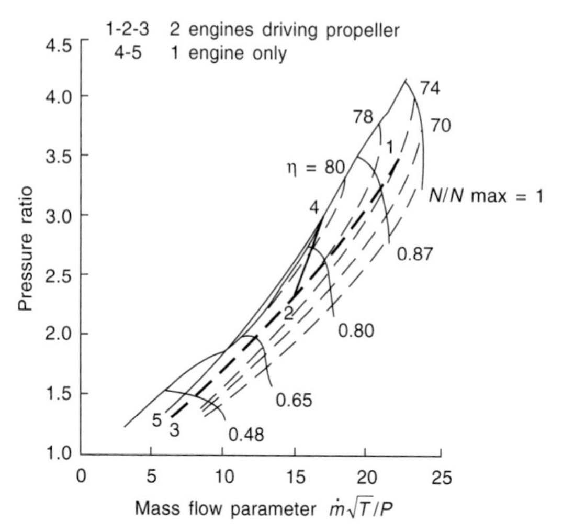

Figure 2.52 Engine mass flow characteristic superimposed on compressor map when one of two engines is driving propeller

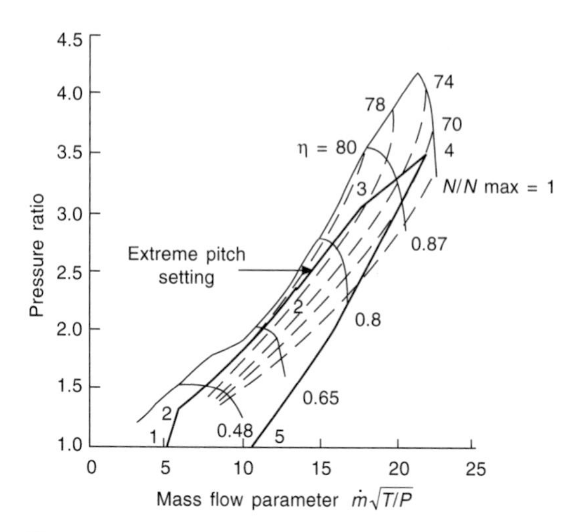

Figure 2.53 Engine mass flow characteristic driving a variable pitch propeller

Frequently, two medium speed engines will be used geared together driving a propeller. The requirement can arise, if one engine fails, of driving the propeller with one engine only. If the engine/turbocharger combination has been matched for optimum performance when producing half of the maximum power required by the propeller, then the match will be unsuitable for operation on one engine. Naturally this engine cannot drive the propeller at full speed (it will not have the power), but it is theoretically capable of turning it at nearly three quarters of its maximum speed due to the power requirement being proportional to the cube of the speed (propeller law). Consider what happens if the engine runs at 70% of full speed, remembering that the power required is double what would be required from each engine if both were running at that speed. If both engines were running, the operating point would be that denoted by point 2, in *Figure 2.52*, but point 4 with one engine. Unless the compressor is matched with a very large surge margin, surge is inevitable. Thus the optimum match with both engines working correctly is compromised.

2.6.4.2 The four-stroke engine with variable pitch propeller

A much broader power versus speed requirement is obtained from a variable pitch propeller. At each pitch setting, the propeller law will apply and hence the characteristic will be an envelope of propeller law curves. When plotting this resultant operating regime on the compressor characteristic (*Figure 2.53*) it can be seen that it is the 'extreme pitch' curve that determines the surge margin, not the maximum speed point (4). The engine can still be matched for optimum performance at full speed, but the compressor will have to be chosen to allow line 2–3 to be well clear of surge. This may result in a small penalty in performance at point 4.

2.6.4.3 The two-stroke engine with fixed pitch propeller

The different types of scavenging systems used by two-stroke engines have been described earlier. In general, only pulse turbocharged, uniflow scavenged engines can operate without a scavenge pump or fan in addition to the turbocharger over the complete operating range required. This class includes the very large opposed piston and 'valve in head' marine two-stroke engines. The power required by the propeller will naturally be governed by the propeller law and it follows that, if the engine is matched at its rated speed, the equilibrium running line will fall on the compressor map in a similar manner to the four-stroke engine. However, the air flow characteristics will be governed by the scavenge period, and may be simulated by two orifices in series (i.e. to represent the intake and exhaust of the engine). This characteristic suits the compressor well, and matching is usually a reasonably straightforward exercise similar to that described for four-stroke engines.

If scavenge pumps are used, for example, on ported cross-scavenge engines, the division of work between them and the turbocharger will influence the matching process. Generally though, the capacity of the scavenge pumps will have been estimated and the turbocharger is matched for optimum performance in conjunction with those pumps. A change in pump capacity might follow, depending on the success of the matching exercise, in which case the whole procedure must be repeated. To ensure that the engine will not stall at low speeds, the scavenge pumps must supply sufficient air to raise the inlet manifold pressure above that in the exhaust. If series pumps are used, the air flow characteristics obtained on the compressor map at low speeds will be highly dependent on the capacity of the scavenge pumps. Thus a range of air flow requirement curves could be plotted on the compressor map. At low speeds the air flow will depend principally on the pump speed, but at higher speeds, the influence of the turbocharger compressor will significantly affect air flow by increasing the charge density at entry to the scavenge pumps.

2.6.5 Matching for diesel-electric traction

Diesel generators are frequently used for rail traction since the characteristics of the basic diesel engine are not ideally suited for direct drive. In particular, the locomotive engine requires high torque at zero speed, to accelerate a heavy train from rest. Sufficient voltage can be generated (by a diesel generator) to produce the excitation necessary at the electric drive motors, the power needed being less than the capability of the engine at speeds below the maximum. If the turbocharger is matched to the engine at its rated speed and load, the equilibrium running lines will usually be quite well positioned on the compressor map at other speeds and loads. A possible area of trouble is surge at low speed, high load and this may mean that the compressor build finally chosen leaves a rather large surge margin at full power.

Since, like the automotive engine, the diesel electric locomotive unit is mobile, it is possible that it may be required to operate at altitude. The turbocharger must be matched to allow sufficient surge margin, and, if it is known that the engine will run at a particularly high altitude, the fuel injection system and turbocharger match will need to be adjusted to suit. The engine will effectively need derating to prevent overloading at altitude (see section 2.7).

2.6.6 Matching for other industrial duties

The power versus speed requirements for other industrial duties will generally fall between that of an automotive engine and the others described above. A typical industrial application might be the drive to a reciprocating compressor.

The compressor can be operated over a reasonably wide speed range, from zero to full load as the control valves open or close. The resultant operating area on the engine and compressor maps is similar to the requirements for an automotive engine (section 2.6.7). The surge margin will be governed by full load operation at low speeds and this will force the full speed and load point well away from surge and possibly into a low efficiency area of compressor operation.

2.6.7 Matching the four-stroke vehicle engine

2.6.7.1 Torque curve

Turbocharger matching for many industrial and marine duties is relatively straightforward due to the limited speed and load ranges required. Matching the turbocharger to an automotive engine is considerably more difficult due to the wide speed and load variations encountered. Although the power required to propel a vehicle increases repidly with speed, a torque curve that rises as speed falls reduces the number of gearbox ratios and gear changes required.

Such a torque curve is said to have good 'torque back-up'. Pulse turbocharging is essential in obtaining good 'torque back-up' at low engine speed. With good torque back-up, the vehicle will benefit from high torque at low speeds to provide a margin for acceleration and to allow the vehicle to lug up very steep hills.

Clearly the turbocharged automotive diesel engine should not be matched at full power and a compromise must be reached between power and a suitably shaped torque curve. *Figure 2.54* shows a typical torque curve for a turbocharged automotive diesel engine.

Torque back-up may be defined as:

$$\frac{\text{maximum torque} - \text{torque at maximum speed}}{\text{torque at maximum speed}}$$

Maximum torque of this engine occurs at 58% of its maximum speed, and the torque back-up is 29%. Since the engine will not normally be required to work below about 40% of the maximum speed, torque rises with reducing speed over half the useful speed range. This characteristic is, fortunately, a reasonable compromise between power and low speed torque, enabling trucks to use five-speed or six-speed gearboxes.

Engine performance will typically be limited (for reliability) by a maximum value of cylinder pressure and possibly exhaust temperature (since the latter is an approximate guide to the thermal loading of four-stroke engines, for example at the exhaust valve). In addition, a limit will be imposed by the acceptable (or legislated) exhaust smoke level, which will be determined largely by the quantity of air delivered by the turbocharger. The turbocharger will be limited by a maximum safe rotational speed and turbine inlet temperature (governed by the creep and scaling properties of the turbine wheel and housing). Depending on the engine rating, some or all of these factors will limit engine

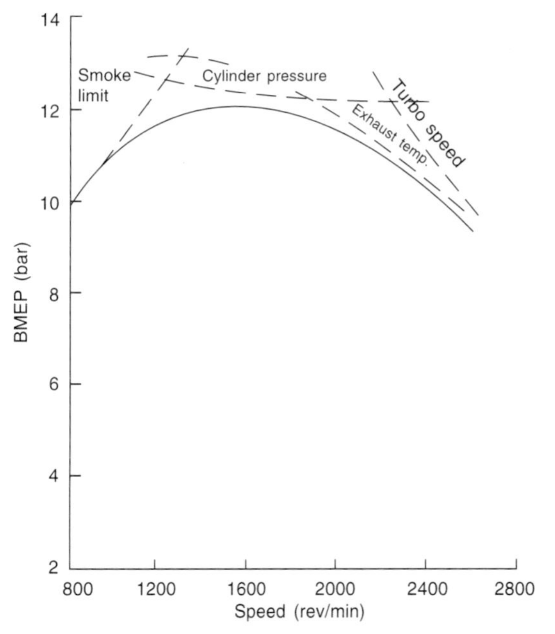

Figure 2.54 Turbocharged truck engine torque curve showing limits to b.m.e.p. set by allowable smoke, cylinder pressure, exhaust temperature and turbocharger speed (Goodlet)

performance. These limits can be superimposed on the torque (or b.m.e.p., since torque $\simeq$ b.m.e.p.) curve as shown in *Figure 2.54*, where the maximum possible torque curve is shown[4]. However, the position of most of these limiting lines will move if the turbocharger match is changed.

It is clear from *Figure 2.54* that the factor that is most restrictive when trying to achieve a desirable torque characteristic is the low speed smoke limit. This comes as no surprise since it was explained that it is normal for boost pressure to rise with engine speed, as a direct result of the air flow characteristics of the turbocharger turbine. For example, the flow characteristics of three radial turbine trims (volutes) are shown in *Figure 2.55*.The smallest area volute will generate the highest turbine inlet pressure and therefore the highest specific available energy at the turbine. Allowing for the variation of turbine efficiency over the pulsating and mean flow range, specific energy recovered by the turbine also increases substantially with engine speed. This characteristic is a consequence of the almost constant effective flow area of a fixed geometry turbine. Thus the natural characteristic of the turbocharged truck engine will be compressor work, and therefore boost pressure, rising with speed. The smoke limit is caused by insufficient boost pressure, and hence air flow, at low engine speeds.

In order to achieve an acceptable torque curve, the fuel delivery (per cycle) is held relatively constant over the speed range and efforts are made to raise boost pressure at low speed. Two techniques are available to achieve this. Either the turbocharger efficiency must be raised at this operating point or the thermodynamic availability of energy delivered to the turbine (i.e. specific available energy, *Figure 2.55*) must be increased. Both techniques are usually adopted, and will be discussed under the headings of turbine and compressor matching.

2.6.7.2 Turbine matching

Figure 2.55 shows that by reducing turbine area, for example from match 1 to match 3, specific available energy at the turbine increases at all speeds. If the fuel delivery schedule is unchanged,

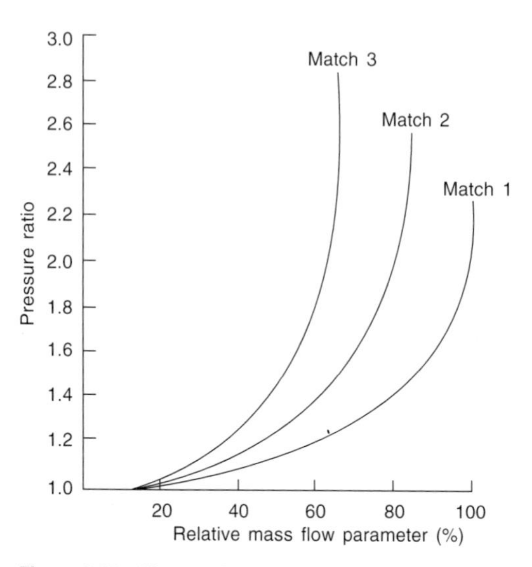

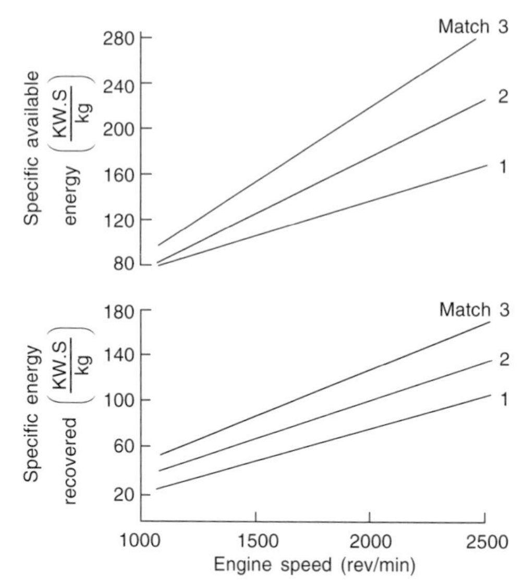

Figure 2.55 Characteristics of three truck turbocharger trims, and their effect on turbine energy on one particular engine

then boost pressure increases as shown in *Figure 2.56*, weakening the air–fuel ratio and reducing low speed smoke. Match 2 is based on a small turbine area reduction, hence the increase in available energy at low speed is small. However, turbine operation is more efficient with this build, hence the benefit in actual turbine work is more substantial.

The benefit of reduced smoke at low speeds does not come without accompanying disadvantages. *Figure 2.55* shows that with the smallest trubine (match 3), the expansion ratio across the turbine will be very high at the maximum engine speed, when air flow is greatest. Thus the piston must pump the exhaust gases out against a high pressure, resulting in poor net power output and fuel consumption. This is seen as low b.m.e.p. and high b.s.f.c. at high engine speed (2000 rev/min) with match 3.

In addition, the engine exceeds the allowable limits of maximum cylinder pressure and turbocharger speed. Thus match 2 is a reasonable compromise, except that maximum power (b.m.e.p. at maximum speed) is marginally less than that achieved with match 1.

Further 'fine-tuning' of the turbine may be achieved by selecting from different components having approximately the same effective area at the mid-engine speed condition, but whose flow and efficiency variations differ over the working range of the engine.

Ideally a variable geometry turbine is required. This would have the effective turbine area of match 1 at full engine speed, but the effective area of match 2 at mid-speed and match 3 at low speed. This would offer the low exhaust smoke, low fuel

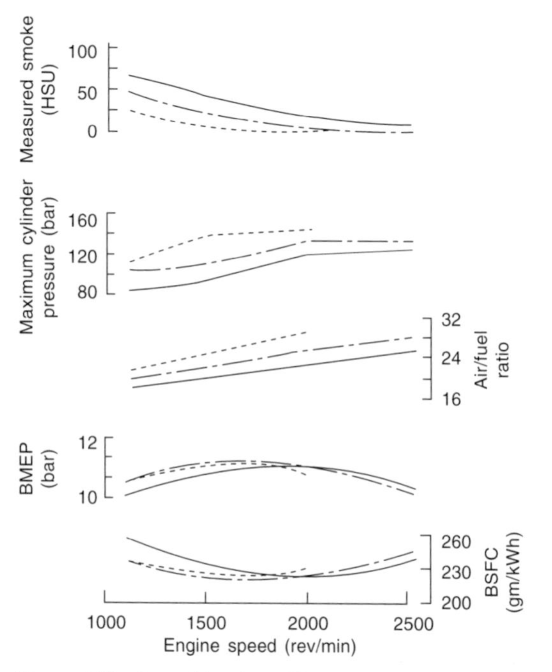

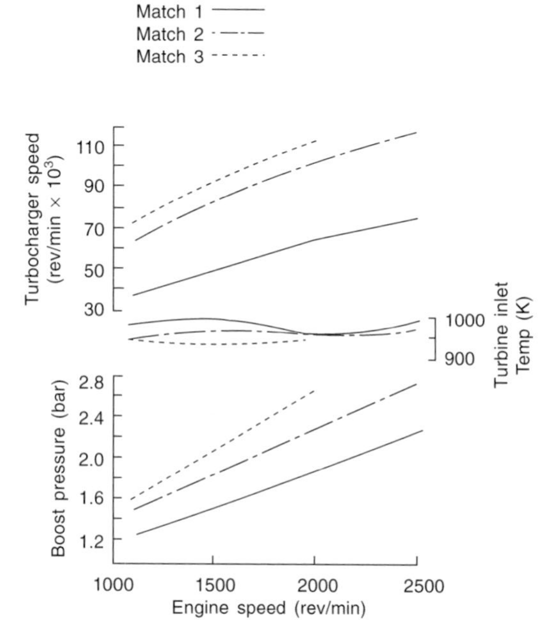

Figure 2.56 Rematching the turbine to improve low speed torque and smoke, with no change in fuel delivery

consumption and high b.m.e.p. of match 3 at 1000 rev/min, with the low cylinder pressure and low fuel consumption of match 1 at 2500 rev/min. This is an ideal that has attracted turbocharger and engine manufacturers for many years, since it reduces the speed dependence of turbine specific available energy shown in *Figure 2.55*. The problem is one of engineering a cheap, reliable and effective system, and has not been solved to date. Prototypes exist, but no system is currently available in mass production (1983).

The problem of overspeeding the turbocharger and coping with high cylinder pressures becomes prominent when engines which operate over a very wide speed range are turbocharged and matched for good torque back-up. The small turbocharged passenger car diesel engine falls into this category. A method of avoiding this problem is to bypass some of the exhaust gas around the turbine (through a waste gate) at high speed and load. Thus, when a small turbine is fitted to achieve good low speed boost, the massive increase in specific available energy at the turbine at high speed is alleviated by increasing the effective flow area out of the exhaust manifold. This has two effects. Firstly, only part of the exhaust gas flow goes through the turbine. Secondly, the increase in flow area reduces the exhaust pressure that would otherwise build up. Both measures reduce turbine work and hence boost pressure. In addition, the second factor reduces pumping work during the exhaust stroke and would, for example, moderate the loss in b.m.e.p. and deterioration in fuel consumption shown in *Figure 2.56* match 3, at high speeds.

The waste gate valve may be built into the turbine casing, and will consist of a spring loaded valve acting in response to the inlet manifold pressure acting on a controlling diaphragm. Different combinations of spring load, diaphragm area and valve area can be used to achieve a wide variety of boost pressure variations with engine speed. Disadvantages are increased cost, potential unreliability and the restriction to a single entry turbine housing.

2.6.7.3 Fuel delivery and engine speed range

Development of fuel injection pumps and associated equipment has introduced additional freedom to vary fuel delivery over the speed range of an engine. The turbocharger matching process must be closely linked with fuel system matching even after optimum injection rates, pressures, nozzle sizes and swirl have been achieved.

Tailoring of the fuel delivery characteristic is a method of achieving good torque back-up within the framework of engine and turbocharger limitations (*Figure 2.54*). For example, maximum fuelling can be restricted at high speeds in order to limit the maximum turbocharger speed with a small area turbine. Thus impressive torque back-up would be achieved, but at the expense of a low maximum power output.

At the other end of the speed range, excessive smoke can be reduced by restricting fuelling until sufficient boost is available to generate a reasonable air-fuel ratio. Thus a spring loaded diaphragm senses boost pressure and allows the maximum fuel stop to open as engine boost increases. Since fuelling is restricted only when the boost pressure is zero or low, torque is only reduced at very low speeds, that is, below that at which maximum torque is achieved. The device is commonly called an 'aneroid' (fuel controller), and is described in Chapter 10.

The difficulty of achieving a satisfactory match over a wide speed range has been explained. In certain circumstances it may be advantageous to reduce the rated speed of an engine whilst increasing b.m.e.p. to achieve the same maximum power output.

By reducing the turbine area and increasing fuelling, high b.m.e.p is obtained. If maximum rated speed is reduced from 2500 rev/min to 2000 rev/min (*Figure 2.56*, but an extreme case), excessive turbocharger speed is avoided. Naturally the final drive gear ratio of the truck must be raised to compensate, hence the engine is working at a higher load than would normally be the case at the same vehicle speed and load. Since specific fuel consumption reduces with load, fuel savings are possible.

2.6.7.4 Compressor matching

Since the truck engine operates over a wide speed and load range, the air flow requirements cover large areas of the compressor map. A typical superimposition of engine air flow on the compressor map is given in *Figure 2.57*, showing lines of constant engine speed (1000, 1500, 1900, 2400 and 2800 rev/min) and load (lower 3.85, 6.17 and 8.48 bar), and the maximum torque curve.

Selection of the correct compressor is largely a matter of ensuring a sufficient surge margin (A/B, *Figure 2.57*) and that the operating points at maximum torque and power (points X and Y, *Figure 2.57*) occur at reasonable compressor speed and efficiencies. Thus the compressor shown in *Figure 2.57* is a satisfactory choice, since the operational area is clear from the surge line and lies in an area of high efficiency. However, a slightly larger compressor might result in more of the operating regime experiencing higher compressor efficiency, with a less generous surge margin. A small improvement in low speed could also be obtained if maximum compressor efficiency occurred at a lower pressure ratio (e.g. 1.6 compared with 1.8 in *Figure 2.57*).

Reducing turbine area will raise the boost pressure, reducing the surge margin (compare *Figures 2.57 and 2.58*) and therefore a smaller compressor trim may be required in some cases. However, the surge margin shown in *Figure 2.58* is adequate, the compressor being well matched.

Figure 2.59 shows a poor compressor match, using too small a compressor trim on an intercooled engine. At maximum speed and load (point Y) the compressor efficiency is low. Boost pressure actually falls as the engine speed increases from 2400 to 2800 rev/min as a result. It may occasionally be convenient to deliberately match into an inefficient region at maximum engine speed and load, in order to hold the boost pressure, and therefore maximum cylinder pressure, down. However, an excessive reduction in efficiency occurs in the extreme case shown, and piston pumping work during gas exchange will suffer.

2.6.8 Matching the two-stroke vehicle engine

Only a few manufacturers now produce automotive two-stroke engines. General Motors (Detroit Diesel) is the only large company involved, and the subject will not be discussed in detail here.

Turbocharging the two-stroke automotive diesel engine involves some additional concepts to those already discussed above. Since the engine must work over a wide speed and load range, and must be capable of starting from a battery, some form of assistance will be required to produce the pressure drop from intake to exhaust manifolds so essential for scavenging at cranking speeds. Normally a Roots blower is used, placed in series after the turbocharger compressor. The Roots blower will ensure that the air flow characteristics are similar to those of a four-stroke engine. However the balance of work between Roots blower and the turbocharger will affect the power output, torque curve and specific fuel consumption.

Similarly to a four-stroke engine, the boost pressure will rise with speed and load. Therefore, the Roots blower will be expected to provide all of the boost pressure at low speeds and loads, yet a reducing proportion of the total as speed and load rises. The capacity of the Roots blower will be governed by the need for acceptable performance at low speeds. The Roots blower absorbs power from the crankshaft of the engine and reduces the power

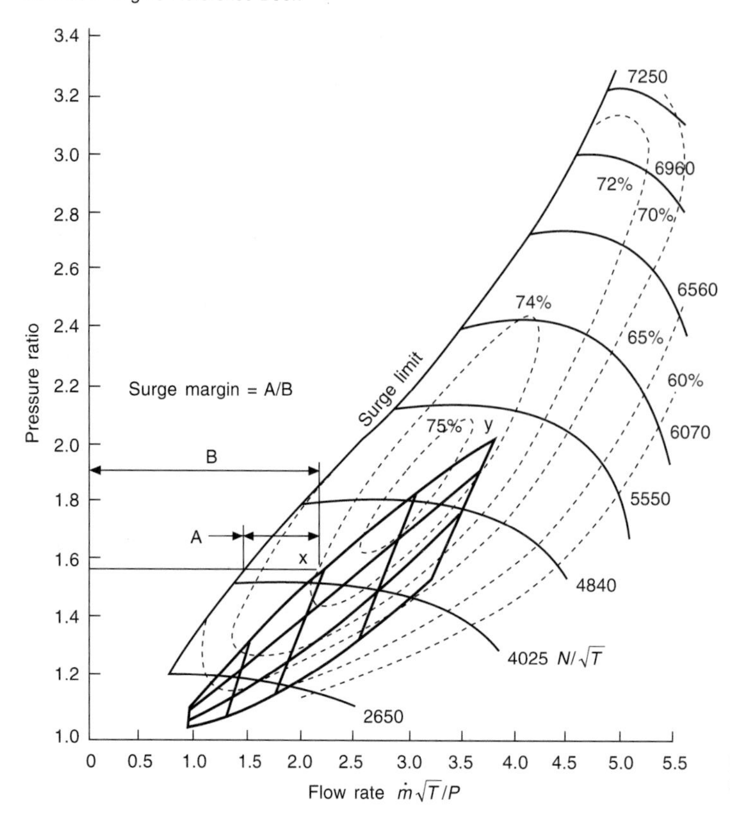

Figure 2.57 Engine operation area superimposed on compressor map, showing surge margin

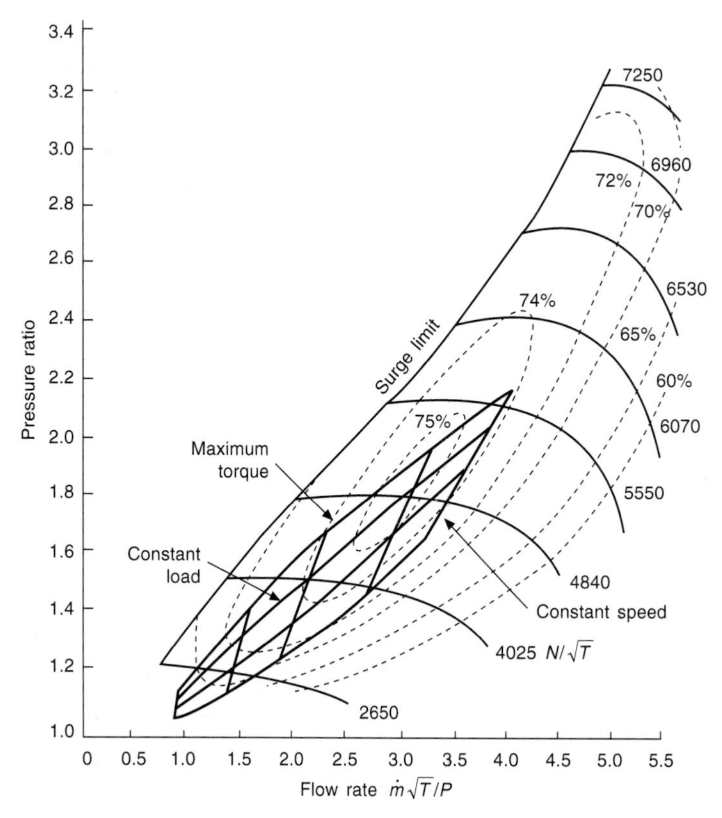

Figure 2.58 Engine operating area superimposed on compressor map, showing surge margin with reduced turbine area

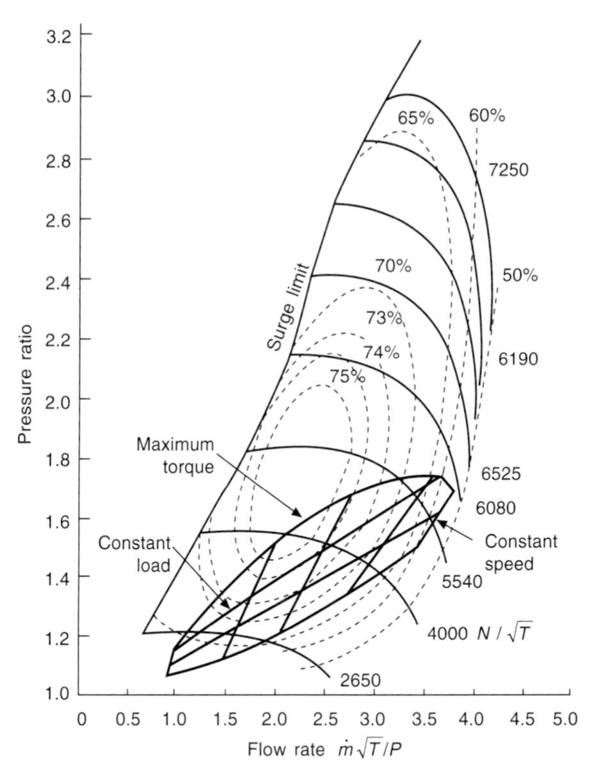

Figure 2.59 A very badly matched compressor

available at the flywheel. The larger the capacity of the Roots blower, the greater the loss in engine power output. Specific fuel consumption deteriorates. If the Roots blower is too small, scavenging at low speeds will be poor. If it is too large, then power will be wasted in developing an excessive boost pressure at high speeds and loads, where the turbocharger alone could provide sufficient boost. A compromise is required.

2.7 Changes in ambient conditions

2.7.1 Introduction

Changes in inlet air density may be caused by ambient temperature and pressure changes at sea level, or operation of the engine at altitude. The change of air flow may readily be predicted if the engine is naturally aspirated, but this is more difficult if the engine is turbocharged.

In mobile applications, such as a truck, the engine may be required to operate at sea-level in a cold winter climate and perhaps at altitude during a hot summer. Thus the final compressor and turbine matches selected will be something of a compromise, particularly if the pressure ratio is high. The match selected will suit the normal operating environment of the engine, but with sufficient margins on surge, turbine inlet temperature and turbocharger speed to cover other conditions.

If the engine is designed for stationary applications, its operating altitude will be known, hence the manufacturer will have the option of rematching the turbocharger to suit the environment. The alternative will be to derate the engine for operation at altitude.

Although operation under changing ambient conditions introduces additional complications for the manufacturer of a turbocharged engine (such as reduction of the surge margin), the turbocharging system does offer partial compensation for reducing air inlet density at altitude. As air density and therefore air mass flow rate reduce, so the turbine inlet temperature will rise due to the richer air-fuel ratio. This means that the ratio of compressor to turbine pressure ratios will be altered in favour of the compressor. Its pressure ratio will increase, partially offsetting the reduction in air inlet density, Also as ambient pressure falls, so the expansion ratio of the turbine increases, raising compressor pressure ratio, provided that the turbine inlet pressure does not fall at the same rate as ambient pressure.

An increase in ambient temperature however, has an undesirable effect on the turbine to compressor energy balance, hence the turbocharger will tend to amplify the effect of such a change on air flow rate. Low ambient temperatures reduce the required compressor power, hence boost rises, sometimes causing compressor surge.

2.7.2 Operation under changing ambient conditions

Wide variations in ambient conditions can lead to problems due to compressor surge, excessive cylinder pressure, turbine inlet temperature, turbocharger speed or smoke emission. The actual performance of an engine under varying ambient conditions will depend on several factors that, for convenience of explanation, were assumed to be constant in the simplified analysis given above. For example, if air mass flow rate and compressor pressure ratio change, movement across the operating map of the compressor will be accompanied by an efficiency change. It follows that engines of similar performance at sea-level will not necessarily perform comparatively at altitude. Techniques have been developed for accurately predicting the effect of varying ambient conditions, but these require detailed turbine and compressor maps. A simpler, but less rigorous approach, is to correlate the performance of existing engines obtained when operating at altitude in very hot and very cold climates.

The parameter that limits engine performance will depend on the design of individual engines. At high ambient temperatures (*Figure 2.60*)[5], the limits are likely to be smoke, due to reducing air flow, then turbine inlet or exhaust valve temperature or thermal loading of the engine. At low ambient temperature, compressor surge (due to high pressure ratio) or maximum cylinder pressure may be a limiting factor. It will be the limitations of a particular engine and turbocharger combination that will govern to what extent fuelling must be reduced, derating the engine for acceptable reliability or smoke emission.

Operation at altitude is usually, but not always, accompanied by a reduction in temperature. The turbocharging system offers partial compensation of the inlet air density reduction at altitude, thus an engine may have to be derated, but not by as much as a naturally aspirated engine.

The effect of altitude on a turbocharged truck engine at full power, with and without intercooling, is shown in *Figure 2.61.* Although the absolute inlet manifold pressure reduces with altitude, the fall-off is slower than that of ambient pressure. Turbocharger speed increases due to the increase in turbine inlet temperature and expansion ratio. It can be seen that thermal limits and the maximum permissible turbocharger speed will be the limiting factors, particularly the latter. Movement towards surge on the compressor map will be greatest for a non-intercooled engine.

If a stationary engine is not rematched for operation at altitude, then initially smoke emission, then turbocharger speed or inlet temperature will be the factors governing the reduction in fuelling and therefore rated power output, required. CIMAC (Conseil International des Machines à Combustion) recommend an empirical formulate for derating at altitude, based on limitation of constant turbine inlet temperature. The formula is

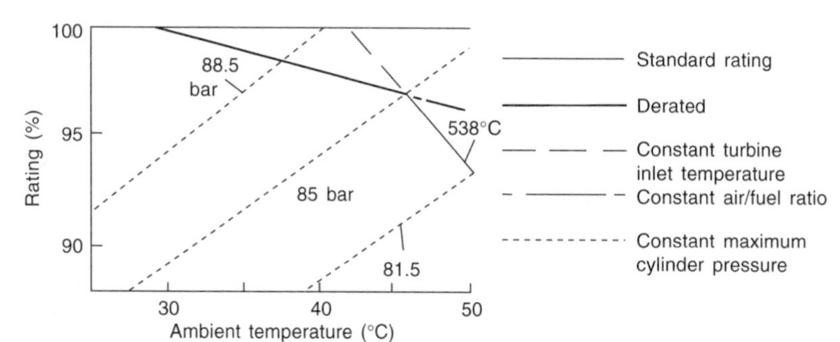

Figure 2.60 Derating chart, for changes in ambient pressure with smoke (air-fuel ratio) and turbine inlet temperature limits (Lowe)

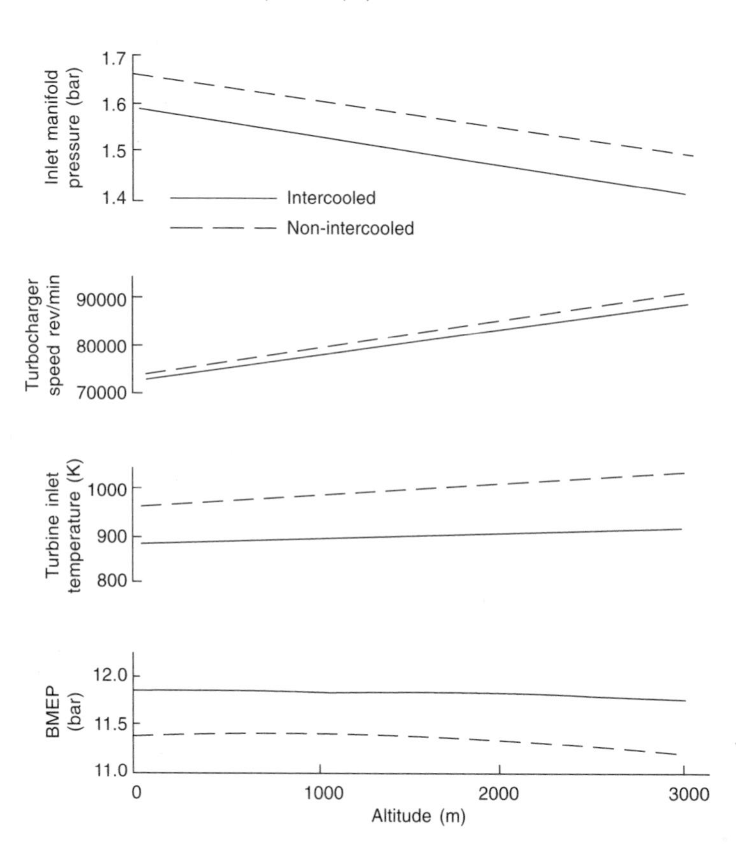

Figure 2.61 Effect of altitude on turbocharged truck engine performance

$$\dot{W}_{\text{alt}} = \dot{W}_{\text{ref}}\left[K - 0.7(1-K)\left(\frac{1}{\eta_{\text{mech}}} - 1\right)\right] \tag{2.32}$$

where

$$K = \frac{\dot{W}_{\text{alt(ind)}}}{\dot{W}_{\text{ref(ind)}}} = \left[\frac{P_{\text{a alt}} - hf\,\phi_{\text{alt}}\,P_{\text{svp}}}{P_{\text{a ref}} - hf\,\phi_{\text{ref}}\,P_{\text{svp ref}}}\right]^m \times \left[\frac{T_{\text{ref}}}{T_{\text{alt}}}\right]^n \left[\frac{T_{\text{cool,ref}}}{T_{\text{cool,alt}}}\right]^q$$

Values of the indices, m, n and q are given in *Table 2.2*. In practice the empirical formula is rather conservative.

Table 2.2 Exponents in CIMAC formula for derating at altitude

Four-stroke turbocharged diesel engines	m	n	q
Without charge cooling	0.7	2.0	—
With charge cooling	0.7	1.2	1.0

2.7.3 Rematching to suit local ambient conditions

If the turbocharger is selected to suit the local ambient condition, additional density compensation can be provided in some cases. By reducing the turbine trim, more work can be extracted from the turbine enabling boost pressure to be raised, offsetting the loss in ambient density. This will, for example, delay the smoke limit or turbine inlet temperature limit to higher altitudes or ambient temperatures (compare *Figures 2.60 and 2.62*)[5]. Thus derating may be avoided altogether or reduced. However, the influence of a more restrictive turbine on exhaust manifold pressure levels must be considered. Some engine performance deficit may occur.

2.8 Closure

Previous sections of this chapter have outlined the principles of turbocharging, as applied to various types and application of diesel engines. Various engine types are described in Chapters

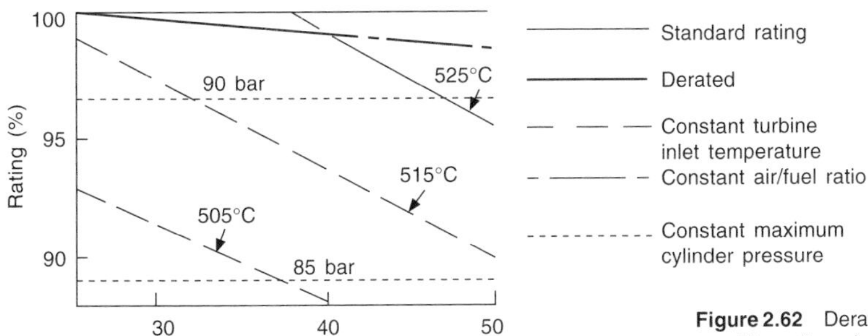

Figure 2.62 Derating chart, with turbocharger rematch to suit ambient conditions (Lowe)

22 to 27 inclusive, including their turbocharging systems. The reader should turn to these chapters to find further examples of turbocharging practice.

N.W.

References

References cited:

1 RYTI, M. and MEIER, E., 'On Selecting the Method of Turbocharging Four-stroke Diesel Engines', *Brown Boveri Review,* **56,** No. 1 (1969)

2 JANOTA, M. S., *Ph.D. thesis, University of London, 1967,* see Ref. 6

3 McLEAN, D. H. and IHNEN, M. H., 'The Design and Development of an Air-to-air Intercooled Engine for Agricultural Applications', *ASME paper 78-DGP-28* (1978)

4 GOODLET, I. W., 'Turbocharging of Small Engines', *Proc. I.Mech.E.,* **188,** No. 3/74 (1974)

5 LOWE, W., 'The Effect of Ambient and Environmental Atmospheric Conditions (on Diesel Engines)', *Proc. I.Mech.E.,* **184,** Pt. 3P (1969/70)

General references

6 WATSON, N. and JANOTA, M. S., *Turbocharging the internal combustion engine,* Macmillan Press (Wiley, USA) (1983)

7 ZINNER, K., *Supercharging to internal combustion engines* (Springer-Verlag (1978))

Acknowledgements

The author would like to acknowledge the help given by the late Professor M. S. Janota (Queen Mary College, London) and the following companies: AiResearch Industrial Division (Garrett Corp.), Brown Boveri & Cie, Holset Engineering Co. Ltd, MAN, Napier Turbochargers Ltd.

Diagrams and parts of the text are reproduced from Reference 6, by kind permission of Macmillan Press.

Nomenclature

a	Velocity of sound
AFR	Air-fuel ratio (by mass)
b.m.e.p.	Brake mean effective pressure
C	Gas velocity after isentropic expansion to turbine exit pressure
C_p	Specific heat at constant pressure
D	Diameter (compressor impeller or turbine rotor)
f	A function of (...)
h	Specific enthalpy
hf	Humidity factor
KE	Kinetic energy
l	Length
$\dot{m}$	Mass flow rate
N	Rotational speed
P	Pressure
PE	Potential energy
$\dot{Q}$	Heat transfer rate
R	Gas constant
s	Specific entropy
T	Temperature
t	Time
ΔT	Temperature change
u	Rotor tip tangential velocity
V	Velocity (gas) or volume
v	Specific volume
$\dot{W}$	Power
γ	Ratio of specific heats (constant pressure/constant volume)
ε	Effectiveness
η	Efficiency
$\Delta\theta$	Change in angle
ν	Viscosity
ρ	Density
ϕ	Relative humidity

Suffices

0	Stagnation
1	Inlet
2	Outlet
3	Charge cooler outlet
a	Air
Alt	Altitude
cool	Coolant
e	Exhaust gas
f	Fuel
ind	Indicated
is	Isentropic
m	Manifold (inlet)
mech	Mechanical
ref	Reference
svp	Saturated vapour pressure
sw	Swept (by pistons)
t	Turbine
TS	Total to static
TT	Total to total
vol	Volumetric

3 Compound and other engine systems

Contents

3.1 Introduction

Chapter 2 deals extensively with many aspects of turbocharging (*Figures 3.1a* and *3.1b*), which has become the preferred means of increasing the output of virtually every type of diesel engine and of improving many other operational factors including specific fuel consumption and emissions.

However, much theoretical and experimental work has been undertaken in addition to that on turbocharging, with a view to assessing the performance potential of other combinations of the diesel engine with compressors and turbines under the headings:

(a) Gas generators (*Figure 3.1c*)
(b) Compound engines (*Figure 3.1d*).

The term 'gas generator' refers to a combination in which the diesel engine drives a reciprocating or rotary supercharging compressor through gearing, the whole of the engine output being absorbed by the compressor, while a mechanically independent turbine, supplied with engine exhaust gas, drives the load. The arrangement has certain advantages particularly for very highly rated two-stroke engines, and will be referred to again in section 3.2.

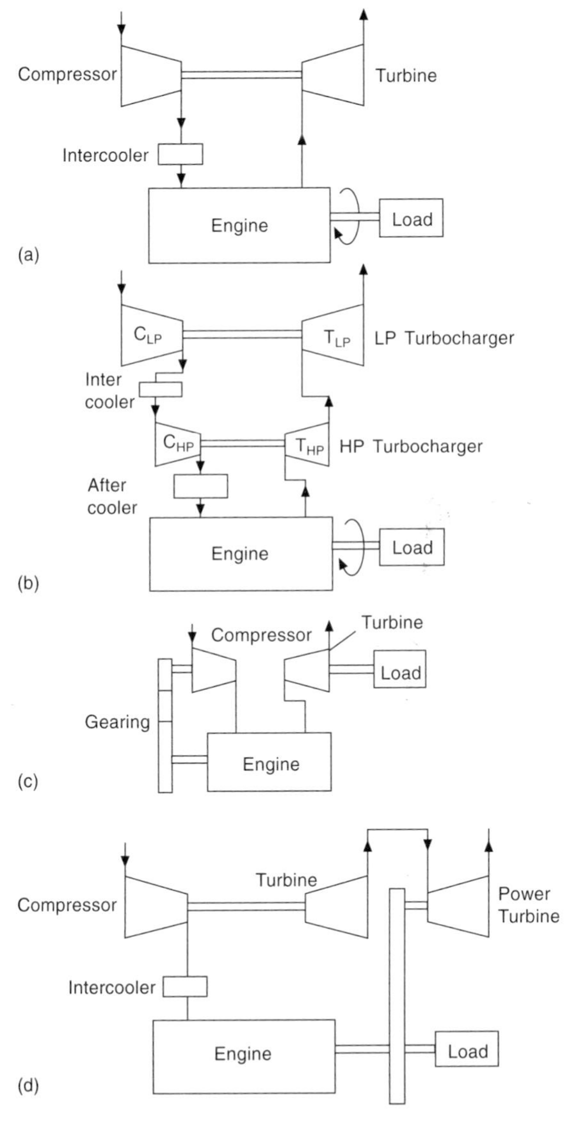

Figure 3.1 Engine turbomachinery systems: (a) turbocharged engine (single-stage); (b) turbocharged engine (two-stage); (c) gas generator; (d) compound engine

The term 'compound engine' is applied to any combination of the diesel engine with one or more compressors and turbines, and with a mechanical connection through gearing or other means between the engine and at least one turbomachine, of which a turbine must be one. Compounding is attractive because it enables any surplus of power developed by the turbine(s) over that absorbed by the compressor(s) to be fed back to the output shaft thereby significantly improving the efficiency and output of the power plant. Various forms of compound engine will be described in a later section. The possibility of power feedback is not available to the turbocharged engine except in very attenuated form through 'positive loop' pumping work.

Both the gas generator and compound schemes involve considerable complication compared with the turbocharged engine and can therefore be justified only if the advantages in terms of specific power output, efficiency, operational flexibility, emissions, etc. are substantial. In practice the turbocharged engine has found almost universal application, but increasingly stringent environmental legislation and mounting pressure for improved fuel economy may well result in renewed interest particularly in compound engines.

3.2 Gas generator and compound schemes compared with the turbocharged engine

Figures 3.1a, *3.1c*, and *3.1d* represent, respectively, the simplest forms of turbocharged engine, gas generator and compound engine. Since in examining the potential of alternative arrangements the design point rating is of crucial importance, it is usual to base such calculations on boost pressure in conjunction with an assumed air–fuel ratio. Using a very simple diesel engine cycle model (see section 1.3, Chapter 1), with appropriate assumptions for compression ratio and maximum cylinder pressure, as well as compressor and turbine efficiency, it is possible to construct curves of specific power $W/\dot{m}$ (where W = power and $\dot{m}$ = rate of air mass flow) for engine, compressor and turbine as a function of boost pressure.

Figure 3.2 represents such a simplified specific power relationship over a wide range of boost ratios for a four-stroke engine with P_{max} = 138 bar, and compression ratio reduced progressively from 15:1 at low boost in order to maintain the set limit of P_{max}. The assumed air–fuel ratio is 25:1. The heat loss to coolant is adjusted so as to give a constant turbine inlet temperature of 600°C. Under these somewhat artificial assumptions engine power W_e exceeds compressor power W_c over the whole of the boost pressure ratio range up to 7:1. Thus power balance between engine and compressor could be achieved only at pressure ratios of 7:1 or above, a highly unrealistic state of affairs. In practice the gas generator concept has been applied only to two-stroke engines in which the specific airflow and the resultant overall air–fuel ratio are considerably higher than for four-stroke engines. Such gas generators have operated at pressure ratios of the order of 4:1. It will also be observed that turbine power consistently exceeds compressor power so that theoretically, since $W_e = W_c$, the useful shaft power output of such a unit, namely that of the turbine, W_t, should exceed engine power, implying an output shaft efficiency greater than that of the engine. This train of reasoning constitutes the theoretical justification for the gas generator cycle. In practice, part load efficiency is strongly dependent on boost, both diminishing rapidly with decreasing load. This inherent disadvantage, coupled

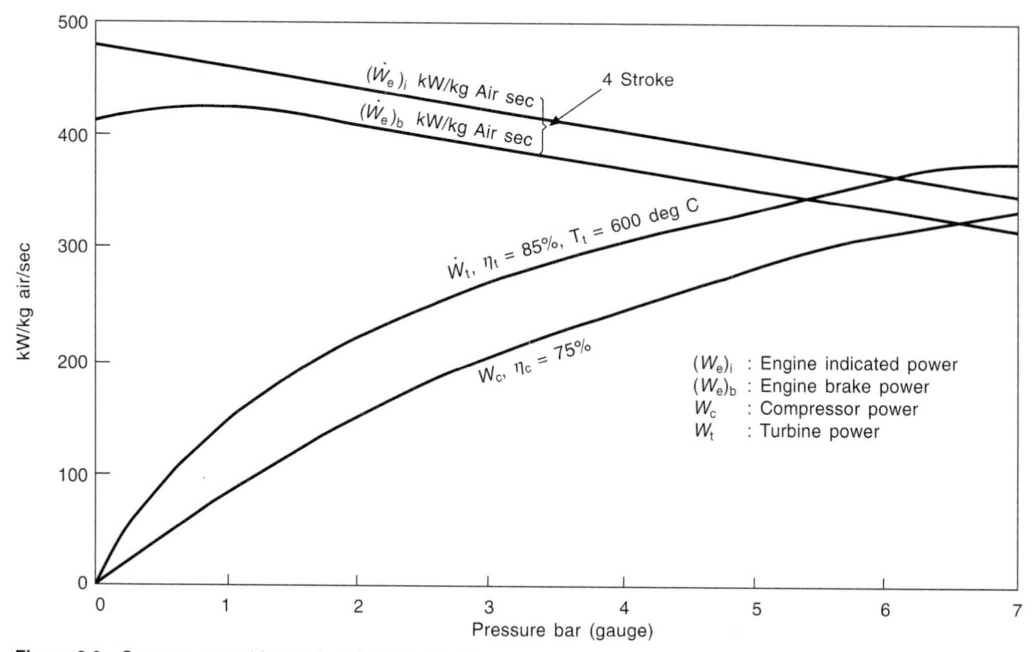

Figure 3.2 Compressor, turbine and engine power curves

with mechanical complexity and unreliability, has led to the abandonment of the diesel gas generator as a practical power plant.

On the other hand compound engines have rather more attractive characteristics, in that the excess of turbine power W_t over compressor power W_c (*Figure 3.2*) becomes an addition to engine power W_e giving augmented shaft power $W_s = W_e + W_t - W_c$, and hence improved efficiency over the whole of the boost range. Compound engines thus undoubtedly have considerable theoretical advantages over turbocharged engines, in which shaft power is necessarily limited to W_e, and should therefore be considered as a means of improving the fuel economy of the diesel engine still further. With indicated or brake thermal engine efficiencies determined by appropriate cycle analysis, the efficiency of the compund cycle, η_{cpd}, may be expressed as a function of engine efficiency, η_e, as follows:

$$\eta_{cpd} = \eta_e \left[\frac{W_e + W_t - W_c}{W_e} \right] = \left[1 + \frac{W_t - W_c}{W_e} \right] \eta_e \qquad (3.1)$$

The above discussion of trends based on rather crude basic assumptions may give a somewhat misleading picture of the likely development potential of the gas generator vis-à-vis that of the turbocharged and compound engine. One practical embodiment of the gas generator, the so-called free piston engine[1], while undoubtedly constituting a most ingenious solution of the basic design layout, nevertheless proved a failure in service.

Compound engines, while undoubtedly attractive in view of their potential for improved thermal efficiency[2], have again not been widely accepted, mainly because of their substantially greater complexity and cost compared with turbocharged engines, the latter themselves having progressed rapidly in recent years in terms of efficiency and specific output.

3.3 Analysis of turbocharged and compound engine systems based on full cycle simulation

Detailed analysis of the many possible types of turbocharged and compound engine schemes requires a flexible method of modelling and linking the various components. This is provided by computer programs in which the constituent elements, i.e. engine, compressor(s), turbine(s), cooler(s), etc. are represented by accurate subroutines linked by a master program ensuring that the various compatibility conditions governing the operation of these elements within the system, i.e. the balances for mass flow, energy, speed and torque, are satisfied[3]. This more refined analysis will be applied in two stages. In the first approach, section 3.3.1, the engine itself is fully modelled, while the compressor(s) and turbine(s) are represented by idealized compression and expansion machines having assumed polytropic efficiencies, in this case 85% and 80%, respectively. This approach is intended to provide efficiency and output trends for turbocharged and compound engine schemes on similar lines to those indicated by *Figure 3.2*, but now based on a much more refined engine submodel.

In the second approach, section 3.3.2, the entire system is fully modelled, using compressor and turbine performance maps, together with the pressure drop characteristics of coolers, filters and silencers. This approach will be applied to turbocharged and compound systems with a single turbocharger (*Figures 3.1a* and *3.1d*) and corresponding systems using two stage turbocharging (*Figure 3.1b*).

3.3.1 Analysis based on compression and expansion machines with fixed polytropic efficiencies of 85% and 80% respectively

This analysis, based on the simulation code ODES (Otto-Diesel-Engine-Simulation)[3], has been applied to an 8 litre, direct injection, 6-cylinder four-stroke engine of 110 mm bore and 140 mm stroke operating in conjunction with the idealized compression and expansion machines as described above. Rated speed and trapped air–fuel ratio are assumed to be 2200 rev/min and 25:1, respectively. The computations are performed at four nominal levels of boost pressure ratio of 3:1, 5:1, 7:1 and 10:1, in each case varying back pressure so as either to satisfy the condition of equality of compressor and turbine power for turbocharged operation, or to maximize the system efficiency for compound operation, as defined by eqn (3.1).

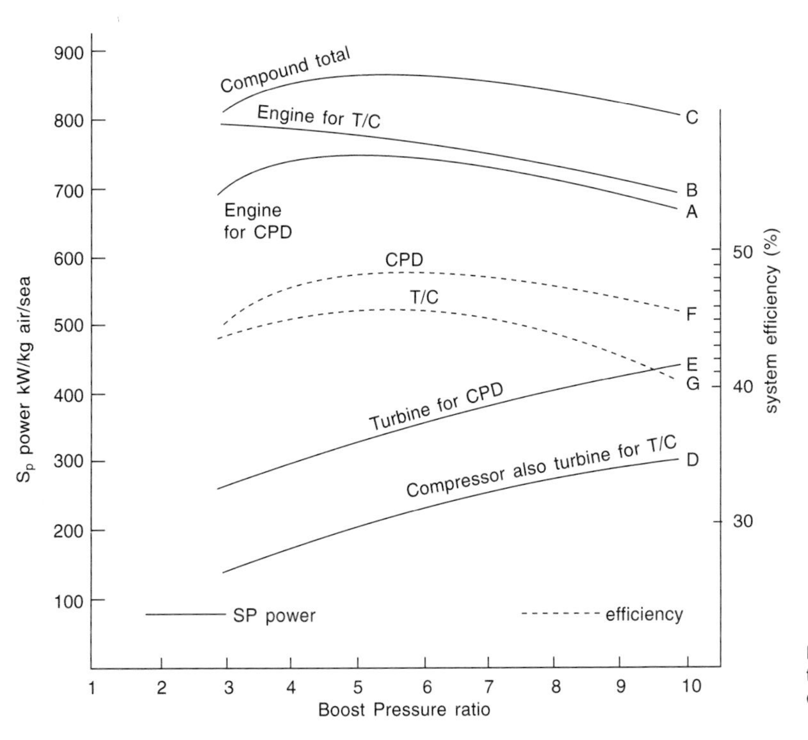

Figure 3.3 Specific power and efficiency for turbocharged and compound engines (by detailed computer model)

Figure 3.3, which should be compared with *Figure 3.2*, again shows specific component power in kW/kg/s of airflow together with turbocharged and compound system efficiency, shows trends which differ substantially from those shown in *Figure 3.2*. In particular, the intercept between the compressor and engine power curves observed in *Figure 3.2* at a boost pressure ratio of approximately 7:1, which represents the gas generator condition, is now absent from the diagram due to the much higher level of engine power, curves A and B in *Figure 3.3* based on the more refined engine model. However, *Figure 3.3* does provide interesting information on turbocharged and compound operation.

(a) *Engine specific power* (curves A and B for compound and turbocharged operation) Curve A for compound systems lies consistently below curve B for turbocharged systems due to the substantial negative pumping loop work implied by the higher exhaust back pressure relative to turbocharged operation (see *Table 3.1*). At boost pressure ratios above 5:1 both curves fall progressively, from 727 $kW/kg_{air}/s$ and 773 $kW/kg_{air}/s$, respectively, to 663 and 686 $kW/kg_{air}/s$, respectively, at a boost pressure ratio of 10:1, due to the inevitable loss of engine efficiency implied by the drastic reduction in compression ratio from 14.5:1 to 8:1, which is dictated by the need to limit maximum cylinder pressure to a maximum of 275 bar, a value reached by some experimental high output engines, but well beyond current practice.

(b) *Compressor and turbine specific power* (curves D and E) The compressor and turbine power curves follow a similar trend to the corresponding curves in *Figure 3.2*. For turbocharged operation, turbine power exactly balances compressor power, both of which are therefore represented by curve D, with curve E applying exclusively to compound operation. The difference between the two curves, some 125 $kW/kg_{air}/s$ represents the excess of turbine over compressor power which is fed back to the output shaft, and is mirrored by the difference between curve A, i.e. engine power under compound operation, and curve C for total compound system power.

(c) *Turbocharged and compound system specific power* (curves B and C) Curve B for turbocharged operation shows a continuous decrease in *specific* power over the boost pressure range from 3:1 to 10:1 due to the reduction in compression ratio already referred to. However, this does not apply to *absolute* power which increases from 328 kW at 3:1 boost to 1021 kW at 10:1 boost due to the massive increase in airflow over this range of boost pressure ratios (see *Table 3.1*). The compound specific power curve C is seen to peak at 865 $kW/kg_{air}/s$ at 5:1 boost, corresponding to 545 kW of *absolute* power, declining to 800 $kW/kg_{air}/s$ at 10:1 boost, corresponding to an *absolute* power level of 1073 kW or 134 kW/litre of swept volume (see *Table 3.1*).

(d) *System efficiency* (curves F and G for compound and turbocharged operation) Curve F for the compound system peaks at 48.19% for a boost pressure ratio of 5:1 (see *Table 3.1*), rising from 44.8% for 3:1 boost. Thereafter it declines to 45.33% at 10:1 boost due to the adverse effect of reducing compression ratio. The efficiency of the turbocharged system, curve G, peaks at 45.73% again at 5:1 boost, and then declines steeply to 40.05% at 10:1 boost. Although curves F and G show reducing trends beyond the optimum boost pressure ratio of 5:1, the difference between them widens significantly due to the increasing excess of *absolute* turbine power over compressor power under compound operation, as a result of the rapid increase in both gas mass flow and turbine pressure ratio.

(e) *Summary of the discussion under (a) to (b) above* It is clear that compounding at current levels of boost pressure ratio, of the order of 3:1, yields only marginal improvements in specific power and efficiency relative to turbocharging. The greatest gains in efficiency are achieved in the region of 5:1 boost, where values approaching 50%, or a specific consumption of 170 g/kW.h may be approached. At the same time specific air consumption, of the order of 4.2 kg/kW.h is particularly good at this boost level.

At boost levels above 5:1 efficiency falls off due to reducing compression ratio, although absolute power continues to rise steeply.

Table 3.1 Turbocharged and compound engine performance as function of boost pressure ratio

	Nominal boost 3:1		*Nominal boost 5:1*		*Nominal boost 7:1*		*Nominal boost 10:1*	
	turbocharged	*cpd*	*turbocharged*	*cpd*	*turbocharged*	*cpd*	*turbocharged*	*cpd*
Actual boost pressure ratio	2.96	2.96	4.93	4.93	6.89	6.89	9.84	9.84
Compression ratio	15:1	15:1	14.5:1	14.5:1	10.5:1	10.5:1	8:1	8:1
Sp. engine output kW/kg. s	799.9	689.5	773.3	727.3	745.6	721.9	686.2	663.0
Sp. compressor input kW/kg. s	136.1	136.1	192.9	192.9	242.0	242.0	298.0	298.0
Sp. turbine output kW/kg. s	136.1	257.0	192.9	330.8	242.0	359.4	298.0	435.4
Sp. compounded output kW/kg. s	—	810.4	—	865.2	—	839.3	—	800.4
Total output kW	328.0	305.2	577.0	545.0	787.6	802.2	1021.0	1073.0
Sp. fuel consumption g/kWh (engine alone)	192.7	217.7	183.9	206.1	190.0	203.9	210.0	223.1
Sp. fuel consumption g/kWh (overall)	192.7	187.7	183.9	174.5	190.0	176.2	210.0	185.5
Max. cylinder press bar	155.9	147.2	265.8	249.2	270.9	262.4	276.4	264.4
Back pressure ratio	2.21	4.09	3.20	6.65	3.97	7.35	5.25	10.54
Trapped air–fuel ratio	25	25	25	25	25	25	25	25
Trapped air mass/cyl kg	0.004	0.0036	0.0067	0.0060	0.0096	0.0089	0.0139	0.0126
System thermal efficiency %	43.64	44.80	45.73	48.19	44.27	47.73	40.05	45.33
Improvement in efficiency % cpd. over turbocharged	2.65		5.37		7.82		13.20	

For practical purposes, therefore, future compound system development should be aimed at boost pressure ratios in the region of 5:1. Such systems would inevitably be more complicated than current turbocharged systems, operating at pressure ratios of 3:1, and would involve two-stage rather than single-stage turbochargers, in conjunction with a geared power turbine, while the engine itself would have to be designed to withstand maximum cylinder pressures in the region of 250 bar, if these high potential efficiencies are to be achieved.

3.3.2 Analysis based on fully modelled system, including compressor, turbine and cooler characteristics

The analysis of section 3.3.1 based on notional compression and expansion machines with given polytropic efficiencies as opposed to compressors and turbines each with their own given characteristics, identified the general trends summarized in *Table 3.1* and *Figure 3.3*.

In the present section the analysis is extended to two specific sets of turbocharged and compounded systems, viz.

System 1: Turbocharged engine operating at 3:1 nominal boost pressure ratio

System 2: Compound engine with single-stage turbocharger and geared power turbine operating at 3:1 nominal boost pressure ratio

System 3: Two-stage turbocharged engine operating at 5:1 nominal boost pressure ratio

System 4: Compound engine with two-stage turbocharger and geared power turbine operating at 5:1 nominal boost pressure ratio

The analysis is applied to the same 8 litre, direct injection, 6 cylinder four-stroke engine as that used in the simplified analysis of section 3.3.1, but now based on manufacturers' compressor and turbine maps, suitably scaled in the case of the compound system to achieve the required increase in back pressure as described in section 3.3.1 Furthermore, to achieve the greatest possible benefit from compounding, efficiency scale factors of 1.1 were applied to both the compressor and turbine maps supplied by the manufacturers, since high turbo machinery efficiencies are a precondition for successful compounding.

The rated engine speed is again taken as 2200 rev/min, but the trapped air–fuel ratio has been increased to 27:1 to provide an adequate fuelling margin for torque back up.

3.3.2.1 Turbocharged and compound engines with single stage turbocharger, systems 1 an (nominal boost pressure ratio 3:1, Figures 3.1a and 3.1d)

The results for rated conditions of both systems are summarized in *Table 3.2* which, in addition, also contains the results for the two-stage turbocharged engine (system 3) and the compound engine with two-stage turbocharger (system 4), both of which are dealt with in section 3.3.2.2.

The rated power for systems 1 and 2 with single-stage turbocharger and operating at a nominal boost pressure ratio of 3:1 was set at approximately 300 kW.

Comparing the compounded system 2 with the turbocharged system 1, the gain in thermal efficiency from 42.76% to 43.88% is very modest, and compares closely with that given by the simple analysis of section 3.3.1, *Table 3.1*, which predicted an improvement of the order of one percentage point.

The introduction of the power turbine raises the exhaust back pressure from 2.87 bar to 3.46 bar and leads to a reduction in engine power from 307.2 kW to 274.2 kW as a result both of negative pumping work and of the reduced mass of trapped air (see *Table 3.2*).

The increase in back pressure has been achieved by the use of a smaller turbocharger turbine (see turbine scale factor in *Table 3.2*) followed by a substantial larger power turbine which contributes 24.6 kW and thus raises the total output of the compound engine to 274.2 + 24.6 = 298.8 kW. The simple treatment of section 3.3.1 predicted a much larger contribution by the power turbine because it exaggerated the difference in back pressure between turbocharged and compound operation, and hence also the difference in engine power. The predicted compounded power was therefore of the same order as with the present analysis.

The relatively small gains in system efficiency achieved at boost pressure ratios of the order of 3:1 agree well with those claimed for the SCANIA compound engine[4] which was introduced in 1994.

3.3.2.2 Turbocharged and compound engines with two stage turbocharger, systems 3 and (nominal boost pressure ratio 5:1 Figure 3.1b)

The results for the rated conditions of both systems are again summarized in Table 3.2. Rated power at this higher level of

Table 3.2 Rated operating conditions of turbocharged and compound systems

			Boost level 3:1		Boost level 5:1	
			Turbocharged system 1	*Compound system 2*	*Turbocharged system 3*	*Compound system 4*
Engine		Actual boost ratio	30.2	2.91	6.23	5.14
		EVO (°ATDC firing)	140	140	155	155
		IVO (°ATDC firing)	350	350	350	350
		EVC (°ATDC firing)	375	375	375	375
		IVC (°ATDC firing)	590	590	636	616
		Compression ratio	15:1	15:1	14:1	15:1
		Output (kW)	307.2	274.2	480.1	436.1
		BMEP (bar)	20.99	18.73	32.83	29.78
		Sp. fuel consumption (g/kW.h)	196.7	191.7	194.3	192.2
		Max. cylinder pressure (bar)	155.2	145.8	208.5	221.4
		Exhaust pressure ratio	2.87	3.46	5.39	5.32
		Trapped A/F	27.00	27.00	24.00	27.00
		Exhaust temp. (K)	766.8	794.1	847.7	832.3
HP turbocharger	compr.	Scale factor	—	—	0.7432	0.7432
		Pressure ratio	—	—	2.48	2.35
		Efficiency (%)	—	—	72.7	79.5
		Speed (rev/min)	—	—	147860	140688
	turb.	Scale factor	—	—	0.860	0.841
		Pressure ratio	—	—	2.13	1.81
		Efficiency (%)	—	—	73.46	73.39
		Speed (rev/min)	—	—	147680	140688
LP turbocharger	compr.	Scale factor	1.0	1.0	1.745	1.680
		Pressure ratio	3.2	3.06	2.76	2.37
		Efficiency (%)	72.1	77.4	72.5	82.3
		Speed (rev/min)	127473	119767	96498	98347
	turb.	Scale factor	1.0	0.762	1.680	1.422
		Pressure ratio	2.79	2.36	2.32	1.77
		Efficiency (%)	74.6	77.1	74.7	77.7
Power turbine		Scale factor	—	1.91	—	2.456
		Pressure ratio	—	1.45	—	1.54
		Efficiency (%)	—	75.61	—	79.81
		Speed (rev/min)	—	62858	—	64663
		Power (kW)	—	24.6	—	45.3
System		Output (kW)	307.2	298.8	480.1	481.4
		Sp. fuel consumption (g/kW.h)	196.7	191.7	194.3	174.6
		Efficiency (%)	42.76	43.88	43.29	48.19
		BMEP (bar)	20.99	20.36	32.83	32.91

boost pressure ratio was set at approximately 489 kW, an increase of 60% over systems 1 and 2 operating at 3:1 boost pressure ratio.

Comparing the compounded system 4 with the turbocharged system 3, the gain in thermal efficiency from 43.29% to 48.19% is now very substantial, compared with the very modest gains observed with systems 1 and 2 with single-stage turbocharging.

The introduction of the power turbine in series with the two turbocharger turbines, both smaller than those for the turbocharged engine, again creates a negative pressure gradient across the engine, thus reducing engine power from 480.1 kW for the turbocharged engine to 436.1 kW for the compound engine. However, with the large power turbine contributing 45.3 kW, the total power of the compound system becomes 436.1 + 45.3 = 481.4 kW, compared with 480.1 kW for the turbocharged engine.

The increase in boost pressure ratio from 3:1 to 5:1 (nominal) inevitably leads to an increase in maximum cylinder pressure from approximately 150 bar to well over 200 bar. Substantial redesign of the engine is therefore required in addition to the more complex turbocharging system.

3.3.2.3 Part load characteristics of systems 1, 2, 3 and 4 (Figures 3.4, 3.5, 3.6 and 3.7)

Figures 3.4 and 3.5 represent the respective performance maps for the turbocharged engine with single-stage turbocharger, system 1 and the compound engine, again with single-stage turbocharger, both operating at nominal boost pressure ratios of 3:1.

Referring to the limiting torque curves, *Figure 3.4* shows a best bsfc of 188.8 g/kW.h (44.55% thermal efficiency) at a peak BMEP of 23.78 bar and 1600 rev/min, the corresponding values for the compound engine, *Figure 3.5* being 187.4 g/kW.h (44.88%) at a BMEP of 22.74 bar and 1900 rev/min. Below this speed the torque characteristic is very flat, with bsfc increasing slightly due to the diminishing contribution of the power turbine. Under part load conditions the bsfc of the compound engine tends to

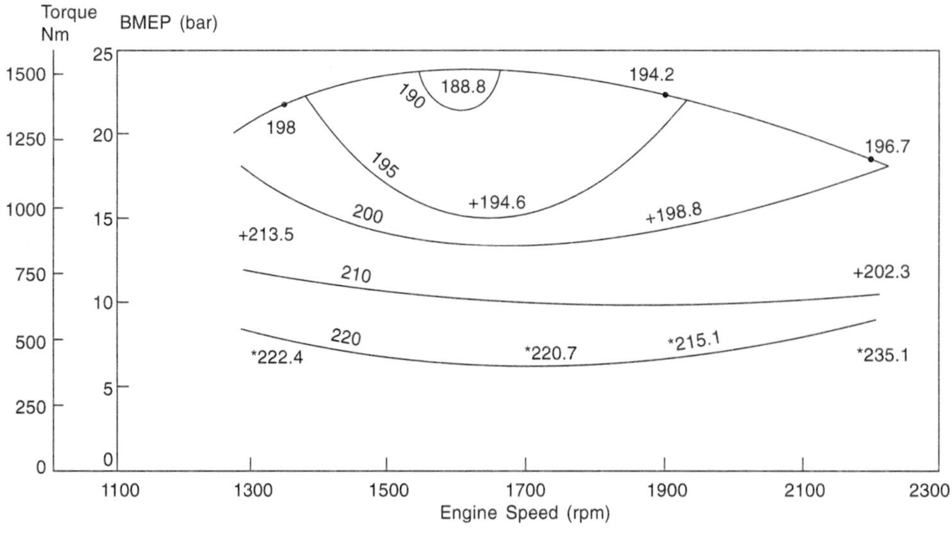

Figure 3.4 Performance map of single-stage turbocharged engine with bsfc contours (g/kWh)

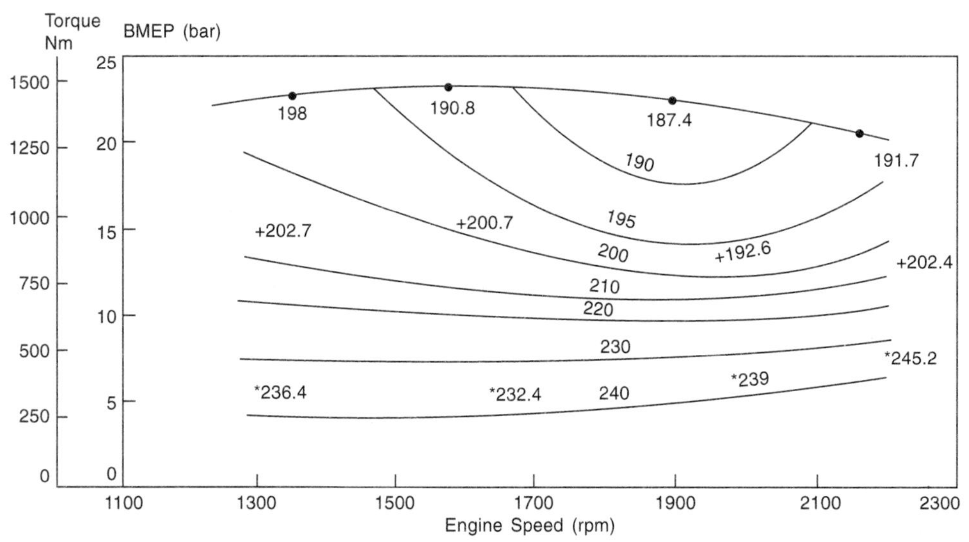

Figure 3.5 Performance map of compound engine with single-stage turbocharger with bsfc contours (g/kWh)

be inferior to that of the turbocharged engine except at very high speeds and loads, where the power turbine makes a significant contribution. This confirms the earlier conclusion that compounding at modest boost pressure ratios yields disappointing results.

Figures 3.6 and *3.7* represent the respective performance maps for the two-stage turbocharged engine, system 3 and the compound engine with two-stage turbocharger, system 4.

Figure 3.6 for the two-stage turbocharged engine, system 3, indicates a best bsfc of 189 g/kW.h (44.5% thermal efficiency) at a peak BMEP of 41.9 bar and 1500 rev/min, compared with rated values of 194.3 g/kW.h with a BMEP of 32.88 bar at 2200 rev/min, a torque rise of 27.6%.

Figure 3.7 for the compound engine system 4 indicates that between 1750 and 2200 rev/min on the limiting torque curve, bsfc remains below 174.6 g/kW.h (48.19%), but increases progressively as speed is reduced due to the diminishing contribution of the power turbine. Nevertheless the bsfc of 178.8 g/kW.h (47.06%) at the peak torque BMEP of 41.9 bar and 1400 rev/min is still substantially lower than that for the turbocharged engine at its torque BMEP which, furthermore, occurs at the higher speed of 1500 rev/min. However, the percentage torque rise is equal to that of the turbocharged engine at 27.6%.

3.3.2.4 The case for compounding at high levels of boost

The analysis of the previous sections clearly shows that the gains achievable by compounding increase substantially as boost pressure ratio is raised from conventional levels of 2.5 to 3 to approximately 5:1.

This is due to the fact that the contribution of the power turbine to total output increases both in absolute and relative terms, as shown in table 3.3 derived from *Tables 3.2* and *3.1*.

The higher power turbine contribution, in turn, leads to higher system efficiency at the higher boost level, both methods predicting system efficiencies of the order of 44% at 3:1 boost, as against 48% at 5:1 boost. At the same time the power density increases dramatically from 37.3 kW/litre at 3:1 boost to 60.0

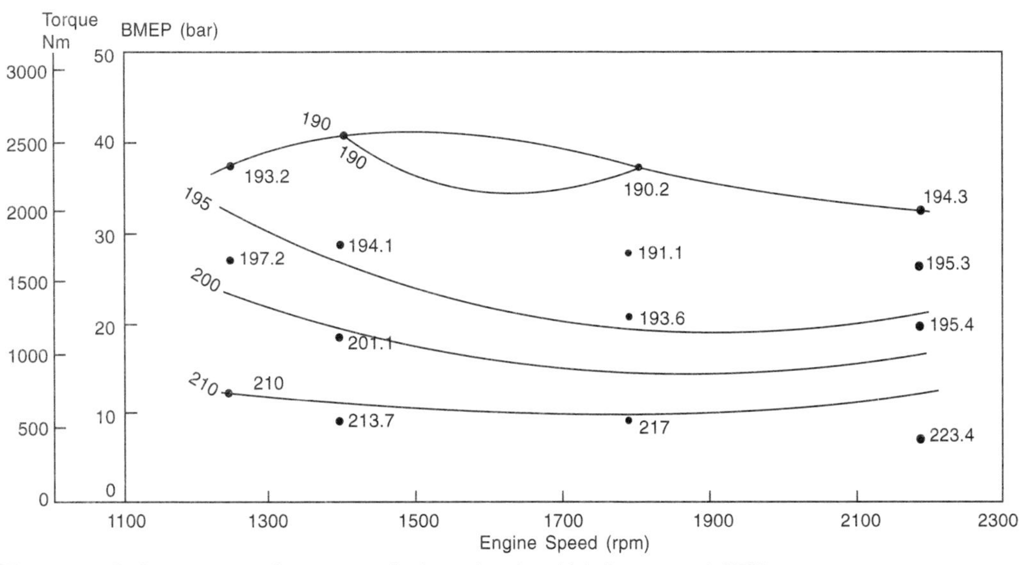

Figure 3.6 Performance map of two-stage turbocharged engine with bsfc contours (g/kWh)

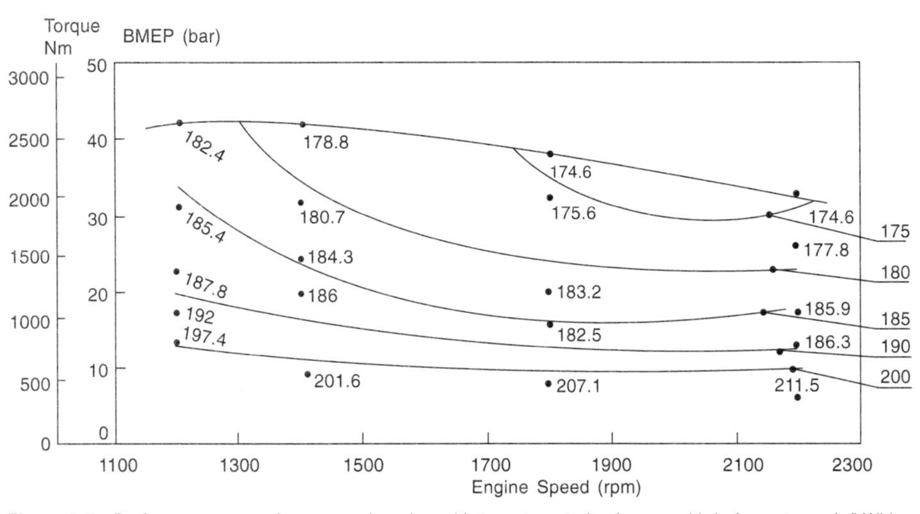

Figure 3.7 Performance map of compound engine with two-stage turbocharger with bsfc contours (g/kWh)

Table 3.3 Contribution of power turbine to total output

	Boost pressure ratio			
	3:1		*5:1*	
	Full prediction	*Simple prediction*	*Full prediction*	*Simple prediction*
Total system output W_{syst} (kW)	298.8	302.0	480.0	540.0
Power turbine power W_{pt} (kW)	24.6	42.0	45.3	82.0
Percentage power turbine contribution $\frac{W_{pt}}{W_{syst}} \times 100.$	8.23	13.9	9.44	15.2

kW/litre at 5:1 boost, albeit with a substantial increase in cylinder pressure from 146 bar to 221 bar.

The increased cost and complexity of the higher boost scheme would therefore appear to be justified by its greatly improved efficiency and power density, especially in military and off-highway applications.

3.4 Other compounded or related engine schemes

3.4.1 The differential compound engine (DCE)

This concept, intended specifically for heavy vehicle applications constitutes an attempt to combine the thermodynamic advantages of compounding as described in section 3.3 with the characteristics of an integrated engine-transmission system of the continuously variable type[5–7]. Like most of the latter it employs a fully floating epicyclic gear train in which the engine is connected to the annulus or ring gear, while the sungear and planet carrier drive, respectively, the supercharging compressor and the output shaft. Connected to the latter through reduction gearing is the exhaust turbine so that the power available at the output shaft becomes the sum of the nett power transmitted through the epicyclic gear train and the turbine power. The scheme is illustrated in *Figure 3.8*, while the torque and efficiency characteristics are shown in *Figure 3.9*.

The scheme has been extensively investigated both experimentally and analytically. The results presented in *Figure 3.9* refer to an 8 litre four-stroke DI diesel engine similar to that used for the comparison between turbocharged and 'conventional' compounded engines in section 3.3.2. Rated engine speed and boost pressure are again 2200 rev/min and 5 bar (approx.), respectively, but rated BMEP is substantially lower in order to meet certain compatibility conditions imposed by the epicyclic gear train.

The turbocharger is now replaced by a two-stage centrifugal compressor driven by the sungear through step-up gearing; the turbine is also of the two-stage type and is connected to the output shaft through reduction gearing.

Table 3.4 summarizes the operating conditions of the various system elements at rated and stall conditions along the limiting output shaft torque curve which is characterized by a massive torque rise from 1374 Nm at 2280 rev/min to 5057 Nm at 570 rev/min, corresponding to a torque ratio of 3.68:1 over a speed ratio of 1:4, implying that output shaft power is maintained almost constant over this very wide speed range. This very high torque rise further implies that such a system could operate in a heavy vehicle without the need for a conventional change-speed gear box, and is illustrated in *Figure 3.9*.

However, while the system succeeds partially in combining the thermodynamic benefits of compounding with those of a continuously variable transmission system, substantial engineering compromises are involved as already indicated (see *Table 3.4*).

First, the epicyclic gear train demands that a fixed ratio be maintained between engine and compressor torque—and between the former and planet carrier torque. In view of the fact that with the epicyclic gear train operating as a differential, any reduction in output shaft speed implies a corresponding increase in compressor speed, the power and hence torque demand of the compressor increases rapidly with decreasing output shaft speed. This demand has to be met by engine torque which therefore increases at the same rate. This can be achieved only by substantially derating the engine at rated output shaft speed. Thus the rated power of the engine running as before at 2200 rev/min is now only 327.2 kW compared with 436.1 kW for the 'conventionally' compounded engine. (see *Tables 3.4 and 3.2*).

Secondly, again as a result of the increase in compressor speed and hence boost with reducing output shaft speed, the air throughput of the system increases, resulting in a substantial excess of total air throughput over that demanded by the engine. Thus, at the stall condition (see *Table 3.4*) the overall air–fuel ratio is 39:1 as against 24:1 for the engine alone, the surplus airflow being bypassed before combining with the engine exhaust flow to form the total turbine flow. This mixing process represents a substantial thermodynamic irreversibility and hence loss of efficiency.

However, the ability of the differential compound engine to deliver virtually constant power over a 4:1 speed range, with the resultant elimination of the change-speed gearbox, constitutes a major advantage over both the turbocharged and the 'conventional' compound engine.

Figure 3.10 compares the limiting torque curves for

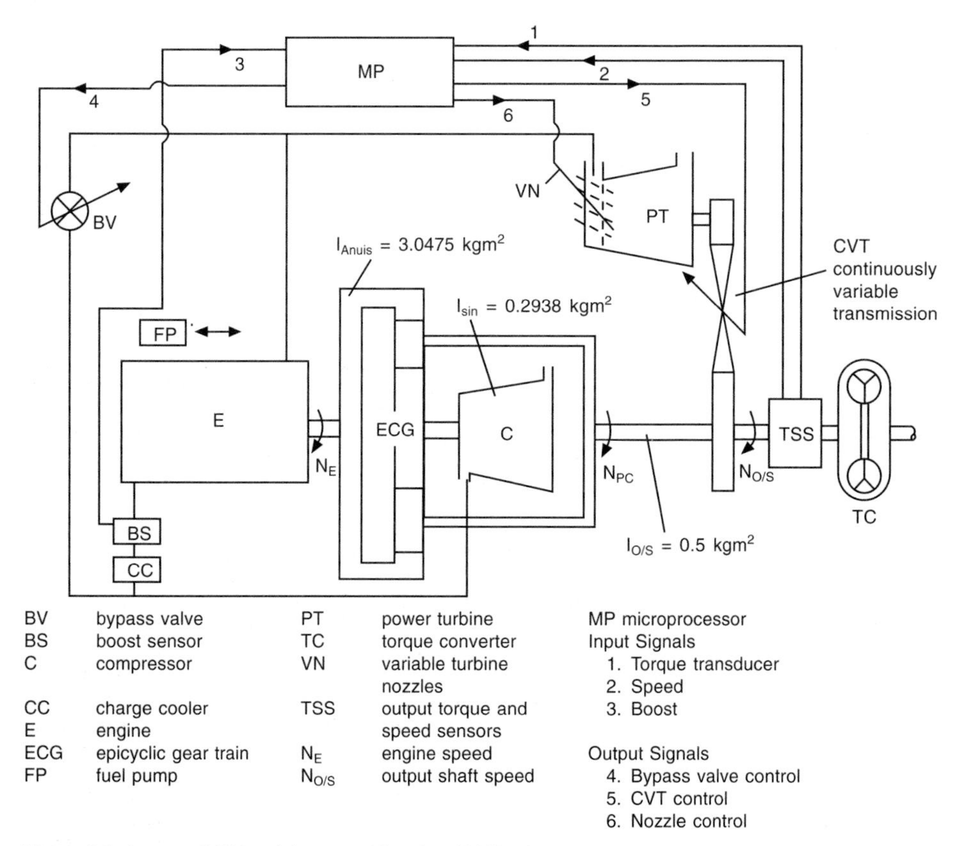

Figure 3.8 Layout of differential compound engine (DCE)

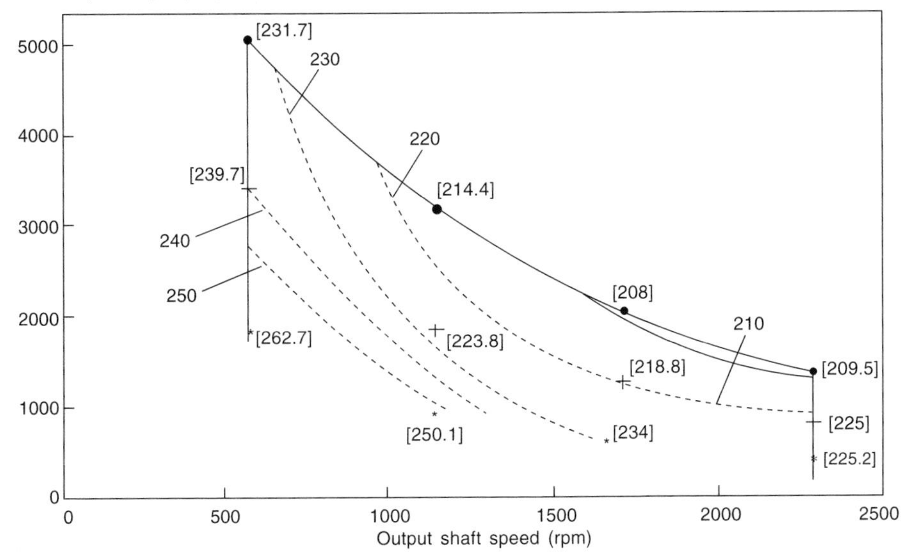

Figure 3.9 Torque characteristics of differential compound engine with bsfc contours (g/kWh)

Table 3.4 Rated and 'stall' operating conditions of differential compound engine

		Rated conditions	*Stall conditions*
Engine	Engine speed (rev/min)	2200	1470
	Compression ratio	12.5:1	12.5:1
	Output (kW)	327.2	345.1
	BMEP (bar)	22.39	35.34
	Sp. fuel consumption (g/kW.h)	205.9	199.8
	Max. cylinder pressure (bar)	156.4	235.0
	Boost pressure (bar)	4.885	6.745
	Exhaust pressure (bar)	4.974	6.621
	Trapped A/F	26.5	24.0
Compressor	Speed (rev/min)	61256	71024
	Pressure ratio	5.14	7.10
	Power (kW)	149.1	275.8
Power turbine	Speed (rev/min)	64000	69712
	Pressure ratio	4.707	6.450
	Power (kW)	152.7	232.0
Output shaft	Output (kW)	327.7	301.9
	Output efficiency (%)	40.9	36.3
	Output sp. fuel consumption (g/kW.h)	205.6	231.7
	Oput speed (rev/min)	2280	570
	Output torque (Nm)	1374	5057
	Output BMEP (bar)	22.40	79.72

(a) the turbocharged engine,
(b) the 'conventional' compound engine,
(c) the differential compound engine.

At 1200 rev/min the conventional compound engine delivers a torque of 2640 Nm as against 3067 Nm for the differential compound engine, with the torque of the latter still rising to 5057 Nm at 570 rev/min, while both the turbocharged and conventional compound engine suffer rapid reduction in torque below 1200 rev/min.

In spite of the engineering compromises already referred to, the differential compound concept clearly offers major advantages over conventional diesel power plant, especially for off-highway applications where a continuously rising torque curve is particularly attractive.

3.4.2 The differentially supercharged diesel engine (DDE)

This scheme, pioneered by G. Dawson[8] in the 1960s, though not strictly a compounded scheme, nevertheless provides an interesting comparison with the differential compound engine.

Figure 3.11 shows a simplified layout. Again an epicyclic geartrain provides a differential connection between engine, compressor and output shaft. However, there is no power turbine, the engine exhaust passing directly to atmosphere. High boost pressure can nevertheless be accommodated by greatly reduced valve overlap. A further difference between the two schemes is that while the differential compound engine preferably uses a centrifugal compressor, the differentially supercharged engine operates with a rotary positive displacement compressor having substantially lower efficiency. On the other hand, the more moderate rotational speed of the latter unit eliminated the need for expensive step-up gearing.

Although the experimental unit employed a 5.8 litre, 6 cylinder DI diesel engine, the following tabulation *Table 3.5* is again based on the same notional 8 litre engine as that used in the previous comparisons between the turbocharged, conventionally and differentially compounded engines.

Table 3.5 compares the rated and 'stall' performance of the differentially supercharged (DDE) and differentially compounded (DCE) engines.

Both the output shaft speed range and the relative output torque rise of the DDE are seen to be much more limited than those of the DCE. This is again a function of the fixed ratio between engine and compressor torque imposed by the epicyclic gear train. With the much higher ratio (rated engine power/rated compressor power) of the DDE (181.0/20.58 = 8.795 cf. 327.2/149.1 = 2.195 for the DCE) it can also be shown that the rate at which compressor speed now increases with reduction in output shaft speed is much greater for the DDE than for the DCE. This in turn imposes more severe limitations on the range of output shaft and engine speeds over which the DDE operates.

In terms of absolute power and torque, the DDE operates at less than half the corresponding values for the DCE as a direct result of the much lower boost levels of the DDE and the absence of a powerturbine which makes its own substantial contribution to output power.

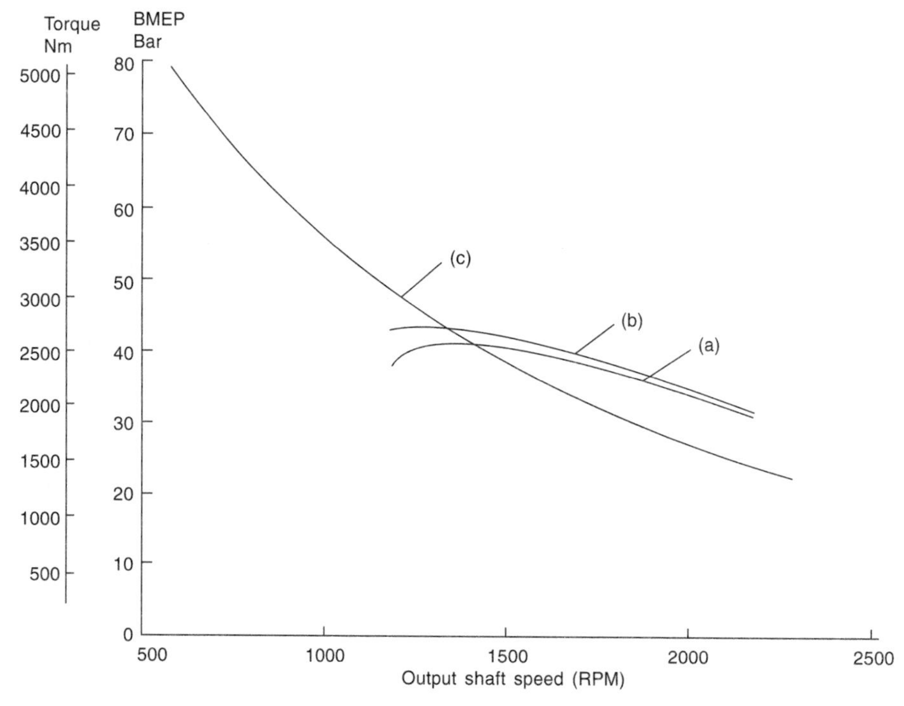

Figure 3.10 Torque characteristics of:
(a) two-stage turbocharged engine;
(b) compound engine with two-stage turbocharger;
(c) differential compound engine

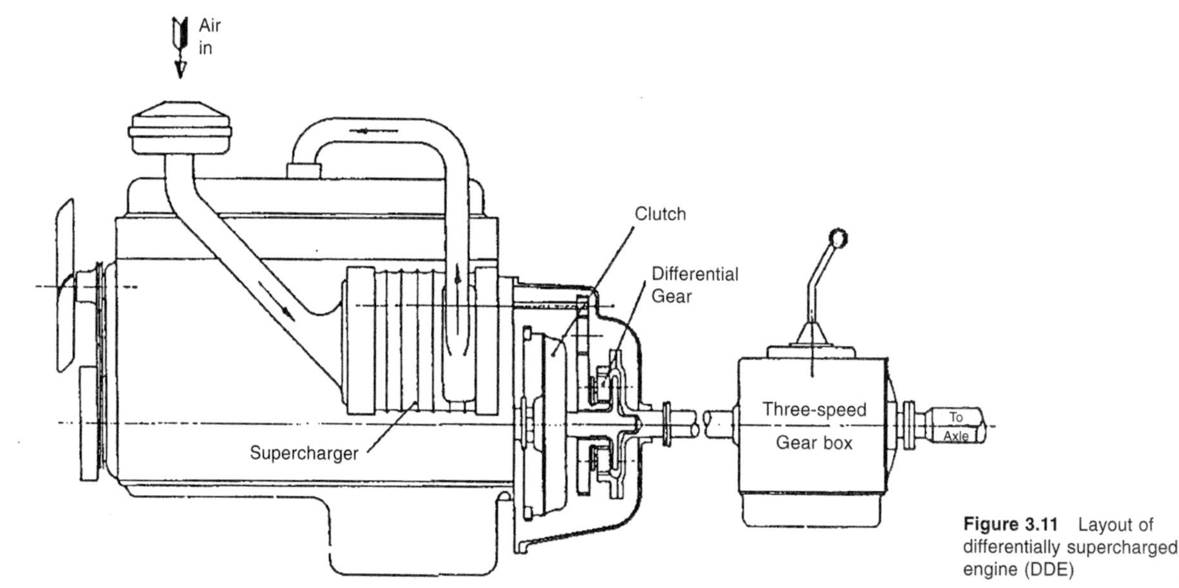

Figure 3.11 Layout of differentially supercharged engine (DDE)

While the DDE was undoubtedly an ingenious method of combining the characteristics of a mechanically supercharged engine with a continuously variable transmission, the DCE which represents a similar concept, but additionally employs a power turbine, clearly exhibits greatly enhanced performance.

3.5 Other turbocharged or pressure charging systems

3.5.1 Two-stage turbocharging (*Figure 3.1b*[a])

Single-stage turbocharging with aftercooling has become very much the norm for truck and bus diesel engines, and more recently has also been adopted for high speed passenger car engines.

Figure 3.6 is typical of the torque and bsfc characteristics of such engines. In general torque backup is superior to that of single-stage turbocharged engines due to the fact that the HP turbocharger maintains a higher speed relative to its design speed over a wider speed range than is the case with a single-stage turbocharger, with beneficial effects in the mid-speed range.

Bsfc is strongly dependent on compression ratio, but allowing for substantially higher maximum cylinder pressures of more than 200 bar compared with 150 bar for single-stage turbocharged engines, a similar or slightly better level of bsfc may be expected.

It is doubtful whether in view of the great increase in complexity associated with two-stage turbocharging, and in spite of increases in rating of the order of 55%, the system will find widespread commercial acceptance.

Table 3.5 Comparative performance of the differentially supercharged and compound

	Engines			
	Rated conditions		*Stall conditions*	
	DDE	*DCE*	*DDE*	*DCE*
Engine				
speed (rev/min)	2800	2200	1350	1470
power (kW)	181.0	327.2	159.0	345.1
Compressor				
flow (kg/s)	0.313	0.635	0.240	0.405
pressure ratio	1.50	5.14	2.40	7.10
speed (rev/min)	7950	61256	7155	71024
power (kW)	20.58	149.1	33.96	275.8
Turbine				
flow (kg/s)	—	0.661	—	0.624
pressure ratio	—	4.707	—	6.450
speed (rev/min)	—	64000	—	69712
power (kW)	—	152.7	—	232.0
Output shaft				
speed (rev/min)	3500	2280	1500	570
power (kW)	160.5	327.7	125.5	301.9
torque (Nm)	440.7	1374	799.0	5057

3.5.2 Variable geometry turbocharging

Variable geometry turbocharging (VGT) involves the use of swivelling nozzle blades—or an equivalent arrangement—in the turbine stator housing as a means of varying the flow capacity of the turbine, with a turn down ratio, defined as

$$\frac{\text{mass flow at maximum flow setting}}{\text{mass flow at minimum flow setting}}$$

of the order of 2:1 (see *Figure 3.12*). By this means the pressure ratio across the turbine, and thus boost pressure can be maintained at high levels down to much lower engine speeds than is possible with fixed geometry turbocharging (FGT)[10, 11]. As a result the limiting torque curve can be drastically re-shaped to give higher torques at lower speeds than is possible with FGT (see *Figure 3.13*) with torque back-up typically increasing from 20% to 35%.

The effect on transient response is also highly beneficial. For a typical fuel or load step from a low speed, low load condition to a high speed, high load condition, the response time for boost pressure from its initial to its final value (typically 0.2 bar gauge to 1.1 bar gauge) can be reduced to less than 50% of the corresponding time with FGT, by closing down the nozzles over the greater part of the transient (see *Figure 3.14*).

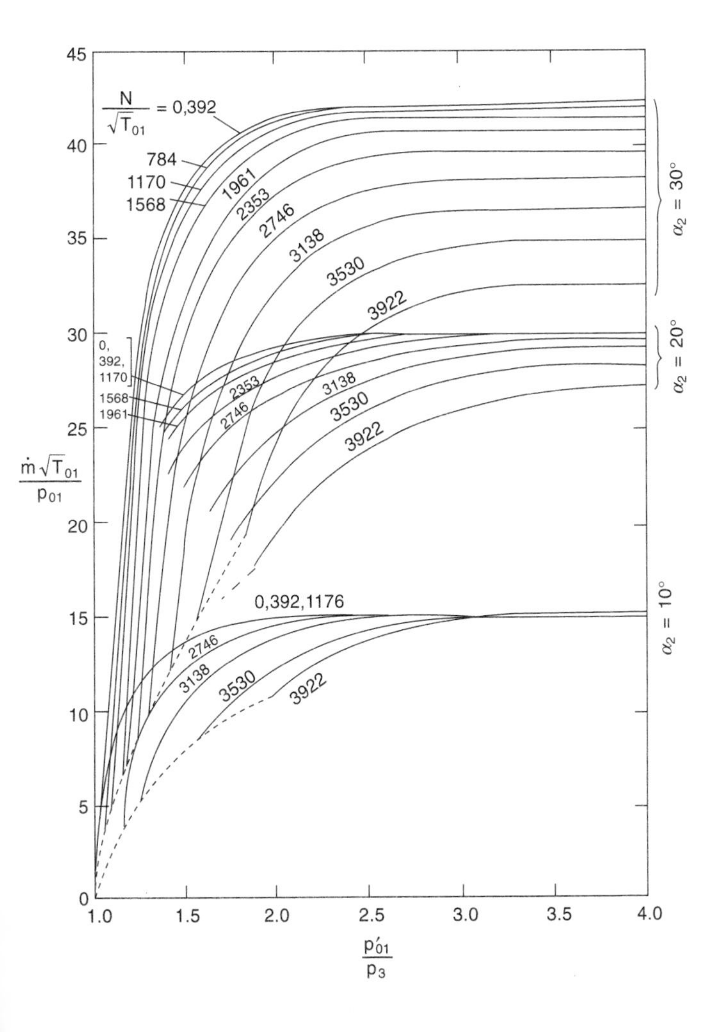

Figure 3.12 Turbine mass flow characteristics with variable nozzle angle

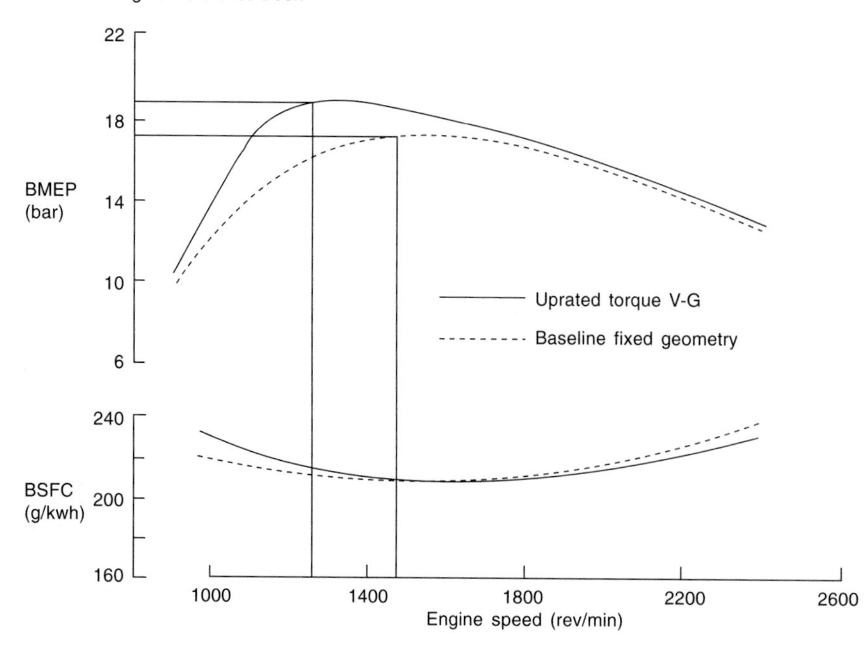

Figure 3.13 Torque characteristics of fixed geometry (FGT) and variable geometry (VGT) turbocharged engine

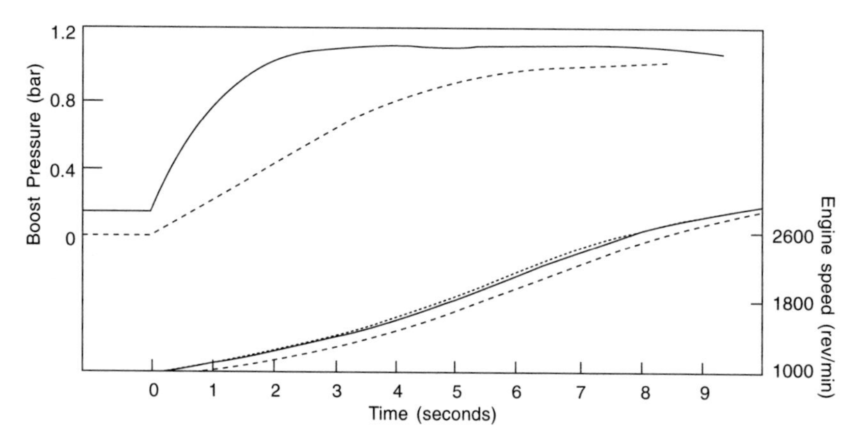

Figure 3.14 Transient response of fixed geometry (FGT) and variable geometry (VGT) turbocharged engine

In practice the VG turbocharger has to be under electronic control, in parallel with the fuel injection system to achieve acceptable results both under steady state and transient conditions[12, 13].

The above remarks apply primarily to truck engines. More recently VGT has also been applied to high speed passenger car diesel engines as a means of improving not only torque back-up and transient response, but also reducing emissions by control, simultaneously, of the VG turbocharger and of the exhaust gas recirculation (EGR) valve, which is now universally fitted to such engines to control the emission of oxides of nitrogen[14].

A further benefit of combined VGT and EGR control is that substantial improvements in fuel consumption are possible over an equivalent FGT system using EGR control only.

3.5.3 The pressure wave supercharger

(*Figures 3.15 and 3.16*[15–17])

The pressure wave supercharger, better known as the COMPREX device, is intended to achieve similar objectives to variable geometry turbocharging, i.e. increased torque back-up and improved transient response. However, the operating principle is totally different from that of turbocharging (*Figure 3.15*).

Compression of the aspirated air is achieved by making use of pressure waves propagated from the exhaust manifold into axial passages on the periphery of the belt-driven rotor when these passages move past the inlet port in the exhaust stator housing. These waves, in their passage towards the inlet stator housing compress air previously at or near atmospheric pressure, to a pressure closely related to that in the exhaust manifold. As the rotating passage subsequently moves past the air delivery port, the compressed air in the passage is discharged into the inlet manifold. Discharge of the exhaust gas, now at reduced pressure, through the outlet port of the exhaust stator housing, followed by admission of air from the opposite side of the passage through the inlet port in the inlet stator housing, leading to a repetition of the events described above, completes the operating cycle.

The pressure wave supercharger has two outstanding advantages over the simple turbocharger: first the compression mechanism, using pressure wave action, is far less severely dependent on engine speed than the turbocharger with its severely limited boost pressure at low engine speeds. Secondly, transient response which is critically dependent on the availability of air when fuelling is suddenly increased, is virtually instantaneous due to the absence of significant inertia in the compression system.

As a result of these characteristics, both low speed torque and response of the engine are substantially improved compared with the conventional single-stage turbocharged system. At the same time smoke levels in the low speed, high torque regime and under severe transients are very much lower than for turbocharged engines. The response of a large truck diesel engine, in either turbocharged or pressure wave supercharged form, to a large fuel step when operating on a propellor law load curve. The much more rapid build-up of boost in the case of the pressure wave supercharged engine is immediately apparent particularly for transient A.

However, the system has not, so far, found general acceptance due to its inherently more complex installation requirements and its higher cost. It may possibly find its most appropriate application in the supercharged passenger car engine, in either diesel or gasoline form, where its relative insensitivity to engine speed and good transient response make it particularly suitable[17].

References

1 WRIGHT, E. J. and WALLACE, F. J., 'Comparative assesment of the Control Characteristics of Outward and Inward Compressing Free Piston Gas Generators', *ASME Diesel and Gas Engine Power Conference*, paper 67-DGP-6 (1967)

2 WALLACE, F. J., 'High Output Compound Diesel Engine Schemes—How do they compare with State-of-the-Art Turbocharged Engines', *SAE Off Highway and Powerplant Congress*, Milwaukee, paper 952099 (1995)

3 WALLACE, F. J., 'ODES—a flexible Simulation Package for Diesel and Gasoline Engines in 2 and 4 stroke form operating in conjunction with Turbomachinery', Universities Internal Combustion Engines Group (UNICEG), London, December (1994)

4 Anon, 'Turbocompound—un grand pas en avant' Scania publicity brochure (1994)

5 WALLACE, F. J., 'The Differential Compound Engine', *SAE International Congress*, Detroit, paper 670110 (1967)

6 WALLACE, F. J. 'The Differential Compound Engine', *Proc. I. Mech. E.*, **187**, 48/73, p. 548 (1973)

7 HALL, J. and WALLACE, F. J., 'Control Design for the Differential Compound Engine' *SAE International Congress*, Detroit, paper 890392 (1989)

8 DAWSON, J. G., HAYWARD, W. J. and GLAMANN, P. W., 'Some experiences with a Differentially Supercharged Diesel Engine', *Proc. I. Mech. E., Automobile Division,* **178**, part 2A No. 5, p. 157 (1964)

9 KELLETT, E., BETTERIDGE, J. F. and MISTOVSKI, M., 'Investigation of Diesel Engine and Turbocharger Interaction', *Proc. I. Mech. E.,* **184**, pt. 1, no. 15 (1967/68)

10 WATSON, N. and BANISOLEIMAN, K., 'Performance of a highly rated Vehicle Diesel Engine with Variable Geometry Turbocharger', *I. Mech. E. Conference on Turbocharging and Turbochargers*, paper C 103/86 (1986)

11 Mc CUTCHEON, A. R. S. and BROWN, M. W. G., 'Evaluation of a Variable Geometry Turbine on a commercial Diesel Engine', *I. Mech. E. Conference on Turbocharging and Turbochargers*, paper C 104/86 (1996)

12 WALLACE, F. J., ROBERTS, E. W. and HOWARD, D., 'Variable Geometry Turbocharging—Optimisation and Control under Steady State Conditions', *I. Mech. E. Conference on Turbocharging and Turbochargers*, paper 97/86 (1986)

13 WALLACE, F. J., ANDERSON, K. and HOWARD D., 'Variable Geometry Turbocharging—Control under Transient Conditions', *I. Mech. E. Conference on Turbocharging and Turbochargers*, paper C 98/86 (1986)

14 CAPOBIANCO, M. and GAMBAROTTA, A., 'Unsteady flow Performance of Radial Flow Turbocharger Turbines', *I. Mech. E. Conference on Turbocharging and Turbochargers*, paper C 405/017 (1990)

15 SCHWARZBAUER, G. E., 'Turbocharging of Tractor Engines with Exhaust Gas Turbochargers and the BBC Comprex', *I. Mech. E. Conference on Turbocharging and Turbochargers*, paper C 69/78 (1978)

16 SUMMERAUER, I., SPINNLER, F. and MAYER, A., 'A Comparative Study of the Acceleration Performance of a Truck Diesel Engine with Exhaust Gas Turbocharger and with Pressure Wave Supercharger Comprex', *I. Mech. E. Conference on Turbocharging and Turbochargers*, paper C 70/78 (1978)

17 SCHRUF, G. M. and MAYER, A., 'Fuel Economy for Diesel Cars by Supercharging, SAE paper 810343 (1981)

4 Diesel combustion and fuels

Contents

The two principal driving forces behind modern diesel engine design are the desire for lower exhaust emissions and improved fuel economy. These goals, often conflicting, are accomplished through the application of turbocharging and electronic controls as well as improvements in combustion chamber design and optimum matching of fuel injection and air motion in the cylinder. This chapter focuses on the in-cylinder combustion processes in diesel engines and how they affect the goals of low emissions and low fuel consumption. Recently, diesel fuel's role in meeting engine emissions and performance objectives has become more widely recognized so diesel fuel properties and characteristics will also be discussed.

4.1 Diesel combustion

The processes that occur during diesel combustion are still somewhat mysterious, in spite of extensive attempts at photography and other, more advanced, optical diagnostics. The environment in the diesel combustion chamber appears designed to resist detailed investigation. It is small, hot, and subject to high pressure and intense vibration. When windows are installed in the chamber to observe the combustion, soot blocks the passage of light and quickly deposits on the windows, obscuring further investigation. The combustion involves gas, liquid, and solid phases as well as complex physical processes and chemical reactions. In spite of this complexity, researchers generally agree about the sequence of processes that occurs in the chamber.

Diesel combustion is the process that occurs when a fuel blend, chosen for its readiness to auto-ignite, is injected into a volume of turbulent air that has been compressed to a high temperature and pressure. The fuel does not ignite immediately. A time period elapses, called the *ignition delay*, during which the fuel must vaporize, mix with air, and undergo preflame chemical reactions that produce the chemical species necessary for spontaneous ignition. Because the air temperature is above the thermodynamic critical point of many of the fuel components, vaporization takes place very quickly. In some engines, vaporization is complete within a few millimetres of the injection nozzle.

After sufficient time has elapsed, ignition will occur spontaneously in regions of fuel–air mixture that have fuel–air ratios close to the *stoichiometric*, or chemically correct mixture. Combustion proceeds very rapidly because of the backlog of prepared or nearly prepared fuel–air mixture formed during the ignition delay period. The rapidly rising temperature and pressure in the cylinder accelerate the combustion in an uncontrolled manner until the backlog is depleted. The fuel in the spray core is still too rich to burn, and the fuel in the periphery of the spray is too lean to burn, so combustion slows down and is controlled by the rate at which the air is entrained and a combustible mixture formed. The first phase of combustion, where prepared fuel burns quickly, is known as the *premixed* phase and the second phase is known as the *diffusion* or *mixing-controlled* phase. The rate of burning during the mixing-controlled phase depends on the air motion and fuel spray momentum. The burning rate starts quite high because there is considerable excess air and the fuel spray entrains air rapidly. After the end of fuel injection, particularly at high loads when there is not as much excess air as with light loads, the burning rate decreases gradually to zero.

Each of the important features of the diesel combustion process will be discussed in this chapter. First, some basic combustion theory will be presented. Then, the ignition delay and fuel–air mixing processes will be discussed. Finally, combustion system design issues will be presented before moving to a discussion of diesel fuels and their effect on diesel combustion, emissions, and performance.

4.1.1 Basic combustion theory

Combustion is the chemical reaction that converts the energy contained in the fuel to the internal energy of product gases. The internal combustion engine serves as a mechanism to convert this internal energy into useful work. This section discusses the basic chemical reactions that relate to diesel combustion and how the reactions associated with chemical equilibrium and chemical kinetics influence combustion. A brief discussion of hydrocarbon combustion is also included.

4.1.1.1 Stoichiometric combustion

Although diesel engines never intentionally run with the chemically correct, or stoichiometric, amount of air, it is useful to compare the actual fuel to air ratio to the stoichiometric amount as a measure of air utilization. Since diesel fuel composition varies considerably, it is desirable to have a laboratory analysis of the fuel that gives its composition. Method D5291 from the American Society for Testing and Materials (ASTM) can give the percentages of hydrogen and carbon in the fuel. If an average molecular weight is also available, an equivalent hydrocarbon molecule can be determined. Universal Oil Products Method 375-86 can be used to estimate the fuel molecular weight using the fuel viscosity, density, and distillation curve[1].

A typical No. 2 diesel fuel will have a molecular weight of 183, a carbon mass fraction of 86.57%, and a hydrogen mass fraction of 13.43%. For a hydrocarbon molecule of the form C_xH_y, x and y need to be determined to match this measured data. Because $MW_{carbon} = 12.0111$ and $MW_{hydrogen} = 1.00797$, then

$$x(12.0111) + y(1.00797) = 183 \tag{4.1}$$

In 1 kg of fuel, there is 0.8657 kg of carbon/12.0111 = 0.0721 kmol of carbon and 0.1343 kg of hydrogen/1.00797 = 0.1332 kmol of hydrogen. Thus,

$$y/x = 0.1332/0.0721 \tag{4.2}$$

This system of two equations and two unknowns can be solved to get $x = 13.2$ and $y = 24.4$. The equivalent diesel fuel molecule is $C_{13.2}H_{24.4}$.

The stoichiometric reaction for this fuel can be obtained by atom balances to be:

$$C_{13.2}H_{24.4} + 91.90\ (0.21\ O_2 + 0.79\ N_2) \Rightarrow 13.2\ CO_2 + 12.2\ H_2O + 72.60\ N_2 \tag{4.3}$$

The molar air–fuel ratio is 91.90 kmol air/kmol fuel, which can be converted to a mass basis as follows:

$$91.90\ \frac{kmol\ air}{kmol\ fuel} \times \frac{28.97\ kg\ air}{kmol\ air} \times \frac{kmol\ fuel}{183\ kg\ fuel}$$

$$= 14.55\ \frac{kg\ air}{kg\ fuel} \tag{4.4}$$

The *equivalence ratio* is defined as the actual fuel–air ratio divided by the stoichiometric fuel–air ratio. If an engine using the fuel described above were running with a 30:1 air–fuel ratio, then its equivalence ratio would be

$$\phi = \frac{F/A)_{actual}}{F/A)_{stoich}} = \frac{1/30}{1/14.55} = 0.485 \tag{4.5}$$

This ratio indicates that the engine is using less than half of the air supplied for combustion. Diesel engine air utilization is generally limited to $\phi < 0.7$. Higher equivalence ratios cause excessive smoke emissions. This can make it difficult for naturally aspirated diesel engines to develop as much power per unit of displacement as spark ignited engines.

4.1.1.2 Complete combustion and equilibrium

When diesel fuel burns with air, the overall process can be expressed by a chemical reaction such as:

$$C_{13.2}H_{24.4} + 91.90/\phi(0.21\ O_2 + 0.79\ N_2) \Rightarrow 13.2\ CO_2 + 12.2\ H_2O + (19.30/\phi - 19.30)\ O_2 + 72.60/\phi\ N_2 \quad (4.6)$$

This equation assumes complete combustion where all of the fuel carbon goes to CO_2 and all of the fuel hydrogen goes to H_2O. Except for the presence of small amounts of pollutant species, this equation accurately describes diesel engine exhaust gas.

At high temperatures, some of the CO_2 will dissociate to CO and O_2, and some of the H_2O will dissociate to H_2, OH, O, and H. Even O_2 and N_2 can dissociate to form O and N atoms. Some of these species are only present at high temperatures and are extremely reactive. They are called *radicals* and they are important participants in auto-ignition and subsequent combustion. The amount of dissociation and the resulting concentrations of radical species can be determined from chemical equilibrium calculations that minimize the thermodynamic Gibb's function. This mixture is known as the *equilibrium* mixture. Several computer programs are available to compute equilibrium compositions[2,3].

4.1.1.3 Equilibrium and chemical kinetics

All chemical reactions tend to move systems closer to equilibrium. However, the rate at which they proceed to equilibrium may be very slow. In general, reaction rates increase with temperature. For example, fuel oxidation is very slow at room temperature, but is very fast at 2500 K.

Just as complete combustion was a simplification, chemical equilibrium can also be a simplification that may not be justified. Equilibrium calculations indicate that pollutants such as nitric oxide and carbon monoxide should not be present in diesel engine exhaust. The reactions that eliminate these species are simply too slow to be completely effective before the exhaust valve opens and releases these pollutants to the environment.

4.1.1.4 Global and elementary reactions

Reactions such as eqn (4.6) are called *global reactions*. They represent the overall chemical process but they do not accurately describe the way individual atomic species interact. Fundamentally, chemical processes occur through *elementary reactions* involving one, two, or, at most, three particles. Many of the participants in elementary reactions are radical species that do not even appear in the global reaction.

All chemical reactions proceed through long chains of these elementary reactions that begin with radical-creating reactions, continue with radical-propagating reactions, and end when radical-consuming reactions become dominant. Computer programs have been developed to model the chemical kinetics of the large numbers of elementary reactions it takes to describe the ignition and combustion of simple fuels such as methane and ethane. These programs can be used to explain a wide variety of combustion phenomena such as ignition, flame propagation and speed, quenching, and flammability limits. Approximate techniques are required for complex fuel mixtures such as diesel fuel.

4.1.1.5 Hydrocarbon combustion

In order for ignition to occur in a low-temperature *radical-poor* environment, before ignition has occurred, preflame reactions must form radical species. One way this can occur is to split C–C bonds. At low temperatures, C–C bonds will open more easily than C–H bonds because they have a lower dissociation energy (80 kcal/mole compared to 90 to 100 kcal/mole). The radicals created then strip hydrogen atoms from other molecules.

If the environment is oxygen rich, large amounts of OH radical will be formed. Paraffinic hydrocarbons tend to react with this OH as follows:

$$OH + C_nH_{2n+2} \Rightarrow H_2O + C_nH_{2n+1} \Rightarrow C_{n-1}H_{2n-2} + CH_3 \quad (4.7)$$

C_nH_{2n+1} radicals are inherently unstable and will break up to form ethylene with some propene, butene, and isobutene. The oxidative attack that eventually converts the hydrocarbon to CO and then CO_2 is against these small olefinic intermediates[4]. These olefinic species become aldehydes and then acetyl or formyl radicals through oxidative attack by other radicals. These acetyl and formyl radicals are then converted to CO by reactions with other species such as

$$R\text{-}C = O + M \Rightarrow R + CO + M \quad (4.8)$$

Finally, CO is oxidized to CO_2 primarily by the reaction

$$CO + OH \Rightarrow CO_2 + O \quad (4.9)$$

To summarize, paraffinic hydrocarbons are generally oxidized through a process that includes an olefinic intermediate, usually ethylene, followed by an aldehyde or oxygenated radical, followed by carbon monoxide, which is finally converted to CO_2 by an OH radical.

With aromatic hydrocarbons, there is some formation of CO early in the reaction scheme in contrast to paraffins where it only occurs near the end. The butadienyl radical is known to be an important participant in the aromatic oxidation process. It can undergo a large number of reactions to produce species such as vinyl acetylene, butadiene, and acetylene, all of which are important intermediates for soot production[4]. It is probably for this reason that higher levels of aromatics in diesel fuel are generally associated with higher particulate emissions.

4.1.2 Ignition delay

Ignition delay is defined as the time period between the start of fuel injection and the start of combustion. As described earlier, the fuel must vaporize, mix with air, and undergo preflame reactions before auto-ignition occurs. The classical notion of ignition delay identifies the vaporization and mixing processes as the *physical delay* and the preflame reactions as the *chemical delay*. This notion can be deceptive because all of the processes are occurring simultaneously. The temperature of the compressed air is the most important variable affecting ignition delay because it accelerates vaporization and the radical-forming preflame reactions. Fuel chemical structure is also important because some fuels resist auto-ignition. The cetane number is a property of the fuel that indicates the fuel's readiness to autoignite. Fuels with high cetane numbers have short ignition delays.

Turbocharged diesel engines have very short ignition delays at full load but naturally aspirated engines and lightly loaded turbocharged engines can have ignition delays of 1–2 milliseconds. Ignition delay can be correlated using equations with the same general form as those used for chemical kinetics calculations. One such equation, developed by Hardenberg and Hase[5], is shown below.

$$ID = [0.36 + 0.22\overline{U}_P] \exp\left[E_A\left(\frac{1}{RT_{im}\, r^{n_c-1}} - \frac{1}{17190}\right) + \left(\frac{21.2}{P_{im}\, r^{n_c} - 12.4}\right)^{0.63}\right] \quad (4.10)$$

where

ID = ignition delay period, degrees of crank rotation
$\bar{U}_P$ = mean piston speed, m/s
$E_A = \dfrac{618\ 840}{CN + 25}$
CN = cetane number of the fuel
R = ideal gas constant, = 8.31434 J/kmol-K
T_{im}, P_{im} = intake manifold temperature and pressure, K and bar
r = compression ratio
n_c = polytropic exponent for compression

Ignition delay is an important variable in diesel combustion because it has a strong correlation to the amount of fuel that is burned in the premixed combustion phase. Longer ignition delays allow more fuel to be injected and prepared for combustion. When ignition finally occurs, it involves more fuel and produces a violent autoignition, sometimes called *diesel knock*. In addition to being a source of undesirable noise, high levels of premixed combustion contribute to high nitric oxide (NO) levels in the exhaust. Experiments that sample the entire cylinder contents at different times during the combustion process have shown that NO is formed early and primarily from the products of the premixed combustion[6]. This product gas is compressed by further combustion and gets to the highest temperature in the cylinder for the longest time. Recent results have shown that in highly turbocharged engines with late injection timing and very short ignition delays, less NO is formed and it is not necessarily associated with the first fuel to burn[7].

4.1.3 Mixing controlled combustion

When auto-ignition finally occurs, the prepared fuel burns very quickly, producing a sudden rise in cylinder pressure and the characteristic *knocking* sound of diesel combustion. The remainder of the fuel burns at a rate determined by the rate at which it mixes with air. While chemical kinetics dominate the ignition delay and premixed combustion periods, the high temperatures and pressures of the post-ignition gases promote very fast reaction rates, which make fuel–air mixing the rate-determining process.

The heterogeneity of the diesel combustion process is responsible for some of the greatest advantages of the diesel engine but also for some of its greatest disadvantages. The injection of fuel into compressed air ensures that a flammable mixture will always exist somewhere in the cylinder regardless of how lean the overall mixture might be, which allows the load of a diesel to be controlled without throttling. The spray also allows low volatility petroleum fractions to be used that would be unacceptable for gasoline.

Unfortunately, when fuel is injected into the air, a portion will be mixed beyond its lean flammability limit before it has a chance to burn. This fuel is likely to be emitted as *unburned hydrocarbon*, particularly at light loads where the probability that the overmixed fuel and air will encounter a high-temperature rich zone is low. When ignition delays are long, there is more time for fuel to be overmixed and hydrocarbon emissions will increase.

Within the core of the spray, there are regions where little air is present. As the temperature of this fuel increases by heat transfer and contact with high-temperature product gases, it undergoes chemical reactions that produce polynuclear aromatic hydrocarbons (PAH) and soot.

While much of the soot will be oxidized before the exhaust valve opens, a portion will remain and be emitted from the cylinder. Late in the expansion process and in the exhaust system, the soot will collect high molecular weight hydrocarbons and sulfates (primarily condensed sulfuric acid and its hydrates). It is this material that the Environmental Protection Agency (EPA) identifies as *particulate*.

Accelerating the fuel–air mixing process can decrease the amount of fuel in the high-temperature rich zones and thus the production of soot. Decreasing soot production can be accomplished by higher injection pressures or higher air swirl levels. However, a byproduct of more rapid mixing and combustion is higher levels of *oxides of nitrogen* (NOx). In general, changes in engine design or operating parameters that reduce particulates will increase NOx and vice versa. This natural tradeoff between particulate emissions and NOx is one of the critical challenges in the design of diesel combustion systems.

4.1.3.1 Spray penetration

To achieve effective mixing during the diffusion burning phase, the spray momentum must be adequate to ensure that the spray penetrates across the combustion chamber. Impingement of the vaporized fuel jet on the piston surface can enhance mixing[8]. Deliberate impingement of liquid fuel on the piston has been used to control the combustion process although most manufacturers find that minimizing liquid impingement facilitates meeting emission regulations. Many correlations for spray penetration have been developed that allow the effect of injection system parameters to be predicted[9]. One such correlation is that of Hiroyasu, Kadota and Arai[10].

Hiroyasu, Kadota and Arai used a high-pressure combustion bomb to simulate the environment in the diesel combustion chamber. They noted that the jet appeared to be divided into two regions, a developing region from the injector tip to the point of transition and a fully-developed region after the transition point. The characteristics of the developing region followed that of a liquid stream and the transition point was where this stream broke up into fine droplets. Other researchers have also reported the presence of these two flow regimes[11]. They provide the following relationships for the spray penetration, the spray angle, and the impact of air swirl on these two phenomena.

The spray tip penetration is as follows:

$$s = 0.39\sqrt{\frac{2}{\rho_l}\Delta P}\ t \qquad \text{for} \quad 0 < t < t_{break} \tag{4.11}$$

$$s = 2.95\left(\frac{\Delta P}{\rho_a}\right)^{1/4}\sqrt{d_o t} \qquad \text{for} \quad t \geq t_{break} \tag{4.12}$$

where

s = spray tip penetration, m
ρ_l = density of liquid fuel, kg/m^3
ΔP = difference between injection pressure and ambient pressure, Pa
t_{break} = transition time for spray to break up, seconds $= 28.65\dfrac{\rho_l d_o}{\sqrt{\rho_a \Delta P}}$
d_o = nozzle orifice diameter, m
ρ_a = density of air, kg/m^3

and the spray divergence angle is as follows:

$$\theta = 0.05\left(\frac{d_o^2 \rho_a \Delta P}{\mu_a^2}\right)^{1/2} \tag{4.13}$$

θ = spray divergence angle
μ_a = viscosity of air, Pa-s

When crossflow is present, these penetration and angle equations are modified as follows:

$$s_s = \frac{s}{\left(1 + \dfrac{\pi r_s n s}{30 u_0}\right)} \tag{4.14}$$

$$\theta_s = \theta\left(1 + \frac{\pi r_s ns}{30u_0}\right)^2 \qquad (4.15)$$

where

s_s = spray tip penetration with crossflow, m
r_s = swirl ratio
n = engine speed, rpm
u_0 = initial jet velocity; $0.39\sqrt{\dfrac{2\Delta P}{\rho_l}}$, m/s
θ_s = spray angle with crossflow

This correlation shows that the fully developed penetration depends on the injection pressure to the 1/4th power. The penetration rate increases as the injection pressure and nozzle orifice diameter increase and decreases as the ambient density increases. At high injection pressures, the momentum of the spray is much higher than the crossflowing air so the deflection of the spray is small. However, the fuel and burned product gases on the periphery of the spray can be swept downstream. Air swirl may also contribute to enhanced mixing when a wall jet is formed as fuel spray contacts the piston surface.

4.1.3.2 Droplet size distribution

An important function of the fuel injection process is to atomize the fuel into very small droplets, typically around 10 μm in size. Atomization greatly increases the surface area of the liquid fuel and accelerates the vaporization process. Measurement of droplet size in diesel sprays is very difficult because the sprays are dense and thus resistant to optical techniques. To be useful, the measurements must be taken under conditions simulating actual diesel combustion. The pressure and temperature must be the same as that in the cylinder, and to accurately reflect the initial and final phases in the spray it must also be transient.

Hiroyasu, Kadota, and Arai[10] measured droplet sizes under simulated diesel conditions. Using a photographic technique, they correlated the droplet sizes using the Sauter mean diameter (SMD), the diameter of a droplet whose ratio of volume to surface area is the same as that of the entire spray. Although the measurements were taken at room temperature, so vaporization would be limited, the effect of the injection parameters should be accurate. Their relationship is as follows:

$$\bar{x}_{32} = A(\Delta P)^{-0.135}(\rho_a)^{0.121}(B)^{0.131} \qquad (4.16)$$

where

$\bar{x}_{32}$ = Sauter mean diameter, m
A = 2.33×10^{-3} for orifice nozzles, = 2.18×10^{-3} for throttling pintle nozzles, = 2.45×10^{-3} for pintle nozzles
ΔP = difference between injection pressure and ambient pressure, Pa
ρ_a = density of air, kg/m^3
B = amount of fuel delivered, m^3/stroke

The most important factor for obtaining small droplets is to have a high relative velocity between the liquid jet and the ambient air. As the injection pressure increases, the jet velocity increases and the correlation correctly predicts a decrease in the SMD. Although the correlation indicates that injection pressure has only a weak influence on droplet size, even small changes in droplet size can dramatically increase the vaporization rate.

An increase in air density will slow the jet and increase the SMD. The amount of fuel delivered is included as a parameter to reflect the increased coalescence of droplets and the resulting increase in SMD that will occur as more fuel is injected. Coalescence is an important effect in the core of the spray where droplet densities are high.

4.1.4 Combustion system design

Designing diesel engines to produce their maximum power without excessive smoke or other pollutants requires careful matching of the combustion chamber geometry, the in-cylinder air motion, and the fuel injection. A large number of successful designs have been developed that involve very different approaches.

Diesel engines can be divided into direct injection (DI) engines and indirect injection (IDI) engines. Direct injection engines have the fuel injected directly into the main chamber above the piston. Indirect injection engines have the fuel injected into a separate chamber that is connected to the main chamber by one or more small passageways. These two types of combustion chambers should be considered as systems since more factors are involved than the chamber shape. The injection pump, injectors, and the air induction system must be properly matched to the chamber design. The design requirements of these two types of combustion systems are described below.

4.1.4.1 Direct injection engines

Figure 4.1 shows several common types of DI combustion chambers. The wide flat chambers are associated with high pressure injection systems and the deeper bowls are used with high swirl, low injection pressure systems. Direct injection engines depend primarily on the kinetic energy of the fuel spray to mix the air and fuel. This dependence increases the importance of the fuel injection system for optimizing the combustion system in DI engines. Increased air swirl can enhance the fuel–air mixing and extend the smoke limiting fuel–air ratio, but it increases NOx at the same time.

Direct injection engines inject fuel directly into a combustion chamber that usually consists of a recess or bowl in the piston as shown in *Figure 4.1*. In two-valve engines (one intake and one exhaust valve per cylinder) the injector is often inclined and is not centred in the bore. An offset bowl is required in the piston, which is not optimum for developing swirl. Four-valve engines have the advantage that they can provide excellent breathing and minimum pumping losses while providing a centred injector, which greatly minimizes particulate emissions.

Small bore engines generally use low injection pressure to limit penetration and rely on high swirl levels to mix the fuel and air. Large bore engines use higher injection pressure and low swirl or quiescent combustion chambers. The reduced need for air swirl in large engines improves efficiency by decreasing the work required to pump air in and out of the engine.

There is a relationship between the number of injector spray holes and the air swirl level. As the number of injector holes increases from 4, the normal minimum, to as high as 10 or 12, the required air swirl level decreases. To maintain the same total flow area, the hole size must be decreased as the number of holes increases. Smaller hole sizes require higher injection pressures to penetrate across the combustion chamber. Successful design of DI combustion systems requires optimum matching of the number of holes, the air swirl level, and the fuel injection pressure.

Increasing the fuel injection pressure can cause a dramatic reduction in the carbon fraction of the particulate in a DI engine[13]. To achieve this reduction with a minimum of NOx increase, it is desirable to have injection rate shaping that allows the injection pressure to increase without increasing the amount of fuel injected during the ignition delay period. Experiments involving sampling of the total cylinder contents at different points during the combustion process have shown that NOx emissions from engines with a significant amount of premixed combustion are formed early[6] and can be minimized by reducing the initial rate of injection into the cylinder or even using pilot injection. This

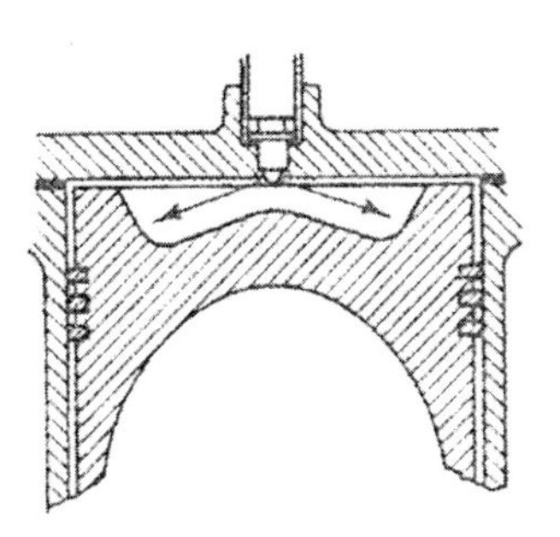
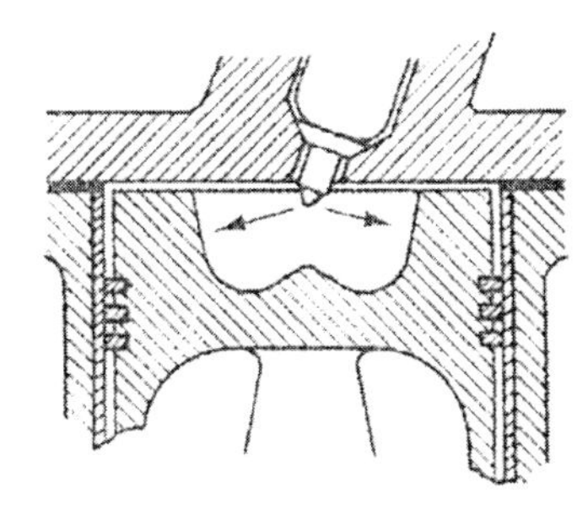
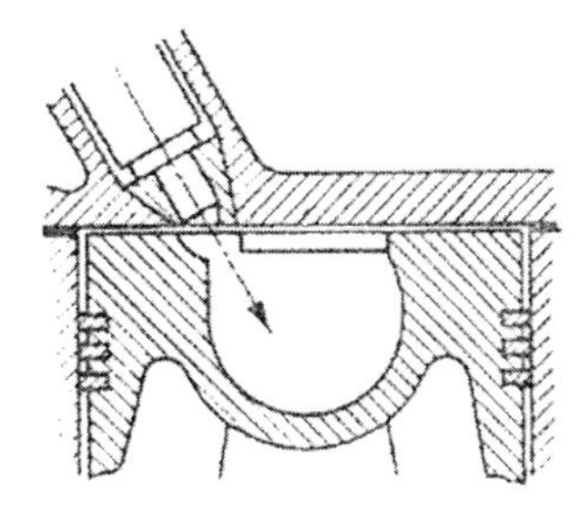

Figure 4.1 Direct injection combustion systems[12]

effect appears to be less important in highly turbocharged engines at full load[7] but should still be important at part load.

To increase the mixing that results from high-pressure injection, without increasing the amount of fuel injected during the ignition delay, it is necessary to: (1) lengthen the duration of the injection by reducing the spray hole diameters, (2) reduce the initial fuel rate by techniques such as needle seat throttling or pilot injection, or (3) control the initial pressure from the pump[14].

The shape of the bowl in the piston must be matched to the spray penetration and the air swirl. Systems utilizing high injection pressure give greater penetration and require less swirl. These systems favour shallow, large-diameter bowls. Smaller, deeper bowls will generate greater swirl as intake port generated swirl is compressed into a small diameter bowl. Because of conservation of angular momentum, the reduction in diameter greatly accelerates the angular velocity of the air. This system can effectively use a lower–pressure injection system. The former approach of high-pressure injection has proven to be more effective in meeting the desire for both low NOx and low particulates. Other bowl shapes have been used with some success. Hino Motors developed a square bowl with rounded corners and a re-entrant lip that has significant advantages over a conventional flat bowl in terms of lower smoke emissions[15].

The contour of the bowl lip has been shown to be important in reducing emissions. Re-entrant bowls lower particulates, in one case by 20%, by producing a shorter, faster combustion event[16]. Sharp-edged bowl lips appear to provide additional benefits. Kawatani *et al.*[17] advocate sharp-edged bowl lips that, on the basis of model predictions, increase the peak turbulent energy in the cylinder by 50% or more. Sharp–edged bowl lips can produce durability concerns in aluminum pistons and are one of the reasons for the recent popularity of steel-topped two-piece articulated pistons.

4.1.4.2 Indirect injection engines

Many different types of combustion chambers have been developed for IDI engines but the most successful recent designs have used some variation of the Ricardo Comet swirl chamber design shown in *Figure 4.2*.

During compression, air is forced at high velocity, from the main chamber, through the narrow connecting passage, and into the *swirl chamber*, or prechamber. As this air enters the prechamber, the chamber shape turns the flow and induces a strong swirl in the chamber. Fuel is injected into the swirling flow and ignited after a brief ignition delay. The fuel–air ratio in the swirl chamber is relatively rich, because only about half of the trapped air is present in the swirl chamber. This rich combustion keeps the NOx emissions low. As the pressure in the swirl chamber rises because of combustion and the pressure in the main chamber falls because of piston motion, the burning gases expand into the main chamber where the CO and unburned hydrocarbons burn with the remaining air.

Indirect injection engines have been popular for light-duty diesel applications because of their lower NOx emissions, wider speed range, and quieter operation. However, the high flow velocities in the swirl chamber and connecting passage produce greater heat transfer losses for IDI engines than for DI engines. These engines generally require compression ratios greater than 20:1 for reliable starting and acceptable fuel economy. Most IDI engines still require some form of starting aid such as a glow plug located in the swirl chamber. Because of the high swirl rates in the prechamber, IDI engines can achieve the fuel–air mixing rates required for high-quality combustion with low-pressure fuel injection. Most IDI engines use relatively inexpensive distributor-type fuel injection pumps and either single-hole or pintle-type injectors.

Many variations of the Ricardo Comet design have been developed. Jones *et al.*[19] found that by varying the orientation of the connecting passage they could decrease the piston heating. They opted for a side outlet from the prechamber as shown in *Figure 4.3*; it directs the hot gases emerging from the prechamber into the valve pockets rather than directly onto the piston crown. Jones also found that the optimum location for the injector was inboard, closer to the cylinder bore axis, with the fuel injected in the direction of the air swirl. This location provided the lowest NOx and noise, high priorities for their programme, but they found it necessary to use a fan-shaped multi-hole nozzle to achieve their power and unburned hydrocarbon goals. Jones, along with other engine developers[20], found that directing the fuel spray toward the centre of the prechamber gave lower unburned hydrocarbon emissions but resulted in large increases in NOx and combustion noise.

Amano *et al.*[20] designed a prechamber with a sharp edge that disrupts the swirling flow, achieving lower NOx emissions. They attributed this effect to lower maximum combustion temperatures resulting from weaker swirl. Swirl chamber design requires optimization of swirl chamber geometry, injection nozzle location, injection rate and timing, the throat area of the connecting passage, and compression ratio.

4.1.4.3 Comparison of IDI and DI engines

As mentioned earlier, the DI engine has up to 20% better fuel efficiency than the IDI engine[21]. Most of this loss is due to greater heat transfer losses in the swirl chamber and connecting passageway and to thermodynamic losses because of the late combustion process that occurs as the rich swirl chamber mixture expands into the main chamber. Throttling losses through the connecting passageway are sometimes cited as a loss term. However, these passages are usually large enough that the pressure difference between the main chamber and the swirl chamber is not large, and this effect is partially offset by the fact that the IDI engine does not have the pumping losses associated with the inlet port generated swirl typical of DI engines.

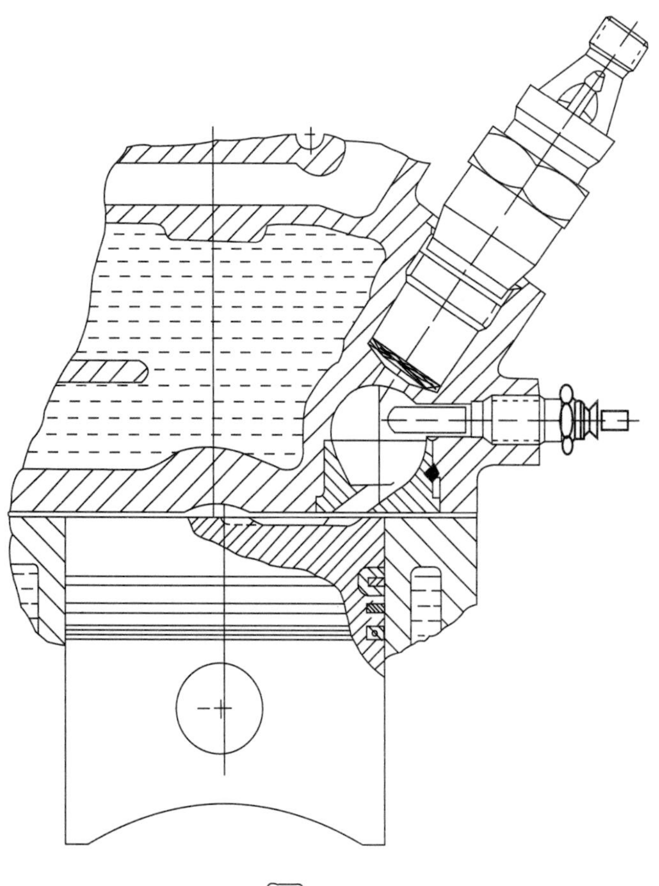

Figure 4.2 Ricardo Comet indirect injection combustion chamber[18]

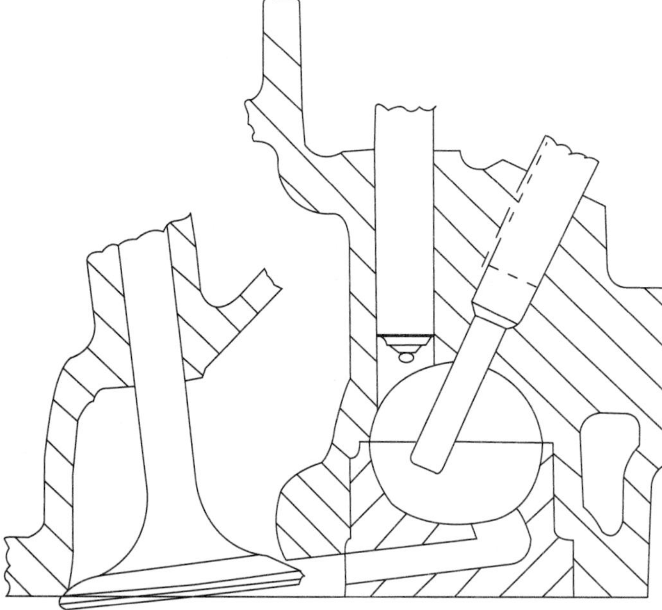

Figure 4.3 Side discharge for prechamber passage

Both IDI and DI engines require small clearances between the piston and the cylinder head. This clearance restricts the timing of exhaust valve closing and intake valve opening unless valve cutouts are provided in the piston. These cutouts increase piston cost and adversely affect the in-cylinder flow field.

4.1.5 Analysis of cylinder pressure data

4.1.5.1 Pressure measurement

Much useful information about the diesel combustion process

can be obtained by measurement and analysis of the cylinder pressure. Specifically, the indicated mean effective pressure and the apparent heat release rate can be computed. Other quantities such as the peak pressure and the peak rate of pressure rise are also useful as indicators of overall stress and noise levels, respectively.

Accurate measurement of cylinder pressure requires three things: a pressure transducer that can survive the engine environment, a means for synchronizing the pressure measurements with the engine, and some way to record the pressure measurements. Modern pressure transducers are piezoelectric devices that use quartz crystals to produce electric charges when mechanically loaded. These transducers are designed to resist errors induced by vibration and thermal radiation. While designs are available that can operate above 300°C, water-cooled transducers are recommended for best accuracy. Lancaster *et al.*[22] have published detailed procedures for collecting and validating cylinder pressure data.

To time the data acquisition process to the engine, an incremental optical encoder is directly coupled to the engine crankshaft. This device consists of a rotating disc with very fine lines etched in a radial pattern. A light source illuminates the disc and an optical sensor measures the alternating signal produced as the disc rotates and the lines block the light. Two output channels are provided: one gives a single output signal per revolution, used for indexing the encoder, and the other gives 1440 pulses per revolution (one every quarter of a degree). Coarser divisions are available but are not recommended for heat release analysis. Pressure data used to compute the heat release rate should contain information at frequencies as high as 10 kHz. To collect information at this frequency, the Nyquist sampling criterion mandates a minimum sampling frequency of 20 kHz. To avoid aliasing, all signal information beyond the frequencies of interest should be electronically low-pass filtered before the analogue-to-digital conversion takes place.

Cylinder pressure data usually requires smoothing before it can be used for heat release rate analysis. This smoothing can be accomplished by averaging a large number of consecutive cycles, polynomial smoothing[23], spline smoothing[24], or digital filtering[25].

4.1.5.2 Apparent heat release rate

Heat release calculations are an attempt to learn something about the combustion process in an engine from measured cylinder pressure data. While based on a model that contains a large number of assumptions, some of which are clearly invalid, heat release calculations can provide useful information. If the assumptions behind the analysis are understood, there is much less likelihood that the results will be misinterpreted. A number of approaches to heat release analysis have been presented in the technical literature but the most widely used is that developed by Krieger and Borman[26].

The apparent heat release rate is the rate that fuel and air would need to react to equilibrium combustion products to give the observed cylinder pressure curve. This rate is calculated from the First Law of Thermodynamics applied to a control volume consisting of the engine cylinder. The intake and exhaust valves are assumed to be closed and blowby is negligible. With these assumptions, the energy equation becomes

$$\frac{dU}{dt} = \dot{Q} - P\frac{dV}{dt} + \dot{m}_f h_f \tag{4.17}$$

where

U = internal energy of the cylinder contents
t = time
$\dot{Q}$ = heat transfer rate
P = cylinder pressure
V = cylinder volume
$\dot{m}_f$ = fuel addition rate (also the rate at which fuel is burned to equilibrium products)
h_f = fuel enthalpy

The contents of the cylinder are characterized by a single pressure, temperature, and composition, and the ideal gas equation is used to relate the pressure, volume, and mass to the temperature. The cylinder volume is calculated from geometry as a function of the crank position. The heat transfer rate from the cylinder can be estimated using empirical correlations that are based on heat transfer measurements from other engines. Because the in-cylinder gases are assumed to be at equilibrium, the internal energy and gas constant values can be computed from the temperature, pressure, and composition. The equations reduce to the following form where *dm/dt* is the calculated fuel burning rate.

$$\frac{dm}{dt} = \frac{-P\dfrac{dV}{dt} - m\dfrac{\partial u}{\partial P}\dfrac{dP}{dt} + \dot{Q} - mCB}{u - h_f + D\dfrac{\partial u}{\partial \phi} - C\left[1 + \dfrac{D}{R}\dfrac{\partial R}{\partial \phi}\right]} \tag{4.18}$$

where

$$B = \frac{1}{P}\frac{dP}{dt} - \frac{1}{R}\frac{\partial R}{\partial P}\frac{dP}{dt} + \frac{1}{V}\frac{dV}{dt}$$

$$C = \frac{T\dfrac{\partial u}{\partial T}}{1 + \dfrac{T}{R}\dfrac{\partial R}{\partial T}}$$

$$D = \frac{1 + f_0 m}{f_s m_0}$$

and where

m = mass in the cylinder
u = internal energy per unit mass
ϕ = equivalence ratio
R = gas constant for the equilibrium gas mixture

The apparent heat release rate is determined by multiplying this calculated fuel burning rate by the heating value of the fuel. Usually the heat release rate is presented as a normalized quantity where the rate is divided by the product of the total mass of fuel injected and the heating value. This makes the area under the normalized curve equal to 1, and simplifies comparison of profiles for different fueling levels.

A mass balance on the cylinder can provide an equation relating the mass and equivalence ratio, and the ideal gas equation can be used to derive an equation for the temperature. To provide the heat release rate, these equations can be integrated with eqn (4.18) and experimentally measured pressure data. *Figure 4.4* shows a typical heat release rate profile for a naturally aspirated diesel engine. Heat release is zero during the compression process until shortly after the start of fuel injection. At this point, there is typically a slight negative dip in the curve due to the fuel vaporization and endothermic preflame reactions. When auto-ignition occurs, the heat release rate rises rapidly as the premixed fuel burns. The premixed combustion gives a characteristic spike to the heat release rate. This premixed spike will be smaller or even nonexistent in highly turbocharged engines. After the premixed fuel is consumed, the combustion rate decreases to the slower rate of the mixing controlled combustion.

The heat release rate can be used to identify the start of combustion, the fraction of fuel burned in the premixed mode, and the differences in combustion rates because of injection system changes or air swirl levels.

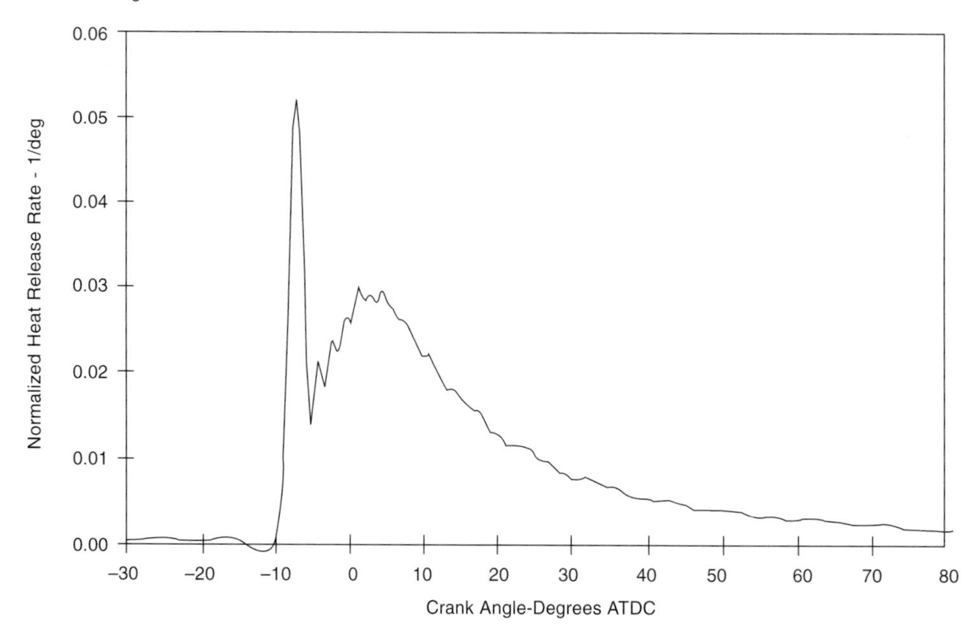

Figure 4.4 Normalized heat release rate

4.2 Diesel fuels

Concern about future fuel quality and availability and recognition of the influence of fuel properties on diesel combustion and emissions provide the justification for including a detailed discussion of fuel chemistry and properties. As diesel engine manufacturers press the limits of performance and emissions improvements that are attainable through engine modifications, the fuel is always identified as an area where substantial changes are still possible. However, because the cost of the fuel burned by a typical diesel engine significantly exceeds the cost of the engine, it is more cost effective to invest in engine technology than in all but the most inexpensive fuel changes.

Petroleum-based fuel composition can be described in many different ways. The most direct way is to identify the specific hydrocarbon species present and their amounts. However, this approach is not practical for diesel fuels because they contain measurable amounts of hundreds of individual fuel species. A more successful approach has been to characterize the fuel by the range of the boiling points of its constituents along with more specialized tests that relate to the properties of the fuel rather than its composition.

Although it is very difficult to identify all of the individual components of the fuel, it is possible to group the types of hydrocarbons into general classes that have similar molecular structures. These general hydrocarbon classes will be discussed in this chapter along with their relation to diesel fuel properties. Diesel fuel properties, quality issues, and specifications will be discussed also.

4.2.1 Hydrocarbon types

A carbon atom can form covalent bonds with other carbon atoms or with other atomic species. Each atomic species has its own preferred number of bonds that it requires to achieve a stable configuration. This preferred number of bonds is called the *covalent valence* of the element. Carbon atoms seek to form four bonds and hydrogen atoms one bond. *Hydrocarbons* are molecules that contain only hydrogen and carbon. The simplest hydrocarbon molecule, methane, is formed when four hydrogen atoms attach to a carbon.

Those atomic species that have a covalent valence greater than one may form double or triple bonds. Hydrocarbons with only single bonds are called *alkanes*, those with one or more double bonds are called *alkenes*, and those with triple bonds are called *alkynes*. *Aromatic* hydrocarbons are another class that involves a special type of bond that is between a single and a double bond.

4.2.1.1. Alkanes

Methane is part of a class of hydrocarbons that contains no double or triple bonds. These hydrocarbons are called alkanes or *paraffins*. Alkanes contain the largest possible number of hydrogens bonded to carbons in the molecule and thus are called *saturated*. Other alkanes can be formed from methane by adding carbons. For example, two carbons can be bonded to each other, with each carbon having three hydrogens attached to it to form C_2H_6, or ethane, As additional carbons are chained together, many familiar molecules such as propane, butane, and pentane are formed. These molecules are sometimes called *straight-chain* hydrocarbons because their carbon atoms are arranged in a single interconnected string. They are also called *normal* paraffins, a term that is sometimes used as a prefix for the molecule name. For example, octane is referred to as *normal-octane* or *n-octane*.

Table 4.1 shows some of the physical properties of selected normal paraffins. It also shows their motored octane number (MON) and cetane number. These two properties are measures of the fuel's readiness to auto-ignite. Some relationships between the size of the molecules and their properties are clear from *Table 4.1*. As the molecules become longer, their boiling points and melting points increase. This is reflected in the normal states where the smaller paraffins are normally gaseous, the molecules in the mid-range are liquid, and the large molecules are solid at room temperature. The tendency toward auto-ignition also shows a clear trend. Fuels with a high resistance to autoignition have high octane numbers and are valuable in spark-ignition engines. Short-chain paraffins have high octane numbers. However, as the chain length increases, the octane number decreases quickly. The cetane number, which is inversely related to the octane number, increases with longer chain lengths, indicating a readiness to auto-ignite.

It is possible to have alkane structures other than straight chains. Two hydrocarbon molecules can have the same number

Table 4.1 Physical properties of some normal paraffin hydrocarbons[27–29]

Name	*Formula*	*M.P., °C*	*B.P., °C*	*Specific gravity*	*Normal state*	*MON*	*Cetane number*
Methane	CH_4	– 182.6	– 161.4	—	gas	120+	—
Ethane	C_2H_6	– 172.0	– 88.3	—	gas	99	—
Propane	C_3H_8	– 187.1	– 44.5	—	gas	97	—
n–Butane	C_4H_{10}	– 135.0	– 0.5	—	gas	90	—
n-Pentane	C_5H_{12}	– 129.7	36.2	0.6264	liquid	61.9	—
n-Hexane	C_6H_{14}	– 94.0	69	0.6594	liquid	26.0	—
n-Heptane	C_7H_{16}	– 90.5	98.4	0.6837	liquid	0	56
n-Octane	C_8H_{18}	– 56.8	124.6	0.7028	liquid	– 15*	—
n-Decane	$C_{10}H_{22}$	– 32	175	0.7300	liquid	– 38*	76
n-Dodecane	$C_{12}H_{26}$	– 9	216	0.7490	liquid	—	80
n-Pentadecane	$C_{15}H_{32}$	10	271	0.7720	liquid	—	95
n-Hexadecane	$C_{16}H_{34}$	18.5	287	0.7730	liquid	—	100
n-Octadecane	$C_{18}H_{38}$	28	308	0.7700	solid	—	110

*Blending MON

of carbon and hydrogen atoms but one molecule might have the carbons connected as a straight chain and the other molecule might have a shorter chain with carbon atoms branching off at intermediate points. These two compounds are called *structural isomers*. Branching has a stabilizing effect on molecules, and large molecules with several branched chains can have very low cetane numbers.

4.2.1.2 Cyclic hydrocarbons

Some hydrocarbon compounds contain three or more carbon atoms arranged in a ring. They are called cyclic compounds. If the ring contains no double bonds, the structure is saturated and is called a *cyclo-paraffin* (also known as *naphthene*). The names of these molecules are derived from the number of carbons in the ring. Most naphthenic hydrocarbons are based on a cyclic structure containing five or six carbons (cyclopentane or cyclohexane) with side chains attached to one of the cyclic carbons.

4.2.1.3 Alkenes and alkynes

Alkenes are compounds that contain a carbon–carbon double bond. Alkenes, also called *olefins*, are unsaturated. They have fewer hydrogens than an alkane would have for the same number of carbon atoms. An *alkyne* is a hydrocarbon with one or more carbon–carbon triple bonds.

4.2.1.4 Aromatics

Aromatic compounds contain bonds that cannot be classified as either single or double but are something halfway between. This phenomenon is called resonance. Aromatic compounds are unsaturated and are most commonly found in hexagonal structures with benzene as the most basic example. Aromatics have higher densities than paraffins and, in spite of having slightly lower energy content per unit mass, have higher energy content per unit volume. Aromatics are powerful solvents and tend to cause swelling of elastomers.

4.2.1.5 Oxygenated hydrocarbons

The oxygenated hydrocarbons of primary interest as fuels are alcohols and ethers, although esters are attracting some attention because they can be produced from renewable sources. Aldehydes and ketones are of interest primarily as exhaust pollutants. They are strong irritants and are highly reactive in photochemical smog formation.

An alcohol consists of a hydrocarbon alkyl group with an OH, or hydroxyl group, connected to one of its carbons. Ethers have an oxygen atom that is connected to two hydrocarbon alkyl groups through carbon–oxygen single bonds. Aldehydes and ketones have an oxygen atom that is connected to a carbon with a double bond. The carbon connected to the oxygen, called the carbonyl carbon, will have either two hydrogens attached to it or two additional carbons. If the carbonyl carbon has at least one hydrogen attached to it, it is an aldehyde. If it has two carbons attached to it, it is a ketone.

4.2.2 Petroleum-derived fuels

Table 4.2 shows the boiling point ranges corresponding to various products produced by petroleum refining. As shown, No. 1 diesel fuel, commonly used as a winter-grade diesel fuel, covers the boiling range from 170°C to 270°C. No. 2 diesel fuel, the most common fuel for medium- and high-speed diesel engines, has a boiling point range from 180°C to 340°C. The terms *distillate fuel* and *residual fuel* were formerly used to designate diesel fuel types. Distillate fuel indicated that the fuel was recovered from a distillation process and thus could be identified with a specific boiling point range. Residual fuel referred to any fuel that contained fractions of residue from distillation or thermal cracking.

While No. 1 diesel fuel is generally a straight-run fraction from the refinery called *light virgin distillate*, No. 2 diesel fuel and heavier fuels are blends of two or more refinery streams including cracked gas oil, heavy, naphtha, light and heavy gas oils, reduced crude, and pitch[31].

When the boiling point ranges shown in *Table 4.2* are compared with *Figure 4.5* which shows the distribution of the various hydrocarbon types throughout the boiling point range of crude petroleum, the dominate hydrocarbon types for each fuel can be

Table 4.2 Typical refinery products[30]

Product	*Boiling range, °C*
LPG	–40–0
Gasoline	30–200
Kerosene, Jet Fuel, No. 1 Diesel	170–270
No. 2 Diesel, Furnace Oil	180–340
Lube Oils	340–540
Residual Oil	340–650
Asphalt	540 +
Petroleum Coke	Solid

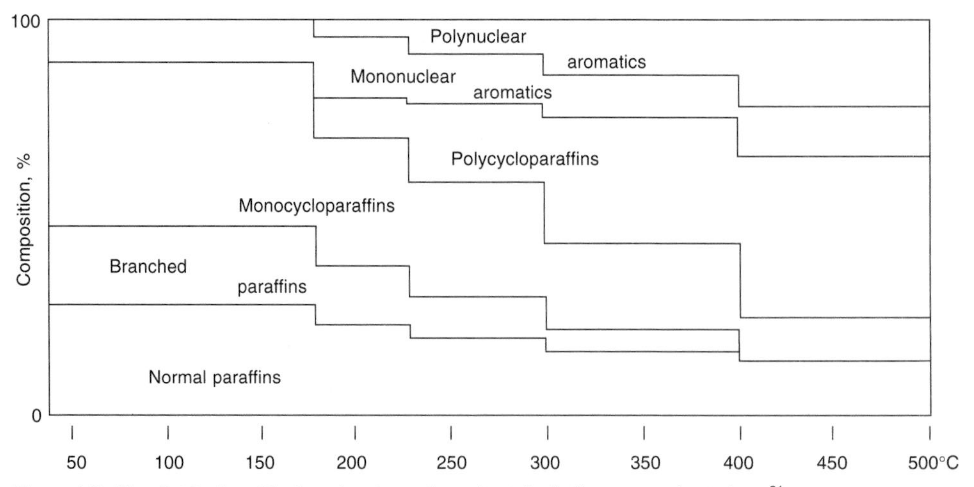

Figure 4.5 The distribution of hydrocarbon types throughout the boiling range of petroleum[31]

Table 4.3 Hydrocarbon composition of three typical No. 2 diesel fuels[32]

Fuel	*T-1*	*T-2*	*T-3*
	Composition, wt%		
Sulfur	0.24	0.25	0.06
Br#	0.61	0.61	0.64
Paraffins	31.5	31.3	35.3
Cycloparaffins	40.3	42.7	32.0
Alkylbenzenes	8.0	8.9	7.4
Indans and tetralins	5.0	4.6	6.5
Indene	1.3	1.2	1.0
Naphthalene	0.2	0.3	0.6
C_{11} + naphthalene	4.2	3.0	7.3
Acenaphthylene	4.0	3.5	3.7
Tricyclicaromatics	2.2	1.9	2.1
Total	100	100	100
	Naphthalene Distribution		
C10	0.20	0.29	0.6
C11	0.65	0.57	1.5
C12	1.19	0.70	2.1
C13	0.80	0.60	1.8
C14+	1.49	1.10	1.9
Total	4.40	3.30	7.9

identified. In the range for No. 2 diesel fuel, one would expect to see many normal paraffins, cycloparaffins, and one- and multi-ring aromatics. The analyses for the No. 2 diesel fuels provided in *Table 4.3* shows this to be true[32].

4.2.3 Diesel fuel properties

Diesel fuel is characterized by a number of properties that provide the basis for fuel specifications. The most important of these properties are discussed below along with the procedures used to measure them.

4.2.3.1 Density

The density of diesel fuel is a secondary indicator of its distillation range. In general, heavier fuels have higher boiling points. The energy contents of hydrocarbon fuels are very similar on a weight basis but most fuels are sold by volume. Higher-density fuels have more energy per unit volume.

Specific gravity. The density of petroleum products is usually expressed as a specific gravity. The *specific gravity* is defined as the ratio of the mass of a volume of the fuel to the mass of the same volume of water. It is dependent on the temperature of both the fuel and the water. It is commonly expressed as *sp gr @ 60°F/60°F* which means that both the fuel and water are at 60°F.

API gravity. The American Petroleum Institute (API) gravity is also a widely used measure of a fuel's density. It is related to the specific gravity of the fuel by the following equation:

$$API\ gravity = \frac{141.5}{specific\ gravity\ @\ 60°F/60°F} - 131.5$$

4.2.3.2 Ignition indices

One of the most important properties of a diesel fuel is its readiness to auto-ignite at the temperatures and pressures present in the cylinder when the fuel is injected. The cetane number is the standard measure of this property although it is difficult to measure precisely and has been criticized in recent years for not accurately reflecting the auto-ignition conditions in modern turbocharged engines.

Cetane Number (ASTM D613). The cetane number is an engine-based test that uses a single-cylinder, indirect-injection diesel engine. The engine speed is fixed at 900 rpm, and while the engine is naturally aspirated, the intake air temperature is held at 150°F. The test is conducted by carefully adjusting the fuel–air ratio and the compression ratio to produce a standard ignition delay of 13° with the test fuel. Then the test fuel is removed from the engine and a blend of reference fuels is added. The primary reference fuels are n-cetane (n-hexadecane), which has a cetane number of 100, and heptamethylnonane (HMN), which has a cetane number of 15. Different combinations of these two fuels are added until a blend is found that restores the ignition delay to 13 degrees. When this occurs, the cetane number is computed from the following relationship:

Cetane number = % n-cetane + 0.15 (% HMN)

Because the price of primary reference fuels is quite high, most commercial cetane testing is done with secondary reference fuels which have been calibrated to known cetane values.

Cetane index. The cetane index is a calculated quantity that is intended to approximate the cetane number. It is much cheaper to determine than the engine-based cetane number but its accuracy is limited to the class of fuels on which it is based. The cetane index generally does not provide an accurate indication of cetane number for non-petroleum-based alternative fuels or if the fuel contains cetane-improving additives. Two methods are available for computing the cetane index. ASTM standard D976 gives the following empirical equation for the cetane index:

$$\text{Cetane index} = 454.74 - 1641.416D + 774.74D^2 - 0.554T_{50} + 97.803\,[\log_{10}(T_{50})]^2$$

where D = fuel density at 15°C in g/ml
T_{50} = temperature corresponding to the 50% point on the distillation curve in °C

ASTM standard D4737 gives the cetane index according to the following four-variable equation:

$$\text{Cetane index} = 45.2 + 0.0892(T_{10}N) + 0.131(T_{50}N) + 0.0523(T_{90}N) + 0.901B(T_{50}N) - 0.420B(T_{90}N) + 4.9 \times 10^{-4}(T_{10}N)^2 - 4.9 \times 10^{-4}(T_{90}N)^2 + 107\text{B} + 60B^2$$

where $T_{10}N = T_{10} - 215$
$T_{50}N = T_{50} - 260$
$T_{90}N = T_{90} - 310$
T_{10}, T_{50}, and T_{90} = temperatures for 10%, 50%, and 90% distilled in °C
$B = [\exp(-3.5\ \text{DN})] - 1$
DN = density at 15°C (kg/liter) – 0.85

Diesel index. The diesel index is an ignition index that roughly correlates to the cetane number. It is calculated from the following equation:

$$\text{Diesel index} = \text{aniline point (°F)} \times \frac{\textit{API gravity}}{100}$$

The diesel index includes information about composition (the aniline point indicates the amount of aromatics) and density. A high aniline point indicates a low aromatic content and therefore a greater likelihood of a high cetane number. A high API gravity indicates a low density, which is indicative of a high paraffin fraction and high cetane number.

The cetane number of a fuel can be increased with fuel additives that are designed to readily decompose, giving precursors to combustion and thus enhancing the rate at which auto-ignition occurs in a diesel engine. Typical compounds used are alkyl nitrates, ether nitrates, dinitrates of polyethylene glycols, and certain peroxides. Because of low cost and ease of handling, alkyl nitrates such as ethylhexyl nitrate are the most widely used *cetane improvers.*

4.2.3.3 Cold flow properties

The heavier diesel fuels, including No. 2 diesel fuel, will crystallize, or gel, at temperatures commonly encountered in the northern United States, Canada, and Europe. Four different tests are used to characterize this problem.

Cloud point (ASTM D2500, IP 219). The cloud point is the temperature at which a cloud of wax crystals first appears in a fuel sample that is cooled under controlled conditions. The cloud point is determined by visually inspecting a normally clear fuel for a haze.

Pour point (ASTM D97, IP 15). The pour point is the lowest temperature at which movement of the fuel sample can be determined when the sample container is tilted. At every 3°C of cooling, the sample is inspected, and when no movement is detected after 5 seconds, the test is stopped. 3°C is added to the temperature where no movement was observed and this is the pour point. Pour points are always expressed in multiples of 3°C.

Low-temperature flow test (ASTM D4539). The low-temperature flow test (LTFT) is designed to evaluate whether a fuel can be expected to pass through an engine fuel filtration system. The test determines the lowest temperature at which 180 ml of fuel can be drawn through a 17 μm screen in 60 seconds or less with 20 kPa of vacuum.

Cold filter plugging point (IP 309). The cold filter plugging point (CFPP) is similar to the LTFT test. It determines the lowest temperature where 20 ml of fuel can be drawn through a 45 μm screen in 60 seconds with 200 mm of water (1.96 kPa) of vacuum. The CFPP provides a close correlation with vehicle operability limits[33].

The cloud point is the highest temperature used for characterizing cold flow and the pour point is the lowest. The LTFT and CFPP temperatures will be somewhere between the cloud and pour points. The cold flow properties of diesel fuels can be modified with pour point depressants. Most pour point depressants, also known as cold flow improvers, work on similar principles. As the fuel sample is cooled, small wax crystals form. The temperature at which this occurs is the cloud point. As the sample is cooled further, the crystals agglomerate and grow in size until the entire sample solidifies. Most pour point depressants do not alter the initial formation of the crystals and thus do not generally affect the cloud point. Rather, they inhibit the crystals from combining and growing to a size large enough to plug filters. The additives are generally waxes that are used in small amounts. They surround the small crystals and provide a barrier to agglomeration. The most common technique for improving the cold flow properties of No. 2 diesel fuel is to add 50% No. 1 diesel fuel, which has a pour point well below –40°C.

4.2.3.4 Volatility

Diesel fuel volatility can be characterized by the distillation curve and the flash point.

Distillation curve (ASTM D86). The distillation curve is determined by measuring the fraction of a fuel sample removed by heating a fuel sample to progressively higher temperatures. Typically, the curve is characterized by the initial point, which is the temperature of the fuel when the first drop of liquid leaves the condenser, the temperatures when each 10% of the liquid is removed; and the temperature for the last drop of fuel which is called the end point. The temperature where 90% of the fuel has been removed is believed to be related to crankcase oil dilution. Higher 90% distillation points increase the likelihood that some fuel will escape combustion and leak past the piston rings to the crankcase.

Flash point (ASTM D93). The flash point is the lowest temperature at which a combustible mixture can be formed above the liquid fuel. It is important for fire safety considerations and depends on both the lean flammability limit of the fuel as well as the vapour pressure of the fuel constituents. The flash point is determined by heating a sample of the fuel in a stirred container and passing a flame over the surface of the liquid. If the temperature is at or above the flash point, the vapour will ignite and an easily detectable flash can be observed. The flash need not correspond to a sustained flame. The *fire point* is sometimes used to designate the fuel temperature that will produce sufficient vapour to maintain a continuous flame.

4.2.3.5 Viscosity

Viscosity is a measure of a fluid's resistance to flow. The greater the viscosity, the less readily the liquid flows. The viscosity of petroleum oils is a strong function of temperature; the viscosity decreases as the temperature increases. ASTM D445 is a standard test procedure for determining the kinematic viscosity of liquids. It provides a measure of the time required for a volume of liquid to flow under gravity through a calibrated glass capillary tube. The kinematic viscosity is then equal to the product of this time and a calibration constant for the tube. The dynamic viscosity can be obtained by multiplying the kinematic viscosity by the density of the fluid.

4.2.3.6 Miscellaneous properties

Heating value, net and gross (ASTM D240). Actually, two heating values are in common use: the higher, or gross, heating value and the lower, or net, heating value. Both quantities are measured using a calorimeter that measures the heat transfer from the hot combustion gases as they are cooled to the initial temperature of the reactants. The higher heating value assumes that all of the water in the products is condensed liquid while the lower value assumes that all of the water is present as vapour. The lower heating value is normally used for engine applications. When calorimeter data is not available, the heating value can be estimated using empirical correlations such as those described in ASTM standards D1405 and D3338.

Ramsbottom carbon residue (ASTM D524). This test involves heating a small sample of fuel to 550°C. The volatile matter evaporates and the heavier fraction undergoes cracking and coking reactions. The amount of residue, as a percentage of the original sample, is used as an indicator of a fuel's tendency to form combustion chamber deposits. The test is sometimes performed on the 10% of the fuel with the highest distillation temperature and reported as such.

Sulfur content (ASTM D2622). Petroleum includes a large number of sulfur-containing compounds. Sulfur can contribute to higher engine wear, can poison catalysts, and can produce sulfates that form part of the exhaust particulate matter. In the United States, the EPA requires diesel fuel used on-highway to contain less than 0.05% sulfur. ASTM D2622 uses x-ray spectrometry to determine the mass fraction of sulfur in the fuel.

Ash (ASTM D482). This test involves heating a small sample of fuel to 775°C until all organic material is either vaporized or burned off. The residue is an ash that usually originates from soluble metallic compounds or contaminants such as dirt and rust.

Water and sediment (ASTM D1796). Water in diesel fuel causes corrosion of storage tanks and fuel injection equipment. The total amount of water and sediment is determined by centrifuging the fuel and reporting the result as a percentage of the fuel.

Copper corrosion (ASTM D130). Fuels can be corrosive, largely due to sulfur compounds. The copper corrosion test is a general indicator of this effect; it involves placing a polished copper strip into a heated fuel sample for 3 hours and comparing the tarnish on the strip to a standard scale. A rating of 1 corresponds to slight tarnish and 4 is corrosion.

4.2.4 Diesel fuel quality issues

4.2.4.1 Contaminants

Diesel fuel contaminants can be divided into three general categories: (1) those originating as foreign materials and introduced into the fuel such as dust, dirt, rust, and water; (2) Those produced in the fuel through biological activity of bacteria, yeast, and fungi; and (3) those produced by fuel degradation such as gums and sediments. This latter category relates to fuel stability and is discussed as a separate section.

Dust, dirt, and rust can be controlled by rigorously enforced standards for cleanliness in fuel transport, storage, and utilization. Regular cleaning of fuel tanks and supply lines can minimize the potential for damage from these sources.

Water can cause corrosion of fuel storage tanks and fuel injection system components. It contributes to fuel degradation and microbial growth. Water most often enters the fuel tank as condensate from moist air entering through vents. Many fuel storage tanks have condensed water at the bottom, and this water should be removed regularly. When the water cannot be removed, the fuel should be treated with a biocide and fuel should not be drawn from the bottom of the tank.

Diesel fuel can be contaminated with microbial growth. Species such as cladosporium resinae and candida tropicalis have been frequently observed in diesel fuel samples. These microbes, consisting of both aerobic and anaerobic bacteria as well as yeast and fungi, require water and are generally found at the fuel–water interface. They are usually controllable by eliminating any free water. Microbes can also be controlled by adding biocides to the fuel or to the water at the bottom of the tank, but good housekeeping practices can minimize water contamination. This latter approach is preferred.

4.2.4.2 Fuel stability

When diesel fuel is subjected to air at elevated temperatures for extended times, it undergoes chemical changes that alter its colour from light yellow to dark brown or black. It also forms gum and other sediments. Gum may be either soluble gum, which is indicated by darkening of the fuel, or insoluble gum, which precipitates from the fuel and deposits on exposed surfaces. The processes that form gum are not well understood but may involve the formation of radicals by heating or catalytic action of transition metals such as Fe, Cu, and Mn. These radicals react with oxygen to form peroxides that combine to form gum particles that can plug fuel filters and cause sticking of fuel pumps, injectors, and governor components[34]. The presence of water can aggravate gum plugging of fuel filters by converting a reduction in fuel flow into a completely plugged filter[34].

Peroxides and hydroperoxides can form in diesel fuel and can actively attack elastomers that are normally compatible with diesel fuel. Antioxidants such as hindered phenols and amines effectively prevent peroxide buildup although they may produce colorbodies as they consume peroxides. Metal deactivators are also sometimes added to diesel fuel. They function by converting any trace metal salts, which could catalyse fuel oxidation, into chelates that have no catalytic effect.

ASTM Test Methods D3241, Thermal Oxidation Stability of Aviation Turbine Fuels (JFTOT Procedure), and D2274, Oxidation Stability of Distillate Fuel Oil, can be used to characterize a fuel's stability and the effectiveness of additives.

4.2.5 Diesel fuel specifications

Several sets of standard specifications have been established for diesel fuels. The most common specifications used in the United States are those established by ASTM Standard D975. This standard covers three grades of diesel fuel oils that vary according to their service applications.

Grade No. 1-D. A light distillate fuel for applications requiring a higher volatility fuel for rapidly fluctuating loads and speeds, as in light trucks and buses. The specification for this gade of diesel fuel overlaps with kerosene and jet fuel, and all three are commonly produced from the same base stock. One major use for No. 1-D diesel fuel is to blend with No. 2-D during winter to provide improved cold flow properties.

Grade No. 2-D. A middle distillate fuel for applications that do not require a high-volatility fuel. Typical applications are high-speed engines that operate for sustained periods at high load.

Grade No. 4-D. A heavy distillate fuel that is viscous and may require fuel heating for proper atomization of the fuel. It is used primarily in low- and medium-speed engines.

Table 4.3 lists the property values for these three grades of diesel fuel.

Table 4.3 ASTM specifications for diesel fuel oils (ASTM D975)

Property	*ASTM test*	*Grade 1-D*	*Grade 2-D*	*Grade 4-D*
Flash point (°C)	D93	38	52	55
Water and sediment(%)	D1796	0.05	0.05	0.50
Distillation temperature	D86			
90% recovered, min		none	282	none
max		288	338	none
Kinematic viscosity	D445			
min		1.3	1.9	5.5
max		2.4	4.1	24.0
Ramsbottom carbon residue on 10% distillation residue	D524	0.15	0.35	none
Ash, % mass, max	D482	0.01	0.01	0.10
Sulfur, % mass, max	D129	0.50*	0.50	2.00
Copper strip corrosion rating, max 3 h at 50°C	D130	No. 3	No. 3	none
Cetane number, min	D613	40	40	30
Cloud point, °C, max	D2500	Guidance only. Should be 6°C higher than the tenth percentile minimum ambient temperature for the region		

*In October 1993, the U.S. EPA mandated that all diesel fuel sold for on-highway use must contain less than 0.05% sulfur. This ruling was made to lower particulate emissions and to allow future use of catalytic converters that can be poisoned by excessive fuel sulfur.

No. 1 diesel fuel is essentially the same distillate cut as (1) No. 1 fuel oil, defined by ASTM standard D396; (2) kerosene, defined by ASTM standard D3699; (3) Jet A, defined by ASTM standard D1655; and (4) JP-8, defined by Mil-T-83133A. These fuels can be produced from the same light virgin distillate base stock but have different additive requirements related to their intended use.

Standards for No. 5 and No. 6 fuel oils are provided in ASTM standard D396. These residual fuels are used mostly for burner fuel but, if heated, they can be used in large, low-speed diesel engines.

In the United States, the EPA sets standards for the fuels to be used for emission certification testing. *Table 4.4* shows the requirements for EPA certification, or *cert*, fuel. This fuel is standardized to ensure that emission measurements can be reproduced.

4.2.6 Alternative fuels

Because of concerns about fossil fuel reserves, alternative fuels originally gained attention in the 1970s as potential substitutes for petroleum-based fuels. These shortages did not materialize but the environmental advantages of alternative fuels have sustained public interest. The fuels of greatest interest are reformulated diesel fuel, compressed natural gas, alcohols, and biodiesel.

4.2.6.1 Reformulated diesel fuel

Reformulated diesel fuel is often not considered to be an alternative fuel because it is still based on nonrenewable

Table 4.4 EPA certification fuel specifications

Item	*ASTM*	*Type 1-D*	*Type 2-D*
Cetane number	D613	40–54	40–48
Cetane index	D975	40–54	40–48
Distillation range	D86	—	—
IBP (°C)	—	165.6–198.9	171.1–204.4
10% (°C)	—	187.8–221.1	204.4–237.8
50% (°C)	—	210.0–248.9	243.3–282.2
90% (°C)	—	237.8–271.1	293.3–332.2
EP (°C)	—	260.0–293.3	321.1–365.6
API gravity	D287	40–44	32–37
Sulfur, %	D2622	0.03–0.05	0.03–0.05
Hydrocarbon composition	D1319	—	—
Aromatics, %	—	8 min	27 min
Paraffins, %	—	—	—
Olefins, %	—	—	—
Flashpoint, min.(°C)	D93	48.9	54.5
Viscosity, cS	D445	1.6–2.0	2.0–3.2

petroleum. This fuel is primarily a result of changes in diesel fuel specifications; the specific intention is to reduce emissions without requiring engine hardware modifications. The most common fuel changes in reformulated diesel fuel are lower sulfur, higher cetane number, lower boiling range, and lower aromatics[35]. Lowering the cetane number, aromatic content, and amount of sulfur reduces particulate emissions and raising the cetane number reduces NOx[36]. Achieving these changes at an acceptable cost is a major challenge for refiners.

Fuel additives can assist in achieving at least some of the goals of reformulated diesel fuel. Cetane additives are in common use today and are quite cost effective. Oxygenates can be very effective in reducing particulates but the EPA has expressed reservations about their use due to increases in NOx. To be an acceptable fuel additive, an oxygenate must fall within accepted norms for volatility, solubility with diesel fuel, material compatibility and cost. All of the current oxygenate options present difficulties in one or more of these areas.

4.2.6.2 Compressed natural gas

Compressed natural gas (CNG) is already used in heavy-duty, spark-ignited engines, although its high octane number renders it unsuitable for direct use in compression-ignition engines. Compressed natural gas is used either with conventional spark-ignition technology or as a secondary fuel in *dual-fuel* engines. Dual-fuel engines inject a small amount of diesel fuel that ignites a mixture of natural gas and air supplied through a carburetor. While these engines can use diesel fuel if the natural gas supply is interrupted and the dual-fuel equipment can be easily retrofitted to conventional diesel engines, they are only a transitional technology for automotive use. The heavy-duty natural gas market is moving to dedicated spark-ignited engines.

4.2.6.3 Alcohols

Alcohol fuels, either methanol or ethanol, can also be used as alternative diesel fuels. Alcohols have low cetane numbers and are difficult to use directly in diesel engines without large amounts of cetane-enhancing additives. By adding cosolvents and emulsifiers, alcohol can be used as a mixture or an emulsion with diesel fuel. This approach can greatly reduce NOx and particulate emissions, but the additives greatly increase the fuel cost and the effect of the fuel on engine durability is still in question. Alcohols can also be *fumigated* or injected into the intake manifold after the turbocharger compressor and ignited with an injection of diesel fuel similar to the dual-fuel engine described earlier. This approach can also reduce NOx and particulates, but it may require a catalytic converter to control unburned hydrocarbons.

4.2.6.4 Biodiesel

Biodiesel has received increasing attention during the last several years because it is perceived to be an environmentally friendly fuel. In Europe, it is primarily produced from rapeseed oil, and in the United States, it is derived from soybean oil. Biodiesel is produced by chemically reacting an animal fat or vegetable oil with an excess of alcohol (usually methanol) in the presence of a catalyst to produce alkyl monoesters from the fatty acids present in the oil. These esters are usually referred to as biodiesel. They may be used neat or in blends with diesel fuel.

Biodiesel is nontoxic and biodegradable, and it tends to reduce soot emissions. When used in blends with diesel fuel, the nontoxicity and biodegradability advantages disappear but it still provides lower emissions. Carbon monoxide, unburned hydrocarbons and the carbon portion of particulate are generally lower with biodiesel but NOx and the soluble portion of the particulate increase.

Biodiesel cost fluctuates with the commodity price of the oilseed feedstock, which represents the primary production expense. In Europe, some countries have provided tax incentives to biodiesel that make it price competitive with diesel fuel. In the United States, these tax incentives are not available and biodiesel is four to six times more expensive than diesel fuel.

Biodiesel can be used in existing diesel engines with a minimum of changes. Some elastomers are not compatible with esters, and concerns have been expressed about interactions with the lubricating oil and the thermal and oxidative stability of biodiesel.

References

1 Universal Oil Products, 'Calculation of UOP Characterization Factor and Estimation of Molecular Weight of Petroleum Oils,' UOP Method 375-86, Universal Oil Products, Des Plaines, Illinois (1986)

2 OLIKARA, C. and BORMAN, G. L. 'A Computer Program for Calculating Properties of Equilibrium Combustion Products with Some Applications to I.C. Engines,' *Society of Automotive Engineers Paper No. 750468*, SAE, Warrendale, PA (1975)

3 GORDON, S. and McBRIDE, B. J. 'Computer Program for Calculation of Complex Chemical Equilibrium Compositions, Rocket Performance, Incident and Reflected Shocks, and Chapman-Jouguet Detonations,' *NASA SP-273* (1971)

4 GLASSMAN, I., *Combustion*, Academic Press, New York (1977)

5 HARDENBERG, H. O. and HASE, F. W. 'An Emperical Formula for Computing the Pressure Rise Delay of a Fuel from its Cetane Number and from the Relevant Parameters of Direct Injection Diesel Engines,' *Society of Automotive Engineers Paper No. 790493*, SAE, Warrendale, PA (1979)

6 CHAN, T. T. and BORMAN, G. L. 'An Experimental Study of Swirl and EGR Effects on Diesel Combustion by Use of the Dumping Method,' *Society of Automotive Engineers Paper No. 820359*, SAE, Warrendale, PA (1982)

7 DONAHUE, R. J., BORMAN, G. L. and BOWER, G. R. 'Cylinder-Averaged Histories of Nitrogen Oxide in a D.I. Diesel with Simulated Turbocharging,' *Society of Automotive Engineers Paper No. 942046*, SAE, Warrendale, PA (1994)

8 MATSUOKA, S., 'Combustion in the Diesel Engine,' *Internal Combustion Engineering: Science and Technology*, edited by J.H. Weaving, Elsevier, New York (1990)

9 HAY, N. and JONES, P. L. 'Comparison of the Various Correlations for Spray Penetration,' *Society of Automotive Engineers Paper No. 720776*, SAE, Warrendale, PA (1972)

10 HIROYASU, H., KADOTA, T. and ARAI, M., 'Supplementary Comments: Fuel Spray Characterization in Diesel Engines,' *Combustion Modeling in Reciprocating Engines*, edited by MATTAVI, J. N. and AMANN, C. A. Plenum (1980)

11 CALLAHAN, T. J., RYAN, T. W. III, DODGE, L. G. and SCHWALB, J. A. 'Effects of Fuel Properties on Diesel Spray Characteristics,' *Society of Automotive Engineers Paper No. 870533*, SAE Warrendale, PA (1987)

12 HEYWOOD, J. B., *Internal Combustion Engine Fundamentals*, McGraw-Hill, New York (1988) p. 494

13 PORTER, B. C., DOYLE, D. M., FAULKNER, S. A., LAMBERT, P., NEEDHAM, J. R., ANDERSSON, S. E., FREDHOLM, S. and FRESTAD, A., 'Engine and Catalyst Strategies for 1994,' *Society of Automotive Engineers Paper No. 910604*, SAE, Warrendale, PA (1991)

14 HERZOG, P. L., BURGLER, L., WINKLHOFER, E., ZELENKA, P. and CARTELLIERI, W., 'NOx Reduction Strategies for DI Diesel Engines,' *Society of Automotive Engineers Paper No. 920470*, SAE, Warrendale, PA (1992)

15 SHIMODA, M., SHIGEMORI, M. and TSURUOKA, S. 'Effect of Combustion Chamber Configuration on In-cylinder Air Motion and Combustion Characteristics of a D.I. Diesel Engine,' *Society of Automotive Engineers Paper No. 850070*, SAE, Warrendale, PA (1985)

16 CARTELLIERI, W. P. and WACHTER, W. F. 'Status Report on a Preliminary Survey of Strategies to Meet US-1991 HD Diesel Emission Standards without Exhaust Gas Aftertreatment,' *Society of Automotive Engineers Paper No. 870342*, SAE, Warrendale, PA (1987)

17 KAWATANI, T., MORI, K., FUKANO, I., SUGAWARA, K. and KOYAMA, T., 'Technology for Meeting the 1994 USA Exhaust Emission Regulations in Heavy-Duty Engine,' *Society of Automotive Engineers Paper No. 932654*, SAE, Warrendale, PA (1993)

18 HOFBAUER, P. and SATOR, K. 'Advanced Automotive Power Systems—Part 2: A Diesel for a Subcompact Car,' *Society of Automotive Engineers Paper No. 770113*, Warrendale, PA (1977)

19 JONES, J. H., KINGBURY, W. L., LYON, H. H., MUTTY, P. R. and THURSTON, K. W. 'Development of a 5.7 Litre V8 Automotive Diesel Engine,' *Society of Automotive Engineers Paper No. 780412*, SAE, Warrendale, PA (1978)

20 AMANO, M., SAMI, H., NAKAGAWA, S. and YOSHIZAKI, H. 'Approaches to Low Emission Levels for Light-Duty Diesel Vehicles,' *Society of Automotive Engineers Paper No. 760211*, SAE, Warrendale, PA (1976)

21 MONAGHAN, M. L., 'The High Speed Direct Injection Diesel for Passenger Cars,' *Society of Automotive Engineers Paper No. 810477*, SAE, Warrendale, PA (1981)

22 LANCASTER, D. R., KRIEGER, R. B. and LIENESCH, J. H., 'Measurement and Analysis of Engine Pressure Data,' *SAE Paper No. 750026* (1975)

23 HAMMING, R. W., *Numerical Methods for Scientists and Engineers*, 2nd edition, McGraw-Hill, New York (1973)

24 CRAVEN, P. and WAHBA, G., 'Smoothing Noisy Data with Spline Functions,' *Numer. Math.* **31**, 377–403 (1979)

25 HAMMING, R. W., *Digital Filters*, 3rd edition, Prentice-Hall, Englewood Cliffs NJ (1989)

26 KRIEGER, R. B. and BORMAN, G. L., 'The Computation of Apparent Heat Release for Internal Combustion Engines,' *ASME Paper No. 66-WA/DGP-4* (1966)

27 ROSE, J. W. and COOPER, J. R. (eds), *Technical Data on Fuel*, 7th edition, Wiley, New York (1977)

28 OBERT, E. F., *Internal Combustion Engines and Air Pollution*, Intext, New York (1973)

29 LINSTROMBERG, W. W., *Organic Chemistry*, 2nd edition, D.C. Heath, Lexington, MA (1970)

30 SCHMIDT, G. K. and FORSTER, E. J., 'Modern Refining for Today's Fuels and Lubricants,' *SAE Paper 861176* (1986)

31 SPEIGHT, J. G., *The Chemistry and Technology of Petroleum*, Marcel Dekker, New York (1980)

32 LONGWELL, J. P., 'Interface between Fuels and Combustion,' in *Fossil Fuel Combustion—A Source Book*, edited by W. Bartok and A.F. Sarofim, Wiley, New York (1991)

33 COLEY, T. R., 'Diesel Fuel Additives Influencing Flow and Storage Properties,' in *Gasoline and Diesel Fuel Additives*, edited by K. Owen, Wiley, New York (1989)

34 REDDY, S. R., 'Fuel Filter Plugging by Insoluble Sediment in Diesel Fuels, *'Distillate Fuel-Contamination, Storage and Handling, ASTM STP 1005*, edited by Chesneau H. L. and Dorris, M. M., ASTM, Philadelphia (1988) pp. 82–94

35 BENNETHUM, J. E. and WINSOR, R. E. 'Toward Improved Diesel Fuel,' *Society of Automotive Engineers Paper No. 912325*, SAE, Warrendale, PA (1991)

36 WALL, J. C. and HOEKMAN, S. K., 'Fuel Composition Effects on Heavy-Duty Diesel Particulate Emissions,' *Society of Automotive Engineers Paper No. 841364*, SAE, Warrendale, PA (1984)

5

Thermal loading

Contents

5.1 Introduction

Thermal loading is a vital factor in the design and operation of diesel engines. The proportion of the total heat rejected to the coolant reflects the level of thermal loads within the engine. The gross heat rejection to the coolant is particularly important in highly rated truck and locomotive engines, due to the difficulty of locating the cooling water radiator within the confined space of the vehicle while still allowing flow area for cooling air.

In all highly rated engines the local heat flows are very important. In certain critical areas of the cylinder head, for example in the bridge between the valves, and in the crown of the piston, the temperature gradient through the metal is high enough to result in thermal stress levels which can ultimately lead to thermal failure. Engine poppet valves can also fail due to excessive thermal stress as can the combustion chamber members of pre and swirl chamber engines. Examples of thermal failures of engine components are shown in *Figure 5.1*.

Other problems, which may be caused by high temperatures resulting from high local heat flows, occur in the cylinder liner/piston ring/piston area where breakdown of the oil film due to the resulting low oil viscosity can lead to ring or piston scuffing, or to chemical degradation of the oil itself. This may lead to ring sticking and/or packing by the products of decomposition of the lubricating oil. It is important therefore to design the engine so that there shall not be excessive metal temperatures in any area where such temperature could lead to operating problems. It is also important to limit thermal stress levels to those which will not lead to failure during the required life of the component.

5.2 Gross heat losses

The gross heat loss to the coolant normally lies in the range of 10–35% of the fuel heat, depending on the size and rating and the detail design of the engine and the type of combustion chamber employed. Heat loss data for various engine types is shown in *Figures 5.2a and b*.

In might be thought that heat losses of this order—which on a naturally aspirated indirect injection engine equal 1.1 or 1.2 times the heat equivalent of the brake power—would impose a considerable cycle efficiency penalty. In fact, however, since much of the heat loss occurs either a considerable way down the expansion stroke, during the exhaust stroke, or from the exhaust tract itself, the gain in efficiency which could be achieved by eliminating the gross heat loss, i.e. by the use of an 'Adiabatic Engine', is only of the order of about 10% of current engine efficiencies. There would of course be a very considerable resulting increase in exhaust temperature, and further gains in efficiency would result from the use of a compound cycle—with a power turbine geared into the engine output shaft—and/or by the use of a Rankine bottoming cycle working off the exhaust heat.

The resulting power plant complication is however formidable and is only worth contemplating with a relatively large engine operating at high power. Due to the surface/volume ratio effect, the larger the engine cylinder swept volume the lower the specific heat rejection. It will be seen that the quiescent direct injection chambers which have the lowest surface areas and the lowest gas velocities give the lowest losses. There is also a reduction in the proportion of heat loss with increase in the level of turbo-charging. These effects can, however, be masked or at any rate reduced by changes in detailed engine design and in particular by changes in the area of the exhaust port in contact with the cooling water.

Heat is of course also lost to the lubricating oil, as shown in *Figures 5.2a and b*.When the engine is fitted with oil cooled pistons, this is increased.

5.3 Prediction of local heat flows

For design purposes it is essential to be able to predict the local heat flows and to calculate the component temperatures. By means of an iterative process it is then possible to modify the design until the temperatures and the resulting thermal stresses are acceptable.

A. Swirl chamber light alloy piston with 'Torch' erosion of metal at the point of outflow from the combustion chamber leading to piston seizure

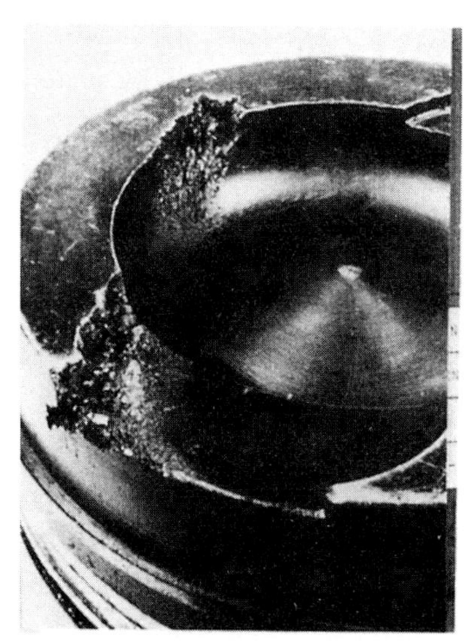

B. Direct injection light alloy piston with severe metal removal due to detonation

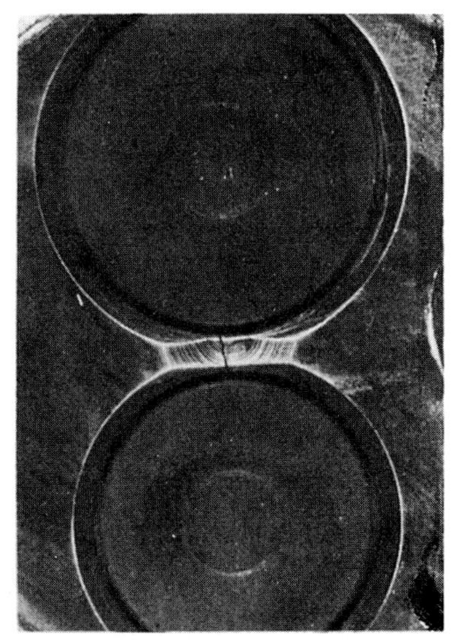

C. Cast iron cylinder head with cracked valve bridge

Figure 5.1 Examples of thermal failure of engine components

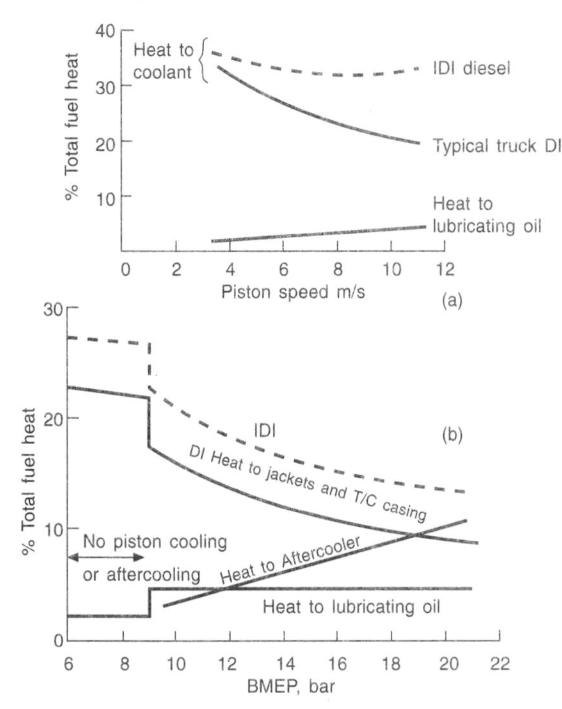

Figure 5.2 Heat to coolant and to lubricating oil: (a) naturally aspirated and lightly turbocharged engines of up to 130 mm bore; (b) effect of increase in the level of turbocharging on engines of 150–250 mm bore with piston cooling and aftercooling

The heat transfer process in an engine is however very complex. Firstly, combustion in the diesel engine is heterogeneous, and at any one time there are wide variations in gas temperatures through the charge. Secondly, radiation is an important contributor to heat flow, but there are wide variations in the radiant energy from various portions of the charge, and finally, there are considerable variations in the local velocity of the various portions of the charge. In recent years, powerful computer programs have been developed to model the complete engine cycle, including combustion. These have made it possible to make more accurate estimates of local heat flows than could be obtained in the past. The computing power required, however, is substantial and there still remains a place for empirical methods especially in the early stages of design.

In theory (the majority of heat passing by convection) if one assumes a mean effective gas temperature concept, the surface temperatures of the components lining the combustion chamber should have an important influence on the local heat flows. In practice the effect is small, and the errors involved in assuming a heat flow independent of metal temperature are found to be small.

The reasons for these effects being small are believed to be twofold. Firstly, radiation plays a part in heat transfer and for this, with a fourth power law, the metal temperature plays an insignificant part. Secondly, as first suggested by Alcock[1], the convective heat is transferred from pockets of flame at temperatures much above the mean gas temperature, the variations in heat flow being due to variations in the contact time of the pockets of flame with the wall rather than from variations in its temperature.

The empirical formulate which have been used for predicting heat flows are of one of two forms. They either calculate a time mean heat flow or, following the availability of the high speed digital computer, they calculate and sum the heat flows at discrete intervals over the engine cycle. Such calculations can of course

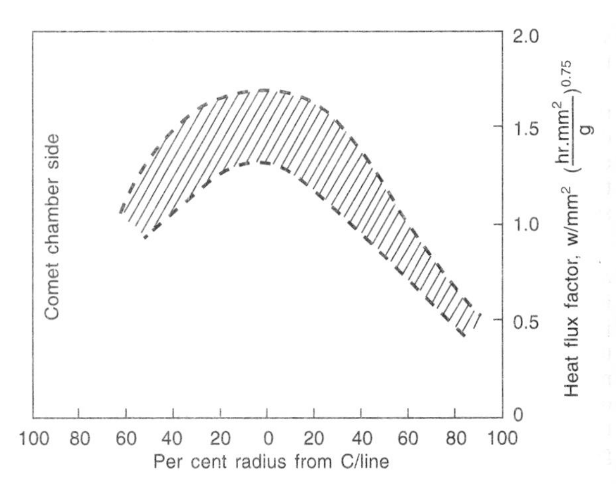

Figure 5.3 Heat flux factor band for indirect injection engine cylinder heads

be carried out as a part of engine performance cycle calculations, although in this case it is more convenient to calculate a mean heat flow rather than the local values. The original studies in this area were those by Nusselt[2] and Eichelberg[3] whose formulae provided the basis for later work, but formulae by Annand[4] and by Woschni[5] are now more generally used.

For simplicity, the *heat flux factor* approach which was devised by Alcock has much to commend it. Values of the factor for a number of engine types together with details of the method of application are given in Reference 6. As an example, *Figure 5.3* shows (in a slightly revised form to that given in Reference 6) the distribution of flux factor, defined as:

$$q = F\left(\frac{G}{A_p}\right)^{0.75}\left(\frac{P_a}{P_b}\right)^{0.3}\cdot\frac{T_b}{T_a}\ \text{W/mm}^2 \qquad (5.1)$$

where
q = Local heat flux (W/mm^2);
F = Heat flux factor (W/mm^2);
G = Gross fuel flow to engine (g/h);
A_P = Total piston area for engine (mm^2);
P_a = Ambient pressure (bar);
P_b = Boost air pressure (bar);
T_b = Boost air temperature (K);
T_a = Ambient temperature (K).

This is for the cylinder head of indirect injection engines. The large variation over the area of the head is very clear. The more powerful computer methods of advanced cycle simulation which are now available overcome the limitations inherent in the use of the calculated "mean heat flux" values[2–5]. One such engine simulation code is IRIS[7]. This code contains sophisticated physically based models for the important in-cylinder processes, including:

- In-cylinder flow model to calculate turbulence and swirl as these are produced or influenced by intake flow, chamber geometry, piston motion, injection, combustion and compression.
- Combustion models for spark ignition or diesel combustion including the effects of two-zone (burned and unburned) thermodynamics, and in addition for spark ignition, the effects of combustion chamber shape, flow velocities, turbulence and fuel properties.
- Convective heat transfer model based on local flow velocities (swirl, squish) and turbulence adjacent to interior surfaces. This model can therefore represent the effects of bowl-in-piston as well as various head geometries on the flow velocities and on combustion. The resulting thermal boundary conditions thus

vary spatially within the combustion chamber as well as in time.

• Radiation heat transfer model based on soot kinetics (generation and burnup) and calculation of radiation flux distribution along the combustion surfaces as a function of chamber shape, piston position, soot density, radiation temperature, wall absorption and reflections from surface to surface.

IRIS performs a simultaneous coupled finite element solution of all the cylinder components accounting for all of their interactions and relative motions, including the piston-ring-liner interface ensuring a full balance of all heat fluxes throughout the entire structure. Thermal conditions under transient conditions may be calculated as may the cyclic swings in the temperature of the surface layers of components. The calculation of piston-ring–cylinder friction and the effects of the resulting heat generation at the piston/cylinder interface is also included.

Such models as IRIS may be used to calculate the heat transfer coefficients and gas temperatures and their distribution over the combustion chamber surfaces. The predicted boundary conditions are thus a function of the engine type, in-cylinder geometry, in–cylinder velocities, operating conditions and of the friction generated heat.

Figures 5.4 and 5.5[17] show comparative experimental and predicted heat fluxes for a Cummins NH engine under both motoring and firing conditions demonstrating the accuracy which can be achieved. *Figure 5.5* also demonstrates quite good agreement in predicting spatial variations.

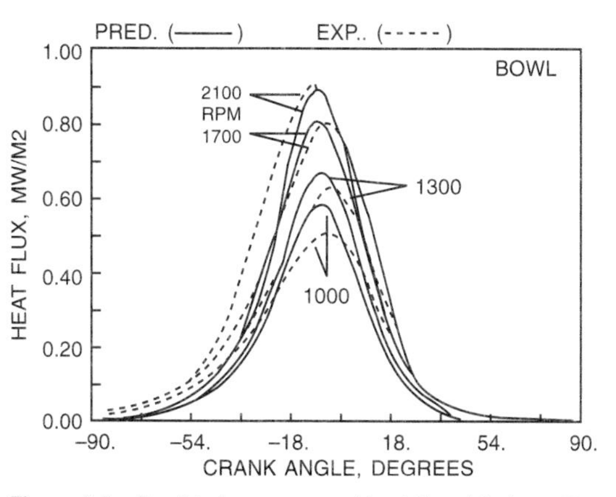

Figure 5.4 Predicted vs measured heat flux data for a Cummins single-cylinder NH engine under motoring operation at different engine speeds.

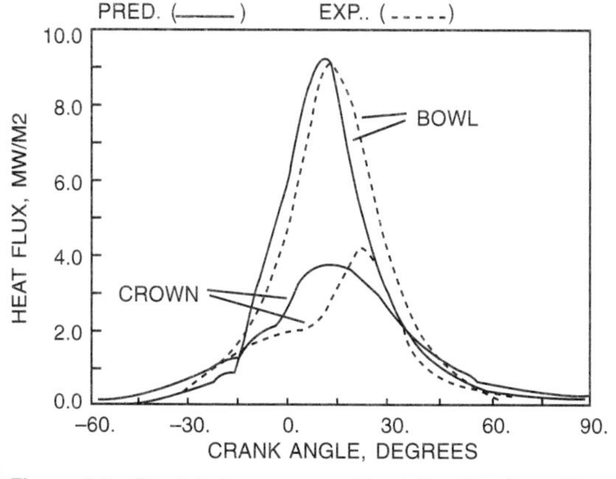

Figure 5.5 Predicted vs measured heat flux data for a Cummins single-cylinder NH engine under rated speed operation at two locations in the head.

5.4 Heat transfer at coolant side

5.4.1 Stationary surfaces—cylinder head and liner

The large majority of highly rated diesel engines employ liquid cooling. While there may be special reasons, for example in military engines, for using air cooling, the problem of cooling local regions with very high heat flows—for example the valve bridges and adjacent to the injector—make air cooling unattractive for highly rated engines.

In the critical regions of the cylinder head, where the heat flux values are high, heat transfer is by means of nucleate boiling. Here, although the bulk temperature of the coolant is below its boiling point, steam bubbles are formed at the metal/coolant interface. Heat transfer therefore involves the latent heat of the liquid, and very high rates of heat transfer can occur without the large temperature steps that would be necessary with forced convection heat transfer.

While boiling heat transfer has received considerable attention in recent years, with resultant voluminous literature, the great majority of this has involved chemically clean surfaces and chemically pure coolants. There can be no piece of heat transfer equipment where the fluids and the surfaces depart further from these ideals. In the engine, the metal surfaces are normally as-cast, cast iron with adhering sand and scale, rust and possibly other deposits from the coolant.

The coolant is water which may well be untreated and with appreciable quantities of hardness. Oil can be present as a result of gasket leaks, even if it is not added in 'soluble' form to suppress cavitation erosion; and other inhibitors, which may have a profound effect on heat transfer, may be added to avoid corrosion of engine coolant system components.

Heat transfer under engine conditions has been investigated by Ricardo in the rig described in Reference 8. In this rig the coolant is circulated past the face of a specimen which is electrically heated. The temperature gradient through the specimen is measured by a traversing thermocouple, as described later.

Investigations on this rig have revealed that a severe barrier to heat transfer can exist on the as-cast surface of cast iron, which results in a substantial increase in the operating temperature of the components. Temperature increases due to the barrier in excess of 55° C have been measured, and while the barrier can be eliminated by machining the surface, this is not normally possible in a cylinder head, and careful control of the foundry process is necessary to prevent the occurrence of the barrier. This barrier is not due to the presence of core sand inclusions in the metal, sand burnt onto the surface of the metal, or to gas voids in the metal, although clearly any of these result in an additional barrier to heat flow. To date the barrier has not been experienced in an inoculated cast iron.

Under forced convection heat transfer conditions and in the absence of a thermal barrier, heat transfer coefficients rising from about 4000 to 12 000 W/m^2. K are obtained as the water velocity rises from 0.25 to 1.0 m/sec. For boiling heat transfer, it is appropriate to define the heat transfer by means of the boiling potential, that is, the metal surface temperature minus the boiling temperature of the coolant. Typical curves for a non-barrier cast iron are shown in *Figures 5.6 a and b*. It will be seen that in the boiling regime, increasing the coolant velocity only reduces the metal temperature if the velocity is raised sufficiently high to suppress the boiling, and this is not normally feasible since the water pressure would be excessive and the pump power too high. Reductions in coolant temperature do not affect metal temperature and may indeed accentuate thermal stress problems in cylinder heads since the outer restraining areas of the cylinder

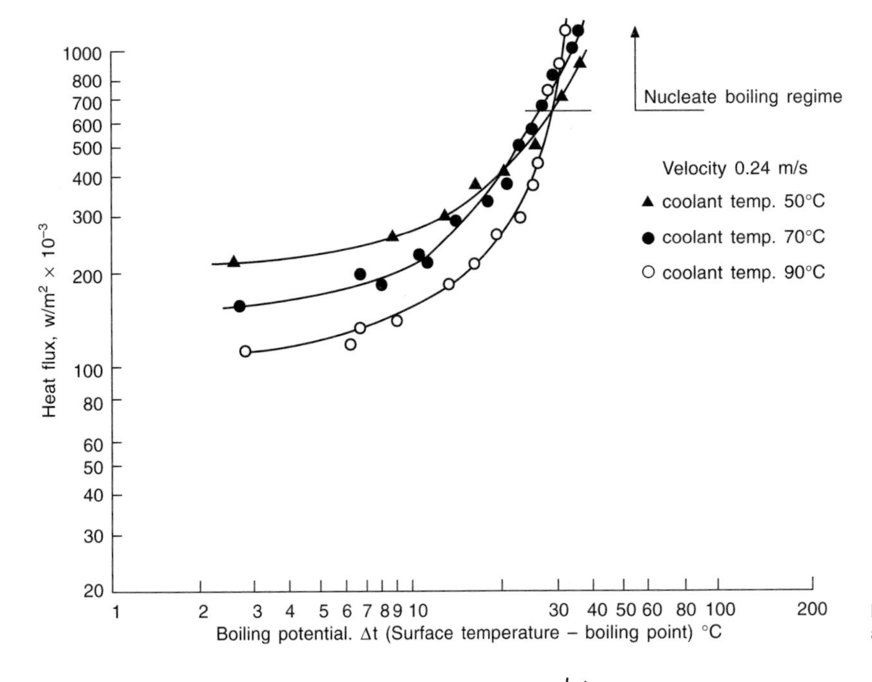

Figure 5.6 (a) Heat flow curves to water at various bulk temperatures

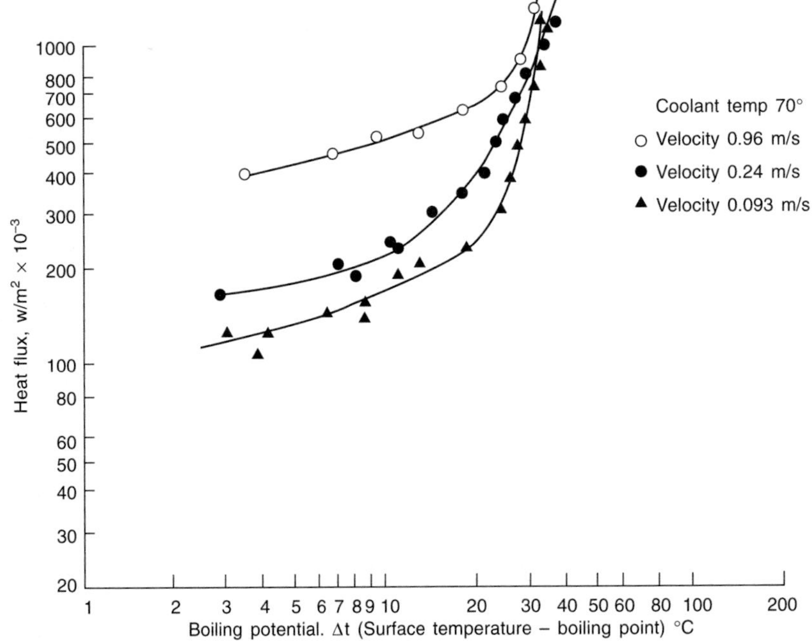

Figure 5.6 (b) Heat flow curves to water at various velocities

head deck are cooled by forced convection and will be reduced in temperature.

An increase of coolant pressure gives a direct increase in metal temperature since the boiling temperature increases as the pressure increases.

The effects of inhibitors and of other coolant additives on heat transfer have been studied on the pool boiling rig described in Reference 8. Here gas heating of the specimen is employed, and the temperature gradients and hence the heat flow are monitored by fixed microthermocouples. Typical results from this rig are given in *Figure 5.7*. It can be seen that with a 'bad' inhibitor—in this case a soluble oil—the metal temperature after a running time of 1700 h rose by 130°C above that of a similar specimen in contact with untreated distilled water. Clearly this would have led to a cylinder head failure in service.

The effects of additions of ethylene glycol as an antifreeze do not lead to a large reduction in heat transfer effectiveness. With forced convection, the heat transfer coefficients for 20% glycol/water mixtures are about 0.75 times those with water. With 100% glycol, the heat transfer coefficients are about 0.66 times those with water.

Under a nucleate boiling regime, with 25% glycol, the boiling potential is unchanged from that with water. Care should of course be taken that the inhibitors or other substances which are present in the antifreeze formulation do not adversely affect heat transfer. *Figure 5.8*[9] shows a comparison of measured heat

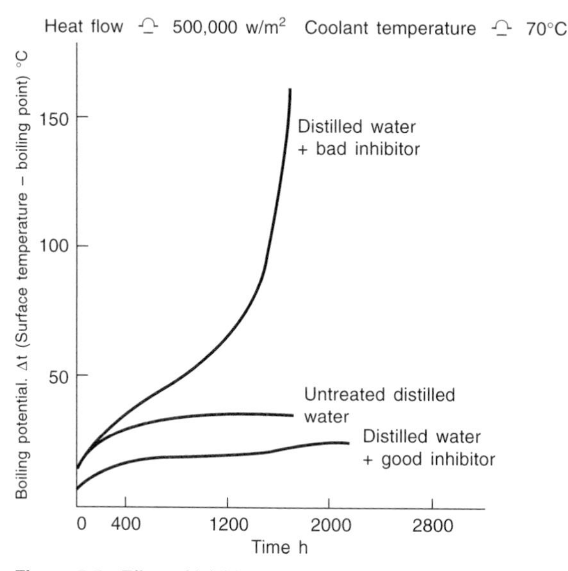

Figure 5.7 Effect of inhibitors on heat transfer

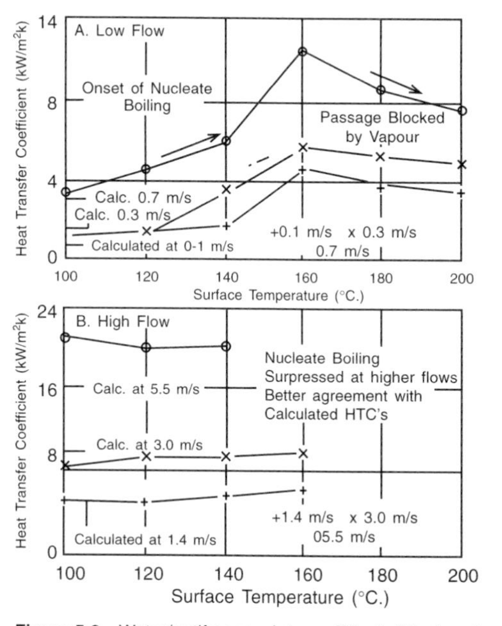

Figure 5.8 Water/antifreeze mixture—Effect of Surface temperature on heat transfer

transfer data from reference 6 and from reference 10 with results calculated using the Dittus/Boelter equation[11], plotted as heat transfer coefficients against surface temperature.

At lower flow rates (upper graph) there is reasonable agreement between the measured and predicated data at low surface temperatures. As the surface temperature is increased the heat transfer coefficient is increased owing to the onset of nucleate boiling but above 160°C a reduction of heat transfer occurs as a result of the presence of excessive vapour. This is not true film boiling, but is due to the accumulation of vapour bubbles in the narrow coolant passage. Similar phenomena can occur in the drilled coolant passages of highly rated engines (particularly IDIs). If the coolant velocity is low, the collapse of the bubbles can lead to cavitation erosion.

At higher flow rates, nucleate boiling is suppressed. For flows over the range 1–3 m/s (which is the range found in most thermally critical areas), agreement between measurement and prediction is excellent.

Coolant flow distribution has a major effect on the engine operating temperature distribution. For engine designs with no organized cooling within the cylinder head it is often acceptable to specify coolant side heat transfer conditions based on knowledge of the coolant properties, bulk flow rate and experience.

For engines with complex flow paths, or where accurate temperature prediction is essential (for example when calculating marginal safety factors), it is necessary to obtain detailed information regarding the coolant-side heat transfer distribution. This distribution is a function of the local flow velocity, the local turbulence level, the metal surface and the metal surface temperature.

The heat transfer coefficient is a weak function of coolant velocity for the conditions experienced in most regions of most engines. This makes it possible to use approximate methods for the specification of coolant-side conditions for many analyses. At high heat flows, nucleate or even film boiling may occur, the former of course beneficial, the latter disastrous.

Overall, the coolant-side heat transfer may be nonlinear and form part of a coupled system comprising the engine structure plus coolant. While the solution for the complete comprehensive system would have to incorporate all these features, this is neither practicable nor beneficial as the potential advantages are more than outweighed by the problems, due to a lack of fundamental data to describe the basic heat transfer phenomena.

In recent years however, it has become possible to calculate the flow within the coolant passages as a stand-alone analysis through solution of the 3-dimensional incompressible form of the Navier–Stokes equation using computational fluid dynamics. While the geometry of the coolant passage is invariably complex, modern CFD programs, such as VECTIS[12] have the ability to automatically generate the computational meshes from definitions of the engine components. It is now therefore economic to employ CFD to optimize the design of engine cooling systems, to eliminate spots of stagnant coolant flow or for example to minimize coolant volume spaces to ensure quicker engine warm-up.

While such flow network models allow an outline analysis which is adequate if backed up by experimental development, it does not properly allow for flow dynamics effects. Often the best approach is one based on experience of heat transfer achieved in similar designs. The presence of a good database of past measurements, as in so many areas of engine thermal analysis, is essential!

A Significant part of the thermal load applied to the cylinder head originates in the exhaust ports. Thus there is a need to prescribe thermal boundary conditions on the gas side of intake and exhaust ports. The flow in the ports is highly unsteady with velocities ranging from essentially stagnant conditions to very high instantaneous values. To calculate the resonances and dynamics involved, the equations governing the compressible flow in the inlet and exhaust tracks must be solved. One-dimensional fluid dynamics methods have been widely used to solve such problems and one example is by the use of the Ricardo WAVE computer code which calculates both the flows and the heat flows these produce.

A highly promising tool for calculation of thermal boundary conditions is computational fluid dynamics (CFD). It provides a full field solution of the equations governing the fluid flow, turbulence and heat transfer, and thus in principle can provide fine resolution of heat fluxes over the combustion chamber surfaces. An example of a typical in-cylinder CFD grid is shown in *Figure 5.9* and the results of calculations of air mass fraction and fuel mass fraction in *Figure 5.10*

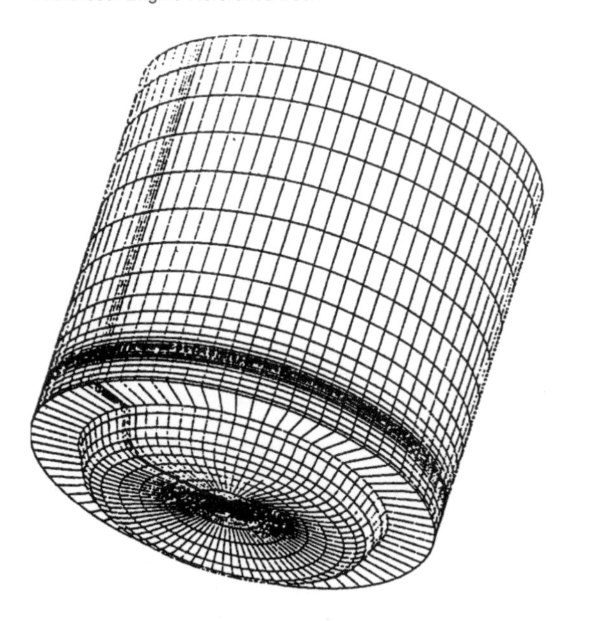

Figure 5.9 Three-dimensional CFD grid of the combustion chamber of a two-stroke diesel engine converted to natural gas operation

At this time however, further refinement of the tool is required as it needs the availability of improved combustion models. These are particularly important since much of the heat transfer occurs during the combustion process and this limits the use of CFD for calculation of gas-side thermal boundary conditions at this time.

5.4.2 Moving components—pistons

While in the cooling of cylinder liners and cylinder heads the relative motion between the coolant and the metal surface is imposed solely by the coolant pump, in piston cooling there may be additional relative velocities due to the acceleration of the piston. Indeed, with a 'cocktail shaker' type of piston, these velocities are the dominant ones.

Investigations into piston cooling have to be carried out either on an engine or on a rig with reciprocating motion of the piston. Ricardo's rig studies have been carried out on the rig described in Reference 13. The piston is electrically heated and the leads from the piston thermocouples are carried away along a swinging link. Coolant is fed to and from the test piston by means of articulated pipes. The rig can operate at speeds of up 8.5 rev/s, but for higher speeds, tests on an actual engine have been carried out, as described in the same Reference.

Very much higher heat transfer rates are obtained with water cooling than with oil cooling. There would therefore be much to be said for employing water colling of the pistons in highly rated engines. Unfortunaely the problems of obtaining a completely liquid-tight system are such that it is not possible to do this in a trunk type piston engine since water leaks into the lubricating oil are inadmissible.

The large slow speed main propulsion marine diesel engines however are of the crosshead type, and they incorporate a diaphragm seal at the crosshead to prevent the lubricating oil from the cylinder bores and the blow-by materials from contaminating the crankcase lubricant. Under these circumstances water cooling of the piston is feasible and is in fact employed on a number of engines. An increase in heat transfer coefficient of 5–10 times is obtained on low speed engines by the use of water cooling.

For trunk piston engines, oil cooling of the piston has to be used. Two basic types of cooling are employed, although the actual design of piston may incorporate both types.

(i) *'Solid' ms' flow systems* In this type the cooling passages run full of oil at all times and there is no cocktail shaking. It is found that the heat transfer coefficient is unaffected by the speed of rotation of the engine but is a function of the normal fluid parameters which affect heat transfer under pipe flow conditions. Under typical engine conditions it is found that the heat transfer coefficient obtained is about 1150 W/m^2 K.

(ii) *'Cocktail shaking'* This is strongly influenced by the engine speed but hardly at all by flow rate. Experimental work carried out by Bush[14] has shown that it is rotational speed rather than piston speed or velocity that is the important parameter, and the author and his colleagues have obtained good correlations using similar non-dimensional functions to those developed by Bush but with somewhat different indices.

The general equation under cocktail shaking conditions for the rig described in Reference 13 with both oil and water coolants is:

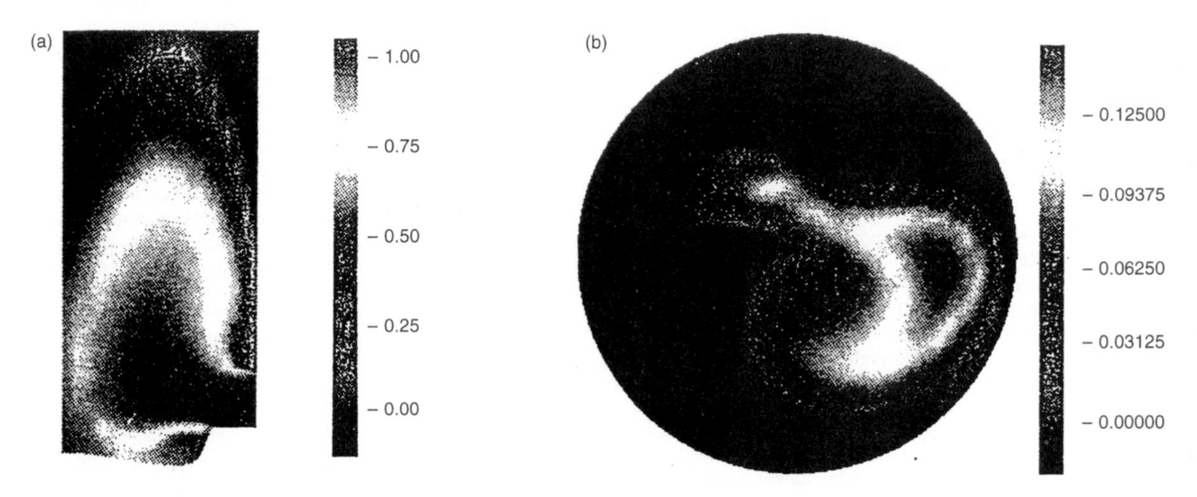

Figure 5.10 Results of CFD analysis of a two-stroke diesel engine converted to natural gas operation. Contour plots of (a) air mass fraction during scavenging in an axial-radical plane cutting across an intake port at BDC, and (b) fuel mass fraction at 240°ATDC (5° after end of injection) in a circumferential-radial plane just above the piston crown.

$$[N_{nu}] \propto [N_{Rb}]^{0.54} \times [N_{pr}]^{0.33} \times \left[\frac{\eta_b}{\eta_s}\right]^{0.14} [D^*]^{0.33} \qquad (5.2)$$

where N_{nu} = Nusselt number = UD/Z;
N_{Rb} = Reynolds number = $(26/t)\ D/\nu$;
N_{pr} = Prandtl number = $\eta C_p/Z$;
b = Piston cavity length (mm);
C_p = Specific heat (J/kg. K);
D = Piston cylindrical cavity diameter (mm);
$D^* = D/b$;
U = Heat transfer coefficient W/m^2 K;
t = Period of piston motion $1/N$ sec;
N = rev/s;
Z = Thermal conductivity W/m. K;
ν = Kinematic viscosity (St);
η = Absolute viscosity (P).

Subscripts
b = Bulk coolant conditions; s = Surface conditions.

From the results of experiments on the rig, supplemented by extensive engine test data, it is found that optimum values for piston heat transfer coefficients with oil cooling are as follows:

1.67 rev/sec	50	285 W/m^{2}K
5.0 rev/sec	150	860 W/m^{2}K
33.3 rev/sec	500	2850 W/m^{2}K

It can be seen therefore that with large slow speed engines, solid flow systems give more effective cooling than do cocktail shaking systems. In the range 5–6 rev/s there is little to choose between the two types, but for high speed engines there is a clear advantage in favour of cocktail shaking.

5.4.3 Establishing temperature maps

Having established the distribution of local heat flux for a particular design of combustion chamber and knowing the rating of the engine and hence the amplitude of the heat fluxes, the coolant side heat transfer coefficients and the metal thicknesses and thermal conductivity, it is possible to predict the temperature distribution in the engine components surrounding the combustion chamber.

This one-dimensional procedure is straightforward where there is an essentially linear temperature gradient through the thickness of the material as applies for example to the cylinder liner and the centre sections of the crown of an oil-cooled piston. In the case of valve bridges, the heatflux from the flame plate is augmented by heat flow from the valve seats and heat transfer to the coolant is increased by the presence of the port walls and corrections must be made to allow for this.

Two-dimensional and axisymmetric FE modelling can give a good compromise between time and accuracy when applied at an intermediate stage in the design process in component areas where the thermal conditions are more complex as for example in the outer areas of the piston crown and complex areas of the cylinder head.

For the final definitive design of the thermally loaded components of any highly rated engine, however, it would be recommended practice to carry out a full 3D FE assessment. The Ricardo approach, in so far as the choice of analytical design techniques are concerned, for various stages of the thermal analysis is summarized in *Table 5.1.*

5.5 Thermal stress

5.5.1 Thermal stress failures

Failures due to excessive thermal stress are one of the common causes of engine breakdown under highly rated conditions. Under extreme conditions, as with burning of exhaust valves following exhaust gas blow-by, or burning of piston crowns or cylinder heads by contact with outflow jets from a pepper-pot prechamber type of combustion chamber, or yet again as a result of 'detonation' when a diesel engine is operated on gasoline, metal may actually be removed in appreciable quantities.

More commonly however, a fatigue crack is formed and this propagates through in time to the coolant space, allowing gas and oil to blow into the coolant under running conditions and/or water to leak into the cylinder when the engine is shut down. The crack normally develops following yield of the metal as described by Fitzgeorge and Pop[15] due to an excessive compressive thermal stress, and under load cycling conditions of operation a tensile fatigue failure occurs.

In low speed engines—c. 2 rev/s—the cyclic fluctuation of temperature at the gas face is of the order of 40°C, which will give an additional cyclic stress at the surface of 74 MPa which, while it is not a dominant consideration, will be a substantial addition to the fatigue conditions.

The cyclic temperature swing is, however, an inverse function of the square root of the operating speed, and hence the effect is much smaller in medium and high speed engines where it is sufficient to consider the load cycle from low load up to full load. The importance of load cycling is clear from the common experience with engines in service or on the test bed where a satisfactory life is obtained under continuous high load operation, but where failure occurs quite rapidly under cycling conditions of operations.

Since the number of these cycles during the life of an engine may be relatively small, for example in ocean going ship service, as shown by Alcock[16], the total life of the engine component such as a liner will only entail operation through 500–1000 such cycles, we are working in the range of low cycle fatigue where quite a high stress may be allowable without failure occurring in service. In city delivery service or in taxi service, the number of cycles will be much larger.

Table 5.1 Thermal analysis—The Ricardo approach

Design process	*Model type*	*Heat input*	*Coolant flow*	*Coolant heat transfer*
Concept study	1-D calculations	Empirical heatflow formula based on database	Experience	Experience
	2-D or axisymmetric FE models			
Design layout	3-D flameface model		Pressure drop calculations	Empirical heat transfer formulae
Definitive design	Full 3-D model	Spatially resolved simulation (IRIS)	Computational fluid dynamics (VECTIS)	

5.5.2 Materials

In considering materials for operation under conditions of thermal fatigue, it is important to remember that in the cylinder head, liner and piston crown one is not working between fixed temperature limits but with approximately fixed local heat fluxes, as explained earlier. A reduction in thermal conductivity therefore increases the temperature limits and increases the peak thermal strain. It was a failure to appreciate this that led to extensive trials, with consequent severe failure rates in many cases, of nodular cast iron for cylinder heads. While nodular cast iron can give an improved life under conditions where the mechanical loads predominate, it is inferior where thermal loads predominate, due to its low thermal conductivity, to flake graphite grey cast iron.

The most useful criterion for comparing materials for operation under thermal conditions is the Eichelberg[17] quality factor:

$$\text{Quality factor} = Z \times \frac{\text{Ultimate tensile stress}}{\alpha E} \quad (5.3)$$

where Z = Thermal conductivity (W/m. K);
Ultimae tensile stress (MPa);
α = Coefficient of linear expansion 1/K;
E = Young's modulus (N/m^2).

Figure 5.11 shows clearly that the relative rating of materials on this basis contrasts sharply with the simple basis of strength versus temperature shown in *Figure 5.12*

For components such as 'hot plugs' in a Comet combustion chamber engine, the pepper-pot in other IDI chambers, or for exhaust valves, where the gas side heat transfer coefficients are high and hence the components closely approach the mean effective gas temperature, the temperature of the component is less affected by its thermal conductivity and also, due to its shape and construction, there is less restraint from cooler areas. Under these circumstances the thermal conductivity of the metal is much less important and an adequate hot strength is all-important.

5.5.3 Strongbacked constructions

The most important method of reducing thermal stress is to reduce the gas side temperature. *Figure 5.13* shows the factors which control this in the valve bridge area of a cylinder head. It is clear that the most important factor is the temperature drop through the metal. In the absence of a thermal barrier at the metal/coolant interface, little can be done to improve the

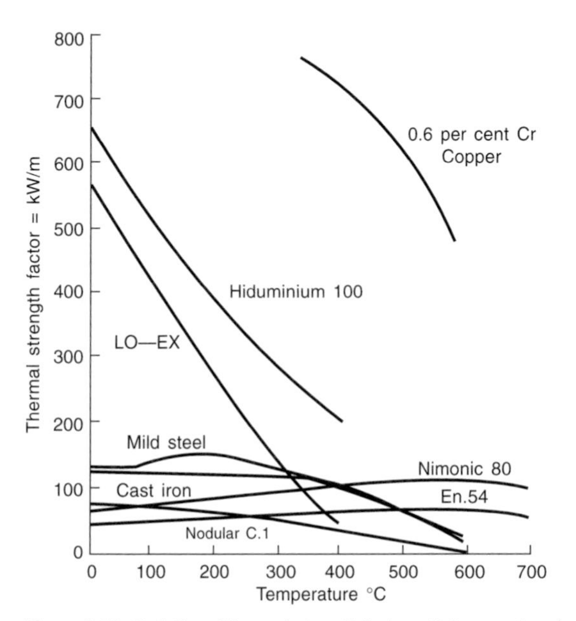

Figure 5.11 Variation of thermal strength factor with temperature for various metals

Thermal strength factor

$$= \frac{\text{thermal conductivity} \times \text{ultimate tensile stress}}{\text{coefficient of linear expansion} \times \text{Young's modulus}} \text{ W/m}$$

effectiveness of the cooling since nucleate boiling exists and, as already shown, increases of coolant velocity or reductions in coolant temperature are ineffective.

We are then faced with a need to reduce metal thickness. While this may be possible in the smaller engines where pressure stress tends to be insignificant, it is not possible in larger engines where the pressure stresses form a significnat part of the total stress level. The obvious answer is to employ a strongback where a well-cooled, relatively heavy member carries the gas loads while the thermal loads are carried by a thin member with a large number of supports transmitting the pressure loads to the strongback. Examples of this form of construction are common in the large marine diesel engines as exemplified by Doxford cylinder liners, MAN cylinder heads and Fiat pistons.

A method for calculating the optimum strongbacking of a 'hoop supported' member, for example a cylinder liner, has been given by Alcock[16] who has demonstrated that in this case it is possible to give excessive support with a resulting tensile thermal fatigue failure at the bore surface, although clearly with optimum proportions a strongback gives considerable advantages.

5.5.4 Calculation of thermal stress

While the simplified method mentioned above is applicable to strongbacked liners, the calculation of thermal stress in other engine components is difficult. The simplest case is where the component and its temperature distribution are axisymmetric. One of he earliest procedures employed for this case was that, due to Fitzgeorge and Pope[15], for a piston which was considered as a disc attached to the top of the cylinder.

More recently, as has been mentioned in the section on temperature prediction, with the development of finite element programs for digital computers, much greater accuracy can be obtained for this axisymmetric case, and indeed provided there is axial symmetry, quite complex shapes can be accommodated. Where the component and/or its temperature distribution depart radically from an axisymmetric case, a fully three-dimensional solution becomes necessary.

Indeed for the final definitive design of engine components a full three-dimensional finite element thermo-mechanical analysis can usually be justified since a number of pre and post-processing packages are now available which greatly shorten the once prohibitively lengthy and expensive calculations.

A flow chart of the procedure from reference 7 is given in *Figure 5.14*. The geometry definition may be through engineering drawings, 2D or 3D CAD files or via a solid modeller. Commercial packages for mesh generation are now available. The method of calculating the thermal boundary conditions is as described earlier. These conditions, together with separately calculated mechanical boundary conditions and modal constraints are then applied to the finite element model and the analysis is performed using a commercially available proprietary FE code. This analysis may be linear or nonlinear, static or dynamic, steady state or transient in nature.

Following analysis the results are post-processed into a form suitable for engineering assessment. The software can additionally be employed to assess the results against criteria of acceptability developed by experience. It is, for example, possible to obtain a direct plot of fatigue safety factors taking account of the minimum and maximum stress levels and the local material properties at each nodal point on the model. The results may be presented in a number of different ways, e.g. colour contour plots, deformed geometry, vectors, animations or graphs. Examples of predicted against measured temperatures for the cylinder head of a locomotive diesel engine are shown in *Figure 5.15*. *Figure 5.16* shows an example of calculated high cycle fatigue factors for a piston of a gasoline engine.

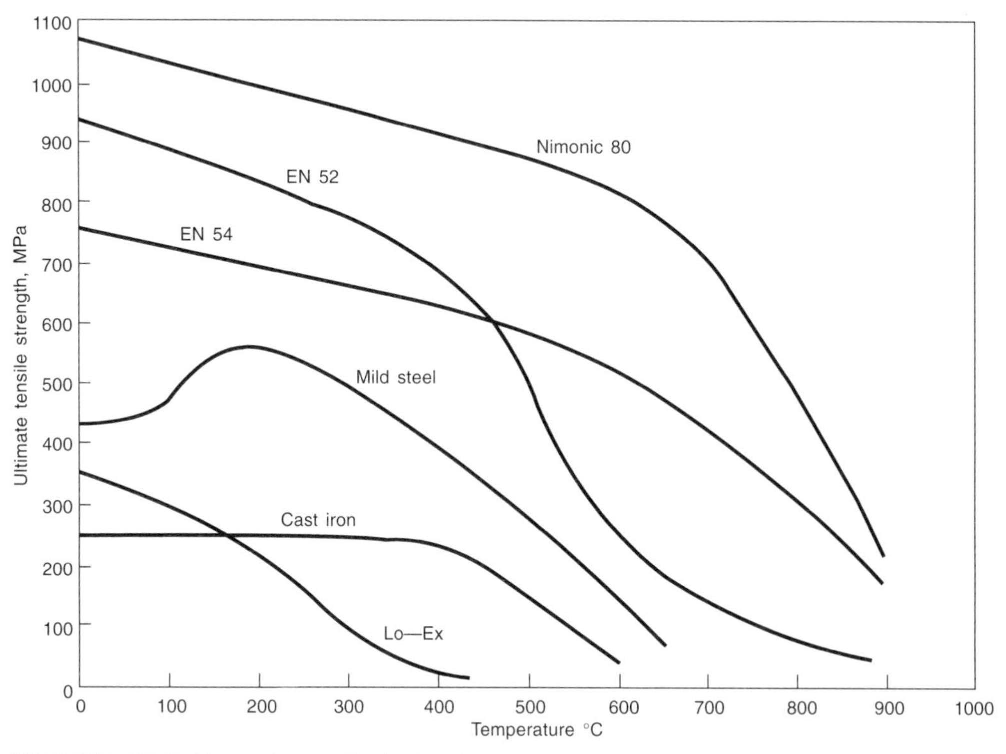

Figure 5.12 Effect of temperature on ultimate strength

5.6 Limiting conditions in operation

5.6.1 To meet lubrication requirements

While piston ring scuffing may also be a function of peak cycle pressure, piston speed lubricant composition, and liner surface finish, there are strong reasons for holding the gas side liner temperature at the upper limit of top ring travel to below 180°C

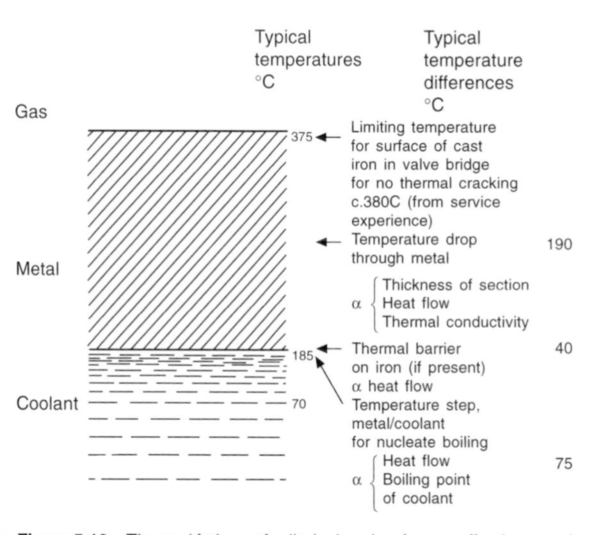

Figure 5.13 Thermal fatigue of cylinder heads—factors affecting metal temperatures

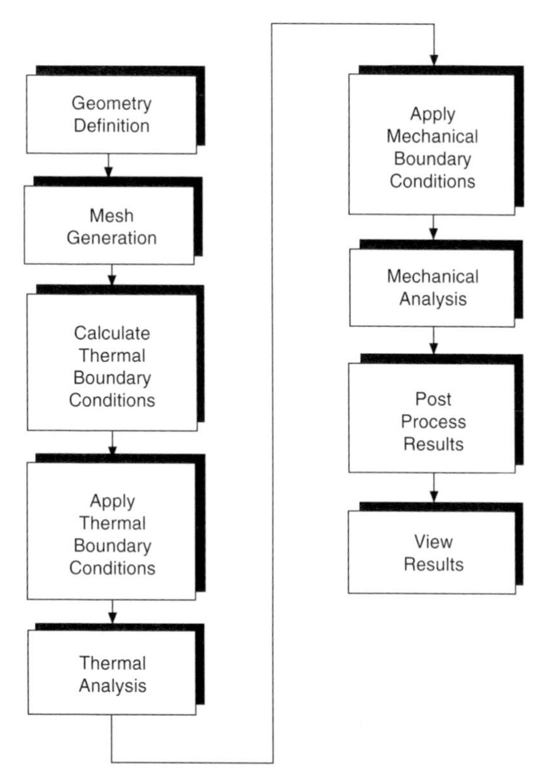

Figure 5.14 Flow chart of thermo-mechanical analysis procedure.

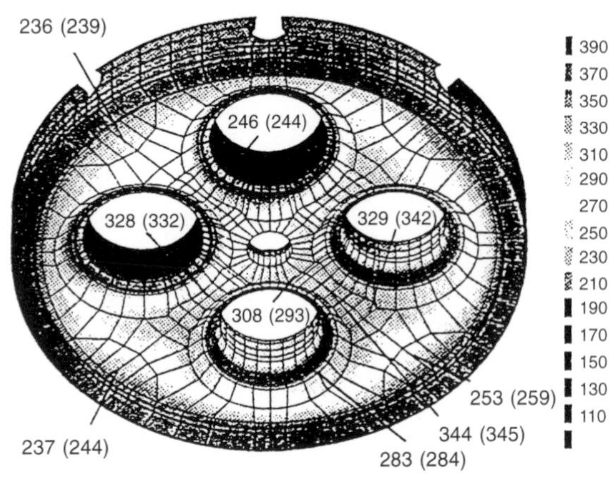

Figure 5.15 Predicted (measured) temperatures in °C or 5000hp V16 diesel locomotive engine cylinder head at 1050rmp ful load

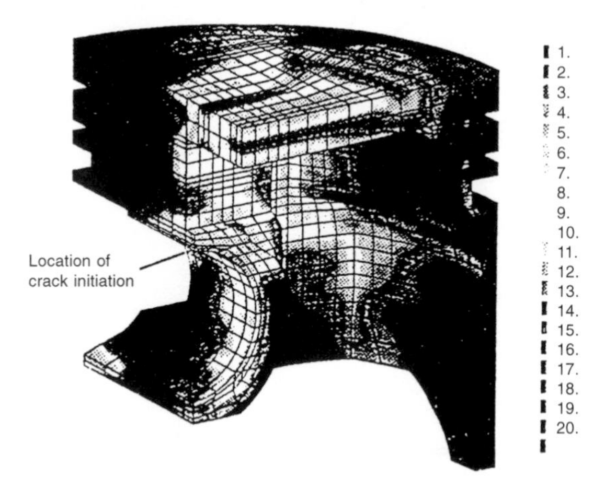

Figure 5.16 Contours of high cycle fatigue safety factor in a gasoline engine piston

if scuffing is to be avoided. In fact, if possible, the author prefers to design to a figure of 160°C, but this may not be possible when a high rating is required.

Top piston ring groove temperatures are more related to ring sticking, but packing behind the ring can also be a major factor contributing to ring scuffing and to high rates of ring wear. The target peak temperatures which the author uses for design and development purposes of automotive size engines are a function of oil type, as set out in *Table 5.2*. Where a keystone type of top ring is fitted, with taper sides, about 20°C may be added to each of these figures. Additional information may be found in Chapter 16.

Table 5.2

Oil	*Duty*	
	Intermittent °C	*Continuous °C*
HD	220	200
Supplement I	230 (Series I)	210
Supplement II	250 (Series II)	220
Series III	260	230

5.6.2 For thermal strength

While, ideally, a full thermal stress analysis would be carried out, it is found that, in practice, relative freedom from thermal problems will be experienced if certain empirical rules are followed. In particular, with conventional cast iron cylinder heads manufactured from good quality unalloyed flake graphite grey cast iron, over a wide range of engine sizes, the gas side temperature of the valve bridge material should be kept below 400°C and preferably below 380°C. With temperatures much above 400°C cracking in service under cycled load conditions will be severe and at even higher temperatures will be catastrophic.

It should be emphasized that these temperature limits may not apply to engines fitted with integral cylinder heads, which will have less restraint due to the absence of the cold outer areas of the cylinder head deck. Here, as with cast iron pistons, somewhat higher temperatures may be acceptable. For freedom from thermal cracking, a peak temperature for the crowns of aluminium alloy pistons of 380°C has been quoted by Law[18] but the author would prefer to see a value of 30°C lower than this.

5.6.3 Fuel injector

Blockage of the spray holes in the fuel injection nozzles is another problem which has its origin in excessive temperature, leading in this case to a partial thermal degradation of the fuel and to lacquer or to carbon formation.

From development experience the following are critical temperatures for hole type nozzles:

1. Maximum metal temperature at nozzle tip—280°C.
2. Corresponding maximum metal temperature at nozzle seat—230°C.
3. Smallest hole recommended for freedom from blocking by lacquer formation—0.25 mm.
4. Detailed information on the measurement of nozzle temperatures, together with some typical values, is given in Reference 19.

5.7 Designing to meet thermal requirements

5.7.1 Cylinder head

5.7.1.1 Materials

Flake graphite cast iron or aluminium alloy are by far the most commonly used for the cylinder head, although steel is also employed, as is nodular cast iron and bronze. If it were not due to difficulties in casting, steel would probably be more widely used since it does offer possibilities of a better fatigue life under engine conditions. With steel it is essential that an inhibitor is added to the coolant since the oxide scale formed gives a serious barrier to heat transfer.

In the author's experience, an improvement in life under thermal stress conditions is found with a flake graphite iron up to an ultimate sress level of 260 MPa and possibly 310 MPa, but above this figure the necessary alloying elements give a progressive reduction in thermal conductivity coupled with an increase in the modulus of elasticity which more than counteracts any gain in strength. For reasons given earlier, an inoculated iron is preferable since it reduces the possibility of a thermal barrier being present at the metal/coolant interface.

The author does not believe that a liquid cooled cylinder head in aluminium alloy can be operated at a higher rating than a well designed liquid cooled one in cast iron. The overall weight with aluminium alloy would, of course, however, be less.

5.7.1.2 Drilled passages

The critical region in a cylinder head is normally the bridge between the valves and, in an indirect injection engine, the bridges between the valves and the combustion chamber. With the thin sections of the cores there are limits to how close one can bring the coolant to the gas face, and the possibility always exists that a damaged core may lead to a faulty casting and hence a poor cooling in this region.

For this reason it is preferable to drill a cooling passage through the bridge in engines for very high ratings. A typical example is shown in the cylinder for a research engine shown in *Figure 5.17*. This engine has operated successfully at 22 bar b.m.e.p., which leads to very high thermal loadings in an indirect injection engine. Comparative temperatures in the valve bridge between this head and an earlier one without the drilling passages at a somewhat lower rating are shown in *Figure 5.18*. Curve 1 shows the temperature gradient in the original cylinder head. The dashed curve shows the temperature gradient which would have existed without a thermal barrier. Curve 2 shows the gradient for the original head thinned by machining away 1 mm of metal. Curve 3 shows the temperature gradient in the final head with a drilled passage.

It can be seen that in this case there was a severe thermal barrier that has been eliminated by the drilling. Furthermore, an additional reduction in temperature of 100°C has been made since it was possible to bring the drilling closer to the gas face than was possible with the cored water passage.

Modern Sulzer large slow speed engines employ cylinder heads which are cooled by closely spaced drilled passages rather than by cored cooling passages. In this way very good cooling close to the hot side metal surfaces is obtained, as shown in *Figures 5.19 a and b.*

5.7.1.3 Flame plate

Probably the ultimate in effectiveness of cylinder head cooling is achieved in the construction patented by Ruston and Hornsby

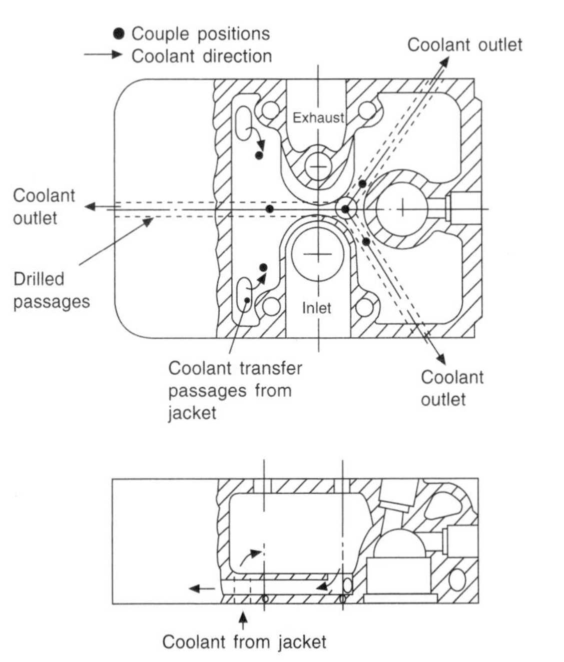

Figure 5.17 Arrangement of drilled coolant passages and traversing thermocouple positions in Comet Mark V swirl chamber diesel engine

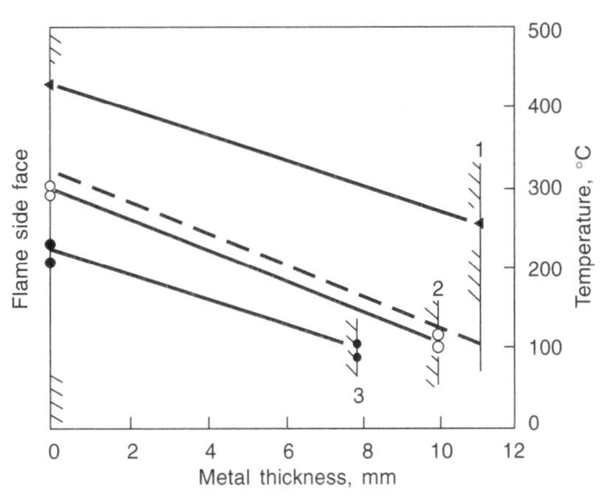

Figure 5.18 Effect of thermal barrier on metal temperature

which was used in the A0 two-cycle engines. *Figure 5.20,* taken from British Patent Specification No. 977678, shows that a thick steel plate is attached to the gas face of the main cylinder

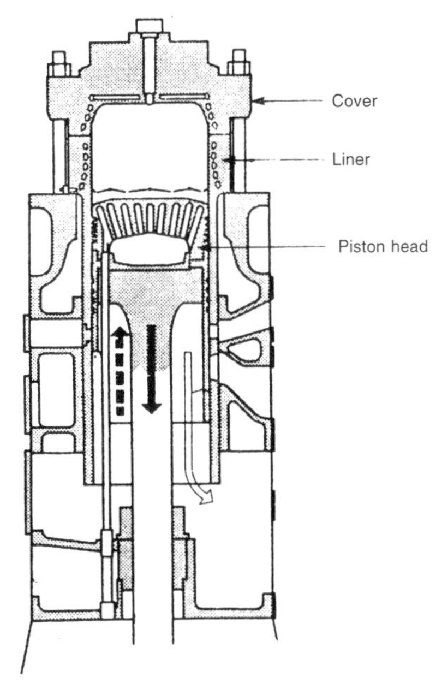

Figure 5.19 (a) Sulzer bore cooled engines-RLA 90

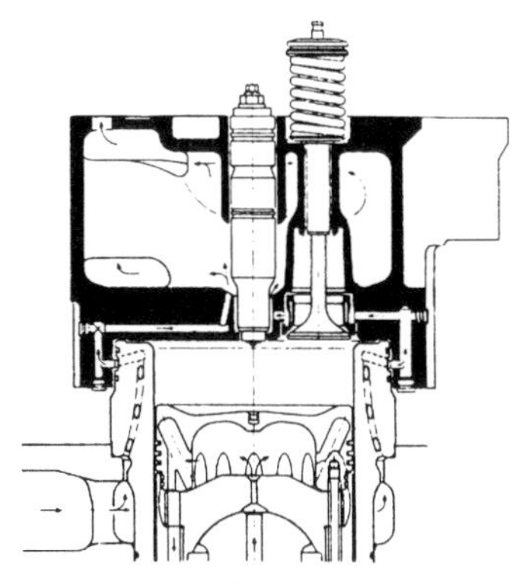

Figure 5.19 (b) Sulzer bore cooled engines-ZA 40

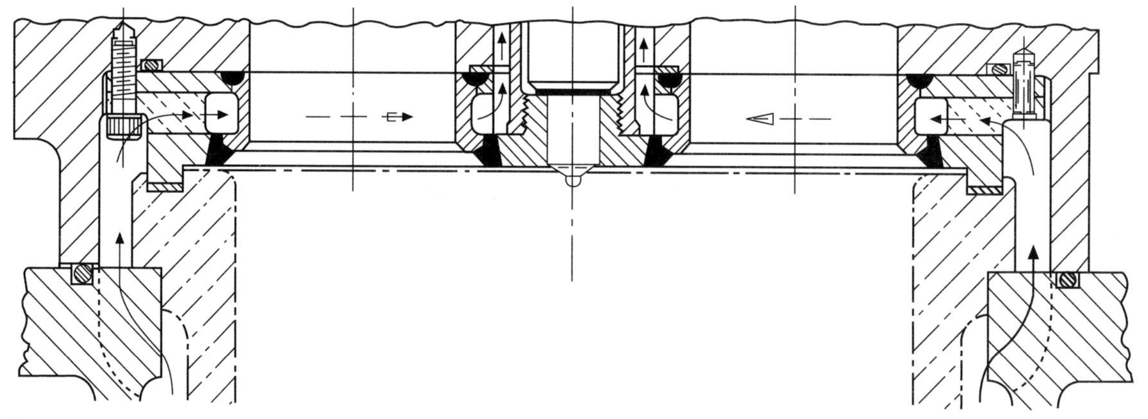

Figure 5.20 Ruston 'Flame plate' cylinder head

head and cooling passages are machined in the plate, giving direct cooling of the valve seat inserts. The cylinder head casting proper is thus free from thermal loading, apart from that arising from the heat flows in the walls of the exhaust port.

5.7.2 Cylinder liner

The use of a separate steel strongback to improve the strength of the liner has already been described, but there is a variant which may be easier to employ in the case of smaller engines. This is the use of drilled cooling passages in the liner or in a very deep liner flange. These drillings may be axial, as in the General Electric locomotive engines, or may be angled somewhat with radial outflow drillings to allow for drilling from the lower side of the flange without the drillings breaking through the top face of the liner. Such drillings are used by Sulzer; see *Figures 5.19a and b.* In this case the liner forms its own strongback, and advantage is taken of having a relatively thick liner to resist mechanical stress and incidentally reduce the risk of cavitation erosion, while giving the effective cooling of a thin liner.

5.7.3 Piston design

The following types of cooling are possible.

5.7.3.1 The Thermoflow

This type has a deep arch undercrown shape to transmit heat to the skirt and hence to the liner. This can either be used uncooled or with jet cooling from the top of the connecting rod, or alternatively from a fixed jet attached to the crankcase.

5.7.3.2 Cocktail shaker

Cooling is by means of a chamber which is partly filled with oil. As the piston is reciprocated, high relative velocities between the cooling oil and the metal surfaces are effected. The oil level is maintained by means of a weir, and oil may be supplied either from the top of the connecting rod or from a fixed jet attached to the crankcase. The undercrown profile is commonly similar to the Thermoflow.

5.7.3.3 Pistons with internal passages in crown

These passages usually take the form of a cooling gallery behind the ring belt which may be formed by electron beam welding a separate machined ring belt member on the crown, or more commonly today by casting, using soluble cores.

5.7.3.4 Pistons with separate crowns

Low alloy steel crowns are normally used, although nickel aluminium bronze crowns were used on the Napier Deltic opposed piston two-cycle engines. The Maybach engine used an austenitic steel crown, the low conductivity being offset by very thin metal sections coupled with good mechanical support to reduce stress.

Some of the results from an extensive series of engine tests carried out in the Ricardo Laboratories have been given in Reference 13. These tests covered the first, second and fourth piston types above, which were fitted to the highly rated $4\frac{3}{4}$ inch bore Ricardo Comet combustion chamber engine mentioned earlier. In general it was found that for this size of engine:

(i) With jet cooling there is little to choose between a jet at the top of the connecting rod and a fixed jet in the crankcase. The conrod jet gives better cooling of the crown, but the fixed jet gives better cooling of the ring pack.
(ii) The cocktail shaker gives better cooling than jet cooling, and this is especially so when there is a limited supply of cooling oil.
(iii) As would be expected, the two part piston shows the most effective cooling. It is interesting to note that this occurs as a result of thinner metal sections between the hot side and the coolant side, as shown earlier. In a high speed engine the heat transfer coefficients between metal and coolant are much higher for cocktail shaking than for solid flow.
(iv) A cocktail shaker of the improved type, with the addition of a cored hole through the piston pin boss struts behind the ring pack is extremely effective for an automotive size of engine at very high ratings. For larger engines however a two part piston offers considerable advantages. Typical modern piston designs for a range of engines are shown in *Figure 5.21.*

5.7.3.5 Piston reinforcement

In pistons with very re-entrant combustion chambers the chamber lips are particularly prone to thermal cracking. This problem may be reduced by the use of ceramic fibre reinforcement of the parent aluminium alloy piston in the vicinity of the lip. Such reinforcement gives a local increase in the strength of the material.

5.7.4 Injector cooling

In engines burning distillate fuels there is normally sufficient cooling from the passage of the cold fuel to prevent the nozzle from overheating. In small engines however, unless pencil type injectors are employed, there is a tendency for the nozzle face area to be large in comparison with the quantity of fuel that is

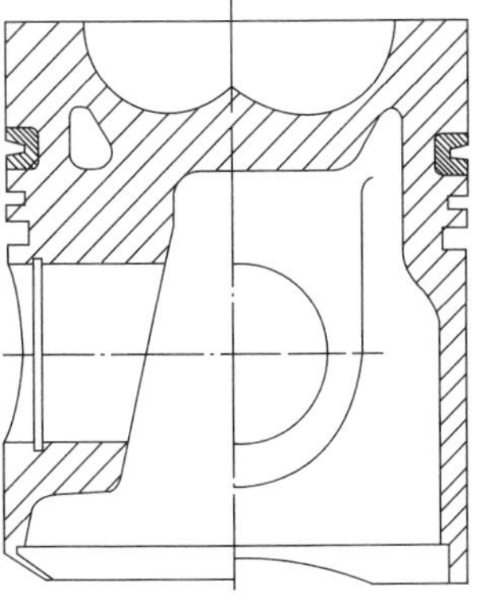

A. 'Soluble Core' oil gallery piston

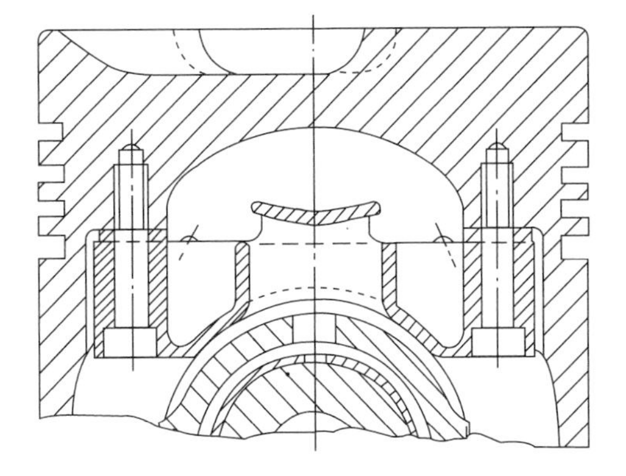

B. 'Cocktail Shaker' piston

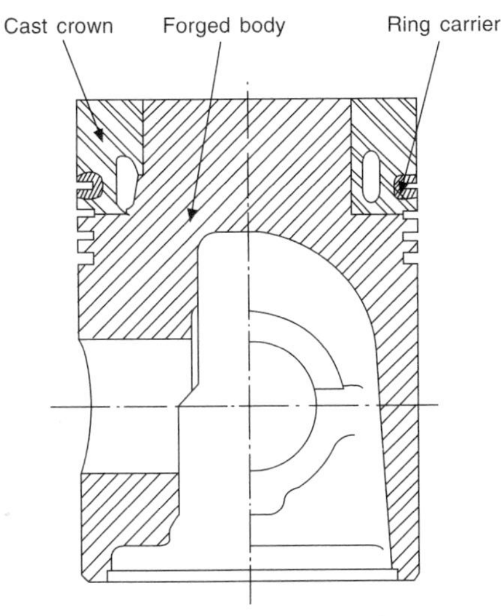

C. Electron beam welded oil gallery piston (right hand form preferred for large pistons)

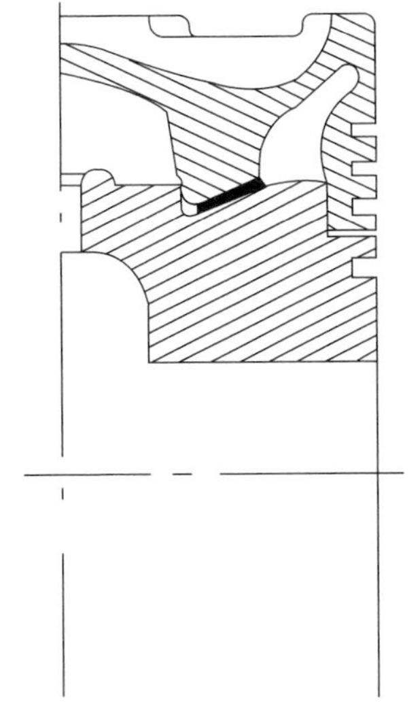

D. Steel crown two-part piston (bolt fastening)

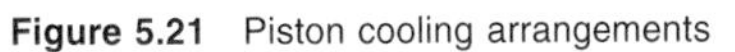

Figure 5.21 Piston cooling arrangements

pumped into the engine. As a consequence the fuel cooling is less effective, and overheating of the nozzle results.

In indirect injection automotive engines for example this can lead to blocking of the auxiliary holes in the Pintaux type of nozzle. The provision of a heat shield in the form shown in *Figure 5.22* has been found to be extremely effective in reducing the nozzle temperature to an acceptable figure.

In larger engines burning the heavier kind of residual fuel oil, the fuel is heated to a temperature of 100-120°C to enable it to be pumped. In this case there is no effective fuel cooling of the nozzle, and it is necessary to supply water cooling of the nozzle either by providing a water jacket around the outside of the nozzle or by providing cooling passages within the nozzle itself.

5.8 Measurement of local temperature gradients and heat fluxes

As a part of engine development it is desirable to establish the magnitude of the temperature gradient in the metal adjacent to the gas face. The following methods are employed for temperature measurement.

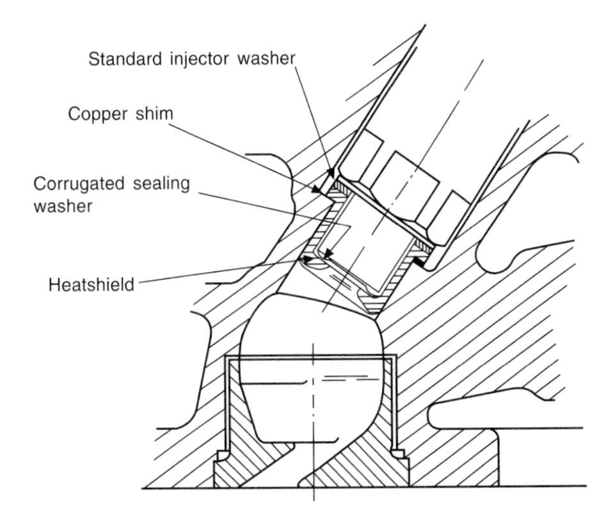

Figure 5.22 Arrangement of nozzle heat shield

5.8.1 Fixed thermocouples

These are generally easy to apply, but only a limited number can usually be fitted, normally one close to the gas face and one close to the coolant face. Errors are therefore introduced if the gradients are not linear.

5.8.2 Traversing thermocouples

As shown in *Figure 5.23,* the thermocouple button is traversed through a drilling from the coolant side of the gas face, a guard tube being fitted to ensure the coolant does not enter the drilling. The author has found these thermocouples to be accurate under all conditions, for cylinder heads and liners, but there are difficulties in installation in some locations.

5.8.3 Hardness recovery methods

As applied to complete aluminium alloy pistons, to complete exhaust valves, and as inserted plugs of material in the case of Shell Templugs which can be fitted to any component. Hardness recovery methods have many uses, particularly since a number of temperature readings may be taken at various distances from the surface at one time and hence the temperature map established from one test.

The disadvantage of using hardness recovery methods on pistons or exhaust valves is that one component has to be destroyed for each test condition unless plugs are used. Furthermore, with Templugs there is some evidence that if there is a large temperature gradient along the length of the plug, the temperatures can read high or low depending on whether the plug is inserted from the hot or from the cold side.

5.8.4 Fusible plugs

The fitting of a range of plugs manufactured in eutectic melting alloys provides a quick and convenient way of measuring piston temperatures. Test conditions must be maintained constant for one hour for each reading but the temperatures can be read immediately upon dismantling the engine in order that the plugs may be inspected.

5.9 Exhaust valves and seats

The exhaust valve on a poppet valve engine has a particularly severe duty cycle. The valve opens with a high pressure ratio across it so that sonic velocities occur in the annulus between the valve head and the seat, giving high rates of heat transfer at the radius behind the valve head, with peak metal temperatures which will be of the order of 800°C.

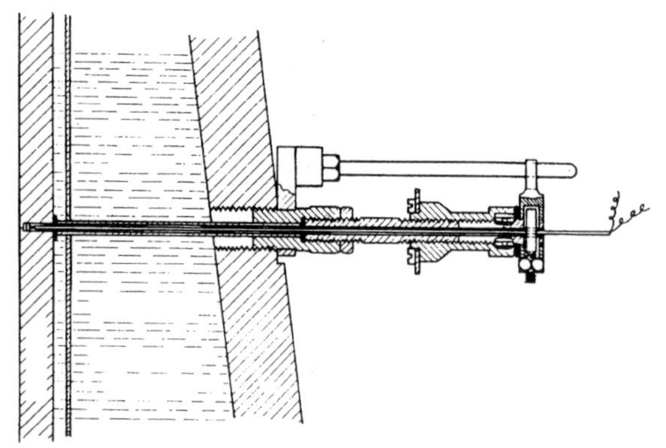

Figure 5.23 Traversing thermocouple

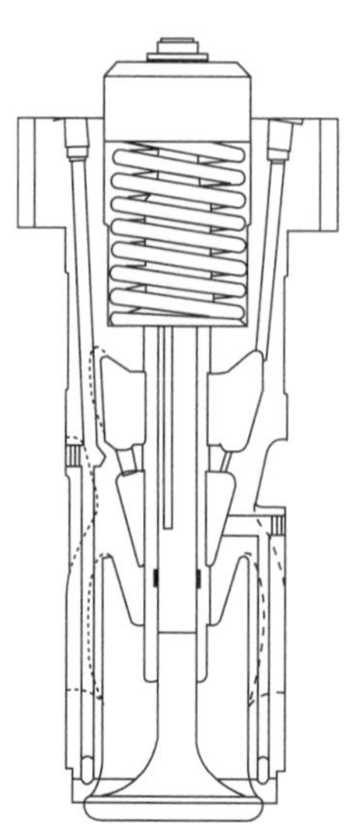

Figure 5.24 Mirrlees electron beam welded water-cooled exhaust valve cage

Austenitic steels such as 21–12N (21% Cr, 12% Ni) or 21-4N (21% Cr, 4% Ni) are normally adequate, but for very high duties it may be necessary to employ Nimonic materials. With such materials, an adequate valve life is obtained unless appreciable seat distortion occurs. With seat distortion, the valve no longer seats, and hence there is a tendency for the high pressure, high temperature combustion gases to blow past the valve and seat. This results in a very high valve temperature and burning of the valve head material. The solution is to improve the cooling of the cylinder head and valve seat in order to give a more uniform temperature and hence less distortion.

Exhaust valve life is a particularly difficult problem when burning residual fuels. These fuels contain vanadium and sodium salts which deposit as 'fluxes' on the valve head and valve seat. This does not lead to problems unless the temperature of the metal of the valve head exceeds 600°C or so. At this temperature, the flux melts locally, leaving a leakage path between the valve head and the cylinder head as a localized 'gutter'. There will then be a very rapid burning of the exhaust valve maerial. The solution is to improve the cooling of the valve head, either directly by means of sodium or water cooling, or indirectly by reducing the temperature of the valve seat in the cylinder head by bringing the coolant closer to the seat.

Engines which run on residual fuel are normally of 250 mm bore and upward. In such engines it is quite common to have exhaust valves fitted in a separate 'cage' which carries the seat. This cage is easily removable for renewal, and by making the cage in two parts with electron beam welding in the critical areas, as shown in *Figure 5.24*, it is possible to bring the coolant very close to the seat.

Acknowledgements

The author must acknowledge his indebtedness to the management of Ricardo Consulting Engineers Ltd. and to his former colleagues there for their assistance in the preparation of portions of this article and for giving permission for the inclusion of material from published articles.

References

1 ALCOCK, J. F., 'Heat Transfer in Diesel Engines', *International Heat Transfer Conference,* Boulder, Colorado, p. 174 (1961)

2 NUSSELT, W., 'Der Wärmeübergang in der Verbrennungs-Kraftmaschine', *V.D.I.*, **27** (1923)

3 EICHELBERG, G., 'Temperaturverlauf und Wärmespannungen in Verbrennungsmotoren', *V.D.I. Forschungsheft 263* (1923)

4 ANNAND , W. J. D., 'Heat Transfer in the Cylinders of Reciprocating Internal Combustion Engines', *Proc. Inst. Mech. Engineers,* London, **177,** p. 973 (1963)

5 WOSCHNI, G., 'Die Berechnung der Wandverluste und der thermischen Belastung der Bauteile von Dieselmotoren', *MTZ,* **31**, No. 19 (1971)

6 FRENCH, C. C. J., 'Taking the Heat Off the Highly Boosted Diesel', *SAE* Paper No. 690463 (1969)

7 LOWE, A. S. H. and MOREL, T., 'A New Generation of Tools for Accurate Thermo-Mechanical Finite Element Analysis of Engine Components', *SAE Paper No. 920681* (1992)

8 FRENCH, C. C. J., 'Problems Arising from the Water Cooling of Engine Components' *Proc. Inst. Mech. Engineers,* London, **184** Part 1 (1969–70)

9 OWEN, N. J., ROBINSON, K. and JACKSON, N. S., 'Quality Assurance for Combustion Chamber Thermal Boundary Conditions—A Combined Experimental and Analytical Approach', *SAE Paper No. 931139* (1993)

10 FINLAY, I. C., BOYLE, R. J., PIRAULT, J. F. and BIDDULPH, T., 'Nucleate and Film Boiling of Engine Coolants Flowing in a Uniformly Heated Duct of Small Cross Section', *SAE Paper No. 870032* (1987)

11 DITTUS, F.W. and BOELTER, L. M. K., *'University of California Publications Engineering'* Vol. 2, (1930) p. 443

12 SMITH, A., 'Another Fine Mesh', *Vehicle Engineering and Design,* June (1994)

13 FRENCH, C. C. J., 'Piston Cooling', *SAE* Paper No. 720024 (1972)

14 BUSH, J. E., 'Heat Transfer in a Reciprocating Hollow Piston Partially Filled with a Liquid', *Stanford University Technical Report No. 56,* Stanford, California (1963)

15 FITZGEORGE, D. and POPE, J. A., 'An Investigation of the Factors Contributing to the Failure of Diesel Engine Piston and Cylinder Covers', *Transactions of North East Coast Institution of Engineers and Shipbuilders,* Newcastle, England, **71**, p. 163 (1955)

16 ALCOCK, J. F., 'Thermal Loading of Diesel Engines', *Transactions of the Institute of Marine Engineers,* **77**, No. 10 (October 1955)

17 EICHELBERG, G., 'Some New Investigations on Old Combustion Engine Problems', *Engineering,* p. 463 (27–10–39)

18 LAW. D. A., 'New Features in Diesel Pistons Above 6 in. Diameter', *Diesel Engineers and Users Association,* Publication No. 308 (1966)

19 'Temperature Measurement of Injection Nozzles', *Hawker Siddeley Technical Journal,* **2,** No 1 (August 1960)

Part 2

Engine design practice

6

Thermodynamic mathematical modelling

Contents

6.1 Introduction

It is convenient to link together the theory of diesel engine performance with an introduction to computer modelling of the diesel engines thermodynamics in engine simulations, as the latter make the most complete use of the former and the use of such models is becoming widespread. Engine modelling is a very large subject, in part because of the range of engine configurations possible and the variety of alternative analytical techniques, or sub-models, which can be applied in overall engine models. The main reason for the growth in engine modelling activity arises from the economic benefits; by using computer models, large savings are possible in expensive experimental work when engine modifications are being considered. Models cannot replace real engine testing but they are able to provide good estimates of performance changes resulting from possible engine modifications and can thus help in selecting the best options for further development, thus reducing the amount of hardware development required. They can also save much expenditure by assisting in the optimization of, for example, control strategies and component matching, an area of increasing importance as engine sub-systems become ever more complex and difficult to optimize on an engine test bed. Finally, they can give considerable insight into particular aspects of engine behaviour which may not be perceived from experimental work.

Engine modelling is a fruitful research area, and as a result many universities have produced their own engine thermodynamic models, of varying degrees of complexity, scope and ease of use. This widespread activity probably arises from a combination of reasons, including the importance of engine modelling and its relevance to almost any aspect of engine research and development. It also has considerable intrinsic interest arising from the fact it directly embodies such a wide range of technology areas (basic engine design, turbomachinery, gas dynamics, chemical reaction kinetics, advanced analytical and computing techniques) and has close links to other disciplines such as computational fluid dynamics and the application of finite element techniques to engine structural and thermal analysis. There are also now available a number of fairly comprehensive 'commercial' models which have a wider, more general purpose use with refined inputs and outputs to facilitate their use by engineers other than their developers; most of these models had their origins in university-developed models. They include WAVE from the US, PROMO from Germany and TRANSENG/ICENG/MERLIN from the UK, among others. A new code covering complete engine systems has emerged recently, GT-suite.

While the more advanced models are extremely large and complex, the basics of an engine thermodynamic model are quite straightforward and easily understood; the complexity arises later in the refinement of the calculation methods, the level of detail of sub-system representation, and the accommodation of a wide variety of alternative engine configurations and control systems. In this section the basics of steady-state engine modelling and diesel engine theory are first presented, focused on direct injection (DI) four-stroke turbocharged diesel engines. The aim is to give an insight into engine cycle simulation models, how they are built and how the different components interact, as most literature (excellent though it may be) does not make this very clear. Hopefully, the section will make an engine model less of a 'black box' and provide the reader with a starting point for understanding the more specialist literature.

The later sections 6.10 and following then briefly discuss transient modelling, more complex and other engine types, and give an introduction to both enhanced gas property/combustion calculations and gas dynamic flow calculations. A number of text books, of which [1 to 4] are extremely useful, provide much more detailed information than can be given in the present overview. Other references given below are to a few key papers on specific aspects, but no attempt has been made to provide a complete bibliography.

6.2 Fundamentals and the energy equation

Figure 6.1 shows a straightforward 6-cylinder diesel engine with a single-stage turbocharger and charge air cooler, with simple connecting pipework, cylinder valves and flow losses (flow restrictions). In the schematic there are nine 'control volumes' (CVs), the six fuelled cylinders, two inlet manifolds and an exhaust manifold, plus two infinite volume (constant pressure) CVs representing the ambient air feed to the engine and the back pressure (which may be ambient).

The essence of modelling the engine is to determine gas conditions right through the engine at successive small intervals of time (normally of the order of one degree of crank angle (CA) rotation). Appropriate summation of these gas conditions over an engine cycle then leads to an estimate of engine performance. By 'conditions through the engine' is basically meant pressures, temperatures, gas composition and mass/energy flows. The core of any model is the energy equation for each CV in the engine. The energy equation and its solution, which are both given below, provide the rate of change of pressure and temperature in each CV at any instant of time from data which is known or can be calculated, and leads to the rate of change of other gas parameters. If the rates of change of gas parameters in the CV are found at successive timesteps, these can be integrated up from timestep to timestep to obtain a time history of the CV parameters. Euler integration is the most straightforward integration method and, for temperature for example, is

$$T_{n+1} = T_n + \tau[(dT/dt)_n + (dT/dt)_{n+1}]/2 \qquad (6.1)$$

where T_{n+1}, $(dT/dt)_{n+1}$ are the temperature and its rate of change at the end of the new step and the n subscripted values those at the end of the previous step. Equivalent equations can be written for other gas properties such as pressure (P), mass (M) and gas composition (ϕ, see below). Other more complex integration methods exist such as the well known Runge–Kutta method (involving calculation of rates at four points through the step) and 'predictor-correction' methods (which use values from the end of several previous steps), but their advantages are questionable.

Equations of the type (6.1) are solved iteratively. For the temperature example, a first estimate of T_{n+1} is obtained from the known values at the end of the previous step (T_n, $(dT/dt)_n$) and assuming $(dT/dt)_{n+1}$ is unchanged from $(dT/dt)_n$. The same process is applied to the other parameters to give a first estimate of all gas parameters at the end of the new step. The energy equation is then solved, using these estimates, to give a new value of $(dT/dt)_{n+1}$ and the other end of step rates of change. Reapplying eqn (1.1) then gives improved estimates of end of step values of T etc. The process can be repeated until the parameters have been obtained to the required accuracy, although in practice for timesteps equivalent to 1° CA only one iteration will be required for much of the time (as the rates do not change dramatically within a timestep, except, for example, at start of combustion).

To model engine performance then, the first essential is the energy equation for a CV, which is derived simply from the First Law of Thermodynamics and the Perfect Gas Law. The first law states that the rate of change of internal energy of the gas volume equals the sum of the net energy flow into the volume from connected gas flows, less any nett heat transfer out through the volume walls and less any work done by the CV gas against its surroundings. Thus

$$\frac{d(uM)}{dt} = \sum_I h_i \dot{m}_i - \frac{dQ}{dt} - P\left(\frac{dV}{dt}\right) \qquad (6.2)$$

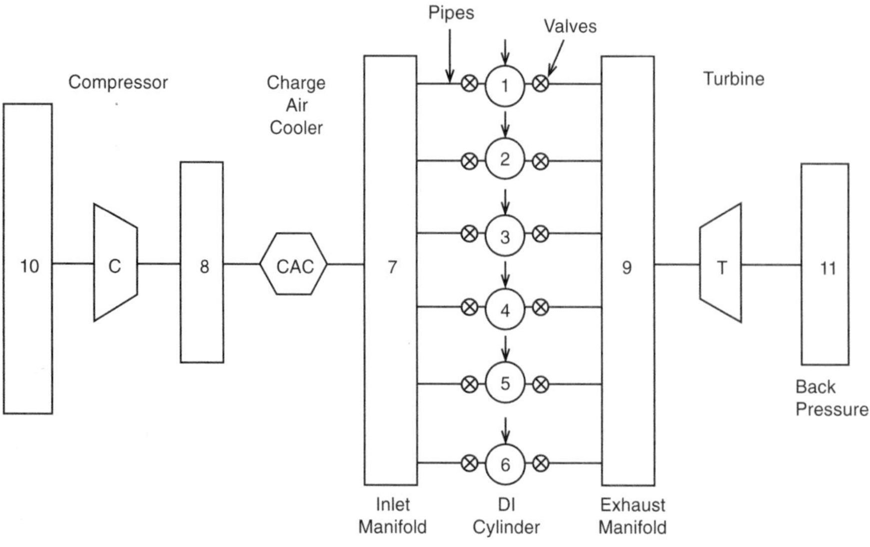

Figure 6.1 A simple single-stage turbocharged diesel engine configuration

where u is the specific internal energy per unit mass of gas in the CV and there are I pipes connecting to the CV, the flow in the ith pipe having specific enthalpy h_i and mass flow rate $\dot{m}_i$ (negative for outflows). Net heat transfer out of the CV is dQ/dt, M is the mass of gas in the CV at pressure P and this gas carries out nett work on its surroundings of PdV/dt, where dV/dt is the CV's current rate of change of volume (zero for manifold CVs, but not for cylinder CVs). To be useful, the left-hand side of eqn (6.2) requires expanding, the simplest expansion being to assume that the gas in the CV is homogenous and that internal energy u depends only on temperatures and gas composition, expressed as equivalence ratio or air–fuel ratio in the CV. The equivalence ratio, ϕ, is defined by

$$\phi = \frac{M_f/M_a}{M_{fs}/M_{as}} = \frac{M_f/M_a}{FAS} \tag{6.3}$$

subscripts f and a denoting fuel and air, subscript s the masses in a stoichiometric mixture (see section 6.3 below) and FAS the stoichiometric fuel–air ratio (the ratio of fuel to air which will exactly give complete combustion of the fuel and air), ϕ thus being unity for a stoichiometric mixture. Values of FAS do not vary much for a wide range of diesel fuels, 0.068 being a typical value (stoichiometric air–fuel ratio 14.71). For use with eqn (6.2) the value of ϕ is that based on M_f being the burnt fuel mass present at any time, rather than total burnt and unburnt fuel. With the assumption that $u = f(T, \phi)$ then the left-hand side of eqn (1.2) expands to

$$\frac{d(uM)}{dt} = M\frac{du}{dt} + u\frac{dM}{dt} = M\frac{\partial u}{\partial T}\frac{dT}{dt} + M\frac{\partial u}{\partial \phi}\frac{d\phi}{dt} + u\frac{dM}{dt}$$

where the partial derivative $\partial u/\partial T$ is the change in u due to a change in T alone, and $\partial u/\partial \phi$ is the change in u due to a change in ϕ alone. Using $\dot{T}$ for (dT/dt) etc., eqn (6.2) becomes:

$$M\left(\frac{\partial u}{\partial T}\right)\dot{T} + M\left(\frac{\partial u}{\partial \phi}\right)\dot{\phi} + u\dot{M} = \sum_I h_i\dot{m}_i - \dot{Q} - P\dot{V} \tag{6.4}$$

which is the energy equation for the simplified case of a homogeneous mixture of air and burnt fuel in a CV, ignoring the effects of dissociation of the gas molecules which can occur at high temperatures (section 6.12). This equation is a reasonable approximation for much basic diesel engine work.

Equation (6.4) can be solved for any CV to give $\dot{T}$, the rate of change of temperature as all other quantities can be calculated using the estimated gas parameters for the end of timestep. The rate of change of CV mass $\dot{M}$ in eqn (6.4) is given by

$$\dot{M} = \sum_I \dot{m}_i \tag{6.5}$$

the sum of current rates of inflow or outflow (negative) in any connected pipe, which can be calculated by the methods of sections 6.4 and 6.5 using the estimated P, T, etc. in the CVs connected by each pipe. Mass M can then be calculated by integration, as in (6.1). $\dot{Q}$ is the current rate of heat transfer which may be zero or calculable by the methods of section 6.8. The $P\dot{V}$ term will be zero for fixed volume CVs, and can be calculated for the cylinder with the equations in section 6.6. Evaluation of ϕ is slightly awkward. Substituting $(M - M_f)$ for M_a in (6.3) and differentiating

$$\dot{\phi} = \frac{d\phi}{dt} = \frac{1}{FAS(M - M_f)^2}\left\{M\frac{dM_f}{dt} - M_f\frac{dM}{dt}\right\} \tag{6.6}$$

where

$$M_f = M\phi FAS/(1 + \phi FAS) \tag{6.6A}$$

by rearranging (6.3). It should be noted that for the assumption that the cylinder contents are homogeneous and that the mixture equivalence ratio ϕ for use in the equations is based on burnt fuel mass, then ϕ is the cylinder will vary from near zero at the start of combustion (only residual burnt fuel present) to its final value corresponding to actual engine fuel–air ratio; the combustion energy release is incorporated in (6.4) through the change in ϕ during combustion. This leaves the four 'gas property' terms h, u, $(\partial u/\partial T)$ and $(\partial u/\partial \phi)$ to be found to solve eqn (6.4).

Some basic gas property equations now need to be stated, namely

$$\text{Specific Internal energy, } u = \int_{T_0}^{T_1} c_v\, dT \qquad c_v = \frac{\partial u}{\partial T} \tag{6.7}$$

$$\text{Specific Enthalpy } h = \int_{T_0}^{T_1} c_p \, dT \qquad c_p = \frac{\partial h}{\partial t} \tag{6.8}$$

$$\text{Gas Constant } R = c_p - c_v \quad \gamma = c_p/c_v \tag{6.9}$$

so

$$h = u + R(T_1 - T_0) \tag{6.10}$$

where T_1 is the temperature of interest and T_0 an arbitrary reference temperature. Quantities c_p, c_v are specific heats at constant pressure and constant volume, respectively, R the gas constant. To deal with R first, this can be taken as 0.287 kJ/kg K for air or lean burnt gas mixture below 2000 K, though it does vary slightly with gas composition. R is related to the Universal Gas Constant R_0 by

$$R = R_0/\overline{MW} \; (= 8.314/\overline{MW}) \text{ kJ/kg K} \tag{6.11}$$

where $\overline{MW}$ is the average molecular weight of the gas (section 6.3). Values of u and h depend on reference temperature T_0 but the value of this does not matter provided it is used consistently; values given below are for a T_0 of 0 K but some texts use values based on a T_0 of 298 K, and they must not be mixed (without a correcting factor). While the important subject of gas properties is dealt with in the next section *Figure 6.2* gives the variation of u, $\partial u/\partial t(c_v)$ and $\partial u/\partial \phi$ for an air/burnt diesel mixture, produced by the Kreiger Borman properties model (section 6.3).

It will be noted that at 2000 K, dissociation occurs and so u has a pressure dependence (higher pressures reducing the amount of dissociation) but the data can be used for preliminary diesel engine work if the cylinder gases are assumed homogeneous. The figure gives u, $\partial u/\partial t$ and $\partial u/\partial \phi$ directly for use in eqn (6.4), specific enthalpy being calculable from (6.10) with R equal to 0.287 kJ/kg K, T_0 zero. For in flow to the CV of course the value of h in (6.4) should be based on inflow T, ϕ, not on their values for the CV itself.

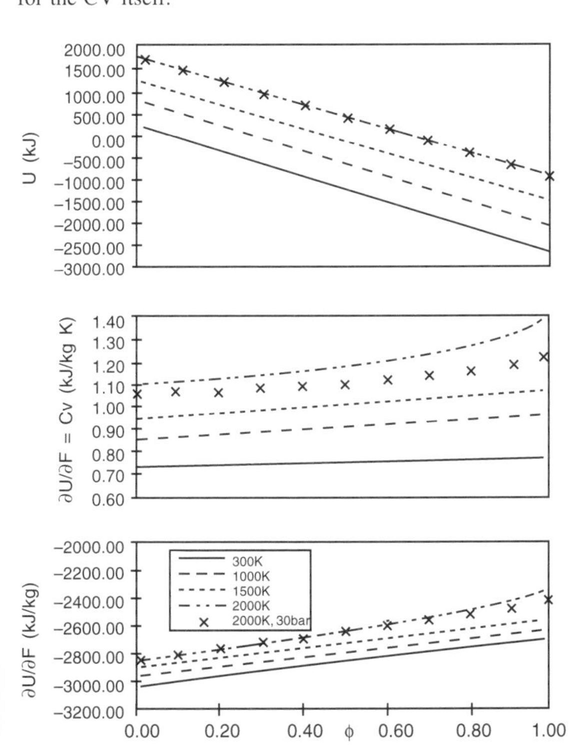

Figure 6.2 Gas properties for lean burnt air/diesel mixture

Thus with the simplifying assumptions of homogeneity and ignoring dissociation, rates of change of temperatures in all CVs can be calculated at any instant from eqn (6.4). Integrating the temperature up from its value at the previous step gives the new end of step temperature. Now for this slightly simplified case the new end of step pressure could be derived from the perfect gas equation directly

$$PV = MRT \tag{6.12}$$

as M, V, R, and T are known. However, as it is necessary when using more complete versions of (6.4), as in section 6.12, it is worth mentioning that $\dot{P}$ the rate of change of P, can be obtained and then integrated up, just like $\dot{T}$. Taking logarithms of (6.12) and differentiating

$$\frac{\dot{P}}{P} + \frac{\dot{V}}{V} = \frac{\dot{M}}{M} + \frac{\dot{T}}{T} \tag{6.13}$$

for the case (as here) where R is assumed to be independent of ϕ and T. While these equations are adequate for much basic engine modelling work, more general forms of eqns (6.4) and (6.13) can be written to cover the case where high temperature dissociation can occur, where the CV is subdivided into different zones and where individual species in the gas (e.g. air unburnt fuel, burnt mixture) are separately considered. This extension is mentioned in section 6.12.

Before discussing pipe flow rates, turbomachinery and combustion it is appropriate to present some further information on gas properties, although eqn (6.4) can be solved using just data of the type given in *Figure 6.2*.

6.3 Gas properties

The properties of any gas mixture depend on the relative proportions of different molecular species in the mixture. These mixture properties (e.g. *Figure 6.2*) are effectively derived by summing the known properties of the individual molecular species, weighted by their proportions in the mixture, the properties of the molecular species being available from standard tables (the JANAF tables[5]) or empirical fits[6,7] to these standard tables. The effective summation can then either be carried out directly, or the results obtained from further empirical relationships, as will be described. Direct summation will be outlined frist as this enables both the underlying principles and a number of fundamental relationships to be presented. The reader is warned that considerable care is needed in using data from the literature as different sources use a wide variety of units, reference temperatures, molecular weights, etc., not to mention variations in assumed air composition, the use of mass and mole based quantities, and numerous polynomial fits to empirical data; it is very easy to get the wrong numbers.

First, it is useful to define mass and molar fractions. The mass fraction of species i in a mixture of mass m is given by

$$x_i = m_i/m \tag{6.14}$$

where m_i is the mass of species i in the mixture. The mass m_i is also given by

$$m_i = N_i M_i \tag{6.15}$$

where M_i is the molecular weight (MW) of species i and N_i is the number of moles of the species in mass m_i. Now the molar fraction is defined by

$$y_i = N_i/N \tag{6.16}$$

where N is the total number of moles of mixture. Both x_i and y_i of course sum to unity for any mixture. Further, the average molecular weight of the mixture is given simply by

$$M = \Sigma\, y_i M_i \tag{6.17}$$

From (6.15) to (6.17) the relationship between mass and molar fractions is

$$x_i = \frac{m_i}{m} = \frac{N_i M_i}{NM} = y_i \left(\frac{M_i}{M}\right) \tag{6.18}$$

that is the mass fraction of species i is given by multiplying the molar fraction by the ratio of species MW to average mixture MW.

Now diesel fuels are hydrocarbons of composition $C_\alpha H_\beta$, α and β giving the proportions of carbon and hydrogen atoms in the fuel. While α, β are integers for any real molecule, fuels are a mixture of many different hydrocarbons so that $C_\alpha H_\beta$ gives the fuel's overall composition, with α, β generally non-integer. One diesel fuel might be $C_{14.1}H_{24.9}$, for example, while in much of the literature a general C_nH_{2n} is used as an approximation. In engine work the types of mixtures of interest are the burnt mixture from combustion of the fuel, and mixtures of air, unburnt fuel and residual burnt gases; the burnt mixture is of course the most difficult to deal with.

The proportions of molecular species in a burnt mixture can be derived from the chemical reaction equation. For the special case where exactly the correct quantity of fuel exists to burn completely with air and to consume all of the available oxygen in the air (the stoichiometric case) the reaction equation at low temperature is

$$\varepsilon\, C_\alpha H_\beta + (0.21\, O_2 + 0.79\, N_2) \rightarrow a_1\, CO_2 + a_2 H_2O + a_3 N_2 \tag{6.19}$$

where ε is the number of moles of fuel required to fully react with one mole of air (assumed composition 0.21 molar fraction of oxygen, 0.79 molar fraction of nitrogen). As atoms must be conserved:

$$\left.\begin{aligned} &\text{For carbon:} && \varepsilon\alpha = a_1 \\ &\text{For hydrogen:} && \varepsilon\beta = 2a_2 \\ &\text{For oxygen:} && 0.42 = 2a_1 + a_2 \\ &\text{For nitrogen:} && 1.58 = 2a_3 \end{aligned}\right\}$$

$$\text{giving} \left\{\begin{aligned} a_1 &= 0.21(1 + \beta/4\alpha) \\ a_2 &= 0.105/(\alpha/\beta + 1/4) \\ a_3 &= 0.79 \\ \varepsilon &= 0.21/(\alpha + \beta/4) \end{aligned}\right. \tag{6.20}$$

by solving the four simultaneous equations. The stoichiometric fuel–air ratio (*FAS*) by mass for the fuel is thus given by

$$FAS = \left(\frac{m_f}{m_a}\right)_s = \left(\frac{N_f M_f}{N_a M_a}\right)_s = \frac{\varepsilon(N_c M_c + N_H M_H)}{(N_{02} M_{02} + N_{N2} M_{N2})}$$

$$= \frac{\varepsilon(12.01\alpha + 1.01\beta)}{(0.21*2*16.00 + 0.79*2*14.01)} = \frac{\varepsilon(12.01\alpha + 1.01\beta)}{28.86} \tag{6.21}$$

where eqn (6.15) has been used and N_f, N_c etc. are the number of moles of fuel of MW M_f, carbon of MW M_c, etc. As stated, ε is the number of moles of fuel reacting with one mole of air, and assumed atomic weights are 12.01 for C, 1.01 for H, 16.00 for O and 14.01 for N. For a diesel fuel of composition $C_{14}H_{28}$ say, then ε is 0.01 (eqn (1.20)) and FAS is 0.0680 (corresponding to a stoichiometric air–fuel ratio of 14.71). The stoichiometric fuel–air ratios do not vary much for most hydrocarbon fuels.

For completeness, the molar fractions of CO_2, H_2O and N_2 for $C_{14}H_{28}$ in the burnt mixture, for fuel burnt stoichiometrically, can then be obtained by dividing a_1, a_2, a_3 from eqn (6.20) by their sum; using (6.16):

$$y_i = N_i/N = a_i/\Sigma\, a_i \tag{6.22}$$

as a_i are the numbers of moles in the burnt mixture for one mole of air. The mass fractions can then be obtained from eqn (6.18). Thus for $C_{14}H_{28}$ burnt stoichiometrically, using (6.20), (6.22) and (6.18):

$$\begin{array}{llll} a_{CO_2} = 0.14 & a_{H_2O} = 0.14 & a_{N_2} = 0.79 & \Sigma\, a_i = 1.07 \\ y_{CO_2} = 0.1308 & y_{H_2O} = 0.1308 & y_{N_2} = 0.7384 & \Sigma\, y_i = 1 \\ x_{CO_2} = 0.200 & x_{H_2O} = 0.082 & x_{N_2} = 0.718 & \Sigma\, x_i = 1 \end{array}$$

where the MWs used for x values are 44.01 for CO_2, 18.02 for H_2O and 28.02 for N_2, giving a mixture MW (eqn (6.18)) of 28.80.

Equation (6.19) is important, both as an introduction to reaction calculations and because it directly relates stoichiometric fuel–air ratio to fuel composition, but fuel is only occasionally burnt stoichiometrically. For an engine at equivalence ratio ϕ then from eqns (6.3) and (6.15)

$$\phi = \frac{(m_f/m_a)}{(m_{fs}/m_{as})} = \frac{N_f M_f / N_a M_a}{N_{fs} M_f / N_{as} M_a}$$

$$= \frac{N_f}{N_a} \Big/ \frac{N_{fs}}{N_{as}} = \frac{N_f}{N_a} \cdot \frac{1}{\varepsilon}$$

where s denotes stoichiometric quantities. Thus the number of fuel molecules relative to air molecules is now

$$(N_f/N_a) = \varepsilon\,\phi \tag{6.23}$$

Now, if the mixture is rich ($\phi > 1$) then the burnt mixture will contain CO and H_2 as there is insufficient air to fully burn the fuel, but this will not be considered further here as diesel engines usually run lean. Conversely if the mixture is lean ($\phi < 1$) then there will be unconsumed oxygen in the burnt mixture. Further, for higher combustion temperatures, dissociation will occur and the burnt mixture will contain numerous other species, the most important of which are uncombined hydrogen and oxygen atoms, hydroxyl radicals and nitric oxides. For the lower temperatures (or up to about 1500 K where dissociation is ignored) eqn (6.19) can be written more generally for lean mixtures by using (6.22) for non-stoichiometric fuelling:

$$\begin{aligned} &\varepsilon\phi\, C_\alpha H_\beta + (0.29\, O_2 + 0.79\, N_2) \\ &\quad = b_1 CO_2 + b_2 H_2O + b_3 N_2 + b_4 O_2 \end{aligned} \tag{6.24}$$

Similarly to eqn (6.20) for stoichiometric mixtures, an atom balance gives

$$b_1 = \phi \in \alpha \quad b_2 = \phi \in \beta/2 \quad b_3 = 0.79 \quad b_4 = 0.21(1 - \phi) \tag{6.25}$$

The b_i values can be converted to molar/mass fractions using eqns (6.22) and (6.18), as before with a_i for a stoichiometric mixture.

For lean, non-dissociated burnt mixture the molar fractions can thus be determined. For air they are simply 0.21 for oxygen and 0.79 for nitrogen, corresponding mass fractions being derivable from (6.18); for unburnt fuel, molar and mass fractions are of course unity. These various fractions can be linked directly to the gas property data given in section 6.2 (e.g. *Figure 6.2*). For a gas with a variety of constituents (whether a mixture of combustion products or of combustion products mixed with air and unburnt fuel), the average value of the gas constant (c.f. eqn (6.11)) is given by

$$\overline{R} = R_0/M = R_0/\Sigma\ y_i M_i \qquad (6.26)$$

Where M is the average MW of the mixture given by (6.17) and y_i are species molar fractions. $\overline{R}$ for a burnt diesel/air mixture will be about 0.287 kJ/kg K as mentioned earlier. To obtain the specific enthalpy (h) for a gas mixture, the specific enthalpies of the species have to be found, and these can be derived from empirical fits to the specific heat variation of each species in the mixture. For a given species the value of c_p depends only on temperature and can be expressed as a polynomial in T:

$$(c_{pi}/R_0) = a_{i1} + a_{i2}T + a_{i3}T^2 + a_{i4}T^3 + a_{i5}T^4 \qquad (6.27)$$

where a_i values vary with species of course. Tables of a_i are readily available for combustion products[1,3] although care is needed in their use. If R_0 in (6.27) is taken as 8.314 J/mole K then c_p values are in mole terms (J/mole K). Using eqn (6.8), species-specific enthalpy is given by integrating (6.27)

$$(h_i/R_0) = a_i T + a_{i2}T^2/2 + a_{i6} \qquad (6.28)$$

where a_{i6} is an integration constant which includes allowance for the chosen reference temperature. The mixture specific enthalpy is then obtainable from

$$h = \Sigma\ y_i h_i \qquad (6.29)$$

if h_i is mole based, and can be converted to a mass based value (e.g. J/kg K) by multiplying the average MW of the mixture. If h is evaluated by the above methods for some mixture temperature T, and the specific internal energy then calculated using eqn (6.10), the resulting value will correspond to that given in *Figure 6.2* (these use a reference temperature 0 K). The derivatives in *Figure 6.2* can similarly be obtained by making use of appropriate differentiation of eqns (6.10) and (6.28) for $\partial u/\partial T$ or (considerably more complex) eqns (6.25) for $\partial u/\partial \phi$.

Before leaving the basic derivation of properties it is worth mentioning the derivation of fuel calorific value, from equation (6.19). This equation gives the combustion products if ε moles of fuel are burnt stoichiometrically with one mole of air. Using the mole values a_i from (6.20) with enthalpy values from eqns (6.28, etc.) gives the total enthalpy of combustion products on the right of (6.19). If the enthalpy of the unburnt fuel and air are calculated using the methods indicated above, the difference between the combustion products enthalpy and the input fuel–air enthalpy will give the combustion heat release per mole of fuel if divided by ε. Dividing by molecular weight will give the heat release per unit mass of fuel. If the evaluation is carried out assuming the fuel, air and combustion products all to be at 298 K then this heat release will be the Lower Calorific Value (Q_{LCV}) of the fuel, the standard measure of fuel heat release. Values of Q_{LCV} do not vary markedly with exact fuel composition for diesel fuels, being around 44 000 kJ/kg. The calculations can be carried out more rapidly if, instead of using enthalpy relationships like (6.28), the species enthalpies are obtained directly from the widely available extracts from JANAF tables[1,3]; the values of a_i and ε will still be needed from (6.20).

While the above methods can be used to obtain gas property data for engine modelling purposes, the computations are relatively slow, particularly if dissociation is to be accounted for. Thus while extensions of equations (6.24) must by solved if actual fractions of combustion product species are required for emission calculations, empirical methods have been developed to obtain thermodynamic properties which are all that are required for engine performance (excluding emission) calculations. Two basic methods have been used, though there are a number of variations on them. The oldest and probably most widely used was developed by Kreiger and Borman[8], the more recent and apparently more flexible is due to Martin and Heywood[9]; both methods rely on data originating from the JANAF tables of properties.

The KB method applies to fuels of composition C_nH_{2n}, which is a reasonable compromise as the C:H ratio is the dominant factor in determining hydrocarbon fuel properties and a ratio of around 2 applies to most hydrocarbon fuels of interest. It consists of a variety of curve fits to JANAF data for combustion products. The basic fit is for lean mixtures ignoring dissociation, two linear variations with equivalence ratio ϕ being used for internal energy and gas constant. For higher temperatures where dissociation can occur corrections are applied which depend on ϕ, T and P. For richer than stoichiometric, which is of less importance for diesel engines, a number of very complex fits were obtained but for specific values of ϕ. For rich mixtures the development of Watson and Mazouk, which is broadly similar in form to the lean KB method, is easier to apply.

The MH method is considerably more flexible than KB in that it can take proper account of actual values of fuel composition, and is somewhat less empirical. While published only for $C_\alpha H_\beta$ fuels it can be extended to include fuels incorporating oxygen and nitrogen. It is a very ingenious method but the derivation of its equations is correspondingly complex. The method is based on a fuel–air reaction equation similar to an extended eqn (6.24) except that the right-hand side is written in terms of numbers of moles of monatomic, diatomic and triatomic molecules rather than in terms of CO_2, H_2O and CO, etc. The method gives good agreement with results calculated both from fundamentals and KB, and is much faster than the former. It appears preferable to KB in that it can cover a wide range of real fuel compositions but is quite complex to program.

6.4 Pipe flows, valves, throttles and flow restrictions

Equations (6.4) and (6.13) which are the basis of the energy equation, require instantaneous mass flows through any pipes and turbomachinery components attached to each CV of interest, and also the entry and exit flow specific enthalpy (h). While turbomachinery is dealt with later, this section will be concerned with evaluating the flow through pipes, including any valves, throttles or flow restrictions attached to the ends of the pipes. In the simple example of *Figure 6.1* the pipes connecting to the cylinders have the cyliner valves on one end. Other pipes not on the figure might have wastegate valves by-passing the turbine, exhaust gas recirculation (EGR) valves controlling flow from the exhaust manifold back to inlet manifold, or other pipes containing a variety of flow control valves or different purposes. All of these flow control devices can be characterized by a (usually varying) constriction of the gas flow area and a loss factor, so that this section is firstly concerned with calculating instantaneous flow through a pipe with some known minimum area and loss factor at that instant. The second part of the section will briefly discuss determining areas and loss factor at any instant. In this section only so-called 'filling-and-emptying' steady flow calculations are considered, where fluctuation of pressure, mass flow, etc. within the pipe are ignored, the more complex gas dynamic calculations being mentioned in section 6.13. The quasi-steady flow calculations are adequate for much diesel engine work.

The basic equation determining flow through a pipe and constriction is the orifice equation, of which three forms are of interest depending on the pressure difference between the CVs joined by the pipe. It is first useful to derive several basic gas flow equations, which then lead on to the orifice equation. The stagnation or total temperature, T_t, at any point in a flow is given by

$$c_pT_t = c_pT + v^2/2 \tag{6.30}$$

where T_t is the temperature that the gas flowing at velocity v (with static temperature T) would reach if it were brought to rest adiabatically; the equation is simply an energy balance, c_pT being a measure of the static energy, $v^2/2$ the kinetic energy (p.u. mass) and c_pT_t the total energy (enthalpy). Using the definition of Mach Number, M_a

$$M_a = v/c = v/\sqrt{(\gamma RT)}$$

where c is the velocity of sound (which can be shown to equal $\sqrt{\gamma RT}$) then

$$T_t/T = 1 + v^2/2c_p = 1 + \gamma RT M_a^2/2c_p$$

By manipulating $\gamma = c_p/c_v$ and $c_p - c_v = R$ then c_p can easily be shown to equal $R\gamma/(\gamma - 1)$ so that, substituting back, the well-known total-to-static temperature relationship is obtained

$$T_t/T = 1 + (\gamma - 1) M_a^2/2 \tag{6.31}$$

If the flow is brought to rest adiabatically then

$$P/\rho^\gamma = \text{constant} \tag{6.32}$$

or, using the gas equation (6.12) with $\rho = M/V$:

$$P_t(RT_t/P_t)^\gamma = P(RT/P)^\gamma$$

which on rearranging and combining with (6.31) gives the well-known total-to-static pressure equation:

$$\frac{P_t}{P} = \left(\frac{T_t}{T}\right)^{\frac{\gamma}{\gamma-1}} = \left[1 + \left(\frac{\gamma-1}{2}\right) M_a^2\right]^{\frac{\gamma}{\gamma-1}} \tag{6.33}$$

The standard orifice equation can be derived directly from (6.32) and (6.33), although the derivation is somewhat contorted however it is carried out. Substituting back $(v^2/\gamma RT)$ for M_a^2 is (6.33) and rearranging, the velocity at any point is given by

$$v^2 = \left(\frac{2\gamma}{\gamma-1}\right) RT \left[\left(\frac{P_t}{P}\right)^{\frac{\gamma-1}{\gamma}} - 1\right] \tag{6.34}$$

Now suppose that the flow is in a pipe coming from a large control volume where pressure is P_1 and the velocity of the gas in the CV can be assumed zero. Equatin (6.34) can now be used to give velocity at the point of smallest area along the pipe (the 'throat'), denoted by subscript 'T' (i.e. conditions at throat v_T, R_T, P_T, P_T, area A_T). If the flow from CV to throat is adiabatic there is no loss of total pressure, that is

$$P_{t_T} = P_{t_1} = P_1$$

the second equality following from the velocity in the control volume being zero. (e.g. (6.33) with M_a zero). Thus (6.34) becomes

$$v_T^2 = \left(\frac{2\gamma}{g-1}\right) R_T T_T \left[\left(\frac{P_1}{P_T}\right)^{\frac{\gamma-1}{\gamma}} - 1\right]$$

$$= \left(\frac{2\gamma}{g-1}\right)\left(\frac{P_T}{\rho_T}\right)\left[\left(\frac{P_1}{P_T}\right)^{\frac{\gamma-1}{\gamma}} - 1\right] \tag{6.35}$$

using the perfect gas equation to replace $R_T T_T$. Now for adiabatic flow

$$(\rho_T/\rho_1) = (P_T/P_1)^{1/\gamma} \tag{6.32A}$$

thus

$$\left(\frac{P_T}{\rho_T}\right) = \frac{P_T}{\rho_1 (P_T/P_1)^{1/\gamma}} = \frac{P_T}{P_1} \cdot \frac{P_1}{\rho_1 (P_T/P_1)^{1/\gamma}}$$

$$= \frac{P_1}{\rho_1}\left(\frac{P_T}{P_1}\right)^{1-\frac{1}{\gamma}} \tag{6.36}$$

$$v^2 = \left(\frac{2\gamma}{\gamma-1}\right)\left(\frac{P_1}{\rho_1}\right)\left(\frac{P_T}{P_1}\right)^{\frac{\gamma-1}{\gamma}}\left[\left(\frac{P_T}{P_1}\right)^{\frac{\gamma-1}{\gamma}} - 1\right] \tag{6.37}$$

where (P_1/P_T) from (6.35) has been inverted. However, the mass flow equation gives mass flow $\dot{m}$ at any point in a flow where flow velocity is v, flow area A and flow density ρ as:

$$\dot{m} = \rho A v \tag{6.38}$$

therefore

$$\dot{m}_T^2 = A_T^2 v_T^2 \rho_T^2$$

$$= A_T^2\left(\frac{2\gamma}{\gamma-1}\right)\left(\frac{P_1}{\rho_1}\right)\left(\frac{P_T}{P_1}\right)^{\frac{\gamma-1}{\gamma}}\left[\left(\frac{P_T}{P_1}\right)^{\frac{1-\gamma}{\gamma}} - 1\right]\rho_1^2\left(\frac{P_T}{P_1}\right)^{\frac{2}{\gamma}}$$

using (6.32A) and (6.37). Multiplying together the (P_T/P_1) terms then:

$$\dot{m}^2 = A_T^2 = \left(\frac{2\gamma}{\gamma-1}\right) P_1\rho_1\left[\left(\frac{P_T}{P_1}\right)^{\frac{2}{\gamma}} - \left(\frac{P_T}{P_1}\right)^{\frac{\gamma+1}{\gamma}}\right] \tag{6.39}$$

The only assumptions thus far are that the flow is adiabatic and originates from a control volume which is large enough for velocity to be assumed effectively zero (and hence CV static pressure equal to total pressure). Equation (6.39) gives mass flow in terms of CV parameters (P_1, ρ_1, γ) and throat area and (static) pressure. The problem is throat pressure, which is normally unknown, and the usual assumption is that throat pressure is equal to the downstream CV pressure (P_2) on the basis that the flow loses its kinetic head ($P_{t_T} - P_T$, equal to $P_{t_T} - P_2$). With this assumption and including a flow loss factor C_D (the 'discharge coefficient') equation (6.39) gives the standard compressible flow orifice equation,

$$\dot{m} = C_D A_T P_1 \sqrt{\left(\frac{2\gamma}{\gamma-1}\right)\frac{1}{RT_1}\left[\left(\frac{P_2}{P_1}\right)^{\frac{2}{\gamma}} - \left(\frac{P_2}{P_1}\right)^{\frac{\gamma+1}{\gamma}}\right]} \tag{6.40}$$

where ρ_1 has been replaced by (P_1/R_{t1}), (P_2/P_1) is the pressure ratio across the orifice between two CVs at pressures P_1 (entry) and P_2. Equation (1.40) is widely used for calculating orifice flows (e.g. flow through valves and throttles) and is a very basic equation in engine modelling. Values of C_D, the discharge coefficient, are normally obtained experimentally. The equation works very well for abrupt orifices such as valves and throttles where the pressure is lost by flow separation after the throat (it is not so good for a smooth venturi where there is pressure recovery and so no complete loss of head, requiring C_D greater than unity).

There are two special cases of eqn (6.40), depending on pressure ratio (P_2/P_1), which are of considerable importance in flow calculations. Firstly if the pressure drop across the orifice is large enough (P_2/P_1 small) the flow becomes fixed at a maximum value ('choked flow') corresponding to the flow velocity at the throat reaching the local speed of sound. Maximum flow occurs when the derivative of $\dot{m}$ (from (6.40) for example) with respect to pressure ratio becomes zero. Differentiating (6.40) with respect to (P_2/P_1), setting the result to zero (maximum flow) and re-arranging, the "critical pressure ratio" for chocked flow is then:

$$\left(\frac{P_2}{P_1}\right)_{crit} = \left(\frac{\gamma+1}{2}\right)^{\frac{\gamma}{1-\gamma}} = \left(\frac{2}{\gamma+1}\right)^{\frac{\gamma}{\gamma-1}} \tag{6.41}$$

Substituting back into (6.40) the corresponding critical (choked) mass flow is given, after some manipulation, by:

$$\dot{m}_{cr.} = C_D A_T P_1 \sqrt{\frac{\gamma}{RT_1}\left(\frac{2}{\gamma+1}\right)^{\frac{\gamma+1}{\gamma-1}}} \tag{6.42}$$

which is independent of actual pressure ratio. As pressure ratio falls (pressure drop rises) then, mass flow increases until the critical pressure ratio is reached, at which point mass flow is given by (6.42). If exit pressure falls further, then the mass flow stays at the value given by (6.42) and the flow is choked; increasing P_1 will, however, still increase flow (from 6.42) because of the higher inlet density, but the throat velocity will remain sonic. For air, where γ is 1.4 the critical pressure ratio is 0.528.

The second special case is at the other extreme of small pressure difference across an orifice. In this case equation (6.40) can be binomially expanded by putting (P_2/P_1) = (1 – $\Delta P/P_1$), where ΔP is the small pressure drop ($\Delta P << P_1$), giving

$$\dot{m} = C_D A_T \sqrt{2\Delta P_{\rho_1}} \tag{6.43}$$

Equation (6.43), using loss factor K in place of C_D, is the well-known hydraulic equation for incompressible flow:

$$\dot{m} = A_T \sqrt{2\rho\Delta P/K} \quad \text{or} \quad \Delta P = (1/2\ \rho v^2)K \tag{6.44}$$

(using 6.38 for $\dot{m}$) where K is ($1/C_D^2$). Equation (6.44) is suitable for flow restrictions in engines where presure losses are small (< ≈ 10%) so that flow can be assumed incompressible and where it is more appropriate to use a loss factor K rather than a discharge coefficient.

To return to the data required by the energy equation, eqns (6.40), (6.42) and (6.44) can be used to determine the instantaneus mass flow into and out of pipes connected to a control volume. If there is no heat transfer through the pipe wall or friction (frequently a reasonably valid assumption) then the total enthalpy p.u. mass of flow can be taken as that in the originating control volume, which can be calculated for known CV conditions. For cylinder valves, throttles and other control valves where pressure ratios can be large, it is essential to use eqns (6.40) and (6.42), as compressibility efects can be large. For open pipes or ones with a small flow restriction (e.g. inlet tracts, exit from turbines to back pressure or ambient) where pressure drops are small, the incompressible flow assumptions of eqn (6.44) are valid. While simply using the above equations to obtain instantaneous flow from timestep to timestep through a pipe/orifice between two control volumes works extremely well for many turbocharged diesel engines, they can cause incorrect results or instability in some cases. The well-known cases are at high engine speeds or for long pipes, where the flow cannot be considered as 'steady', that is varying slowly enough relative to engine events for the above steady-flow equations to be valid. In these cases the speed of propagation of pressure changes in control volumes down the pipes begins to determine flow, and this propagation speed is determined by the local speed of sound. Thus for high speed engines or ones with long pipes non-steady flow equations have to be used, and these are briefly discussed in section 6.13. At low engine speeds and loads, instability can result because gas inertia effects become important when flow rates are very low. Nevertheless, for much diesel engine work flows can be (and are) calculated by using the above orifice equations.

In order to use the orifice equations, the area and discharge coefficient (or loss factor) must be known for the constricting part of the pipe, normally a cylinder valve, some form of examples are cylinder valves and control valves, where areas vary with time and losses vary with area.

For cylinder valves the variation of flow area is determined by the cam profile and for a particular engine the variation of area will normally be available from measurements in the form of area versus crank angle curves, for inlet and exhaust valves (*Figure 6.3*). Often the curves give effective area (AC_D) as this can easily be derived from combining test rig measurements of flow with the use of eqns (6.40) and (6.42) in reverse.

Deriving effective area curves in this way avoids difficult problems of separately determining actual flow area through the valve and the discharge coefficient. Some organizations prefer to express valve profiles by valve lift versus crank angle, with associated area-lift and C_D-lift curves, the data for these still coming from flow rig measurements. In this case, care has to be taken in defining 'flow area' as actual flow area through a valve varies with lift in a complex way. Frequently an approximation to flow area is used, the associated C_D values then accounting for the approximation. One common approach is to use the 'curtain area', simply

$$\text{Curtain Area} = \pi DL \tag{6.45}$$

where D is valve head diameter and L the valve lift from flow measurements, C_D values can be obtained again using the orifice equations in reverse. However, it is important to realize that these C_D values go with the curtain areas of (6.45) and must not be used with other definitions of flow area, which will need their own set of C_D–lift curves. In using eqns (6.40, 6.42) in a model the effective flow area is obtained each timestep from look-up tables of either effective flow area versus crank angle, or lift versus crank angle with an associated C_D–lift table, using linear interpolation. The pressure ratio is determined by the pressures in the cylinder and either inlet or exhaust manifold at the timestep.

Other types of control valves can be represented quite easily in a model but additional program will be needed to represent the valve's control system. While there are many possible types of control valves in engines, from wastegates to throttles, they all operate by changing flow areas; if their flow area and discharge coefficient are known at any instant then flow can be calculated from (6.40, 6.42). However, the flow area at any instant will be determined by the characteristics of the valve and its control system, and these must be represented in the model. Taking a simple throttle which is closed as engine speed increases as an example, curves (or look-up tables) of area versus angle and C_D versus area will be needed, together with the algorithm for determining throttle angle, in this case possibly a simple algorithm representing the throttle fully closed up to some engine speed, opening over an engine speed band and being fully open above some specified higher speed.

At any instant, knowing engine speed would then give throttle effective angle, which in turn would give flow area and C_D, allowing (6.40, 6.42) to provide mass flow for the current pressures in the control volumes on either end of the pipe containing the throttle. For an exhaust wastegate, the valve lift might depend simply on the ratio of exhaust manifold pressure to back pressure.

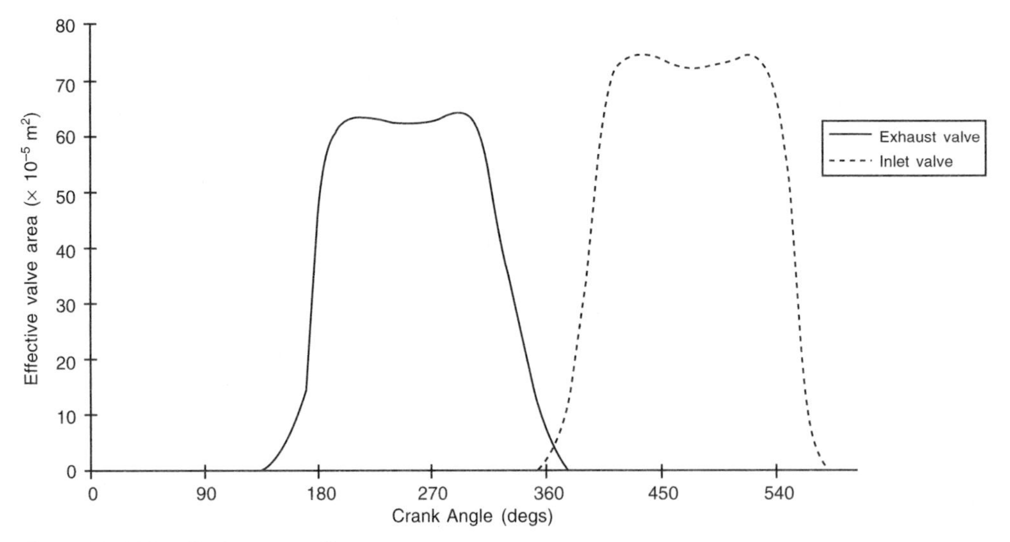

Figure 6.3 Valve effective area profiles

In this case a curve of look-up table of lift versus pressure ratio will be required so that at any instant the model can determine instantaneous lift which, combined with lift versus area and area–C_D curves will allow (6.40, 6.42) to again determine instantaneous flow. For steady state operation this is all of the data that is often required, but for transient modelling (section 6.9) additional information will be needed for the valve's response rate as a real valve will not respond instantaneously to changing conditions. Such information will also be needed for steady state modelling if the valve has a sufficiently fast response to respond to in-cycle changes in engine conditions. This will not usually apply for throttles and wastegates of the types mentioned (they will not respond fast enough to 'see' in-cycle changes) but it will apply, for example, to reed-valves which respond very rapidly to variations in the pressure ratio across them. For these fast-responding valves it is still adequate to assume quasi-steay flow, and thus to use eqns (6.40, 6.42), but the valve area must be calculated each timestep from the 'control algorithm'; in the particular case of reed-valve of course the 'control algorithm' is the mechanical characteristics of the valve rather than a separate control system. The division between steady state and transient modelling is not as distinct as may be thought.

Using these methods then instantaneous flow through pipes and valves can be determined from the conditions in the adjoining control volumes, and form part of the input data required by the energy equation for these CVs. However, there are other components for which the flow is required, notably turbo-machinery components.

6.5 Turbomachinery and charge air coolers

Compressors and turbines are normally represented in engine models by their characteristic curves, which can come in several forms relating mass flow and efficiency to pressure ratio across them, inlet temperature and rotational speed. Usually they are expressed in terms of 'non-dimensional' parameters as this limits the variation between curves for different types, but it is not essential that they be in this form. Compressors and turbines are treated elsewhere in considerable detail, so the following is merely introductory and aimed at the representation of these components in engine models.

Compressor characteristics (*Figure 6.4*) are normally given in terms of so-called non-dimensional (ND) quantities (they are not all actually ND, but behave as if they were):

Speed	$N/\sqrt{T_1}$	rpm/(K)$^{1/2}$
Pressure Ratio	P_1/P_2	
Mass flow	$\dot{m}\sqrt{T_1}/P_1$	kg/s(K)$^{1/2}$/(kN/m^2)
Efficiency	η	

where T_1, P_1 are inlet temperature and pressure, P_2 is exit pressure and N is compressor rotational speed. These characteristics are input into models by two large look-up tables, one for ND mass flow as a function of pressure ratio and ND speed, the other for efficiency in terms of the same two parameters. While they differ considerably in detail, similar forms of characteristics are obtained for both radial flow (centrifugal) and axial compressors, and both can be handled similarly in an engine model.

The energy equation for the compressor is given by balancing heat input into the compressor less the work output, to the enthalpy change across the compressor. Thus ignoring any potential energy change:

$$\dot{Q} - \dot{W} = \dot{m}[(h_2 + KE_2) - (h_1 + KE_1)] = \dot{m}[h_{t1} - h_{t2}] \tag{6.46}$$

where $\dot{m}$ is mass flow, h is static specific enthalpy, KE is specific kinetic energy and h_t total specific enthalpy, with subscript 2 being for exit conditions. $\dot{Q}$ is normally taken to be zero (heat transfer is very small compared to flow energy changes) and as specific heat at constant pressure (c_p) is constant for a given gas

$$-\dot{W} = \dot{m}(h_{t2} - h_{t1}) = \dot{m}c_p(T_{t2} - T_{t1}) \tag{6.47}$$

T_t being total temperature. $\dot{W}$ is negative as work is done on the gas. The efficiency is defined by the ratio of work required for adiabatic compression to that actually required, thus

$$\eta_{TT} = \frac{\dot{m}c_p(T'_{t2} - T_{t1})}{\dot{m}c_p(T_{t2} - T_{t1})} = \left(\frac{T'_{t2} - T_{t1}}{T_{t2} - T_{t1}}\right) \tag{6.48}$$

where T'_{t2} is the temperature at the exit were the compression to be adiabatic. However, because kinetic energy from the compressor is not very useful and tends anyway to be lost in the following manifold it is usual for efficiency to be expressed on a total-to-static basis:

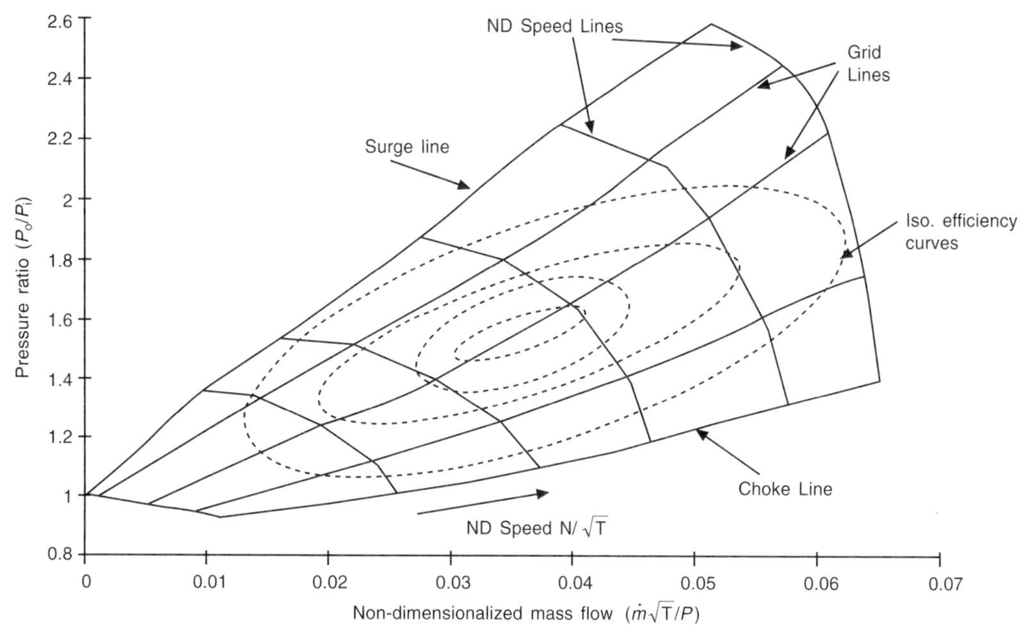

Figure 6.4 Compressor characteristics

$$\eta_{TS} = (T_2' - T_{t1})/(T_{t2} - T_{t1}) \tag{6.49}$$

where T_2' is the static temperature at exit for adiabatic compression. From eqn (6.33), for flow brought to rest adiabatically:

$$(T_t/T) = (P_t/P)^{(\gamma-1)/\gamma}$$

and so for an adiabatic flow

$$(T_2/T_{t1}) = (P_2/P_{t1})^{(\gamma-1)/\gamma}$$

giving

$$\eta_{TS} = T_{t1}[(P_2/P_{t1})^{(\gamma-1)/\gamma} - 1]/(T_{t2} - T_{t1}) \tag{6.50}$$

For a compressor between two control vlumes, P_{t1}, T_{t1} are pressure and temperature in the inlet volume, P_2, T_{t2} the pressure and temperature in the exit volume as velocity in each volume is assumed to be zero. Returning to eqns (6.47, 6.50) then

$$\dot{W} = -\frac{\dot{m}c_p}{\eta_{TS}}(T_2' - T_{t1}) = -\frac{\dot{m}c_p T_1}{\eta_{TS}}[(P_2/P_1)^{(\lambda-1)/\lambda} - 1] \tag{6.51}$$

gives the power absorption of the compressor, the 't' subscript being dropped as total equals static in the CVs. The torque input required by the compressor is then

$$TQ = \dot{W}/\omega = 30\dot{W}/(N\pi) \tag{6.52}$$

for N the compressor rotational speed in rpm (ω rad/s).

The use of the compressor characteristics in modelling is quite straightforward. Knowing the (predicted) pressures in the two adojining CVs (and temperature in the inlet one) at a timestep, and the compressor rotational speed, the pressure ratio and ND speed are calculated. A double linear interpolation (*Figure 6.5*) of the mass flow characteristic (in look-up table form) gives the ND mass flow, from which actual mass flow can be calculated.

Efficiency is similarly calculated from a double interpolation of its look-up table. Now while exit pressure P_2 is known, exit temperature is not (it is *not* necessarily equal to the exit CV

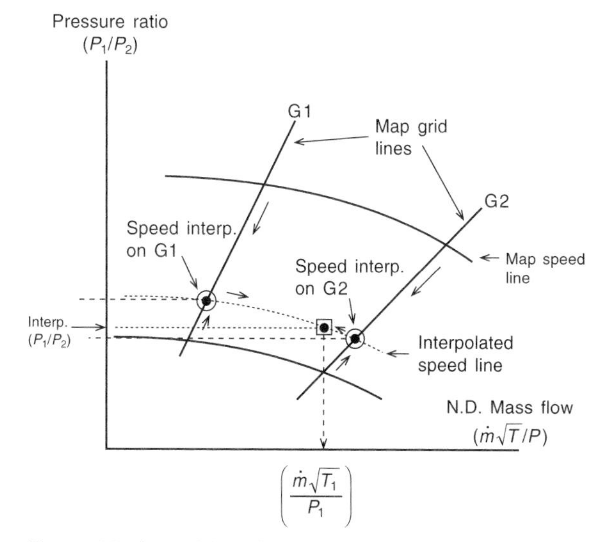

Figure 6.5 Interpolation of compressor mass flow

temperature) an this is required to obtain the flow specific enthalpy from the compressor, which will of course be much higher than the entry CV specific enthalpy as work is done on the gas by the compressor. The exit temperature is given by re-arranging (6.50)

$$T_{t2} = T_1\{[(P_2/P_1)^{(\gamma-1)/\gamma} - 1]/\eta_{TS} + 1\} \tag{6.53}$$

which of course can be evaluated from known pressure ratio, T_1 and η_{TS} just evaluated. For use in matching compressor/turbine performance at any instant (see below) the power and torque input required by the compressor are needed, and can be derived from (6.51) and (6.52).

Turbine performance is dealt with similarly, using turbine characteristic curves which relate similar parameters to those for the compressor but are of slightly different form. The mass flow curves give ND mass flow ($\dot{m}\sqrt{T}/P$, where T and P are turbine inlet manifold conditions) to ND speed $(N/\sqrt{T})$ and pressure ratio (P_1/P_2). The speed variations are not so large as for compressors and as pressure ratio increases the speed lines converge (*Figure 6.6*), mass flow reaching a limiting value, this corresponding to choking (sonic) flow in the turbine. Efficiency can be plotted in several ways, but the most convenient is to plot it against turbine blade speed ratio (*u/c*) as the curves vary little with speed and so a single all-speed efficiency curve is adequate. The blade speed ratio is defined to be the ratio of rotor tip velocity to gas velocity if the gas were isentropically expanded through the turbine and is derived below.

As for the compressor

$$\dot{Q} - \dot{W} = \dot{m}[(h_2 + KE_2) - (h_1 + KE_1)] = m[h_{t1} - h_{t2}] \quad (6.54)$$

Again ignoring $\dot{Q}$ (heat transfer through walls negligible compared with flow energy)

$$\dot{W} = \dot{m}(h_{t1} + h_{t2}) = \dot{m}c_p(T_{t1} - T_{t2}) \quad (6.55)$$

which is now positive, the turbine outputing work. The efficiency is defined on a total-to-total basis as

$$\eta_{TS} = \frac{\text{Actual work}}{\text{Work for isentropic expansion}} = \frac{h_{t1} - h_{t2}}{h_{t1} - h'_{t2}} = \frac{T_{t1} - T_{t2}}{T_{t1} - T'_{t2}} \quad (6.56)$$

where T'_{t2} is the exit temperature if expansion through the turbine were isentropic. As for compressors, *KE* from the turbine is not useful and will tend to be lost in the exit CV so it is usual to define efficiency on a total-to-static basis, so that

$$\eta_{TS} = \frac{T_{t1} - T_{t2}}{T_{t1} - T'_2} = \frac{(T_{t1} - T_{t2})}{T_{t1}[1 - (P_2/P_{t1})^{(\gamma-1)/\gamma}]} \quad (6.57)$$

analogous to equation (6.50) for a compressor. Using (6.55) the work output of the turbine is

$$\dot{W} = \dot{m}c_p\eta_{TS}T_1[1 - (P_2/P_1)^{(\gamma-1)/\gamma}] \quad (6.58)$$

where 't' subscripts have been dropped as total and static conditions are equal in the CVs. In using these equations, c_p and γ must of course be for the hot exhaust gases, not for air. As for compressors, turbine troque output is given by

$$TQ = 30\,\dot{W}/(N\pi)$$

for *N* rotational speed (rpm). The *u/c* ratio for use with the efficiency map is given by

$$u/c = \frac{\text{Rotational tip speed}}{\text{(Isentropic gas velocity)}} = \frac{(\pi DN/60)}{v}$$

where *D* is the turbine rotor tip diameter for a radial turbine (mean blade diameter for an axial turbine) and

$$v^2/2 = (h_{t1} - h'_{t2}) = c_p(T_{t1} - T_{t2}) = c_pT_{t1}[1 - (P_2/P_1)^{(\gamma-1)/\gamma}]$$

giving

$$u/c = \frac{\pi DN/60}{\sqrt{2c_PT_1[1 - (P_2/P_1)^{(\gamma-1)/\gamma}]}} \quad (6.59)$$

again dropping 't' subscripts.

Thus knowing predicted pressures in the adjoining control volumes, inlet temperature and rotational speed, then non-dimensional speed and pressure ratio can be calculated. Interpolating the mass flow map (LUT) then gives ND mass flow and hence actual mass flow. Knowing turbine rotor diameter (6.59) gives *u/c* allowing turbine efficiency to be found from the efficiency map. Turbine exit temperature can then be found by re-arranging (6.57)

$$T_{t2} = T_1\{1 - \eta_{TS}[1 - (P_2/P_1)^{(\gamma-1)/\gamma}\} \quad (6.60)$$

Turbine power and torque can then be found from (6.58) and (6.52).

If a compressor and turbine are linked in a turbocharger then the rotational speeds must be identical, the turbocharger speed automatically adjusting until turbine power input (less any mechanical losses) equals compressor power output requirement, i.e.

$$\eta_m\dot{W}_T = \dot{W}_c \quad (6.61)$$

where η_m is usually very nearly unity and $\dot{W}_T$, $\dot{W}_c$ given by (6.58) and (6.51), respectively. More usefully turbocharger acceleration $(\dot{N})$ is given by

$$\dot{N} = \eta_m[TQ_T - TQ_c]/I \quad (6.62)$$

where *I* is the total turbocharger moment of inertia. At steady state, turbocharger speed does vary somewhat through the engine cycle as a result of the pulsating flow, and eqn (6.62) allows this to be represented by providing $\dot{N}$ for the turbocharger at each timestep, integration of which gives the (varying) turbocharger speed. For transient engine modelling of course, turbocharger speed will vary widely and (6.62) must be used to determine this. In steady state modelling, it is usual to start a run off with an estimated turbocharger speed and then correct this as the model 'converges' to the correct steady state conditions. One

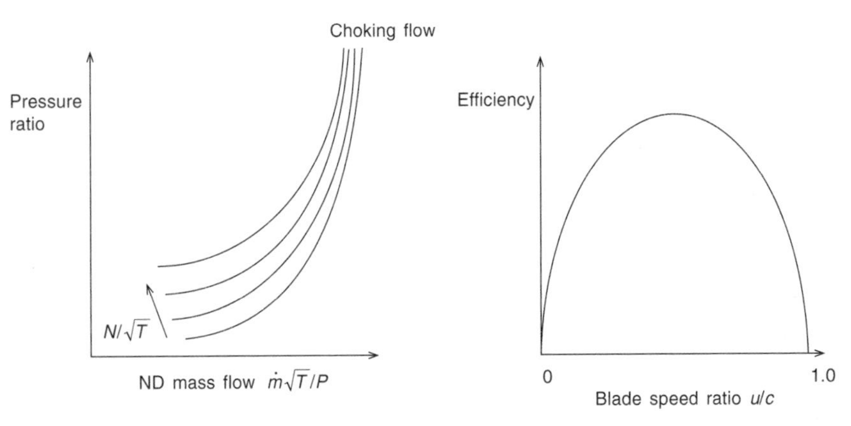

Figure 6.6 Turbine characteristics

way of doing this is to calculate TQ_T, TQ_C at each timestep from the above equations and obtain an artificially high acceleration by using an artificially low inertia in eqn (6.62), to speed up convergence.

While the above has been represented in terms of a simple turbocharger, a properly designed model should allow any configuration of compressors and turbines, including multiple shafts, more than one compressor or turbine on a shaft, or direct connection to engine crankshaft (superchargers and compound turbines). By proper model structuring any such arrangement can be accommodated and the calculation principles are exactly as given above, except that of course the acceleration eqn (6.62) will vary to reflect the connections. In the case of superchargers and compound turbines the speeds will be related directly to crankshaft speed of course and power will add to or subtract from the engine shaft power.

Before leaving turbomachinery, it is appropriate to deal with an important and frequently associated component, the charge air cooler (CAC), or intercooler. The compressor raises gas temperature significantly (eqn (6.53)), for example for a compression ratio of 3:1 and 70% efficiency the exit temperature is about 455 K for 298 K inlet temperature. As the high temperature gives low density and thus low cylinder charge the benefits of using a heat exchanger to cool the air before induction into the cylinder are considerable. This is normally carried out using some variety of air or water-to-air heat exchanger, and the benefits frequently outweigh the disadvantages of some loss of pressure, cost, additional space requirements and the need for a source of coolant. Whatever the type of CAC it can be represented for steady state engine performance modelling purposes by a heat exchanger effectiveness and a flow resistance. The CAC effectiveness is defined by:

$$\eta = \frac{\text{Actual heat transfer}}{\text{Max. possible H.T.}} = \left(\frac{T_i - T_o}{T_i - T_c}\right) \tag{6.63}$$

where T_i, T_o are the inlet and exit temperatures of the engine air flow through the CAC and T_c the coolant temperature. Frequently η can be assumed constant but if appropriate a variation of η with, for example, mass flow could be accommodated by a look-up table; η values can vary widely with the particular installation. The pressure loss through the CAC is given by eqn (6.44) for incompressible flow (which is valid as pressure differences are small):

$$\Delta P = (1/2\rho v^2)K$$

where K is the pressure loss factor. For modelling purposes the CAC mass flow is required. which is given by the alternative version of (6.44)

$$\dot{m} = A\sqrt{2\rho\Delta P/K}$$

where A is the engine air flow cross-sectional area in the CAC. In practice K may vary with mass flow so that a LUT or K is then required. In using these equations the coolant temperature (T_c) and the predicted CAC entry control volume temperature (T_i) are used with the known effectiveness in eqn (6.63) to obtain the CAC exit temperature. The predicted pressures in the CVs on either side of the CAC give ΔP so that (6.44) gives mass flow rate, using known flow area, K and density in the inlet CV. If K and η are functions of mass flow it will usually be adequate to take the previous timestep mass flow to evaluate K and η; if more accuracy is required the calculation can be iterated, the calculated mass flow being used to determine new values of K and η. With exit temperature known the exit enthalpy can be found using the property routines, thus giving all of the data from the CAC that is required to solve the adjoining CV energy equations.

6.6 The cylinder

In order to solve the energy equation for the cylinder, the rate of change of volume is required (e.g. eqn (64)), as well as the heat transfer rate which will depend on the combustion chamber surface area. It is appropriate first then to deal with cylinder geometry The cylinder compression ratio is given by

$$CR = (V_s + V_c)/V_c \text{ so } V_c = V_s/(CR - 1) \tag{6.64}$$

where V_s is the swept volume and V_c the clearance volume, the combustion chamber volume when the piston is at top dead centre (TDC). V_s is of course given by

$$V_s = \pi B^2 S/4 \tag{6.65}$$

for bore B and stroke S, the stroke being twice the crankshaft throw (r)

Referring to *Figure 6.7*, the instantaneous combustion chamber volume V and its rate of change, are given by

$$V = V_c + (\pi B^2/4)(l + r - x) \qquad \dot{V} = -(\pi B^2/4)\dot{x} \tag{6.66}$$

where l is the connecting rod length. Similarly the combustion chamber surface area is given by

$$A = A_P = A_h = A_l = A_P + A_h + \pi B(l + r - x) \tag{6.67}$$

where subscripts p, h, l refer to areas of the piston crown, cylinder head and exposed liner area, respectively. To solve (6.66 and 6.67) when crank angle is θ requires only the valves of x and $\dot{x}$. From Figure 6.7,

$$x = r\cos\theta \ (l^2 - r^2\sin^2\theta)^{1/2} \tag{6.68}$$

and differentiating:

$$\dot{x} = [r\sin\theta + (l^2 - r^2\sin^2\theta)^{-1/2} r^2\sin\theta\cos\theta]\,(d\theta\ dt) \tag{6.69}$$

where ($d\theta/dt$) is related to engine speed N (rpm) by

$$d\theta/dt = \pi N/30 \text{ rad/s} \tag{6.70}$$

so that chamber volume, its rate of change and exposed surface area can be calculated for any engine speed and crank angle.

The solution of the energy equation for the cylinder, and the data required for it, depends on the point that the cylinder is at in the engine cycle, which is illustrated in *Figure 6.8* for a 720° crank angle four-stroke cycle. The inlet closes shortly after BDC during the compression stroke, air and any residual burnt gases then being compressed and fuel injection starting towards the end of the compression stroke. After a short delay, combustion is initiated, usually towards the end of the compression stroke and continuing well into the expansion stroke. Late in the expansion stroke the exhaust valve opens, remaining open usually right through the exhaust stroke and closing shortly after TDC in the induction or intake stroke. Towards the end of the exhaust stroke the inlet valve is normally opened, so that both inlet and exhaust valves are simultaneously open around TDC the scavenge period. The inlet valve then stays open throughout the induction stroke through into the early part of the compression stroke. Solution of the energy equation for the cylinder will briefly be described for each of the cycle phases before dealing with the combustion phase in more detail.

To illustrate the form of calculations it will be assumed that the simple forms of the energy eqns (6.4 to 6.13) are being used. Thus the contents of control volumes (including the cylinder) are taken to be homogeneous and the equivalence ratio (eqn (6.3)) in any volume is the ratio of total burnt fuel present to total mass, divided by FAS, and thus can vary between near zero (air/residuals only) to the engine's operating equivalence ratio ϕ_e determined by the ratio of fuel flow to air flow.

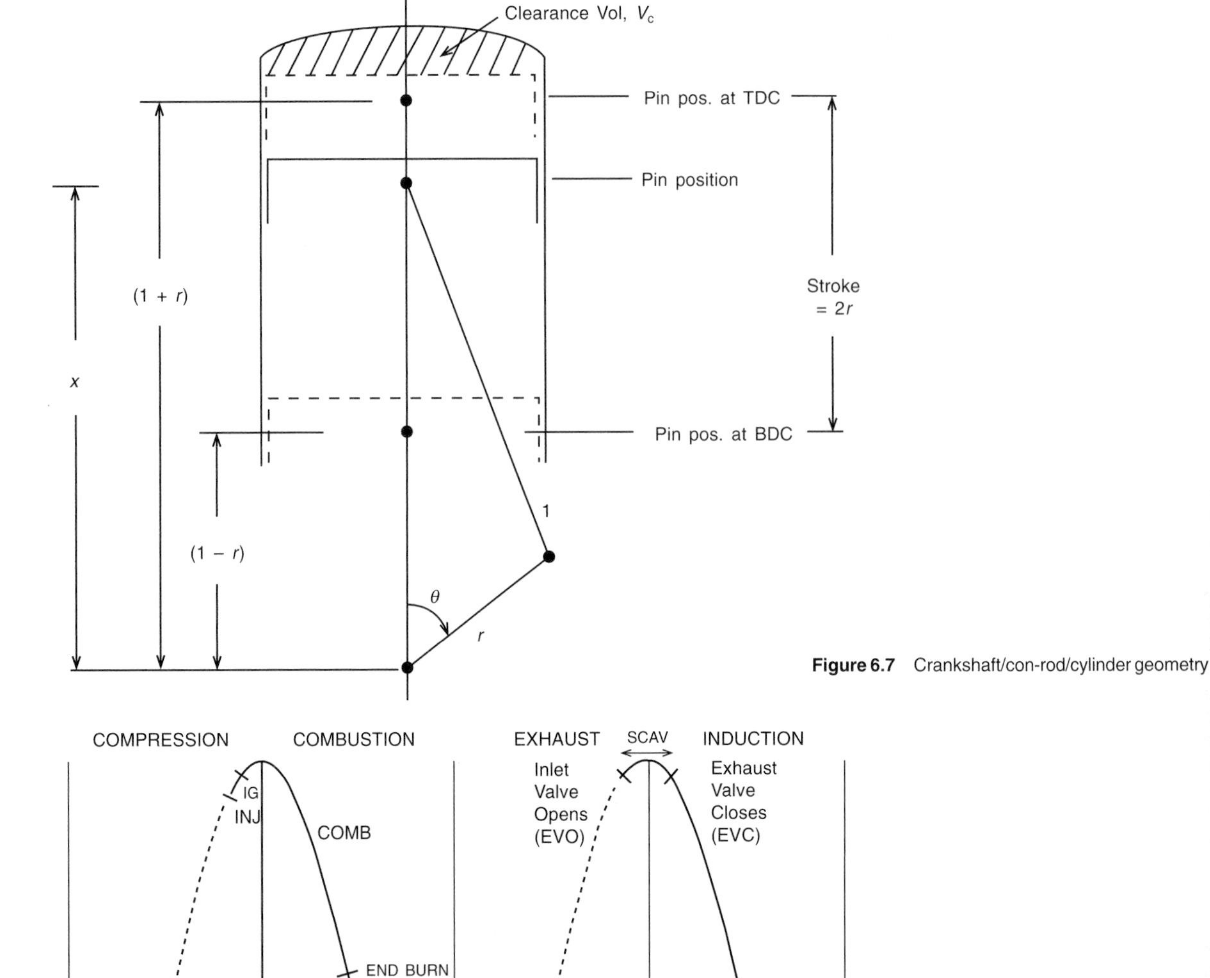

Figure 6.7 Crankshaft/con-rod/cylinder geometry

Figure 6.8 Four-stroke cycle

Considering the inlet-valve-open period first (*Figure 6.9*), at the time the inlet valve starts to open the cylinder pressure may exceed inlet manifold pressure, giving reverse flow of burnt gas from cylinder to inlet manifold; this period is discussed later in connection with scavenging. After a short time, cylinder pressure drops and normal flow is established, the flow rate (dm/dt) being determined by the methods of section 6.4; the flow will be mainly air, plus some returning residual gas. The specific enthalpy for the flow will be that of the inlet manifold, as given by the property routine (or *Figure 6.2*, eqn (6.10)), which will be close to that for air at manifold temperature and pressure but will include allowance for any small amount of residual burnt gas having reached the inlet manifold by earlier reverse flow. During this normal flow, the fresh air charge will mix in the cylinder with the residual burnt exhaust gas, so that the overall equivalence ratio ϕ in the cylinder will steadily decrease due to fresh air dilution ($d\phi/dt$) being given at any instant by eqn (6.6). If the period after EVC (i.e. after end of scavenging) is being considered, the enthalpy given by the product of this mass flow and specific enthalpy will be the only contribution to the $\Sigma\, \dot{m}h_i$ term in (6.4). As this is the only flow $\dot{M}$, the rate of change of cylinder mass, will be equal to the inflow (dm/dt). The current cylinder mass can be obtained either from integrating up the inlet mass flow or from the perfect gas equation, using current

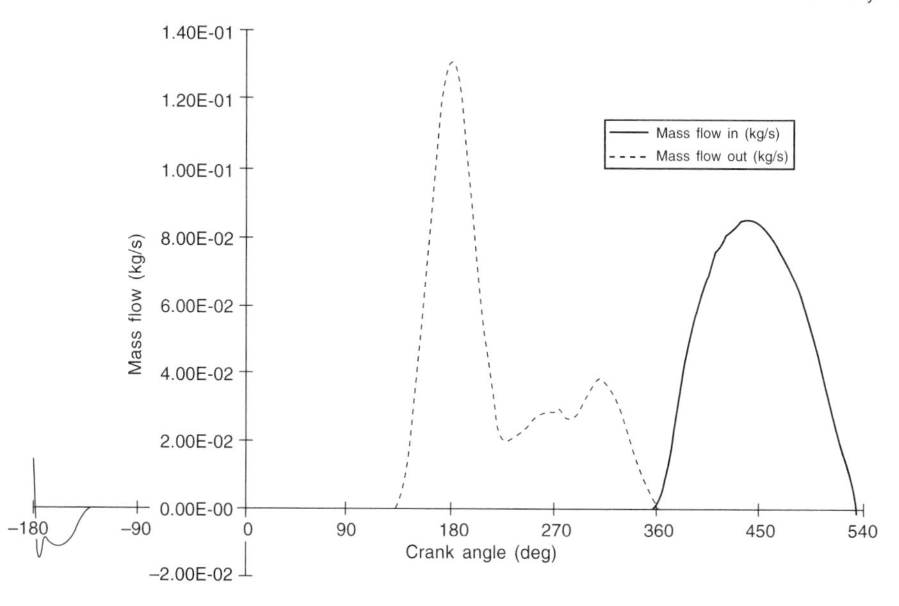

Figure 6.9 Cylinder flows

predicted cylinder pressure temperature and volume from (6.66). Similarly PV can be evaluated from predicted pressure and V from (6.66). $(\partial u/\partial T)$ and $(\partial u/\partial \phi)$ are obtained from the property routines (or Figure 6.2) for the current pressure, temperature and equivalence ratio in the cylinder. The two remaining terms required to evaluate $\dot{T}$ from (6.4) are heat transfer rate $\dot{Q}$ and rate of change of equivalence ratio $\dot{\phi}$. Discussion of heat transfer is referred to section 6.8 where it will be shown that $\dot{Q}$ can be readily evaluated. The value of $\dot{\phi}$ is determined from eqn (6.6) as follows. The ratio of burnt fuel mass to total mass at any point in the engine is given, using eqn (6.3), by

$$\frac{M_f}{M} = \frac{M_f}{M_a + M_f} = \frac{\phi FAS}{1 + \phi FAS} \tag{6.71}$$

where ϕ is the local equivalence ratio. Thus M_f, the burnt fuel mass in the cylinder, can be obtained from current cylinder mass M, FAS and the currently estimated in-cylinder equivalence ratio. The remaining term required is (dM_f/dt). During this inkate period the only flow is from the inlet manifold and so (dM_f/dt) must equal the fuel flow rate through the inlet valve, i.e.

$$dM_f/dt = dm_f/dt = (dm/dt)\left(\frac{\phi_m FAS}{1 + \phi_m FAS}\right) \tag{6.72}$$

where (dm/dt) is inlet flow rate and ϕ_m is the value of equivalence ratio in the inlet manifold. For the intake flow ϕ_m is low so (dM_f/dt) is low, being due only to any residual exhaust gas which has reached the manifold. In this case then the $(dM_f/dt$ term in (6.6) is a minor contribution $(d\phi/dt)$ itself from (6.6) will be negative and fairly large, as ϕ in the cylinder decreases rapidly from the initial residual gas value to the mainly air value (~ zero) With this data then (and the heat transfer term to be discussed eqn (6.4) can be solved for any timestep in the intake period to give $\dot{T}$ and thus a better estimate of cylinder temperature, T. Equation (6.12) (or 6.13, for $\dot{P}$) then gives a better estimate of pressure P.

The procedure for the exhaust-valve-open period after end of scavenge is analogous to the inlet valve period, but actual numbers are of course very different. Mass flow from the methods of section 6.6 for the exhaust valve, and the specific enthalpy for the cylinder contents (burnt gas) from the property routine (or Figure 6.2), give the $\Sigma \dot{m}h$ term. $\dot{M}$ will equal the outflow, while $(d\phi/dt)$ can be evaluated from (6.6) again, using

$$dM_f/dt = (dm_f/df)_{ev} = (dm/dt)_{ev}\ (\phi_c\ FAS/(1 + \phi_c\ FAS)) \tag{6.73}$$

where now $(dm/dt)_{ev}$ is the exhaust valve low and ϕ_c is the equivalence ratio in the cylinder (from where the exhaust gas originates).

For the scavenge period when both inlet and exhaust valves are open, the procedure is similar but two flows have to considered rather than just one. Thus there are two contributions to the $\Sigma \dot{m}h$ term, one evaluated as in the inlet only case, the other as in the exhaust only case. $(d\phi/dt)$ can again be calculated from (6.6) but now (dM_f/dt) has two components, a large one from (6.73) for the exhaust flow, a small one from (6.72) for the inlet flow. It may be noted that in this model it is simply assumed that during scavenge the inflow gas mixes completely and immediately with the residual burnt gas, as cylinder gas properties are derived for the mixture. More complex scavenging sub-models are possible where a mixing rate is introduced so that exhaust gas can be richer than given by the simple model.

Thus the energy quation can be solved or inlet, exhaust and scavenge periods. Solution during the compression period between IVC and fuel injection and from end of combustion to EVO are carried out in the same way but of course the external flows are zero. Solution of the remaining part of the cycle from start of injection to end of combustion is considered next.

6.7 Injection and combustion

It is not the purpose here to describe the combustion process in detail as this is covered elsewhere, but to illustrate how this can be included in an engine model. For the present purposes, it can be assumed that following start of injection of fuel there is a delay period while some of the injected fuel evaporates and mixes with air in the cylinder to the point where some mixture reaches roughly stoichiometric composition where combustion can take place. In the diesel engine then, combustion starts at a number of points, more or less simultaneously, around the fuel spray from each injector hole. As the fuel and air around these centres of combustion is pre-mixed, there is usually a short period of very rapid burning (pre-mixed burning), followed by a rather lower rate of burning as the remaining fuel and air diffuse into the already burning regions and provide further mixture of a burnable composition (diffusion burning). This general burn pattern is as in the schematic *Figure 6.10* which shows burn rate ($\dot{x}$) in terms of crank angle.

The division between pre-mix and diffusion burning varies considerably with combustion chamber design and engine operating conditions. In quiescent chambers the pre-mix region may be almost absent, whereas in engines with high swirl, giving rapid fuel–air mixing, the pre-mixed burning region will tend to be more significant. Similarly long ignition delays (see below) caused by low P and T after injection give more time for mixing before ignition and hence a relatively longer pre-mix burning phase. A variety of sub-models have been produced to represent combustion but even the most detailed fluid dynamic models have not been able to fully represent diesel engine combustion because of the number of complex processes involved. These include fuel jet pattern and fuel vaporization, gas motion in the cylinder, flame speeds, air mixing, chemical reaction kinetics, wall interaction and crevice effects, the often quite large influence of small changes in chamber design, etc. There is only a limited understanding of many of these processes. In complete engine cycle simulation there is also a run-time problem with the more sophisticated sub-models.

To model the combustion process for cycle simulation purposes, two key parameters are required, the ignition delay from start of injection and the variation of fraction of fuel burnt with crank angle after ignition has taken place. If these are known then the use of the burnt gas property routines and the energy equation in the usual timestep process allows the calculation of cylinder pressure and temperature variation through the combustion process. In the simplest models, which are adequate for much diesel cycle simulation work, the combustion chamber is treated as a single homogeneous zone, pressure, temperature and burnt fuel equivalence ratio being assumed uniform across the chamber at any time. To illustrate the broad procedure, assume that P, T and ϕ are known at the end of a timestep during combustion, and that predicted (estimated) P and T are known for the end of the next timestep. If the fuel burning rate is known in the new timestep then eqn (6.6) will give the rate of change of burnt fuel equivalence ratio in the next step, allowing a new ϕ value to be determined. With this and the predicted P/T, the gas property routines (or *Figure 6.2*) will give new values of internal energy and other properties of the cylinder gas at the end of the timestep; this property data will reflect the fact that during the timestep a small mass of air and fuel has been converted to burnt gas, with energy release. Application of the energy eqn (6.4) will then give $\dot{T}$ and $\dot{P}$ at the end of the step, thus leading to new estimates of end of step pressure and temperature. Thus the cylinder conditions can be determined through combustion provided the variation of fuel burning rate is known; this can be expressed in terms of $\dot{x}$ the fractional burning rate (actual rate equals $\dot{x}$ times total fuel delivery to the cylinder). First, it is necessary to determine the ignition delay from start of injection.

The ignition delay time τ_I is given reasonably accurately by an empirical expression due to Wolfer:

$$\tau_I = 3.454 \times 10^{-3} \exp(2.1008 \times 10^3/\bar{T})/\bar{P}^{1.028}\,\mathrm{s} \qquad (6.74)$$

where $\bar{T}$ (K) and $\bar{P}$ (kN/m^2) are the average cylinder temperature and pressure during the delay period. To use (6.74), $\bar{T}$ and $\bar{P}$ are calculated each step after start of injection and τ_D values calculated; ignition is assumed to occur on that step where τ_D is less than time since start of injection.

For the burning rate variation, a widely used diesel combustion sub-model is the semi-empirical one due to Watson, Pilley and Marzouk[2], which separately models the pre-mix and diffusion phases and uses a further empirical factor (β) to determine the relative proportions of the two phases. The overall fuel fraction burnt at any instant is then given by:

$$\left.\begin{aligned} x &= \beta x_P + (1-\beta)\,x_D \\ \beta &= 1.0 - 1000a\,\phi_e^b/\tau_I^c \end{aligned}\right\} \qquad (6.75)$$

where a, b, c are empirical constants, τ_I the ignition delay and x_P, x_D (see below) the fractions of fuel burnt in the pre-mix and diffusion phases respectively at the time of interest. All of x, x_p, x_D vary 0 to 1 through combustion (*Figure 6.10*). ϕ_e is the engine operating equivalence ratio, not ϕ based on burnt fuel which varies through combustion.

For the diffusion burning phase a standard Wiebe function is used for x_D:

$$x_D = 1 - \exp(-A\,\Omega^B) \qquad \Omega = (CA - CAI)/BDUR \qquad (6.76)$$

where A and B are shape factors given by:

$$A = e/\phi_e^f \qquad B = \dot{g}A^h$$

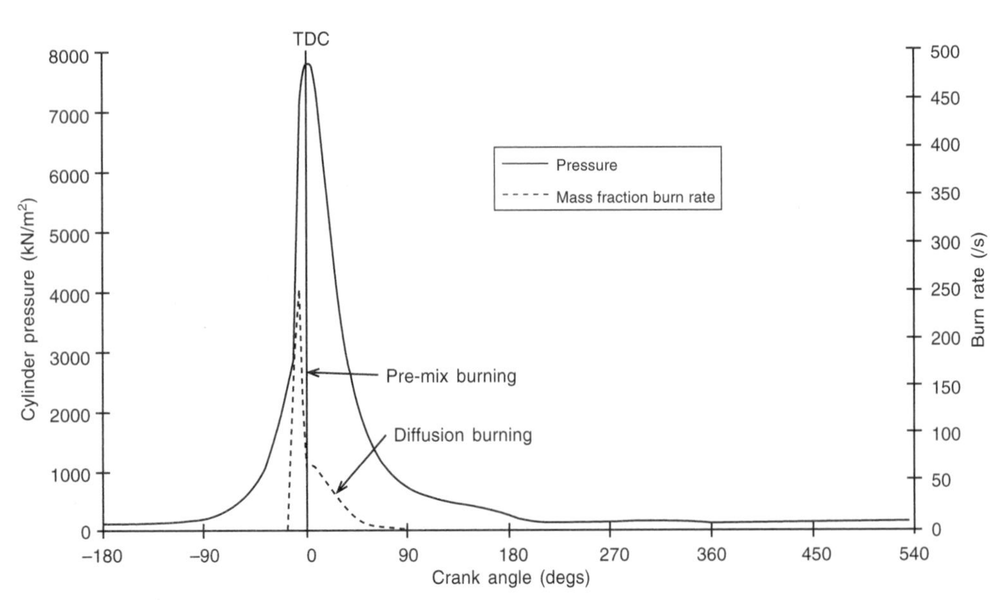

Figure 6.10 Schematic cylinder pressure and burn rate profiles

where e, f, g, h are further empirical constants, and Ω is the non-dimensional crank angle, measured from ignition (*CAI*), normalized by burn duration (*BDUR*). For pre-mix burning a function fairly similar to Wiebe is used:

$$x_p = 1 - (1 - \Omega^n)^{5000.0} \tag{6.77}$$

where

$$n = 2.0 + 0.002703\ (\tau_I\ 6N)^{2.40}$$

where the '6*N*' factor is introduced to convert τ_I from seconds to deg *CA*. The eqns (6.75), (6.76) and (6.77), with (6.74) for the ignition delay thus give the mass fraction burnt at any instant, although care is needed in using them to avoid numerical problems in some cases. It is also necessary to know the values of empirical 'constants' *a* to *h* and the burn duration. The latter vary to some extent with the engine but a typical set for a turbocharged truck engine, and their likely ranges for other engines, are as follows:

	Eqn (6.75) – β			Eqn (6.76) – x_D			
	a	*b*	*c*	*e*	*f*	*g*	*h*
Typical	0.93	0.37	0.26	14.2	0.644	0.79	0.25
Range	0.8–0.95	0.25–0.45	0.25–0.55	7–17	0.6–0.7	0.7–1.2	0.13–0.25

With the fuel fraction burnt, *x*, known at any instant the total fuel burnt can be obtained simply by multiplication by the cylinder's fuel charge per cycle (M_{ft}). To solve eqn (6.6) the rate of burning fuel (dM_f/dt) is required and this can be obtained by either differencing (xM_{ft}) values at successive timesteps or by differentiating eqns (6.76) and (6.77):

$$\dot{x}_D = AB\Omega \exp(-A\Omega^B)\dot{\theta}$$

$$\dot{x}_P = 5000C\Omega^{n-1}(1 - \Omega^n)^{5000-1}\ \dot{\theta} \tag{6.78}$$

where $\dot{\theta}$ is the (non-dimensional) rate of change of crank angle (6*N*/*BDUR* for *N* rpm, *BDUR* in deg. *CA*). The value of *BDUR* in the equation is not critical as long as it is at least as long as actual combustion duration and a value of 125 degrees was used originally.

To a reasonable approximation for much work (dM/dt) in eqn (6.6), the rate of change of total cylinder mass at any instant, can be taken to be the same as (dM_f/dt) and the actual injection pattern ignored; this is equivalent to the assumption that the fuel is injected as in is burnt. Alternatively an injection time profile can be input, either giving the variation in rate of fuel per unit time or the injector open area; in the latter case fuel flow rate at any instant can be calculated from the hydraulic eqn (6.44), using an appropriate discharge coefficient and an instantaneous pressure difference across the injector taken as fuel rail pressure less current cylinder pressure. If injection is represented explicitly then the fuelling rate is included in the (dM/dt) term of (6.6); (dMf/dt) is still the fuel burning rate as the form of model described uses a burnt-fuel based equivalence ratio. Whatever method is used to represent injection/combustion it will be necessary to provide the model with data on start of injection time for each cylinder, normally speed dependent. For steady state modelling it is usually sufficient to specify input fuel per cycle per cylinder as an input variable but alternatives such as specifying fuel–air ratio and letting the model determine fuelling are fairly straightforward. For transient modelling (section 6.4) it will be necessary to in corporate a fuel pump/governor sub-model to determine fuelling as engine speed and load change; there will usually also need to be an aneroid control override to limit fuelling at low boost pressure. For more modern fuelling systems fuel control 'maps' (look-up tables) will be required (again generally specified in terms of engine current speed and load) which can be interpolated to obtain current injection timing, pattern, fuel quantity, etc.

Before leaving combustion it is worth giving some further information on heat release. First, the approximate heat release up to any instant is given by

$$Q_t = x\ M_{ft}\ Q_{LCV} \tag{6.79}$$

where *x* is the current fraction burnt from (6.75, 6.76, 6.77), M_{ft} is the fuel charge and Q_{LCV}, the fuel's lower calorific value. The heat released of course is split approximately between raising the charge temperature, doing work on the piston and heat transfer through the chamber walls. It is also worth mentioning that when a fuel–air mix burns, there is no change in specific enthalpy, that is

$$h_u = h_b \quad \text{or} \quad u_u + RT_u = u_b + RT_b \tag{6.80}$$

with obvious subscripts. If the specific internal energies of the unburnt mix and burnt gas are known, with the unburnt mix temperature immediately before combustion, the equation gives the adiabatic flame temperature. Finally, while it should be apparent that by deriving fuel burn data from eqns (6.74)–(6.78) and combining this with the energy equation it is possible to compute the time variation of *P* and *T* in the cylinder, it may not be apparent that one can do the reverse. That is, using a measured cylinder pressure diagram one can reverse the procedure to calculate burnt fuel fraction variation and, by repeating the exercise for alternative engine speeds and loads, go further back and generate values of the various coefficients in eqns (6.74)–(6.78) for the specific engine of interest. However, this is quite a lengthy exercise analytically, and it is important to have very accurate cylinder pressure measurements as any small ripple in the pressure curve (real or otherwise) will cause a large bump in the resulting burn rate curve. Normally, empirical adjustment of the coefficients in the equations is adequate and quicker.

6.8 Heat transfer and friction

To avoid distraction in the above sections, heat transfer (HT) has been omitted, but as mentioned it must be evaluated to solve the energy equation. It is important not only for the cylinder but also for manifolds and pipes, at least on the exhaust side of the engine. Fortunately as far as the cylinder is concerned the heat transfer calculations, while they have to be included, are not too critical for engine performance analysis as HT accounts for a relatively small part of the total combustion energy; this is fortuitous as the HT process is quite complex and cannot be represented with great precision. Of course the HT calculations become more important if the results are to be used in cooling system or engine thermal stress work rather than simply engine performance. Some of the data required for HT calculations will be difficult to obtain with any accuracy but at least for performance work, fairly crude estimates usually suffice.

For convection and conduction heat transfer, the heat flow rate through a material layer is given by

$$\dot{Q} = hA(T_h - T_c) \tag{6.81}$$

where *h* is the layer heat transfer coefficient per unit area, *A* the HT surface area and T_h, T_c the hot and cold side surface temperatures. The same equation can be used for a multiple-component layer, for example a gas convection layer, a cylinder wall conduction layer and a coolant layer, using a composite heat transfer coefficient (h_{TOT}). Further, if steady state heat flow can be assumed then such an equation can be used for each individual sub-layer. The heat transfer rate through all layers must be equal so, using the notation of *Figure 6.11*:

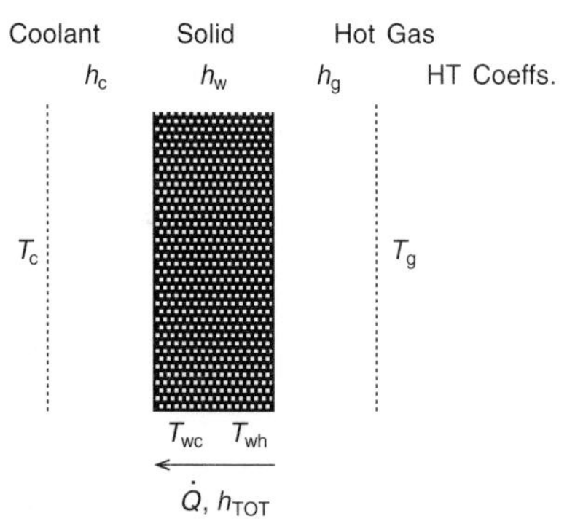

Figure 6.11 Wall heat transfer

$$\dot{Q} = h_{TOT}A(T_g - T_c) = h_gA(T_{hg} - T_{wh}) = h_wA(T_{wh} - T_{wc}) = h_cA(T_{wc} - T_c) \quad (6.82)$$

Replacing the HT coefficients by thermal resistances:

$$\dot{Q} = (T_{hg} - T_c)/R_{TOT} = (T_{hg} - T_{wh})/R_g = (T_{wh} - T_{wc})/R_w = (T_{wc} - T_c)/R_c \quad (6.83)$$

where

$R = 1/hA$ for convection $R = 1/hA$

$= L/kA$ for conduction (6.84)

k being the conducting material conductivity (e.g. kW/(mK)) and L the material thickness. By manipulation of (6.83):

$$R_{TOT} = R_g + R_w + R_c \quad (6.85)$$

for this 'in-series' heat transfer path. The resistance R_i can be handled exactly as electrical resistance; equations for parallel HT paths of HT networks can be derived similarly to the above and the resistances in these can be manipulated exactly as their electrical analogues.

For an engine cylinder the HT rate varies rapidly thrugh the engine cycle as cylinder gas temperature changes. For the cylinder gas–wall boundary layer at any instant

$$\dot{Q} = h_gA(T_g - T_w) \quad (6.86)$$

where T_g is the cylinder gas temperature, obtained from the cylinder energy equation which of course requires this value of $\dot{Q}$! This is not as difficult as it may appear because $\dot{Q}$ can effectively be estimated (from 6.86) at any instant using the predicted cylinder gas temperature (provided h_g, A and T_W are known), and then used in the cylinder energy equation to obtain the correct cylinder gas temperature, which will be near the predicted one, or the 'predicted' value can be improved by iteration. The problem is to find h_g and T_w (and A). The calculation of h_g is discussed later, to avoid a large distraction at this point. T_w is obtained from a solution of the heat transfer network from gas to engine coolant. Now while the instantaneous $\dot{Q}$ values vary rapidly, the HT network has considerable thermal inertia so that its temperatures (other than T_g) do not vary very quickly and in practice a time-averaged value of wall temperature can be used from the previous engine cycle for all timesteps in the next cycle. If T_w is assumed known for a cycle then $\dot{Q}$ at any instant can be obtained from (6.86) using the current timestep prediction of T_g and calculated h_g (below) values. At the end of the cycle, cycle-averaged values of θ, h_g and T_g can be determined and used in thermal resistance network equations to make a new evaluation of T_w for use on the next cycle, the whole process being started off by an initial estimate of T_w. The cycle-averaged values are simply:

$$\bar{\dot{Q}} = \Sigma \dot{Q}/CYCDUR;\ \bar{h}_g = \Sigma h_g/CYCDUR;\ \bar{T}_g = T_w + \bar{\dot{Q}}/\bar{h}_g A \quad (6.87)$$

the last being a re-arrangement of (6.86) using cycle-averaged parameters.

At the simplest level one could assume an average thickness and conductivity for the cylinder 'wall' and coolant boundary layer. Then:

$$\bar{\dot{Q}} = \bar{h}_g A(\bar{T}_g - T_w) = (1/R_w)(T_w - T_c) = (k_wA/L_w)(T_w - T_c) \quad (6.88)$$

from which T_w can readily be evaluated. In practice this is too simplistic and it is sensible to divide the cylinder 'wall' into at least three parts, the liner (L), piston (P) and head (H), in this case using three sets of equations (6.87), three wall temperatures T_{wL}, etc. However, the principles of determining the T_w values are still as in (6.88), but a HT network must be analysed. A simple thermal resistance network for an engine cylinder might be as in *Figure 6.12* (though the network could be considerably more complex), for HT from cylinder gas through L, P, H components to main coolant and oil.

The network need only be solved at the end of each cycle as mentioned earlier, not each timestep. A to H are resistances (R_i in 6.83, 6.85), T_{GL}, etc. are cycle-average gas temperatures just outside the cylinder gas boundary layer for the liner, piston and head (the T_g of eqn 6.87), T_c and T_{oil} are the two specified coolant temperatures and T_{NODE} is the temperature at some point in the piston to allow piston-to-liner HT calculation; it is assumed that the head and liner are thermally isolated. Resistances *A–C* are for the convection from cylinder gas to L, P, H walls, respectively; *F–H* are combined resistances for both the P, L, H conduction and convection from their cold side surfaces through the relevant coolant boundary layer. *D* and *E* are resistances from node to liner and piston hot side, respectively. T_{wL} etc., are the three wall temperatutres. For the isolated cylinder head, the wall temperature is given by eqn (6.88), with appropriately changed subscripts. For the connected liner and piston, three simple simultaneous equations have to be solved, and an additional heat input $\dot{Q}_f$ from piston ring friction into the liner allowed for. Carrying out a heat flow balance at the 'T_{wL}' point:

$$\dot{Q}_{out} = \left(\frac{T_{wL} - T_c}{F}\right) = \dot{Q}_f + \left(\frac{T_{gL} - T_{wL}}{A}\right) + \left(\frac{T_{NODE} - T_{wL}}{D}\right) + \dot{Q}_f \quad (6.89)$$

where *F*, *A* and *D* are resistances. Similarly heat flow balances at T_{NODE} and T_{wP} give:

$$\left(\frac{T_{NODE} - T_{wL}}{D}\right) + \left(\frac{T_{NODE} - T_{oil}}{G}\right) = \left(\frac{T_{gP} - T_{NODE}}{E + B}\right) \quad (6.90)$$

$$\left(\frac{T_{NODE} - T_{wP}}{E}\right) = \left(\frac{T_{gP} - T_{wP}}{B}\right) \quad (6.91)$$

Appropriate straightforward manipulation of the three equations then give the three temperatures T_{NODE}, T_{wP} and T_{wL} provided the other quantities are known, which they will be. The value of $\dot{Q}_f$ is normally taken as a fraction (e.g. 20%) of the friction power (see below). The thermal resistances *D* to *H* are all calculated in a similar way, but are likely to be fairly approximate estimates. The example of *F*, the liner wall to coolant resistance will illustrate the procedure.

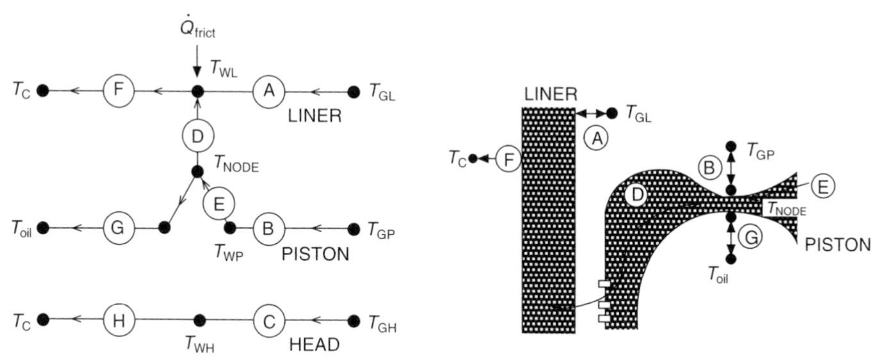

Figure 6.12 Heat transfer network

$$F = \frac{1}{h_1 A_L} + \frac{1}{h_2 A_L}$$

where A_L is the (cycle-average) liner area exposed to the cylinder gas, h_1 is the HT coefficient for the wall (material conductivity/thickeness) and h_2 is the coefficient for the coolant boundary layer. For water coolant an expression for h_2 which has been used is

$$h = (\bar{\dot{Q}}/A_L)^{0.644}/7.88 \quad \text{kW/(m}^2\text{ K)} \tag{6.92}$$

for $\dot{Q}$ in kW, A_L in m^2. The other resistances D to H are calculated similarly but for some it is difficult to know exactly what flow paths should be used (e.g. D from piston node to liner). Fortunately, as mentioned, precise numbers are not usually required.

To complete the HT calculations for the cylinders, the resistances A to C and the related cylinder gas heat transfer coefficient h_g are required, time average $\bar{h}_g$ being related to resistance by

$$B = 1/(\bar{h}_g A_P)$$

where A_P is piston HT surface area. A number empirical equations exist for h_g in the cylinder at any instant, perhaps the most commonly used being that by Woschni[10]. Now dimensional analysis indicates that heat transfer in gases can be expressed by a relationship between three dimensionless parameters, the Nusselt, Reynolds and Prandtl numbers, that is

$$\text{Nu} = c_1 \text{Re}^{c_2} \text{Pr}^{c_3} \text{ or } \left(\frac{hl}{k}\right) = c_1 \left(\frac{\rho v l}{\mu}\right)^{c_2} \left(\frac{c_p \mu}{k}\right)^{c_3} \tag{6.93}$$

where the parameters are gas heat transfer coefficient (h), conductivity (k), density (ρ), and viscosity (μ), specific heat (c_p), while v and l are, respectively, a characteristic velocity and length for the application; c_i are empirical constants. Prandtl number varies very little and can often be ignored. Equation (6.93) is a fundamental equation in gas heat transfer work, and Woschni's equation for cylinder HT, as well as others, was based on it, supplemented by measurement on a variety of engines:

$$h = \frac{0.003264\, P^{0.8} v^{0.8}}{B^{0.2} T^{0.53}} \text{ kW/m}^2\text{ K} \tag{6.94}$$

where P is pressure (kN/m^2), B cylinder bore (m), T is temperature (K) and v is an average cylinder gas velocity given by

$$v = a_1 v_p + a_2[(V_s T_r / P_r V_r)(P - P_m)] \tag{6.95}$$

where V_s is swept volume, P_r, T_r and V_r are cylinder parameter values at start of combustion, P is actual cylinder pressure at the timestep of interest, P_m is the corresponding pressure for a 'motored' (unfired) engine, v_p is mean piston speed (by averaging eqn 6.69 over the cycle) and a_1, a_2 are constants depending on the part of the cycle that the cylinder is in at the timestep. All of the parameter values in the fairly complex equations are quite easily obtained except P_m the current timestep pressure for a motored engine. However, to reasonable approximation, isentropic compression and expansion can be assumed so that:

$$(P_m/P_r) = (V_r/V)^{1.32} \tag{6.96}$$

Regarding a_1, a_2, the value of a_1 is 6.18 for the valve open periods but 2.28 for other (compression, combustion, expansion) periods. a_2 is zero except for combustion and expansion (to EVO) when it is 3.24×10^{-3}. For engines with swirl, a_1 is increased by ($0.417\, v_s/v_p$) for valve open periods or by ($0.308\, v_s//v_p$) at other times, where v_s is the swirl velocity.

In practice radiation plays some part in cylinder heat transfer but it is difficult to represent with precision and is normally taken to be incorporated approximately in (6.94). Radiation originates from the CO_2 and H_2O in the cylinder, but these are 'band' radiators, radiating in only part of the spectrum and are not too important (or easy to represent). The other radiation source is carbon particles, which radiate continuously across the spectrum, but again precise methods of calculation are not available.

Turning to components other than cylinders (pipes and manifolds), heat transfer can be represented in a precisely similar way, using internal gas temperature, external coolant temperature (normally ambient temperature for components cooled only by ambient air) and a thermal resistance. The latter will include an external convection heat transfer coefficient, a wall conduction term and an internal heat transfer coefficient usually based on eqn (6.93) with differing c_1–c_3 values depending on application. As Prandt number varies very little it is usual to use the simpler form

$$\text{Nu} = hl/k = c_1 \text{Re}^{c_2} = c_1 (\rho v l/\mu)^{c_2} \tag{6.97}$$

For pipes, characteristic length l is normally taken as pipe diameter and c_2 is usually 0.8. The value of c_1 varies[3] from about 0.02 for turbulent flow in a smooth pipe to 0.05 in an exhaust pipe to as high as 0.16 in an exhaust port, presumably due to the more complex structure and turbulent flow near the exhaust valve.

Friction in the engine is obviously very important but is difficult to calculate with precision from the various individual friction sources. Fortunately, there are empirical calculation methods available which give a good representation of total friction for diesel engines. The most widely used is probably that due to Chen and Flyn[2] which gives friction mean effective pressure (FMEP) as a function of maximum cylinder pressure and engine speed:

$$FMEP = 13.79 + 0.005\, P_{max} + 1.086\, rN \text{ kN/m}^2 \tag{6.98}$$

where P_{max} (kN/m^2) is maximum cylinder pressure, r the crankshaft throw (m) and N engine speed (rpm). While the constants give a reasonable fit to many turbocharged diesel engines, alternative values of these can be derived for specific engines if measurements of both indicated and brake power are available (see Section 6.9). Other variants of eqn (6.98) are possible. For example or SI engines a term proportional to $(rN)^2$ is often used without the P_{max} term but for turbocharged diesel engines the high cylinder pressures do give a significant increase in friction whereas engine speeds are (c.f. SI) relatively low. Regarding sources of friction, around 60% is attributable to the piston and piston rings, other sources being valves, camshafts, bearings and agitation of engine oil by the crankshaft. Although there is considerable literature on calculating the contributions of individual sources, there is some difficulty in getting them to sum up properly and it is better for most modelling purposes to use a relationship of the (6.98) type, preferably with constants from measurements on the engine of interest.

6.9 Model results and engine performance

Using the methods outlined in the previous sub-sections it is possible to build a basic steady-state diesel engine model. While it would require enhancement or detailed performance analysis it would contain the main elements and be sufficiently accurate for preliminary investigations. The model would course require input routines to provide the data needed for the various calculations described and output routines to provide the results of the form indicated below. For steady-state operation it would be necessary to specify the required fuelling rate and engine speed, together with initial estimates of P, T and ϕ in each control volume and of turbocharger speed. The model would then run for several cycles to achieve 'convergence' that is the correct P, T, ϕ and turbocharger speed for the specified fuelling and engine speed. The following describes the type of information which can be obtained from such a model and serves to introduce a number of basic engine performance equations.

The model can provide two basic types of results, those giving the variation of engine parameters through the 720° (for a four-stroke engine) of the cycle and time-averaged performance data for the cycle. Dealing with in-cycle data first, some examples are shown in *Figure 6.13* for a 6-cylinder turbocharged diesel; while these were provided by the ICENG model, similar results could be obtained from a simpler model of the type described. Outputs which can be obtained from this type of model include:

(a) For all control volumes (including cylinders) the variation of P, T and ϕ through the cycle, along with mass of gas in the CV and its specific enthalpy if required.
(b) Mass flow rate variation through pipes and valves, through turbomachinery, and into and out of CVs; if required, variations in valve areas can be obtained.

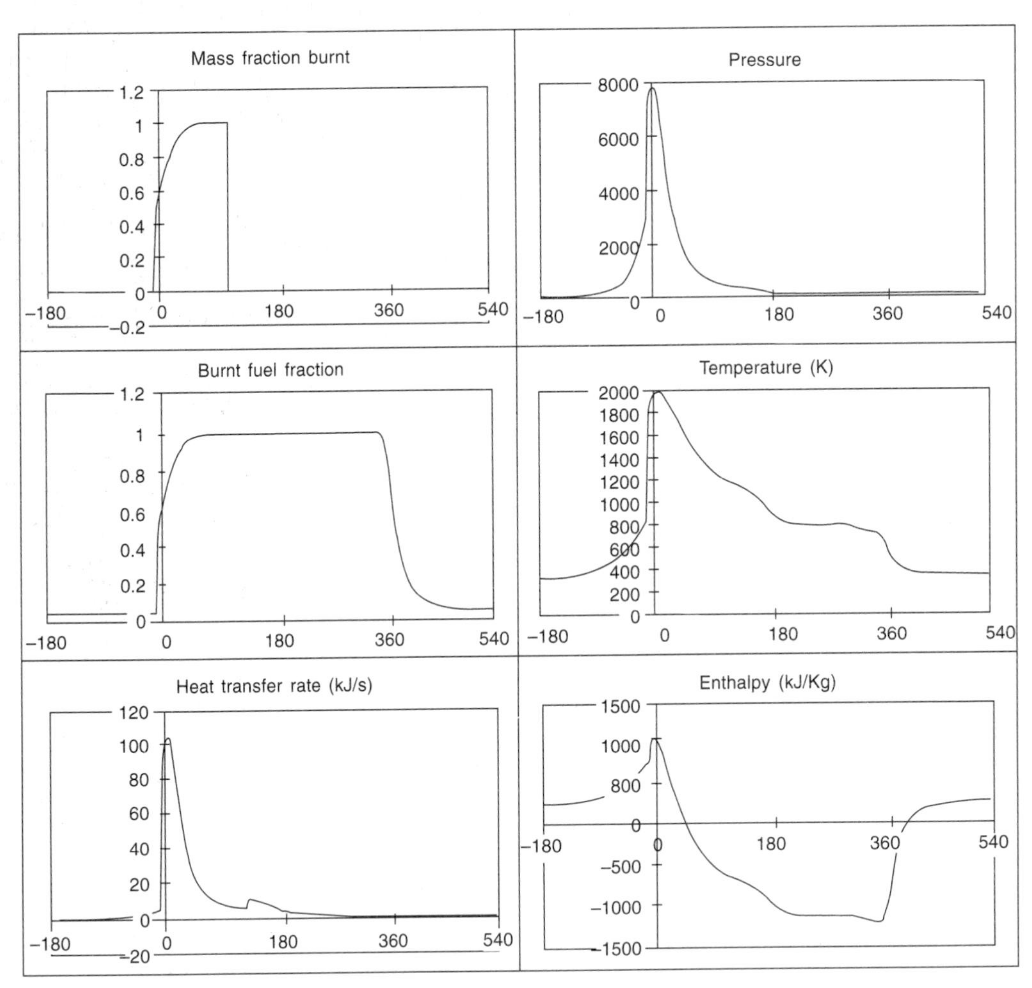

Figure 6.13 Example in-cycle results of cylinder

(c) For the cylinder, variation of mass fraction burnt, burning rate, indication of start of injection/combustion and end of combustion, variation of cylinder volume. This cylinder data allows determination of the P–V diagram and identification of the important maximum P, T, and the crank angles at which they occur.
(d) For the turbomachinery, the variation of pressure ratio, mass flow, efficiency, speed through the cycle, together with instantaneus power absorbed (eqn (6.51) for compressor) or produced (eqn (6.55) for turbine).

A more convenient summary of important results, together with time-averaged data can be produced based on this extensive in-cycle data. Such a summary would include fuel and air flow rates, trapped fuel and total mass in the cylinders, peak cylinder pressures and temperatures, injection and other timings, cycle-averaged heat transfer rates and turbomachinery power inputs/outputs, efficiencies, mass flows, etc. Overall performance data can also be calculated by summation and time averaging, the important basic quantities being the total airflow into each cylinder and work done by the gases in each cylinder. Cylinder airflow during a cycle is

$$M_a = \int_{CYC} \dot{m}_{ai}\, dt = \sum_{CYC} \dot{m}_{ai}\, \tau \qquad (6.99)$$

where $\dot{m}_{ai}$ is the instantaneous airflow rate into the cylinder and τ the timestep. The work done by the gases in a cycle, for one cylinder, is given by

$$W = \int_{CYC} P\, dV = \sum_{CYC} P(dV/dt)\,\tau = \sum_{CYC} P\,\delta V \qquad (6.100)$$

where δV is the change in cylinder volume in a timestep τ, P the corresponding instantaneous pressure (W in kJ for PkN/m^2 and δV m^3). The indicated mean effective pressure (IMEP) for the cylinder is then given by

$$IMEP = W/V_s \qquad (6.101)$$

where V_s is the swept volume; IMEP values are normally in the range 1000–1600 kN/m^2 for turbocharged diesel engines. Calculated in this way, IMEP is the net value, allowing for pumping losses through intake and exhaust. The friction mean effective pressure (FMEP) is given directly by eqn (6.98) and the brake mean effective pressure (BMEP) is then simply, for no engine auxilliaries fitted.

$$BMEP = IMEP - FMEP \qquad (6.102)$$

The indicated power from a single cylinder is simply the work done in a cycle (W) divided by cycle duration (t_c), thus for a four-stroke engine

$$\text{Indicated power} = \frac{W}{t_c} = IMEP{*}V_s$$

$$= \frac{W{*}N}{120}\left(= \frac{IMEP{*}V_s}{t_c}\right) \text{ (kW, for WkJ)} \qquad (6.103)$$

where N is engine speed in rpm. Similarly

Friction power = $FMEP{*}Vs{*}N/120$
Brake power = $BMEP{*}Vs\ {*}N/120$

To give total engine power, these figures have to be summed up over the cylinders (or, if all cylinders are identical, multiplied by the number of cylinders). The delivered or brake torque is derived from brake power using

$$\text{Torque} = \text{Brake power} * (30/\pi N) \quad \text{(kNm, for kW power and N rpm)} \qquad (6.104)$$

Specific fuel consumption can be calculated from the fuel mass flow rate divided by power. Thus brake s.f.c. given by

$$\text{B.s.f.c.} = \text{Fuel flow rate/(Brake power)} \qquad (6.105)$$

and similarly for indicated s.f.c. using indicated power. Regarding efficiencies the most important is the volumetric efficiency given by

$$\eta_{vol} = M_a/(\rho_a V_s) = M_a RT/PV_s \qquad (6.106)$$

the mass of air actually inducted into the cylinder divided by the mass which would be inducted if the cylinder were filled at inlet pressure. If T and P are ambient conditions, η_{vol} will be much greater than unity for a turbocharged engine (η_{vol} referred to ambient), but if inlet manifold conditions then η_{vol} will reflect engine port and valve efficiency. Mechanical efficiency is simply

$$\eta_{mech} = BMEP/IMEP \qquad (6.107)$$

while thermal efficiency is

$$\eta_{therm} = W/m_f\, Q_{LHV} \qquad (6.108)$$

where W is the cylinder work per cycle (1.100), m_f is the fuel injected per cycle per cylinder and Q_{LHV} the fuel lower calorific value. Finally the scavenging efficiency is given by

$$\eta_{scav} = M_a/M_{tr} \qquad (6.109)$$

where M_a is total air inducted into the cylinder and M_{tr} the total trapped mass in the cylinder at the end induction.

6.10 Transient modelling

The above sections have largely described the content of a steady-state model, but this can relatively easily be extended to a transient model, to investigate engine performance as load, speed or rack position change. Additional data and program are required for this of course.

First, while a fixed fuelling rate suffices for a basic steady-state model, a transient model will require fuelling 'maps' giving the fuelling for different rack positions (*RP*) and speeds (*N*), which might be of the from of *Figure 6.14*; for the current engine values of *N*, *RP* the model interpolates the map to obtain fuelling rate. Other maps may be required as well, depending on the fuel system design, for example an aneroid map to limit fuelling at low boost pressures (giving maximum fuelling rate for variations of boost pressure and engine speed), or a maximum fuelling rate map (varying with engine speed); in these cases actual fuelling is taken to be the minimum given by the three calculations. Another possibility is to model an engine governor. Modern digital fuelling systems may require a variety of other maps depending on their design, but these all lead to a value of fuelling which can be derived from other known parameters in the engine (e.g. speed, rack, CV pressures). As well as fuelling, injection timing is also likely to vary with engine condition and in this case injection timing and burn duration will have to be derived from a map rather than being a simple input value; one form of map would give injection timing variation with N and the fuelling rate, but others are possible. A more sophisticated engine model will allow for a variety of alternative methods of determining fuelling, injection timing and other variables for transient simulation, but they will be on similar lines.

As well as this type of additional basic engine data, further information is required to specify the transient of interest. This is largely in the form of time profiles (schedules of load, speed and rack setting, or some of these. The load schedule is most

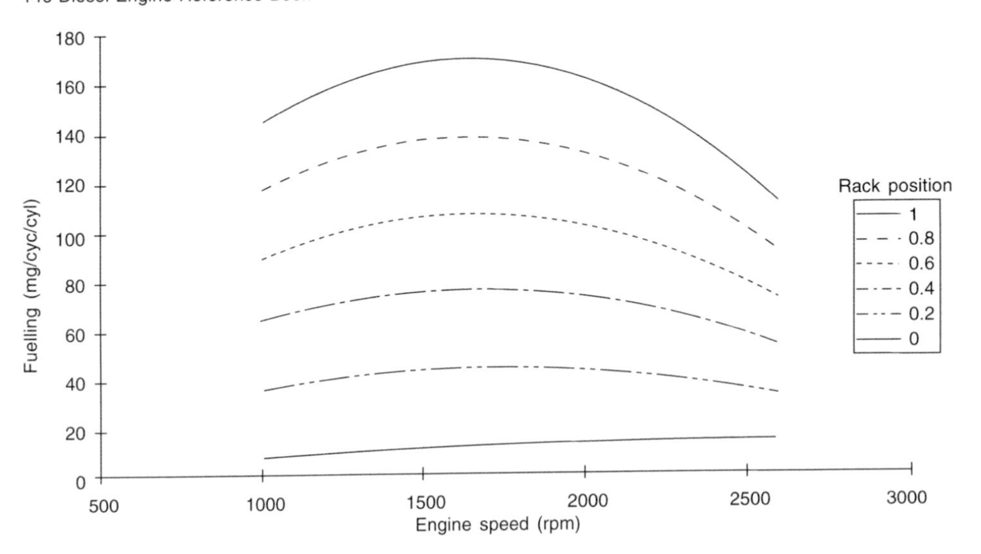

Figure 6.14 Fuelling curves

easily dealt with, being the variation of applied load with time and from which engine acceleration can be calculated simply using

$$\ddot{\theta} = (T_e - T_l)/I \tag{6.110}$$

where T_e is instantaneous developed torque from eqn (6.104), T_l the current load torque (interpolated from schedule) and I the engine total inertia (input). Integration of $\ddot{\theta}$ gives the current engine speed. One possible transient is simply to vary load at constant *RP* and observe speed and developed torque profiles. Alternatively, the load schedule (which could be a constant load) could be coupled with a rack schedule, allowing examination of the engine response to simultaneous load and rack variations.

Following a specified speed schedule (or trying to hold a fixed speed, as may be of interest for generators) under varying load is another type of transient and will require inclusion of a rack control system in the model. This control might be a proportional-integral-differential controller (PID), though other types are possible. These controllers sense the change in actual engine speed (from integration of 6.110), compare this with required speed from the speed schedule, and adjust *RP* accordingly to try to null the speed error. For a PID controller, the demanded rack position (RP_D) in response to a speed error ΔN is given by:

$$RP_D = K_1\Delta N + K_2(\Sigma\tau\Delta N) + K_3(\Delta N - \Delta N_p)/t \tag{6.111}$$

where $\Sigma\tau\Delta N$ is the integral of all previous errors, ΔN_p is the previous timestep error and K_1 to K_3 are the proportional, integral and differential PID constants respectively. In practice, the controller will not respond instantly. If the controller is assumed to be a simple exponentially lagged system with time constant *TCON* then to good approximation the actual rack setting, using x for *RP*, is given by

$$x = x_P + (x_D - x_P)(1 - \exp(-\tau/TCON))$$
$$\approx x_P + (x_D - x_P)\ \tau/TCON \tag{6.112}$$

where x_P and x_D are, respectively, the *RP* value at the end of the previous step and the demanded RP_D. Alternatively, the controller can be modelled in detail to provide the response.

Figure 6.15 shows an example of a required applied load/required speed schedule together with the output speed and developed engine torque. A model should of course be able to produce all of the data that a steady-state mdel can produce for each timestep through the transient, but care is needed to avoid too much output!

6.11 Other engine components

The above sections have dealt with a basic diesel engine, but real engines are now getting much more complex and the basic type of model described, while being of considerable use, needs to be extended to deal with these. It is not appropriate to go into these extensions in detail, but it is worth briefly indicating some of the more important areas of extension. Alternative fuelling representation has already been mentioned, and sections 6.12 and 6.13 separately deal with combustion calculation extensions and gas dynamics respectively.

6.11.1 Turbomachinery

There are a number of additional options here which a comprehensive model should cover. First there could be more than one stage of turbocharging, a low pressure compressor preceding a high pressure one, the LP compressor normally being on a common shaft with a LP turbine, the HP compressor similarly being linked to a HP turbine. Other combinations are possible, such as two compressors linked to a single turbine or vice versa. Expander turbines can be inserted before the engine to reduce inlet temperature, hence increasing air density and volumetric efficiency. Sequentially turbocharged engines have multiple turbochargers, switched in and out by control valves to optimally match the engine at differnt speed/load points. All of these can be modelled using the principles given earlier, though of course the model structure will have to be more flexible than for a single turbocharger, and for sequential turbocharging algorithms or input data will be required to handle switching between turbochargers. Similarly, a compressor can be a supercharger, or a turbine can be a compound turbine, geared to the engine crankshaft in both cases; these can be introduced quite simply. Multiple entry turbines are frequently used, where two or more exhaust manifolds feed a single turbine, leading to interactions between the separate flows into the turbine as the manifold pressures are normally out of phase because of their different cylinder connections; additional modelling is required to deal adequately with this. Finally, variable geometry compressors and turbines can be used where the gas flow through these components is controlled (usually by vanes) to match the turbocharger performance more optimally to current engine condition. Such VG turbomachinery can be represented by using multiple characteristic curves, covering the range of VG settings, and interpolating between curves for the current VG settings.

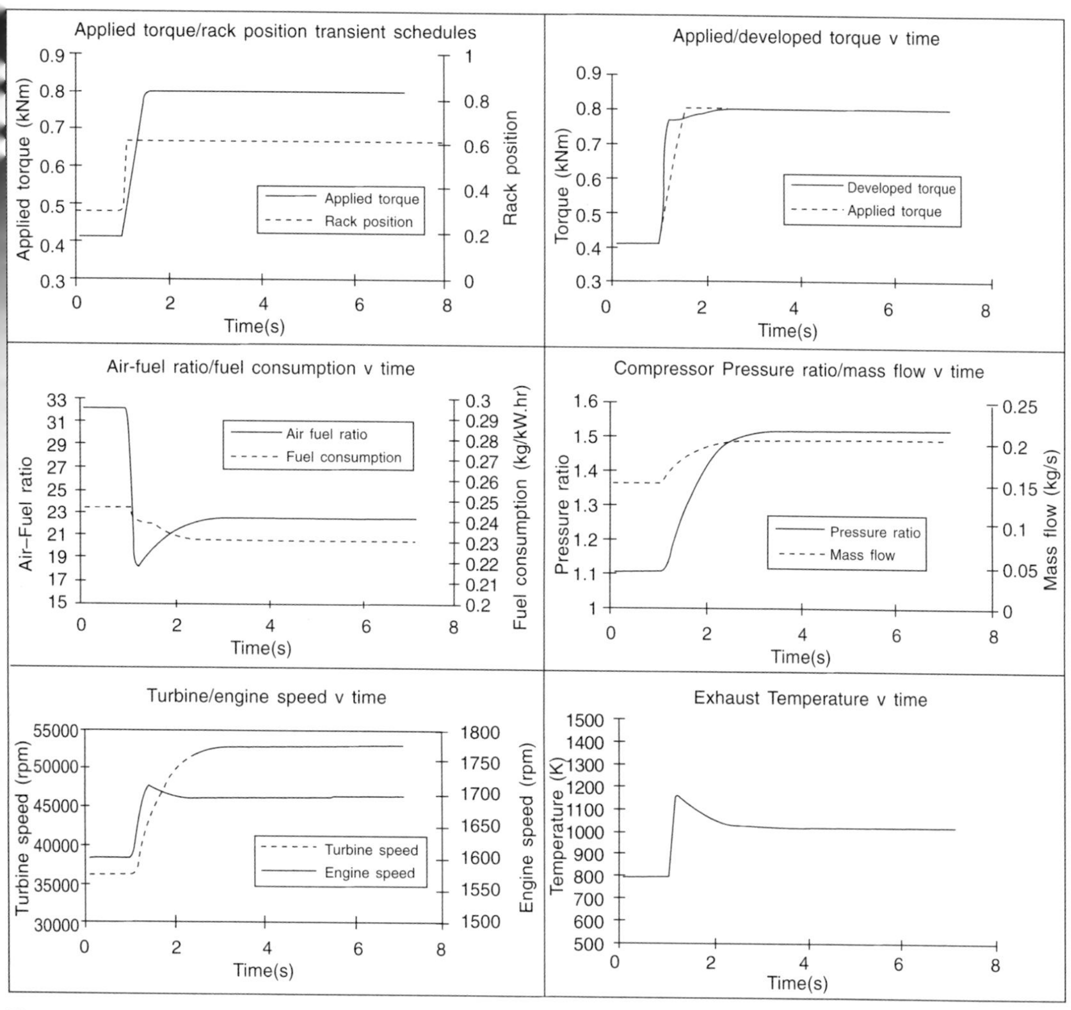

Figure 6.15 Transient result examples

They thus also require a controller to be represented to determine the current VG setting, which might be represented by maps relating VG setting to boost and exhaust manifold pressures, engine speed, etc.; the controller could be a PID and will have a finite response time, so could be represented similarly to the engine speed controller in section 6.10.

6.11.2 Control valves

A wide variety of gas flow control valves are possible in modern diesel engines, usually opened and closed using algorithms relating their area to other parameters in the engine. The most common examples are wastegates and exhaust gas recirculation valves. The turbine wastegate performs some of the functions of the alternative VG turbine, reducing turbine flow and hence power output at high engine speeds to give a better match to engine requirements; the wastegate by-passes flow around the turbine. It can be represented in a model by a variable area valve in a pipe in parallel with the turbine, the actual valve area being determined in general by an algorithm using other sensed parameters in the engine (typically boost and exhaust manifold pressures) and its controller characteristics. The latter can be modelled similarly to a VG turbine controller, or a simple mechanical wastegate can be modelled as the actual mechanical system. EGR can be represented simply by a fraction of the gas flow being recycled from exhaust to inlet manifolds or again by a more complex control system algorithm. Other flow control valves are becoming more common on engines in order to provide ever more close matching of flows to requirements, and these can usually be modelled in similar ways provided the basic model structure is flexible enough.

6.11.3 Indirect injection and other fuelling methods

A common class of diesel engines, particularly for cars, are indirect injection (IDI) engines where most of all of the fuel is injected into a small pre-chamber connected by a nozzle to the main cylinder chamber. The pre-chamber is usually of fixed volume, although some designs have variable volumes. Such engines can be fairly simply represented as single chamber engines, using appropriately selected combustion parameters, but for more accurate analysis they have to be explicitly modelled. The latter introduces a number of new features relating to the early combustion of fuel in the pre-chamber, flow through the connecting nozzle and complete combustion in the main cylinder. This is beyond the present scope and the reader is referred to Reference 1 for basic information on such modelling. Other fuelling methods are, for example, pilot injection into the cylinder (of a DI engine) where a small amount of fuel is burnt to raise cylinder temperature before the main injection burn, and fumigation where some fuel is premixed with air in the inlet manifold rather like an SI engine, to give a lean-mix charge in the cylinder (rather than just air) prior to the main in-cylinder direct fuel injection.

6.11.4 Two-stroke engines

The basic model outlined can be applied to two-stroke engines with fairly minimal modification relating to a 360° rather than a 720° crank angle cycle. However, while this may be adequate for preliminary work special features of two-stroke engines really require specific attention. In particular, scavenging needs to be more accurately modelled than in a basic four stroke model, as it has a large effect on performance and varies with the type of scavenging, while port flows and other features need some special attention. Two-stroke engines and their modelling is covered in detail in Reference 11.

6.12 Energy equation, gas properties and combustion extensions

More general versions of eqns (6.4) and (6.13) can be written to cover disassociation, where the CV (particularly the cylinder) is split into a number of zones and where the gas is a mixture of several species (either different chemical molecules or, for example, separate air, unburnt fuel and burnt mixture). If emission calculations are to be carried out then at least two zones (burnt gas and unburnt gas) are required in the cylinder as the burnt gas is at higher temperatures than the average homogeneous mixture dealt with in earlier sections, and emissions are critically dependent on temperature. The use of a separate burnt mix is also one way of dealing with problems which may arise with rich mixtures ($\phi > 1$) which, while not usual in diesel engines, can occur in transients.

To deal with the energy equation first, for each of J zones in a CV, there are J equations corresponding to each of (6.4) and (6.13), a pair or each zone. Also at higher temperatures the internal energy u, enthalpy h and gas constant R are all functions of P, T and ϕ. Assuming that all zones in a CV are at the same pressure then (6.4) for the jth zone is now replaced by the more general form:

$$\dot{M}_j\bar{u}_j + M_j\left[\dot{P}\sum_K\left(\alpha_{kj}\frac{\partial u_{kj}}{\partial P}\right) + \dot{T}_j\sum_K\left(\alpha_{kj}\frac{\partial u_{kj}}{\partial T_j}\right)\right.$$
$$\left. + \sum_K\left(\dot{\phi}_{kj}\alpha_{kj}\frac{\partial u_{kj}}{\partial \phi_{kj}}\right) + \sum_K(\dot{\alpha}_{kj}u_{kj})\right]$$
$$= \sum_{I,J}\bar{h}_{ij}\dot{m}_{ij} - \dot{Q}_j - P\dot{V}_j \qquad (6.113)$$

where α_{kj} is the mass fraction of species k in this jth zone, u_{kj} is the specific internal energy of the kth species in the zone, M_j is the total mass of the zone and $\bar{u}_j$ is the average internal energy of the gas in the zone, mass-averaged across the different species. $\dot{Q}$ and $\dot{V}_j$ now refer to zone heat transfer and volume change respectively, while $\bar{h}_{ij}$, $\dot{m}_{ij}$ refer to mass-average specific enthalpy and mass flow rate (summed across the species) for flow to or from the jth zone from the ith source (which may be another zone). The term involving $\dot{\alpha}_{kj}$ is to allow for the variation with time of the fraction of the kth species in the zone. An equation corresponding to (6.13) can similarly be obtained from the perfect gas flow for each zone:

$$\frac{\dot{P}}{P}\left[1 - \frac{P}{R_j}\sum_K\left(\alpha_{kj}\frac{\partial R_{kj}}{\partial P}\right)\right] + \frac{\dot{V}_j}{V_j}$$
$$= \frac{\dot{M}_j}{M_j} + \frac{\dot{T}_j}{T_j}\left[1 + \frac{T_j}{R_j}\sum_K\left(\alpha_{kj}\frac{\partial R_{kj}}{\partial T_j}\right)\right]$$
$$+ \frac{1}{\bar{R}_j}\left[\sum_K\left(\dot{\phi}_{kj}\alpha_{kj}\frac{\partial R_{kj}}{\partial \phi_{kj}}\right) + \sum(\dot{\alpha}_{kj}R_{kj})\right] \qquad (6.114)$$

where $\bar{R}_j$ is the average gas constant for the zone, mass-averaged across the different species, and $\bar{R}_{kj}$ is the gas constant for the kth species in zone j. The mass-averaged values of the various 'bar' terms are from, for example,

$$\bar{R}_j = \sum_K(\alpha_{kj}M_jR_{kj})/\sum M_j\text{, etc. and }\sum_K\alpha_k$$
$$= 1 \text{ for each of the zones} \qquad (6.115)$$

While eqns (6.113) and (6.114) appear complex all of the terms in them can be evaluated. Further, they can be manipulated into a form suitable for rapid computer evaluation of $\dot{P}$ and the J zone values of $\dot{T}_j$. The manipulation is straightforward but lengthy, and is not given here. Basically, eqns (6.114), are multiplied through by V_j and rearranged to give $\dot{V}_j$ which is substituted back into eqns (6.113), further rearrangement giving J equations of the form:

$$\dot{T}_j = (a_j - b_j\dot{P})/c_j \qquad (6.116)$$

where a, b, and c are combinations of calculable terms in (6.113) and (6.114). Again multiplying the J equations (6.114) by P_j and adding them together gives a single equation of the form:

$$\alpha_j\dot{P} - \sum_J(\beta_j\dot{T}_j) = \gamma_j - P\sum_J\dot{V}_j = \gamma_j - P\dot{V}$$

$$= \varepsilon_j \text{ say, as } \sum_J\dot{V}_j = \dot{V}_{CV}$$

where again α, β, γ and ε are combinations of calculable terms from (6.113) and (6.114). Thus using (6.116)

$$\left.\begin{aligned}&\alpha_j\dot{P} = \varepsilon_j + \sum_J\beta_j(a_j - b_j\dot{P})/c_j\\ &\text{or}\\ &\dot{P} = \frac{\varepsilon_j + \sum(\beta_ja_j/c_j)}{\alpha_j + \sum(\beta_jb_j/c_j)}\end{aligned}\right\} \qquad (6.117)$$

dP/dt can thus be evaluated from (6.117), and substitution back into (6.116) yields J values of dT_j/dt, one for each zone. The equations require (as well as all of the data discussed below) values of individual zone volumes V_j and these can be assumed proportional to the zone masses if better knowledge is not available.

To solve the above equations requires values of u, h and R, and their partial derivatives with respect to P, T and ϕ, beyond those in, for example, Figure 6.2. These are given by the empirical methods or property routines, mentioned in section 6.3 due to Kreiger and Borman (KB) and Martin and Heywood (MH), or can be generated from an equilibrium model based on extension of the reactions eqn (6.24). Any enhanced model will have to use some form of property routine, and while the empirical KB and MH methods are sufficient for a thermodynamic model it is necessary to use an equilibrium model, with other extensions, if emissions are to be calculated as the KB and MH methods do not represent individual molecular species.

For high temperatures where dissociation occurs, eqn (6.24) can be extended to represent dissociation species. Normally 10 combustion products are dealt with and (6.24) becomes

$$\varepsilon\phi\ C_{\alpha}H_{\beta} + (0.29\ O_2 + 0.79\ N_2) = b_1\ CO_2 + b_2H_2O + b_3N_2 + b_4\ O_2 + b_5\ CO + b_6\ H_2 + b_7H + b_8O + b_9\ OH + b_{10}\ NO \tag{6.118}$$

There are now 10 unknown b_i values and to determine these one uses the four atom balances used with (6.24) together with six dissociation reaction equations. The latter relate reaction equilibrium constants to the concentrations of the different molecules. As the equilibrium constants can be obtained from tables and expressed as polynomials in temperature, these equations can be used with the atom balance to determine the concentrations of the different molecules[3,7]. Solving 10 simultaneous equations requires quite complex matrix manipulation but it does give the concentrations of the 10 species in an equilibrium mixture for any given P, T, ϕ set. With the concentrations known, the mixture molecular weight, gas constant R, enthalpy/internal energy, and derivatives of R and u can be found by the same methods as outlined in section 6.3 for the simpler non-dissociated gas mixture. Such an equilibrium model is slow running compared to the empirical methods, but is necessary if emission calculations are to be carried out.

While an equilibrium model is satisfactory for hydrocarbon products of combustion, and hence for their contributions for emissions, it is not suitable for NO_x as the NO_x reactions are too slow to reach equilibrium. Accordingly for calculation of NO_x emissions, the equilibrium model is supplemented with a non-equilibrium NO_x calculation, usually based on the extended Zeldovich reaction equations[1,3]. While the CO reactions are faster than those for NO_x, the CO concentration can also differ somewhat from equilibrium and similar techniques have sometimes been used to deal with this rather than assume CO to be at equilibrium concentration.

Before leaving gas properties it is worth mentioning air, unburnt fuel and burnt gas mixtures. For such a mixture the average gas constant is given by

$$\bar{R} = \Sigma\ x_i R_i = x_a R_a = x_u R_u + x_b R_b \tag{6.119}$$

where R_a, R_u, R_b are gas constants for air, unburnt fuel and burnt gas respectively, while the x values are corresponding mass fractions. Precisely analogous expressions apply to mixture average enthalpy and internal energy (with mass based h_b, u_b, etc.). Derivatives for the mixture are then, for example,

$$\frac{\partial \bar{R}}{\partial T} = \Sigma x_i \left(\frac{\partial R_i}{\partial T}\right) = x_a \frac{\partial R_a}{\partial T} + x_u \frac{\partial R_u}{\partial T} + x_b \frac{\partial R_b}{\partial T} \tag{6.120}$$

with corresponding equations for derivatives w.r.t. P and ϕ and for u. For air and unburnt fuel, derivatives with respect to P and ϕ are zero, while the temperature derivatives can be found similarly to those for combustion products (but much simpler).

The need for at least two zones in the cylinder has been mentioned as being required to carry out emission calculations, and it is appropriate to briefly introduce more complex sub-models of combustion. At least two zones are required because of the very different composition and temperature of the burnt and unburnt parts of the cylinder contents at any instant during combustion. The empirical methods of section 6.7 can in fact be used for a simple two-zone model, as they give the masses of burnt/unburnt gases, but because of the complex mechanism of mixing and burning in a diesel engine in particular, a larger number of zones are really required. Models exist with several hundred zones, though in practice, for run-time reasons these are normally used in a single-cylinder combustion model rather than a complete cycle simulation. It may be noted that eqns (6.113) to (6.117) can be used to determine cylinder pressure and temperatures in any number of zones provided information is available on the contents and properties of each zone.

The normal modelling procedure[1,12,13] is to model the time behaviour of each injector spray and its surrounding air/residual gas. The spray, including mixing regions at the edges, is divided into a number of zones, new zones being formed as the spray develops and existing zones expanding as the spray moves away from the nozzle and entrains the cylinder air. The fuel starts as a liquid, forms droplets and eventually evaporates, these processes being modelled. Empirical relationships for spray penetration provided the velocity of zones on the spray centreline at any instant, assumptions about jet mixing giving the relative velocities of off-centre zones. With zone velocities known, zone mass and hence the amount of air entrained in a timestep can be calculated by assuming conservation of momentum. Ignition delay for each zone can be calculated by an equation similar to (6.74) Combustion calculations can be based either on achieving a stoichiometric mixture in the zone or on a reaction rate equation of the standard Arrhenius type,

$$\dot{m}_b = a\rho^2[\text{Fuel vapour}]\ [\text{Oxygen}]^b \exp(-E/RT) \tag{6.121}$$

The burn rate in the zone is thus dependent on concentrations of fuel vapour and oxygen, temperature, density (ρ) and on empirical reaction activation energy (E); a and b are empirical constants. As well as mass transfer between zones, from the entrainment model, heat transfer also has to be allowed for. With all of this information, the energy equation leads to cylinder pressure, and temperatures of the individual zones. Use of an equilibrium model, with additional calculations for NO_x and solids, will then give the relative quantities of species in each zone which can be summed over the cylinder to give total emission products at each timestep. The calculations need to be modified somewhat later in the combustion process both as the spray impinges on the cylinder walls and when all the air/residual gas in entrained, but the principles are similar.

Such models are obviously quite complex but do give a reasonable representation of combustion and allow estimation of emissions, but the complexity of the in-cylinder processes are such that small changes in, for example, cylinder geometry can give quite large changes in emissions. There are unfortunately not many simpler intermediate sub-models currently available lying between two-zone and many-zone sub-models, as these may be more suitable for complete engine simulations.

6.13 Gas dynamics

In section 6.4 pipe flow were calculated by the orifice equation which does in fact give good results for mid-speed range diesel engines. However, at the higher engine speeds, particularly for large engines, significant errors can occur, while at low speeds and loads use of the orifice equation can lead to instability if implemented as in 6.4. Both of these problems can be overcome by using gas dynamics calculations which properly account for mass, energy and momentum transfer between finite elements of the pipe flow. The penalty for this increased precision is a significant increase in program complexity and model run time. As gas dynamics is of considerable general interest, is widely used in engine models and has generated a large volume of literature, it is appropriate to present the basic equations and to indicate the alternative methods of solution. The following is restricted to one-dimensional gas dynamics, the form normally used in complete engine cycle simulation. The three-dimensional methods of computational fluid dynamics (CFD) are widely used in engine research, but for looking at a small part of the engine for a very limited time period. CFD methods take considerable effort to set up (days) and have relatively very long run times (hours) but can give very detailed flow information; they are not further discussed here.

To develop the one-dimensional gas dynamic equations a pipe is divided into a number of finite elements or meshes, usually of equal length (*Figure 6.16*), and the mass, momentum and energy conservation laws applied to each element. There are then a variety of ways of manipulating the resulting equations. In the following, two equations are given for each conservation law, a basic equation (A) and a derived form (B).

Mass conservation. Using subscripts 1 and 2 to denote values at entry and exit to the element (unsubscripted are element centre values) then

Rate of change element mass = Rate of mass inflow − Rate of mass outflow

$$\partial M/\partial t = \dot{m}_1 - \dot{m}_2 \qquad (6.122A)$$

where M is element mass. Using

$\dot{m}_2 = \dot{m}_1 + [\partial(\rho A v)/\partial x]\delta x$; $M = \rho A \delta x$ then leads to

$$\frac{\partial \rho}{\partial t} + \frac{\partial}{\partial x}(\rho v) + \frac{\rho v}{A}\frac{\partial A}{\partial x} = 0 \qquad (6.122B)$$

where ρ, v and A are element centre density, velocity and cross-sectional area.

Momentum conservation. This requires that

Rate of change of momentum + Rate efflux of momentum − Rate influx of momentum

= Pressure force + Friction force

The pressure force has to include the force on the walls of the element. Thus

$$\frac{\partial}{\partial t}(Mv) + \dot{m}_2 v_2 - \dot{m}_1 v_1 = \frac{\partial}{\partial t}(Mv) + \frac{\partial}{\partial x}(\rho A v \cdot v)\,\delta x$$

$$= (P_1 - P_2)A + P(A_2 - A_1) + F\left[-\frac{\partial}{\partial x}(PA) + P\frac{\partial A}{\partial x}\right]\delta x \qquad (6.123A)$$

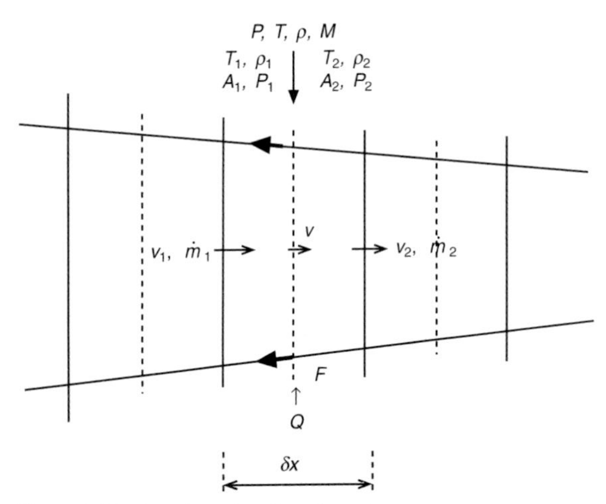

Figure 6.16 Gas dynamics pipe element

where P is pressure, F the friction farce at the walls and δx the element length. This equation can be manipulated to give

$$\frac{\partial v}{\partial t} + v\frac{\partial v}{\partial x} + \frac{1}{\rho}\frac{\partial \rho}{\partial x} + F = 0 \qquad (6.123B)$$

where the mass conservation equation has been used in the manipulation.

Energy equation This is

Rate of change of element energy = Energy flow in − Energy flow out

+ Work done on element + Heat flow through walls

$$\frac{\partial}{\partial t}\left[M\left(u + \frac{v^2}{2}\right)\right] = \dot{m}_1\left(u_1 + \frac{v_1^2}{2}\right)$$

$$- \dot{m}_2\left(u_2 + \frac{v_2^2}{2}\right) + P_1 A_1 v_1 - P_2 A_2 v_2 + Q \qquad (6.124A)$$

where u is specific internal energy and Q is heat flow rate in through the walls. The v^2 terms are of course kinetic energy. This again can be manipulated to give an alternative form, using $\dot{m} = \rho A v$, $P = \rho R T$ so $PAv = \dot{m}RT$; $u = c_v T$ and the general relation $Y_2 = Y_1 + (\partial Y/\partial x)\,\delta x$, then

$$Q = A\,\delta x\left[\rho c_v T + \frac{\rho v^2}{2}\right]$$

$$+ \frac{\partial}{\partial x}\left[vA\left(\frac{\rho v^2}{2} + \rho c_v T + P\right)\right]\delta x \qquad (6.124B)$$

The heat transfer rate Q can be calculated as for any pipe flow, using element temperature (eqn 6.81 etc.), while pipe friction force is given by

$$F = fA_{sur}(\rho v \mid v \mid /2) \qquad (6.125)$$

where f is the friction factor for the pipe, A_{sur} the element surface area of pipe ($\pi D\,\delta x$ for diameter D) and $|v|$ has been used to ensure the force opposes gas motion. The above equations are the full so-called 'non-homentropic' equations, the simpler 'homentropic' equations being obtained by putting Q and F to zero.

The '*B*' set of three equations are efectively the basis of the Method of Characteristics (MoC) and the Lax–Wendroff (LW) method. The earliest work on pipe flows used MoC, Benson[14] having written the standard text on the subject, and it is still possibly the most widely used method. For the MoC, Benson further manipulated the energy equation by expanding and using the mass and momentum equations (6.123B, 6.124B) to give

$$\left(\frac{\partial P}{\partial t} + v\frac{\partial P}{\partial x}\right) - c^2\left(\frac{\partial \rho}{\partial t} + v\frac{\partial \rho}{\partial x}\right) - (\gamma - 1)\,\rho\left[\frac{Q}{M} + vF\right] = 0 \qquad (6.126)$$

where γ is the ratio of specific heats and c the velocity of sound ($\sqrt{\gamma RT}$). By further combining (6.123B), (6.124B) and (6.126) a new set of equations with two new variables (the Reimann variables) can be derived and solutions to these quations obtained by MoC. The calculations method is cumbersome and limited, involving 'reference conditions' which can cause difficulties. Whilst widely used in the past it is now being superceded by other methods. The velocity of sound c plays a central role in the MoC and leads to much consideration of pressure waves in the pipe; while these are real enough there is no need to explicitly consider either c or the waves as these fall out naturally from the basic equations. The energy equation includes c implicitly, and indeed it is also implicitly in the simple 'energy equation' (6.30) relating T_t in terms of T and v^2, as indicated in deriving (6.31).

The basic paper on the LW method by Takizawa *et al.*[15] uses exactly the three '*B*' equations given above (and does not explicitly use c at all), and they are restated in the vector form

$$\frac{\partial X}{\partial t} + \frac{\partial Y}{\partial x} = D$$

where X, Y and D are the vector functions indicated by the '*B*' equations above. The LW method is a mathematical technique for solving these equations using a finite difference approach which is not discussed further here. While the method has numerous advantages it does have stability problems, leading to the use of artificial 'viscosity' terms to slow down rapid flow changes.

While the LW method is now widely used another group of methods, variously described as finite difference, finite volume, or upwind methods, have been developed working with equations more closely related to the 'A' equations given above. One of earliest papers on these methods was by Chapman *et al.*[16] who used a complex finite difference approach employing Lagrangians and artificial viscosity (more limited than LW) to deal with instabilities caused by flow discontinuities. The methods essentially involve calculating mass flows across the element boundaries and then using these to solve the momentum and energy equations for the element. The 'upwind' group of methods involve assuming gas densities at the element boundaries to be those at the centre of the upwind element, rather than interpolated values, at locations where the gas properties in the pipe change rapidly, again to try to introduce stability. Despite the problems of dealing with instability, these methods and LW do give good results, with good experimental agreement.

While the above has outlined the basis of methods for GD calculations within a pipe, a separate, related, set of calculations have to be carried out at the pipe 'boundaries'. These might be entries to a manifold (CV), at a throttle, or valve, or at a junction between several pipes. The gas state at the boundaries is again obtained by solving mass, momentum and energy equations assuming quasi steady flow[1,14,15]. While it is not appropriate to go into this in depth here it is worth giving the equations relating to one case, that of a throttle in a pipe (*Figure 6.17*). The relevant equations are:

Mass conservation

$$\rho_1 A_1 v_1 = \rho_2 A_2 v_2 = \rho_3 A_3 v_3 \tag{6.127}$$

Energy conservation

$$c_p T_1 + v_1^2/2 = c_p T_2 + v_2^2/2 = c_p T_3 + v_3^2/2 \tag{6.128}$$

Pseudo momentum equation

$$A_3 (P_2 - P_3) = A_3 \rho_3 v_3^2 - A_2 \rho_2 v_2^2 \tag{6.129}$$

The third equation needs some explanation. The RHS of the equation is the increase in momentum across the throttle, which will equal the force accelerating the gas. The accelerating force is $(P_3A_2 + P'(A_3 - A_2) - P_3A_3)$, where P' in the centre term is the pressure acting on the annulus between the pipes and this can be taken to equal P_3 on the grounds that there is no large transverse acceleration at exit from the throat; the equation then follows. The assumption regarding P' is a standard one in hydraulics and leads to the equation for losses at entry to a large volume. Similar equations are used for other boundaries but for exit to a large volume, throat pressure can be taken as the volume pressure (see discussion with (6.40), section 6.4), the adiabatic condition can be used for entry to a throttle (P/ρ^{γ} = const. and, for junctions a pressure loss can be assigned to each pipe end.

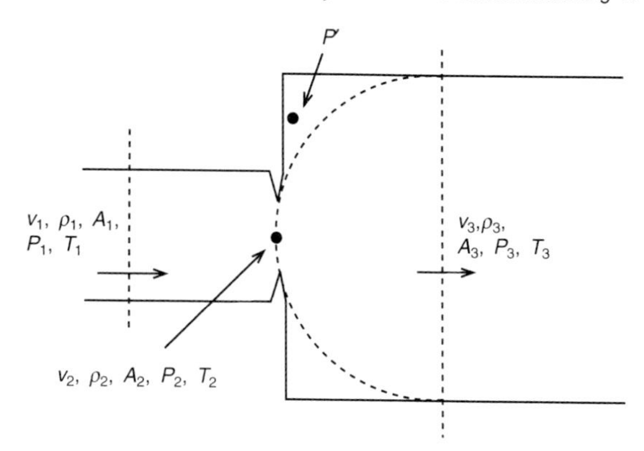

Figure 6.17 Pipe boundary example

Using these gas dynamic calculation techniques then quite accurate modelling of the pipe flow in an engine can be carried out, allowing for the real effects of pressure waves travelling up and down the pipes. A comparison of the alternative methods is given in Reference 17. While these effects can be significant (and extremely so in high speed SI engines) the simpler orifice equation solutions do give quite good results for medium speed diesel engines, though GD calculations are now widely used.

References

1 HEYWOOD, J. B., *Internal Combustion Engine Fundamentals*, McGraw Hill (1988)

2 WATSON, N. and JANOTA, M. S., *Turbocharging the Internal Combustion Engine*, Macmillan (1982)

3 FERGUSON, C. R., *Internal Combustion Engines*, Wiley & Sons (1985)

4 RAMOS, J. J., *Internal Combustion Engine Modelling,*, Hemisphere (1989)

5 NAT. U.S. BUREAU STANDARDS, *JANEF Thermochemical Tables*, NSRDS-NB537 (1971)

6 GORDON, S. and McBRIDE; B.,'Computer Program for the Calculation of Complex Chemical Equilibrium Composition', *NASA SP-273* (1971)

7 'OLIKARA, C. and BORMAN, G. L., A Computer Program for Calculating Properties of Equilibrium Combustion Products with Some Application to I.C. Engines', *SAE 750468* (1975)

8 'KREIGER, R. B. and BORMAN, G. L., The Computation of Apparent Heat Release for Internal Combustion Engines', *ASME Paper 66-WA/DGP-4* (1966)

9 MARTIN, M.K. and HEYWOOD, J. B. 'Approximate Relationships for the Thermodynamic Properties of Hydrocarbon-Air Combustion Products', *Combustion Science Tech.*, **15** (1977)

10 WOSCHNI, G., 'A Universally Applicable Equation for the Instantaneous Heat Transfer Coefficient in the Internal Combustion Engine', *SAE 670931* (1967)

11 BLAIR, G. P., *The Basic Design of Two-stroke Engines*, SAE (1995)

12 HIROYASU, H., KADOTA, T. and ARAI, M.,'Development and Use of Spray Combustion Modelling to Predict Diesel Engine Efficiency and Pollutant Emission', *Bull. JSME, Paper 214–12* (1983)

13 BAZARI, Z.,'A DI Diesel Combustion and Emission Predictive Capability for Use in Cycle Simulation', *SAE 920462* (1992)

14 BENSON, R. S., *The Thermodynamics and Gas Dynamics of Internal Combustion Engines*, Vol 1., Clarendon Press (1982)

15 TAKIZAWA, M., UNO, T., OUE, T. and YURA, T., 'A Study of Gas Exchange Process Simulation of an Automotive Multi-Cylinder Internal Combustion Engine', *SAE 820410* (1982)

16 CHAPMAN, M., NOVAK J. and STEIN, R.,'Numerical Modelling of Inlet and Exhaust Flows in Multi-Cylinder I.C. Engines', *ASME Conf. on Flow in IC Engines*, Phoenix (1982)

17 CHEN, C. VESHAGH, A. and WALLACE, F. J., 'A Comparison Between Alternative Methods for Gas Flow and Performance Prediction of I.C. Engines', *SAE 921734* (1992)

7 Computational fluid dynamics

Contents

7.1 Introduction

Engine manufacturers are facing increasingly stringent federally mandated emission standards due to rising environmental awareness. This has resulted in the need for new combustion analysis tools such as Computational Fluid Dynamics (CFD) for predicting engine performance. The internal combustion engine represents a challenging fluid mechanics problem because the flow is compressible (with large density variations, but with low Mach number (typically $M < 0.4$)), turbulent, unsteady, cyclic and non-stationary, both spatially and temporally. The combustion characteristics are greatly influenced by the details of the fuel injection process, which introduces the complexity of describing the physics of vaporizing, dense two-phase flows. Pollutant emissions are controlled by the details of the turbulent fuel–air mixing and combustion processes, and a detailed understanding of these processes is required to improve engine performance and reduce emissions while not compromising fuel economy.

Much progress has been made in multidimensional modelling of engines in recent years[1]. However, in spite of the detailed nature of the most comprehensive engine models, they will not be entirely predictive for the foreseeable future due to the wide range of length and time scales needed to describe engine fluid mechanics. Modelling of unresolved physical processes is required, and this introduces empiricism in computations. For example, a modern heavy-duty truck diesel engine operates with injection pressures as high as 200 MPa and uses injector nozzles with holes that are less than 200 μm in diameter. At these high pressures, the fuel jet enters the combustion chamber at velocities that are close to sonic (around 600 m/s) and breaks up into droplets with diameters of the order of 10 μm in times of the order of microseconds. To begin to resolve the details of this process in an engine with a combustion chamber bore of 100 mm it is seen that more than $(10^4)^3 = 10^{12}$ grid points would be needed in a 3-D CFD simulation. Similar grid resolution requirements are needed to be able to resolve the heat transfer process in the thin boundary layers on the combustion chamber walls. These requirements by far exceed current practical super-computer storage and run times which are limited to about 10^5 or 10^6 grid points.

Once atomized, the injected fuel vaporizes and mixes with the air in the combustion chamber. At some point in the process of fuel–air mixing, ignition takes place, followed by a fast flame spreading process. The burning that takes place shortly after ignition rapidly consumes the premixed mixture, while the subsequent burning is thought to be mixing-controlled or diffusion-type combustion. A description of these processes requires an understanding of spray dynamics and mixing, ignition kinetics, and post-ignition combustion. Submodels are required for the turbulence, spray injection, atomization, drop breakup and coalescence, vaporization, ignition, premixed combustion, diffusion combustion, wall heat transfer, and emissions (soot and NOx). All of these submodels must work together in a turbulent flow field[1].

The complexity of diesel combustion has prevented widespread success in diesel modelling. However, recent progress in submodel development and application has shown considerable promise for predicting engine performance with sufficient accuracy to provide directions for engine design[1–4]. The purpose of the present article is to review diesel spray, combustion, and emission models, and to indicate some areas where further research is needed. This review is not intended to be a comprehensive guide to the CFD modelling literature. Introductory reviews of engine flow modelling are available elsewhere[5,6]. Instead, this work emphasizes recent advances in spray, combustion and emissions applications and modelling.

This chapter is organized as follows. After describing spray, combustion and emissions models, examples are given of their application to heavy-duty diesel engines. Validation results are reviewed from a study in which the CFD models were applied over the entire operating range of a heavy-duty diesel engine. Then a discussion is given of the use of engine CFD models to explain the anomalous effect of reduced injection pressure on particulate emissions at low loads, and the reasons for emissions reductions with the use of multiple injections. Finally, an example is given showing the usefulness of CFD combustion modelling in the diesel engine design process.

7.2 Model description

Many published multidimensional engine combustion modelling studies are based on the Los Alamos National Laboratory KIVA codes[7–9]. Often the codes are modified by the user to include advanced submodels for engine subprocesses[1,2]. For example, a list of some of the submodels added at the ERC is given in *Table 7.1*. The present review outlines the implementation of such submodels in the KIVA codes for modelling engine spray combustion.

Table 7.1 Submodel implementation in the KIVA codes at the ERC

Submodel	*KIVA*	*UW-updated*	*References*
Heat transfer	Law-of-the-wall	Compressible,	SAE Paper 960633
turbulence	standard k–ε	unsteady RNG k–ε/compressible	Combust. Sci. Tech. 106, p. 267, 1995
Atomization	Drop Taylor analogy	Surface-wave-growth	SAE Paper 960633
& drop drag	rigid sphere	drop distortion	SAE Paper 960861
wall impinge	none	rebound-slide model	SAE Paper 880107
Ignition	Arrhenius	Shell autoignition model	SAE Paper 950278
combustion	Arrhenius	laminar-turbulent char time	SAE Paper 950278
NOx	Zeldo'vich	extended Zeldo'vich	SAE Paper 940523
soot	none	Hiroyasu and Surovkin Nagle Strickland oxidation	SAE Paper 960633
Vaporization	Single component low pressure	Multicomponent high pressure	SAE Paper 952425 SAE Paper 952431
Intake flow	Assumed initial flow	Compute intake flow	SAE Paper 951200

7.2.1 Gas-phase modelling

The gas-phase equations are solved using an Eulerian finite difference or finite volume approach. Good introductions to numerical methods for CFD are given by Roache[10] or Warsi[11], and the methods used in the KIVA codes are described by Amsden *et al.*[7–9]. The three-dimensional computational domain is divided into a number of small hexahedral cells, or in some applications, tetrahedrons. Typical computational meshes are either structured or unstructured (i.e. the cells are joined regularly or in an 'arbitrary' manner to fill any volume, respectively), and the cells can distort in a prescribed time-varying fashion to accommodate boundary motion.

The governing equations are the mass, momentum and energy equations coupled with the turbulence model equations. Source terms account for the effects of chemical reaction (which effects the concentration of participating gas-phase species) and interaction between the spray droplets and the gas phase. Thus, the mass conservation equation for species m has source terms arising from the vaporizing spray and chemical reactions, i.e.

$$\frac{\partial \rho_m}{\partial t} + \nabla \cdot (\rho_m \mathbf{u}) = \nabla \cdot \left[\rho D \nabla \left(\frac{\rho_m}{\rho} \right) \right] + \dot{\rho}_m^c + \dot{\rho}^s \delta_{m1} \quad (7.1)$$

where ρ_m is the mass density of species m, ρ is the total gas mass density, $\mathbf{u}$ is the gas velocity, D is the (turbulent) diffusion coefficient of Fick's Law, and $\dot{\rho}_m^c$ and $\dot{\rho}^s \delta_{m1}$ are source terms that refer to particular chemical reactions and spray conditions. The species 1 corresponds to the fuel and δ is the Kronecker delta function. The chemical reaction source terms depend on the combustion model, as described below. The gas-phase momentum equation is

$$\frac{\partial \rho \mathbf{u}}{\partial t} + \nabla \cdot (\rho \mathbf{u}\mathbf{u}) = -\nabla P - \nabla \left(\frac{2}{3} \rho k \right) + \nabla \sigma + \mathbf{F}^s + \rho \mathbf{g} \quad (7.2)$$

where P is the gas pressure, k is the turbulent kinetic energy, σ is the (turbulent) viscous stress tensor in Newtonian form, $\mathbf{F}^s$ is the rate of momentum gain per unit volume due to the spray, and $\mathbf{g}$ is the gravity body force. The internal energy equation is

$$\frac{\partial \rho I}{\partial t} + \nabla \cdot (\rho \mathbf{u} I) = -P \nabla \mathbf{u} - \nabla \mathbf{J} + \rho \varepsilon + \dot{Q}^c + \dot{Q}^s \quad (7.3)$$

where I is the specific internal energy, exclusive of chemical energy, $\mathbf{J}$ is the heat flux vector which includes turbulent heat conduction and enthalpy diffusion effects, and $\dot{Q}^c$ and $\dot{Q}^s$ are the source terms due to the chemical heat release and spray interactions. In the KIVA codes, turbulence is modelled by the k–ε model with the transport equations for turbulent kinetic energy k and its dissipation rate ε as follows,

$$\frac{\partial \rho k}{\partial t} + \nabla \cdot (\rho \mathbf{u} k) = -\frac{2}{3} \rho k \nabla \mathbf{u} + \sigma : \mathbf{u}$$

$$+ \nabla \cdot \left[\left(\frac{\mu}{Pr_k} \right) \nabla k \right] - \rho \varepsilon + \dot{W}^s \quad (7.4)$$

$$\frac{\partial \rho \varepsilon}{\partial t} + \nabla \cdot (\rho \mathbf{u} \varepsilon) = -\left(\frac{2}{3} C_{\varepsilon 1} - C_{\varepsilon 3} \right) \rho \varepsilon \nabla \mathbf{u} + \nabla \cdot \left[\left(\frac{\mu}{Pr_\varepsilon} \right) \nabla \varepsilon \right]$$

$$+ \frac{\varepsilon}{k} [C_{\varepsilon 1} \sigma : \nabla \mathbf{u} - C_{\varepsilon 2} \rho \varepsilon + C_s \dot{W}^s] - R \quad (7.5)$$

The source term $-\left(\frac{2}{3} C_{\varepsilon 1} - C_{\varepsilon 3} \right) \rho \varepsilon \nabla \mathbf{u}$ accounts for length scale changes due to compressibility. Source terms involving the quantity $\dot{W}^s$ arise due to interactions with the spray. The turbulent transport of mass (due to diffusion), momentum (due to viscous stresses) and energy (due to heat conduction) is controlled by diffusion coefficients of the form $D = C_\mu k^2/\varepsilon$ and $C_\mu = 0.09$, with specified turbulent Schmidt and Prandtl numbers.

The term R in eqn (7.5) was investigated by Han and Reitz[12] in an improved turbulence model that is based on the Renormalization Group model of Orzag *et al.*[13]. In this case,

$$R = \frac{C_\mu \eta^3 (1 - \eta/\eta_0)}{1 + \beta \eta^3} \frac{\varepsilon^2}{k} \quad (7.6)$$

where $\eta = S\bar{k}/\bar{\varepsilon}$ is the ratio of the turbulent-to-mean-strain timescale, $S = (2S_{ij}S_{ij})^{1/2}$ is the magnitude of the mean strain $S_{ij} = \frac{1}{2}(\partial u_i/\partial x_j + \partial u_j/\partial x_i)$, and $\nu = \nu_0 + \nu_T$. The model constants are $C_\mu = 0.0845$, $C_{\varepsilon 1} = 1.42$, $C_{\varepsilon 2} = 1.68$, $Pr_k = Pr_e = 1.39$, $\eta_0 = 4.38$, and $\beta = 0.012$. Compared with the standard k–ε model[9], the k-equation remains the same in the RNG version. The R term is small for weakly strained turbulence, such as a homogenous shear flow, and is large in the rapid distortion limit when η approaches infinity.

The equations of state are assumed to be those of an ideal gas mixture and the corresponding thermodynamic properties are interpolated from the JANAF tables. Rigid wall boundaries are used together with the turbulent law-of-the-wall conditions. Recent improvements to the wall boundary conditions have been made to account for the fact that engine processes are unsteady and compressible[14]. Under the equilibrium assumption in the log layer of a turbulent boundary layer, it can be shown that

$$k = C_\mu^{-1/2} \mu^{*2} \quad (7.7)$$

$$\varepsilon = \frac{C_\mu^{3/4} k^{3/2}}{\kappa y} \quad (7.8)$$

where $\kappa = 0.41$ is the von Kármán constant. A zero gradient of k (i.e., $\nabla k \cdot \mathbf{n} = 0$) which follows from differentiating eqn (7.7) is used for the k equation and eqn (7.8) is used as the boundary condition for the ε equation, where y is the distance normal the wall to the nearest cell. The velocity wall function of Launder and Spalding[15] is used to calculate the momentum flux

$$\tau_w = \rho U \hat{U} \quad (7.9)$$

where τ_w is the wall shear stress, U is the magnitude of the gas velocity in the wall cell and

$$\hat{U} = \frac{u^*}{y^+} \qquad y^+ \leq 10.18 \quad (7.10)$$

$$\hat{U} = \frac{\kappa u^*}{\ln(y^+)} \qquad y^+ > 10.18$$

The friction velocity u^* is calculated using eqn (7.7), and y^+ is the dimensions distance from the wall, u^*y/v. In the heat transfer model the wall heat flux is computed by

$$q_w = \frac{\rho c_p u^* T \ln(T/T_w)}{2.1 \ln(y^+) + 2.513} \quad (7.11)$$

where ρ, T and c_p are the gas density, temperature and specific heat, respectively, and T_w is the wall temperature. This differs from the incompressible flow formulation in which the wall heat flux is predicted to vary linearly with the temperature difference (T–T_w). The approach adopted here gives more realistic flame structures along the chamber wall and a better prediction of wall heat flux[14]. Determinaion of the combustion chamber wall temperatures requires consideration of heat conduction in the piston, liner and head components, as described by Liu and Reitz[16].

7.2.2 Liquid-phase modelling

A spray equation is solved to describe the spray dynamics, viz.

$$\frac{\partial f}{\partial t} + \nabla_x \cdot (f\mathbf{v}) + \nabla_v \cdot (f\mathbf{F}) + \frac{\partial}{\partial r}(fR) + \frac{\partial}{\partial T_d}(f\dot{T}_d)$$

$$+ \frac{\partial}{\partial y}(f\dot{y}) + \frac{\partial}{\partial \dot{y}}(f\dot{y}) = f_{coll} + f_{bu} \quad (7.12)$$

The droplet distribution function f has eleven independent variables including three droplet position components $\mathbf{x}$, three velocity components $\mathbf{v}$, radius, r, temperature T_d (usually assumed to be uniform within the drop), distortion from sphericity y, the time rate of change $dy/dt = \dot{y}$, and time t. Therefore, $f(x, v, r, T_d,$

$y, \dot{y}, t)\, dv\, dr\, dT_d\, dy\, d\dot{y}$ is the probable number of droplets per unit volume at position x and time t with velocities in the interval $(v, v + dv)$, radii in the interval $(r, r + dr)$, temperatures in the interval $(T_d, T_d + dT_d)$ and displacement parameters in the intervals $(y, y + dy)$ and $(\dot{y}, \dot{y} + d\dot{y})$. In the spray equation, the quantities $\mathbf{F}$, R, $\dot{T}_d$, and $\ddot{y}$ are the time rates of changes, following an individual drop, of its velocity, radius, temperature, and oscillation velocity $\dot{y}$. The terms $\dot{f}_{coll}$ and $\dot{f}_{bu}$ are the sources due to droplet collisions and breakup.

The spray equation is solved using the method of characteristics in a Lagrangian formulation[7]. The trajectories of spray drops are traced in the eleven dimensional phase space once they are injected. The spray model considers the drop interactions with turbulence and walls, and calculates drop momentum changes due to drag, drop breakup, collision and evaporation. These processes control the drop locations, sizes, velocities, temperatures and distortions. The status of drops, i.e. the function f, is then updated and the contribution of fuel spray to the gas phase is also be obtained since mass, momentum and energy is transferred between the phases.

The interaction between the spray and the gas-phase occurs through the exchange functions $\dot{\rho}^s$, $\mathbf{F}^s$, $\dot{Q}^s$, and $\dot{W}^s$ in eqns (1–5). These terms are obtained by summing the rate of change of mass, momentum, and energy of all droplets at position x and time t., i.e.

$$\dot{\rho}^s = -\int f\rho_d\, 4\pi r^2 R\, d\mathbf{v}\, dr\, dT_d\, dy\, d\dot{y} \tag{7.13}$$

$$\mathbf{F}^s = -\int f\rho_d\,(4/3\pi r^3 \mathbf{F}' + 4\pi r^2 R\mathbf{v})\, d\mathbf{v}\, dr\, dT_d\, dy\, d\dot{y} \tag{7.14}$$

$$\dot{\mathbf{Q}}^s = -\int f\rho_d \left\{4\pi r^2 R\left[I_l + \frac{1}{2}(\mathbf{v} - \mathbf{u})^2\right] + 4/3\pi r^3\, [c_l \dot{T}_d + \mathbf{F}' \cdot (\mathbf{v} - \mathbf{u} - \mathbf{u}')]\right\} d\mathbf{v}\, dr\, dT_d\, dy\, d\dot{y} \tag{7.15}$$

$$\dot{\mathbf{W}}^s = -\int f\rho_d + 4/3\pi r^3\, \mathbf{F}'\, \mathbf{u}'\, d\mathbf{v}\, dr\, dT_d\, dy\, d\dot{y} \tag{7.16}$$

where $\mathbf{F}' = \mathbf{F} - \mathbf{g}$, $(\mathbf{v} - \mathbf{u})$ is the drop-gas relative mean velocity, $\mathbf{u}'$ is the turbulence velocity which is assumed to be normally distributed with a variance of $2/3k$[17]. This Monte Carlo approach accounts for the drop dispersion by the turbulence. The drop equations of motion are integrated with a gas velocity field $\mathbf{u} + \mathbf{u}'$ until the drop enters a new eddy. The drop-eddy interaction time is determined by the eddy life time, or by the time taken by the drop to pass through an eddy[17]. The term $\dot{W}^s$ is the negative of the rate at which the turbulent eddies do work in dispersing the droplets. I_l and c_l are the internal energy and specific heat of liquid drops, respectively.

Additional consideration is needed in eqn (7.12) to describe the injection, atomization, distortion, breakup, collision and coalescence of spray drops. An efficient injection modelling approach is to introduce the liquid into the combustion chamber as computational parcels containing large numbers of identical drops or 'blobs' with a characteristic size equal to the injector nozzle diameter[18]. The number and velocity of the injected drop parcels is determined from the fuel flow rate and knowledge of the nozzle discharge coefficient[19,20].

Modelling liquid atomization represents a particularly difficult challenge since there is still much uncertainty about the fundamental mechanisms of atomization[21]. Models have been proposed which ascribe atomization to the turbulent and/or cavitation flow processes within the nozzle passage, and to aerodynamic effects outside the nozzle, and to other mechanisms[20,21]. Alternatively, the initial atomization of the injected blobs, as well as the subsequent breakup of the drops produced from the atomization process can be modelled using drop breakup models. This procedure removes the requirement of having to specify drop sizes at the nozzle exit, and it is based on the reasonable assumption that the atomization of the injected liquid and the fragmentation of drops or liquid 'blobs' are indistinguishable processes within the dense liquid core region near the injector nozzle exit[18].

Two drop breakup models have been widely used: the Taylor Analogy Breakup (TAB) model[22] and a 'wave' breakup model[23]. The TAB model compares an oscillating-distorting drop to a spring-mass system where the aerodynamic force on the drop, the liquid surface tension force, and the liquid viscosity force are analogous, respectively, to the external force acting on a mass, the restoring force of a spring, and the damping force. The distortion parameter y is calculated by solving a spring-mass equation of the form

$$\ddot{y} = \frac{2}{3}\frac{\rho}{\rho_d}\frac{\mathbf{w}^2}{r^2} - \frac{8\sigma}{\rho_d r^3}y - \frac{5\mu}{\rho_d r^2}\dot{y} \tag{7.17}$$

where ρ is the gas density; ρ_d, σ, μ are the droplet density, surface tension, and viscosity, respectively; and $\mathbf{w} = \mathbf{v} - \mathbf{u}$ is the local relative velocity between the droplet and the surrounding gas. If the value of y exceeds unity, the droplet breaks up into smaller droplets with radius chosen randomly from a χ-squared[9] or other specified distribution (e.g. a Rosin Rammler distribution[24], or a modelled distribution derived from entropy minimization arguments[20] with a specified Sauter mean radius that is obtained from considerations of surface energy conservation[22].

The 'wave' breakup model considers the unstable growth of Kelvin–Helmholtz waves at a liquid–gas interface. A stability analysis leads to a dispersion equation relating the growth rate, Ω, of an initial perturbation on a liquid surface of infinitesimal amplitude to its wavelength, Λ, and to other physical and dynamic parameters of both the liquid and the ambient gas. Curve fits of the numerical solutions for the maximum growth rate and its wavelength are[23]

$$\frac{\Lambda}{a} = 9.02\,\frac{(1 + 0.45Z^{0.5})(1 + 0.4T^{0.7})}{(1 + 0.87We_2^{1.67})^{0.6}} \tag{7.18}$$

$$\Omega\left(\frac{\rho_1 a^3}{\sigma}\right)^{0.5} = \frac{0.34 + 0.38\,We_2^{1.5}}{(1 + Z)(1 + 1.4T^{0.6})} \tag{7.19}$$

where

$$Z = \frac{We_1^{0.5}}{Re_1};\quad T = ZWe_2^{0.5};\quad We_1 = \frac{\rho_1 U^2 a}{\sigma};$$

$$We_2 = \frac{\rho_2 U^2 a}{\sigma};\quad Re_1 = \frac{Ua}{\upsilon_1}$$

and the subscripts 1 and 2 refer to the liquid and gas phases, respectively.

The liquid breakup is modelled by postulating that new drops are formed (with drop radius r) from a parent drop or blob (with radius a) with[23].

$$r = B_0\Lambda \qquad (B_0\Lambda \le a) \tag{7.20a}$$

$$r = \min\begin{cases}(3\pi a^2 U/2\Omega)^{0.33}\\(3a^2\Lambda/4)^{0.33}\end{cases} \qquad (B_0\Lambda > a,\ \text{one time only}) \tag{7.20b}$$

where $B_0 = 0.61$. In eqn (7.20a), it is assumed that small drops are formed with drop sizes proportional to the wavelength of the fastest-growing or most probable unstable surface wave;

eqn (7.20b) applies to drops larger than the jet (low speed breakup) and assumes that the jet disturbance has frequency $\Omega/2\pi$ (a drop is formed each period) or that drop size is determined from the volume of liquid contained under one surface wave. The mass of new droplets due to breakup is subtracted from the parent drops. The change of the radius of a parent drop is assumed to follow the rate equation

$$da/dt = -(a - r)\tau \quad (r \le a) \tag{7.21}$$

where τ is the breakup time

$$\tau = 3.726 B_1 a/\Lambda\Omega \tag{7.22}$$

where B_1 is the breakup time constant with the suggested value $B_1 = 60$[19]. However, this value depends on the injector characteristics.

Figure 7.1 shows details of a spray computed using the 'blob' injection method and the 'wave' breakup model in a heavy-duty Cummins diesel engine[25]. The engine had a centrally located injector with an 8-hole nozzle tip. The computations considered one of the eight spray plumes and the mesh consisted of a 45 degree sector of the engine combustion chamber with the assumption of sector symmetry. In *Figure 7.1*, the injector nozzle is located at the upper left and the spray is directed downwards into the piston bowl. The locations of the spray drops are represented by the circles which are the computed spray parcel positions. The predicted spray is shown at 8 degrees before top-dead-centre (BTDC) and can be seen to be very dense in the core region near the nozzle exit. The injection started at 18 degrees BTDC.

Drop collisions are important in dense sprays. For a binary collision between drops, the outcome may result in a coalescence or separation. depending on the operating conditions, as described by Ashgriz and Poo[26]. In the O'Rourke and Bracco model[27] used in the KIVA code, the process is modelled by computing the collision frequency ν_{12} between drops in parcels 1 (containing larger drops) and 2 in each computational cell, where

$$\nu_{12} = N_2 \pi (r_1 + r_2)^2 \mid \mathbf{v}_1 - \mathbf{v}_2 \mid /Vol \tag{7.23}$$

and N_2 is the number of drops in parcel 2, $\mathbf{v}$ is the drop velocity vector and Vol is the volume of the cell. The probable number of collisions, n, within the computational timestep Δt is then equal to $\nu_{12}\Delta t$. The probability of no collisions is $p(n) = e^{-\nu_{12}\Delta t}$ so that $0 < p(n) < 1$. A collision event is assumed if $p(n)$ is less than a random number generated in the interval (0, 1). Coalescence of colliding drops results if the collision impact parameter b is less than a critical value b_{crit}, where

$$b^2 = q(r_1 + r_2)^2 \tag{7.24a}$$

and

$$b_{crit}^2 = (r_1 + r_2)^2 \ \mathrm{Min}\,(1.0, 2.4\,(\gamma^3 - 2.4\,\gamma^2 + 2.7\,\gamma)/We_1) \tag{7.24b}$$

If b exceeds b_{crit}, the droplets maintain their sizes and temperatures but undergo velocity changes. In eqn (7.24a), q is a random number in the interval of (0, 1) and $\gamma = r_1/r_2$. If coalescence is predicted, n drops are removed from parcel 2 and the size, velocity, and temperature of drops in parcel 1 are modified appropriately[9].

The momentum exchange between the drops and the gas is due to drop drag. The acceleration term, $\mathbf{F}$ in eqn (7.12), is obtained from the equation of motion of a drop moving at a relative velocity, $\mathbf{w}$, in the gas, i.e.

$$\rho_1 V\,\mathbf{F} = C_D\,A_f\,\rho_2\,\mathbf{w}^2/2 \tag{7.25}$$

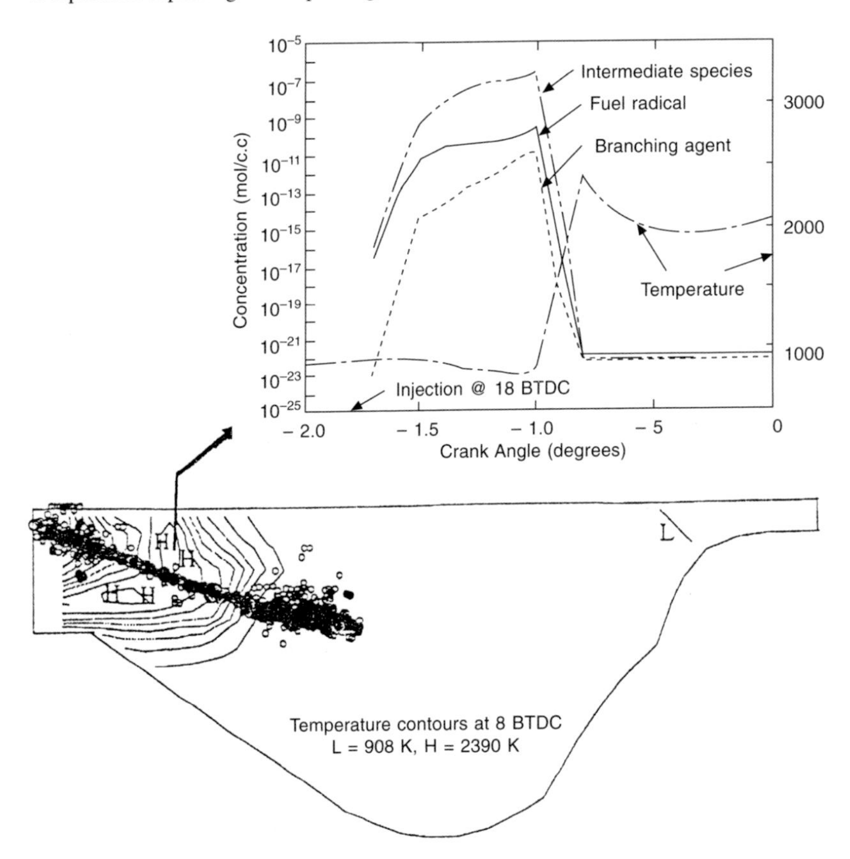

Figure 7.1 Predicted spray drop locations and temperature contours in the plane of the spray in a Cummins heavy-duty diesel engine at 8 degrees BTDC. Inset plot shows variation of gas temperature and ignition species concentrations with crank angle in the ignition cell[25]

where V and A_f are the drop volume and frontal area, respectively. To calculate the drop drag, in KIVA the drop is taken to be a sphere with drag coefficient[9].

$$C_D = \begin{cases} \frac{24}{Re_d}\left(1 + \frac{1}{6} Re_d^{2/3}\right) & Re_d \leq 1000 \\ 0.424 & Re_d > 1000 \end{cases} \tag{7.26}$$

where Re_d is the drop Reynolds number. However, in diesel sprays the drops undergo high distortion due to the high injection velocity and the drag coefficient changes as a drop departs from the spherical shape. To account for this, the distortion of a drop can be calculated from the TAB model, i.e. eqn (7.17)[22]. The distortion parameter lies between the limits of a sphere ($y = 0$) and a flattened drop or disk ($y = 1$) which has a drag coefficient $C_D = 1.54$. A simple expression for drag coefficient has been formulated to recover those limits for high speed drops, as follows[28]:

$$C_D = C_{D,sphere}\,(1 + 2.632y) \tag{7.27}$$

Consideration of drop distortion effects on drag coefficients has been found to be important also in engines that have spray/wall impingement since impingement destabilizes drops. Indeed, the impact of a drop on a heated surface may lead to its instantaneous breakup, sudden vaporization, or to the development of a thin liquid film on the surface[29]. Results from single-drop wall impingement experiments of Wachters and Westerling[30] show that the drop Weber number, $We = \rho_1 v^2 d/\sigma$, where v and d are the drop's normal velocity component and diameter, respectively, is an important parameter. For $We \leq 80$ the drop rebounds from the wall while for $We > 80$ the drop may disintegrate into small drops that move away from the impingement site parallel to the surface, depending on the surface conditions and temperature[30]. In the model of Naber and Reitz[31], at high Weber numbers ($We > 80$) the impinging drop is assumed to slide along the wall surface. This model has also been extended to include the rebounding drop case for $We < 80$. In this case, the tangential velocity component of the rebounding drop is assumed not to change during the collision and the normal velocity component is evaluated using a correlation between the arrival and departure Weber numbers of the form[32].

$$We_o = 0.678\, We_i\, e^{-0.04415\, We_i} \tag{7.28}$$

where the subscripts i and o refer to the incoming and outgoing rebounding drop, respectively. The subsequent disintegration of the drop depends on the relative velocity between the drop and the gas. Wall impingement represents a sudden disturbance acting on a drop. Therefore, the drop breakup time constant has been assigned a different value, $B_1 = 1.73$, after wall impingement, based on experimental data[33]. The effect of liquid films and wall heat transfer have also been considered[34,35].

The final term that must be modelled in eqn (7.12) is the rate of drop radius change, **R**, due to vaporization. The KIVA code uses the Frossling correlation[9]

$$\mathbf{R} = dr/dt = -\rho_2\, DBSh/(2\rho_1 r) \tag{7.29}$$

where D is the (laminar) mass diffusivity of fuel vapour in air, B is the mass transfer number, and Sh is the Sherwood number. The fuel mass fraction at the drop surface (which appears in B) is obtained by assuming the partial pressure of fuel vapour equal to the equilibrium vapour pressure at drop temperature. The rate of change in drop temperature is calculated from an energy balance involving the latent heat of vaporizaton and the heat conduction from the gas. The rate of heat conduction to the drop is

$$Q = \alpha(T_2 - T_1)\, Nu/(2r) \tag{7.30}$$

where α is the (laminar) thermal diffusivity, T_2 and T_1 are the gas and drop temperatures, respectively, and Nu is the Nusselt number. Other details about drop vaporization are described by Amsden *et al.*[9].

7.2.3 Ignition, combustion and emissions

The ignition delay is an important parameter in the operation of diesel engines since it influences hydrocarbon and NO_x emissions. During the delay period, the injected fuel undergoes complex physical and chemical processes including atomization, evaporation, mixing and preliminary chemical reaction. Ignition takes place after the preparation and reaction of the fuel–air mixture and leads to fast exothermic reaction. The multistep 'Shell' ignition model has been developed for the autoignition of hydrocarbon fuels at high pressures and temperatures by Halstead *et al*[30]. The model accounts for multistage ignition and 'negative temperature' coefficient phenomena. The model has been applied to diesel combustion by Theobald and Cheng[37], and Kong *et al.*[38].

The reactions and species involved in the Shell kinetic model are as follows:

$$RH + O_2 \rightarrow 2R^* \qquad K_q \tag{7.31a}$$
$$R^* \rightarrow R^* + P + \text{Heat} \qquad K_p \tag{7.31b}$$
$$R^* \rightarrow R^* + B \qquad f_1 K_p \tag{7.31c}$$
$$R^* \rightarrow R^* + Q \qquad f_4 K_p \tag{7.31d}$$
$$R^* + Q \rightarrow R^* + B \qquad f_2 K_p \tag{7.31e}$$
$$B \rightarrow 2R^* \qquad K_b \tag{7.31f}$$
$$R^* \rightarrow \text{termination} \qquad f_3 K_p \tag{7.31g}$$
$$2R^* \rightarrow \text{termination} \qquad K_t \tag{7.31h}$$

where RH is the hydrocarbon fuel (C_nH_{2m}), R* is the radical formed from the fuel, B is the branching agent, Q is a labile intermediate species, and P is oxidized products, consisting of CO, CO_2, and H_2O in specified proportions. The expressions for K_q, K_p, K_b, K_t, f_1, f_2, f_3, f_4, etc. are those given by Halstead *et al.*[36]. In addition, the local concentrations of O_2 and N_2 are needed to compute the reaction rates.

The premise of the Shell model is that degenerate branching plays an important role in determining the cool flame and two-stage ignition phenomena that are observed during the auto-ignition of hydrocarbon fuels. A chain-propagation cycle is formulated to describe the history of the branching agent, eqns (7.31b–f), together with one initiation, eqn (7.31a), and two termination reaction, eqns (7.31g–h). Some interpretations of these generic species have been proposed. The branching agent B is related to hydroperoxides (RO_2H) at low temperatures and to hydrogen peroxides (H_2O_2) at high temperatures. The intermediate species Q is related to aldehydes (RCHO) during the first induction period and to the alkylperoxy radical (HO_2) and its isomerization products during the second induction period. The formation of intermediate species eqn (7.31d) is the crucial reaction leading to the production of branching agent, which in turn induces hot ignition. The Shell model specifies different values for the kinetic parameters of this reaction for different fuels, and values are given by Kong *et al.*[38].

Figure 7.2 shows a comparison between spray ignition experiments and computations of ignition delay times using the Shell model[25]. The experiments were conducted in a constant volume bomb by Edwards *et al.*[39], and used diesel fuel (simulated in the computations as hexadecane) injected into compressed air at 30 atmospheres. Good levels of agreement between measured and predicted ignition delay times are seen over a wide range of operating temperatures. Another application of the ignition model is given in *Figure 7.1* which shows the predicted location and time-history of ignition in the Cummins diesel engine combustion chamber[25]. Fuel injection started at 18 degrees BTDC. Ignition takes place some distance downstream of the

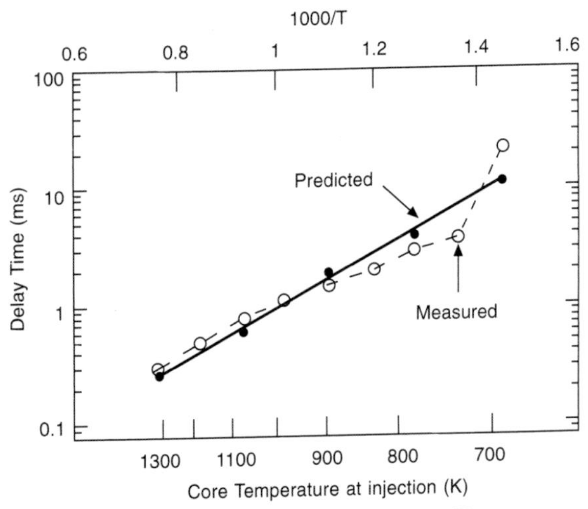

Figure 7.2 Comparison of measured[39] and predicted[38] ignition delay times for constant volume bomb at different chamber gas temperatures

nozzle at the spray edge at about 9 degrees BTDC. This is evidenced by the high temperature contours which are seen to surround the computed spray. The build-up and subsequent rapid consumption of the intermediate, branching and radical species that control the ignition process are also shown as a function of crank angle in the ignition cell in *Figure 7.1*. After the ignition, the gas temperature variation in the ignition cell reflects the balance between the energy released due to combustion and the energy required to vaporize the liquid fuel.

The Shell model has been modified to account also for fuel composition (Cetane number)[40] and for the effect of residual gases[41]. Since the formation of the branching species is crucial to the ignition process, the approach was to adjust the rate of reaction of reaction (7.31d). The formation rate of Q is f_4K_p, where $f_4 = A_{f04} \exp(-E_{f4}/RT) [O_2]^{x_4} [RH]^{y_4}$ with $E_{f4} = 3.0 \times 10^4$, $x_4 = -1$, $y_4 = 0.35$, and the rate K_p is given by Kong *et al.*[38]. To account for the effect of the residual gas the constant A_{f04} was adjusted as a function of the cylinder gas temperature at intake valve closure, T_{IVC}, as[41]

$$A_{f04} = (160.5 - 0.424T_{IVC}) \times 10^4 \quad (7.32)$$

Equation (7.32) shows that the autoignition reaction becomes slower as the initial temperature increases (i.e. as the amount of residual gas increases). This is consistent with expectation since an increase in the amount of residual gas leads to a reduction in the concentration of the reactants. To account for fuel effects the activation energy of reaction (7.31d), E_{f4}, was modified by the factor $65/(CN + 25)$[40]. In other words, for a Cetane number, CN, of 40, which is typical for diesel fuels, the factor is unity and for higher values of CN the activation energy decreases, resulting in shorter ignition delays and vice versa, consistent with experimental data[5].

In practical combustors, once ignition has occurred, the majority of combustion is thought to be mixing-controlled, and the interactions between turbulence and chemical reactions have to be considered. In recent years several turbulent combustion models have been proposed which include the effects of turbulence on mean reaction rates based on flamelet concepts. The model of Pitsch *et al.*[2] combines a detailed elementary chemical kinetics model with the flamelet model concept, while the models of Dillies *et al.*[42] and Musculus and Rutland[43] use simpler reduced chemistry models.

A simpler combustion model that has proved to be successful is based on the characteristic-time combustion model of Reitz and Bracco[44], which was originally applied to spark-ignition engine combustion by Abraham *et al.*[45]. The diesel combustion model used by Kong *et al.*[25,38] and Xin *et al.*[41] combines the 'Shell' ignition model and the characteristic time combustion model. In the combustion model, the time rate of change of the partial density of species i due to conversion from one chemical species to another, is given by

$$dY_i/dt = -(Y_i - Y_i^*)/\tau_c \quad (7.33)$$

where Y_i is the mass fraction of species i, and the * indicates local and instantaneous thermodynamic equilibrium values. $\tau_c = \tau_l + \tau_t$ is the characteristic time for reaction which is formulated as a sum of timescales such that the longest timescale controls the combustion rate. τ_l is a laminar Arrhenius (high temperature) chemistry time, and $\tau_t = f\, C_2\, k/\varepsilon$ is the turbulence mixing time. k and ε are the turbulence kinetic energy and its dissipation rate, and $C_2 = 0.1$ when k and ε are computed using the RNG model of Han and Reitz[12]. f is a 'delay coefficient' that accounts for the fact that turbulence does not influence early flame growth, $f = (1 - e^{-r})/0.632$, where r is the ratio of the amount of products to that of total reactive species. Its value varies from 0 (no combustion yet) to 1 (complete consumption of fuel and oxygen). Accordingly, the delay coefficient f changes from 0 to 1 depending on local conditions.

Xin *et al.*[41] accounted for the effect of residual gas concentration on combustion through the laminar timescale

$$\tau_l = A[RH]^{0.75} [O_2]^{-1.5} \exp(E/RT) \quad (7.34)$$

by introducing the residual gas concentration in the pre-exponential constant with $A = (1 + 3.3x_r) \times 3.24 \times 10^{-12}$, where x_r is the residual gas mass fraction[13]. This ensures that the laminar characteristic time increases with increasing residual gas concentration, consistent with experimental data on laminar flame speeds, such as that of Metghalchi and Keck[46]. This concept of modifying the laminar timescale to account for residual gas effects was also used successfully by Kuo and Reitz[47].

With the combustion model in place, the chemical source term in the species eqn (7.1) and the chemical heat release in the energy eqn (7.5) are found from

$$\Delta\rho_m = -\rho(Y_m - Y_m^*)(1 - e^{-\Delta t/\tau_c}) \quad (7.35)$$

$$\Delta Q = -\sum_m \frac{\Delta\rho_m}{W_m} (\Delta h_f^0)_m \quad (7.36)$$

where Δt is the numerical timestep, and $\dot{\rho}_m^c = \Delta\rho_m/\Delta t$ and $\dot{Q}^c = \Delta Q/\Delta t$. Further details of the combustion model are given by Kong and Reitz[33] and Xin *et al.*[41].

Modelling of engine emissions is of critical importance for the engine industry, which is facing stringent emission regulations. Unburned hydrocarbon emissions (HC) are predicted by the characteristic time combustion model in regions where the temperatures become low enough that combustion times become very large. For NO, the extended Zeldovich mechanism[48] has been widely used:

$$\begin{aligned} O + N_2 &\leftrightarrow NO + N \\ N + O_2 &\leftrightarrow NO + O \\ N + OH &\leftrightarrow NO + H \end{aligned} \quad (7.37)$$

The reaction equations are solved by assuming a steady state population of N and equilibrium for $O + OH \leftrightarrow O_2 + H$. The resulting NO formation rate is

$$\frac{d[NO]}{dt} = \frac{2R_1(1 - \alpha^2)}{1 + (\alpha R_1/(R_2 + R_3))} \beta_{NO} \quad (7.38)$$

where α is the ratio of the kinetic [NO] to that which would be in equilibrium and reaction rates, R_i, are given by Patterson *et al.*[49]. To close the solution, N_2, O, O_2, and OH are assumed to be in equilibrium at local conditions. The factor β_{NO} in eqn (7.38) is a constant calibration factor adjusted to allow NOx predicted from the NO model to match the engine-out NOx data. Other more comprehensive NO models have also been proposed (e.g. Miller and Bowman[50]). However, adequately accurate results have been obtained with the extended Zeldovich model.

Similarly, detailed soot models have been proposed (e.g., Pitsch *et al.*[2], and Fusco *et al.*[5]. However, good results have been obtained using the soot formation model of Hiroyasu[52] with the Nagle-Strickland Constable (NSC) oxidation model[53]. The Hiroyasu soot formation model uses an Arrhenius rate expression to compute the rate of soot formation, i.e.,

$$\frac{dM_{soot}}{dt} = \frac{dM_{form}}{dt} - \frac{dM_{oxid}}{dt} \qquad (7.39)$$

where

$$\frac{dM_{form}}{dt} = A_f M_{fv} P^{0.5} \exp(-E_f/RT)$$

and M_{fv}, M_{form}, M_{oxid}, M_{soot} are the fuel vapour, formed, oxidized and net soot masses, respectively, P is the pressure, and the Arrhenius pre-exponential and activation terms are $A_f = 100$, $E_f = 12500$ cal/mole, respectively. In the NSC oxidation model, carbon oxidation occurs by two mechanisms whose rates depend on surface chemistry involving more reactive A sites and the less reactive B sites. The net reaction rate is

$$R_{Total} = \left(\frac{K_A P_{O_2}}{1 + K_Z P_{O_2}}\right) x + K_B P_{O_2} (1 - x)$$

where x, the proportion of A sites, is given by

$$x = \frac{P_{O_2}}{P_{O_2} + (K_T/K_B)}$$

P_{O_2} is the oxygen partial pressure, and the rate constants, K_A, K_B, K_Z and K_T, are given by Patterson *et al.*[49]. The soot mass oxidation rate is given by

$$\frac{dM_{oxid}}{dt} = \frac{6M_c}{\rho_s D_s} M_{soot} R_{Total} \qquad (7.40)$$

where M_c is the carbon molecular weight, ρ_s is the soot density (2.0 g/cm^3), D_s is the soot diameter (3×10^{-6} cm).

7.3 Applications

An integrated numerical model has been developed for diesel engine computations based on the KIVA codes, as described by Kong *et al.*[25,38], Han *et al.*[12,19], Xin *et al.*[41], and Senecal *et al.*[54]. The improved submodels include models for piston ring-pack crevice flow, wall heat transfer, spray, ignition, combustion, soot and NOx emissions, as described in the previous section, in the references, and in *Table 7.1*. The 'blob' injection and wave breakup model was used because it removes the need to specify an assumed initial drop size distribution at the nozzle. The drop drag submodel accounts for the effects of drop distortion and oscillation. The Shell multistep ignition kinetics model was used to simulate the low temperature chemistry during ignition delay period. The laminar-and-turbulent characteristic-time combustion model was used for modelling high temperature combustion. The extended Zeldovich mechanism was implemented for predicting NOx formation. The soot emission model used the formation model of Hiroyasu, and the NSC soot oxidation model.

7.3.1 Modelling the gas-exchange process

The accuracy of simulating combustion processes depends on the initial conditions within the combustion chamber at intake valve closure. Residual gas left in the combustion chamber from the previous engine cycle effects the combustion process through its influence on the charge mass, charge temperature and dilution. These parameters, in turn, influence the ignition delay and the combustion rate, as mentioned previously. Three-dimensional computations of the flow field set up during the gas exchange process were conducted by Senecal *et al.*[54] using a modified version of the KIVA-3 code. Details of the engine geometry simulated are given in *Table 7.2* and in *Figure 7.3*. The code employed a 'snapping' procedure for the intake and exhaust valves similar to that used to move the piston[55]. As a result, the valves move realistically according to the defined valve lift curves which are given as input. The grid is made up of over 120 000 cells which allows for accurate representation of the valve region and the combustion chamber with piston bowl. The port and valve dimensions used are also given in *Table 7.2* and details of the results and the grid generation methods used are given by Senecal *et al.*[54].

The simulations began at exhaust valve opening and continued until intake valve closure. *Figure 7.4* presents computed velocity vectors showing the flow and the residual gas mass distribution in the engine for the baseline 1600 rev/min, 75% load, zero exhaust gas recirculation (EGR) case of *Table 7.3* at 144° ATDC. As can be seen, the highest mixing of the incoming fresh charge and the combustion products is predicted to occur where the flow velocities are the largest. Constant pressure boundary conditions were set at the ends of both the intake and exhaust ports corresponding to the values given in *Table 7.3* for the various engine operating modes of the experiments of Montgomery and Reitz[56]. Combustion simulation results were used to supply an initial estimate of the cylinder conditions at exhaust valve opening. After the gas exchange process was modelled, revised initial conditions (at intake valve opening) were used for a new combustion simulation, and the process

Table 7.2 Test engine and fuel system details

Engine details (Caterpillar 3401)	
Cylinder bore	137.2 mm
Stroke	165.1 mm
Compression ratio	14.5
Displacement	2.44 litres
Simulated turbocharge to 4 atm.	
Cooled EGR	0–15%
Fuel injection system	
Injectors 6 holes	0.26 mm diameter
Injection pressure (nominal)	90 MPa
Single injections	
Up to 4 multiple injections	
Port/valve data	
Exhaust valve opening	–217° ATDC
Intake valve opening	220° ATDC
Intake valve diameter	45 mm
Exhaust valve diameter	41.8 mm
Maximum valve lift	11.0 mm
Intake port diameter	40.38 mm
Exhaust port diameter	37.21 mm
Port lengths	130 mm

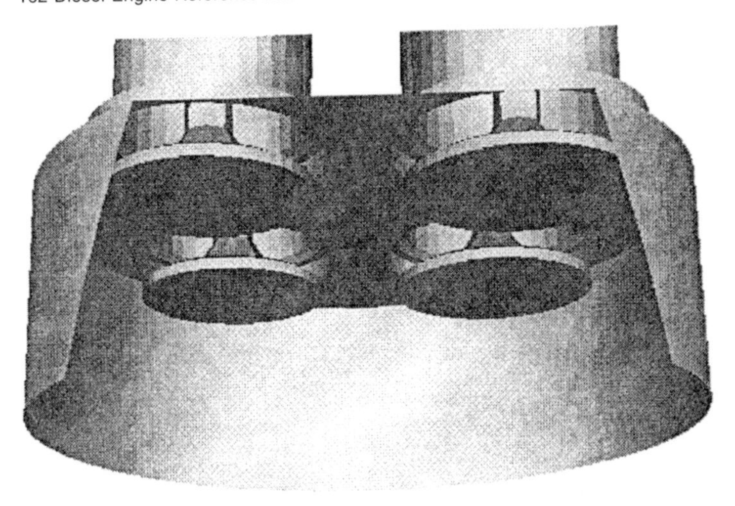

Figure 7.3 Intake and exhaust valve arrangements for modelled diesel engine[54]

was repeated until convergence was achieved. For computational efficiency the combustion computations were made in a domain that considered only one of the six spray plumes (see *Table 7.2*), and 60 degree sector symmetry was assumed in the engine. In this case, the results from the gas exchange computations (which considered the entire engine combustion chamber) were spatially averaged to provide initial conditions for the combustion computations.

The results of the gas exchange model predictions are summarized in *Table 7.4* for the zero EGR engine operating conditions of *Table 7.3*. As can be seen, the predicted residual gas amount varies from about 1 or 2% of the cylinder charge mass for high and medium load cases, to over 5% of the charge mass for light load cases. The residual gas level is seen to also significantly influence the average gas temperature in the combustion chamber at intake valve closure T_{IVC}.

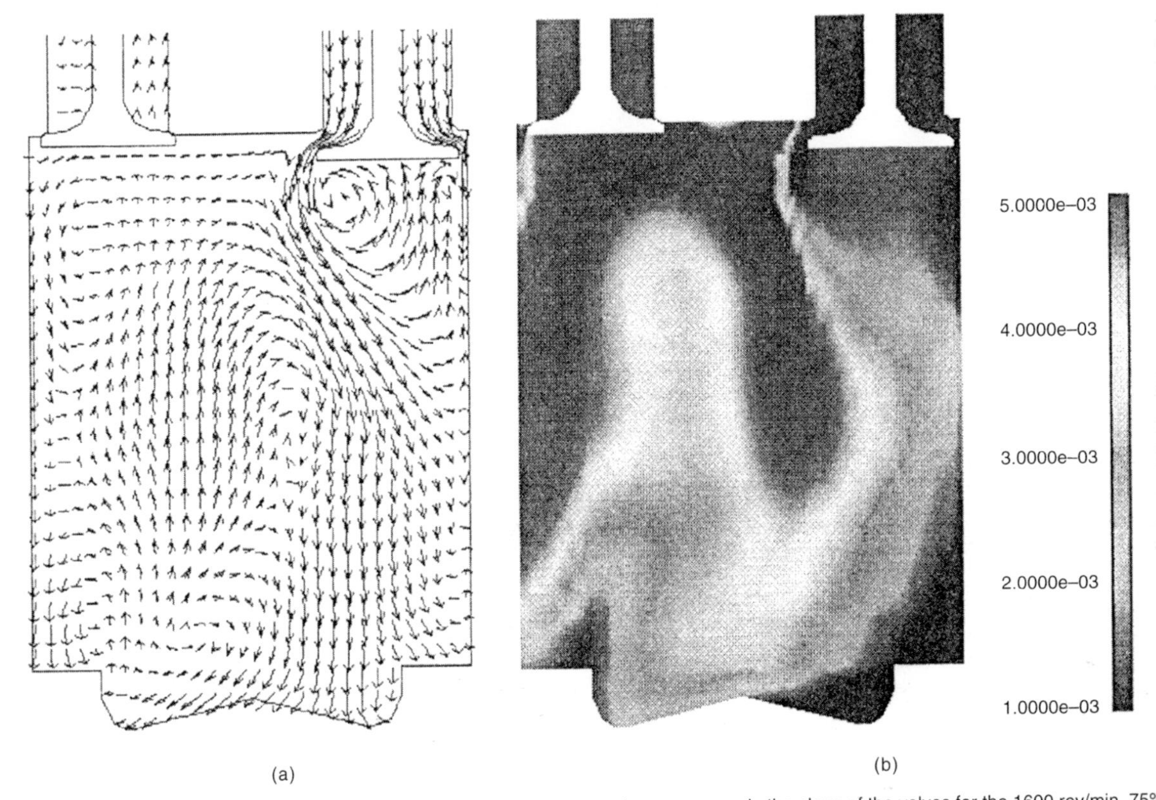

Figure 7.4 Computed flow details at 144 degrees ATDC during the gas exchange process in the plane of the valves for the 1600 rev/min, 75% load baseline case of *Table 7.3*. Intake valve is on the right. CO_2 density, normalized by the total density is used to represent the residual gas mass fraction: (a) flow velocity vectors; (b) residual mass distribution[54]

Table 7.3 Six-mode engine test conditions[56]

Running Condition	*Mode 1*	*Mode 2*	*Mode 3*	*Mode 4*	*Mode 5*	*Mode 6*	*1600 rev/min 75% Load*
Speed (rev/min)	750	953	1074	1657	1668	1690	1600
Fuel consumption rate (kg/hr)	0.5	2.0	6.4	10.2	5.4	3.2	7.7
Intake (°C) temperature	26	29 (40 at 20% EGR)	31	40	32	29	36
Intake pressure (kPa)	100	108	168	239	164	132	183
Baseline exhaust pressure (kPa)	100	112	144	220	164	143	159
Exhaust pressure at 6% EGR (kPa)	—	—	181	286	197	—	216
Exhaust pressure at 10% EGR (kPa)	—	—	—	—	—	—	219
Exhaust pressure at 20% EGR (kPa)	—	116	—	—	—	—	—

Test Conditions (nominal)

Table 7.4 Predicted residual gas levels at intake valve closure[41]

Mode	*Residual%*	$T_{IVC}(K)$
1	5.3	330
2	4.3	345
3	0.73	310
4	0.2	337
5	3.5	376
6	5.2	350
Baseline	1.4	312

7.3.2 Combustion and emissions model validation

The experiments for the tests conditions of *Table 7.3* are summarized in *Figure 7.5*. The engine is described in *Table 7.2*, and details of the measurements and instrumentation are given by Montgomery and Reitz[56]. In this case a common rail fuel injector was used with single injections, and an injection pressure of 90 MPa was used in each case. The total cycle emissions (obtained using a weighted sum of the results from each of the six modes to represent the federal transient emissions cycle[56] gave 5.15 g/bhp-h NOx and 0.44 g/bhp-h particulate. The overall cycle specific fuel consumption was 206 g/bhp-h. As can be seen in *Figure 7.5*, the experimental results show that mode 1 (idle point) contributes as much as 41% of the total particulates in the six-mode cycle. However, by reducing the injection pressure from 90 MPa to 30 MPa for Mode 1, the particulate emission from Mode 1 was found to be significantly reduced to 6% of the total, as shown in *Figure 7.6*. In this case the total cycle emissions were 5.5 g/bhp-h NOx and 0.26 g/bhp-h particulate. The overall cycle specific fuel consumption was also reduced to 199 g/bhp-h.

It is usually thought that particulate emissions increase with reduced injection pressure. The reasons for the dramatic reduction in particulate with decreased injection pressure at idle were explored using the computational models[41]. The simulation results for the two injection pressures (30 MPa and 90 MPa) are shown in *Figure 7.7* to *7.9*. Comparisons between the measured and predicted cylinder pressures are shown in *Figure 7.7a* and *b*. In this case, the nozzle discharge coefficient was also reduced

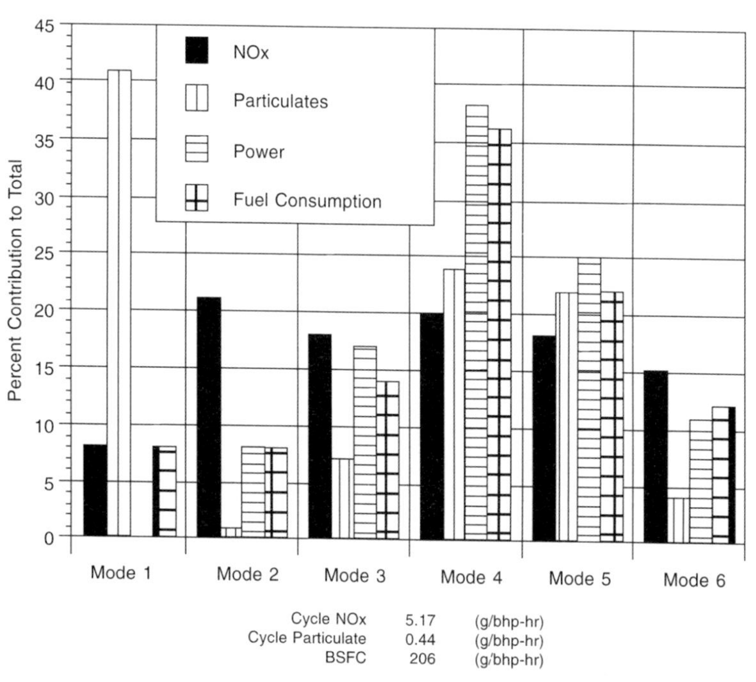

Figure 7.5 Engine measurements for six-mode analysis with 90 MPa injections[56]

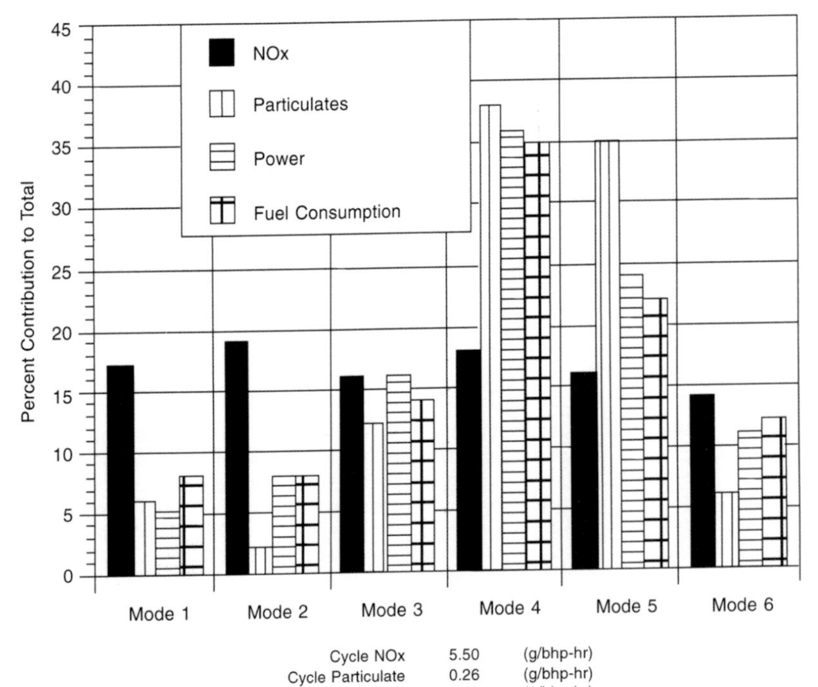

Results of six mode analysis with a 30 MPa single injection at mode 1 and 90 MPa single injections at all other modes

Figure 7.6 Engine measurements for six-mode analysis with 90 MPa injections except at Mode 1 (idle) which uses 30 MPa injection pressure[56]

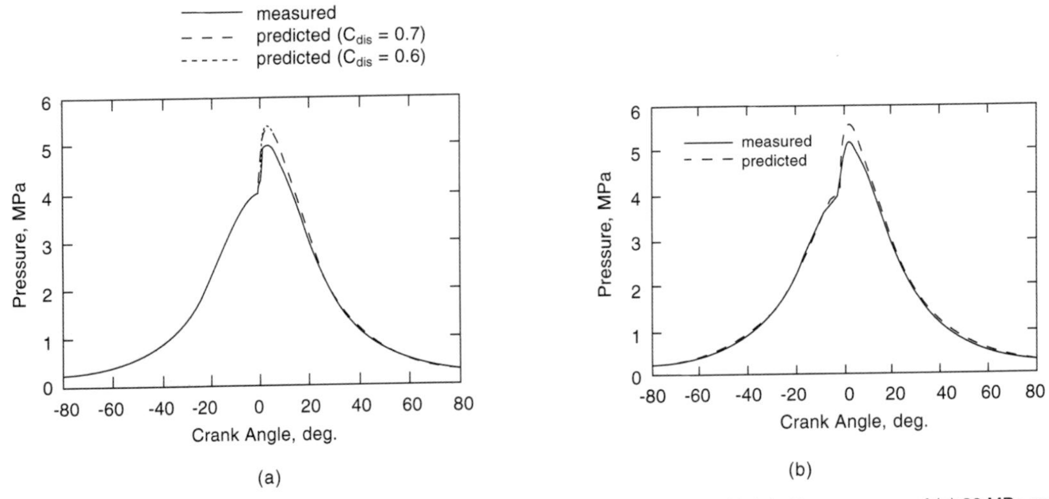

Figure 7.7 Predicted and measured cylinder pressure history for mode 1 (idle) cases with injection pressures of (a) 30 MPa and (b) 90 MPa[41]

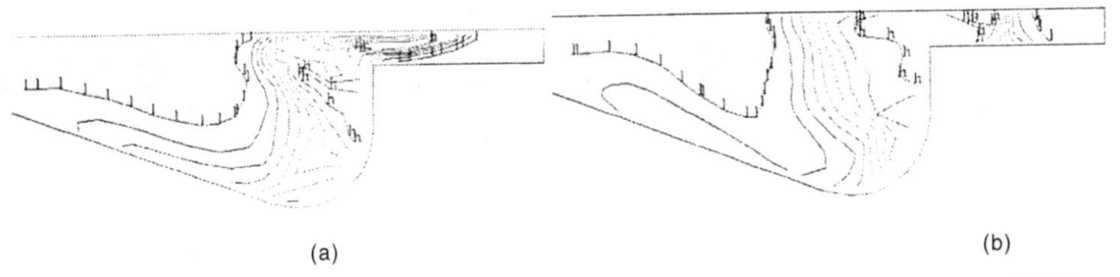

Figure 7.8 Predicted temperature distributions in the combustion chamber on the spray axis at 2 degrees ATDC for mode 1 (idle). (a) 30 MPa injection pressure 1 = 1050K, h = 2560 K; (b) 90 MPa injection pressure 1 = 1050 K, h = 2510 K[41]

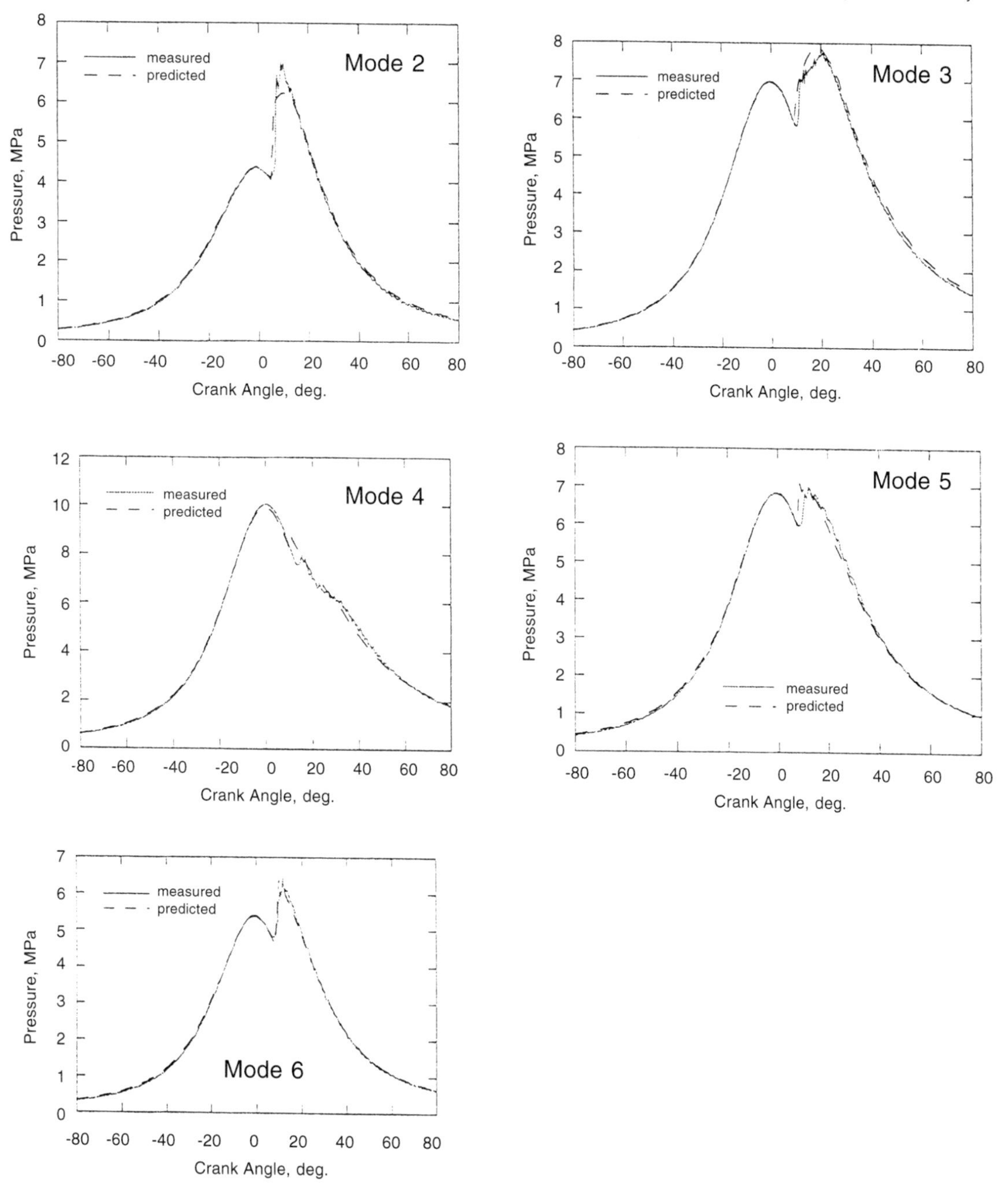

Figure 7.9 Predicted and measured cylinder pressure histories for modes 2 to 6[41]

from 0.7 to 0.6 when the injection pressure was reduced from 90 MPa to 30 MPa to explore its effect on engine-out emissions. For the same flow rate and nozzle exit area, decreasing the discharge coefficient increases the injection velocity which improves atomization and mixing. *Figure 7.7* shows that the predicted cylinder pressures agree reasonably well with the experimental results for both injection pressures, and the effect of the discharge coefficient is actually quite small.

The emissions are influenced by the distribution of the fuel–air mixture in the combustion chamber. The local mixture temperature is affected by mixing (through the equivalence ratio) and the ignition delay, which are different when the injection pressure is reduced, as shown in *Figure 7.8*. When the injection pressure is reduced, the injection duration increases, and the initial droplet velocity decreases. Consequently, the 30 MPa spray penetration becomes short compared with the 90 MPa case. The reduced penetration prevents fuel from over-mixing in this case of very low overall fuel–air ratio (see *Table 7.3*).

For the 90 MPa injection pressure case, initial droplet velocities are high. Since the in-cylinder gas density is low at idle, the spray penetrates further during the ignition delay. The over-mixed fuel is too lean to burn and is likely to be exhausted as unburned hydrocarbons and as high soluble organic fraction (SOF) of the particulate. Also, because of the short injection duration, the droplets cluster together, and this reduces evaporation rates. By the time the mixture auto-ignites, some of the fuel has reached the squish region, and the flammable mixture continues to move into the squish region. The high temperature zones (combustion zones) are distributed along the combustion chamber head surface, as shown in *Figure 7.8b* (shown at 2 degrees ATDC). For the 30 MPa case, the spray is well distributed spatially within the piston bowl at the time of ignition. By 2 degrees ATDC, the high temperature zones are still located within the piston bowl, as seen in *Figure 7.8a*. Therefore, for the 30 MPa injection pressure case, the combustion is confined, to the piston bowl and there is less heat loss through the head surface, which tends to increase NOx emissions somewhat, but the soot formation is suppressed significantly.

Figure 7.9 shows comparisons of predicted and measured in-cylinder pressure histories for modes 2–6, while *Figure 7.10* summarizes the engine-out emissions comparisons for the six modes and the baseline case. Mode 2 is characteristic of low load (25%), low speed (953 rev/min) engine operation. Like the mode 1 idle case, the predicted results show good agreement with the experimental pressure results. However, the predicted peak pressure is somewhat lower than that measured for mode 2. The NOx prediction agrees with the engine-out data very well for both modes 1 and 2.

It should be noted that current soot models only consider the 'dry' carbon part of the particulate. Because of the significant contribution of SOF to the measured particulate at low loads, it is expected that the present soot predictions should be low, and this is indeed found (see *Figure 7.10b*). For mode 2, it is about 0.18 g/kgfuel compared with the experimental particulate measurement 0.39 g/kgfuel. For mode 1, most of the measured particulate is SOF, and this is in qualitative agreement with the predicted low soot level.

Mode 3 represents a high load (75%) and low speed (1074 rev/min) engine operating condition. For this mode, the predicted in-cylinder pressure history has excellent agreement with the experimental result, as shown in *Figure 7.9*. From the NOx comparison with the experimental engine-out result shown in *Figure 7.10a*, it is seen that the simulation predicts slightly higher NOx than that measured. The soot prediction is higher than that measured (measured particulate emission 1.09 g/kgfuel, predicted soot 1.5 g/kgfuel). It is not known at this time why this discrepancy exists, but it is expected that the effect of overmixing on SOF should be smaller than that found for modes 1 and 2. It is known that the percentage of SOF in diesel particulates decreases as engine loads rise.

In the mode 4 case, the engine is operated at full load and high speed. The pressure traces of mode 4 are shown in *Figure 7.9*, which indicates that the pressure history is predicted quite well. The numerical NOx results show excellent agreement with the experimental engine-out results. The predicted engine-out soot emission is equal to 1.39 g/kgfuel which is reasonably close to the measured particulate value (1.41 g/kgfuel). Mode 5 is characteristic of high engine speed and moderate load (50%). *Figure 7.9* shows that there is very good agreement between the experimental and predicted pressure results. The predicted NOx also agrees with the experimental engine-out result very well. The soot comparisons of the predicted and measured results show fair agreement with the experimental engine-out data, again possibly because of the increased contribution of SOF to the total particulates at the reduced load.

Finally, mode 6 has a similar engine speed to mode 5, but the load is only 25%. As listed in *Table 7.3*, the exhaust pressure is higher than the intake pressure in this mode, and it is expected that a significant amount of combustion products flow into the intake manifold during the valve overlap period. Therefore, the initial temperature for the simulation at IVC is estimated to be 350 K, which is higher than those in other low load modes (see *Table 7.4*). The predicted pressure history of the mode 6 case has good agreement with the measured results, as shown in *Figure 7.9*. The predicted engine-out NOx is also very close to that measured (see *Figure 7.10a*). *Figure 7.10b* also shows that the predicted soot is lower than the measured particulate emissions. This is expected because SOF contributes a significant portion of particulates at low load conditions. This is in agreement with measured SOF levels of Choi *et al.*[57] for the same engine operating conditions.

From *Figure 7.10a*, it is seen that the overall NOx prediction accuracy is satisfactory in all cases. However, there exist larger discrepancies between the predicted soot and the measured particulate, probably because of the role of unburned fuel and lubricating oil in the SOF portion of the particulates. The results also show that the soot prediction accuracy for high and moderate load engine conditions is better than that for low load engine conditions. The low load cases feature longer ignition delays and dominant premixed-mode combustion. Therefore, mixing and evaporation play a more important role in low load modes, and the present results indicate that improvements may be needed in those models. This conclusion has also been reached by comparing model predictions with endoscope combustion visualization images. At light loads, discrepancies have also been found between model and experimental results by Ricart and Reitz[58]

7.3.3 Effect of multiple injections

As described by Pierpont *et al.*[59], and Montgomery and Reitz[56], split injections used in conjunction with EGR can significantly reduce engine-out NOx and particulate emissions. This is also shown in *Figure 7.11* which presents overall cycle results for

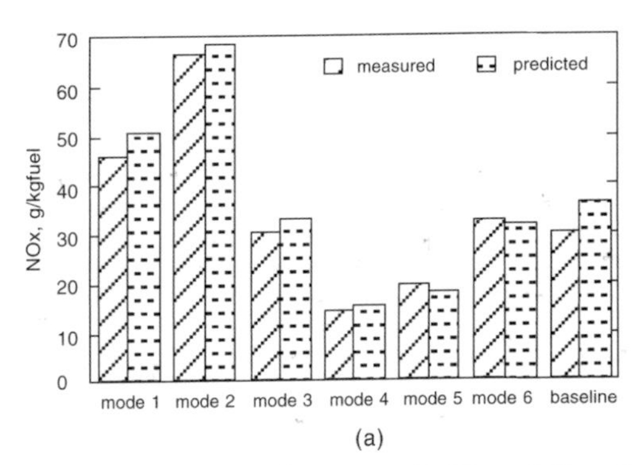

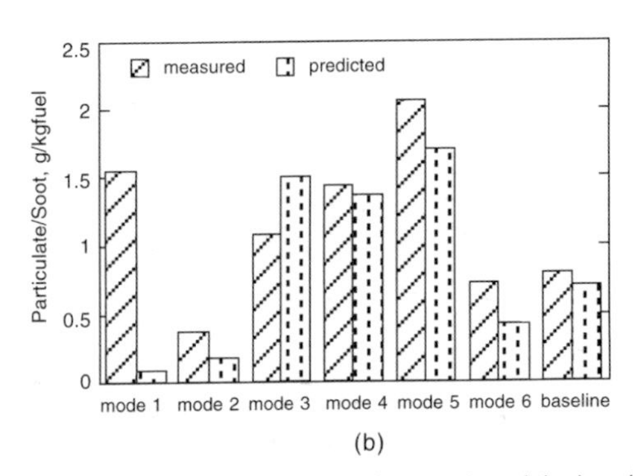

Figure 7.10 Predicted and measured (a) NOx and (b) soot/particulate histories (g/kgfuel) for modes 1 to 6, and the baseline case (see *Table 7.3*)[41]

the present six mode simulation when the engine was operated with the optimized double injection/EGR combinations shown in *Table 7.5*. The total cycle NOx is now 3.7 g/bhp-h, the particulate has been reduced to 0.11 g/bhp-h and the specific fuel consumption is reduced to 184 g/bhp-h. Comparison with the 90 MPa single injection, zero EGR results of *Figure 7.5* show that the cycle NOx, particulate and fuel consumption have been reduced by 32%, 75% and 12%, respectively, through the use of double injections and EGR. It is worth noting that these low emission levels were achieved with the use of relatively low injection pressures, relatively low boost pressures, and a fuel injector nozzle with a relatively large exit orifice diameter, and a high nozzle discharge coefficient. Also, no correction was applied to account for the increased friction of the single cylinder test engine.

The mechanism of emissions reduction accompanying the use of multiple injections has been studied computationally by Han *et al.*[19]. Computer animations of the results are summarized schematically in *Figure 7.12*. Soot is formed and it accumulates in the tip region of the spray jet, consistent with experimental observations in optically accessible DI diesel engines[60]. In single injection combustion, the high-momentum injected fuel penetrates to the fuel-rich, relatively low temperature region at the jet tip and continuously replenishes this rich region as long as the injection continues, producing soot. In a split-injection, however, the replenishment is interrupted allowing the soot at the tip to oxidize. The second-pulse-injected fuel enters into a relatively fuel-lean and high-temperature region which is left over from the combustion of the first pulse. Soot formation is therefore significantly reduced because the injected fuel is rapidly consumed by combustion before a new rich soot-producing region can accumulate. As a result, the net production of soot in split-injection combustion can be reduced substantially, particularly if the dwell between the two injections is optimized—long enough so that the soot formation region of the first injection is not replenished with fresh fuel, but short enough that the in-cylinder gas temperature environment seen by the second pulse remains high enough to prompt fast combustion, reducing soot formation.

The CFD predictions of the effect of multiple injections on emissions are presented in *Figure 7.13a* for a double injection case where 75% of the fuel is injected in the first pulse, and the remaining 25% is injected after an 8 crank angle degree delay [19]. The predictions agree well with measured trends for a similar double injection scheme on the same engine of Tow *et al.*[61], shown in *Figure 7.13b*. The experiments and the model predictions both demonstrate that significant particulate reductions are possible with the use of a double injection, without a penalty in NOx. The CFD model results explain the mechanism of the emissions reduction. This information is invaluable to provide guidelines for further emissions reductions.

Triple and quadruple injections have also been investigated experimentally by Montgomery and Reitz[56] and have been shown to give further reductions in particulate emissions. In fact, the combination of multiple injections and EGR has been shown to be a very promising strategy for the control of both NOx and particulate emissions[56]. Work is in progress using the simulation models to study the mechanisms of emissions reductions for these cases and to extend the study to the entire engine operating range.

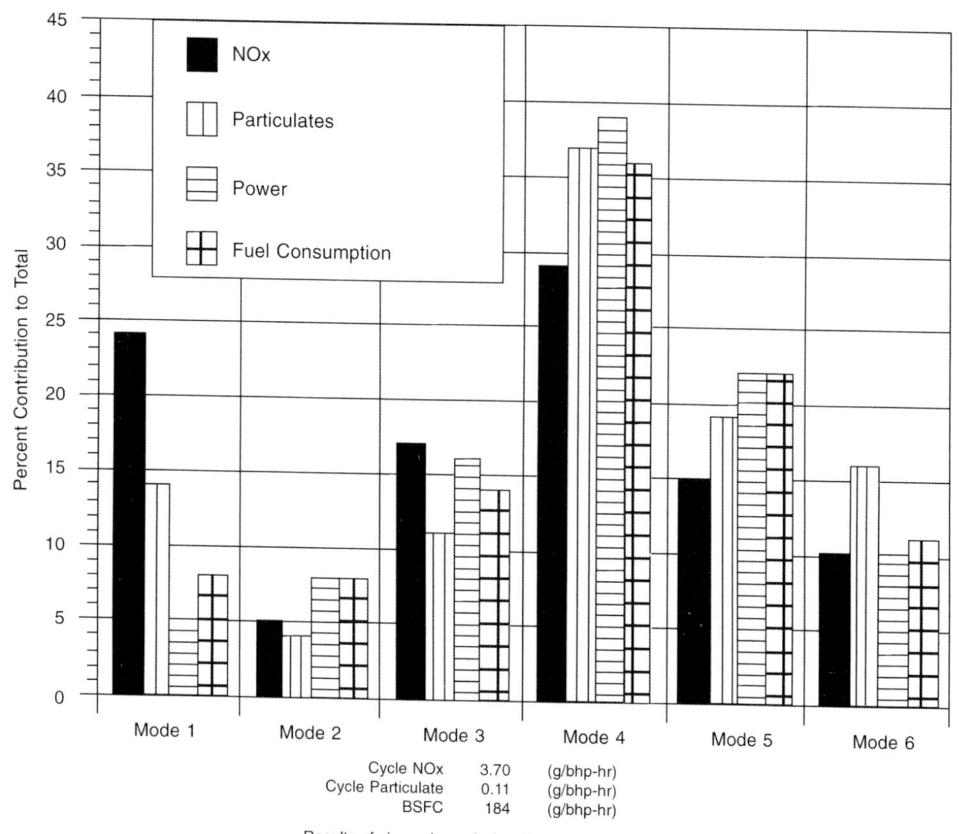

Figure 7.11 Engine measurements for six mode analysis for the double injections and EGR conditions shown in Table 7.5[56]

Table 7.5 Optimized 6-mode test operating conditions[56]

Mode	Conditions
1	0% EGR 30 MPa single injection
2	20% EGR 90 MPa double injection
3	6% EGR 90 MPa double injection
4	6% EGR 90 MPa double injection
5	6% EGR 90 MPa double injection
6	0% EGR 90 MPa double injection

7.3.4 Use of CFD in engine design

Recently novel engine designs have been investigated that use a multiple injector combustion system (MICS) to enhance air usage in a direct injection diesel engine[3,4]. *Figure 7.14a* illustrates the MICS concept, which includes six, single hole injectors equally spaced around the edge of the piston bowl. This enhances the swirl motion and results in significantly improved air–fuel mixing. As a result, soot oxidation is increased and hence soot production is reduced dramatically, as shown in the predicted soot–NOx tradeoff results of *Figure 7.14b*. While the practicality of such an engine is questionable, the purpose of the work was to explore the potential for emissions reductions using alternative engine concepts. As this example illustrates, engine CFD can be an invaluable tool to an engine designer since it can provide insight into the highly complex physical and chemical processes in the engine. Novel design ideas can be introduced and analyzed much faster and at lower cost than with conventional methods. For example, a dual injector concept has been explored by Senecal *et al.*[4] that exploits the same physical mechanisms that cause the predicted dramatic soot reduction in the original six-injector design. The computational results show that this design is also very promising[4].

7.4 Summary and conclusions

Detailed comparisons between measured and predicted engine cylinder pressures, and soot and nitric oxide (NOx) emissions have been made, and the predicted results are seen to be in very good agreement with engine experiments. This indicates that engine CFD spray, combustion and emissions models are now available to the engine industry for use as predictive design tools to provide directions in engine design.

The importance of sub-grid scale models is emphasized. In particular, drop breakup and drop drag effects govern the penetration and mixing of vaporizing diesel sprays. The effects of spray wall impingement are also important. The Shell model gives good predictions of diesel ignition, and combustion can be modelled adequately when both laminar chemistry and turbulent mixing are incorporated into the combustion model. The present soot and NOx predictions are also encouraging. However, further testing of the spray, combustion and emissions models is needed for engines other than those studied here.

CFD and grid generation codes are now available for modelling the intake and exhaust flow processes. This permits accurate prediction of the residual gas concentrations which set the stage for the combustion process. Current spray, combustion and emissions models have been shown to be able to predict engine data over the entire engine operating range. Cylinder pressure and NOx predictions are satisfactorily accurate, and 'soot' is also predicted at high loads with no model tuning. However, soot models need modification to account for the observed high percentage of SOF in particulate at low loads.

The present results show that computational models can be used to explain engine data, and for the evaluation of new engine design concepts. The computations show that low injection pressure reduces particulate emissions at idle and light loads, in

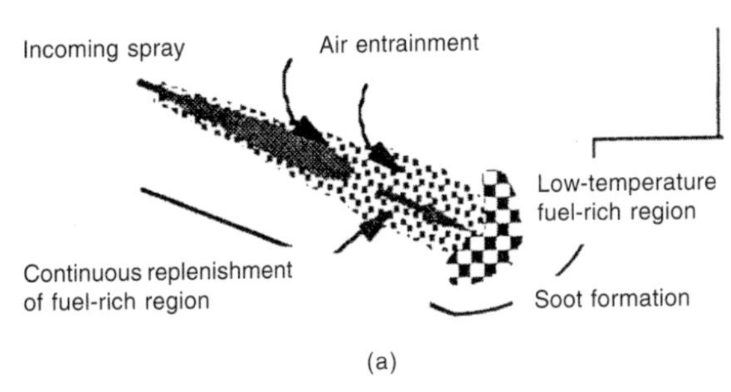

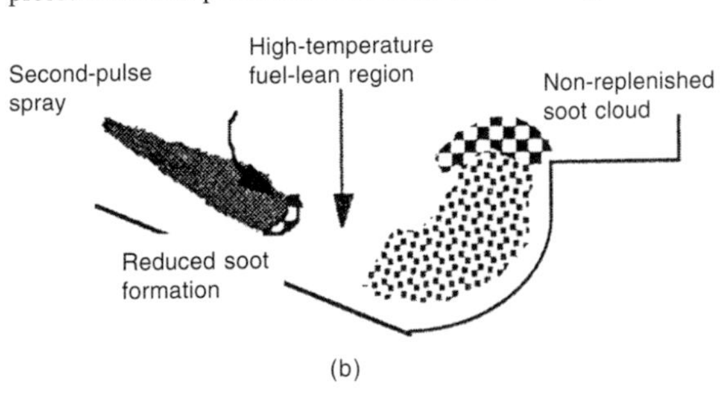

Figure 7.12 Schematic diagram showing soot-reduction mechanisms of split injections: (a) single injection; (b) split injection[19]

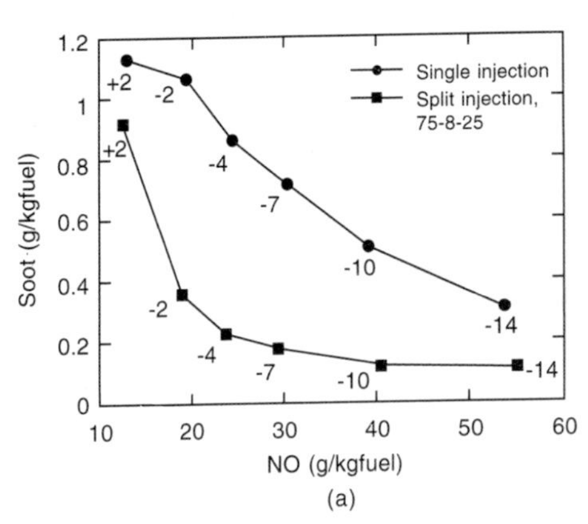

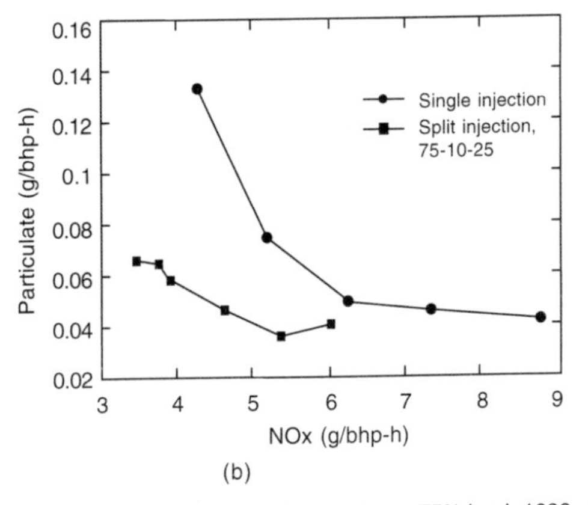

Figure 7.13 Trade-off curves showing emission reductions using split injections for the Caterpillar engine at 75% load, 1600 rev/min. Injection timings were varied from – 12 to + 1 ATDC. (a) Computed soot-NO trade-off[19] (b) Measured particulate-NOx trade-off[61]

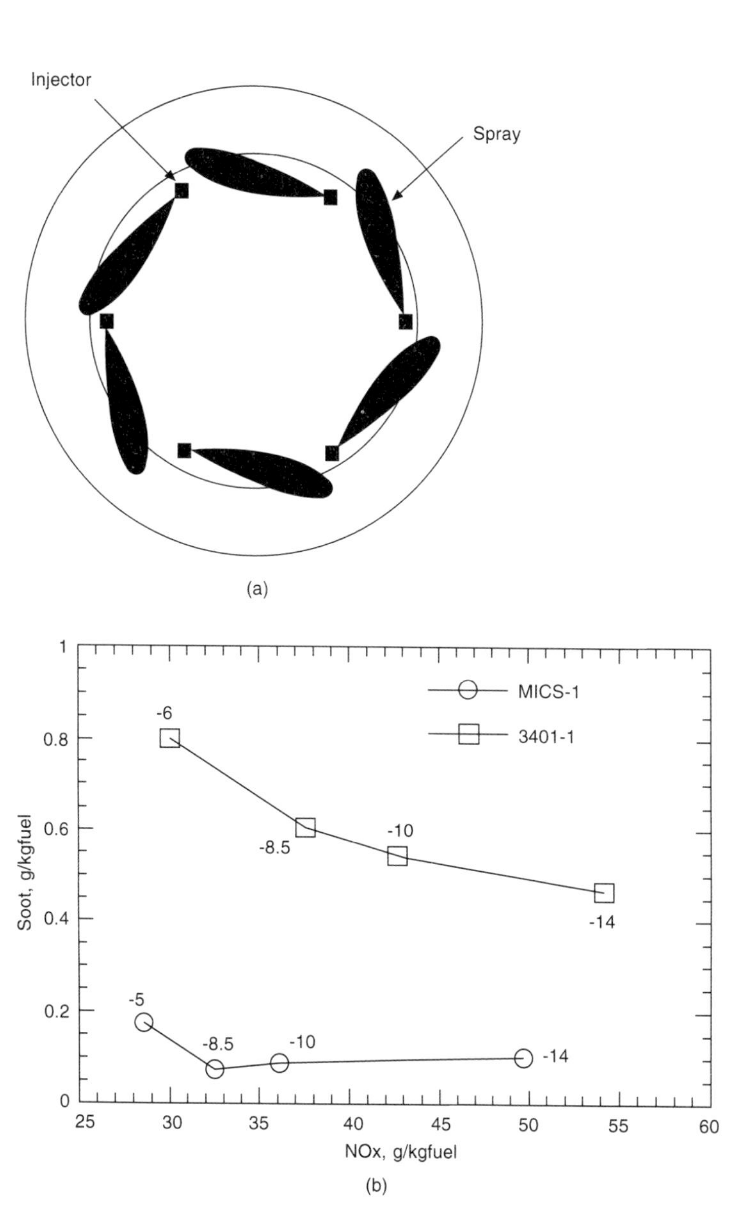

Figure 7.14 (a) Schematic diagram of the multiple injector combustion system (MICS) chamber (top view) and (b) predicted soot-NOx tradeoff curves for the MICS and a baseline single centrally located six-hole injector design (Caterpillar 3401 engine)[3]. Numbers indicate start of injection timings

agreement with experiments. The explanation for this trend and the mechanism of over-mixing with the use of high injection pressures was explained with the simulations. Significant NOx and particulate emissions reductions have been achieved, together with improved fuel economy, by using split injections combined with Exhaust Gas Recirculation (EGR). The present CFD spray, combustion and emissions models have been used to explain the mechanism of particulate emissions reductions with split injections.

Acknowledgements

This work is the outcome of collaborations with many colleagues at the Engine Research Center whose names appear in the list of references. Major funding for the work was provided by the Army Research Office, Caterpillar, Ford, Cray Research and the U.S. Department of Energy from NASA-Lewis.

References

1 REITZ, R. D. and RUTLAND, C. J., "Development and Testing of Diesel Engine CFD Models," *Progress in Energy and Combustion Science,* **21,** 173–196 (1995)

2 PITSCH, H., BARTHS, H. and PETERS, N., "Three-Dimensional Modelling of NOx and Soot Formation in DI-Diesel engines Using Detailed Chemistry Based on the Interactive Flamelet Approach," *SAE paper* 962057 (1996)

3 ULUDOGAN, A., XIN, J. and REITZ, "Exploring the Use of Multiple Injectors and Split Injection to Reduce DI Diesel Engine Emissions," *SAE Paper 962058* (1996)

4 SENECAL, P. K., ULUDOGAN, A. and REITZ, R. D., "Development

of Novel Direct-Injection Diesel Engine Combustion Chamber Designs using Computational Fluid Dynamics", *SAE Fuels and Lubes Meeting*, May (1997)
5 HEYWOOD, J. B., *Internal Combustion Engine Fundamentals*, McGraw-Hill, New York (1988)
6 RAMOS, J. I., *Internal Combustion Engine Modelling*, Hemisphere, Taylor and Francis, New York (1989)
7 AMSDEN, A. A., Butler, T. D. and O'ROURKE, P. J. "The KIVA-II Computer Program for Transient Multi-Dimensional Chemically Reactive Flows with Sprays," *SAE Paper 872072* (1987)
8 AMSDEN, A. A., O'ROURKE, P. J., BUTLER, T. D., MEINTJES, K. and FANSLER, T. D. "Comparisons of Computed and Measured Three-Dimensional Velocity Fields in a Motored Two-Stroke Engine," *SAE Paper 920418* (1992)
9 AMSDEN, A. A., "KIVA-3: A KIVA Program with Block Structured Mesh for Complex Geometries", Los Alamos National Laboratory Report, LA-12503-MS (1993)
10 ROACHE, P. J., *Computational Fluid Dynamics*, Hermosa Publishers, New Mexico (1976)
11 WARSI, Z. U. A., *Fluid Dynamics—Theoretical and Computational Approaches*, CRC Press, Florida (1993)
12 HAN, Z. and REITZ, R. D., "Turbulence Modelling of Internal Combustion Engines Using RNG k-ε Models," *Combust. Sci. and Tech.*, 106 (4–6), 267 (1995)
13 YAKHOT, V., ORSZAG, S. A., THANGAM, S., GATSKI, T. B. and SPEZIALE, C. G., "Development of Turbulence Models for Shear Flows by a Double Expansion Technique," *Phys. Fluids A*, **4**, 1510 (1992)
14 HAN, Z. and REITZ, R. D., "A Temperature Wall Function Formulation for Variable Density Turbulent Flows with Application of Engine Convective Heat Transfer Modelling," *International Journal of Heat and Mass Transfer*, **40**, No. 3, 613–625 (1997)
15 LAUNDER, B. E. and SPALDING, D. B., The Numerical Computation of Turbulent Flows, *Computer Methods in Applied Mechanics and Engineering*, **3**, 269 (1974)
16 LIU, Y. and REITZ, R. D., "Modelling of Heat Conduction Within Chamber Walls for Multidimensional Internal Combustion Engine Simulations," Submitted to *International Journal of Heat and Mass Transfer* (1997)
17 GOSMAN, A. D. and IOANNIDES, E., "Aspects of Computer Simulation of Liquid-Fueled Combustors," *AIAA Paper No. 81–0323* (1981)
18 REITZ, R. D. and DIWAKAR, R., "Structure of High-Pressure Fuel Sprays," *SAE Paper 870598*, *SAE Transactions*, **96**, Sect. 5, 492–509 (1987)
19 HAN, Z., ULUDOGAN, A., HAMPSON, G. and REITZ, R. D., "Mechanisms of Soot and NOx Emission Reduction Using Multiple-Injection in a Diesel Engine," *SAE Paper 960633* (1996)
20 ARCOUMANIS, C. and GAVAISES, M., "Effects of Fuel Injection Processes on the Structure of Diesel Sprays," *SAE Paper 970799* (1997)
21 CHIGIER, N. and REITZ, R. D., "Regimes of Jet Breakup", AIAA *Progress in Astronautics and Aeronautics*, Recent Advances in Spray Combustion, K. Kuo (ed.), Vol. I, pp. 109–136 (1996)
22 O'ROURKE, P. J. and AMSDEN, A. A., "The TAB Method for Numerical Calculation of Spray Droplet Breakup," *SAE Paper 872089* (1987)
23 REITZ, R. D. "Modelling Atomization Processes in High-Pressure Vaporizing Sprays," *Atomisation and Spray Technology*, **3**, 309–337 (1987)
24 HAN, Z., PARRISH, S. E., FARRELL, P. V. and REITZ, R. D., "Modelling Atomization Processes of Pressure-Swirl Hollow-Cone Fuel Sprays," Submitted, *Atomization and Sprays* (1997)
25 KONG, S.-C. and REITZ, R. D. "Multidimensional Modelling of Diesel Ignition and Combustion Using Multistep Kinetics Models," Paper 93-ICE-22, ASME Transactions, *Journal of Engineering for Gas Turbines and Power*, **115**, No. 4, 781–789 (1993)
26 ASHGRIZ, N. and POO, J. Y., "Coalescence and Separation in Binary Collisions of Liquid Drops," *J. Fluid Mechanics*, **221**, 183–204 (1990)
27 O'ROURKE, P. J. and BRACCO, F. V., "Modelling Drop Interactions in Thick Sprays and a Comparison with Experiments," Stratified Charge Automotive Engines, I. Mech. E. Conference Publications 1980–9, pp. 101–116 (1980)
28 LIU, A. B., MATHER, D. and REITZ, R. D., "Modelling the Effects of Drop Drag and Breakup on Fuel Sprays," *SAE Paper 930072* (1993)
29 SENDA, J., KOBAYASHI, M., IWASHITA, S. and FUJIMOTO, H., "Modeling of Diesel Spray Impingement on a Flat Wall," *SAE Paper 941894*
30 WACHTERS, L. H. J. and WESTERLING, N. A. J., "The Heat Transfer from a Hot Wall to Impinging Water Drops in the Spheroidal State," *Chem. Eng. Sci.*, **21**, 1047–1056 (1966)
31 NABER, J. D. and REITZ, R. D., "Modelling Engine Spray/Wall Impingement," *SAE Paper 880107* (1988)
32 GONZALEZ, D., M. A., BORMAN, G. L. and REITZ, R. D., "A Study of Diesel Cold Starting using both Cycle Analysis and Multidimensional Calculations," *SAE Paper 910180* (1991)
33 KONG, S.-C., and REITZ, R. D., "Modelling Engine Spray Combustion Processes," *Progress in Astronautics and Aeronautics* Recent Advances in Spray Combustion, K. Kuo (ed.), Vol. II, pp. 395–424 (1996)
34 ECKHAUSE, J. E. and REITZ, R. D., "Modelling Heat Transfer to Impinging Fuel Sprays in Direct Injection Engines," *Atomization and Sprays*, **5**, 213–242 (1995)
35 STANTON, D. and RUTLAND, C., "Modelling Fuel Film Formation and Wall Interaction in Diesel Engines," *SAE paper 960628* (1996)
36 HALSTEAD, M., KIRSH, L. and QUINN, C., "The Autoignition of Hydrocarbon Fuels at High Temperatures and Pressures—Fitting of a Mathematical Model," *Combustion and Flame*, **30** (1977)
37 THEOBALD, M. A. and CHENG, W. K., "A Numerical Study of Diesel Ignition," *ASME Paper 87-EF-2* (1987)
38 KONG, S.-C., HAN, Z. and REITZ, R. D., "The Development and Application of a Diesel Ignition and Combustion Model for Multidimensional Engine Simulations," *SAE Paper 950278* (1995)
39 EDWARDS, C. F., SIEBERS, D. L. and HOSKIN, D. H., "A Study of the Autoignition Process of a Diesel Spray via High Speed Visualization," *SAE Paper 920108* (1992)
40 AYOUB, N. S. and REITZ, R. D., "Multidimensional Modelling of Fuel Composition Effects and Split Injections on Diesel Engine Cold-starting," Accepted, AIAA *Journal of Propulsion and Power* (1997)
41 XIN, J., MONTGOMERY, D., HAN, Z. and REITZ, R. D., "Modelling Diesel Engine NOx and Particulate over a Six-Mode Emissions Cycle," Accepted, ASME *Journal of Gas Turbines and Power* (1997)
42 DILLIES, B., MARX, K., DEC., J. and ESPEY, C., "Diesel Engine Combustion Modelling Using the Coherent Flame Model in Kiva-II," *SAE Paper 930074* (1993)
43 MUSCULUS, M. P. and RUTLAND, C. J., "Coherent Flamelet Modelling of Diesel Engine Combustion," *Combustion Science and Technology*, **104**, 295–337 (1995)
44 REITZ, R.D. and BRACCO, F. V., "Global Kinetics and Lack of Thermodynamic Equilibrium," *Combustion and Flame*, **53**, 141 (1983)
45 ABRAHAM, J., BRACCO, F. V. and Reitz, R. D., "Comparisons of Computed and Measured Premixed Charge Engine Combustion," *Combust. Flame*, **60**, 309–322 (1985)
46 METGHALCHI, M. and KECK, J., "Burning Velocities of Mixtures of Air with Methanol, Isooctane, and Indolene at High Pressure and Temperature," *Combustion and Flame*, **48**, 191–210 (1982)
47 KUO, T. -W. and REITZ, R. D., "Three-Dimensional Computations of Combustion in Premixed-Charge and Fuel-Injected Two-Stroke Engines," *SAE Paper 920425* (1992)
48 WESTENBERG, A., A., "Kinetics of NO and CO in Lean, Premixed Hydrocarbon-Air Flames," *Combustion Science and Technology*, **4**, 59–64 (1971)
49 PATTERSON, M. A., KONG, S.-C., HAMPSON, G. J. and REITZ, R. D., "Modelling the Effects of Fuel Injection Characteristics on Diesel Engine Soot and NOx Emissions," *SAE Paper 940523* (1994)
50 MILLER, J. A. and BOWMAN, C. T., "Mechanism and Modelling of Nitrogen Chemistry in Combustion," *Prog. Energy Comb. Sci.*, **15**, 287–338 (1989)
51 FUSCO, A., KNOX-KELECY, A. L. and FOSTER, D. E., "Application of a Phenomenological Soot Model to Diesel Engine Combustion," *International Symposium COMODIA 94*, pp. 571–576 (1994)
52 HIROYASU, H. and NISHIDA, K., "Simplified Three-Dimensional Modelling of Mixture Formation and Combustion in a D. I. Diesel Engine," *SAE Paper 890269* (1989)
53 NAGLE, J. STRICKLAND-CONSTABLE, R. F., "Oxidation of Carbon between 1000–2000 C," *Proc. of the Fifth Carbon Conf.*, Vol. 1, Pergamon Press, p. 154 (1962)
54 SENECAL, P. K., Xin, J. and REITZ, R. D., "Predictions of Residual Gas Fraction in IC Engines," *SAE Paper 962052* (1996)
55 HESSEL, R. P. and RUTLAND, C. J., "Intake Flow Modelling in a Four-Stroke Diesel Using KIVA-3", *Journal of Propulsion and Power*, **11**, 378–384 (1995)

56 MONTGOMERY, D. T. and REITZ, R. D., "Six-mode Cycle Evaluation of the Effect of EGR and Multiple Injections on Particulate and NOx Emissions from a D.I. Diesel Engine," *SAE Paper 960316* (1996)

57 CHOI, C., BOWER, G. and REITZ, R. D., "Effects of Biodiesel Blended Fuels and Multiple Injections on DI Diesel Engine Emissions, *SAE Paper 970218* (1997)

58 RICART, L. M., and REITZ, R. D., "Visualization and Modelling of Pilot Injection and Combustion in Diesel Engines," *SAE Paper 960833* (1996)

59 PIERPONT, D. A., MONTGOMERY, D. and REITZ, R. D., "Reduction of Diesel Soot and NOx emissions using EGR and Multiple Injections," *SAE Paper 950217* (1995)

60 DEC, J. E. and ESPEY, C., "Ignition and Early Soot Formation in a DI Diesel Engine Using Multiple 2-D Imaging Diagnostics," *SAE Paper 950456* (1995)

61 TOW, T., PIERPONT, A. and REITZ, R. D., "Reducing Particulates and NOx Emissions by Using Multiple Injections in a Heavy Duty D.I. Diesel Engine," *SAE Paper 940897* (1994)

Nomenclature

a	parent drop radius
A	pre-exponential constant of reaction rate; combustion model constant
A_f	frontal area of drops, pre-exponential constant
b	drop collision impact parameter
B	mass transfer number; branching agents in ignition model; combustion model constant
B_0	drop breakup size constant
B_1	drop breakup time constant
c	specific heat
C_2	0.1, combustion model constant
C_D	drag coefficient
CN	fuel cetane number
$C_{\varepsilon 1,2,3}$,	C_μ, $C_{\mu\varepsilon}$, C_s turbulence model constants
D	diffusivity
E	activation energy
f	spray model function; combustion model delay coefficient
f_1, f_2, f_3, f_4	kinetic parameters in ignition model
$\dot{f}_{coll}$	drop collision source term
$\dot{f}_{bu}$	drop breakup source term
$\mathbf{F}$	force; time rate of change of drop velocity
$\mathbf{g}$	specific body force
Δh_f^0	enthalpy of formation at 0 K
I	specific internal energy
$\mathbf{J}$	heat flux vector
k	turbulent kinetic energy
K_p, K_t, K_q, K_b	kinetic parameters in ignition model
K_A, K_Z, K_B, K_T	kinetic parameters in soot model
M	molecular mass
n	number of drop collisions
$\mathbf{n}$	surface normal vector
N	number of drops
Nu	Nusselt number
p	probability function
P	fluid pressure; cylinder pressure; ignition model products
Pr	Prandtl number
q	random number, heat flux
Q	energy; intermediate species in ignition model
r	drop radius; combustion products fraction
R	gas constant; rate of change of drop radius; RNG turbulence model source term; reaction rate
R^*	radical in chemical reaction
Re	Reynolds number ($= \rho V d/\mu$)
RH	hydrocarbon fuel
S	strain rate
Sh	Sherwood number for mass transfer
t	time
T	temperature; Taylor parameter
$\mathbf{u}$	gas velocity
$\mathbf{u}'$	turbulent velocity
u^*	wall friction velocity
U	relative velocity between drops and gas, gas velocity
$\mathbf{v}$	drop velocity
V	drop volume
Vol	volume
x	parameter in NSC soot oxidation model
x_r	residual gas mass fraction
$\mathbf{x}$	drop position vector
y	distance from walls; drop distortion parameter
Y_m	species mass fraction
$\mathbf{w}$	relative velocity of drops and gas, $\mathbf{u}$-$\mathbf{v}$
W	work; molecular weight; reaction rate
We	Weber number ($= \rho V^2 d/\sigma$)
Z	Ohnesorge number
α	thermal diffusivity
β	RNG turbulence model parameter, NO model constant
δ	Kronecker delta function
Δ	incremental amount
∇	gradient operator
ε	turbulent kinetic energy dissipation rate
γ	drop radius ratio ($= r_1/r_2$)
η	surface wave amplitude, turbulence timescale ratio
κ	von Karman constant
λ	wave length
Λ	wavelength of unstable surface wave
μ	gas dynamic viscosity
ν	gas kinematic viscosity
ν_{12}	collision frequency
ρ	species density
σ	surface tension; viscous stress tensor
τ	liquid breakup time
τ_c, τ_1, τ_t	characteristic timescales
τ	wall shear stress
Ω	growth rate of unstable surface wave

Subscripts

1, d, l	liquid drop; drop number 1,
d	based on diameter
2, g	gas; drop number 2
bu	drop breakup
c	combustion
coll	drop collision
crit	critical quantity
ε	dissipation of turbulent kinetic energy
	form soot formation
fv	fuel vapor
i	incoming drop, running index
j	running index
k	turbulent kinetic energy
l	laminar
m	species index
o	outgoing drop; oxygen atom
oxid	soot oxidation
sphere	spherical drop
t	turbulent

Superscripts

*	equilibrium state; radical species, wall friction value
.	first order derivative
..	second order derivative
c	chemical reaction
s	spray

8 Modern control in diesel engine management

Contents

8.1 What is the purpose of control?

The purpose of an engine control system is to keep the performance of the engine within specified limits. Historically for most engines this performance objective is a speed which must be kept at a certain nominal value and within an acceptable tolerance band. On vehicles the engine control system, often now electronic, is responsible for the achievement of the legislated emissions performance as well as efficiency while keeping driveability at an acceptable level throughout the operating range of the engine.

8.1.1 Fundamental components

A control system always has three components (*Figure 8.1*).

A *sensor* obtains a measurement of a physical variable through a direct measurement or a combination of measurement and computation. A 'soft sensor' delivers a value through an intermediate computation. Sensors can be specified to measure a range of physical and chemical quantities at the speed of response needed even by a high speed diesel engine. Several factors limit those that prove acceptable in a production engine including cost and durability.

A *processor* calculates a control action which will keep system performance at the required level. The processor does not have to be electronic and indeed the most widely known form of speed control in engines uses a purely mechanical system. The 'centrifugal' governor uses the position of fly weights to set the required fuelling and was first applied to engines by James Watt, although the principle came from earlier windmill technology. The majority of control systems fitted to today's engines are electronic and are usually programmable.

An *actuator* which is set by the processor to effect the required control action. Fuel injection equipment is the most fundamental actuator on the engine and is the means of supplying energy to the cylinders. Historically the control system has adjusted the fuel rack which has adjusted fuelling in parallel to all the cylinders. Modern FIE systems permit fully flexible timing control and where there are unit injectors control by individual cylinders. Other actuators include the EGR valve, variable geometry devices, and the turbocharger wastegate.

8.1.2 The structure of a control system

The primary structural distinction that can be drawn between control systems is between *closed loop* and *open loop*. In a closed loop controller a measurement of the quantity being controlled is explicitly compared with the desired value. The difference is used as the basis of the control action.

Speed governing illustrates closed loop control, while the type of control utilized in first generation diesel controls is a scheduled or open loop control.

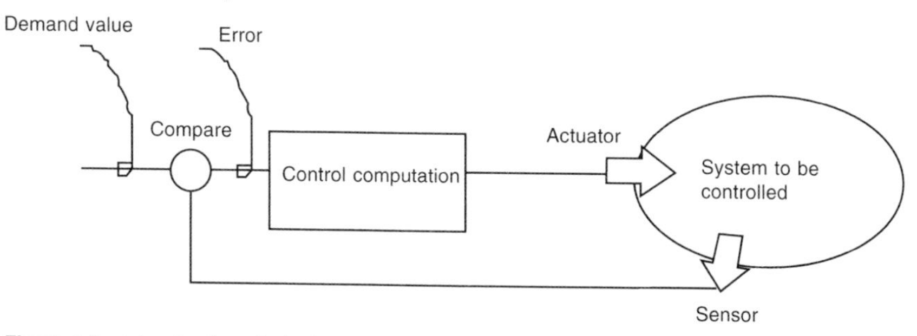

Figure 8.1 A feedback control scheme

8.1.2.1 Closed loop control (Figure 8.2)

A closed loop controller calculates a control action on the basis of the error between a demand signal and the value measured from the system under control. An all speed governor is a common example of such a feedback control system. The form of control computation decides the dynamic behaviour of the controlled system, and may be as simple as the multiplication of the error by a constant.

Feedforward is the technique of adding to the control action to compensate for a measured disturbance. If a measurement is available and its effect on the system to be controlled can be calculated, then feedforward is an excellent way of ensuring a rapid response. It is important to retain a feedback element in the control system to compensate for the inevitable inaccuracies that exist even in the best executed feedforward scheme.

In *Figure 8.2* a closed loop is formed when a measurement, **m** is compared with a demand, **r**. The error, **e** is calculated as **r-m** and then included in a simple calculation to form the control signal **u**. In a simple feedback system **u** is sent direct to an actuator whose position is set accordingly. For feedforward a measurement, **d** is in effect advance warning of a coming change. A computation of the effect of this variable is done and delivers a value **f**, that is the amount by which the actuator must be moved to compensate for the change implied by the measurement **d**.

Feedback provides a clear advantage by adding some useful properties to the control system. An example is the reduction of the effect of a disturbance to the output. For a governor such an output disturbance would be caused by a change of load. Where there is simple feedback the effect is diminished by a factor:

1+ (gain of the controller) * (gain of the engine)

More sophisticated feedback will remove the disturbance altogether. An open loop controller would pass the disturbance through unmodified until a regular inspection of the speed led to manual control correction.

If the controller contains a calculation in which the error is integrated and the resulting integral scaled and applied as a control action, the controller is said to include integral action. This feature is enough to eliminate steady state error in the control response. This feedback is the same as applying infinite DC gain. Integral action can result in both instability and a phenomenon known as *reset windup*. Both aspects will be covered in detail later in the chapter.

8.1.2.2 Open loop control

A purely open loop control system has no embedded feedback mechanism, and relies on external agents, normally people, to make corrections. Scheduled control is part way between pure open loop control and closed loop control. The example shown in *Figure 8.3* is illustrative.

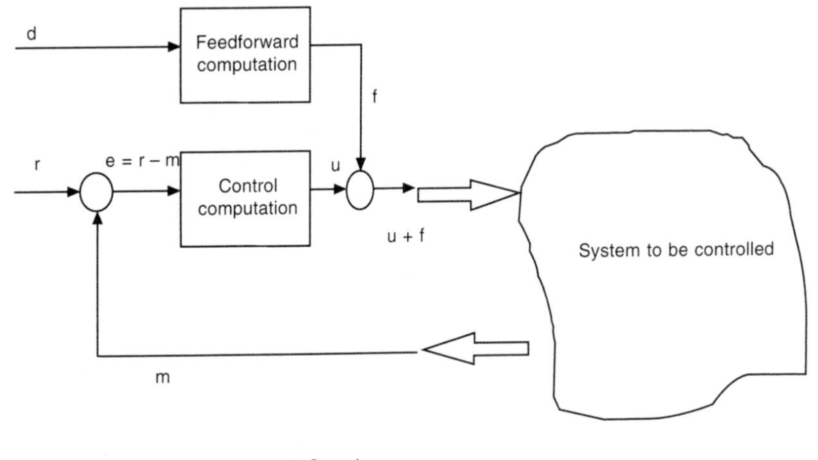

Figure 8.2 A closed loop control system

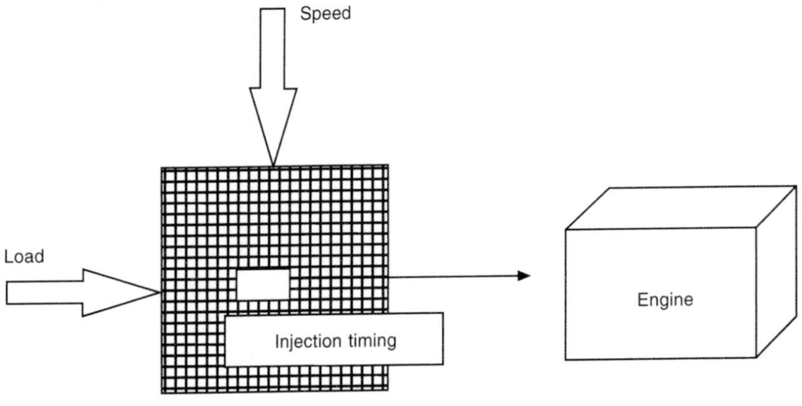

Figure 8.3 A scheduled control system

Within the engine control system, measured values of speed and torque are used to index a table of injection timing values. The value is calculated using an interpolation algorithm and translated into the output value to set the injection timing. It can be understood quite easily that if the timing values change in the area of the table where indexing is taking place, the injection timing will change as the speed and load undergo small changes. The speed and load will change as a consequence of an injection timing change and those changes will in turn feed back to modify the injection timing and so on. Limit cycling where the injection timing hunts around a small area of the lookup table is possible, and is a feature which calibration engineers must eliminate.

Modern engine control systems have a large number of tables and demand substantial effort to fill in the lookup tables ('calibrate'). While full closed loop control looks like an attractive alternative, its scope is severely limited in the absence of reliable sensors for closed loop control. Scheduled control permits control, while the sensors can be used to deduce the engine's operating condition. The prognosis is for the availability of sensors to progressively shift control more to closed loop.

8.1.3 The shape of the future—model-based control

In the example of the previous section, the desired control behaviour is represented as a lookup table. There are other possible representations. For example, an EGR control scheme may be organized as a simple table lookup as shown in the previous section. The product of the lookup process will be a position set point for the EGR valve spindle. Instead consider a *model-based* approach. The root of such a model-based approach is a system model which is used during design and development of the control system, or for diagnosis purposes is carried in the target system itself and acts as a reference for 'normal' system behaviour.

The EGR system can be considered to be made up of a series of components. (*Figure 8.4*):

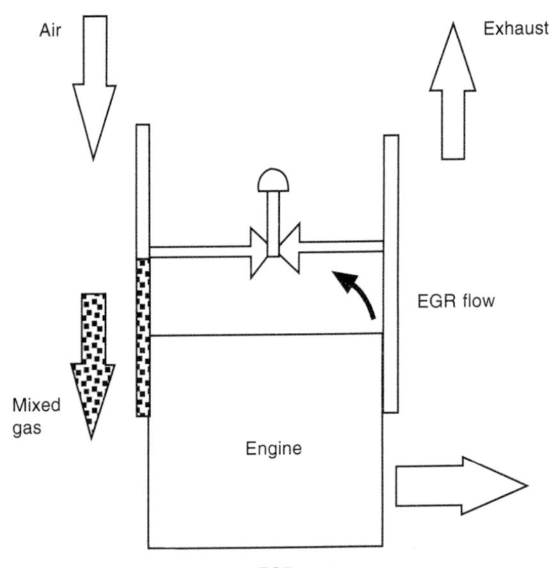

Figure 8.4 A high pressure EGR system

Air filter
Intake plenum
Mixing chamber (for EGR)
Turbocharger compressor
Intake manifold
Exhaust manifold
EGR valve.

The dynamics of each element, together with the pumping effect of the engine contribute to the overall dynamic behaviour. For a similar step by step consideration of the dynamics of various engine components see Reference 1. A successful control design and development exercise should start with the modelling of components which contribute to the dynamics of the system.

8.1.3.1 EGR system elements

Air filter: simply drops pressure, producing a flow dependent pressure drop

Intake plenum: acts as buffer but does not store gas, rather mixes the properties of incoming gases with those already there. Such mixing may vary from perfect to completely separate (plug flow).

Mixing chamber: in practice this may be a join in two pipes, but may be considered to be a plenum with mixing properties varying from complete to 'plug flow'.

Turbocharger compressor: has a complex static behaviour, and dynamics related to the inertia of mechanical parts, and viscous friction.

Intake manifold: The intake manifold behaves as a plenum but now is subject to the pressure fluctuations corresponding to intake valve events. A *mean value* model will not consider such variations whereas a full engine model will.

Exhaust manifold: is a plenum into which cylinders discharge in turn. Like the intake manifold, the exhaust manifold is subject to valve 'events'. Mixing of each cylinder's contents follows the pattern described for the intake plenum. The exhaust manifold will discharge into a turbocharger turbine.

EGR valve: is a valve with carefully designed discharge characteristics which will have a positioning time constant due to the actuation mechanism. It may also have some flow hysterisis which must be captured for any modelling work.

Turbocharger turbine: has dynamics due to rotor intertia and the gas flow. A modern diesel engine may use a variable geometry turbine where there will be dynamics associated with the moving stator vanes.

We will consider the variable geometry turbocharger in detail from a modelling point of view but cover several general principles first.

8.1.3.2 Some principles of modelling

For controls analysis purposes the most useful models are linear models which may include a separate non-linear element such as hysterisis. In practice such linear models are relatively easy to represent and combine and can be accurate over a wide range of engine operations.

In building a model for control design purposes, the level of detail needs to be chosen carefully. Too much detail will lead to an unwidely model of limited utility. Too little and the model will lead to a poor control design. The detail in the model will be a matter of engineering judgement and some numerical tests which are both fast and straightforward to perform. It is quite legitimate to lump various model elements together to produce an overall system model which is quite representative.

A model is some way of representing system behaviour in a way that helps the development of a controller. Modelling has been done for a considerable time even in the engine world, although the results have only recently been widely known. Until the 1960s the most common form of model was a *non-parametric* model, that is one where the system behaviour is captured in a non-numeric form such as a Bode or Nyquist diagram. Starting with a Bode diagram it is quite straightforward to designing a controller of the lead-lag type to give an effective control over the frequency range of interest.

Parametric models are those where the model is captured in the form of parameters which are probably the parameters of a differential or difference equation which describes the system behaviour. Such models became more popular with the availability of computers able to derive them. The most common model *identification* method, is a variation on the least squares method widely used in statistics to fit a relationship between independent and dependent variables logged from an experiment.

Further information on the use of techniques for calculating models can be found in the general literature[2].

8.1.3.3 Discrete and continuous time models

The engineering analysis of the dynamics of engines normally starts with a continuous time analysis. The use of continuous time representations and Laplace transforms results in the familiar s domain representation. A thorough analysis of systems and controls can be done in continuous time and is thoroughly presented in introductory texts on control systems design[3].

Modern controls are almost exclusively implemented in a digital form with a microprocessor or micro-controller as the key element. However digital systems require a different design approach and one which is again well documented in a number of standard texts[4,5]. One important aspect of the modelling of reciprocating engines is highlighted in Reference 5, and in turn is drawn from Reference 6. The engine is inherently sampled. The impulsive nature of injection and combustion leads to a behaviour which has a sampled character. Early control studies of engines showed erratic results at high speeds. Results could not be repeated. Once it was understood that at high engine speeds, the engine events were close to the Nyquist frequency and there was consequent interference from sidebands (that is aliased signals with a beat frequency of f-f_s, where f is the frequency of the observed measurement, and f_s is the sample frequency). The solution is to synchronize the sampling with the engine speed, and avoid such interference.

The inherently sampled nature of engine operation was developed in two important papers[7,8] which make an important contribution to the controls analysis of reciprocating engines. The papers were both written specifically for compression ignition engines. The authors' analysis is extensive but contain several useful concepts which we can briefly review here.

- Torque development in an engine is inherently impulsive. The development of torque begins after the injection, and continues sometimes into the next injection/torque event.
- This continuing torque development can be represented by a partial first-order hold (*Figure 8.5*). The torque can be approximated as constant in the first sample interval, then as a diminished but constant value for the first part of the next interval.
- The impulses are distinct in a four-cylinder engine. There is some overlap in a six-cylinder engine, while there is substantial overlap in an eight-cylinder engine. The net effect as that the eight-cylinder engine is inherently less stable than the six-cylinder engine, a phenomenon that can be observed in

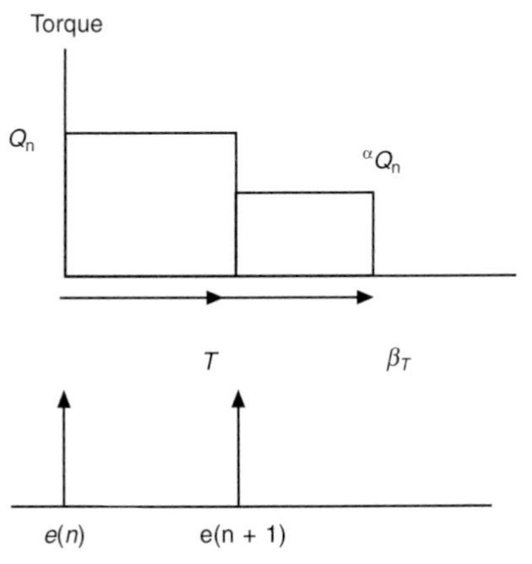

Figure 8.5 Torque development in a diesel engine[7,8].

the respective root locus diagrams of the two engine configurations.

The diagrams show the concept of the partial first-order hold which captures the torque development of each cylinder. The torque contribution in the next cycle continues at a level of alpha times the original, and continues for a proportion B of the cycle.

8.1.3.4 The variable geometry turbocharger as an example

Figure 8.6 shows a schematic diagram of a variable geometry turbocharger which illustrates the main dynamic elements.

The three main components interact with the gas flow. The vane movement will be dominated by the actuator dynamics which will be second order with respect to the input signal (or as viewed by ECU). As a result of the vane movement, the stator flow area will change. A larger area will allow a larger flow rate but the change will be fast compared with the initial movement of the vanes.

As the pressure changes and the flow rate changes the rotor will experience a greater internal torque. The acceleration of the rotor will be the result of the difference between the developed torque of the rotor and the current compressor torque. The current rotor speed will dictate the pressure ratio developed.

In general the gas dynamics are sufficiently fast that we could regard the turbine characteristics as representing all the states between 'steady state' conditions. The mechanical parts have longer time constants which will govern the response of the turbocharger under transient conditions.

The two principal components—the vanes and the rotor—are subject to experimentation to determine the important dynamics. Taking a very simplified view the rotor is subject to torque due to gas flow, and is resisted by bearing friction and the compressor torque. The corresponding model, expressed mathematically is:

$$J\ddot{\theta} + m\dot{\theta} + T_c = T_t$$

where: J is the moment of inertia of the turbine rotor,
θ is the position (hence $\dot{\theta}$ the rotational velocity)
T_c is the compressor torque, and
T_t is the torque developed across the turbine by the change of gas velocity.

The vanes are subject to friction and a resistance due to pressure. Taking this into account, the equation for vane movement can be expressed:

$$I\ddot{\phi} + p\dot{\phi} + q\phi = Kv$$

where ϕ is the position of the vanes, assumed to be with reference to a neutral position where the torque due to pressure forces is zero.

It may be possible to calculate the parameters of this model—but the accuracy of a calculation, particularly of friction torque could be limited. Instead an experiment can be performed to identify the parameters of a model like this. What follows is a brief overview of the process. In practice the model will be calculated from an experimental record using a technique known as parameter estimation. This is part of class of techniques known as system identification. For a detailed account of system identification see Reference 2.

Consider the experiment shown in *Figure 8.7*. Suppose that the model which best represents the physical model is first order, that is one that can be represented as

$$y(t) = ay(t - 1) + bu(t - 1)$$

This is a very common formulation in *discrete time*. Note how the current value of the output ($y(t)$) is related to previous values of y and u. The equation has the appearance of a linear regression and it is solved in a similar way. Just as a linear regression makes certain assumptions about the type of randomness that will appear in the data so does this type of analysis. Complex iterative procedures are needed where the noise is coloured and could create bias in the estimates of parameters.

For a detailed description of the applications of analysis and system identification to a turbocharged diesel engine see Reference 9.

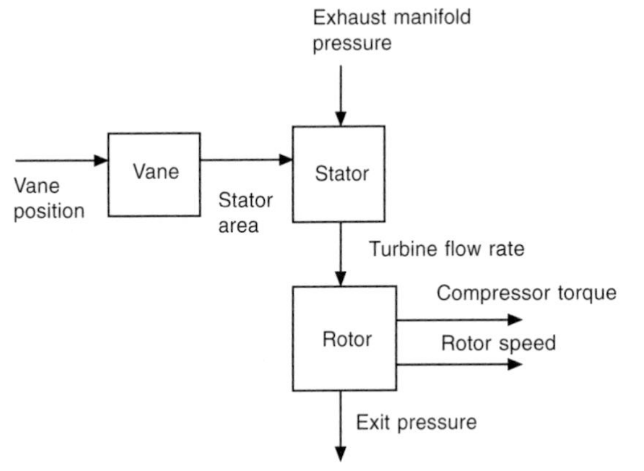

Figure 8.6 The main dynamic elements of a variable geometry turbocharger

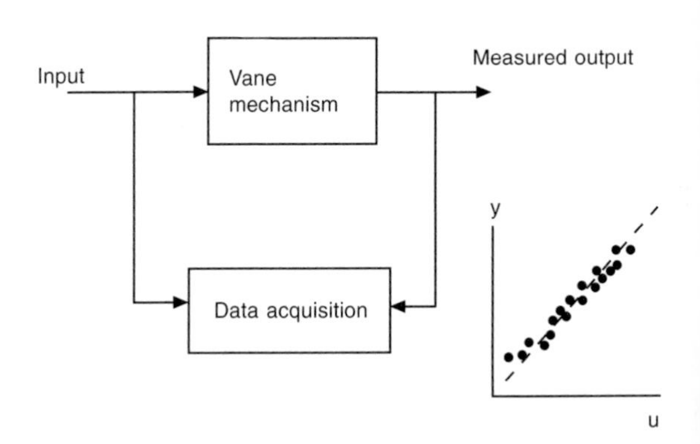

Figure 8.7 System identification example: vane dynamics

8.1.3.5 Models for control

Models are used in the design of control algorithms. Even the widely used PID controller assumes that the system to be controlled is second order without a time delay.

A good example comes from the discipline of robust control, which uses a model for design which explicitly represents the uncertainty in a model. In the model shown in *Figure 8.8*[10], the behaviour of a rotary fuel pump is represented by a model with a multiplicative uncertainty.

8.1.3.6 Models for data transformation and diagnosis

Such models may also be used for diagnosis, although a wider range of models can be employed in, for example, knowledge based diagnosis. In this chapter we will restrict comment to a class of diagnosis procedures which are closely related to controls—the so-called analytical redundancy techniques. Analytical redundancy simply refers to the property of a system that in the nature of the way its measurements change gives away details of the internal structure of the system.

Figure 8.9 shows the structure of a non-linear observer[11]. The function of the observer is to reconstruct certain unmeasurable quantities which though not measured can be inferred from other measurements. One pertinent example is the reconstruction of manifold pressure in throttled engines. Though this example is most relevant to spark ignition engines it illustrates some important aspects of observer technology. The controlled device in the example is the inlet sub-system of the engine, and the output x is the pressure. The real measurement is inherently slow and usually noisy and to be useful must be filtered. The result is that the measurement is available too late to be effectively included in air–fuel ratio control.

The 'model' is that of the manifold, and is formulated so that when it receives an input from the throttle it computes the manifold pressure and makes the estimate available to the control system. In the short term the accuracy can be made very high but in the longer term, the estimate may drift. Feedback will compensate for such longer term changes.

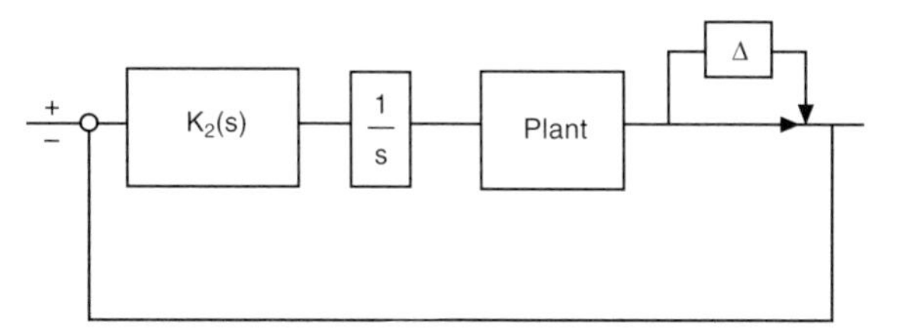

Figure 8.8 Model of the fuelling quantity control in a rotary fuel pump

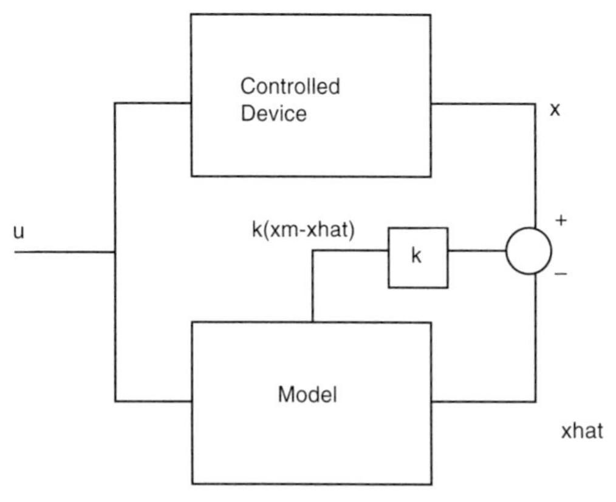

Figure 8.9 A non-linear observer for engines

8.2 The context of engine control

The engine control system fits into the context of a wider system in just the same way as an engine fits into the context of a vehicle, a ship or a power generation system. The co-ordination of the engine with other items of plant and equipment is managed by another level of control activity. Some examples will illustrate the influence the context of the engine has on the control 'problem'.

- A diesel electric locomotive must co-ordinate the generator excitation to match the maximum power available from the engine. The driver will demand engine speed through the engine governor, while the governor and generator controller will co-ordinate to ensure the generator matches the engine's capability to supply torque.
- In a power plant a supervisory controller will select which engines to operate to ensure a plant-wide fuel economy target. Speed is given once a generating set is connected to the supply, but power output will vary according to demand placed on each engine generating set.
- In a truck the primary input to the engine is the driver's demand which historically has been a fuelling demand. That is changing with the advent of reliable drive-by-wire systems and the increasing use of cruise control. 'Intelligent' transmissions and engine brakes need to be co-ordinated with the engine and will exchange control information. The truck engine finds itself between a cab computer supervising cruise control, a transmission controller, an engine brake with perhaps a link to an anti-skid system.

These examples illustrate the wider picture that engine control fits into. It is no longer sufficient to think of engine control in isolation.

A feature list of a modern ECU may contain the following:

- Control—the functions required to regulate the engine operation including the control of fuel injection, EGR control, supercharging devices, regulation of maximum and minimum engine speeds.
- Communications: ISO9141, CAN, J1850, J1939—a range of physical and logical interfaces to allow the engine controller to be integrated into other vehicle systems.
- Diagnosis: both self diagnosis and for some road vehicles on-board diagnosis of emissions control equipment.
- Vehicle related functions such as cruise control, traction control, vehicle dynamics controls, and air conditioning.

8.3 What a control system does

When confronted with a control problem, the designer needs to answer a series of fundamental questions about the system to be controlled. In practice a control problem is likely to be solved with a 'standard' control solution but this does not absolve the designer from the need to understand the system to be controlled. Indeed the choice of a standard solution (from a library of possible solutions) is dictated by the underlying system.

The most common technical solution to a control requirement is the three-term controller (PID = proportional, integral and derivative, referring to the three elements of the calculation). This form of control has been widely and successfully applied in the process industry over the last 50 years. The flyweight engine governor is one implementation of the three-term controller.

The three-term controller contains three essential components for an effective control system:

- the proportional term offers speed of response: a high proportional gain will give a fast response to a developing error but will risk making the controller unstable;
- the integral term ensures that there is no control error in the steady state: a proportional term alone can never finally correct a steady state error;
- a derivative term ensures that the control signal is modified if the error changes quickly: this term has a damping effect on the control response.

The three-term controller has a certain intuitive appeal. Modern control systems have been conceived since the three-term controller. A surprising result is that in general they can be reduced to a three-term form. Why is it so successful?

The three-term controller 'matches' a second-order plant well, that is a system to be controlled whose behaviour is defined by a second-order differential equation. In practice most physical systems are truly second order or behave approximately second order. Three-term control begins to break down when for example a time delay is introduced. The structure of the three-term controller makes no assumption about time delay, and a significant value tends to make the controller become unstable.

A control system must be customized to the engine's behaviour, while its tolerance to changes in the engine is a critical factor in deciding the type of control algorithm. All controllers are tolerant to some extent of the variation in the underlying system. It is only recently, however, that control algorithms which make an explicit allowance for model variations have been developed and proven. Such a *robust* controller is discussed later in the chapter.

Figure 8.10 shows in a simplified form a control algorithm that may be used for pressure control in a fuel injection system. The measured values of engine coolant temperature and boost pressure are used to index a table of pressures for a common rail or the oil pressure in a servo-assisted injection system (such as Caterpillar's HEUI system). The desired pressure (or set-point) is compared with the measured pressure, and the resulting error is processed through the PID algorithm which sets the actuator value, which in this case is the position of a valve in the hydraulic circuit. Such a PID controller is shown in block diagram form in *Figure 8.10*.

The performance specification of the controller includes speed of response, stability and accuracy. The PID controller offers these features in each of its three 'terms'. These terms are calculations:

- The *P* term produces a multiple of the current error. The greater the factor, the faster the control action, and the more accurate the control. However a *P* term alone cannot completely eliminate a control error since an error is needed in order to generate the control action.
- The *I* term forms a sum of the control error over time. If there is a persistent control error, this term keeps accumulating the error producing a stronger corrective action.
- The *D* term differentiates the measurement and subtracts the result from the control output. This tends to damp the control response, and acts as a stabilizing influence on the control activity.

There are important practical requirements for such a controller:

- The *D* term can amplify the effect of high frequency noise and must be accompanied by measures of filter noise from the measurement.
- The *I* term can saturate if the control error persists for a long period. There must be measures applied to prevent this so-called 'intergral windup'.
- The controller is a linear controller—so that it will behave in the same way whether the control demand is low or high. If the system under control is non-linear, and engine systems are severely non-linear, the parameters must be changed as the operating condition is changed. This technique of gain scheduling is widely applied in engine control.
- Tuning is a matter of selecting the correct parameters for each of the three terms. There are rules for tuning which allow the parameters to be chosen following some simple tests.

Practical aspects of applying PID control

The PID controller continues to be almost universally applied in engine control. This is primarily because it is the right solution for most practical control problems. Where the system to be controlled is second order and includes only small transport delays (delay connecting the actuator and measurement—usually due to delays in the properties of flowing fluids)—PID is the best solution. It fulfils one of the key rules for control engineers—that integral action should be included to ensure good steady state accuracy.

Where a controller's performance must match a plant whose characteristics vary over its operating range (an engine is characterized by its varying characteristics), it may be possible to design a controller which anticipates all possible conditions. In practice this approach often results in a controller which is stable at the most demanding conditions of the plant. At less demanding conditions, the controller is less effective and results in an overall performance which may be weak.

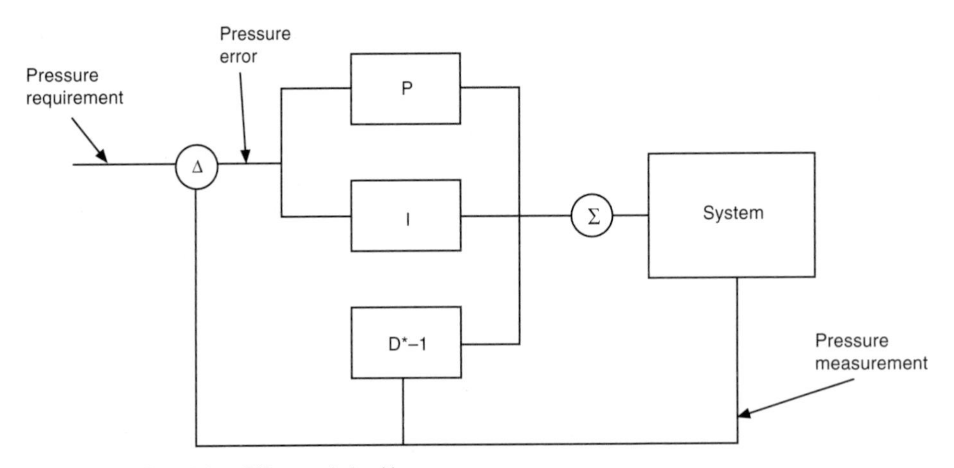

Figure 8.10 Signals in a PID control algorithm

There are two principal approaches:

- In a gain scheduling controller, the gains (corresponding to the P, I and D parameters in a three-term controller) are adjusted based on some observable parameters. In an idle speed controller where fuelling is adjusted to reach a set speed, the behaviour of the *speed loop* is dependent on engine temperature. The temperature may be the scheduling parameter and as the temperature varies, the controller will select a different set of parameters.
- In an adaptive controller, the controller is subject to fundamental changes depending on a observation of the plant being controlled. In a later section we will consider the different types of adaptive controller types.

The selection of gains for such a gain scheduling controller is a matter of tuning, an exercise in which the control loop is exercised according to certain criteria. The gains identified in such an exercise offer a known level of performance. The exercise would be repeated at different engine conditions. The number of different conditions is dependent on the quality of control needed, and would be established in experiments.

Implementation needs care, as there are no firm theoretical underpinnings for gain scheduling. Limit cycling and instability are conceivable fault conditions, and the designer should be careful about the conditions under which gains are switched and how they are switched.

8.3.1 Sensors for control

In this section we will consider the types of sensor that are used to provide the inputs needed by a control system. Most sensors employed in current control schemes are simple in nature primarily because the sensors needed to implement advanced strategies are not yet available either cheaply enough of sufficiently robust. In the following section we will distinguish between sensors which are current and accessible technology and those which are promising but not yet commercial. Useful overviews of sensor technology in engine application can be found in References 12 and 13.

While actuators are essential for the transformation of a control requirement into a change in the state of the engine, it is sensors which dictate the type of control which can be offered. For example, there is a clear trend towards control strategies in diesel engines which modify in-cylinder conditions in real time. This is not possible until sensors are available which have the required bandwidth within cost and reliability criteria.

8.3.1.1 Current technology

Pressure

Pressure sensors have reached a high level of reliability for low bandwidth applications such as boost pressure. High pressure measurement—in the common rail and in the combustion chamber itself are recent developments. Cylinder pressure measurement for large engines is well established for monitoring purposes but for all power ranges there is the opportunity offered by control based on cylinder pressure measurement.

Light and heavy duty needs are likely to be met by a low cost pressure sensor using a modulating transmission from a deflecting diaphragm. Medium and low speed applications may be met by more sophisticated technologies such as interferometric methods based around the change in dimensions of a Fabry Perot cavity.

Temperature

Temperature needs include fuel, intake air, coolant, and oil. All need to be monitored for their effect on the control system. In general low cost, low bandwidth methods are used and the thermistor has proved popular while thermocouples are occasionally used in high temperature environments such as in the exhaust system.

Temperature will be used for adjusting calibration parameters and in the near future will provide control feedback for temperature control of elements of the engine structure.

Injection parameters

Needle lift is an indirect measure of injection timing and can be measured inductively. An extension to the injector needle will change the flux linkage in a coil held stationary in the injector body. In light duty systems, only one cylinder is in general equipped and is used as an indirect measurement of injection timing.

Flow in diesel fuel injection systems indicates the start and end of injection as well as the actual quantity delivered. The timing of injection may also be inferred from the injection events in both common rail and in unit injector systems.

Pump position

Position measurement is needed as a feedback parameter in the control of injection pumps. In early conversions of rotary pumps, linear potentiometers provided position feedback for timing and maximum fuelling control.

Engine speed and crankshaft position

Magnetic and Hall effect sensors are fitted to both crankshaft and camshaft. The camshaft position gives details of the 'revolution', while the crankshaft sensor identifies both the incremental and absolute position. The absolute position is given by a 'missing tooth'. Within the control system, the engine position can be interpolated based on the position and instantaneous speed information.

8.3.1.2 Sensors for advanced functions

Pressure measurement for modern diesel systems

Common rail fuel injection systems require rail pressure measurement for regulation. Conditions are clean, and the pressure consistent at the operating pressure. Here piezo devices are well suited to both temperature and pressure requirements.

The performance of such systems will be enhanced by the application of pressure measurement to fuel injection, for fast detection of start and end of injection, and to cylinder pressure. Pressure feedback during combustion will permit the computation of peak pressure and its timing as well as heat release and the timing of events.

Both types of pressure measurement require miniature, low cost transducers. Some progress has already been made with Fabry-Perot techniques. Simple optical displacement/modulation-type sensors have reached a sufficiently low cost per unit to be considered. For in-cylinder pressure sensing, Texas Instruments[14] and Robert Bosch[15] have offered a sensor in which the pressure sensing diaphragm is connected to a measurement device. In the TI system, a piezo element is used to sense movement and hence pressure. In the Robert Bosch system the measurement is done using a micro-machined strain gauge bridge.

Optrand[16] have pioneered the use of the optical displacement/modulation device which has already been integrated with a spark plug for SI engine applications. The challenge remains to integrate the sensor with a diesel fuel injector to avoid more complexity in the cylinder head design.

Engine torque

Engine torque remains a vital measurement, but where sensing at a reasonable cost has not proved possible. For test stands, in-line torquemeters are practical.

Engine speed variations give some insight into torque contributions from each cylinder although a shaft torque measurement remains the most direct. There are several successful torque transducers already applied to EPAS (electrically power assisted steering) systems. Surface acoustic wave (SAW) remains a distinct possibility for engines.

Exhaust gas measurement

Both NOx and particulates remain the key elements of diesel exhaust and viable real time sensors will permit new diagnosis and control regimes. (One such control scheme based on NOx sensing is described later in the chapter.)

NOx measurement has been tried with a number of different phenomena. Pure chemical techniques have used reactions on the surface of thin films to generate a resistance change. Other chemical methods rely on a reversible reaction on a NOx sensitive elements which can be detected optically or electrically.

The NOx sensor from NSK[17] consists of two elements which act successively on the exhaust gas. In the first any NOx in the exhaust gas is reduced over a catalyst resulting in a small increase in the oxygen partial pressure. In the second a zirconia-based device is used to equilibrate the oxygen partial pressures between a reduced and unreduced sample. The difference in oxygen partial pressure is a measure of the NOx concentration.

Particulates represent a significant challenge for real time measurement. For current particulate sizes (typically 0.5 micron to 10 microns), obscuration and scattering techniques are possible. However, as particulate sizes fall, new techniques will be needed which allow particles whose length is of the same order of magnitude as the wavelength of visible light to be detected.

8.3.1.3 Sensors for future systems

Future sensors are likely to focus on the combustion process itself. Current sensing techniques focus on measurable quantities such as components of the exhaust gas, or pressure and temperatures. Such quantities are sometimes surrogates for the quantities to be controlled which are generally concerned with the combustion of fuel. There is a tendency as sensing techniques continue to develop that they will get closer to cylinder events and feed information about the fundamentals of combustion.

Two-colour sensing

Optical measurement inside the cylinder has been the subject of a number of studies[18]. They describe an approach to measuring cylinder temperature based on the 'colour' of the gases during combustion. Spectral content in the range 400 nm to 1100 nm gives a strong indication of temperature and correlates well with NOx emissions. In practice the two-colour method[19] needs to be used because the emissions from the burning gases do not exhibit true black body behaviour.

Electromagnetic methods

Electromagnetic techniques may also be used to analyse the contents of the cylinder. At close range the benefits of measuring ionisation have already been demonstrated in SI engines. A number of measurements may be extracted including temperature, pressure, and the characteristics of detonation[20,21].

In a diesel the challenge is to obtain a spatial measurement so that the start of combustion can be observed. This may be done using an antenna embedded in the cylinder and observing changes in the electromagnetic properties of the burning gases. Both optical and electrical properties will be much more responsive to the start of combustion than will an in-cylinder pressure sensor.

8.3.2 Actuators for control

For diesel engines there are three prime categories of actuators which we shall briefly review. The design and engineering of these actuators is covered in other chapters, but here we will briefly review the control aspects.

8.3.2.1 Exhaust gas recirculation (EGR)

EGR valves are designed to connect exhaust and inlet manifolds on either a high pressure or low pressure sides of the system. In light duty systems the valve may simply operate on the pressure difference created between inlet and outlet manifolds. Where the engine is turbocharged, the matching and the EGR value operation must be designed to give a high enough EGR rate.

EGR valves have traditionally been operated on air pressure, but electric valves will offer more flexibility. Such a valve will also carry a position feedback device so that the valve will be cascade controlled—a technique frequently used in the process industry.

EGR control is likely to be done by the valve working in concert with a variable geometry boost system. The importance for co-design of such a system is underlined by the complexity of a combined EGR, VGT and VG intake system.

8.3.2.2 Turbocharger wastegate

A wastegate is a device designed simply to avoid over-boosting the engine. It diverts some proportion of the exhaust gas flow around the turbine thereby increasing the flow cross sectional area. At present waste gate devices are actuated pneumatically but in the future electric operation will offer more flexibility.

8.3.2.3 Fuel injection equipment

The details of fuel injection equipment (FIE) are covered in a separate chapter. However, we will briefly review the control implications of fuel injection equipment, and how different architectures make different demands on the control of the engine. We will consider FIE in these distinct categories:

- rotary pumps,
- in-line pumps,
- rotary pumps with local control (VP44 and ESR10 are typical),
- unit injectors and the variation of pump-line-injector (PLD) systems,
- common rail.

Rotary pumps typically control timing and injection quantity. In some early conversions of rotary pumps, timing was controlled by means of a small stepper motor which regulated the position of the timing ring. Quantity was still controlled directly by the driver although maximum fuelling was regulated by an adjustable fuel stop itself controlled by a linear positioner.

In later development such as VP44, the control of both quantity and timing is executed at the pump itself. The injection is controlled by a solenoid, and as with unit injectors, the solenoid controls the start and duration of the injection. Both VP44 and ESR10 are characterized by pump mounted electronics which offload the detailed monitoring and control of the injection process. Such drive by wire systems allow more control flexibility, since the control system has direct control of fuelling rather than simply following a given fuelling demand.

In-line pumps dominated the heavy duty market up to the introduction of EURO2 emissions standards. In-line pumps are amenable to both quantity and timing control although local closed loop control of the actuator is needed for fast and accurate operation.

The majority of future diesel engines will use either unit injectors or common rail fuel injection. Both pose their own set of control issues, and common rail allows more degrees of freedom than does any other fuel injection system. The unit injector, because the pressure profile is generated by a cam, is still subject to that pressure cycle. The unit injector will start the injection during this pressurization period and will stop when the required quantity is injected. Multiple injections are possible. The control of the injection process is potentially quite sophisticated, and the injection current must be monitored for key events such as valve closure.

Common rail systems provide more degrees of freedom in the injection process than either pump or unit injector systems. Multiple injections at different injection rates allow:

- initiation of combustion with a pilot injection;
- a main injection which may in turn be 'rate controlled' by multiple injections;
- a final injection which will leave hydrocarbons in the exhaust gas for a DeNOx catalyst.

For a description of the control potential in a common rail system see Reference 22. For ESR10 see Reference 23.

8.4 Current engine control technology

Modern engine ECMs are largely based on RISC industrial micro-controllers with a simple I/O and memory architecture. With the emphasis on low cost the design is generally simple. The programs are written in a mixture of assembly language and a high level language, probably 'C'. Choice of control algorithms is kept deliberately simple although this has led to an increasing calibration load.

The key features of current technology control systems might be summarized as below.

- Light duty—compact light weight unit borrowing extensively from SI units developed for passenger cars. Simple controls in general with a sophisticated idle and load management strategy.
- Heavy duty (truck) engine—generally larger units to accommodate more interfaces, larger memory and injector drive circuits for high pressure injection. More emphasis than in passenger cars on data collection and diagnosis functions.
- Medium speed for power generation—probably 16 or 32 bit controller with strong emphasis on the quality of speed regulation. Interfaces to other 'industrial control' equipment. Diagnosis functions and communications links for standby units.
- Medium speed for rail and marine—similar to a heavy duty control system and like heavy duty uses a system based on a 16 or 32 bit micro-controller. 'Features' generally offered in software will differ between applications but will include diagnosis, prognosis and communication.

Throughout all current applications, the control algorithms applied are usually quite simple. Until now simple linear controls typified by the three-term (PID) controller have proved adequate, although gain scheduling must generally be used to achieve acceptable performance over the complete engine operating regime. In spite of significant advances in the hardware implementation of engine control, the software solution which incorporates control algorithms remains rudimentary.

8.5 Algorithms for control

The algorithm adopted and implemented by a control systems designer lies at the heart of the control systems solution. A good control algorithm choice can make a substantial difference to both the quality of control and the speed with which the product can be brought to the market.

Earlier in the chapter we reviewed the three-term (PID) controller which is widely applied throughout engine control. However, in recent years there have been a number of new algorithms emerge which offer considerable promise both in controls and in diagnosis. The following brief descriptions highlight the principal techniques which have been offered as potential solutions for engine control and some possible application areas.

8.5.0.1 Multivariable techniques

Multivariable is a general term for control systems which accommodate several input and several output variables. An example is shown in *Figure 8.11* which shows the diesel engine with a boost control device (variable geometry turbocharger) and fuel quantity control (through a fuel rack).

The engine is a multivariable system, in that it is required to simultaneously achieve certain performance criteria with a number of control variables. The diagram shows a multivariable compensator which is designed to 'unravel' the complex internal interactions. In practice such multivariable compensators are rarely used and instead engineers use single loop controls and try to 'decouple' those controls.

Multivariable controls are used widely in other domains, and design tools are available. Again it is the non-linear nature of the diesel engine which discourages the widespread application of true multivariable control methods. Reference 24 provides a useful insight into the design of multivariable control systems.

8.5.0.2 Model-based control systems technology

Control and diagnosis systems which use a model of the target system to compute control actions or help in the diagnosis are often referred to as *model based.* The term is very widely used and applicable to range of control and diagnosis algorithms and to development techniques. Some examples will help illustrate the term.

- An adaptive controller must follow changes in the system it is controlling and adapt its actions. This is generally done by maintaining a model of the system internally to the controller which is used to periodically re-design the control algorithm.
- In a diagnosis system, the model provides a reference for

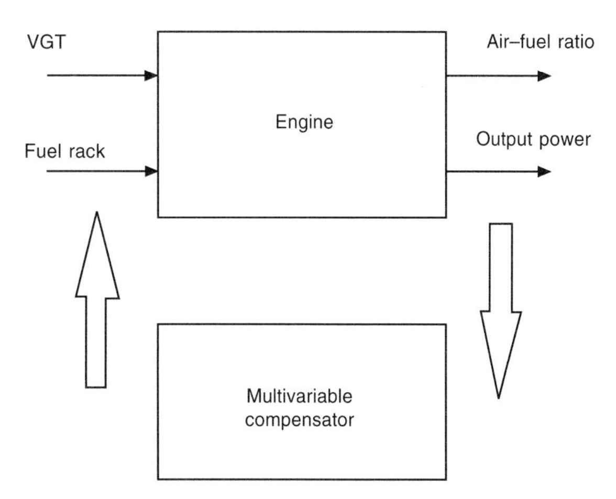

Figure 8.11 A multivariable compensator for EGR and speed in a diesel engine

system behaviour. The outputs from the real system are compared with the computed outputs from the model. Certain types of deviation provide evidence of a fault.

- During development, an engine component (such as the intake system) may be represented implicitly (as lookup tables) or explicitly as a physical model. The use of a physical model allows the control parameters to be pre-calculated rather than decided through experiment.

8.5.0.3 Time delay compensation: an example of model-based control

Measures to compensate for time delay in a control loop have been well established but are not widespread in engine control. The most common, the so-called Smith predictor is used widely in the process control industry. While it is quite difficult to apply, it nevertheless illustrates an early use of model-based techniques, where a model of the process to be controlled is included in the controller itself. Such a control is likely to be needed in the management of exhaust after-treatment in larger devices, and in particular to the flow of reagent in SCR systems. *Figure 8.12* illustrates the operation.

The control algorithm, probably PI or PID controls the plant directly. The feedback loop has two parts. In one branch, the control output is connected to a model of the process from which the output is delayed by the transport delay of the process. The controller is in fact controlling both the real process and the model process in parallel, but the feedback it sees is from the model. Because of this configuration, the controller can be made quite simple, since it is in fact controlling a process without time delay.

The output from the two branches is compared. Any difference is due for example to differences between the real plant and the plant model. Such differences when fed back reduce the effect of the difference.

8.5.0.4 Adaptive control

An adaptive controller is one that is able to change its design—either structure or parameters—to suit new conditions under which it is operating. Adaptive control covers a large number of different algorithms and we can consider them in two broad classes:

- Model reference adaptive controllers (MRACs), keep an internal model of the process under control: if the model changes the controller adapts its design to suit. One of the earliest and probably best known MRAC laws is the so called MIT law. The structure of an MRAC controller is shown in *Figure 8.13*. An example of the application of MRAC to engine control can be found in Reference 25.
- Self-tuning adaptive controllers monitor the behaviour of the system they control by refitting a model at every control sample. The structure of a typical self-tuning adaptive controller is shown in *Figure 8.14*.

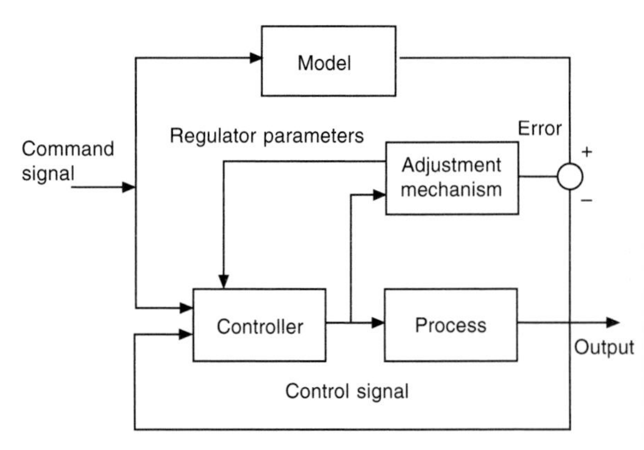

Figure 8.13 MRAC structure

The MRAC controller has an inner loop consisting of the controller, the process and the feedback of the process output. An outer loop is responsible for the adjustment of the controller parameters, and is driven by the error between the actual and predicted process output. If the MRAC controller was responsible for the speed control of a diesel engine, the reference model would represent the 'ideal' dynamics between the load demand and the speed output.

The error, e drives the adaptation of the controller. The MIT rule determines the change in the control parameters, θ:

$$\dot{\theta} = -\alpha * \text{error} * \text{grad}_{\theta}\, e$$

α determines the adaptation rate, and the direction is determined by how the error varies with the control parameters.

In a self-tuning controller, the two main functional blocks are the controller and the parameter estimator. The structure of the controller has many similarities with MRAC but differences in detail. Like all feedback controllers there is an inner loop consisting of a controller and regulator in a feedback configuration. The parameter estimator forms a model of the process building an explicit model which in turn is used by a design algorithm to update the regulator.

The ethos is different in that a control law is redesigned by the tuning of parameters rather than adjustment using a performance-based rule.

8.5.0.5 Predictive control

As its name suggests, a predictive controller will use a knowledge of the system to be controlled, to both make a prediction and to calculate future control actions.

A predictive controller is model based: at every control sample,

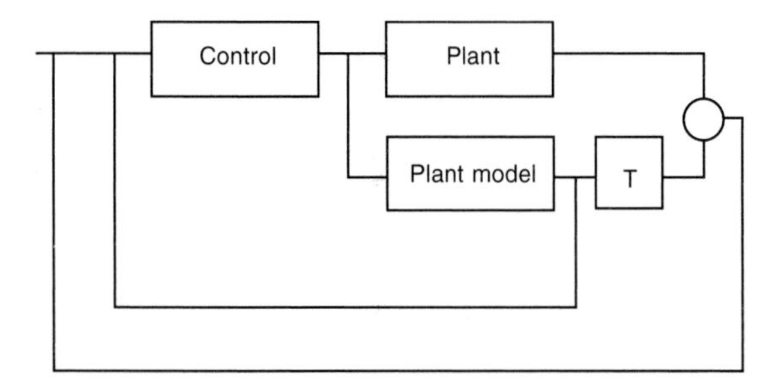

Figure 8.12 A simple Smith predictor or time delay compensator.

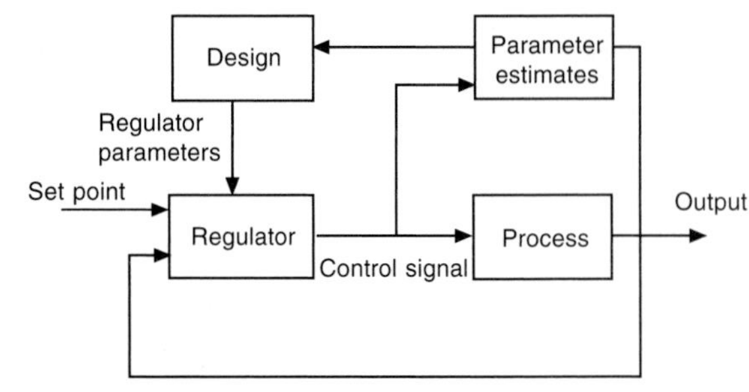

Figure 8.14 Structure of a self tuning controller

the control algorithm calculates the best sequence of control actions based on the model of the system to be controlled. The computation optimizes the future series of control actions based on future demand values, and a 'cost function' which guides the optimization.

Such predictive controls have excellent properties and have been demonstrated in the automotive environment[26]. However they require a knowledge of future demand values which makes them particularly appropriate where there is a 'forward view' of demand values. Power generation and some plant applications of engines admit this form of load prediction.

See Reference 27 for an excellent introduction to Generalised Predictive Control, a particular development of predictive control.

8.5.0.6 Optimal control

An optimal controller is one which is designed to select control signals which optimize a cost function. Typical of optimal controllers is the LQG (linear quadratic Gaussian) which uses a cost function made up of:

- deviation between the demand, the achieved output of the controlled system, and
- the size of the control signal.

The weighting placed on each of the two aspects of cost determine how the controller will behave in operation. If control is expensive—for example if a control action requires the use of additional reducing agent in a selective catalytic reduction (SCR) system—then the weighting on control will be high. If error is expensive—for example if an emissions error is costly—then the weighting on error will be high. Such an optimal controller is designed from a cost function qualitatively defined as

$$J = \Sigma\ (Qx^2 + Ru^2)$$

where the sum is formed over a long period the solution for the feedback law converges to a constant value. This approach is generally followed in practice rather than re-computing the feedback law from time to time, although it is possible to formulate the LQG in a self-tuning form.

Figure 8.15 illustrates the structure of such an LQG controller. The controller uses the state feedback matrix, **L** which is applied to a state estimate, $\hat{x}$. That state estimate is calculated by means of an observer which uses the control signal and the measurement, y from the process. In this implementation there is an additional function which checks for and corrects the control output for saturation. Note how the observer uses the corrected control signal.

8.5.0.7 Robust control

Robust control is a relatively recent development in the field of control theory and with the first significant papers emerging in the early 1980s (see References 28 and 29 as good examples).

In a robust control system, the design of the controller is such that it will be both stable and deliver the required performance whatever the plant behaviour is within a quantified range of error. The plant is modelled as a nominal model and an uncertainty. The uncertainty can represent a change which is difficult to model but which can be characterised. *Figure 8.16* shows one

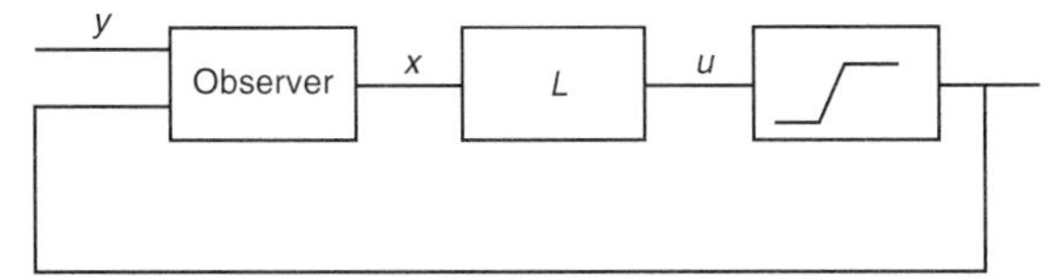

Figure 8.15 The structure of an LQG controller

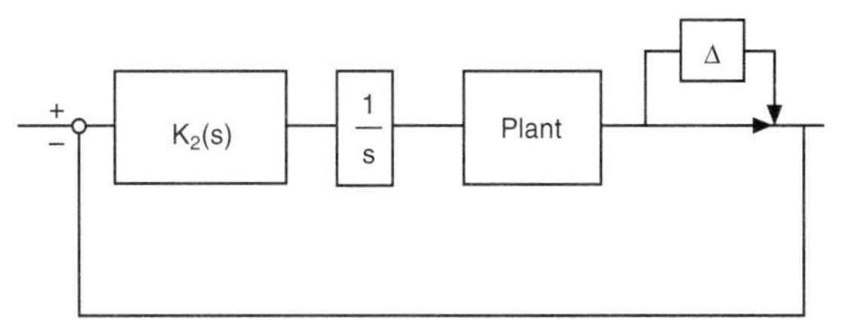

Figure 8.16 Structure of a fuel pump controller with a multiplicative uncertainty

type of uncertainty, a multiplicative uncertainty. The uncertainty may represent for example the fuel temperature in a fuel injection system, whose effect can be modelled off-line and included in the model for control design purposes.

A simple example will illustrate the principles and how in particular (delta) is used to design a stable controller, K. In this example, the controller K is in series with an integrator. Such an integrator is always present where the control actuator is an electric motor. The current input is related to torque where we are usually interested in a velocity or position. The plant can be considered as

$P = 1/s\ *$ the plant transfer function

T is now defined as $PK/(1 + PK)$, that is the nominal transfer function. L is the open loop transfer function PK. S is defined as $1/(1 + L)$ or $1 - T$. T is the sensitivity function associated with stability while S is associated with performance.

The controller will be stable if the following condition is satisfied.

$$\| W_2 \cdot T \|_\infty < 1$$

that is if the least upper bound of $W_2 \cdot T$ is less than one over all frequencies.

In a robust control design, the two sensitivity functions S and T are, respectively, weighted to give the controller performance and stability. In the final analysis, it is this relative weighting which lies at the heart of robust design.

8.5.0.8 Neural and fuzzy control

Both neural and fuzzy control are vast fields and well served by a research community and a number of industrial applications. Here we will consider the broad principles of neural and fuzzy systems and look at some of the applications to diesel engines. The reader is recommended to read one of the textbooks which focuses on the application of neural networks and fuzzy logic to control: Reference 30 is a good example and also provides some useful insights into the commonality between fuzzy logic and neural networks.

There is a strong relationship between fuzzy logic and neural networks (NN) for control. The most commonly used NN architectures can also be implemented using a fuzzy approach. For such systems, the neural network is an excellent learning tool, while a fuzzy system can be implemented to operate efficiently in real time.

A fuzzy controller is based on two principal components: a set of membership functions which relate measurements to fuzzy sets, and a set of rules which allow the controller to deduce control actions. The particular arrangement of these components is shown in *Figure 8.17*.

The first step, fuzzification, is a misnomer since its purpose is to change the frame of reference of the measurement rather than to lose information. In fuzzification the measurement is related to the 'degree of membership' of a fuzzy set. For example a temperature which is low might be judged to a member of the set *temperature low* to the degree 0.8.

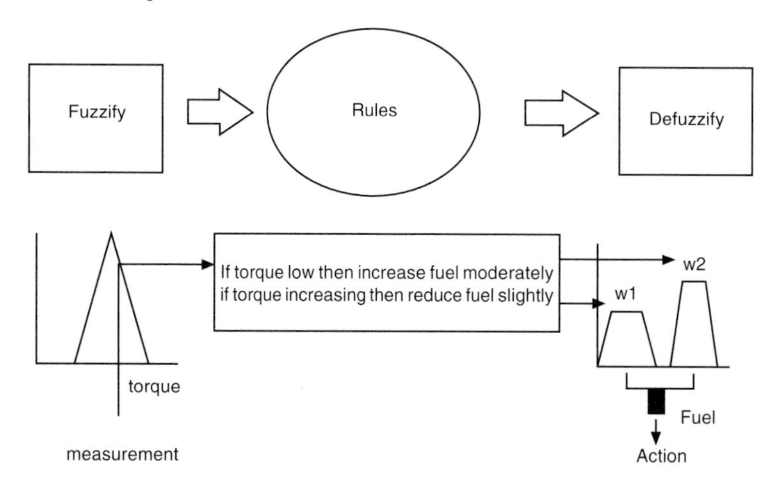

Figure 8.17 Principles of operation of a fuzzy logic based controller

Each membership of such a set can relate to the firing of a rule. Any rule which is based on a temperature low criterion would be applied to this measurement. However, the recommended output of such a rule would only be applied according to the degree of the input, that is 0.8. This application of rules *by degree* leads to a collection of outputs which themselves are membership functions but truncated. The defuzzification process 'balances' these outputs—by effectively taking a centre of area of the truncated functions. Again defuzzification is a change in a frame of reference.

Neural control shares many characteristics with fuzzy control. Both are useful if the model of the system to be controlled is unclear. Fuzzy control is most useful where there are some control heuristics which already exist or can be easily derived from 'experts'. A neural network acts as a means of learning a function and replaying it for controls or diagnostics purposes. The neural network will:

- learn and imitate the characteristics of a dynamic system to replay the 'desired' or model response of a control system, or
- be trained with the behaviour of a system whose behaviour we want to capture and replay for diagnosis purposes.

The example[31] illustrates how fuzzy techniques might be used to optimize the fuel efficiency of a diesel engine. The authors propose three rules which summarize a heuristic approach to identifying optimum conditions. In summary,

- Rule 1 states that if the measurement of successive torques is significant, then continue to follow the sign, otherwise re-measure and if still lacking confident information continue in the same direction as before.
- Rule 2 states that if the gradient sign changes the search is close to the optimum, then use a regression fit to recent data to identify the optimum point.
- Rule 3 is concerned with comparing the identified optimum point with the designers expectation and resolving the difference.

Application of such simple rules led to the typical search time for an optimum point in a large marine diesel engine falling from several days to less than 2 hours.

In a neural system computing is done by a large number of simple computational elements which can be understood to receive an input $x(t)$ and which generates an output $S(x(t))$. The neural network's topology then determines its properties 'in the large'. A simple form of S is the function,

$$S(x) = 1/(1 + \exp(-cx))$$

In one of its simplest forms, the network has just three layers, *Figure 8.18*. The network is supplied with data—where each of the input nodes receives one data item. At each node the processing scheme is broadly

1. take each input weighted according to the connection scheme,
2. form a sum,
3. process the sum through the function S,
4. pass the result on to the next layer.

This process is illustrated in *Figure 8.19*.

There are many different architectures each of which offers slightly different properties. One of the most significant distinctions in architectures is the difference between feedforward

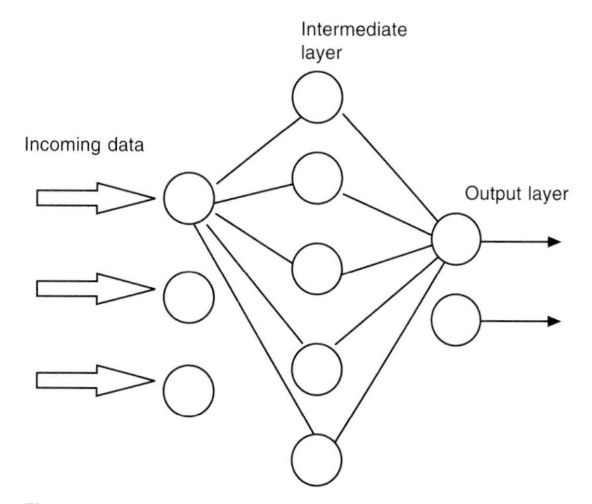

Figure 8.18 A simple neural network with a single 'hidden' layer

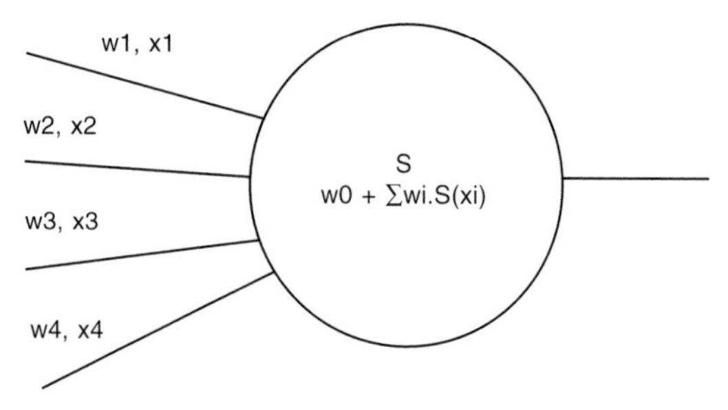

Figure 8.19 The operation at a single neuron

and feedback schemes. In a pure feedforward scheme, signals propagate forward through the network. Such schemes have been applied widely for identification of objects in vision systems. In a feedback scheme the outputs of layers are fed back to preceding layers. In general, control algorithms can adequately use a feedforward structure.

In the learning or training phase, the weightings are learnt through one of a number of processes. In supervised learning of which the process of back-propagation is an example, the network is presented with both the incoming data and the pattern or output which corresponds to that input. Such supervised learning may require a large number of iterations, and must find global rather than local minima in the 'search space'

The neural network makes an excellent classifier, relating complex incoming data to a limited set of outputs. The particular form of classification is determined by the structure of the network and in particular, the number of layers. Increasing the layers improves the classification resolution but costs more in computing power both in learning and replay modes.

If S (t) is a radial basis function (RBF) which has a Gaussian form

$$S(x) = \exp \{- 1/2s^2\ \sigma, j = 1 \text{ to } n\ (x_i - c_j)\}$$

the learning process is relatively simpler and can take place in real time, using algorithms which can be made robust and compact.

With RBF, the aim is to model a dynamic relationship in a set of data, where typically a set of data $\{[x_i,y_i] : i = 1, \ldots, n\}$ needs to be modelled. The task in learning from a set of data is to compute a set of weights so that the relationship

$$\hat{y}(x) = w_0 + \sigma, j = 1 \text{ to } N,\ \sigma(w_j \cdot S(\| x_j - c_j \|))$$

converges—that is $\hat{y}(x)$ reproduces to an acceptable accuracy the behaviour of the real system.

There is a strong analogy between neural networks and fuzzy systems. This analogy enables the linkage of neural network techniques to fuzzy logic. A neural network equipped with a learning algorithm can readily model the required control behaviour. A translation algorithm can transform the weightings into a set of fuzzy rules and membership functions. The fuzzy system is computationally more efficient than the equivalent neural network and such an approach offers a good development route.

8.5.0.9 Variable structure

Variable structure control takes on a number of guises but in general describes a type of control which changes its form according to the particular situation of the control algorithm. VSS was first developed at the University of Moscow by Professor Utkin. In principle it is simple and described by the following example.

An EGR valve located on the high pressure side of a turbocharged engine. The position controller is set for different engine operating conditions by position. The valve is subject to small disturbances which must be removed as quickly and smoothly as possible. The valve's position is x and velocity $\dot{x}$.

The position of the valve is detected by a sensor, and the controller derives the velocity by differencing between samples. The objective of the controller is to force the valve back to its nominal position as a first-order response. The function of the VSS controller is to force the state (described by x and $\dot{x}$) along a surface which corresponds to a first order response, that is

$$x + T\dot{x} = 0,$$

or graphically as shown in *Figure 8.20*.

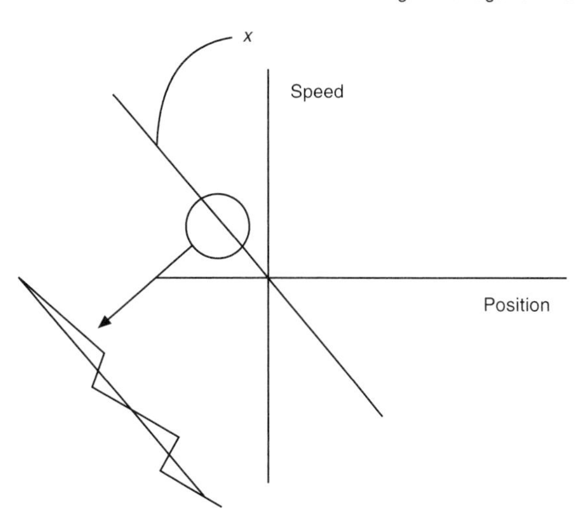

Figure 8.20 Illustrating the operation of a simple variable structure controller

The surface is designed to 'catch' the free trajectory of the system. Once x and $\dot{x}$ conicide with the surface the controller cuts in and forces the state trajectory. This is done by switching from one side of the surface to the other producing a chattering effect while at the same time controlling the position and velocity of the valve.

For the original discourse on VSS control see Professor Utkin's book[32], and for a modern automotive control example see Reference 33. For use of sliding mode *observers* see Reference 34. It is this later example that shows the improvement made to VSS by the use of boundary layers. The state rajectory is drawn to the surface and kept close by an appropriate choice of algorithm. This results in a smooth response (without the chattering) at the cost of greater computational complexity.

8.5.0.10 Extremum control

Extremum control describes a class of algorithms in which the objective of the controller is finding a peak in the system response. The early work on extremum control was done by C S Draper[35] on optimizing ignition control for SI engines. Other applications have included an adaptive peak seeking ignition controller for SI engines developed by Scotson[36]. The principle is simple but the implementation requires care to avoid for example limit cycling around the peak of the response.

The following example, *Figure 8.21* illustrates the operation of an extremum controller.

The objective of the controller is to find the optimum injection timing at a particular engine condition. The optimum is a peak but the response is not necessarily symmetrical. The function of the controller can be summarized.

- the timing is adjusted stepwise while the torque contribution increases;
- at or close to the maximum the controller finds a decrease in torque and will step back so as to maintain a position close to the optimum;
- in practice the response curve is different for each engine condition although the shape of the curve is constant;
- the curve can be characterized and used as the basis of an adaptive algorithm which continuously identifies the curve and allows the controller to reach the peak response within a few samples.

This apparently simple problem is amenable to modern control

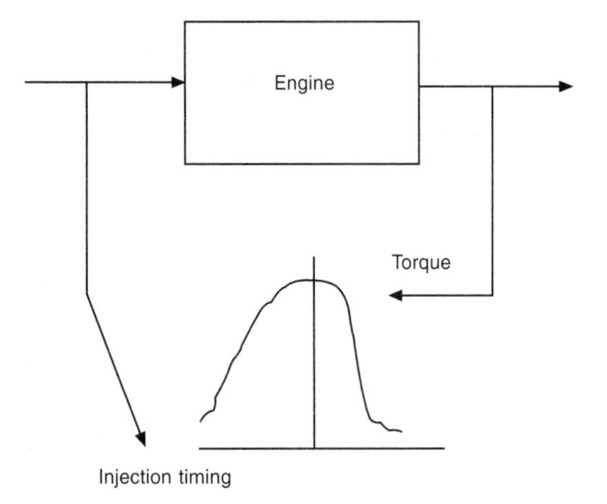

Figure 8.21 The extremum controller identifying optimum injection timing

techniques. Without the adaptive approach the performance would always be slow and probably inaccurate.

8.5.1 Predictors and filters

Conventionally, a *predictor* makes a forward projection of a signal, while a *filter* gives an estimate of the signal now based on a current observation and the immediate past history. One of the more famous types of predictor is the Kalman filter which makes a single sample prediction of the states of a system. States may be described as the internal measures of system behaviour. In a mechanical system where acceleration and position are measured, velocity will be one of the system states, and could be predicted by such a filter.

Indeed if we were to design a controller which required a measure of speed to stabilize the system, we would need such a filter to generate a 'measurement'.

8.5.1.1 Observers as advanced sensors

The observer is a concept introduced earlier in the chapter and applied to the estimation of hidden parameters. Where there is noise in the sensor readings and the measurement information is available periodically the Kalman filter provides a route to making state estimates. The processing route for a Kalman filter can be roughly described as follows:

New state estimate = old state estimate + gain * estimation error

This simple statement stands to reason in that in such a system, feedback is an effective way to correct an error. The gain is non-linear and depends on an estimate of the covariance of the state estimates. In a general form of the Kalman filter, the covariance increases every sample time whenever a measurement (for any reason) is not received. This tends to make the gain larger (closer to unity). When a measurement is received the tendency is for the covariance to fall and for the state estimates to converge.

Such Kalman filters are based on models of the underlying process. The filter uses the internal model to estimate the hidden states from the received but noisy measurements. The states are reconstructed on the assumption of a healthy model. *Figure 8.22* is an illustration of the algorithm. The states could equally be re-constructed on the assumption of a certain type of fault. Where there is a restricted family of faults such a filter bank is feasible, *Figure 8.23*. Such an algorithm has been demonstrated to provide good estimates of speed and acceleration from a position sensor[37]. This scheme avoids the pitfalls associated with the direct estimate of a differential.

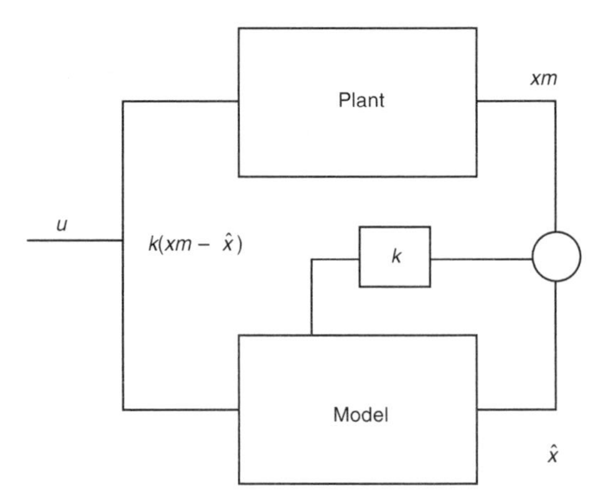

Figure 8.22 The structure of a non-linear observer

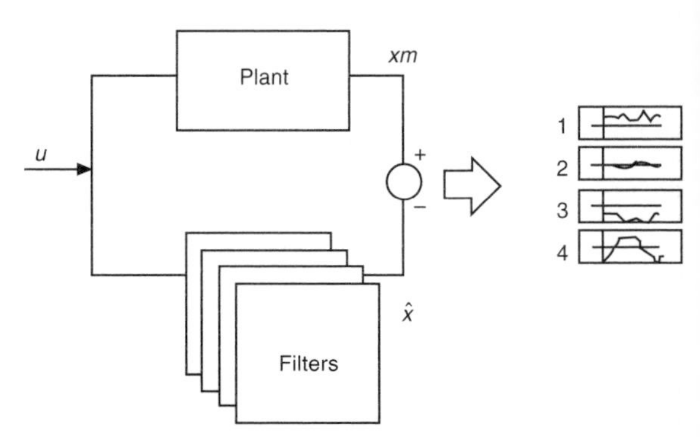

Figure 8.23 A diagnosis process using a bank of filters

The *residuals* between the predicted and actual plant output can be generated by a number of fault filters each tuned to a particular fault condition. If one filter generates a residual which is close to zero while the others are non-zero and there is confidence at a sufficiently high level, then one of the faults will be suspected. The filters may be tuned to represent classes of faults, such as in individual engine sub-systems—fuelling, induction, and exhaust. Algorithms for this type of diagnosis are introduced by Willsky[38].

8.5.2 The future

With many of these new types of control, there are already a number of applications in research within major engine companies. A prediction made in Reference 39 of the application of modern control techniques is still applicable to the vision and opportunity that such techniques offer.

Robust control offers substantial benefits in the design of practical control systems which must meet their performance criteria in spite of variations in the system components. Research is proceeding[40] but as yet there is no practical scheme for applying robust control to production engines.

8.5.3 Modern control—an example

The example (*Figure 8.24*) considers the use of two sensors which are currently only under consideration for engine control—cylinder pressure and NOx. The first problem is topical while the second is rather more visionary, posing questions like 'if we

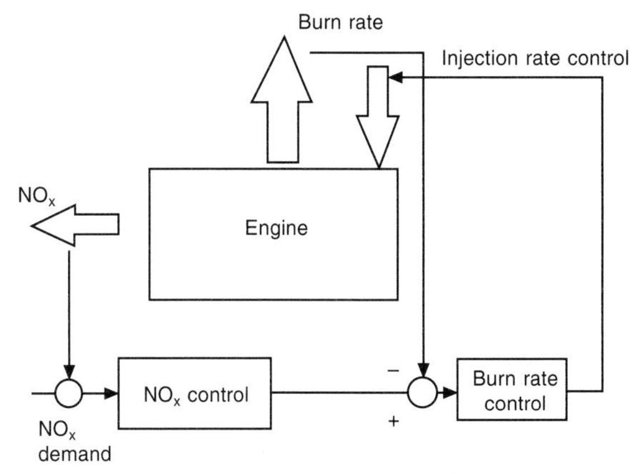

Figure 8.24 A conceptual control scheme to limit NOx emissions

had an NOx sensor how could we best apply it in closed loop control?'.

Given the scheme shown in *Figure 8.24* the next step in the design would be to consider a control block diagram that could be implemented for prototype purposes in a design environment such as *Simulink. Figure 8.25* shows such a block diagram with the main control elements highlighted. The links are those that would need to be made to an external engine simulation.

The choice of control algorithm for both the overall NOx control as well as the burn rate control depends strongly on the characteristics of those variables. They will be non-linear, although for initial tests a simple PI controller will be sufficient.

The next step will require a series of system identification tests by means of which the underlying dynamics will be recognized. This step could lead to a number of conclusions:

- If the dynamics are simple and vary little over the operating range of the engine, a simple linear controller such as PI may be enough and certainly the best direction for initial tests.
- If the models vary over the operating range and are dependent on only one or a few parameters then a gain scheduled PI controller may work.
- If there is a substantial time delay in the NOx feedback which will depend on where the sensor is located, there may be a need for time delay compensation.
- If the models vary and there are many parameters influencing the change, then a robust controller may provide the secure operation that is needed.

The need for a fast prototype may be served by a fuzzy controller whose behaviour can be developed according to a set of rules. The rules represent a very direct way for an engineer to specify the details of control. Finally, the algorithm may be implemented in different form but a fuzzy prototype may help create the flexible environment needed early in the development process.

8.6 Designer's guide

In this section we will take a brief review of the design process. As far as possible we have chosen topics which are generic and which are not reliant on current styles of engineering tools. The engineering methodologies are described in detail to convey the philosophy of control design. The reader will interpret these according to specific policy and practice in their place of work.

We assume that the designer is principally involved with the specification and testing of control or monitoring algorithms. Such algorithms will be integrated into an overall hardware and software solution. In many cases the engineer will be using much existing code, but this environment does not override the need to robust engineering practice in algorithm development.

8.6.1 Developing control systems

The control system is an information system so that its development will follow techniques adopted for information systems development. Such models were first proposed and followed in the early years of computer systems development. The essential elements of such a model are:

- requirements capture: understanding what the system needs to do;
- specification: specifying the system behaviour and the details of the environment in which the system is going to operate;
- design: identifying the architecture and components of the proposed system;
- implementation: putting the system together;
- testing: verifying that the system meets the requirements.

The waterfall model of system development was first proposed in 1956[41]. At this time it was understood that computer systems developments would be impossible without an organized approach. The waterfall model described a process of step-wise refinement which followed the path of specification, design, implementation and testing.

The V model[42] is a development of the waterfall model puts

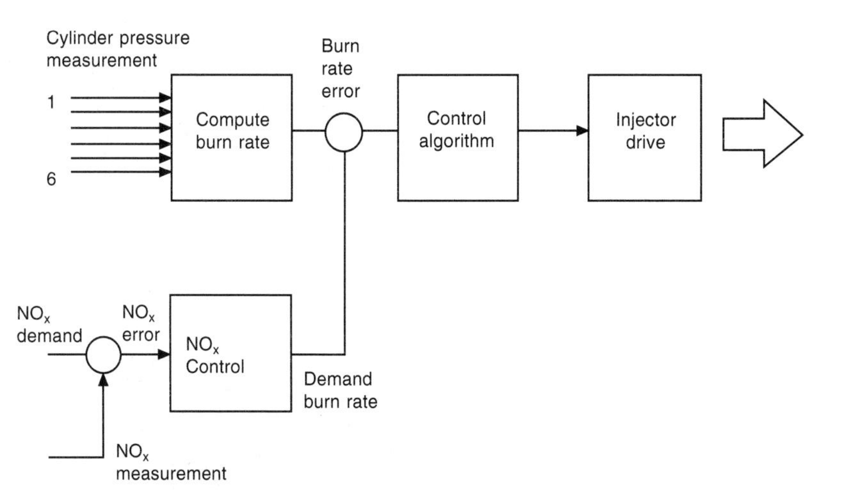

Figure 8.25 Control block diagram for some initial investigations

a stronger emphasis on validation and verification. The V model requires that a full set of documents are complete at each stage of the development process—a situation which is rare in practice. A more adaptive approach is needed for most systems development.

Validation is the step of checking that a step such as implementation has been faithful to the goals. Verification compares the results of implementation with the design objectives.

The Spiral Model[43] accommodates each of the previous models as special cases. In the spiral model, the system design is progressively refined through a series of cycles. In each cycle some aspect of the design is developed. Risk assessment and planning are done before entering into the next cycle of refinement. It allows development to proceed where neither requirements nor specification have been well thought out.

Given this history it is worth refining nomenclature before proceeding to a full description of the development process.

- A development model is *a description of the process whereby the product is developed*—the choice of model is fundamental to the quality of the engineering product developed. A thorough development model includes a specification model and techniques for extracting and using user requirements.
- A system or component model used in the development of an algorithm is a description of the component *at a level of detail which will support the development of a control algorithm.*
- The control strategy can be defined as *the functional statement of the control policy to be applied to a particular engine product.* A functional statement is independent of any implementation detail and includes statements of control policy with regards to different aspects of the engine systems.
- The development path to a control system is well documented in general texts on systems development (for example: [5]) and in the specific engine literature [44].

8.6.2 General comments about system development

System development is concerned with delivering a solution which meets a set of customer requirements including functional requirements (performs certain types of control) and non-functional requirements (will run on several different hardware platforms). Other non-functional requirements include maintainability: how can the system be maintained at a reasonable cost?

An important aspect of maintainability is keeping versions under control both before and after release. Version control is an essential component of fault tracking and will allow previous configurations of software and hardware to be exactly reassembled. Version control will start even during the development of algorithms where a cut-and -try approach which is needed in some development proceeds in a fashion where it can be 'wound back'.

We will assume that system development is supported by engineering tools—probably a systems development method accompanied by a software toolkit. For the purposes of illustration we will use the *DeMarco* structured analysis method using the real time modifications proposed by Hatley and Pirbhai[45]. Such a structured method is concerned with data flows—that is how data is passed between processes which transform that data. For a controls specialist, data received from sensors is generally the incoming data flow, while the data flow out of the process is a signal to an acutator.

We intend to answer the question—what does the controls specialist need to understand of the system development methods? A formal approach to systems development as we will describe it does not preclude rapid prototyping—indeed some of the newer ideas in systems development embrace the need for rapid prototyping to progressively remove risk during systems development.

8.6.3 Specifying functions

The specification of functions is done in a top down way. The overall system is bounded in the form of a context diagram which shows the system input and output dataflows. The analysis proceeds through a number of levels where the data flows required to fulfil the required functions are progressively analysed and refined until the content of each process 'bubble' are no longer divisible only be data flows. There is some more detailed specification needed in the form of a description of how that process is performed. Such a description, normally called a P-Spec (short for *program specification*) may well be a description of the control algorithm which fits into the process bubble.

Figure 8.26 shows the process for an example in which a controller for intake manifold pressure is identified and isolated from the rest of the design.

The context, *Figure 8.26a* shows the whole system of which we have highlighted the incoming raw pressure signal from the inlet manifold, and the drive signal to the wastegate. Nothing more is specified at this stage. The context is very useful in defining boundaries and indeed would provide the opportunity to debate the merits of smart or 'dumb' sensors for pressure measurement in this application.

The designer will proceed to develop the requirement around these two signals. Conceptually the two signals are connected by a process which we now understand to be the calculation of the actuator signal, *Figure 8.26b*. There is now an additional detail in that the pressure controller must know about the demanded manifold pressure before embarking on a calculation to set the wastegate. It may need other signals, and it is at this point that those signal needs should be identified.

There are still conceptual processes 'hidden' in the bubble of *Figure 8.26b* and the designer must continue the refinement to isolate the fundamental processes. One interpretation is shown in *Figure 8.26c*. There are now three processes, with one, *calculate control* now clearly identified with performing a classical control task, that of taking demand and actual measurements, and of then calculating the required wastegate position.

This design sets the context for the controls engineer who will then proceed to develop a control algorithm to perform the required duty. As the task proceeds, the controls development may require some new data. This can be proposed as a subsequent modification to the diagram, and the *data dictionary.*

8.6.3.1 Developing the control system

Function and constraints

The selection of a control algorithm is possible once the requirement is understood in detail. The starting point will be the various data flows to and from the controller. There will also be constraints primarily in the form of resources: processor capacity to be used and the memory budget the algorithm may use.

Developing a system model

The system model will be one primary factor bearing on the choice of algorithm. In practice most algorithms will be the regular PID, but other types of algorithm are now feasible and have superior qualities. Those functional qualities must be balanced with the costs of implementing such advanced control schemes. It may require specialist staff and computer tools.

The engineer should construct the plant model using one of the techniques we described earlier. Such *system identification* is an experimental process which must be carefully planned otherwise it will give misleading results. There are some clear steps in this process:

- define the plant to be controlled;

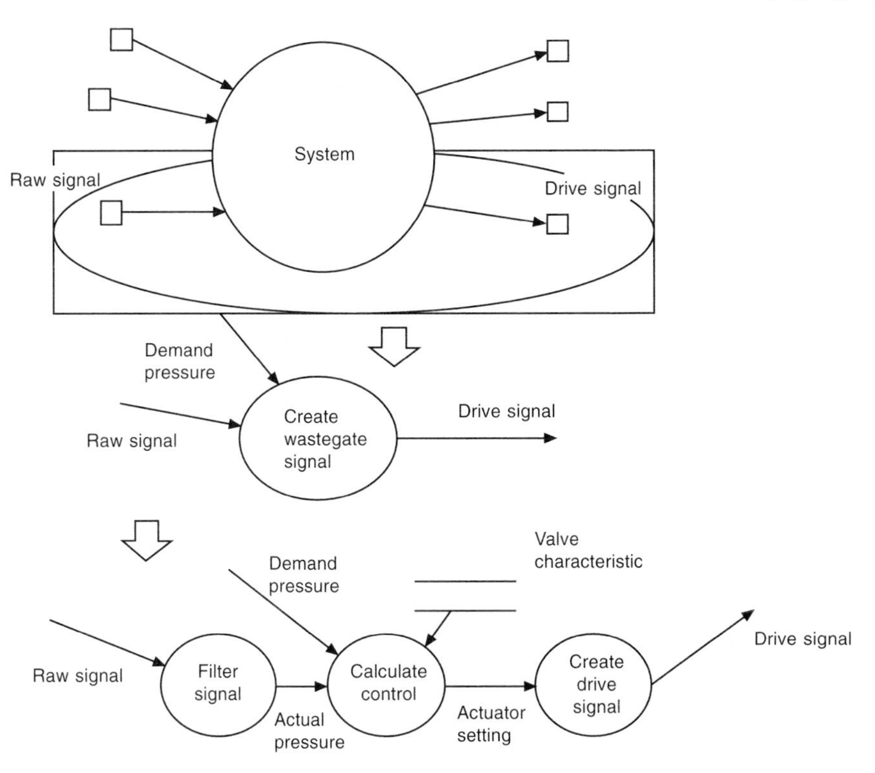

Figure 8.26 The development of a wastegate control module starting from the context diagram in a structured analysis

- if the dynamics are simple, derive them from first principles; or otherwise
- plan an experiment to capture the dynamics in the form of a system model.

The details of such an experiment and how to process the data can be understood from a standard text on the subject of *system identification*. The experimenter must avoid certain pitfalls and be vigilant even when the experiment is underway. Very often when conducting this exercise on an engine[9], the best approach is to instrument the engine so that the data streams are representative of actual operating conditions. In the context of the engine, the principal 'rules' are:

- sample at a rate about 10% of the rise time of the system: faster than this and the numerical algorithms may not converge;
- be careful where there is a closed loop control in place — sample'inside' the controller;
- assume that noise sources (measurement noise) are not white and model the system accordingly;
- the model is likely to vary across the operating regime of the engine
- at high engine speeds sampling should be synchronized with the firing events.

The control engineer must then select the algorithm on the basis of the expected performance and the resources that can be allocated. The pressures include:

- resource allowance (influences the choice of algorithm and the sampling time);
- rise time and disturbance rejection;
- through life stability.

Some of the aspects of this decision will be considered in the next section.

Selecting a Control Algorithm

Algorithm selection is based on a balance between performance and implementation cost.

In the following table the algorithm which has been introduced in earlier sections is shown against the following measures:

- Complexity: how difficult is the algorithm to implement and adjust? ● = most difficult)
- Resources: how much memory and processor resource does the algorithm take? (● = most resources)
- Tuning: how much effort does the algorithm take to get it working to specification? (● = most resources)
- The characteristics of processes the algorithm is 'good for'.
- The extent of readily available design support tools. (● = difficult to obtain)
  (See *Table 8.1* on page 192.)

Assuring safety through the system development process

The algorithm must be safe by design, but is only a single component in a larger system which too must be safe by design. The safety level to be applied to development is a matter of the role of the system under development. In general an engine management unit is not *safety critical* although is related to safety—it is conceivable for example that an under-fuelling of the passenger car engine when the driver is demanding torque could result in a dangerous situation. However, when the engine management unit is linked to traction control, and in modern vehicle dynamics control systems there is a need to link braking, traction and steering. The role of the engine management computer is now firmly a safety related component, and will need to be engineered with that aspect in mind.

Safety management is a substantial subject and is covered well in Reference 46. We will take a brief look at the process used in modern systems management and investigate the impact on the work of the control systems designer.

Table 8.1 Criteria against which controller algorithms will be compared.

	Complexity	*Resources*	*Tuning*	*Good for....*	*Design support*
PID	●●●	●●●	●●	Ordinary dynamic systems	●●
Smith predictor	●●	●●	●	Systems with marked transport delay	●●
Self tuning regulator	●●	●	●●	Changing complex processes	●
Robust controller	●●	●●	●	Processes with uncertain parameters	●●●
Sliding mode controller	●●	●●	●	Uncertain processes where control power is not a concern	●
Fuzzy logic	●●	●●	●●	Processes where the model is difficult to find	●●
Neural networks	●●	●	●●●	Processes with typical performance data	●●
Extremum control	●●	●●	●●	Processes with an optimum which changes	●

The safety is rooted in the systems development process, which needs to be both formalized and integrated with safety management tools. We can identify three phases of the safety management process and underlying is a documentation process which centers on the 'hazard log', a record of the proceedings which allows auditing of all safety-related comments on the system.

- Preliminary hazard analysis (PHA) is performed in the early stages of the programme, just after technology choices and a broad system architecture have been selected. The PHA identifies hazards and their potential causes. It tends to be focused on function rather than the details of technology and sets the agenda for the next phases where resources must be deployed.
- The main hazard analysis phase is where design documentation is used to support a detailed but high level analysis of the design. This step can be applied at any stage during structured analysis and design and indeed can be repeatedly applied to ensure that hazards are not introduced.
- The detailed analysis of components is done using FMEA and fault tree analysis. FMEA is widely practised and is well supported by manuals and documentation[47]. Fault trees are generated from top level faults—that is system behaviour which is hazardous and the designer will build up a tree showing how those faults are related to failures in system components. The contrast between FMEA and fault trees is the usual one between a top-down and a bottom-up approach. Both have their place in the design process.

Using prototypes in the development process

One view of the system development process suggests that it is cyclic—that is development consists of a series of refinement cycles where prototyping plays a significant role. *Figure 8.27* shows this concept diagrammatically. All systems development contains an element of risk, and a project manager must progressively reduce risk as the design evolves. One key element of risk is the performance of algorithms and this explains the need to develop prototype versions.

Typically, during a prototype development, the project team will answer questions concerning choice of algorithm, how it is implemented and how fast it must run.

The engineering tools used for the testing of prototypes cover a spectrum of activities and include:

- mean value modelling where engine and algorithm are all modelled by computer; the engine model is an approximation but good for an initial assessment of an algorithm;
- full phenomenological models, which are now being used for controls assessment and offer more accuracy and the possibility of more elaborate development and test facilities than a mean value model would permit;
- *hardware in the loop* (HIL) testing in which an item of hardware representing either the controller or the engine is included in the simulation. HIL is suitable for the late stages of testing, and for example allows a design verification step to be taken in a controllable environment. Too early use of HIL testing can lead to expense and an over emphasis on development at the expense of design.

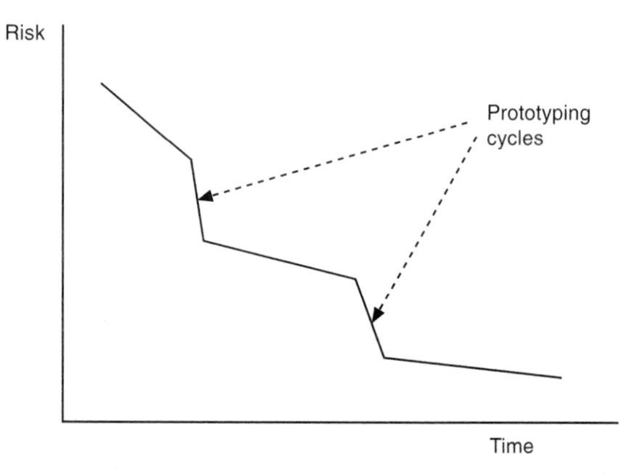

Figure 8.27 Progressively reducing risk through the development cycle

We will explain each of these types of tool in more detail:

8.6.3.2 Mean value modelling

This approach is illustrated in Reference 48 where a model of a spark ignition engine is developed using the block structure of the Simulink system modelling tool. There are numerous other diesel and spark ignition examples of which Reference 49 is a good example of a diesel mean value model.

In general, the mean value model is based on mixture of relationships derived from first principles and empirical correlations. The timescale in such models ignores individual cylinder events to maintain simplicity.

For example, in Reference 49 the authors propose a model of indicated efficiency:

$$\eta_{\text{ind}} = f_1 \cdot (1 - k_1 \cdot \lambda^{k_2})$$

where f_1 is a second order polynomial in engine speed
k_1 and k_2 are constants
λ is the theoretical air ratio

which is an empricial correlation, whereas the model of engine speed development,

$$\dot{N}_e = (K_I \cdot (\dot{m}_f H_u - P_{fr} - P_b))/I_e \cdot N_e$$

where $\dot{N}_e$ is engine speed
K_I is a constant
$\dot{m}_f$ is the fuel flow rate
H_u is the lower heating value of fuel
P_{fr} is friction power
P_b is the brake power
I_e is the engine inertia

is based on the engine dynamics.

In a mean value engine model, the engine will be built up from a series of components which can be tested and validated individually. The modular approach[48] allows component models to be swapped and offers flexibility in building options into the model.

The particular functions and advantages cited[48] include:

- the off-line testing of control algorithms—by using the model as an engine simulation in the controls prototyping environment;
- as a real-time engine model—which can be run on special high speed computers which will allow the model to function in real time. In this context the model behaves like an engine and helps avoid the cost and complexity of a real engine in a preliminary test of a new control system;
- as an embedded model for diagnosis in engine management systems;
- testing of sensor and algorithm concepts—which may be tested on the model engine as a first step to selection for production units;
- the engine model as a sub-system in a powertrain or vehicle model.

These are generic functions, which apply irrespective of the type of model employed.

Mean value models are relatively easy to configure and can be made to run quickly on special purpose hardware. However, their precision is ultimately limited by the kinds of physical relationships they can represent based as they are on empirical correlations and simple physical formulas.

Another type of model is illustrated in Reference 50 where a non-linear form of input–output model is used to capture diesel engine dynamics.

8.6.3.3 Phenomenological modelling

Increasing computer power has allowed a full phenomenological model to be run quickly enough to form part of a prototype system. *Figure 8.28* shows a screen dump from a typical session where the GT-Power code[51], is integrated with the control simulation environment, Simulink.

The *C*-Power environment which is illustrated here is a combination of the GTPower* engine modelling code and the Simulink† controls simulation package. The objective in bringing these tools together is to allow controls development to be set in the context of an accurate engine model.

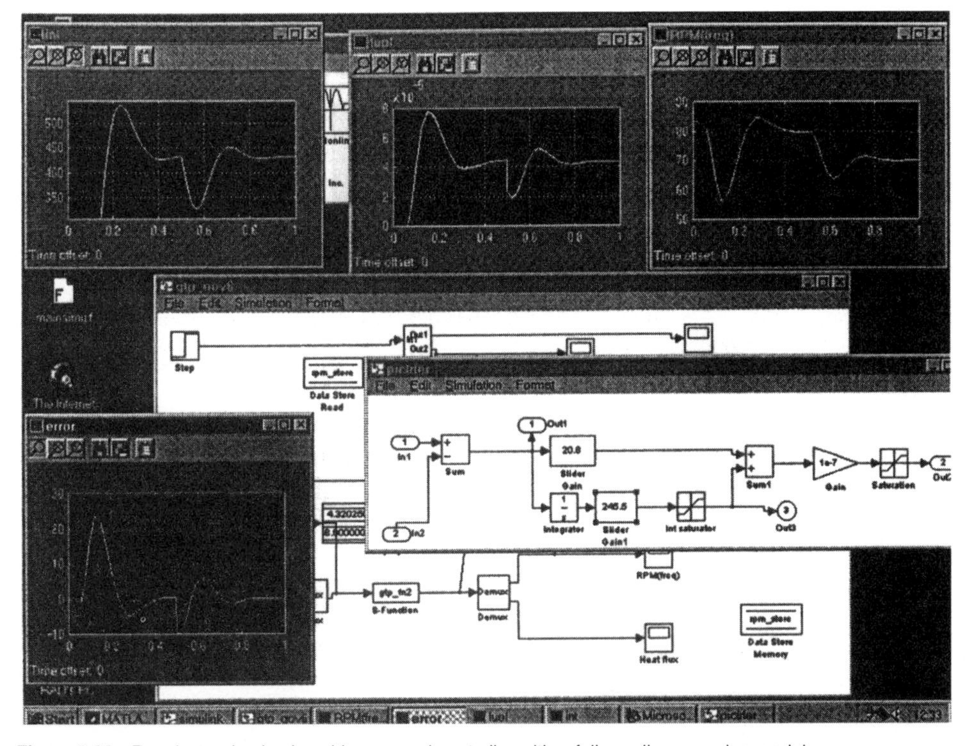

Figure 8.28 Running a simple closed loop speed controller with a full non-linear engine model

*GTPower is a trademark of Gamma Technologies Inc, Westmont, IL
† Simulink is a trademark of The Mathworks, Natick, MA

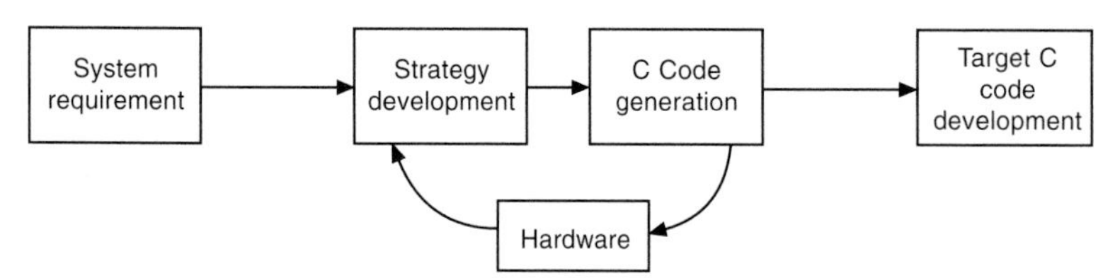

Figure 8.29 The flow from requirements to tested software

By making this combination the engineer is offered significant advantages over mean value engine models for control systems development. In particular:

- the representation of most engine non-linearities;
- system identification of engine components before hardware is available;
- the testing and evaluation of engine strategies which rely on a novel or untested component (a sensor currently in development for example);
- a rapid evaluation of proposed algorithms and strategy elements before engine hardware is available.

The engine designer will prepare a model of the engine based on the system components. Such a model is readily constructed but must be validated. For an entirely new design, a detailed validation may not be possible, but some component level validation may be possible. During this process, the engine designer, working with the controls engineer, will define the sensors and actuators. Connections are made by the engine designer to a 'wiring harness' which forms the link between the engine simulation and the controls simulation. By detaching the details of the engine simulation from the controls simulation, the task of creating a simulation is eased. The controls specialist need only connect signals for sensors and actuators while regarding the engine as a 'black box'.

8.6.3.4 Hardware in the loop

A useful example of the use of hardware in the loop is discussed in Reference 52. The objective of hardware in the loop (HIL) testing, is to quickly assess and modify control and signal processing algorithms. *Figure 8.29,* shows one application of HIL testing.

The rapid prototyping computer (RPC) shown in *Figure 8.30.* is a special item of computer hardware optimised for high speed execution of complex algorithms. It is usually based on advanced DSP components although high performance general purpose processors are the basis for the system described in Reference 51. The RPC is in turn connected to the engine management hardware through a high speed link, resulting in the arrangement illustrated in *Figure 8.30.*

The engineer will develop one or a small number of specialist algorithms in the fast hardware system. Previously built in a general purpose controls environment such as Simulink, ASCET-SD, or Systembuild, the algorithm will be pre-processed and converted directly into machine code for the RPC to execute. The RPC by-passes the default control algorithms contained in the engine ECM and both its input data are directly obtained from the ECM interfaces. Such a structure enables a very fast turnaround on algorithm assessment and allows options to be tested and analysed in rapid succession.

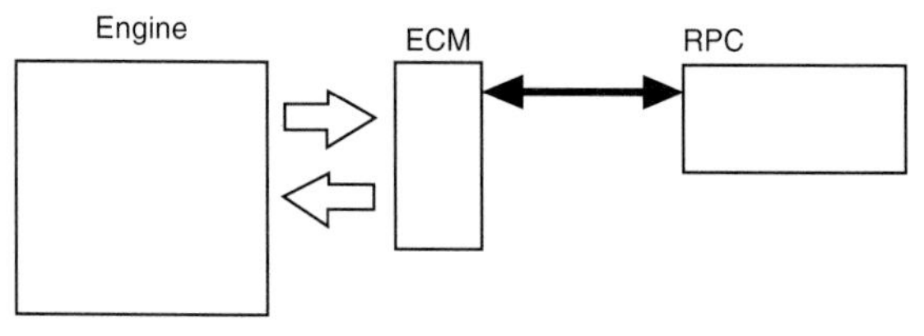

Figure 8.30 The links between engine, ECM and the rapid prototyping computer

The HIL approach requires a corresponding discipline including planning of the test program to optimise experiment time, and a configuration management that matches the algorithm with the test results.

References

1 COOK, J. A. and POWELL, B. K., "Modelling of an Internal Combustion Engine for Control Analysis", *IEEE Control Systems Magazine,* August 20–26 (1988)

2 NORTON, J. P., *An Introduction to Identification,* Academic Press (1986)

3 RICHARDS, R. J., *An Introduction to Dyanamics and Control,* Longman (1979)

4 FRANKLIN, G. F., POWELL, J. D. and EMAMI-NAEINI, A., *Feedback Control of Dynamic Systems,* ADDISON-WESLEY (1986)

5 ASTROM, K. J. and WITTENMARK, B., *Computer Controlled Systems,* Prentice-Hall (1984) 24 November 1997

6 BOWNS, D. E., "The Dynamic Transfer Characteristics of Reciprocating Engines", *Proc. Mech. Engrs,* **185** (1971)

7 HAZELL, P. A. and FLOWER, J. O., "Sampled-data theory applied to the modelling and control analysis of compression ignition engines—Part 1", *Int. J. Control,* 13, No3, 549–562 (1971)

8 HAZELL, P. A. and FLOWER, J. O., "Sampled-data theory applied to the modelling and control analysis of compression ignition engines—Part 2", *Int. J. Control,* 13, No4, 609–623 (1971)

9 WELLSTEAD, P. E., THIRUAROORAN, C. and WINTERBONE, D. E., "Identification of a Turbocharged Diesel Engine", *Proceedings of the 7th IFAC World Congress,* pp. 361–377 (1978)

10 KURAOKA, H. OHKA, N., OHBA, M., HOSOE, S. and ZHANG, F., Application of H-Infinity Design to Automotive Fuel Control, *IEEE Control Systems Magazine,* April, 102–106 (1990)

11 HENDRICKS, E., VESTERHOLM, T. and SORENSON. S., "Nonlinear, Closed Loop, SI Engine Observers", *SAE 920237* (1992)

12 WESTBROOK, M.H., "Automotive transducers: an overview", *Proc. IEE,* **135,** Pt D, No 5, September, 339–347 (1988)

13 CHALLEN, B. J., "Some Diesel Engine Sensors", *SAE 871628* (1987)

14 ANASTASIA, C. N. and PESTANA, G. W., "A cylinder pressure sensor for closed loop engine control", *SAE 870288* (1987)

15 HERDEN, W. and KÜSSEL, M., "A New Combustion Pressure Sensor for Advanced Engine Management", *SAE 940379,* in SP-1029 (1994)

16 WLODARCZYK, M. T., "Enabling Controlled Combustion Engines: Embedded In-Cylinder Fiber Optic Pressure Sensor", *Engine Technology International 1997*

17 KATO, N. NAGAKAKI, K. and INA, N., "Thick Film, ZrO_2 NOx Sensor", *SAE 960334,* in SP-1149, (1996)

18 NUTTON, D. and PINNOCK, R. A., "Closed Loop Ignition and Fuelling Control Using Optical Combustion Sensors", *SAE 900486* (1990)

19 HOTTEL, H. C. and BROUGHTON F. P., "Determination of true flame temperature and total radiation from luminous gas flames", *Industrial and Engineering Chemistry, Analytical Edition,* Vol 4 (1932)

20 ERKISSON, LARS, NIELSEN, L., and NYTOMT, J., "Ignition Control by Ionisation Current Control", *SAE 960045,* in SP-1149 (1996)

21 REINMANN R., SAITZKOFF, A. and MAUSS, F., "Local Air Fuel Ratio Measurements Using the Spark Plug as an Ionisation Sensor", *SAE 97856,* in SP-1236 (1997)

22 BOEHNER, W. and HUMMEL, K., "Common Rail Injection System for Commercial Diesel Vehicles", *SAE 970345,* in SP-1219, Fuel Spray Studies

23 FELTON, G. N., "High Pressure Rotary Spill Pump with Electronic

Control, *Proceedings of a Seminar on Diesel Fuel Injection Systems,* Institution of Mechanical Engineers, London (1985)
24 PATEL, R. V. and MUNRO, N., *Multi-variable System Theory and Design,* Pergamon Press (1982)
25 HARLAND, G. E. and GILL K. F., Design of a Model-Reference Adaptive Control for an Internal Combustion Engine, *Transactions of the Institute of Measurement and Control,* **6,** 167–173 (1973)
26 NOBLE, A. D., BEAUMONT, A. J. and MERCER, A. S., "Predictive Control Applied to Transient Engine Testbeds, *SAE 880487* (1988)
27 CLARKE, D. W., MOHTADI, C. and TUFFS, P. S., Generalized Predictive Control, Parts 1 and 2, *Automatica,* **23**, 137–160
28 FRANCIS, B. A., "A course in H-infinity Control Theory", Vol 88 in *Lecture Notes in Control and Information Sciences* (1987)
29 BOYLE, J. C., FRANCIS, B. A. and TANNENBAUM, A . R., *Feedback Control Theory,* Macmillan (1992)
30 KOSKO, B., *Neural Networks and Fuzzy Systems*, Prentice-Hall International (1992)
31 MURYAMA, Y., TERANO, T., MASUI, S. and AKIYAMA, N., "Optimising Control of a Diesel Engine", from *Industrial Applications of Fuzzy Control,* edited by M. SUGENO, Elsevier Science Publications (1985)
32 UTKIN, V. I., *Sliding Modes and their Application in Variable Structure Systems,* MIR, Moscow, (1978)
33 WANG, Y., KRISHNASWANI, V. and RIZZONI, G., "Event Based Estimation of Indicated Torque for IC Engines using Sliding Mode Observers", *Proceeding of the 1996 IFAC World Congress,* Vol Q, pp. 105–110, SAN FRANCISCO, July (1996)
34 CARNEVALE, C., COIN, D. SECCO, M. and TUBETTIM P., "A/F ratio control with sliding mode technique", *SAE 950838,* in SP-1086 (1995)
35 DRAPER, C. S. and LI, Y., *Principles of Optimising Control Systems,* ASME Publications (1954).
36 SCOTSON, P. G. and WELLSTEAD, P. E., "Self Tuning Optimisation of Spark Ignition Automotive Engines", *IEEE Control Systems Magazine,* April 94–101 (1990)
37 HEBBALE, K. V. and GHOEIM, Y. A., "A Speed and Acceleration Algorithm for Powertrain Control, *Proceeding of the American Control Conference,* Vol. 1, pp. 415–420, Boston (1991)
38 WILLSKY, A. S., "A Survey of Design Methods for Failure Detection in Dynamic Systems", *Automatica,* **12,** 601–611 (1978)
39 SWEET, L. M., "Automotive Applications of Modern Control Theory", *SAE 820913* (1982)
40 BRANDSTETTER, M., "Robust Air-Fuel Ratio Control for Combustion Engines", PhD thesis (1996)
41 BENNINGTON, H. D., and Production of Large Computer Systems, *Proceedings ONR, Symposium on Advanced Programming Methods for Digital Computers,* June (1956)
42 *The STARTS Guide to methods and software tools for the construction of large real time systems,* NCC Publications, Hobbs Southampton (1987)
43 BOEHM, B. W., A Spiral Model of Software Development and Enhancement, *IEEE Computer,* May (1988)
44 HENDRICKS, E., JENSEN, M., CHEVALIER, A. and WESTERHOLM T., "Conventional Event Based Engine Control", *SAE 940377,* in SP-1029 (1994)
45 HATLEY, D. and PIRBHAI, I., *Strategies for Real Time Systems Development,* Dorset House (1988)
46 STOREY, N., *Safety-Critical Computer Systems,* ADDISON-WESLEY, London (1996)
47 *FMEA "Cookbook"*
48 WEEKS, R. W. and MOSKWA, J. J., "Automotive Engine Modelling for Real Time Control Using Matlab/Simulin", *SAE 950417* (1995)
49 JENSEN, J-P., KRISTENSEN, A. F., SORENSEN, S. C., HOUBAK, N. and HENDRICKS, E., "Mean value Modelling or a Small Turbocharged Diesel Engine", *SAE 910074* (1991)
50 RACHID, A., LIAZID, A. and CHAMPOUSSIN, J. C., "Non-linear Modelling of a Turbocharged Diesel Engine"
51 CHALLEN, B., STOBART, R., MAY, A. and MOREL, T., Modelling for Diesel Engine Control: the CPower Environment. Paper to be presented at the SAE Congress, (1998)
52 LUKICH, MICHAEL, S., "Rapid Prototyping of Embedded Systems: 1997 Update", *1997 Earthmoving Industry Conference,* Peoria, IL, SAE 97EIC-17 (1997)

Part 3

Engine sub-systems

9 Inlet and exhaust systems

Contents

9.1 Introduction

Since air breathing capacity and utilization determine the output of the diesel engine, the flow characteristics of the intake and exhaust systems are crucial in the achievement of good performance. In addition, the amplitude of the air pressure pulsations of the reciprocating engine must be attenuated in order to achieve reasonable levels of intake and exhaust noise.

The design and development of effective intake and exhaust systems remains an engineering challenge. Routinely intake and exhaust external flow systems are now designed and optimized using various design analysis software systems. These are based on various mathematical approaches, including complete one-dimensional solution of the wave equation for engine simulation. Complete three-dimensional calculations (computational fluid dynamics solutions) are used locally for optimization of elements such as exhaust downpipes, catalyst connections and also to determine optimum designs for internal components such as ports.

The intake ports of a diesel engine may have to generate air motion ('swirl') in the cylinder so that the air charge retains significant angular motion when the fuel is later injected, improving air–fuel mixing. Most engines rely on some swirl but modern high-speed direct injection engines require only low levels of swirl, much of the air–fuel mixing energy coming from the fuel spray.

Volumetric efficiency is determined by the complete flow system calling for care in selection and application of all components including intake and exhaust silencers.

Four-stroke engines have employed sleeve valves and rotary valves but it is the poppet valve that is most common. Two-stroke engines use a variety of airflow controls mostly piston-covered ports for scavenging, with valve-actuated exhaust systems in higher ratings.

This chapter covers mainly the design of poppet valve systems and piston-covered ports for two-stroke engines. Many of the design calculations covered here are often computed today, but the fundamentals remain unchanged. A section on factors affecting silencer design and selection is included.

9.2 Gas flow

Assuming an incompressible fluid the fundamental equation for fluid flow, derived from Bernoulli's equation is

$$v = c\sqrt{2gh} \tag{9.1}$$

where v = Velocity of flow in m/sec;
c = Coefficient of discharge;
g = Gravitation acceleration = 9.80665 m/sec^2;
h = Head of fluid causing the flow in metres.

Further if
W = Mass of fluid flowing kg/sec;
A = Area through which the fluid flows (m^2);
ρ = Density of the flowing fluid (kg/m^3).

$$\text{Then } W = \rho A v = \rho A c\sqrt{2gh} \tag{9.2}$$

For the present discussion the fluids involved in measurements are air, water, mercury. For these the densities at room temperature taken to be 20°C (293 K) are:

Air (dry)	1.205 kg/m^3
Water	998.2 kg/m^3
Mercury	13554 kg/m^3

In many measurements of air flow it is convenient to balance the pressure, or in some cases the pressure drop across an orifice, against a head of water, sometimes mercury for larger pressures, contained in a manometer.

Since pressure = ρh

$$h_{air} = h_w \frac{\rho_w}{\rho_{air}} = \frac{998.2}{1.205} h_w = 828.38\, h_w$$

If for convenience h_w is measured in mm (H_2O). Then, h_{air}(m) = 828.38 × $10^{-3} h_w$. So eqn (9.1) becomes

$$v = c\sqrt{2 \times 9.807 \times 828.38 \times 10^{-3}\, h_w}$$

$$\text{i.e. } v = 4.031\, c\sqrt{h_w\,(\text{mm})}\ \text{m/s} \tag{9.3}$$

In many cases the flowing air may not be at room temperature, or even pressure so its true pressure head must be balanced against a head of water at 20°C.

$$\rho_{air(t,p.)} = \rho_{air}\ (20°\text{C, 760 mm Hg}) \times \frac{p(\text{mm Hg})}{760} \times \frac{293}{(t + 273)}$$

$$= 0.3855\,\rho\,(20°\text{C, 760 mm Hg}) \times \frac{p(\text{mm Hg})}{(t + 273)} \tag{9.4}$$

The above is valid for small pressure differences such as the pressure drop across the inlet valve of a naturally aspirated diesel engine. Where higher pressures are involved a more complicated expression allowing for compressibility must be used.

In the four-stroke engine the pressure drop across the inlet valve is created by the section generated by the descending piston in the case of the naturally aspirated version. When the engine is supercharged the boost pressure is added usually by a turbocharger.

By contrast, the piston in relation to the ports gives a varying port area only in the two-stroke engine. With the selection of suitable timings the cylinder pressure will have dropped to scavenge belt pressure or below by the time the scavenge ports open The positive pressure supplied to the scavenge belt will cause the air to flow into the cylinder displacing the residual exhaust products from the previous cycle into the exhaust. Inevitably, the scavenging process is not fully efficient so it is usual to supply more than one cylinder swept volume, usually about 1.2 SV, of scavenge air per cycle. Due to the different processes involved the four-stroke and two-stroke cases will be dealt with separately.

9.3 Four-stroke engines

In the case of the diesel engine two general types of inlet ports need to be considered. One type is used on various prechamber engines in which swirl required for the mixing of air and fuel to obtain good combustion is generated during the compression stroke by forcing the trapped air charge through a throat or series of holes into the combustion chamber. In this case the inlet valve and port is designed to give the freest possible entry to the air, and timings are chosen to obtain the highest practical volumetric efficiency.

Another type of inlet port is used on the direct injection engine, where the fuel is sprayed directly into the cylinder, the piston having a central circular combustion chamber formed in it. The inlet port form must be designed to cause the ingoing air to rotate or 'swirl' about the cylinder axis. The degree of swirl needed varies with the size and type of combustion chamber employed which is interrelated to the number of nozzle sprays used. The problem is therefore to generate the required degree of swirl needed for efficient combustion with a minimum loss of pressure through the inlet port and valve. Combustion performance levels have increased over the years largely through increased injection pressures. This allows a reduction in the swirl level of the intake air charge. The reduced pressure losses in the intake port that this allows also aid improved performance.

The first step is to decide the largest valve sizes that can be accommodated whilst maintaining adequate cooling between them and core thicknesses which permit sound castings. Practically all diesel engines use vertical valves due to the need to have very close poston-to-head clearances (about 10% of the stroke), in order to have a compact combustion chamber and the necessary high compression ratio needed for good cold starting and light load high speed operation. It is usual to make exhaust valves slightly smaller in diameter than the inlet ones since a slightly higher pressure drop is permissible without penalizing engine breathing and output.

For two vertical valves which are able to operate within the cylinder bore D without interference the maximum inlet port diameter, i.e. the throat diameter at the minimum cylinder head port diameter, is from 0.43–0.46D diameter and the corresponding exhaust port size 0.35–0.375D diameter.

Four valve heads were once the prerogative of larger engines, but are now used for well below 1 litre/cylinder high-speed direct injection engines, primarily to preserve the central location for the fuel injector. In these the maximum intake throat typically is around 0.32D, while for the exhaust up to 0.30D can be used. With the use 45° valve seats the valve head diameter is about 1.09 times the throat diameter d, compared with 1.11d when 30° seats are used.

For many years the mean inlet gas velocity has been used as a 'rule-of-thumb' assessment of maximum operating speed. Typical values at maximum power output being 76 m/s for automotive types and up to about 52 m/s for the larger stationary engine.

If the stroke is L metres and the engine crankshaft speed N rev/s then mean piston speed

$$v_p = 2LN \text{ m/s} \tag{9.5}$$

So with an inlet throat diameter of dm and a cylinder bore diameter of D the mean inlet valve gas velocity

$$v_g = 2LN\frac{A_p}{A_v} = 2LN\left(\frac{D}{d}\right)^2 \text{ m/s} \tag{9.6}$$

Ideally the area represented by the valve stem diameter should be subtracted from the valve throat area but as its value is small it is usually neglected when making these empirical design comparisons. The valve stem diameter is typically 0.20–0.22 times the valve throat diameter.

To have the same nominal area as the valve throat, the valve should be lifted by a quarter of the valve throat diameter (d/4). In practice an attempt is made to make the maximum valve lift at least equal to 0.26d. This can usually be done on low speed engines, but valve train dynamics may make this difficult for push rod operated systems in small high speed automotive types.

Cams for modern engines almost entirely use some form of smooth acceleration curve, often a polynomial form in which there are no abrupt changes, or discontinuities. In this, the lift curve is assumed to have the form

$$y = c_o + c_p x_p + c_q x_q + c_r x_r + c_s x_s + \tag{9.7}$$

by successive differentiation the velocity, acceleration, jerk and jounce expressions are obtained which are then equated to the required end conditions. For example at full lift $x = 0$ and $y = L_T$ (lift above ramp); y' = velocity = 0; y'' = acceleration = A_N = maximum nose deceleration; and y''' = jerk = 0. When $x = 1$; $y = 0$, y' = ramp velocity; $y''= 0$; and $y''' = 0$. From these equations a range of constants can be calculated based on a series of values for the exponents. The selection of suitable exponents is bewildering at first but with experience various trends which assist selection become obvious. $x = 0$ at the full lift condition.

Having evaluated the constants corresponding to the exponents selected the lift ordinates are calculated from:

$$L_\theta = L_T + c_p\left(\frac{\theta}{\theta_r}\right)^p + c_q\left(\frac{\theta}{\theta_r}\right)^q + c_r\left(\frac{\theta}{\theta_r}\right)^r + \tag{9.8}$$

where θ_T is the half cam period, assuming a symmetrical cam.
θ is the angle within the half period for which the lift is being calculated.

The cam ordinates are obtained by dividing the valve lift by the rocker ratio.

For cams such as this the master cam must be generated by indexing and grinding to the required lit curve. In practice it is easier to start at the cam nose position and then rotate and machine to the drop from full lift figure.

A full treatment of valve gear and cam design is a very large subject beyond the scope of this chapter but references are appended for those wishing to study the subject.

Figure 9.1 shows a plot of relative valve opening area against cam angle for a valve opening to 0.25d maximum lift, with two ratios of maximum flank acceleration/maximum nose deceleration (A_F/A_N). The advantage of the higher opening flank acceleration can be seen. The lesson is that the valve should be opened as quickly as other aspects, usually dynamic problems, of the valve train design permit. Dotted on the same *Figure 9.1* is a lift curve for a valve having a 33% greater maximum lift. Obviously the lift in excess of 0.25d is valueless but the increase of area under the curve at lower lifts is valuable when valve gear dynamics permit.

9.3.1 Valve timings

At least a preliminary choice of valve timings must be made to design the space required to accommodate the cam tip to centre of rotation radius, the camshaft bearing diameter where the camshaft is inserted axially from one end of the engine, and the

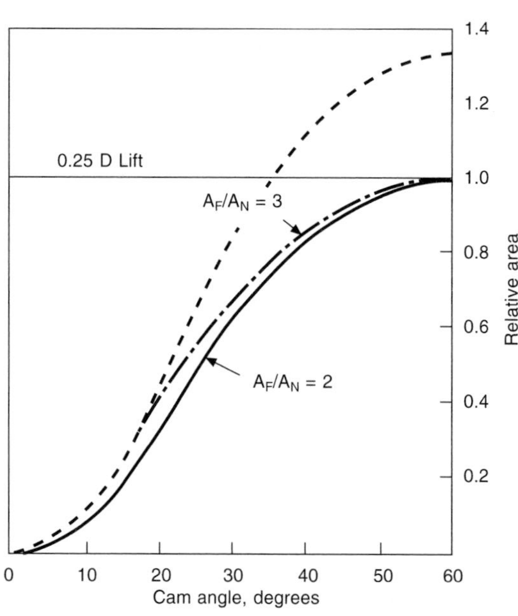

Figure 9.1 Area–lift curve-acceleration ratio and maximum lift

size of the cam followers. These sizes follow from an early cam design based on the desired valve period and maximum lift.

For naturally aspirated engines the piston-to-head clearance is only about 1.0% of the stroke in order to keep the 'lost volume' small at firing TDC. For this reason the small high speed diesel cannot use as much valve overlap in either period or area as is conventional for petrol engines. As a result the inlet valve cannot be opened earlier than about 12° BTDC (exhaust) and the exhaust valve closed later than 12° ATDC without the risk of the valves hitting the piston with flat topped pistons. The machining of cavities in the piston crown to accommodate the valve heads with wider timings distorts the combustion chamber too much to be acceptable in small engines.

Large direct injection engines use a larger diameter shallower combustion chamber cavity in the piston crown, partly because lower compression ratios are adequate for cold starting and partly because they use a lower swirl ratio if conjunction with a larger number of fuel injection sprays to obtain satisfactory combustion. The effect of machining room for the valves during the overlap period is less disastrous for combustion. Indeed with large turbocharged engines wide overlap periods of up to 150° crank are used to pass scavenge air through the cylinder giving some cooling to the cylinder head and valves but in particular keeping the mean operating temperature of the turbocharger down to reduce thermal stresses and creep in the turbine rotor.

In small high speed automotive engines the selection of inlet closing timing is a compromise between making as much use of inertia ram in the intake system as possible at the top speeds, and maintaining a high effective compression ratio at low starting speeds to assist cold starting. In general the inlet valve closing will occur at about 40–45° ABDC for this type of engine. In the larger slower speed engines the inlet closing tends to be rather earlier, around 35° ABDC.

The exhaust valve opening should ideally occur at a timing such that atmospheric pressure is reached within the cylinder at an equal number of degrees after BDC as the exhaust opening occurred before BDC. If this is done the work lost on the expansion stroke and the negative work done during the early part of the exhaust stroke are minimized. For an all speed engine such as used in road vehicles a compromise timing must be used anyway. In practice the exhaust opening for the small high speed engine is usually 55–45° BBDC.

Similar timings for the exhaust opening are usually adopted for engines which are moderately turbocharged. The actual timings are often modified to get the best effect from exhaust pulses used to improve the exhaust turbine work. As the boost pressure is increased, more work is required for the exhaust turbine of the turbocharger. The tendency is to open the exhaust valve earlier to increase the pressure drop across the exhaust turbine. Whilst some engine cylinder expansion work is sacrificed the higher boost pressure obtainable as a consequence benefits the overall engine output. This is discussed in considerable detail in Chapter 2. Alternative exhaust pipe arrangements to suit pulse turbocharging, the use of pulse converters and constant pressure with the highest boosts are also described.

When turbocharging is used it is advantageous to use valve overlap partly to reduce the turbine working temperature and partly to help in the cooling of the exhaust valve(s) and the cylinder head. The greater the boost used the more important this becomes. As previously stated, it is difficult to provide significant valve overlap in the smallest high speed engines, but the use of lower compression ratios with a larger volume direct injection piston cavity combustion chamber, permissible in larger engines, allows cutaways to be machined in the piston crown without seriously affecting combustion. Some overlap can be provided in truck engines, but overlap periods of 140–160° are often used for large stationary and locomotive engiens, see *Figure 9.2*.

Cams have been designed in which the exhaust valve almost closes at exhaust TDC and then partly reopens again before finally closing in order to minimize the machining of the piston crown to avoid valve/piston interference. With this approach the inlet valve is opened early, partly closed until the piston starts to descend and then finally lifted to its normal full lift.

As explained in other chapters, much effort has been put into modelling the complete engine working cycle including the complete gas flow phenomena in inlet and exhaust systems, using computers. Whilst not yet completely satisfactory, fair agreement is now possible between predicted and measured pressures within the cylinder and for the pressure fluctuations in the inlet and exhaust systems. In any case it is still well worth taking cylinder pressure diagrams during prototype testing to optimize valve timings and hence engine performance.

All the above timings are the nominal top of the quietening ramp values. Ramps are used to ensure that valve opening and closing takes place at a known, usually constant, velocity. They allow some margin for wear, for wind-up in the valve train due to its flexibility and for changing clearances due to relative expansions or contractions during transient speed and load changes. Ramp heights should not be excessive to avoid possible valve burning at exhaust opening, and excessive gas leakage at the inlet valve closure during cold starting.

9.3.2 Valve areas

The sizes of valves which can be accommodated and the fact that 45° seat angles are normally used has already been discussed. A few cases use a 30° valve seat angle. Before dealing with valve port and valve flow coefficients a few words must be said about methods in use for defining the valve area.

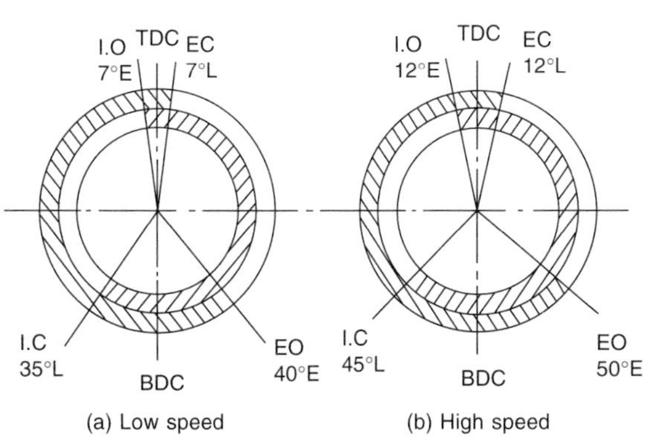

(a) Low speed (b) High speed

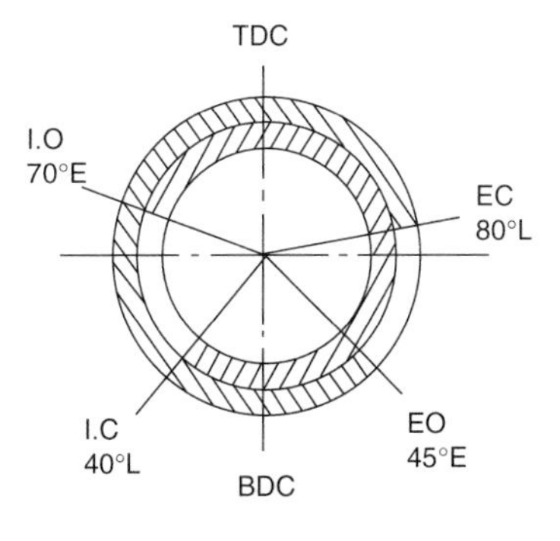

(c) Medium speed—turbocharged

Figure 9.2 Typical valve timings: (a) Low speed; (b) High speed; (c) Large engine—turbocharged

Figure 9.3 shows a poppet valve lifted from its seat. At low lifts, shown on the right, the flow area normal to the seat faces is that of the surface for part of a frustrum of a cone. The diameters for the cone are d_1 and d_L. Then XY is the slant height

$$XY = L \cos \alpha = S \text{ (say)} \tag{9.9}$$

where L = Valve lift;
α = Angle of valve seat.

Now

$$d_L = d_1 + 2XZ = d_1 + 2XY \sin \alpha$$
$$= d_1 + 2L \cos \alpha \sin \alpha \tag{9.10}$$

The surface area of a right circular cone is

$$A = \frac{\pi}{2} S(d_1 + d_L) \tag{9.11}$$

This gives the valve opening area per valve

$$A_v = \pi L \cos \alpha(d_1 + L \cos \alpha \sin \alpha) \tag{9.12}$$

$$= \pi d_1^2 \frac{L}{d_1} \cos \alpha \left(1 + \frac{L}{d_1} \cos \alpha \sin \alpha\right)$$

$$= \pi d_1^2 \cos \alpha \left(1 + \frac{L}{d_1} \cos \alpha \sin \alpha\right) \frac{L}{d_1} \tag{9.12a}$$

This last equation has been derived since it is useful when comparing flows on an L/d_1 basis, as a non-dimensional lift, which is convenient when comparing flow characteristics for differing valves and ports.

For valves having a 45° seat angle eqn (9.12a) becomes:

$$A_v = \pi d_1^2 \left(0.7071 + 0.35354 \frac{L}{d_1}\right) \frac{L}{d_1} \tag{9.12b}$$

For a 30° seat angle

$$A_v = \pi d_1^2 \left(0.866 + 0.375 \frac{L}{d_1}\right) \frac{L}{d_1} \tag{9.12c}$$

If a valve stem diameter of $0.21d$ is used the net port area without the valve head is

$$\frac{\pi}{4} d_1^2 - (0.21d_1)^2 = 0.2390 n d_1^2 \tag{9.13}$$

Equating this to eqn (9.12b)

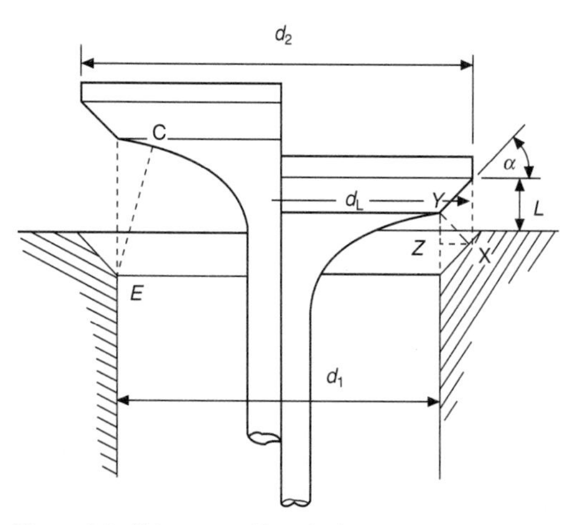

Figure 9.3 Valve area with conical seats

$$\pi d_1^2 \left(0.7071 + 0.3536 \frac{L}{d_1}\right) \frac{L}{d_1} = 0.239 \pi d_1^2$$

This occurs when $L/D = 0.2946$ with 45° seats
and $L/D = 0.2491$ with 30° seats

Often the valve stem area is neglected in which case eqn (9.13) becomes equal to $\pi/4d_1^2$.

So for 45° seat angle this occurs when $L/D = 0.307d_1$
and for 30° seat angle this occurs when $L/D = 0.259d_1$

However when the valve lift becomes such that the perpendicular from the inner corner of the valve head seat, Y, falls outside of the outer diameter of the stationary head seat the opening area is no longer that of the frustrum surface and, as indicated in the left hand side *Figure 9.3*, a line EC must be found for which the flow area at the lift concerned is the minimum.

In practice when flow coefficients are being determined, the frustrum area is often used beyond the lift for which its truth is strictly accurate, the associated coefficients being calculated on this assumption.

Two other approaches to nominal area are in use. The first is to assume that the nominal flow area is the vertical collar area based on the valve throat diameter, i.e. $\pi d_1 L_v$ for the lift L_v being considered. The second, usually adopted in the USA, is a similar vertical collar area based on the valve head diameter d_2, i.e. the area is $\pi d_2 L_v$. (Yet another approach is merely to use the valve throat area as the reference area.) For this reason the experimental details and the basis for the nominal area adopted should be checked before any published values are assumed.

9.3.3 Determination of flow coefficients

Measurements may be made on existing cylinder heads and ports, but more often they are made on accurately made wooden or plastic models at the design stage when the coefficients can be determined and experimental changes made to improve the flow characteristics. This is particularly important for direct injection engines when the ports are shaped to generate the required degree of swirl within the cylinder. Swirl generation will be discussed later; but flow tests are required to obtain the necessary degree of swirl for minimum pressure loss, since this will penalize the engine's volumetric efficiency and hence power output.

Figure 9.4 shows the schematic arrangement of the flow rig used by Ricardo for both flow measurement and for the measurement of swirl. Essentially it consists of a centrifugal fan, shown in the lower half of the figure, whose speed can be varied. Arrangements are made so that either the fan entry can be used to suck air through the model, when required with the air sucked through a viscous flow air meter which is equipped with the necessary manometers and flow trimming valves fitted on its downstream side before the fan eye is reached.

Alternatively, and in fact the method normally used, the fan delivers air, using flow trimming valves, through an air cleaner into a viscus flow air meter and then on through a plastic convoluted pipe to an air box. Within this air box are eight layers of expanded metal to eliminate any directional effects from the supply pipe. This pressure air box is fitted with a manometer downstream of the expanded metal so that the required pressure for blowing through the port to be tested can be monitored.

The cylinder head or wooden model is fitted to the outlet of the pressure box usually with a bell mouthed block interposed having the smaller diameter arranged to line up with and match accurately the entry end of the head port. At times an actual inlet manifold may be used instead when its shape is such that it could affect swirl generation. Normally the cylinder head is

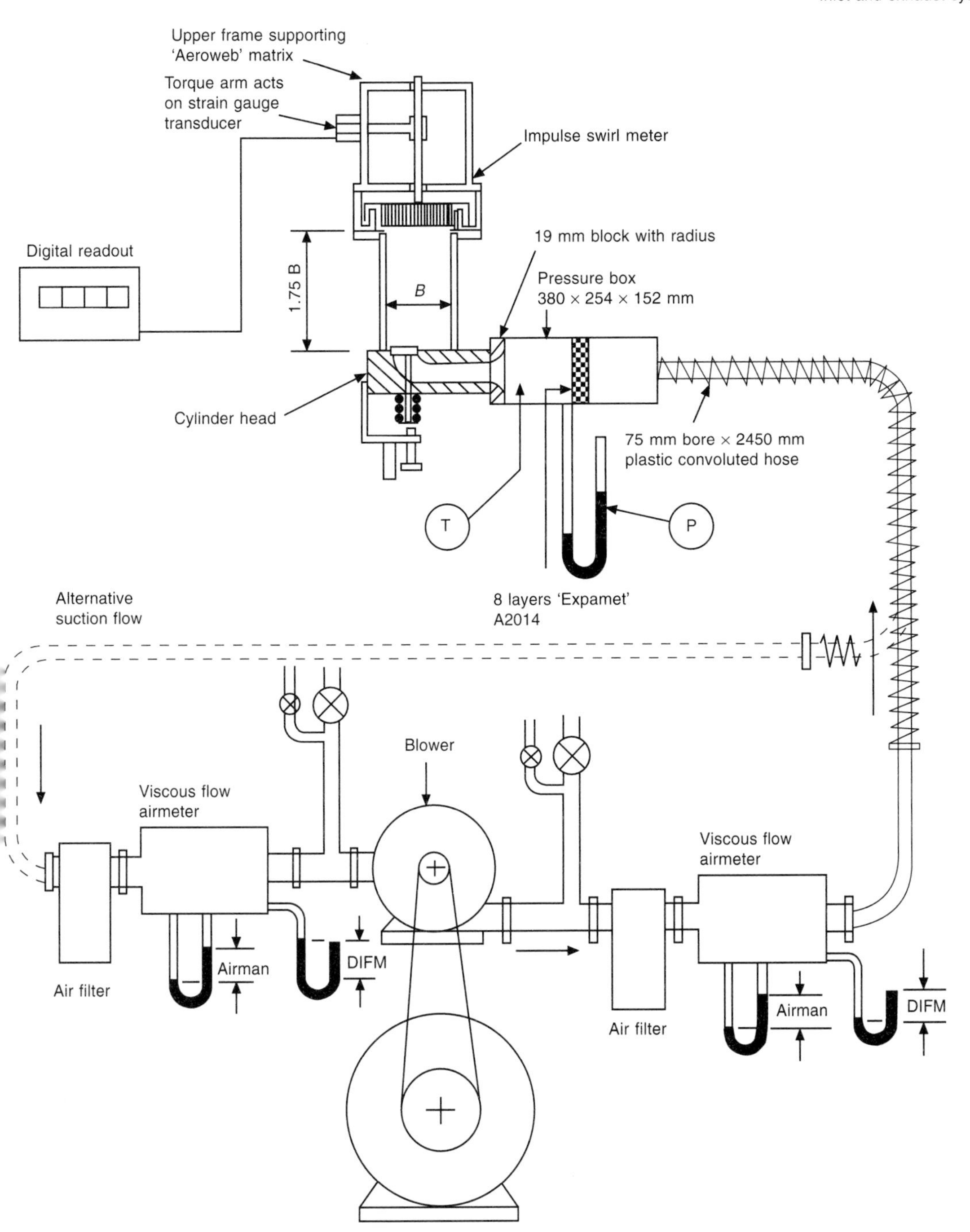

Figure 9.4 Ricardo flow and swirl measuring rig

inverted for convenience, and a cylinder liner of the correct diameter for the engine for which the head has been designed clamped to its top and correctly orientated with respect to the valve(s). The test liner is made 1.75 times the cylinder bore long. When merely determining flow resistance the impulse swirl meter is removed.

When testing inlet ports, air is blown from the pressure box through the port and valve into the cylinder liner. When flowing exhaust valve and port configurations, air is sucked from the cylinder through the valve into the pressure box. Means are provided whereby the valve under test, fitted with a light spring, is lifted in a series of accurately determined lifts from its seat up to the maximum designed lift. At each lift the flow from the fan, or to it for exhaust valves, is adjusted so that the pressure box pressure is maintained at 254 mm of water above atmospheric for an inlet port, and below in the case of an exhaust valve and port. With the flow steady the air meter pressure drop, downstream pressure and temperature are measured. Thus the flow is determined. Of course any other accurate means for measuring the air flow can be used as an alternative to the viscous flow air meter whose chief advantage is that the pressure drop is linearly proportional to the air flow. By contrast an orifice plate has a square law variation with flow. The volume flows are corrected to the pressure air box conditions.

For a given valve and cylinder head port design it is possible to measure the air passing through the whole port/valve combination at a range of lifts from zero to the full designed lift. In order to make comparisons for differing designs and engine sizes these lift/flow figures are often converted into flow coefficients which are plotted against a non-dimensional lift, which is merely the actual lift divided by the valve throat diameter.

The flow coefficients given on the typical curves shown here are based on the valve throat area.

$$\text{Flow Coefficient} = \frac{Q}{AV_0} = C_F \tag{9.14}$$

where A = Valve throat area = $\pi/4D^2$ (m^2)
Q = Measured quantity of air flow (m^3/s)
V_0 = Theoretical air velocity = $\sqrt{2\Delta_p/\rho}$ (m/s)
Δ_p = Pressure drop across valve (N/m^2)
Tests were normally with Δ_p = 2491 N/m^2 (≡ 254 mm water gauge) which makes V_0 = 64.4 m/s (approx.)
ρ = Air density at pressure box conditions

Where more than one valve is used the total throat area should be used, i.e.

$$A = n\frac{\pi}{d}D^2 \text{ (m}^2\text{)}$$

Where the Discharge Coefficient, C_D, is referred to, this is defined as:

$$\text{Discharge Coefficient} = C_D = \frac{Q}{A_v V_0} \tag{9.15}$$

Where A_v = Conical frustrum flow area as given by eqn (9.12).

Assuming, for the ideal case, that C_D = 1, then at an L/D ratio of 0.125 C_F becomes 0.456 for 30° valve seats or 0.376 for 45° valve seats.

Figure 9.5 shows a typical curve for flow coefficient against L/D ratio for an exhaust valve and port whilst *Figure 9.6* shows a similar curve for a non-swirling inlet port.

For the flow velocity formula given in eqn (9.15) to be true the air flow must be fully turbulent. *Figure 9.7* indicates the required velocity head against inner seat diameter to ensure that the air flow is in fact turbulent.

Again the simple formula for non-compressible flow has so far been given (eqn 6.1). This can be shown to be equal to

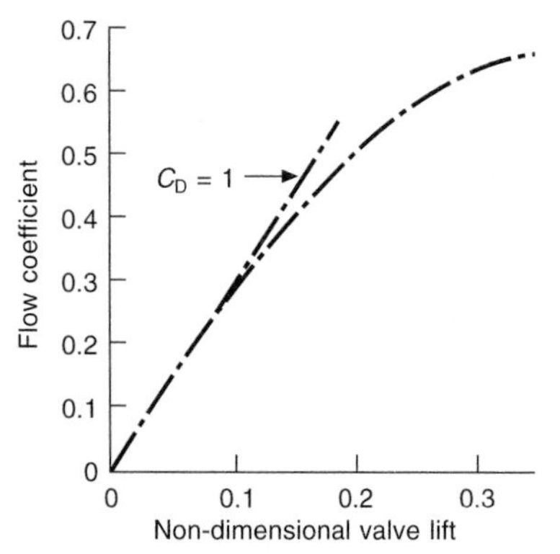

Figure 9.5 Flow coefficient *vs L/D* ratio for an exhaust valve and port

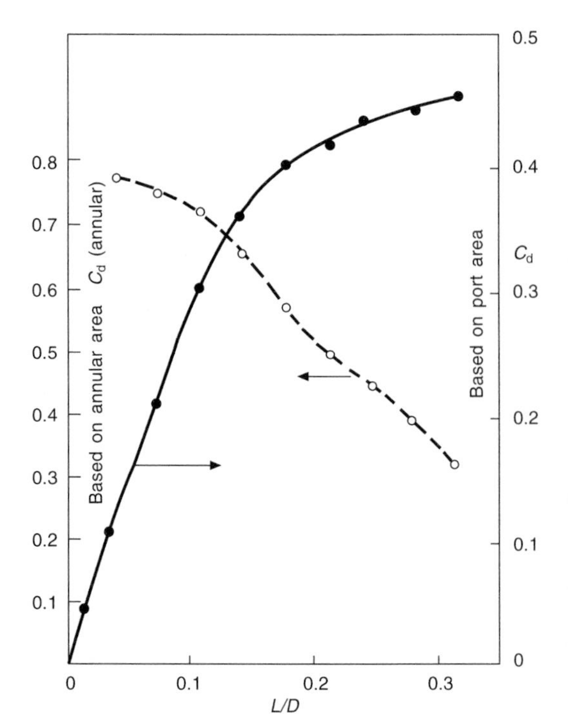

Figure 9.6 Flow coefficient *vs L/D* ratio or a non-swirling inlet valve and port

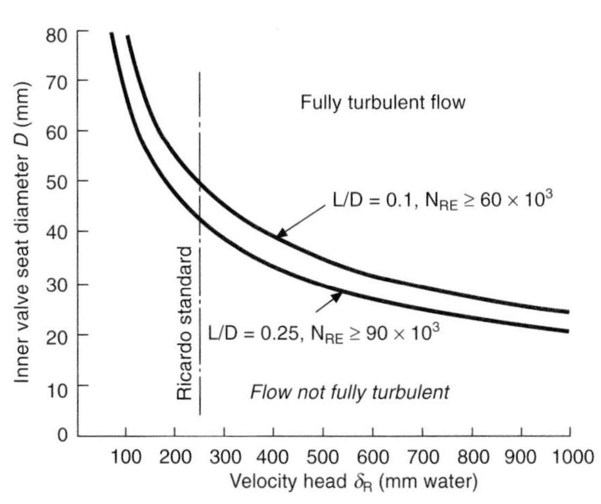

Figure 9.7 Limits of full turbulent flow in IC engine ports

$$V_0 = \sqrt{2RT_0\frac{(r-1)}{r}} \tag{9.16}$$

For compressible, frictionless, flow through a simple orifice

$$V_2^1 = \sqrt{\frac{2\gamma RT_0}{\gamma - 1}\left\{1 - \left(\frac{P_2}{P_1}\right)\frac{\gamma-1}{\gamma}\right\}} \tag{9.17}$$

where T_0 = Upstream temperature (stagnation) (K)
P_2 = Downstream static pressure (N/m^2)
P_1 = Upstream stagnation pressure (N/m^2)
R = Gas constant (J/kg K)
γ = Ratio of specific heats = 1.4 for air
$r = P_1/P_2$

It can be shown that for the 254 mm pressure drop used for tests the use of the simple formula underestimates the true value of C_F by 1.3%.

9.3.4 Engine breathing demands

As mentioned earlier the size of inlet valve has been used to determine whether the gas speed will permit successful operation up to the desired maximum output speed. This oversimplifies the problem by overlooking the valve lift characteristics, the fact that the air temperature at the inlet port may not be atmospheric, and that swirl ports have a higher resistance to flow. For this reason it is becoming usual to use actual flow coefficients in a more sophisticated formula. This basic approach was originally used by Livergood, Rogowski and Taylor.

The first step is to replot the flow coefficients determined on the blowing rig to create a curve of flow coefficient against the valve opening diagram in crank degrees or which a cam has been designed. The area under this curve divided by the base length will give a mean value for the flow coefficient for the valve (see *Figure 9.8*). Stated mathematically.

$$\text{Mean Flow Coefficient} = C_{F\text{mean}} = \frac{\int_{\theta_1}^{\theta_2} C_F d\theta}{\theta_2 - \theta_1} \tag{9.18}$$

This is then used to give a more realistic mean gas velocity and the answer is divided by the velocity of sound at the intake port temperature concerned.

Thus from eqn (9.6) the mean inlet gas velocity assuming the valve at full port area during the whole inlet stroke is given by

$$v = 2LN\left(\frac{D}{d}\right)^2 \text{ m/s}$$

Using the mean flow coefficient a better mean flow velocity becomes

$$v = \frac{2LN\left(\frac{D}{d}\right)^2}{nC_{F\text{mean}}}$$

here n is the number of inlet valves.

In practice it was found that better correlation of engine volumetric efficiencies was obtained if this improved inlet mean gas velocity is divided by the speed of sound in air. This answer, which is a species of Mach Number, has become known as the 'Gulp Factor'. If 'a' is the speed of sound at valve inlet conditions:

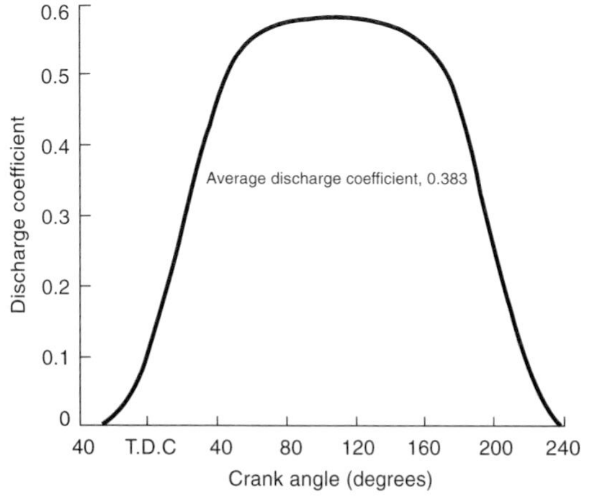

Figure 9.8 Flow coefficient *vs* crank angle curve

$$\text{Gulp factor} = \frac{2LN}{naC_{F\text{mean}}}\left(\frac{D}{d}\right)^2 \tag{9.19}$$

Whilst not a firm definitive figure, experience suggests that the Gulp Factor at rated speed should not be greater than 0.6.

9.3.5 Actual non-swirling port shapes

Exact details for inlet and exhaust ports depend greatly on details of the overall engine design and its size. However *Figure 9.9* indicates general details for a non-swirl inlet port for a naturally aspirated high speed diesel engine using a single valve, whilst *Figure 9.10* shows a typical exhaust port.

Generally the ports in the cylinder head should be as short as is practicable. This avoids undue pick-up of heat from the coolant in the case of the inlet or the transfer of too much exhaust heat to the coolant in the case of the exhaust port. Inlet manifolds should have as generous an area as is possible, and the branches flared into the manifold as smoothly as possible to avoid any 'vena contracta' effects due to a sharp cornered sudden change in flow area.

The actual lengths of the branches from the manifold, or plenum to the inlet valve can have a marked effect on the shape of the volumetric efficiency curve over the speed range. *Figure 9.11* shows one set of test results for a naturally aspirated automotive diesel engine. The tuning of both inlet and exhaust pipe lengths to augment output particularly over a narrow speed range is well known and has been exploited in motor racing. It has been used to a limited extent on diesel engines and at least one patent exists for using tuned inlet pipes with a turbocharged engine.

9.3.6 Swirl producing ports

Swirl is used in direct injection engines to assist combustion by causing air to rotate past the fuel injection spray jets. Practical limitations of hole size, both from the production angle and liability to blockage in service, cause a smaller number of fuel jets to be used in the smaller sized engines than in large engines. Automotive and truck sized engines usually use 4-hole nozzles, whilst large marine engines use eight or even more.

As a consequence, the smaller engines are found to require a higher rate of air rotation in the combustion chamber than is the case for large ones. This is in spite of using small diameter piston cavities to speed up the rotation on the principle of a forced vortex during compression. As small engines require higher compression ratios to improve cold starting and light load operation, the use of the smaller piston cavity diameter is in fact a great help in obtaining the smaller combustion chamber volume required. In general then small engines must have a higher degree of swirl than large engines.

Originally the swirl was obtained by some form of obstruction at the valve end of the port such that the air only entered the cylinder through part of the valve circumference. One well-known means was the use of a mask on part of the inlet valve circumference which virtually stopped the air flow over that angle of the valve circumference it occluded. This required greater pressure drops for air flow and penalized volumetric efficiency at the higher speeds. In addition fatigue failures of the valve head often occurred as the result of the local change in head stiffness.

As engine speeds rose and greater output with clean exhaust was required, efforts were made to find means other than directional obstruction to generate swirl. This has mostly been achieved by careful detailed modifications to the whole inlet port tract. Broadly two general types are now used, the so-called 'directed port' and the vortex or helical type of port.

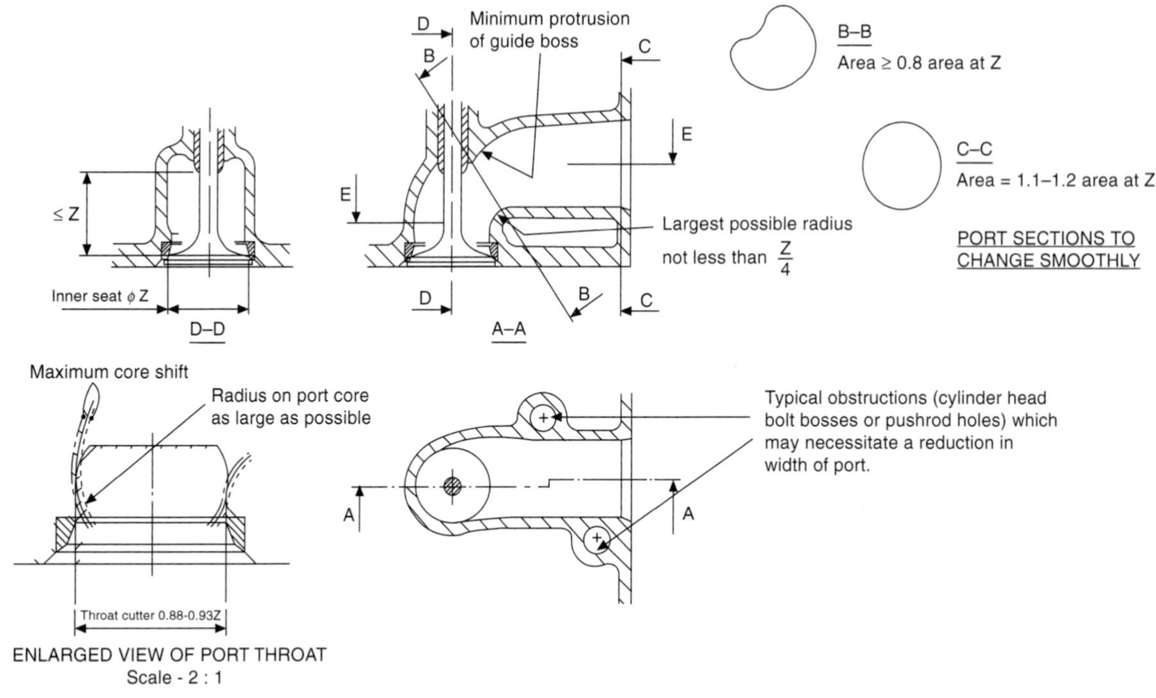

Figure 9.9 Non-swirling inlet port

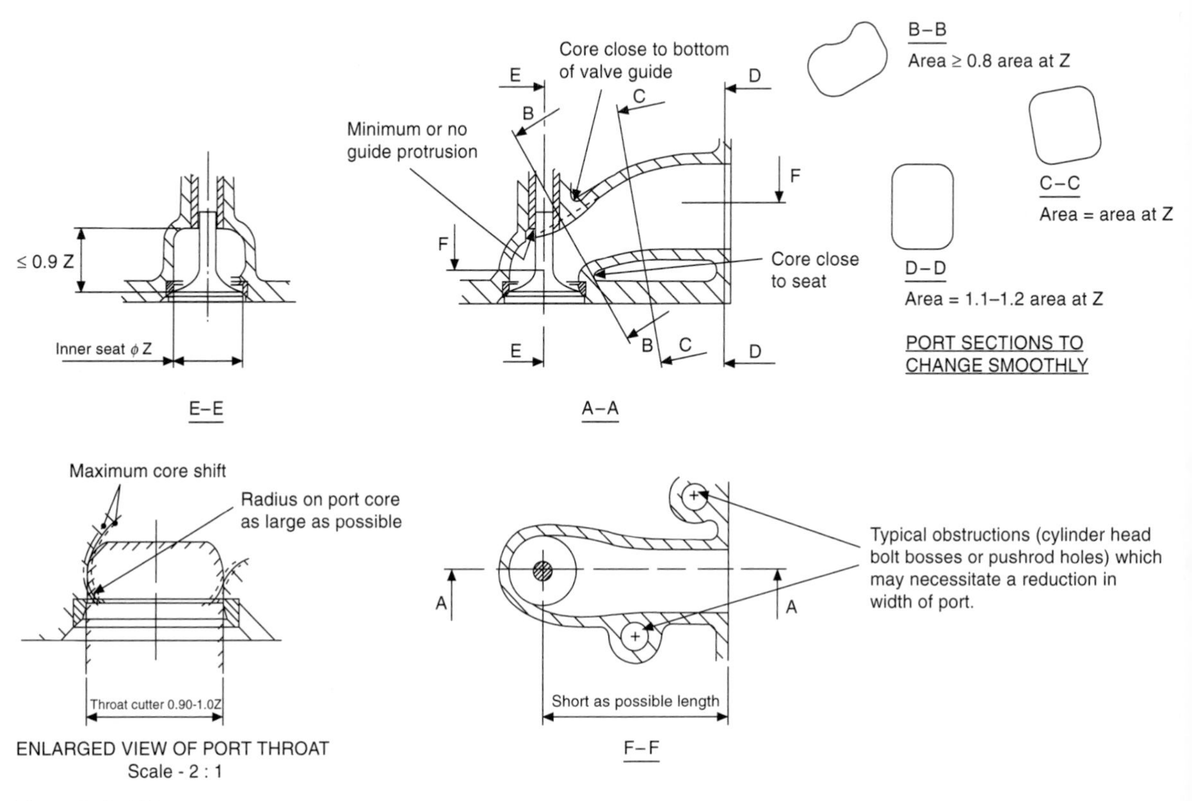

Figure 9.10 Exhaust port

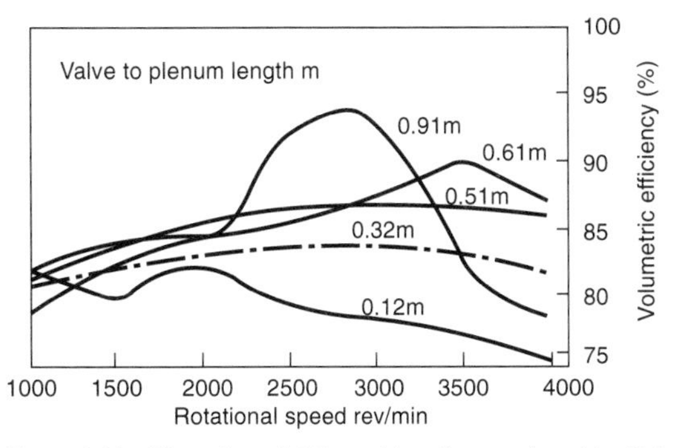

Figure 9.11 Effect of manifold branch lengths on volumetric efficiency over speed range

The directed port is arranged to cause the air to enter the cylinder at a shallow angle more or less symmetrically about the inlet valve stem and roughly normal to the radius joining the inlet valve and cylinder centrelines. The directed port is more readily used for a twin inlet valve head than the helical port. *Figure 9.12* shows semi-diagrammatically one form for a single inlet valve.

Although the detailed form of a helical port may vary quite widely it is usually offset from the valve axis so that a rotary air movement is superimposed on the linear velocity as it enters the cylinder. *Figure 9.13* illustrates one form of helical port. Drawings for such ports are at first a little difficult to understand so a perspective view is added to show the tangential approach to that part of the port immediately above the valve, and the rapid helical descent to the back of the valve such that the air largely enters the cylinder with a rotary motion at the outer cylinder radius.

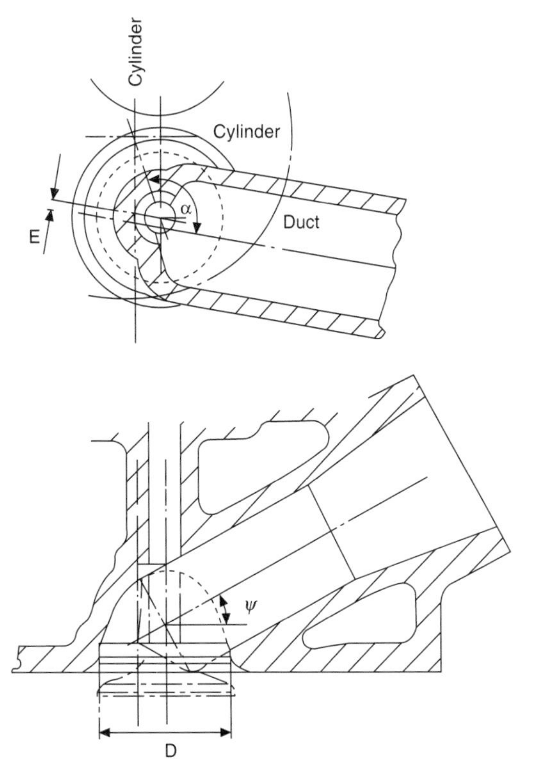

Figure 9.12 Directed port

It is necessary to assess the swirling ability of a swirl producing port prior to building the prototype engine. This is done by making a wooden or plastic full scale model of the port as designed, and blowing through the inlet port and valve in a precisely similar way to that described earlier for measuring the flow coefficients for non-swirl ports.

To measure the swirl intensity within the cylinder a swirl meter must be used. Formerly a light vane was fitted at the upper, open end, of the cylinder liner one bore diameter from the valve which was made to rotate by the swirling air and the speed of rotation measured. Since the air may have a spiralling motion down the cylinder bore, there is some doubt about the measurement of the air's angular momentum.

As a consequence, the recently developed impulse meter has now been adopted. This consists of a light aluminium foil honeycomb which is freely suspended on precision ball races with its periphery sealed by an oil filled trap. The suspended honeycomb is prevented from rotation by a light torque arm bearing on a fixed spring at its outer end. The movement of the restraining spring is measured by an electronic transducer. Thus with a swirling air charge moving up the cylinder, the axial holes in the honeycomb, through which the air must pass to escape, destroy the swirl causing the destroyed total angular momentum flux to produce a torque reaction on the suspended honeycomb. Measurement of this reaction torque is a direct measure of the swirl's angular momentum. Comparative tests indicate that the impulse swirl meter generally gives a higher swirl reading by a factor of 1.3 than the vanes swirl meter. This impulse swirl meter is fitted 1.75 bore diameters from the head face.

As for the flow testing with the simple inlet port, readings are taken at a series of valve lifts from zero to full lift, the swirl meter torque and the actual air flow through the port and valve being measured at each lift. In the light of early tests, modifications may be made to the port shape by cutting material away or adding plasticine until the best swirl and air flow compromise is obtained. Before finishing these swirl tests it is wise to check that the shape of the engine manifold to be used does not have any ill effects. Prototype heads should also be checked so that the required swirl is actually obtained.

The following is a brief derivation of the Ricardo Momentum Summation Method for predicting engine swirl. It should be noted that other workers use differing methods.

It is assumed that (i) the flow through the port is incompressible and adiabatic on both the engine and the steady state flow rig; (ii) the port retains the some characteristics (i.e. C_F and N_R) under the transient conditions in an engine as it does under steady flow rig conditions; (iii) the pressure drop, Δ_p, across the port is assumed constant during induction; (iv) angular momentum is conserved, and skin friction does not impede swirl. (v) Volumetric efficiency is 100^{λ}; (vi) flow occurs between inletvalve opening and closing so that flow is dependent on valve lift.

$$\text{Swirl ratio} = \frac{\text{Charge swirl speed at end of induction}}{\text{Engine crankshaft speed}}$$

$$= R_s = \frac{\omega_c}{\omega_e} \qquad (9.20)$$

$$\text{Impulse meter torque} = G = I\omega_R = \dot{m}\frac{B^2}{8}\omega_R$$

$$\text{So non-dimensional rig swirl} = N_R = \frac{\omega_R B}{V_0} = \frac{8G}{\dot{m}BV_0} \qquad (9.21)$$

where m is mass flow rate;
B is cylinder bore diameter;
ω_R is rig swirl as measured at any valve lift.

The swirl ratio is determined by equating the angular momentum

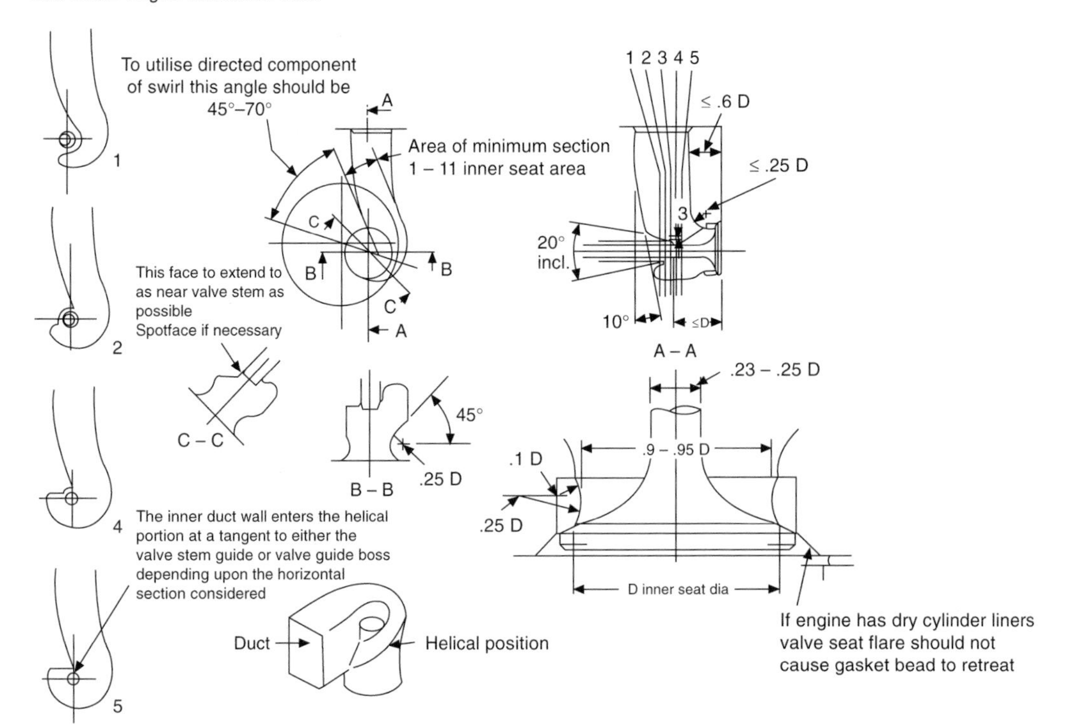

Figure 9.13 Helical port

of the charge at the end of induction to the sum of the angular momentum flux generated by the inlet port over the induction period. The following refers to conditions in the engine.

Assuming constant Δ_p and incompressible flow

$$\dot{m} = \rho A V_0 C_F \tag{9.22}$$

Assuming the same port properties on the rig as in the engine then from eqn (9.21)

$$\text{Angulor momentum flux} = G = \frac{\dot{m} B V_0 N_R}{8} \tag{9.23}$$

Angular momentum at end of induction = $I_c\omega_c$

$$\text{By conservation of momentum } I_c\omega_c = \int_{t_1}^{t_2} G\,dt \tag{9.24}$$

$$\text{Engine speed } \omega_E = \frac{dx}{dt}$$

$$\text{So } I_c\omega_c = \frac{\rho A B V_0^2}{8\omega_E} \int_{\alpha_1}^{\alpha_2} C_F N_R\,dx \tag{9.25}$$

Assuming a forced vortex in the cylinder, $I_c = \dfrac{MB^2}{8}$ where M is the cylinder mass flow rate.

Ignoring the residual gases $M = \displaystyle\int_{t_1}^{t_2} \dot{m}\,dt$

$$\text{So } M = \frac{\rho A V_0}{\omega_E} \int_{\alpha_1}^{\alpha_2} C_F\,dx \tag{9.26}$$

By the definition of volumetric efficiency

$$M = \frac{\rho \pi B^2 s}{4} \eta_v$$

If for simplicity η_v is taken as unity, then from eqns (9.25) to (9.27)

$$\omega_c \left[\int_{\alpha_1}^{\alpha_2} C_F\,d\alpha\right]^2 = \omega_E \frac{BS}{nD^2} \int_{\alpha_1}^{\alpha_2} C_F N_R\,d\alpha \tag{9.28}$$

Ricardo swirl ratio,

$$R_s = \frac{\omega_c}{\omega_E} = \frac{L_D \displaystyle\int_{\alpha_1}^{\alpha_2} C_F N_R\,d\alpha}{\left[\displaystyle\int_{\alpha_1}^{\alpha_2} C_F\,d\alpha\right]^2} \tag{9.29}$$

where L_D = Engine shape factor = BS/nD^2;
n being the number of inlet valves;
D the inlet valve throat diameter.

The integrals can be evaluated either graphically or using a computer, utilizing the C_F values determined from the rig flows and N_R values determined from the impulse meter torque at each valve lift measured on the rig. The rig lift values are of course associated with the valve lift corresponding to a certain angle during the lift period as determind by the cam design.

Typical rig swirl ratio values are, for direct injection engines,

75–90 mm bore 4.5–3.5,
90–110 mm bore 3.5–3.0,
110–150 mm bore 3.0–2.0.

These values assume the swirl is compromised over the engine's speed range and that normal combustion chamber proportions are used.

Figure 9.14 illustrates some swirl rig results showing the variation of flow coefficient and non-dimensional swirl with valve lift.

It is usually found that a 10% allowance must be made for losses incurred in translating results from a model to a cylinder head casting. With care this can be reduced to about 5%. To avoid this and the cylinder to cylinder variations which can occur in production due to small core shifts, etc., there is a tendency for manufacturers to machine the ports.

9.4 Turbocharging

The discussion so far might suggest that most engines are naturally aspirated. This is far from true since virtually all large diesel engines are turbocharged often to high pressure ratios. Even the small high speed automotive diesel engine is increasingly turbocharged now that suitable small flow rate turbochargers, running at very high speeds with small diameter rotors, have been developed. In automotive sizes the Brown Boveri 'Comprex' pressure exchanger is now applicable.

So far as breathing is concerned port design is very similar to that for naturally aspirated engines and only slightly lower swirl ratios seem to be required to optimize performance. The main difference is in large engines where quite wide overlaps are used, as referred to under section 9.3.1, to scavenge the clearance volume and help to cool the cylinder head, exhaust valve(s) and the turbocharger turbine at the higher boosts and loads.

9.5 Two-stroke engine scavenging

Unlike the four-stroke engine the two-stroke engine does not sweep out its exhaust residuals by the piston. This must be done by the current of air passed through the engine during the scavenging period, i.e. when the exhaust and scavenge ports are simultaneously open.

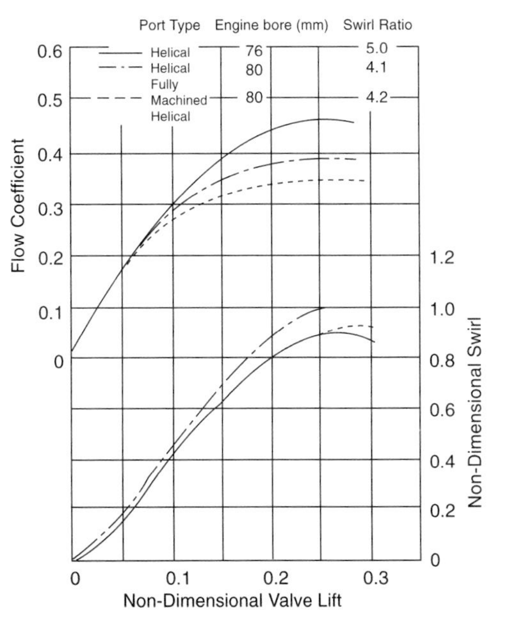

Figure 9.14 Typical swirl rig results

In the ideal engine, the cylinder is scavenged of all residual products from the previous cycle, compression of the new charge is that of pure air, and at the same time no air is lost to the exhaust. This is far from the truth on real engines. The residual gases cannot be swept out without mixing with some of the new air charge. As a consequence it is necessary to supply a larger volume of air than that corresponding to one engine swept volume. This is usually called the 'scavenge ratio' which is the volume of air supplied, usually at scavenge belt conditions. Typical values lie between 1.2 and 1.3 swept volumes.

Three main types of air–flow scavenging have been used. These are cross scavenging, loop scavenging and uniflow scavenging.

9.5.1 Cross scavenging

This is almost entirely found in small petrol engines. The arrangement is illustrated diagrammatically in *Figure 9.15.* To assist scavenging the piston crown is usually shaped to provide a deflector for the incoming charge.

The scavenge efficiency is not high and it is not possible to get large port areas except by the use of extended timings. For this reason their output is low. Their chief merit is that of simplicity and cheapness for industrial applications where a high specific output is not required.

9.5.2 Loop scavenging

There are three main types of the loop scavenging system. These are shown in *Figures 9.16, 9.17, 9.18.*

Again the problem is to obtain enough port area for high speed operation since the cylinder periphery must be shared by two sets of ports. However, they have served well for many years in slow speed and medium speed engines. As specific outputs have risen it has become increasingly difficult to overcome liner wear and piston ring and groove wear in those designs using trunk pistons. In this size range the highly turbocharged four-cycle engine seems to be taking over largely due to the reduced liner and piston problems.

In the type of loop scavenge engine shown in *Figure 9.16,*

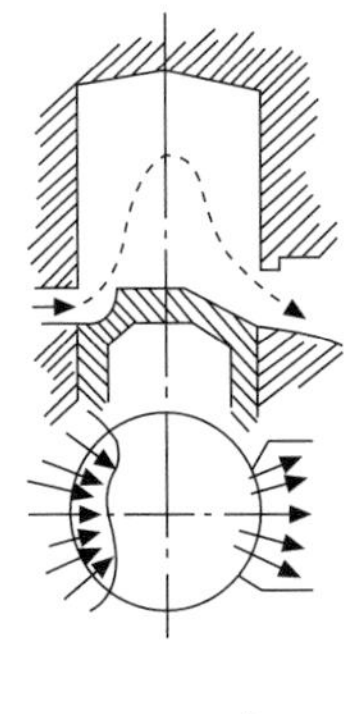

Figure 9.15 Cross scavenging

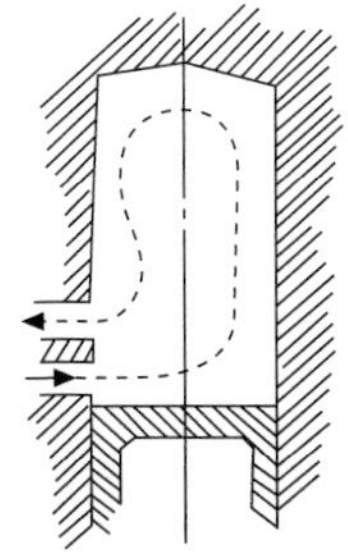

Figure 9.16 MAN scavenging

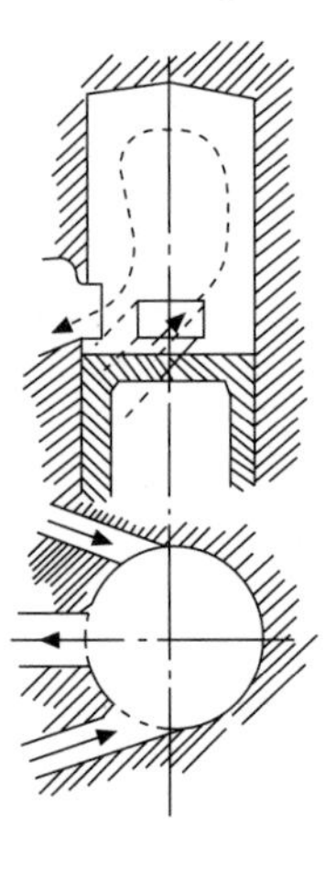

Figure 9.17 Schnuerle scavenging

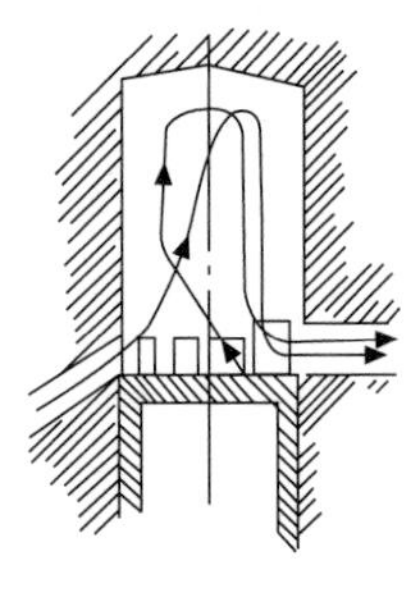

Figure 9.18 Curtis scavenging

the exhust ports are above the scavenge ports so that after exhaust 'blow-down' the scavenge air moves horizontally across the piston crown, then loops upwards and over to sweep most of the residual gases out of the exhaust ports. In the Schnuerle type shown in *Figure 9.17* the scavenge air passes into the cylinder through ports on two sides of the cylinder which are inclined upwards. The exhaust port(s) are in between the two sets of scavenge ports. The result is that the scavenge air sweeps the bore of the cylinder as it enters on two sides in a rising direction and then loops over towards the exhaust port(s). The third variety, the Curtis type, is shown in *Figure 9.18*. This has the exhaust ports on one side of the cylinder with the scavenge ports on the other three sides, the individual scavenge ports having differing horizontal and vertical inclinations which are greatest for the ports most remote from the exhaust ports.

To obtain adequate area most loop scavenge engines have their scavenge ports as high as the exhaust ports and in some cases even higher, i.e. they open at least as early as the exhaust ports. To avoid the flow of exhaust gases into the scavenge belt and contaminating the new air charge some form of reed, rotary or other kind of non-return valves is fitted in the scavenge belt near to the ports.

Where the exhaust ports open earlier than the scavenge ports, a rotary valve has been used in the exhaust belt so that the scavenge period is cut off by the rotary valve. This permits the air pressue in the cylinder to build up to full scavenge belt pressure.

9.5.3 Uniflow scavenging

For high output two-stroke engines and those operating at high speed the uniflow scavenging system is to be preferred. In this type, the exhaust is arranged at one end of the cylinder and the scavenge ports at the other, i.e. the scavenge air sweeps from one end of the cylinder to the other leading to much improved scavenge efficiency.

Two main types are to be distinguished. The first, shown diagrammatically in *Figure 9.19*, has a single piston which overruns the scavenge ports controlling their timing which must be symmetrical about BDC; one or more poppet exhaust valves are fitted to a cylinder head operated by cams. The second type, *Figure 9.20*, has two opposed pistons working in the one cylinder bore. One piston controls the scavenge port timing and the other the exhaust ports.

Apart from the improved scavenging the uniflow type of engine enables much larger port areas to be obtained since the scavenge ports alone can use as much of the cylinder periphery as is permitted by the strength of the port bars, sufficient support for the piston and rings and for the passage of adequate oil for piston and ring lubrication when trunk pistons are used. Up to about 60% of the bore periphery may be available for port width. If swirl is required to assist good combustion, the scavenge ports can be machined tangentially to a certain pitch circle so that the scavenge air enters with a tangential swirling velocity.

In the case of one or more poppet exhaust valves in the cylinder head ample area can be obtained. The major advantage is that their timings can be optimized with the exhaust opening before the scavenge ports so that the pressure in the cylinder has dropped to, or almost to, the scavenge belt pressure by the time the piston opens the scavenge ports. Similarly the exhaust can be closed prior to the scavenge ports.

When the opposed piston arrangement is used ample areas are available for both scavenge and exhaust ports. The exhaust ports can be opened earlier and closed earlier than the scavenge ports by a suitable exhaust port period combined with giving the exhaust piston a suitable lead angle, so that it reaches its inner TDC before that of the scavenge piston. True combustion TDC is of course the point of nearest approach of the two pistons and occurs at half the lead angle, i.e. the exhaust piston is half the lead angle after its own TDC position and the scavenge piston still has to move half the lead angle to reach its own inner TDC.

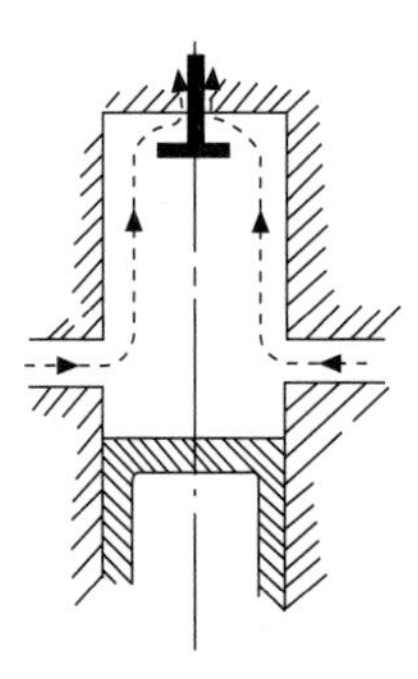

Figure 9.19 Uniflow scavengin with poppet valve

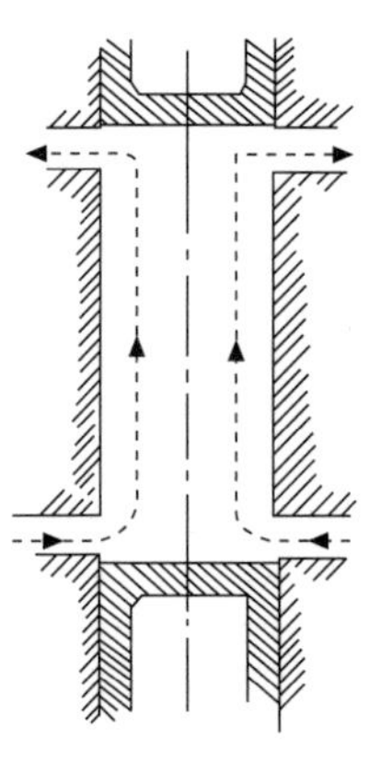

Figure 9.20 Uniflow scavenging opposed piston

The two pistons may be connected together as follows:

(1) via their crankshafts and gearing, e.g. Fairbanks Morse;
(2) by means of two side rods from the exhaust piston to two crankshaft throws, or even eccentrics, immediately adjacent to that for the scavenge piston as is done in the Doxford engines;
(3) by means of the pistons working on a piston rod to operate a rocker arm at each end of the cylinder, the other end of each 'dog-legged' rocker arm being connected to a common crankshaft by means of a connecting rod, e.g. Tilling-Stevens TS3 engine.

The use of rocker arms has mostly been confined to small high speed engines. Apart from high stress levels in the rocker arms and their fulcra, the lubrication of the always loaded oscillating bearings is a very difficult problem.

Due to the exhaust piston being ahead of its own TDC geometrically it has a crank angle past its own TDC when the point of minimum compression volume occurs. As a result the effective crank radius for the exaust piston is greater than that for the scavenge piston so that the instantaneous torque, being the product of the gas pressure and the effective crank radius, is greater on the exhaust crank than the scavenge piston crank. This means that a greater power output is given by the exhaust piston. For this reason the stroke for the exhaust piston is sometimes made shorter than that for the scavenge piston.

9.5.4 Port areas and timings

The treatment in this section of port areas and timings can only be approximate since most two-stroke engines are now supplied with pressurized air by turbochargers. As mentioned earlier for turbocharged four-stroke engines, the exhaust and scavenge systems are likely to have marked wave effects in them, initiated in part by the sudden port openings. Depending on the arrangement of the exhaust pipes used to connect with the exhaust turbine of the turbocharger, their diameter and length and the nozzle ring area in the turbine, significant dynamic wave-like pressure variations can occur. In the case of the two-stroke engine, particularly, these waves can influence the scavenging process.

Considerable effort has been devoted by Benson and others to develop calculation methods to predict the likely pressure diagrams in exhaust and scavenge systems and their inter-reaction with engine performance. These methods require the use of a large digital computer. As these methods are dealt with in detail in other chapters only the more elementary and cruder approach will be used here.

In the case of piston controlled ports their area is the product of the piston crown movement from the port opening position times the combined width of the ports around the bore. In the case of radial ports when used for scavenge ports in uniflow engines up to about 70% of the bore periphery has been used as port width. In practice it is usual to incline the ports tangentially to a pitch circle so as to provide some swirl to assist good combustion, particularly in smaller sizes. The effective peripheral length is the normal to the air flow and up to about the equivalent to 60% of the bore periphery can usually be obtained. Ports direct the air better if they are not too wide compared with their maximum height. In any case they must not be made too wide to avoid the piston rings bulging into them leading to excessive wear and possibly ring breakage. Pegging of rings so that their gap always passes up a port bar is unacceptable in large high output engines.

Details for loop scavenge engines depend very much on the configuration and number of exhaust and scavenge ports it is elected to use. As large a flow area as is practicable should be provided in the scavenge port belt, and care taken to see that the flow direction does not detract from the swirl the scavenge ports may have been designed to generate.

Flow and swirl generating characteristics can be experimentally determined, when desired, by methods very similar to those employed for four-stroke inlet and exhaust ports. As a rough guide the mean flow coefficients for poppet valve exhausts will be about 0.44 whilst piston controlled ports will be higher, up to 0.65.

Figure 9.21 shows a diagram of port opening, plotted vertically, against crank angle for a relatively small high speed two-stroke engine. Timings are with respect to piston BDC. During the period marked A the exhaust ports alone are open and the cylinder contents discharge to exhaust. Depending on the pressures prevailing, the pressure difference may be above critical, i.e. when the exhaust pressure is about 0.53 times that in the cylinder. As the pressure drops a diffierent flow formula is required.

Fundamentally it is necessary to decide how early the exhaust port (or valve(s)) should be opened to allow the cylinder pressure to drop to scavenge belt pressure at or immediately after the scavenge ports start to open. Making a first guess for the exhaust opening timing it is possible to plot the opening area as it increases, plotting this area against the angle from opening. If the pressure within the cylinder is known together with port flow coefficient it is possible to calculate the progressive drop in cylinder pressure with time, i.e. crank angle. For the initial approximate selection of exhaust lead, the following method, which although empirical includes all the variables, may be used.

It has been found that the required time-area integral in cm^3 degrees before the scavenge ports open should be:
For port controlled exhaust:

$$\frac{\text{Swept volume of 1 cylinder (cm}^3\text{)} \times \text{rev/min}}{5600} = \int A d\theta \tag{9.30}$$

where A is the port opening area in cm^2;
θ is angle from port opening.

For poppet valve exhaust:

$$\frac{\text{Swept volume (cm}^3\text{)} \times \text{rev/min}}{7700} = \int A d\theta \tag{9.31}$$

If the scale of the plotted diagram is x degrees to 1 cm horizontally and y cm^2 for 1 cm on the vertical axis then the required time-area integral is represented by $\int A d\theta / xy$ cm^2 on the diagram. Using a planimeter this area can be found under the curve and the angle required noted. This represents the exhaust lead required. Next if the scavenge port timings are assumed, a preliminary diagram showing both the exhaust and scavenge areas can be drawn.

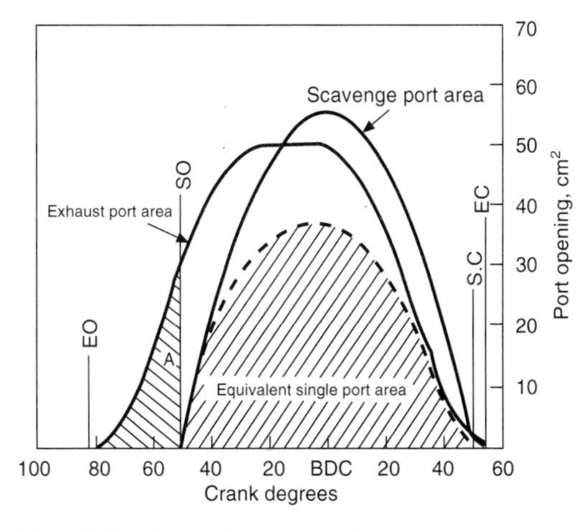

Figure 9.21 Two-stroke port area diagram

If, as the simplest case, it is assumed that the discharge coefficients for the exhaust and scavenge ports are equal, which is approximately true for opposed piston engines, it can be shown that the area of a single port having the same resistance to flow as the scavenge and exhaust ports in series as they are during the scavenging period is

$$A_{\text{equiv}} = \frac{se}{\sqrt{s^2 + e^2}} \tag{9.32}$$

Using this expression at various angles during the scavenge period allows the dotted 'equivalent single port' area to be plotted. The area under this curve is the time-area integral for the scavenge period. If this is B cm^2 degrees, the area of a constantly open hole which will pass the same quantity of air with the same pressure differences would be $B/360$ cm^2. It is now possible to estimate the likely flow during the scavenge period with the proposed nominally constant pressure difference between the scavenge and exhaust pressures. In the light of the quantity estimated, adjustments can be made to the timings and areas to obtain the required scavenge flow.

For modest pressure differences the mean gas velocity through the constantly open hole will be approximately $0.69\sqrt{(P/\rho)}$ where P = pressure difference across the engine in N/m^2 and ρ = relative air density compared with atmospheric temperature and pressure.

9.6 Silencers

Three sources of noise are important: pressure pulsations, flow-induced and shell noise from the surfaces of the silencer. The assistance provided by modern mathematical modelling tools is substantial, especially for the pressure pulsation source, where engine models can be extended to the silencers.

Attenuation of the pressure pulsations in the intake and exhaust of diesel engines is achieved through the use of flow path expansions and contractions, where acoustic energy is lost as heat, the use of reflections to provide internal cancellation and also by the use of acoustic absorption materials. Design factors include the intake pipe diameter, the main body diameter (which therefore controls the area expansion ratio), the lengths of the internal pipes, as well as the positioning of internal baffles to form multiple chambers. Calculations of silencer attenuation in realistic systems quickly become complex and computational tools are required to produce rapid and reliable designs. *Figure 9.22* illustrates the simulation of a simple muffler using a current conventional engine performance software tool (GT-Power).

The overall optimization of silencer systems, must take account of the flow loss (back pressure), acoustic performance, size, weight and life. Additional acoustic performance can be achieved by the use of side-branched resonators, such as quarter-wave pipes and Helmholz resonators.

In practice, it is often preferable to use perforated pipes in place of stub pipes. Perforation levels up to 40% are employed. Also the addition of some form of acoustic absorption material can significantly improve specific acoustic performance results. Such systems can be accurately designed using advanced software packages[8]. Such absorption materials must be carefully chosen for the planned lifetime of the silencer,

Figures 9.23 illustrate both the overall form of the attenuation that can be obtained as well as the good agreement that can be achieved with modelling. For simple exhaust or intake systems, where the design is not critical, approximations can be used. For designs where cost/weight/performance issues are important, more complex designs necessitate a modelling approach to achieve an efficient optimization.

Similar calculations apply to intake systems, where the complexities associated with high thermal gradients are often exchanged for filtration challenges where large filter bodies can exceed the limits of 1-D modelling assumptions, so that other tools are used.

Physical construction of the silencer determines the shell noise, which is radiated from the surface as a result of internal

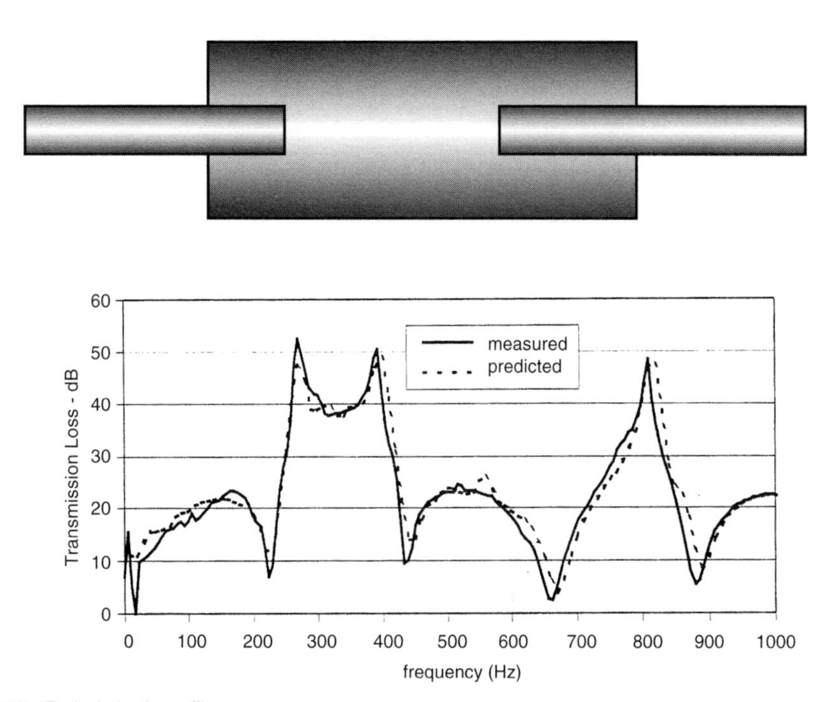

Figure 9.22 Typical simple muffler

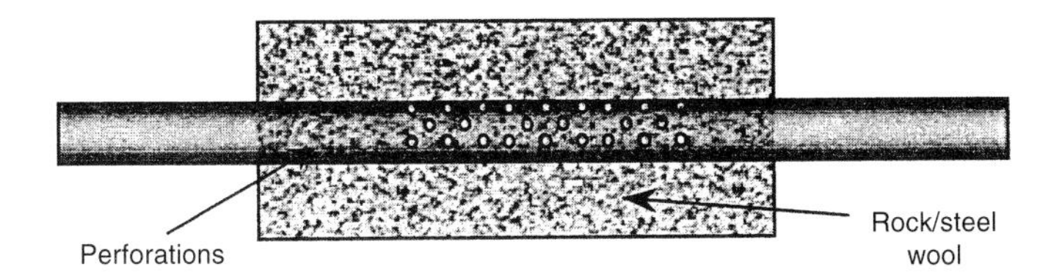

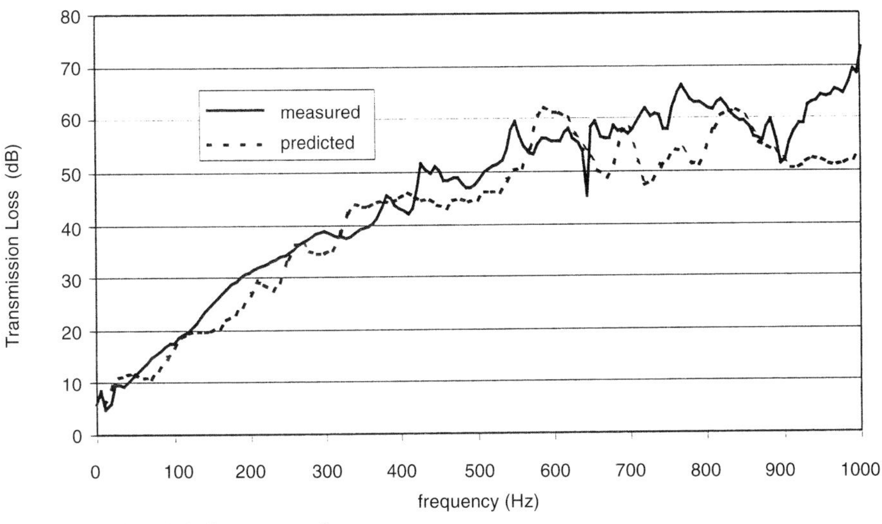

Figure 9.23 Practical silencer example

acoustic excitation and mechanical coupling to the engine. To minimize this, the outer walls and end-caps of the body are constructed of double-skin material. This introduces frictional damping in the surface layers, reducing the vibration and so the radiated noise.

References

1 TAYLOR, Professor C.F., *The Internal Combustion Engine in Theory and Practice,* Vol. I, M.I.T. Press (USA)

2 SCHWEITZER, Professor, P. H., *Scavenging of Two-Stroke Cycle Diesel Engines,* Macmillan, New York (1949). (Out of print but a good fundamental reference book)

3 ROTHBART, H.A., *Cam: Design, Dynamics and Accuracy,* Wiley (1956)

4 Society of Automotive Engineers Sub-Committee, *Application of Computers in Valve Gear Design.* Collection of four papers dealing with Mechanics of Valve Gear Design. Distributed by Macmillan Co., New York and Pergamon Press, England (1963)

5 BENSON, R.S. *Thermodynamics and Gas Dynamics of I.C. Engines,* Vol. I. Fundamentals, Vol. II. Clarendon, Oxford (1982)

6 WATSON, N. and JANOTA, M. S., *Turbocharging the Internal Combustion Engine,* Wiley-Interscience (1982)

7 BAXA, D. E. (ed) *Noise Control in Internal-Combustion Engines,* Wiley-Interscience

8 PILO, GAMBA and CHALLEN, SAE, 971985 'Prediction of Vehicle Radiated Noise'

10 Design layout options

Contents

10.1 Introduction

For a new engine design it has been observed that the decisions that are most costly—and often most regretted—are made in the first few weeks of the project. Current project management approaches, where the concept phase is highly intensive, aim to carry out a considerable amount of analysis work at the 'front end' of the work. This makes the observation even truer. There is little spare time to correct errors so that the initial design decisions are critical. The objective is to have just one prototype—confirming the design correct choices. The application of computer-aided design tools allows many fundamental design options to be studied and the optimum determined for the target specification.

This part of the book deals with some of the most crucial choices in a new engine—the overall layout, its balancing, living with torsional vibrations, and the overall compromises that constitute design. Charles Beard (to whom one of the editors, BJC, owes much of his early design analysis appreciation) wrote some sections for the first edition of the book. They are retained in this edition since they deal very effectively with the topics. Much of the analytical work is today carried out through numerical analysis and often with the use of finite element models. The outline approaches covered here remain good initial starting points for any new design or substantial upgrade of an existing design.

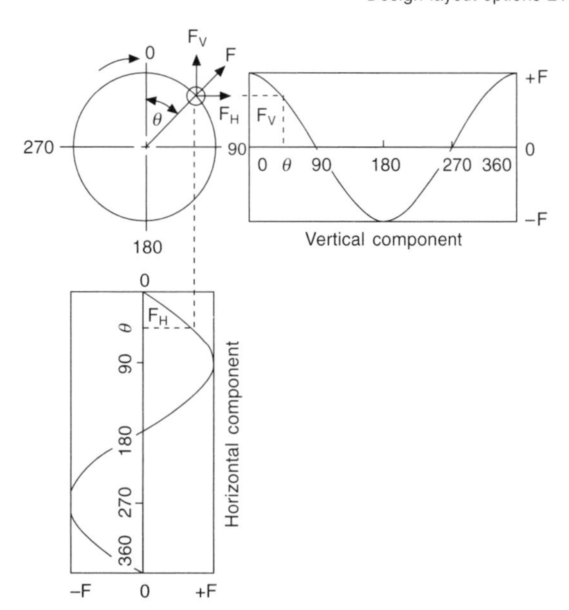

Figure 10.1 Vertical and horizontal forces produced by a rotating mass

10.2 The balancing of engines

10.2.1 Consideration of the forces involved

10.2.1.1 Revolving forces

If a single concentrated mass attached by a thin arm such that it maintains a fixed radius is rotated, a centripetal acceleration is imposed of the mass to cause it to remain at the fixed distance from the axis of rotation. The correspoding equal and opposite force—known as centrifugal force—can be shown to be

$$F = Ma = M\frac{v^2}{r} = M\omega^2 r \qquad (10.1)$$

where F = centrifugal force (N);
a = centripetal acceleration (m/s^2);
v = tangential velocity of mass as it rotates (m/s);
r = radius of rotation (m);
ω = angular velocity (radian/s)

Since it is usual to measure engine rotational velocity in revolutions per second eqn (10.1) may be rewritten:

$$F = \omega^2 Mr = (2\pi n)^2 Mr = 39.48n^2 Mr \qquad (10.1a)$$

It should be noted that for a given speed this is a fixed vector which rotates at the same angular velocity as the rotating shaft (*Figure 10.1*). Now if the light arm, referred to above, holding the concentrated mass at a fixed radius is likened to a crankthrow of an engine it is convenient to refer the angle through which the mass turns to a datum corresponding to the mass being vertically above the axis of rotation. In the parlance of internal combustion engines this corresponds to Top Dead Centre (TDC). Suppose now the mass rotates in a clockwise direction. Then at any angle θ from the TDC position the constant centrifugal force produces a vertical component $F_V = F\cos\theta$ and a horizontal component $F_H = F\sin\theta$. These relationships will be of value later but, for the moment, it is worth getting a clear mental picture of the way in which the rotating weight produces vertical and horizontal forces varying with crank angle, i.e. with time.

Now if a balancing mass is attached directly opposite the first disturbing mass, the out-of-balance force will be eliminated provided the masses are co-planar and produce equal and opposite disturbing forces. If M_B is used for the balancing mass it is necessary that:

$$\omega^2 M_B r_B = \omega^2 Mr$$

Since ω^2 is common this reduces to:

$$M_B r_B = Mr \qquad (10.2)$$

Thus the mass of the balance weight(s) used may be different from the mass causing the out-of-balance provided their radii of action are such that eqn (10.2) is complied with. It should be noted that the force produced by the balance mass must, for the ideal case just quoted, be directly opposite the out-of-balance mass. This is not usually practical in actual engines but the methods used will be explained later.

Next suppose there are two rotating masses in the same plane (*Figure 10.2*). The first mass M_1 produces a force F_1 when rotating at radius r_1 and an angular velocity ω radian/sec. The second mass M_2 is phased at an angle of ϕ to the first mass and being at a radius r_2 and rotating at the same angular velocity produces a force F_2.

In the position drawn F_1 acts vertically upwards and F_2 acts downwardly and to the right. Resolving F_2 into these components the downward component is:

$$F_2 \cos(180 - \phi) = -F_2 \cos\phi$$

and the horizontal component is:

$$F_2 \sin(180 - \phi) = F_2 \sin\phi$$

Adding the vertical component

$F_1 - (-F_2 \cos\phi)$ is obtained (*Figure 10.2(b)*).

As F_1 is vertical it has no horizontal component so the sum of the horizontal forces is $F_2 \sin\phi$. The resultant of these two net forces at right angles is:

$$F_R^2 = \{F_1 - (-F_2\cos\phi)\}^2 + \{F_2\sin\phi)^2$$

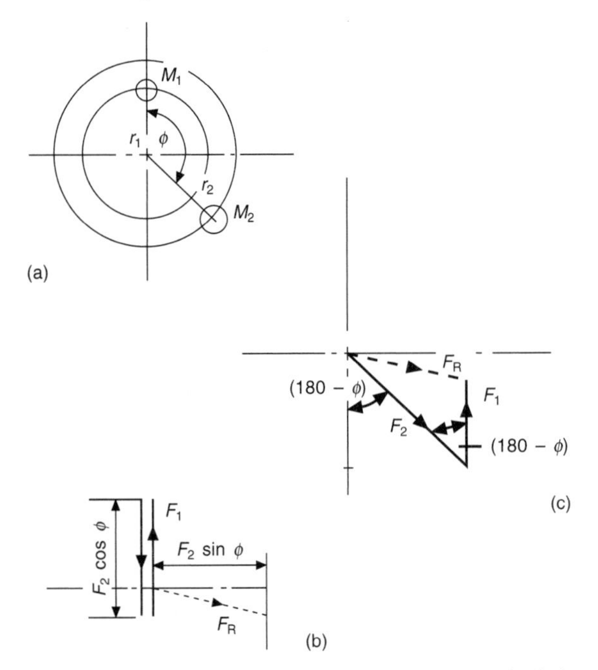

Figure 10.2 Resultant force produced by two masses rotating in the same direction

$$= F_1^2 + 2F_1F_2 \cos\phi + F_2^2 \cos^2\phi + F_2 \sin^2\phi$$

$$= F_1^2 + F_2^2 + 2F_1F_2 \cos\phi \qquad (10.3)$$

or

$$F_R = [F_1^2 + F_2^2 + 2F_1F_2 \cos\phi]^{1/2} \qquad (10.3a)$$

This resultant's position is

$$\theta_R = \tan^{-1}\frac{(F_1 - F_2 \cos\phi)}{F_2 \sin\phi} \qquad (10.4)$$

below the horizontal. If M_1 is regarded as the reference mass the resultant trails F_1 by the angle $(90 + \theta_R)$, as drawn.

Traditionally the resultant would be found by drawing a triangle, or maybe a parallelogram of forces. This is shown in *Figure 10.2c*. Such vector diagrams are slow in use and, depending on the angle between the forces, may be difficult to draw accurately. The advent of the small electronic calculator able to generate its own trigonometrical functions means that taking and summing vertical and horizontal components of each force, and these may be more than two in practice, is quicker and more accurate once this approach is mastered. This, of course, leads on to the use of a digital computer for complicated multi-mass systems or when bearing load diagrams are required around the operating cycle.

From the classical triangle of forces shown in *Figure 10.2c* the resultant by using the Cosine Rule, i.e.:

$$F_R^2 = F_1^2 + F_2^2 - 2F_1F_2 \cos(180 - \phi)$$

$$= F_1^2 + F_2^2 + 2F_1F_2 \cos\phi$$

which is identical to eqn (10.3).

10.2.1.2 Couples

Next suppose two equal masses both attached at a radius R are positioned opposite to one another but not in the same plane (*Figure 10.3a*). Each produce equal and opposite forces, i.e. there is no force out-of-balance. However as these equal and opposite forces are positioned at a longitudinal distance L apart they produce a couple $M\omega^2RL$ which may be rewritten ω^2MRL. This out-of-balance couple must be balanced by attaching balance weights which must each have the same Mr value but which need not be the same Mr value as that for the original out-of-balance masses, when spaced at a distance L_B apart. These attached balance weights must produce the same magnitude of couple but of opposite sense to the out-of-balance couple. Expressed mathematically

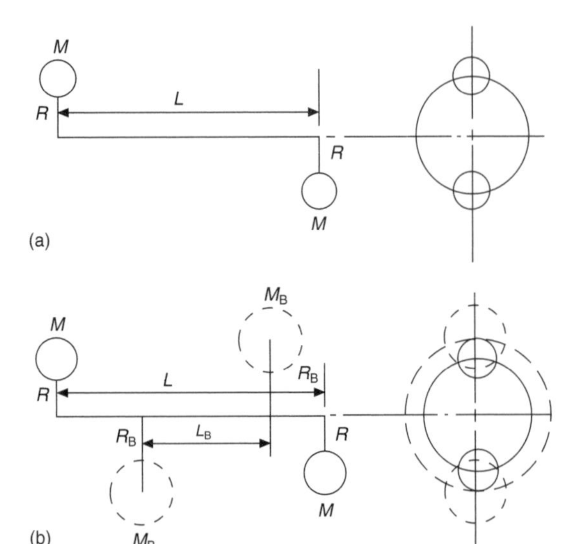

Figure 10.3 Out-of-balance rotating couple prodced by two opposing equal masses in the same plane and its balancing

$$\omega^2MrL = \omega^2M_Br_BL_B \qquad (10.4)$$

or

$$MrL = M_Br_BL_B$$

When two uneqal masses are present on the same shaft (*Figure 10.4*), it is necessary to arrange for force balance as well as couple balance. If $M_1r_1 > M_2r_2$, force balance can be established by fitting a balance weight having a M_Fr_F value equal to $(M_1r_1 - M_2r_2)$ opposite to M_1. This leaves a clockwise couple of M_2r_2L to be eliminated by the fitting of two weights to produce an anticlockwise balancing couple each having a moment M_cr_c spaced at longitudinal distance L_c (*Figure 10.4b*).

This leads on to the point that a sigle out-of-balance mass need not be balanced by a single mass having the same Mr placed directly opposite at 180°. It can be split into two equal values if this happens to be convenient; for example fitting equal balance weights on each crankweb of a crankthrow, provided they are equally spaced each side of the out-of-balance force (*Figure 10.5a*). If this is not so, a couple will be produced and must be separately dealt with. In *Figure 10.5b* the system is in force balance but L_1 being larger than L_2 means that an anticlockwise moment of $Mr/2(L_1 - L_2)$ exists and must be balanced by the provision of a suitable pair of balance weights producing an opposing clockwise moment.

This diagram may also be used to demonstrate that an out-of-balance couple can be found by taking longitudinal moments about any convenient point. If moments are taken about A and clockwise moments are considered positive, we have a total couple of

$$-MrL_1 + \frac{Mr}{2}(L_1 + L_2) = \frac{Mr}{2}(L_2 - L_1)$$

Taking moments about C a total couple of

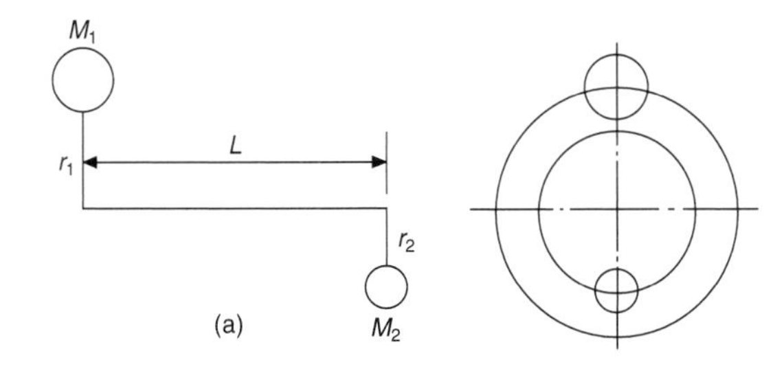

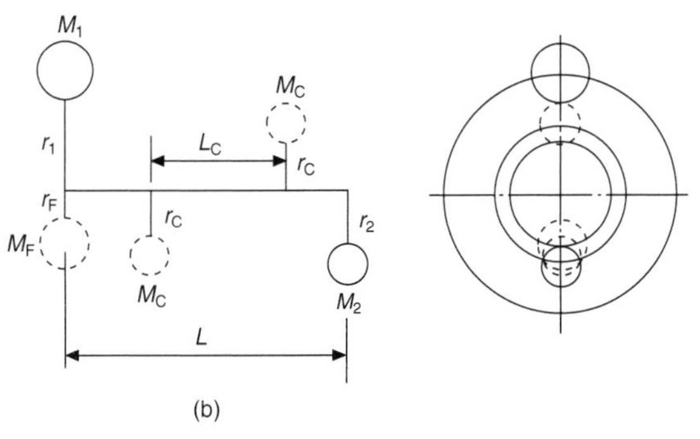

Figure 10.4 Balacing of two opposing unequal masses and moments not in the same plane

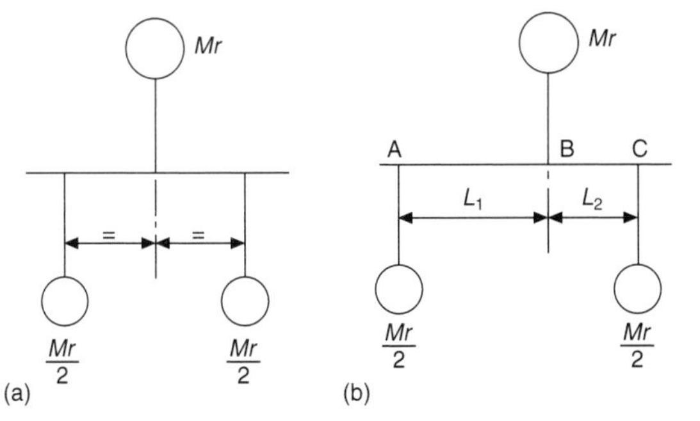

Figure 10.5 Balance of a single out-of-balance mass by two masses: (a) Equally spaced opposite mass to be balanced; (b) Production of couple if two equal balancing masses are not equally spaced

$$MrL_2 - \frac{Mr}{2}(L_1 + L_2) = \frac{Mr}{2}(L_2 - L_1) \text{ again.}$$

10.2.1.3 Multiple masses

Where multiple masses are attached to a shaft at various angles, it is necessary to consider both force balance and couple balance. This is most conveniently done by resolving the individual disturbing forces into two components at right angles with reference to standard reference axes. In the case of internal combustion engines this is usually with reference a 'vertical' axis through the centre-line of the cylinders for in-line engines. The inverted commas have been used since in-line engines are sometimes installed with the cylinder axes inclined at an angle to the true vertical. In the case of vee engines, the vertical axis is usually taken as the bisector of the vee angle.

For the engine to be in balance the algebraic sum of the vertical forces and horizontal forces must be equal to zero or a suitable balance weight(s) fitted. Couple balance is checked by taking moments of each horizontal and vertical force component about a suitable longitudinal point. For perfect balance the sum of the clockwise and anticlockwise moments should be equal to zero for both vertical and horizontal components. Expressed mathematically the requirements are:

$$\left.\begin{aligned}\Sigma\, Mr\cos\theta &= 0\\ \Sigma\, Mr\sin\theta &= 0\end{aligned}\right\}\text{ Force balance} \qquad (10.5)$$

$$\left.\begin{aligned}\Sigma\, MrL\cos\theta &= 0\\ \Sigma\, MrL\sin\theta &= 0\end{aligned}\right\}\text{ Couple balance} \qquad (10.6)$$

When most internal combustion engines are considered the task is simplified by the fact that the Mr value for every crankthrow is the same, cylinder longitudinal spacing is usually equal, and to obtain equal firing intervals multiples of a fixed angle occur. This often permits the use of a simpler semigraphical approach as will be shown later.

10.2.1.4 Reciprocating masses

So far reference has only been made to the balancing of rotating masses. In internal combustion engines the piston assembly, the gudgeon pin and that part of the connecting rod whose mass may be regarded as moving with the gudgeon pin reciprocate up and down the cylinder bore, and are accelerated and decelerated at the top and bottom dead centres thereby producing disturbing forces along the axes of the cylinder bores.

Referring to *Figure 10.6*, suppose the total reciprocating mass is M_r, L is the centre-to-centre length of the connecting rod, R

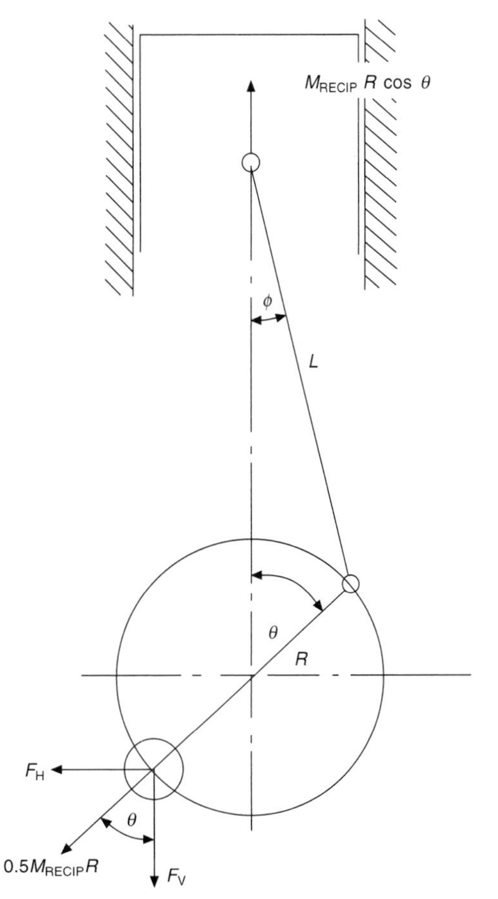

Figure 10.6 Diagram to show basic balancing principles for single throw crank

is the crankthrow radius, y the displacement of the piston from its top dead centre position, θ the angular movement of the crank from TDC, and ϕ the connecting rod obliquity. Then:

$$y = R + L - (L \cos \phi + R \cos \phi)$$

but $R \sin \theta = L \sin \phi$

$$\text{or } \sin \phi = \frac{R}{L} \sin \theta$$

$$\text{making } \cos \phi = \left\{1 - \left(\frac{R}{L} \sin \theta\right)^2\right\}^{1/2}$$

$$y = R + L - R \cos \theta - L \left\{1 - \left(\frac{R}{L} \sin \theta\right)^2\right\}^{1/2} \quad (10.7)$$

Putting $R/L = \rho$ and expanding the bracketed term equal to cos ϕ we obtain:

$$\cos \phi = 1 - \tfrac{1}{2}\rho^2 \sin^2 \theta - \tfrac{1}{8}\rho^4 \sin^4 \theta - \tfrac{1}{16}\rho^6 \sin^6 \theta - \ldots$$

using the relationships

$$\sin^2 \theta = \tfrac{1}{2} - \cos 2\theta$$

$$\sin^4 \theta = \tfrac{3}{8} - \tfrac{1}{2} \cos 2\theta + \tfrac{1}{8} \cos 4\theta$$

$$\sin^6 \theta = \tfrac{5}{16} - \tfrac{15}{32} \cos 2\theta + \tfrac{3}{16} \cos 4\theta - \tfrac{1}{32} \cos 6\theta, \text{ etc}.$$

we obtain

$$y = R + L - R \cos \theta - L[1 - \tfrac{1}{4}\rho^2 - \tfrac{3}{64}\rho^4 - \tfrac{5}{256}\rho^6 - \ldots]$$

$$- L \cos 2\theta [\tfrac{1}{4}\rho^2 + \tfrac{1}{16}\rho^4 + \tfrac{15}{512}\rho^6 + \ldots]$$

$$- L \cos 4\theta [- \tfrac{1}{64}\rho^4 - \tfrac{3}{256}\rho^6 - \ldots]$$

$$- L \cos 6\theta [\tfrac{1}{512}\rho^6 + \ldots]$$

which on rearranging becomes:

$$y = R\left(1 + \frac{1}{\rho}\right) + K + A \cos \theta + B \cos 2\theta + C \cos 4\theta + D \cos 6\theta + \text{etc.} \quad (10.8)$$

where

$$K = - R\left(\frac{1}{\rho} - \tfrac{1}{4}\rho - \tfrac{3}{64}\rho^3 - \tfrac{5}{256}\rho^5 - \ldots\right)$$

$$A = - R$$

$$B = - R(\tfrac{1}{4}\rho + \tfrac{1}{16}\rho^3 + \tfrac{15}{512}\rho^5 + \ldots)$$

$$C = + R(\tfrac{1}{64}\rho^3 + \tfrac{3}{256}\rho^5 + \ldots)$$

$$D = - R(\tfrac{1}{512}\rho^5 + \ldots)$$

$$\text{Piston velocity } = \dot{y} = v = \frac{dy}{dt} = \frac{d\theta}{dt}\frac{dy}{d\theta}$$

$$= - \omega [A \sin \theta + 2B \sin 2\theta + 4C \sin 4\theta + 6D \sin 6\theta + \ldots] \quad (10.9)$$

$$\text{Piston acceleration } = \ddot{y} = \frac{d\theta}{dt}\frac{dv}{d\theta}$$

$$= - \omega^2[A \cos \theta + 4B \cos 2\theta + 16C \cos 4\theta + 36D \cos 6\theta + \ldots]$$

$$= + \omega^2 R[\cos \theta + B_1 \cos 2\theta + C_1 \cos 4\theta + D_1 \cos 6\theta + \ldots] \quad (10.10)$$

where

$$B_1 = \rho + \tfrac{1}{4}\rho^3 + \tfrac{15}{128}\rho^5 + \ldots$$

$$C_1 = - \tfrac{1}{4}\rho^3 - \tfrac{3}{16}\rho^5 - \ldots$$

$$D_1 = \tfrac{9}{128}\rho^5$$

Evaluating these for a range of possible practical values:

$\frac{L}{R} = \frac{1}{\rho}$	3	3.5	4	4.5
B_1	0.3431	0.2918	0.2540	0.2250
C_1	– 0.0100	– 0.0062	– 0.0041	– 0.0028
D_1	0.0003	0.0001	0.0001	0.0000

Examining eqn (10.10) it will be seen that the piston acceleration has terms in cos θB_1 cos 2θ, C_1 cos 4θ, D_1 cos 6θ, etc. the constants for which rapidly diminish with the number of the term. In fact after tne third term the values become negligibly small. The expression also shows that the total acceleration can be represented as made up of three vectors of decreasing magnitude rotating at once, twice and four times crankshaft speed. These are normally known as the primary, secondary and fourth order components. Being accelerations they generate corresponding disturbing forces when multiplied by the total reciprocating mass.

These disturbing forces generated by the accelerations of the total reciprocating masses act entirely along the line of the cylinder bore. Thus for the usual engines having their cylinder axes vertical the disturbances at each individual cylinder act in a vertical plane only.

The reciprocating mass per cylinder consists of the piston plus its rings and gudgeon pin together with the small end mass of the connecting rod. The reciprocating mass of the connecting rod is approximately one third of its total mass, but for accurate work it should be determined by weighting with the rod axis horizontal and the small and large end bearings supported on knife edges.

10.2.2 Balance of a single-cylinder engine

Referring to *Figure 10.6* the rotating out-of-balance mass M_{Rot} consists of the equivalent out-of-balance mass at crankpin radius plus the rotating mass of the connecting rod complete with its big-end bearing. This makes the mass out-of-balance moment equal to $M_{Rot}R$ where R is the crankthrow. In practice when working from drawings it is easier to calculate the individual Mr values for the crankpin, the webs, etc. separately may be breaking the webs into convenient geometric shapes for ease of calculation and then summing all the Mr values. The out-of-balance moments are all calculated with reference to the line joining the crankpin centre to the journal centre. The moment of any material on the remote side of the journal centre line is negative in value. This rotating out-of-balance is readily balanced by two balance weights, each having a moment equal to $0.5M_{Rot}R$, with their centres of gravity opposite to that of the crankpin.

Merely balancing the rotating weight leaves the reciprocating forces acting along the cylinder axis untoched. Suppose the Mr value for the reciprocating masses is $M_{Recip}R$. This leads to a primary out-of-balance force proportional to $M_{Recip}R \cos \theta$ where θ is the instantaneous crank rotation from its TDC position. Suppose it is arranged to supplement the Mr value already provided to balance the rotational out-of-balance moment by an amount equal to $0.5M_{Recip}R$, then from *Figure 10.1* this extra moment produces vertical and horizontal components each having maximum values proportional to $0.5\ M_{Recip}R$.

Referring now to *Figure 10.6* the reciprocating mass produces an upward primary force proportional to $M_{Recip}R \cos \theta$ at a crank angle θ from *TDC*. The additional balance mass over and above that provided to balance the rotating mass of the crank unit of $0.5M_{Recip}R$ will produce a downward component of $0.5M_{Recip}R \cos \theta$, as drawn, and a horizontal component $0.5M_{Recip}R \sin \theta$.

Since the downward component due to the balance weight opposes the upward force due to the reciprocating mass the net out-of-balance along the cylinder axis is

$$M_{Recip}R \cos \theta - 0.5M_{Recip} R \cos \theta = 0.5M_{Recip}R \cos \theta$$

Thus the vertical out-of-balance primary disturbance has been reduced to one half its original value but at the cost of introducing a harmonically varying horisontal force of $5.0M_{Recip}R \sin \theta$.

From this it can be seen that if balance weights are fitted to reduce the primary reciprocating out-of-balance by a proportion x a horizotal out-of-balance $xM_{Recip}R \sin \theta$ is simultaneously introduced. In practice half the reciprocating primary out-of-balance is usually balanced as a compromise, but larger or smaller proportions may be chosen depending on the rigidity of the engine mounting.

10.2.2.1 Primary balancers

Now suppose two masses having the same out-of-balance of $0.5MR$ are simultaneously arranged to rotate in opposite directions at the same speed. Then each of them produces (*Figure 10.7*) an upward force $0.5\omega^2 MR \cos \theta$ at a rotational speed of ω radian/sec. Sticking to the use of moments, but remembering they must be multiplied by ω^2 when actual forces are required, it can be seen from *Figure 10.7a* that the vertical components of the two 0.5 MR values are in phase and always have equal magnitudes. Simultaneously the horizontal components $0.5\ MR \sin \theta$ have equal values but opposite signs due to the counter-rotation as shown in the lower part of *Figure 10.7a*. If now the vertical components are added it is seen that they add to one another producing a sinusoidally varying moment of $MR \cos \theta$ whilst the horizotal components cancel one another, *Figure 10.7b*.

This concept is a useful theoretical one since it permits replacing a single harmonically varying moment (or force) acting in one plane only by two counter-rotating vectors each having half the original magnitude. When used for theoretical purposes this device is known as the 'Reverse crank concept'.

This principle can be applied in the balance of primary and secondary disturbing forces in the case of a single-cylinder engine and for dealing with primary and secondary out-of-balance forces and couples where they occur in multi-cylinder engines. Practical use of contrarotating masses for balancing primary and secondary forces was invented by Dr Lanchester.

Referring to *Figure 10.8* an ideal arrangement is shown for balancing the primary forces of a single-cylinder engine. Here the balance weights shown on the crankshaft eliminate the rotating out-of-balance forces only but two balancer shafts are arranged to be rotated in opposite directions at crankshaft speed. If the axes of rotation of the two balancer shafts are equidistant from the vertical cylinder axis and the centre of mass for the two balancer masses are arranged to be on the longitudinal cylinder axis the primary reciprocating out-of-balance is eliminated. If the reciprocating out-of-balance maximum moment is $M_{Recip}R$ each of the two balancer masses must be arranged to have a moment of $0.5\ M_{Recip}R$ and phased to be acting vertically downwards when the piston is at TDC. If it is more convenient each of the balancer masses may be made as an eccentric cylinder, or broken into two separate masses spaced apart longitudinally, providing their centre of mass is on the cylinder's longitudinal axis.

In practice the primary balancer weight rotating in the same

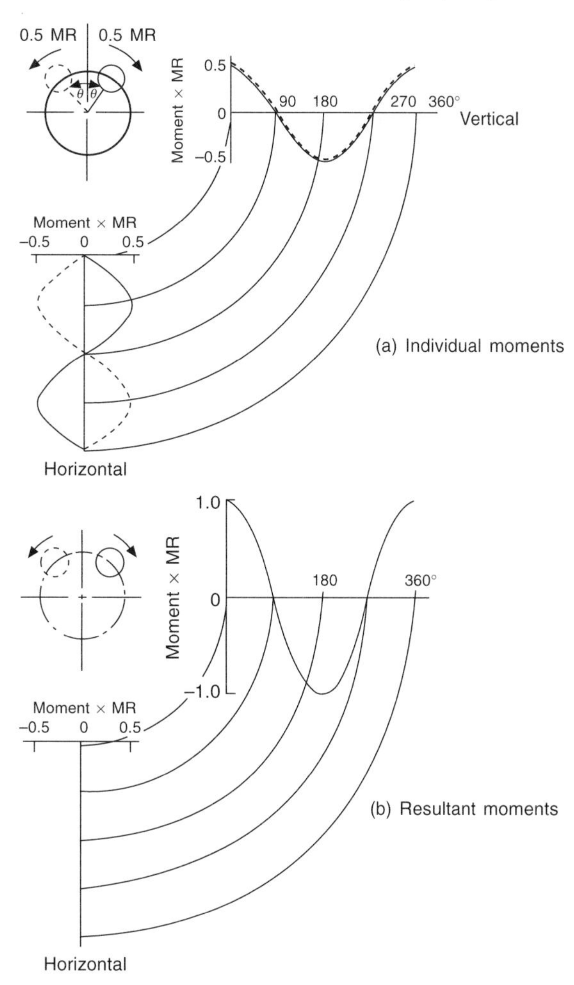

Figure 10.7 Vertical and horizontal moments produced by two equal moment reverse rotation masses (Principle of reverse cranks)

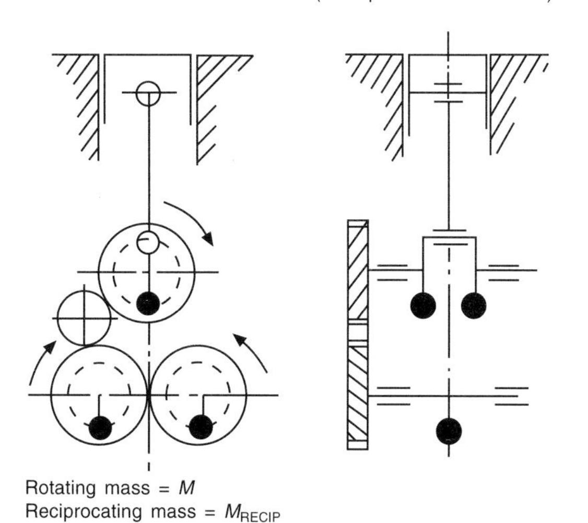

Figure 10.8 Primary balancing of single-cylinder engine using two contrarotating balancer shafts

direction as the crankshaft is often incorporated in the crankshaft balance weight to reduce size and cost. However using the arrangement shown in *Figure 10.9* the primary out-of-balance force is eliminated but at the cost of introducing a harmonic rolling couple eqal to p.$5M_{\text{Recip}}RL$ due to the vertical distance between the two balancer shafts, in effect.

It should be mentioned that it is possible to arrange to have appropriate reciprocating masses working in the opposite direction to the disturbing ones so that the total disturbance becomes zero. This approach is not normally used, since it tends to be more complex and costly than the use of the contrarotating masses principle, and tends to have higher friction losses.

10.2.2.2 Secondary balancers

When seeking to cancel secondary forces (and couples) it must be remembered the value of the maximum secondary disturbing force, from eqn (10.10), is given by:

$$F_2 = \omega^2 M_{\text{Recip}} R B_1 \cos 2\theta$$

$$= \omega^2 M_{\text{Recip}} R \frac{R}{L} \cos 2\theta$$

$$= \omega^2 M_{\text{Recip}} R \frac{R}{L} \text{ when } \theta = 0$$

Here ω is the angular velocity of the crankshaft.

This secondary disturbing force can only be balanced by a force rotating at twice crankshaft speed. The secondary force acts along the line of the cylinder axis so two balancer shafts rotating in opposite directions must be used in a similar way to primary balancers. If the moment to be provided on each balancer shaft is $M_1\, r$, then for maximum force balance

$$= \omega^2 M_{\text{Recip}} R \frac{R}{L} = (2\omega)^2 (2 M_1 r)$$

Making $M_1 r$ for each secondary balancer shaft

$$= \frac{M_{\text{Recip}} R \frac{R}{L}}{8}$$

The balancer masses must be phased to act downwards when the piston is at TDC.

10.2.3 Two-cylinder engines

Three main arrangements of two-cylinder engines are to be found. These are (a) cranks at 180°, (b) cranks at 360° and (c) horizontally opposed. *Figure 10.10* indicates in (a) that number 2 cylinder (No 2 crank) must fire 180° after cylinder number 1 with the crank arrangement shown in *Figure 10.10b* and that 540° crankshaft occurs before cylinder No 1 fires again. In *Figure 10.10b* the vertical crankthrows shown on the left have been projected to the right. If the rotating mass per crank is M and the radius of the throw R then a rotating couple MR exists producing a rotating force vector at the centre line of each crankthrow equal to $\omega^2 MR$, where ω is the angular velocity of the crankshaft.

Now as the two out-of-balance forces are equal and opposite they produce force balance for the crankshaft as a whole. However they are not co-planar, but spaced apart by the cylinder center distance L, and produce a rotating force couple of $\omega^2\, MRL$ in a clockwise direction as shown. This is resisted, if no balance weights are fitted to the crankshaft, by the two engine main

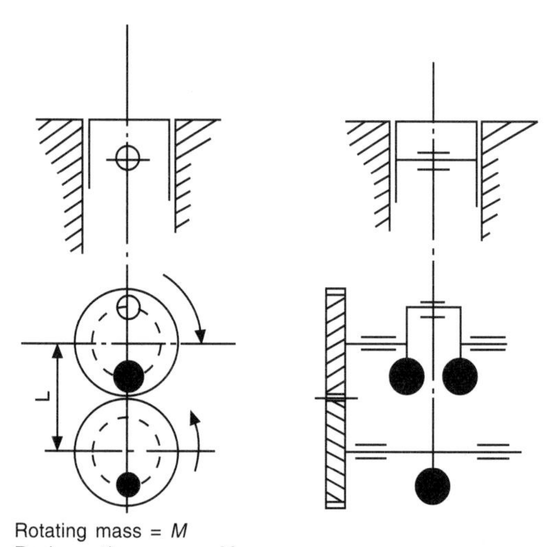

Rotating mass = M
Reciprocating mass = M_R
Crank radius = R

Total moment of crankshaft balance weight

$$= mr = \left(M + \frac{M_R}{2}\right) R$$

Moment of each balancer mass = $m_1 r_1 = 0.5\ M_R R$
Rolling couple introduced = $0.5 M_R RL$

Figure 10.9 Compromise single-cylinder engine primary balance using single balancer shaft

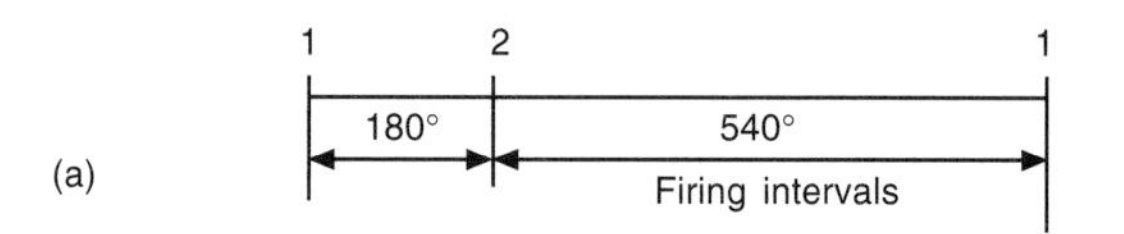

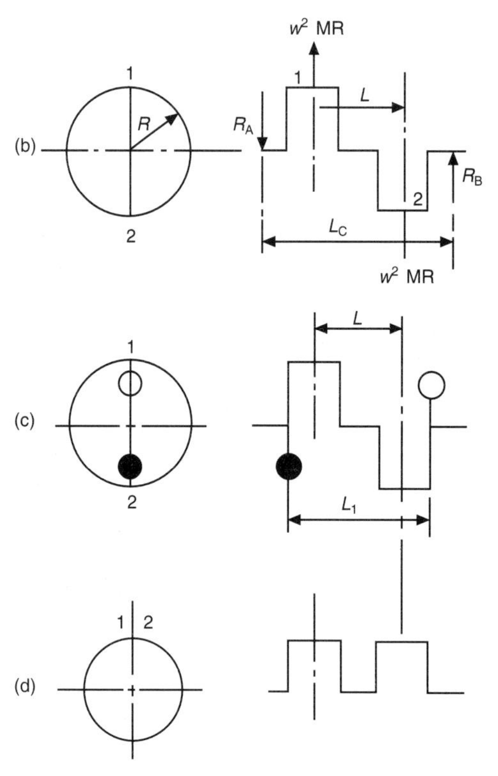

Figure 10.10 Balance of 180° crank two-cylinder engine

bearings which if spaced a distance L_c apart results in a constant rotating load equal to $\omega^2 MRL/L_c$ being applied at a crankshaft angular velocity ω radian/sec. This unbalanced rotating couple is transmitted through the bearings to the crankcase and its mountings leading to engine vibration.

To eliminate this rotating couple balance weights must be fitted to oppose it. If two balance weights are fitted on the two end webs as shown in *Figure 10.10c* each balance weight must have a moment mr equal to MRL/L_1, to eliminate the rotating couple completely.

If four balance weights are fitted, one to each crank web, they each must have a moment mr equal to $MRL/2L = MR/2$. Since the two inner weights are at a longitudinal distance apart which is less than L, whereas the two outer weights are greater than L, their mass is used less efficiently in the reduction of the longitudinal rotating couple. Also, since the disturbing force acting at the centre bearing is zero due to the forces from the adjacent throws being equal and opposite, balance weights are only used on the two inner webs when the outside radius for the balance weights has to be minimized.

With a reciprocating mass M_{Recip} acting along the centre line of each cylinder a primary out-of-balance moment of $\pm M_{\text{Recip}}R \cos\theta$ acts along each cylinder axis producing a primary out-of-balance force equal to $\pm\,\omega^2 M_{\text{Recip}}R\cos\theta$. Now the two primary forces are phased apart by the crankthrow angle of 180° so that in their maximum value position crankthrow 1 is up and crankthrow 2 is down the configuration being similar to that shown for the rotating forces in *Figure 10.10b*, it will be seen that these primary forces always balance one another. However, as for the rotating forces, the spacing apart of the cylinders by a distance L means that, for an engine having its cylinders vertical, a vertical pitching couple of $\pm\,\omega^2 M_{\text{Recip}}RL$ will exist.

As for the single-cylinder engine, the fitting of any extra mr value to the rotating crankshaft weights will introduce a horizontal yawing harmonically varying moment in the horizontal plane, and equal to that proportion of the pitching couple removed from the vertical plane. Where it is wished to eliminate the vertical pitching couple, it is necessary to arrange for two contra-rotating balancer shafts rotating at crankshaft speed each having two longitudinally spaced balance weights, so as to produce an opposing reciprocating couple to the pitching moment due to the primary forces. The principle is identical to that used in the single cylinder except that an opposing primary couple must be produced instead of a force. If the contrarotating weights each have an mr value of $m_B r_B$ and are spaced distance L_B apart logitudinally on their shafts, then

$$2m_B r_B L_B = M_{\text{Recip}}RL$$

Now as previously mentioned in connection with eqn (10.10) the motion of a piston on its connecting rod produces a secondary out-of-balance force equal to $\omega^2 M_{\text{Recip}}R(R/L)\cos 2\theta$. Here ω and θ both refer to the crankshaft movement. This harmonically varying force has a vector which rotates at twice crankshaft speed. So referring to *Figure 10.10d* with the secondary vector for No 1 cylinder in its vertical position the vector for No 2 cylinder is 2 × 180° = 360° behind that for No 1; this is akin to the diagrammatic cranks being drawn as shown. It will be seen that these are unbalanced harmonically varying forces, no couple being present for this two cylinder. So the engine has in addition to the primary pitching couple a secondary vertical disturbing force of

$$2\omega^2 M_{\text{Recip}} R \frac{R}{L} \cos 2\theta$$

With the two cranks for a two-cylinder engine at 360°, with the two cylinders firing alternately, equal firing intervals are obtained. However with the crankthrows up together the balancing is really the same as for a large single-cylinder engine, the forces being twice those for a single-cylinder engine having the same bore and stroke.

The horizontally opposed twin engine is really a case of a 180° vee engine with the big-end bearings attached to separate crankthrows. The general layout is diagrammatically shown in *Figure 10.11*.

It will be seen that the rotating out-of-balance is similar to that of he 180° crankshaft arrangement with two veritical cylinders and can be eliminated by balance weights on the two end crankwebs in a similar way. The pistons are at their respective TDCs at the same time but are always moving in opposite directions. This means that both primary and secondary forces are cancelled. However the distance between the cylinder centres results in both primary and secondary couples being unbalanced. With the conventional mounting of the cylinder axes horizontally —hence 'horizontally opposed'—these harmonically varying primary and secondary couples exist in the horizontal plane.

10.2.4 Four-cylinder in-line engines

The form of crankshaft normally used in four-cylinder four-cycle engines is shown in *Figure 10.12*. It is easy to manufacture and gives equal firing intervals, the usual firing order being 1–3–4–2–1. As can be seen, it is really two 180° twin cylinder shafts end to end but with one half tuned end to end. The result, as shown in the upper part of the figure, is to produce couples in each half of the crankshaft of opposite sign, which means that for the whole shaft there is complete rotating and primary balance.

However if no balance weights are fitted, the rotating couple of MRL in each half of the shaft will result in a rotating force of MRL/L_1 being imposed at the centre of bearings 1, 3 and 5. In the case of bearing No 3 the force is due to two couples and amounts to $2\omega^2 MRL/L_1$, being the resultant Mr value multiplied by ω^2 to produce the force at the prevailing crankshaft speed.

These internal rotating couples react on the crankcase tending to bend it. So the crankcase must be made strong enough to resist them. It is usual to fit balance weights to crankwebs 1, 4, 5 and 8 to reduce the loads on the bearings, particularly No 3 bearing, and to minimize crankcase deflections.

The lower diagram in *Figure 10.12* shows the relationship for the secondary forces, the vector angle between them being twice that for the primaries. It will be seen that they all have their maximum value at the same time so that the secondaries are completely unbalanced. No secondary couples are produced. The total secondary force unbalance has a maximum value of

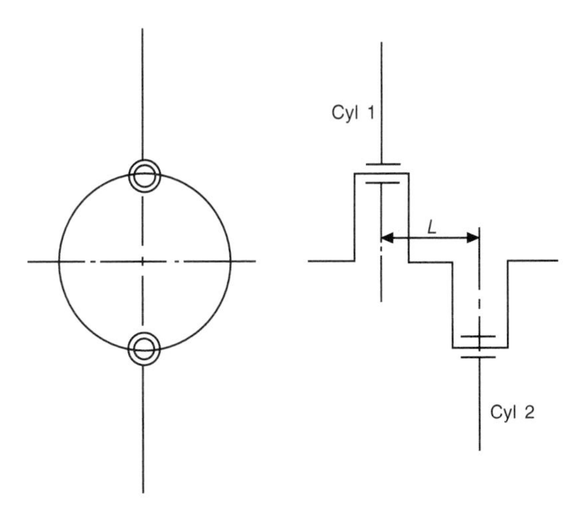

Figure 10.11 Arrangement of two-cylinder horizontally opposed engine

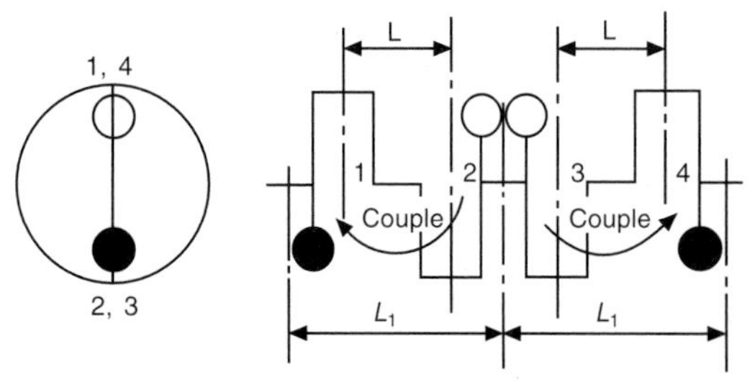

Rotating and primary forces and couples

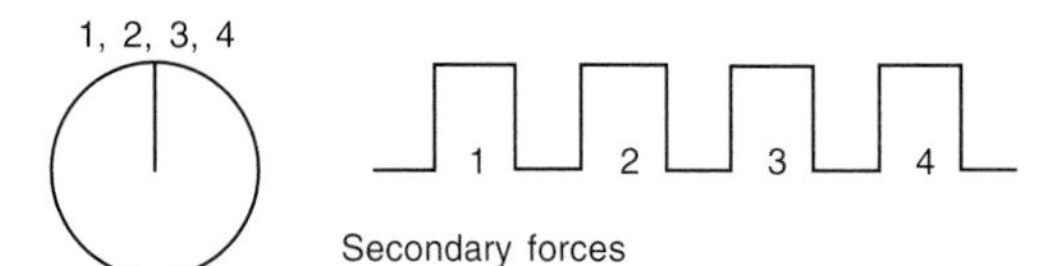

Secondary forces

Figure 10.12 Balance of four-cylinder in-line engine

$$\pm 4\omega^2 M_{\text{Recip}} R \frac{R}{L} \approx \pm \omega^2 M_{\text{Recip}} R$$

In most automotive engines the use of flexible mountings for the engine having a low stiffness, i.e. the engine has a low natural frequency on its mountings, usually reduces the secondary harmonically varying vertical forces reaching the chassis to a value which is acceptable. When it is required to eliminate these secondaries, it is necessary to fit two contrarotating shafts running at twice engine speed fitted with suitable balance weights, arranged so that their cancelling secondary harmonic force is at the centre line of the centre main bearing.

10.2.5 Three-cylinder engines

As seen in *Figure 10.13* the three crankthrows are arranged at 120° apart, although the firing interval for a four-cycle engine will be 240°.

Consider now the rotating balance of the shaft. With a rotating moment *MR* on each crankpin, we have in the vertical projection in relation to No 1 crankthrow, shown to the right, a vertical upward force at No 1 pin proportional to *MR*, whilst each of pins 2 and 3 have a downward value of *MR* cos 60° = 0.5*MR*. Thus the rotating forces are in balance. Similarly in the horizontal projection below the rotating forces are in balance.

Taking moments about point *X*, which for convenience has been taken at a distance of *L*/2 from the centre line of No 1 crankthrow, or cylinder (this in fact approximates to the centre line of No 1 main bearing but any point could have been chosen), the moments about *X*, calling clockwise couples positive, are from the vertical projection

Total Moment

$$= -MR\frac{L}{2} + MR\tfrac{3}{2}L\cos 60^\circ + MR\tfrac{5}{2}L\cos 60^\circ$$

$$= MR\left[-\frac{L}{2} + \tfrac{3}{4}L + \tfrac{5}{4}L\right]$$

$$= +\tfrac{3}{2}MRL \text{ (clockwise)}$$

From the horizontal projection:

Total Couple = *MR* sin 60°*L* = 0.866*MRL*

or working it out at length taking moments about *X*

$$\text{Total Moment} = 0\frac{L}{2} - MR\sin 60^\circ\tfrac{3}{2}L + MR\sin 60^\circ\tfrac{5}{2}L$$

$$= [-\tfrac{3}{2} + \tfrac{5}{2}]\,MRL\sin 60^\circ$$

$$= 0.866\,MRL \text{ (clockwise)}$$

The Resultant Couple = $\{1.5^2 + 0.866^2\}^{1/2}$ = 1.732*MRL*

whose angle from No 1 crank is $\tan^{-1}\dfrac{0.866}{1.5} = 30^\circ$

which in fact is at right angles to crank No 2.

Now suppose *Figure 10.13* is redrawn with crank No 2 vertical as shown in *Figure 10.14*. It is immediately obvious that in the vertical projection both the forces and couples are in balance whereas in the horizontal projection the forces are in balance but a maximum couple exists having the magnitude *MR* sin 60°2*L* = 1.732*MRL* as found above. In many cases it is possible to select the most appropriate planes for the projections and hence to simplify the work.

Knowing the plane in which the maximum couple occurs permits the selection of appropriate balance weights and their line of action. For example if four balance weights are fitted as indicated on *Figure 10.14* their *mr* value will each be

1.732*MRL*/4*x*2*L* = 0.217*MR*

Considering the primary out-of-balance analytically, first of all

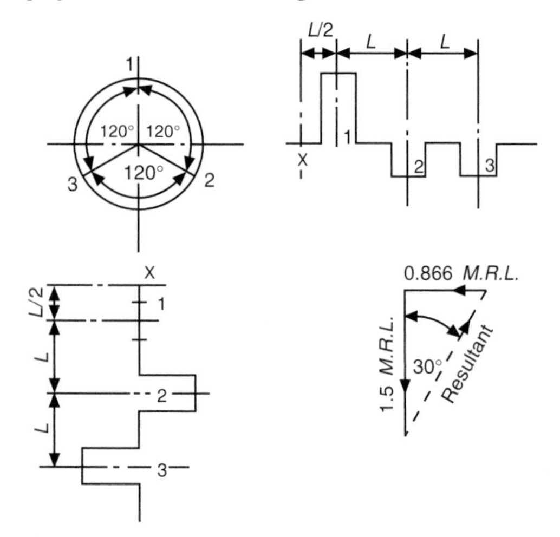

Figure 10.13 Rotating out-of-balance for three-cylinder crankshaft

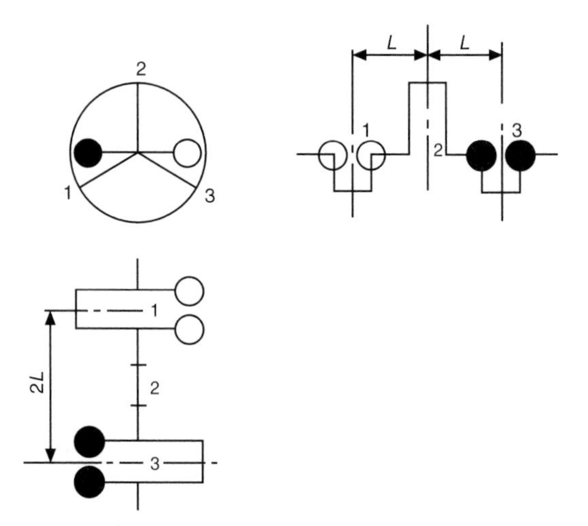

Figure 10.14 Rotating balance for three-cylinder crankshaft

the disturbance is along each cylinder axis and varies harmonically from a maximum at TDC. This maximum value has a moment of $M_{Recip}R$ and the value at any crank angle θ is given by $M_{Recip}R \cos\theta$.

Using No 1 cylinder as the reference cylinder No 2 cylinder lags 120° behind and No 3 cylinder 240° behind. So the total primary moment for all three cylinders is

$$[\cos\theta + \cos(\theta - 120°) + \cos(\theta - 240°)]M_{Recip}R$$

$$= [\cos\theta + \cos\theta\cos 120° + \sin\theta\sin 120° + \cos\theta\cos 240° + \sin\theta\sin 240°]M_{Recip}R$$

$$= [\cos\theta - 0.5\cos\theta + 0.866\sin\theta - 0.5\cos\theta - 0.866\sin\theta]M_{Recip}R$$

$$= 0$$

Considering the primary couple this is, clockwise positive, about X:

$$-M_{Recip}R\left[\frac{L}{2}\cos\theta + \tfrac{3}{2}L\cos(\theta - 120°) + \tfrac{5}{2}L\cos(\theta - 120°)\right]$$

$$= -M_{Recip}RL[0.5\cos\theta + 1.5(\cos\theta\cos 120° + \sin\theta\sin 120°) + 2.5(\cos\theta\cos 240° + \sin\theta\sin 240°)]$$

$$= -M_{Recip}RL[0.5\cos\theta + 1.5(-0.5\cos\theta + 0.866\sin\theta) + 2.5(-0.5\cos\theta - 0.866\sin\theta)]$$

$$= -M_{Recip}RL[(0.5 - 0.75 - 1.25)\cos\theta + 0.866(1.5 - 2.5)\sin\theta]$$

$$= -M_{Recip}RL[-1.5\cos\theta - 0.866\sin\theta]$$

Differentiating the expression within the bracket

$$\frac{dy}{d\theta} = 1.5\sin\theta - 0.866\theta$$

$$= 0 \text{ for maximum or minimum}$$

So $1.5\sin\theta = 0.866\cos\theta$

i.e. $= \tan^{-1}\dfrac{0.866}{1.5} = 30°$ or 210° from TDC for No 1 cylinder.

So the maximum value for the primary reciprocating couple becomes

$$= M_{Recip}RL + 1.5\cos 30° + 0.866\sin 30°$$

$$= 1.732M_{Recip}RL\text{(clockwise)}$$

This is similar in form and position to that for the rotating out-of-balance couple resultant. It is of course the maximum value of the primary pitching couple acting on the whole engine.

It is also possible to derive the primary inertia balance by using the concept of 'reverse cranks' referred to earlier. In this case each crank has a moment of 0.5 $M_{Recip}R$ considered attached at each crankpin. The direct crank arrangement will be identical to that shown for the rotating out-of-balance in *Figure 10.13*. That for the reverse crank will also be similar but rotating in the reverse direction. It follows that the resultant for the two cranks will be as determined mathematically above. The secondary forces will be in balance but again a secondary pitching couple will be present for the whole engine.

Figure 10.15 uses the principle of the reverse cranks to eliminate the primary pitching couple. Here balance weights are provided on the crankshaft to eliminate all the rotating out-of-balance plus half the pitching moment as the 'direct crank' whilst the 'reverse crank'—the reverse rotating balancer shaft—deals with the other half of the primary pitching couple.

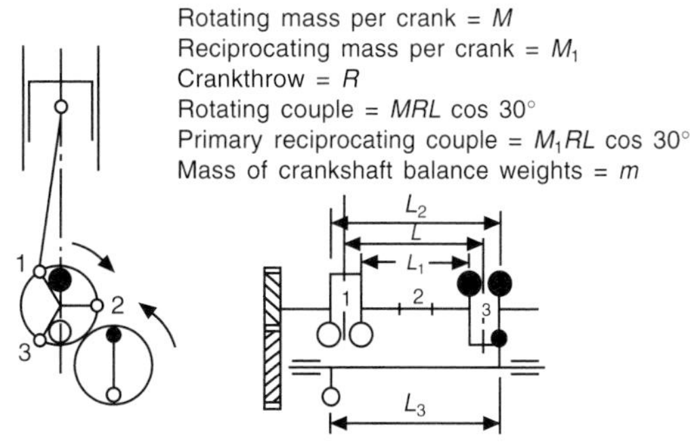

For balance wts on crankshaft $mr(L_1 + L_2) = RL\cos 30°\left(M + \dfrac{M_1}{2}\right)$

i.e. moment required for balance weights on crankshaft $= Mr = \dfrac{\left(M + \dfrac{M_1}{2}\right)RL\cos 30°}{(L_1 + L_2)}$

Moment required for countershaft balance weights $= \dfrac{M_1RL\cos 30°}{2L_3}$

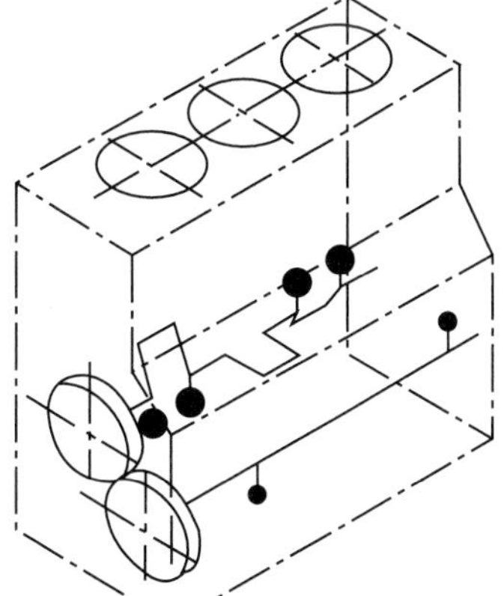

Figure 10.15 Rotating and primary balance of three-cylinder engine using one primary balancer shaft

10.2.6 Six-cylinder engines

The crankshaft arrangement normally used for an in-line six-cylinder engine is shown in *Figure 10.16*. It will be seen that this is two three-cylinder shafts placed end to end, but with one half reversed. As shown, the shaft as a whole is in rotating force and couple balance, together with the primary forces and couples. To reduce the magnitude of the two opposing internal couples, and hence the loads on Nos 1, 4 and 7 main bearings, four balance weights are indicated placed angularly at right-angles to Nos 2 and 5 crankpins.

A number of other ways of positioning balance weights are possible with a view to relieving the between throws rotating bearing load. The one indicated is the minimum number of weights required to eliminate or reduce, if smaller weights are used, the magnitudes of the two opposing internal couples. *Figure 10.17* shows the arrangement of the secondary forces and couples. These, like the primary ones, are innately balanced for the whole engine.

10.2.7 Vee engines

10.2.7.1 The general case—single throw

Figure 10.18 shows a diagrammatic section through a vee engine having an angle 2α between the two banks normally referred to as the bank angle.

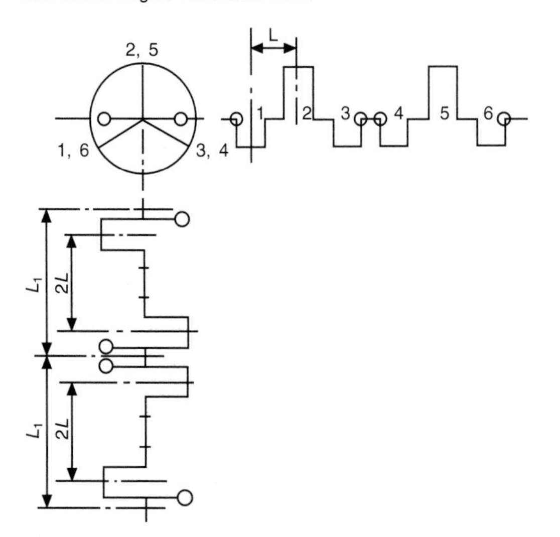

Figure 10.16 Balance of six-cylinder shaft. Rotating and primary balance

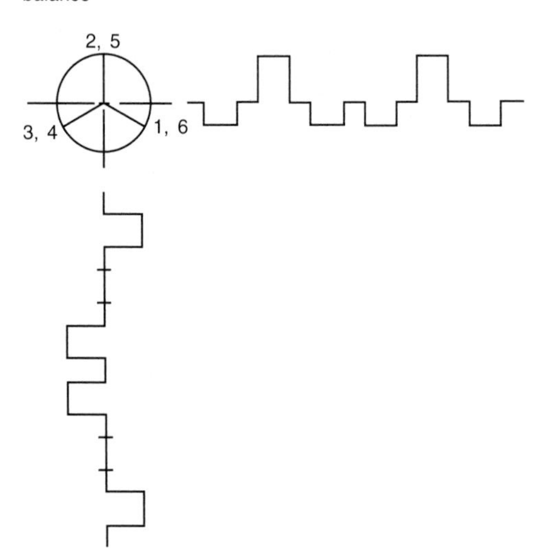

Figure 10.17 Diagram for secondary out-of-balance for six-cylinder engine

Primary out-of-balance forces

If the total reciprocating mass for each cylinder is M_{Recip}, the crankthrow is R, the connecting rod length L, then if the crank angle turned through from TDC of the left-hand cylinder A is θ the primary inertia force acting along $0A$ is given by:

$$F_1 = \omega^2 M_{Recip} R \cos\theta$$

$$= K \cos\theta \qquad (10.11)$$

where $K = \omega^2 M_{Recip} R, \omega$ being the crankshaft angular velocity in radian/sec.

This may be resolved into

$K \cos\theta \cos\alpha$ vertically

and $K \cos\theta \sin\alpha$ horizontally

Considering bank B its TDC occurs at an angle 2α later than that of bank A.

So when the crank has turned through θ from TDC of A bank it has only turned $(\theta - 2\alpha)$ from TDC B bank.

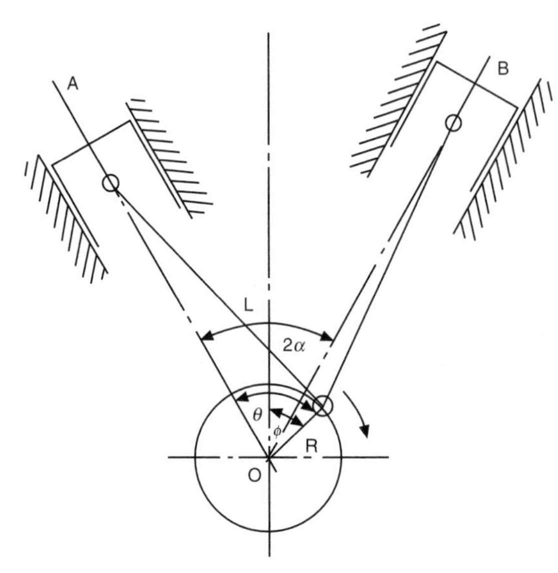

Figure 10.18 Layout of one crank unit for a vee engine

So the total vertical primary force due to both banks

$= K[\cos\theta \cos\alpha + \cos(\theta - 2\alpha)\cos\alpha]$

$= K \cos\alpha[\cos\theta + \cos(\theta - 2\alpha)]$

If crank angle is measured from the vertical centre line—the axis of symmetry

then, $\phi = (\theta - \alpha)$

i.e. $\theta = (\phi + \alpha)$

so that the total vertical primary force becomes

$= K \cos\alpha[\cos(\phi + \alpha) + \cos(\phi + \alpha - 2\alpha)]$

$= K \cos\alpha[\cos(\phi + \alpha) + \cos(\phi - \alpha)]$

$= K \cos\alpha\,[2\cos\phi \cos\alpha]$

$= 2K \cos^2\alpha \cos\phi$

$$= K(1 + \cos 2\alpha)\cos\phi \qquad (10.12)$$

Horizontal primary forces

Forces to right considered as positive.

Total Horizontal Force $= K[\cos(\theta - 2\alpha)\sin\alpha - \cos\theta \sin\alpha]$

$= K \sin\alpha[\cos(\theta - 2\alpha) - \cos\theta]$

$= K \sin\alpha[\cos(\phi + \alpha - 2\alpha) - \cos(\phi + \alpha)]$

$= K \sin\alpha[\cos(\phi - \alpha) - \cos(\phi + \alpha)]$

$= K \sin\alpha\,[-2\sin\phi \sin(-\alpha)]$

$= K2 \sin^2\alpha \sin\phi$

$$= K(1 - \cos 2\alpha)\sin\phi \qquad (10.13)$$

Secondary out-of-balance forces

Secondary out-of-balance force along axis of bank A

$$F_\pi = \omega^2 M_{Recip} R \frac{R}{L} \cos 2\theta$$

$$= k \cos 2\theta \text{ where } k = \omega^2 M_{Recip} R \frac{R}{L}$$

So in a similar manner to that for the primary forces the total vertical secondary force for both banks

$k \cos \alpha[\cos 2\theta + \cos 2(\theta - 2\alpha)]$

hich by manipulation becomes

$k(\cos 3\alpha + \cos \alpha) \cos 2\phi$ (10.14)

he total horizontal secondary force

$k(\cos \alpha - \cos 3\alpha) \sin 2\phi$ (10.15)

valuating the bracketed terms in these expressions for V angles ormally met with, the values tabulated in *Table 7.1* are obtained.

Figure 10.19 shows the full state of affairs for one crankthrow a 60° Vee engine. When the crank is vertical, i.e. on the sector of the Vee angle, the maximum primary out-of-balance oment of $1.5M_{Recip}R$ is present in an upward direction, whilst horizontal disturbance occurs. As the crankshaft rotates the ertical disturbance varies as cos ϕ, since the bracketed term is onstant as such as K in eqn (10.12). Similarly the horizontal orce force from eqn (10.13) varies as sin ϕ. This means that the adius vector swept out by the co-ordinate disturbing forces ollows an elliptical path, as shown by the dotted locus. The eader should be reminded that whilst it is convenient to use the oment as a constant time $M_{Recip}R$, when actual forces are equired for any crankshaft angular velocity ω rad/sec, the moment xpression must be multiplied by ω^2 to give the force disturbance n newtons.

The disturbing forces leading to the elliptical locus for their ector mgnitude as the crankthrow rotates can be simulated by wo masses, one equal to the reciprocating mass moment for ne cylinder revolving in the same direction as the crank, and he other equal to one half the reciprocating mass moment for ne cylinder rotating in the opposite direction. Both masses are t crankpin radius. With the two equivalent masses vertical they roduce an upward force proportional to $1.5M_{Recip}R$, 180° later he downward force is again proportional to $1.5M_{Recip}R$. In the 0° and 270° positions the large equivalent mass is in opposition o the small one, so that the horizontal vector is proportional to $.5M_{Recip}R$.

0.2.7.2 *Six-cylinder 60° Vee engine*

Figure 10.20 shows diagrammatically the balancing for the otating and primary out-of-balance for a six-cylinder 60° Vee ngine. The crankthrow arrangement is similar to that for a hree-cylinder in-line engine. The upper diagram (a) shows alance weights fitted on the crankshaft at right angles to No 2 hrow, to eliminate the rotating out-of-balance couple due to the rankthrow's own *MR* value plus that due to two big-end masses n each crankpin. Two contrarotating balancer shafts are used, he one rotating in the same direction as the crankshaft deals with the 'direct acting crank' fo the primary out-of-balance, i.e. $M_{Recip}R$, whilst the reverse rotation shaft deals with that for the 'reverse crank', i.e $0.5M_{Recip}R$.

The lower part (b) of *Figure 10.20* shows that the 'direct acting' crank balance weights can be incorporated with those dealing with the rotating out-of-balance of the crankshaft whilst

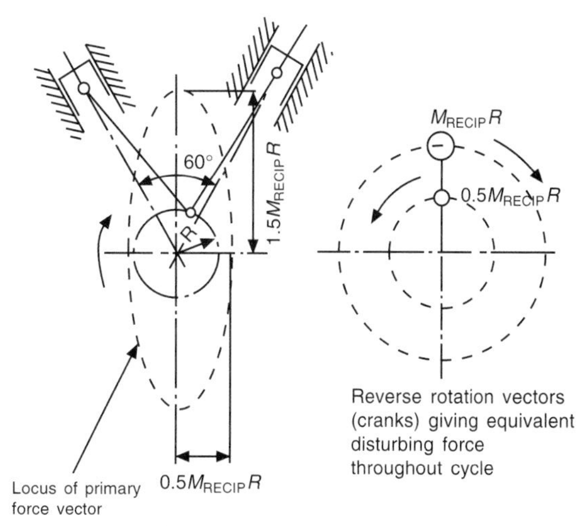

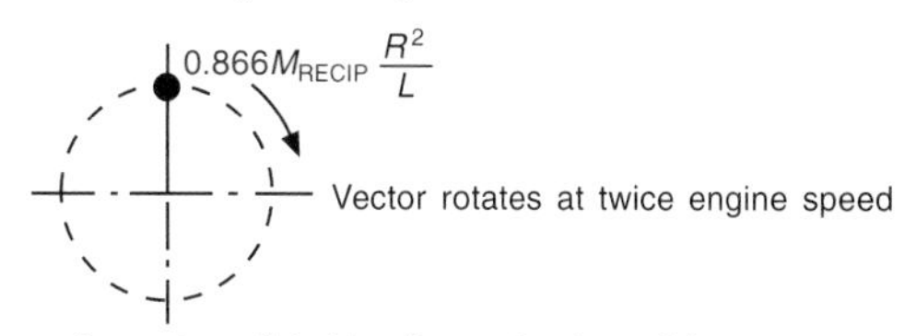

Figure 10.19 Primary and secondary out-of-balance forces for one throw of a 60° vee engine

the single contrarotating balancer shaft deals with the couple arising from the 'reverse crank' primary forces.

10.2.7.3 *Eight-cylinder 90° vee engine*

For one throw of a 90° vee engine the vertical component for the resultant primary out-of-balance force is $\omega^2 M_{Recip}R \cos \phi$, and the horizontal disturbing component is $\omega^2 M_{Recip}R \sin \phi$. The vector locus is therefore a circle. This is equivalent to adding the reciprocating mass for one piston to the crankpin. The out-of-balance per crankthrow is therefore the sum of one reciprocating mass plus two rotating masses at crankpin radius. These add to the *mr* value for the crankshaft throw alone. If this is called $M_{equiv}R$, use can be made of it for balancing the crankshaft.

The usual crankshaft layout used to obtain equal firing intervals is shown in *Figure 10.21*. From this it can be seen that a vertical couple of magnitude $3M_{equiv}RL$ exists and a horizontal one of $M_{equiv}RL$. These have a resultant of $3.162M_{equiv}RL$ acting at an angle of approximately $\tan^{-1}(\frac{1}{3})$ or 18.5° from the line of No 1 crankthrow. This out-of-balance rotating couple is normally eliminated by the use of balance weights fitted as indicated.

Table 10.1

	Primary force		*Secondary force*	
V angle 2α	*Vertical* $(1 + \cos 2\alpha)$	*Horizontal* $(1 - \cos 2\alpha)$	*Vertical* $(\cos 3\alpha + \cos \alpha)$	*Horizontal* $(\cos \alpha - \cos 3\alpha)$
30	1.866	0.134	1.673	0.259
45	1.701	0.293	1.306	0.541
60	1.500	0.500	0.866	0.866
72	1.309	0.691	0.500	1.118
90	1.000	1.000	0	1.414
120	0.500	1.500	– 0.500	1.500
180	0	2.000	0	0

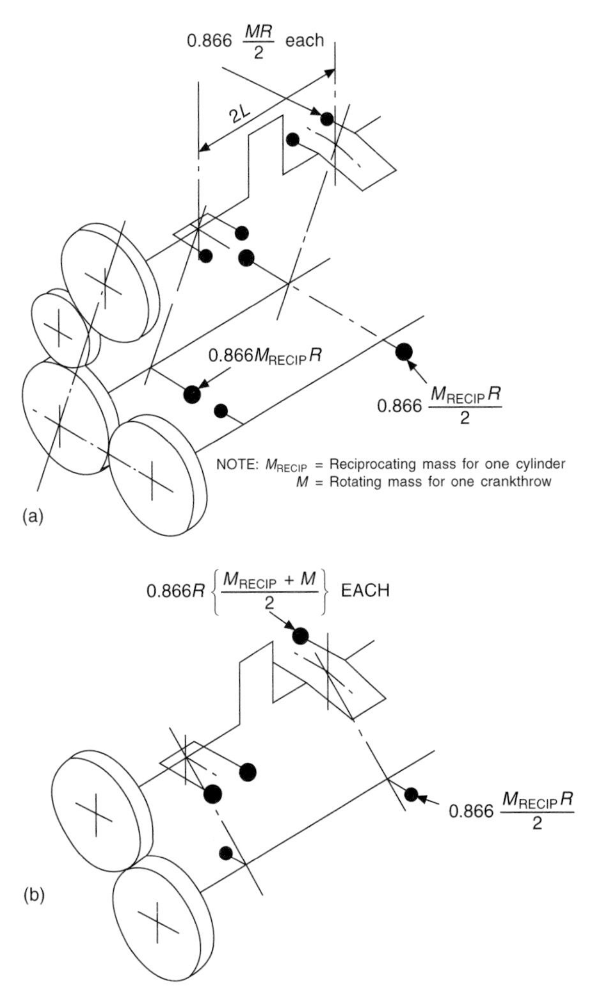

Figure 10.20 Balancing of six-cylinder 60° vee engine

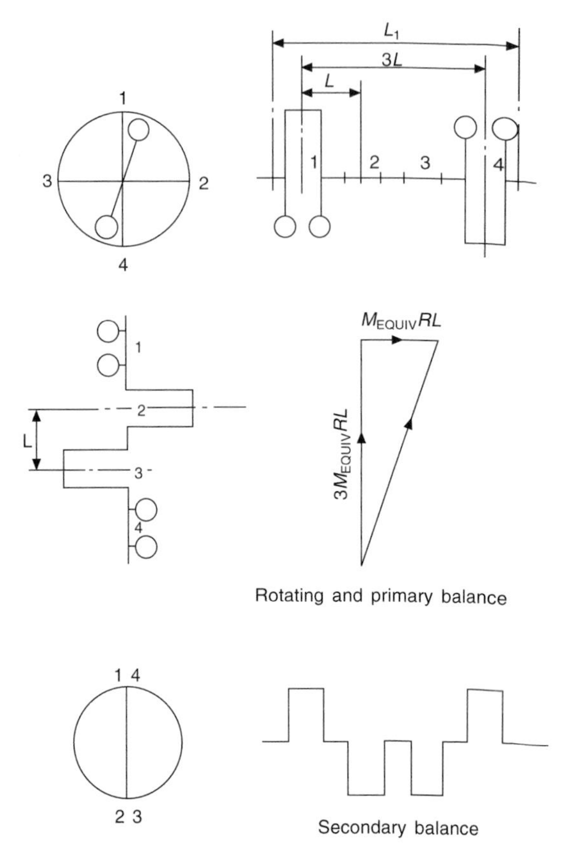

Figure 10.21 Balance of crankshaft for eight-cylinder 90° vee engine

So far as secondary balance is concerned the resultant secondary out-of-balance per crankthrow represents a harmonically varying vector in the horizontal direction only. However when the phasing and longitudinal position of these vectors is considered (see lower half of *Figure 10.21*) it can be seen that for the whole shaft both the secondary forces and couples are balanced.

Where dynamic balancing in production of eight-cyinder 90° vee engine crankshafts is done it is necessary to clamp a concentric weight around each crankpin whose mass is equal to two big-end masses plus the mass of one reciprocating mass. It is also necessary to maintain the absolute mass, or weight for the big-ends, small-ends and fully equipped pistons to close tolerances in production.

10.2.8 Two-stroke engines

Similar methods may be used to balance two-stroke engines as for four-stroke ones. However due to the fact that the full sequence of cylinders must fire every crankshaft revolution it is not possible to arrange cranks as an exact mirror image in the two longitudinal halves of the crankshaft. As has been shown this can, for the four-stroke engine shaft, allow the shaft as a whole to be in balance due to two opposing couples cancelling out. In general this cannot be so readily done for two-stroke engine shafts.

In the case of four-cycle engines having even numbers of cylinders there are at least several firing sequences which can be used for a given crank arrangement. In the case of the two-cycle engine, if equal firing intervals are to be maintained each firing order requires its own crankshaft arrangement.

Taking, for example, a six-cylinder two-stroke engine, *Figure 10.22* indicates three possible crankthrow sequences with their firing order. *Figure 10.23* shows the crankshaft (*Figure 10.22a*) laid out for balancing considerations in a similar way to that shown for four-stroke engines. With this shaft arrangement there are two opposing couples each of magnitude $MRL \sin 30°$ in the vertical projection to the right. In the case of the horizontal projection drawn below there is a clockwise couple of MRL due to cranks 3 and 4 opposed by two anticlockwise couples each of magnitude $MRL \sin 30° = 0.5MRL$, so that the primary forces and couples are in balance.

Figure 10.24 shows the crank arrangement given in *Figure 10.22c* laid out to show its balance. In the vertical projection to the right it can be seen that the forces are in balance and no couples exist. On the other hand the horizontal projection downwards is the plane in which the maximum couple out-of-balance occurs. As drawn, two clockwise couples exist each having a value of $0.866MR2L$ making the total for the whole shaft $2 \times 0.866MR2L = 3.464MRL$. One possible arrangement of balance weights to eliminate the rotating out-of-balance in the shaft as a whole is shown dotted. The primary out-of-balance couple will also have a maximum value of numerically the same value but, of course, occurs in the engine's vertical plane causing a pitching couple.

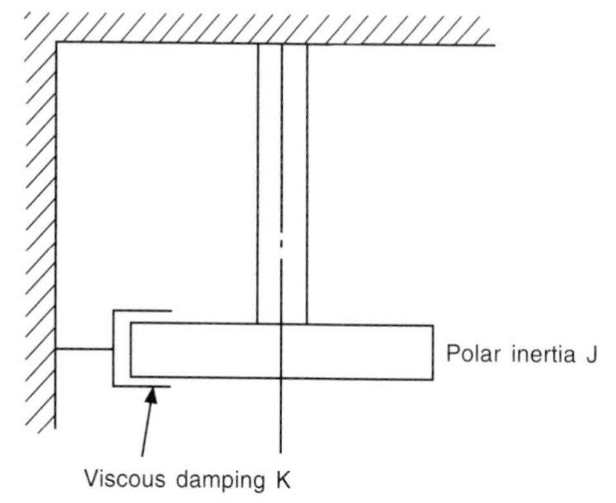

Figure 10.27 Simple torsional pendulum with damping

or

$$J\frac{d^2\theta}{dt^2} + k\frac{d\theta}{dt} + C\theta - T_0 \sin \omega t = 0$$

or

$$J\ddot{\theta} + k\dot{\theta} + C\theta - T_0 \sin \omega t = 0 \quad (10.19)$$

Assuming the particular solution to be of the form

$$\theta = \theta_{max} \sin (\omega t - \phi) \quad (10.20)$$

where θ_{max} = maximum amplitude of vibration;
ϕ = phase angle.

Then eqn (10.20) becomes

$$J\omega^2\theta_{max} \sin (\omega t - \phi) - k\,\omega\theta_{max} \sin \left(\omega t - \phi + \frac{\pi}{2}\right)$$

$$- C\theta_{max} \sin (\omega t - \phi) + T_0 \sin \omega t = 0$$

From the vector diagrams *Figure 10.28*, considering the right hand triangle

$$T_0^2 = [(C - J\omega^2)^2 + (k\omega)^2]\,\theta_{max}^2$$

$$\text{So } \theta_{max} = \frac{T_0}{\sqrt{(C - J\omega^2)^2 + (k\omega)^2}} \quad (10.21)$$

and

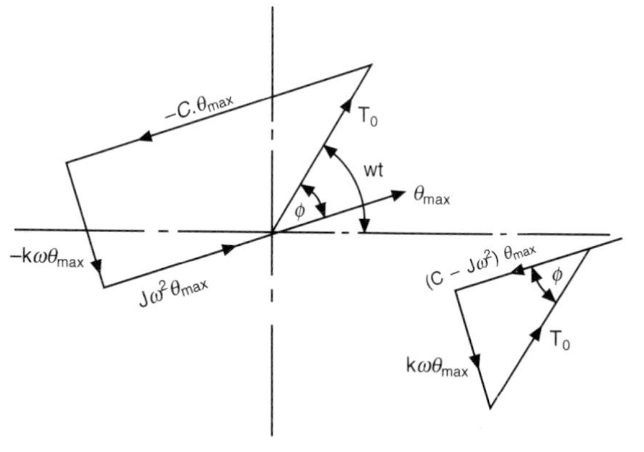

Figure 10.28 Vector diagram for torsional pendulum with forced vibration and damping

$$\tan \phi = \frac{k\omega}{(C - J\omega^2)} \quad (10.22)$$

Dividing numerators and denominators by C, gives

$$\theta_{max} = \frac{T_0/C}{\left[\left(1 - \frac{J\omega^2}{C}\right)^2 + \left(\frac{k\omega}{C}\right)^2\right]^{1/2}} \quad (10.21)$$

$$\tan \phi = \frac{(k\omega/C)}{\left(1 - \frac{J\omega^2}{C}\right)} \quad (10.22)$$

These may be further reduced for practical use by adopting the following relationships:

$\omega_n = \sqrt{\frac{C}{J}}$ = natural frequency of undamped oscillation in rad/sec;
$\zeta = k/k_c$ = damping factor;
$k_c = 2J\omega_n$ = critical damping coefficient;
$\theta_{st} = T_0/C$ = deflection of spring mass system if a steady torque of T_0 were applied.

Then

$$\frac{\theta_{max}}{\theta_{st}} = \frac{1}{\sqrt{\left[1 - \left(\frac{\omega}{\omega_n}\right)^2\right]^2 + \left[2\zeta\frac{\omega}{\omega_n}\right]^2}} \quad (10.21b)$$

$$\tan \phi = \frac{2\zeta\left(\frac{\omega}{\omega_n}\right)}{\left[1 - \left(\frac{\omega}{\omega_n}\right)^2\right]} \quad (10.22b)$$

The term θ_{max}/θ_{st} is called the Dynamic Magnifier. This is the factor by which the deflection of the system under a steady torque T_0 must be multiplied to find the actual maximum deflection under vibratory conditions with damping. Both θ_{max}/θ_{st} and ϕ, it should be noted, are only dependent on the applied frequency ratio ω/ω_n and the damping factor ζ.

For most considerations of torsional vibration problems little note is taken of phase angle due to damping. The 'dynamic magnifier' is used considerably when estimating amplitude and stresses. Measurements of torsional vibration amplitudes on actual engines produce a series of amplitude against engine speed curves, corresponding to various order numbers, over the speed range very similar to those plotted in *Figure10.29*. The reason for this will be dealt with later.

10.3.1.3 Vibration of two-mass system

So far, for simplicity, reference has been made to a single mass vibrating against a fixed point. It is necessary in practice to consider two or more masses connected by shafts having stiffness but no mass, or polar inertia, with the whole shaft nominally rotating at a constant mean angular velocity which is what occurs in crankshaft systems of engines. Such a system is generally known as a free-free system since from a torsional vibration point of view there are no external constraints on the system.

Two-mass system

In *Figure 10.30a* two masses J_1 and J_2 are shown connected by

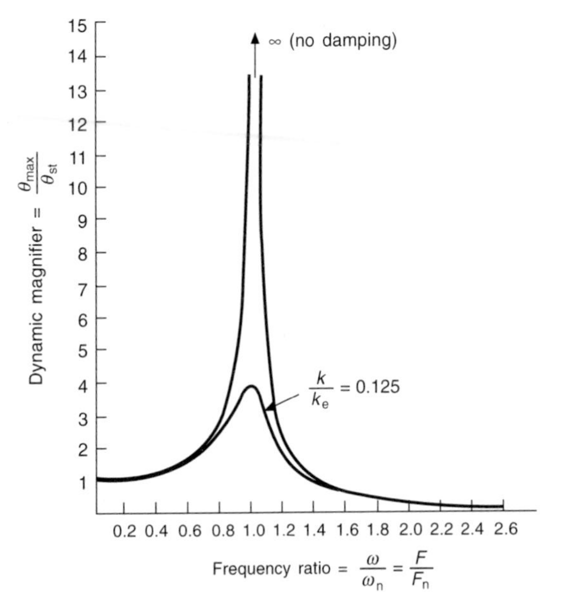

Figure 10.29 Variation of dynamic magnifier with frequency ratio and constant forcing torque

a length of shaft L. If it is imagined that equal and opposite torques T are applied simultaneously to the two ends of the shaft a total windup of T/C radians will occur in the shaft between the two masses. If now the applied torques are instantaneously removed the strain energy in the shaft is released causing the masses to move in opposite directions to one another. At a point to be determined there is no movement. This is called the node. The system can be regarded therefore as equivalent to that shown in *Figure 10.30b* when the left and right hand sections are equivalent to that given in *Figure 10.26*.

The natural frequency of the left hand system has the value

$$F_1 = \frac{1}{2\pi}\sqrt{\frac{C_1}{J_1}} \text{ from eqn (10.18)}$$

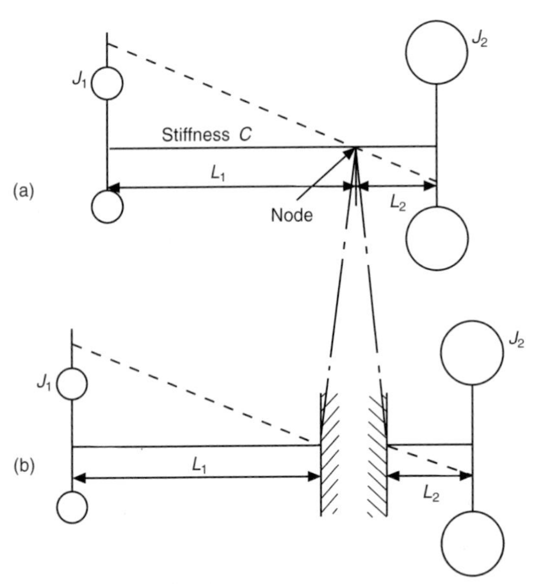

Figure 10.30 Two-mass system

$$= \frac{1}{2\pi}\sqrt{\frac{IG}{J_1 L_1}} \qquad (10.23)$$

For the natural frequency of the right hand system, similarly,

$$F_2 = \frac{1}{2\pi}\sqrt{\frac{C_2}{J_1}} = \frac{1}{2\pi}\sqrt{\frac{IG}{J_2 L_2}} \qquad (10.23a)$$

These frequencies must have the same value as in reality they belong to the same shaft-mass system. So the system natural frequency

$$F = F_1 = F_2$$

$$\text{So } F = \frac{1}{2\pi}\sqrt{\frac{C_1}{J_1}} = \frac{1}{2\pi}\sqrt{\frac{C_2}{J_2}}$$

$$= \frac{1}{2\pi}\sqrt{\frac{IG}{J_1 L_1}} = \frac{1}{2\pi}\sqrt{\frac{IG}{J_2 L_2}}$$

Now for a given system π, G and I are constants so

$$J_1 L_1 = J_2 L_2$$

$$\text{or } \frac{L_1}{L_2} = \frac{J_2}{J_1} \qquad (10.24)$$

i.e. the node divides the length of shaft inversely as the polar inertias of the end discs or masses. Now

$$L = L_1 + L_2 = L_1\left[1 + \frac{J_1}{J_2}\right]$$

making

$$L_1 = \left[\frac{J_2 L}{J_1 + J_2}\right]$$

Substitution of this value in eqn (10.23) gives

$$F = \frac{1}{2\pi}\sqrt{\frac{IG(J_1 + J_2)}{J_1 J_2 L}}$$

$$= \frac{1}{2\pi}\sqrt{\frac{C(J_1 + J_2)}{J_1 J_2}} \text{ vib/sec or Hertz} \qquad (10.25)$$

since $IG/L = C$ the torsional rigidity for the whole shaft between the two masses.

If, in a practical system, the shaft length between the two masses is significantly long its rotational mass is not negligible. It can be shown that this can be corrected for by adding one third of the polar inertia of the shaft length L_1 between the node and the left hand mass to J_1 and one third the polar inertia of the right hand length L_2 to J_2.

10.3.1.4 Three-mass system

Figure 10.31 shows a simple three-mass system having three heavy masses J_1, J_2 and J_3 connected together with shafting having a stiffness C_1 between masses J_1 and J_1 and a stiffness C_2 between masses J_2 and J_3.

In such a free-free system having no exterior restraining torques the torques due to the vibratory movements of the masses must exactly equal the elastic resisting torque due to the maximum twist of the connecting shafts. Remembering that the torque T required to vibrate a mass J through an amplitude $\pm\,\theta$ radians at a frequency of ω radian/second is $T = \pm\, J\omega^2\theta$, then starting at the left hand of the system given in *Figure 10.31* and supposing the maximum amplitude at mass A is $\pm\, a_1$ we have.

Torque to left of Mass $A = 0$

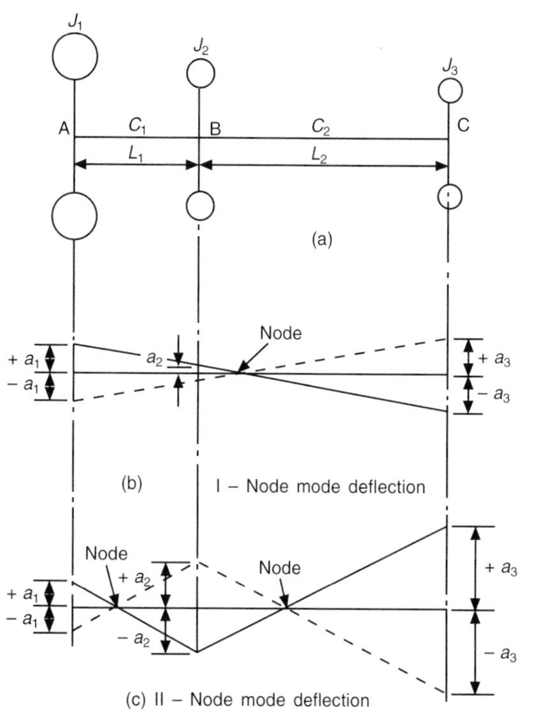

Figure 10.31 Three-mass system

Torque to right of Mass $A = M_2 = 0 + J_1\omega^2 a_1$

$$\text{Angle of twist between } J_1 \text{ and } J_2 = \frac{\text{Torque}}{\text{Stiffness}} = \frac{J_1\omega^2 a_1}{C_1}$$

$$\text{So deflection of Mass } B = a_2 = \left[a_1 - \frac{J_1\omega^2 a_1}{C_1}\right]$$

so torque to right of Mass B

$$= M_3 = M_2 + \text{Torque to accelerate mass } J_2$$

$$= J_1\omega^2 a_1 + J_2\omega^2\left[a_1 - \frac{J_1\omega^2 a_1}{C_1}\right]$$

$$= \omega^2(J_1 a_1 + J_2 a_1) - \frac{\omega^4 J_1 J_2 a_1}{C_1}$$

So angle of deflection at Mass C

$$a_3 = a_1 - \frac{J_1\omega^2 a_1}{C_1} - \left[\frac{\omega^2(J_1 a_1 + J_2 a_1) - \dfrac{\omega^4 J_1 J_2 a_1}{C_1}}{C_2}\right]$$

$$= a_1 - \omega^2\left[\frac{J_1 a_1}{C_1} + \frac{J_1 a_1}{C_2} + \frac{J_2 a_1}{C_2}\right] + \frac{\omega^4 J_1 J_2 a_1}{C_1 C_2}$$

So torque to right of Mass C

$$= M_4 = \omega^2(J_1 a_1 + J_2 a_1) - \frac{\omega^4 J_1 J_2 a_1}{C_1}$$

$$+ J_3\omega^2 a_1 - \omega^2\left[\frac{J_1 J_3 a_1}{C_1} + \frac{J_1 J_3 a_1}{C_2} + \frac{J_2 J_3 a_1}{C_2}\right]$$

$$+ \frac{\omega^6 J_1 J_2 J_3 a_1}{C_1 C_2} \qquad (10.26)$$

Now as there are no external restraints to the right of Mass C, $M_4 = 0$. Equating eqn (10.26) to zero and dividing both sides by $\omega^2 a_1$ we obtain

$$(J_1 + J_2 + J_3) - \omega^2\left[\frac{J_1 J_2}{C_1} + \frac{J_1 J_3}{C_1} + \frac{J_1 J_3}{C_2} + \frac{J_2 J_3}{C_2}\right]$$

$$+ \frac{\omega^4 J_1 J_2 J_3}{C_1 C_2} = 0 \qquad (10.27)$$

This equation has two real roots so that a simple three-mass system has two principal modes of vibration, i.e. two natural frequencies.

At the lower frequency or fundamental mode of vibration there is a single node in the system when one of the end masses, dependent on the actual system values, moves in one direction whilst the other two masses move in the opposite direction. This is usually called the I-node mode of vibration of the system (*Figure 10.31b*). At the higher frequency there are two nodes (*Figure 10.31c*) one on either side of the central mass B so that masses A and C move in the same direction but opposite to that of mass B. This is called the II-node mode of vibration for the system.

With systems having more masses the same process can be used to derive expressions from which to obtain the natural frequencies. This however leads to more unwieldy expressions and greater analytical difficulty in solving them for the increasing number of roots or natural frequencies.

Fortunately this problem can be overcome by the use of iterative procedures which the digital computer can handle easily and increasingly economically as the number of masses involved gets larger. Two broad approaches are used. The first is based on matrix methods when mass and stiffnesses matrices are drawn up from the details of the reduced system under consideration when the natural frequencies or eigenvalues are determined using standard programs within the computer. After this the relative movement of the various masses are calculated for each of the natural frequencies. The second approach, which it is proposed to use here, is that due to Holzer since it is easier to follow for those not skilled in the art. Classically, Holzer's method is a tabular iterative method in which successive approximations to the frequencies are made but the method can readily be programmed for computer solutions.

10.3.2 The solution of multi-cylinder crankshaft systems

10.3.2.1 Holzer tabulation method

This method is based on the approach used to derive eqn (10.27). For a free-free vibration system the total torque at both free ends of the system must be zero. So at the end of the shaft where mass A is located a maximum torque of $J_1\omega^2\theta_1$ is required to vibrate mass J_1 at a frequency of ω radian/sec through an angle of $\pm\,\theta_1$ radians. So organizing the calculations in tabular form *Table 10.2* is obtained for the three-mass system shown in *Figure 10.31*.

Displayed in symbols this looks complex but once the system is understood it is easy and readily done. For systems having a large number of masses it becomes time consuming and tedious, particularly as more than one frequency has to be determined, so a computer is then used.

The basis of the columns is as follows:

Columns 1 and 2 merely put on record an identification of the mass and its polar inertia.
Column 3 The iterative process used is to assume a frequency $\omega = 2\pi F$ (Hertz) redian/s which it is hoped may be the correct frequency for the mode of vibration being considered. If ω has

Table 10.2

1	2	3	4	5	6	7	8
Mass	*Polar Inertia*	*Torque per unit deflection*	*Deflection in plane of mass*	*Torque in plane of mass*	*Total torque*	*Shaft stiffness*	*Change in deflection*
	J	$J\omega^2$	$\pm\,\theta$	$J\omega^2\theta$	$\Sigma\, J\omega^2\theta$	C	$\Delta\theta$
	kg m^2	N m	± rad	N m	N m	N m/rad	rad
A	J_1	$J_1\omega^2$	1	$J_1\omega^2$	$J_1\omega^2$	C_1	$J_1\omega^{2/C}{}_1$
B	J_2	$J_2\omega^2$	$\theta_2 = 1 - \dfrac{J_1\omega^2}{C}$	$J_2\omega^2\theta_2$	$T_2 = J_1\omega^2 + J_2\omega^2\theta_2$	C_2	T_2/C_2
C	J_3	$J_3\omega^2$	$\theta_3 = \theta_2 - \dfrac{T_2}{C_2}$	$J_3\omega^2\theta_3$	0	—	—

Assumed Frequency ω rad/sec

been correctly assumed, then the total torque in column 6 at the last mass will be zero. If not, another value for ω is taken and the table repeated. For the currently assumed value for ω column 3 then becomes the values for J_n times ω^2.
Column 4 It is assumed that the amplitude of movement in the plane of the first mass is ±1 radian. So when the table is finally balanced column 4 gives the relative amplitudes at all the masses. In following rows the value for θ is the value for the preceding row less the change $\Delta\theta$ for the preceding row obtained in column 8.
Column 5 is the product of $J_n\omega^2\theta_n$ for the mass J_n being considered which gives the torque required to vibrate that mass.
Column 6 This column really starts with torque 0 prior to mass A so the total torque in the plane of the first mass A is $0 + J_1\omega^2\theta_1 = J_1\omega^2$ as θ_1 has been made equal to unity in column 4. In successive lines the total torque is qual to that in the previous line plus the vibration torque in the line being considered. The final value when ω has been correctly assumed to be that of a natural frequency for the system should be zero. It should be noted that the total torque may become negative depending on the position of the node or nodes in the system corresponding to the mode whose frequency is being sought.
Column 7 Lists the stiffnesses of the shaft between the masses.
Column 8 This is the shaft twist between the two adjacent masses being considered and is obtained by dividing the total torque by the shaft stiffness between the two masses. So $\Delta\theta = \Sigma\, J\omega^2\theta/C$. The value so obtained is then subtracted from the deflection θ_n in that row to give the value for θ_{n+1} in column 4 for the deflection in the plane of the next mass J_{n+1}.

If the value for ω is incorrect a positive or negative residual torque will be left at the bottom of column 6. Assuming a number of frequencies will give a number of residual torques which, if plotted, result in a curve. Such a plot will help the location of the correct values of the required frequencies.

This process can be speeded up by using a method suggested by Mahalingam (*Proc. R.Ae. S.* Oct. 1966 p. 953). If, since the system is free at both ends, the system is considered as a rigid body, a more accurate approximation for ω^2 can be calculated by adding a compensating term for the resulting change in the maximum kinetic energy. This results in a second order term $R\bar{\theta}$ being added to the 'Rayleigh quotient'. It can be shown that the maximum strain energy for the vibrating system is

$$V = \frac{1}{2}\left[\sum_{i=1}^{n} J_i\omega^2_{k(1)}\theta^2 - R\theta_n\right]$$

Since R should be zero a new value for ω would be

$$\omega^2_{k(2)} = \omega^2_{k(1)}[\Sigma^n_{i=1}\, J_i\omega^2_{k(1)}\theta_i^2 - R\theta_n]/[\Sigma^n_{i=1}\, J_i\omega^2_{k(1)}\theta_i^2] \tag{10.28}$$

This is Rayleigh's quotient.

Here R is the amount the total torque at the end of the Holzer Table differs from zero;
θ_n = the deflection of the last mass n;
$\omega^2_{k(1)}$ = initial assumption for frequency (rad/sec) squared;
$\omega^2_{k(2)}$ = better assumption for ω^2;

Mahalingam's method first derives an ampitude

$$\bar{\theta} = \frac{R}{[\Sigma^n_{i=1}\, J_i]\,\omega^2_{k(1)}}$$

which is then used in a modified formula for a better estimate for ω^2.

$$\omega^2_{k(2)} = \omega^2_{k(1)}\frac{[\Sigma^n_{i=1}\, J_i\omega^2_{k(1)}\theta_i^2 - R\theta_n]}{[\Sigma^n_{i=1}\, J_i\omega^2_{k(1)}\theta_i - R\bar{\theta}]} \tag{10.29}$$

This form of correction gives a quadratic iteration which gives rapid convergence if the first approximation is reasonably good.

10.3.2.2 Gas torque and harmonic components

An internal combustion engine produces its instantaneous torques, and the resultant mean output torque, as a result of the instantaneous gas pressures acting on each piston throughout the cycle. The actual gas pressures will depend on a number of factors such as whether the engine inhales its charge at atmospheric pressure or at a higher value due to its being pressure-charged, whether it is a spark ignited engine or a diesel engine, the actual compression ratio used, and the spark advance used at a particular speed and load for a spark ignited engine or the injection advance in use in the case of a diesel engine.

For a given set of engine conditions a diagram can be obtained giving the variation of the gas pressure within the cylinder against crank angle throughout the whole cycle. This is repeated every 360 crank degrees for a two-cycle engine, or 720° (two revolutions) in the case of a four-cycle engine.

Since the direction in which the load due to the gas pressure within the cylinder acts on the crank arm varies as the crankshaft rotates, the gas load at any angle must be multiplied by the effective crank radius at that angle.

The instantaneous crank radius is derived by the instantaneous torque rate of work to the force work rate acting on the piston. Thus at any crank angle

$$T\omega = FV = pAV = pA\omega R\left[\sin\theta + \frac{\sin 2\theta}{2\sqrt{n^2 - \sin^2\theta}}\right]$$

So $$T = pAR\left[\sin\theta + \frac{\sin 2\theta}{2\sqrt{n^2 - \sin^2\theta}}\right] \qquad (10.30)$$

Here T = instantaneous torque;
p = prevailing gas pressure at crank angle θ;
A = piston area;
R = crank radius
θ = crank angle from TDC (firing);
L = connecting rod length;
$n = L/R$;
V = piston velocity at angle θ
ω = crankshaft angular velocity (radian/s);
F = force (N).

The term within the brackets is dependent on the crank angle nd the connecting rod to crank ratio, which does not vary videly for a given type of engine, and may be called the 'effective rank radius factor'. In view of this it is possible to construct n instantaneous torque curve for unity piston area and unity rank radius by multiplying the pressure at each crank angle by he effective crank radius factor. This curve is known as the Tangential Effort' or 'Tangential Pressure Curve' since the nits are those of a pressure (N/m^2 m) for unity crank radius. Such a curve can be used with reasonable accuracy to derive quickly the torque curve for other sizes of cylinders working nder similar operating conditions and indicated mean effective ressure by multiplying by the piston area and crank radius for he size being considered. Since the mean height of this tangential ressure against crank angle corresponds to the indicated mean ffective pressure the phasing of TDC must be as accurate as ossible when the original pressure diagram is obtained.

Figure 10.32 shows an example of such a tangential pressure urve. As is well known, such a periodically repeating curve nay be harmonically analysed by Fourier's analysis and epresented by a constant and a series of harmonically (sinusoidal) arying curves of different amplitudes and phase occurring once, wice, three times, etc. in one complete engine cycle. This is hown for the first seven components of the cycle.

Since it is convenient to relate these harmonics to the engine peed it is usual to refer to the number of complete harmonic ycles occurring in one crank revolution and to refer to them as he order numbers of the harmonics. For the four-cycle diagram hown the 1, 2, 3, 4, etc. harmonic cycles occur during two rankshaft revolutions, so that when referred to a crankshaft evolution these become respectively the $\frac{1}{2}$, 1, 1 $\frac{1}{2}$, 2, etc. armonic orders.

In the case of two-cycle engines the order numbers and the armonics of the cycle are always integers of the same value, s one engine cycle occurs every crank revolution.

Table 10.3 shows typical values for the harmonic gas torque omponents for three indicated mean effective pressures. It will e noticed that the values increase with load. Although it is sual to use such tabulated figures, it is the writer's opinion that here is a growing need for further studies, particularly since the nore sophisticated computerized studies often incorporate amping at every mass. There are indications that variations ccur with changes in compression ratio and injection advance, s well as with the load.

0.3.2.3 Effect of reciprocating masses

he mass of the reciprocating parts result in inertia forces which ary with the square of the speed. This, multiplied by the effective rank radius, produces an inertia torque diagram which, when

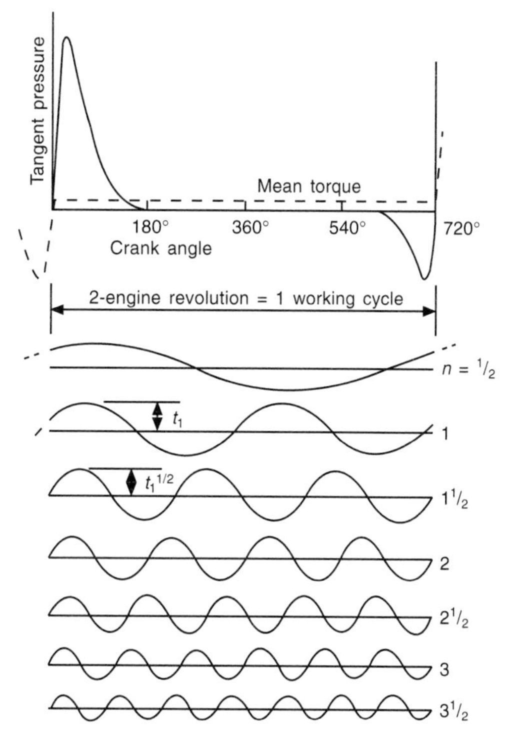

Figure 10.32 Typical diesel engine single-cylinder instantaneous tangential pressure (torque) curve throughout one cycle and its harmonic components

Table 10.3 Values for harmonic torque coefficients, t_n

Order No.	*IMEP* (bar) →	t_n(N/mm^2-m) 7	10	15
$\frac{1}{2}$		0.290	0.380	0.537
1		0.345	0.435	0.585
$1\frac{1}{2}$		0.345	0.425	0.575
2		0.295	0.357	0.463
$2\frac{1}{2}$		0.248	0.300	0.385
3		0.207	0.250	0.320
$3\frac{1}{2}$		0.170	0.200	0.258
4		0.135	0.160	0.203
$4\frac{1}{2}$		0.105	0.125	0.158
5		0.082	0.095	0.120
$5\frac{1}{2}$		0.063	0.075	0.090
6		0.050	0.057	0.070
$6\frac{1}{2}$		0.040	0.046	0.057
7		0.034	0.037	0.046
$7\frac{1}{2}$		0.026	0.030	0.037
8		0.021	0.024	0.029
$8\frac{1}{2}$		0.017	0.020	0.025
9		0.014	0.017	0.021

harmonically analysed, produces a series of orders. The expressions can be analytically derived and expanded when it is found that integral order numbers only exist which are all symmetrically disposed about top dead centre, i.e. have zero phase angle, and consist of sine components only.

By contrast the gas torque components have varying phase. For this reason the inertia and gas torque components cannot be directly added. As well as this, those due to gas forces vary slightly with IMEP in amplitude and phase, whereas those arising from the inertia forces vary as square of the operating speed (or angular velocity).

The approximate values of the first four harmonic orders are given in *Table 10.4a*. It will be noted that sine and cosine values are given for the gas components and sine values only for the inertia torque components (the cosine values are zero).

To find the combined value for the harmonic torque components for a particular order it is necessary to proceed as follows:

Table 10.4a Sine and cosine values for gas harmonic torques for diesel engines (N/mm^2-m)

Order No.	*Sin or Cos*	*IMEP—bar* 7	10	15
1	sin	0.285	0.364	0.472
	cos	0.105	0.150	0.225
2	sin	0.293	0.351	0.446
	cos	– 0.014	– 0.020	– 0.030
3	sin	0.195	0.226	0.277
	cos	– 0.021	– 0.040	– 0.060
4	sin	0.114	0.128	0.150
	cos	– 0.035	– 0.050	– 0.075

Table 10.4b *N*-th order inertia torque component

Con rod / *Crank* Ratio		3	3.5	4	4.2	4.4
Order No.	*Sign*	Harmonic component constant (*sine only*)				
1	+	0.086	0.073	0.064	0.060	0.058
2	–	0.500	0.500	0.500	0.500	0.500
3	–	0.261	0.220	0.192	0.182	0.174
4	–	0.030	0.022	0.016	0.015	0.013

Estimate or determine the mass m in kg of the reciprocating parts and divide by the piston area to give the mass per unit area (kg/m^2). This is then multiplied by the inertia torque coefficient t from *Table 10.4b* for the n-th order, by the crank radius (m) and the crankshaft angular velocity square $(rad/sec)^2$. So the inertia torque component for the n-th order

$$S_{I_n} = t_{n_I} mR\omega^2 \text{ Nm/m}^2\text{m} \quad (10.31)$$

Note that t_{n_I} may have a negative value thus cancelling part of the gas sine component. Calling the gas sine component S_{G_n} and the gas cosine component C_{G_n} the resultant for the combined inertia and gas components for the n-th order for the engine being considered T_n is

$$T_n = [(S_{G_n} + S_{I_n})^2 + C_{G_n}^2]^{0.5} \text{ Nm for 1 m}^2 \text{ piston}$$

and 1 m crank throw.

10.3.2.4 Determination of crankshaft stiffness

The first stage in the determination of the natural torsional frequency of an engine system is to estimate the torsional stiffness of the crankshaft units. Initially it is convenient to express each section as an equivalent length of journal of equal stiffness.

It is impossible to derive formulae for the stiffness of a crankthrow from first principles due to its complexity although estimates are now being made using 3-D finite element methods. It is usual to use one of a number of empirical formulae evolved from practical tests on a series of shafts but, as various investigators have used shafts of differing designs and sizes, none is completely satisfactory for universal application to crankshafts of all proportions. The author has had satisfactory results using a combination of two formulae as suggested by Ker Wilson in his book. The first due to Major Carter applies mainly to shafts having relatively flexible webs and stiff crankpins and journals. The second due to Ker Wilson is for shafts having relatively flexible pins and journals combined with stiff webs. Using the symbols shown in *Figure 10.33*.

Carter's Formula:

$$L_e = D^4\left[\left(\frac{2a + 0.8b}{D^4 - D_1^4}\right) + \left(\frac{0.75c}{d^4 - d_1^4}\right) + \left(\frac{1.5R}{bw^3}\right)\right] \quad (10.32)$$

Ker Wilson's Formula:

$$L_e = D^4\left[\left\{\frac{2a + 0.4D}{D^4 - D_1^4}\right\} + \left\{\frac{c + 0.4d}{d^4 - d_1^4}\right\} + \left\{\frac{R - 0.2(D+d)}{bw^3}\right\}\right] \quad (10.33)$$

i.e. the length of soild shaft of journal diameter D having the same stiffness as the crank unit from centre to centre of the bearings.

After the evaluation of these formulae, if the values are within 10% of one another a mean of the two values is adopted. If greater than 10% difference is found the longer length is adopted. Where longer journal lengths exist, such as the centre main bearing at times and the front and rear bearings, the appropriate extra length is added to the equivalent length for that shaft section. Usually there is one or more step downs in diameter at the front end of the crankshaft where timing gears, or a heavy pulley is fitted. Their equivalent lengths when converted to crankshaft journal diameter is their length from the centreline of the keyway to the point of shaft step down multiplied by $(D/D_x)^4$, where D_x is the diameter of the actual shaft and D the journal diameter being used as standard.

Another method for reducing a crankthrow to an equivalent length of uniform shaft was evolved by the British Internal Combustion Engine Research Association. The method uses factors derived from a series of curves for various web and shaft proportions and design features. Space does not permit its inclusion here. Although originally devised for medium speed engines it appears to give good results for smaller high speed engines as well.

Once the equivalent lengths of uniform shaft have been determined the stiffness of each section can be calculated from

$$C = \frac{I_p G}{L} \text{ Nm/radian}$$

where I_p = shaft section polar moment of inertia $= \pi D^4/32$ (m^4);
D = journal diameter (m);
G = shaft material modulus of rigidity (N/m^2);
L = reduced shaft equivalent length for one crankthrow, or other length (m).

10.3.2.5 Polar inertias

Next, it is necessary to calculate the total polar inertia of each crankthrow together with those for the flywheel and any masses attached to the free end of the crankshaft. If the engine is coupled to, and drives any further masses the polar inertias in the coupled system and the stiffnesses between them must also be calculated.

For a crankthrow it is assumed that the rotating mass of the connecting rod is attached to the crankpin together with half the total mass of the reciprocating masses per cylinder. These are then multiplied by the square of the crankthrow. The rest of the

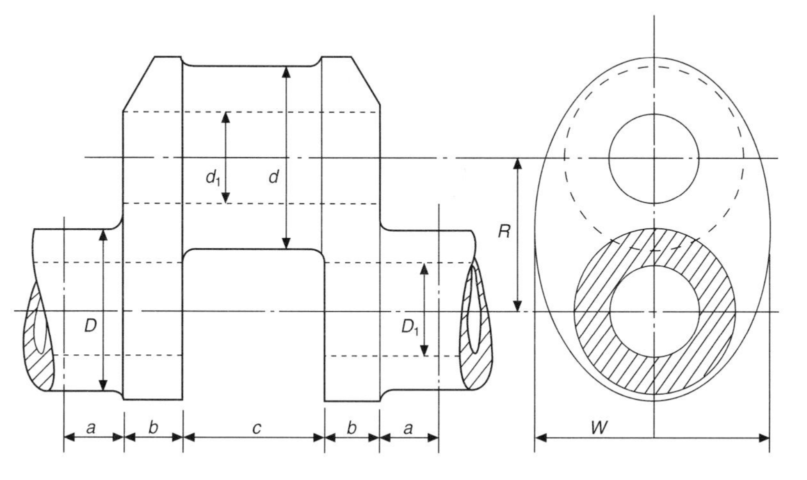

L_0 = equivalent length of shaft, D outside diameter and D_1 inside diameter
D = outside diameter of shaft journal
D_1 = inside diameter of shaft journal
d = diameter of crankpin
d_1 = diameter of hole in crankpin
a = half journal length
b = web thickness
c = crankpin length
W = width of crank webs
R = crank radius

Figure 10.33 Symbols used for the reduction of a crankthrow to its equivalent length of uniform diameter shaft

crankthrow is broken into a number of simple geometric shapes and the polar inertia of each section about the journal axis (axis for crankshaft rotation) calculated. This should include the two half journal lengths up to their centre of length. The sum of all these parts gives the total polar inertia for each throw which is assumed to act at the cylinder centre line.

Similarly the polar inertia for the flywheel and any other masses such as front end pulleys are calculated. Where torsional vibration dampers are fitted their polar inertias are either calculated or taken from manufacturers' figures.

10.3.2.6 Total equivalent torsional system and determination of system's frequency

Once the various individual stiffnesses and polar inertias have been determined the whole equivalent system can be assembled. *Figure 10.34* shows a system for a six-cylinder automotive type engine.

It is now possible to calculate the natural frequency or frequencies of the system using the Holzer Tabulation method as already described, which also gives the relative deflections at each mass corresponding to the assumed unity angular deflection at the free end of the crankshaft. This can be done using a computer.

If a desk calculator is used the amount of work can be reduced by first making an approximation to the natural frequency. This may be done by assuming all the crank masses are concentrated at the centre of the equivalent shaft length and vibrate as a two-mass system against the flywheel.

An alternative approach assumes the 'swinging form' varies

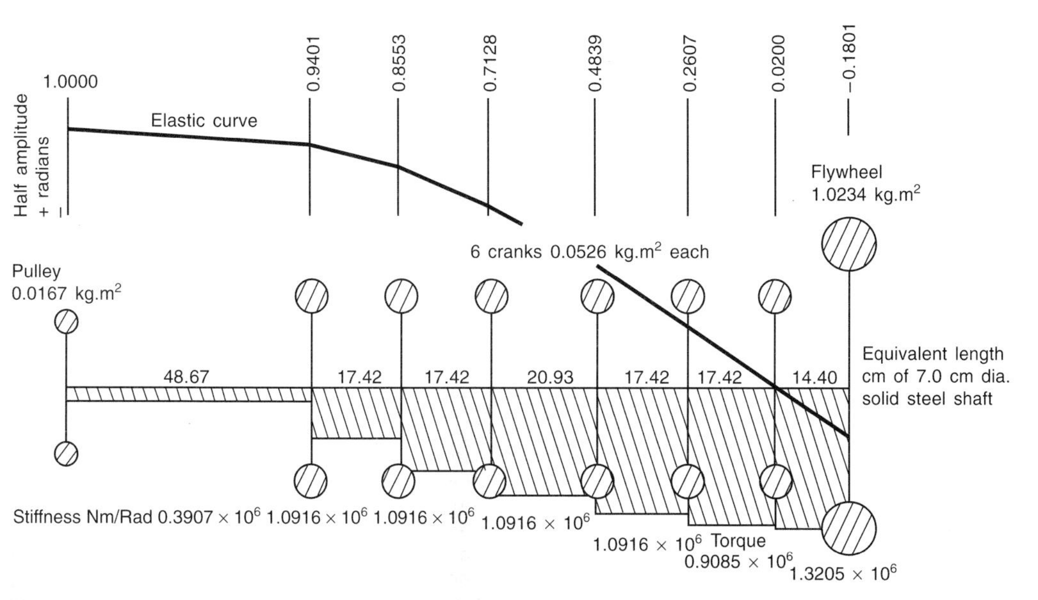

Figure 10.34 Typiclal equivalent torsional system for a six-cylinder engine

linearly along the shaft length. Then treating the crank inertias as masses and the equivalent length from the flywheel as fixing their position on a beam, a point of balance can be found in a manner analogous to finding the centre of mass. This is done by using the flywheel as the reference point then multiplying the inertia masses by their equivalent length from the flywheel, adding their products and dividing the sum by the total of the masses, including that of the flywheel, to find the point of balance. This approximates to the node with the crank masses vibrating against the flywheel. The approximate natural frequency can then be found by regarding the flywheel as being a simple torsional pendulum on a shaft having the length just determined from eqn (8.3).

$$f = \frac{1}{2\pi}\sqrt{\frac{C}{J}} = \frac{1}{2\pi}\sqrt{\frac{GI}{LJ}} = \frac{1}{2\pi}\sqrt{\frac{\pi}{32}\frac{GD^4}{JL}}$$

$$= 0.0499\sqrt{\frac{GD^4}{JL}}\ \text{vib/s} \qquad (10.34)$$

Since the swinging form is not in fact a straight line this frequency should be divided by 0.92 to give a reasonable approximation of frequency with which to start the Holzer Tabulation. It will normally be necessary to make some adjustment to the assumed frequency until the total torque at the end of the system becomes satisfactorily near to zero. *Table 10.5* gives the Holzer Table corresponding to the system shown in *Figure 10.34* for the *I*-node mode of vibration.

10.3.2.7 Estimation of torsional vibration amplitudes

The first step in estimating the likely torsional vibration amplitudes and stresses is to calculate the 'Equilibrium Amplitude'. This is done by equating the forcing work done on the shaft by the cylinders to the maximum kinetic energy in all the vibrating masses.

The forcing work applied to the shaft equals $0.5t_nAR\Sigma\theta$, where t_n is the harmonic component for the n-th order under consideration in Nm/m^{2}m, A is the area of one piston (m^2) and R the crank radius (m). $\Sigma\theta$ is the vector sum for the order number being considered with the firing order proposed, where θ is the deflection in the plane of each cylinder given in *Table 10.5* column 4.

The maximum kinetic energy in the system is given by $\Sigma 0.5J\omega_c^2\theta^2$ for all the masses. ω_c is the angular vibration velocity, also known as the phase velocity, for the natural frequency already determined.

Remembering that the values of θ determined in the Holzer Table assume unit deflection at the free end of the shaft, and calling the deflection due to the work put in, but without any magnification due to resonance, θ_e or the 'Equilibrium Amplitude', then equating the above two expressions gives:

$$0.5t_nAR\Sigma\theta\theta_e = 0.5\Sigma J\omega_c^2\theta^2\theta_e^2$$

$$\text{From which,}\quad \theta_e = \frac{t_nAR\Sigma\theta}{\omega_e^2\Sigma(J\theta^2)} \qquad (10.35)$$

For the mode of vibration being considered, ω_c has just been determined from the Holzer Table, $\Sigma(J\theta^2)$ can be evaluated from the Holzer Table by multiplying *all* the masses shown in column 2 by the square of the figure in column 4 for the deflection in the plane of the mass and then totalling all the figures so obtained. $\Sigma\theta$ is the vector sum for the crank masses only.

Remembering that the engine rev/s multiplied by the order number equals the natural frequency (vib/s) just determined, the order number occurring within and just above the normal operating speed range can be determined. Once these are known the corresponding values for the harmonic torque component t_n for each order at the IMEP involved can be looked up which, when multiplied by the piston area and the crank radius, gives the forcing torque $T_n = t_nAR$ acting on each piston.

This leaves only the vector sum corresponding to the vibration amplitude at each cylinder to be calculated based on the figures in column 4 of the Holzer Table which correspond to the assumed ± 1 radian at the free end of the crankshaft. This vector sum is dependent on the crankshaft arrangement and the firing order chosen.

For the six-cylinder engine used as an example the crankshaft has crankthrows spaced 120° apart. The firing sequence often chosen is 1–5–3–6–2–4–1. Remembering that the order number is the number of complete cycles occurring during one crankshaft revolution, then the vectors for the 1-st order will be spaced 120° apart in the sequence corresponding to the firing order. 2-nd order vectors rotate at twice crank intervals so each succeeding 2-nd order vector will be spaced at 2 × 120° = 240° apart in their correct sequence and so on.

Figure 10.35a shows the spacing and magnitude of the vectors for the first order—the spacing is the same as the crank angles

Table 10.5 I-node mode frequency tabulation (Holzer Table) with no damper fitted

1	*2*	*3*	*4*	*5*	*6*	*7*	*8*
	Moment of inertia J	*Torque per unit deflection* $J\omega^2$	*Deflection in plane of mass* θ	*Torque in plane of mass* $J\omega^2\theta$	*Total torque* $\Sigma J\omega^2\theta$	*Shaft stiffness* C	*Change in deflection* $\Delta\theta$
	kg m^2	N m × 10^6	± rad	N m × 10^6	N m × 10^6	N m/rad × 10^6	rad
Pulley	0.0167	0.0234	1.0000	0.0234	0.0234	0.3907	0.0599
Cyl. 1	0.0526	0.0736	0.9401	0.0692	0.0926	1.0916	0.0848
Cyl. 2	0.0526	0.0736	0.8553	0.0692	0.1555	1.0916	0.1425
Cyl. 3	0.0526	0.0736	0.7128	0.0525	0.2080	0.9085	0.2289
Cyl. 4	0.0526	0.0736	0.4839	0.0356	0.2436	1.0916	0.2232
Cyl. 5	0.0526	0.0736	0.2607	0.0192	0.2628	1.0916	0.2407
Cyl. 6	0.0526	0.0736	0.0200	0.0015	0.2643	1.3205	0.2001
Flywheel	1.0234	1.4328	– 0.1801	– 0.2581	0.0062	—	—

F_1 = 11300 vpm
= 188.33 vps
$\omega^2 = 1.400 \times 10^6$ (rad/s)^2

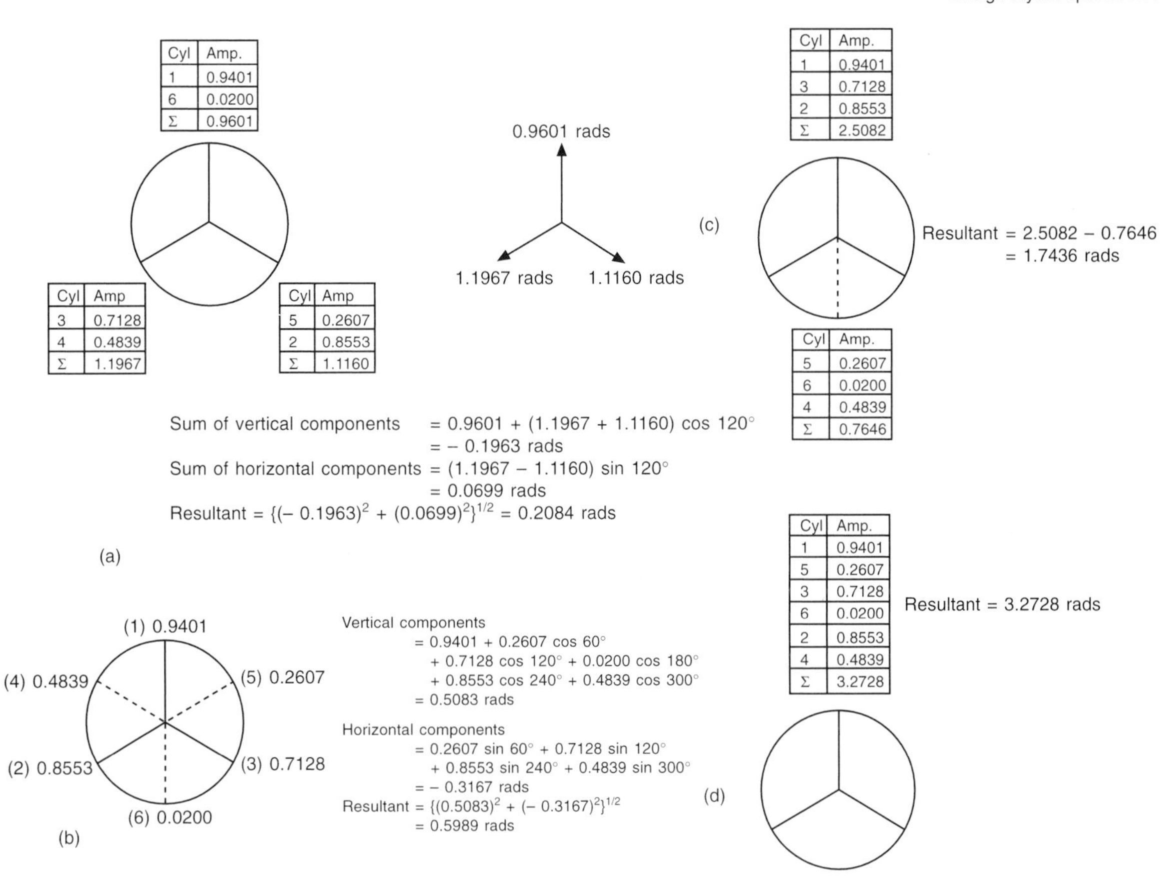

Figure 10.35 Vector sums for six-cylinder engine example given in *Figure 10.34*: (a) Vector sum for orders Nos. 1, 4, 7, 10, etc. (b) Vector sum for order Nos. $\frac{1}{2}$, $2\frac{1}{2}$, $3\frac{1}{2}$, $5\frac{1}{2}$, $6\frac{1}{2}$, etc. (c) Vector sum for orders Nos. $1\frac{1}{2}$, $4\frac{1}{2}$, $7\frac{1}{2}$, etc. (d) Vector sum for orders Nos. 3, 6, 9, etc.

= 120°. The cylinder number and the amplitude corresponding to it is shown. It is found that more than one vector has the same angular oposition, so these have been arithmetically added making the diagram to the right. Then by taking cosine components and algebraically adding the total cosine (vertical) component for the resultant vector is found, and by taking the sine components the resultant sine component is similarly found. The resultant vector magnitude is then

$$\{(\text{sine component})^2 + (\text{cosine component})^2\}^{1/2}$$

It will be noticed that the 4-th order successive vectors are spaced at 4 × 120 = 480° = (360 + 120°) apart, and therefore have the same pattern as for the 1-st order. Similarly for the 7-th and 10-th orders.

Figure 10.35b gives the diagram for the $\frac{1}{2}$, $2\frac{1}{2}$, $3\frac{1}{2}$, $5\frac{1}{2}$, $6\frac{1}{2}$, etc. orders. The $1\frac{1}{2}$, $4\frac{1}{2}$, $7\frac{1}{2}$ orders shown in *Figure 10.35c* all act up and down, as drawn, and have a moderately large resultant.

In the case of order numbers which are multiples of half the number of cylinders, i.e. 3, 6, 9, etc. for the six-cylinder engine used as an example, all the vectors act in the same direction (*Figure 10.34d*). For this reason, they are called 'Major Orders'. These, for engines in which the node is close to the flywheel, are always equal to multiples of half the number of cylinders. Where the flywheel is light compared with the crank masses, or a heavy mass is attached to the free end of the shaft, the node will move towards the centre of the crankshaft, and this rule does not hold. Hence a full investigation is necessary.

In the case of vee engines the firing sequence is usually the same for each bank. It is possible to make the vector sums for each bank as indicated and then vectorially combine them, remembering that the vector spacing is the order number times the firing interval between corresponding cylinders in the two banks. Remember also that the firing angle between banks may be either the bank angle or (360° + bank angle) depending on which has been chosen.

Some variation in the vector sums for the different orders can be obtained by altering the firing order. For example, changing the firing order used on the illustrative six-cylinder engine to 6, 5, 3,1, 2, 4 reduces the vector sum for the $1\frac{1}{2}$, $4\frac{1}{2}$, $7\frac{1}{2}$ group of orders but raises the value for the $\frac{1}{2}$, $2\frac{1}{2}$, $3\frac{1}{2}$, $5\frac{1}{2}$, $6\frac{1}{2}$ orders, the major orders stay the same. Limits to the variations in firing orders are set by the crankshaft arrangement required for good engine balance and manifold arrangements required for good breathing.

It is now possible to evaluate the expression (10.35) for the equilibrium amplitudes for the various orders. This represents in the absence of any resonance effects the amplitude at the free end of the crankshaft arising from the forcing torques. However, as previously stated, the repeated application of the forcing torques at or near the resonant speed involved, leads to the building up of amplitudes considerably larger than the equilibrium amplitude. The actual magnitue is dependent on damping in the whole system arising from piston and bearing viscous losses, shaft material, hystersis, etc.

The ratio of the actual shaft deflection to the equilibrium amplitude is called the 'Dynamic Magnifier'. The actual value for crankshafts not deliberately fitted with amplitude limiting devices, dampers, must be based on previous experience with similar engines. Values of 20–25 may be taken as a rough guide for low alloy and carbon steels and 40–50 for high alloy steels for major orders and for the minor orders with low alloy shafts. In the case of nodular iron shafts the dynamic magnifier will be about 20.

With estimates now available for the maximum amplitudes of vibration at the front end of the shaft, the ± 1 radian movement on which the Holzer Table was based can be reduced to the estimated value and the total torque in each shaft section modified accordingly. Then by dividing by the minimum section modulus $\pi/16d^3$, usually that of the crankpin, the stress levels are determined. It is now possible to sketch the variation of the torsional stresses for the various orders over the operating speed range using the peak values, as just determined, faired into flank values which are closely similar to those of the mathematically derived curve as shown in *Figure 10.29*. It must be remembered that significantly high lower flank amplitudes and stresses may be present in the running speed range for at least some of the low order numbers, e.g. the 3-rd major order in a six-cylinder engine, although their peak values occur well above the maximum operating speed.

Since thes evarious orders have different phases with regard to one another it is usually assumed for safety that the actual stress level at each operating speed is the sum of the stresses due to all the order numbers present at that speed. It should also be remembered for more complicated systems that there may be more than one mode of vibration occurring simultaneously. Each will have its own frequency and associated order numbers.

Wherever practical, actual measurements of torsional vibration amplitudes should be made on a prototype engine to confirm the various assumptions made. This will be briefly referred to later.

If an unacceptably high amplitude occurs in the speed range steps must be taken at the design stage to reduce it. Whilst the frequency may be raised by increasing the shaft diameter where the objectionable order is close to the maximum operating speed, this is often not possible for other reasons. In most cases, therefore, some form of damping or detuning device must be fitted to prevent the build-up of large amplitudes at resonance.

10.3.3 Vibration dampers

These may be divided into a number of broad categories although in some cases the end result may be due to an interaction of the approaches. These are:

(1) means for altering the frequency of the system when vibration occurs;

(2) devices which absorb work when vibration occurs thereby reducing the energy available to deflect the shaft;

(3) attachments which go into antiphase when vibration occurs so that the vibratory torque is opposed.

Wherever possible damping and detuning devices should be fitted where maximum movement occurs so that their size can be minimal.

10.3.3.1 Frequency varying devices

In its simplest form when used for couplings it is possible to use a two-rate spring system such that, in normal operation and with small vibration amplitudes, the coupling has a low stiffness and the system a low natural frequency. However when significant vibration amplitudes occur within the coupling a second set of springs, often rubber buffer blocks, are contacted. The natural frequency is raised as a result of the increased stiffness and amplitudes cease to build up since the forcing torque being applied is no longer in resonance.

An early form of a more sophisticated type was the Wellman-Bibby coupling. In this the two coupling halves have curved axial slots machined around their outer circumference. A series of grid-iron like steel springs are fitted in the slots and connect the two halves together. They are retained against centrifugal force by a cover. In operation any increase in transmitted torque particularly that due to vibration deflects the springs tangentially against the sides of the grooves. As the grooves have a diminishing radius of curvature as the junction between the two halves is approached the springs progressively sit on the sides of the grooves, their active length is reduced and the coupling stiffness increases. Thus a continuously varying spring rate stiffening with angular deflection of the coupling is obtained. Another well-known type using readial steel spring packs between the hub and the outer member is the Geislinger coupling.

Although not normally used with high speed engines the Bibby principle is sometimes employed as a damper-detuner at the free end of the crankshaft in medium speed and slow speed engines. Usually, if the change in spring rate between extremes of movement is some 30%, resulting in roughly 15% change in frequency, the detuning is sufficient to eliminate dangerous torsional amplitudes.

10.3.3.2 Work absorbing devices

These are friction and viscous friction dampers. The solid friction damper originally devised by Lanchester consists of a disc rigidly fitted to the front end of the damper, whose inertia is made as small as possible consistent with handling the imposed torques, with a pair of circular masses lined with a suitable friction material arranged to clamp on each side of the disc by a series of springs. At small vibration amplitudes the masses move back wards and forwards with the rigidly attached disc. However when the vibratory angular acceleration $\omega^2\theta$ produces a torque $J_d\omega^2\theta$ to oscillate the clamped masses which is greater than the clamping torque, slip occurs between the masses and the driving disc and vibratory energy is converted into heat. Here J_d is the polar inertia of the damper flywheel masses. It can be shown that the optimum friction torque setting for a multi-cylinder engine is given by

$$T_d = \pm 1.11\ T_n\Sigma\theta \tag{10.36}$$

Due to wear in long-term operation, leading to a reduction in the torque setting, solid friction dampers are rarely used today, the viscous friction damper, usually simply called a viscous damper, being preferred.

10.3.3.3 Viscous dampers

These consist of a light metal casing rigidly attached to the front end of the crankshaft in which a relatively heavy annular mass is enclosed having centring bearings and a relatively small clearance between its sides and its circumference and the enclosing casing (*Figure 10.36*). The clearances between the mass and the casing are filled with a silicone fluid. The silicones used have a high viscosity the actual value of which is selected to suit a particular damper design. They also have a very flat viscosity index curve, much less than for hydrocarbon oils used for normal lubrication purposes.

At small angular accelerations and movements the viscous drag is sufficient to cause the internal mass to move with the casing but as vibratory amplitudes grow the mass starts to slip at the peaks of angular cyclical acccelerations and energy is dissipated as heat. This type of damper was originally devised by the Houdaille-Hershey Corporation in 1946. Nowadays, viscous dampers are obtainable in a wide number of standard sizes from manufacturers specializing in their supply. Finer details of the design and the silicone viscosity used are selected by the

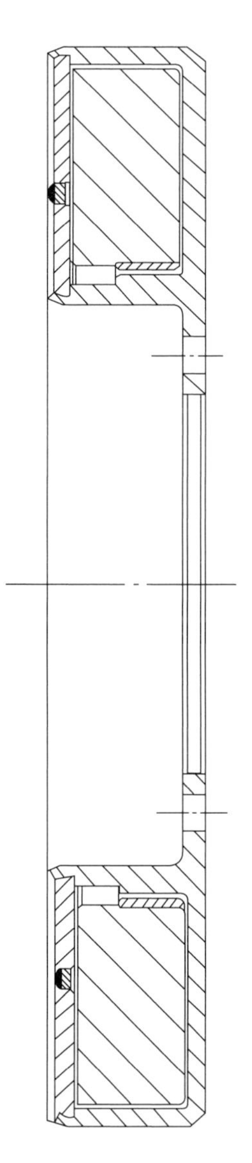

Figure 10.36 Cross-section of a typical viscous damper

suppliers to provide optimum damping for the engine system to which they are to be fitted.

In action, two distinct extreme vibratory regimes exist. At low vibration amplitude the inertial mass moves as one with the casing, and therefore reduces the natural frequency of the crankshaft to which it is fitted. At very large vibration amplitudes the accelerations are such that the damper mass is always sliding, i.e. only the relatively light casing can be regarded as rigidly attached to the forward end of the crankshaft so only a small drop in shaft frequency occurs.

These two extremes are shown diagrammatically in *Figure 10.37*. With the damper mass locked the lower frequency is obtained and theoreticaly infinite amplitudes are obtained at the resonant frequency (A). With zero viscous drag, i.e. the damper mass loose, the upper frequency is obtained again with infinite magnitude (B) in the absence of damping. In practice as the viscous drag is reduced some slip occurs at the lower frequency and the dynamic magnifier is reduced from infinity at resonance as sketched in curve C. With low viscous drag torques little movement occurs at the lower frequency but damping work is done at the higher frequency reducing the dynamic magnifier curve D. Between these two extemes an optimum damper torque exists which results in curve E which has a maximum for the dynamic magnifier at the point where the upper flank for the damper locked frequency crosses the lower flank of the casing only resonance curve. The frequency at which this maximum danamic magnifier for the optimum damping occurs approximates to the mean of the upper and lower frequencies.

For this reason an estimate for the vibration amplitude assuming optimum damping torque is provided by the damper is obtained by carrying out a Holzer Table for the original system having an inertia mass equal to that of the damper casing plus one half that of the damper's flywheel inertial mass. These values are supplied by damper manufacturers.

If the resultant vibration amplitude at the free end of the crankshaft is $\pm\,\theta$ radians then the input harmonic work at the pistons is:

$$\pi\,|T_{\mathrm{m}}|\Sigma a\theta = \pm\,\pi t_{\mathrm{n}} AR\Sigma a\theta \text{ joule/cycle} \tag{10.37}$$

where Σa is the vector sum of the amplitudes at the cylinder masses. With an optimum damper torque $= \pm\,2|T_{\mathrm{n}}|\,\Sigma a$

$$\text{The damper work } = \pm\,\pi J_{\mathrm{d}}\omega_{\mathrm{c}}^{2}\,\frac{\theta^{2}}{2} \text{ joule/cycle} \tag{10.38}$$

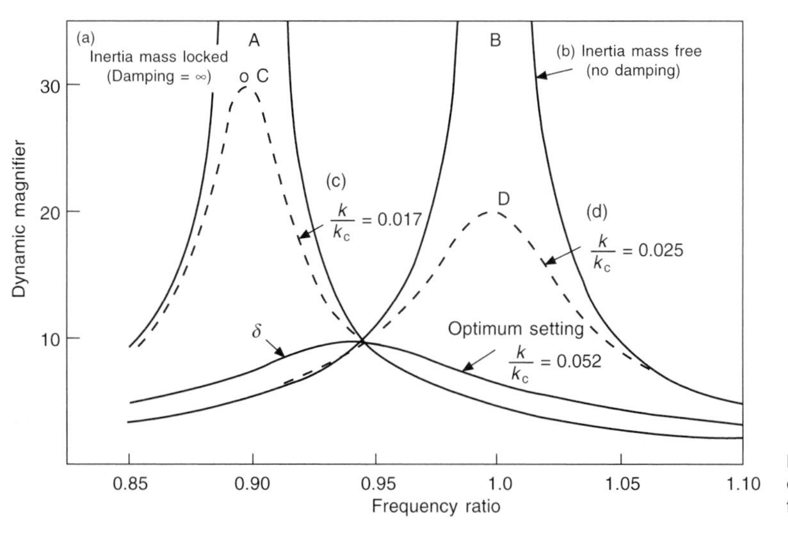

Figure 10.37 Response curves for a viscous damper—dynamic magnifier pltted against frequency ratio

where J_d = polar inertia of damper flywheel mass;
ω_c = phase velocity at resonance determined from the Holzer Table;

Equating the amplitude at the point of damper application

$$\theta = \frac{2|T_n|\Sigma a}{J_d \omega_c^2} \text{ radian} \qquad (10.39)$$

This is in fact a pessimistic value since damping within the rest of the system will result in a lower value.

Now that an estimated amplitude has been obtained, the stresses in the various parts of the system may be determined, as before, using the Holzer Table. With a damper fitted, the maximum crankshaft stress may be found to exist in the front end of the shaft when a marked reduction from journal diameter has been made. This also has the undesirable effect due to its flexibility of unduly dropping the crankshaft frequency. In general therefore the length of the crankshaft front end should be as short as possible to the point of damper attachment and as large a diameter as can be tolerated.

10.3.3.4 Pendulum dampers

These are masses attached by some form of link(s) to one or more balance weights at the forward end of a crankshaft. When vibration occurs for a particular engine order number the pendulum which has been suitably proportioned vibrates in antiphase to the crankshaft movement thereby resisting the shaft's vibratory movements.

The pendulum is designed so that the ratio of the radius from the crankshaft axis to the point of pendulum attachment (R) divided by the length of the pendulum (L) is equal to the square of the order number to be controlled. This form of damper is only occasionally used today. It requires accurate control of the dimensions, and wear, due to the movement under fairly high contact pressures at the points of suspension, must be small or the dampers will go off tune in service. More than one pendulum damper will be required if there exists more than one high amplitude order number in the running speed range.

10.3.3.5 Rubber dampers

These can be relatively cheap to make and are very popular for large production automotive type engines. They consist of a relatively light hub attached to the front end of the crankshaft surrounded by an annular ring of rubber on the outside of which is fitted a comparatively heavy inertia ring. The rubber ring is usually inserted so that it is radially compressed and the working torques transmitted as a result of the interference loads acting on the inside and outside of the rubber ring. *Figure 10.38* shows a typical construction.

Often the axial section of the rubber ring is curved so as to avoid the possibility of the mass working off if any crankshaft axial movements or even vibration occur. Sometimes the rubber is chemically bonded to the hub and mass usually when a different design is used from that illustrated. Vee-belt grooves provided to drive engine auxiliaries may be formed in the hub but also may be formed in the outer diameter of the damper mass.

The damping action is complex which makes the design and performance difficult to predict accurately. In manufacture, care must be taken to maintain the correct rubber mix to give the required stiffness and to have consistent damping (hysteresis) properties.

The fitting of a rubber damper, in fact, adds another mass and elasticity to the original system and therefore introduces another natural frequency to the original system. It should be noted that in the initial stages of selection both the damper inertial mass and the stiffness of the rubber ring can be varied as well as the damping factor of the rubber to some degree. Depending on the type and grade of rubber employed it is found that the working stiffness is greater than the static stiffness. Fortunately the dynamic stiffness does not itself vary very much with frequency.

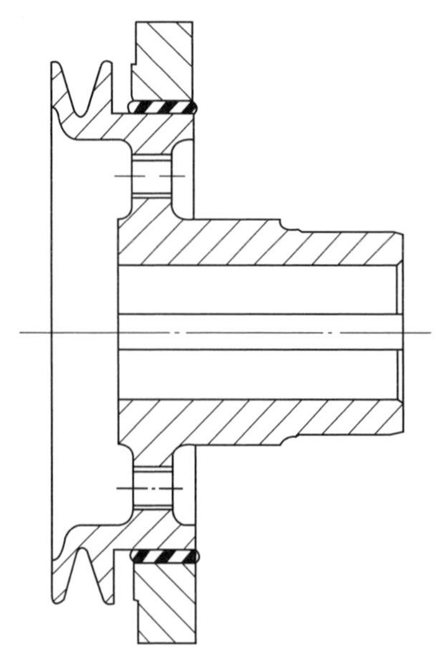

Figure 10.38 Cross-section of a rubber damper

Related to the parameters chosen three natural frequencies can be calculated which for a particular order number, say the 6-th major order for a six-cylinder engine, gives the result shown in *Figure 10.39*. The dynamic magnifier against frequency ratio for the original system is shown in the centre chain dotted curve. With the additional damper mass regarded as clamped to its hub, i.e. the rubber has infinite damping, the slightly lower frequency is obtained. If the rubber is next considered to have no damping two frequencies are found to exist, one higher and one lower than the original frequency.

With no damping, the amplitudes for the damper mass and crankshaft will be infinite as shown in the two outer dotted curves, but with damping present due to hysteresis losses in the rubber, this state is never realized. It is found that the actual curve for the dynamic magnifier always passes through the circled points where the flanks for the no damping and infinite damping resonance curves intersect. The relative dynamic magnifiers at these points depend on the damper mass and stiffness in relation to the rest of the system. However the dynamic magnifiers under the resonance peaks are dependent on the rubber damping.

According to the mutual adjustments made, the height of the lower resonant peak may be made greater or less than the higher resonant peak. This feature may be deliberately used to reduce a particular order occurring just above the maximum engine operating speed, when the peak for the lower resonance curve is made smaller than that for the higher one which would be in tune at a yet greater speed above the maximum operating speed.

In practice, the damper manufacturer usually designs two or three experimental dampers based on the reduced torsional equivalent system for the engine and installation, selecting the best damper for use by carrying out torsional vibration tests on the engine test bed. The calculations will also cross check on the maximum deflections and torque acting on the rubber, so that the stresses are such as to ensure a long working life; and check that the hysteresis work does not heat the rubber to a point where its properties deteriorate.

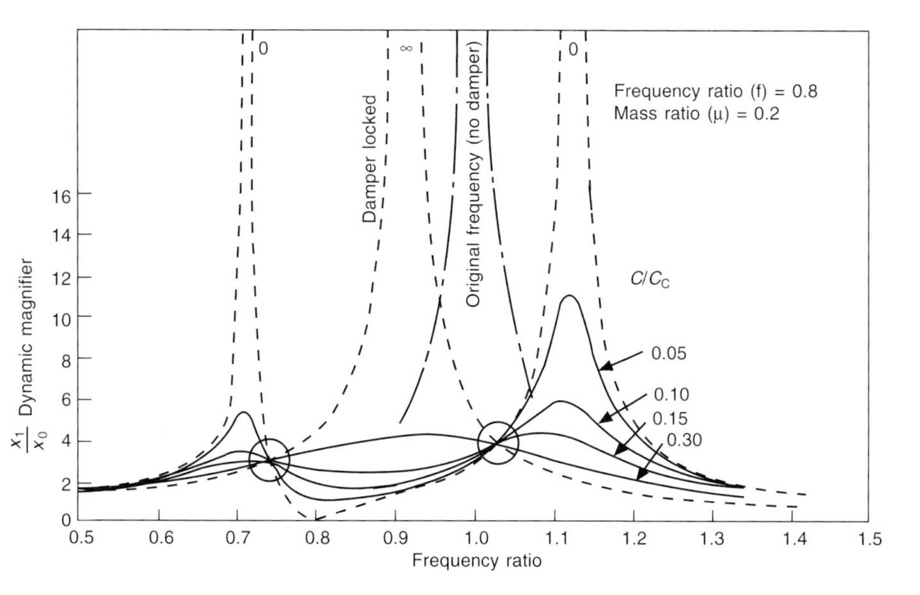

Figure 10.39 Dynamic magnifiers against forced frequency ratio for rubber damper and various damping ratios

10.3.4 Torsiographs and torsional vibration tests

The presence of any torsional vibration results in a positive and negative vibratory movement superimposed on the nominally constant angular velocity of the whole crankshaft, the movements being greatest at a position remote from the node(s). This point being at the front end of the crankshaft in the majority of installations, means are required to measure these vibratory movements.

Until recently, this has been done by attaching a light rotating mass to the front end of the crankshaft by means of a torsionally rigid coupling and driving a heavey mass by the light one through very flexible springs. The natural frequency of the heavy mass is very low so that at high frequencies the heavy mass rotates at constant angular velocity independent of the angular excursions of the light mass. By fitting means to measure the relative movements between the heavy driven mass and the light one, the variations of the crankshaft movement from constant angular velocity can be measured.

One of the earliest torsiographs was that by Geiger, which used a light bell crank with its fulcrum attached to one mass and one arm to the other. Relative angular movement was translated to an axial linear movement with a light spring load and feeler pin which operated a recording pen marking the movements on paper tape passing under the pen at uniform speed. The addition of time and cycle markings permits the numbers of peaks per cycle and the frequencies present to be measured and analysed.

As engine speeds rose and crankshaft frequncies were designed to be higher, the purely mechanical recording means were no longer able to operate fast enough to give accurate readings. A series of small lighter instruments, directly mounted on the crankshaft end, were then devised using changes in electrical characteristics with relative movement between the light and heavy masses the signals being taken away by means of some form of sliprings to appropriate amplifiers and electronic analysing equipment, usually with a cathode ray oscillograph as a monitoring unit.

Amongst such units were: the MIT-Sperry and Sunbury-Stansfield generating a signal proportional to the relative velocity between the masses so that integration and dynamic calibration was required to obtain amplitudes; the Carter and SLM types using variations of capacity to modulate a frequency sensitive system, and the Southern Instruments one using variation of inductance with frequency modulation. These instruments gave good service on medium and low speed engines, but when used on modern high speed petrol and diesel engines often needed a good deal of maintenance due to the general roughness of the engine and shaft which imposed high axial and flexural accelerations on the whole torsiograph pick-up unit.

Today the preferred instruments are based on the use of a light toothed wheel rigidly attached to the crankshaft rotating past a magnetic transducer attached as rigidly as possible to the crankcase. Examples of this type of torsiograph are marketed by Associated Engineering Developments Limited in the United Kingdom and Scientific Atlanta in the United States. As the crankshaft rotates a roughly sinusoidal voltage is induced by the passage of the teeth past the trasducer, the frequency of which is proportional to the product of the number of teeth and the shaft speed. This sinusoidal voltage is fed to a frequency modulated discriminator which, having a high frequency, forms the system's carrier frequency. Then torsional vibrations are superimposed, the carrier waveform is frequency modulated.

At constant shaft speed with no vibration present the carrier frequency is greater than the cut-off frequency of a low pass filter, and the filter output is a steady d.c. voltage. With torsional vibration present at a frequency lower than the filter cut-off frequency the filter output varies as a function of the torsional vibration waveform. After passage through an integrating amplifier and the removal of the d.c. component an output voltage is obtained directly proportional to the torsional vibration amplitude regardless of frequency. This can then be recorded or harmonically analysed to determine the order numbers present and their magnitudes as required. The narrow width of the toothed wheel used makes it possible to attach it at other points in the system if so desired. Dynamic calibration is required.

The measured amplitudes for the various order numbers present are then used in conjucnction with the Holzer Table to determine the actual stress levels present in the operating speed range, the tests having been carried out at a number of engine speeds. The maximum stress level in the crankshaft at any point and speed is usually assumed to be the sum of the mean output torque stress and the sum of the individual stresses present due to the various orders present. In complicated systems there may be order numbers present due to more than one mode of vibration.

10.4 General design practice and use of materials

10.4.1 Introduction

The process of design is a complex problem of integrating many difficult, often conflicting, technical, manufacturing, and cost problems. Aesthetically the completed design should look attractive. Manufacturing, including assembly and subsequent servicing costs, must be as low as possible. The whole package size must usually be minimal, but in many cases also conform to some fixed size to allow it to be installed in, say, a motor car; or, in larger sizes, in a locomotive which must itself conform to standard loading gauges and overall weight. Usually a high specific output is required with a high degree of thermal and mechanical reliability.

Clearly a single chapter cannot being to deal adequately with the subject of design. Some aspects are already covered in their specialist details in other sections of ths book. It is therefore proposed to discuss general principles with particular reference to materials and to the manufacturing processes normally used.

In recent years there have been rapid advances in all the individual disciplines involved in the successful design and manufacture of an internal compbustion engine. This means that there must be close co-operation of a whole team of specialists. Such co-operation demands increasingly that each specialist should have some understanding of other disciplines in order that effective discussions can take place.

10.4.2 The design process

The starting point for an engine design is the obvious one of deciding the power output and speed required to meet the desired application. In the case of a naturally-aspirated engine the bore and stroke can usually readily be determined by assuming a certain number of cylinders and a stoke/bore ratio. These days even automotive diesel engines may be turbocharged whilst truck, locomotive and marine engines are almost invariably turbocharged. In the highest specific output engines boost pressures of three times atmospheric or even higher may be used.

At the same time, the decision must be made as to whether a precombustion chamber is to be used or whether an open or direct injection chamber will be used. Truck size, and upwards, engines are virtually always direct injection. At the moment the smaller engines operating to higher speeds, and covering a wider speed range, almost always use some form of high swirl precombustion chamber. The lower fuel economy and easier cold starting offered by the direct injection forms of combustion chamber is resulting in intensive efforts to operate the direct injection engines successfully at higher speeds so that they can compete with the precombustion chamber engines.

Having decided a preliminary cylinder size and number of cylinders, the type of combustion chamber to be used, and the degree of boost to be used, it is then possible to decide, in the light of previous experience, the compression ratio and the probable maximum cylinder pressures. The designer can now begin to draw in the centrelines for the cylinders, based on empirical values from previous experience for engines of a similar type, and the cylinder bores followed by the outline pistons, connecting rods, crankshaft bearings, etc. Similarly the initial transverse section of the engine can be started. At this preliminary stage the overall design will be very fluid while the requirements for any novel design features, the timing drives and the accommodation of auxiliaries including the turbocharger(s), and maybe aftercooler, are investigated to make the most compact overall arrangement.

Having a preliminary design available it becomes possible to begin quantifying stress levels in the various components on the one hand and to consider manufacturing details on the other. These deliberations will lead to many detail changes to the first schemes. For example, an evaluation of the bending stresses and the torsional vibration stresses in the crankshaft may lead to some adjustment of sizes and occasionally a change in the firing order to be selected. It may be that the evaluation of the torsional vibration stresses will point to the need to fit some form of damper to the front end of the crankshaft. At the same time the total reciprocating mass and the rotating mass acting at each crankthrow are calculated, along with the out-of-balance moment of each crankthrow, when preliminary values of bearing loadings can be made at both low and high speeds.

In the light of these values the size and disposition of balance weights can be decided. It may be thought necessary to modify bearing sizes at this point. The crankshaft arrangement and balance weight sizes and disposition now permit the value of any internal rotating couples to be evaluated with a wiew to ensuring that the crankcase scantlings, particularly those in the horizontal plane, are made sufficiently robust to oppose the couples without excessive deflection. Case are known where inadequate tansverse crankcase stiffness have led to bending fatigue failure in one of the crankshaft's central crankwebs.

Where a range of engines having different numbers of cylinders is being designed, the crnkshaft and bearings should be designed for the maximum number of cylinders. At the same time some thought must be given to likely future uprating of the basic engine design whether by increasing the bore and/or stroke or by increasing the inlet boost pressure. In the interest of light weight it is usual to work to the highest practical stress levels. This means that where crankshaft regrinds take place, to overcome ovality and wear in the course of a long life, allowance must be made for the reduced crankpin and journal diameters which result as well as the type and depth of any crankpin or journal hardening used.

In the case of automotive type engines, some allowance must be made for overspeeding on downgrades, due to the vehicle's weight driving the engine, when evaluating inertia forces and stresses. In hilly areas exhust gas braking is increasingly called for in the interest of safety to avoid such overspeeding but still the possibility should be considered at the engine design stage.

At an early stage of the desing at least some preliminary calculations must be carried out on the cam and valve train. This is to ensure that the cam height from the centre of rotation is such that with the acceleration characteristics chosen, particularly the decelerations at or near maximum lift, when associated with the working valve train mass, referred to the cam follower, and initial valve spring design results in an acceptable contact (Hertzian) stress to avoid pitting or excessive wear in service.

The size of the cam follower should be made adequate to cover the maximum cam offset, which for nominally flat-faced tappets is 57.3 times the maximum cam velocity in mm/cam degree, and the cam width used together with any axial displacement of the tappet with respect to the cam used to promote tappet rotation. At the valve end the initial cam calculations will permit the required valve springs carriers, etc. to be drawn in. Final, more accurate, calculations may well be carried out later, as well as dynamic valve train responses over the speed range using a computer model of the whole equivalent valve train but early calculations are required to determine that adequate, but not excessive, space has been provided.

At this stage, prticularly for the larger multicylinder engines at the top end of the power ranges, it is wise to check that the camshaft diameter is sufficiently large to avoid undue bending deflections under the operating loads. This is really necessary where a fuel pump cam is positioned within the camshaft bearing spans. The natural torsional frequency should also be calculated for the camshaft since with long camshafts it may be necessary to fit a suitable damper at the free end. Any significant torsional vibration will not only lead to high camshaft stresses but will invalidate the tacit assumption of constant angular camshaft velocity when calculating lifts and accelerations. Extra torque of a vibratory nature may well be added to those prevailing in the camshaft drive train leading to overloading of the gears, chain, or toothed belt, used to drive the camshaft.

As the design develops by steps, akin to successive approximations, various other aspects must be considered including materials, casting and other metal forming processes, even assembly and servicing requirements. It becomes possible to consider the inter-reaction of all these things and to define and stress the various components in even greater detail.

Recent developments in digital computers make possible calculations, and calculation techniques, which could not have been contemplated a few years ago. One such technique is that of 'finite elements', increasingly used for evaluating stresses and deflections of components, even major components, under prescribed loading conditions. The method can cope with temperature changes within the component under consideration and will accommodate non-linearity and creep in the stress-strain relationships. The size of storage now available permits the recall and cross-linking of many programs, so that they can be used interactively to optimise the design of systems such as valve train details and predict their dynamic behaviour very quickly and thoroughly.

The whole field of the use of computers in design is currently developing and expanding at an ever increasing rate. This covers not only complete design and stress methods, but also the speeding up of the making of detail and general arrangement drawings and in suitable cases, the production of tapes for numerically controlled machine tools.

Having very briefly touched on the general approach to engine design it is necessary to consider factors which determine the selection of materials and the permissible stress levels which may be used.

10.4.3 General properties of materials

The average engineer is primarily interested in the mechanical and physical properties of the materials he uses and tends to leave the details of their production to the metallurgist. However, since the ultimate tensile strength, the fatigue strength, the hardness obtainable, machineability coefficient of expansion, thermal and electrical conductivity, and other properties of interest depend on the metallic constituents used to make the alloy and the details of melting, pouring, cooling rate and mechanical shaping methods, considerable additional knowledge is required. Whilst much detailed information is now available in published form, every endeavour should be made to discuss all the aspects and requirements during the design stage not only with a metallurgist but also with the pattern makers, the foundry, the forging manufacturer, etc. This is because the final properties will be related to the effect of all the manufacturing processes which, themselves, are related to the production quantities involved and the total economics of the production methods used.

Before considering the typical materials used for engine parts, it is necessary to discuss briefly what is meant by certain simple terms which are usually quoted very glibly.

Amongst the properties first taught to students is 'ultimate tensile strength'. This is, of course, obtained by applying as truly axial tensile load as possible to a specimen of the material under consideration, usually machined accurately to some recognised standard size and proportions, noting the load at which the specimen breaks. This load divided by the original cross sectional area gives the 'ultimate tensile stength'. In the case of very ductile materials good deal of elongation occurs accompanied by a local waisting or 'necking'. This means that the true local breaking stress is higher than the nominal value normally quoted.

At failure, the diameter of the fracture is measured and the reduction of area calculated.

$$\text{Reduction of area } \% = \left(\frac{\text{original area}-\text{final area}}{\text{original area}}\right) \times 100 \quad (10.40)$$

Also, the
Elongation at failure, % on gauge length

$$= \frac{\text{extension}}{\text{original gauge length}} \times 100 \quad (10.41)$$

It is essential to quote the gauge length due to the local larger elongation when necking occurs. The gauge length used for ductile materials is made $5.65\sqrt{\text{area}} = 5d$ in the UK following ISO, $4.47\sqrt{\text{area}} = 3.96d$ in the USA, and $11.3\sqrt{\text{area}} = 10d$ in Germany. Many quoted English values are based on the old gauge length of $3.54d$.

The values for the reduction of area and elongation at failure are a measure of the material's ductility. In the case of steels the initial stress-strain curve is linear at low values, the ratio of stress/strain being Young's Modulus of Elasticity. Depending on the ultimate strength of the material this linear reltionship is ended by a distinct yield point, where the elongation increases quite suddenly for low stength low alloy steels, but becomes gradually curvilinear for high strength steels which also have much smaller elongation at fracture.

Similar conditions apply to cast irons and light alloys. Reference should be made to the relevant British Standard Specifications for details of the diameter, gauge length and the whole specimen form.

Engineers tend to refer to the ultimate tensile strength (UTS) when considering materials and their allowable stress levels. For steels, an approximation exists between the hardness and ultimate strength in the absence of surface layer heat treatments such as carburising, induction and flame hardening, nitriding, etc. These are carried out ot reduce bearing wear and enhance the fatigue properties of the material in either bending or torsion when the maximum stress levels occur at the surface.

An important point to remember is that the basic properties of materials are dependent on their analysis and the rate of cooling which occurs during cooling for castings and after heat treatment. Large and/or thick sections cool more slowly than thin ones and, as a result, have lower strengths than small sections. *Figure 10.40*, which is based on BS 1452:1977, illustrates this effect for grey cast irons. The Grade number is the ultimate tensile stregth of a 20 mm diameter test piece mchined from a 30 mm cast bar. Similar effects occur with nodular cast irons.

In one series of tests a nodular iron, having a mainly pearlitic structure as cast, the UTS fell from 660 N/mm^2 with a 30 mm cast section size to 450 N/mm^2 with a section size of 300 mm. Steels too are subject to size effects, the precise effect being dependent on the analysis and the heat treatment used. By way of example a steel having an analysis of C = 0.3/0.4%; Si = 0.1/0.35%; Mn = 1.3/1.8%; Mo = 0.20/0.35%, oil quenched from 845°C and tempered from 600°C, can have an UTS of 940 N/mm^2 in sizes up to 22 mm, but falls to 725 N/mm^2 with a 100 mm section.

Since there may be a considerable degree of scatter in the properties achieved due to minor variations in analysis, heat treatment, metallurgical structure, degree of rolling or forging involved, etc., most material specifications usually quote the minimum properties required.

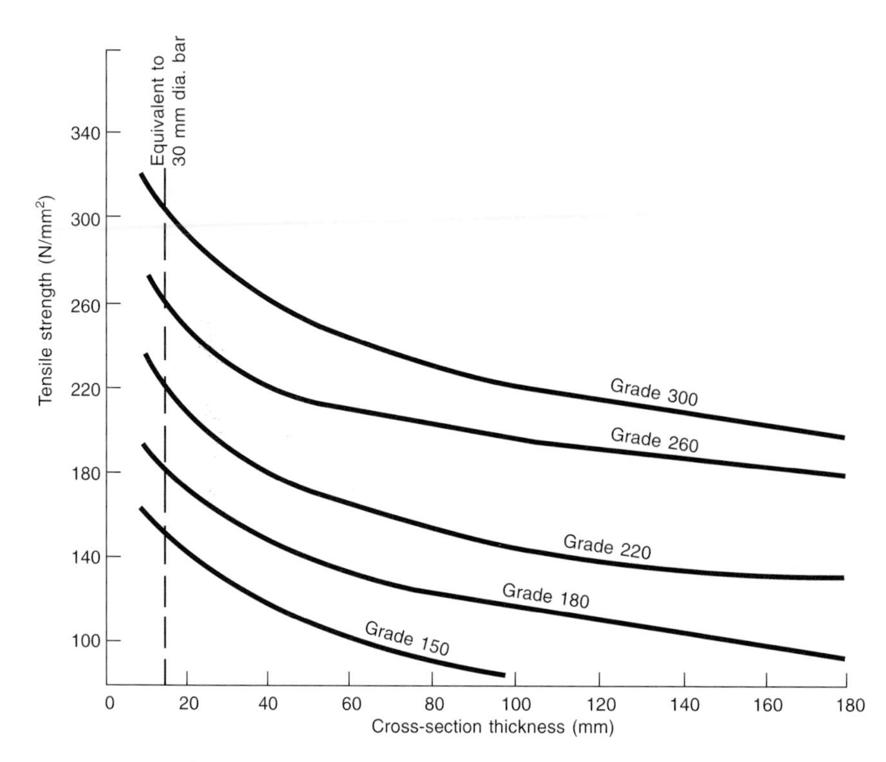

Figure 10.40 Variation of ultimate tensile strength with section size for grey cast irons (from BS 1452:1977)

10.4.4 Behaviour of materials under repeated loads—fatigue

In internal combustion engines most parts are subjected to loads and stresses which are repeated every cycle and may vary with load and speed. The various stresses may be due to direct tension, bending and torsion or combinations of these. The signs may alternate between positive and negative at various times. Some parts such as cylinder head and big end bolts are tightened to quite high stress levels—just below their yield point. This stress remains a major stress for any one tightening, although additional small cyclical stresses will be added to the main stresses when the engine is operating. With a correctly designed and tightened joint these superimposed fluctuating stresses should be small.

Test machines have been devised to subject materials to varying stress levels and, by carrying out tests with various loads, determine how the stress level to produce cracking or complete failure varies with the number of cycles of load repetitions. One of the earliest and simplest test devices is that originally devised by Wohler.

Essentially, a specimen is held in a suitable chuck at one end, whch is supported by rigid bearings and rotated by an electric motor, whilst the cantilevered specimen has the experimental load attached at its free end through a ball bearing. The specimen is therefore subject to a bending moment and the stress fluctuates sinusoidally as it rotates. A revolution counter records the number of cycles to which the specimen has been subjected before failure occurrred.

Usually a microswitch or similar device is set to switch off the machine automatically when the specimen breaks. This permits operation without supervision. A series of tests at various loads permits the plotting of a curve of stress level against the number of cycles before failure. Due to the large number of cycles required before no failures occur, for most ferrous metals, such curves are usually plotted with logarithmic or semi-logarithmic co-ordinates and are usually referred to as S-N curves i.e stress against number of cycles.

Figure 10.41 shows diagrammatically the trend for a steel specimen. As plotted, the stress producing failure drops as a straight line until about 2×10^6 or more cycles when it becomes horizontal. The stress level at the horizontal line is known as the 'fatigue-limit stress' or 'endurance limit'.

The specimens for a smooth endurance limit test are accurately made and finished by careful grinding and polishing. In spite of this, slight imperfections in the material such as inclusions,

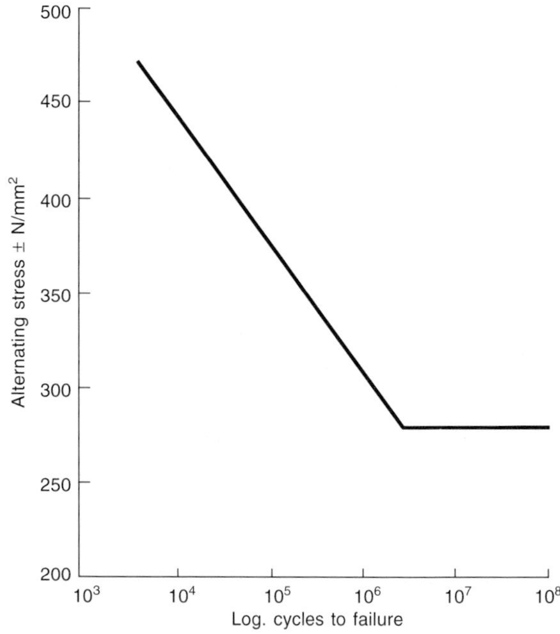

Figure 10.41 Typical S-N curve for a steel specimen

minor differences in analysis or structure and slight experimental irregularities, result in a significant scatter of test points. For this reason; the largest practical number of specimens should be tested.

Tests carried out in push-pull or pure tension require a more sophisticated machine to impose the axial loads on the specimen but also to ensure that the loads are applied truly axially on the specimen to avoid the simultaneous imposition of a bending stress as well as the required axial value. Since with bending stresses their value is maximum at the surface and theoretically drop to zero at the centre the importance of any imperfections in causing cracking is greatest at or near the outer surface. In truly axial tension the stress level is uniform across the whole section and failure may start from an imperfection anywhere in the section. Statistically, the chance of having an imperfection leading to failure is greater for axial loading than for bending. For this reason fatigue strengths in push-pull fatigue are lower than for those in bending fatigue.

In practical engineering components a truly smooth specimen is rarely found. Changes in section occur, notches such as keyways are cut, holes are drilled, whilst screwed parts have a spiral groove formed in them. Again scars from corrosion and fretting lead at the very least to surface blemishes. It is found that the introduction of such irregularities results in a significant lowering of the fatigue strength.

In some cases the increase in local stress levels at the tip of a notch can be estimated by mathematical methods and, more recently, a considerable number of studies are being made using finite element methods. It is found that the local stresses may be considerably higher than the nominal values. The ratio of the theoretical peak value to the nominal is known as the *theoretical stress factor* (K_t). In the case of 60° vee notches with a small tip radius values of up to 3.6 have been estimated.

Actual fatigue tests with notched samples often show a reduction which is less than such theoretically derived values. The ratio of the fatigue strength of specimens with no stress concentration present to that with a deliberate stress concentration present is known as the *fatigue strength reduction factor*, usually denoted by the symbol K_f.

Attempts have been made to correlate K_t and K_f using the term *notch sensitivity* (q) defined as

$$q = \frac{k_f - 1}{K_t - 1} \tag{10.42}$$

where q usually has a value between 0 and 1. Curves have been published such as *Figure 10.42* suggesting relationships but these must be regarded as a rough approximation only, since q is not only dependent on the material in use but also varies with the specimen size, the stress conditions appertaining and the number of life cycles being considered.

In general, the notched fatigue strength of steels improves up to static strengths of about 1000 N/mm^2 and then deteriorates slightly at greater strengths. Flake graphite grey cast irons having relatively sharp edges to the graphite flakes, i.e. have in-built multiple notches, are less notch sensitive than the recently introduced compacted graphite cast irons having flakes with rounded edges, and both are less notch sensitive than nodular or spheroidal irons having significant ductility and in which the graphite is in the form of substantially spherical nodules.

Non-ferrous materials, particularly aluminium alloys, tend not to have a distinct fatigue limit. In the unnotched form the S-N curve never becomes truly horizontal although the rate of fall within the number of cycles decreases. For this reason unless low stress levels are used in design, infinite life cannot be guaranteed, *Figure 10.43*. In the notched form practically no increase in fatigue stresses occur with increased ultimate strength. In internal combustion engines, note must be taken of the operating temperature with aluminium alloy components, since at relatively low temperatures compared with ferrous materials i.e. above 200°C, their ultimate strength drops off and creep can occur. Creep is a continuation of extension, or compression, with time when the specimen is subject to a given stress.

10.4.4.1 Modified Goodman diagram

So far reference has been made to fatigue tests in either rotating bending or push-pull. In the usual relatively simple testing this

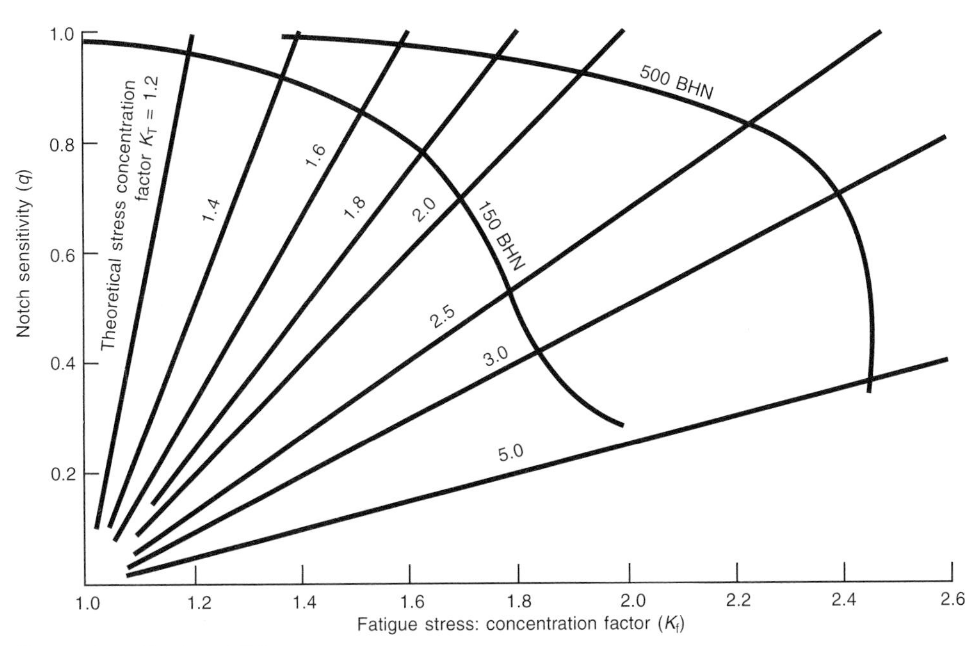

Figure 10.42 Notch sensitivity curves

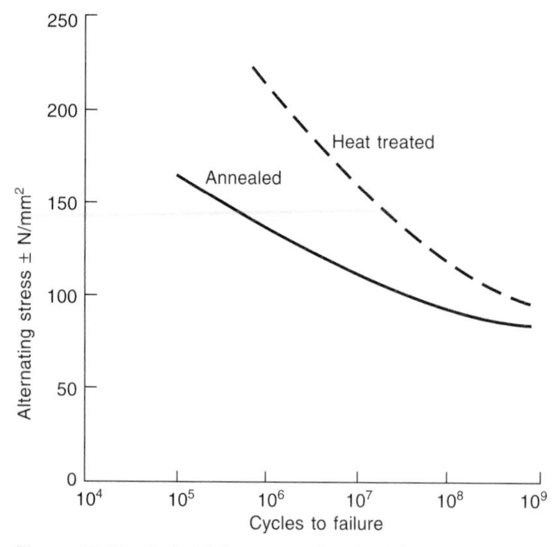

Figure 10.43 Typical fatigue curve for aluminium alloys

means that alternating loads and stresses varying sinusoidally with time are applied, the maximum cyclical tensile stress being equal to the maximum compressive stress. This alternating stress with time is indicated in *Figure 10.44A* . It will be noted that the mean cyclic stress is zero. In practice this state of affairs rarely occurs and other cases may be distinguished.

It the mean stress is tensile as indicated in *Figure 10.44B* with a sinusoidally varying stress superimposed, a fluctuating tensile stress is said to be imposed. If the mean stress imposed is equal to the maximum value of the cyclically imposed stress, the minimum stress imposed is zero as indicated in *Figure 10.44C* and this pattern is known as a pulsating stress.

When the mean stress is greater than the applied half amplitude of the cyclically varying stress, the applied stress is that of fluctuating tension stress as shown in *Figure 10.44D*. In most practical cases one of these patterns will occur albeit often in a more complex form.

By way of example the stresses in the shank of a connecting rod for a naturally-aspirated diesel engine at low speed will almost be that of a pulsating compressive stress. To a first rough approximation the applied peak gas load will remain constant over the load and speed range of the engine so that the resulting gas compressive stress range in the rod shank will also remain substantially constant. However as the speed rises the upward inertia force at TDC due to the reciprocating weight will reduce the gas load applied to the rod at firing TDC but will apply a tensile force to the rod at idle TDC, in the case of a four cycle engine. As a result, the applied stresses become a case of fluctuating compressive stress.

In practice, many diesel engines are now turbocharged, so the actual maximum cylinder pressures will vary with speed and load in a manner dependent on details of turbocharger matching and the variations of fuel injection timing with load and speed, so these pressure changes and their effect on the rod stresses must be taken into account.

It has been found, from series of fatigue tests in which varying ranges of fluctuating stress are superimposed on various mean stress levels, that a diagram as shown in *Figure 10.45* in which for positive values of mean stress, i.e. tensile stress, straight lines joining the experimentally determined alternating fatigue stress to the ultimate tensile strength gives safe estimates of the expected fatigue strengths for various mean stress levels. This type of diagram is known as the 'modified Goodman diagram'.

Strictly the Goodman diagram applies to that part of the diagram

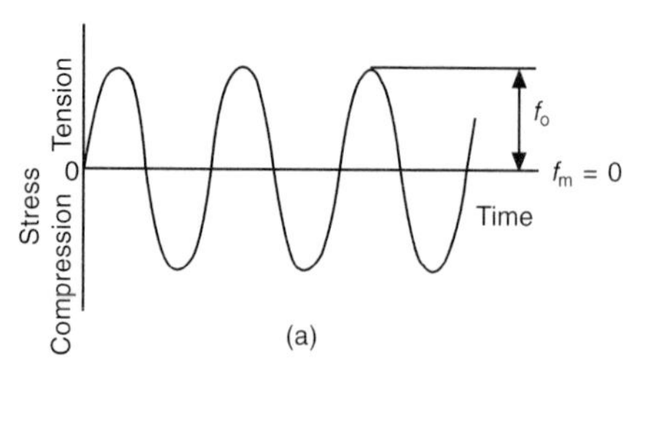

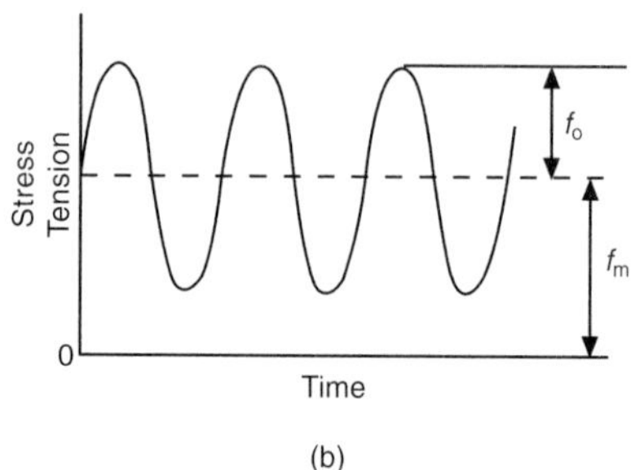

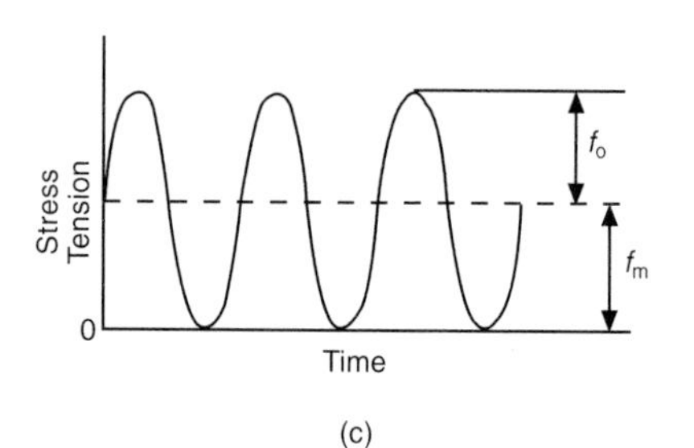

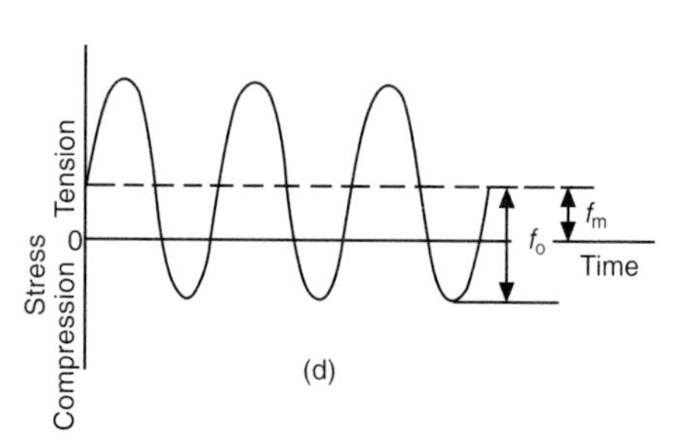

Figure 10.44 Various types of fatigue stress. (a) Alternating; (b) Fluctuating tensile; (c) Pulsating; (d) Fluctuating

in which the mean stress is positive but to cover compressive mean stresses it is usual to extend the diagram as shown. Since a part in tension will extend plastically if the yield point is exceeded maximum cyclical stresses in excess of the yield point value must be avoided. For this reason the top of the diagram is cut off as shown.

Reference has been made to the fact that simple laboratory fatigue specimens are carefully manufactured and finished off with a polished surface. In most engine componensts, parts are subject to complex stresses, which may be tension and/or compression or even shear, in three dimensions. The solution to dealing with these theoretically in relation to fatigue is not yet in sight, but the calculation of such complex stresses in complicated shapes is now tractable using finite element methods. In practice tests are carried out on complete components, often

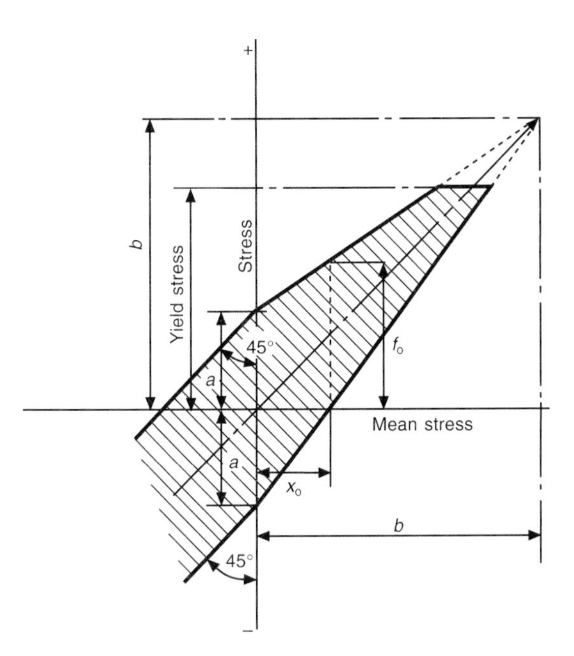

f_o = Maximum fatigue stress when minimum stress = 0
a = Alternating fatigue stress
b = Ultimate tensile strength of material

$$f_o = \frac{2a.b}{(a+b)} \qquad x_o = \frac{a.b}{(a+b)}$$

Figure 10.45 Modified Goodman diagram

within the engine, to check that satisfactory service will be given. This is a vast field which is still under investigation. Here, only a very elementary study is possible and more detailed information should be sought in the references given and in reports published in various magazines and the proceedings of the various professional institutions.

In simple parts subjected to unidirectional stresses it is found that a sudden change in section, the existence of a notch such as a groove particularly one with sharp root, or even a rough finish will significantly lower the stress at which fatigue failure occurs. To obtain an estimate of the lower fatigue value, tests are carried out as for the smooth specimens already previously referred to but with a suitable notch machined in each. Often this is a 60° vee with a small radius at the bottom. From a series of tests a fatigue S-N curve is obtained as for the plain specimens when it will be found that the fatigue limit known as the *notched fatigue limit* is significantly lower than that for the unnotched specimen. In the case of steels the curves become horizontal at about 10^7 cycles as for the unnotched specimens. As previously mentioned, the term *fatigue strength reduction factor* is given to the ratio given by the fatigue strength for the plain specimens divided by that for the notched specimens. This is usually denoted by K_f.

In simple cases it is possible to calculate the maximum local stress prevailing theoretically in the notch, and the term (theoretical) *stress concentration factor,* K_t, is given to ratio of the maximum local stress level to the nominal local stress prevailing in the vicinity of the discontinuity as evaluated by simple theory.

The theoretical stress concentration factor and the practically determined fatigue strength reduction factor rarely agree and vary from material to material. A very approximate relationship has been obtained by introducing a notch sensitivity factor—q defined in eqn. (10.42).

If $k_f = K_t$, $q = 1$ the material is said to be fully notch sensitive. On the other hand should the presence of a notch not affect the fatigue strength $K_f = 1$ making $q = 0$ and the material is described as notch insensitive. In practice q has a value between these two extremes. Unfortunately published values, or indeed experimentally determined ones, for q must be regarded as very approximate since it not only depends on the material in use but the size, metallurgical details, and actual prevailing stress in the part or specimen. It cannot be regarded, therefore, as a material constant.

In design every endeavour must be made to avoid sudden changes in section, and if these must exist, to use the largest practical radius which in production must be as perfectly formed and blended as possible to keep stress concentrations factors to a minimum. The surface finish on highly stressed components must be good as surface 'jogs', even part number stamps or etching, represent notches and will lower the part's fatigue strength if present in highly stressed zones.

During design and manufacture every attempt must be made to avoid locked in tensile stresses such as can occur in cast or heat-treated components having widely different sections and, in consequence, cooling rates; during grinding with too high a feed rate or an unsuitable wheel; or as a result of cold straightening. Such locked in tensile stresses mean that the real mean stress is higher than that due to the applied working loads. Conversely, deliberate steps can sometimes be taken to cause locked in compressive stresses to be present in highly loaded parts. Examples are cold rolling of fillet radii, shot peening with spherical shot and in the case of helical springs, scragging.

Compressive stresses at the surface of parts are often induced as the result of material phase changes during hardening and quenching, particularly with induction and flame hardening and the more recent laser beam hardening, and the introduction of additional elements combined with quenching as in carburising, cyaniding, nitriding, tufftriding, etc.

It must be remembered that such induced surface compressive strains will be balanced by tensile stresses at, or near, the junction of the hardened layer and the core material. For parts subject to tensile loads and for shafts in bending or torsion, when the bending or torsional stresses have not fallen off significantly at the radius of the junction of core and hardened layer, it may be found that fatigue failure starts at, or immediately below, the hardened zone junction.

10.4.4.2 Low cycle fatigue

Most applied stresses are repeated every engine cycle although their actual magnitudes may vary with load and speed. Normal design procedure should ensure that their magnitude is below the fatigue limit corresponding to infinite life.

However certain cases occur where the cyclical stresses are low, or comparatively low, but are superimposed on quite high static stress. Two examples are connecting rod bolts and cylinder head bolts or studs. These are normally tightened to about 90% of their yield stress to provide the required bolting load. In operation, with correctly proportioned abutments, the working stress range will be small. However if the bolts are tightened and untightened many times, they are subject to a stress range from zero to the tightening stress which being quite high will lead to fatigue failure, usually in the first thread, in a relatively small number of cycles. This is known as *low cycle fatigue.*

A similar state of affairs might occur in inlet and exhaust manifolds when high inlet and exhaust pressures are used, unless this has been foreseen during design. The major stress range, neglecting thermal stresses, is that due to the operating pressure within. This varies between atmospheric and the maximum value, with lesser changes due to operating changes in boost with load and speed.

10.4.4.3 Thermal fatigue

As discussed in chapter 5, the metal towards the centre of the cylinder bore can reach quite high temperatures when operating at maximum load and speed particularly when turbocharging is used. Typical values and information concerning their interdependence with operating conditions, materials and head deck thickness can also be found in chapter 5.

In a typical cylinder head the gas side metal temperature will be greatest at or near the centre of the cylinder bore decreasing as the cylinder bore is approached. Thus the metal in the centre wants to expand but is constrained by the cooler metal surrounding it including the comparatively cool ring of metal outside the cylinder-to-head joint line. As consequence the central metal is stressed in compression. At the same time the metal temperature decreases vertically through the head deck from the hot gas face to that prevailing at the coolant side.

The temperature difference involved depends on the local heat flux, the thermal conductivity of the head material and the local metal thickness. Assuming this temperature changes linearly with the distance from the gas face, the hot metal at the gas face will be held in compression relative to the mean temperature existing at mid-thickness whilst the cooler metal at the coolant side will be strained in tension. Again, a further horizontal temperature difference will occur in the bridge(s) between the inlet and exhaust valves and also between the exhaust value(s) and central injector in the case of direct injection engines.

Clearly the thermal strains and stresses prevailing within the gas face of a cylinder head are very complex, and these in turn may have additional stresses imposed by interference fit valve seat inserts and combustion chamber hotplugs, in the case of indirect injection engines. Furthermore the cyclically imposed stresses due to the operating gas pressures will be additive to these thermal strains and stresses.

For the present discussion, and for the sake of simplicity, consider the radial stresses occurring at mid-metal thickness of a head assumed to have a uniform head deck thickness. Starting with a new head, taken as stress free, and supposing it to be operated up to full load on starting, the metal temperatures will rise with time until they reach their stable values corresponding to the power output involved, when they will stay at those values with time until a load change is made. As already pointed out, the hotter metal in the centre of the bore will be constrained by the cooler metal surrounding it so the central material will be compressively stressed and the cool metal outside the cylinder bore will be stressed in tension. Due to the gas side temperature being hotter than the mean, the compressive stresses will be greater than the mean values. On the coolant side of the deck the temperatures will be less than the mean so that the stresses will be tensile relative to those at the mean deck thickness.

Now the ultimate strength of ferrous materials tends to decrease slightly as temperatures rise whilst that for light alloys can drop significantly at temperatures above about 200°C. This means that creep, or a slow rate of yielding, of the hot gas face material will occur if the temperature and thermally induced compressive stresses are high enough. If creep occurs in compression when hot, the face material will no longer have the zero stress level, assumed to exist in the new head, when the engine operating temperatures fall at lighter loads or as an extreme when the engine is stopped and the temperatures fall to ambient values. The stress-time history is, therefore, similar to that shown diagrammatically in *Figure 10.46*. As indicated, the major stress levels σ_{th} are compressive, or tensile, due to temperature with relatively smaller high frequency cyclical changes in stress σ_p superimposed due to the gas forces prevailing. The major stress cycle is that due to warming up and cooling down. Road vehicles tend to run for relatively short periods of time, so many cycles of thermal stress will occur in a year. Their smaller sizes and metal thicknesses tend to result in smaller thermal stresses and

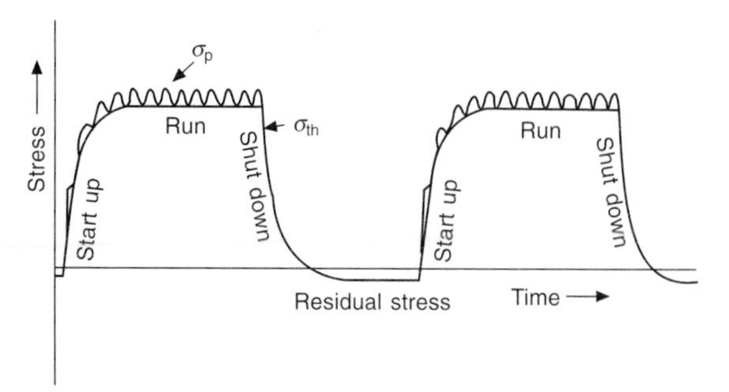

Figure 10.46 Idealized stress cycle for a combustin chamber component

less creep will occur; also the time for creep to take place is comparatively short. By contrast a large marine engine will work most of the time at a high load level and for periods of weeks.

Consequently, the higher thermal stresses will be maintained for longer periods, and the amount of creep will be larger even if equivalent temperatures and stress levels were to occur as in the small engines. When such a large engine does cool to ambient temperatures the residual stresses will be relatively larger. This will occur only three or four times a year depending on voyage times.

Failure due to thermal fatigue is due to low cycle fatigue, the cracks usually being manifest on the gas side metal when cold. Even when visible, their propagation time may be comparatively low.

No universally agreed factor has been derived for the selection of the best material to use for cylinder heads. The aim is to keep metal thicknesses in regions of high heat flow as thin as possible, and to use the highest practical thermal conductivity for the material used. This keeps the temperature difference prevailing to the lowest practical value. When associated with a low coefficient of material thermal expansion and a low Young's Modulus, the lowest practical stress levels are obtained. This, ideally, should be combined with a high ultimate and fatigue strength for the material chosen. Unfortunately high strength materials need to be alloyed which lowers the thermal conductivity and increases Young's Modulus. As a result the final selection is usually based on a good deal of practical, and experimental experience.

As will now be realised, the whole subject of stressing and selection of materials is a vast one and it has only been possible to give the basic guidelines in this chapter. However, the references given at the conclusion of the chapter will permit studying the subject in much greater depth. Information about the materials in most general use for the main components now follows.

10.4.5 Typical materials used in production

10.4.5.1 Bedplates and crankcases

In the largest engines, bedplates and crankcases are fabricated by welding from low carbon steels. These may have small quantities of niobium added to improve their strength without affecting their other properties, prticularly that of ease of welding. An ultimate strength of about 540 N/mm^2 is obtained, slightly higher than mild steel but with the high ductility being maintained. Every care must be taken in the preparation of the plate for welding and on the actual welding techniques, used. The fatigue strength of welds is less than for the parent metal, so stress levels at welds should be kept low.

Medium-sized engines usually have cast-iron crankcases and bedplates, when the latter are used. Vast improvements have been made in the past twenty years in the production of higher quality cast irons. Grey flake graphite cast irons have been improved mainly by removing phosphorus and sulphur. At the same time it has been realised that quite small quantities of elements such as lead and antimony, often introduced from mixed steel scrap, can markedly reduce the strength obtainable. Eutectic cell size have been reduced in many cases by 'innoculating', adding finely divided calcium silicide as the molten cast iron is run into the ladle. Grey cast irons having strengths of up to 400 N/mm^2 are now obtainable although, as already indicated, this is interrelated with section size.

Starting with cast iron having less than about 0.02% of sulphur and phosphorus an addition of either magnesium, cerium, or magnesium plus cerium is made which has the effect of transforming the graphite flakes typical of grey irons into spheroids of graphite. Such irons are known as spheroidal or nodular irons. These irons will give ultimate strengths up to 800 N/mm^2 as cast with an elongation of about 2%. Although care is required to obtain a fully and truly nodular structure, some large engines are now using nodular iron for their crankcases as foundries with sufficiently large pouring capacity become available.

In truck and automotive engines a high proportion use grey flake graphite irons for their combined crankcase and cylinder block castings. Where integral dry cylinder bores are used, a relatively high phosphorus iron is employed giving a network of hard iron phosphide relatively resistant to wear. A large proportion of engines are now fitted with wet cylinder liners, the necessary coolant seal at the bottom being by rubber O-rings.

In the interest of lighter engines, particularly for the smaller sizes, there is a growing tendency to use aluminium alloys. A number of high volume production engines are to be found having crankcases ingeniously designed to provide the necessary relative uniformity of wall thicknesses, so permitting successful manufacture by aluminium alloy die-casting. Liners are usually of cast iron, whether wet or dry types, but in a few cases special high silicon alloys are used which after machining and etching provide a satisfactory hard surface compatible with the pistons and rings. Further information about liner materials may be found by reference to chapter 12.

10.4.5.2 Cylinder heads

In the largest sizes, steel castings may be used but more often a good quality flake cast iron is preferred usually with minimal alloy content in the interest of maintaining a good thermal conductivity.

Attempts have been made to employ nodular irons, but although they have higher strengths their thermal conductivity is about two-thirds that of flake graphite irons and their modulus of elasticity some 30% greater. As a result nodular irons are more subject to thermal cracking problems.

Currently experimental heads are being tried in 'compacted cast iron'. These are low sulphur and phosphorus irons containing small amounts of magnesium and titanium, which gives a metallurgical structure intermediate between flake and nodular irons. In compacted irons the flakes typical of grey irons have rounded ends. These irons give strengths and thermal conductivities intermediate between flake and nodular irons.

In truck and automotive engines pearlitic flake cast irons are mostly used but there is a growing trend in spite of higher cost to prefer aluminium alloy to reduce weight. These, of course, require the use of value seat inserts. Typical aluminium alloys include BS 1490 alloys LM8, LM23P, while LM24 and LM25 are often employed particularly for pressure die cast components.

10.4.5.3 Crankshafts

For very large engines low alloy forged steel is used. In general the crankthrows are individually forged and machined, the required shaft arrangement being assembled by shrinking on to the journals, although the webs may be made separately and shrunk on to both the crankpins and journals.

Medium speed engines use steels, the whole shaft being forged as one and the machined. A wide variety of steels may be chosen depending on the crankshaft size, since the ruling section determines the ultimate strength obtainable. For sizes up to about 230 mm journal diameter a '40' carbon (En 8) steel may be used in its hardened and tempered condition giving an UTS of about 620 N/mm^2. By the use of heat treated low alloy steels such as En 14 or 15 (carbon manganese), En 12(1% nickel) and En 18 (1% chromium) a similar ultimate strength can be obtained but with improved yield point strengths.

As sizes reduce, a wider range of steels is available and this by the use of appropriate heat treatments. For example En 16 (Mn-Cr) will reach 695 N/mm^2 UTS in 150 mm section and 850 N/mm^2 with a 65 mm ruling section. Other often used steels include En 19 (1% Cr-Mo), En 24 ($1\frac{1}{2}$% Ni-Cr-Mo), and En 40 (3% Cr-Mo). With high bearing loadings relatively hard bearing materials are required to avoid their fatigue; and high loadings also lead to reduced oil film thickness with consequent greater wear of the crankpins and journals. To avoid this some form of surface hardening is used for the smaller shafts. This is frequently done by flame-hardening or induction-hardening in which the surfaces to be hardened are rapidly heated and then water quenched.

The hardness obtainable by these methods is dependent on the carbon content of the steel. For this reason the carbon content should preferably be between 0.4 and 0.55%. In the case of induction hardening the hardened layer thickness is dependent on the heating time, i.e. power used, and the frequency used for the inductor blocks. Where induction blocks are employed the hardening is ended before the start of the fillet radii to avoid complicating the stressing conditions at the radii. Where it is desired to harden the fillet radii and part of the crankweb faces as well, a single inductor 'wire' is used for heating and scanned around the pin and journal.

In highly stressed shafts gas nitriding or carbonitriding is often adopted. Here the shafts, made from steels containing elements which form nitrides, are exposed to ammonia and hydrocarbon gases at around 500–600°C for periods of up to 100 hours. The details depend on the actual steel used and the thickness of hardened layer required. The result is an extremely hard (700–800 DHN) layer about 0.25 mm thick which is also in compressive stress leading to an improved fatigue strength. A similar result using a bath of liquid salts is also available known as 'Tufftriding'.

The development of reliable and high strength nodular irons, which can be shell-moulded closely to size and shape, are now increasingly used for the high volume production automotive size engines. Such cast shafts behave well with bearing materials so long as the surface finish of the pins and journals is good. Induction hardening can also be used if required.

Fillet radii are often rolled to improve their form and to induce compressive stresses, both of which will enhance the fatigue strength. This treatment can be used with nodular iron shafts.

10.4.5.4 Connecting rods

These are almost invariably made from forged high tensile steels. A few small engines are beginning to use cast nodular iron rods where the stress levels are reasonable.

10.4.5.5 Cams and tappets

In large engines roller type cam followers are used made of a through hardened high carbon steel working in conjunction with case hardened cams.

As sizes reduce flat tappets or curved finger followers are employed sometimes in steel but increasingly in chill-cast iron. In the smaller sizes chilled, or flame or induction hardened, followers and cams are almost invariably used. Hardnesses of 45 Rc or more are achieved. Apart from the details of cam design in which maximum Hertzian or contact stresses are calculated, and these days also the minimum oil film thicknesses estimated, experience indicates that good wear and fitting behaviour depends on obtaining good surface finishes on both the cams and the tappets. On highly stressed systems finishes as good as 10 microinches(c.l.a) {0.25 micrometres} or less for the tappet face and 0.51 micrometres for the cams have been found necessary. Usually some form of anti-scuff surface treatment such as 'Parco-Lubrising' is given after manufacture.

10.4.5.6 Pistons

These are dealt with in chapter 12, but for completeness the following general remarks are made here.

In the largest sizes pistons are usually cast steel having a complex internal design of cavities and drillings to permit their cooling by circulated oil or water in order to dissipate the large heat fluxes which occur. Medium size engines use thin nodular iron or somewhat thicker aluminium pistons. Again in both cases arrangements are made to provide internal oil cooling to limit the maximum operating temperatures.

In truck and automotive sizes aluminium alloys are almost always employed. Because of their slightly lower coefficient of expansion and somewhat better high temperature strength many pistons are made in alloys such as 'Lo-Ex' containing about 12% of silicon, the eutectic value, and sometimes even greater amounts up to 22% i.e. the so-called hyper-eutectic alloys. With high outputs as obtained by turbocharging such pistons are directly oil cooled using either an oil jet impinging on the underside of the crown or circulated in a plain or even spiral internal gallery cast into the piston body, using salt cores which are dissolved out when the casting cools.

10.4.5.7 Valves

Engine valves have an arduous life. They have to work at high temperatures, particularly the exhaust valve, and maintain a high tensile strength and hardness being free from creep at working temperatures. In the case of the larger engines which often operate on poor quality bunker fuels likely to contain sulphur and significant proportions of vanadium and sodium as well as a relatively large ash content the steels used must have a high resistance to corrosion.

The complexes of vanadium and sodium oxides form glass like deposits on the valve head sealing surfaces if the valve head temperatures exceed their relatively low melting points. If deposited they prevent the exhaust valve from seating properly and, breaking away locally, lead on to locally guttering and burning. Even if burning does not occur the valve head temperatures are raised and distortion and fatigue failure can ensue.

Larger engines use valve cages which permit the valves and their seats to be removed without lifting the complete cylinder head for servicing. Particular care is now taken to ensure that the valve seating in the cages is well cooled by circulated water which, when combined with a large excess combustion air factor and scavenge air, keeps the valve head seat temperature below about 500°C so that the vanadium-sodium complexes do not adhere to the sealing zones.

Exhaust valve heads are frequently made of a highly heat resistant material such as Nimonic 80 fraction-welded to a less highly alloyed stem. Apart from reducing material costs, the stem material is selected to be compatible with the valve guide material so avoiding the need for chromium flashing and similar treatments when an integral austenitic valve stem is used.

To minimise corrosion and wear the valve head seats are protected by welding on a facing of stellite or similar high nickel-cobalt-chromium material. To help equalize the peripheral temperature valve rotators are often fitted.

In smaller engines the burning of a larger proportion of the available air at full load coupled with the higher operational speeds leads to higher exhaust valve temperatures—up to about 800°C. On the other hand the use of distillate fuels which contain only very small amounts of metallic salts and, usually, only small amounts of sulphur considerably reduces the problems of corrosion.

For small diesel engines 21-4N is often preferred but if corrosion occurs 21-12N having a lower manganese content may be more suitable. For highly rated engines Nimonic 80A or 81 may be used.

In the case of inlet valves En 24, 52, or 59 may be used for naturally aspirated engines. With turbocharging, even automotive sizes, inlet valve temperatures up to nearly 700°C have been observed in indirect injection engines at full load and speed. At these temperatures the ultimate strength of the named materials has fallen to about 185 N/mm^2 so it becomes necessary to adopt 21-4N or 21-12N with their greater strength at these high temperatures.

10.4.5.8 General remarks

In referring to steels, reference has been made to En number specifications largely because these are still empoyed in practice. It should be pointed out that these numbers used in BS 970: 1955 are now regarded as obsolete, and have officially been replaced in BS 970: 1970. The new standard specifies a wider range of steel compositions using numbers and letters to identify their composition, whether to be supplied only to a chemical specification, to specified mechanical properties, etc. It resembles in many ways systems used in the USA and elsewhere. In the attached list of materials and properties the corresponding new designation is given where a correspondence exists.

A good deal of interest is currently being shown in the possible use of new materials such as ceramics on the one hand, which are heat resistant but brittle, or plastics on the other, which are heat sensitive and subject to creep at relatively low stresses and temperatures. Considerable problems have still to be overcome before they can be used on a large scale in the internal combustion engine field but if the difficulties can be overcome they offer potential thermal advantages, lower engine weights and possibly lower prices in years to come as the richest metallic ores, particularly of alloying elements, become exhausted leading to higher prices if not actual shortages.

Table 10.6 Summary of material properties

Steels

	Heat treatment °C	*Size mm*	*PS 0.1%*	*PS 0.2%*	*PS 0.5%*	*YP N/mm^2*	*UTS N/mm^2*	*Elong. %*	*RA %*	*IZOD ft. lb.*	*E N/mm^2 $\times 10^3$*	*Density kg/m^3*	*Therm cond. W/m°C*	*Coeff. exp. $\times 10^6/°K$*	*Spec.heat J/kgK*
En2	N 930	28.6	—	—	—	263	368	42	74	—	205	7861	60.7	12.6	0.427
	WQ 900	28.6	—	—	—	347	347	37.5	75.5	—	—	—	—	—	—
080M40															
En8	N 860	38.1	—	—	—	340	608	30	68	70	207	7833	45.6	12.2	0.481
	N 860	203	—	—	—	334	633	28	—	—	—	—	—	—	—
	OQ 840 T610	28.6	—	—	—	578	771	25	—	71	—	—	—	—	—
	OQ 850 T500	28.6	—	—	—	862	940	16.5	52	33	—	—	—	—	—
605M36															
En 16	OQ 845 T600	28.6	—	—	—	734	846	21.5	66	89	210	7861	48.1	13.2	0.461
		100 SQ	—	—	—	587	729	23.5	62	90	—	—	—	—	—
	N840	111	—	—	—	670	754	14.5	22	47	—	—	—	—	—
	OQ 850 T 630	28.6	—	—	—	772	927	20	—	77	—	—	—	—	
	OQ 850 T 630	100	—	—	—	664	834	22	—	50	—	—	—	—	—
709 M 40															
En 19	OQ 550 T 550	14.3	—	—	—	1124	1174	17	59	46	213	7833	42.7	13.2	0.473
	OQ 550 T 550	63.5	—	—	—	976	1059	18	57	40	—	—	—	—	—
	OQ 550 T 550 (O)	82.6	—	—	—	988	1103	17	55	35	—	—	—	—	—
	OQ 550 T 550 (C)	82.6	—	—	—	988	1092	19.5	59	46	—	—	—	—	—
817 M 40															
En 24	OQ 840 T 200	28.6	—	—	—	1655	2000	10.5	27.5	20	208	7833	38.1	12.4	0.481
	OQ 840 T 650	41.3	797	—	—	—	951	22	59.5	73	—	—	—	—	—
	OQ 830 T 650	63.5	—	—	—	911	1004	21	—	29	—	—	—	—	—
	OQ 830 T 650	100	—	—	—	905	985	23	—	68	—	—	—	—	—
826 M 31															
En 25	OQ 840 T660	38.1	72.3	—	—	—	871	23	66	77	204	7861	35.6	12.2	0.494
	OQ 840 T660	108	—	—	—	788	927	20	57	65	—	—	—	—	—
	OQ 860 T200	19	1283	—	—	—	1912	11	—	15.5	—	—	—	—	—
535 A 99															
En 31	OQ 850 T 430	13 SQ	1603	1640	1668	—	1764	9.5	20	7	213	7805	33.5	13.1	0.481
En 32	BC, WQ900 WQ780	25	—	—	—	346	553	35	—	85	207	7861	50.2	13.0	0.461
		51	—	—	—	332	493	35	—	84	—	—	—	—	—
		100	—	—	—	297	477	40	—	80	—	—	—	—	—
722M24															
En 40B	OQ 900 T 650	89	—	—	—	795	914	23	70	92	208	7861	35.6	12.2	0.486
897 M 39															
En40C	OQ 930 T 600	76 SQ	—	—	1392	1473	14.5	47	28	—	—	—	—	—	
401S45															
En 52	OQ 950 T 650	19	—	—	—	68.6	937	25	33	8	212	7612	12.6	12.6	0.481
En 55	WQ 1100	28.5	340	374	419	—	826	49	56	35	196	7889	15.9	16	0.502
321S20															
En 58C	WQ 1100	25	—	—	—	284	636	54	64	103	193	8000	15.5	17	0.502

Table 10.6 Contd

Grey iron casting (BS 1452 1977)

Grade	Heat treatment °C	Size cast mm	PS N/mm² 0.1%	PS N/mm² 0.2%	PS 0.5%	YP N/mm²	UTS N/mm²	Elong. %	Elastic stratin %	IZOD ft. lb.	E N/mm² × 10³	Density kg/m³	Therm cond. W/m °C	Coeff. exp. × 10⁶/°K	Spec. heat J/kgK
150	—	30	42	98	—	—	150	0.60/0.75	0.15	—	100	7050	52.2	10	265
180	—	30	50	117	—	—	180	0.50/0.70	0.17	—	109	7100	51.5	—	330
220	—	30	62	143	—	—	220	0.39/0.63	0.18	—	120	7150	50.1	—	420
260	—	30	73	169	—	—	260	0.57	0.20	—	128	7200	48.8	—	460
300	—	30	84	195	—	—	300	0.50	0.22	—	135	7250	47.4	—	460
350	—	30	98	228	—	—	350	0.50	0.25	—	140	7300	45.7	—	460
400	—	30	112	260	—	—	400	0.50	0.28	—	145	7300	44.0	—	460
Nodular iron castings (BS 2789: 1973)															
370/17	(Specification)	25		230	—	—	370	17	—	—	—	—	—	—	
380/17	BCIRA		230	242	256	—	380	17	—	—	169	7100	36.5	11.0	
420/12	BCIRA		266	278	292	—	420	12	—	—	169	7100	35.5	11.0	
500/7	BCIRA		323	339	356	—	500	7	—	—	169	7100	35.5	11.0	
600/2	BCIRA		346	372	409	—	600	2	—	—	174	7170	32.8	11.0	
700/2	As cast (N)		385	416	462	—	700	2	—	—	176	7200	31.0	11.0	
800/2	As cast (N)		440	471	517	—	800	2	—	—	176	7200	31.0	11.0	
Vermicular iron (compacted graphite iron) (BCIRA lab. tests)															
No. 1			246	272	304	—	380	2	—	—	147	—	—	—	
5			307	335	372	—	473	2	—	—	161	—	41.0	—	

Pure metals

Type of metal	Heat treatment °C	Size mm	Melting point °C	PS 0.2%	PS 0.5%	YP N/mm²	UTS N/mm²	Elong. %	RA %	IZOD ft. lb	E N/mm² × 10³	Density kg/m³	Therm. cond. W./m °K 100°C	Coeff. exp. × 10⁶/°K	Spec. heat J/kgK
Gold	Cold rolled 50%	—	1064	—	—	207	221	4	—	—	82.7	19321	24.8	14.2	—
Iridium	Annealed	—	2447	—	—	—	551.6	—	—	—	517.1	22504	4.9	6.84	—
Osmium	—	—	3056	—	—	—	—	8	—	—	558.5	22698	—	6.12	—
Palladium	Annealed	—	1554	—	—	34.5	138/ 193	24/40	—	—	124.1	12013	5.9	11.7	—
Platinum	Annealed	—	1772	—	—	13.8/ 37.9	126/ 186	30/40	—	—	172.4	21452	6.1	9.2	—
Rhodium	Annealed	—	1963	—	—	—	503	—	—	—	324.0	12373	7.2	7.9	—
Ruthenium	Annealed	—	2314	—	—	—	379	—	—	—	413.7	12207	—	9.2	—
Silver	Annealed	—	962	—	—	55.2	151.7	48	—	—	110.0	10491	34.9	19.6	—
Columbium	—	—	2468	—	—	137.9	241.3	—	—	—	103.4	8581	4.54	6.84	—
Molybdenum	—	—	2610	—	—	689.5	758.4	—	—	—	324.1	10214	12.18	4.86	—
Tantalum	—	—	2996	—	—	241.3	310.3	—	—	—	186.2	16608	4.54	6.48	—
Tungsten	—	—	3410	—	—	1379.0	1516.8	—	—	—	406.8	19293	13.93	4.50	—

Table 10.6 Contd

Aluminium alloys—sand cast

	Heat treatment °C	*Size mm*	*PS N/mm² 0.1%*	*PS 0.2%*	*PS 0.5%*	*YP N/mm²*	*UTS N/mm²*	*Elong. %*	*RA %*	*IZOD ft. lb*	*E N/mm² × 10³*	*Density kg/m³*	*Therm. cond. W.m°K*	*Coeff. exp. × 10⁶/°K*	*Spec. heat J/kgK*	*Hardness BHN*
LM2-M	None		93	—	—	—	148.2	2	—	—	68.9	2700	—	20	0.92	70
LM4-M	—		62	—	—	—	139.1	2	—	1.0	68.9	2750	130.0	21	0.92	60
LM4-WP	—		185	—	—	—	279.2	5	—	0.5	68.9	2750	—	21	0.92	100
LM5-M										5.8		2649	100.5	23		60
LM6-M	—		55.2	—	—	—	179.2	5	—	4.5	68.9	2650	160.5	20	0.92	50
LM9WP	—		201	—	—	—	239	—	—	0.9	68.9	2657	146.5	20	0.92	95
LM13WP	—		154	—	—	—	170	—	—	0.5	68.9	2713	121.4	19	0.92	125
LM14WP	—		—	—	—	—	185.3	2.5	—	0.6	68.9	2823	163.3	23	0.92	100
Wrought aluminium alloys																
Duralumin	—		232	—	—	—	386	12	—	—	73.8	2800	134	23.4	—	—
RR56	—		334	—	—	—	432	12	—	—	71.7	2740	159	22	—	—
RR59	—		263	—	—	—	371	6	—	—	71.7	2740	142	23	—	—
Y-alloy	—		216	—	—	—	371	20	—	—	71.7	2740	175.8	22.4	—	—

Miscellaneous metallic alloys

Type of alloy	*PS 0.1% N/mm²*	*UTS N/mm²*	*Elong. %*	*IZOD ft. lbf*	*E N/MM² × 10³*	*Density kg/m³*	*Thermal cond. W/mm°C*	*Coeff. exp. × 10⁶°C*	*Spec. heat J/kg-K*	*Hardness BHN*
Magnesium alloy mag. 1. W	69.5	247.1	4.5	4.0	44.8	1799	79.5	28	1.00	55
Magnesium alloy mag. 7. WP	97.3	210.0	2.0	0.5	44.8	1799	83.7	27.2	1.00	75
Titanium alloy 679	970	1300	8	—	110	4429	7.27	9.5	—	—
18/8 stainless steel En 58	270	642	54	109	193.1	7960	15.5	17.0	0.50	—
Nimonic 80A (room temp.)	602.3	1066	44	30/75	186.2	8170	11.3	11.9	0.43	250/350
Nimonic 105 (room temp.)	787.7	788	7	4/20	220.6	7990	10.9	12.9	0.46	330/400

Refractory hard materials

Material	*Density kg/m³*	*UTS N/mm²*	*Comp. strength N/mm²*	*E (tension) N/mm²*	*Hardness R_A*	*Coeff. thermal exp. × 10⁻⁶°K*	*Thermal cond. W/m°K*
Tungsten carbide + 7.8% Co matrix	14.670	1103	4619	606	89.5	5.76	7.86
Tungsten carbide + 12.2% Co matrix	14,117	814	3861	567	88.0	6.12	8.16
Tantalum carbide/tungsten carbide + Cr-Co matrix	14,117	—	4688	533	92.0	6.48	450
Tungsten-titanium carbide + Co matrix	11,072	862	4206	483	91.7	7.02	2.36

Table 10.6 Contd

Ceramics

		UTS flexural N/mm^2	UTS comp. N/mm^2	Melting point °C	$E_{tension} \times 10^3$ N/mm^2	Density k/m^3	Thermal cond. 21°C W/m°K	Coeff. exp. $\times 10^{-6}$/°K	Spec. heat J/kg-K
Alumina (dense sintered)	21°C	317/345			379/		17.3/		
	1232°C	145/303	2344	—	310.3	3737	32.4	7.65	—
Boron carbide	21°C	303	2896	—	448.1	2408	26.7	5.76	—
Beryllium carbide		—	724	—	—	—	20.9	10.62	—
Boron nitride		—	276	—	—	2104	27.8	7.74	—
Molybdenum chromium alumina		—	1655	—	—	6089	—	9.36	—
Silicon nitride (sintered)	21°C	896							
	1232°C	310	3477	—	310.3	3200	4.2	3.06	—
Silicon nitride (reaction bonded)	21°C	224							
	1232°C	293	689	—	172.4	2600	4.2	3.06	—
	21°C	758							0.670
	1332°C	552	3447	—	427.4	3160	72.1	4.86	1.398

Plastics, unreinforced

Heat treatment °C	Size mm	Deflection temperature at 1.7 N/mm^2 °C	Deflection temperature at 0.46 N/mm^2 °C	PS 0.5%	YP N/mm^2	UTS N/mm^2	Elong. %	RA %	IZOD ft. lb	E N/mm^2 $\times 10^3$	Density kg/m^3	Therm. cond. W/m°K	Coeff. exp. $\times 10^6$/°K	Spec. heat J/kgK
ABS, general purpose	—	87	89	—	—	41.4	5.20	—	6.5	2.27	1010/	—	95.4	
											1040			
ABS, high impact	—	95	98	—	—	34.5/	5/50	—	7/7.5	1.72/	1020/	—	90/	
						41.4				2.21	1040		108	
Acetals homopolymer	—	124	170	—	—	68.9	—	—	1.4	3.59	1420	0.231	135	
Acrylics, standard	—	92.2	101	—	—	72.4	6	—	0.4	2.96	1190	—	64.8	
Acrylics, high impact	—	77	86	—	—	37.2	75	—	2.0	1.38	1150	—	100.8	
Acrylics, high temperature	—	105	115	—	—	68.9	3	—	0.3	3.24	1160	—	54.0	
Epoxy resin (no filler)	—	—	—	—	—	27.5/	3/6	—	0.2/1.0	2.41	1110/	0.167/	81/	
						89.6					1400	0.209	108	
Polytetrafluoroethylene	—	56	121	—	—	23.1	300	—	3.0	—	2130/	0.245	135/	
											2200		151	
Nylon 6/6	—	66	182	—	—	81.4	60	—	0.9	2.90	1130/	0.245	145.8	
											1150			
Phenolics (heat resistant)	—	166/	—	—	—	34.5	—	—	0.26	9.65	1410/	—	50.4	
		193									1840			
Polycarbonate (G.P.)	—	132	138	—	—	62.1/	110/	—	12/16	2.34	1200	0.195	118.8/	
						72.4	125						126	
Polyethylene (high density)	—	—	—	—	—	21.3/	20/	—	0.5/	0.41/	—	—	928/	
						37.9	1000		20	1.24			940	
Polypropylene		51/	93	—	—	34.5	10/20	—	0.5/	1.10	905	0.117	104/	1.69
		60	121						2.2				184	
Polyphenylene sulphide	—	135	—	—	—	65.5	1.6	—	0.5	3.31	1300	0.288	86.4	—
Polysulfone	—	174	—	—	—	70.3	50/100	—	1.3	2.48	1240	0.117	97.2	—
Torlon (poly-amideimide)	—	260	—	—	—	189 (room)		—	—	4.60	1400	—	320	—
						48 (230°C)								

Table 10.6 Contd

Fibre materials

Material	*Density kg/m^3*	*UTS N/mm^2*	*E N/mm^2 × 10^3*
E glass	2550	3447	72.4
S glass	2500	4482	86.9
Boron	2360	2758	379.2
Carbon type I	1950	1896	399.9
Carbon type II	1780	2758	241.3
Tungsten	1940	3999	406.8
Molybdenum	1020	2206	358.5
Steel	774	4137	199.9
Whisker materials			
Alumina	3960	20682	427.5
Boron carbide	2520	13790	482.6
Silicon nitride	3180	13790	379.2
Silicon carbide	3120	20684	482.6

Cement

Type	*Tensile strength (flex.) N/mm^2*	*Comp. strength N/mm^2*	E *N/mm^2* × 10^3	*Density kg/m^3*	*Coeff. thermal exp.* × 10^{-6}°*K*
Hard cement paste (0.3 W/C)	15	60	40	2400	10/20
Plain concrete	6	50	30	2400	10/12
High strength concrete	10	100	45	2400	10/12
Reinforced concrete	30	—	35	2500	10/12
Polymer impregnated concrete	25	150	40	2500	10/12
Glass reinforced cement	30	—	30	2200	10/12
Steel fibre concrete	10	—	30	2400	10/12

References

Design

1 HOWARTH, M. H., *Design of High Speed Diesel Engines,* Constable (1960)
2 SCHWEITZER, P. H., *Scavenging of Two-Stroke Cycle Diesel Engines,* Macmillan, New York (1949)
3 SHAW, M. C. and MACKS, E. F. *Analysis and Lubrication of Bearings,* McGraw-Hill (1949)
4 WILCOCK, D. F. and BOOSER, E. R., *Bearing Design and Application,* McGraw-Hill (1957)
5 CARLSON, H., *Spring Designer's Handbook,* Marcel Dekker (1978)
6 SOCIETY OF AUTOMOTIVE ENGINEERS, Ed. ABELL, R. F., *Application of Computers in Valve Gear Design* (1963) Distributed by Macmillan, New York and Pergamon Press, Oxford
7 ROTHBART, H. A., *Cams: Design, Dynamics and Accuracy,* Wiley (1956)
8 CHEN, F. Y., *Mechanics and Design of Cam Mechanisms,* Pergamon Press (1982)
9 STODDART, D. A., 'Polydyne Cam Design', Machine Design No. 1–3, pp 121, 146, 149 (1953)
10 MERRITT, DR. H. E. *Gear Engineering,* Pitman (1971)
11 BARNES-MOSS, H. W., 'A Designer's Viewpoint', IMechE Paper No. C343/73 (1973)
12 BARNES-MOSS, H. W., *Engine Design for the Future*', SAE Paper No. 741130 (1974)
13 BICKFORD, J. H., *An Introduction to the Design and Behaviour of Bolted Joints,* Marcel Dekker Inc, New York (1981)
14 NEALE, M. J., Ed., *Tribology Handbook,* Butterworth (1973)

Stressing and fatigue

15 DAVIES, A. J., *The Finite Element Method: A First Approach,* Clarendon Press, Oxford (1980)
16 ZIENKIEWICZ, O. C., *The Finite Element Method,* McGraw-Hill (1977)
17 ROARK, R. J. and YOUNG, W. C., *Formulas for Stress and Strain,* McGraw-Hill (1975)
18 PETERSON, R. E., *Stress Concentration Factors,* Wiley-Interscience (1953)
19 LIPSON, C., NOLL, G. C. and CLOCK, L. S., *Stress and Strength of Manufactured Parts,* McGraw-Hill (1950)
20 LIPSON, C. and JUVINALL, R. C., *Handbook of Stress and Strength,* Macmillan, New York (1963)
21 *Fatigue Design Handbook,* S.A.E. Advances in Engineering Series No. 4 (1968)
22 MANSON, S. S., *Thermal Stress and Low Cycle Fatigue,* McGraw-Hill (1966)
23 FORREST, P. G., *Fatigue of Metals,* Pergamon (1962)
24 FUCHS, H. O. and STEPHENS, R. I., *Metal Fatigue in Engineering,* Wiley (1980)
25 GURNEY, T. R., *Fatigue of Welded Structures,* CUP (1968)
26 WILSON, W. Ker., *Practical Solution of Torsional Vibration Problems,* Vol. 3, 3rd Ed.: 'Strength Calculations,' Chapman and Hall (1965)
27 BERTODO, R. and SIVAKUMARAN, S., 'An Assessment of Diesel Engine Poppet Valves', IMechE Proc. **187** Paper No. 2/7 (1973)
28 ROMBAKIS, S., 'Design to Minimize Stress Concentration', SAE Paper No. 720361 (1972)

Materials

29 ROLLASON, E. C., *Metallurgy for Engineers,* Arnold (1973)
30 WOOLMAN, J. and MOTTRAM, R. A. (Compilers) *The Mechanical and Physical Properties of the British Standard Engineering Steels(BS 970: 1955),* 3 Vols. Pergamon (1964–69)
31 AMERICAN SOCIETY FOR METALS, *Metals Handbook* (5 Vols) Vol 1. Properties and Selection: Irons and Steels. Vol 2. Properties and Selection: Nonferrous Alloys and Pure Metals. Vol 3. Properties and Selection: Stainless Steels, Tool Materials and Special Purpose Metals. Vol 4. Heat Treating. Vol 5. Surface Cleaning, Finishing and Coating.
32 GILBERT, G. N. J., *Engineering Data on Cast Iron,* (Summary of Typical Properties) British Cast Iron Research Association (1974)
33 ANGUS, DR. H. T., *Physical and Engineering Properties of Cast Iron,* Butterworth (1976)
34 *Conference on Engineering Properties and Performance of Modern Iron Castings,* BCIARA (1972)
35 NECHTELBERGER, E., *The Properties of Cast Iron up to 500°C,* English Edition Technology Ltd., Stonehouse, Gloucester (1980)
36 *Society of Automotive Engineers Handbook (2 vols),* Published Yearly. (Contains various automotive standards and has section on materials)
37 BERTODO, R., 'Grey Cast Irons for Thermal-Stress Applications', IMechE, Journal of strain Analysis Vol. 5 No. 2 98 (1970)
38 HABIG, K-H., 'Wear Protection of Steels by Boriding, Vanadising, Nitriding, Carburising. and Hardening', *Materials in Engineering,* **2** p. 83 (December 1980)
39 TAYLOR, B. J. and EYER, T. S., 'A Review of Piston Ring and Cylinder Liner Materials', *Tribology International* p. 79 (April 1979)
40 GODFREY, D. J.,'The Use of Ceramics for Engines', *Materials and Design* **4** p. 759 (June/July 1983)

Balancing

41 KER WILSON, W., *Practical Solution of Torsional Vibration Problems, Vol. II, Amplitude Calculations,* 3rd end, Chapman & Hall (1963)
42 KER WILSON, W., *Balancing of Oil Engines,* Griffin (1929)
43 CORMAC, P., *A Treatise on Engine Balance Using Exponentials,* Chapman & Hall (1923)
44 KER WILSON, W., *Vibration Engineering,* Griffin (1959)
45 ANON, 'Balancing the Chevrolet V-8 Engine', *General Motors Engineering Journal,* 3rd Quarter, p. 14 (1962)
46 HARKNESS, J. R., 'Methods of Balancing Single Cylinder Engines', *SAE Paper No. 680571* (1968)
47 CLINK, R., 'Balancing of High Spped Four Stroke Engines', *I. Mech. E. Symposium on the dynamics of the IC engine,* Automobile Division, 1958–59, No. 2, p. 73
48 SEINO, T. and SHIMAMOTO, S., 'The Control of the Primary Inertia Force and Moments in Engines with Three Cylinders or Less', *SAE Paper No, 680023* (1968)
49 KER WILSON, W., 'The Fundamentals of Engine Balancing', *Gas and Oil Power,* 1955; July pp. 207–210; August pp. 233–238; September pp. 261–264; October pp. 286–288

Torsional vibration

50 KER WILSON, W., *Practical Solution of Torsional Vibration Problems,* (5 Volumes), Chapman & Hall, 3rd edn. Vol I. Frequency Calculations (1956); Vol. II. Amplitude Calculations (1963); Vol. III. Strength Calculations (1965); Vol. IV. Devices for Controlling Amplitude (1968); Vol. V. Vibration Measurement and Analysis (1969); This massive treatise contains an extensive bibliography under their various headings together with a long list of selected British Patents relating to torsional vibration damping devices etc.
51 BICERA, *A Handbook of Torsional Vibration,* compiled by E. J. Nestorides, C.U.P. (1958). A very concise but thorough review of all aspects of calculations
52 HATTER, D. J., *Matrix Computer Methods of Vibration Analysis,* Butterwoth (1973). A general introductory text
53 DEN HARTOG, J. P., *Mechanical Vibrations,* McGraw-Hill (1940). A classic theoretical text on vibration generally, including dampers

11 Fuel injection systems

Contents

11.1 Introduction

Often called 'the heart of the engine', the fuel injection system is without any doubt one of the most important systems. It meters the fuel delivery according to engine requirements, it generates the high injection pressure required for fuel atomization, for air–fuel mixing and for combustion and it contributes to the fuel distribution in the combustion system–hence it significantly affects engine performance emissions and noise.

The components of the fuel injection system require accurate design standards, proper selection of materials and high precision manufacturing processes. They lend themselves to mass production techniques and they become complex and costly.

As the applications of diesel engines diversified so did the fuel injection systems. Along with the conventional pump-line-nozzle systems new concepts evolved such as distributor pumps, common-rail systems, accumulator systems, unit pumps, unit injectors, etc. In addition, the 'intelligence' of electronics enhanced the capability of the 'muscle power' of hydraulics making the combustion system much more flexible and responsive to new parameters: pressure, temperature, engine speed, etc. Combustion can be thus optimized for best performance, emissions, smooth operation etc., according to the needs of the application. Through electronics, the fuel injection system can interface with other systems; automatic transmissions, cruise control, turbocharger operation, traction control, exhaust aftertreatment, etc. in an integrated system approach.

The net result of this integration is an advanced diesel engine with high power density, very low emissions, low noise and superior drivability. Probably the most dynamic application of advances in fuel injection and electronic management is in the area of light-duty vehicles (passenger cars, light trucks, sport utilities) where constraints of high performance, low emissions, low noise, low cost, etc., render optimization very challenging.

The research and development in fuel injection systems continues at a very sustained pace. The following sections illustrate several examples of state-of-the-art and near term prospects in fuel injection–from the perspective of major manufacturers.

11.2 Diesel fuel injection systems—Lucas Diesel Systems

11.2.1 Compression ignition combustion processes

11.2.1.1 Introduction of fuel

In indirect injection engines, the fuel is introduced via pintle nozzles into an anti-chamber to the space above the piston, which is called a 'prechamber' or a 'swirl chamber', depending upon the intensity of rotary air motion in this chamber. A swirl chamber layout is shown below in *Figure 11.1* together with a much magnified section through a pintle nozzle. In IDI engines, much of the energy to mix the fuel with the air comes from the air motion, including swirl before ignition and combustion-driven mixing in the swirl-chamber throat and above the piston after ignition.

When fuel pressure is applied to the differential area between the guide and the seat, the needle lifts when the force developed exceeds the preload in the spring that holds the needle valve closed. If the pressure is applied progressively, an 'obturator' in the nozzle hole restricts flow during the first part of the valve. This controls the initial rate of injection to reduce noise. At the bottom of the needle is a 'pintle' which forms the spray into a hollow cone

Indirect injection (IDI) engines are being replaced progressively by direct injection (DI) engines, from the larger sizes downwards, and all European truck and larger engines have been direct injection diesel engines for some time. In most of these combustion systems, more of the energy to mix the fuel with the air comes from the momentum imparted to the fuel as it leaves the nozzle.

In small high speed direct injection engines for van and passenger car applications, the combustion chamber has a high swirl re-entrant bowl (also known as having a 'squish lip') to promote turbulence and hence faster mixing towards the end of combustion. In the intermediate medium truck (1 litre/cylinder) size, both quiescent and swirl-assisted combustion have their champions. The quiescent chamber requires more nozzle holes as shown in *Figure 11.2*.

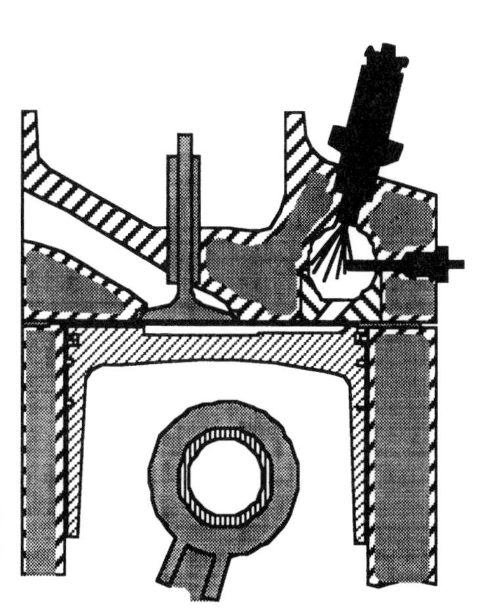

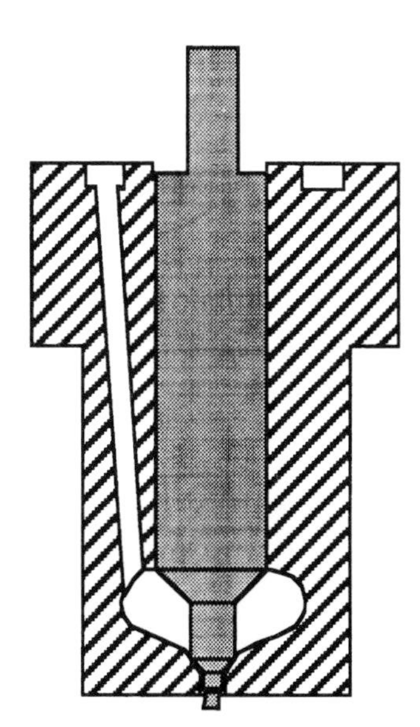

Figure 11.1 Cross-sections of an IDI combustion system and a pintle nozzle

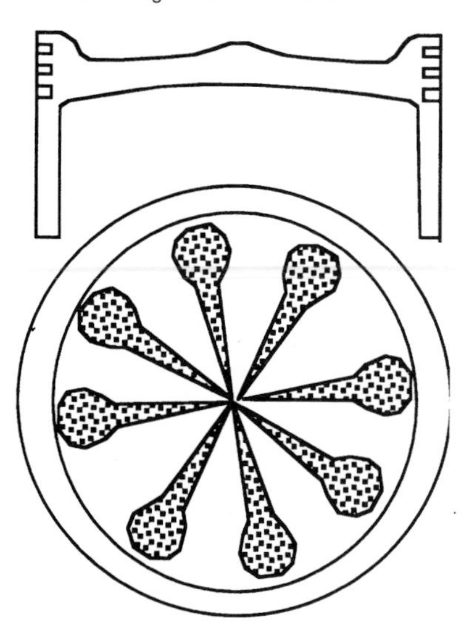

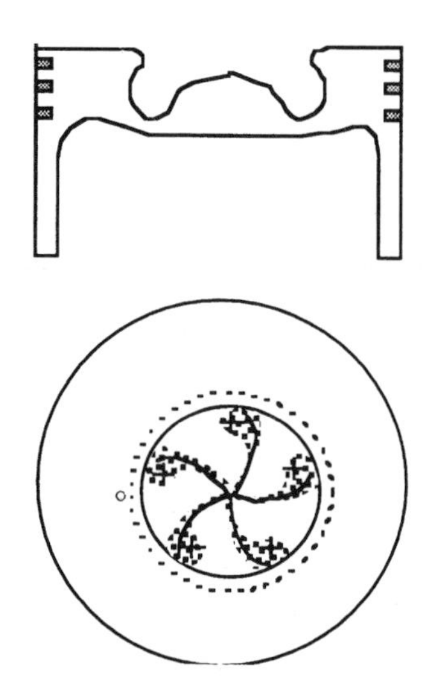

Figure 11.2 Quiescent chamber for a truck DI engine and a swirl-assisted HSDI engine chamber

The arrangement of the hole-type injector in a swirl-assisted 2-valve HSDI combustion chamber is shown in *Figure 11.3*. A cross section of the complete injector is shown in *Figure 11.4*.

The piston is close to the top of its travel, and autoignition can occur, between 20° crank angle before top dead centre (btdc) to 40° after top dead centre (atdc), approximately, with 18:1 compression ratio. At 2400 rev/min this represents only 4.2 ms and in the case of passenger car engines rotating at up to 4500 rev/min, this is only 2.2 ms.

11.2.1.2 Sprays

The literature on spray formation is very extensive and covers over 20 years of intensive developmen of the spray combustion

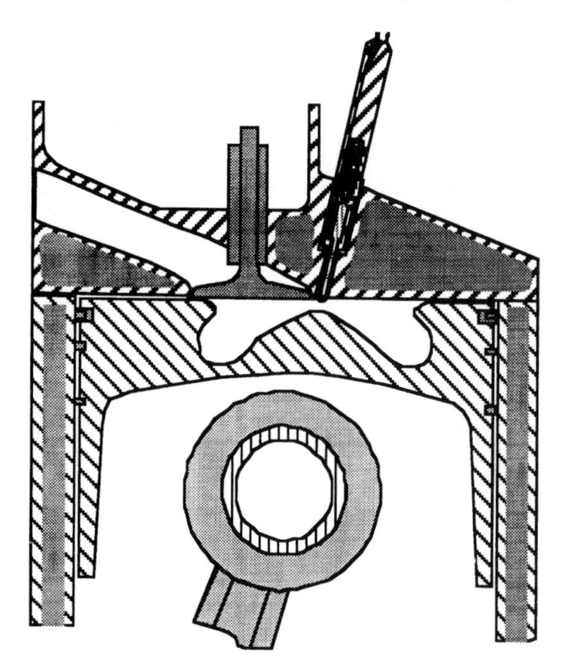

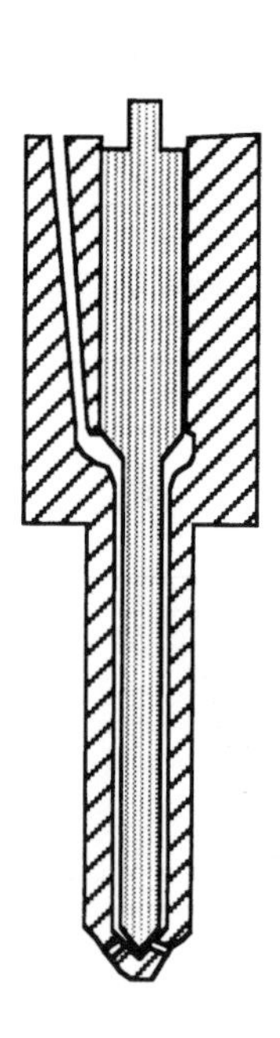

Figure 11.3 Diagram of 2-valve HSDI and hole-type nozzle to inject fuel into the toroidal combustion chamber

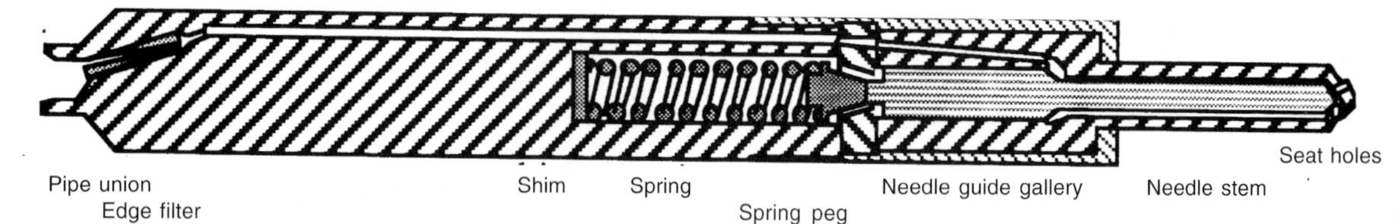

Figure 11. 4 Hole-type nozzle mounted in a low-spring injector

processes. During this period the fuel injection equipment has changed considerably. Early papers[1–3] reflect the performance of injection systems which provided pressures across the nozzle holes that rarely exceeded 500 bars. Some of the injected fuel may be spread along the wall by its own momentum and that of the air swirl. The careful observations by many academic and industrial researchers provided empirical relationships[3,4] and insight into the break-up of liquid sprays into ligaments, droplets and daughter droplets. Several alternative models of the various combustion processes were developed[5].

If the injector nozzle holes are reduced in diameter, and the fuel pressure across them is increased to obtain the same penetration in the compressed air charge, several workers found that the soot and particulate matter in the exhaust were reduced[6–10]. In North America, high injection pressures have been common for many years in injection systems for truck engines which have been equipped with shallow quiescent combustion chambers. More recent work to apply mixing models to guide further reduction in soot generation has shown that these chambers respond well to further improvements in atomization[11]. In addition, the large surface area of the fuel droplets expedites mixing and evaporation. In the 1990s injection pressures in European fuel injection systems rose to over 1000 bars at the nozzle and the upward trend continues through 2000 bars in the late 1990s to reduce the mass of particulate matter in the exhaust.

A typical diesel injection spray leaves each nozzle hole with a narrow included angle, and develops a head vortex where spray momentum is transferred to the compressed air[12, 13]. Each successive element of fuel seems to pass through the head vortex of the previous element, to form a new head vortex further across the bowl until the combustion chamber wall is reached, or the injection is terminated. The spray entrains air as it moves through the air in the combustion chamber[6]. The air entrainment and mixing models, some of which have been extensively validated against experimental results, show that when fuel is introduced into the combustion chamber as finely atomized sprays, the air entrainment increases providing a mixture which is closer to stoichiometric near the centre of each spray.

The generation of diesel sprays by the injector nozzle has been studied with large-scale models[14,15]. Above a critical pressure ratio, cavitation in the nozzle hole occurs which finely divides the fuel before it leaves the hole. In consequence the spray angle is larger and more air is entrained into the spray.

11.2.1.3 Ingnition

Compression ignition combustion occurs only after a delay of approximately 0.0002 to 0.002 seconds after the start of injection. This delay occurs because:

- The fuel has to travel into the combustion chamber.
- It has to mix with the air sufficiently to form a near-stoichiometric mixture.
- The fuel has to evaporate by taking heat from the compressed air.
- The mixture has to heat up to the auto-ignition temperature (*Figure 11.5*).
- Certain chemical reactions must take place in which unstable hydrocarbon-oxygenate species form which will ignite spontaneously.

11.2.1.4 Pre-mixed burn

Fuel which has been injected into the combustion chamber leaves the nozzle at 150 to 500 m/s, so the 20 to 30 mm radius of an HSDI combustion chamber can be traversed by a liquid jet in 0.04 to 0.2 ms, and somewhat longer by the evaporating droplets in a finely atomized spray tip. A significant proportion of the fuel injected during the ignition delay period will have mixed with the air in the combustion chamber when the first element ignites spontaneously (autoignition). Thus virtually all of the fuel which is injected in the ignition delay period (less between one and two crank degrees depending upon the injection pressure and the engine speed) is consumed in the premixed burn, unless the injection is specifically configured to avoid this (for example the M-combustion system[16].

The premixed burn provides the rapid initial heat release which is typical of unrefined direct injection combustion systems, and causes a rapid increase in cylinder pressure that is the origin of combustion noise and in extreme cases piston failures. Fuel injection system developments such as pilot injection (pilot)

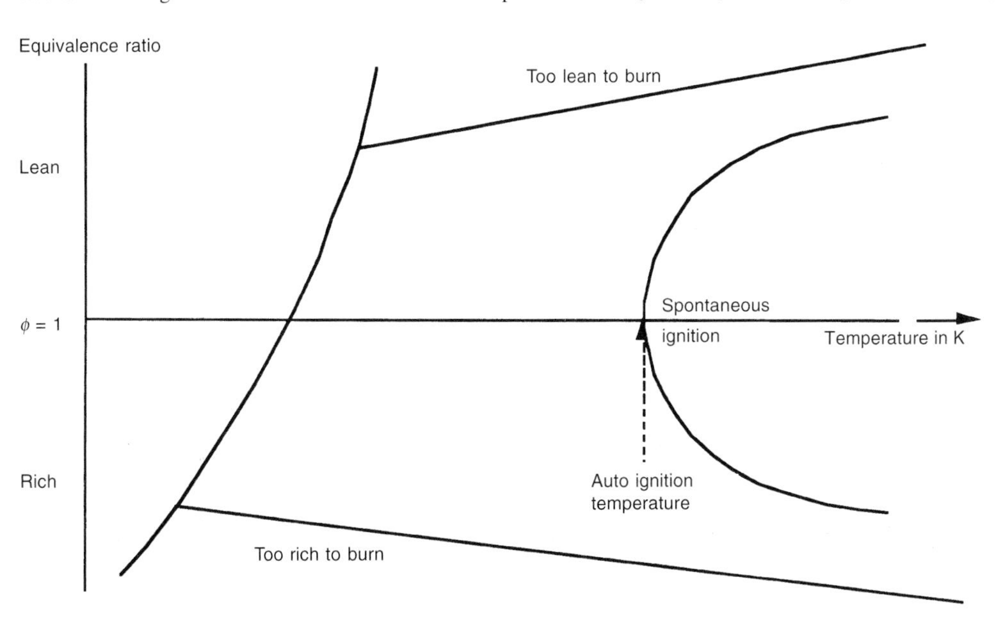

Figure 11.5 Effect of mixture strength and temperature on ignition and combustion of a hydrocarbon fuel

and initial rate controls (IRC) have evolved to control the initial rate of heat release[17].

11.2.1.5 Diffusion burn

Fuel that is injected after ignition meets very different temperature and pressure conditions to that injected before ignition. The conditions may exceed the critical temperature and pressure so that the fuel flashes into vapour as it gains heat. The temperature in the combustion chamber is sufficient to pyrolize fuel if insufficient oxygen is available to burn it, especially near the centre-line of a spray. The rate of consumption of the fuel governed mainly by the rate at which it is mixed with sufficient oxygen and the exhaut emissions reflect this dependence.

When the piston moves down from top dead centre, the vortices generated by the movement of gas force the burning mixture over the lip of the combustion chamber (reverse squish). If the lip is provided with a sharp edge, the turbulence caused will promote even more intimate mixing between fuel and air and will reduce much of the soot formed previously. This is very beneficial for high speed direct injection engines; but calculation of the remaining soot is not easy.

11.2.2 Formation of nitric oxide by lean combustion

Nitrogen and oxygen will combine together to form nitric oxide when heated to temperatures above 1500 K, primarily by the Zeldovich mechanism, although several other chemical reactions have been mentioned in the literature. Under steady state conditions, the rate of formation of nitric oxide increase rapidly with temperature as shown in *Figure 11.6*.

The abscissa of *Figure 11.6* is the equivalence ratio of the mixture which is unity when the mixture is stoichiometric. When the equivalence ratio is greater than 1, very little nitric oxide is formed, as the fuel consumes all the oxygen available at the flame front. As the mixture becomes leaner, increasing proportions of nitric oxide are formed in the mixture. If the temperature is increased at an equivalence ratio of say 0.8, the initial step from 1500 K to 2000 K results in only 0.3% increase in nitric oxide formation as a result of a new equilibrium between O_2, N_2 and NO. The next 300 K leads to a similar 0.3% increase in nitric oxide formation. However, the increase of 300 K between the top two curves yields over 0.5% increase in the equilibrium value of nitric oxide.

Two courses of action are available to reduce nitric oxide formation:

(1) Reduction of the volume of lean mixture in the combustion process.
(2) Reduction of the peak local temperatures of lean burn combustion.

From these two basic options, a number of alternative treatments have emerged which are effective in reducing nitric oxide formation:

(1) Retarding the injection timing, to reduce the peak cycle temperature. (However the thermodynamic efficiency, and hence the fuel efficiency suffers.)
(2) Recirculating some exhaust gas in controlled proportions to dilute the oxygen available at part load conditions and slow the chemical reaction rate. If the recirculated exhaust gas is cooled, the temperature of combustion is reduced.
(3) Injecting water through the same spray nozzle as the fuel (the evaporation of the water reduces the peak cycle temperature).
(4) Reducing the oxide of nitrogen with catalytic combustion of extra fuel or ammonia injected into the exhaust system.

The fuel injection system and electronic control unit are central to all these treatments; either directly, or indirectly since even the EGR and aftertreatments require a precise measurement of the fuelling to be effective without making other emissions worse.

11.2.3 Unburned hydrocarbons

A compression ignition engine emits far less unburned hydrocarbons at normal operating temperatures than intake- and port-injected spark ignition engines, simply because less fuel comes into contact with the film of lubricating oil in compression ignition engines. However, there are about six independent sources of unburned hydrocarbons in a diesel engine[18]. Of these, four are directly controlled by the fuel injection equipment:

(1) If fuel is injected at high velocity into the combustion chamber, before 20° crank btdc, the air motion mixes the fuel so effectively before ignition so that a mixture forms that becomes too lean to auto-ignite and too lean to sustain

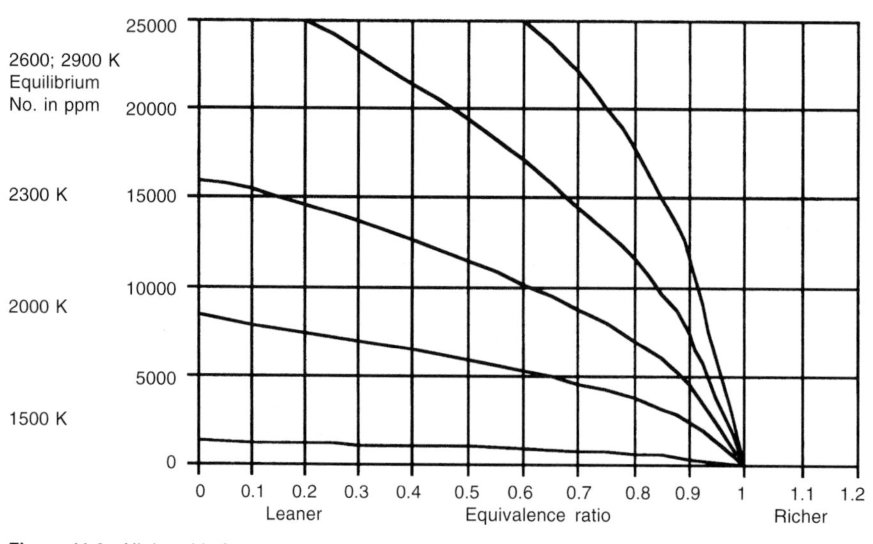

Figure 11.6 Nitric oxide formation in a hydrocarbon–air mixture at high combustion temperatures

a flame. This is the 'lean-limit source' described by Greeves et al.[19].

(2) Hydrocarbons trapped in the sac and holes downstream of the needle seat join the burning mixture late in the cycle, when air motion draws this fuel out[19]. Injector nozzles with smaller sacs or with valve-covers-orifice (VCO) configurations reduce this volume as shown in *Figure 11.7.*

(3) Fuel that is injected late in the engine cycle, will find little oxygen in which to burn. As the piston descends, cylinder temperatures drop below the auto-ignition temperature.

(4) Fuel that is sprayed onto the combustion chamber walls, can conribute to unburned hydrocarbon emissions where the quantity of fuel on the wall exceeds the capacity of the air motion to evaporate it at some particular operating conditions or temperatures. Excess smoke may arise from such conditions also.

11.2.4 Origins of noise in diesel combustion processes

The rapid consumption of most of the fuel injected during the ignition delay period in the pre-mixed burn release heat very rapidly, and the cylinder pressure rises almost instantaneously. This imparts a large and steep-fronted force pulse to the sructure which excites most of the mechanical resonances and causes the surfaces of the engine to vibrate[20]. Acoustic radiation from the vibrating surfaces completes the transmission to the ear. The sound generated by traditional diesel combustion has a characteristic 'knock'; however the structure responds in the same way to mechanical impacts as the pistons move in their bores and in the timing drive. Subjectively the noise sounds very similar to combustion knock if it has a mechanical origin.

Much of the literature suggests that combustion noise depends almost entirely upon the peak rate of pressure rise caused in turn by an initial peak in the rate of heat release. For individual engines, quite good relationships exist between peak rate of rise in cylinder pressure and the noise that originates directly in the combustion processes. However, when such relationships for several different engines are compared, large discrepancies appear[21].

The shape of the cylinder pressure curve can be related to the Fourier Analysis (or spectrum) of the cylinder pressure quite simply:

(1) The compression ratio and turbocharger boost ratio directly influence the level of components up to 500 Hz (as well as indirectly influencing the peak rate of pressure rise via the compression temperature and hence the ignition delay).

(2) The peak cylinder pressure influences the average level of low frequency spectrum components up to between 500 and 900 Hz.

(3) The peak rate of pressure rise influences the components between 500/900 Hz and 3/5 kHz depending upon the engine speed, turbocharger boost (if any) and the rate of injection diagram.

(4) The second derivative of pressure with respect to time can influence very high frequency components (above 5 kHz).

The first option to reduce combustion noise is to reduce the ignition delay and hence the quantity of fuel that contributes to the premixed burn. Increasing the compression ratio, heating the intake air, turbocharging or supercharging, fumigating and reducing heat transfer into the piston and cylinder head may be deployed to this end.

Pilot injection reduces the fuel injected during the ignition delay, and hence the heat released from the premixed burn. Control of the initial rate of injection (IRC) and 'boot-shaped' injection rate diagrams are effective for the same reasons, but only if the control extend over most of the ignition delay; hence pilot is more appropriate for cool combustion conditions encountered in urban traffic and cold start/cold idle. Models of fuel injection systems which reduce combustion noise have been used to explore the design freedoms in such systems[23,24].

Exhaust gas recirculation will reduce the rate of the premixed burn and hence the peak rate of rise in the cylinder pressure.

The modulation of the initial rate of fuel injection, either by pilot or IRC or combinations of these has become an area where the fuel injection equipment manufacturers can add value in terms of refinement in passenger car and public service vehicle applications.

11.2.5 Particulate emissions

Particulate emissions from compression ignition engines are mostly microscopic pieces of carbon, bound together with unburned hydrocarbons into particles which range in size from a few hundredths of a micron to over ten microns. Sulphate particles and particles formed from any metals in the fuel may add to the fine particulate emissions. The fuel injection equipment has a controlling influence upon the generation of soot-based particulates via the air entrainment in the sprays.

If more oxygen can be introduced into the centre of the sprays by increasing atomization and spray velocity and hence air entrainment, less soot particles form in the centre of the sprays. This is the basis of soot and particulate reduction by reducing the nozzle hole diameter and increasing injection pressure.

If the combustion system is designed to work with appreciable wall wetting, then the injection equipment is required to control the proportion of the fuel that is deposited upon the wall.

11.2.5.1 Soot reduction during subsequent combustion

As the piston of a HSDI engine descends, the burning mass of

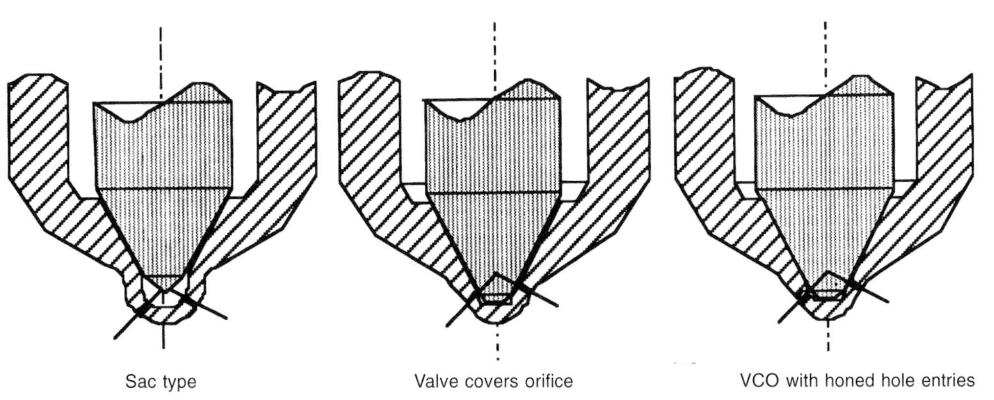

Figure 11.7 Reduced sac volumes and valve-covers-orifice (VCO) to reduce unburned hydrocarbon emissions

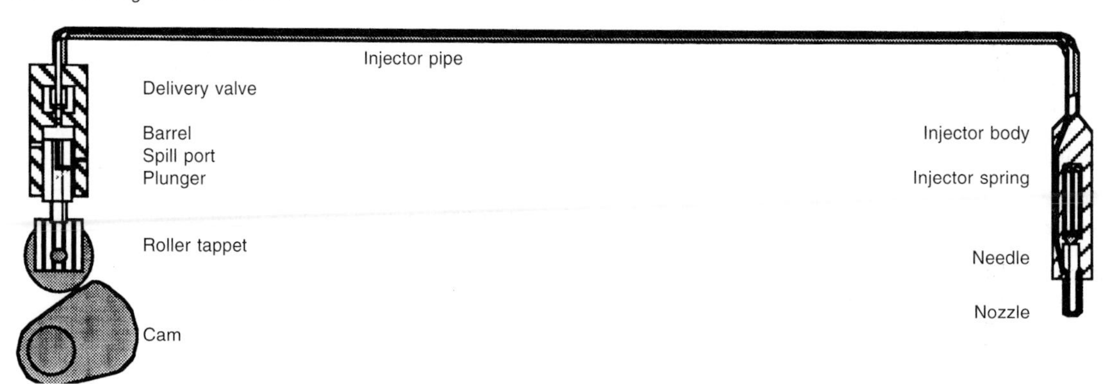

Figure 11.8 Components of a "traditional" pump pipe-nozzle diesel fuel injection system

fuel and air is forced over the rim of the chamber, which generates considerable turbulence and mixes the remaining unburned hydrocarbons with the air above the piston crown. Incandescent particles of carbon are bought into contact with the remaining oxygen and up to 95% of them are burned completely. The detail of the complex motion involving swirl and squish giving rise to a moving toroid of air, into which the fuel is injected; followed by the expansion and spilling of this burning mass of air and fuel over a lip into the space above the piston as the piston starts to descend is difficult to model accurately. Furthermore, as the soot in the exhaust is the difference between that generated during the earlier parts of the diffusion burn, and consumed in almost equal amounts by later diffusion burn processes, its computation involves the subtraction of two large quantities to predict a small difference.

Research with a variety of mechanical fuel injection equipment devices[22] indicates that the main injection must terminate as abruptly as is mechanically feasible, as well as meeting certain injection pressure criteria in order to minimize soot generation.

11.2.6 Traditional jerk pump

The traditional jerk pump was developed to give the prime example of a variable-delivery, positive displacement hydraulic pump. *Figure 11.8* shows the components of a traditional pump-pipe-nozzle system.

The chamber above the plunger within the barrel, in which high pressure is generated by upward movement of the plunger, is connected by a central drilling to a helical or angled groove cut into the side of the plunger. There are two drillings in the barrel wall, one to allow fuel to flow into the chamber when the plunger is withdrawn, which is known as the 'filling port' and a second drilling which is known as the 'spill port'.

In operation, the plunger is withdrawn down the barrel, opening the filling port, allowing fuel to enter the chamber above the plunger from a low pressure supply. As the camshaft rotates and the cam bears upon the roller of the tappet, the plunger is driven upwards until it seals the filling port. The fuel trapped above it is pressurized rapidly, and the pressure opens the delivery valve. When the delivery valve opens, the rapid motion of the plunger creates a high pressure hydraulic wave in the injector pipe.

The high pressure hydraulic wave travels along the pipe until it meets the seat of the needle valve in the injector. While the needle valve is closed, the wave is reflected, causing an additional pressure at the seat of the needle valve from the combined pressure of the incident and reflected waves. When the total hydraulic pressure on the differential area around the valve seat exceeds the nozzle opening pressure (NOP), the needle valve in the injector rises off its seat. The needle opens rapidly when the whole area of the needle is exposed to this pressure. Subsequent injection ensures that the valve remains open, held at its lift stop, despite any temporary fall in pressure caused by the needle displacement.

Inside the injection pump, the plunger continues to travel upward until the helical groove in the plunger engages with the spill port. When this occurs, fuel flows through the central drilling to the helical groove and out through the spill port causing the fuel pressure in the chamber to collapse. This terminates the effective stroke of the pump and generates an expansion wave in the injector pipe. When this expansion wave reaches the drilling within the injector, it causes the pressure under the needle at the seat to collapse below the nozzle closing pressure. (NCP is the pressure which acts on the whole needle area to provide a force equal to that of the spring.) The spring above the needle then begins to accelerate the needle towards its seat with a force that is equal to the spring force minus the force due to any remaining pressure acting upon the bottom of the needle. The needle displacement is a modified sinusoidal movement as the spring relaxes and drives the needle onto its seat.

As the expansion wave passes through the delivery valve, its spring begins to close the valve and the collar on the valve (*Figure 11.9*) enters the valve guide, thus separating fuel within the injection pump from fuel within the pumping chamber. As the delivery valve continues to close, the displacement of the collar multiplied by the area of the guide 'unloads' the system downstream to control the residual pressure within the pipe and hence reduces the risk of secondary injections (injections after the main injection which cannot burn completely as they appear near the end of the combustion process).

Jerk pump operation is reliable and repeatable. The plunger and barrel are made to tight manufacturing tolerances with a very small clearance between them to contain the high injection pressures inside the pumping chamber. The rate of spill controls the collapse of line pressure and hence the rate of the expansion wave that terminates injection. This can be improved by machining the spill port to be oval or even spark eroding it to a parallelogram section to match the helix angle, in order to increase the rate of area increase as the port is opened.

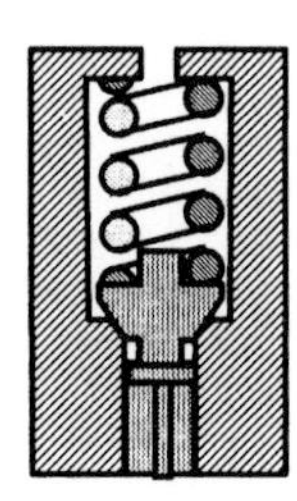

Figure 11.9 Volume unloading valve, which is designed to reduce secondary injection

These pumps are used on large truck engines with separate control of the timing of the start and end of injection. The cam rate, plunger diameter and nozzle characteristics are chosen to control soot generation by injecting at high pressures through small nozzle holes with an injection rate which increases as the injection timing is retarded.

11.2.7 Unit injectors

Several manufacturers of diesel engines have produced their own fuel injection equipment, usually in the form of a 'unit injector'. These injectors are driven from the engine camshaft and comprise the pumping element and the nozzle in the same assembly. Fuel control in usually by rotation of the pumping element, as in the basic jerk pump, but other ingenious mechanical means have been introduced to control fuel delivery. The engine camshaft needs to be somewhat stiffer in torsional deflection in order to maintain accurate injection timings despite variations in loading due to valve and injection events. The juxtaposition of nozzle and pumping element reduces the hydraulic wave travel time, however, the mechanism of operation still generates waves in the drillings within the unit injector, which are difficult to eradicate when high pressures are generated.

11.2.7.1 Electronic unit injectors

The requirement for fully flexible control of injection timing and injection delivery gave rise to a new form of unit injector called the 'electronic unit injector' (EUI). In the EUI, the traditional jerk pump principle has been replaced with a plain plunger and an electromagnetically operated spill valve.

The cross-section of an EUI is shown in *Figure 11.10*. The low pressure fuel supply is direct to each unit injector via drillings in the cylinder head. The connections to the EUI are sealed by high temperature rubber O-rings. The pumps has a conventional filling port to allow fuel into the chamber when the cam-operated plunger is raised by its return spring.

As the cam on the engine camshaft drives the plunger downwards, via a rocker or a roller tappet, the filling port is covered and the pump begins to deliver fuel into a passage which is connected to both the nozzle below it and to a spill valve to one side. To start the injection, a current is passed through an actuator coil, which pulls the spill valve closed against the action of a compression spring. Fuel pressure in the chamber rises rapidly; and an hydraulic wave is developed which passes down the drillings into the nozzle body to the needle seat. When the current through the actuator coil is cut off, spill resumes and the injection terminates.

At low speed and light load conditions, the actuator is fast enough to be operated twice per injection, giving a pilot ahead of the main injection ('electronic pilot'). Very small pilots can be produce down to less than 1 mm^3 which is sufficient to minimize combustion noise when using fuel with a Cetane Number greater than 50.

Many innovative features have been incorporated into experimental EUIs, including combined pilot and initial rate control and control of rate of termination of injection.

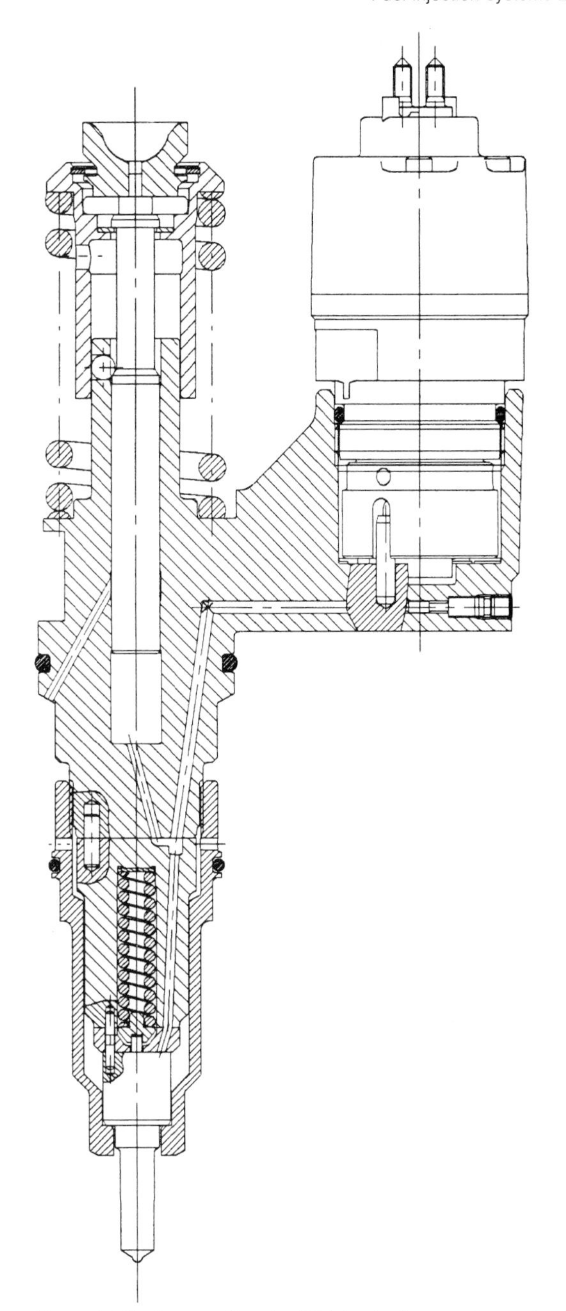

Figure 11.10 Cross-section of an EUI showing the pumping mechanism and spill valve actuator near the top of the equipment and the conventional nozzle at the bottom of the equipment.

11.2.8 DP rotary distributor pumps

The rotary distributor pump design has proved to be very successful, since its introduction in the mid-1950s. This pump design incorporates a totally different pumping mechanism, as shown in cross-section in *Figure 11.11*.

The first practical 'Over the Nose' rotary distributor pump was designed by Roosa, who designed a compact and effective pumping mechanism which drives two plungers in a common bore rotating inside an enlarged shaft. The plungers are driven inwards by an internally-lobed cam ring, acting through rollers and shoes, as shown in *Figure 11.11*. The injection is terminated by allowing the rollers to roll over the 'nose' of the cam and down the reverse flank of the cam. The reverse cam flank has a plateau to unload the injector pipe and drillings before the discharge port in the rotor is sealed from the pump outlet as the rotor rotates. The outlet port of the pumping mechanism is arranged to connect with each injector pipe in turn by carefully-mechined drillings in the rotor and its sleeve. Pumps have three, four or six outlets as appropriate for the engine application.

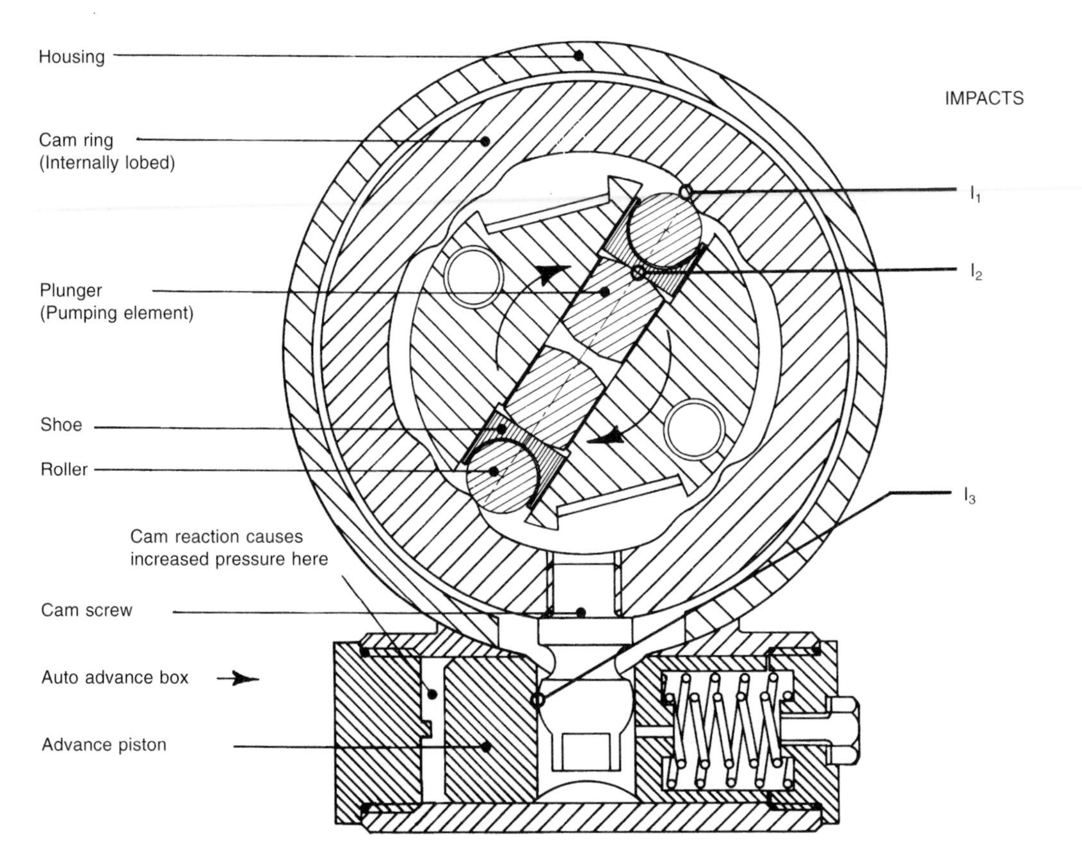

Figure 11.11 Cross-section of DP rotary pump mechanism showing internally-lobed cam ring and two plungers in a common bore which pump via rollers each equipped with a shoe. The cam ring is rotated to set injection timing

The pumping mechanism has to be filled between each pumping stroke. The volume between the plungers and the central drilling in the rotor are connected alternately to the intermediate pressure supply to fill the pump and the high pressure outlets via separate ports in the distributor rotor. The high pressure containment is maintained by a close fit between the distributor rotor and its sleeve.

The intermediate pressure required to fill the pump is generated by a transfer pump, which is outboard of the high pressure pump in the DP range of pumps. The transfer pump consists of a vane pumps with an eccenric stator, to transport fuel from low pressure at the fuel inlet to the filling pressure. The fuel in the cam box is at transfer pressure. The transfer pump provides a consistent pressure that increases progressively with rotation speed up to 7–8 bars, depending upon the pump type.

The transfer pressure is applied to a timing-advance piston operating in a transverse bore and connected to the cam ring to rotate the cam ring. As the transfer pressure rises with speed, the advance piston moves against the action of a return spring, moving the position of the cam between injections, via a ball-ended screw fitted to the cam. When pumping starts, the cam reaction load is taken by hydraulic pressure in the advance box and a non-return valve ensures hydraulic lock so that the cam remains in the position at which it has been set.

The fuel delivery to each engine cylinder is controlled by a metering valve upstream of the pumping mechanism. The metering valve is a throttle in the filling passage to limit the fuel taken into the pumping chamber per stroke. The angular position of the metering valve, and hence the fuel quantity entering the space between the plungers before each pumping stroke, is controlled by driver demand which can be overruled by a mechanical governor.

When it was introduced in the mid-1950s, the first rotary distributor pumps in volume production (the DPA) revolutionized agricultural tractor engines, by providing higher injection pressures to improve specific output and fuel consumption without excessive smoke. The new design spread rapidly to truck engine applications. The pump stroke started earlier on the cam flank as the engine load and hence fuelling was increased, so the natural characteristic of the pump was a constant end of injection. The start of injection advanced as the load was increased, which characteristic gave a useful reduction in visible smoke at appreciably lower cost than with the contemporary in line pumps.

The original medium duty rotary distributor pump has expanded into the modular DP200 range, for truck, agricultural tractor, industrial and marine applications. *Figure 11.12* shows a view of the current development of this type of rotary distributor pump, the DP203 pump. The cam ring is near the centre of the drawing and to the right of it, the rotor which distributes fuel to each injector pipe in turn as it rotates.

The mechanical governor shown in Figure 11.12 was originally developed for agricultural tractor applications which require precise governing at all speeds. This governor has been adapted to suit all the applications from vehicle to generator sets. Six weights are assembled in a pressed-steel housing which rotates with the drive shaft. Extended 'toes' on each weight bear on a collar around the driveshaft which operates the metering valve via a lever and spring-loaded rod. The driver of the vehicle

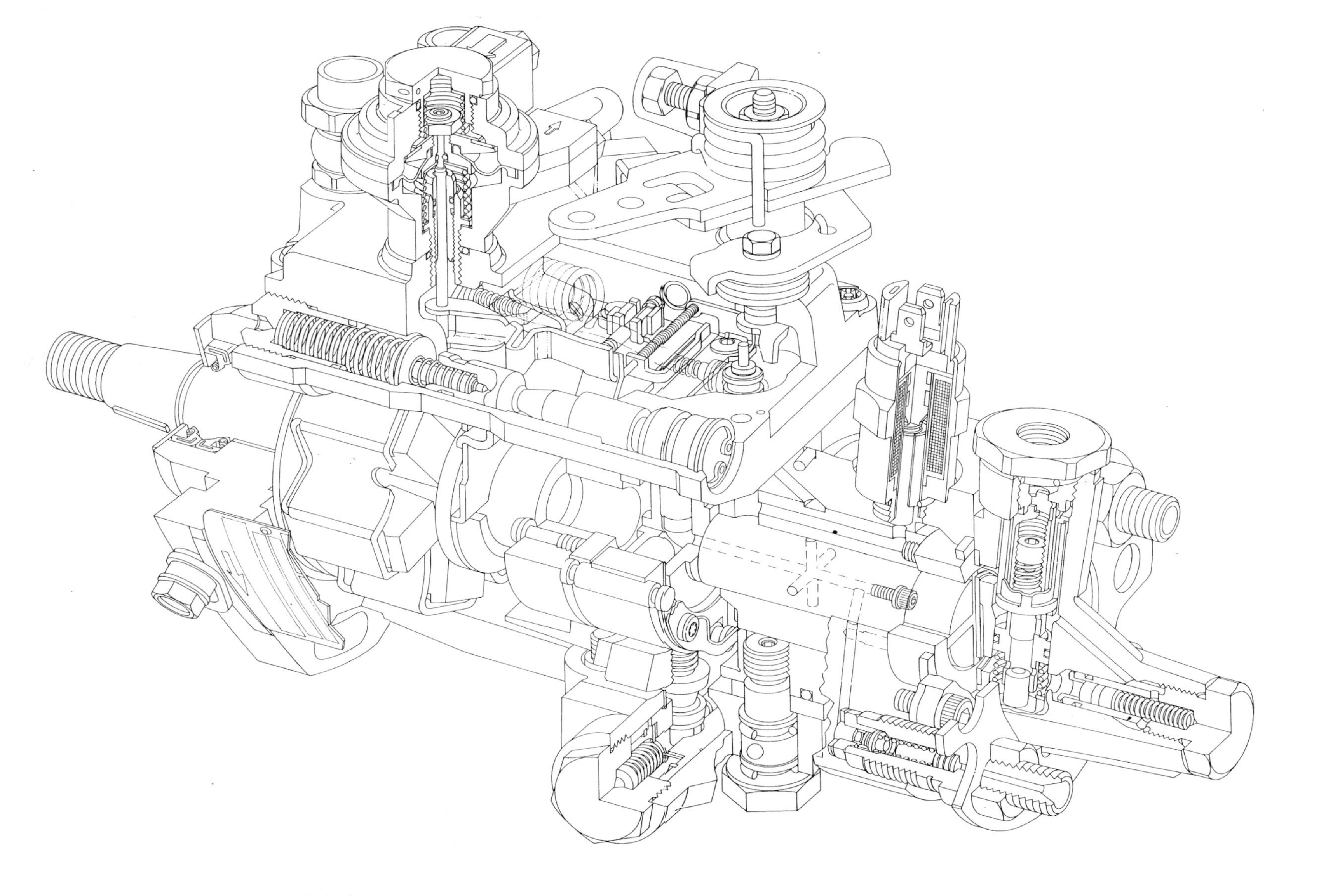

Figure 11.12 DP203 rotary distributor fuel injection pump

applies a fuelling 'demand' to the metering valve via a spring in response to driver demand. At high speeds, this can be overruled by the governor.

The Torque Trimmer control sets the maximum fuel at each speed was via scroll-rings that limit the roller radial travel as shown in *Figure 11.13*. The scroll rings are mounted on either side of the cam ring. Pins extend from each roller that bear on the scroll rings to limit the maximum fuel volume between the pumping plungers according to the angle of rotation of the scroll rings. Both scroll rings are moved by a sliding carriage at the top of the pump, which can be moved even further to provide excess fuel for cold starting. Thus the maximum fuelling and hence the engine torque curve can be set without compromise to the high pressure hydraulic performance.

This pump range has a common hydraulic head with axial outlets, and incorporates a number of detail design improvements; the options include, as well as torque control:

- high pressure rating (over 800 bars at the injector) to minimize soot generation and fuel consumption for low emissions engines;
- minimum dead volume in the high pressure hydraulics, to maximize pressure;
- zero backlash drive to maintain maximum rate of injection;
- stiff advance system to maximize emissions margins;
- progressive light load advance;
- start advance device;
- excess fuel for cold starts;
- enhanced mechanical governor with precise speed control and capable of maintaining repeatable governing throughout the design life of the pump;
- fuel viscosity compensation;
- boost control to increase fuel delivery as the turbocharger compressor delivers pressurized charge air;
- electric ('key-off') engine shut-off.

An electronically-governed option is available for isochronous governing for generator sets.

This design principle has been adapted separately to passenger car applications to form the DPC pump range that has been applied to numerous passenger car engines. This pump range has some of the same features as the medium duty DP200 range, but engineered specifically for passenger car engines. This range includes the DPCN with electronic control of injection timing and a protection device to deter thieves.

11.2.9 Electronically conrolled rotary pumps (EPIC)

As emissions limits become more stringent, much more flexibility is required in the control of injection timing and delivery quantity. A radical redesign of the ring-cam principle employed in earlier Lucus rotary distributor pumps provides precise fuel metering and injection timing under the control of digital electrical signals. This pump, known as EPIC, incorporates magnetically-operated valves to control hydraulic servo-mechanisms to fulfill this requirement.

The pump delivery is controlled by a unique sliding rotor and ramps which are arranged to limit the travel of each plunger during the filling stroke as shown in *Figure 11.14*. Stroke-control ramps are machined into the inner surface of the drive shaft opposite the pumping mechanism. Roller-shoes fit into slots in these ramps and thus the rollers are driven around the internally-lobed cam ring, transmitting the drive torque to the pumping mechanism. The roller-shoes carry substantial 'ears' which rest against the control ramps. These limit the outward travel of shoes, rollers and plungers. When the rollers come into contact with the cam lobes, the shoes move radially to force two (or more) plungers inwards to compress fuel in the volume between

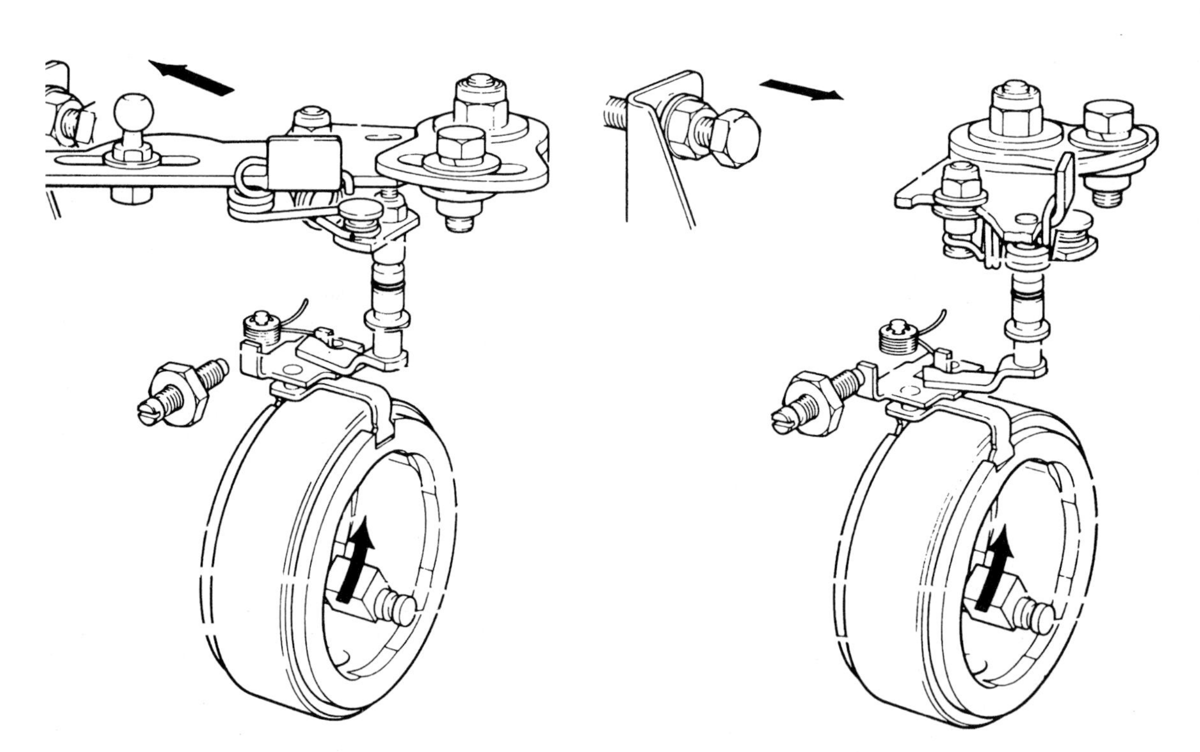

Figure 11.13 Scroll plate control of maximum fuelling in a Lucas Diesel Systems DP200 pump. The scroll rings control the maximum throw of the rollers, thus controlling the maximum fuelling

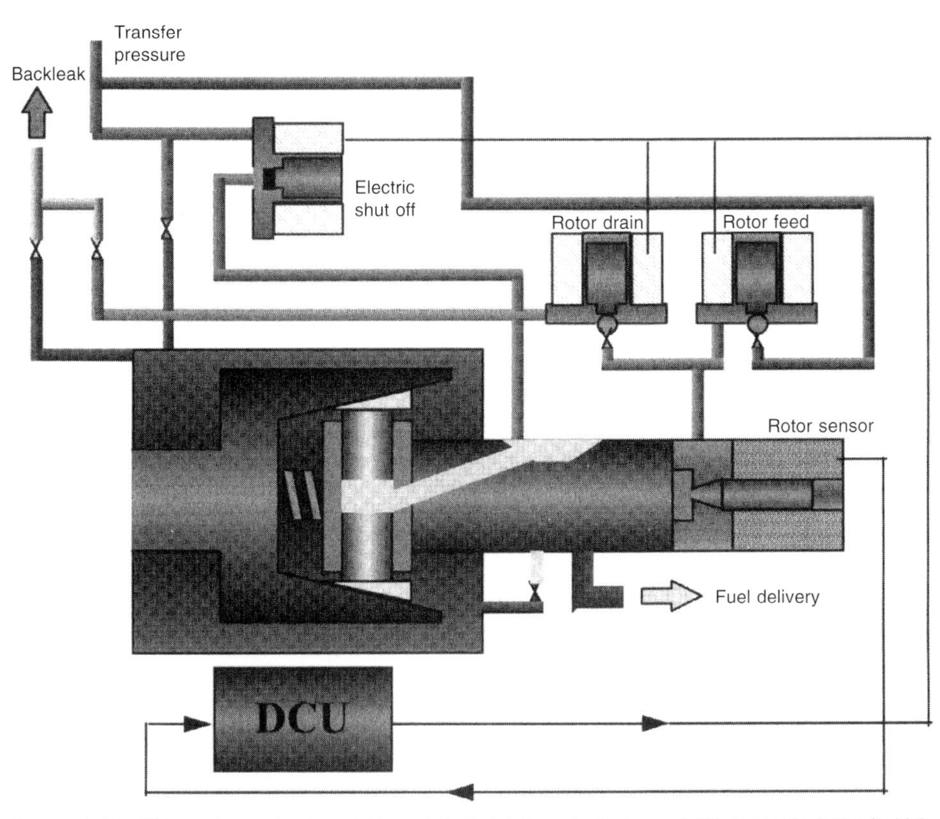

Figure 11.14 Sliding rotor mechanism which controls fuel delivery in the Lucas EPIC rotary distributor fuel injection system

the plungers and pump it through a delivery port. The rotor distributes the fuel to the appropriate outlet during each pumping stroke and connects the pumping chamber to a filling port between pumping strokes.

The pump is equipped with a transfer pump which provides higher pressure than earlier DP type pumps to both fill the pumping mechanism between pumping stokes and to drive the hydraulic servomechanisms. When the rotor is driven to the left by hydraulic pressure applied to its end face, the ears on each shoe prevent the plungers from moving outwards during filling, which restricts the filling and therefore the delivery. As the rotor moves to the right in *Figure 11.14*, the plungers are allowed to move outwards and the pump fills with more fuel, which is delivered as the rotor rotates further. Excess fuel can be provided, to start direct injection engines from cold without starting aids, if the rotor is moved further to the right.

The timing of the injection is adjusted, as in earlier rotary distributor pumps, by rotating the cam ring with a plunger subjected to hydraulic pressure on one end against a spring force applied to the other end. In this case the hydraulic pressure is provided and drained via electromagnetically-operated valves using a variable mark–space ratio to control flow into and out of the hydraulic chamber. Light load advance and separate timing setting for cold starts can obtained with suitable settings in the software map in the electronic control unit.

The idle speed and maximum speed governing is performed by digital electronics, and therefore a speed sensor is required. This is mounted on the engine flywheel (four flags triggering a magneto-resistive sensor). Further sensors are provided to detect coolant temperature, ambient air temperature, boost pressure in turbocharged engines, etc. The maximum fuelling available from the pump is continuously adjusted and positively limited to the suit the engine condition and provide a precise control of torque without excessive soot generation.

A medium duty direct injection engine pump and a lighter duty indirect injection engine pump were developed from the same basic design principles. Both are in volume production for low emissions engines.

11.2.10 Advanced rotary distributor pumps

Lucas have developed to the prototype stage and advanced high pressure rotary distributor spill pump incorporating the internally-lobed cam ring and radical pistons which they pioneered. A servo-assisted central spill valve is incorporated into the rotor, which causes an expansion wave to travel towards the nozzle when it is opened. The amplitude of this wave can approach the sum of the injection over-pressure and the residual line pressure with a correspondingly high local fluid velocity. The direction of this local velocity is inward, or towards the pump, so the wave de-pressurizes the pipe as it travels towards the nozzle. When this wave reaches the nozzle, fuel between the needle valve and the nozzle holes is abstracted back into the feed drilling, and so less fuel is left under the needle to provide resistance to needle closure. This feature of the pump design terminates the injection rapidly; minimizing the injection of fuel at low velocity as the needle closes and consequently minimizing any soot produced by less effective mixing at the end of injection. High pressure volumes have been reduced to minimize the dissipation of hydraulic wave energy both during the injection and at termination. This pump has a larger cam ring to accommodate the faster pumping required to generate up to 1400 bars injection pressure at the nozzle.

The electromagnetic trigger valve for the servo mechanism which spills the fuel pressure to terminate injection is mounted in the hydraulic head and connected to the drive piston in the rotor by drillings and ports to provide a better environment for this key component.

A major problem with high pressure spill pumps is that considerable energy is thrown away each time the fuel is spilled to terminate the injection. In this pump, an accumulator is placed within the rotor to conserve this energy and so minimize the power required to drive the fuel injection system. This has the added advantage that the minimum of heat is released when the fuel is spilled thus reducing thermal distortion of the rotor, which is required to run with very small clearances to distribute fuel to each injector without excessive leakage.

These fundamental design features, plus a more robust construction make a major contribution towards soot reduction, some of which can be traded for reductions in nitrogen oxides without significantly affecting fuel economy by controlled, cooled, EGR. The Lucas advanced rotary spill pump is shown in *Figure 11.15*.

The injection timing is precisely adjusted by a hydraulic servo mechanism which rotates the cam ring by a piston under the control of an electromagnetic valve.

The electronic control unit (ECU), which controls all the pump functions, is mounted on top of the pump where it can be cooled by fuel flow. The ECU provides local control and interfaces with the vehicle controls via a standard interface bus. This enables a high level of functionality and better management of EMR. A more complete description of this pump was published earlier[25].

11.2.11 Control of rate of injection with conventional FIE

In earlier sections, the importance has been stressed of rapid termination of injection to reduce soot. In the more advanced electronic unit injectors and the common rail systems, the needle is closed by pressure applied to the upper face of the needle, which both accelerates the needle onto its seat, and also generates pressure underneath the needle seat so that the last remaining elements of fuel are injected with high momentum and small droplet size.

The effects of modulating the initial rate of injection have been studied for many years at universities and by FIE system manufacturers who have published their work to reduce cobustion noise extensively[26–33]. With conventional fuels with a cetane number of 48 or above, a single pilot is effective at all but the coolest conditions in naturally aspirated engines, and for turbocharged engines at low speed light load conditions. However in some areas of the world, fuels have cetane numbers well below this, at 40–45 for example. For quiet operation with low cetane fuels, more complex modulation of the initial rate of injection is required.

The optimum pilot quantity seems to reduce with the effectiveness of atomization. With current FIE, the useful range seems to be from 0.5 mm^3 to 2 mm^3 for most automotive engines. For the optimum reduction of noise, the pilot should ignite the main injection; however, this requires more fuel to be burned in the diffusion burn, so some manufacturers compromise by injecting the pilot either earlier, or later, to achieve non-optimum noise reduction and consequently less to achieve to reduce particulate emissions.

For turbocharged engines at medium speeds and medium loads, the ignition delays are so short that the pilot timing needs to be very close to that of the main, and a stepped or 'boot-shaped' initial rate control (IRC) is preferable. This is also the case for engines which have a high compression ratio, for example exceeding 19:1.

A large number of practical mechanical devices have been developed which will produce either pilot or initial rate control (IRC). In some cases devices have been developed which will produce a pilot at low speeds and IRC at high speeds. Of these three are shown in *Figure 11.16* which are particularly useful, or in production in some form.

Two-spring injectors have been produced to throttle the flow into the cylinder at the needle seat during the first stage of injection[17,24]. When injection pressure is applied to the two-

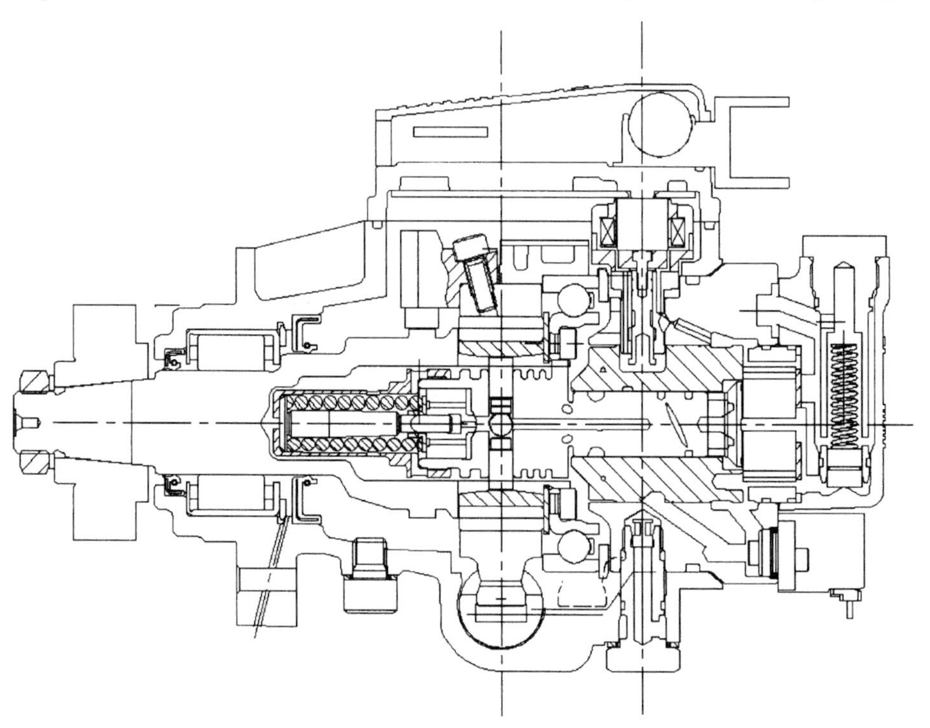

Figure 11.15 Lucas ESR 10 rotary fuel injection pump with servo-spill. The termination of injection is controlled by the servo spill valve which is driven is by an intermediate hydraulic pressure, controlled in turn by the trigger valve on the top-right of the drawing. Ports are cut into the rotor and connected to the spill actuation piston by drillings to actuate the valve over a small angle

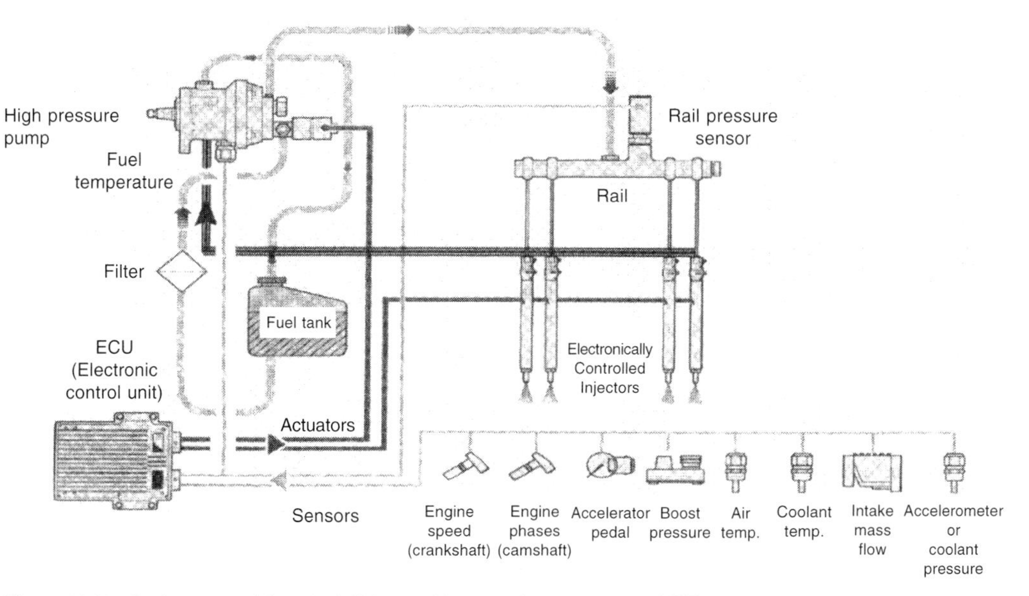

Figure 11.16 Devices to modulate the initial rate of injection from conventional FIE

spring device, illustrated above, the needle lifts by between 20 and 50 microns when the combined incident and reflected wave pressure exceeds the first nozzle opening pressure (NOP 1). The upper spring controls the first nozzle opening pressure, and the needle opens against the action of this spring until the triangular-shaped lower spring cap picks up the lower spring. This spring is pre-loaded to a much higher opening pressure (acting together with the upper spring) to achieve 350 to 400 bars which is known as the second nozzle opening pressure (NOP 2). Thus the needle only opens fully, to inject at the normal rate, when the fluid pressure under the needle exceeds the value of NOP 2.

In production, these devices are set with very tight tolerances on the flow rates in the first stage of injection. These injectors are matched to a rotary distributor pump to provide optimum performance across the speed and load range and they have proved to be an effective control for combustion noise.

It is vital with this type of device to ensure that the termination of pumping is very rapid, otherwise the injector will hang up in the first stage of lift at the end of injection and the consequent low rate of injection late in the combustion period will generate soot and visible smoke. The rate of pumping may be low at high speed part load conditions also, and two-spring injectors have remained in the first stage of lift at these operating conditions with similar effect.

11.2.11.1 Spilt injection device (SID)

This device has been called many things including ‘retraction piston’ and it is often confused with the shuttle pilot device. In its most common form, it consists of an additional piston with a differential seat on its upper face which is mounted above the injector spring[17,23]. A push rod and end cap over the injector spring ensures that the injector spring holds this additional piston (accumulator piston) onto its seat as well as the needle. The accumulator seat diameter is chosen to open at a slightly higher pressure than the nozzle opening pressure.

At low speeds, the needle is forced back onto its seat by the combined action of the hydraulic expansion wave and the spring compression wave. At high speeds, the wave speed in the spring prevents the spring force from being applied to the needle in time to close the needle, and the effect is to cause the needle to hover providing an initial rate of injection. However, this rate needs to be well-controlled which makes severe demands on the production process controls. At low speeds, the size of the pilot depends critically upon the ratio of the seat diameters (needle/accumulator piston). This makes the device very difficult to manufacture in volume, and difficult to use if wave causes differential changes in seat diameter[23].

If the SID is used in a fuel injection system, it is necessary to increase the unloading at the pump by an amount equal to the volume displaced by the accumulator piston.

11.2.11.2 Shuttle pilot device

The shuttle spill device, which has been patented by Lucas Diesel Systems, produces pilot at low speeds and IRC at high speeds. In each injector there is a shuttle which has a internal blind drilling connecting two grooves cut in the shuttle surface to the injector drilling[17,24].

In operation, the compression wave from the pump drives the shuttle piston downwards and starts pumping fuel from the underside of the shuttle. This pumping action causes a compression wave in the injector drilling between it and the needle seat which opens the needle valve to inject a pilot. After a predetermined shuttle travel, the first gallery opens to the spill port (spring back-leak chamber) which causes the pressure to collapse in the drilling between the shuttle and the needle seat. The shuttle continues to be driven forward by the compression wave from the pump. At some later shuttle travel, a second gallery opens a by-pass drilling from the high pressure line and the line pressure is connected directly to the injector drilling and hence to the needle seat. After each injection, the shuttle retracts under the action of a light spring, and fuel is drawn in via the non-return valve to fill the drilling between the shuttle and the needle seat.

At low speeds, the shuttle moves relatively slowly, so that there is time for the needle to close when fuel in the drilling is spilled via the shutle. A high speeds, the effect is to produce a step in the hydraulic wave downstream of the shuttle, which provides an initial rate of injection of controlled size. The shuttle device provides precise control of the pilot injection quantity.

All the devices benefit from a fairly gradual increase of initial rate of pumping, but the shuttle device is capable of working with a wider range of pumps than the other two.

11.2.11.3 Effects of initial rate modulation upon combustion noise

The primary reason for developing the rate control devices was to control 'diesel knock'. Small pilot deliveries are best for combustion noise control; but the mixing of the main injection needs to be improved to counteract the shift in fuel from premixed burn to diffusion burn. In the case of the two-spring injector, the higher NOP 2 gives some measure of improvement in soot generation. In the case of the split injection device, the increase in nozzle opening pressure helps, but the large accumulator volume significantly softens the injection and is a second reason for its current unpopularity. The shuttle pilot device includes some extra volume but this can be minimized and an increase in nozzle opening pressure 'NOP 1' is sufficient to recover any smoke penalty in the first generation of HSDI engines. The improvement in combustion noise can be seen in the *Figure 11.17*.

11.2.12 Lubrication of fuel injection components

Fuel injection pumps and injectors contain many parts which are lubricated by the fuel itself. Examples are the barrels and plungers inside inline pumps and electronic unit injectors, the needle valve and guide in almost all injectors and the rotor and sleeve inside a rotary distributor and its sleeve. Fuels from crude oils with high sulphur content are hydrogenated to reduce their sulphur content which reduces the oxides of sulphur in the exhaust emissions. The hydrogenation process also removes the components in the fuel which provide the lubrication for the FIE. When such fuel first became available, a number of fuel injection systems failed in service due to extremely rapid wear. Fuel injection manufacturers have tested a number of fuels to find the effect of fuel quality on the durability of FIE[35].

A test procedure was developed to assess the lubricity of fuels based on the High Frequency Reciprocating Rig (HFRR). This test procedure has been standardized (ISO 12156 part 1).

The European oil companies have agreed to ensure that enough lubricity additive is put into the fuel that such rapid wear should not recur and that fuels for sale will not cause a wear scar greater than 460 μm which provides protection for the 50 million or so distributor pumps which have been put into service so far, which were designed to make the best use of the natural lubricity of middle distillate fuel.

11.2.13 Common rail systems

With the advent of variable geometry turbochargers (VGTs), the air charge is much more dense at the top dead centre of the compression stroke at low and mid-range engine rotation speeds. Higher injection pressures are required to exploit the potential improvement in mid-range torque and hence vehicle driveability. This characteristic can be met in part by advanced cam-operated systems, such as recent developments of unit injectors for passenger cars. However, the drive torque required poses some challenges in terms of mechanical noise and torque fluctuations, due to the high rate cams required.

An alternative scheme, which offers added flexibility, is to supply the requirements from each cylinder from a common high pressure supply (a common rail) via a valve in each injector. In the Lucas system, the valve is the needle valve of the injector; and it is opened and shut by hydraulic pressure under the control of a small electro-magnetically operated valve immediately above it. The advantage of such a system is that the response of the servo mechanism is very fast. An overview of the Lucas common rail system is shown in *Figure 11.18*.

The Common Rail is charged by a variable-delivery high pressure pump to between 400 and 1600 bars, depending upon the engine operating conditions and the boost pressure available. Lucas produce a high pressure pump which incorporates the internally-lobed cam ring and radial plungers principle, which has proved successful over many years in their rotary distributor pumps. In this pump an inlet metering valve is fitted to minimize the volume of fuel pumped per stroke and so minimize the work required to drive the fuel injection system. These high pressure pumps are equipped also with a pressure relief valve, to control the rail pressure precisely and rapidly as required for each operating condition. There is a further mechanical pressure relief valve which activates above a certain maximum pressure, to prevent damage to the system.

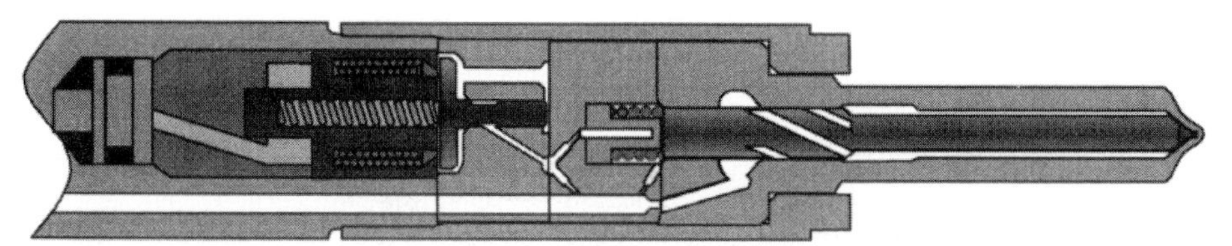

Figure 11.17 Lucas Diesel Systems common rail injector showing the servomechanism which controls injection timing and delivery, which is fast enough to deliver close pilot and post injection

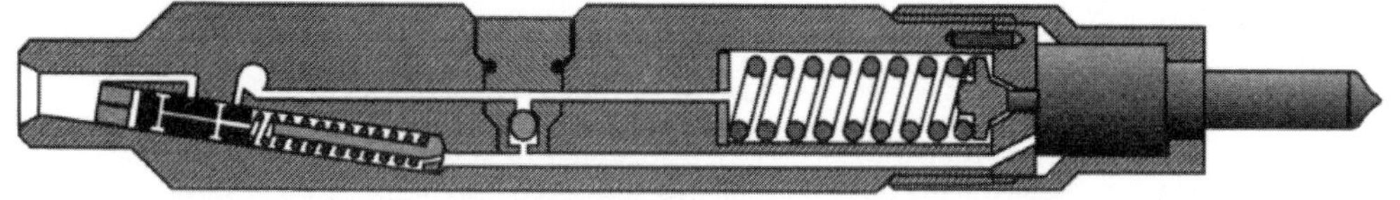

Figure 11.18a Devices to modulate the initial rate of injection from conventional FIE: Shuttle Pilot Device

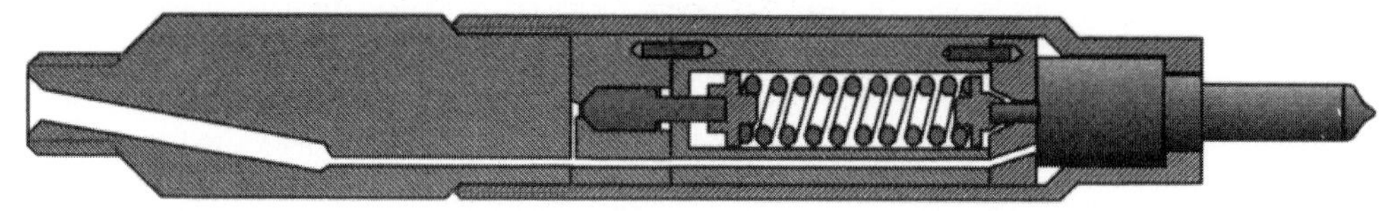

Figure 11.18b Devices to modulate the initial rate of injection from conventional FIE: Split Injection Device (SID)

Each injector has a conventional needle valve, which is operated as a servo mechanism by altering the pressure differential between the underside and the top of the valve. A light spring provides additional force to hold the needle shut. Mounted above the needle in each injector is a control valve. This system breaks new ground by incorporating the servo mechanism within the confines of a 17 mm injector, which is very space efficient for four-valve passenger car engines; as shown in *Figure 11.19*. The hydraulic servo mechanism needs to operate in a small fraction of a millisecond, so large forces and small components masses are incorporated in the design.

A variety of injection strategies have been proposed, for emissions and noise control, in which two or more injections are made into each cylinder in each engine cycle. This common rail system has the flexibility to inject with these with a very short timing separation which is an important attribute. Combustion noise, which can be very noticeable when high rates of injection are used at low engine speeds, can be controlled by injecting a pilot ahead of the main injection. For noise reduction the first pilot needs to be very small, and the second can be somewhat larger in the range 0.5 to 2 mm^3 in quantity for each stroke.

One of the strategies which have been proposed to reduce nitrogen oxide emissions involves splitting the main injection into two, or more, to meter the rate at which fuel is injected into the engine, with the aim of reducing peak cycle temperature and therefore nitric oxide formation.

Some engine manufacturers are advocating a 'post injection' (or secondary injection) to deliver fuel to a catalyst system in the exhaust. The exact timing for a post injection is critical; too early, and some of the fuel is ignited by the end of the diffusion burn. However, if it is injected too late in the engine cycle, then the sprays may impinge upon the cylinder wall with consequent risks in the loss of lubrication between the piston and cylinder bore, dilution of the oil in the sump and ultimately engine seizure. However, if a suitable window exists in the engine cycle, this common rail system is fast enough to provide an additional injection to reduce nitrogen oxides when a suitable storage deNOx catalyst has been proved to be reliable.

11.2.14 Integrated fuel injection systems

11.2.14.1 Matching FIE to engine requirements

Fuel injection equipment has to be precisely matched to a combustion system and therefore must incorporate several important detail changes or each variation for chamber diameter, chamber depth, lip profile, etc. Hence, to obtain the best torque and specific power output for a given level of particulate emissions, noise and gaseous emissions, a substantial development programme must be completed for each and every combustion chamber.

To expedite FIE design and development, a fuel injection simulation has been written and validated extensively with many experimental results. The first predictive program was written in the 1950s[2]. This program incorporated cavitation almost from the outset and many fuel injection systems were run on test rigs to check that the mathematical formulation of the underlying physics was rigorous and sufficient. This program grew with the advances in FIE, firstly by becoming better at predicting inline FIE at high speed operating conditions, and then by predicting the performance of EUI, rotary distributor and common rail FIE as these were developed. Each step involved some additions to the total description of the physics involved, and each step had to be validated against test results rigorously. This computer program has been developed to be a precise simulation of the performance of FIE products[36]. With increasing injection pressures, non-linear terms have had to be incorporated into the bulk modulus formulation.

11.2.14.2 Containment of high pressure fluids

Fuel injection pumps have always incorporated metal-to-metal seals between the moving parts but the manufacturing technologies have been refined to meet each new challenge as injection pressures have risen. Thermal distortion and fine debris are the major concerns with such seals, which rely on radial clearances of the order of 2 microns to seal ports in 19 mm diameter rotors

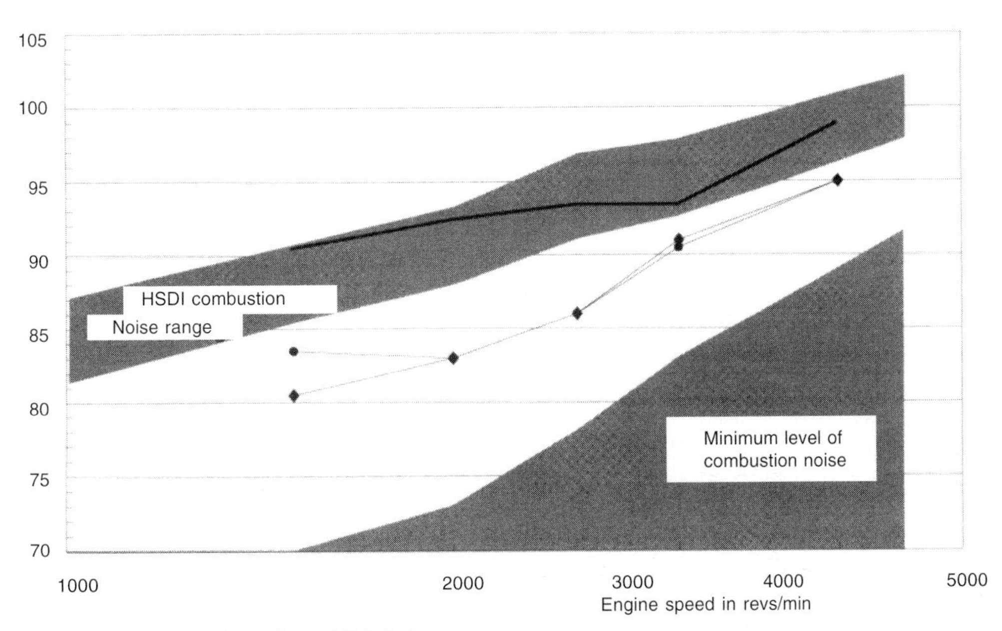

Figure 11.19 Noise reduction from pilot and IRC devices

in their sleeves (locked up stresses are removed during the production processes). Thermal distortion is minimized by careful design.

Fine debris from manufacturing processes is removed by sophisticated deburring and cleaning techniques. The pumps are run with filtered oil during calibration, which removes the debris from assembly. Protecting the components upstream of the nozzle from damage in service by debris, particles and water in the fuel has been a major concern as injection pressures have risen. Current FIE systems all include a filter and most include a water separator. Thus a range of low pressure fuel system components has evolved to provide such protection for the rest of the FIE system.

Debris may be introduced from pipe unions assembly to the engine and a study showed that this operation was responsible for a surprisingly large number of particles which were large enough to block nozzle holes unless great care was taken. Edge filters have been introduced into injectors to reduce such debris to such a size that they will pass through the nozzle holes.

11.2.14.3 Low pressure system

The low pressure supply, filter and backleak components of fuel injection systems are probably the source of more day-to-day problems than the precision-manufactured high pressure components. The high pressure components have to be made from high strength materials, and steel is used almost exclusively; hence these must be protected against water ingress and excessive contamination of fuel by water. Most vehicles are therefore fitted with water traps as part of the filter assembly.

The filter was added to the low pressure as clearances between rotor and sleeve shrank in order to seal the outlet pressure adequately when the rotary distributor pumps were introduced. Clearly if particles were to find their way into the clearance between the rotor and its sleeve the heat generated locally would be liable to cause the metal to distort and perhaps weld. This has been shown to cause seizure and catastrophic failure. The solution to the seizure problem with fuel which contained particles was solved by interposing a fine filter made from a special grade of paper between the lift pump or tank and the transfer pump. A photograph of the range of filters and water separators is shown in *Figure 11.20*. Any water collected has to be drained from the sysem as soon as practicable. The filters have replaceable elements and periodic replacement is essential to ensure a long service life for the precision FIE components.

11.2.14.4 Control systems

In the early pumps simple mechanical governors were linked to the rack that rotated the pumping elements to maintain a constant speed at idle and to limit the maximum engine speed. The design of rotating weights and linkage had to account for the limited force available at low speeds since the force available varied with the square of rotation speed. Pneumatic and hydraulic governors were developed to provide better control. The early rotary distributor pumps used the increase in pressure from a vane pump to operate a governor and to control injection timing. As the high pressure pump was developed, a mechanical governor was incorporated, which could control the engine speed more precisely to any desired speed between idle and rated speed. This governor is still very cost effective and is used for most rotary distributor pump systems. Further mechanisms have been added to prevent undershoot, which will cause the engine to stall, as the speed is reduced rapidly from high idle to idle. The design of these mechanisms includes compensation for wear to ensure that the FIE system stayed within its calibrated range over a long service life.

With the introduction of electronic control, the mechanical governor was replaced by a magnetic 'pickup' which sensed projections ('flags') attached to the engine flywheel and a suitable algorithm in the microprocessor. The delivery and injection timing was controlled to match a table ('map') of predetermined settings. These were held in a non-volatile digital memory within the Engine Control Unit (ECU). The values read from the map depended upon air flow, air temperature and coolant temperature as well as speed and fuel demand. As the limits on exhaust

Figure 11.20 Various fuel filters and water separator assemblies

emissions became more stringent, the map expanded to accommodate special cold start and warm start procedures to minimize emissions. The changes in fuelling and injection timing can be made to vary with speed and load in any way to meet emissions simply by adjusting the values stored in the map. Thus if a particular operating condition contributes more to some exhaust emission than others, the map can be set to minimize the effect at this condition.

The maps in the ECU have become very complex, with 'patches' of values inserted to cope with the particular requirements at each operating condition. However, when the insight from years of research is incorporated into algorithms based on mathematical formulations of the dominant physical effects (models) of the engine behaviour, each algorithm can cover a larger domain in the load–speed range. The move to electronic control has bound together the FIE and the engine by controlling the FIE on data derived from sensors attached to the engine. The system of sensors and actuators which is associated with typical electronic control of FIE is shown in *Figure 11.21*. (PATS is the passive anti-theft system and AMF is air mass flow.)

Improvements in vehicle refinement demanded better control of vehicle acceleration from rest and smoother governing performance. With digital electronic control it is possible to develop algorithms to eliminate any sudden application of torque and so to improve the feeling of refinement. For passenger car applications, a drivetrain model is used to predict the vehicle response to any drivetrain input. The control algorithms are developed to ensure a smooth transfer of torque to the driven wheels. Thus the advanced fuel injection system provides not only better performance within the narrowing envelope of emissions legislation, but also improved refinement in combustion noise, vibration (cylinder balance) and driveability.

11.2.15 Summary

Diesel fuel injection equipment has been developed to provide precise and reliable control over the fuel delivery into each engine cycle. This delivery is adjusted at idle to maintain the idle speed and in some applications to maintain engine speed irrespective of load up to the torque limit for that engine. The delivery is gradually reduced at speeds above the rated speed to protect the engine from the damage which would result from running at excessive speed.

The timing of the injection is adjusted rapidly and automatically by the FIE to minimize fuel consumption within an envelope of timing (and other) constraints imposed by the legal requirement to comply with emissions limits. During a cold start, the timing is automatically advanced to increase the time available to burn the fuel injected. At other operating conditions, the injection timing may be retarded to reduce cycle efficiency and hence emissions of nitrogen oxides at predetermined operating conditions.

The formation of the injection sprays by the nozzles has developed to divide the fuel into fine droplets with enough collective momentum to engage all the oxygen available in the combustion chamber. This has been achieved by reducing the size of the nozzle holes, increasing their number and increasing injection pressures. The finely-divided droplets and vapour form a spray with a wider included angle than early liquid sprays. The wider spray entrains more oxygen. The fuel near the centre of each spray burns instead of pyrolizing as the outer elements of the spray burn. This reduces the soot particles generated in the first part of the diffusion burn and hence the number of soot particles which need to be consumed as the reverse squish accelerates mixing in the later diffusion burn.

In the most recent FIE developments (for example advanced EUI and common rail systems) the injection pressure is modulated to match the density of the gas charge in the cylinder. (The gas charge comprises air, which may be compressed before induction by a turbocharger, and recirculated and cooled exhaust gas.) The fuel sprays have to mix with almost all of the gas in the combustion chamber to burn in the oxygen available, to minimize nitric oxide formation.

At the end of injection, the flow of fuel into the combustion chamber is terminated abruptly to avoid placing any fuel into regions of the combustion chamber in which the gas charge has been depleted of oxygen.

At the start of injection a variety of schemes have been developed to modulate the initial rate of injection. Initial rate control devices, such as two-spring injectors which inject fuel at a low rate for the initial period of the injection. These reduce

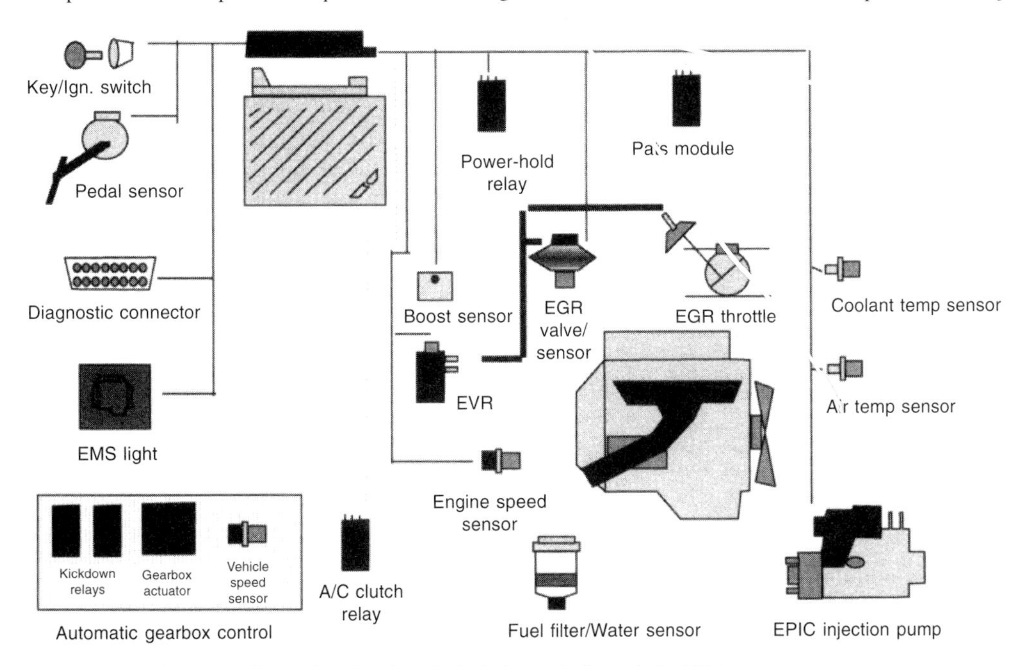

Figure 11.21 Sensors and control system for a typical electronically controlled FIE

combustion noise adequately when the cetane number of the fuel is 48 or more. At cold idle, when driving in urban traffic and when accelerating a vehicle after a protracted period when little heat was released into the combustion chambers, excessive combustion noise ('diesel knock') may be produced even with such devices. A more flexible noise control device has been introduced in pilot injection. The heat release from one or more pilots approximates to the ideal rate of heat release to minimize combustion noise without adverse fuel consumption effects. Pilot injection is effective also in reducing unburned hydrocarbons from the lean limit source. However, pilot injection is so effective in removing fuel from the diffusion burn, that additional measures are required to maintain the soot-limited power output (increases to the injection pressure in the main injection, for example). Recent FIE developments feature combinations of pilot and initial rate control, with the option of adjusting the pilot quantity and timing over the load and speed range.

The containment of high pressure fuel within the FIE system has been a major concern and is effected by close control of clearances between the moving parts. As pressures rise to reduce soot in the exhaust, this containment demands more refined engineering and production methods. The fuel filtering has been improved to prevent damage to the containment surfaces or even seizure due to local heat generation if a particle should find its way into a high pressure seal.

The fuel injection system incorporates control functions which are increasingly performed by digital electronics and appropriate software. These control functions demand high performance from the microprocessor(s) in the Control Unit. Currently control is based on'maps' of delivery, timing, pilot quantity(ies), pilot timing(s) and exhaust gas recirculation flow.

11.2.16 Acknowledgement

The publishers gratefully thank the Directors of Lucas Diesel Systems who authorized the publication of this section and are pleased to have this opportunity to thank the many Lucas Diesel Systems engineers who helped to prepare the material. The figures in this section remain the copyright if Lucas Diesel Systems and shall not be reproduced in any form, in an electronic retrieval system or otherwise, without the prior written authorization from Lucas Diesel Systems.

11.3 Diesel fuel injection systems–Robert Bosch Corp.

11.3.1 Fuel-injection systems

11.3.1.1 Assignments

The fuel-injection system is responsible for supplying the diesel engine with fuel. To do so, the injection pump generates the pressure required for fuel injection. The fuel under pressure is forced through the high-pressure fuel-injection tubing to the injection nozzle which then injects it into the combustion chamber.

The fuel-injection system includes the following components and assemblies: the fuel tank, the fuel filter, the fuel supply pump, the injection nozzles, the high-pressure injection tubing, the governor, and the timing device (if required).

The combustion process in the diesel engine depends to a large degree upon the quantity of fuel which is injected and upon the method of introducing this fuel to the combustion chamber. The most important criteria in this respect are the fuel-injection timing and the duration of injection, the fuel's distribution in the combustion chamber, the moment in time when combution starts, the amount of fuel metered to the engine per degree cankshaft, and the total injected fuel quantity in accordance with the engine loading. The optimum interplay of all these parameters is decisive for the faultless functioning of the diesel engine.

11.3.1.2 Types

To keep pace with the ever-increasing demands placed upon the diesel fuel-injection system, it has been necessary to continually develop and improve the fuel-injection pump. The result is that today we have available an extensive range of in-line pumps, distributor pumps, and single-plunger pumps in a wide variety of sizes and types.

The following fuel-injection systems are in line with the present state-of-the art:

- In-line injection pump (PE) with mechanical (flyweight) or electronic governor and, if required, timing device.
- Control-sleeve injection pump (PE) with electronic governor and infinitely variable port closing (start of delivery). Without fitted timing device.
- Single-plunger injection pump (PF).
- Distributor injection pump (VE) with mechanical or electronic governor and integral timing device.
- Unit injector (PDE), in the form of a compact system.
- Unit pump (PLD), a modular fuel-injection system.

Table 11.1 provides an overview of diesel fuel-injection pumps.

11.3.2 Fuel-injection techniques

11.3.2.1 Fuel metering

Depending upon the diesel combustion process in question, to ensure a good air–fuel mixture, the injection pumps must inject the fuel at very high pressures, as well as metering it with maximum possible precision. An accuracy of approx, 1°cks (crankshaft) for start of injection is necessary if optimal matching is to be reached between fuel economy, exhaust-gas emissions, and running noise.

With the standard in-line injection pump, a timing device is used to control start of injection and to compensate for pressure-wave propagation time in the high-pressure fuel-injection line. This adjusts the pump's start of delivery (port closing) in the

Table 11.1 Overview

Characteristics	*Diesel fuel-injection pumps*			
	VE	*PE*	*PF*	*PDE/PLD*
Injection pressure in bar (pump-side)	up to 700	up to 1150	up to 1500	up to 1500
Application	High-speed passenger car and commercial vehicle engines	Commercial vehicles, special vehicles, stationary engines	Marine engines, construction machinery	Commercial vehicles, passenger cars
Output per cylinder in kW/cylinder	up to 25	up to 70	up to 1000	up to 70

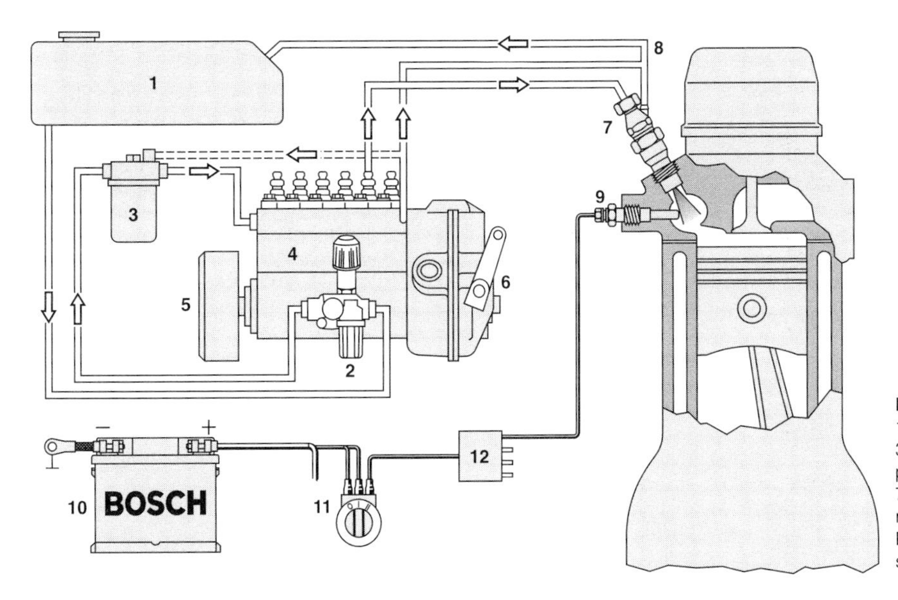

Figure 11.22 Fuel injection system
1 Fuel tank, 2 Fuel supply pump, 3 Fuel filter, 4 In-line fuel injection pump, 5 Timing device, 6 Governor, 7 Nozzle holder with nozzle, 8 Fuel return line, 9 Glow plug (GSK), 10 Battery, 11 Glow plug and starter switch, 12 Glow control unit (GZS).

'advance' direction as pump speed increases. In special cases, load-dependent control is provided. The diesel engine's load and speed control are determined by the injected fuel quantity (correct terms for fuel quantity are: injected fuel volume [mm^3/stroke] and injected fuel mass [mg/stroke]) without intake-air throttling (*Figures 11.22* and *11.23*)

11.3.2.2 Plunger stroke-phase sequence (Figure 11.24)

With the pump plunger at BDC, the pump-barrel inlet ports are open. Through them, the fuel which is under supply-pump pressure flows from the pump's fuel gallery and into the high-pressure chamber. As it moves up in the barrel, the pump plunger passes the inlet ports and closes them. This plunger stroke is termed the prestroke. During the further course of the plunger stroke, fuel pressure is increased so that the delivery valve opens. If a constant-volume valve is used the plunger travels through a further stroke known as the retraction stroke. During the effective (working) stroke, the fuel is forced through the high-pressure line to the nozzle. The effective stroke is terminated as soon as the plunger's helix opens the spill port (or inlet port). From this instant, no more fuel is delivered to the nozzle because during the remaining plunger stroke (residual stroke), the fuel is forced back through the plunger's vertical slot and into the fuel gallery.

Following reversal of plunger travel at TDC, the fuel continues to flow through the vertical slot into the barrel until the plunger helix closes the spill port (or inlet port) again. During the plunger's continuing return stroke, a vacuum is generated in the pump barrel, and fuel flows into the high-pressure chamber as soon as the plunger opens the inlet port. The cycle can begin again.

Among other things, the power delivered by a diesel engine depends upon the injected fuel quantity. It is the injection pump's job to always meter the appropriate quantity of fuel to the engine in accordance with its loading.

The injected fuel quantity is varied by changing the plunger's effective stroke. To do so, the control rack turns the pump plunger in the barrel so that the helix, which runs diagonally around the plunger circumference, can open the inlet port sooner or later and in doing so change the end-of-delivery point and with it the injected fuel quantity. In the case of maximum delivery, port opening cannot take place until the maximum effective stroke has been reached, in other words at the maximum possible delivery quantity. With partial delivery, port opening occurs sooner depending upon the exact position of the pump plunger. At the zero-delivery position, the vertical slot in the plunger is directly opposite to the inlet port. This means that for the complete plunger stroke the high-pressure chamber is connected to the fuel gallery through the plunger's vertical slot. No fuel is delivered. This is the position to which the plungers are rotated when the engine is switched off (*Figure 11.25*). The PE.. A in-line injection pump uses a toothed rack to turn the plungers and vary the injected fuel quantity (*Figure 11.26*).

11.3.2.3 Cam shapes

Different combustion systems and combustion chambers necessitate individually tailored conditions of injection, in other words, the injection process must be specifically matched to the particular engine type. The plunger speed, and therefore the duration of injection, depend upon the plunger actuating cam's lift relative to the angle of cam rotation. This is why a wide variety of different cam contours are required for everyday operation. In order to imporve the conditions of injection, such as rate-of-discharge curve and pressure loading, special cam shapes can be generated by calculation.

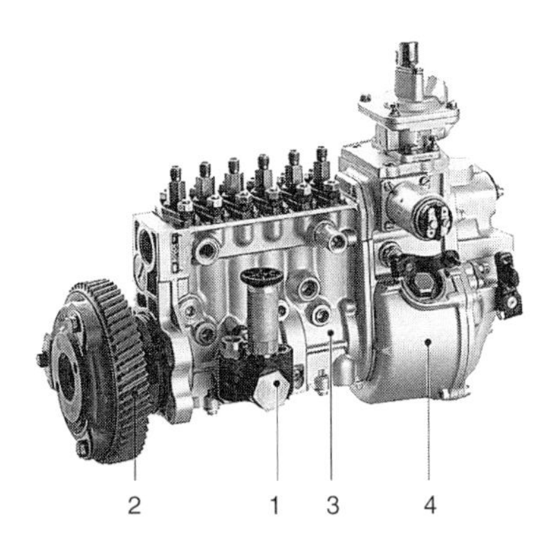

Figure 11.23 Injection pump assembly
1 Fuel injection pump, 2 Governor, 3 Fuel supply pump, 4 Timing device.

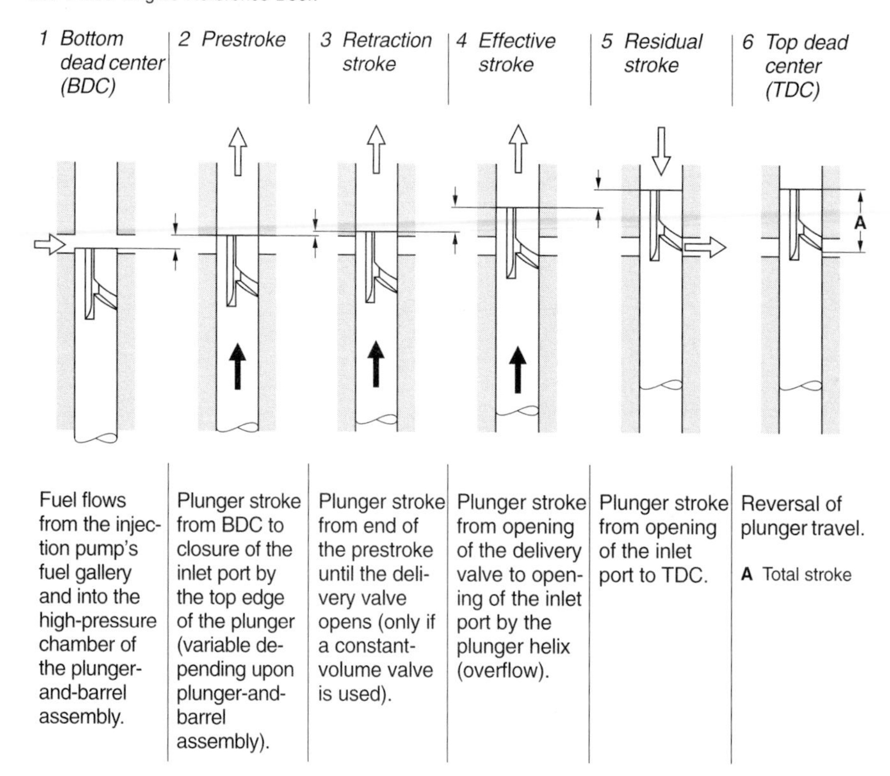

Figure 11.24 Plunger-stroke phases

Use is made of symmetrical cams, eccentric cams, and cams (back-kick cam) which prevent the engine being started in the false direction of rotation (*Figure 11.27*)

11.3.3 Pump-and-barrel assemblies (pumping elements)

Plunger-and-barrel assembly: basic version. The pump plunger together with the pump barrel form the plunger-and-barrel assembly. This utilizes the overflow principle in conjunction with port and helix control.

The pump plunger has been so finely lapped into the plunger barrel that it provides an adequate seal even at high pressures and low speeds, and no additional sealing elements are required.

In addition to its vertical slot, the pump plunger has also been machined on the side and the resulting diagonal edge cut in the plunger's wall is termed the control helix (*Figure 11.28* a and b).

A single helix suffices for injection pressures up to 600 bar, but above this figure the plunger needs two diametrically opposite helixes. This measure serves to prevent plunger 'seizure' because the plunger is no longer forced against the barrel wall by the injection pressure. The barrel is provided with one or two inlet ports for fuel inlet and end of delivery.

Due to the precise matching of the plunger to the barrel, it is essential that only complete plunger-and-barrel assemblies are exchanged and never the plunger or the barrel alone.

Plunger-and-barrel assembly with leakage-return duct

If the fuel-injection pump is connected to the engine's lube-oil circuit, under certain circumstances leak fuel can result in lube-oil dilution. This is avoided to a great extent by plunger-and-barrel assemblies with a leakage-return duct to the pump's fuel gallery. In such cases, the barrel is provided with a ring-shaped groove which is connected to the fuel gallery through a separate passage, or the leak oil is collected in a ring-shaped plunger groove and then returned to the fuel gallery through appropriate slots in the plunger (*Figure 11.29*).

Versions

Special requirements calling for the reduction of noise or exhaust emissions for instance, dictate that some form of load-dependent start-of-delivery (port closing) is provided. Plunger versions which in addition to the lower helix also have an upper helix permit the start of delivery to be adjusted as a function of load (*Figure 11.30*). In order to improve the starting performance of

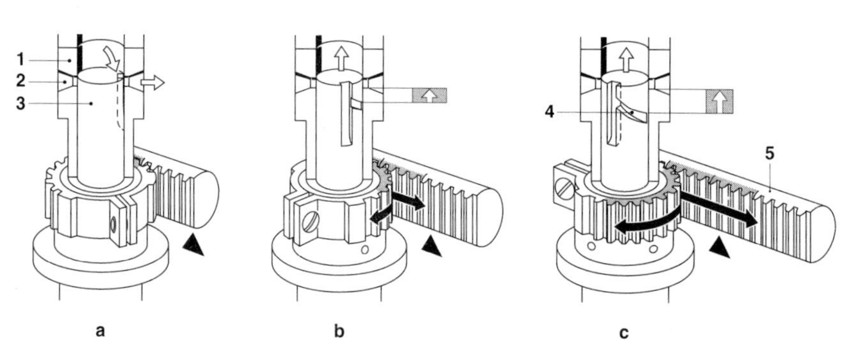

Figure. 11.25 Fuel delivery control Using a toothed control rack. (a) Zero delivery, (b) Partial delivery, (c) Maximum delivery. 1 Pump barrel, 2 Inlet port, 3 Pump plunger, 4 Helix, 6 Control rack.

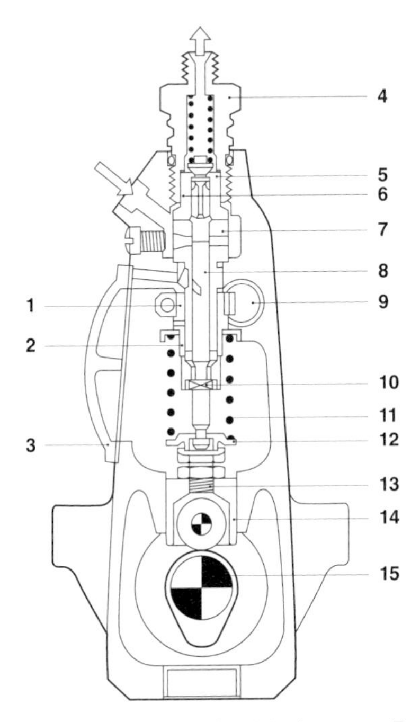

Figure 11.26 In-line fuel-injection pump: Type PE..A.
1 Control-sleeve gear, 2 Control sleeve, 3 Spring-chamber cover, 4 Delivery valve holder, 5 Valve holder, 6 Delivery valve, 7 Pump barrel, 8 Pump plunger, 9 Control rack, 10 Plunger control arm, 11 Plunger return spring, 12 Spring seat, 13 Adjusting screw, 14 Roller tappet, 15 Camshaft.

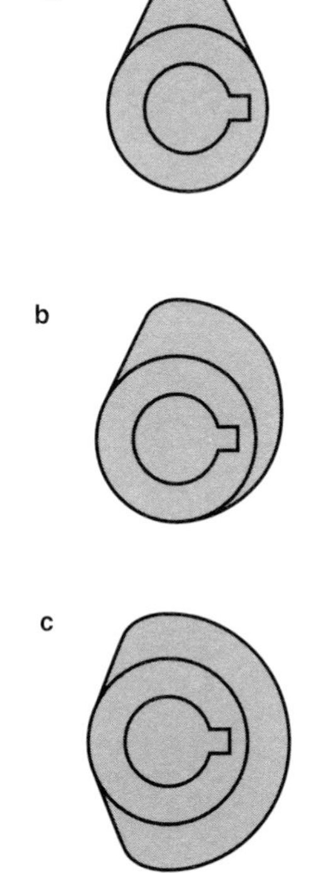

Figure 11.27 Cam shapes. Versions:
(a) Symmetrical cam, (b) Asymmetrical cam, (c) Back-kick cam.

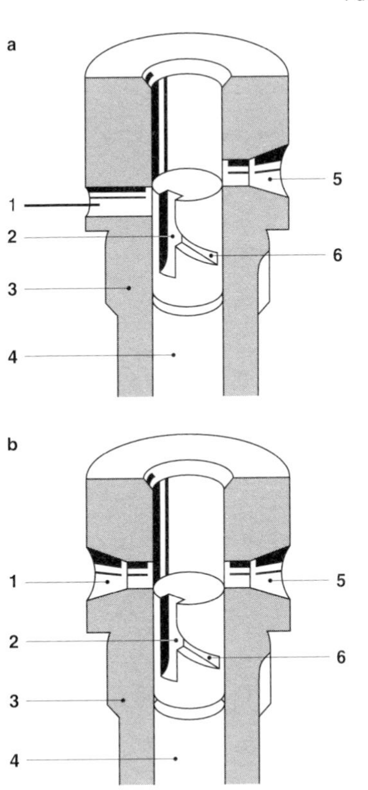

Figure 11.28 Plunger-and-barrel assemblies (pumping elements)
(a) Single-port plunger and barrel assembly, (b) Two-port plunger and barrel assembly. 1 Inlet port, 2 Vertical slot, 3 Barrel, 4 Plunger, 5 Spill port (inlet and return), 6 Control helix.

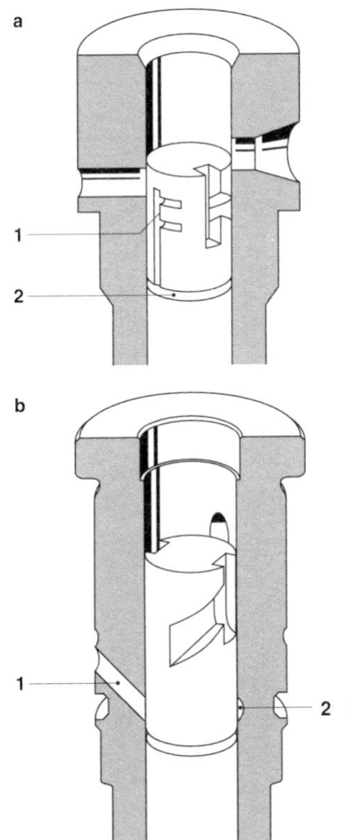

Figure 11.29 Plunger-and-barrel assembly with leakage-return duct
(a) 1 Leakage-return slot, 2 Ring shaped plunger groove,
(b) 1 Leakage-return duct, 2 Ring shaped barrel groove.

certain engine types, special plungers are used which have a special starting groove. This starting groove is machined into the plunger's top edge and is effective only when the plunger is in the start position. It results in a start of delivery which is retarded by 5. . .10° relative to the crankshaft setting.

11.3.3.1 Delivery valves (Figure 11.31)

It is the job of the delivery valve to interrupt the high-pressure circuit between the high-pressure fuel-injection lines and the pump plunger, as well as to relieve the high-pressure lines and the nozzle space by reducing pressure to a given static level. This reduction in pressure causes the nozzle to close rapidly and precisely, as well as preventing undesirable fuel dribble. During the actual injection process, the pressure which is generated in the nozzle's high-pressure chamber causes the delivery valve's cone to lift from its seat in the valve holder, and the pressurized fuel is forced through the delivery-valve holder and the high-pressure lines to the injection nozzle. As soon as the plunger's helix terminates the fuel delivery, pressure drops in the nozzle's high-pressure chamber, and the delivery-valve spring forces the valve cone back onto its seat. This separates the chamber above the pump plunger from the high-pressure circuit until the plunger's next effective (working) stroke.

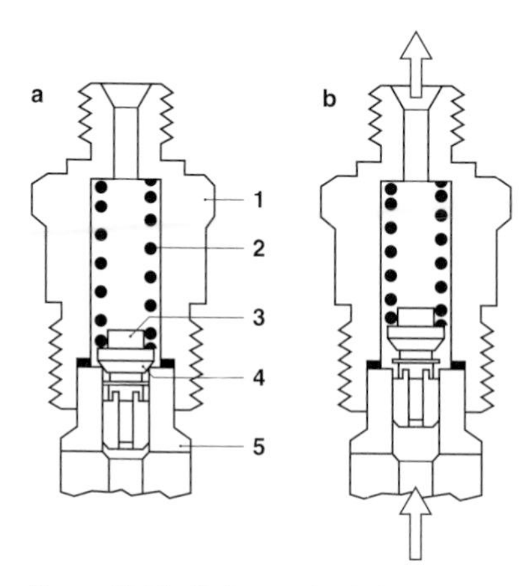

Figure 11.31 Delivery-valve holder with delivery valve
(a) Closed (b) During fuel delivery.
1 Delivery-valve holder, 2 Delivery-valve spring, 3 Delivery valve, 4 Valve seat, 5 Valve holder.

Constant-volume valve without return-flow restriction (Figure 11.32)

In the constant-volume valve part of the valve-element stem is shaped like a piston (retraction piston) and is precisely lapped into the valve-steam guide. When the plunger's helix terminates the fuel delivery, and the spring closes the delivery valve, the piston enters the valve-stem guide and closes off the high-pressure line from the nozzle's high-pressure chamber. This means that the volume available to the fuel in the high-pressure line is increased by the retraction piston's stroke volume. This retraction volume is matched to the length of the high-pressure line, which means that the line length is not to be changed. In order to achieve specific fuel-delivery characteristics, compensation valves can be used in special cases. These have an extra machined section on the retraction piston.

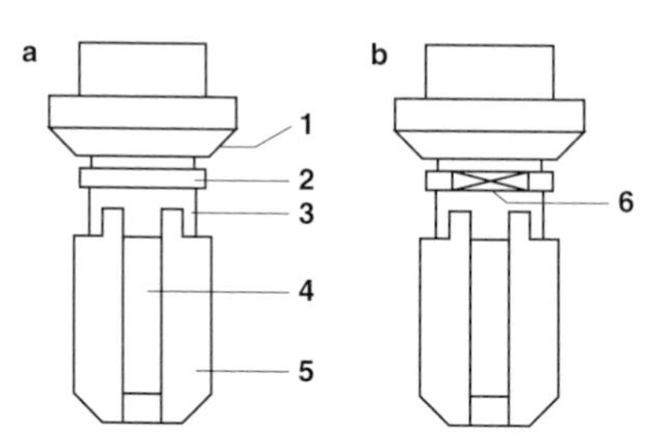

Figure 11.32 Constant-volume delivery-valve element
(a) Normal, (b) With compensation
1 Valve seat, 2 Retraction position, 3 Ring-shaped groove, 4 Valve stem, 5 Vertical slot, 6 Mechined section.

Constant-volume valve with return-flow restriction (Figure 11.33)

The return-flow restriction can be used in addition to the constant-volume valve. The reverse pressure waves which are generated when the injection nozzle closes can cause wear and cavitation in the delivery-valve's high-pressure chamber. This can be reduced or even prevented completely by the damping effect of the return-flow restriction which is located in the upper section of the delivery-valve holder, in other words between the constant-volume valve and the nozzle. There is a narrow restriction passage in the valve body which on the one hand provides the required throttling effect and on the other for the most part prevents pressure-wave reflection. The valve opens in the delivery direction and there is no resriction or throttling effect. A plate is used as the valve body for pressures up to 800 bar, and for higher pressures a guided cone.

Constant-pressure valve

The constant-pressure valve (*Figure 11.34*) is used with high-

a

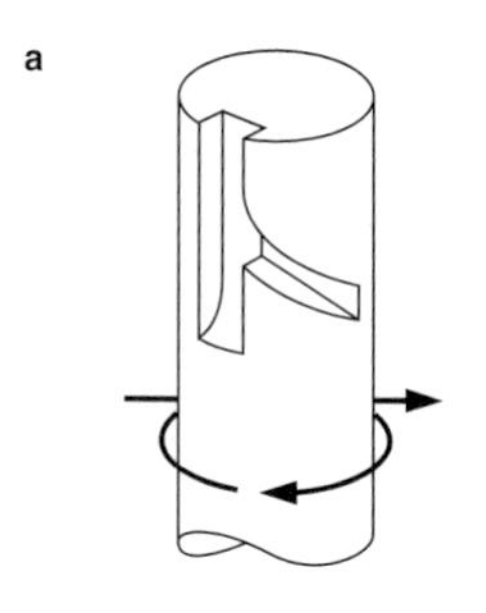

b

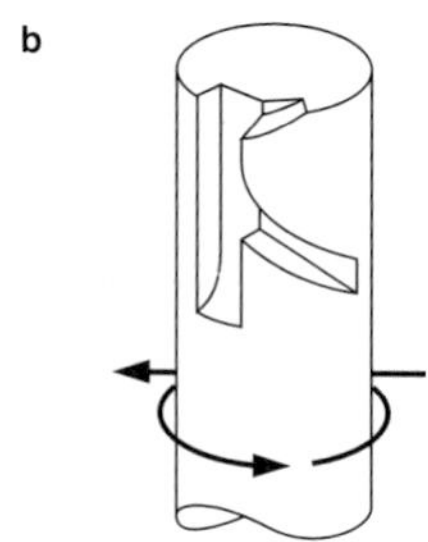

c

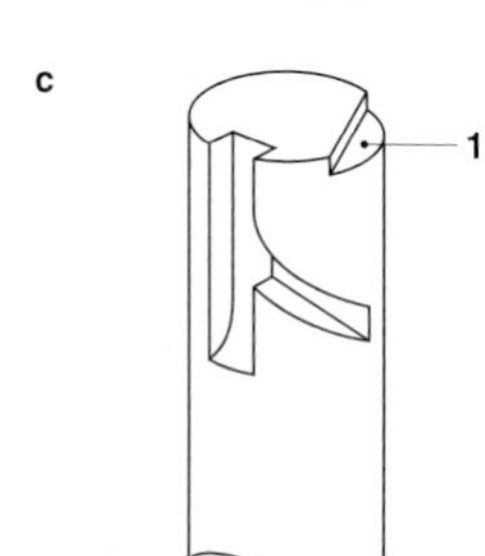

Figure 11.30 Pump plunger versions
(a) Lower helix,
(b) Upper and lower Helix,
(c) Lower helix with starting groove (1).

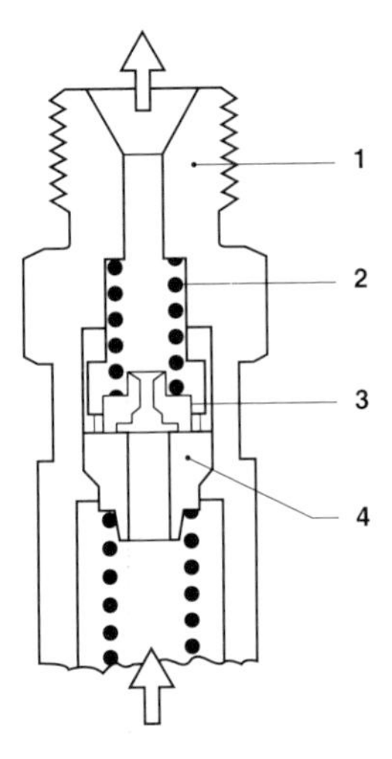

Figure 11.33 Delivery-valve holder with return flow restriction
1 Delivery-valve holder, 2 Delivery-valve spring, 3 Valve plate, 4 Valve holder

pressure fuel-injection pumps. It comprises a forward-delivery valve in the delivery direction, and a pressure-holding valve in the return-flow direction. Between injections, the latter maintains the static line pressure as constant as possible under all operating conditions. The advantage of the constant-pressure valve lies in the avoidance of cavitation and in improved hydraulic stability. If the constant-pressure valve is to be employed efficiently, this necessitates more precise adjustments and governor modifications. The constant-pressure valve is used for high-pressure fuel-injection pumps (pressures above approx. 800 bar), and for small, high-speed direct-injection (DI) engines.

11.3.4 Standard PE in-line injection pumps

11.3.4.1 Design and construction

The standard PE in-line injection pumps incorporate their own camshaft and a plunger and barrel assembly (pumping element) for each engine cylinder (*Figure 11.35*).

The complete fuel injection system is comprised of:

- A fuel injection pump;
- a mechanical (flyweight) or electronic governor for control of engine speed and injected fuel quantity;

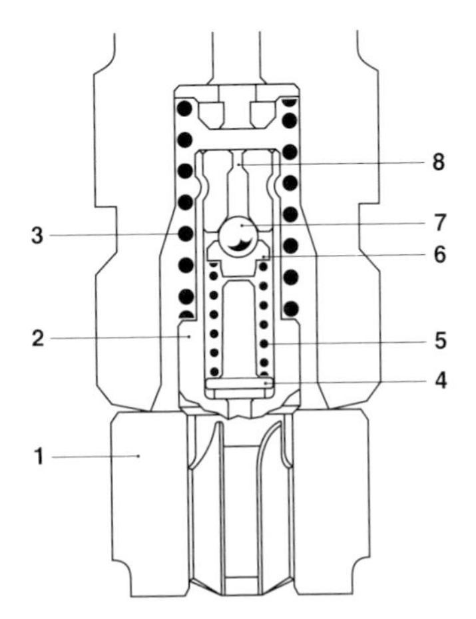

Figure 11.34 Constant-pressure valve
1 Valve holder, 2 Valve element, 3 Valve spring, 4 Filler piece, 5 Compression spring, 6 Spring seat, 7 Ball, 8 Restriction passage.

- a timing device (if required) for varying the start of delivery as a function of engine speed;
- a fuel supply pump for delivering the fuel from the fuel tank, through the fuel filter and the fuel line, to the injection pump;
- a number of high pressure fuel injection lines, corresponding to the number of engine cylinders, connecting the injection pump and the injection nozzles;
- the injection nozzles.

The injection pump's camshaft is driven by the diesel engine. Injection pump speed and crankshaft speed are identical for two-stroke engines. For four-stroke engines, pump speed is the same as engine camshaft speed, in other words half crankshaft speed.

The drive between injection pumps and engine must be as torsionally rigid as possible if today's high injection pressures are to be generated.

There are a number of different sizes of in-line injection pumps for the various engine outputs.

The injected fuel quantity depends upon the swept volume of the injection pump barrel, and maximum (pump side) injection pressures are between 400 and 1150 bar.

To lubricate the moving injection pump components (e.g. camshaft, roller tappets, etc.) there must be a certain amount of oil in the injection pump. The injection pump is connected to the diesel engine's lube-oil circuit, and oil circulates through the pump during operation.

Each pump type is allocated to a given type series, which in some cases overlap with respect to their power ranges. These will be described in the following chapters.

Two different construction principles are used for in-line injection pumps: The principle for the M and A pumps, and that for the MW and P pumps.

The power output of diesel engines equipped with in-line injection pumps ranges from 10 to 70 kW per cylinder. This broad power output range is made possible by the availability of a wide variety of different pump versions. The pump sizes A, M, MW and P are manufactured in large batches (*Figure 11.36*).

The pump sizes ZW, P9, and P10 are available for even higher cylinder power outputs.

11.3.4.2 Method of operation

Interaction between the components

The camshaft of the PE in-line injection pump is integrated in the aluminum pump housing. It is connected to the diesel engine either through a timing device, through a coupling element, or directly. A roller tappet with spring seat is located above each camshaft cam. The spring seat provides a positive drive connection between pump plunger and roller tappet. The pump plunger moves up and down in the pump barrel, and together these two components form the plunger and barrel assembly (pumping element).

The injection pump barrel has either one or two inlet ports which lead from the pumps's fuel gallery into the pump barrel. The delivery valve holder complete with delivery valve is located at the top of the plunger and barrel assembly.

The control sleeve is the connection between the pump plunger and the control rack. The control rack, which is free to move lengthwise in the pump housing as dictated by the governor, engages with the control sleeve gear or with a linkage lever to turn the 'control sleeve/pump plunger unit' in accordance with the governor output as described in the Section 'Engine speed Governing'. This permits the precise control of injected fuel quantity.

Injection pump drive

In the in-line injection pump, camshaft rotation is converted

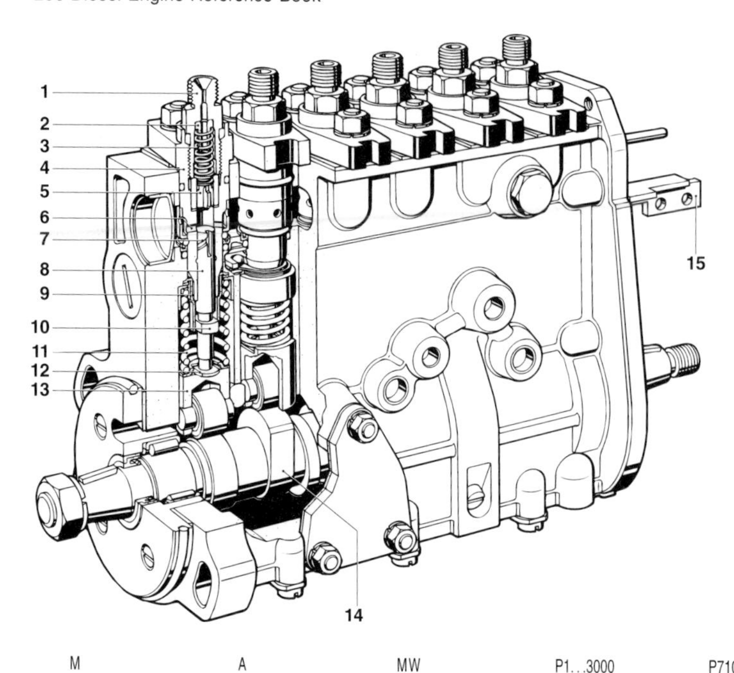

Figure 11.35 PES in-line fuel injection pump
1 Delivery-valve holder, 2 Filler piece, 3 Delivery-valve spring, 4 Pump barrel, 5 Delivery valve, 6 Inlet port and spill port, 7 Control helix, 8 Pump plunger, 9 Control sleeve, 10 Plunger control arm, 11 Plunger return spring, 12 Spring seat, 13 Roller tappet, 14 Cam, 15 Control rack.

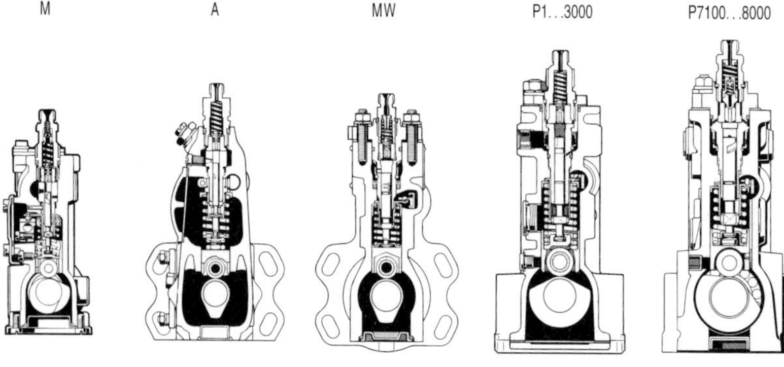

Figure 11.36 Inline injection pumps
Size comparison (looking onto camshaft end)

directly to the vertical lift of the roller tappet which results in the reciprocating plunger movement (*Figure 11.37*). The pump plunger's total lift cannot be varied, although the effective stroke, and with it the delivery quantity, can be changed by using the control rack to rotate the pump plunger. The plunger is forced up to TDC by the cam, and back down again to BDC by the plunger return spring. This spring must be selected so that even at maximum pump speed the roller tappet cannot jump off of the cam. This must be avoided at all costs because the impact caused when the roller 'hits' the cam again is bound to lead to cam or roller damage. The angular offset between adjoining cams ensures that the injection sequence agrees exactly with the engine's firing sequence and firing interval.

11.3.4.3 Additional components

Engine speed control (governing)

The governor's main task is to limit the engine's maximum speed (maximum no load speed). It must limit engine speed to the maximum permitted by the engine manufacturer, because otherwise the unloaded diesel engine will speed up out of control until it destroys itself. It must also be possible to maintain specific engine speeds inside a given engine speed range or within the complete range. Depending upon governor design, this can apply for instance to idle speed and maximum speed.

The governor also has a number of other functions: changing full load delivery as a function of engine speed (torque control), or as a function of atmospheric pressure or charge air pressure, or provision of injected fuel quantity needed for starting. To do so, the governor shifts the control rack so that the pump plunger is rotated to the appropriate setting for the required delivery quantity (*Figure 11.38*). For governing on in-line injection pumps, mechanical (flyweight) governors, or Electronic Diesel Control (EDC), are used. Pneumatic governors are no longer used because they cannot comply with the severe requirements made on a modern diesel engine. The above governors are described in the Technical Instruction manual 'Governors for Diesel In-Line Fuel injection Pumps'.

Mechanical engine speed control

There are a variety of different mechanical governor types in use:

- Maximum speed governor, for limiting the maximum speed (high idle speed).
- Minimum maximum speed governor (mainly for automotive applications), governs only at the upper and lower limits of the engine speed range, but not in between. The driver changes the injected fuel quantity by means of the accelerator pedal.

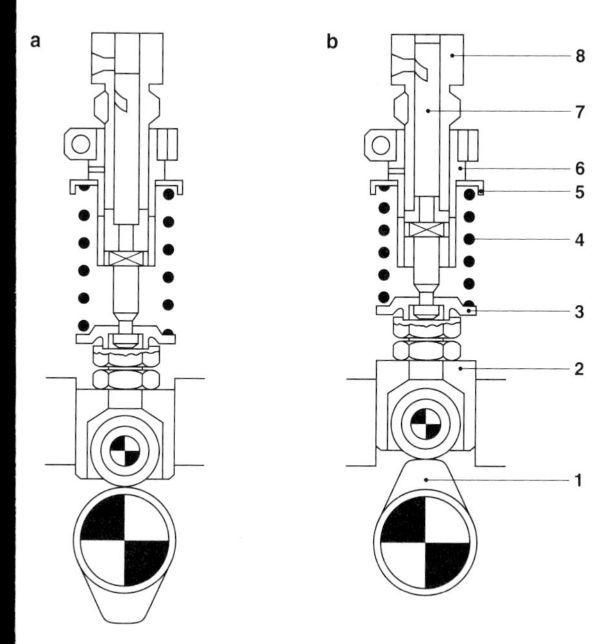

Figure 11.37 Plunger and barrel assembly (pumping element). Drive (a) BDC position, (b) TDC position
1 Cam, 2 Roller tappet, 3 Lower spring seat, 4 Plunger return spring, 5 Upper spring seat, 6 Control sleeve, 7 Pump plunger, 8 Pump barrel.

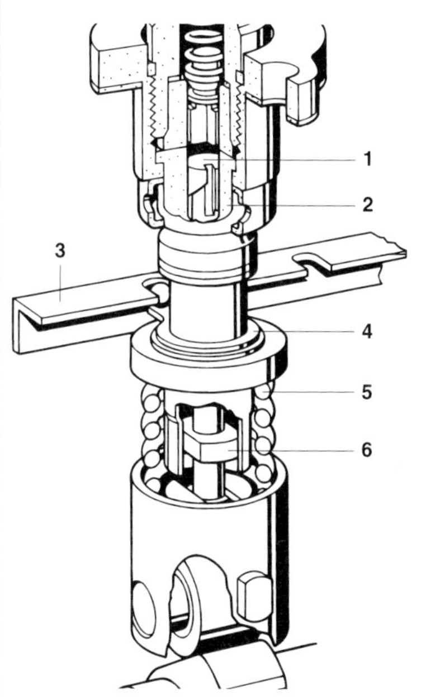

Figure 11.38 Fuel-delivery control
Using a control-sleeve lever.
1 Plunger, 2 Barrel, 3 Control rack, 4 Control sleeve, 5 Plunger return spring, 6 Plunger control arm.

- Variable speed governor, governs throughout the complete speed range in addition to the maximum (high idle) and idle speeds.

Developments in diesel fuel injection techniques are today determined by the steadily increasing demands made on exhaust gas 'quality' fuel economy, derivability and comfort, and engine power. Consequently, the demands made on the fuel injection system, and particularly on the governor, are also rising.

Electronic engine speed control

The Electronic Diesel Control (EDC) complies in full with the increasing demands made upon the engine speed governing system. As well as permitting electrical measurement and electronic data processing, EDC incorporates control loops with electrical actuators which compared to the mechanical governor provide for more functions as well as improving existing functions. The EDC comprises:

- a variety of sensors;
- the electronic control unit (ECU); and
- the actuator fitted to the injection pump.

Injection timing

The most important criteria for diesel engine optimization are:

- low exhaust emissions;
- low combustion noise; and
- low specific fuel consumption.

The moment at which the injection pump starts to deliver fuel is known as the 'start of delivery' (or port closing). This moment in time is selected according to injection delay and ignition delay. These are variable parameters which are a function of the particular operating point. Injection delay is defined as the delay between start of delivery and start of injection. Ignition delay as the delay between start of injection and start of combustion. Start of injection is defined as the crankshaft angle in the TDC area at which the nozzle injects fuel into the combustion chamber. Start of combustion is defined as the moment of A/F mixture ignition which can be influenced by start of injection. With the PE fuel injection pump, the speed dependent adjustment of the start of delivery (port closing) is best performed using a timing device.

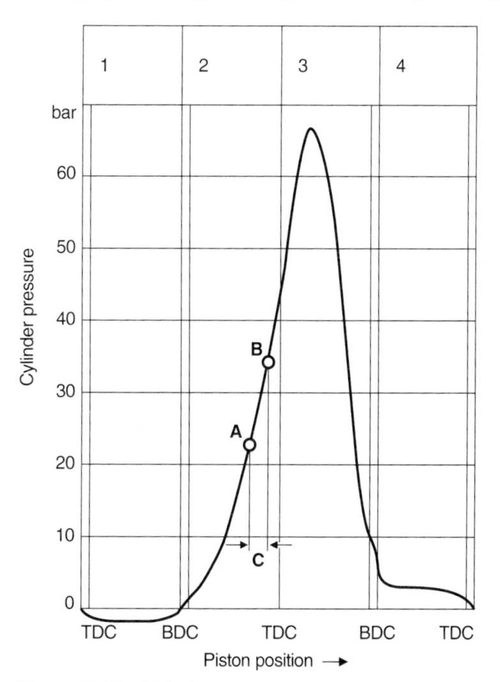

Figure 11.39 Cylinder pressure
A Start of injection, B Start of combustion, C Ignition lag
1 Induction stroke, 2 Compression stroke, 3 Power stroke, 4 Exhaust stroke.
OT=TDC, UT = BDC

Assignments

Because it directly changes the start of delivery point, the timing device should be termed a start of delivery adjuster. The timing device (eccentric type) transfers the drive torque to the injection pump while at the same time performing its timing function. The drive torque required by the injection pump depends upon pump size, number of barrels, injected fuel quantity, injection pressure, plunger diameter, the cam contour. The fact that drive torque has a direct effect upon the timing characteristic, must be taken into account during design work, as well as the power capability.

Design and construction

The timing device for the in-line injection pump is mounted directly on the end of the pump's camshaft. Basically, one differentiates between the open type and the closed type of timing device.

The 'closed' timing device has its own lube-oil reservoir which makes it independent of the engine's lube-oil circuit. The 'open' design on the other hand is connected directly to the engine's lube-oil circuit. Its housing is screwed to a toothed gear, and the compensating and adjusting eccentrics are mounted in the housing so that they are free to pivot. The compensating and adjusting eccentrics area guided by a pin which is rigidly connected to the housing. Apart from costing less, the 'open' type has the advantage of needing less room and of being more efficiently lubricated.

Operating principle

The timing device is driven by a toothed gear which is accommodated in the engine timing case. The connection between the input and drive output (hub) is through interlocking pairs of eccentric elements (*Figure 11.40*). The largest of these, the adjusting eccentric elements (4), are located in holes in the backing disk (8) which in turn is bolted to the drive element (1). The compensating eccentric elements (5) fit in the adjusting eccentric elements (4) and are guided by them and the hub bolt (6). On the other hand the hub bolt is directly connected with the hub (2). The flyweights (7) engage with the adjusting eccentric element (4) and are held in their starting positions by progressive springs (*Figure 11.41*).

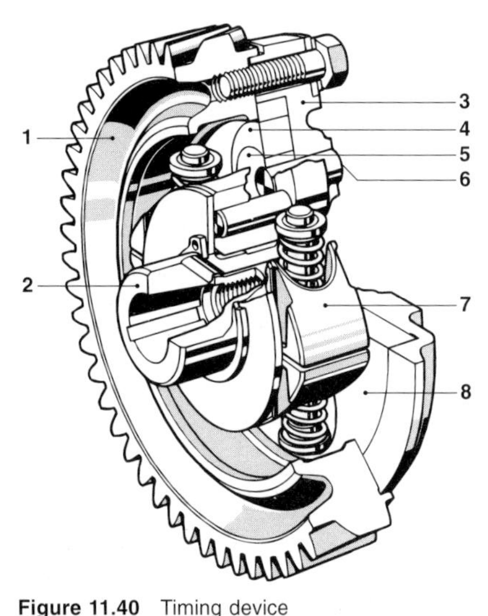

Figure 11.40 Timing device
Design and construction.
1 Drive element, 2 Hub, 3 Housing 4 Adjusting eccentric element, 5 Compensating eccentric element, 6 Hub bolt, 7 Flyweights, 8 Backing disc.

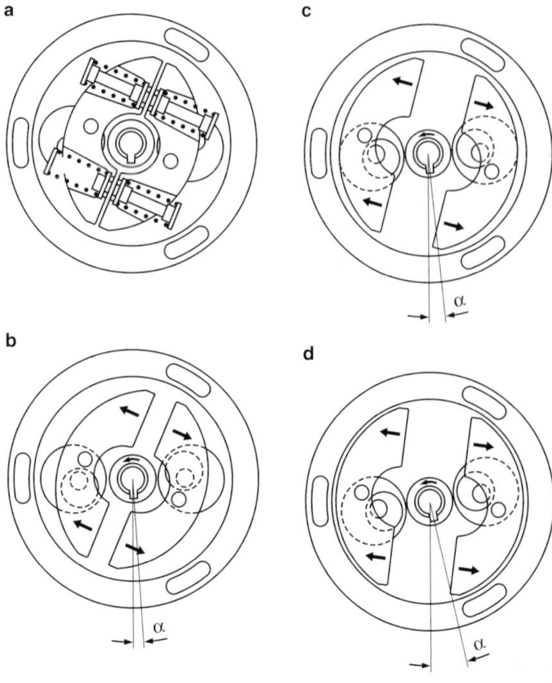

Figure 11.41 Timing device
Function.
(a) In initial position
(b) Setting at low speeds
(c) Setting at medium speeds
(d) Final setting at high speeds
α = Advance angle

Sizes

The size of the timing device as defined by its outside diameter and depth, determines the installable flyweight mass, the distance between centres of gravity, and the possible flyweight travel. These three factors also define the timing device's power capacity and the application range.

11.3.4.4 Pump sizes

Bosch–Size A pump

The size A in-line injection pumps (*11.42* and *11.43*) with their larger delivery ranges follow directly after the size M pumps. This pump also has a light metal housing and can be attached to the engine by flange or cradle mounting. The A pump is also of the 'open' design, and the pump barrels are inserted directly

Figure 11.42 A-type injection pump

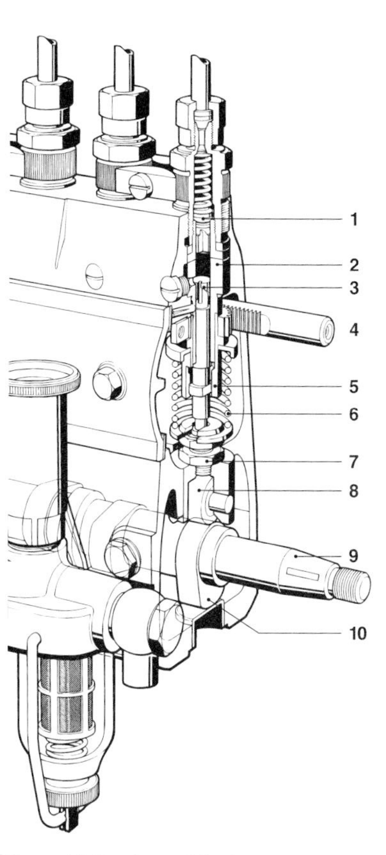

Figure 11.43 A-type injection pump
1 Delivery valve, 2 Pump barrel, 3 Pump plunger, 4 Control rack, 5 Control sleeve, 6 Plunger return spring, 7 Adjusting screw, 8 Roller tappet, 9 Camshaft, 10 Cam lobe.

into the aluminum housing from above, whereby the delivery valve assembly is pressed against the pump housing by the delivery valve holder. The sealing pressures, which are far in excess of the hydraulic delivery pressures, must be absorbed by the pump housing. For this reason, the A pump peak injection pressure is limited to 600 bar.

For adjusting the delivery quantity by means of the control rack, the A pump is equipped with a pinion control. A gear segment clamped on the plunger's control sleeve engages with the control rack, and to adjust the plunger and barrel assemblies for equal delivery quantities, the locking screws must be released and the control sleeve turned relative to the gear segment and thus relative to the control rack.

All adjustment work on this type of pump must be carried out with the pump at standstill and the housing open. The A pump has a side mounted spring chamber cover which must be removed in order to gain access to the pump's interior.

For lubrication, the pump is connected to the engine's lube-oil circuit.

The A pump is available in versions with up to 12 cylinders, and is suitable for operation.

Bosch—Size MW pump

The MW in-line injection pump was developed to satisfy the need for higher peak injection pressures (*Figure 11.44* and *11.45*). The MW pump is a 'closed type' in-line injection pump, and its peak injection pressure is limited to 900 bar. It also has a light

Figure 11.44 MW-type injection pump

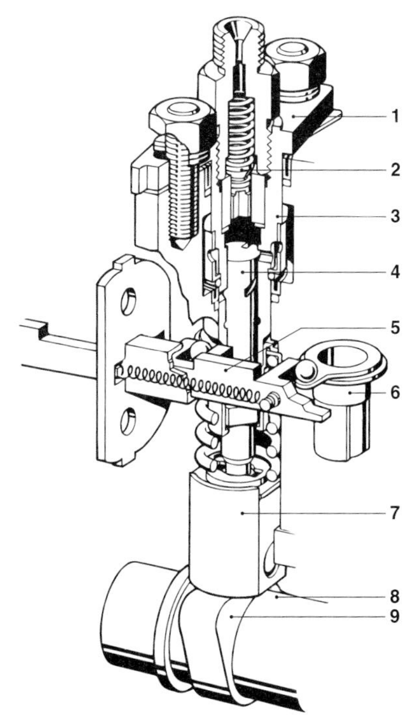

Figure 11.45 MW-type injection pump
1 Fastening flange for the plunger-and-barrel assembly, 2 Delivery valve, 3 Pump barrel, 4 Pump plunger, 5 Control rack, 6 Control sleeve, 7 Roller tappet, 8 Camshaft, 9 Cam lobe.

metal housing and is fastened to the engine using either cradle-, flatbed-, or flange mounting.

The design of the MW pump differs considerably to that of the A pump. The major difference being the use of a plunger and barrel assembly comprising the pump barrel, delivery valve, and delivery valve holder. This is assembled outside the pump and inserted into the pump housing from above. On the MW pump, the delivery valve holder is screwed directly into the pump barrel which has been extended upwards. The prestroke is adjusted by shims of varying thickness which are inserted between the housing and the barrel and valve assembly. The adjustment of uniform delivery between the individual plunger barrels is carried out from outside the pump by turning the barrel and valve assemblies. The assembly fastening flanges are provided with slots for this purpose.

The pump plunger's position remains unchanged when the barrel and valve assembly is turned.

The MW pump is available in versions with up to max. 8 barrels (8 cylinders) and is suitable for a variety of different mounting methods. It operates with diesel fuel, and lubrication is through the engine's lube-oil circuit.

Bosch–Size P pump

The Size P in-line injection pump was also developed to provide higher injection pressures (*Figures 11.46 and 11.47*). Similar to the MW-pump, it is of the 'closed' type and is fastened to the engine using a base or flange mounting. In the case of P-pumps designed for a peak injection of 850 bar, the pump barrel is inserted in a flange bushing which is already provided with threads for the delivery-valve holder. With this version of pump-barrel installation, the sealing forces put no load on the pump housing. Prestroke adjustment is performed the same as with the MW-pump.

Low-pressure in-line injection pumps use conventional fuel-gallery flushing. Here, the fuel flows through the fuel galleries of the individual barrels one after another, and in the direction of the pump's longitudinal axis. Fuel enters the galleries at the fuel inlet and leaves at the fuel return. Taking the P8000 version of the P-pump, which is designed for pump-side injection pressures of up to 1150 bar, as an example, this flushing method would lead to excessive fuel-temperature gradient (of up to 40°C) inside the pump between the first and last barrel. Since the fuel's energy density decreases along with its increasing temperature and its resulting increase in volume, this would lead to the injection of differing 'quantities of energy' into the engine's combustion chambers. These injection pumps therefore use 'cross-flushing', a method whereby the fuel galleries of the individual pump barrels are separated from each other by means of tattling points. This means they can be flushed parallel to each other (at right angles to the pump's longitudinal axis) under practically identical temperature conditions.

This injection pump is also connected to the engine's lube-oil circuit for lubrication. The P-pump is also available in versions for up to 12 barrels (cylinders), and is suitable for diesel fuel as well as for multi-fuel operation.

Figure 11.46 P-type injection pump

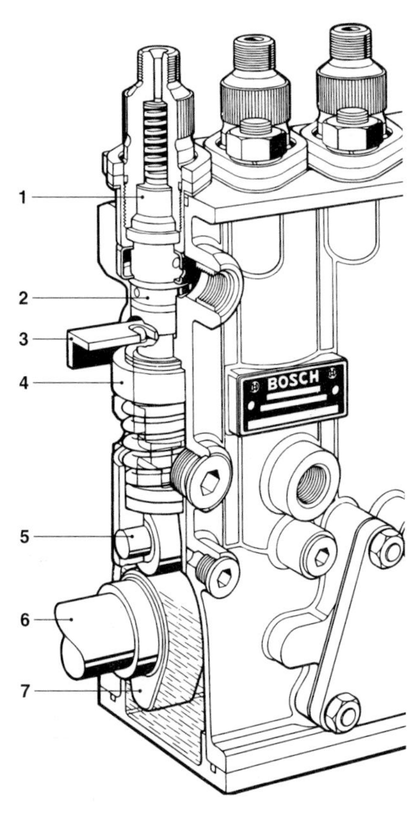

Figure 11.47 P-type injection pump
1 Delivery valve, 2 Pump barrel, 3 Control rod, 4 Control sleeve, Roller tappet, 6 Camshaft, 7 Camshaft lobe

11.3.5 PE in-line injection pumps for alternative fuels

11.3.5.1 Application, design, and construction

Specially designed diesel engines can operate on alternative fuels, whereby one differentiates between:

- multi-fuel engines which in addition to diesel fuel, can also run on gasoline (petrol) or kerosene;
- alcohol engines which operate with methanol or ethanol (ethyl alcohol);
- engines which run on biomass fuels.

The change from one type of fuel to another necessitates adaptive measures in the fuel metering in order to avoid excessive variations in power output. The most important fuel characteristics are: viscosity, boiling point, lubricity, density, and self ignition point. Constructional measures are necessary at the fuel injection equipment and at the engine in order that these characteristics can be optimally matched to each other. If diesel fuels with a considerably reduced sulphur content are used with Bosch in-line fuel injection pumps no negative effects are to be expected. Alternative fuels though have a lower boiling point which means that the injection pump's fuel gallery must be flushed more intensively and at higher pressures. A special fuel supply pump is available for this purpose.

11.3.5.2 Subassemblies

Injection pumps for multi-fuel engines

In the case of low density fuels (gasoline/petrol), the full load delivery is increased by means of a switchable control rod stop. On the other hand, in order to prevent fuel losses when low viscosity fuels are used, the pump's plunger and barrel assemblies are provided with an oil block in the form of two ring-shaped grooves in the pump barrel, the upper groove being connected with the pump's fuel gallery through a passage. During the effective (working) stroke, the fuel that leaks between the plunger and the barrel wall, flows through this passage and back to the fuel gallery.

There is an inlet passage in the bottom groove through which engine lube-oil is forced into the groove under pressure via a fine filter. At normal engine operating speeds, this pressure exceeds the fuel pressure in the fuel gallery so that the plunger and barrel assembly is sealed off effectively. A non-return valve prevents fuel entering the lube-oil circuit when the oil pressure drops below a certain level at idle.

Injection pumps for alcohol engines

The in-line injection pumps are also suitable for operation with methanol and ethanol (ethyl alcohol) provided they have been modified beforehand. Such modifications include:

- fitting special seals and gaskets;
- providing special protection for surfaces in contact with alcohol;
- fitting rustproof springs;
- operating with special lubricants.

In order that the engine receives an amount of energy equivalent to that provided by diesel fuel, the delivery quantity for methanol is 2.3 times higher and that for ethanol 1.7 times higher. And compared to diesel fuel, considerably more wear must be expected at the delivery valve seats and nozzle/needle seats.

Injection pumps for biomass fuels

Serious complications in the diesel fuel injection installation are not to be expected when an engine is run with rape oil (RME). Although, depending upon the specifications of the fuel in use special measures may be required.

11.3.6 In-line control sleeve fuel-injection pumps

Increasing attention is being paid to reducing the toxic content of commercial vehicle exhaust gases, and engine designers are concentrating on preventing the generation of toxic substances at the source, or at least on reducing it.

With commercial vehicle diesel engines, high injection pressures and optimum start of injection timing are making major contributions towards achieving this target. A direct result of these endeavours has been the development of a new generation of high pressure injection pumps, the in-line control sleeve injection pumps (*Figure 11.48*).

11.3.6.1 Design and construction

Due to its 'control sleeve' which can be moved up and down the pump plunger, the control sleeve injection pump differs from the conventional in-line injection pump both in operating principle and in design and construction. The control sleeve used with this pump supersedes the timing device as described in 'Injection timing'. Technically speaking the design and construction of the rest of the pump has remained unchanged. The in-line control sleeve injection pump operates with injection pressures of approx. 1150 bar. Thanks to the technology used, it is possible to freely programme the start of delivery timing. The essential feature of this pump principle is that the start of delivery timing is practically independent of injected fuel quantity, and is implemented by adjustments carried out simultaneously at all control sleeve elements.

This means that the former 'rigid' timing device attached to the end of the camshaft and designed to handle the injection pump's high driving torques is no longer necessary (*Figure 11.49*).

11.3.6.2 Operating principle

In order to change the start of delivery (port closing) or start of injection, the prestroke is adjusted by means of a control sleeve which can be moved up and down on the pump plunger. This means that compared to the conventional in-line injection pump, electronics has been introduced as a second method to govern the injection pump.

Each pump plunger is equipped with a control sleeve which incorporates the spill port. All sleeves are adjusted simultaneously by the control sleeve levers which engage with them through 'windows' in the barrel (*Figure 11.36*). The levers are fastened to the control sleeve shaft. Depending upon the control sleeve's vertical position, start of delivery (port closing) begins sooner or later relative to the camshaft lobe. The injected fuel quantity is metered using the port and helix control familiar from the conventional in-line injection pump. The in-line control sleeve injection pump is an element in the electrical servo system with which start of in injection and injected fuel quantity can be programmed as a function of a variety of different influencing variables. This form of control permits pollutant emissions to be reduced to a minimum (important for instance, with respect to the severe US limits), optimization of fuel economy in all operating states, precise fuel metering, and further improvement of the starting phase and in particular of the warm-up phase.

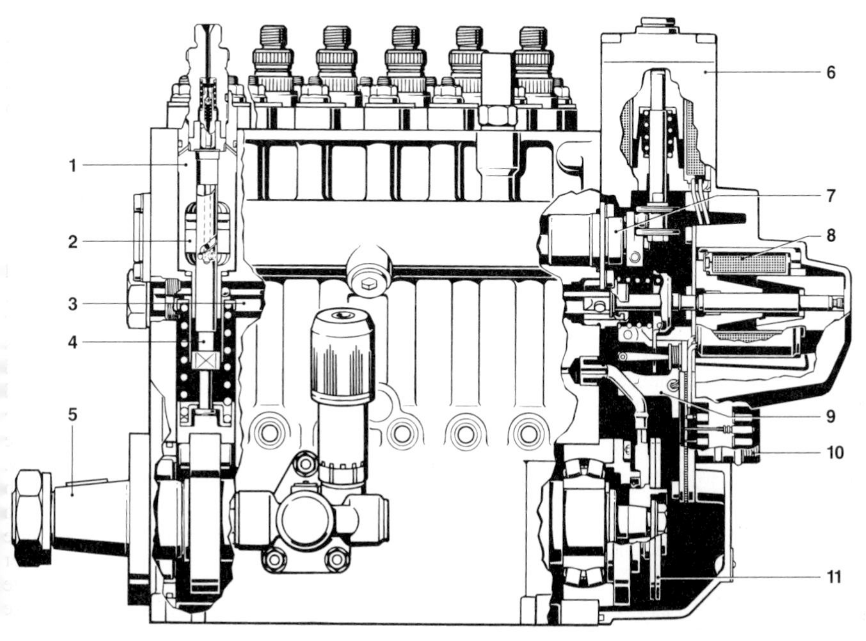

Figure 11.48 In-line control sleeve injection pump
1 Pump barrel, 2 Control sleeve, 3 Control rod, 4 Pump plunger, 5 Camshaft, 6 Port-closing actuator solenoid, 7 Control-sleeve setting shaft, 8 Rod-travel actuator solenoid, 9 Inductive rod-travel sensor, 10 Plug-in connection, 11 Disk for port-closing block and for oil return pump.

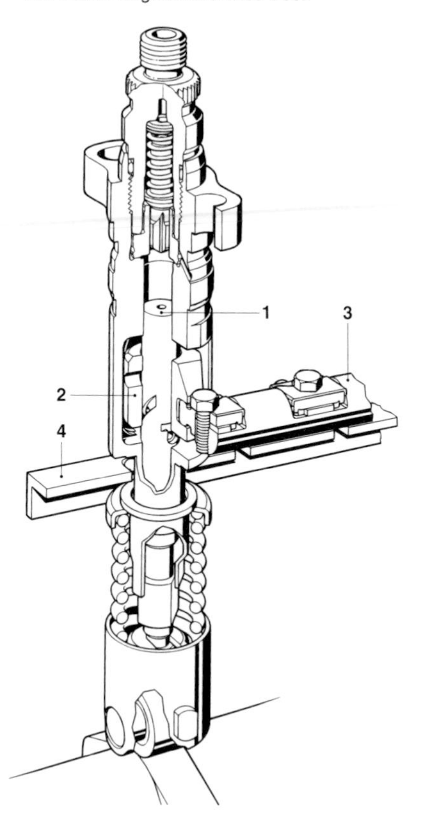

Figure 11.49 Control-sleeve adjustment mechanism
1 Pump plunger, 2 Control sleeve, 3 Control-sleeve shaft, 4 Control rod

Electronic engine speed control (governing)

Operating data acquisition
A number of sensors are attached at various points on the control sleeve in-line injection pump itself, on the vehicle's engine, and on the vehicle. These measure the temperature of the air, the fuel, and the engine, as well as the air pressure, the intake air quantity, the engine speed, and the accelerator pedal position (driver input). The sensors convert these environmental and operating parameters into electrical signals which are inputted into the ECU.

Operating data processing
A computer-based electronic control unit (ECU) uses the incoming input data to generate the desired value for the pump setting. The ECU then forwards this information to the pump in the form of electrical signals through the various solenoid actuators in the pump's actuator system. The set point injected fuel quantity outputted by the ECU is set by means of the rod position control loop. The ECU specifies a set point rod travel and receives the feedback of the actual rod position via the rod travel sensor. To complete the closed control loop, the ECU repeatedly calculates the current which must be input to the injection pump's actuator system in order for the set point value and the actual value to conform with each other. Safety considerations dictate that a return spring is fitted to bring the control rod back to the 'zero delivery' position when no current is applied to the actuator system.

Start of delivery (port closing) is also adjusted using a closed control loop. One of the nozzle holders is equipped with a needle motion sensor which signals the start of injection to the computer-based ECU which then calculates the actual value for the moment of injection, taking into account the camshaft position. This is then compared with a set point value, and by means of current control the computer-controlled port closing actuator system is adjusted so that actual and set point values are identical. Since the port closing actuator system is 'structurally stable', a special position check back signalling unit is not required. Structurally stable means that the lines of force of the solenoid actuator and of the return spring always have a clear intersection point, so that the travel of the solenoid actuator is proportional to the applied current. This is equivalent to closing the control loop.

Fuel delivery

Port closing (start of delivery)
As soon as the pump plunger has travelled a certain distance towards TDC, the bottom edge of the control sleeve closes the pump plunger's spill port. Pressure can now build up in the high pressure chamber above the plunger and fuel delivery commences.

Port opening (end of delivery)
Following the plunger's remainder stroke to TDC, the pump plunger helix and the spill port in the control sleeve terminate the fuel delivery. Port opening and with it the injected fuel quantity can be varied by rotating the plunger by means of the control rod.

Port closing (start of delivery) adjustment
The control sleeve must be moved in the direction of TDC (or BDC) in order to adjust the port closing and with it the start of injection. Whereas moving the control sleeve to a position nearer to TDC results in increased prestroke and therefore in delayed start of delivery (injection), moving it nearer to BDC reduces the prestroke and advances the start of delivery (injection).

Depending upon the shape of the camshaft lobe, not only the delivery velocity is changed but also the delivery rate (theoretical quantity of fuel delivered per degree camshaft) and the injection pressure.

11.3.7 Electronic Diesel Control (EDC)

11.3.7.1 Application

The development of the automotive diesel engine is governed primarily by requirements for clean exhaust, improved fuel economy, and the optimization of derivability. These stipulations are placing increasingly stringent demands upon the fuel-injection system, namely:

- sensitive controls;
- ability to process additional parameters;
- tighter tolerances and increased accuracy even over very long periods of operation.

These demands are fulfilled by the Electronic Diesel Control (EDC). This system provides for electronic measurement, as well as flexible data processing, and closed control loops with electrical actuators. In comparison to the conventional mechanical governor therefore, EDC implements new and improved control functions.

In the diesel engine, operating characteristics and combustion are influenced by:

- injected fuel quantity;
- start of injection;
- exhaust-gas re-circulation (EGR);
- charge-air pressure.

These controlled variables must be optimally adjusted for every working mode in order to ensure efficient diesel engine operation. To this end, EDC incorporates automatic control loops for the main parameters.

11.3.7.2 System blocks

The electronic control is divided into three system blocks (*Figure 11.50*):

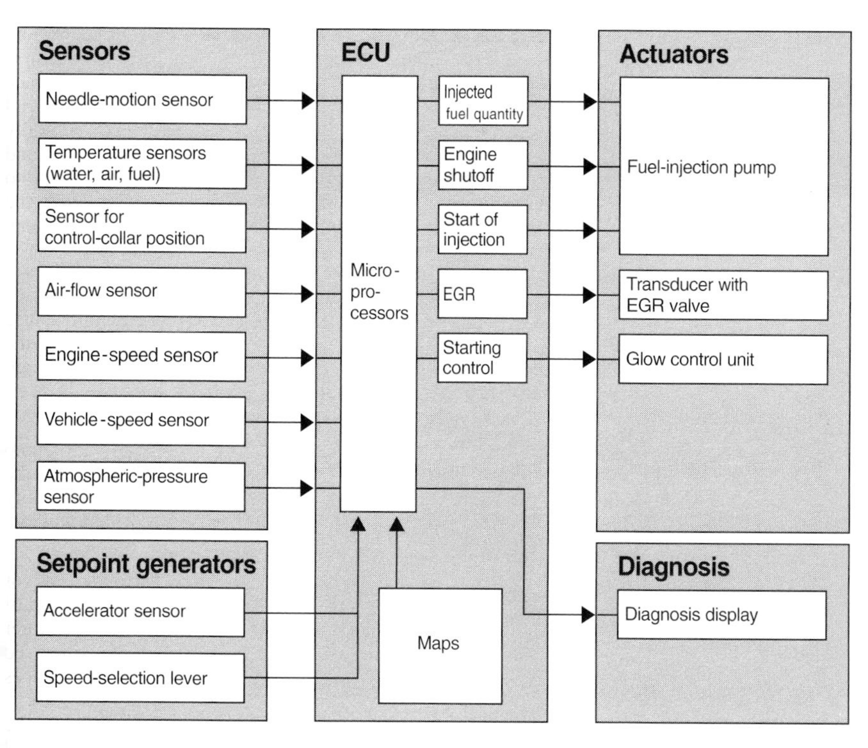

Figure 11.50 Electronic diesel control (EDC): System blocks

1. Sensors for registering operating conditions. A wide variety of physical quantities are converted into electrical signals.
2. Electronic control unit (ECU) with microprocessors which processes the information in accordance with specific control algorithms, and outputs corresponding electrical signals.
3. Actuators which convert the ECU's electrical output signals into mechanical quantities.

11.3.7.3 Components

Sensors

The positions of the accelerator and the control collar in the injection pump are registered by the angle sensors. These use either contacting or non-contacting methods. Engine speed and TDC are registered by inductive sensors. Sensors with high measuring accuracy and long-term stability are used for pressure and temperature measurements. The start of injection is registered by a sensor which is directly integrated in the nozzle holder and which detects the start of injection by sensing the needle movement (*Figure 11.51* and *11.52*).

Electronic control unit (ECU)

The ECU employs digital technology. The microprocessors with their input and output interface circuits form the heart of the ECU. The circuitry is completed by the memory units and devices for the conversion of the sensor signals into computer-compatible quantities. The ECU is installed in the passenger compartment to protect it from external influences.

There are a number of different maps stored in the ECU, and these come into effect as a function of such parameters as: load, engine speed, coolant temperature, air quantity, etc. Exacting demands are made upon interference immunity. Inputs and outputs are short-circuit-proof and protected against spurious pulses from the vehicle electrical system. Protective circuitry and mechanical shielding provide a high level of EMC (Electromagnetic Compatibility) against outside interference.

Solenoid actuator for injected fuel-quantity control

The solenoid actuator (rotary actuator) engages with the control collar through a shaft (*Figure 11.53*). Similar to the mechanically governed fuel-injection pump, the cutoff ports are opened or closed depending upon the control collar's position. The injected fuel quantity can be infinitely varied between zero and maximum (e.g. for cold starting). Using an angle sensor (e.g. potentiometer), the rotary actuator's angle of rotation, and thus the position of the control collar, are reported back to the ECU and used to determine the injected fuel quantity as a function of engine

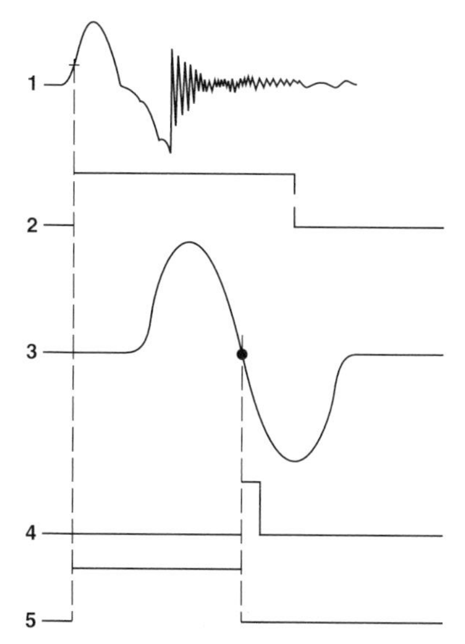

Figure 11.51 Sensor signals 1 Untreated signal from the sensor (NBF), 2 Signal derived from the NBF signal, 3 Untreated signal from the engine-speed sensor 4 Signal derived from untreated engine-speed signal, 5 Evaluated start-of-injection signal.

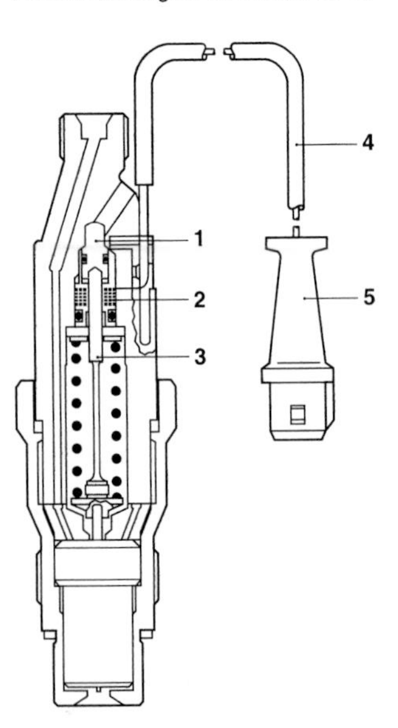

Figure 11.52 Nozzle-and holder assembly with needle-movement sensor (NBF)
1 Setting pin, 2 Sensor winding, 3 Pressure pin, 4 Cable, 5 Pin

speed. When no voltage is applied to the actuator, its return springs reduce the injected fuel quantity to zero.

Solenoid valve for start-of-injection control

The pump interior pressure is dependent upon pump speed. Similar to the mechanical timing device, this pressure is applied to the timing-device piston (*Figure 11.53*). This pressure on the timing device pressure side is modulated by a clocked solenoid valve.

With the solenoid valve permanently opened (pressure reduction), start of injection is retarded, and with it fully closed (pressure increase), start of injection is advanced. In the inter-mediate range, the on/off ratio (the ratio of solenoid valve open to solenoid valve closed) can be infinitely varied by the ECU.

11.3.8 Bosch—Single-plunger fuel-injection pumps

11.3.8.1 Bosch—PF single-plunger fuel-injection pumps

Design and construction

The PF single-cylinder (or single-plunger) injection pump has no integral camshaft of its own (the 'F' in PF indicates external drive). The PF injection pump and the PE in-line (multi-cylinder) injection pump operate according to the same principles. The PE pumps are suitable for use with small, medium-size, and large engines (*Figure 11.54*), to which they are flange-mounted.

Since each engine cylinder is allocated its own injection pump, the use of single-plunger injection pumps with multi-cylinder engines make it possible to use very short high-pressure delivery lines. This means that when single-cylinder pumps are used with multi-cylinder engines, only one type of pump and high-pressure line is employed. The PF injection pump is equipped with a locking device when it leaves the factory. This keeps the pump in its full-load delivery setting so that additional adjustment work is not needed when mounting it on the engine.

Engine-speed control

On large engines, the governor is fitted directly to the engine housing. The adjustment of injected fuel quantity needed for controlling engine speed is defined by the governor and transmitted to the individual pumps through a linkage integrated in the engine. Mechanical-hydraulic, electronic, and purely mechanical governors are in use, although the latter are rarely encountered. A sprung intermediate element in the transmission linkage to each pump ensures that governing can still take place even if the pump's adjustment mechanism should block.

Fuel supply and delivery

With regard to fuel supply and delivery, fuel filtering, and bleeding of the injection system, the same stipulations apply for the PF pump as they do for the PE in-line injection pump. The gear-type fuel-supply pump generates a pressure of 3–10 bar, and its delivered fuel quantity is 3–5 times the injected fuel quantity. Dirt is to be kept out of the injection system by using fine filtration with 10–30 μm aperture sizes. PF injection pumps

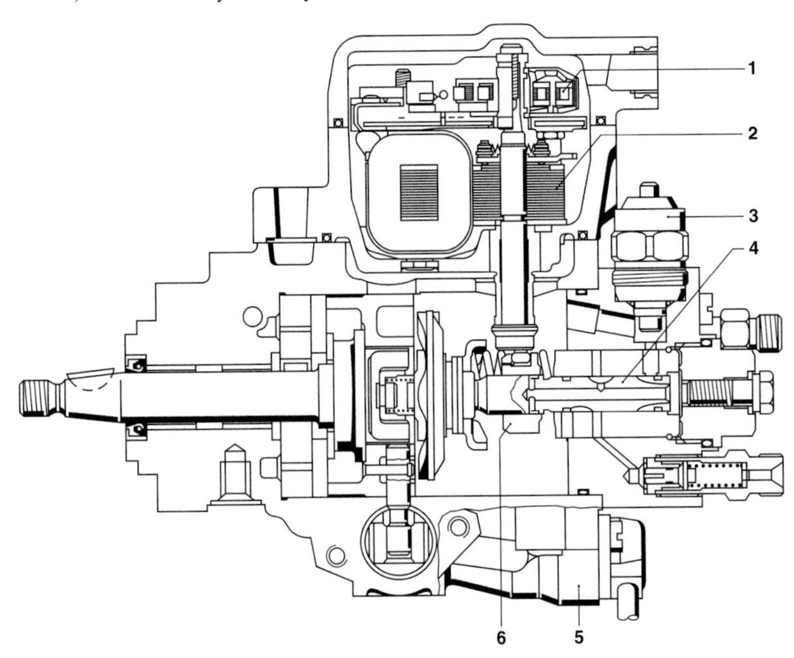

Figure 11.53 Distributor injection-pump fo electronic diesel control
1 Control-collar-position sensor, 2 Solenoid actuator for the injected fuel quantity, 3 Electromagnetic shutoff valve, 4 Delivery plunger, 5 Solenoid valve for start-of-injection timing, 6 Control collar.

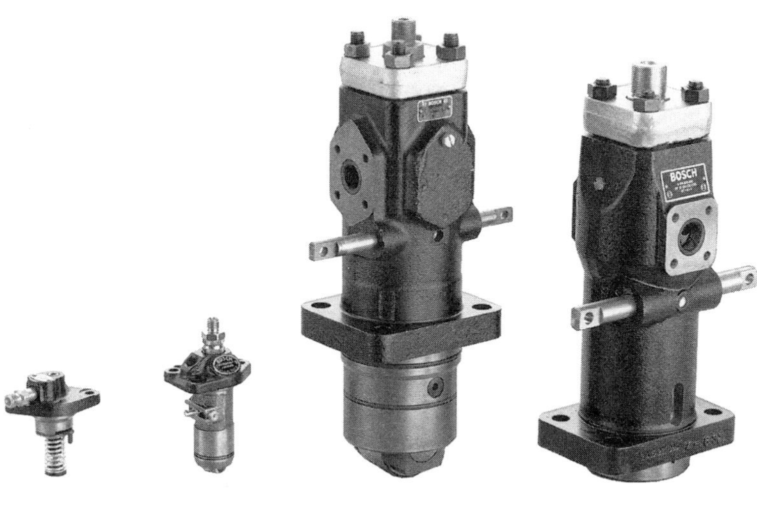

Figure 11.54 PF fuel-injection pump
Size comparison.
(a) Type PFE 1 Q, (b) Type PFR 1 K,
(c) Type PFR 1 W, Type PF 1 D

developing powers of up to 100 kW/cylinder are used not only for diesel fuel injection, but also for injecting heavy-oil with viscosities up to 700 mm^2/s at 50°C. In order to be able to deliver and inject this oil it must be heated to 150°C before reaching the fuel-supply pump so that it has the necessary injection viscosity of approx. 10–20 mm^2/s.

Injection timing

The injection cams for the individual PF injection pumps are on the engine's timing-gear camshaft. This means that turning the camshaft relative to the timing gears is out of the question for injection-timing purposes. Instead, an intermediate member is used, for instance a rocker fitted between roller tappet and camshaft (*Figure 11.55*), which provides for an advance angle of several degrees. This permits not only the optimization of fuel consumption and exhaust-gas emissions, but also the adaptation to the ignition quality of different fuels.

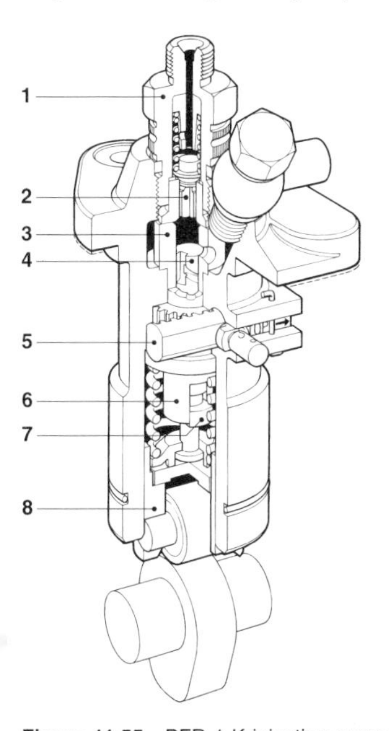

Figure 11.55 PFR 1 K injection pump
1 Delivery-valve holder, 2 Delivery-valve, 3 Pump barrel, 4 Pump plunger, 5 Control rack, 6 Control sleeve, 7 Plunger control arm, 8 Roller tappet

11.3.8.2 Sizes

Bosch—PE-pumps up to 50 kW/cylinder

These single-plunger injection pumps are used in diesel engines for powering small construction machines, pumps, agricultural tractors, and engine-generator sets.

The pump types PFE 1 A.. and PFE 1Q.. are of the single-cylinder type without integral roller tappet, which instead is integrated in the engine block. Control of fuel delivery on both pump ranges is by means of a control-sleeve lever which engages with the control rack running in the engine block. The PFR..K pump types with integrated roller tappet are fitted to 1-, 2-, 3-, or 4-cylinder engines. On all versions, the plunger is rotated by means of a toothed control sleeve which engages in a control rack mounted in the pump housing (*Figure 11.55*).

The small PF pumps have a maximum pump speed of about 1800 min^{-1}. Depending upon plunger diameter (5–9 mm), full-load injected fuel quantity is max. 95 mm^3 per plunger stroke, and maximum permissible peak injection pressure in the high-pressure lines (pump end) is 600 bar. PF fuel-injection pumps are equipped with constant-volume valves (with or without return-flow restriction). Constant-pressure valves are used in applications characterized by high pump loading and increased demands on the stability of the injected fuel quantity.

PE pumps above 50 kW/cylinder

Single-plunger injection pumps (*Figure 11.56*) are installed in diesel engines with output powers of up to 1000 kW per cylinder. They operate with diesel fuel and heavy oils of various viscosities. With (pump end) peak injection pressures of up to approx. 1200 bar, through-holes are drilled in the pump barrels. When peak pressures up to 1500 bar are concerned, blind-hole plunger-and-barrel assemblies are used to keep the deformation in the area of the plunger-and-barrel head to a minimum.

To prevent pump-housing damage due to the high-energy jet of fuel generated at the end of delivery (port closing), baffle screws are fitted in the immediate vicinity of the pump-barrel spill ports.

In order to seal the delivery valve against very high pressures, the surfaces between it and the flange, as well as between it and the pump barrel are lapped and flat. Pressure compensation at the pump plunger prevents asymmetrical loads. The pump-plunger control helixes are arranged similarly to those on the in-line injection pump. The delivery quantity is adjusted by rotating the plunger via the control rack. Rack-travel indication is provided.

The plunger-and-barrel assembly is provided with leak-oil

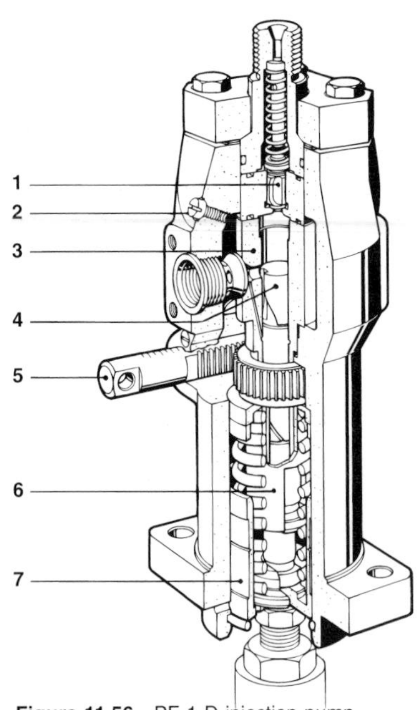

Figure 11.56 PF 1 D injection pump
1 Delivery valve, 2 Vent screw, 3 Pump barrel, 4 Pump plunger, 5 Control rack, 6 Control sleeve, 7 Guide sleeve.

return and an extra oil block for leak fuel. A second ring-shaped groove together with a blocking-oil inlet passage is machined into the pump barrel. Filtered oil is forced into this groove at a pressure of 3–5 bar, a pressure which under normal circumstances exceeds that in the pump's fuel gallery, and suffices to form an oil block to prevent passage of the leak fuel. The extremely small leakage quantity concerned (a mixture of fuel and lube-oil) is drawn off through a separate outlet and led into a collecting tank.

In the case of heavy-oil operation, the roller tappet of the PFR pump, or the guide sleeve of the PF pump, are lubricated with engine lube-oil through a separate connection.

11.3.9 Innovative fuel-injection systems

11.3.9.1 Bosch—Unit injector (PDE)

The unit injector (*Figure 11.57*) is screwed directly into the engine's cylinder head. This design combines the injection pump and the injection nozzle in a single unit which is driven by the engine camshaft. Each unit injector has its own high-speed solenoid valve which controls start and end of injection. With the solenoid valve open the unit injector forces fuel into the return line, and when the solenoid valve closes, into the engine cylinder. The start of injection is defined by the solenoid closing point, and the injected fuel quantity by the closing time (length of time the solenoid remains closed). The solenoid valve is triggered by an ECU with map-based control, which means that start and end of injection are programmable and therefore independent of piston position in the engine cylinder. Compared to the injection valve/injector in electronically controlled gasoline injection systems (Jetronic/EFI) the diesel solenoid valve must be able to control pressures which are 300–500 times higher, and must also be able to switch 10–20 times faster.

In today's conventional fuel-injection systems, the maximum injection pressure is limited by the physical characteristics of the high-pressure lines between injection pump and injection nozzle. The unit injector makes such lines superfluous, which

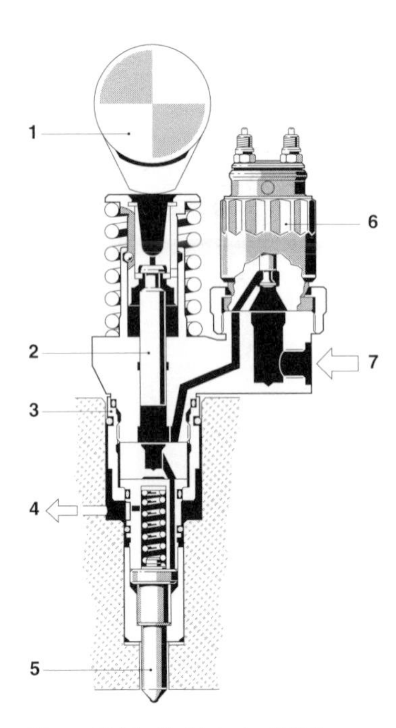

Figure 11.57 Unit injector (PDE)
1 Cam, 2 Pump plunger, 3 Engine, 4 Return, 5 Injection nozzle, 6 Solenoid valve, 7 Inlet

means that injection pressures of up to 1500 bar are possible. Pressures of this magnitude, together with map-based control of start of injection and duration of injection (injected fuel quantity), lead to a considerable reduction in the diesel engine's pollutants emission. Using electronic control concepts, special functions such as temperature-controlled start of injection, engine smooth-running control, anti-buck damping, and in future pilot fuel injection for even further reductions in noise, become feasible. In addition, the use of unit injectors makes it possible to switch off individual engine cylinders during part-load operation (*Figure 11.57*).

11.3.9.2 Bosch—Unit pump (PLD)

The unit-pump system (*Figure 11.58*) is a modular high-pressure injection system. From the control engineering viewpoint, it is closely related to the unit injector. Both systems use an individual injection pump for each engine cylinder which is driven by an extra cam (injection cam) on the engine's camshaft. The use of an electronically triggered high-speed solenoid valve enables the moment of injection and the injected fuel quantity to be precisely adjusted for each cylinder. This permits:

- fuel delivery to the injection nozzle;
- interruption of fuel delivery; and
- return of excess fuel to the fuel tank.

Similar to the PDE system, the PLD unit-pump system registers the most important engine and environmental parameters, and converts them into the optimal start of injection and optimal injected fuel quantity for the given operating conditions. The system comprises the following modules:

- high-pressure pump with attached solenoid valve;
- short high-pressure delivery line; and
- nozzle-and-holder assembly.

With this modular design, which contrasts with the compact

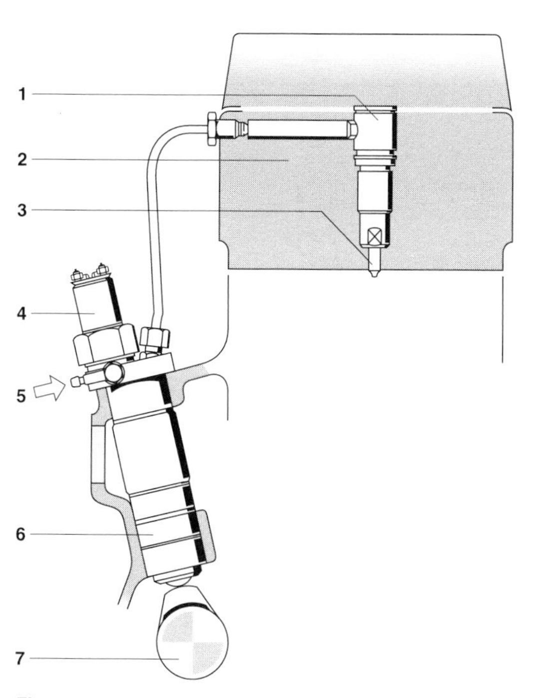

Figure 11.58 Unit pump (PLD)
1 Nozzle holder, 2 Engine, 3 Injection nozzle, 4 Solenoid valve, 5 Inlet, 6 High-pressure pump, 7 Cam

design of the PDE system, the unit-pump system represents a directly controlled high-pressure injection system which is suitable for a wide range of different installation requirements.

This system is further characterized by a fault-recognition facility, the possibility of emergency operation and self-diagnosis, as well as the option of communicating with other control systems via already existing interfaces.

11.3.10 Peripheral equipment for diesel fuel-injection systems

11.3.10.1 Bosch Nozzles and nozzle holders

Assignments

In the diesel engine's fuel-injection system, the nozzles in their respective nozzle holders are an important link between injection pump and engine. Their assignments are:

- meter the injected fuel;
- manage and prepare the fuel spray;
- define the rate-of-discharge curve; and
- seal off the injection system from the combustion chamber.

Diesel fuel is injected at high pressures, with peak pressures as high as 1200 bar, which in future will be even higher. At such high pressures, the diesel fuel no longer behaves like a rigid liquid but becomes compressible. During the brief delivery period (approx. 1 ms), the high pressure causes the injection system to 'expand' at certain points, whereby the nozzle cross-section defines the quantity of fuel which is injected into the combustion chamber. The nozzle's spray-hole length and diameter, and (to a limited extent) its spray-orifice shape, have an influence upon fuel-spray management and, as a result, upon the engine's output power, its fuel consumption and its exhaust emissions.

Within certain limits, the rate-of-discharge curve can be tailored to requirements by 'correct' control of the nozzle fuel-flow section (as a function of needle lift) and by controlling the nozzle-needle motion. And finally, the nozzle must be able to seal off the fuel-injection system against the hot, highly compressed gases from the combustion chamber (up to approx. 1000°C). In order to avoid blow-back of these gases when the injection nozzle opens, the pressure in the nozzle's pressure chamber must always be higher than the combustion pressures. This requirement is particularly difficult to comply with at the end of the injection process (when injection pressure has already dropped while combustion pressure is increasing rapidly), and it requires careful matching of the injection pump, the nozzle, and the pressure spring.

Designs

According to whether they have a divided combustion chamber (pre-chamber, turbulence, for swirl-chamber engine) or a non-divided combustion chamber (direct-injection (DI)), each design requires its own special nozzle.

The throttling pintle nozzle is used on pre-chamber, turbulence or swirl-chamber engines with divided combustion chamber. This nozzle injects a coaxial shaped jet of fuel and the needle normally opens inwards. On the other hand, hole-type nozzles are used for direct-injection (DI) engines with non-divided combustion chambers.

Throttling pintle nozzles
The standard nozzle-and-holder assembly for pre-chamber and turbulence (swirl) chamber engines is composed of the injection nozzle (Type DN.. SD..) together with the nozzle holder (Type KCA with screw-in thread). The normal version of this nozzle holder has an M24×2 screw-in thread and an A/F dimension of 27 mm. Usually, DN O SD.. injection nozzles are used which have a needle diameter of 6 mm and a spray angle of 0° (pencil spray). It is much rarer to find an injector nozzle with a defined spray angle (e.g. 12° in the DN 12 SD..). For restricted cylinder-head space, more compact nozzle-holder versions (e.g. KCE) are available.

One of the characteristic features of the throttling pintle nozzle is the control of its discharge cross-section, in other words the flow quantity, as a direct function of the needle lift. Whereas in the case of the hole-type nozzle the cross-section increases sharply as soon as the needle opens, the throttling pintle nozzle features a very flat cross-section characteristic in the range of small needle strokes. In this range, the throttling pintle, a pin-shaped extension of the nozzle needle, remains inside the spray hole and only the small ring-shaped area between the spray hole and the pintle remains available as the flow cross-section. When large needle strokes take place, the pintle lifts out of the spray hole completely and the flow cross-section increases rapidly (*Figure 11.59*).

To a certain degree, this change in cross-section as a function of needle stroke controls the rate-of-injection curve, in other words the injected fuel quantity per unit of time. At the start of injection only a small quantity of fuel can leave the nozzle, while a large quantity emerges at the end of the injection process. Above all, this characteristic has a positive effect on engine combustion noise.

It must be noted that if the cross-section values are too small there is insufficient needle lift, the injection pump accelerates the needle in the 'open' direction more quickly than would otherwise be the case, and the pintle leaves the spray hole sooner, thus terminating the throttling action more quickly. The injected fuel quantity per unit of time climbs rapidly as a result and combustion noise increases. Excessively small cross-sections at the end of injection have a similar negative effect because when the needle closes again the displaced fuel has difficulty in leaving through the restricted cross-section, with the attendant end-of-injection delay. It is therefore important to match the

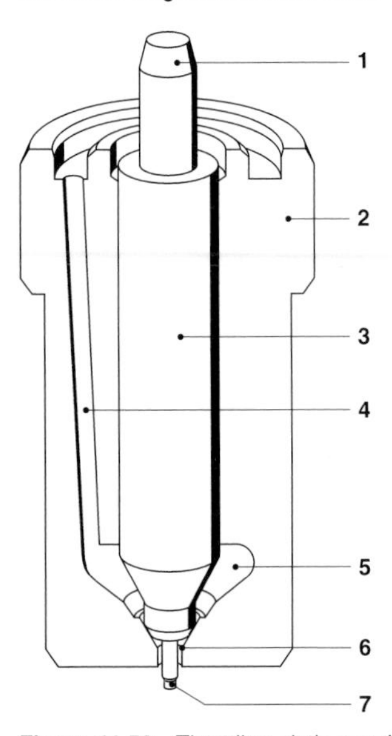

Figure 11.59 Throttling pintle nozzle
1 Pressure pin, 2 Nozzle body, 3 Nozzle needle, 4 Inlet passage, 5 Pressure chamber, 6 Spray hole, 7 Pintle

cross-section characteristic to the rate-of-discharge curve and to the particular combustion process.

For the spray holes, appropriate manufacturing processes must be applied in order to comply with the close-tolerance dimensions concerned.

During operation, the throttling gap cokes up rather heavily and very unevenly. The degree of coking is determined by the fuel quality and by the engine's operating mode. Only about 30% of the original fuel-flow section remains free of coke.

The so-called flat-pintle injection nozzle, is a special version of the throttling pintle nozzle. The ring gap between the nozzle's spray hole and its throttling pintle is practically zero and apart from coking up less than the throttling pintle nozzles, coking is more evenly distributed (*Figure 11.60*). The pintle of this injection nozzle is provided with a ground surface which opens the fuel-flow section when the needle is lifted (*Figure 11.60, 2a and 2b*). A flow channel is then generated whose total surface, referred to as the fuel-flow section, is less whereby the self-cleaning effect is greater. The pintle's ground surface is often parallel to the nozzle needle's axis. If the ground angle is increased, the flat section of the through-flow curve rises faster and this results in a gentler transition to the fully open state. This has a positive effect on the vehicle's part-load noise and upon its derivability (*Figure 11.60*). Because temperatures at the nozzles above 220°C also result in pronounced coking, thermal protection plates and caps are available which dissipate the combustion-chamber heat away from nozzles and into the cylinder head.

Hole-type nozzles

There are a very wide variety of different nozzle-and-holder assemblies for hole-type nozzles on the market. In contrast to the throttling-pintle nozzles, the hole-type nozzles must be installed in a given position. The spray holes are at different angles in the nozzle body and must be correctly aligned with regard to the combustion chamber. The nozzle and holder assembly is therefore fastened to the cylinder head with hollow screws or claws. A special mount is used to lock the nozzle in the correct position.

The hole-type nozzles (*11.60* and *11.61*) have needle diameters of 4 mm (Size P) and between 5 and 6 mm (Size S). The seat-hole nozzle is only available as a Size P version. The nozzle

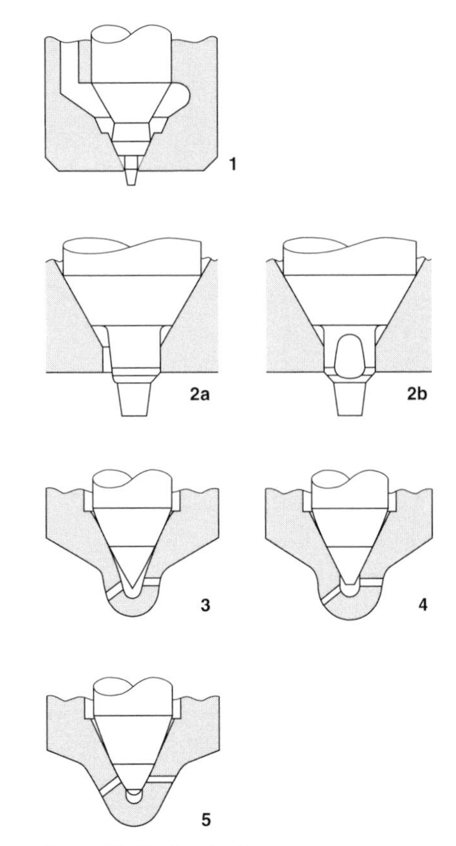

Figure 11.60 Nozzle shapes
1 Throttling pintle nozzle, 2 Throttling pintle nozzle with flat-cut pintle, 2a Side view, 2 b Front view, 3 Hole-type nozzle with conical blind hole, 4 Hole-type nozzle with cylindrical blind hole. 5 Seat-hole nozzle

pressure springs must be matched to the needle diameters and to the high opening pressures which are usually above 180 bar. The nozzle-sealing function is particularly important at the end of injection because there is a risk of the combustion gases blowing back into the nozzle and in the long run destroying it and causing hydraulic instability. Precision matching of the pressure spring and the needle diameter ensures efficient sealing. In certain cases, it may be necessary to take into account the oscillations of the pressure spring. There are three designs for the arrangement of the spray holes in the nozzle cone (*Figure 11.60*). These three designs also differ from each other with respect to the amount of fuel which remains inside the injector and which can evaporate into the combustion chamber when injection has finished. Versions with cylindrical blind hole, conical blind hole, and seat hole, have decreasing fuel quantities in this order. Furthermore, the less fuel that can evaporate from the nozzle, the lower are the engine's hydrocarbon emissions. The levels of these emissions therefore also correspond to the (nozzle) order given above.

The nozzle cone's mechanical integrity is the limiting factor in the length of the spray hole. At present, spray-hole length is 0.6–0.8 mm in the case of the cylindrical and conical blind holes. With the seat-hole nozzle, the minimum spray-hole length is 1 mm, whereby special hole-making techniques must be applied.

Developments are proceeding towards shorter hole length, because as a rule the shorter the hole, the better the engine's smoke values. In the case of hole-type nozzles, when the spray hole is bored this results in through-flow tolerances of ±3.5%. If rounding is also carried out (hydro-erosive machining), the

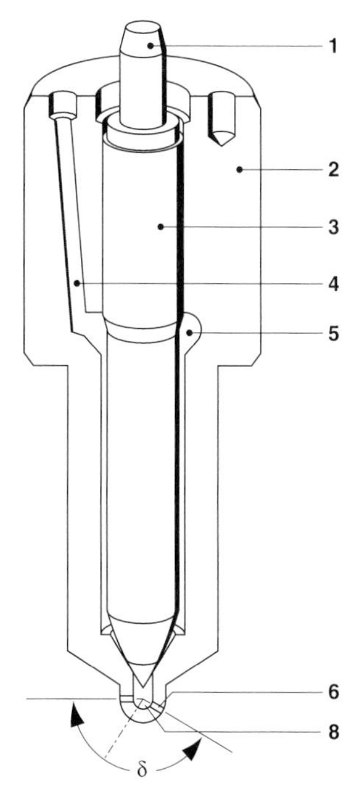

Figure 11.61 Hole-type nozzle
1 Pressure pin, 2 Nozzle body, 3 Nozzle needle, 4 Inlet passage, 5 Pressure chamber, 6 Spray hole, 7 Blind hole, δ Spray-hole cone angle.

through-flow tolerances can be reduced to ±2%. Due to the thermal stability of the materials used in the hole-type nozzle, its upper temperature limit is in the area of 270°C. Particularly difficult applications can necessitate the use of thermal protection sleeves, and cooled injection nozzles are available for large engines.

Standard nozzle holders
Figures 11.62 and *11.63* show the basic design of two nozzle-and holder assemblies each consisting of an injection nozzle and nozzle holder. The nozzle itself comprises the nozzle body and the nozzle needle which moves freely within the nozzle body's guide bore, while at the same time sealing against high injection pressures. At the combustion-chamber end, the nozzle needle has a sealing cone which is forced by the pressure spring against the nozzle body's conical sealing surface when the nozzle is closed. The sealing cone's opening angle is slightly different to that of the nozzle body's sealing cone, with the result that linear contact with high compression and good sealing properties is formed between them.

The needle-guide diameter is larger than the seat diameter. The injection pump's hydraulic pressure is applied to the differential area between the needle cross-section and the area covered by the seat. As soon as the product of sealing surface and pressure exceeds the force of the pressure spring, the nozzle opens. Since the area of nozzle which is subjected to pressure increases abruptly by the area of the seat as soon as the needle starts to lift, the injection nozzle 'snaps' open very quickly provided the injection pump's delivery rate is sufficient. The nozzle closes again only when the pressure falls below the closing pressure (which is lower than the opening pressure). In the design of injection systems, this hysteresis effect is particularly important with regard to hydraulic stability.

The opening pressure of a nozzle-and-holder assembly (approx. 110–140 bar for the throttling pintle nozzle, and 150–250 bar

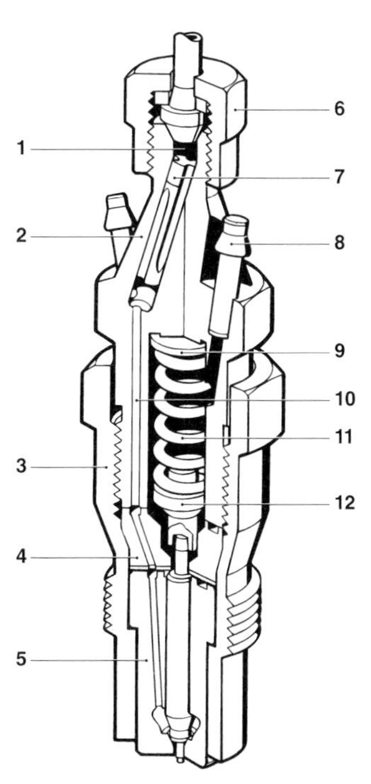

Figure 11.62 Nozzle-and-holder assembly
With throttling pin nozzle.
1 Inlet, 2 Nozzle-holder body, 3 Nozzle-retaining nut, 4 Intermediate element, 5 Injection nozzle, 6 Union nut with fuel-injection tubing, 7 Edge filter, 8 Leak-off connection, 9 Pressure-adjusting shims, 10 Pressure passage, 11 Pressure spring, 12 Pressure pin.

for the hole-type nozzle) is adjusted by inserting shims underneath the pressure spring. The closing pressure then results from the injection-nozzle geometry, that is, the ratio of needle-guide diameter to seat diameter, the so-called pressure stage.

2-spring nozzle holders
These are mainly used with direct-injection (DI) engines, in which pilot injection is the most important measure for minimizing the combustion noise. Since pilot injection ensures a relatively gentle pressure rise, engine idle is both quiet and stable, and combustion noises are reduced. The 2-spring nozzle holder (*Figure 11.64*) achieves this effect by improving the shape of the rate-of-discharge curve. The improvement is the result of the adjustment and matching of

- opening pressure 1,
- opening pressure 2,
- prestroke, and
- total needle lift.

Opening-pressure adjustment is the same as for the single-spring nozzle holder. Opening pressure 2 is given by the pretension of spring 1 together with that of spring 2 which is supported on a stop sleeve into which the prestroke dimension has been machined. When injection takes place, the nozzle needle opens at first in the prestroke range, whereby common lifts in this range are 0.03–0.06 mm. The further rise of the pressure in the nozzle causes the stop sleeve to lift and the nozzle needle can open fully.

There are also special nozzles available for the 2-spring nozzle holder in which the nozzle needle has no pintle and the needle shoulder is level with the end of the nozzle body.

To put it more simply: the springs of the 2-spring nozzle holder are so calibrated that first of all only a small amount of fuel is injected into the combustion chamber with a resulting

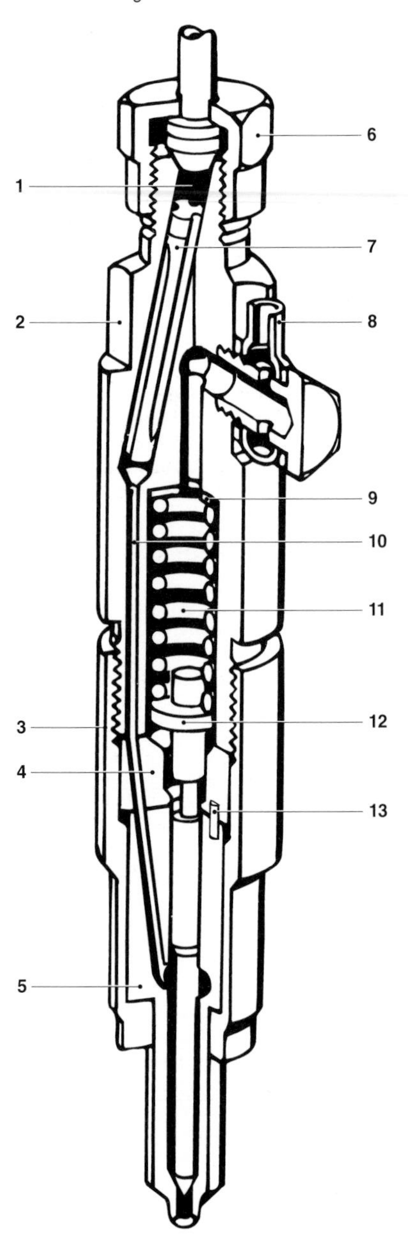

Figure 11.63 Nozzle-and-holder assembly
With hole-type nozzle.
1 Inlet, 2 Nozzle-holder body, 3 Nozzle-retaining nut, 4 Intermediate element, 5 Injection nozzle, 6 Union nut with fuel-injection tubing, 7 Edge filter, 8 Leak-off connection, 9 Pressure-adjusting shims, 10 Pressure passage, 11 Pressure spring, 12 Pressure pin, 13 Locating pins.

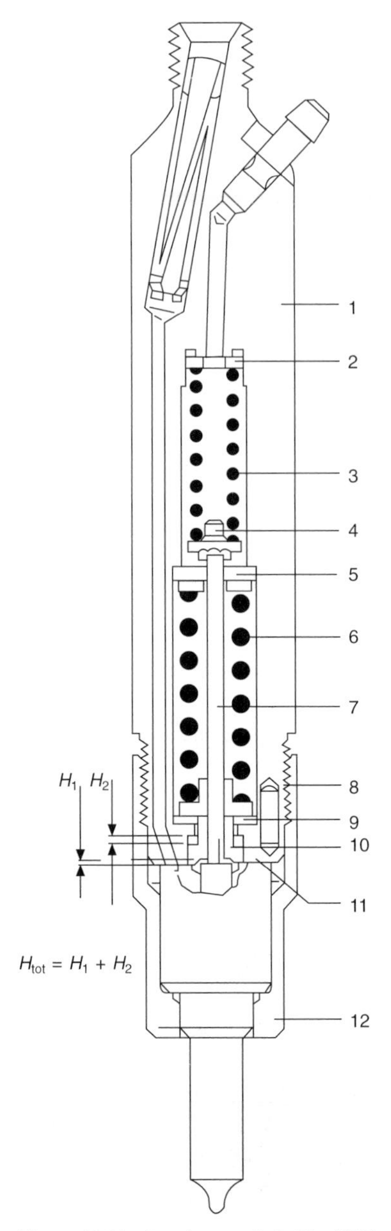

Figure 11.64 2-spring nozzle holder KBEL..P...
H_1 Pre-stroke, H_2 Main needle stroke, $H_{tot} = H_1 + H_2$ Main needle stroke.
1 Nozzle-holder body, 2 Shim, 3 Pressure spring, 4 Pressure pin, 5 Guide element, 6 Pressure spring, 7 Pressure pin, 8 Spring seat, 9 Shim, 10 Stop sleeve, 11 Intermediate element, 12 Nozzle-retainer nut.

slight rise in combustion pressure. This extends the duration of injection and together with the following increased residual fuel quantity leads to the desired softer combustion.

There are also 2-spring nozzle holders available for pre-chamber and turbulence-chamber engines. Their settings are matched to the injection system in question. The various opening pressures are around 130/180 bar and the prestroke approx. 0.1mm.

11.3.11 Bosch—Distributor injection pumps VE

11.3.11.1 Applications

Today's small, high speed diesel engine demands a lightweight and compact installation. The VE distributor pump fulfills these stipulations by combining fuel-supply pump, high pressure pump, governor, and timing device in a small, compact unit. The diesel engine's rated speed, its power output, and its configuration determine the parameters for the particular distributor pump.

Distributor pumps are used in passenger cars, commercial vehicles, agricultural and industrial tractors and stationary engines.

11.3.11.2 Subassemblies

In contrast to the in-line pump, the VE distributor pump has only *one* pump cylinder and *one* plunger regardless of the number of cylinders in the engine (*Figure 11.65*). The fuel delivery by the pump plunger is apportioned by a distributor groove to the outlet ports as determined by the engine's number of cylinders. The distributor pump's closed housing contains the following functional groups:

- high pressure pump with distributor;
- mechanical (flyweight) governor;
- hydraulic timing device;
- vane type fuel supply pump;
- shutoff device; and
- engine specific add-on modules.

Figure 11.66 shows the functional groups and their assignments. The add-on modules facilitate adaptation to the specific requirements of the diesel engines in question.

11.3.11.3 Design and construction

The distributor pump's drive shaft runs in bearings in the pump housing and drives the vane-type fuel-supply pump. The roller ring is located inside the pump at the end of the drive shaft although it is not connected to it. A rotating-reciprocating movement is imparted to the distributor plunger by way of the cam plate—which is driven by the input shaft and rides on the rollers of the roller ring. The plunger moves inside the distributor head which is bolted to the pump housing. Installed in the distributor head are the electrical fuel shutoff device, the screw plug with vent screw, and the delivery valves with their holders. If the distributor pump is also equipped with a mechanical fuel shutoff device this is mounted in the governor cover.

The governor assembly comprising the flyweights and the control sleeve is driven by the drive shaft (gear with rubber damper) via a gear pair. The governor linkage mechanism which consists of the control, starting, and tensioning levers, can pivot in the housing. The governor shifts the position of the control collar on the pump plunger. On the governor mechanism's top side is the governor spring which engages with the external control lever through the control-lever shaft which is held in bearings in the governor cover. The control lever is used to control pump function. The governor cover forms the top of the distributor pump, and also contains the full-load adjusting screw, the overflow restriction or the overflow valve, and the engine-speed adjusting screw. The hydraulic injection timing device is located at the bottom of the pump at right angles to the pump's longitudinal axis. Its operation is influenced by the pump's internal pressure which in turn is defined by the vane-type fuel-supply pump and by the pressure-regulating valve. The timing device is closed off by a cover on each side of the pump (*Figure 11.66*).

11.4 Diesel fuel injection systems–Caterpillar Inc.

The 'common rail' fuel system has been reintroduced into the diesel engine industry. What characterizes the common rail fuel system from current, more popular fuel systems is that all of the injectors are supplied by a common, pressurized fuel supply line, or manifold, known as the rail. This separates the fuel delivery system and its operation from being solely dependent on engine load and speed. In general, common rail fuel systems are electronically controlled and capable of managing all aspects of the injection process.

A fuel system based on the common rail principle was originally introduced by Vickers in 1913 for Atlas Imperial Diesel Company.

Figure 11.65 VE distributor pump fitted to a 4-cylinder diesel engine

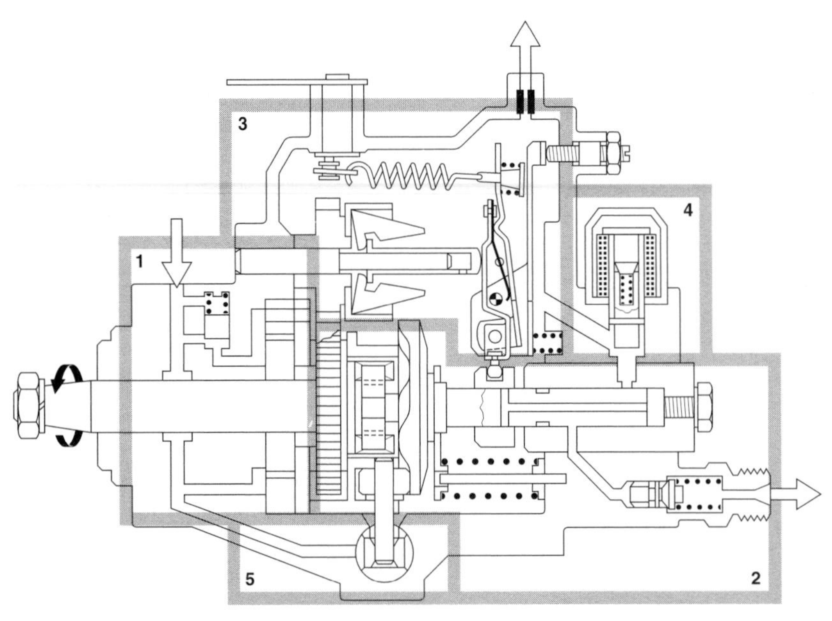

Figure 11.66 VE distributor pump. The subassemblies and their functions
1 Vane-type fuel supply pump with pressure regulation valve: Draws in fuel and generates pressure inside the pump.
2 High pressure pump with distributor: Generates injection pressure, delivers and distributes fuel.
3 Mechanical (flyweight) governor: Controls the pump speed and varies the delivery quantity within the control range.
4 Electromagnetic fuel shutoff valve: Interrupts fuel supply.
5 Timing device: Adjusts the start of delivery (port closing) as a function of the pump speed and in part as a function of the load.

This system employed a multi-cylinder pump and an accumulator operating at an approximate pressure of 350 bar with mechanically controlled injector nozzles. It is interesting to note that electronically controlled injector nozzles were also introduced by the Atlas Imperial Diesel Company. The Vickers common rail system with its accumulator as part of the fuel manifold was an attempt to increase the injection pressure by the expansion of the compressed fuel in the accumulator volume.

The driving motivation behind common rail is adherence to ever-increasing emission regulations while maintaining, if not improving upon performance, economy, durability, reliability, cost and sociability. The single most influential system on today's diesel engine, which also possesses the capability of meeting these stringent emission regulations, is the fuel injection system. Many improvements have been made to conventional fuel injection systems in an effort to meet these requirements. However, the more recently introduced common rail fuel injection systems have demonstrated their ability to meet or exceed emission and manufacturing requirements.

Today, there are several different designs based upon the common rail theory: however, fewer than a handful are in full production.

11.4.1 Caterpillar's Hydraulically-actuated Electronic Unit Injector (HEUI) fuel system

Caterpillar introduced a dual line common rail fuel system in 1993. Unlike most conventional fuel systems, the HEUI injector's actuation is accomplished through hydraulics (thus the dual line), not the engine camshaft. This permits the HEUI injector to operate independent of engine speed.

The HEUI injector sends fuel directly into the cylinder under high pressure. Because this pressure is created hydraulically, the fuel remains at low pressure until its injection. The hydraulic oil supply pump provides the muscle to power the fuel injection. Upon receiving a signal from the Electronic Control Module (ECM), the injector poppet valve opens to allow high pressure oil to enter the injector and start injection. This is unlike the traditional common rail system, where the entire fuel line remains under high pressure in order to power fuel injection.

Inherent features of the HEUI fuel injection system include fuel injection pressure control independent of engine load or speed, flexible injection timing, and full electronic control of fuel injection parameters.

The HEUI fuel injection system consists of six main components. They are as follows:

- High pressure oil pump
- Rail pressure control valve (RPCV)
- Hydraulically actuated Electronic Unit Injector (HEUI)
- Sensors
- Electronic control module (ECM)
- Fuel transfer pump

Figure 11.67 illustrates a typical HEUI fuel system. The ECM contains information defining optimum fuel system operating parameters and controls key system components. Multiple sensor signals, such as engine speed, timing, rail and boost pressures are transmitted to the ECM to identify the engine's current operating conditions. The ECM uses these input signals to control the operation of the fuel system.

The high pressure oil pump is a gear driven, axial piston pump. The rail pressure control valve (RPCV) electronically controls the output pressure of the oil supply pump.

As can be seen, the HEUI system incorporates two fluid circuits–fuel and oil. The fuel circuit provides fuel to the injectors. The fuel transfer pump draws fuel from the fuel tank and delivers it through a filter to the injectors via a passage located in the cylinder head; thus, the common rail design. Typical system pressure is approximately 450 kPa and is controlled by a regulating valve. Unused fuel is returned to the fuel tank. Typically one part fuel is returned to the tank for every two parts used.

The oil circuit consists of a low pressure section and a high pressure section. The low pressure section typically operates at a pressure of 3 bar. Its function is to provide filtered oil to the high pressure oil supply pump as well as meet the demands of the engine's lubrication system. Lubrication oil is drawn from the engine oil sump and supplied through the oil cooler and filter to both the engine and the high pressure supply pump.

Two types of high pressure supply pumps are currently in production–fixed and variable displacement axial pumps. These are adaptations of axial piston pump technology commonly found on today's vehicular hydraulic systems. The fixed displacement pump incorporates a 7-piston rotating group that operates at a pressure range from 35 to 230 bar. Output pressure is controlled by the RPCV that directs excess flow back to the return circuit. The variable displacement pump features a 9-piston rotating group and a pivoting yoke for variable displacement control. Flow control is accomplished by a closed-loop electro-hydraulic system that pivots the yoke from a minimum angle of 0 degrees to a maximum angle of 15.5 degrees and a RPCV. Depending on the engine's rating, operating conditions and the engine's mapping characteristics, output pressure is controlled between 80 and 230 bar.

The oil delivered from the supply pump provides actuation oil to the injector via the high pressure circuit. This high pressure oil flows through lines into an oil manifold either integral to the head or in a rail located near the injectors. The manifold stores the oil at actuation pressure ready for injection operation. Oil is discharged from the injector under the valve cover so that no return lines are required.

The ECM controls the pressure in the high pressure oil circuit and thus injection pressure, Operational maps stored in the module's memory identify the optimum rail pressure for best engine performance.

A HEUI injector consists of three main subassemblies as illustrated in *Figure 11.68*. These are the poppet valve, intensifier plunger and barrel and nozzle.

The poppet valve's purpose is to initiate and finalize the injection process. It is composed of a poppet valve, armature and solenoid. High pressure actuating oil is supplied to the valve's lower seat at all times. To begin injection, the solenoid is energized by the ECM, moving the poppet valve off the lower seat to the upper seat. This action admits high pressure actuation oil to the spring cavity and the passage to the intensifier, Injection continues until the solenoid is de-energized and the poppet valve moves from the upper seat back to the lower seat. Actuation oil and fuel pressures decrease as spent oil is ejected from the injector through the open upper seat to the valve cover area.

The middle segment of the injector consists of the hydraulic intensifier piston, the plunger and barrel, and the plunger return spring. Pressure intensification of the fuel to the desired injection levels is accomplished by the ratio of areas between the intensifier piston and the plunger. The intensification ratio can be tailored to achieve desired injection characteristics. Injection begins as high pressure actuating oil is supplied to the top of the intensifier piston. As the piston and plunger move downward, the pressure of the fuel below the plunger rises. The piston continues to move downward until the solenoid is de-energized causing the poppet to return to the lower seat, blocking oil flow. The plunger return spring returns the piston and plunger to their initial positions. As the plunger returns, it draws replenishing fuel into the plunger chamber across a springless ball check-valve.

The nozzle is typical of other diesel fuel system nozzles; although, a mini-sac version of the tip is also available. Fuel is supplied to the nozzle through internal passages. As fuel pressure increases, the nozzle check lifts from the lower seat allowing injection to occur. As pressure decreases at the end of injection, the spring returns the needle to its lower seat.

The HEUI fuel injection system provides new and unique injection characteristics that were not achievable with typical mechanically driven fuel injection systems. Chief among these is injection pressure control over the engine's entire operating range. Typical mechanical injection systems increase injection pressure proportionally with engine speed in a linear fashion; however, HEUI is electronically controlled and completely independent of engine speed. The ability for independent electronic control of injection pressure has proven advantages in smoke and particulate reduction and in greatly improved low speed engine response. System pressure response is quite fast: 300 to 1200 bar in 30 ms (1 to 2 engine revolutions).

Another benefit of a time-based injection system is the ability of tailoring the injection duration for any engine operating speed. Injection duration with the HEUI system decreases proportionally with engine speed and can also be controlled through electronic control of the oil supply pressure, or rail pressure. This characteristic is beneficial in improving engine performance and fuel consumption.

In addition to injection pressure and duration control, the time based HEUI fuel injection system has the flexibility of controlling injection timing. Timing can be optimized without concern for cam profile limitations. This flexibility provides proven advantages in lower emissions, reduced noise, reduced smoke, improved hot and cold starting, white smoke cleanup, and high altitude operation.

The HEUI fuel injection system also has rate shaping capability. Rate shaping is the process of tailoring the initial portion of fuel delivery to control the amount of fuel delivered during ignition delay and main injection. This process modifies the heat release characteristics and is beneficial in achieving low emissions and noise levels. Although the effects of rate shaping are documented,

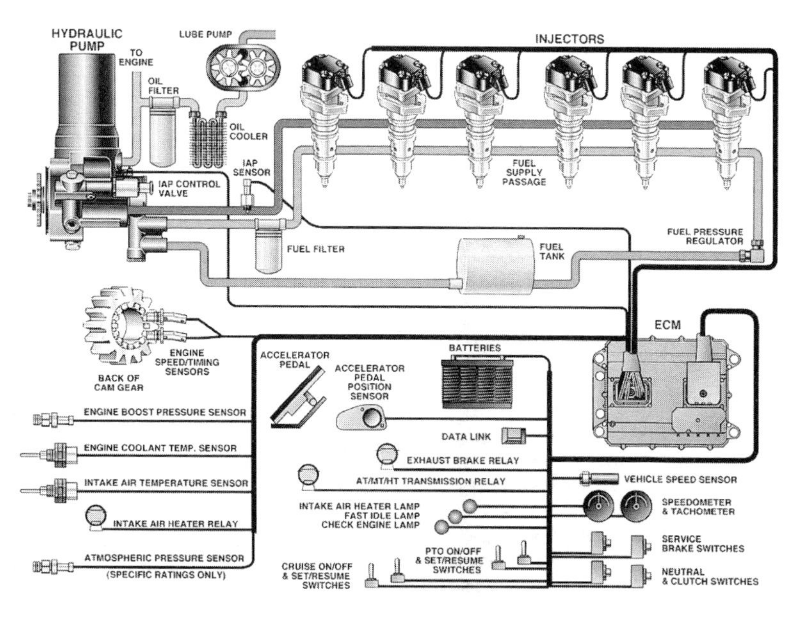

Figure 11.67 HEUI fuel system schematic

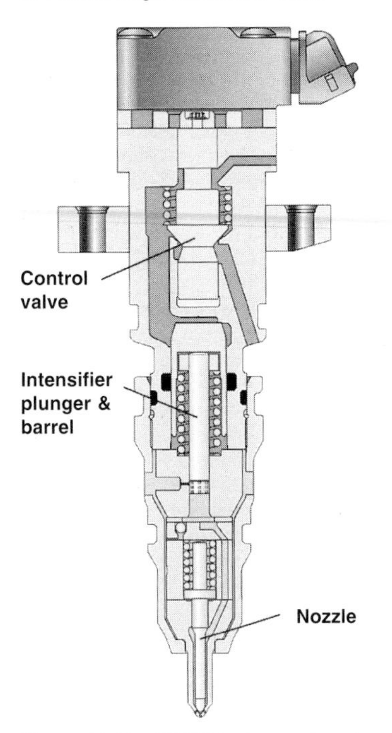

Figure 11.68 HEUI injector—main subassemblies

the tradeoffs with overall engine performance are not well understood. This is due in part to the fact that conventional fuel injection system rate shaping characteristics are limited in operating range; they are dependent on engine load and speed. The rate shaping developed with the HEUI fuel injection system operates over the entire load and speed range, and can be readily tailored to the engine combustion system.

HEUI's rate shaping device, called pre-injection metering (PRIME), is a precision-ported spill-control device in the plunger and barrel that adds no additional parts. In combination with the independent injection pressure control of the HEUI system, engine performance can be optimized by varying the idle and light load rate characteristics, independent of rated and high load conditions. The resulting benefits to performance, noise, and emissions generally will depend on the engine and the objectives of the manufacturer.

11.4.2 Next generation: HEUI-B

Increasingly stringent global regulations, whether they are noise, particulate, or gaseous emissions, are driving the diesel engine designs of the next generation. Caterpillar has addressed the global challenges with a next-generation fuel system appropriately called HEUI-B (see *Figure 11.69*). Based on its predecessor's technology, the HEUI-B fuel injection system raises the bar of innovation several notches above the standard set by HEUI.

Both HEUI and HEUI-B fuel injection systems use a single solenoid with two wires. The HEUI-B injector has a unique valve concept where the single solenoid controls the oil pressure to the intensifier (as in the HEUI system) and the oil pressure to control the nozzle check. This is accomplished by precise hydraulic and event control of two valves, one acting as a pilot to the other. Also it should be pointed out that discrete valve events control the injection process. Neither valve depends on any sort of modulation during its travel to accomplish injection control. With the direct control of the nozzle check for both opening and closing, the various injection rate shapes are realized by control of opening and closing independently.

Figure 11.69 Next generator. HEUI B—outside view

References

1 AUSTEN, A. E. W., and LYN, W-T. 'Relation between fuel injection and heat release in a direct injection engine and the nature of the combustion process', *I Mech E symposium 'Four Papers on Diesel Engine Fuel Injection, Combustion and Noise'* Proc. Auto. Div. I Mech E London 1960–61 No. 1 pp. 47–62 (1960)

2 KNIGHT, B. A., 'Fuel-injection system calculations', *I Mech E symposium 'Four Papers on Diesel Engine Fuel Injection, Combustion and Noise'* Proc. Auto. Div. I Mech E London 1960–61 No. 1 pp. 25–33 (1960)

3 HIROYASU, H. and ARAI, M., 'Structures of fuel sprays in diesel engines', *SAE paper No. 900475 Soc. Auto. Eng. Congress*, Detroit (1990)

4 DENT, J. C. and MEHTA, P. S., 'A phenomenological combustion model for a quiescent chamber diesel engine', *SAE paper No. 811235 Soc. Auto. Eng. Congress*, Detroit (1981)

5 RAMOS, J. I., 'Internal combustion engine modelling', Hemisphere, Washington (1989)

6 KHAN, I. M., GREEVES, G. and WANG, C. H. T. 'Factors affecting smoke and gaseous emissions from direct injection engines and a method of calculation', *SAE paper No. 730169 Soc. Auto. Engrs. Congress*, Detroit (1973).

7 DOYLE, D. M., NEEDHAM, J. R., FAULKNER, S. A., and FREESE, R. G., 'Application of an advanced inline injection system to a heavy duty diesel engine', *SAE paper No. 891847 Soc. Auto. Eng. Int. Off-Highway and powerplant Congress*, Milwaukee Sept. (1989)

8 HERZOG, P. 'The ideal rate of injection for swirl-supported HSDI diesel

engines' *Proc. I Mech E seminar 'Diesel Fuel Injection Systems'*, Shirley, Birmingham Oct. 10–11 1989 ISBN 0 85298 708 0 (1989)

9 DEC, J. E., 'A conceptual model of diesel combustion based on laser-sheet imaging', *SAE paper No. 970873, Soc. Auto. Engrs. Congress,* Detroit (1997)

10 GREEVES, G. 'Response of diesel combustion systems to increase of fuel injection rate', *SAE paper 790037 Soc. Auto. Engrs. Congress* Detroit Feb 26–Mar 2 (1979)

11 GREEVES, G. and TULLIS, S., 'Contribution of EUI 200 and quiescent combustion towards US 1994 emissions', *SAE paper 930274 Soc. Auto. Engrs. Int. Congress,* Detroit (1993)

12 RAO, K. K., WINTERBONE, D. E., and CLOUGH, E., 'Laser illuminated photographic studies of the spray and combustion phenomenon a small high speed DI diesel engine', *SAE paper No. 922203, Soc. Auto. Engrs* (1993)

13 SHIOZAKI, T., NAKAJIMA, H., YOKOTA, H. and MIYASHITA, A. 'The visualisation and its analysis of combustion flame in a diesel engine', *SAE paper No. 980141* (1998)

14 SOTERIOU, C., ANDREWS, R., and SMITH, M 'Direct injection sprays and the effect of cavitation and hydraulic flip on atomisation', *SAE paper 950080 Soc. Auto. Engrs. Congress,* Detroit (1995)

15 ARCOUMANIS, C., GERVAISES, M. and NOURI, J. M., 'Effect of fuel injection processes on diesel injection sprays', *SAE paper No. 970080 Soc. Auto. Eng. Congress, Detroit* Feb. (1997)

16 MEURER, J. S., 'Evaluation of reaction kinetics eliminates diesel knock', *SAE Trans.* **64** (1956)

17 RUSSELL M. F., NICOL, S. and YOUNG, C. D., 'Rate modulation for direct injection diesel engines', *Proc. I Mech E seminar 'Diesel Fuel Injection Systems',* Shirley, Birmingham Oct. 1989 ISBN 0 85298 7080, pp. 79–86 (1989)

18 SUGIHARA, K. MATSUI, Y. and 'Origins of hydrocarbons in small direct injection diesel engines', *SAE paper No. 851213*

19 GREEVES, G KHAN, I. M., WANG, C. H. T. and FENNE, I., 'Origins of hydrocarbon emissions from diesel engines', *SAE paper No. 770259 Soc. Auto. Engrs. Congress,* Detroit, Feb. 28–Mar 4 (1977)

20 AUSTEN, A. E. W., and PRIEDE T. 'Origins of diesel engine noise', *Proc Symposium on Engine Noise and Noise Suppression,* London 24 Oct. pp. 19–32 *Proc. I. Mech. E.* **173,** 19 (1958)

21 RUSSELL, M. F., and YOUNG, C. D., 'Measurement of Diesel combustion noise', *Proc I Mech. E. Autotech 85 Seminar* 2S 326 (1985)

22 RUSSELL, M. F., 'The dependence of diesel combustion on injection rate' *Proc I Mech. E seminar Future engine and fuel injection technology: the Euro IV challenge,* Dec. 3–4 London (1997)

23 RUSSELL, M. F., and LEE, H. K., 'Modelling diesel fuel injection equipment to control combustion noise', *paper No. C487/27 Proc. I Mech E conference vehicle NVH and Refinement,* Birmingham May 2–4 ISBN 085 298 9059 pp. 19–30 (1993)

24 RUSSELL, M. F., and LEE, H. K. 'Modelling injection rate control devices', *Proc I Mech. E Conference Diesel Fuel Injection Systems,* Sept. 28–29, pp. 115–132 (1995)

25 FELTON, G. N., 'High pressure rotary spill pump with electronic control', *Proc I Mech E Conference Diesel Fuel Injection Systems,* Sept. 28–29, London (1995)

26 RUSSELL, M. F., 'Reducing noise emissions from diesel engine surfaces', *SAE paper 720135 SAE Int. Auto. Eng. Congress,* Detroit Jan. (1972)

27 RUSSELL, M. F., 'Automotive diesel engine noise and its control', *SAE paper 730243 SAE Int. Auto. Eng. Congress,* Detroit Jan. (1973) and *SAE Transactions,* **82** (1973)

28 RUSSELL, M. F., 'Recent C. A. V. research into noise, emissions and fuel economy of diesel engines', *SAE paper No. 770257 Soc Auto. Eng Congress,* Detroit Feb. (1977), and SAE book PT-79/17 ISBN 0 89883 105 9 (1979)

29 HEAD, H. E., and WAKE, J. D., 'Noise of diesel engines under transient conditions', *SAE paper No. 800404 Soc. Auto. Engrs. Congress,* Detroit Feb. (1980)

30 RUSSELL, M. F., 'Diesel engine noise: control at source', *SAE paper No. 820238 Soc. Auto. Engrs. Int. Congress,* Detroit Feb. (1982)

31 RUSSELL, M. F., PALMER, D. C., and YOUNG, C. D., 'Measurement of noise at source with a view to its control', *paper C290/35 Proc I Mech. E. Conference,* London, June (1984)

32 RUSSELL, M. F., and HAWORTH, R., 'Combustion noise from high speed direct injection diesel engines' *Soc. Auto. Engrs. SAE paper 850973 in P-161 Proc Surface Vehicle Noise and Vibration Conference* in Traverse City, May (1985)

33 RUSSELL, M. F., 'Effect of fuel ignition quality on diesel combustion noise', *Proc. I Mech. E. Auto. Div. seminar Current Operational Problems wih Diesel Fuels and Future Prospects,* 14 June (1985)

34 GUERRASSI, N. and DUPRAZ, P., 'A common rail injection system for high speed direct injection diesel engines', *SAE paper 980803 in SP - 1316 Soc. Auto. Engrs. Congress,* Detroit Feb. (1998)

35 HALES, B. and RUSSELL, M. F., 'Effect of fuel quality on diesel pump performance', *paper S496/008/97 Proc. I Mech E seminar Automotive Fuels for the 21st Century,* London Jan. 21 pp. 119–139 (1997)

36 SMITH, M. and SOTERIOU, C., 'From concept to end product—computer simulation in the development of the EUI 200', *SAE paper No. 960866 Soc. Auto. Engrs. Congress,* Detroit Feb. (1996)

12 Lubrication and lubricating oils

Contents

12.1 Introduction

When one solid surfce is made to slide over another the motion is resisted by friction between the surfaces, and heat is generated to an extent depending on the speed and the load. Extreme examples are the primitive method of producing fire by rubbing together two pieces of wood, and the overheating and eventual seizure of a dry metal bearing.

Friction can be reduced, motion facilitated, and overheating prevented by interposing a fluid film between the two surfaces; the fluid is a *lubricant* and its application provides *lubrication*. This is true whatever the solid—wood, plastic, or metal—and whatever the fluid, e.g. water, oil, air or other gas. This chapter, however, dealing with the lubrication of diesel engines, is concerned primarily with metal–metal contacts and mineral oil lubricants.

A metal surface, however well machined and however smooth it may appear to be, is not so smooth in microscopic or molecular dimensions; it presents an irregular pattern of asperities and intervening troughs. When two such surfaces come together, the contact is not over the whole area, but is only where asperities meet. Under load asperities deform and adhere together. If sliding takes place, force is required to overcome the adhesion, and this force is equal to the sliding friction. The mechanical energy supplied will be converted to heat, and wear will occur whenever the rupture of an asperity adhesion results in transfer of material from one surface to the other, or formation of a separate particle.

By applying a fluid film between the surfaces they are kept apart, provided the thickness of the film is greater than the height of the asperities and the load is insufficient to disrupt the film. Resistance to motion is then due solely to resistance within the film which, by definition, is the viscosity of the fluid. Friction is low, heat generation and wear are negligible.

This is the ideal state of lubrication, and is known as 'full-film lubrication' or 'fluid-film lubrication'. The film may be provided and maintained by supplying the lubricant under sufficient pressure to withstand the load ('hydrostatic lubrication') or, more usually, by the lubricant being drawn in between the surfaces as a result of the design of the system to form a wedge-shaped film in which sufficient pressure is generated to support the load ('hydrodynamic lubrication').

In recent years it has been realized that at high pressures two other factors, distortion of the metal and increase in oil viscosity have to be taken into account to explain the maintenance of lubrication under high loads that would be precluded by conventional hydrodynamic theory. These phenomena may occur under high pressure, resulting in an increase in film thickness and load-bearing capacity. Lubrication under such conditions is designated 'elastohydrodynamic lubrication'.

Under high loads the film may be so reduced in thickness that it is easily ruptured, and contact between asperities may occur. Resistance to motion (friction) will then be due no longer solely to the viscosity of the lubricant, but to deformation of asperities and rupture of welds, with consequent increase in friction and onset of wear. Effective lubrication can still be maintained if the lubricant contains polar molecules that adhere to or are adsorbed onto the metal surfaces to form a film that shears more readily than does the metal substrate. Resistance to motion is then due to shearing of the adsorbed layer rather than to deformation of asperities. Lubrication under these conditions is called *boundary lubrication*. Mineral oils are deficient in polar molecules and are therefore poor boundary lubricants unless reinforced by the incorporation of suitable additives.

At still higher loads boundary conditions may be surpassed and the adsorbed film of polar material will no longer prevent metal–metal contact, high friction, and wear. Again lubrication may be maintained by the use of additives—extreme pressure (EP) additives—that contain compounds of sulphur, chlorine or phosphorus which do not readily react with metals at moderate temperatures but do so at the hot spots between asperities to form films that shear more readily than does the metal.

In a diesel engine full-film lubrication exists between the liners and the rings and pistons, except at the extremities of piston travel, but not until the engine has been 'run in'. During running-in the major asperities are worn off the surfaces of the liner and rings and the rings acquire a curved profile which promotes hydrodynamic lubrication. This running-in process, which may be accelerated by the use of a special running-in lubricant, gives a controlled but not too fine surface finish to liner and rings, and is the best method of obtaining a good working fit between them.

Bearings and journals do not normally need running-in; the accurate finish of bearing shells, crankpins and journals achieved by modern machining techniques makes it unnecessary.

In the lubrication of diesel engines somewhat different conditions exist as between high- and medium-speed trunk-piston engines, in which the crankcase oil lubricates both the cylinders and the bearings, and low-speed crosshead engines in which the cylinder and crankcase systems are separated. As will be discussed later (Section 12.6.2), crosshead engines require a special type of cylinder oil.

12.2 Lubricating oils

The primary function of a lubricating oil is to provide a continuous fluid film between surfaces in relative motion so as to reduce friction and prevent wear. Secondary, but nevertheless important, functions are to cool the working parts, to protect metal surfaces against corrosion, to flush away or prevent ingress of contaminants and to keep the engine reasonably free from deposits.

Early lubricants were animal and vegetable oils (fatty oils) which satisfactorily fulfilled the lubrication requirements of contemporary engines and machinery. As conditions became more severe and as loads and speeds increased, these oils became less suitable, because, although excellent lubricants, they are readily oxidized, especially at high temperatures, resulting in increase in viscosity and the formation of acids and of resinous or gummy deposits. The advent of petroleum provided an alternative—mineral oils—whose chemical and physical properties were more suitable to the new requirements.

12.2.1 Mineral oils

Lubricating oils are now manufactured mainly from petroleum—crude oil—which consists of a mixture of hydrocarbons (compounds of carbon and hydrogen) together with small amounts of compounds of sulphur, nitrogen, oxygen and metals such as vanadium, nickel and iron. The hydrocarbons vary in chemical properties according to their molecular structure, and in physical properties according to their molecular size, ranging from gas, through liquid, to soild. Crude oils vary in properties according to their relative contents of paraffinic, naphthenic (cycloparaffinic) and aromatic hydrocarbons; and the properties of the products manufactured from them similarly depend to some extent on the type of crude, although modem methods of refining permit considerable flexibility in the selection of the crude for the manufacture of a particular type of product.

In the refinery crude oil is split by distillation into a limited number of fractions that are further refined to produce a range of fuels, lubricants, bitumens and waxes. The crude is first distilled under atmospheric pressure, whereby components boiling below 370°C are driven off, and are separated into arbitrary fractions of progressively higher boiling range that are the bases of commercial products—petroleum gases, petroleum spirits, gasoline, kerosine, gas oil, diesel fuel and heating oil. The residue from this distillation ('long residue' or 'topped crude') may be used as residual fuel or it may be distilled under vacuum to yield a number of distillates from which heavy fuels, distillate lubricating oils and paraffin waxes are manufactured. The residue from the vacuum distillation ('short residue') yields residual fuels, residual lubricating oils ('cylinder stock' or 'bright stock') microwaxes and bitumens according to the type of crude. The borderline between successive primary fractions is arbitrary, depending on commercial and technical requirements and not on chemical constitution; except for gases and some lower liquid hydrocarbons, no attempt is made to isolate individual chemical compounds and the fractions consist of mixtures of hydrocarbons. (See also Chapter 4).

The primary fractions produced by distillation are further refined by a variety of physical and chemical processes to obtain commercial products suitable for specific applications.

At one time the lubricating oil fractions were refined by treating with sulphuric acid to remove unstable and unsuitable compounds, and then with clay (earth) to still further improve stability and colour (acid/earth treatment). Nowadays most lubricating oils are refined by solvent extraction whereby unsuitable compounds are selectively dissolved and removed. The solvent-extracted oil is then dewaxed by solvent treatment, refrigeration and filtration, and the dewaxed oil is finally clarified by treatment with earth or hydrogen.

Treatment with hydrogen in the presence of a catalyst (hydroprocessing) is increasingly used to refine lubricating oils. It can be carried out with varying degrees of severity according to the temperature and pressure used. Hydrofining is a mild teatment used instead of acid/earth of improve stability and colour; hydrotreating is a more severe treatment to replace solvent extraction, either wholly or in part, and to improve stability and viscosity/temperature properties.

Only a limited number of refined lubricating oils are manufactured in the refinery and these are then blended in various proportions, together with additive (see Section 12.4), to produce the many different grades that are required to meet the diverse needs of engines and machinery.

12.2.2 Synthetic oils

A number of synthetic chemicals, some derived from petroleum, are used as lubricants but on account of their high cost and certain technical difficulties—such as incompatibility with the normal types of oil seals—they are not widely used except for some special industrial applications. At present relatively small quantities are used in diesel engine lubricants, but the proportion is increasing as requirements become more demanding. The main types used in diesel lubricants are synthetic hydrocarbons (polyalphaolefines, polybutenes and alkylated aromatics) and organic esters (di-esters and polyol esters).

Some synthetic lubricants are compatible with mineral oils and may be blended with them to obtain desired properties. They have the advantage of high viscosity index, in excess of 120 (see Section 12.3), without degradation by mechanical shearing.

12.3 Viscosity—its significance in lubrication

Viscosity is the most important property of a lubricating oil. It is a measure of the resistance to flow; a thin oil, an oil of high fluidity, has a low viscosity; a thick oil, a viscous oil, has a high viscosity. Simple liquids, known as Newtonian liquids, cannot withstand the slightest shearing stress, and relative motion between layers of liquid, i.e. flow, continues as long as stress is applied. This leads to the scientific definition of viscosity as the shearing stress divided by the velocity gradient produced within the fluid. Most lubricating oils are Newtonian liquids provided the temperature is not so low as to cause separation of wax crystals and provided that no high-molecular-weight additives such as viscosity-index-improvers are present. Viscosity is markedly affected by temperature, and all mineral oils decrease in viscosity more or less rapidly with rise in temperature (see page 311).

12.3.1 Viscosity and coefficient of friction

The coefficient of friction between lubricated surfaces depends on the dimensionless parameter ZN/P where Z = dynamic viscosity; N = relative sliding velocity of the surfaces; and P = pressure between the surfaces. *Figure 12.1* shows the typical relationship between the coefficient of friction and ZN/P. Above a limiting value of ZN/P the coefficient of friction increases in direct proportion to it, and fluid-film lubrication prevails, a condition that applies at most lubricated surfaces in an engine. At values of ZN/P below this limiting value the coefficient again increases and the lubrication, designated mixed and then 'boundary', depends on properties of the lubricant other than viscosity. Two oils with identical fluid-film behaviour may have very different coefficients of friction in the the boundary regime, depending on the quality of the oil and the efficiency of the

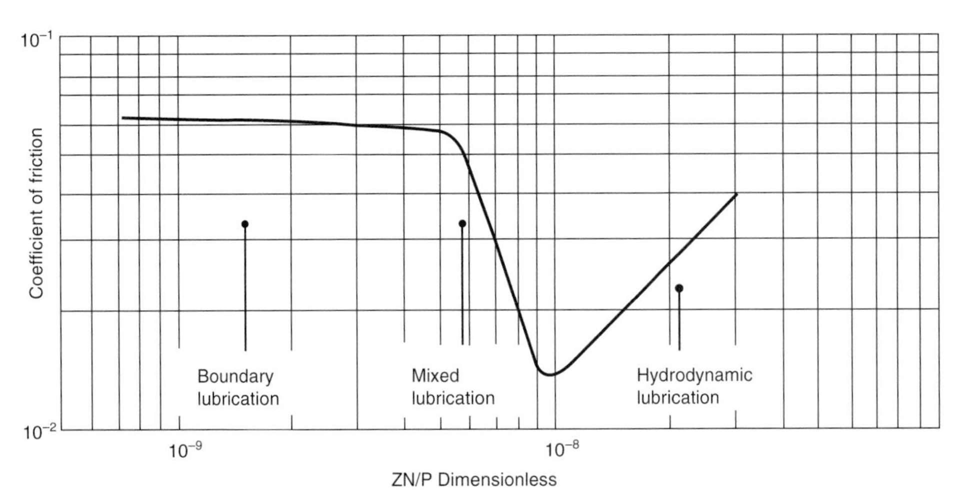

Figure 12.1 Effect of lubricating conditions on coefficient of friction

additives it contains to improve performance under boundary conditions.

In the mixed regime, metal-to-metal contact through the oil film may occur, with an increase in wear as ZN/P decreases. Eventually, in the boundary regime, the loads are carried entirely on solid–solid contacts. If lubrication is inadequate in the mixed or boundary regimes, frictional heating can become excessive and scuffing or seizure can occur. Whether or not lubrication breaks down depends not only on the properties of the lubricant but also on the nature of the surfaces. Top-ring/liner lubrication is close to this condition when running-in.

The way in which viscosity operates in fluid-film lubrication is illustrated by the behaviour of a plain journal bearing. As the journal starts to rotate it tends to roll forward on the bearing, but as the motion continues oil is dragged down between the journal and the bearing at a rate that depends on its viscosity. A wedge shaped oil film is established beneath the journal, and the pressure developed causes the shaft to lift and float freely on a cushion of oil. There is no contact between the metal surfaces, and wear is negligible. If the speed is too low or the load is too high, the wedge may become too thin to be effective, boundary conditions are established, and friction and wear increase. The formation of a wedge-shaped oil film is therefore most desirable on all lubricated surfaces in an engine, and is achieved by correct design.

12.3.2 Viscosity measurement and units

The dynamic viscosity of a fluid is the tangential force on unit area of either of two parallel plates at unit distance apart when the space between them is filled with the fluid and one plate moves relatively to the other with unit velocity in its own plane. The unit of dynamic viscosity most in use is the poise (P). For most practical purposes it is more convenient to use the kinematic viscosity—the dynamic viscosity divided by the density—the unit of which is the Stoke (St). These two units are large in relation to the viscosity of lubricating oils and it is customary to use the one-hundredth part of each unit, namely the centipoise (cP) and the centistoke (cSt). The SI unit of dynamic viscosity is the pascal second (Pa s) and 1 cP = 1 mPa s.

The kinematic viscosity is determined by measuring the time taken for a given volume of oil to flow through a standard capillary tube under controlled conditions of temperature and pressure, and multiplying this time by a factor determined by calibrating the tube against water, the viscosity of which is accurately known.

In such a *viscometer,* as the apparatus is called, the head of oil is fixed and the pressure at the capillary is therefore proportional to the density of the oil. Thus for a given dynamic viscosity oils of high density flow faster through the capillary than do oils of low density, but oils of the same kinematic viscosity flow through the capillary at the same rate.

The dynamic viscosity is established by determining the kinematic viscosity as described above and multiplying it by the density, except in the case of determination at low temperatures in the ASTM Cold Cranking Simulator (see Section 16.3.4). The dynamic viscosity is nearly always the appropriate one in lubrication calculations but the kinematic viscosity is the one most commonly used in characterizing oils and in specifications. Because the viscosity of a lubricating oil changes significantly with change in temperature the specification of viscosity must always include the temperature.

Before the determination and use of kinematic viscosity became almost universal, as they now are, it was customary to determine and specify viscosity in arbitrary units in terms of the time taken for a given volume of oil to flow at a given temperature from a standared orifice in the bottom of a jacketed metal cup. Several such viscometers were used, the Redwood I and II (British), the Saybolt Universal and Furol (American) and the Engler (European). The expression of viscosity in this way was so deeply rooted in the oil industry that oils continued to be so characterized long after mesurement of kinematic viscosity became standard laboratory practice, the kinematic viscosity being converted to arbitrary units by the use of conversion tables. Nowadays the use of kinematic or dynamic viscosity is almost universal in lubricating oil specifications although the old arbitrary units may still be encountered. Another modern development is the general use of 40° and 100°C instead of the erstwhile 70°, 100° and 210°F.

12.3.3 Change in viscosity with temperature and pressure

The viscosity of lubricating oils decreases with rise in temperature and vice versa. The rate of change with temperature depends on the chemical nature of the oil; the viscosity of aromatic and naphthenic oils changes more rapidly than that of paraffinic oils. The rate of change is customarily indicated by the viscosity index (VI), an empirical number calculated from equations involving the determined kinematic viscosities at 40° and 100°C (previously 100° and 210°F) and the use of tables that give the viscosities at 40°C of 0–VI and 100–VI reference oils that have the same viscosity at 100°C as that of the oil under test. The higher the VI the less the change in viscosity with temperature. *Figure 12.2* shows viscosity/temperature relationships for typical oils.

The original Dean and Davis VI system catered for oils with VI ranging from 0 to 100, but oils with VI greater than 100 are now manufactured by more elaborate refining and the use of VI improvers (see Section 12.4), and the VI system has been extended to include such oils. The significance of a VI above 140 is, however, doubtful since oils with the same VI in this region may differ markedly in their viscosity/temperature characteristics.

The viscosity of lubricating oils increases with increase in pressure, and oils that change least in viscosity with temperature tend to change least with pressure. The increase in viscosity with pressure is relatively small until large increases in pressure occur, as in gear trains and rolling bearings where the oil-film pressure may be many times the projected load.

12.3.4 Viscosity classification

The Society of Automotive Engineers (SAE) in the USA has

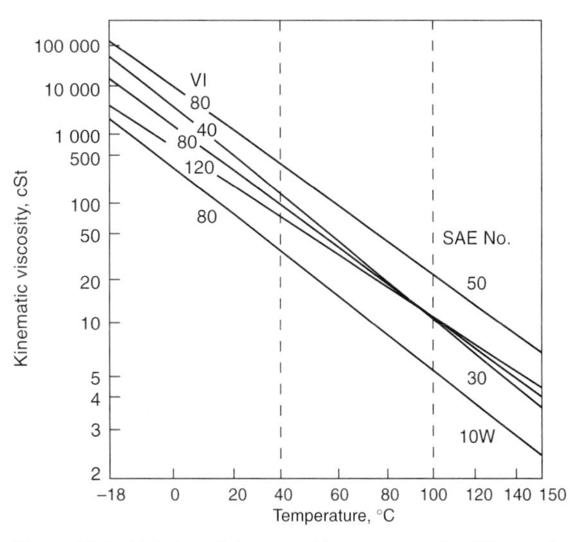

Figure 12.2 Variation of viscosity with temperature for different oils

classified engine lubricating oils into grades according to their viscosity, thus simplifying manufacturers' recommendations and the marketing of oils of the different viscosities required in practice. This SAE system is now used almost universally for the classification of engine oils. There are now six W (for winter) grades (SAE 0W, 5W, 10W, 15W, 20W and 25W) defined by their viscosities at low temperatures particular to each grade. There are five grades (SAE 20, 30, 40, 50 and 60) classified in accordance with their viscosities at 100°C. *Table 12.1* shows the 1995 revision of the classification. It also shows that the added requirement of pumpability down to a specified maximum sub-zero temperature is demanded of the W grades to avoid a start without oil pressure.

In this classification the viscosity at 100°C is the kinematic viscosity (in centistokes) determined by flow through a capillary tube and the viscosity at sub-zero tempeatures is the dynamic viscosity (in centipoises) determined the Cold Cranking Simulator (ASTM D 2602, multi-temperature version—see appendix to SAE J 300).

By the use of VI improvers engine oils can be formulated to fall into more than one SAE grade, and such oils are known as multigrade oils. They are designated by the two extreme SAE numbers, e.g. 10W/30, meaning that they have a low-temperature viscosity appropriate to the W grade and a high-temperature viscosity appropriate to the non-W grade. They have a VI of the order of 130–140. SAE 10W/30, 20W/40 and 20W/50 grades have been those mostly available, but now such combinations as 5W/20 and 15W/40 and others are marketed. These multigrade oils are used more often in gasoline engines, but multigrades are also becoming increasingly used in diesel engines. The use of synthetic oils can enable multigrade performance to be achieved without the use of VI improvers.

Lubricating oils are also classified according to viscosity by the International Standardization Organization (ISO) into eighteen grades numbered from 2 to 1500 according to the mid-point of a range of viscosities at 40°C, and corresponding classifications have been issued by the ASTM and the BSI, but these classifications are not generally applied to engine oils.

Oils may also be described according to their viscosity index as LVI, MVI, HVI, VHVI or UHVI (low, medium, high, very high or ultra-high VI). There is no precise definition of these terms, but in general LVI oils have a VI not greater than 30, MVI from about 35 to 75, HVI from 85 to 115, VHVI from 120 to 140, and UHVI greater than 140. The last two are obtained by a very high or severe degree of refining.

12.3.5 Low-temperature viscosity and ease of starting

Ease of starting depends on many factors but the viscosity of the lubricating oil at the starting temperature is one of the most important. For a diesel engine using a given fuel and lubricant there is a temperature below which it is impossible to start it, mainly because the final compression temperature is too low to ignite the fuel. This is partly because of the low ambient temperature as such, and partly because the lower cranking speed brought about by the increase in oil viscosity at this temperature raises the heat losses from the charge air as well as increasing gas leakage.

The viscosity of the oil at the starting temperature is thus an important factor in controlling the unsticking torque and the subsequent resistance to turning. An indication of startability is given by the dynamic viscosity of the oil as measured in the Cold Cranking Simulator. At very low temperatures wax separation from the oil could still further increase the starting torque, and to avoid this the pour point of the oil should be at least 9°C below the lowest temperature at which the oil is expected to be used. *Table 12.2* shows the lowest reliable starting temperatures for W grade oils of a given viscosity, and the corresponding maximum pour points. It is interesting to note that these temperatures are generally within 5°C of the borderline pumping temperature requirements of the 1995 SAE viscosity classification in *Table 12.1*.

Table 12.1 1995 SAE viscosity grades for engine oils

SAE viscosity grade	*Viscosity (cP) at temperature (°C) Max*	*Pumpability Max*	*Viscosity at 100°C Min*	*Viscosity at 100°C Max*	*High temperature, high shear viscosity (cP) Min*
0W	3250 at –30	60,000 at –45	3.8	—	—
5W	3500 at –25	60,000 at –35	3.8	—	—
10W	3500 at –20	60,000 at –30	4.1	—	—
15W	3500 at –15	60,000 at –25	5.6	—	—
20W	4500 at –10	60,000 at –20	5.6	—	—
25W	6000 at –5	60,000 at –15	9.3	—	—
20			5.6	< 9.3	2.6
30			9.3	< 12.5	2.9
40			12.5	< 16.3	2.9
50			16.3	< 21.9	3.7
60			21.9	< 26.1	3.7

Table 12.2 Starting temperature, viscosity and pour point

SAE grade	*Maximum viscosity (cP) at temperature (°C)*	*Lowest reliable starting temperature (°C)*	*Maximum pour point appropriate to lowest starting temperature (°C)*
5W	3500 at –25	–30	–40
5W/20	3500 at –25	–30	–40
10 W	3500 at –20	–27	–33
10W/30	3500 at –20	–27	–33
15W/40	3500 at –15	–20	–27
20W	4500 at –10	–15	–18
20W/40	4500 at –10	–15	–18

Good low-temperature viscosity characteristics are desirable in the oil even if the lowest possible starting temperature is never reached. Under all conditions lower viscosity means quicker starting and more rapid oil circulation. With electric starting it also means less wear and tear of the battery. An SAE 20W/40 oil is suitable for most engine starting requirements down to an ambient temperature of – 10°C.

12.3.6 Viscosity at running temperatures; friction losses and oil consumption

The overall thermal efficiency and specific fuel consumption of an engine are influenced by the internal friction or mechanical losses, upon which the viscosity of the oil at the running temperature has a pronounced effect. Piston and piston-ring losses account for most of the mechanical losses. Cylinder wall temperatures vary considerably according to engine design and operating conditions, and although 120°C may be taken as an average figure, temperatures of 200°C or more may be reached in road-vehicle engines. The SAE control viscosity of 100°C is undoubtedly well below average cylinder temperature but it may be used as a reasonable guide to oil selection. A viscosity control at 150°C would provide a better guide.

In general the lower the viscosity of the lubricating oil the lower is the fuel consumption, but a limit is imposed on reduction in oil viscosity by the need for adequate lubrication and the avoidance of excessive oil consumption. *Figure 12.3* shows the effect of lubricating oil viscosity on the specific fuel consumption of a 10-litre vehicle engine. The use of an SAE 10W oil in place of an SAE 30W oil gives a worthwhile reduction in specific fuel consumption of around 4 per cent.

Friction reducers' can be added to the lubricating oil to improve fuel consumption, by reducing viscosity and improving 'slipperiness'.

12.4 Additives

Under modern operating conditions straight mineral oils are no longer able to meet the needs of internal combustion engines, and additives are used to improve the quality and performance of lubricating oils beyond what can be achieved by refining. The principal types of additive used in engine oils are shown in *Table 12.3*; they are used in concentrations ranging from a few parts per million to 10% weight, or sometimes more.

Two of the most important functions of additives are to keep the engine clean and to prevent wear. Cleanliness, i.e. freedom from deposits, is achieved jointly by preventing, as far as possible, oxidation of the oil and the consequent formation of oil-insoluble matter, and by keeping in suspension oil-insoluble matter, whether derived from fuel combustion (the major source), oil oxidation, or ingress of adventitious matter. The ability to keep material in suspension in the oil is known as *dispersancy*, and hence the associated terms *dispersant additives and dispersant oils*—terms that are preferable to *detergency/detergent*, which imply removal of dirt from a dirty surface as opposed to prevention of deposition of dirt on a clean one.

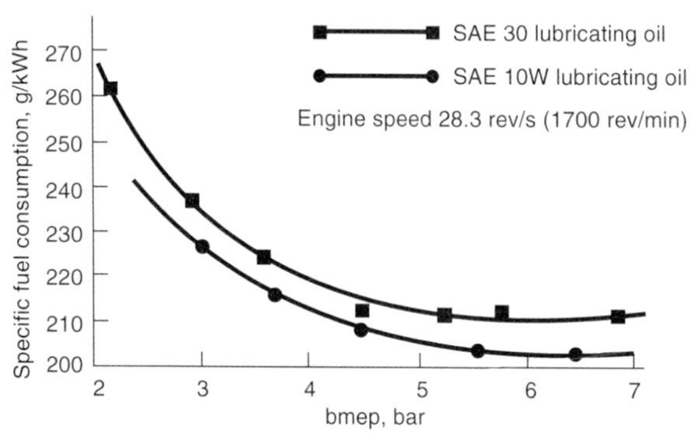

Figure 12.3 Effect of lubricating oil viscosity on specific fuel consumption; 10 litre vehicle engine; oil temperature in 85°C; water temperature out, 75°C

Dispersant oils may present difficulties with fine-particle filters, which may become choked by finely divided matter, but the normal type of filter used in diesel engines to remove particles down to 5 microns is unlikely to give trouble.

After a short period of use a dispersant oil may look dirty to an extent that would be disconcerting in a straight oil, but the dirt is harmlessly suspended in the oil instead of being deposited on engine parts, and the oil can continue to be used until the recommended oil-change period is reached. The oil-change period is materially increased by the use of dispersant oils and this, together with the greater resistance to oxidation conferred by anti-oxidants, is why much longer oil-change periods can be recommended.

Viscosity index improvers are used in the formulation of multigrade oils to lessen the decrease in viscosity with increase in temperature. *Figure 12.4* shows how they function by increasing the viscosity at higher temperatures more than they do at lower temperatures, thus flattening the viscosity/temperature curve. An important aspect of VI improvers is that they should not lose their viscosity-increasing effect as a result of continuous shearing in the engine—they should be shear-stable.

Shear stability can be tested in an Orbahn shear rig in which the oil is circulated thirty times through a diesel fuel pump and injector. A reasonably good correlation exists between the viscosity loss in this test and the loss occurring after 5000 km service in automotive engines. *Figure 12.5* shows the performance in this test of SAE 20W/40 diesel oils formulated with the same base oil but different VI improvers differing in their shear stability, all oils having an initial viscosity of 14.0 cSt at 100°C. The styrene copolymer, gives the best results. Multigrade oils are used in diesel engines on account of less wear, lower oil consumption and longer engine life. These benefits stem from the higher viscosity and hence thicker oil film at ring-belt temperatures, as illustrated in *Figure 12.6*, where it can be seen that a shear-stable SAE 20W/40 oil with the same viscosity at 100°C as that of an SAE 40 oil is about 15 per cent thicker at 180°C, i.e. at ring-belt temperatures. Moreover, an SAE 20W/40 oil is suitable for most engine-starting requirements down to −10°C.

Alkaline additives are an important feature in diesel engine lubricants in order to neutralize inorganic acids derived from the combustion of sulphur-containing fuels that are a potent cause of cylinder wear. They are vitally necessary in marine diesels runing on high-sulphur fuel.

12.5 Oil deterioration

Diesel engine lubricants may deteriorate in a variety of ways. They may become contaminated by carbonaceous particles from incomplete combustion of the fuel, unburnt fuel, acidic water from blow-by gases, sea water (marine diesels), oxidation products from the lubricating oil, ash from lubricating oil additives, metal particles from wear of metal parts, or from adventitious matter such as road dust. Their viscosity may increase owing to oil oxidation, and will certainly increase as their content of suspended oil-insolubles increases. Their additives will be depleted in the normal course of use.

Oil-insoluble contaminants are kept in suspension by disper

Table 12.3 Principal types of additive for engine oils

Type	*Chemical nature*	*Function*
Basic dispersant	Calcium, barium or magnesium phenate or salicylate	1. Neutralization of acids 2. Prevention of lacquer and deposit formation 3. Dispersion of oil-insolubles
Ashless dispersant	Poly-isobutenyl succinimide	1. Dispersion of soot and oxidation products 2. Prevention of lacquer and deposit formation
Anti-oxidant	Zinc dithiophosphate, hindered phenol, phosphosulphurized olefin, metal salicylate	Prevention of oxidation and thickening
Extreme pressure additive	Zinc dithiophosphate, organic phosphate, organic sulphur and chlorine compounds	Prevention of wear (cams and tappets)
Anti-corrosion additive	Calcium, barium or sodium sulphonate, organic amine	Prevention of corrosion
Viscosity index improver	Polymeric compound, polyester or hydrocarbon	Reduction of viscosity loss as temperature increases
Pour point depressant	Polymeric methacrylate	Improves flow properties at low temperatures
Anti-foam additive	Silicone compounds	Prevention of foaming under conditions of vigorous agitation

sant additives. What is not kept in suspension is removed by oil filters or centrifuges. There comes a time, however, when the burden of suspended matter causes too great an increase in oil

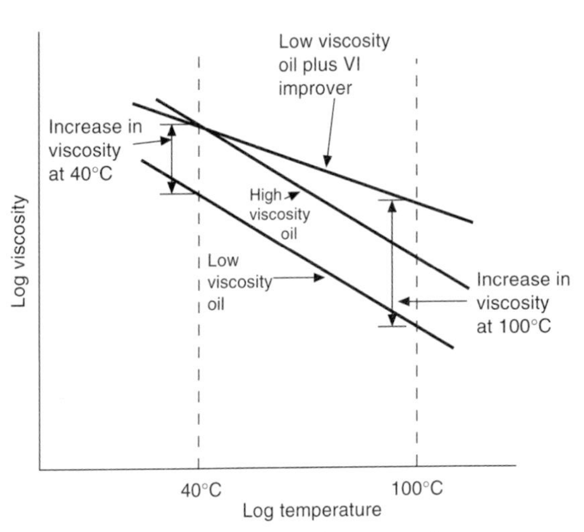

Figure 12.4 How a viscosity index improver works

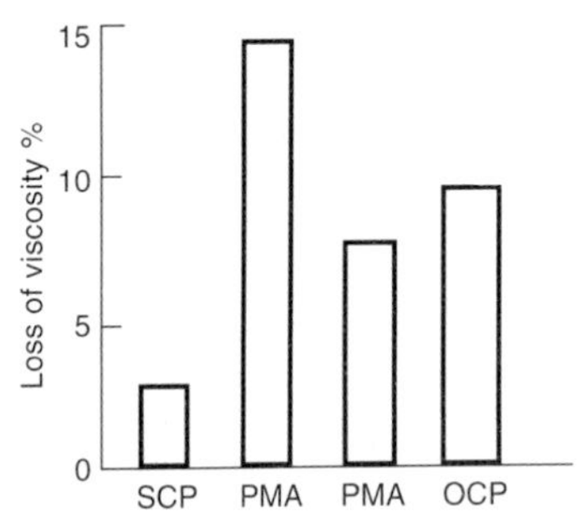

Figure 12.5 Orbahn shear-rig tests on SAE 20W/40 diesel engine oils formulated with various VI improvers. All oils initially 14.0 cSt at 100°C
SCP = Styrene copolymer
PMA = Polymethacrylate
OCP = Olefin copolymer

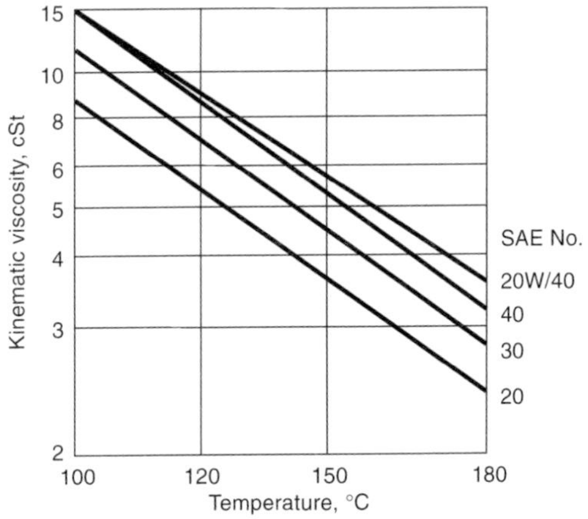

Figure 12.6 Oil viscosities at ring-belt temperatures

viscosity, and the oil must be changed. The observance of recommended oil-change periods takes care of this aspect.

Dilution by unburnt fuel rarely reaches significant proportions in modern engines provided they are well maintained. Dilution by up to 5 per cent by volume of gas oil may be tolerated although it reduces the viscosity of an SAE 30 oil almost to that of an SAE 20 oil.

Leakage of water, particularly sea water, into the crankcase should be avoided. If acidic combustion products get into the crankcase corrosion of the crankshaft and other steel parts could occur, but this is prevented by the use of alkaline additives. In large marine installations continuous centrifugation of the crankcase oil is customary but the earlier practice of water washing is inadvisable with dispersant oils because of the likelihood of emulsification. Some dispersant oils, however, are tolerant of a certain amout of water washing, which may be necessary in an emergency.

Used oils may show an increase in viscosity owing to oil oxidation or suspended material, or may show a decrease in viscosity owing to fuel dilution. All these factors are usually operating simultaneously and interpretation of used oil viscosity is difficult. As stated in Section 12.4, multigrade oils may suffer a reduction in viscosity at running temperatures owing to polymer degradation caused by mechanical shearing.

Additive depletion is a potential cause of deterioration since additives are naturally consumed in the normal course of use, and could become depleted by an excessive period between oil changes. This is unlikely to occur in normal circumstances since ample additives are usually provided for the type of operation for which the oil is designed, and for the expected life of the oil.

12.6 Operational problems

The main problems that are likely to arise in the lubrication of diesel engines come from the deposits arising from the combustion of the fuel and the lubricating oil, and from the partial oxidation of the lubricating oil. The extent to which these problems arise depends on the ability of the lubricant to cope with these phenomena, and to keep wear within acceptable limits. As might be expected, the progressive improvement in the quality of lubricating oils over the last 40 years, to enable them to meet the demands of engines with ever increasing output, has created the need for a means of classifying lubricating oils according to their performance, and for appropriate engine tests to assess and control their quality. The American Petroleum Institute has devised such a classification (see Section 12.7) and appropriate engine tests are described in Section 12.8.

12.6.1 Piston deposits

Deposits are liable to form on the crown, the top and intermediate lands, in the ring grooves, on the skirt surfaces and on the undercrown.

The deposits on the crown are derived mainly from the fuel (carbonaceous combustion products) with a contribution from lubricating oil additives other than ashless additives. Top land deposits are probable from both fuel and lubricant; they occur under both moderate- and high-temperature conditions. Top land deposits can be very hard, and since adjacent deposits also occur on the liner above the top ring travel these rub against each other and often rub each other off so that on removal of the piston the top land may appear clean though scored over portions of the circumference. The liner deposit will tend to vary, the liner being clean where the piston is black and black where the piston is clean. The top lands of light-alloy pistons often become considerably worn by this process if temperatures are high.

Undercrown deposits occur mainly in high-output engines operating under severe conditions. They do not have any serious effects, though carbon can flake off and collect in the crankcase.

Crown deposits can sometimes build up to the extent of causing the piston crown to hit the head (on the idle stroke in four-stroke engines), mostly without any serious consequences.

Lacquer formation on the skirt can reduce the clearance between piston and cylinder, and in severe cases can lead to scoring or seizure of the piston. It may occur in high-output engines operating under severe conditions.

Carbon deposits form particularly in the top groove and generally to a lesser extent in the second groove of high-output engines operating at high temperatures, and these can lead to sluggish ring movement when the side clearance is taken up, to ring sticking and to possible leakage. If the rings remain free, as will occur where tapered rings are used, deposits may completely fill the clearance space behind the ring, a condition known as *ring packing*. Ring packing can lead to rapid ring wear or scuffing of the ring surface since with aluminium pistons a reduction of load after a period of full load may leave the rings projecting beyond the piston and so carrying most of the thrust load. This same effect may occur at some local place on the periphery as a result of ring rotation when packing is present, due to ring radial depth variations. High-quality lubricating oils can help but not if ring-groove temperatures are excessive, e.g. above 250°C in vehicle engines. Stationary engines or marine engines operating at continuous loads have ring temperatures not exceeding 220°C, and under these conditions ring-groove deposits need be no problem.

Piston deposits are held in check as far as possible by antioxidant additives that retard oxidation of the lubricating oil, and dispersant additives that keep insoluble matter in suspension, but if temperatures become too high this is no longer possible.

12.6.2 Engine wear

Wear can occur on any of the surfaces in relative motion, i.e. piston bearing surfaces, bore, rings, grooves valve train, main and big-end bearings. However, the important wear that controls how long the engine can run between overhauls is mainly that of the bore at the upper limit of top ring travel, the radial wear of the top ring with resultant increase of ring gap and consequent blowby, and the increase of side clearance of the top ring due to ring and groove wear. Such wear may be due to abrasion by engine deposits or by foreign matter that enters through the air intake or the crankcase breathers. Abrasion by engine deposits usually increases with increase in the severity of operating conditions, but that due to adventitious matter is usually independent of operating conditions.

Provided air filtration is satisfactory, the main cause of wear of cylinder bores or liners is the corrosive action of inorganic acids derived from the sulphur in the fuel. In vehicle engine fuel (DERV) this is quite low (0.5%w) but in stationary engines running on B grade fuels it can rise to 1.8%w, and in marine engines running on residual fuels or blends of residual and distillate fuels it can be up to 4.5%w. In vehicle engines the thermostat ensures rapid warming up of the engine from cold, so eliminating much of the condensation of water produced during combustion and containing inorganic, mainly sulphuric, acid. However, even so, the dewpoint of the acids can be higher than the liner temperature at light loads, and will increase with gas pressure so that turbocharging makes matters worse. Furthermore, condensate on the lower portions of the liner can be swept upwards by the top ring where conditions of lubrication are least favourable and conditions for corrosion the most favourable. However, this type of wear can be kept within reasonable limits by the use of highly alkaline additives which neutralize the acids on the liner wall. In turbocharged vehicle engines running on DERV a TBN value of the oil of 9 mg KOH/*g* is adequate. In medium-speed four-stroke marine engines running on residual fuels this figure needs to be increased to between 20 and 30 mg KOH/*g*.

In crosshead marine engines, which have separate lubrication for the liner from some eight quill points around the circumference much greater alkalinity and dispersancy are necessary than are required for trunk piston engines in which the cylinders are lubricated by the crankcase oil, and TBN values of 70 mg KOH/g are quite common. This is because the oil film on the cylinder wall is not continuously intrerchanged with oil from the crankcase, as in a trunk piston engine, and because of difficulties with getting the oil to spread sideways from the quill points which can be some 30 cm apart. A greater reserve of additive is therefore necessary in the oil if wear is to be kept within bounds. This results in the use of cylinder lubricants with additive contents high enough to give TBN values of 70 mg KOH/*g* when using residual fuels of high sulphur content. Such a value is necessary to enable the oil to cope with the large amounts of acid condensate derived from the fuel and the greater quantity of partially burnt combustion products, both of which would lead to uneconomically high cylinder wear. The development of such highly alkaline, superbasic lubricating oils around 1956 opened a new era in the operation of marine diesel engines, permitting the use of residual fuels in place of the more expensive distilate fuels with no greater and even less wear, and considerable saving in fuel and maintenance costs. *Figure 12.7* shows the corrosion of a

Figure 12.7 Corrosion of chromium-plated cylinder liner due to inadequate alkalinity of lubricating oil

chromium-plated cylinder liner resulting from inadequate alkalinity of the lubricating oil.

12.6.3 Bearing corrosion

Bearing corrosion can be caused by organic acids produced by the oxidation of the lubricating oil when engine speeds and operating temperatures are high, but it rarely occurs nowadays, oxidation being held in check by anti-oxidant additives and corrosion prevented by anti-corrosion additives. Freedom from bearing corrosion is, however, an important criterion in specification engine tests. *Figure 12.8* shows the type of bearing corrosion that may occur with an unsatisfactory lubricating oil.

12.6.4 Sludge

Sludge may deposit on cool surfaces, where it causes little trouble, but it can cause trouble by blocking oil-ways or by baking to a hard deposit in scraper-ring grooves and oil drain holes, thereby reducing their effectiveness and increasing oil consumption. Sludge formation is accentuated by inefficient combustion, particularly in association with piston blow-by and low temperature operation. Deposition of sludge is retarded, however, by dispersant additives.

12.7 API classification

A classification for diesel engine lubricants based on their performance characteristics in relation to the type of use for which each is intended is shown in *Table 12.4*. The associated engine tests are also named and described further in the next section.

12.8 Engine tests and associated specifications

Engine tests of lubricating oils are so closely associated with specifications devised for their approval or quality control that they can best be discussed in the light of the historical development of such specifications.

Until the late 1930s the primary function of an engine lubricating oil, i.e. to ensure efficient functioning of the moving parts with the minimum of friction and wear, was satisfactorily achieved by straight mineral oils. Advances in engine design and increasing severity of operating conditions were catered for by improvements in the quality of lubricating oils obtained by selection of the crude from which they were manufactured, and by advances in methods of refining, notably the greater use of solvent extraction instead of acid/earth treatment. During this period laboratory tests for chemical and physical properties, such as those described in Section 12.9, sufficed to determine and control the quality of the oil.

By the late 1930s improvements in engine design had led to much greater efficiencies and higher operating temperatures that made greater demands on the lubricating oil. Moreover, in addition to its primary function the oil had to cool the piston, prevent deposits and corrosion, and be generally less susceptible

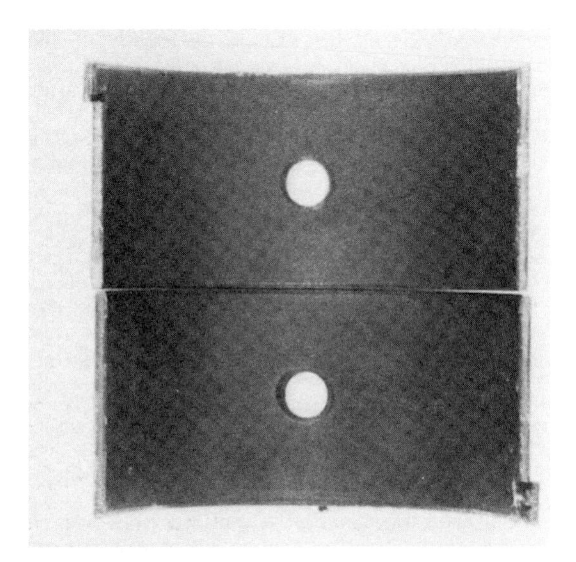

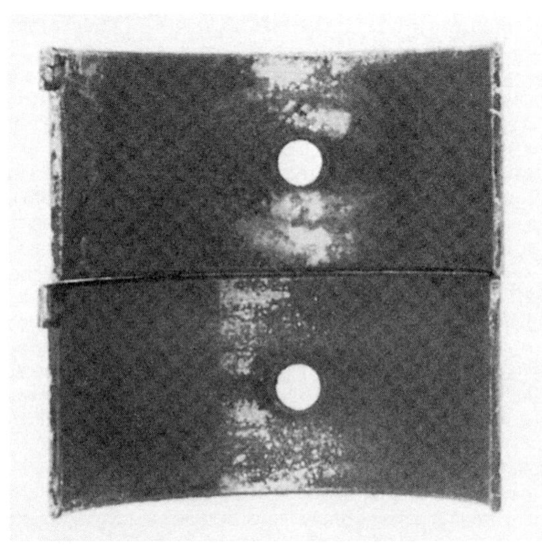

Figure 12.8 Bearings from a Petter W1 corrosion test. Left: satisfactory lubricating oil. Right: unsatisfactory lubricating oil

Table 12.4 API classification for diesel engine lubricating oils

Letter Designation	*API Engine Service Description*	*ASTM Engine Oil Description*
CE	1983 Diesel Engine Service API Service Category CE denotes service typical of certain turbocharged or supercharged heavy duty diesel engines manufactured since 1983 and operated under high load for both low and high speed operation. Oils designed for this service may also be used when API engine service category CD is recommended for diesel engines.	Oil meeting the performance requirements of the following diesel and gasoline engine tests: The 1G2 diesel engine test has been correlated with indirect injection engines used in heavy-duty service, particularly with regard to piston and ring groove deposits. The T-6, T-7 and NTC-400 are direct injection diesel/ engine tests. The T-6 has been correlated with vehicles equipped with engines used in high speed operation prior to 1980, particularly with regard to deposits oil consumption and ring wear. The F-7 test has been correlated with vehicles equipped with engines operated largely under lugging conditions prior to 1984, particularly with regard to oil thickening. Test Method D5290, the NTC-400 diesel engine test has been correlated with vehicles equipped with engines in highway operation prior to 1983. particularly with regard to oil consumption control, deposits and wear. Test Method D5119, the L-38 gasoline test requirement is used to measure copper-lead bearing weight loss.
CF-4	API Service Category CF-4 describes oils for use in high-speed, four-stroke cycle diesel engines. CF-4 oils exceed the requirements of the Service Category CE, are designed to replace CE oils, and provide improved control of oil consumption and piston deposits. CF-4 oils may be used in place of CC and CD oils. They are particularly suited for on-highway, heavy duty truck applications.	Oil meeting the performance requirements in the following diesel and gasoline engine tests. The 1K diesel engine test has been correlated with vehicles equipped with engines used in high-speed operation prior to 1980, particularly with regard to deposits, oil consumption and ring wear. The T6 has been correlated with vehicles equipped with engines used in high speed operation prior to 1980, particularly with regard to deposits, oil consumption and, ring wear. T-7 test has been correlated with vehicles equipped with engines operated largely under lugging conditions prior to 1984, particularly with regard to oil thickening. The Bench Corrosion Test has been shown to predict corrosion of engine oil-lubricated copper, lead, or tin-containing components used in diesel engines. Test Method D5119, the L-38 gasoline engine test is used to measure copper-lead bearing weight loss.
CF	API Service Category CF denotes service typical of off-road indirect injected diesel engines and other diesel engines which use a broad range of fuel types including those using fuel with higher sulphur content for example over 0.5% wt. Effective control of piston deposits, wear, and corrosion of copper-containing bearings is essential for these engines, which may be naturally aspirated turbocharged or supercharged. Oils designated for this service have been in existence since 1994. Oils designated for this service may also be used when API service category CD is recommended. Engine oils that meet the API Service Category CF designation have been tested in accordance with the CMA Code, may use the API Base Oil Interchangeability Guidelines and the API Guidelines for SAE Viscosity Grade Engine Testing.	Oil meeting the performance requirements in the following diesel and gasoline engine tests. The IMPC diesel engine test has been shown to provide correlation with engine oil performance when used in naturally aspirated, turbocharged, or supercharged indirect injection engines. Test method D5119, the L-38 gasoline engine test is used to measure copper lead bearing weight loss.
CF-2	API service category CF-2 denotes service typical of two stroke cycle engines requiring highly effective control over cylinder and ring-face scuffing and deposits. Oil designated for this service have been in existence since 1994 and may also be used when API Service Category CD II is recommended. These oils do not necessarily meet the requirements of CF or CF-4 unless the oils have specifically met the performance requirements of these Categories. Engine oils in the two-stroke cycle DD 6V92TA engine test since January 1, 1992, may be considered for this Service Category provided the tests were conducted in accordance with the test procedure as published in ASTM Research Report RR:DO2-1319 or as revised by the ASTM Test Monitoring Centre. All tests conducted since January 1, 1992, must be done in accordance with the most current test procedures. Engine oils that meet the API Service Category CF-2 designation have been tested in accordance with the CMA Code, may use the API Base Oil Interchangeability Guidelines and the API Guidelines for Viscosity Grade Engine Testing.	Oil meeting the performance requirements in the following diesel and gasoline engine tests. The IMPC diesel engine test has been shown to provide correlation to engine oil performance when used in naturally aspirated, turbocharged, or supercharged indirect injection diesel engines with modified piston deposit rating methodology to relate to effective piston and ring groove deposit control for two stroke cycle diesel engines. The 6V92IA diesel engine test has been correlated with two-stroke cycle diesel engines in heavy-duty service, particularly with regard to ring face distress and liner scuffing.
CG-4	API Service Category CG-4 describes oils for use in high-speed, four-stroke cycle diesel engines used in highway and off-road applications where the fuel sulphur content may vary from less than 0.05% weight to less than 0.5% weight. CG-4 oils provide effective control over high temperature piston deposits wear corrosion foaming, oxidation and soot accumulation. These oils are especially effective in engines designed to meet 1994 exhaust emission standards and may also be used in engines requiring API Service Categories CD, CE and CF-4. Oils designated for API Service Category CG-4 have been in existence since 1995. Engine oils that meet the API Service Category CG-4 designation have been tested in accordance with the CMA code, may use the API Base Oil Interchangeability Guidelines and the API Guidelines for SAE Viscosity Grade Engine Testing.	Oil meeting the performance requirements in the following diesel and gasoline engine tests: The III diesel engine test has been used to predict piston deposit formation in four-stroke cycle, direct injection, diesel engines which have been calibrated to meet 1994 US Federal Exhaust Emissions requirements for heavy-duty engines operated on fuel containing less than 0.05% weight sulphur. The T-8 diesel engine test which has been shown to generate soot-related oil thickening in a manner similar to 1992 emission controlled heavy-duty diesel engines using mechanical injection control systems. Test method D5533, the Sequence IIIE gasoline engine test has been correlated with vehicles used in high temperature service prior to 1988, particularly with regard to oil thickening. Test Method D5119, the L-38 gasoline engine test, is used to measure copper-lead bearing weight loss and viscosity loss due to shearing. The 6.2L diesel engine test has been correlated with hydraulic roller cam follower pin wear in medium duty indirect injection diesel engines used in broadly based field operations. The ASTM D892 foaming test. Sequences I, II and III, and a higher temperature foaming test (under development) have been shown to predict foaming of engine oils used in diesel engines. The Bench Corrosion Test has been shown to predict corrosion of engine oil lubricated copper, lead, or tin containing components used in diesel engines.

to oxidation. Even the best quality straight mineral oils no longer gave satisfactory performance, and there seemed little prospect of any significant improvement in quality by advances in methods of refining. Thus began the use of functional additives to enhance the quality and improve the performance of the lubricating oil and the development of heavy-duty (HD) oils. This increasing demand on the lubricant accelerated dramatically during World War II, and has continued ever since with corresponding increase in the use of additives until today the use of straight mineral oils as engine lubricants is virtually extinct.

Traditional laboratory tests hitherto satisfactory for the evaluation of straight mineral oils proved incapable of predicting or controlling the performace of additive-containing oils, and engine tests became essential for their development and evaluation. No single engine test can be expected to reproduce all operational problems, and several different engine tests are required to evaluate the overall performance of such lubricants. Engine tests are designed to highlight one or more aspects of performance, and operating conditions are adjusted accordingly. Exaggerated conditions are used to reduce the test time to a reasonable duration, e.g. to evaluate performance under high-temperature conditions tests are run continuously at high power outputs, while tests to evaluate sludge and rust are run intermittently at low temperatures.

Engine tests are carried out on engine test-bed installations ranging from single-purpose beds for one type of engine running under steady-state conditions up to multi-purpose beds capable of accommodating a range of engines running under various automatically controlled cycling conditions of speed, load and temperature. Both specially designed laboratory test-engines and commercially available engines are used, but they have to be installed with extreme care and precision to ensure that the test is valid to simulate the condition that is a field problem, and that it is repeatable and reproducible.

The Caterpillar Tractor Company and General Motors in the USA were the earliest to devise engine tests for the evaluation and approval of lubricants for use in their equipment. Five of their tests were standardized by the US Co-ordinating Research Council (CRC) as a series numbered L1 to L5. During World War II these were incorporated by the US Army Ordnance into their 2–104 specification for HD lubricating oils, and it is from this specification that most subsequent military specifications have evolved. These military specifications were designed primarily for the qualification of lubricants for use in military vehicles but they became widely accepted as criteria for commercial diesel engine lubricants, and so were soon an integral part of the vocabulary of engine manufacturers and oil suppliers.

The 2–104B specification that was subsequently introduced. employed the Caterpillar single-cylinder diesel engine running on a 0.4%w sulphur fuel to evaluate ring sticking, engine deposits and wear; and a Chevrolet gasoline engine to evaluate oil oxidation and copper-lead bearing corrosion. Definite acceptance limits were laid down for all these properties.

In 1949 the 2–104 BSupplement 1 specification was introduced to cater for lubricating oils for use in diesel engines operating with fuels of higher sulphur content. The same Caterpillar engine test procedure was used, but severity was increased by the use of a 1.0%w sulphur fuel. Although this specification has been officially obsolete for many years it continues to be an accepted criterion of performance, and oils are still sometimes referred to as of Supplement 1 quality. Such oils fall within the API/CB designation.

In 1954 the designation of these specifications was changed to MIL–L–2104, and a series beginning with MIL–L–2104A (now obsolete) for light-duty diesel engine service was issued. This was followed in 1964 by MIL–L–2104B, specifying lubricating oils for both diesel and gasoline engines operating under severe conditions such as stop-start operations, where low-temperature sludging and rusting could be a problem.

The MIL-L2104B specification incorporated four engine tests: the Caterpillar 1–H test with a supercharged diesel engine to assess piston cleanliness, ring sticking and wear; the CRC L–38 test (replacing the Chevrolet L–4 test using a CLR single cylinder gasoline engine to assess oil oxidation and bearing corrosion; the CLR LTD test) using the same CLR engine, to evaluate low-temperature deposits of sludge and varnish; and a test in an Oldsmobile 8-cylinder gasoline engine to evaluate rust formation.

Another specification, MIL–L–45199B,was used concurrently with MIL–L–2104B to define lubricants for supercharge and non-supercharged highly rated diesel engines. This specification was based on one devised by the Caterpillar Tractor Company for lubricating oils for use in their high-output engines, oils generally known as Caterpillar Superior Lubricants, Series 3 (superseding a previous Series 2). Two tests in single-cylinder diesel engines were specified, the Caterpillar 1–D test and the more highly rated Caterpillar 1–G test to evaluate ring sticking, deposits and wear under supercharged operating conditions. The CRC L–38 test was also included.

Two new specifications were issued in 1970, MIL–L–2104C and MIL–L–46152, the former superseding the MIL–L–2104B and MIL–L–45199B specifications, the latter specifying lubricants for both gasoline and diesel engines in passenger cars and light trucks. The specifications embraced yet more engine tests and required oils of higher quality to meet them. Seven engine tests were specified: CRC L–38 oxidation test, Oldsmobile rusting test, Oldsmobile high-temperature oil thickening test, Ford 302in^3 sludging and deposition test, and the Caterpillar 1–D, 1–G and 1–H tests.

Since 1970 many new specifications and tests have been introduced. Three new versions of MIL-L 2104 have appeared, the latest being MIL–L–2104F, and new American engine tests include Oldsmobile Sequence IIIE, VE, VI and VIA, Caterpillar 1K, 1M–PC, 1N and 1P, Mack T–6, T–7, T–8 and T–9, Cummins NTC400 and M11, and General Motors 6.2 and 6V–92TA. *Table 12.5* summarizes the main varieties of the Caterpillar engine tests alone. Satisfactory and unsatisfactory pistons from a Caterpillar 1–G2 test are shown in *Figure 12.9*.

The US military specifications and tests have been and still are widely used outside the USA to qualify oils for both military and commercial use. From 1951 to 1963 they were incorporated in the DFF 2101 series of specifications issued by the UK Ministry of Supply, which have since been replaced by DEF STAN 41–43. However, the requirements and operating conditions of

Table 12.5 Caterpillar diesel engine tests

	L-1	*L-1 mod*	*1-D*	*1-G*	*1-G2*	*1-H*	*1-H2*	*1M-PC*	*1-K*	*1-N*
RPM	1000	1000	1200	1800	1800	1800	1800	1800	2100	2100
Displacement (ml)	3400	3400	3400	2200	2200	2200	2200	2200	2400	2400
Test duration, h	480	480	480	480	480	480	480	120	252	252
Fuel sulphur, %wt	0.35	±1.0	±1.0	0.35	0.37	0.35	0.37	0.4	0.4	0.04
	min	0.05	0.05	min	0.43	min	0.43			
API category	CA	CB	CD	CD	CD	CC	CC	CF	CF-4	CG-4

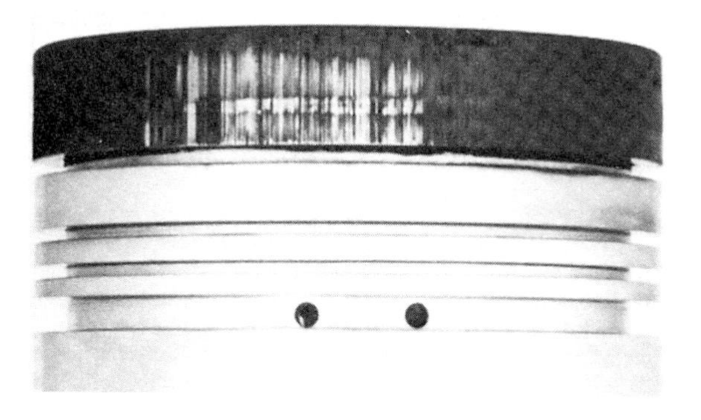

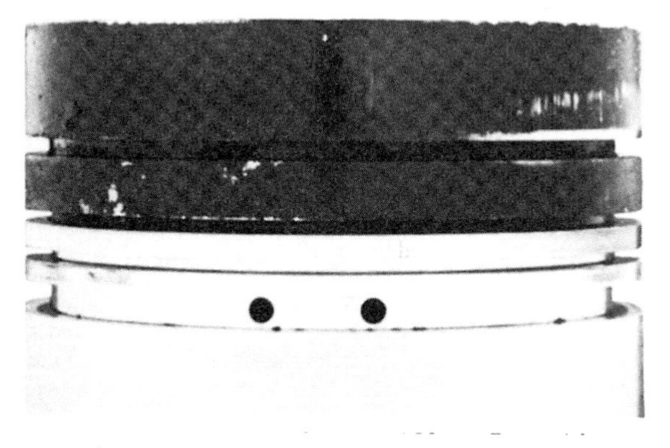

Figure 12.9 Pistons from a Caterpillar 1G2 test. Top: satisfactory lubricating oil. Bottom: unsatisfactory lubricating oil

European engines differ appreciably from those of American engines and it was eventually realized that the American specifications and engine tests were not entirely relevant for European equipment and operating conditions. As a result, European test procedures, using European engines, were developed by the Co-ordinating European Council (CEC) with the co-operation of The Institute of Petroleum (IP), and the Institut Français du Pétrole (IFP). Using these tests the Committee of Common Market Automobile Constructors (CCMC) introduced European sequences for service-fill oils for both passenger cars and diesel-powered commercial vehicles. The CCMC has now been superseded by the Association des Constructeurs Européens d'Automobile (ACEA) and the ACEA has introduced sequences covering gasoline engines, passenger car diesel engines and other diesel engines. The diesel engine oil sequences are reproduced in *Tables 12.6* and *12.7*.

The UK Ministry of Supply also amended their DEF 2101 specification to introduce European engines, firstly in 1963 (DEF 2101C) by replacing the Chevrolet L–4 bearing corrosion and oil oxidation test by a similar one using the British Petter W1 gasoline engine, and subsequently in 1965 when the DEF 2101D specification contained a piston cleanliness test in the Petter AV1 diesel engine operating on 1%w sulphur fuel, and the Petter W1 bearing corrosion and oil oxidation test. The DEF 2101D specification defined an oil of approximately 2–104B, Supplement 1 quality, falling within the API/CB designation. In 1979 DEF 2101D was superseded by DEF STAN 91–43/1, the UK Ministry of Defence Specification which includes the same engine tests, but in addition incorporates an FZG gear test to determine load-carrying capacity. Tests to evaluate elastomer compatibility and foaming characteristics are also included. Details of these tests are given in *Table 12.8*.

Other specifications include those of many vehicle manufacturers. The specification situation is therefore extremely complicated, and changes often occur several times a year. It is therefore difficult for even specialists to keep up to date.

Furthermore, while oils meeting these various specifications have gained wide acceptance and have indeed performed well in commercial service, such oils do not necessarily meet the needs of all diesel engines, and excessive oil thickening, piston deposits and ring sticking have been experienced with some of them, particularly in turbocharged engines. Commercially available branded oils may prove to be superior to oils that merely meet one or other of these specifications and the user should follow the engine manufacturer's recommendations regarding their use in his particular engine and under his operating conditions. It follows also that continued research and development is necessary to keep abreast of the ever increasing demands on the lubricant imposed by new generations of diesel engines.

Table 12.6 API service category chart*

Gasoline Engines	*Diesel Engines*
SA—Obsolete–For older engines, no performance requirement. Use only when specifically recommended by manufacturer.	**CA**—Obsolete–For light duty engines (1940s and 1950s)
SB—Obsolete–For older engines. Use only when specifically recommended by manufacture.	**CB**—Obsolete–Moderate duty engines for 1949 to 1960
SC—Obsolete–For 1967 and older engines.	**CC**—Obsolete–For engines introduced in 1961
SD—Obsolete–For 1971 and older engines.	**CD**—Obsolete–Introduced in 1955. For certain naturally aspirated and turbocharged engines.
SE—Obsolete–For 1979 and older engines.	**CD-II**—Obsolete–Introduced in 1987. For two-stroke cycle engines.
SF—Obsolete–For 1988 and older engines.	**CE**—Obsolete–Introduced in 1987. For high-speed, four-stroke, naturally aspirated and turbocharged engines. Can be used in place of CC and CD oils.
SG—Obsolete–For model-year 1993 and older engines.	**CF-4**—Current–Introduced in 1990. For high-speed, four-stroke, naturally aspirated and turbocharged engines. Can be used in place of CE oils.
SH—Current–Introduced in1993. Will be discontinued in the API Service Symbol, August 1,1997, except when used in combination with certain C categories.	**CF**—Current–Introduced in 1994. For off-road, indirect-injected and other diesel engines including those using fuel with over 0.5% weight sulfur. Can be used in place of CD oils.
SJ—Current–Introduced in the API Service Symbol in October 1996. For all engines presently in use.	**CF-2**—Current–Introduced in 1994. For severe duty, two-stroke cycle engines. Can be used in place of CD-II oils.
Note: Each gasoline engine category above exceeds the performance properties of all the previous categories and can be used in place of the lower one. For example, an SJ oil can be used for any previous category.	**CG-4**—Current–Introduced in 1995. For severe duty, four-stroke cycle engines using fuel with less than 0.5% weight sulfur. Can be used in place of CD, CE, and CF-4 oils.

*Reprinted from API publication 1551 (1996) Courtesy of the American Petroleum Institute, 1220 L Street, NW, Washington, DC 20005.

Table 12.7 ACEA 1998 European oil sequence for service-fill oils for light and heavy duty diesel engines (March 1998)

This document details the ACEA 1998 European Oil Sequences for Service-fill Oils for Gasoline engines, for Light Duty Diesel engines, and for Heavy Duty Diesel engines. These sequences define the minimum quality level of a product for presentation to ACEA members. Performance parameters other than those covered by the tests shown or more stringent limits may be indicated by individual member companies.

These sequences will replace the ACEA 1996 sequences as a means of defining engine lubricant quality from 1 March 1998.

Conditions for use of performance claims against the ACEA oil sequences

ACEA requires that any claims for Oil performance to meet these sequences must be based on credible data and controlled tests in accredited test laboratories.

All engine performance testing used to support a claim of compliance with these ACEA sequences must be generated according to the European Engine Lubricants Quality Management System (EELQMS). This system, which is described in the ATIEL Code of Practice[1], addresses product development testing and product performance documentation, and involves the registration of all candidate and reference oil testing and defines the compliance process. Compliance with the ATIEL Code of Practice is mandatory for any claim to meet the requirements of the 1998 issue of these ACEA sequences.

First allowable use of the designations and performance levels defined in the 1998 issue of the ACEA sequences for performance claims will be from 1st March 1998, and from 1st March 1999 all new claims to meet ACEA sequences must be to the 1998 issue. The 1996 issue will be withdrawn on 1st March 2000, after which no claims to meet those requirements shall be made.

The marketer of an oil claiming to meet ACEA performance requirements is responsible for all aspects of product liability.

Where limits are shown relative to a reference oil, then these must be compared to the last valid Reference Result on that test stand prior to the candidate and using the same hardware. Further details will be in the ATIEL Code of Practice.

The ACEA 1998 European Oil Sequences for Service-fill Oils comprise 3 sets (classes) of sequences: one for Gasoline engines; one for Light Duty Diesel engines; and one for Heavy Duty Diesel engines. Within each of these sets there are categories which reflect different performance requirements—three (A1, A2 and A3) for gasoline engines; four (B1, B2, B3 and B4) for light duty diesel engines; and four (E1, E2, E3 and E4) for heavy duty diesel engines. Typical applications for each sequence are described below for guidance only. Specific applications of each sequence are the responsibility of individual motor manufacturers for their own vehicles/engines.

The sequences define the minimum quality level of a product for self-certificatiion to EELQMS and presentation to ACEA members. Performance parameters other than those covered by the tests shown or more stringent limits may be indicated by individual ACEA member companies.

Nomenclature and ACEA Process

Each set of sequences is designated for consumer use by a 2 part code comprising a letter to define the CLASS (e.g. A), and a number to define the CATEGORY (e.g. A1).

In addition, for industry use, each sequence has a two-digit number to identify the YEAR of implementation of that severity level (e.g. A1-98). An ISSUE number may also be included where requirements have been updated without a change in severity. (e.g. A2-96 Issue 2).

The CLASS indicates oil intended for a general type of engine—currently A = gasoline engines; B = light duty diesel engines; E = heavy duty diesel engines. Other classes may be added in future if, for example, Natural Gas engines prove to require oil characteristics which cannot readily be incorporated into existing classes.

The CATEGORY indicates oils for different purposes or applications within that general class, related to some aspect or aspects of the performance level of the oil. Typical applications for each sequence are described below for guidance only. Specific applications of each sequence are the responsibility of the individual motor manufacturer for his own vehicles and engines. Oils within a category may also meet the requirements of another category, but some engines may only be satisfied by oils of one category within a class.

The YEAR numbers are intended only for industry use and indicate the year of implementation of that severity level for the particular category. A new year number will indicate, for example, that a new test, parameter or limit has been incorporated for the category to meet new/uprated performance requirements whilst remaining compatible with existing applications. An update must always satisfy the applications of the previous issue. If this is not the case, then a new category is required.

An administrative ISSUE Number is added for industry use where it is necessary to update the technical requirements of a sequence without the intention to increase severity (e.g. when a CEC test engine is updated to the latest version whilst maintaining equivalent severity; or where a severity shift in the test requires modification of the specified limits.).

The ACEA Development Decision Guidelines for the update process are shown in Appendix A

This sequence defines the minimum quality level of a product for self-certification to EELQMS and for presentation to ACEA members. Performance parameters other than those covered by the tests shown or more stringent limits may be indicated by individual member companies.

Light Duty Diesel Oils

Requirements	*Test Method*	*Properties*	*Unit*	*Limits*			
				ACEA: B1-98	*ACEA: B2-98*	*ACEA: B3-98*	*ACEA: B4-98*
1. Laboratory Tests							
1.1 Viscosity grades		SAE J300 Latest active issue		No restriction except as defined by shear stability and HT/HS requirements. Manufacturers may indicate specific viscosity requirements related to ambient temperature.			
1.2 Shear stability	CEC-L-14-A-93 (Bosch Injector)	Viscosity after 30 cycles measured at 100°C.	mm^2/s	xW-20: stay in grade xW-30 ≥ 8.6 xW-40 ≥ 12.0	xW-30 ≥ 9.0 xW-40 ≥ 12.0 xW-50 ≥ 15.0	stay in grade	xW-30 ≥ 9.0 xW-40 ≥ 12.0 xW-50 ≥ 15.0
1.3 Viscosity-high temperature high shear rate	CEC-L-36-A-97 (Ravenfield)	Viscosity at 150°C and $10^6 s^{-1}$ shear rate	mPa.s	min 2.9 max. 3.5	> 3.5	> 3.5	> 3.5

(Contd.)

[1]The ATIEL Code of Practice is the sole property of ATIEL and is available from ATIEL (Association Technique de l'Industrie Européenne des Lubrifiants), Madou Plaza, 25th floor, Place Madou 1, B-1210 Brussels, Belgium.

Table 12.7 Contd.

Requirements	*Test Method*	*Properties*	*Unit*	*Limits*			
				ACEA: B1-98	*ACEA: B2-98*	*ACEA: B3-98*	*ACEA: B4-98*
1.4 Evaporative loss	CEC-L-40-A-93 (Noack)	Max. weight loss after 1 h at 250°C	%	≤ 15	≤ 15 for 10W-x or lower. ≤ 13 for others	≤ 13	≤ 15 for 10W-x or lower. ≤ 13 for others
1.5 Sulfated ash	ASTM D874		% m/m			≤ 1.8	
1.6 Oil/elastomer compatibility	CEC-L-39-T-96	Max. variation of characteristics after immersion for 7 days in fresh oil without pre-ageing		RE1	Elastomer RE2	type RE3	RE4
		Hardness DIDC	points	–1/+5	– 5/+5	– 25/+1	– 5/+5
		Tensile strength	%	–50/+10	–15/+10	– 45/+10	– 20/+10
		Elongation rupture	%	– 60/+10	– 35/+10	– 20/+10	– 50/+10
		Volume variation	%	– 1/+5	– 5/+5	– 1/+30	– 5/+5
1.7 Foaming tendency	ASTM D892 without option A	Tendency-stability	ml		Sequence I (24°C) 10 – nil Sequence II (94°C) 50 – nil Sequence III (24°C) 10 – nil		
1.8 High temperature foaming tendency	ASTM D6082 High temperature foam test	Tendency-stability	ml		Sequence IV (150°C) 100 – nil		
2. ENGINE TESTS							
2.1 Ring sticking and Piston cleanliness	CEC L-46-T-93 (VW 1.6 TC D)	Ring sticking	merit	≥ RL 148	≥ RL 148	≥ RL 148	—
		Piston cleanliness	merit	≥ RL 148	≥ RL 148	≥ RL 148	—
2.2 Medium temperature dispersivity	CEC-L-56-T-95 (XUD11ATE) See Note (1) then CEC-L-56-T-9x) (XUD11BTE) (when T status is granted)	Absolute viscosity increases at 100°C and 3% soot (measurement with CEC L-83-A-97 method)	mm²/s	≤ 0.90 × Vk increase with RL 197	≤ 0.90 × Vk increase with RL 197	≤ 0.50 × Vk increase with RL 197	≤ 0.90 × Vk increase with RL 197
		Piston merit (5 elements) (average for 4 pistons)	merit	≥ (RL 197 minus 6 pts)	≥ (RL 197 minus 6 pts)	≥ (RL 197)	≥ (RL 197 minus 6 pts)
2.3 Wear, Viscosity stability & Oil consumption	CEC-L-51-T-95 (OM602A)	Cam wear. average	µm	≤ RL 148	≤ RL 148	50% of ≤ RL 148	≤ RL 148
		Viscosity increase at 40°C	%	FFPL.≤ 90[2)]	FFPL.≤ 90[2)]	FFPL.≤ 90[2)]	FFPL.≤ 90[2)]
		Bore polishing	%	FFPL.≤7.0[2)]	FFPL.≤7.0[2)]	FFPL.≤7.0[2)]	FFPL.≤7.0[2)]
		Piston cleanliness	merit	Rate & Report	Rate & Report	Rate & Report	Rate & Report
		Average engine sludge	merit	Rate & Report	Rate & Report	Rate & Report	Rate & Report
		Cylinder wear. average	µm	FFPL. ≤ 15.0[2)]	FFPL. ≤ 15.0[2)]	FFPL. ≤ 15.0[2)]	FFPL. ≤ 15.0[2)]
		Oil consumption	kg/test	FFPL. ≤ 10.0[2)]	FFPL. ≤ 10.0[2)]	FFPL. ≤ 10.0[2)]	FFPL. ≤ 10.0[2)]
2.4 DI diesel Piston cleanliness & Ring sticking	CEC-L78-T-97 (VW DI)	Piston cleanliness	merit	—	—	—	≥ 65
		Ring sticking (Rings 1 & 2)					
		Average of all 8 rings	ASF	—	—	—	≤0.7
		Max. for 1 individual ring	ASF	—	—	—	≤2.5
		Viscosity increase at 40°C	%	—	—	—	Rate & Report
2.5 Fuel economy See Notes (3)	CEC-L-54-T-96 (M111E)	Fuel economy improvement vs. Reference oil RL 191 (15W-40)	%	≥ 2.5	-	-	-

(1) Data from an existing test to the 1996 issue of these sequences may be reassessed to demonstrate compliance with the requirement, as detailed in the Equivalency Guidelines in Appendix E of the ATIEL Code of Practice.
(2) FFPL: Fit-for-purpose limits determined according to the principles outlined in Appendix E of the ATIEL Code of Practice, and subject to regular review.
(3) ACEA considers the CEC L-54-T-96 test the only valid comparator against which claims of lubricant fuel economy improvement should be made.

Heavy Duty Diesel Oils

Requirements	*Test Method*	*Properties*	*Unit*	*Limits*			
				ACEA: E1-96 Issue 2	*ACEA: E2-96 Issue 2*	*ACEA: E3-96 Issue 2*	*ACEA: E4-98*
1. LABORATORY TESTS							
1.1 Viscosity		SAE J300 Latest active issue		No restriction except as defined by shear stability and HT/HS requirements. Manufacturers may indicate specific viscosity requirements related to ambient temperature.			

(Contd.)

Table 12.7 Contd.

Requirements	*Test Method*	*Properties*	*Unit*	*Limits*			
				ACEA: E1-96 Issue 2	*ACEA: E2-96 Issue 2*	*ACEA: E3-96 Issue 2*	*ACEA: E4-98*
1.2 Shear stability	CEC-L-14-A-93 (Bosch Injector)	Viscosity after 30 cycles measured at 100°C.	mm^2/s	xW-30 ≥ 9.0 xW-40 ≥ 12.0 xW-50 ≥ 15.0 No requirements for single grades			Stay in grade
1.3 Viscosity-high temperature high shear rate	CEC-L-36-A-97 (Ravenfield)	Viscosity at 150°C and $10^6\ s^{-1}$ shear rate	mPa.s			> 3.5	
1.4 Evaporative loss	CEC-L-40-A-93 (Noack)	Max. weight loss after 1 h at 250°C	%			≤ 13	
1.5 Sulfated ash	ASTM D 874		% m/m			≤ 2.0	
1.6 Oil/elastomer compatibility	CEC-L-39-T-96	Max. variation of characteristics after immersion for 7 days in fresh oil without pre-ageing		RE1	Elastomer RE2	type RE3	RE4
		Hardness DIDC	points	–1/+5	– 5/+5	– 25/+1	– 5/+5
		Tensile strength	%	–50/+10	–15/+10	– 45/+10	– 20/+10
		Elongation rupture	%	– 60/+10	– 35/+10	– 20/+10	– 50/+10
		Volume variation	%	– 1/+5	– 5/+5	– 1/+30	– 5/+5
1.7 Foaming tendency	ASTM D892 without option A	Tendency-stability	ml		Sequence I (24°C) 10 – nil Sequence II (94°C) 50 – nil Sequence III (24°C) 10 – nil		
1.8 High temperature foaming tendency	ASTM D6082 High temp. foam test	Tendency-stability	ml		Sequence IV (150°C) 100 – nil		
2. ENGINE TESTS							
2.1 Bore polishing/ Piston cleanliness	CEC L-42-A-92 (OM 364 A) See Note (1). or CEC L-42-T-9X (OM364LA) when T status is granted	Bore polishing	%	≤ RL 134	≤ A	≤ RL 133	—
		Piston cleanliness	merit	≥ RL 134	≥ A	≥ RL 133	—
		Average cylinder wear	µm	≤ RL 134	≤ A	≤ RL 133	—
		Sludge	merit	≥ RL 134	≥ A	≥ RL 133	—
		Oil consumption	kg/test	≤ RL 134	≤ A	≤ RL 133	—
					where A = 0.5 × (RL 133 + RL 134)		
2.2 Wear, Viscosity stability & Oil consumption	CEC-L-51-T-95 (OM602A)	Cam wear. average	µm	≤ RL 148	≤ RL 148	≤ RL 148	≤ 50% of RL 148
		Viscosity increase at 40°C	%	Rate & report	Rate & report	Rate & report	FFPL. ≤ 90[2]
		Bore polishing	%	Rate & report	Rate & report	Rate & report	FFPL. ≤ 7.0[2]
		Piston cleanliness	merit	Rate & report	Rate & report	Rate & report	Rate & report
		Average engine sludge	merit	Rate & report	Rate & report	Rate & report	Rate & report
		Cylinder wear. average	µm	Rate & report	Rate & report	Rate & report	FFPL. ≤ 15.0[2]
		Oil consumption	kg/test	Rate & report	Rate & report	Rate & report	FFPL. ≤ 10.0[2]
2.3 Soot in oil	ASTM D5967 (Mack T8E)	Test duration	Hours				300
		Relative viscosity at		—	—	—	4.8% soot
		1st test		—	—	—	2.1%
		2 test average.		—	—	—	2.2%
		3 test average.		—	—	—	2.3%
	ASTM D4485 (Mack T-8)	Test duration:	Hours	—	250	250	—
		Viscosity increase at:				3.8% soot	3.8% soot
		1st test	cSt	—	Rate & report	≤ 11.5	≤ 11.5
		2 test average.	cSt	—	Rate & report	≤ 12.5	≤ 12.5
		3 test average.	cSt	—	Rate & report	≤ 13.0	≤ 13.0
		Filter plugging, Diff. pressure	kPa	—	Rate & report	≤ 138	≤ 138
		Oil consumption	g/kWh	—	Rate & report	≤ 0.304	≤ 0.304
2.4 Bore polishing	CEC L-52-T-97	Bore polishing	%	—	—	—	FFPL. ≤ 2.0[2]
Piston cleanliness		Piston cleanliness	merit	—	—	—	≥ 40.0
Turbocharger deposits	(OM441LA)	Turbocharger deposits	mg	—	—	—	Rate & report

(1) Data from an existing test to the first 1996 issue of the E1, E2 and E3 sequences may be reassessed to demonstrate compliance with the requirements as detailed in the Equivalency Guidelines in Appendix E of the ATIEL Code of Practice.

(2) FFPL: Fit-for-purpose limits calculated according to the principles outlined in Appendix E of the ATIEL Code of Practice, and subject to regular review.

Table 12.8 UK Ministry of Defence specification DEF STAN 91–43/1 (1979)

API engine service description
Moderate duty diesel engine service. Service typical of diesel engines operated in mild to moderate duty.

ASTM engine oil description
Oil providing protection from piston deposits and bearing corrosion when using high-sulphur fuel.

Engine tests	*Assessment*
Petter AV–1; 120h; Petter AVI diesel engine	Ring sticking; engine deposits
Petter W–1; 36h; Petter W1 gasoline engine	Bearing weight loss; ring sticking; piston skirt and interior ratings; increase in oil viscosity, 40°C

In addition the specification incorporates an FZG gear test to determine load-carrying capacity, and also tests to evaluate elastomer compatibility and foaming characteristics.

12.8.1 Engine test rating

On completion of an engine test the engine is stripped down and individual components are examined to assess the performance of the oil, which is judged by the cleanliness of the components and the amount of wear, the results being compared with those from a reference oil of known performance.

When two engines using different oils are examined side by side it is usually fairly easy to decide which is the better oil, but it is not easy to do this when comparing engines from tests run at other times or in other locations, except for wear, since this is assessed quantitatively whereas cleanliness, often the main criterion, is a subjective judgement. To facilitate such comparisons a 'merit rating' system is used in which a rating from 0 to 10 is given to each component in terms of area, thickness and colour of deposit, and an overall rating is obtained by a summation of the individual ratings weighted according to their importance. Consideration is given, however, to the individual ratings since an extremely poor rating on one component may fail an oil despite a reasonable overall rating.

Two rating conventions are in use: a 'merit' rating where 0 is the maximum deposit and 10 is the cleanest condition, and a 'demerit' rating where 0 is the cleanest and 10 is the dirtiest condition. Rating systems have been developed by the Coordinating Research Council (CRC), The Institute of Petroleum (IP) and the Co-ordinating European Council (CEC). These differ in detail but are all based on the same principle, and give equivalent results.

12.9 Laboratory inspection tests

Apart from engine tests that assess the performance of a lubricating oil, as discussed in Section 12.8, the quality of an oil is assessed by a variety of tests for chemical and physical properties that are applied at various stages in the life of the oil, from manufacture to ultimate use, including tests on the used oil. Some of these tests have a direct bearing on performance, e.g. viscosity and foaming; others have little or no relevance to performance, e.g. carbon residue, flash point and density, but are useful as quality controls.

Some properties, e.g. kinematic viscosity, sulphur content, are absolute quantities and can be determined by any scientifically sound method, but many, e.g. pour point and flash point, are arbitrarily defined and the results obtained in their determination depend on the apparatus and procedure employed. In order to obtain universally significant and comprehensible results it is essential that the arbitrary methods are standardized, and it is desirable that even the absolute methods should also be standardized.

Most of the methods applied to lubricating oils have been standardized by organizations such as the American Society for Testing and Materials (ASTM), the Institute of Petroleum (IP), the British Standards Institution (BSI), the Deutsches Institut für Normung (DIN) and the International Standardization Organization (ISO). The ASTM and the IP standard methods are the most commonly and widely used. Prior to 1963 there were significant differences between ASTM and IP methods for the determination of nominally the same property but since then these differences have been eliminated by joint ASTM/IP methods. ASTM and IP methods are kept up-to-date by constant review and are published annually; the current volume should be consulted for details. The most important of these methods are briefly reviewed in this section in alphabetical order.

Acidity/Alkalinity *See Neutralization value (p. 325).*

Ash Ash is the percentage of mineral matter remaining after burning a sample of oil and further heating the residue to free it from carbon. Well refined, unused straight mineral oils yield only a trace of ash unless they have been contaminated during storage and handling. Unused oils containing other than ashless additives yield a relatively high ash depending on the nature and quantity of the additive. The ash of used oils will contain, in addition to material derived from additives, metal particles derived from engine wear and inorganic matter derived from contamination.

For straight mineral oils ash determination by combustion and subsequent heating (ASTM D482; IP 4) will suffice but for additive-containing oils the residue from combustion is treated with sulphuric acid prior to further heating, thereby converting volatile compounds into less volatile sulphates or oxides. The 'sulphated ash' so determined (ASTM D874; IP 163) is higher than the unsulphated ash both because of the prevention of loss from volatile compounds and because of the extra sulphur and oxygen combined in the ash.

The sulphated ash may be qualitatively or quantitatively analysed, either chemically or spectrographically, for barium, calcium, magnesium, zinc, phosphorus, etc., in order to assess the nature and concentration of the additives.

The ash from an oil can contribute in some measure to combustion chamber deposits and deposits on exhaust valves although the main contribution to these is usually from the fuel.

Asphaltenes *See Insolubles (p. 325)*

Base number *See Neutralization value (p. 325)*

Carbon residue When a straight mineral oil is heated in the absence of air, or in the presence of insufficient air for complete combustion, the volatile fractions are driven off but the nonvolatile, more complex compounds decompose and form carbonaceous deposits. The residue of oils containing ash-forming additives will, of course, also contain inorganic matter. Two methods are used to determine the 'carbon residue' or 'coke number'.

Ramsbottom method (ASTM D524; IP 14) A weighed quantity of the oil is sucked into a glass bulb through its capillary neck and is heated to 550°C by immersing the bulb in a bath of molten metal. The residual coke is weighed and expressed as a weight percentage of the original sample.

Conradson method (ASTM D189; IP 13) A weighed quantity of the oil is heated in a crucible under controlled conditions with virtual exclusion of air until a carbonaceous residue remains. The amount of residue is expressed as a weight percentage of the original sample.

Carbon residue has little significance in relation to engine deposits. It has been thought to indicate the tendency of the oil to form combustion chamber deposits but this is very dubious; often enough some of the best-performing lubricants have high carbon residues.

Cloud point (ASTM D2500; IP 15) The cloud point is the temperature at which a cloud or haze appears in the oil when cooled under the conditions of the test. This is the temperature at which wax begins to crystallize from the oil; it has no practical significance as regards performance.

Colour Colour is measured in arbitrary units by a variety of methods (ASTM D156, ASTM D1500; IP 17). It has no significance in lubrication.

Demulsification number (ASTM D1401; IP 219) The demulsification number is a measure of the ability of an oil to separate from an emulsion with water; it is the number of seconds required for the oil to separate under defined conditions from an emulsion formed either by stirring (ASTM) or by steam blowing (IP). Unused oils usually have a low demulsification number but some additive-containing oils have a high number. Contamination of the used oil by combustion products increases its tendency to form emulsions.

Density (ASTM D287, D1298, D941; IP 59, 160) Density is the mass per unit volume of oil at a specified temperature, usually 15°C for lubricating oils. Specific gravity is the ratio of the weight of a given volume of oil to the weight of the same volume of water, both at specified temperatures, either 15/15°C or 15/4°C being usual for lubricating oils. Density is now the preferred term.

Density or specific gravity has no practical significance in lubrication but is useful for costing purposes or for converting weights to volumes or vice versa. The density or specific gravity multiplied by ten gives the pounds weight per UK gallon. In conjunction with viscosity it can give an indication of the source of the oil or its method of refining.

Diesel fuel diluent The degree of dilution of the lubricating oil by diesel fuel cannot be readily determined by distillation because the distillation ranges of the fuel and the lubricant may overlap. Dilution may sometimes be roughly estimated by the reduction in viscosity of the oil, provided, of course, that the fuel is less viscous than the oil. But this is a rather unreliable method since oxidation of the oil and suspended insoluble material will increase the viscosity of the oil. A new method of determining diesel fuel dilutiion (ASTM D3524) is now available, using gas chromatography to compare the boiling point distributions of the contaminated oil and of the unused oil.

Engines in good condition and operated at normal temperatures usually show little dilution of the crankcase oil but it may occur in vehicle and even locomotive engines where high-speed idling and over-running are frequent. Incomplete combustion or misfiring of the very small amounts of fuel injected per cycle may occur and result in dilution. Such misfiring is promoted by low ambient temperatures and will continue below perhaps 0°C, depending on the engine. Dilution can also occur, apart from that due to leakage into the crankcase, by loss of fuel from the sac of multi-hole injection nozzles if this volume is not kept as small as possible. This can happen in both normally aspirated and turbocharged engines, particularly if there is overlap, since after the inlet valve opens the air flow across the nozzle may cause sufficient pressure difference to force fuel out of the sac. The best safeguard against undue dilution is to maintain the fuel injection system in good order.

Flash point (Pensky-Martens ASTM D93; IP34. Cleveland ASTM D92; IP 36) The flash point is the temperature to which the oil must be heated under prescribed conditions to give off sufficient vapour to form a mixture with air that can be ignited momentarily by a specified flame. The closed flash point is the temperature at which inflammable vapour is produced in a closed space above the oil, the open flash point is the temperature at which inflammable vapour is produced when the surface of the oil is exposed to the atmosphere. The fire point is the temperature at which the oil continues to burn in an open cup after withdrawal of the test flame. The flash point depends on the apparatus used, of which there are several, and it is therefore always necessary to state which instrument is used.

Lubricating oils usually have closed flash points above 150°C: a low flash point indicates contamination with a more volatile material and may then be significant in relation to fire hazard, but otherwise flash point has no relation to performance in an engine.

Foaming (ASTM D892; IP 146) The amount and stability of foam produced by aeration of the oil under standard conditions of air-blowing are determined by measuring the volume of foam produced immediately on cessation of blowing and again after settling for ten minutes. The test is repeated on another sample of oil at a higher temperature. The test indicates the extent to which foaming trouble, such as loss of oil from crankcase vents, is likely.

Inorganic constituents Chemical or spectrographic analysis of either unused or used oil may be carried out to determine the content of various elements that may be derived from additives or from contaminants. Comparison of results from used and unused oils can indicate the nature and amount of contamination or the extent of additive depletion. Analysis of the unused oil will indicate the nature and amount of additives present.

Periodic spectrographic analysis of the crankcase oil has been successfully used by some operators, notably of locomotive and marine engines, to monitor engine conditions. Any sudden increase in copper, iron or aluminium will give timely warning of rapid or abnormal wear in a bearing or a cylinder—and permit the engine to be taken out of service before more serious damage occurs.

Insolubles (ASTM D893; IP 143) The material insoluble in or precipitated by *n*-pentane (ASTM) or *n*-heptane (IP) is first determined. It consists of inorganic matter, carbonaceous matter and asphaltic matter. The asphaltic matter is extracted by solution in benzene or toluene and is weighed after evaporation of the solvent. (The methods specify extraction with benzene but toluene is more generally used because of the toxicity of benzene.) This asphaltic matter is sometimes referred to as *asphaltenes*. The residue left after solvent extraction is inorganic and carbonaceous matter, the former consisting of metallic wear debris and corrosion products, metallic compounds from the decomposition of additives, and inorganic contaminants from external sources. The inorganic matter is determined as sulphated ash from the benzene residue and may be further analysed by chemical or spectrographic methods.

These tests are applied only to used oils and indicate the amount of contamination.

Neutralization value (ASTM D664, D974, D2896; IP 1, 139, 177, 182, 276) Neutralization value is a measure of the acidity or alkalinity of the oil and is determined by titration of the oil with potassium hydroxide (for acidic oils) and hydrochloric acid or perchloric acid (for alkaline oils), the result being expressed as the amount of potassium hydroxide (KOH) required to neutralize one gram of acid oil or to be equivalent to the acid required to neutralize one gram of an alkaline oil. The amount of KOH is expressed in milligrams and the neutralization value is given as mg KOH/*g* oil. Alternative names for neutralization value are *acid number* and *base number* for acid and alkaline oils respectively. It is important to specify which method is used since results differ according to the method.

The total acidity TAN (total acid number) is a measure of the combined organic and inorganic acids; the strong acid number (SAN) is a measure of the inorganic acidity, and the difference is the organic acidity. Similarly the TBN (total base number) is a measure of the alkalinity of the oil.

Acidity in an unused oil indicates that it may be corrosive to metals; inorganic acidity of used oils indicates contamination by acidic products of combustion; organic acidity indicates oil oxidation.

Unused oils containing dispersant additives are normally alkaline. The extent to which alkalinity is reduced in a used oil may be a measure of the loss of dispersancy or alkalinity by additive depletion.

Oxidation stability (ASTM D943; IP 48,280) Lubricating oil in an engine necessarily comes into contact with air (and therefore oxygen) at high temperatures, and sometimes when in a finely divided state. Oxidation is inevitable under these conditions, depending on the chemical constitution of the oil and the temperature to which it is exposed, but engine lubricants nowadays containing anti-oxidant additives that prevent undue oxidation under normal conditions during the expected life of the oil.

There are many laboratory oxidation tests, differing in detail but following the same general principles—exposure for a prescribed time to air or oxygen in the presence of a catalyst at a temperature commensurable with that to which it is exposed in practice, or somewhat higher to expedite the process. The stability of the oil is assessed by the amount of oxygen absorbed, the quantity and quality of the oxidation products, or both.

The correlation between the results of such tests and oxidation stability in practice in the engine is somewhat tenuous and more reliance is nowadays placed on engine tests.

Pour point (ASTM D97; IP 15) The pour point is the lowest temperature at which the oil can be poured under specified conditions. It is determined by cooling the oil in a glass jar until it no longer flows when the jar is tilted. The pour point should always be below the lowest temperature at which it is expected to start the engine but it has no significance under running conditions.

Sulphur content (ASTM D129; IP 61) Sulphur content is determined by igniting the oil in oxygen in a bomb, absorbing the combustion products in aqueous alkali and analysing the resulting solution for sulphur content. Sulphur content of the lubricating oil has no significance in relation to lubrication as such but may be useful indicating the source of the oil. The presence of some additives may increase the sulphur content.

Total acid/base number *See Neutralization value (p. 325).*

Viscosity (ASTM D445; IP 71) The determination of viscosity has already been discussed in Section 16.3.2. Engines are not critical to comparatively large changes in viscosity, and specification of the SAE number usually suffices for the selection of an appropriate grade of lubricating oil. Used oils may show an increase in viscosity owing to oxidation or the presence of suspended matter or a decrease in viscosity owing to fuel dilution. These features tend to compensate each other, and which predominates depends on operating conditions.

Viscosity Index (ASTM D2270; IP 226) Viscosity index (VI)

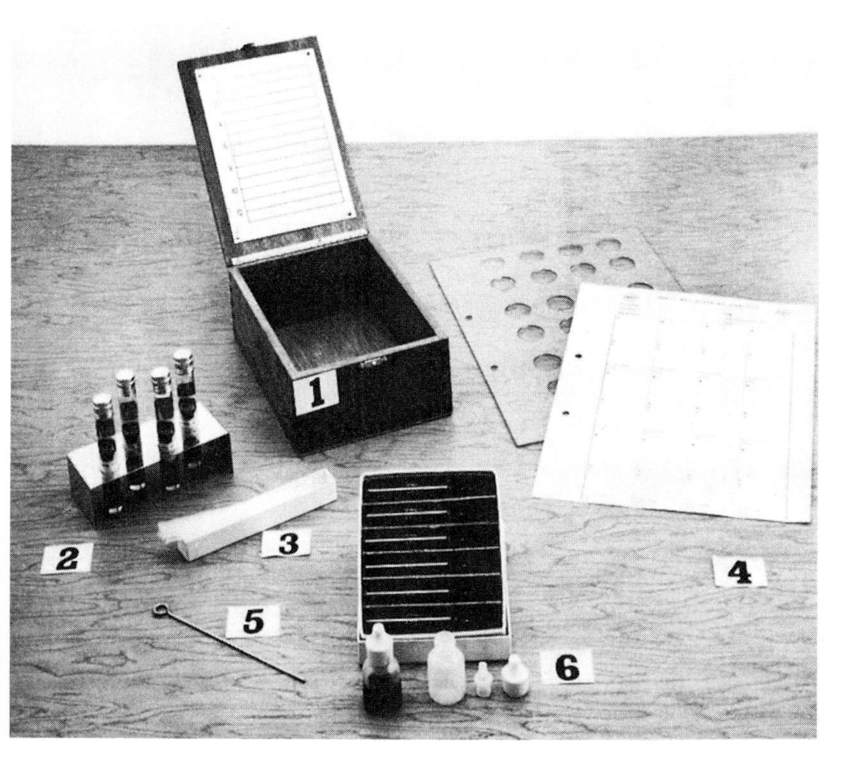

Figure 12.10 A blotter-spot and viscosity test kit

1. Box with polished aluminium plate, marked with centimetre scale
2. Reference oils
3. Glass dropper tubes
4. Blotter sheet and backing card
5. Stainless steel dropper rod
6. Ampoules of indicator solution and dropper bottles

has already been discussed (see Section 12.3.3). The higher the VI the less is the change in viscosity with change in temperature. A high VI is desirable to ensure that the viscosity remains suitable over a wide range of operating temperatures. Apart from this it has no other practical significance in lubrication. It is an indication of the degree of solvent extraction or the incorporation of VI improvers. HVI oils are predominately composed of paraffinic hydrocarbons whereas LVI oils consist mainly of naphthenic and aromatic hydrocarbons. There is no relation between VI and dispersancy.

Water content (ASTM D95; IP 74) Water is removed from 100 ml of oil by distillation and is condensed into a graduated glass tube in which the volume is measured. The water content is expressed as %v. Unused oils should be free from water. Small quantities of water are frequently found in used oils, derived from combustion products, leakage from the cooling system or condensation from the atmosphere.

12.10 Spot tests

Laboratory examination of used oils inevitably entails considerable delay before results are available to the engine operator. More immediate results can be obtained by the use of 'do-it-yourself' test kits whereby engine operators in the field can keep a check on the quality of the lubricating oil, with special reference to oil-change period.

The simplest type of kit, illustrated in *Figure 12.10* is one in which a spot of used oil is placed on an absorbent paper. Oil insolubles and dispersion are judged by the density of the oil spot and the rate at which relatively clean oil spreads out from it. Acidity, neutrality, alkalinity can be gauged by the colour of the spot when treated with a given reagent. Viscosity is judged by the rate of flow of a drop of oil on an inclined metal plate in comparison with the rates of flow of similar drops of two reference oils of known viscosity. Alternatively viscosity may be judged by the rate of fall of a metal ball in a tube of the oil in comparison with the rates of fall in two similar tubes containing reference oils of known viscosity.

A more elaborate type of test kit is shown in *Figure 12.11*. It contains all the necessary solvents, reagents and test vessels to permit a particular brand of oil to be examined for alkalinity, dispersancy, contamination, viscosity, water content, total acid number of total base number.

These spot tests are usually specific to a particular brand of oil, the test kit being supplied by the oil supplier. For example, the indicator solution for the determination of alkalinity, acid and base numbers in the kit illustrated in *Figure 12.11* is specific to the particular engine oil in use and has to be provided by the oil supplier.

Acknowledgements

The author gratefully acknowledges the assistance of Shell International Petroleum Company Ltd and Shell Research Ltd in the provision of tables and figures.

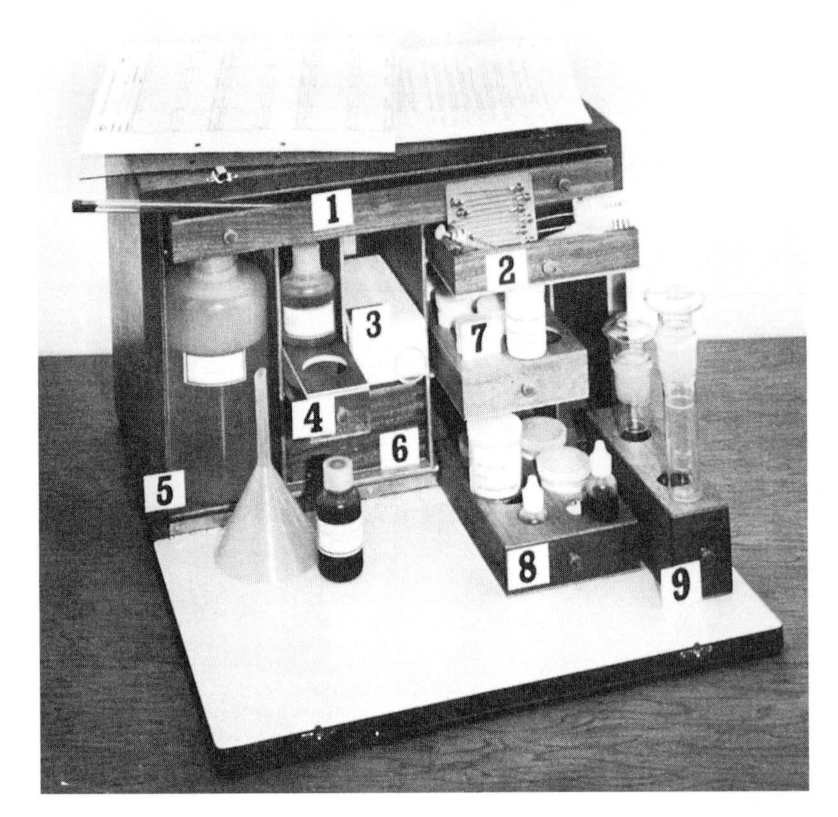

Figure 12.11 A comprehensive test kit
1. Drawer containing instructions, parts list, viscosity graphs, blotter sheets, thermometer
2. Drawer containing syringe and needles, pipette, cleaning brushes, dropper bottles and vials
3. Water content test vessel and holder
4. Holder for alkali indicator and hydrochloric acid
5. Titration solvent and funnel
6. A blotter-spot/viscosity kit as illustrated in *Figure 12.10*
7. Jars of water-content reagent
8. Jars of TAN indicator capsules and dropper bottles for item 6
9. Stand with TAN test vessels

References

BOWDEN, F.P. and TABOR, D., *Friction and Lubrication,* Methuen (1967)
BRAITHWAITE, E.R., *Lubrication and Lubricants,* Elsevier (1967)
ELLIS, E.G., *Fundamentals of Lubrication,* Scientific Publications (G.B.), Broseley, Shropshire (1968)
DOWSON, D. and HIGGINSON, G.R., *Elastohydrodynamic Lubrication,* Pergamon (1969)
PUGH, B., *Practical Lubrication,* Newnes-Butterworth (1970)
PUGH, B., *Friction and Wear,* Newnes-Butterworth (1973)
CAMERON, A., *Basic Lubrication Theory,* Ellis Horwood (1976)
GUNDERSON, R. C. and HART, A. W., *Synthetic Lubricants,* Reinhold (1962)
RANNEY, M. W., *Synthetic Oils and Greases,* Noyes Data Corporation, New Jersey (1979)
RANNEY, M. W., *Lubricant Additives,* Noyes Data Corporation, New Jersey (1978)
HOBSON, G. D. and POHL, W. (Eds), *Modern Petroleum Technology,* Applied Science Publishers (1975)
SELL, G. and WELLS, M. J. (Eds), *Engine Testing of Crankcase Lubricating Oils,* The Institute of Petroleum (1962)
Shear Stability of Multigrade Crankcase Oils, ASTM Data Series Publication DS49 (1973)
McCUE, C. F., CREE, J. C. G. and TOURRET, R. (Eds), *Performance Testing of Lubricants for Automotive Engines and Transmissions,* Applied Science Publishers (1974)
Significance of ASTM Tests for Petroleum Products, ASTM Special Technical Publication C (1977)
ASTM Standards on Petroleum Products and Lubricants, American Society for Testing and Materials (Annual publication)
IP Standards for Petroleum and its Products, The Institute of Petroleum (Annual publication)
SAE Handbook, Section 13, Fuels and Lubricants, Society of Automotive Engineers (Annual publication)
CEC Standard Test Procedures and Codes of Practice; Test Method Summary Sheet (*TMSS*), 1st Edn, Co-ordinating European Council, London (1978)

Abbreviations

ACEA	Association des constructors Europeens d'Automobile
API	American Petroleum Institute
ASTM	American Society for Testing and Materials
BSI	British Standards Institution
CCMC	Committee of Common Market Automobile Constructors
CEC	Co-ordinating European Council
CRC	Co-ordinating Research Council
DIN	Deutsche Industrienormen
	Deutsches Institut für Normung
HVI	High Viscosity Index
IFP	Institut Français du Pétrole
IP	Institute of Petroleum
ISO	International Standardization Organization
LVI	Low Viscosity Index
MVI	Medium Viscosity Index
NV	Neutralization Value
SAE	Society of Automotive Engineers
SAN	Strong Acid Number
TAN	Total Acid Number
TBN	Total Base Number
VHVI	Very High Viscosity Index
VI	Viscosity Index
UHVI	Ultra High Viscosity Index

13 Bearings and bearing metals

Contents

13.1 Introduction

Any bearing can be simply defined as a support or guide which at the same time allows relative movement to take place between two bodies. In a diesel engine, the principal bearings are those which allow rotation of the crankshaft about its own longitudinal axis in the main engine entablature, and those between the connecting rod and crankshaft. In modern engine terminology a bearing has come to mean the component fitted between the journal and either the main bearing housing or the connecting rod, the shaft being supported in the bearing via an oil film. *Figure 13.1* indicates the interaction between the various components in any engine bearing assembly, and the environment within which the bearing system must operate[1].

The co-operating surface, the lubricant and the environment all place constraints on the bearing, the resulting design usually being a compromise between various conflicting requirements. In relation to the bearing itself, the choice is usually restricted to material and geometry. However, environmental factors have to be taken into serious consideration at the design stage if a reliable system is to be achieved.

13.2 Bearing design

13.2.1 Wall thickness

To achieve optimum utilization of space, reliability and repeatability of the bearing assembly, thin shell bearings have been almost universally adopted in modern diesel engines, and fitted retrospectively into many older engines. Such a bearing allows the use of a wider range of surface lining materials than would otherwise be possible.

The thin shell bearing is a precision made pre-finished component consisting of a steel backing with a thin coating of the most appropriate lining material necessary to provide sufficient strength for the applied loading and compatibility with the general environment. The overall thickness/diameter ratio is not critical, but typically varies from 0.05 at 40 mm diameter to 0.02 at 400 mm as shown by *Figure 13.2*.

13.2.2 Interference fit[2]

The assembled form of a thin shell bearing is dictated by the tolerance to which it can be manufactured, and by the housing into which it fits. To ensure conformity of the bearing shell to its housing, an accurate interference fit has to be provided. This is obtained by means of an excess peripheral length in each half-bearing which has to be closely prescribed to allow interchangeability, requiring precise quality control of every half-bearing in a checking fixture (*Figure 13.3*). On assembly the excess peripheral length creates a circumferential stress around the bearing, and a radial contact pressure between the bearing back and the housing bore. This contact pressure resists relative movement, thus preventing fretting. Unfortunately, there is no theoretically correct level; housings with high flexibility require more contact pressure than stiffer ones. On early engines, having thin shell bearings, a contact pressure as low as 2MPa was usually sufficient to resist fretting, but as engine ratings increased, and housing stress analysis became more sophisticated, higher pressures became necessary, often reaching 8–10 MPa today. In these very high interference fit assemblies, particular care has to be taken to ensure that the joint face clamping bolts have sufficient capacity to assemble the bearing, and to resist dynamic separating forces generated by engine operation.

As the contact pressure is increased for any given bearing size, the circumferential stress increases to the point where the

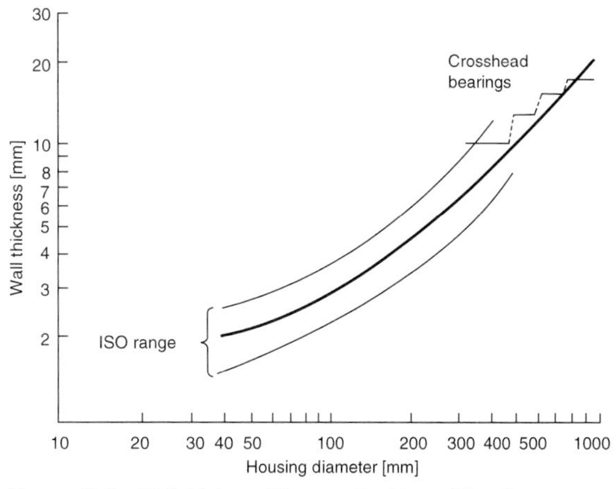

Figure 13.2 Wall thickness/diameter for thin wall bearings

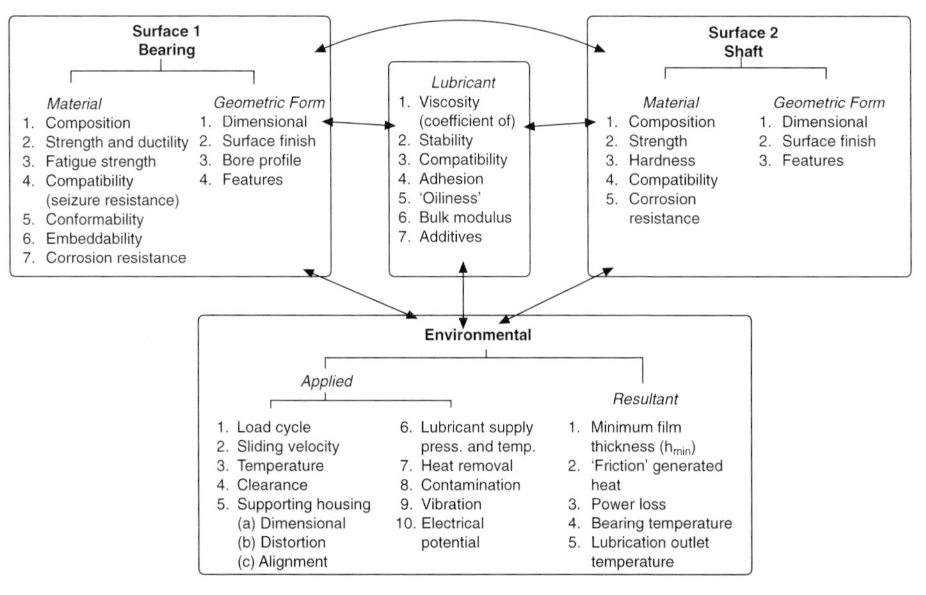

Figure 13.1 The bearing system-interaction between components and environment

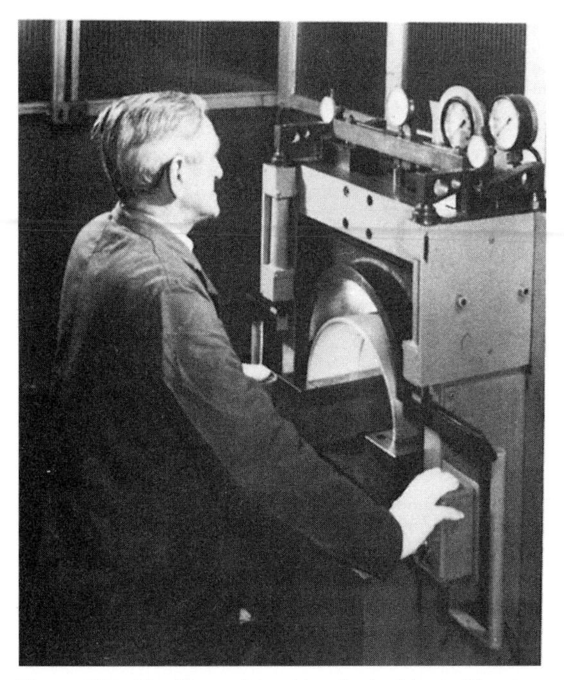

Figure 13.3 Checking peripheral length of a thin shell bearing

steel backing begins to yield adjacent to the joint faces. Knowing the combined effect of bearing steel yield strength and the friction force for bearing assembly, a wall thickness can be determined which will avoid yield, although in some automotive-type engines this may not be feasible.

Increased contact pressure requires a greater bolt tension for fitting bearing caps to their opposite half-housings, and as higher values are incorporated, measurement checks should be carried out on prototype assemblies to ensure that satisfactory closure of the housing joints occurs. *Figure 13.4* shows a convenient method of checking this closure, the micrometer readings being taken as the bolt load is gradually increased. Insufficient bolt pre-load would result in fretting between the bearing and housing, and could allow the housing joints to separate, giving rise to high dynamic loading of the bolts, with possible fatigue fracture.

To reduce the tendency to fretting, it is recommended that the centre line average (CLA) of the housing bore surface finish should not exceed 1.6 μm. Bearing backs are typically 0.8 *μm* CLA surface finish and, in highly loaded zones, should not be unsupported. Cyclic variation of hydrodynamic oil pressure on

Figure 13.4 Method of checking housing joint face closure

the bearing surface will cause the bearing to flex in these unsupported regions. If, for example, there are grooves or oil holes behind the bearing surface, fretting will almost certainly occur. If flexure is sufficiently great, the lining material on the bearing surface can also suffer fatigue damage. *Figure 13.5* shows fretting damage on the back caused by a combination of insufficient interference fit and lack of support. The resulting flexure has also caused the lining material on the bore surface to fatigue.

Provided a bearing is manufactured to the defined length and wall thickness limits, and the housing and bolting arrangements are satisfactory, correct assembly should be achieved every time, with bearing shells being interchangeable. However, locating tangs and free spread greatly assist correct assembly of thin shell bearings, particularly replacement *in situ*.

13.2.3 Locating tangs

Locating tangs allow correct axial positioning of bearing shells thereby ensuring alignment of oil transfer passages and clearance between the ends of the bearing and crankshaft fillets. The tang is located in a corresponding recess in the housing and must be a clearance fit, otherwise the bearing surface may distort. It is also advantageous to relieve the tang below the level of the bearing joint face, thus reducing localized pressure on the tang during assembly.

Tangs are not intended to resist rotation of the bearing shell; that is achieved by the contact pressure. If, under seizure conditions, this contact pressure cannot prevent rotation, the tangs are either sheared or flattened.

13.2.4 Free spread

Free spread is the term given to the excess dimension across the bearing joints with the bearing in its free state, compared with the housing diameter. This allows the bearing to be positively located radially and should generate sufficient friction to hold the shell in its housing even when inverted.

Following operation in an engine, bearings may show a loss of free spread on removal for the following reasons:

(i) The lining of a steel-backed bearing, because of its higher thermal expansion, can yield in compression at operating temperature producing a residual tensile stress when cooled to room temperature. This effect becomes greater with higher lining to steel thickness ratio.
(ii) A high radial temperature gradient across the steel backing, due to malfunction, can cause differential yield and therefore additional free spread loss.
(iii) Relief of strains induced during fitting or by the manufacturing and machining processes.

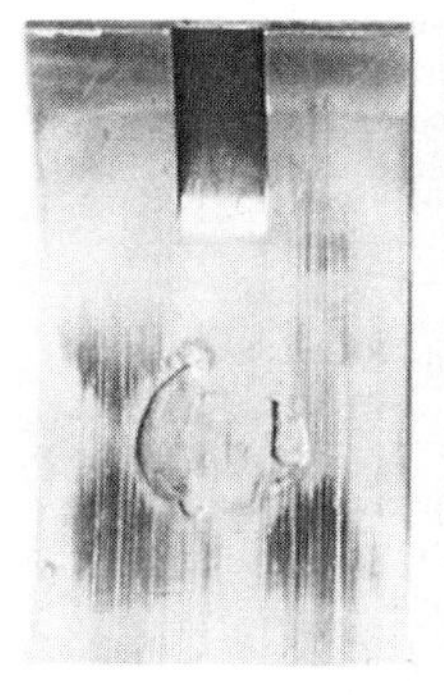
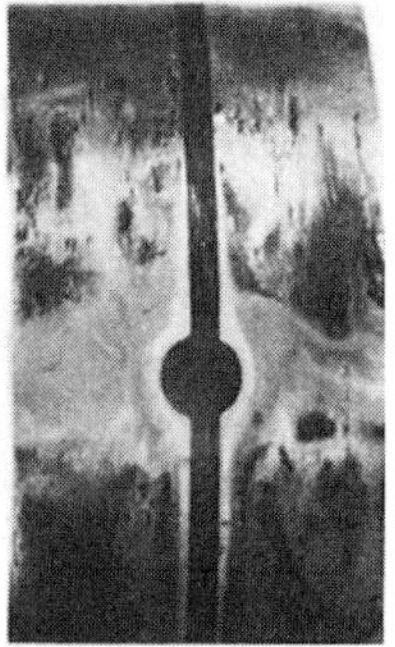

Figure 13.5 Fatigue and fretting damage due to insufficient interference fit and unsupported region of the bearing back

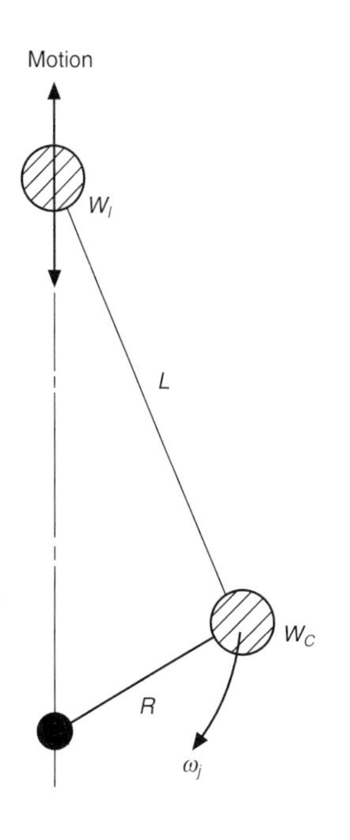

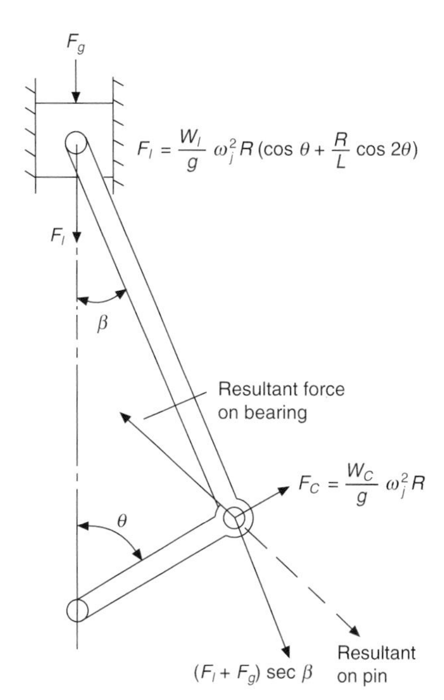

Figure 13.6 Components of load acting on large end bearing

The forces which result in free spread loss after releasing the bearing from its housing are of relatively small magnitude and will not cause malfunction of the bearing whilst in operation. Problems only arise upon reassembly, particularly if one half has negative and the other positive free spread. The half with the negative free spread may become trapped against the shaft and lead to overheating or seizure. Therefore bearings with negative free spread should not be refitted.

13.2.5 Loading on crankpin and main bearings[3]

The load on both crankpin and main bearings varies in magnitude and direction with time. When studying the performance of such bearings, it is necessary to determine these loads, over each complete loading cycle. Such data are usually presented as polar load diagrams.

The loads on the connecting rod big end bearing can be attributed to three components; reciprocating inertia forces, rotating inertia forces and gas forces, as shown by the two diagrams of *Figure 13.6*. Vector addition of these forces at regular intervals of crank angle results in the typical polar load diagram of bearing forces for one loading cycle shown in *Figure 13.7*.

The loads on main bearings are partly due to force reactions from the big end bearings and partly due to out of balance of the crankshaft. The out of balance of the crankshaft is usually reduced by the use of balance weights. When considering the forces from the big end bearing, it is necessary to orientate them to the same non-rotating datum as the main bearing (i.e. relative to a cylinder axis).

In a multi-cylinder engine, which generally in the modern engine has a main bearing between each crank pin, the force on that main bearing is usually assumed to consist only of forces from components in the two crank bays immediately adjacent to that bearing. The resultant force is obtained by treating the crankshaft as a series of simply supported beams resting on supports at the main bearings (statically determinate method). By this means, component forces closest to any one main bearing have the largest effect on that bearing. Then, in much the same way as the polar load diagram of the big end bearing is built up over a complete loading cycle, the overall dynamic load pattern on each main bearing is obtained by vector addition of all the forces at each interval of crank angle.

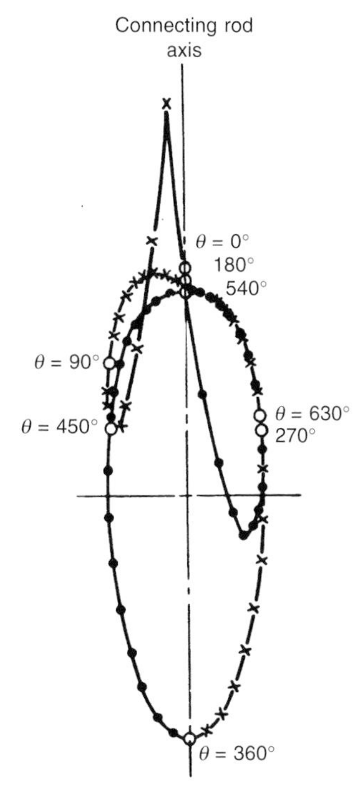

Figure 13.7 Typical large end polar load diagram

13.2.6 Prediction of oil film thickness

Once the dynamic load pattern as a function of time has been etablished for any bearings in the engine, it is possible to predict the minimum oil film thickness, and its position on the bearing surface, at each position of the load diagram. Several different procedures exist for this prediction, varying in complexity and hence in validity with the real case. Each method makes some attempt to solve Reynolds equation governing the performance of hydrodynamic bearings. Some of the methods available are:

(a) Hand calculation, using an equivalent speed concept[4,5,6]
(b) Graphical procedures for the complete locus, using mobility methods[7].
(c) Approximate method to predict the minimum film thickness in the cycle where 'squeeze intervals' predominate[8].
(d) Computer methods which predict the complete journal locus[7,9,10].
(e) Inertia load studies[11].
(f) Experimental results[8].

All these methods are thoroughly reviewed by Campbell *et al.*[8], but whilst each technique has specific advantages in particular circumstances, those most generally employed involve the use of computer methods to predict the complete journal locus over the whole of each loading cycle. The advantage of rapid, low-cost computer programs is that predicted results can be achieved and compared for a large number of different types and sizes of engines at various operating conditions. With the greater flexibility afforded by the use of computers, bearing analysis will continue to advance, but whilst it is already possible to give some consideration to housing, crankshaft and bedplate distortion, the most common computer techniques still treat the crankshaft as a series of simply supported beams. For medium- and slow-speed engines, this is probably realistic due to the high crankshaft stiffnesses demanded by stress limits, and the high standard of bearing alignment enabled by current machine tools and production methods.

Other basic assumptions usually include: perfect alignment of shaft and bearing through the loading cycle; truly circular shaft and bearing; infinitely rigid bearing housing and journal; viscosity unaffected by variations of temperature and pressure around the bearing and through the cycle; and that negative oil film pressures can be neglected.

From the numerous bearing calculations carried out by the Glacier Metal Co., with use of computer calculation procedures incorporating the above assumptions, it has been possible to prepare a graph of computed minimum film thicknesses for different shaft diameters (*Figure 13.8*). This facilitates comparison of calculated values for any other engine against the norm, but does not allow absolute limits to be set. Deduced levels of 0.000 008 per unit diameter for large ends, and 0.000 01 per unit of diameter for main bearings have been superimposed. Further experience, improved production methods, filtration, etc., will presumably allow these levels to be reduced in the future.

A similar exercise was carried out to compare the calculated maximum specific bearing loads (*Figure 13.9*) for the same engines as with the minimum film thicknesses of *Figure 13.8*. The wide scatter band shown by these results is to be expected, but as with the film thickness values, they allow direct comparison with existing practice for any engine design, particularly at the design stage.

13.2.7 Grooving configuration

The simplest grooving arrangement is to have a central circumferential groove of 360° extent in all bearings. This allows unimpeded oil transfer from one area to another. Hydrodynamically, the absence of all grooving is the optimum, and so a compromise between the hydrodynamic requirement and the physical requirement of supplying oil has to be made. This often results in the so-called partially grooved bearing in which there is a central circumferential groove of less than 360° extent. Such a partially grooved bearing usually allows an increase of the order of 100–200 per cent in the oil film thickness in the original thinnest film region.

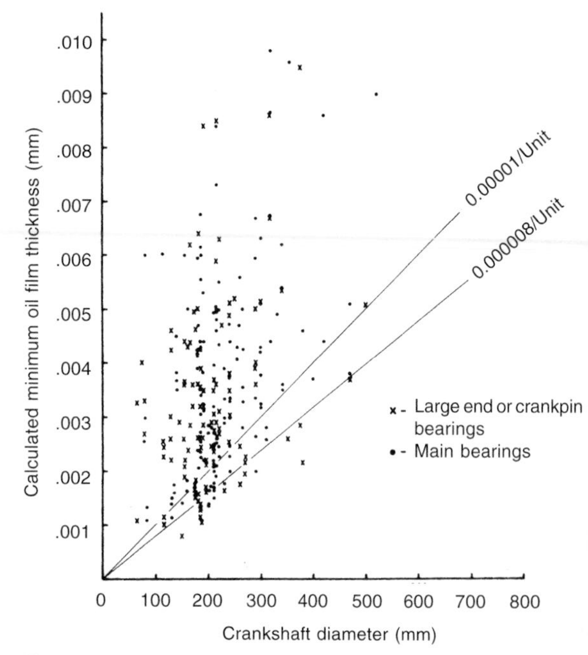

Figure 13.8 Predicted minimum oil film thickness as a function of diameter

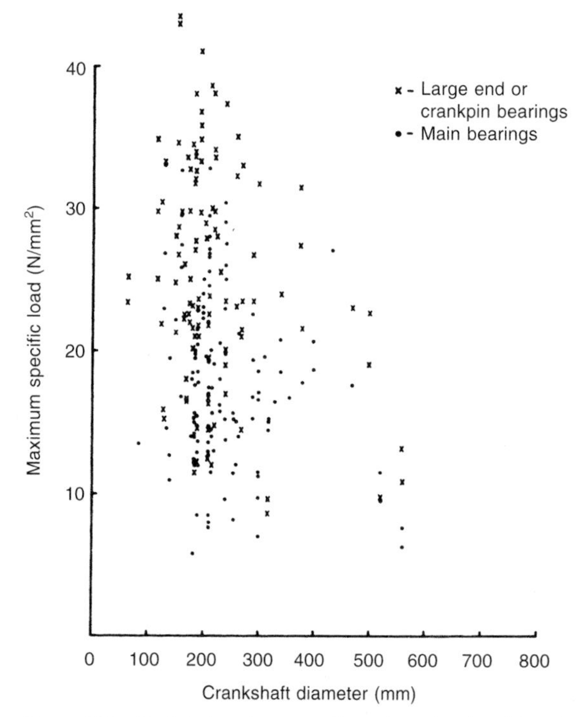

Figure 13.9 Predicted maximum specific load as a function of diameter

When partially grooved bearings are adopted, a system of interconnected cross drillings in the crankshaft is necessary to allow a constant supply of oil to each bearing, or to the piston.

Positioning of, and oil supply passages to, partial grooves can be critical. Predicted oil film thicknesses are based on the assumption that oil is available at all times within the clearance space to allow a film to be generated. Thus any arrangement of partial grooving must allow sufficient oil to be supplied to the required region of the bearing surface.

Two further distinct disadvantages of the partial groove design of bearing are described in the section on bearing damage. These are differential wear of the crank journal, and cavitation erosion of the bearing surface[12].

Thus one should not automatically design partially grooved bearings purely on the basis that increased oil film thickness will be achieved. Alternative methods of improving the film thickness should also be investigated if improvement is considered necessary. This includes increase of bearing dimensions, adjustment of engine masses and modification of firing orders.

13.2.8 Clearance[13]

There are no precise limits on what clearance there should be between bearing bore and journal surface, and it is often determined by experience. As clearance is increased, the theoretical load-carrying capacity is reduced, but this assumes that bearing temperature and oil viscosity are constant. However, an increased quantity of oil can be pumped through the bearing, thus keeping its temperature lower. Too high a value of clearance in any plane around the bearing may also result in cavitation erosion damage[12]. Thus there is a range of clearance over which optimum bearing performance can be expected. *Figure 13.10* presents curves of a semi-empirical nature based on a theoretical heat balance for steadily loaded bearings. These curves can be considerably simplified, by the superposition of a practical band of shaft sizes over a given speed range. Comparison of these two sets of curves results in a generalized value for minimum diametral clearance of 0.00075 × bearing diameter. Manufacturing tolerances of shaft, housing diameter and bearing wall thickness then result in a maximum design clearance somewhat greater than 0.001 × bearing diameter.

When oil film thickness computations are carried out on a comparative basis for various values of clearance, allowing for the lower temperature and higher viscosity corresponding to the larger clearances, it is often found for the medium-speed range of engines that optimum oil film conditions occur at a clearance of approximately 0.001 × bearing diameter, but over a region 0.000 75–0.0015 × bearing diameter there is usually only a small change in the predicted conditions. This upper limit of 0.0015 × bearing diameter can be used as an indicator for limiting wear during engine operation, although absolute limits for any particular engine type can be determined only by experience with that engine type. An additional practical limit can also be imposed by oil pump capacity.

13.3 Bearing damage

The main difficulty of establishing separate categories of damage is that they are often related, or that damage initiated by one cause can result in another effect. However, the commonest types of bearing damage likely to be experienced by diesel engine operators can be approximately classified under the following 10 categories (Sections 13.3.1 to 13.3.10).

13.3.1 Abrasion

This is still probably the most common form of bearing damage, even though filtration standards are generally very high. Long term operation with very fine debris in the oil, or short term operation with coarser contaminant can result in abrasion and scoring of the surface of a bearing *(Figures 13.11 and 13.12)* roughening it to such an extent that overheating can occur due to the surface roughness penetrating the thin oil film. Alternatively, ferrous particles can become embedded in the bearing lining or soft overlay and become work hardened to a level higher than that of the crankshaft, resulting in wear and scoring of the shaft. The presence and degree of ferrous contamination in bearing surfaces can be detected by the technique known as iron printing. In essence, this consists of obtaining a chemical reaction between the embedded particles and a weak solution of potassium ferrocyanide held in a filter paper placed on the bearing surface. Generally speaking, a worn bearing should be renewed, when discovered, if approximately one-third of the projected area of copper-lead or nickel interlayer is exposed. Wear through to expose underlying tin-aluminium is not necessarily a cause for renewal, unless the bearing shows signs of overheating, scuffing, or entrapment of particles proud of the bore surface.

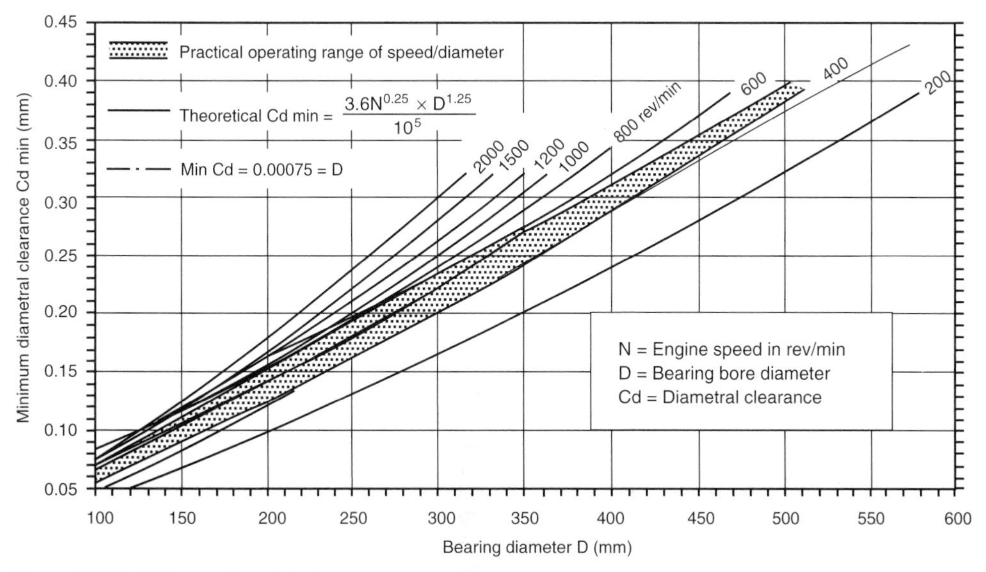

Figure 13.10 Minimum diametral clearance

Figure 13.11 Abrasion by small contaminant over a long period

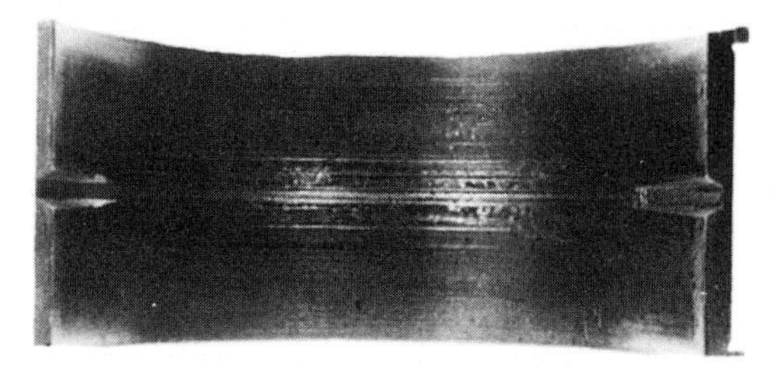

Figure 13.12 Severe scoring by a few large particles

13.3.2 Fatigue

With the general introduction of stronger bearing linings, and improved calculation procedures, fatigue due to design overload is uncommon. Typical fatigue damage of white metal is shown in *Figure 13.13* due to a combination of load, high temperature, and the inconsistency of white metal structure resulting from the direct lining of a large variable-cross-section housing. Fatigue damage of the stronger lining of the modern diesel engine is usually a consequence of some other problem, as was shown in *Figure 13.5*.

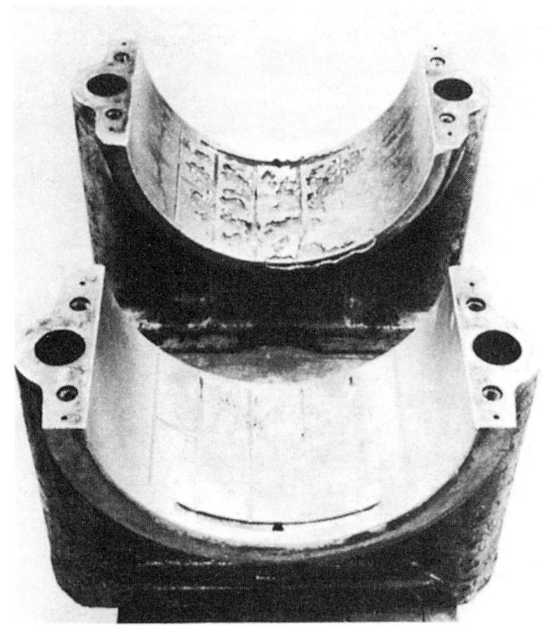

Figure 13.13 Fatigue damage in direct-lined white metal crosshead bearings

Fatigue of the galvanically applied overlay plate is shown in *Figure 13.14*. This layer is applied to the majority of medium-speed engine crankshaft bearings to accommodate oil-borne debris, misalignment and operating deflections.

13.3.3 Corrosion

The different bearing alloys suffer corrosion under different conditions.

Tin-base white metals, particularly in the slow-speed diesel engine, can form a hard, dark (almost black) brittle surface deposit when water is present in the lubricating oil. This has been found to be tin oxide, formed by conversion of the tin matrix of the lining by an electrochemical process[14], and becomes thicker with time. Layers as thick as 0.25 mm have been found. Under these conditions, the bearing becomes prone to overheating and seizure. Alternatively, due to the brittle nature of the layer, it can flake off the surface, leaving a generally pitted appearance.

Copper-lead lining, if exposed to a lubricating oil which has thermally degraded to form organic acids and peroxides, or become contaminated by sulphur containing fuel oils, blow-by of products of combustion of such fuels, or cooling water with antifreeze additives, is prone to corrosion of the lead phase of the lining, be this either of the cast or sinter type. Such attack results in a porous, extremely weak, copper matrix which is easily fatigued by the dynamic loads applied to the surface (*Figure 13.15*).

Aluminium-based linings are completely resistant to engine oils, and to their high-temperature degradation products. However, in direct contact with water, a film of aluminium oxide can form in the bore, together with corrosion of the steel backing and

Figure 13.14 Fatigue of an overlay-plated surface

Figure 13.15 Loss of surface material caused by corrosion of the lead phase

possible lifting of the lining material, although in an oil environment in an engine, this is virtually impossible.

13.3.4 Wiping

This type of damage (*Figure 13.16*) can occur with any lining material, and is caused by insufficient lubricating and cooling oil on the bearing surface. This results in overheating and eventually melting of the lowest melting-point phase of the lining alloy. Potential causes of insufficient oil on the bearing surface include inadequate generated oil film thickness, insufficient clearance, housing distortion, restriction in oil supply system, excessively worn bearings in other locations and inadequate oil pump capacity.

13.3.5 Cavitation

Cavitation erosion damage to bearing surfaces is a form of micro fatigue cracking, initiated by the collapse of vapour cavities, and is thoroughly discussed in Reference 12. Typical damage to a large end bearing is shown in *Figure 13.17* and to a main bearing in *Figure 13.18*.

Figure 13.16 Wiping damage

Figure 13.17 Cavitation erosion of a large end bearing

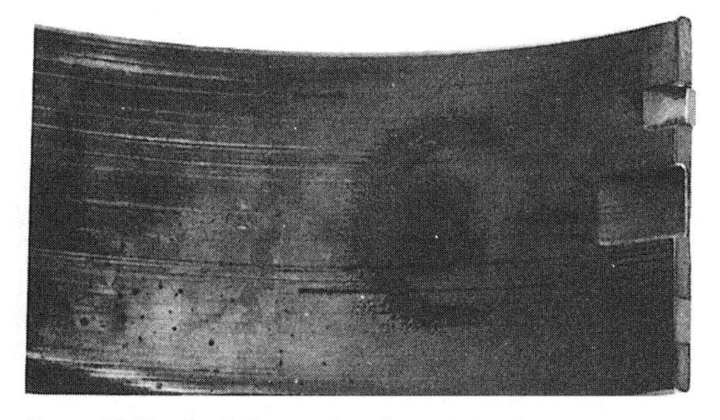

Figure 13.18 Cavitation erosion of a main bearing

13.3.6 Fretting

Figure 13.19 shows a severe case of fretting of a bearing due to insufficient contact pressure, local welding and tearing having taken place between the bearing back and housing bore, resulting in transfer of metal from one to the other. Where fretting of the joint faces of the bearing occurs, as shown in *Figure13.20*, this usually implies inadequate joint face clamping force. Fretting of the housing joints can also be expected under these conditions.

Before fitting new bearings where fretting has occurred, all trace of fretting build-up in the housing must be removed to avoid premature catastrophic damage of the new bearing.

13.3.7 Design faults

Major design faults will not generally be evident in production engines because they would have been found out very quickly during pre-production development testing. Some faults however, do not necessarily cause problems, but require certain other conditions to exist also. An example of this would be blind location features, such as a dowel fitting in a hole in the bearing back. If positioned away from the heaviest loaded region of the bearing, this would not cause problems, unless during assembly or re-fitting the dowel was not located correctly in the hole.

A bearing load diagram which results in an almost pure rotating load pattern, i.e. where the load vector on the bearing surface travels at approximately shaft speed, can result in wiping and overheating even though the theoretical oil film thickness is

Figure 13.19 Fretting of a bearing back due to insufficient contact pressure

Figure 13.20 Fretting of a bearing joint face due to inadequate bolt load

greater on that bearing than on others in the same engine. Such a condition is more likely to exist in an engine running on very light load and full speed. The solution is to fit larger balance weights to counteract the inertia forces of the crankshaft and connecting rods. Adverse load patterns can be further aggravated by incorrect positioning of oil holes in journal surfaces, and damage due to this cause is typified by *Figure 13.21*.

13.3.8 Incorrect assembly

The commonest causes of incorrect assembly are associated with locating devices. Incorrect positioning will mean that oil feed connections are misplaced and can block off the oil feed. Incorrect location of a tang into its recess means that the bearing back will not be in contact with the housing, and the clearance between the bearing bore and the crankshaft can be lost, resulting in local overheating of the bearing surface, and possibly seizure.

Having located the bearing correctly into its housing, care must be taken to ensure that the housing bolts are correctly tensioned. Insufficient bolt load due to incorrect tightening can result in the damage shown in *Figure 13.20* and also excessive dynamic stressing of the bolts and ultimate fatigue fracture. The bolts should also be tensioned in the prescribed sequence given in the engine manual. Failure to follow the same sequence may result in a modified bore shape and unsatisfactory bearing performance.

13.3.9 Environmental factors

13.3.9.1 Electrical discharge

On generator installations, with inadequate earthing, a discharge of current can occur through the oil film between the journal surface and the bearing bore, resulting in fine, cleanly defined pitting of the bearing surface *(Figure 13.22)*. Damage such as this has been found to occur with an electrical potential in excess of only 50 mV.

13.3.9.2 Journal wear ridge

Wear of the journal surface can occur due to embedded hard contaminant in the bearing surface. Less wear occurs in the region associated with grooves, gradually producing a ridge on the journal surface, which in turn causes wear of the bearing surface (*Figure 13.23*) of partially grooved bearings.

13.3.9.3 Static fretting

The damage shown in *Figure 13.24* is caused by vibration of a crankshaft within the bearing clearance. The term 'static' comes from the fact that such damage only occurs when the engine is not running. The damage can occur during transport of assembled engines to site. In marine installations, generating sets mounted on flexible tank tops are prone to similar damage. In such cases, the simplest palliative is to ensure that oil is pumped through the bearing clearance every few hours while the engine is inoperative, although in very severe instances it may also be necessary to redesign the mountings.

Figure 13.21 Surface damage due to incorrectly positioned oil hole

Figure 13.22 Pitted surface caused by electrical discharge

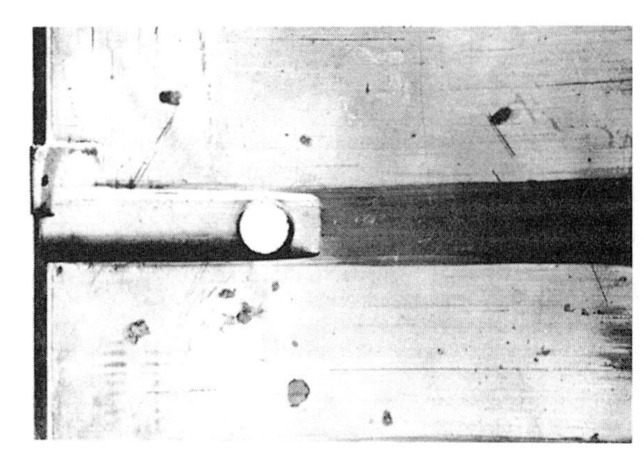

Figure 13.23 Wiping and overloading of a local region of the bore surface

Figure 13.24 Damage to bore surface caused by crankshaft vibration

13.3.9.4 Nitride damage

The specific form of damage shown in *Figure 13.25* characterized by the vee chevrons is caused by iron nitride particles from a

Figure 13.25 Damage caused by iron nitride particles from a nitrided journal surface

nitride-hardened shaft which has not had the thin friable layer ground off from the surface. To be sure of avoiding such damage, up to 0.02 mm needs to be removed.

13.3.9.5 Misalignment

Misalignment through whatever cause is typified by a 'D' pattern of wear, with the longest side on one end of the bearing, as *Figure 13.26*. Causes can be taper in the bearing housing or journal surface, debris trapped between shell and housing, bruising of the steel backing of the bearing shell or housing bore during assembly, taper through the bore of the bearing, or a variation in alignment of a series of main bearing bores along the engine.

13.3.10 Geometric factors

Severe wear of localized regions of the bearing surface are generally caused by specific geometric inaccuracies. Heavy wear in the axial centre region is usually caused by a 'barrelled' journal, whereas heavy wear at both ends of the bearing results from an 'hour glass' profile.

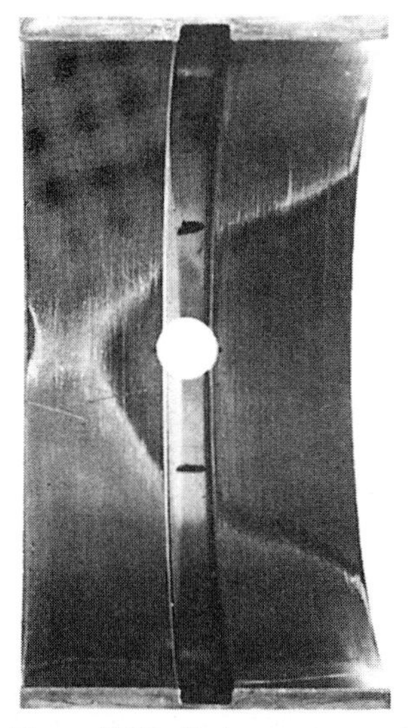

Figure 13.26 Surface wear associated with shaft misalignment

13.4 Slow-speed engine crosshead bearings

Thin shell bearings for probably the most arduously loaded of all engine bearings, the two-stroke crosshead assembly, were tested by the major engine builders in the early 1960s. Following such trials, they were gradually introduced into general production, and as loadings and conditions became even more arduous, the thin shell pre-finished bearing has become almost universally adopted in crossheads, and in several instances, in the bottom end and main bearings also.

The basic problem encountered in direct-lined white metal bearings was lining fatigue (as shown in *Figure 13.13*). Adoption of thin shell bearings permitted an improvement in metallurgical structure and strength, particularly at the steel/white-metal bond line by the use of centrifugal casting and improved uniform cooling of the constant cross-section.

Moreover, once the pre-finished thin shell is adopted, it allows advantage to be taken of the better fatigue properties of the stronger lining materials. One engine range has had 40 per cent tin-aluminium-lined bearings incorporated from inception[15]. The thin shell bearing has other inherent advantages. Carrying of onboard spares is much more convenient, as is the re-fitting of replacements. Spares, when required urgently, are much more easily freighted and ship turn-round time is greatly reduced compared with that for the assembly shown in *Figure 13.13*.

The bearing geometry can be much more closely controlled as a result of specialist manufacture and absence of hand fitting. Optimum performance of the crosshead bearing has been obtained with the so-called bedded-are profile. This is a machinable form, repeatable within fine tolerance, that replaces the hand-scraped bedded-in form. The general configuration is shown in *Figure 13.27*. Over an arc of approximately 120°, symmetrical about the vertical centre line, the radial clearance between pin and bearing is almost zero. This is optimum because the crosshead bearing relies basically upon a squeeze film for lubrication. Axial grooves pitched at the swing of the crosshead pin allow oil to be swept onto the bearing surface. Small bleed grooves at

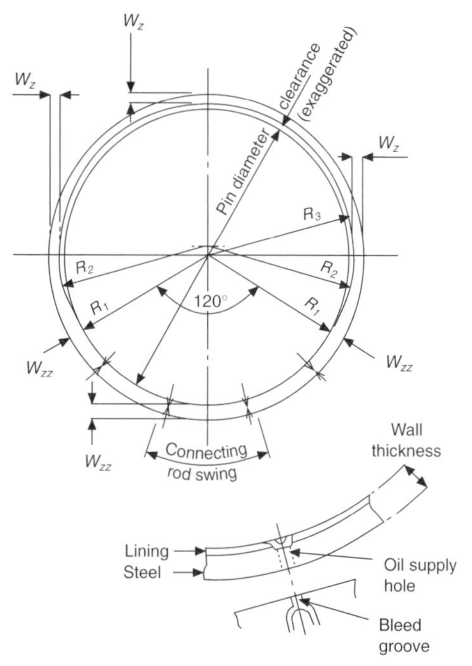

Figure 13.27 General configuration of a 'bedded arc' crosshead bearing

the ends of the main axial grooves induce a flow of oil for added cooling, and also allow lubrication of thrust faces.

Manufacturing tolerances of bearing, pin and housing create a positive clearance between pin and bearing surface, but the application of a precision electroplated overlay, usually 0.02–0.03 mm thick, of a soft lead-tin on the precision-formed bearing bore profile, allows very rapid bedding-in without localized overheating and consequential fatigue cracking, and as such is extremely beneficial, even on white metal.

A different approach has been taken by one engine builder[16], using an eccentric arrangement of bearings and pins where load is transferred from one bearing land to another during the total swing of the rod and pin. The adoption of separate shell bearings for each land, with eccentricity in the wall thickness, simplifies the achievement of the required assembly.

13.5 Bearing metals

The purpose of the remaining sections of this chapter is to describe the materials used as linings for thin shell bearings, with particular emphasis again on crankshaft bearings. The most commonly used alloys are listed in *Table 13.1*.

To begin with, an outline is given of the properties of materials most relevant to bearing performance.

13.5.1 Fatigue strength

Lining fatigue occurs because of the cyclic nature of the loads applied to the bearings. Fatigue ratings for a range of bearing alloys are given in *Figure 13.28*. These results were obtained on a test rig in which a dynamic load is applied to a bearing via hydraulic resistance against an eccentric test shaft[14]. Results from test rigs cannot be used directly for design purposes. However, it is possible to rank materials in an order that correlates with their performance in engines, as illustrated by comparison of *Figure 13.28* with *Table 13.2*, which contains fatigue limits of bearing materials derived from engine experience.

13.5.2 Scuff resistance

Scuffing, also known as seizure or scoring, is a result of local solid-phase welding ('pick-up') between the shaft and bearing surfaces. In extreme cases, scuffing leads to wiping and possibly to complete bearing seizure.

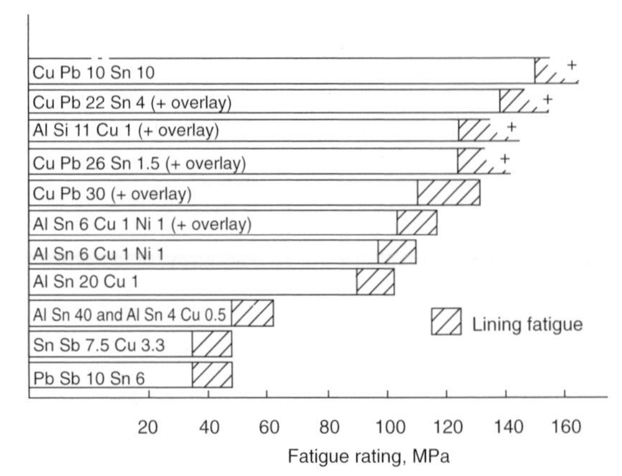

Figure 13.28 Fatigue ratings for a range of commercial bearing alloys.

Table 13.2 Recommended maximum loadings (fatigue limit) for bearing materials in slow- and medium-speed diesel engines

Material	*Maximum loading (MPa)*
AlSn6Cu1Ni1	38
CuPb26Sn1.5	38
AlSn20Cu1	35
AlSn40Cu0.5	20
SnSb8.5Cu3.5Cd1Cr0.1	14
SnSb7.5Cu3.3	12
PbSb 10Sn6	12

The ability of a bearing material to resist scuffing depends on three factors—compatibility, conformability and embeddability. *Compatibility* is an inherent tendency of a material to resist solid-phase welding. *A conformable material.* is able to deform under conditions of shaft misalignment, thereby reducing local load concentrations and maintaining an adequate oil film thickness. A material with good *embeddability* allows hard particles to embed in the surface of the bearing, thus reducing any abrasive damage which such particles can cause both to the bearing and the shaft.

With metallic bearing materials, *both* conformability and embeddability are related inversely to hardness. In *Table 13.1*

Table 13.1 Composition and hardness of some commercial bearing alloys

Nominal composition (% by weight)		*Hardness* (HV 2.5)*
White-metal alloys:		
Tin/7–8% antimony/3–3.5% copper	SnSb7.5Cu3.3	27–29
Tin/8–9% antimony/3–4% copper/0.5–1.0% cadmium/0.05–0.15% chromium	SnSb8.5Cu3.5Cd1Cr	28–33
Lead/9–11% antimony/5–7% tin	PbSb10Sn6	16–20
Lead-bronze alloys:		
Copper/28–32% lead	CuPb30	46–51
Copper/24.5–27.5% lead 1–2% tin	CuPb26Sn1.5	78–82
Copper/20–24% lead/3.5–5% tin	CuPb22Sn4	98–105
Copper/9–11% lead/9–11% tin	CuPb10Sn10	124–140
Aluminium alloys:		
Aluminium/37–43% tin	AlSn40	23–25
Aluminium/37–43% tin/0.3–0.5% copper	AlSn40Cu0.5	29–31.
Aluminium/18.5–24% tin/0.8–1.2% copper	AlSn20Cu1	34–41
Aluminium/5.5–7.0% tin/0.7–1.3% Copper/0.7–1.3% nickel	AlSn6Cu1 Ni1	38–43
Aluminium/10–11% silicon/0.8–1.2% copper	AlSi11 Cu1	59–62

This is the hardness of a 0.25–0.50 mm thick lining on steel, in the form of a 50 mm diameter half-bearing. Flat strip or larger bearings have lower hardness.

typical hardness values are given for the range of bearing alloys. Compatibility is less easy to quantify; in a test used in the Glacier Metal Co., a shaft runs in a bush in a stop-start cycle with the load maintained during the stop part of the cycle. Lubrication is limited to approximately one drop of oil per minute. The load is increased until scuffing occurs, indicated by an increase in running temperature. Alignment between shaft and bearing surface is good so that resistance to scuffing depends primarily on the compatibility of the bearing material. The relative performance of bearing alloys on this test rig is shown in *Figure 13.29.*

Comparison of *Figures 13.28 and 13.29* and *Table 13.1* shows that, in general, bearing alloys with high fatigue strength have high values of hardness as well as relatively poor compatibility. Thus, selection of a bearing material is always a compromise between fatigue strength and scuff resistance. As a general guide, the least hard material is chosen which has sufficient fatigue strength for the application.

13.5.3 Wear resistance

Many wear mechanisms have been proposed for different materials[17], but for engine bearings only two need be considered:

(i) 'severe abrasion' of the bearing and shaft by hard particles in the oil,
(ii) 'mild abrasion' of the bearing by asperities on the shaft.

Wear rates in the two regimes may differ by a factor of up to 10^6. Mild wear without scuffing can be beneficial if it improves conformability between shaft and bearing.

The wear rate of a material is a highly system-dependent property, and test rigs provide an unreliable guide to performance in an application. In general, a harder material has greater resistance to 'severe abrasion' but a correlation between material properties and 'mild abrasion' resistance has not been established.

13.5.4 Cavitation erosion resistance

A simple test rig for producing cavitation erosion of bearing linings has been described[12], in which cavitation is produced close to a bearing surface by an ultrasonically vibrating probe. Some results from this rig are shown in *Figure 13.30* and suggest that cavitation erosion resistance depends on material type rather than mechanical properties. For each type of alloy, however, e.g. lead-bronze, increased hardness and strength produce increased cavitation erosion resistance. It should be noted that

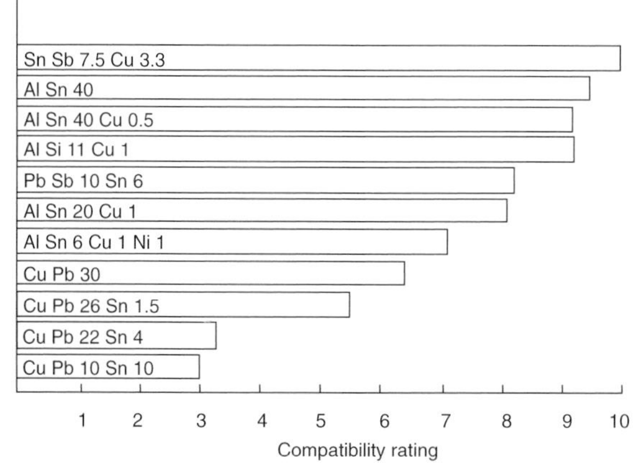

Figure 13.29 Relative compatibility ratings (Sn-based white metal=10) of bearing alloys

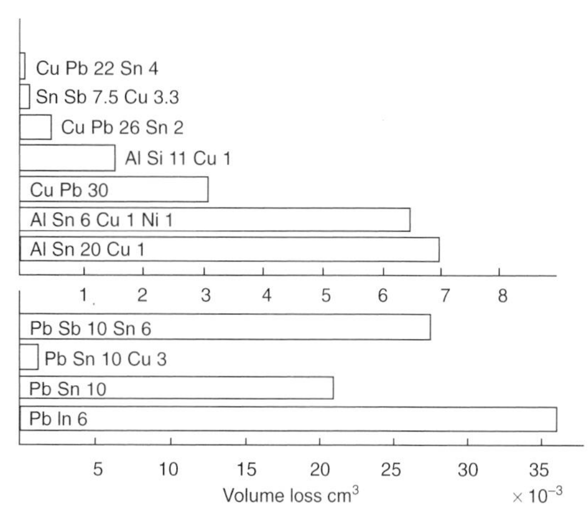

Figure 13.30 Cavitation erosion resistance of bearing alloys and overlays. Volume loss measured after 10 min. Note different scale for lead-based alloys

the results in *Figure 13.30* for the overlay compositions (PbSn10Cu3, PbSn10 and PbIn6) were obtained under less severe conditions than for the other alloys. The two sets of results are not, therefore, directly comparable.

13.5.5 Overlays

Overlays are thin coatings (0.02–0.05 mm) of a soft alloy, usually lead-based. Commonly used compositions are given in *Table 13.3*. Their original purpose was to improve the corrosion resistance and compatibility of lead-bronze bearings although, being soft, they also have good dirt embeddability and conformability. With cast lead-bronze linings a nickel interlayer, 0.002 mm thick, is normally applied to ensure the long-term integrity of the bond between overlay and substrate. Overlays can also be applied to the full range of aluminium alloy bearings. A nickel interlayer is again used, to protect the aluminium from chemical attack during deposition of the overlay.

Overlays are electrodeposited onto the bearing surface, normally after final boring. Lead/tin and lead/tin/copper are co-deposited directly, whereas lead and indium are deposited consecutively and diffused together by a heat treatment. Typical hardness figures are given in *Table 13.3*. In bulk these overlay compositions are weak, but are significantly stronger in the form of thin coatings. *Figure 13.31* shows how fatigue strength increases as thickness is reduced. For a given thickness, lead/indium overlays give slightly higher fatigue ratings than lead/tin or lead/tin/copper.

Overlay fatigue usually occurs at a lower load than for the substrate material, as shown in *Figure 13.28*. However, even a fatigued overlay provides some protection for the substrate and the fatigue strength of a bearing is always higher when overlay plated, as illustrated in *Figure 13.28* for aluminium/6% tin. This does not imply, though, that overlay fatigue can be ignored. With lead-bronze bearings for example, fatigue cracks in the overlay expose lead in the substrate to corrosion.

Table 13.3 Composition and hardness of overlays

Nominal composition		*Hardness (Hv 0.025)*
Lead/10% tin	PbSn10	9–10
Lead/10% tin/3% copper	PbSn10Cu3	12–15
Lead/6% indium	PbIn6	8–9

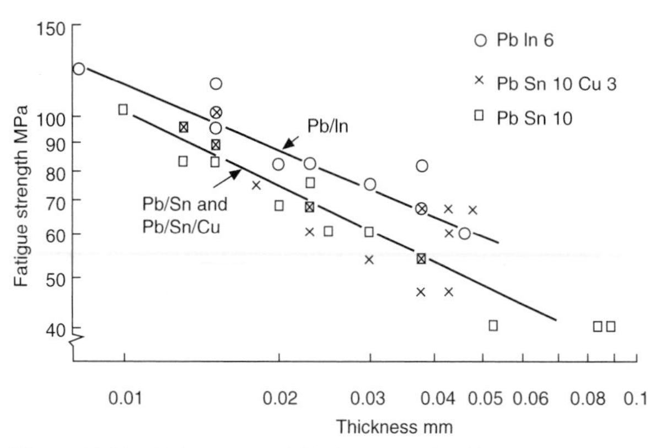

Figure 13.31 Fatigue-strength/thickness relationship for overlays

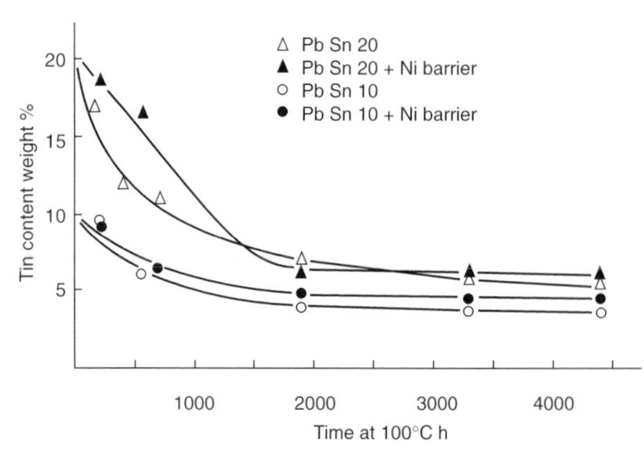

Figure 13.33 Diffusion of Sn from Pb/Sn overlays on Cu/Pb with and without Ni barrier

Relative wear resistance is demonstrated by measurements obtained using the dynamically loaded bearing test rig, and results are shown in *Figure 13.32.* Lead/tin/copper has a higher wear resistance than lead/tin, which is in turn superior to lead/indium. These results concur with those reported by Schaefer[19]. The relatively poor wear resistance of lead/indium negates to some extent its high fatigue rating, since for a given wear life a thicker overlay is required which leads to a lower fatigue strength.

Relative cavitation erosion rates are given in *Figure 13.30.* Lead/tin/copper has the highest resistance to cavitation, followed by lead/tin.

Tin or indium is necessary to improve corrosion resistance of the lead. The minimum levels required for adequate corrosion resistance are 3 per cent tin and 5 per cent indium[20]. Although these levels are below those in the as-plated overlays, in service, tin and indium are lost by diffusion into the substrate[18]. The result is formation of intermetallic compounds with copper or nickel in the substrate or interlayer[21], and corrosion of the overlay if the tin or indium level falls below the safety limit. Loss of tin is demonstrated by results shown in *Figure 13.33,* which were obtained by heating sintered copper-lead bearings plated with lead/10% tin. It is sometimes claimed that a nickel interlayer eliminates tin loss. However, this is not so, as shown in *Figure 13.33,* although the rate of loss of tin is significantly reduced by the presence of a nickel interlayer. Increasing the tin content of the overlay is not a complete solution to the tin diffusion problem, since the rate of diffusion is correspondingly increased.

Lead/tin/copper overlays are more corrosion resistant than lead/tin, possibly because the rate of tin loss is lower with lead/tin/copper as a result of the formation of copper-tin compounds within the overlay. Thus, the combination of high wear resistance, cavitation erosion resistance and corrosion resistance makes lead/tin/copper the preferred overlay for the modern highly rated medium-speed diesel engine, particularly when these operate on residual fuels.

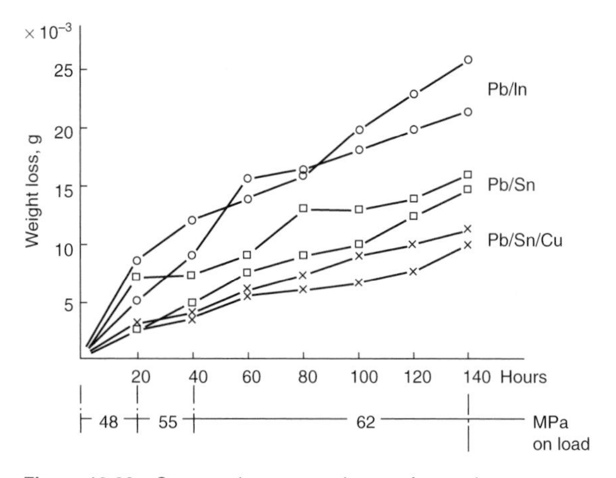

Figure 13.32 Comparative wear resistance for overlays

13.5.6 White metals

Tin-based white metals are alloys of tin, antimony and copper, the most commonly used composition being tin/7.5% antimony/3.3% copper. The metallurgical structure (*Figure 13.34*) consists of a tin-antimony matrix which is strengthened by needles of copper-tin compound.

The effects of composition and method of manufacture on the mechanical and bearing properties of white metals have been described in detail by Pratt[14]. The effect of cooling conditions during production of bearing linings, referred to earlier, is demonstrated by the results in *Figure 13.35.* Rapidly quenched material, obtained during production of thin shell bearings, has a finer copper-tin compound distribution and a higher fatigue strength than slow-cooled linings obtained when white metal is cast directly into large bearing structures.

The fatigue strength of the basic white metal composition

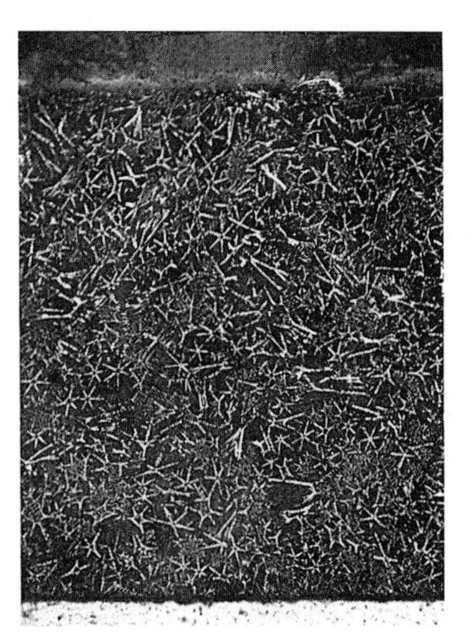

Figure 13.34 Typical structure of SnSb7.5Cu3.3, Showing star-shaped Cu-Sn needles in Sn-Sb matrix

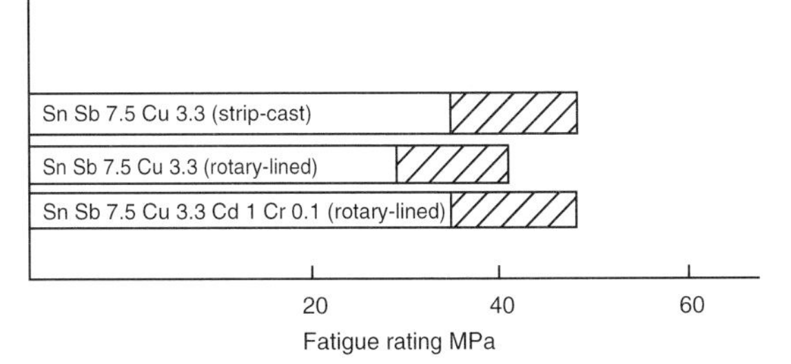

Figure 13.35 Effect of casting method and composition on fatigue strength of Sn-based white metals

can be increased by adding cadmium, which goes into solution in the matrix, and by adding trace quantities (0.1 per cent) of chromium, which refines the distribution of the copper-tin compound. By this means, it is possible to produce commercially a slow-cooled white-metal-lined bearing with a fatigue strength as high as a rapidly cooled or quenched lining (*Figure 13.35*).

Conformability and embeddability of tin-based white metals are very good, as would be expected for a material of low hardness. Corrosion resistance is good, except when the lubricating oil is contaminated by water as described earlier. However, the outstanding characteristic of tin-based white metal is its compatibility, scuffing in service being virtually unknown. In conditions of oil starvation linings suffer wiping rather than scuffing and the wiped surfaces are capable of continued operation, which is not the case with most other bearing alloys.

Lead-based white metals are basically alloys of lead, tin and antimony, containing 6–12 per cent tin and 10–15 per cent antimony. The structure of the 10 per cent antimony/6 per cent tin composition consists of a single-phase lead-tin-antimony matrix. At higher antimony contents, tin-antimony cuboids are formed. Another version contains 15 per cent antimony and 1 per cent arsenic, the structure in this case being a fine antimony-arsenic precipitate, in a lead-antimony matrix.

Cooling conditions during bearing manufacture affect mechanical properties similarly to tin-based white metal, although rapidly cooled lead-based linings must be annealed to provide adequate ductility. Likewise, minor additions of certain metals—arsenic, copper, cadmium or nickel—can increase the fatigue strength of a slow-cooled lining by refining the alloy structure.

In comparison with tin-based white metals, the lead-based alloys are softer (*Table 13.1*) and hence have even better conformability and dirt embeddability. Fatigue strength is similar (*Figure 13.28*). Tin and antimony contents are generally high enough to avoid corrosion problems in most applications. Compatibility, as assessed on the stop-start test rig described above, is not as good as tin-based white metal (*Figure 13.29*) although in service, scuffing of lead-based white metal is generally not a problem. Wear and cavitation erosion resistance of the lead-based alloys are, however, inferior.

White metals were the first alloys developed specifically for plain bearings. Although their fatigue strength is relatively low, they are still widely used on account of their surface properties. Because of the rather better corrosion and scuff resistance of tin-based white metal, it is preferred to lead-based alloys in certain critical applications, e.g. crosshead bearings. For general applications, lead-based white metal bearings are significantly cheaper than their tin-based equivalents, but even so, the tin-based alloys are, by tradition, used more widely.

13.5.7 Copper-lead and lead-bronze alloys

Historically, lead-bronze bearing linings were introduced when fatigue of white metal became a serious problem, as a result of increased engine ratings. Modern thin shell bearings are made either by casting on to steel strip or by a sinter route, in which pre-alloyed powder is applied to steel and consolidated by sintering and rolling. All of the lead-bronze alloys in *Table 13.1* can be made into bearings by either method; copper-lead linings with 30 per cent or more lead are made only by the sinter route.

The most widely used compositions lie within the range 22–26% lead/1–2% tin. The tin goes almost exclusively into the copper, and so the structure consists of a bronze matrix containing lead islands. However, the detailed structure depends on the method of manufacture, as illustrated in *Figure 13.36 (a) and (b)*. Sintered linings tend to have a finer and more uniform lead distribution than cast alloys. In general, strip-cast linings are harder than sintered ones of the same composition.

The 22–26% lead/1–2% tin alloy has a high fatigue strength and is suitable for most medium- and high-speed diesel engines. It is always overlay-plated, primarily to protect the lead phase from corrosion but also because the alloy is relatively hard and incompatible (*Figure 13.29*). For medium-speed applications the overlay thickness may be as high as 0.05 mm, but where higher load-carrying capacity is required a thickness of 0.025 mm is used (see *Figure 13.31*.) Hardened shafts, of minimum hardness 300 HV, are recommended.

An improvement in surface properties and dirt embeddability, at the expense of fatigue strength, is obtained with copper/30% lead. Even so, this alloy is overlay-plated for corrosion protection. Stronger linings can be made from copper/22% lead /4% tin. Such alloys can be used in engines which are too highly rated

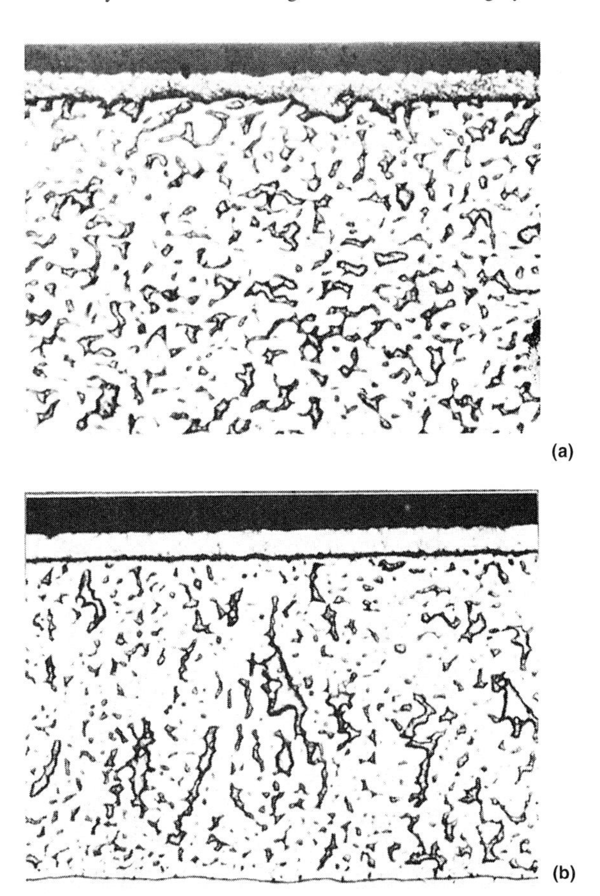

Figure 13.36 Typical structures of CuPb26Sn1.5; (a) sintered; (b) cast. The different lead distributions and matrix structures are shown

for the 26% lead/1.5% tin alloy; a thin overlay and a hardened shaft are required.

The copper/10% lead/10% tin alloy is very strong, but is too hard and prone to scuffing to be used as a crankshaft bearing. However, it is widely used for small end bushes, which are very highly loaded. In order to avoid scuffing, a hardened gudgeon pin (700 HV) is used, with a very fine surface finish.

13.5.8 Aluminium-tin alloys

The major impetus for the development of aluminium-tin bearings was their superior corrosion resistance relative to lead-bronze. Four aluminium-tin alloys are used in diesel engines, and their nominal compositions are given in *Table 13.1*. Bearings are made by roll bonding the alloy to steel strip via an interlayer (for aluminium-tin alloys) which is usually aluminium.

Aluminium/6% tin is a long-established bearing material[22], developed originally for cast bearings. Copper and nickel are added to increase strength; copper is a solution hardener and the nickel forms a fine dispersion of nickel-aluminium compounds, as shown in *Figure 13.37 (a)*.

Fatigue strength is slightly lower than that of the lead-bronzes (*Figure 13.28*), but is more than adequate for most engine applications. Compatibility is better (*Figure 13.29*), but the alloy is relatively hard and so is normally overlay-plated. Overlay thickness is normally in the range 0.02–0.05 mm, and shafts hardened to 250–300 HV are recommended.

Aluminium/20% tin with a 'reticular' tin structure was developed as a more compatible version of the 6% tin alloy. The structure (*Figure 13.37 (b)*) consists of an aluminium/1% copper matrix containing an interconnecting network of tin. This structure is essential if the alloy is to have adequate high-temperature strength, and is obtained by controlled cold working and heat treatment of the as-cast alloy. According to results from the stop/start test rig (*Figure 13.29*), the compatibility of aluminium/20% tin is superior to the 6% tin alloy, and in practice it has proved to be sufficiently scuff resistant to operate successfully without an overlay. Although its fatigue strength is marginally lower than aluminium/6% tin it is high enough to use without an overlay in the majority of high-speed engines; for medium-speed diesel engines it is usually overlay-plated and run against a shaft of minimum hardness 250 HV.

High-tin aluminium alloys, containing 40 per cent or more tin, were developed as a higher strength replacement for white metal in slow-speed diesel engines. Their use in such applications was not possible until thin shell bearings were adopted, because cast-in linings cannot be produced with aluminium alloys. Two forms are available, aluminium/40% tin which is used unplated, and, for crosshead applications, overlay-plated aluminium/40% tin/0.5% copper.

The 'reticular' tin structure and continuous aluminium matrix are also present in these high-tin alloys (*Figure 13.37 (c)*). As a result, aluminium/40% tin maintains its fatigue strength at elevated temperature, whereas white metal becomes significantly weaker, as shown in *Figure 13.38*. At typical engine operating temperatures aluminium/40% tin is approximately 30% stronger than white metal.

The compatibility of aluminium/40% tin is not as good as

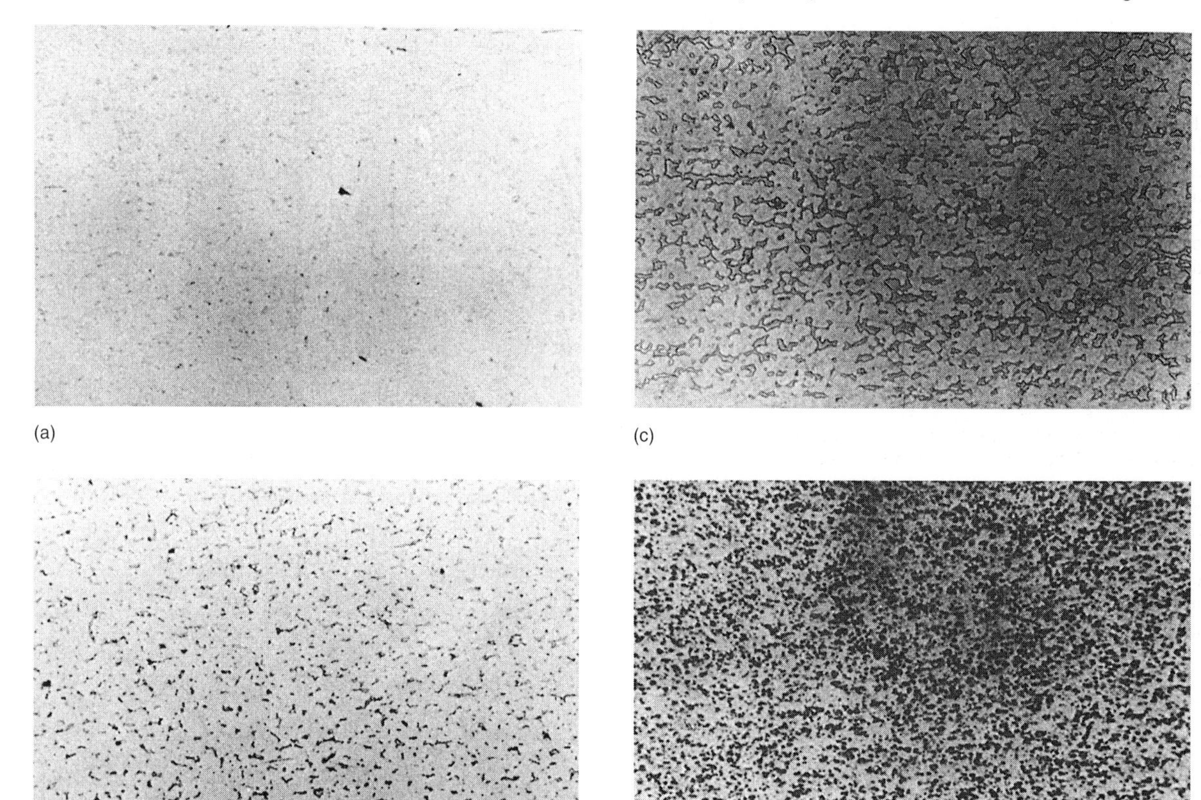

Figure 13.37 Typical structures of (a) AlSn6Cu1Ni1; (b) AlSn20Cu1; (c) AlSn40; (d) AlSi11 Cu1

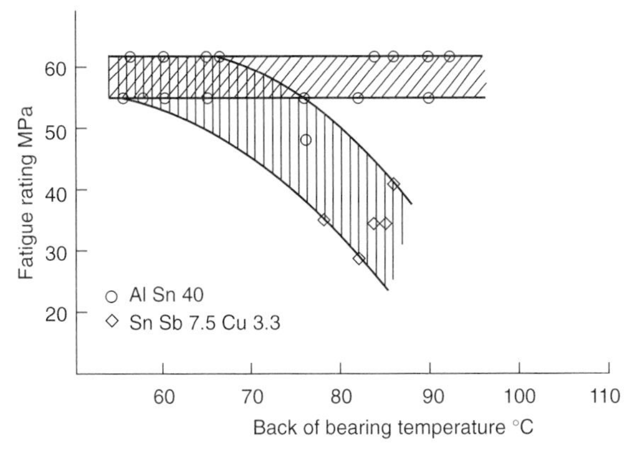

Figure 13.38 Fatigue-strength/temperature relationships for AlSn40 and SnSb 7.5Cu3.3

tin-based white metal, as assessed on the stop/start test rig. Nevertheless, the copper-free alloy does have the ability to 'wipe' under incipient scuffing conditions in the same way as white metal, and has been tested successfully without an overlay in crosshead bearings.

Aluminium containing up to 70 per cent tin has been produced experimentally, in an attempt to further improve compatibility and reduce hardness. However, at the 60 per cent level and above, tin tends to form the matrix in the finished lining, with a consequent reduction in high-temperature fatigue strength.

13.5.9 Aluminium-silicon alloys

Two aluminium-silicon compositions are used as high-strength bearing materials. The first, aluminium/4% silicon/1% cadmium, consists of a dispersion of silicon particles in an aluminium matrix; cadmium provides a soft phase to improve scuff resistance. Copper and magnesium are added, approximately 0.1 per cent of each. The magnesium forms magnesium-silicon compound during heat treatment of the alloy and a heat-treated high-strength version is available.

The second composition is aluminium/11% silicon/1% copper. The structure of this alloy is shown in *Figure 13.37* (*d*). For maximum strength and compatibility the silicon must be finely dispersed throughout the aluminium/1% copper matrix, and this is achieved by cold working the alloy during manufacture. Bimetal is produced by roll bonding directly to steel strip. The material was developed specifically for the turbocharged high-speed diesel and its strength equates to that of the strong lead-bronze alloys also used in this application. However, the superior corrosion resistance of the aluminium-based alloy offers a major advantage over lead-bronze. In very highly loaded high-speed engines the operating temperatures of the oil may be up to 170°C, and under these conditions breakdown of oil can occur very rapidly, especially when catalysed by a copper-based bearing alloy.

In general, the compatibility of aluminium-silicon is very good considering that there is no major soft phase, and on the stop/start test rig aluminium/11% silicon/1% copper is more compatible than lead-based white metal (*Figure 13.29*). This good compatibility is attributed to the non-metallic nature of the silicon which does not exhibit solid-phase welding to the shaft under scuffing conditions[23]. However, the alloy is relatively hard, and must be overlay-plated. Overlay thickness is normally restricted to 0.02–0.03 mm because of the high operating loads. Hard shafts, of minimum hardness 300 Hv, are also required.

References

1 HILL, A., 'Modern bearing design and practice', *Trans. IMarE*, **88** (1976)
2 WARRINER, J. F., 'Factors affecting the design and operation of thin shell bearings for the modern diesel engine', *Diesel Engines for the World 1977/78*, Whitehall Press
3 MARTIN, F. A., 'Design procedures for dynamically loaded bearings', Glacier Metal Co. Internal Publication GC 48/74
4 STONE, J. M. and UNDERWOOD, A. F., 'Load carrying capacity of journal bearings', *Qt. Trans. Soc. Automot. Engrs.*, **1**, 56 (1947)
5 WILCOCK, D. F. and BOOSER, E. R., *Bearing Design and Application*, McGraw-Hill (1957)
6 SHAW, M. C. and MACKS, E. F., *Analysis and Lubrication of Bearings*, McGraw-Hill (1949)
7 BOOKER, J. F., 'Dynamically loaded journal bearings—mobility method of solution', *J. Basic Eng. Trans. ASME Series D*, **187**, September (1965)
8 CAMPBELL, J., LOVE, P. P., MARTIN, F. A. and RAFIQUE, S. O., 'Bearings for reciprocating machinery: a review of the present state of theoretical, experimental and service knowledge', Conf. Lub. and Wear, Paper 4, IMechE, London, September (1967)
9 HORSNELL, R., 'Journal bearing performance', Ph D Thesis, University of Nottingham, May (1963)
10 LLOYD, T., 'Dynamically loaded journal bearings', Ph D Thesis, University of Nottingham, May (1966)
11 MARTIN, F. A. and BOOKER, J. F., 'Influence of engine inertia forces on minimum film thickness in connecting rod big end bearings', *Proc. I Mech E*, **181** (Pt 1) (1966–67)
12 GARNER, D. R., JAMES, R. D. and WARRINER, J. F., 'Cavitation erosion in engine bearings—theory and practice', 13th CIMAC Conference, Vienna (1979)
13 WARRINER, J. F., 'Thin shell bearings for medium speed diesel engines', Diesel Engineers and Users Association, Publication 364, February (1975)
14 PRATT, G. C., 'Materials for plain bearings', Review 174, *International Metallurgical Review*, **18** (1973)
15 BRINER, M. J., 'Development of the Sulzer RN-M type diesel engine range', 11th CIMAC Conference, Barcelona (1975)
16 'GMT Eccentric Crosshead Bearing', *Shipbuilding and Marine Engineering International*, September (1975)
17 SCOTT, D. (Ed.), 'Wear', *Treatise on Materials Science* (edited by Hermann, D.), Vol. 13, Academic Press (1979)
18 PERRIN, H., 'Bearing problems in internal combustion four-stroke rail traction engines', 7th CIMAC Conference, London (1965)
19 SCHAEFER, R. A., 'Electroplated bearings', *Sleeve Bearing Materials* (edited by Dayton, R. W.), ASM (1949)
20 WILSON, R. W. and SHONE, E. B., 'The corrosion of overlay bearings', *Anti-Corrosion*, 9–14 August (1970)
21 SEMLITSCH, M., 'Comparative micro-analysis investigations on multi-layer bearings', *Mikrochim. Acta*, Suppl. IV, 157–169 (1970)
22 HUNSICKER, H. Y., 'Aluminium alloy bearings', *Sleeve Bearing Materials* (edited by Dayton, R. W.), ASM (1949)
23 CONWAY-JONES, J. M. and PRATT, G.C., 'Recent experience and developments in bearings for diesel engines', 12th CIMAC Conference, Tokyo (1977)

14 Pistons, rings and liners

Contents

14.1 Introduction

The performances of pistons, rings and plain or linered bores are so interdependent and interrelated that it is natural to consider them in one section. A consequence of such dependence is that the design of individual components of the cylinder assembly cannot be treated in isolation, and this will be highlighted in the data which follow.

The duty cycle, the thermal and mechanical loading, and the requirements of long life and high reliability of the piston assembly represent one of the most arduous sets of conditions for any mechanical component in engineering today.

A historical review shows few major design landmarks of piston, ring and liner over the last 40 years. Superficially they have always looked similar, but many detailed changes have taken place—in the earlier years mainly by empirical optimization, which has led to the well-defined, precisely machined and more, scientifically based components of today.

Two innovations, however, stand out significantly. The life of an aluminium alloy piston was dramatically increased with the introduction of the Alfin* process top-groove cast-iron insert reinforcement in the early 1940s. At about the same time the introduction of chromium plating to rings produced major improvements in the life before overhaul. Chromium plating combined with improved base material and manufacturing accuracy has led to a reduction in the number of rings on each piston.

As engine ratings have increased and emissions regulations have become more demanding the technologies associated with the piston, ring and liner combination have developed to meet the challenges of all new requirements while at the same time producing much extended life.

14.2 Pistons

14.2.1 Introduction

Somewhat in contrast to the piston ring, piston design today is based on a strongly analytical approach, although empiricism still has some part to play. The need for good piston design principles is emphasized by the words of Sir Harry Ricardo spoken many years ago but still as appropriate today:

'In all internal combustion engines, large or small, the piston is invariably the most vulnerable single member'.

The piston serves three main functions:

(i) It provides the means whereby gas loads are transmitted to the connecting-rod/crankshaft system.
(ii) It acts as a crosshead to react cylinder wall side loads in the connecting-rod/crankshaft system.
(iii) It is a carrier for the gas and oil sealing elements—the piston rings.

14.2.2 Piston loading

It is necessary to understand the loading applied to pistons during engine operation, since this provides the basis for design selection, the reasons for oil cooling and the causes of structural failure.

The loading can be considered to consist of two parts, mechanical loading and thermal loading.

14.2.2.1 Mechanical loading

Mechanical loading is due to the gas pressure in the cylinder and the reactions from the gudgeon pin and the cylinder wall. Inertia loading is not usually significant from a structural viewpoint for diesel engine pistons. However the wider use of high speed engines in light duty applications such as passenger cars and light vans and the need to reduce reciprocating mass for noise, vibration and response features has made consideration of inertia, as with gasoline pistons, of some significance in such cases. In the main, however, the level of mechanical loading can be judged simply in terms of the maximum cylinder pressure (P_{max}).

14.2.2.2 Thermal loading

Thermal loading is due to the temperature and heat transfer conditions in the cylinder and at other boundaries to the piston. No simple or universally applicable relationship is known to exist for assessing thermal loading in terms of engine performance data. Instead two alternative approaches may be used, one of which is very simple and is based on engine rating alone while the other is more detailed, requiring solution by computer, and takes account of more engine variables.

The simpler approach gives an upper-bound measure of thermal loading or heat flux (H.F.) as

$$\text{H.F.} \propto \frac{\text{Power}}{\text{Bore area}}$$

Thus H.F. $\propto$ b.m.e.p. $\times$ mean piston speed and H.F. $\propto$ b.m.e.p. at constant mean piston speed.

This approach has obvious shortcomings since it takes no account of the air intake conditions (boost pressure and temperature), which are known to influence piston temperatures. This was particularly true with engines using quiescent combustion chambers but improvements in turbochargers and intercooling systems have extended this influence to swirl types also.

The other approach uses engine performance and other data, and involves a thermodynamic simulation of the engine cycle by computer. Heat transfer in the cylinder is described by an empirical expression. This approach gives a more accurate indication of thermal loading, and analyses of four-stroke medium-speed engines show that at constant trapped air/fuel ratio and mean piston speed, thermal loading can be related to engine rating by the expression

$$\begin{aligned}\text{H.F.} &\propto (\text{b.m.e.p.})^n \\ &\propto (\text{Power})^n\end{aligned}$$

where n has values between 0.4 and 0.7, as shown in *Figure 14.1*.

The assessment of the thermal loading by the computer approach is preferred in each case where a full design analysis

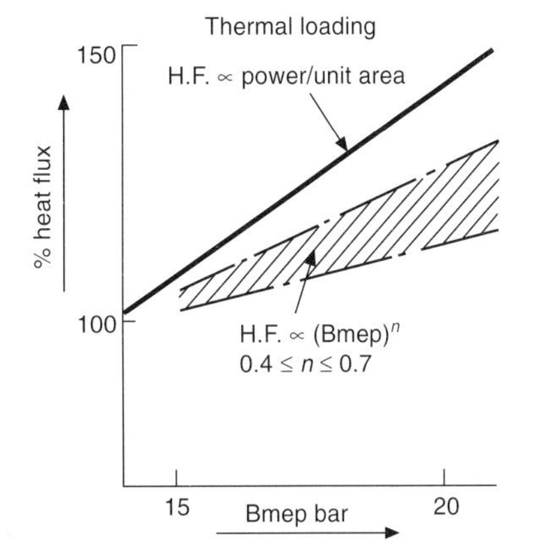

Figure 14.1 Piston thermal loading

*Wellworthy trademark.

is involved, but for comparative purposes the power per unit area is acceptable and is widely used.

The safety of the piston under the combined factors of mechanical and thermal loading can be assessed by analytical methods such as the finite element technique, and this is considered in more detail later.

14.2.3 Piston design

In addition to the engine rating and duty cycle, the maximum cylinder pressure, the combustion chamber shape, emission requirements, the maximum allowable weight and the fuel quality are the essential technical elements in the selection of a particular piston design. The position of the gudgeon pin, the length of the piston, the small-end width and the position of the piston rings are important variables which affect the detail rather than the overall design choice.

Some of the more important piston dimensional and surface finish data are shown in *Table 14.1*.

14.2.3.1 Skirt and land form

The main feature of the piston development process is the derivation of the optimum skirt and land size and shape under operating conditions. Usually a procedure of marking test and piston refinement is carried out until a satisfactory bedding and carbon pattern is achieved. In general, an even piston marking with no hard areas is sought. The skirt thus derived typically has an oval form and a longitudinal barrelled shape. Lands may also have ovality. Skirt lands and cold clearance are finally checked with cold and hot scuff tests when there should be no harsh marking which could lead to seizure.

While skirt shapes were at one time produced mainly by grinding methods, single-point diamond-tool cam turning is now the preferred method, giving flexibility and accuracy of form and controlled surface finish.

Table 14.1 Piston design data

Minimum depth of piston ring grooves

Compression grooves
= Maximum ring radial depth + 0.25 mm
Oil control grooves with drainage
= Maximum ring radial depth + 0.25 mm

Minimum ring side clearances

Ring diameter	Clearance
< 50 mm	0.02 mm (0.001 in)
50–100 mm	0.04 mm (0.0015 in)
100–180 mm	0.064 mm (0.0025 in)
> 180 mm	0.08 mm (0.003 in)

Piston minimum clearances (not controlled expansion type) per unit bore diameter

Material	Skirt	Top land
Lo-ex Aluminium-Silicon alloy (LM 13 type)	0.0012	0.006
Grey cast iron	0.0007	0.0035

Surface finish of piston (Aluminium alloy)

	R_a, μm	CLA, μin
Ring groove side faces	0.4 max	16 max
Skirt bearing surface	2–3	75–125
Gudgeon pin bore	0.2 max	8 max

R_a = roughness average.
CLA = centre line average.

14.2.3.2 Material

The aluminium-silicon eutectic alloy (11/12% Si), known as Lo-ex, is the most widely used material for pistons. In this application it has favourable properties in respect of density, thermal conductivity, wear resistance, expansion, strength and machineability, and offers the option of either casting or forging. Its principal disadvantage is the pronounced temperature dependence of its mechanical properties, which fall off rapidly at temperatures above 200°C (*Figure 14.2*). In practice this means that the temperatures in the highly loaded strut and boss regions should be limited to 200°C, while on the piston crown the operating stresses and duty cycle may allow temperatures up to 350°C to be reached without distress.

Aluminium–copper alloys have better high strength properties, especially, UTS, fatigue strength and creep, providing improved resistance to crown cracking. However, despite the fact that other important properties such as thermal expansion, groove wear, scuff resistance and castability are less acceptable their use in highly rated pistons is growing.

The original material for pistons was cast iron in its grey form, and this was subsequently superseded by aluminium alloy. Recently the much stronger spheroidal graphite cast iron has been applied with success to pistons, particularly for medium-speed engine sizes. The main advantages of cast iron are low expansion, good groove wear resistance and high strength. Apart from weight, its main disadvantage lies in the difficult production of thin-section castings free from defects.

As ratings have increased and, in particular, peak cylinder pressures have exceeded 180 bar, aluminium alloy in any of its forms shows limitations of strength and composite designs with forged steel sections carrying the main loads have been found necessary. Such designs take the form of steel crowns bolted on to aluminium alloy or cast iron skirts or steel crowns with aluminium skirts joined at the pin/pin boss region to form an articulated feature. Monometal thin section steel pistons are in development and as with cast iron the main problem is the casting of sound material. These designs are shown and discussed in more detail later. Despite the considerable increase in hot strength provided by steel it is important that defects such as inclusions or porosity are absent from the highly loaded regions of the piston or cracking will occur.

14.2.3.3 Oil cooling of pistons

As a general guide some 3–5 per cent of the heat available in the fuel is absorbed by the piston in combustion chamber forms

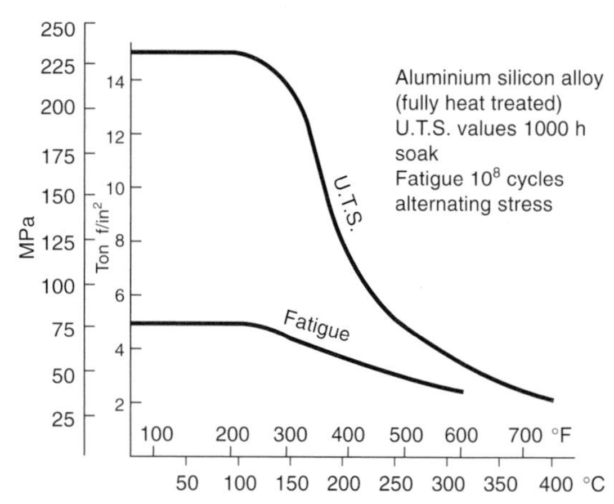

Figure 14.2 Strength of aluminium-silicon alloy at elevated temperatures

of the quiescent type (i.e. mainly above 150 mm (5.9 in)) while approximately 6–8 per cent is absorbed by the swirl-type chambers in the smaller high-speed range.

A pistion has inherent cooling via the rings, land and skrit to the water jacket, and to the crankcase oil by splash/mist. Without additional cooling increased heat flow to the piston results in higher piston temperatures. These can result in increased crown, land and groove carbon deposits. With the continuous improvement in oils, as a general rule, top groove temperatures greater than 270°C (518°F) are considered a potential source of carbon problems.

It can be clearly seen from the factors discussed above, that at a certain rating some form of oil cooling of the piston becomes essential to ensure satisfactory operation. The exact level of this rating and the type of cooling cannot be precisely defined for every case and it varies from engine to engine as a result of such factors as size, duty, combustion chamber form, fuel system, and turbocharger pressure ratio and intercooling arrangements. Efficient intercooling in the turbocharger system results in a very marked reduction in combustion temperatures and permits much higher ratings for a given stress condition. *Figure 14.3* gives a guide to the rating level based on power per unit piston area for engines in the size range 100–400 mm (4–16 in) drawn from a survey of many different designs of engine. It shows that the power rating when oil cooling becomes necessary is lower as the piston diameter increases. The size range has been split into a high-speed range, 100–180 mm (4–7 in) and a medium-speed range, 200–400 mm (8–16 in). The two ranges also recognize that high-speed engines typically have a deep swirl-type combustion chanmber, whereas medium-speed engines normally have a quiescent shallow combustion bowl form. The decision to provide oil cooling, entailing special oil feed/jet arrangements, specific piston designs and often additional oil pump and oil cooling capacity, can only be taken after careful technical and economic assessment. Several types of oil feed arrangements are in regular use today, and are illustrated in *Figure 14.4*.

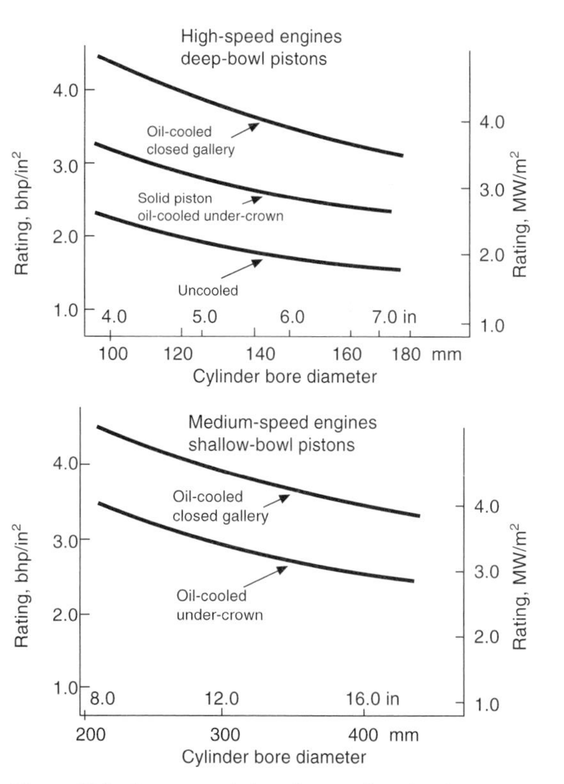

Figure 14.3 Recommended maximum ratings for piston types and diameters

In an aluminium piston some 60 per cent of the heat to the piston is removed by the oil cooling–splash and gallery–the rest being dissipated through lands, skirt and rings. In the case of a ferrous design practically all the heat is removed by the oil cooling because of the thinner sections and lower thermal conductivity of the material. It is seen that effective oil cooling of ferrous types is essential to avoid overheating with its many adverse consequences.

14.2.3.4 Crown cracking

The combined effect of mechanical and thermal loading in severe cases can produce *crown cracking*. The mechanical loading due to gas pressure is a high cycle effect while thermal loading produces a combination of steady state and low cycle stress, the latter due to the transient no-load to rapid full load type of loading as experienced by truck, locomotive and marine trawler applications. Hard anodizing of the crown gives some protection to those regions which are under compressive stress by producing a beneficial tensile stress condition in the sub-aluminium alloy but must be excluded from the pin axis where the stress is already at a high tensile value. A greater improvement is given by oil cooling, particularly of the galleried type. This cooling effectively reduces the stress range by lowering the material temperature and increasing its strength.

In very difficult cases of crown cracking as experienced with deep combustion bowls of the re-entrant form, a more complex and costly solution may be necessary. One such approach used in production is the use of ceramic fibre reinforcement of the bowl edge or indeed the whole crown. The fibrous inserts are infiltrated by the aluminium alloy in the squeeze casting process which also produces more consistently high strength properties in the non-reinforced regions. A typical example is shown in *Figure 14.5*.

14.2.3.5 Skirt coatings

In cases of high speed engine pistons where cold and/or hot scuff of the shirt occurs and cannot be cured by shape modification or clearance or, indeed, if a measure of robustness is required

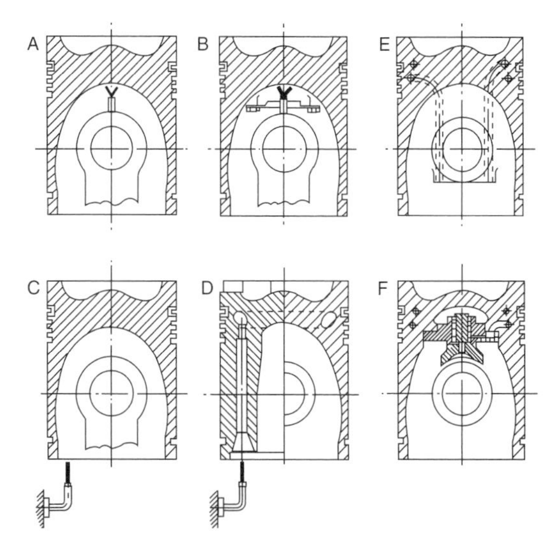

Figure 14.4 Types of oil cooling

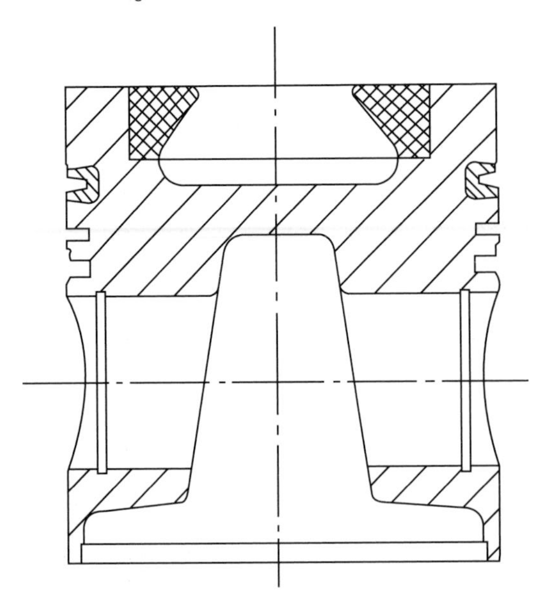

Figure 14.5 Ceramic fibre piston combustion bowl re-enforcement

a coating can be applied to the skirt. For many years this has taken the form of a sprayed graphite with perhaps additions of molybdenum. This is not easy to apply efficiently with good control and does not last the life of the piston. Recently, new resin bonded graphite/molybdenum coatings with wear enhancement additions have become available which can be applied cleanly, uniformly and effectively in mass production conditions by means of silk screen printing. Such coatings are reported to have long life and in some cases reduce frictional losses by up to 3 per cent. It is felt that is more likely to be due to the effect of the coating surface reducing oil film thickness and hence viscous losses, rather than direct boundary friction improvement.

14.2.3.6 Emissions—oil consumption and top ring location

Apart from the combustion system, the piston assembly has a significant contribution to make in the reduction of emissions. Oil consumption is a major source of particulates and for this reason lower target levels down to 0.05% full load fuel are being sought in development.

Detail design of the lands and skirt in combination with the ring pack is required to minimize oil consumption. Also, oil drainage from the oil ring groove by means of holes drilled through into the inside of the piston are being replaced with external drainage slots from the groove on the minor (pin) axis (*Figure 14.6*). This is because more oil can flow from the inside of the piston to the ring groove especially in the case of oil cooled designs. It is also a saving in manufacturing cost.

In addition, the position of the top ring is being raised from a guideline top land height of 20% to 10% bore. This reduces the 'dead volume' formed between the top ring, bore and top land which is a source of inefficient combustion contributing to both hydrocarbon emissions and higher fuel consumption. Clearly, moving the ring upwards into a higher temperature regime, has its disadvantages in terms of groove carbon, ring wear and scuffing. However, ring coating and oil improvements to meet this harsher environment have made this transition possible without major problems. Two further disadvantages of a higher top ring are (1) the tensile stress on the crown is increased especially on deep combustion bowl designs, and (2) when retarded fuel injection timing is used in low emission engines more carbon forms on the bore walls. The higher ring scrapes more of this carbon from the walls and the oil becomes more loaded. In some instances a revision to a top ring location of 15% bore has been introduced.

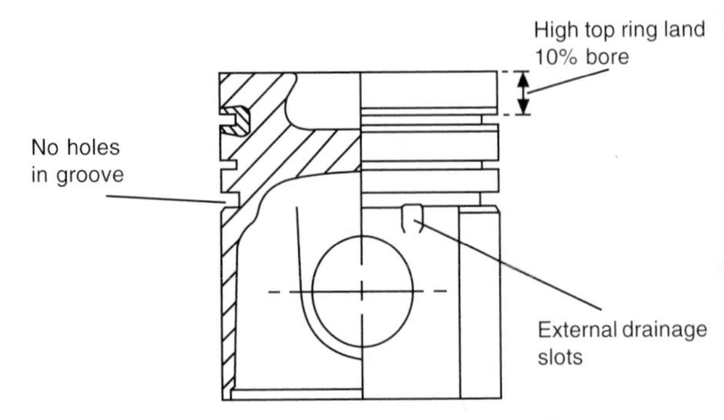

Figure 14.6 High top ring and external drainage piston

14.2.4 Piston types

The unique character of different engine makers' combustion chambers and the many individual features which have been developed over the years for particular applications have resulted in a bewildering range of piston designs. It is possible however to represent these in broad categories, mainly based on engine size and rating. For convenience these are considered under two bore size ranges:

(a) High-speed engines, bore size 80–180 mm (3–7 in)
(b) Medium-speed engines, bore size 180–600 mm (7–24 in)

14.2.4.1 High-speed engine types

Plain aluminium alloy piston (*Figure 14.7*). This is the simplest form of piston, used mainly for naturally aspirated ratings and where long life is not a prime requirement. In automotive applications, with mileages above 100 000, wear of the top groove can be expected to cause high oil consumption and possibly ring breakage. In pistons up to 100 mm in diameter for speeds greater than 3500 rev/min a rolled-in steel rail on the upper face of the top groove (see *Figure 14.8a*) extends useful life by at least 50 per cent. Another development successfully applied in production is the use of ceramic or metal fibres in the ring groove region (*Figure 14.8b*). As with the bowl reinforcement by this method squeeze casting is required.

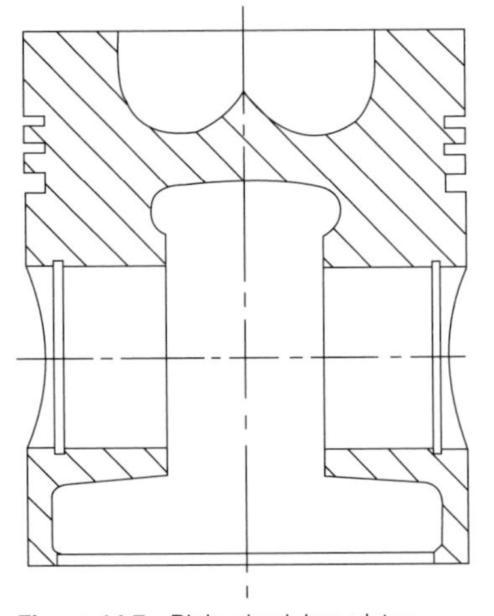

Figure 14.7 Plain aluminium piston

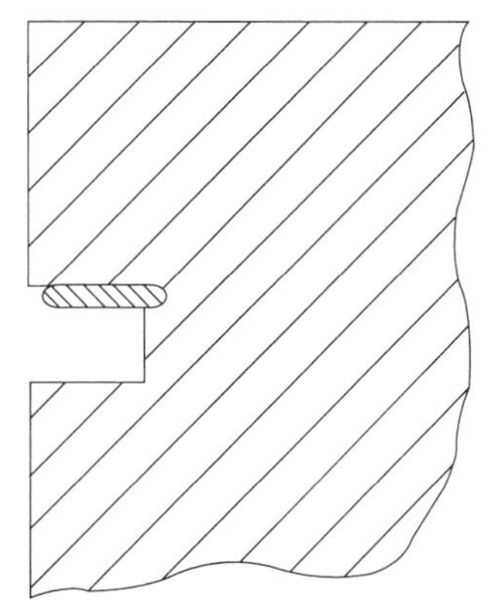

Figure 14.8 Steel rail top groove reinforcement

Aluminium piston with cast-in Alfin insert (Figure 14.9). This is the standard high-speed diesel piston, which in its various forms is used over the complete range of engine applications at both naturally aspirated and turbocharged ratings up to approximately 3.5 MW/m^2, in both cooled and under-crown-cooled designs.

The main feature of the piston is the iron insert in the top groove cast in by means of the long-established Alfin* bonding process. In some cases a double insert encompassing both top and second ring grooves is used (see *Figure 14.11c*) but this is more common towards the top end of the size range. Good bond quality is essential to avoid loosening of the insert and serious failure and effective non–destructive inspection has been made possible by the development of ultrasonic methods. The life of the piston groove can be expected to increase by at least a factor of 3 over the plain design. A wedge or trapezoidal boss form to increase the effective bearing area is used to avoid possible fatigue failure under high cylinder pressure conditions.

Aluminium piston with cast-in cooling gallery (Figure 14.10). At ratings of above approximately 3.5 MW/m^2, additional cooling is necessary in the form of an enclosed gallery in the crown, particularly to avoid carbon formation in the top groove and crown thermal cracking. Such galleries are conveniently produced by means of soluble cores.

In this size range the galleries are usually fed by the standing-jet method, with the flow regulated to allow the gallery to run

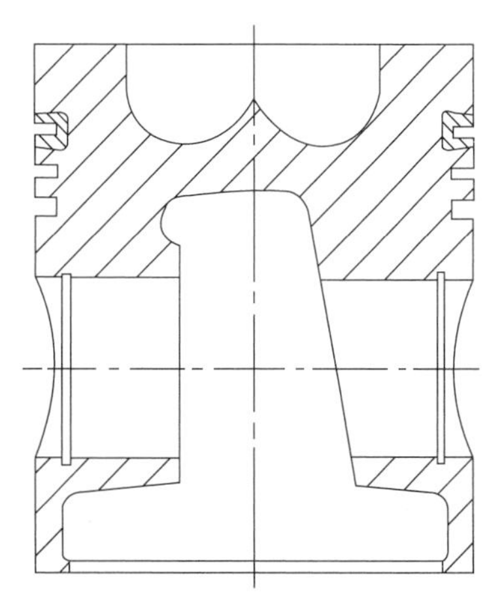

Figure 14.9 Aluminium alloy piston with cast-in Alfin insert

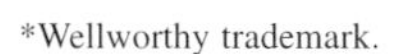

*Wellworthy trademark.

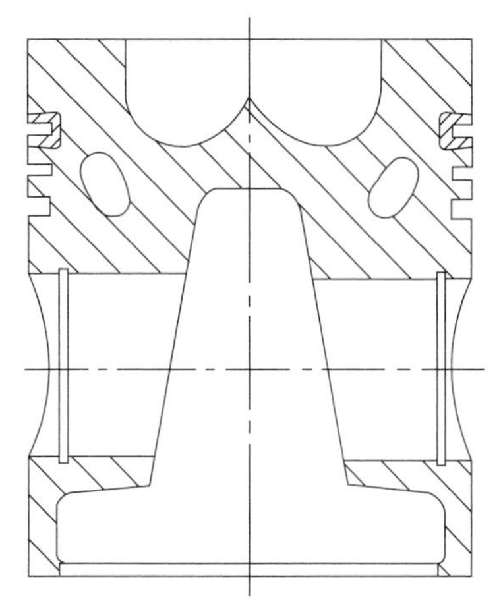

Figure 14.10 Aluminium alloy piston with cast-in cooling gallery

partially full 30–80%, thus introducing a very effective 'cocktail' shaking cooling action.

Articulated pistons (Figure 14.11a and 14.11b). As maximum cylinder pressures have increased above 170 bar due to power density, emission and fuel economy requirements, aluminium alloy, even with bushes, reaches a limit to withstand the resultant high pin boss pressures without cracking. The answer has been in the development of a composite design—a forged steel crown/boss section and an aluminium alloy skirt—joined at the pin hole by means of the gudgeon pin with each part free to rotate about the pin; hence the name 'articulated'. The effect is that the main driving load to the con-rod and the bore side load are separated and the crown section remains more stable from transverse motion. The steel pin bosses are bushed to give compatibility with the steel pin. The steel crown must be adequately oil cooled and '*open*' galleries formed with pockets in the skirt and '*closed*' galleries formed by means of friction welding both feature in current production designs and are typically shown in *Figure 14.11a* and *Figure 14.11b*, respectively.

The strength and reliability of such designs, despite their extra cost, have shown a major market penetration in bores over 120 mm, in both truck and off-highway fields and this penetration is extending into the Medium-speed range.

Monometal ferrous designs (Figures 14.12a and 14.12b). Both cast iron and steel material offer good high strength properties but have limitations in respect of density and manufacture, particularly casting. Despite this, thin walled nodular iron monometal designs are successfully running in production and a typical example is shown in *Figure 14.12a.* An investment cast 17–4 PH steel development version is shown in *Figure 14.12b.*

Close clearance pistons. The need for low noise levels in petrol engines at one time established a prime place for pistons with a cast-in steel insert controlling the expansion of the skrit, to give close running clearances over the complete load range. The high speed diesel engine at a more restricted production level also followed this technology with advantages in subjective noise level, consistency of oil and blow-by control and suppression of liner cavitation damage.

A typical design of the continuous steel band type is shown in *Figure 14.13,* the band terminating in the oil ring groove to give a more effective expansion control.

Today the use of expansion controlling insert pistons in both petrol and diesel engines has reduced very significantly. Weight and strength limitations and a better knowledge and control of

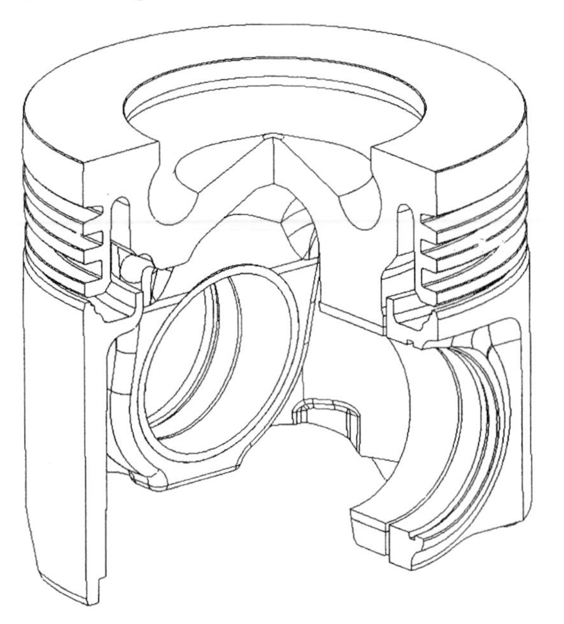

Figure 14.11 (a) 105 mm 'Open' gallery articulated piston for diesel engine

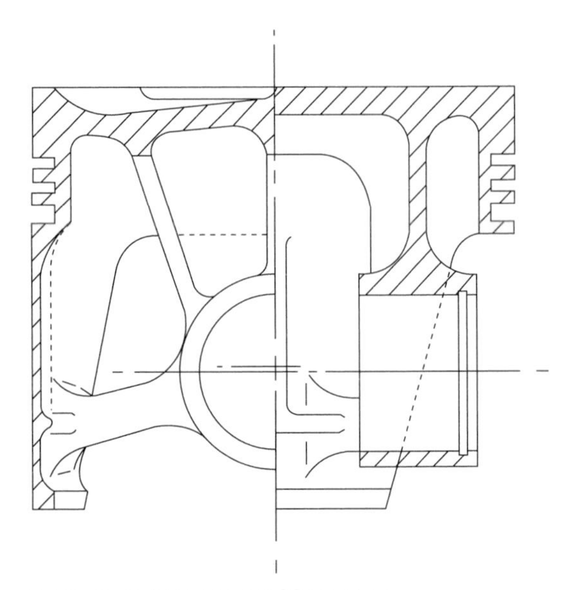

Figure 14.12 (a) Lightweight SG iron piston

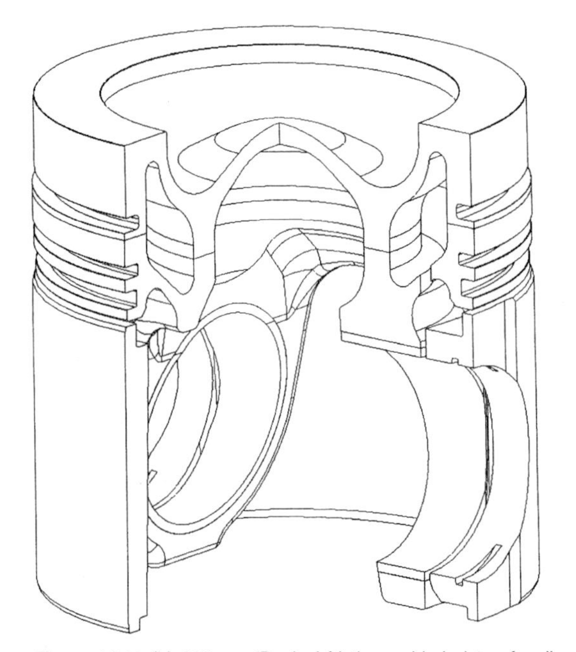

Figure 14.11 (b) 210 mm 'Duplex' friction welded piston for diesel engine

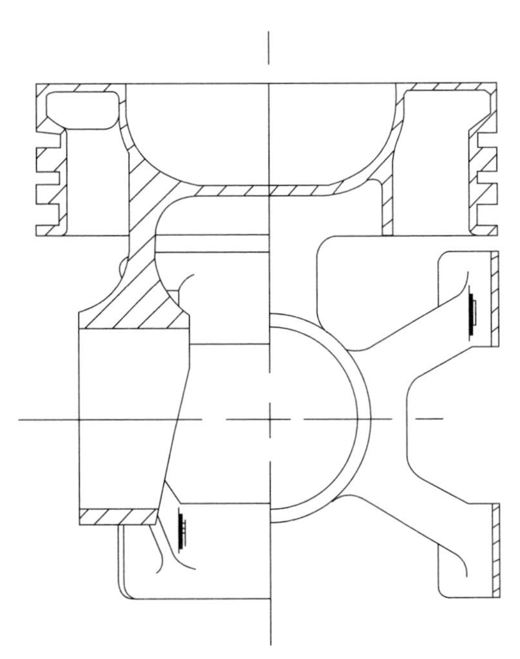

Figure 14.12 (b) Cast steel piston design

skrit shape and stiffness to give low cold clearance have been the main factors leading to this change.

*14.2.4.2 Medium-speed engine types (*Figure 14.14*)*

Plain aluminium piston (*Figure 14.14(a)*). This is the most basic piston. The main structural features are the absence of top ring groove reinforcement, the re-entrant strut above the piston-pin boss, the small gudgeon pin diameter and the low crown thickness. Such a piston generally will be used without oil cooling in engines with only two valves per cylinder and at ratings up to 10 bar b.m.e.p. (or 2.5 MW/m^2 of piston area—the rating here and thereafter in this section refers to a typical piston of 250 mm diameter). This limit is largely set by the maximum cylinder pressure, which can cause structural failures in the re-entrant strut and high wear rates in the top ring groove. The top ring groove would be re-machined at 10 000/20 000 h service intervals.

Aluminium piston with Alfin insert (*Figure 14.14(b)* and (*c*)). This piston type has a cast-in ring groove insert shown in single and double form, in *Figure 14.14(b)* and (*c*) respectively. The double-groove insert is used particularly for heavy fuel applications. This type also includes the stronger D-strut and larger gudgeon pin diameter. It is widely used in engines with four

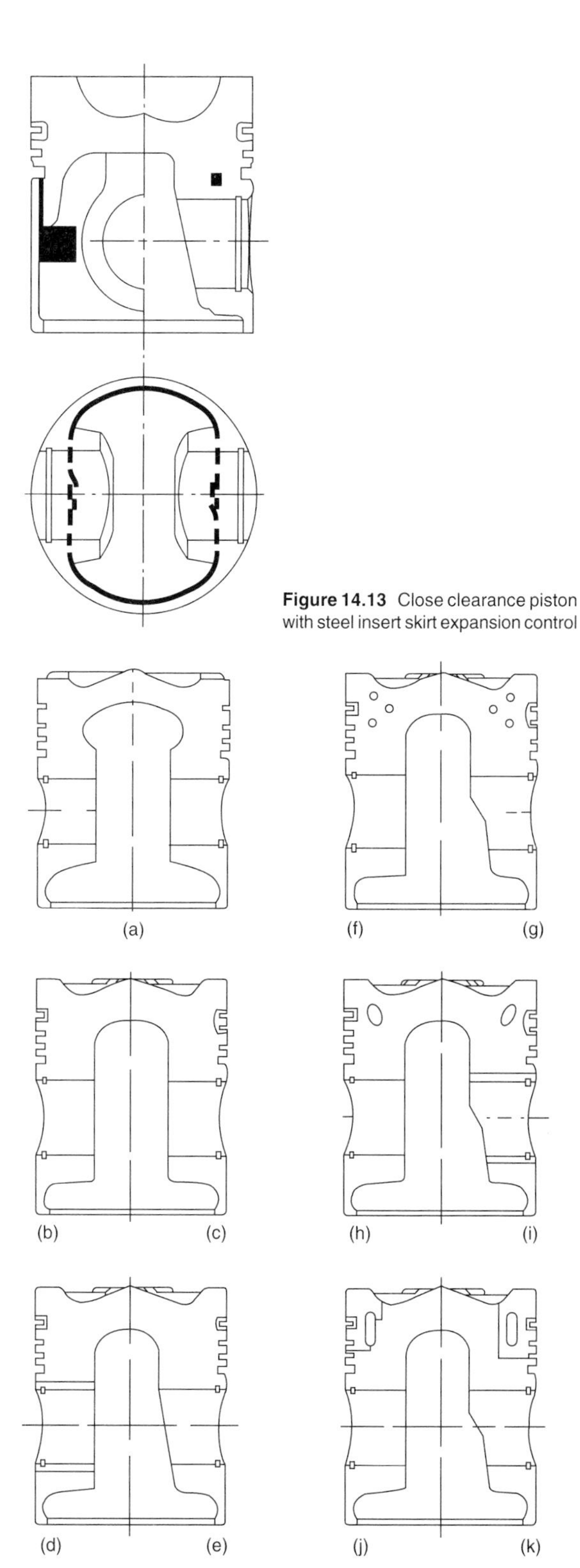

Figure 14.13 Close clearance piston with steel insert skirt expansion control

Figure 14.14 Medium-speed engine piston types

valves per cylinder, with under-crown oil cooling and at engine ratings up to 13 bar (3.3 MW/m^2). The limit is largely set by metal temperature considerations, which can cause structural failures on the crown and oil pyrolysis in the top ring groove. Structural failure in the bore of the cross-hole or boss may be the limiting factor in engines which have either high cylinder pressures or undersized gudgeon pins, and the alternative designs, incorporating a shrunk-in bush or trapezoidal boss (*Figure 14.14(d)* and (*e*)) would be used in such cases. The top ring groove life in Alfin pistons would normally be expected to increase threefold before re-machining.

Aluminium piston with internal cooling gallery (*Figure 14.14(f)*, (*g*), (*h*), (*i*), (*j*) and (*k*)). The provision of an internal cooling gallery allows a larger surface area for cooling and shorter heat flow paths than are possible with under-crown cooling. The internal gallery is formed by one of three methods and each design will be considered separately.

(i) Cast-in steel pipe or coil, with the oil inlet at one end and the exit at the other (*Figure 14.14(f)* and (*g*)). Full flow convection cooling action is employed and high flow rates are required to generate moderate heat transfer conditions. This design is limited by metal temperature considerations and potential boss and strut failures up to an engine rating of 16 bar (4.0 MW/m^2).
(ii) Cast aluminium piston with the cooling gallery formed by means of a water-soluble salt core as shown in *Figure 14.14(h)* and (*i*). The gallery is arranged to run partially filled with oil to promote turbulent cooling action and to make efficient use of the oil. Considerable design freedom exists with this manufacturing method, permitting the optimum gallery shape, inclination and position to be employed. The design, incorporating an oval gallery, will generally be used at ratings up to 16 bar (4.0 MW/ m^2), the limit being set by cross-hole and strut fatigue strength.
(iii) Electron-beam welded piston with cooling gallery (*Figure 14.14(j)* and (*k*)). An electron-beam-welded (e.b.w.) piston consists of two parts jointed by a welding process. The advantage of this construction is that the main structure of the piston is made from higher strength and more ductile forged material while the outer crown annulus is made from cast material containing an Alfin insert.

In *Figure 14.14(j)* the sides of the cooling gallery are machined into each part before assembly in order that a complete gallery is formed between the parts after assembly. The two parts are joined by vertical and horizontal welds which must terminate in solid material and not break into the gallery. A stepped joint is therefore used but has the disadvantage that the joint faces which intersect the gallery remain unwelded and can act as built-in cracks and notches. In the alternative form (*Figure 14.14(k)*) the annulus contains a complete gallery made by the soluble core process and the problem of the built-in notch effect does not arise. The cooling performance of this piston is identical with a soluble core piston, provided the gallery wetted area and position are similar. The design freedom in the gallery shape and inclination is less advantageous than with the cast piston.

Many pistons are in service at ratings up to 16 bar (4.0 MW/ m^2). This limit in the (*j*) form is set by structural failures from the unwelded horizontal joint face on the inner side of the gallery.

Two-piece pistons with steel crown (*Figures 14.15(a)*, (*b*) *and* (*c*)). Two-piece pistons consist of a steel crown member bolted to either an aluminium or a cast iron body. The primary advantage of this construction is that different materials can be used for the crown and the body which better suit their separate functional requirements. In addition, a large gallery for oil cooling purposes is formed between the two parts. This design normally incurs some weight penalty and relies on the performance of tension bolts.

As an assembly with precise fitting parts the two-piece piston is more expensive than an aluminium alloy design of the same size. It is important therefore to establish that the technical needs support the higher cost.

Two-piece pistons with a steel crown are used in place of aluminium gallery pistons when:

(i) The temperature limits are exceeded on the crown or in the ring belt.
(ii) The safety margin in the cooling gallery reaches an unacceptable level.
(iii) Certain heavy fuels are used which cause high groove wear in aluminium grooves and top land abrasive wear.

The rating limit has not been established but is over 20 bar (5.0 MW/m^2). The main areas of vulnerability are in the outer cooling gallery and in the boss.

The crown material should possess good high temperature properties with regard to strength, ductility and corrosion resistance, and good wear properties for the compression ring grooves and top land. Steel forgings, conforming to En 52 (valve steel), En 20B and En 19, hardened and tempered to give an ultimate tensile strength of 900 MN/m^2, are generally used for piston crowns, with En 19 being the most popular choice.

Two or more of the compression ring grooves are contained in the crown, and the grooves may be hardened or chromium plated for longer life. The surface of the crown support is also sometimes chrome plated to resist freeting damage.

The required properties for the body material are mechanical strength, wear resistance and low weight. The maximum operating temperatures are unlikely to exceed 100°C approximately and the temperature gradients will be low. Aluminium-silicon alloy is preferred, although cast irons are used in some engines.

Aluminium bodies will generally be forged and have stepped bosses for the high mechanical loading. Alternatively cast aluminium bodies with stepped bosses and shrunk-in forged bushes are used. The combined weight of a steel-aluminium piston is typically 10–20 per cent heavier than an equivalent aluminium piston. Grey cast iron and spheroidal graphite cast iron are materials also used for the piston body. Both result in increased piston weight and the steel-iron piston is typically 30–60 per cent heavier than an aluminium piston.

The crown and body are bolted together using per-loaded bolts. The pre-load is required to prevent slip and gapping at the joint, and to protect the bolts from loosening and from the full alternating load. The bolts are tightened to a prescribed torque or angular measurement.

Three different bolting arrangements are shown in *Figure 14.15*:

(a) Bolts in the crown and screwed into the body—favoured in high-speed engines (1500 rev/min).
(b) Parallel bolts in the body and screwed into the crown—the most widely used design.
(c) Inclined bolts in the body and screwed into the crown giving greater design freedom—this has been used at ratings up to 35 bar b.m.e.p.

Cast iron pistons (*Figure 14.16*). The use of thin-walled spheroidal graphite (SG) iron pistons for medium-speed engines has developed more widely in the last few years. Good crown and groove wear resistance in heavy fuel applications and the normal advantages of close running clearances are claimed for this type. The main problem seems to be that of casting thin sections consistently to the required high quality free from porosity.

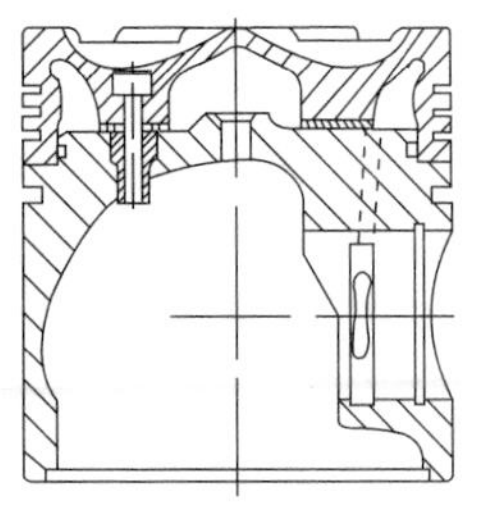
(a) Crown bolts

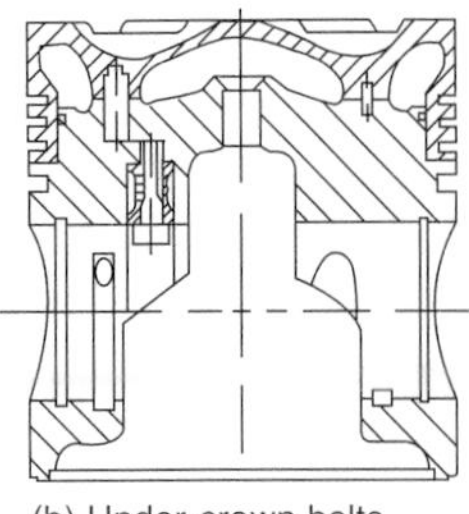
(b) Under-crown bolts

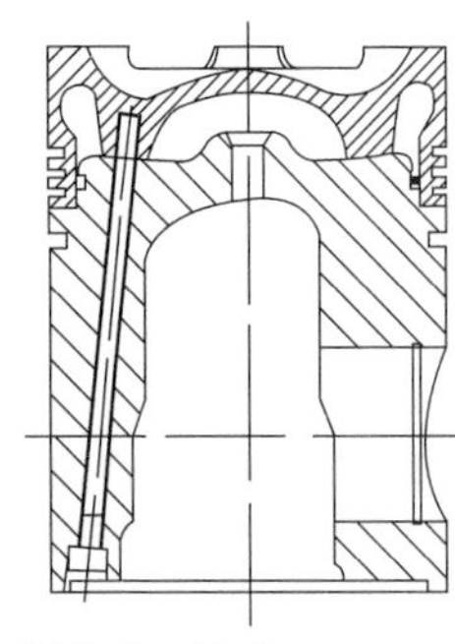
(c) Inclined bolts

Figure 14.15 Two-piece steel crown pistons

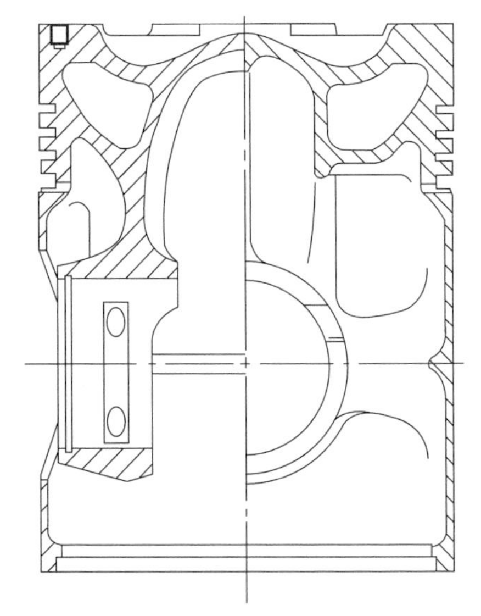

Figure 14.16 Cast iron piston

14.2.5 Gudgeon pins

As an integral major load bearing part of the piston assembly, the gudgeon pin has an important influence on piston strength and, in particular, boss or pin hole strength. This is because bending and ovalization of the pin under load redistributes the pressure on the pin hole from an ideal uniform condition to one which can cause pin scuff or piston cracking of pin hole or crown. The continued increases in peak firing pressure have made the pin and associated pin hole design, the basic starting point in the piston design, a very critical factor for aluminium alloy pistons.

The pin diameter is selected up to an optimum of about 40 per cent piston diameter on the basis of a maximum bearing pressure of 60 MN/m^2 (8700 lbf/in^2) in aluminium and 69 MN/m^2 (10,000 lbf/in^2) in cast iron. The relevant projected area is shown in *Figure 14.17* and the pressure is assumed constant which, as explained above, is not the case in practice. The available projected boss bearing area is limited by the bearing requirements of the con-rod small end which in its typical bushed form has a higher limit of bearing pressure with a guideline of 83 MN/m^2 (12,000 lbf/in^2). Higher allowable boss bearing loads up to

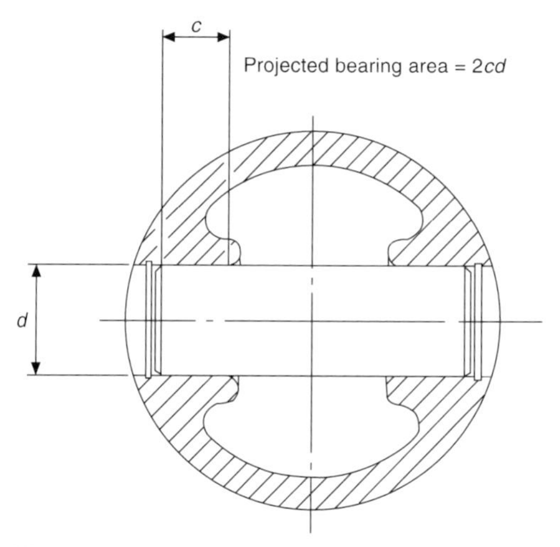

Figure 14.17 Gudgeon pin bearing area

limit of 70 MN/m (10,150 lbf/in^2) can be accommodated by modifying the pin hole shape to alter the pressure distribution by tapering or introducing ovalization relieval slots but such changes increase the crown stress up to a factor of 50 per cent which in some cases can be critical. The ultimate solution is to use a bush which will raise the allowable bearing load to 78 MN/m^2 (11,300 lbf/in^2) but increases cost significantly.

Two key factors–pin ovalization and minimum reciprocating mass–must be considered in the choice of pin diameter and wall thickness. The wall thickness is calculated from a semi-empirical expression based on allowable limits of ovalization and is shown in *Figure 14.18* the pin strength itself arising from such criteria introduces no concerns as long as the manufacturing and particularly the heat treatment is properly controlled. Apart from small high speed diesel engines pin bending, normally, is not taken into account in the design.

A case-hardened steel En 32 to a minimum specification (BS 970) is recommended with good heat treatment to provide the correct case depth and core strength. A tempering temperature of 250/260°C should be used since pin growth in service at the higher temperatures being experienced today can result in pin scuff and seizure. Pins are normally hardened overall and the outside diameter ground to the finish shown in *Table 14.2*.

14.2.6 Piston design analysis

The power and capacity of today's computers and the development of specific software programs has greatly extended the range of predictive analysis of pistons to the point where the need for empiricism is fast diminishing. Current programs cover transverse movement, oil cooling rate, skirt and land contact analysis and temperature, strain and stress prediction, the latter four in association with the 3-D finite element method and combustion models which are subject to constant improvement.

Figure 14.19 shows a typical piston 3-D finite element mesh which has been used to predict the operating thermal and

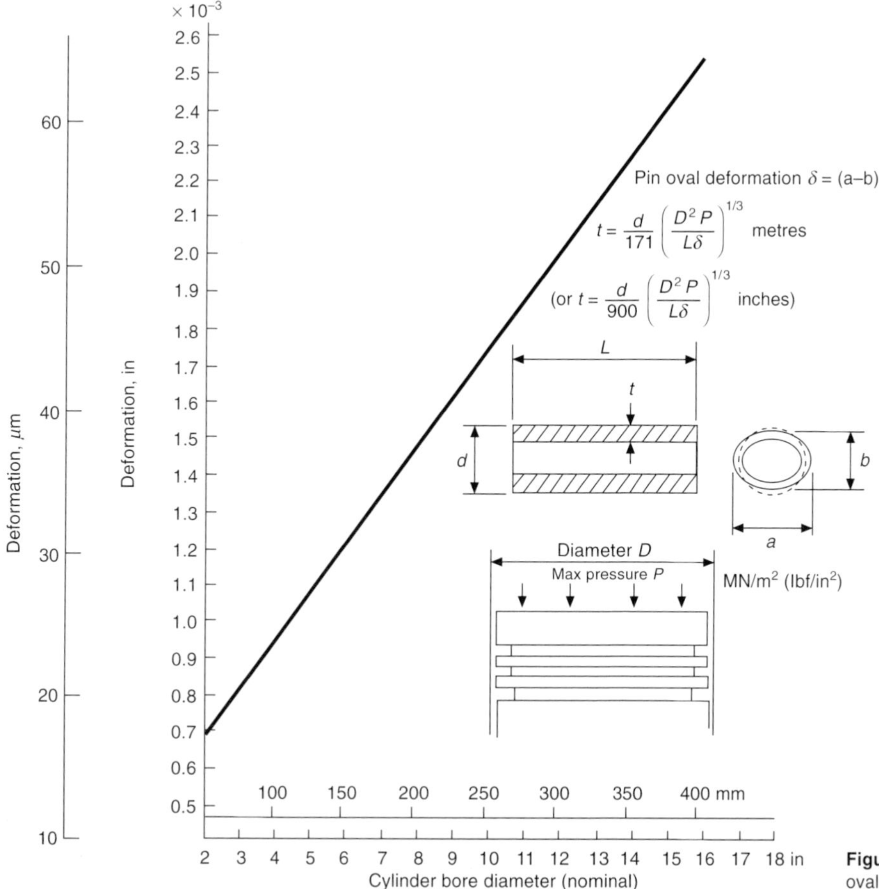

Figure 14.18 Gudgeon pin allowable oval deformation

Table 14.2 Gudgeon pin surface finish

Gudgeon pin	R_a, μm	*CLA*, μin
Outside diameter (O.D.)		
< 75 mm (3 in)	0.1 max	4 max
75–140 mm (3–5.5 in)	0.2 max	8 max

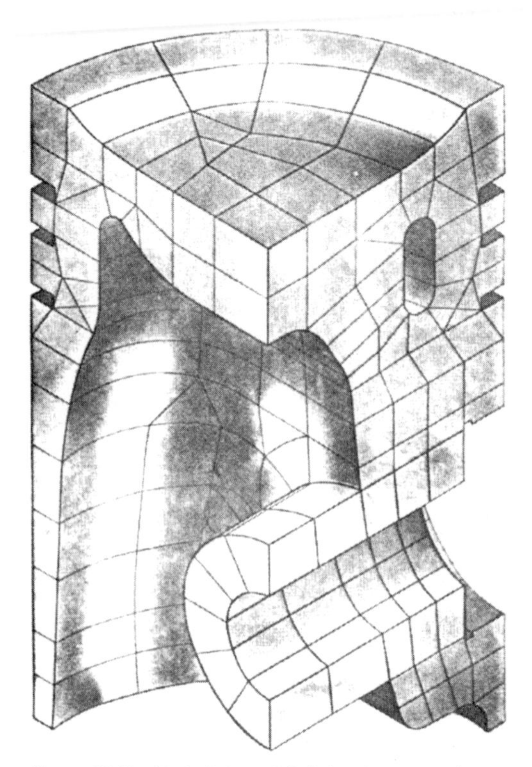

Figure 14.19 Typical piston 3-D finite element mesh

mechanical stresses combined in a factor-of-safety diagram in *Figure 14.20(a)*. The crack shown in *Figure 14.20(b)* occurred after a short period of full load engine running and coincides precisely with the minimum factor-of-safety point, confirming the accuracy of the technique. The piston life prediction problem is an extremely complex one since it requires the combination of steady state thermal, high cycle mechanical and low cycle thermal stresses in a material, in the case of aluminium alloy, whose strength is very temperature dependent.

14.3 Rings

14.3.1 Introduction

The piston ring is a deceptively simple looking component, whose performance is of fundamental importance to the successful operation of the total engine system. The combination of rings carried by each piston and known as the 'ring pack' has two prime functions: (1) to provide a reciprocating seal between the piston and bore limiting the flow of combustion and compression gas into the crankcase (blow-by control); and (2) to control the flow of oil from the crankcase to the combustion chamber space to an acceptable level while giving adequate ring/bore surface lubrication (oil control).

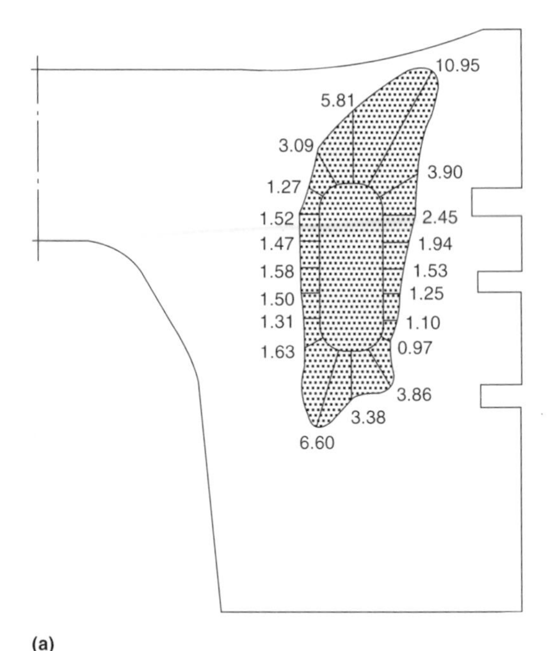

(a)

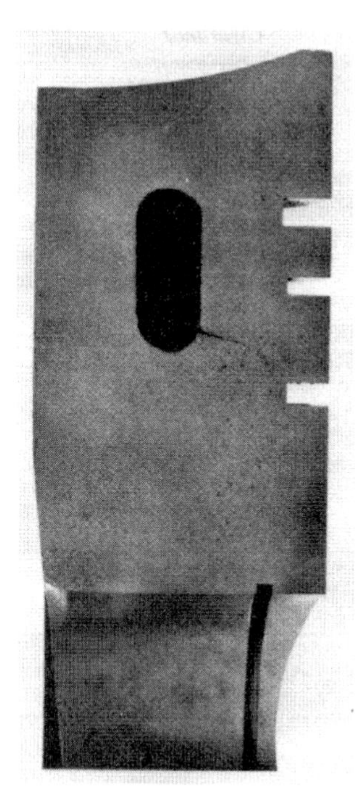

(b)

Figure 14.20 (a) Calculated factor of safety; (b) Piston failure condition

The ring pack must provide these controls for long periods of operation without serious performance drop-off. For example a modern heavy-duty truck engine can have a target life to first overhaul of more than 750000 miles, or more than 20000 hours' duty with no more than a threefold increase in oil consumption. The ring components of today which have been developed to meet such requirements are precisely defined and accurately manufactured elements embodying advanced material and surface technologies.

14.3.2 Ring design

Ring design uses well-established elastic bending theory and is based on a careful compromise between opening stress when fitting the ring onto the piston, and working stress when the ring is in the cylinder. Within the optimum ratio of these stresses and in accordance with the strength and elastic modulus of the available ring materials, current practice defines a free gap which lies between $2\frac{1}{2}$ and $3\frac{1}{2}$ times the radial thickness of the ring, and a ratio of diameter/radial-thickness of between 22 and 34.

Consideration of life, oil consumption, blow-by control, space and freedom from stick, has led, through wide engine running experience, to a range of recommended design data for compression and oil control rings and their grooves. A summary of these data is presented in *Tables 14.3–14.6* and provides a general guideline to design. Each application requires detailed assessment and variations outside the quoted figures are not uncommon. In addition to surface finish control of the ring side, flatness is also important and should be maintained to within 0.013 mm (0.0005 in) and circumferentially within 0.02 mm (0.0008 in) on small rings and 0.07 mm (0.0028 in) on large rings.

Radial clearance is based on an empirical expression for groove root diameter and is as follows:

Compressing ring

Groove diameter $= D - (2t + 0.006D + 0.50)$ mm

or $D - (2t + 0.006D + 0.020)$ in

Oil control ring

Groove diameter $= D - (2t + 0.006D + 1.50)$ mm

or $D - (2t + 0.006D + 1.060)$ in

where t = ring radial thickness (max)
D = cylinder bore diameter

With few exceptions, rings are manufactured to be light tight in a round bore at room temperature. The presssure pattern under such conditions is not necessarily constant since account has to be taken of the thermal effects under running conditions. The ring itself becomes hotter than the bore and the heat flow through the ring from inside to outside surface causes the ring to open out. The net result is a variation in pressure with high values at the gap. As long as this condition does not produce scuff it may be acceptable since as wear takes place the pressure at the gap becomes less giving a ring of longer life. The summary requirement in ring design is to provide a cold free shape which gives a pressure pattern in service shown by practical tests to be needed for the particular application.

The uniform elastic wall pressure P is given by:

$$P = \frac{E_n L}{7.07D\,(D/t - 1)^3}$$

where E_n = nominal modulus of elasticity
L = gap closure = free-gap–fitted-gap
D = external ring diameter
t = radial thickness

$$\frac{\text{Fitting stress}}{E_n} = \frac{4(8t - L + 0.004D)^2}{3\pi t\left(\dfrac{D}{t} - 1\right)}$$

$$\frac{\text{Working stress}}{E_n} = \frac{4(L - 0.004D)}{3\pi t\left(\dfrac{D}{t} - 1\right)^2}$$

$$\frac{\text{Fitting + working stress}}{E_n} = \frac{32}{3\pi\left(\dfrac{D}{t} - 1\right)}^2$$

Table 14.3 Side clearance-compression rings

Ring size	*Clearance*	
	Top ring	*2nd and 3rd rings*
From 3 in to 7 in dia. 76 to 178 mm	0.0025/0.0045 in 0.064/0.114 mm	0.0015/0.0035 in 0.038/0.089 mm
Above 7 in to 9.75 in dia. 178 to 250 mm	0.003/0.005 in 0.076/0.127 mm	0.0025/0.0045 in 0.064/0.114 mm
Above 9.75 in to 16 in dia. 250 to 405 mm	0.004/0.006 in 0.102/0.152 mm	0.003/0.005 in 0.076/0.127 mm
Above 16 in to 23.5 in dia. 405 to 600 mm	0.006/0.0085 in 0.152/0.216 mm	0.005/0.0075 in 0.127/0.191 mm
Above 23.5 in dia. 600 mm	0.006/0.009 in 0.152/0.229 mm	0.005/0.008 in 0.127/0.203 mm

Table 14.4 Side clearance-oil control rings

Ring size	*Clearance*
From 3 in to 7 in dia. 76 to 178 mm	0.0015/0.0035 in 0.038/0.089 mm
Above 7 in to 9.75 in dia. 178 to 250 mm	0.0025/0.0045 in 0.064/0.114 mm
Above 9.75 in to 16 in dia. 250 to 405 mm	0.003/0.005 in 0.076/0.127mm
Above 16 in to 23.5 in dia. 405 to 600 mm	0.005/0.0075 in 0.127/0.191 mm
Above 23.5 dia. 600 mm	0.005/0.008 in 0.127/0.203 mm

Table 14.5 Fitted gap

Engine type	*Gap per unit bore diameter*
Water cooled	0.003/0.004
Air cooled and two-stroke	0.004/0.005

Table 14.6 Side surface finish

Ring size	*Side surface finish CLA*
Up to 7 in 178 mm	16 μin max 0.4 μm max
Above 7 in to 16 in 178 to 405 mm	32 μin max 0.8 μm max
Above 16 in to 36 in 405 to 920 mm	63 μin max 1.6 μm max

which is independent of gap closure L.

To obtain a uniform pressure distribution, the ring free shape is given by:

Radial ordinate

$R_c = R + U + \delta U$ (see *Figure 14.21*)

where

$$U = \frac{FR^4}{E_n I}\left(1 - \cos\alpha + \frac{1}{2}\alpha\sin\alpha\right)$$

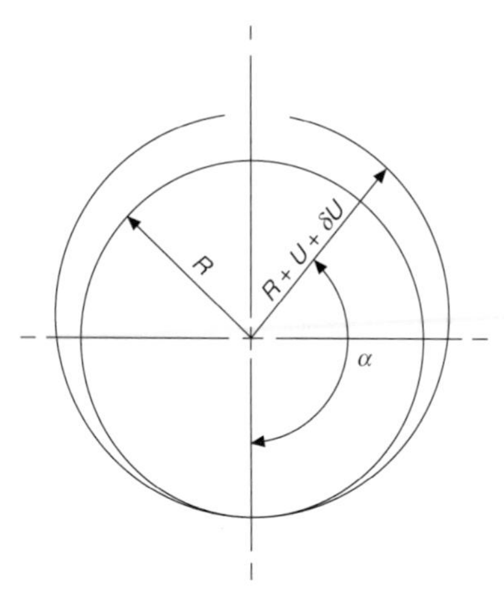

Figure 14.21 Ring free shape

$$\frac{R}{2}\left(\frac{FR^3}{E_n l}\right)^2 \quad (\alpha - \frac{1}{2}\alpha \cos \alpha - \frac{1}{2} \sin \alpha)\ (3 \sin \alpha + \alpha \cos \alpha)$$

F = mean wall pressure × ring axial width
R = radius at neutral axis, when in the cylinder
l = first moment of area

The mathematics of the ring free shape to produce a pressure pattern other than the simple constant form is extremely complex but has been made practical by the modern computer. It is now possible to specify the ring geometry to provide any pressure pattern which theory or practice has indicated to be optimum for any particular engine. Perhaps of more importance, the use of cam turning of rings, as against heat forming, sometimes on to cams, has made it possible to manufacture rings to this required shape.

As a general indication, a test using a flexible band placed round the ring and closed to the bore diameter is commonly employed to described pressure pattern in terms of ovality of the closed ring. The ovality can be positive or negative, positive ovality being defined when the gap diameter exceeds the diameter at 90° to the gap axis.

14.3.3 Ring types

There are a large number of ring types (> 75), albeit with minor variations, in use in current diesel engine practice; also ring packs based on these many types do not fall into a general pattern, and the result is that one seldom encounters two engine designs with the same ring equipment. These wide differences are based on (1) a lack of detailed understanding of the mechanism and control of ring behaviour and performance, and, in fact, that of the piston on which it is assembled and (2) the fact that each engine, by the very nature of its design, has unique characteristics which require individual tailoring of components. More recently, particularly because of the progress made in (1), the ring designer has been able to operate within a much reduced range of individual ring types and complete packs.

A selection of the basic ring types commonly met in current practice is shown in *Figure 14.22*. These illustrate principal design features and are divided by main function into compression and oil control types. It should be understood, however, that all compression rings have an oil control function in addition to sealing gas and the finesse of ring pack design is to allow oil to reach the top ring for tribology reasons while not permitting it to enter the combustion space. It is also important to realize that several of these design features can be combined in one ring design.

The application and operation of the rings shown are as follows:

14.3.3.1 Compression rings

(a) Plain standard The most simple ring, it is used whenever performance conditions do not demand a more complex and expensive design. Its use in high-speed engines, except in chromium bores has diminished considerably, but in medium-speed engines its use has diminished less; it is still common in slow-speed crosshead engines.

(b) Coated barrel-faced This coated barrel periphery form is widely used in parallel and taper-sided top compression rings, giving long life and good scuff resistance. The barrel form ensures quick bed-in and prevents high edge loading due to piston movement and groove distortion and assists in the development of a hydrodynamic oil film. (An added witness lapping operation is sometimes included in the design.) The ring has a 'neutral' oil control behaviour, i.e. it has no upward or downward scraping action. A non-symmetrical barrel form with the high point of the barrel below the ring centreline is now commonly specified but is more difficult to manufacture. This optimizes oil film generation and gives a more positive downward scraping action and hence better oil control.

The coating used most regularly has been chromium plate with sprayed molybdenum in cases where scuff has been a problem. A clean durable edge condition is difficult to achieve with molybdenum which also does not have as long a life as chromium plate. Alternative sprayed ceramic mixtures and ceramic-loaded chromium plate are finding increasing use in applications where the environment is difficult and extremely low levels of oil control essential. (see later under *ring coatings*).

(c) Inlaid plain Various low wear, scuff-resistant materials such as electroplated chrome or one of the sprayed materials detailed later are set into single- or multi-grooves in the ring periphery. The outer land of the base ring material offers edge protection to the deposited coating and also gives good oil control when located in a piston groove with upward run.

A semi-inland form of this ring is popular in high-speed applications. The cast iron land is retained on the lower edge, while the top edge has a full coating with generous radius, thus avoiding any tendency for upward scraping.

(d) Taper-side wedge Most commonly a top ring, but also used in lower grooves. Can have one (half keystone) or both (keystone) sides tapered, with included angles of 0.105 rad, 0.175 rad or 0.262 rad (6°, 10°, or 15°) and is fitted to a groove with a similar taper. The detailed angles and tolerances of the groove and ring are chosen to give an inner or outer lower face bedding for optimization of blow-by or oil control respectively. In pistons where carbon in the top ring groove is severe due to high temperatures, combustion conditions or oil consumption, the relative movement between ring and groove both prevents ring stick and resists the carbon forming. Blow-by can be higher with a keystone ring than with parallel-sided rings. The minimum angle required by the application is used to minimize groove wear although the 0.262 rad (15°) form is now predominant.

As a top ring, this ring nearly always includes a barrel-faced or inlaid coating and may also have an internal step or bevel to promote upward twist. As engine ratings have increased this has become the most widely used and important top ring. In its single taper form (*half keystone*) the upper side is tapered and the lower side provides blow-by similar to that of a parallel ring. Such rings are growing in prominence in small high speed engines up to about 110 mm bore.

(e) Taper faced The face taper is normally between 8.7 and 26.2 mrad (0.5°–1.5°) and is used in plated or unplated forms. When it is unplated either a full-width taper or witness lap is

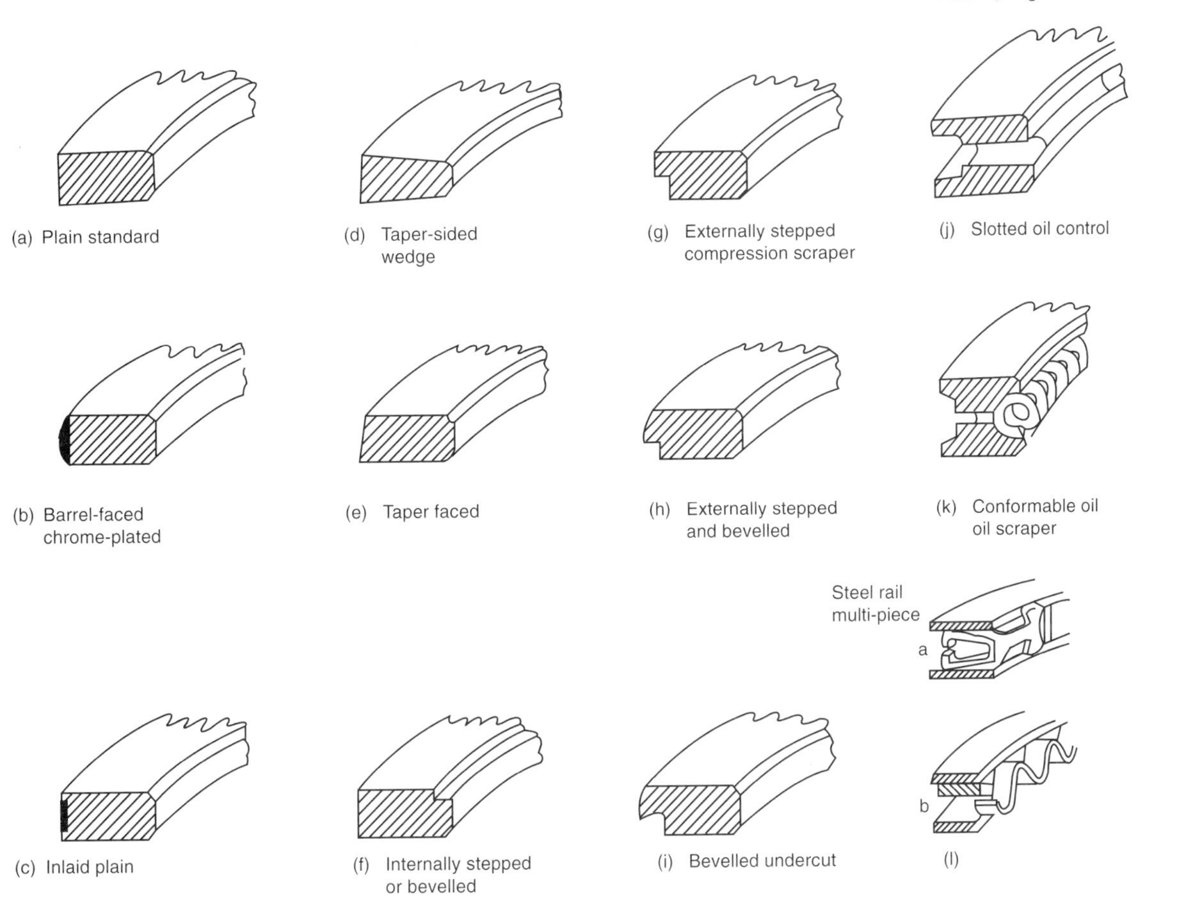

Figure 14.22 Ring types

used, but with the plated form a witness lap is always added. The ring has quick bed-in characteristics and combines the function of a gas sealing and downward scraping oil control ring. Is most often used as an intermediate ring, although its use is not now uncommon in the top groove.

(f) Internally stepped or bevelled A relief on the inner top edge in the form of a step or bevel causes the ring to dish when fitted into the bore giving a bottom edge contact and good oil control. For the very stringent oil control various combinations of features *(e)*, *(f)* *(h)* or *(j)* are sometimes used.

(g) Externally stepped compression scraper Combines gas sealing and oil control functions, the step giving the ring a torsional twist when fitted. According to the size of the step, this ring can perform as a relieved compression ring, or with a maximum step as a scraper ring for fitment immediately above a ventilated oil scraper ring. As such, it can be regarded as a dual-purpose ring design to suit a particular engine duty. Its use on high-speed engines has decreased since the advent of the chrome conformable scraper, but in conjunction with the next ring *(i)* has a useful performance level in the larger engine field.

(h) Externally stepped and bevelled A downward scraping and oil spreading control ring rather than a compression ring. Is used most commonly on the larger bore sizes.

(i) Bevelled undercut A variation of *(i)* is a strongly downward passing oil scraper for medium-speed engines. The hooked relief assures the retention of sharp scraping edges and consistent land contact width.

14.3.3.2 Oil control rings

(j) Slotted oil control A common form of bulk oil scraper, especially for large engines, having two scraping lands separated by ventilation slots or holes. The wall pressure, and hence scraping effectiveness, can be varied by control of the land width, depending on the needs of the application. Is manufactured in both chrome-plated and unplated forms.

(k) Conformable oil scraper The most widely used oil ring in high-speed engines of all types providing consistent oil control and long life. Also being usefully applied to larger engines but only of the 'trunk' piston type.

The ring itself can be inert or self-loaded and the total or additional wall loading is applied by means of a butting helical expander spring, located in a groove on the inside of the ring. The expander springs are made in both round- and square-section wire. The latter spreads the contact load and is preferred in those cases, particularly in high-speed engines, where high spring loads and relative movement may cause wear in the groove. A less frequently used alternative type of spring is the crimped expander, which gives maximum oil drainage with high wall loadings. In some long life applications, where embedment of the spring in the groove can cause lock-up and failure, the back of the ring is chrome plated.

The radial section of the ring is kept to a minimum to provide the maximum conformability to the cylinder bore, and, despite the backing spring, it is completely interchangeable with the slotted oil control type (*k*). Can be used unplated below wall pressures of 0.689 MN/m^2 (100 lbf/in^2) but is normally recommended in chrome plate form for long life and extended intervals between engine overhaul up to a pressure of 1.72 MN/m^2 (250 lbf/in^2), although pressures of 2.41 MN/m^2 (350 lbf/in^2) are not unknown. For the best consistency of oil control the scraping lands are profile ground and such land widths can be as small as 0.025 mm.

The steel form of conformable is becoming standard for small high speed engines and rapidly extending into the larger sizes. The strength and conformability is excellent and can be used with chrome plated or nitrided lands.

(*l*) *Steel rail multi-piece* This design is shown in two forms, one (*a*) using a combined spacer and expander, the other (*b*) a separate spacer and expander. Many different patented designs are in use, providing an inexpensive oil ring with good conformability to bore distortion and high wall pressures. The ring is not suitable for very long life application although rails are nearly always chrome plated, and it tends to be restricted to the type of small high-speed passenger-car engine. However, even here, the conformable types especially the steel design, is dominating in all new engines.

14.3.4 Ring packs

The variety of designs available and their combination on a piston make it difficult to present general pack designs for particular applications. In high speed engines a three-ring combination all above the piston skirt, with surface coated top and conformable oil ring for long life, is almost universally accepted practice. The second ring may be plain cast iron but is normally coated. A two-ring piston has been introduced into production, but there is no general move to follow this pattern.

Figure 14.23(*a*) shows some of the variants in typical high speed light vehicle engine ring packs. The trend is towards steel for the top ring and conformable oil ring, nitrided or plated, and the former progressively moving towards the half keystone type.

The four heavy duty automotive ring packs (*Figures 14.23*(*b*) and (*c*)) show typical European and North American practice. The variations arise, mainly, from the different service conditions and notably the North American packs have higher loaded oil rings.

In medium-speed engines, where varieties of fuel and sulphur content are met world-wide, a pack of four rings all surface treated (see *Figure 14.23*(*d*)) is common practice although, even here, the trend is to three rings. The coating is usually chromium plate, except for the top ring which may have a special coating such as chrome-ceramic because of high corrosive and abrasive wear. The side faces of the top ring and to a less extent the second ring are often chrome plated to resist axial wear and operate in induction or laser hardened grooves. Also shown in *Figure 14.23*(*d*) is an illustration of the way the number of rings on Medium-speed engines has dramatically reduced over the last twenty years.

When developing a new ring pack a standard arrangement based on the experience of a similar size of engine is normally chosen initially. Thereafter if blow-by or oil control performance are not on target a careful examination of run components and the introduction of well-established design change steps will usually lead to a solution. For example, to reduce a case of high blow-by good side face and peripheral sealing are essential, and tell-tale carbon marks will usually provide the necessary clue. In regard to oil control an even and continuous peripheral contact on oil and compression rings is vital, with the latter having a down scraping form.

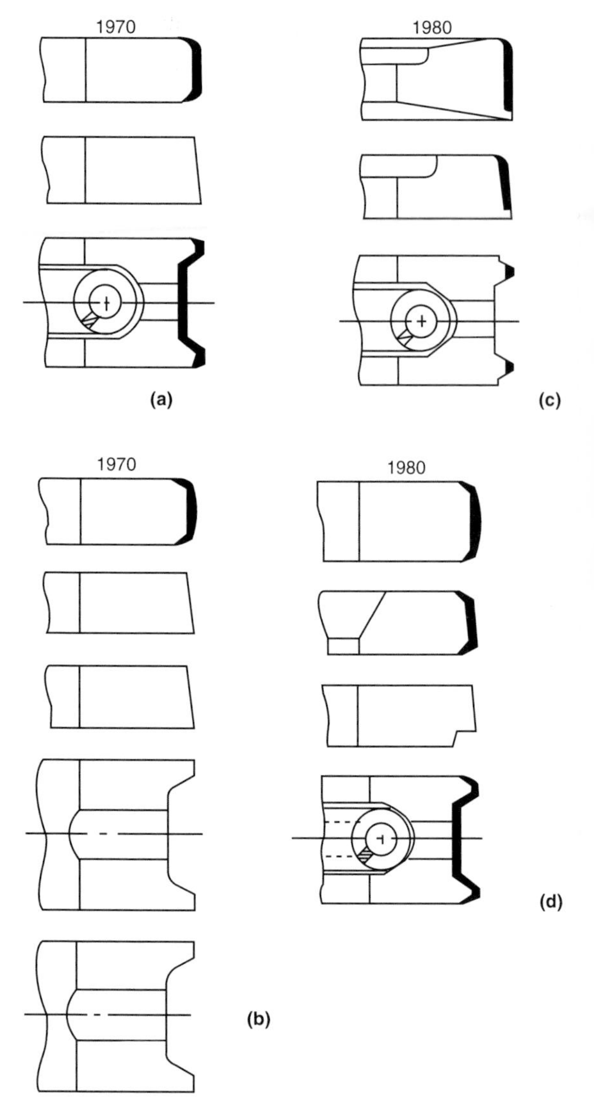

Figure 14.23 (a) High-speed engine ring packs; (b), (c) Medium-speed heavy duty engine ring packs, and (d) Medium-speed engine ring pack

The influence of the piston and liner in such development procedures is most significant. Such features as an unstable piston, inadequate drainage, incorrect land or groove clearance, distorted bores or incorrect bore finish can make the development of a suitable pack extremely difficult. In some cases good oil control is not accompanied by good blow-by control and vice versa, and a compromise in ring pack design has to be found.

Such procedures are typical of empirical design methods based on experience and the need for more accurate predictive technology is well recognized and is being seriously addressed as discussed in Section 14.3.9 Ring research.

14.3.5 Ring materials

Cast iron within its wide variations of chemistry and structure, giving a range of strength, wear and scuff resistance is the most commonly used ring material. However, over the last 10 years, the use of carbon and stainless steels have advanced to the point where they have become a very serious alternative and there are predictions that in the next 10 years they will dominate. The main applications are as top compression rings and conformable oil scrapers when they are coated, mainly with chromium plate, or hardened, usually by gas nitriding. *Table 14.8* shows the range of steels in use today for various ring types and the surface coatings which are being applied.

Table 14.9 details the relevant properties of a representative range of ferrous materials used in current diesel entines. It is seen that as the material strength increases scuff performance drops off and for this reason the high strength irons and steel are usually given a surface coating. As engine ratings and firing pressures have increased the use of high strength material in the top groove has become a necessity to avoid ring breakage. Each ring manufacturer normally has his own particular developed range of cast irons to met the variations in conditions encountered in service. Rings are either manufactured from spun cast sleeves or single cast blanks. Each method has its particular advantages and material characteristics and the choice of a particular route is bound up in a complex set of factors. Production economics and the need to reproduce exactly to an approved specification are usually the controlling factors.

However, in the case of steels the restricted world-wide availability of suitable round and preformed wire has meant that the range of material in use is also restricted. At present the main source is from Japan which has led the revolution in the change from cast iron. Steel has the advantage that it can be rolled to near net shape in a continuous roll and in the case of conformable rings the drainage holes or slots can also be preformed in the I section shaped wire.

14.3.6 Ring coatings

Table 14.8 lists some of the coatings which have been developed in recent years to give longer ring and bore life and improved scuff performance.

Electroplated chromium is still the most widely used coating and provides the best compromise of ring and bore wear. Chromium plate under marginal lubrication conditions tends to scuff and with the very low oil control requirements imposed by emission restrictions today, alternative coatings have had to be developed. One such coating consists of layers of chromium plate with ceramic particles interposed. This gives excellent wear and scuff resistance under very marginal lubrication conditions. Gas nitrided steel has also shown good wear and scuff resistance if due care is taken in regard to the basic brittle 'white layer' which forms on the surface during nitriding.

Molybdenum is used where scuff resistance is a particular problem but does not have such good wear life of bore or ring as does chromium plate.

As with base materials, each ring company has its own range of special proprietary coatings developed in the main for better scuff resistance than chromium plate. Variations in world-wide

Table 14.7 Properties of typical ring materials

	Modulus of elasticity E_n, GN/m²	*Tensile strength, MN/m²*	*Hardness, BHN*	*Fatigue rating*	*Wear rating*	*Scuff compatibility rating*
Grey irons	83–124	230–310	210–310	Fair/good	Good	Vergy good/excellent Good on chrome
Carbidic malleable irons	140–160	400–580	250/320	Good/very good	Excellent	Good. Very good on chrome
Malleable/nodular irons	155–165	540–820	200/440	Excellent	Poor. Usually chromium-plated	Poor. Usually chromium-plated
Sintered irons	120	250–390	130/150	Good	Good	Very good
Carbon/stainless steels or Chromium plated	230	500–(>1000)	130/440	Excellent (not hardened)	Poor usually hardened or chromium plated	Poor. Usually hardened

Table 14.8 Steel rings

Ring type		*Base material*	*Surface coating*
Top compression ring		SAE 9254 (≤ 1% Chromium)	Chrome-plated Chrome Ceramic Plasma sprayed
		13% Chromium 18% Chromium	Nitrided Nitrided Chrome-plated Plasma-sprayed
2-piece oil ring-I section		Carbon steel 13% Chromium 6% Chromium	Chrome-plated Nitrided Nitrided 2 mm axial width
3-piece oil ring	Rails	Carbon steel	Chrome-plated
		13%, 18% Chromium	Nitrided
	Expander	Carbon steel	Plain
		18% Chromium	Plain Nitrided

Table 14.9 Ring coatings

Coating	*Ring wear*	*Scuff/compatibility*	*Bore wear*	*Comment*
Chromium-electroplated	Excellent	Very good	Very good	Most widely used coating
Chrome sprayed	Very good	Very good	Fair	Wide variation in bore wear performance
Molybdenum sprayed	Fair	Excellent	Fair	Suffers from temperature/ time break-up, especially above 250°C
Tungsten-carbide sprayed	Excellent	Good	Good	—
Iron oxide (ferrox)	Fair	Very Good	Good	—
Phosphates (Parko-lubrising)	Fair	Running-in scuff very good	—	Mainly used for running-in
Copper plating	Poor	Running-in scuff excellent	Very good	Mainly used for running-in. Can also be applied to chromium-plate

fuel sulphur levels, longer life requirements and the need for lower oil consumption, already mentioned, have all added impetus to research on new coatings. In this work the technique of plasma spraying has provided a means of depositing single- and multi-constituent metal and ceramic materials ususally fed to the spraying gun in the form of powders.

14.3.7 Oil consumption and blow-by

Over the years there has been a steady reduction in the oil consumption targets for the complete range of diesel engines. In part this is for economic operating cost reasons and also to support sales on a technical excellence basis. In recent years, emission regulations and particularly the aspect of particulates has placed much more emphasis on good oil control in automotive applications. This has meant testing and study of oil consumption and particulates over the complete range of operating conditions. Thus full load oil consumption might be high but particulates low because of the high in-cylinder temperatures. The opposite may apply at part load. Despite this, there is still a general tendency to specify and compare oil control at full load which the engine maker, through experience, can relate to field performance.

In high speed engines the targets for oil consumption fall in the range of 0.05 to 0.25 per cent of full load fuel.

In the medium speed field the target range tends to be higher and currently stands at 0.25–0.5 per cent of full load fuel.

Blow-by, like oil consumption, is a characteristic on which technical excellence of an engine is based. If oil consumption is satisfactory, blow-by will normally not be a problem although there are cases where excessive blow-by can improve oil consumption by effectively blowing oil down through the ring pack into the crankcase. This is one area where analytical studies have assisted the ring pack designer to ensure that ring gaps and inter-ring volumes are such as to give a controlled acceptable flow level of combustion gas (*blow-by*) *through* the compression rings and preventing reverse flow into the combustion space, which increases oil consumption.

Generally, excessive blow-by seems to be of main concern in high speed engines and even here the end user takes a widely different view of acceptable levels. For example the farmers of India expect very low levels from their tractor engines since they can see from a distance the haze which it can create over the vehicle. Current targets are in the range 0.4–0.6 feet^3/bhph (0.015–0.23 m^3/kWh).

14.3.8 Scuffing

One of the main ring problems encountered during engine development, production pass-off or service is 'scuff', the term given to the condition described as 'gross damage characterized by the formation of local welds between sliding surfaces'. The typical effect is a streaky wear pattern on rings and liners which can have varying degrees of intensity and in very severe cases may lead to complete piston seizure. Examples of typical cases on two levels of severity are shown in *Figures 14.24(a)* and *(b)*.

Scuffing can start at any ring in the pack, but in general the top ring is the most susceptible. It is possible to list over sixty factors which can contribute to the scuffing condition, and these can have a most complex interrelationship. Apart from consideration of material compatibility under marginal lubrication, the factors in the main are those which affect the oil film between ring and liner. The major influences on this oil film are:

(1) Temperature
(2) Pressure
(3) Speed
(4) Lubricant
(5) Surface finish
(6) Ring pack design

Within these general categories excess top groove temperature (> 250°C), non plateau-honed bore finish, and incorrect ring peripheral form and edge condition, are highlighted as common causes of scuffing problems.

14.3.9 Ring research

The empirical nature of ring design has already been noted. The need for improved understanding leading to optimization and shorter development time has resulted in major research programmes in this field. The work has been made possible by the development of multi-channel telemetry systems, particularly of the linkage type (see *Figure 14.25*), and also of miniature pressure, movement and oil film transducers. Theoretical computer models developed to predict such features as, ring movement, inter-ring pressure, oil film thickness, blow-by and oil consumption have continued to be refined to increase the

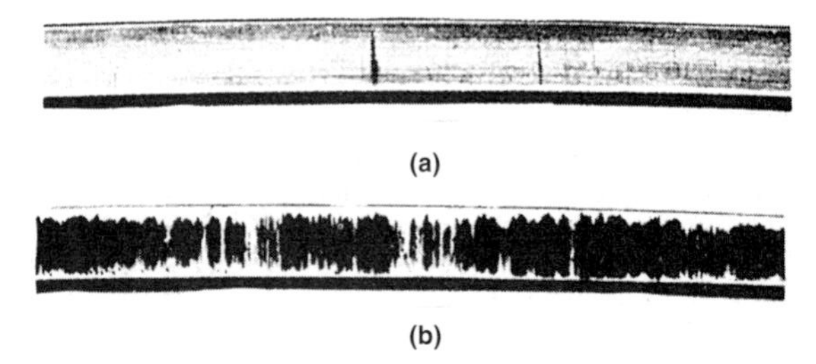

Figure 14.24 (a) Minor ring scuffing; (b) Severe ring scuffing

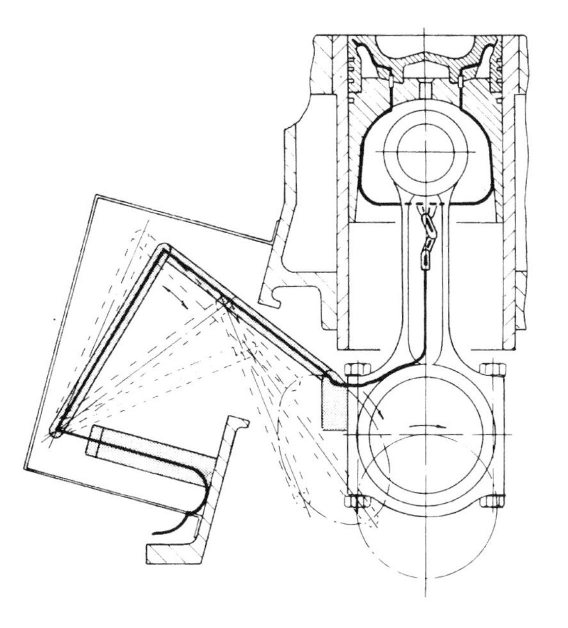

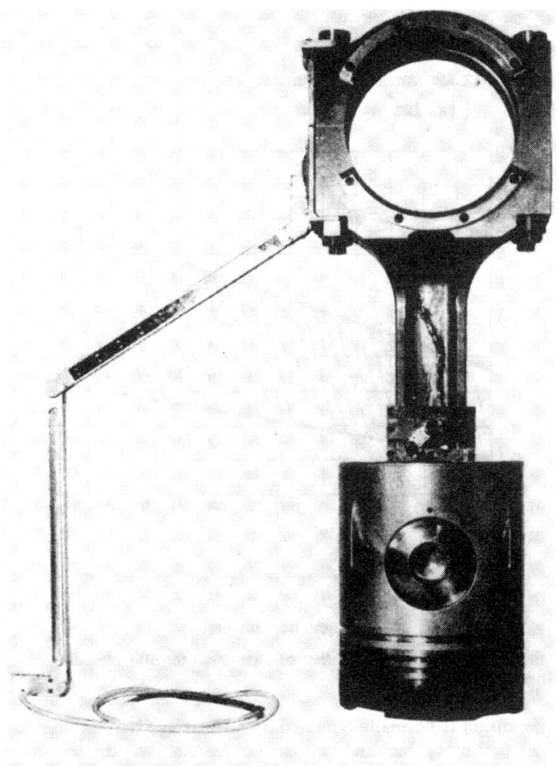

Figure 14.25 Multi-channel linkage telemetry system

agreement with measured figures and while accurate oil consumption prediction, as against trend, with so many factors involved, remains a problem, good progress has been made in many aspects. Thus, based on such research significant detail changes to the piston and ring pack have been suggested including such features as land clearances, inter-ring volumes, oil drainage, ring gaps, ring peripheral form and others.

The empirical character of ring design is certainly diminishing.

14.4 Liners

14.4.1 Introduction

The pistons and rings of many low-cost high-speed engines are run directly in bores machined in the cylinder block. Such designs are termed 'parent bore'.

However, the majority of engines have bores formed by liners fitted into the cylinder block. The material of such liners is not limited to that of the cylinder block type and can be optimized to give longer life and increased scuff resistance.

Liners can conveniently be considered under two type headings, 'dry'—so termed as they do not come into contact with the cooling system, and 'wet'—in direct contact with the engine coolant and an integral part of the cooling system. It is not unknown for some designs to combine the basic features of both wet and dry liners.

For bore sizes up to 150 mm (6 in) both wet and dry liners are widely used. Above 150 mm (6 in) the wet type is almost invariably selected.

14.4.2 Dry liners

Dry liners are fitted directly into the cylinder block, either by means of an interference fit or more recently a push fit and bonded in place with an epoxy compound. Minimum recommended interference fits vary with bore size and wall thickness. Typical figures would be between 0.001 and 0.0015 mm per mm diameter. The dry liner is most commonly used on high-speed engines, particularly of the automotive type, and wall thicknesses are generally low—in extreme cases down to 1–1.5 mm (0.04–0.06 in). It is important for good heat transfer and avoidance of 'hot spots' to ensure full contact of the liner in the block. High accuracy of manufacturing is therefore essential.

While it is commonly assembled in a pre-finished condition, the press fitting inevitably produces some bore distortion, and an *in situ* finishing operation is preferred for optimum piston/ring performance.

Dry liners are made in flange and flangeless forms controlled mainly by the space available and the engine designer's preferred concept. Good gas sealing is essential to avoid the formation of carbon behind the liner, causing bore distortion and possibly serious failure. Although the block casting to accommodate dry liners tends to be more difficult to manufacture and more costly service to accommodate dry liners, the resultant cylinder block structure is stiffer with minimum distance between bore centres particularly with the flangeless forms.

The liner itself is cheaper for service replacement but the removal and fitting are more difficult than for the wet type. Some typical dry liners are shown in *Figure 14.26*.

14.4.3 Wet liners

Always flanged, the wet liner being in contact with the cooling water has to have a suitable sealing arrangement for gas and water. Such arrangements fall into three types:

(a) Flanged and sealed at the top, the lower seal being achieved by elastomeric rings in grooves machined in the liner
(b) As (a) but with the lower grooves machined in the cylinder block
(c) A double flange arrangement with sealing at the flanges by gasket and/or metal-to-metal contact.

Flange sealing of the coolant at the top of the liner is usually achieved without problems by means of metal-to-metal contact, although a liquid sealing compound is sometimes included for additional security.

Wet liners are almost invariably fitted in a pre-finished condition, and typical examples are shown in *Figure 14.27*. The main structural problems of the liner rest in the top flange region, and careful design of the flange, flange undercut and head sealing are necessary, particularly to accommodate the higher firing pressures now being used.

The other main problem in wet liners is cavitation erosion on the outside diameter in contact with the cooling water. The phenomenon of cavitation is caused by vibration of the liner either due to combustion shock or piston impact and in very severe cases complete breakdown of the liner material from cylinder to water can occur in less than 100 hours' operation.

The use of close clearance pistons and inhibitors in the cooling system alleviate the condition and the use of special coatings such as chromium plate or ceramic spray also give beneficial results.

In the main, wet liners use heavier sections than their 'dry' counterparts and a wall thickness minimum of 7 per cent of the bore diameter is recommended to minimize cavitation effects.

Generally the as-fitted and operating bore distortion of wet liners is less than with the dry type. This arises from the more severe fitting stresses of the latter and greater influence of block distortion. Also, although the liner costs are greater than for the dry design, service replacement is a much more simple task. However, because of its flange arrangement cylinder spacings are wider than with the dry type and apart from this the block structure tends to be less rigid.

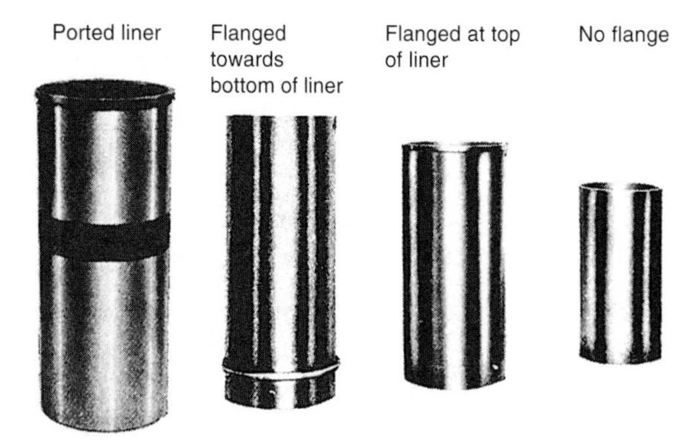

Figure 14.26 Typical dry liners

14.4.4 Liner shape and surface finish

Liner size, shape and surface finish are all critical to good short and long-term bore, ring and piston performance, particularly in respect of scuff and wear.

Rings have limited conformability to non-circular bores and the resultant uneven wall pressure can lead to scuff, high blow-by and excessive oil consumption. While manufacturing accuracy is therefore important this can be overlaid by mechanical and thermal distortion. In sizes up to 150 mm (6 in) a 'good' bore would have no more than 0.05 mm (0.002 in) non-circularity under operating conditions. In general a bore distortion of more than 0.1–0.15 mm (0.004–0.006 in) in this size range of bore will lead to ring problems, although to some extent this depends on the character of the non-circularity.

The importance of having the correct bore surface character cannot be over-emphasized, and the development of precise computer-supported surface measuring equipment has enabled a range of features to be investigated and defined. These are:

(1) Cross hatch angle: 120° +/– 20° (see *Figure 14.28*)
(2) Surface finish (bore up to 300 mm (12 in))
    (a) Roughness average (Ra): 0.4 to 1.0 micron
    (b) Max roughness depth (Rt): 12.5 micron
    (c) Mean roughness depth (Rtm): 5/10 micron
    (d) Mean peak-valley height (Rtm/Rz): 3.0 to 8.0 micron
    (e) Bearing area (tp): 60/80% at 30% depth
    (f) High spot count at mean line-4 mm assessment length: 30/100
    (g) Peak count at 2 micron zone width: > 30/cm
    (h) peak to valley height (Rv/Rp): 2/4
(3) Good surface cleanliness with no loose particles or folded metal. Clean cut—not sharp and uniform in both directions. No burnish or glaze.

Improvements in bore finishing have done much to eliminate early life ring scuffing and reduce running-in times which have serious production cost implications. More recently there has been a need to achieve good oil control at the very early stages of running to minimize emissions at pass-off. The result is a trend to specify average roughness at the lower end of the limits specified and generally create a surface finish as close as possible to the run-in state without inducing early life scuff or wear.

The peak-free plateau condition is normally obtained by a two-stage honing process, the first using diamond or silicon carbide stones and the second, carbide stones sometimes in a flexible matrix which both creates a plateau condition and cleans the surface. *Figure 14.27* shows a typical surface finish trace before and after plateau honing.

For bores more than 300 mm (12 in) a cutting tool finish rather than a honed one is generally preferred with a CLA value of between 2 to 3 micron.

14.4.5 Material

The large majority of liners are made of grey cast iron in one of its many forms. A notable exception is that of the thin walled steel liner, chromium plated in the bore and used in some high-speed applications.

In this text detailed consideration will only be given to grey iron with its inherent properties determined by graphite form

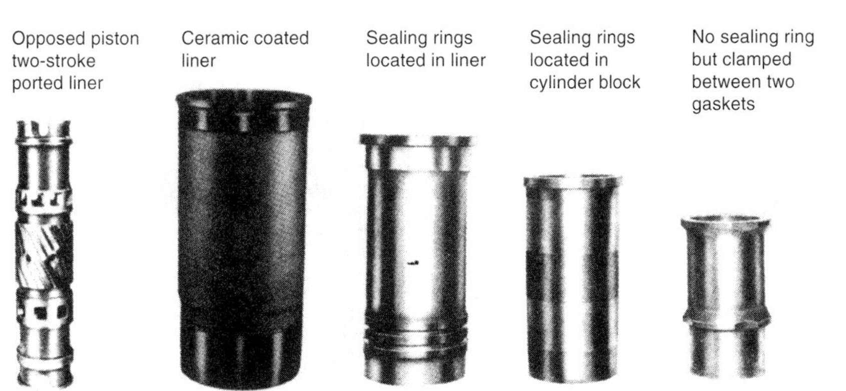

Figure 14.27 Typical wet liners

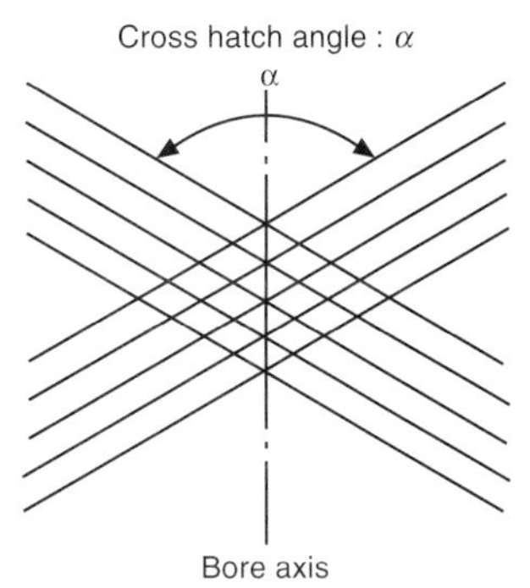

Figure 14.28 Cross hatch angle

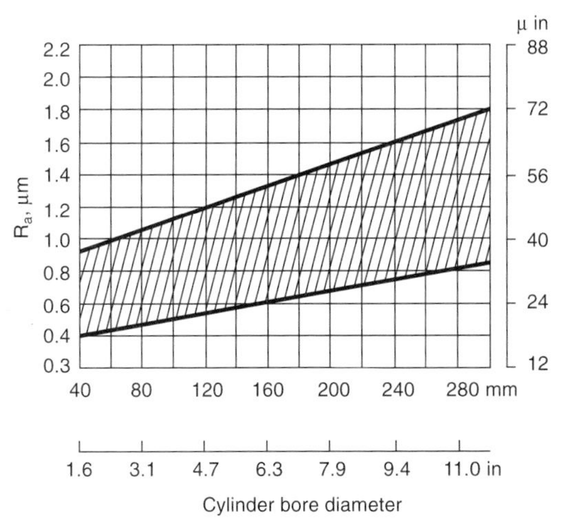

Figure 14.29 Recommended surface finish for nor-plateau honed surface

and size, matrix structure and presence of metallo-compounds. The effect of grey iron chemistry and structure on the basic material properties and characteristics important in the design selection process are summarized in *Table 14.10*. The choice of material is seen to be a compromise based on the essential needs of the particular engine application.

It is seen that A-type graphite offers the best wear, scuff and machining properties at the expense of strength and fatigue resistance. In regard to matrix form in the high-speed diesel engine the choice lies usually between the lowest cost and life (as cast-pearlitic) and increased cost and life (heat treated-martensitic). In field service the latter structure has been known to give a life of up to 1.6×10^6 km (10^6 miles), outlasting both rings and pistons.

Material recommendations for the different types of liner are shown in *Table 14.11*.

14.4.6 Bore polish

One of the most serious and difficult-to-overcome problems regularly encountered in diesel engines today is bore polish leading to early life loss of oil control. The polish arises from the formation of hard carbon on the top land of the piston which builds up to the point where it rubs against the bore under the natural transverse motion of the piston and removes the honing pattern leaving a polished region usually on the non-thrust side. Although the local bore wear is small the rings seem to have difficulty in controlling the flow oil in this region, perhaps due to lack of conformability or character of the surface, and the oil consumption begains to rise. This can happen after a period of from 50 to 500 hours depending on the severity of the carbon. If the top land is cut back by the order of 1 mm, the carbon is of a soft form and readily flakes or breaks off but such cut back, much practised in the USA in the past, is not acceptable today because of the increased 'dead volume' and consequent higher emissions.

Apart from the use of articulated pistons which do not have the same high order of side load on the crown, two solutions are in use in production engines today:

(1) A steel or cast iron ring is located at the top of the bore projecting into it slightly in the region of the top land at TDC. The piston top land is cut back to compensate for this projection and there is no increase in 'dead volume'. The 'cuff' or 'carbon cutting' ring as it is called prevents a build up of hard carbon beyond its inner diameter and no polish can take place as the carbon is always clear of the bore as it moves away from the TDC position.
(2) Ceramic particles are pressed into the bore surface in what is termed the 'Laystall Process' and the resultant very highly wear resistant layer is not subject to polish when rubbed with the hard carbon on the land.

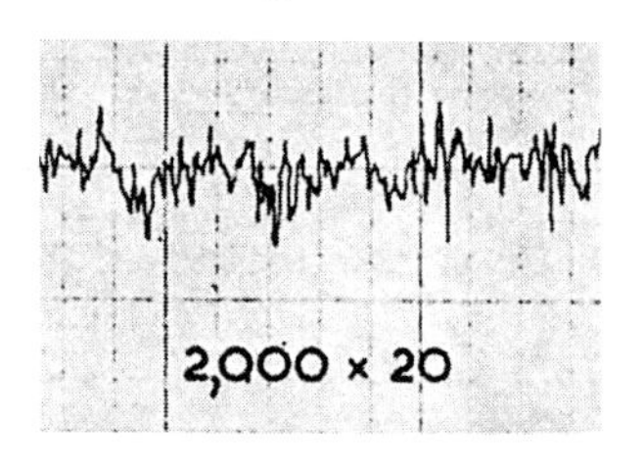

(a)

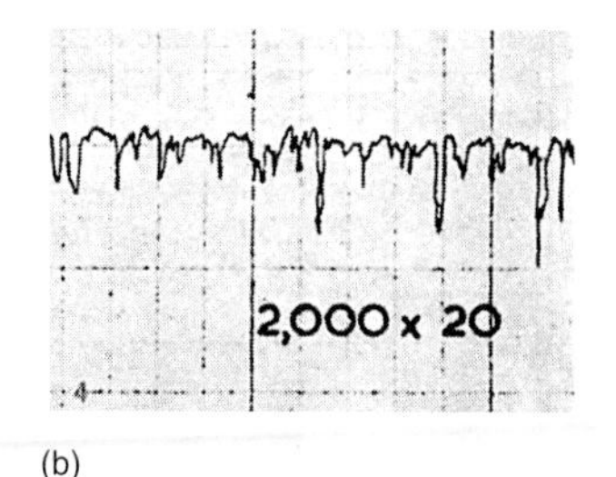

(b)

Figure 14.30 Surface records of (a) Non-plateau honed surface; (b) Plateau honed surface

Table 14.10 Effects of structure upon essential properties

Effect on		*Best* *Worst*		*Feature*
Foundry costs	Mixed A/B-D Medium P Uncontrolled pearlitic	Type A Temper C Low P Carbidic	S.G. High phosphorus Controlled pearlitic	Graphite form Compounds Matrix
Machining costs	Type A A/B-D Medium P Pearlitic	Temper C + Carbide High/Low P	Temper C S.G. Carbidic Martensitic/Bainitic	Graphite form Compounds Matrix
Scuff	Type A Mixed A/B-D Medium/High P Pearlitic Bainitic (Controlled)	Temper C + Carbide Carbidic Martensitic Uncontrolled (free ferrite)	Temper C S.G. Low P Pearlitic	Graphite form Compounds Matrix
Wear	Type A Temper C + Carbide Phosphocarbides Bainitic	Mixed A/B-D Complex carbides Martensitic	Temper C S.G. High/medium P Low P Pearlitic	Graphite form Compounds Matrix
Strength	S.G. Temper C Low P Carbidic Martensitic	Temper C + Carbide Phosphocarbide Bainitic	Type A Mixed A/B-D Medium P High P Pearlitic	Graphite form Compounds Matrix
Cavitation resistance	Typer A (Low modulus) Carbidic Martensitic	Mixed A/B-D	S.G. (High modulus) Others Pearlitic	Graphite form Compounds Matrix
Fatigue resistance	S.G. Temper C Low Phos Pearlitic	Temper C + Carbide Fine carbides Tempered martensitic bainite	Type A Mixed A/B-D Others Martensitic	Graphite form Compounds Matrix

Table 14.11 Marterial application recommendations

(i)	*(ii)*	*(iii)*	*(iv)*	*(v)*
Wet liner High duty	Wet liner Medium duty	Dry liner Normal wall	Dry liner Thin wall or Thin flange	Nitride hardened liners for special duties
Low phos, spun cast alloyed grey iron with predominately type A graphite. Bore surface hardened or differentially hardened. (If used 'as cast' then 5% max. free ferrite).	as (i) in 'as cast' condition alternatively medium phos spun cast unalloyed grey iron, 5% max. free ferrite.	Medium phos spun cast unalloyed grey iron 10% max. free ferrite as cast.	Low phos spun cast unalloyed grey iron predominately Type A graphite 5% max. free ferritte, as cast.	Specially alloyed (Mo/Al) spun cast grey iron, surface nitrides to HV 750 min.

References

The following references provide more detailed information on the subject of pistons, rings and liners. In several cases extracts from these references have been used in the text.

Pistons

1 LONGFOOT, G., 'Piston design data', *Tribology Handbook,* reprint by permission of Newnes-Butterworth

2 MUNRO, R., 'Some diesel piston features in design analysis and experiment', SAE Off-Highway Meeting, Paper 790858, September (1979)

3 DAY, R.A. and SILVESTER, V. C. W., 'Improved piston performance by selective reinforcement or treatment', AE Symposium, Pages 6 (1978)

4 LAW, D.A., 'New features in diesel pistons above six inches diameter', *DEUA,* 17 February (1966)

5 MUNRO, R. and GRIFFITHS, W.J., 'The application of predictive techniques in the design and development of medium speed diesel engine pistons'. International Congress on Combustion Engines, Vienna, Paper D–46 (1979)

6 MUNRO, R. and GRIFFITHS, W. J., 'Diesel piston design and performance prediction', 11th International Congress on Combustion Engines, Barcelona, April–May (1975)

7 MUNRO, R., GRIFFITHS, W. J. and INGHAM, A.P., 'Open gallery

pistons for highly rated high speed diesel engines', 10th International Congress on Combustion Engines, Washington, April (1973)
8 MUNRO, R., GRIFFITHS, W. J., LONGFOOT, G., TRAVAILLE, J. and AVEZOU, J.C., 'Oil cooled and electron beam welded pistons for diesel engines', AE Symposium, Paper 7 (1978)
9 MUNRO, R. and GRAFFITHS, W.J. 'Development and operating experience of pistons for medium speed diesel engines', DEUA General Meeting, March (1979)
10 MORGAN, W. J., 'Finite element predictive techniques applied to crown cracking problems in medium speed diesel engines', Motorsymposium, Czechoslovakia (1979)
11 RHODES, M.L.P., TRAVAILLE, J., PIFFAULT, J., BRUNI, L. and RAGGI, L., 'Pistons for petrol and small diesel engines', AE Symposium, Paper 5 (1978)
12 BACON, M. L., DAY, R.A. and FLETCHER-JONES, D., 'Development of piston materials with special reference to fracture mechanics', AE Symposium, Paper 9 (1978)

Rings

13 LAW, D.A., 'Diesel engine piston ring design factors and application', I.Mech.E. Symposium on Critical Factors in the Application of Diesel Engines, Southampton University, September (1970)
14 MUNRO, R., 'Piston ring design data', *Tribology Handbook,* reprint by permission of Newnes-Butterworth
15 MURRAY, E. J. and SILVESTER, V.C.W., 'Piston ring design trends', AE Symposium, Paper 23 (1978)
16 MUNRO, R. and HUGHES, G. H., 'Current piston and ring practice and the problem of scuffing in diesel engines', DEUA General Meeting (March 1970)
17 MUNRO, R., 'Diesel engine ring scuff—is there a major problem?', I.Mech.E. Conference on Piston Ring Scuffing, Paper C 64/75, 9–14 May (1975)
18 PARKER, D.A., ADAMS, D.R. and MUNRO, R., 'Progress in understanding and control of ring lubrication', Energy and Technology Conference and Exhibition, Houston, ASME Paper 78DGP-25, 5–9 November (1978)
19 GAZZARD, S. T., 'New materials and sprayed coatings for diesel engine piston rings', AE Symposium, Paper 21 (1978)

Liners

20 LAW, D.A., 'Further developments in cylinder bore finishes', SAE National Powerplant Meeting, Cleveland, October (1969)
21 TRAVAILLE, J., DAY, R.A., BRUNI, L., IGUERA, P., AVEZOU, J.C. and COULONDRE, M., 'Developments in cylinder liners', AE Symposium, Paper 20 (1978)

Surface Finish Assessment

BS 1134, *Assessment of Surface Texture, 1972 (with amendment in 1976). Part I, Method and Instrumentation; Part 2, General Information and Guidance.* Related International Standard is ISO/R468 entitled 'Surface Roughness'
I.Mech.E. International Conference on 'Properties and Metrology of Surfaces'; *Proceedings 1967–98*, **182**, Part 3K

15 Auxiliaries

Contents

15.1 Governors and governor gear

15.1.1 Introduction

The three section under this title: Governors, Starting Gear and Heat Exchangers refer to auxiliary systems indispensible to proper diesel engine operation. However, the rate of new developments for these systems is slower; therefore the material of the previous edition was kept almost unchanged–with only minor editing as required. A governor can be defined as a mechanical or electro-mechanical device for automatically controlling the speed of an engine by regulating the intake of fuel. More broadly it can be regarded as equipment which controls some parameter such as speed, load, pressure or temperature and which automatically maintains the desired level of the parameter. The governors described in this chapter are speed governors. These governors are controllers or controlling systems which automatically control the energy inflow to engines that are subject to changes in load or speed so as to maintain a given speed characteristic in accordance with certain requirements and within given limits.

15.1.1.1 System characteristics

The majority of diesel engines are fitted with a speed governor. There are a few exceptions, e.g. a marine engine direct-coupled to a propeller can be controlled by fixing the fuel pump quantity adjustment at any desired value; if the engine speed starts to rise, the propeller torque rises repidly and so allows only a small change in speed. Another example is a road vehicle diesel engine which is controlled by the accelerator pedal, an (idle-maximum) governor controls the idle speed and the maximum speed.

It may be of interest to point out at this stage that the self-limiting characteristics of the unsupercharged or normally aspirated petrol engine will, in most cases, establish a balance between the torque required to overcome friction and windage losses and the torque developed in the unloaded engine at full throttle at a speed sufficiently low to prevent self-destruction. However, it is not so in the case of the diesel engine. In applications where there is a possibility of sudden loss of load, a governor, at least as a maximum-speed limiting device, becomes a necessity for the diesel.

15.1.1.2 Function

Governing control must be performed accurately and stably to be satisfactory. Accuracy is required for steady-state performance while stability is necessary for transient performance.

15.1.2 Basic principles

15.1.2.1 The simple governor

The mechanism shown in *Figure 15.1* is probably the oldest and certainly one of the simplest of speed sensing devices. It depends for its operation upon the fact that a force is required to compel a mass to follow a circular path. This force is proportional to the square of the speed of rotation and to the first power of the distance of the mass from the axis of rotation. In its best known form, the ballhead consists of a pair of weights, usually spherical, at the ends of two arms pivoted near the axis of rotation in such a way that the flyweights can move radially in a plane through the axis. Additional links are attached to the arms and a collar about the axis to form a parallelogram configuration. Thus, when the weights move outwardly, the collar moves up (*Figure 15.1*).

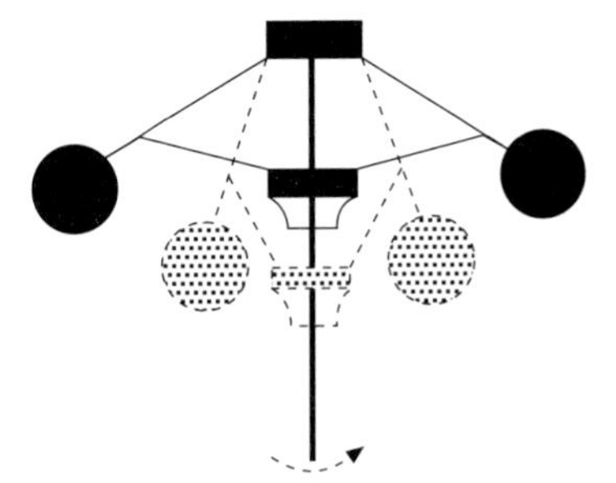

Figure 15.1 Simple governor

Since the centrifugal force always acts at right angles to the axis of rotation, it exerts a torque about its pivot equal to the product of the force times the vertical distance of the ball below the pivot. This torque is opposed and, if no other forces are present, must be balanced at equilibrium by the torque of gravity, which is equal to the weight of the ball times the horizontal distance to the pivot (*Figure 15.2*). Thus as the speed increases the centrifugal force increases and the ball moves outward, decreasing the centrifugal force torque arm and increasing the gravity torque arm until equilibrium is reached. This results in a unique equilibrium position of flyweights and collar for each speed of rotation. In the direct mechanical governor this collar is connected to the throttle so as to close it as the flyweights move outward. It should be noted that the unique relationship between speed and position of ballhead and collar no longer exists if friction is added to the system. This is for the reason that as the speed increases from an equilibrium condition the centrifugal force torque must reach a value equal to the gravity torque plus that due to friction before movement results. Similarly on decreasing speed the centrifugal torque must go down to a value equal to the gravity torque minus that due to firction. The result is a so-called 'dead-band' or region in which the speed may wander without producing a corrective motion of the throttle. Efforts to minimize this 'dead-band' resulted in the huge cast iron flyweights common to early mechanical governors.

15.1.2.2 Mechanical-hydraulic (servo) governors

Figure 15.3 shows a simple from of hydraulic governor in which the ballhead and pilot valve control a simple reciprocating piston servo. An adequate oil supply is assumed to be available. The ballhead is driven at a speed proportional to that of the engine and the servo is connected to operate the fuel racks. It should be noted that the simple combination of ballhead and directly connected pilot valve has only one equilibrium position: the

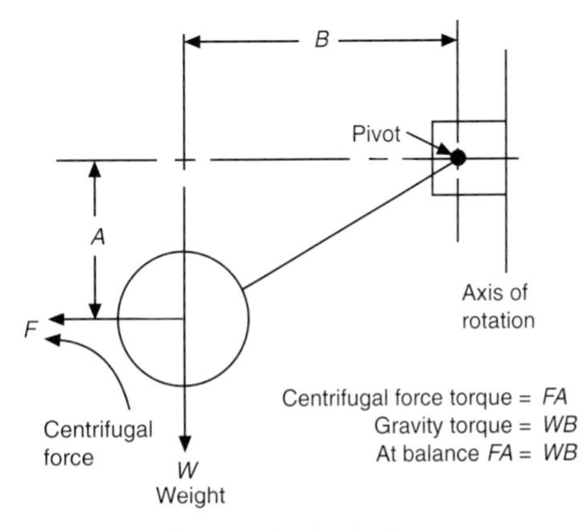

Figure 15.2 Forces acting in simple governor

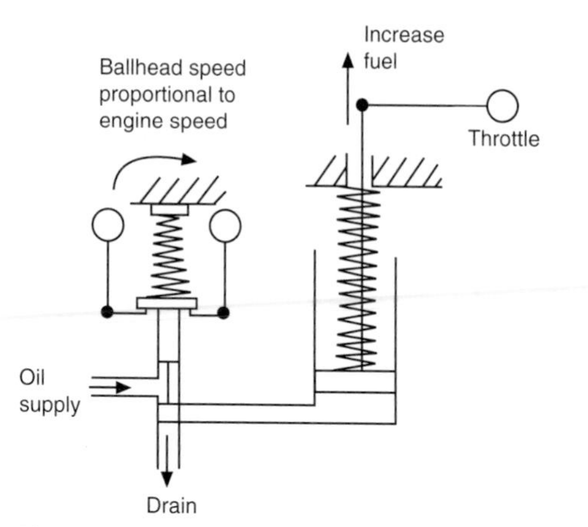

Figure 15.3 Simple hydraulic governor

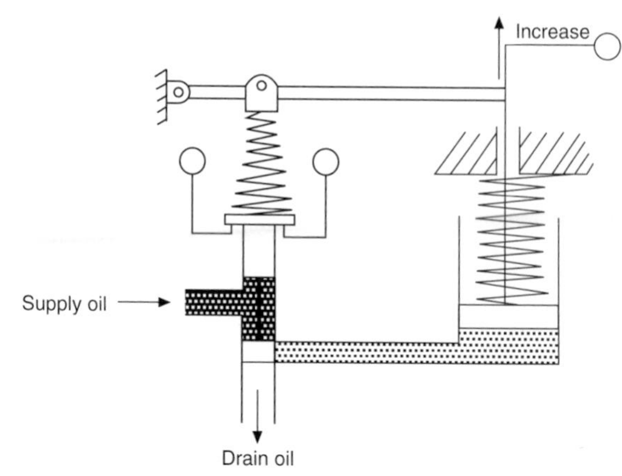

Figure 15.4 Speed droop mechanism in simple hydraulic governor

position in which the valve is colsed, neither admitting oil to nor discharging oil from the servo cylinder. For a given setting of the top end of the speeder spring, the ballhead and valve will take this position at only one speed; in other words, such a speed-sensitive device is inherently isochronous. Unfortunately, such a system is also inherently unstable. This is because the engine speed does not instantly assume a value proportional to the rack position due to the inertia of the rotating mass. Therefore, if the engine is below the governor speed setting, the pilot valve is positioned to move the servo to increase the fuel. By the time the speed has increased to the setting so that the valve is centred and the servo stopped, the fuel has already been increased too much and the engine continues to speed up. This opens the pilot valve the other way and fuel begins to decrease. As before, when the speed gets to the right value, the fuel control has travelled too far, the engine underspeeds, and the whole cycle continues to repeat. Some means for stabilizing such a system must obviously be added to the two components we have described to secure a satisfactory governor.

The simplest method of securing stability in the system described is to add means which will provide *speed droop* in the governor which in turn results in *regulation* in the governed system. The distinction between these terms is explained later.

15.1.2.3 Speed droop

Speed droop in a simple governor can be provided by a mechanical interconnection between servo (and therefore throttle) movement and governor speed setting such that, as fuel is increased, the speed setting is decreased. Such a device may consist simply of a lever of suitable ratio between servo and speeder spring (*Figure 15.4*). The equilibrium relationship between speed setting and servo position for such a system may be represented by a line sloping or 'drooping' downward to indicate decreased speed setting with movement of the servo in the 'increase fuel' direction.

15.1.2.4 Isochronous governor

It is sometimes desirable to have an isolated prime mover run *isochronously* (speed constant regardless of load within the capacity of the prime mover). In such cases we resort to transient speed droop or, as it has come to be generally called, *compensation*. This calls for the introduction of a temporary readjustment of speed setting with servo movement to produce the stabilizing speed droop characteristic followed by a relatively slow return of speed to its original value.

This can be accomplished in a number of ways. One method (*Figure 15.5*) involves the direct application of pressure to the pilot valve plunger, adding to or subtracting from the speeder spring force in order to effect a change in speed setting. The oil actuating the servo is required to deflect a 'buffer' piston against a centring spring load which produces a pressure differential across a receiving piston rigidly attached to the pilot valve plunger. A needle valve permits equalization of pressure across the pilot valve receiving piston to restore the initial speed setting. In operation as oil flows to the servo, the buffer piston is moved against the force of its centring spring, resulting in a higher pressure on the lower side of the receiving piston which produces an upward force on the pilot valve. This in effect decreases the force which the flyweights must balance, resulting in centring of the pilot valve at a lower speed, thus providing speed droop. As the displaced oil is permitted to leak through the needle valve, the buffer piston returns to its equilibrium position, the differential pressure disappears and the speed setting reverts to its original value.

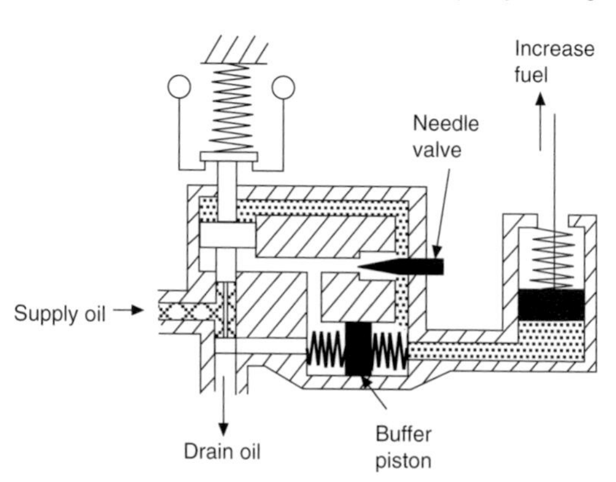

Figure 15.5 Means to enable isochronous operation with a hydraulic governor

15.1.2.5 The electronic governor

Basic electronic governor

The basic electronic gevernor consists of two main parts: (1) an electronic control which senses speed and, in some cases, load; and (2) an actuator to position the prime mover fuel mechanism.

Engine speed can be sensed using one of two methods. One method uses a magnetic pick-up mounted in the close proximity of a ferrous metal gear (such as the starter ring gear of the engine). The second sometimes used on generator set applications measures the generator output frequency. Either method produces a speed signal.

The speed signal is converted to a d.c. signal proportional to speed. This signal is then compared with a preset voltage representing the desired speed. If an error exists a signal is provided from the control amplifier to the actuator for corrective action (either increasing or decreasing fuel to maintain speed).

Basic electronic control

The basic electronic control compares the electrically measured speed signal with the preset/adjustable speed setting reference signal. Any error between the two signals is amplified and fed to the actuator coil (*Figures 15.6 and 15.7*).

Electrohydraulic actuators

The electrohydraulic actuator (*Figure 15.7*) converts the electric signal applied to it into a mechanical position of the actuator power piston. The actuator does no sensing but only provides the mechanical means of operating the control linkage.

The basic electronic control and actuator (*Figure 15.8*) is an alternative to the simple mechanical governor.

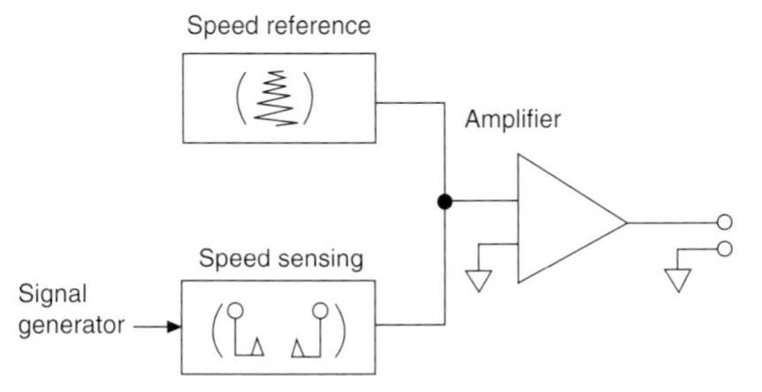

Figure 15.6 Basic electronic speed control

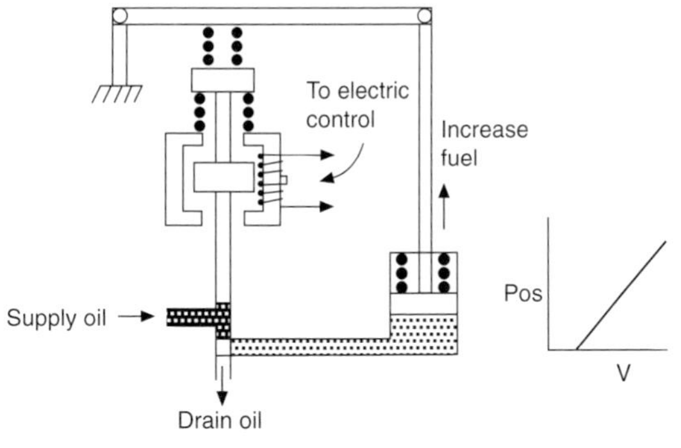

Figure 15.7 Basic electrohydraulic actuator. Right hand diagram shows actuator output position *vs* voltage

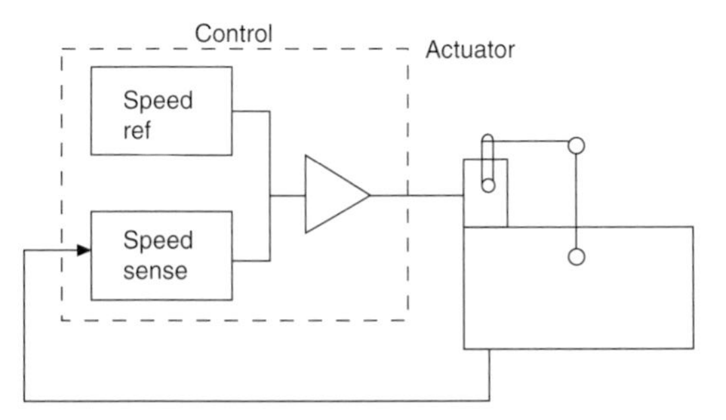

Figure 15.8 Basic electronic governor

15.1.3 Basic governing terms

15.1.3.1 Isochronous operation

Isochronous operation maintains the governed speed the same regardless of total load within the capabilities of the engine except during transient load changes. As stated previously this is a desirable feature for many applications. *Figure 15.9* shows a family of isochronous characteristic curves. Each curve has a different speed setting. The dotted lines illustrate effects of load change.

15.1.3.2 Speed droop

Speed droop is a governing speed characteristic which requires a decrease in speed to produce an increase in servo position. Since an increase in servo position is required if the engine is to carry more load, it follows that increased load means decreased speed. It is expressed as the percentage change in speed corresponding to the full travel of the servo. Speed droop provides stability and is also necessary when two or more engines fitted with speed-sensing governors are operating in parallel, mechanically or electrically, to share load proportionally.

Figure 15.10 (a) shows the speed droop characteristics with different droop settings and illustrates the effect of a speed rise due to load decrease on paralleled unit A (2 per cent droop) and unit B (6 per cent droop). Unit A output position decreases three times more than Unit B. *Figure 15.10* (b) shows the speed droop characteristics with the same droop settings but with different speed settings and illustrates the effects of a speed decrease due to load increase on paralleled units A and B (4 per cent droop). The increase in output positions are the same for both units.

15.1.3.3 Steady-state speed regulation

Steady-state speed regulation is a governing speed characteristic which decreases with increase in power output of the prime mover or of the load driven. It is expressed as the percentage change in speed corresponding to the change from zero to full power output. It also provides stability and load division. *Figures 15.11 (a)* and *(b)* show the speed regulation characteristics with different speed regulation settings and different speed settings. The effects of speed rise or decrease on paralleled units A and B are illustrated.

Speed droop is a phrase often used as synonymous with *steady state speed regulation*. Briefly speed droop is the

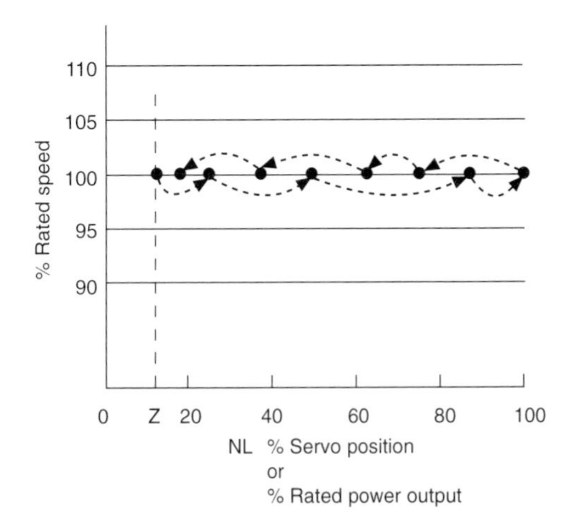

Figure 15.9 Load changes with isochronous governing

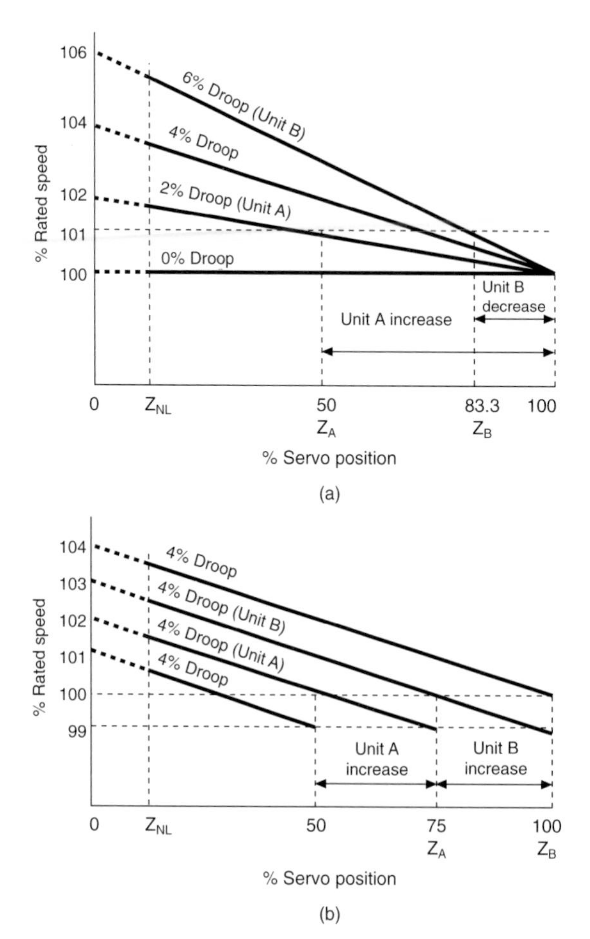

Figure 15.10 Speed droop characteristics

percentage change in speed resulting from full travel of the governor output shaft. Steady-state speed regulation is the percentage change in speed corresponding to the change from zero to full power output of the engine. If a governor was connected to the engine fuel rack so that only 50 per cent of the governor output shaft travel was required to move the fuel rack from no load to full load, the regulation would be 2 per cent although the speed droop is 4 per cent.

15.1.3.4 Steady-state governing speed-band

A steady-state governing speed-band is a speed-time envelope having upper and lower speed values represented by the maximum and minimum instantaneous speeds which occur under steady-state operation with the governing system controlling. *Figure 15.12* illustrates a magnified steady-state governing speed-band.

15.1.3.5 Transient perfromance

The quality of transient performance is often measured on the basis of speed deviation, momentary overspeed, recovery time, response time and damping factor in response to a step change. *Figures 15.13* (*a*) and (*b*) illustrate the meaning of speed deviation, momentary overspeed and underspeed and recovery time for a droop governor. *Figure 15.13* (*c*) illustrates the meaning of response time and damping factor.

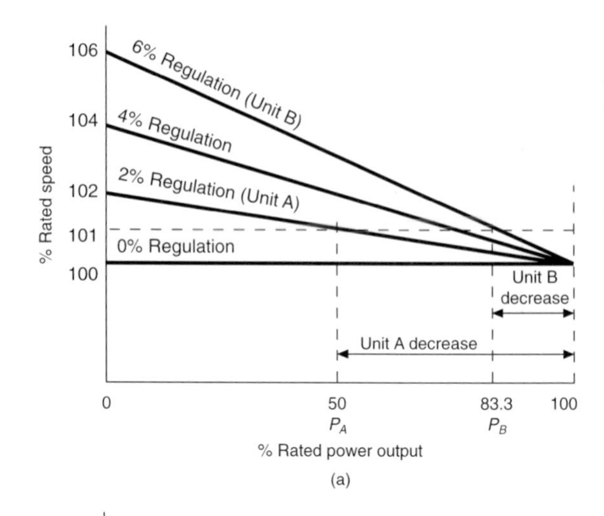

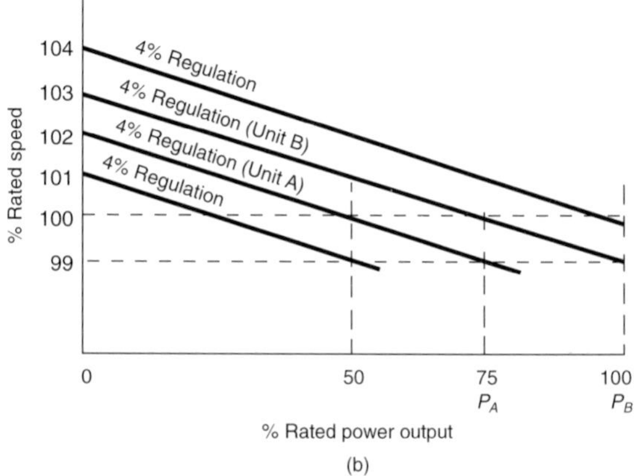

Figure 15.11 Speed regulation characteristics

15.1.4 Typical governors

15.1.4.1 Direct mechanical type

There are numerous ways of constructing governors corresponding to *Figure 15.14*. The weights can be pivoted in various ways, the restoring force can be provided by a spring as shown or by springs connecting the weights together, etc. The hand control can be fitted on the governor casing or led to a convenient position by a linkage.

Droop is inherent in purely mechanical governors in so far as the governor weights (flyweights) take up a position relative to the fuel pump setting, with a corresponding change in speed.

Combined mechanical governor and fuel injection pump

The unit shown in *Figure 15.15* was designed and developed for use on high-speed diesel engines (6 or 8 cylinders) with a capacity of about $1\frac{1}{2}$–$3\frac{1}{2}$ litres per cylinder. This variable range governor consists basically of two or four flyweights (49 in *Figure 15.15*) on a hub fitted on the end of the camshaft (71). The toes of the flyweights about a sliding sleeve (50) on the hub. The sliding sleeve incorporates a fork which engages a pivoted crank level (47); the fork is supported by a ball bearing in the sleeve. The longer arm of the crank lever is connected by a telescopic link (134) to a bridge link (135) on the end of the control rod (17); the shorter arm is connected by the governor springs (159 and 167) to the speed lever shaft (122). The engine speed is controlled by the spring tension applied through the

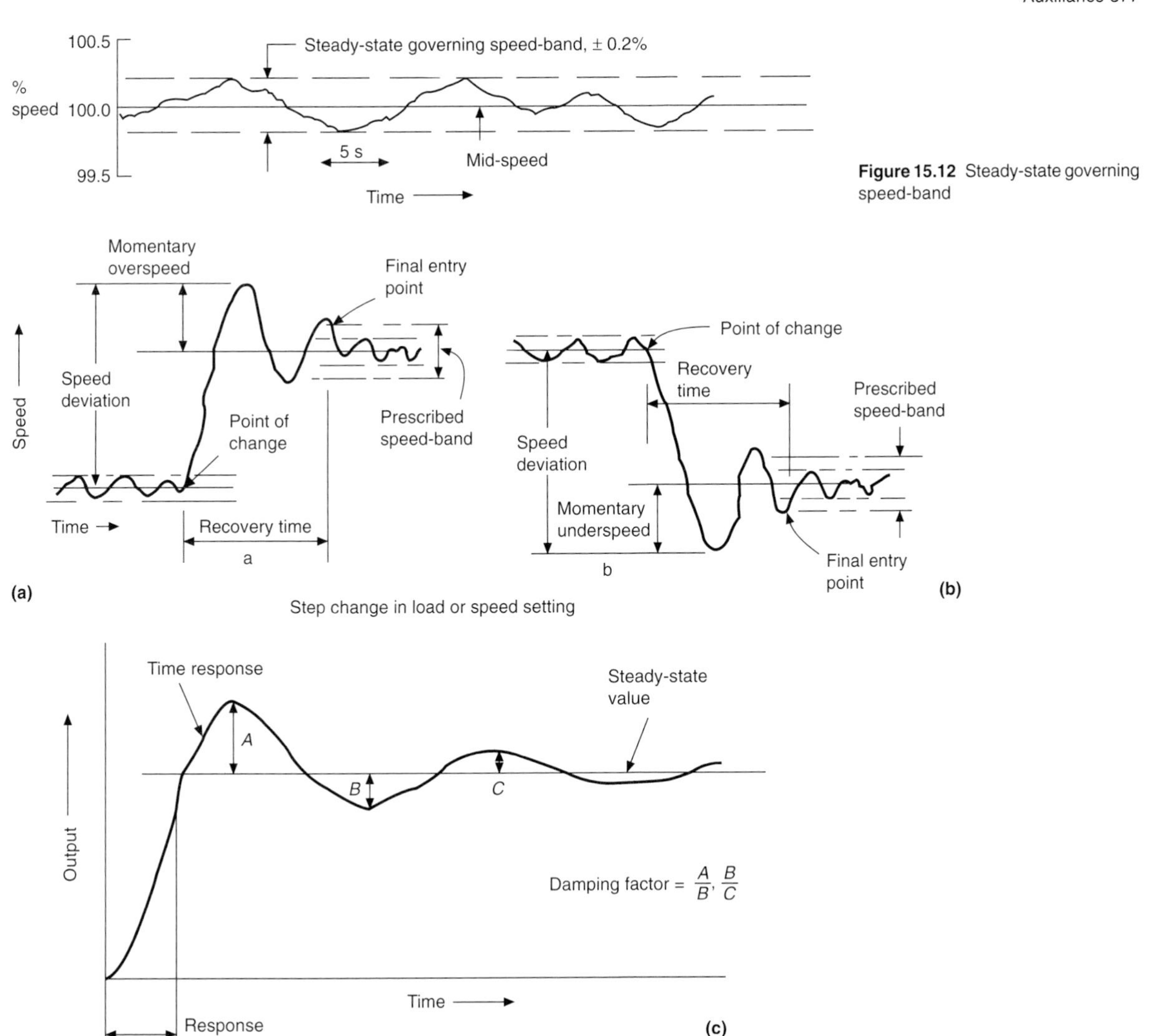

Figure 15.12 Steady-state governing speed-band

Figure 15.13 Transient performance

operation of the speed lever (120). When the speed lever is moved to a higher speed position, the control rod is moved accordingly towards the maximum fuel delivery position while, at the same time, the sliding sleeve is moved against the toes of the flyweights to close the flyweights.

When the engine speed increases just above the selected speed the centrifugal force to the flyweights, acting on the crank lever, moves the control rod towards the minimum fuel position, thereby reducing the fuel delivery and engine speed. Consequently, as the setting of the speed lever remains unchanged, the tension in the governor springs moves the control rod to the selected speed position. This completes the cycle of operation.

15.1.4.2 Mechanical hydraulic governors

The governors of Woodward design Type UG8 shown in *Figures 15.16 and 15.17* are in common use on diesel engines. They have a stalled work capacity of 8 ft lb, and the useful work output over the full available output shaft travel of 42° is 5 ft lb.The two basic versions of this type of governor are shown.

Figure 15.16 has dial speed setting; the manually adjustable dials include synchronizer-speed adustment for single engines, or changes of load when engines are paralleled. The speed droop is externally adjustable for 42° of terminal shaft rotation. The load limit is used to limit terminal shaft travel (engine fuel). An indicator dial shows terminal shaft limit position.

Figure 15.17 shows the governor with lever-type speed adjustment, which is accomplished with a lever mounted on the control shaft connected to a remote speed adjustment linkage. This version is mainly used on marine engines, compressors or similar applications where the possibility of changing speed rapidly over a wide speed range is desirable.

Operation of the UG8 dial governor

Figure 15.18 on page 380 shows a schematic diagram of the dial speed setting UG8 governor. This schematic is of a basic design and does not include any auxiliary equipment. Changes in governor speed setting produce the same governor movements as do changes in load on the engine.

Decrease in load Assume the prime mover is running on speed, i.e. the control land of the pilot valve plunger is centred over the control port of the rotating bushing. The flyweights are in a vertical position for normal steady-state operation. A reduction

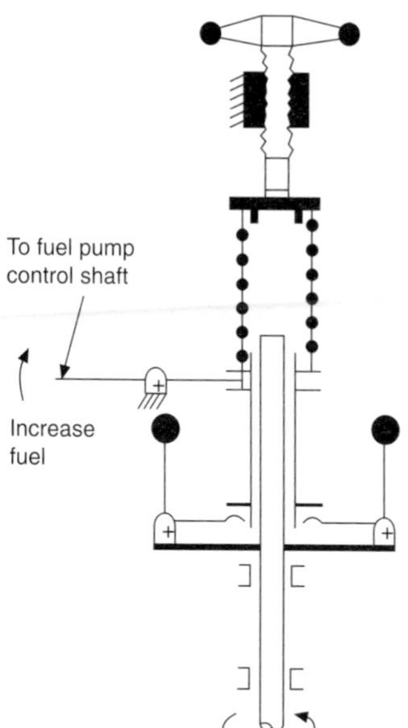

Figure 15.14 Diagrammatic arrangement of mechanical-type governor

in load creates an increase in speed. As speed increases the centrifugal force of the flyweights increases, overcoming the opposing speeder spring force. The flyweights tip outward raising the speeder rod and the right-hand end of the floating lever. This raises the pilot valve plunger, dumping oil from under the power piston. As the power piston moves downwards the terminal shaft is rotated in the decrease fuel direction.

As the power piston moves downward the actuating compensation piston moves upward. This applies suction to the small compensation piston, pulling it downward. The left end of the floating lever is pulled down, forcing the pilot valve plunger downward and closing off the control port. The compensation system, in esssence, anticipates the amount of fuel required to accept the new load change. The amount of movement (compensation) of the actuating compensation piston is controlled by the compensation adjustment and fulcrum. The terminal shaft and power piston's movement are stopped in the new decreased fuel position required to run the engine at normal speed with the decrease in load. As oil dissipates through the needle valve from the compensation system, the receiving compensation piston is returned to normal at the same rate as the speeder rod. This keeps the pilot valve plunger in its centred position.

Increase in load Again assume the prime mover is running on speed. The pilot valve plunger is centred and the flyweights are in a vertical position. An increase in load creates a decrease in speed. As speed decreases, centrifugal force of the flyweights decreases, allowing them to tip inward and decrease the upward force on the speeder spring. The speeder spring forces the speeder rod downward, which forces the pilot valve plunger down. Pressured oil is released through the control port into the lower cylinder of the power piston. The power piston is forced upward by the pressured oil acting on the larger lower surface area of the power piston. The terminal shaft is rotated in the increase

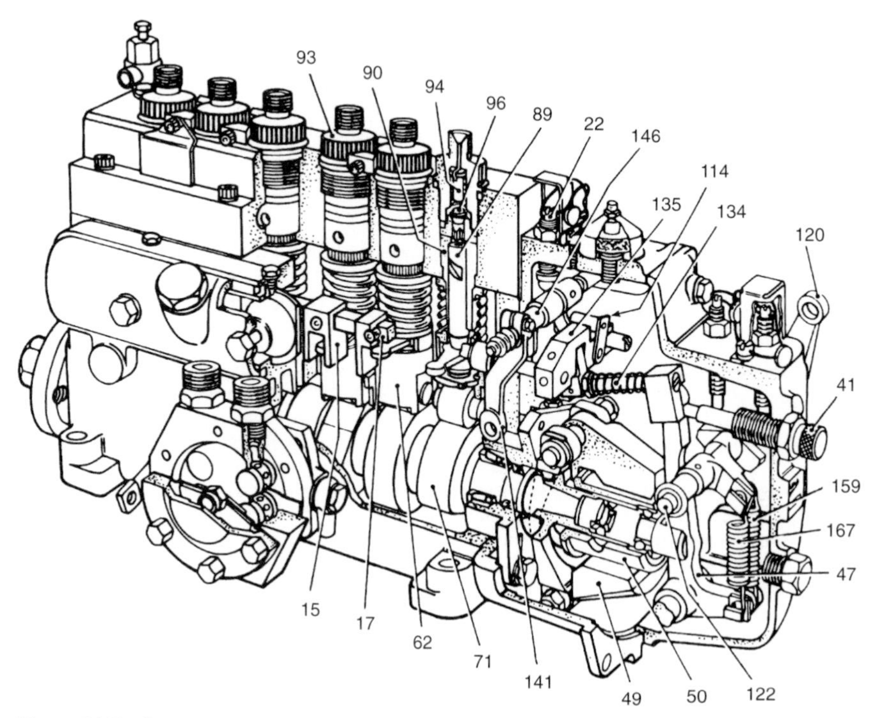

Figure 15.15 Cutaway section of typical 6-cylinder Majormec pump (Lucas CAV Ltd)

15 Control fork
17 Control rod
22 Max fuel stop screw
41 Damper
47 Crank lever
49 Governor flyweight
50 Governor sleeve
62 Tappet assembly
71 Camshaft
89 Plunger
90 Barrel
93 Delivery valve holder
94 Volume reducer
96 Delivery valve
114 Trip lever
120 Speed control lever
122 Speed lever shaft
134 Telescopic link
135 Bridge link
141 Sleeve control lever
146 Excess fuel device
159 Governor idling spring
167 Governor main spring

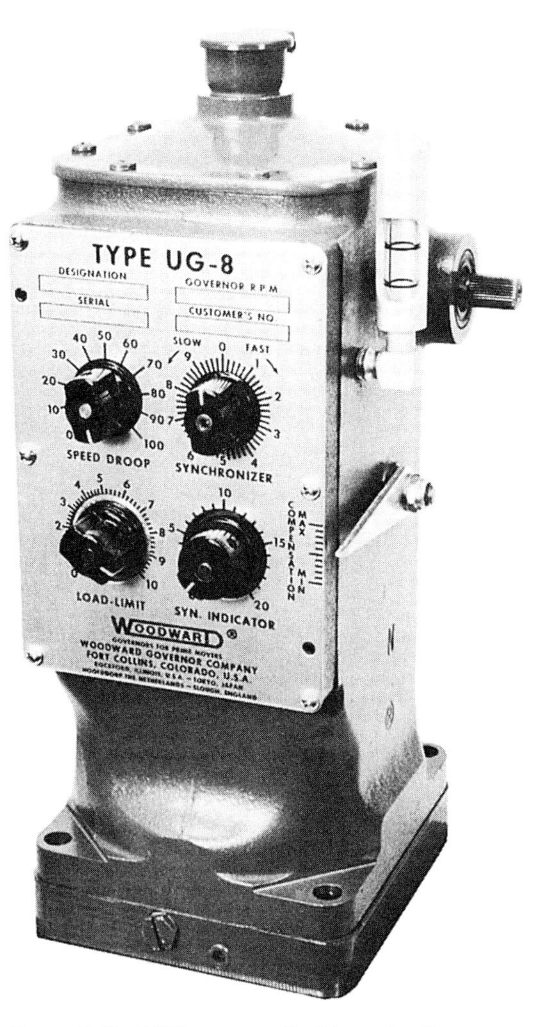

Figure 15.16 UG8D governor with dial speed setting

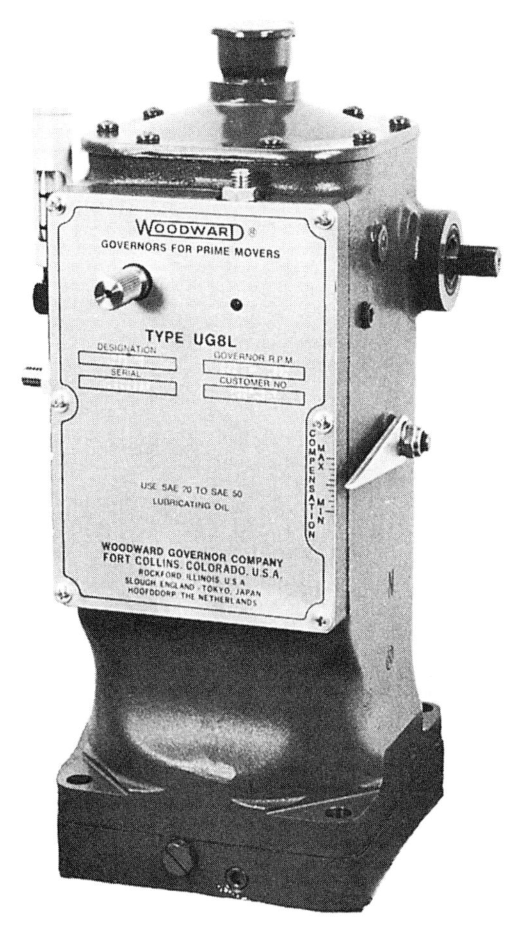

Figure 15.17 UG8L governor with lever speed setting

fuel direction. Linkage from the output shaft lifts the compensation adjusting lever which pivots at the fulcrum, pushing down on the actuating compensation piston. Oil is forced into the chamber of the smaller receiving compensation piston, raising the floating lever and in turn the pilot valve plunger. When correctly adjusted the compensation system effectively anticipates the amount of fuel necessary to bring the engine up to the proper output to accept the increased load.

The pilot valve plunger is re-centred, the speeder spring and flyweight forces are rebalanced, and the terminal shaft is in position to provide the new fuel requirements.

Type PGA mechanical hydraulic governor

The governor shown in *Figure 15.19* is of a type designed to fill the needs of a wide variety of applications from electric power generation to pipe-line pumping stations and from diesel locomotives to marine propulsion engines.

A number of optional auxiliary features and devices are available to permit the governor to perform other secondary functions such as limiting engine load, controlling engine load to maintain a constant power output for each speed setting, minimizing the tendency to over-fuel when starting, permitting temporary overloads, emergency shut-down in the event of ancillary equipment failure or loss of lubricating oil pressure, etc.

Speed setting can be by an air pressure signal from a pneumatic air transmitter or controller, dial, lever or solenoids (stepped).

The elements of a basic version of this governor are illustrated in *Figure 15.20.*

The governor drive shaft passes through the governor base into the pump dirve gear, which is direct connected to the rotating pilot valve bushing. The flyweight head is secured to the upper end of the pilot valve bushing, thus providing a direct drive from the engine to the flyweights. At any speed setting of the governor, when the engine is on speed, the centrifugal force of the flyweights will balance the opposing force of the speeder spring with the flyweights in the vertical position, and the control land of the pilot valve plunger will be covering the regulating ports in the rotating pilot valve bushing.

The governor pump supplies pressure oil to the accumulators and rotating pilot valve bushing, with excess oil (at maximum pressure) by-passing from the accumulators to the governor oil sump. Duplicate suction and discharge ball check valves at the governor pump permit rotation of the governor in either direction.

Movements of the power piston are transmitted by the piston rod to the engine fuel linkage. Regulated oil pressure under the power piston is used to raise the power piston—to increase fuel—and the power spring above the power piston is used to lower the power piston to decrease fuel.

Located between the pilot valve bushing and the power piston is the buffer compensating system, consisting of the buffer cylinder and piston, the buffer springs and the compensating needle valve. Lowering the pilot valve plunger permits a flow

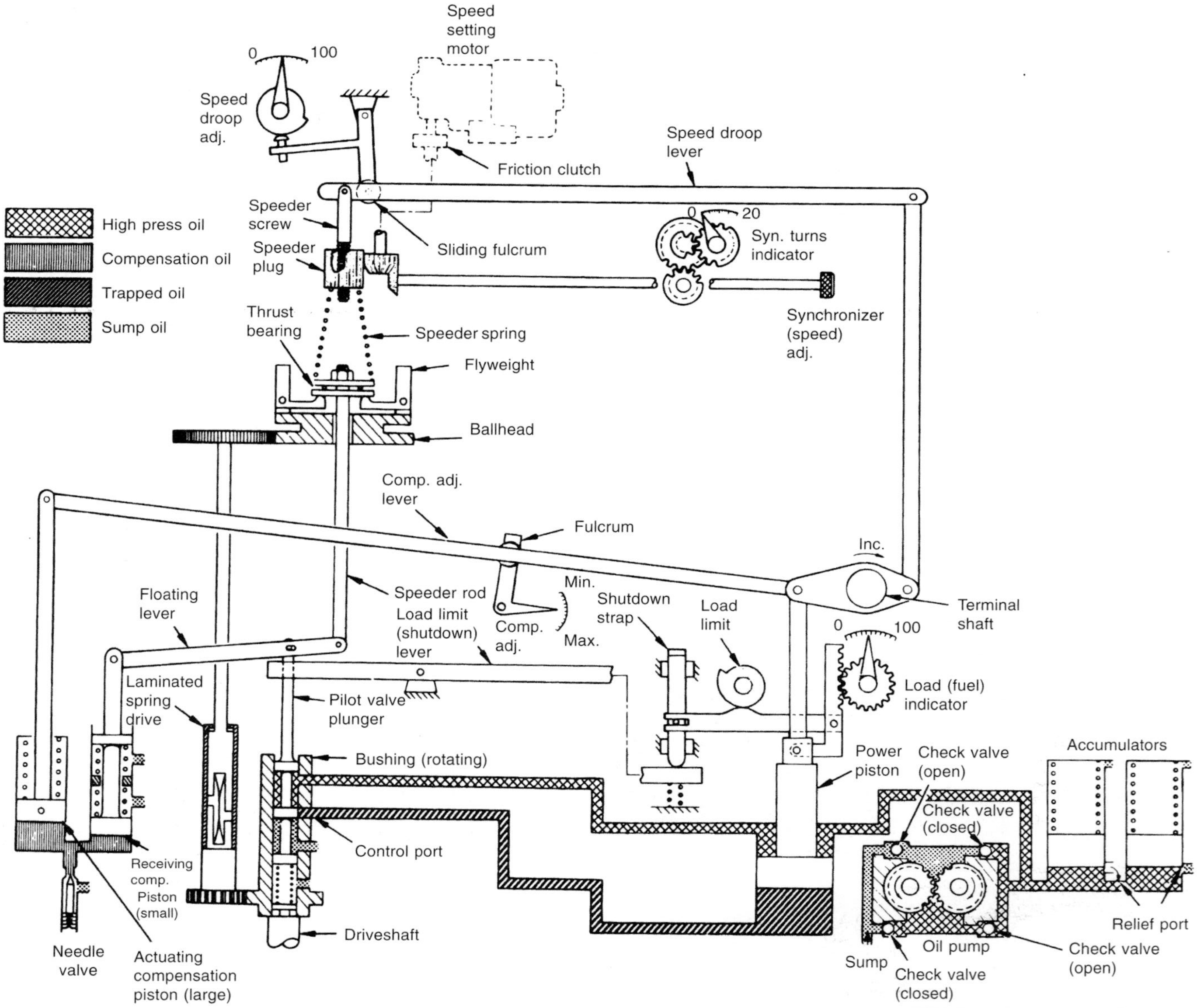

Figure 15.18 Schematic diagram of the dial speed setting UG8 governor (Woodword Governor Co)

of pressure oil from the pilot valve bushing into the buffer system and power cylinder to raise the power piston and increase fuel. Raising the pilot valve results in a flow of oil from the power cylinder and buffer system to the governor sump, and the power spring moves the power piston down to decrease fuel to the engine.

This flow of oil in the buffer system—in either direction—carries the buffer piston in the direction of flow, compressing one of the buffer springs and releasing the other. The action creates a slight differential in the pressures of the oil on opposite sides of the buffer piston, with the higher pressure on the side opposite the spring which is compressed. These differential oil pressures are transmitted to the areas above and below the compensating land on the pilot valve plunger, producing an upward or downward force on the compensating land which assists in re-centring the pilot valve plunger whenever a fuel correction is made.

The vertical position of the flyweights with the control land of the pilot valve covering the regulating port indicates that the engine is on speed.

15.1.4.3 Electric governors

Advantages

The applications and types of electric governors currently available are wide and varied, and range from electromechanical to electrohydraulic actuators. Whilst to date they have been mainly used on generating sets they are being used in increasing numbers on engines for pipe-lines, compressor stations and marine propulsion.

Governing for isolated, remote semi-automatic or completely automatic, attended or unattended power plants, is more easily integrated into the system if the governor is electric. Electric governors lend themselves to automation very readily and with related auxiliaries provide complete control in the area of synchronizing and loading.

Basic advantages in power generation applications where all units are equipped with electric governors are: (1) smaller transient speed changes; (2) isochronous operation; (3) automatic load division; and (4) adaptability to complex control requirements.

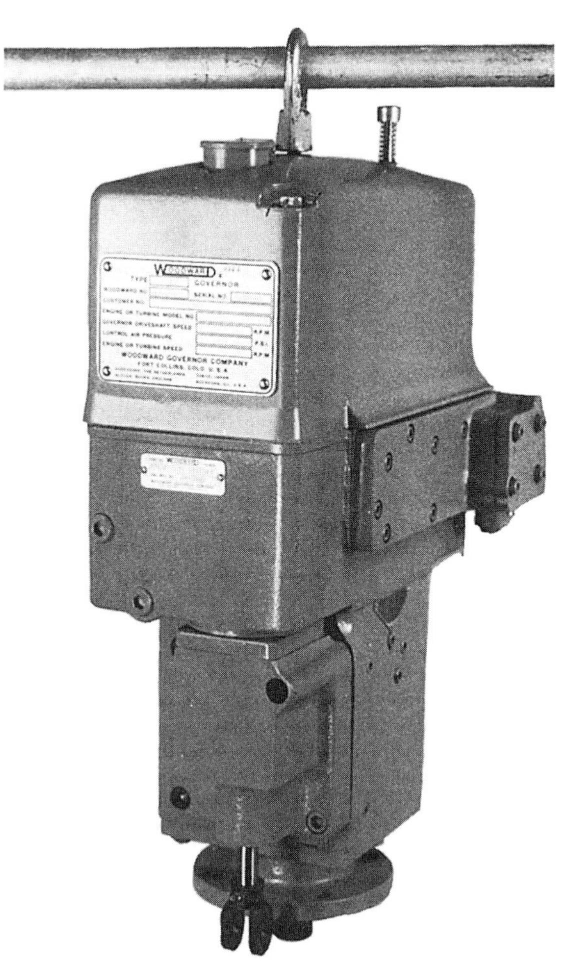

Figure 15.19 Type PGA mechanical hydraulic governor

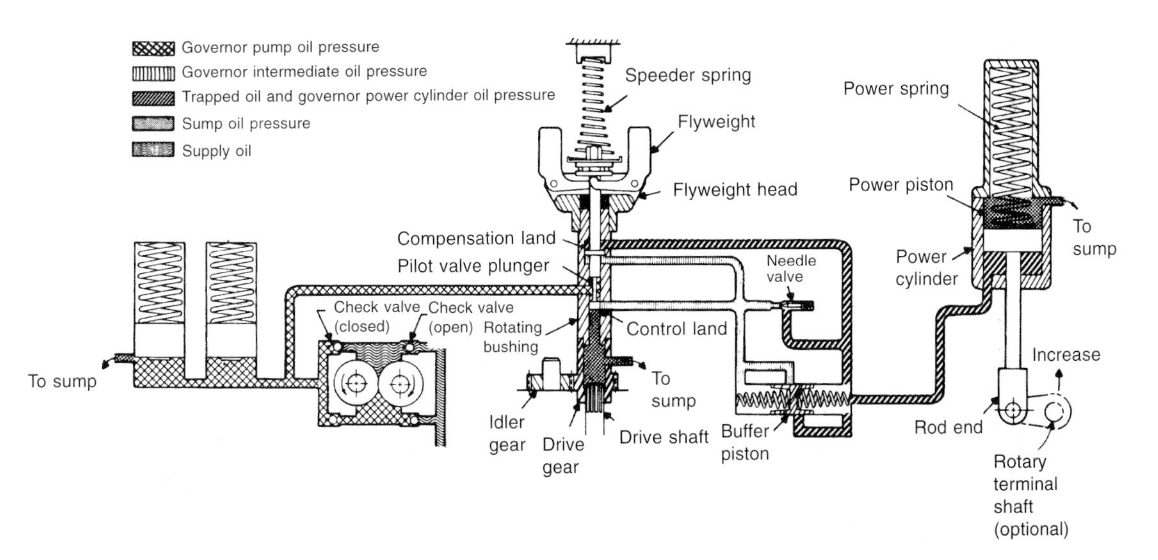

Figure 15.20 Schematic diagram of a PG governor

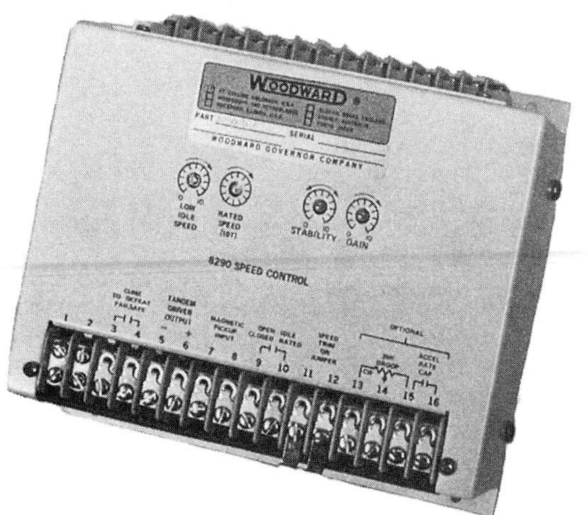

Figure 15.21 An electric speed control

Another advantage of electric governors is that it is not necessary for them to be driven by the engine. The electrohydraulic actuators can be driven by the engine, by an electric motor or by an oil motor. The electromechanical actuators require no drive whatsoever. The exception to the foregoing are actuators that contain a mechanical hydraulic governor in addition to the electric governor.

Clearly the precise form and design of electric governors will depend on the manufacturer. Examples of Woodward products are given below.

Speed control

The speed control shown in *Figure 15.21* is designed for isochronous control of a generator set. It provides a high degree of accuracy over a wide speed range. External wiring can provide remote speed trim, acceleration rate control and droop if parallel operation is required with engines fitted with speed-sensing governors.

Load sharing and speed control

Figure 15.22 shows an isochronous speed control and load sensor for paralleled generator sets. Proportional load division is automatic. Droop operation may also be used for paralleling to an infinite bus. For tandem applications such as two engines driving a single alternator, one of the controls can be used with two actuators.

Modular integrated electronic controls

Figure 15.23 shows an example of an integrated electronic control system with multi-function capability. Most control functions are isolated and not interdependent. An isolated power supply protects all circuits and the functions are designed to failsafe.

All functions are mounted on double-sided printed circuit boards which plug into card edge connectors. Relays are the hermetically sealed crystal-can type. Packaging is usually for front panel mounting and includes light emitting diodes to identify numerous functions such as controlling mode, switch operated, failsafe and over-limit shut-down.

With minor differences, these systems are applicable to generator sets, pumping sets, mechanical loads and marine applications.

Governor/actuator

Combined electrohydraulic actuator with a complete hydraulic ballhead governor. It is used with electric control systems if it is desirable to have manually adjustable back-up governor operation. An optional speed setting motor can privide remote speed setting for the ballhead governor section (*Figure 15.24*).

Electrically powered governors

A complete electric governor comprises a speed control and an actuator which requires neither mechanical engine drive nor hydraulic supply (*Figure 15.25*). The governor is useful for isochronous control of small single-unit generators and a load sensor can be added for paralleling generator sets. The electrically powered actuator positions the fuel control.

Electrohydraulic actuators

An electrohydraulic actuator (*Figures 15.26 and 15.27*) is powered by a mechanical drive from the prime mover or by an optional electric motor or oil motor. A separate, pressure oil supply is

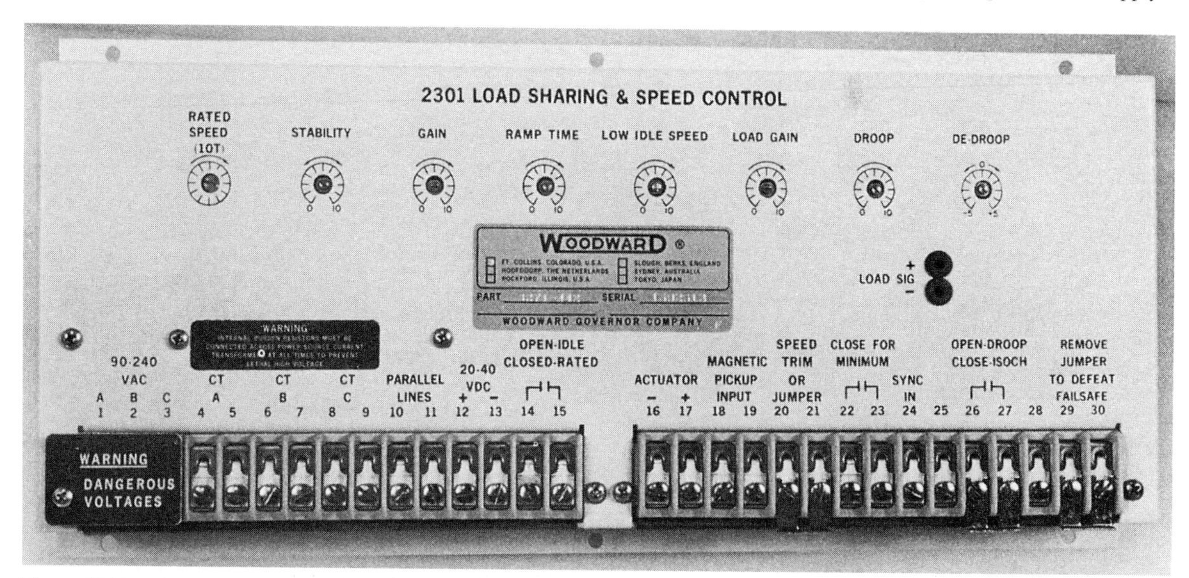

Figure 15.22 A load sharing and speed control

Figure 15.23 An integrated electronic control

required. It can be used in tandem applications of two prime movers driving a common load. The oil pump model of this actuator must be driven by the prime mover or by some other means such as an electric motor. Oil from the esternal supply source enters the suction side of the oil pump. The pump gear and bushing gear carries the oil to the pressure side of the pump, first filling the oil passages and then increasing the hydraulic pressure. Internal pressure is maintained by the relief-valve oil-pump system.

The engine fuel pump linkage is connected to the terminal shaft and the terminal shaft is connected through linkage to the power piston. Constant hydraulic pressure on top of the power piston tends to rotate the terminal shaft in the 'decrease' direction. However, the power piston cannot move down unless control oil is released to the sump.

The pilot valve plunger controls the flow of control oil to and from the power piston. The plunger is centred when its control land covers the control port in the pilot valve bushing. *Figure 15.27* shows this bushing centred.

The pilot valve plunger is connected to a permanent magnet that is spring-suspended in the field of a two-coil solenoid. The output signal from the electric control box is applied to the solenoid coils to produce a force which is proportional to the current in the coils. This force always moves the magnet and pilot valve plunger downward. The centring spring force always moves the pilot valve plunger and magnet upward. The restoring force spring exerts a downward force on the pilot valve plunger. This downward force is dependent upon the position of the restoring lever. The restoring spring lever moves upward to decrease the restoring spring force as the terminal shaft rotates in the 'increase' fuel direction. The resultant force from the combined output of the centring spring and restoring spring is always urging the pilot valve plunger in the upward direction. This resultant force increases as the terminal shaft moves in the 'increase' fuel direction. With the unit running on-speed under steady-state conditions, the resultant spring force and the force from the current in the solenoid coils are equal but opposite. This keeps the terminal shaft at a position to maintain the engine at the required speed.

The oil motor model operates in the same way as the oil pump model except for the oil supply system. When equipped with an oil motor the actuator does not require a drive shaft. An outside source of pressured oil supplies the working pressure for the actuator. This pressured oil also operates the oil motor, which rotates the pilot valve bushing.

Figure 15.24 A combined electrohydraulic actuator with hydraulic ballhead governor

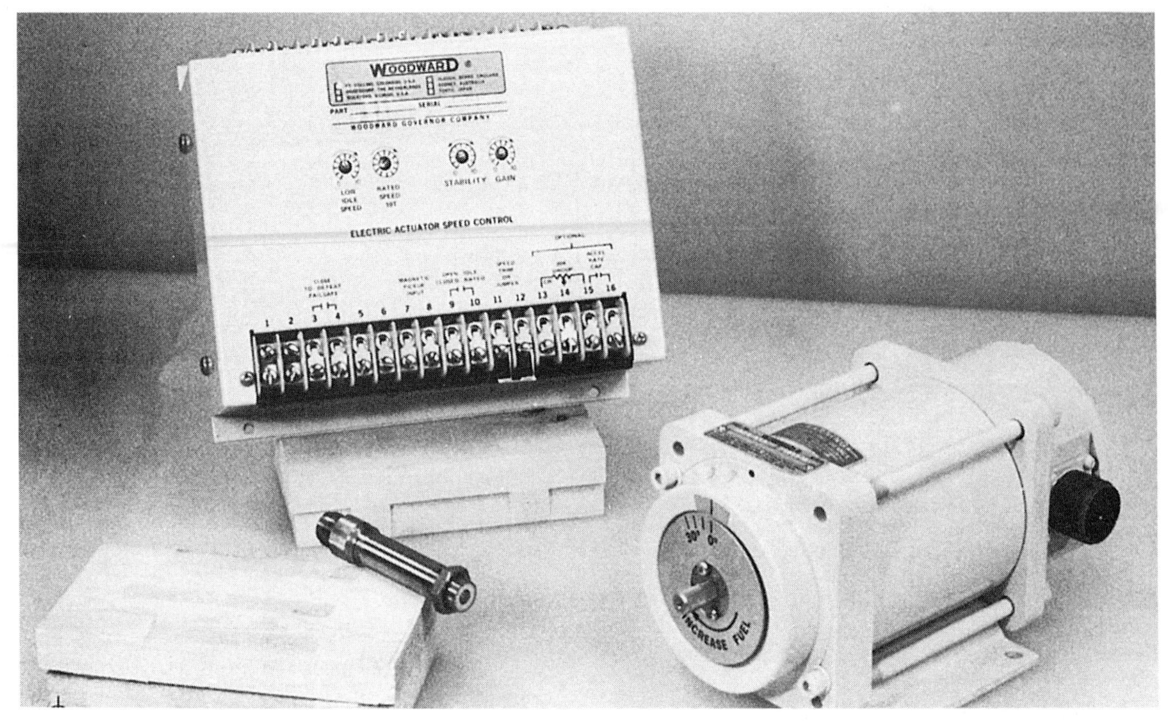

Figure 15.25 A complete electric governor with electrically powered actuator

15.1.5 Application requirements and governor selection

For the majority of engine applications governing performance requirements are quoted and these are used in selecting a governor. A number of organizations, institutions and government agencies such as the British Standards Institution, American Institute of Electrical Engineers, American Institute of Mechanical Engineers, etc., issue governing specifications. The governing performance is determined by the engine-driven machinery-governor combination.

Governor selection involves consideration of two prinicpal factors: (1) application requirements; and (2) prime mover and requirements. The most common considerations are:

1. Requirements of the application
(i) Steady-state speed control.
(ii) Steady-state speed regulation.
(iii) Transient response.
(iv) Parallel and automatic operation.

2. Factors related to requirements of the prime mover
(i) Work required to operate fuel metering mechanism. The governor must produce an output force sufficient to move the fuel metering input throughout its full range. This governor output, force times distance, is calculated and specified as a work rating of the governor. Work requirements for the fuel system are most often defined by test and experience. Elements contributing to the work required include the inertia of linkage, reaction forces in the fuel system, and friction.
(ii) Type of mechanical drive available.
(iii) If a suitable drive is not available an electrohydraulic or an all-electric governor system is required.

15.1.5.1 Governor drives

The basic requirement for a governor drive is to provide an accurate reflection of crankshaft speed. Care must be taken in the manufacture of the gearing between the crankshaft and the governor and also to avoid any distortion of true crankshaft speed. In general the ideal drive is one which is taken from the flywheel end of the crankshaft and goes through a minimum of gears in reaching the governor. It is still quite common however for accessory drives to be taken from the free end of the crankshaft. In addition since direct connection is seldom feasible a gear train is commonly used. Any eccentricities of gear pitch lines or errors in tooth form superimpose further unwanted speed deviations on the drive. If the gearing also drives the camshaft and especially if the governor is driven from the free end of this shaft, the periodic load variations due to operation of the valves produce additional errors.

The problem presented to the governor manufacturer by a rough drive is that whilst the governor must be sensitive and fast to respond satisfactorily to the speed deviations that represent changes in average engine speed it should not respond to the deviations which are due to a rough drive.

Torsional vibrations at the governor input at frequencies below one hundred per minute are practically impossible to damp inside the governor and result in excessive terminal shaft movement with consequent fuel pump and linkage wear. Higher frequencies may be damped by use of a special governor head drive (internal) but only with some sacrifice in quality of performance. Very high frequencies or high amplitudes may result in actual failure of the mechanical parts of the governor or governor drive.

15.1.5.2 Fuel control linkage

The governor output should be transmitted to the fuel pumps or other fuel control mechanism with as little lost motion as possible. There should be less than one-quarter of one per cent of the total travel wasted in lost motion. Friction in this linkage should be held to a minimum and there should be a linear relationship between engine torque and governor terminal shaft position.

In most cases it is desirable to use about 60 per cent of the

Figure 15.26 EG-10P electrohydraulic actuator

governor stroke to move the fuel control linkage between full load and no load positions. Also, the governor should not be able to move the fuel control linkage much below the point at which the engine will stop. Work used over the full governor stroke should not exceed about 60 per cent of the rated governor capacity, which is usually expressed in foot-pounds.

15.1.5.3 Linkages and hand controls

A typical system is shown in *Figure 15.28*. The governor A is connected to the fuel-pump control shaft B through the link C, which is normally extended to its full length by the spring compressed inside it. If the hand lever D is moved so as to bring the fuel pumps to the 'Stop' position, the link is shortened. The force required at the hand lever is therefore not dependent on the construction of the governor. When using a large, powerful governor, it would be very difficult to exert enough effort to override the governor.

By moving lever D to a position between 'Run' and "Stop', the output of the engine can be limited to any desired extent. The governor can always move in the direction which reduces fuel, so that the engine is always under control.

It will be seen that shaft B is connected to each fuel pump through two levers, one fixed on the shaft and the other free, the levers being held together by a spring. This is a safety device which acts in the event of a fuel-pump seizure. If the shaft B were rigidly connected to the pumps, such an accident would prevent the governor from operating in either direction and the engine might run away if the load were reduced. With the device shown, the governor can always reduce fuel on all lines except the one which has seized, so that this danger is avoided. Fuel cannot be increased and an increase in load would stall the engine; this is obviously not so dangerous.

All linkages must operate with the minimum amount of friction to avoid interference with the smooth action of the governor. The various joints should be lubricated periodically and the mechanism should be checked from time to time. It may be easier to do this if one of the joints is uncoupled, so that the linkage and the pump control shaft can be moved separately.

A screw E is provided at each pump to adjust the relative fuel quantities delivered by the pumps. A screwed stop F limits the fuel to any desired maximum; it would normally be set to correspond to the overload rating of the engine.

15.1.5.4 Overspeed protection

A diesel engine should be equipped with a separate overspeed shut-down device to protect against runaway or damage to the engine with possible injury to loss of life to personnel should the engine governor, driving mechanism, fuel control or the linkages fail.

To achieve the greatest degree of protection for the engine, the overspeed governor or sensor should be entirely independent of the normal governing mechanism. It should, therefore, be separately driven from a convenient shaft on the engine, so as to remain operative in the event of a breakage of the normal governor drive. It is also possible to arrange for it to cut off fuel by some means distinct from the normal control gear. For example, it can operate a shut-off valve in the fule supply line; or it can release mechanism which prevents the fuel pumps from being reciprocated by the camshaft. An arrangement of this kind enables the overspeed governor to act, even though the fuel pump control gear has jammed or broken, putting the normal governor out of action.

It is necessary to set the emergency governor operating speed a reasonable amount above the maximum speed which the engine is likely to reach in service. A suitable value for a generating set is, say 15 per cent above normal speed.

As with all other protective devices, it is very desirable to test the device at intervals, to ensure that it is maintained in perfect working order.

15.1.6 Typical applications

15.1.6.1. Parallel operation of generating sets

When two or more alternator sets are run in parallel, although the operation is relatively simple it raises a number of problems which must be successfully solved in order to obtain satisfactory performance. Whilst the problems involved are beyond the scope of this chapter, they must be carefully considered by the manufacturers of the diesel engines and alternators to avoid resonance.

For two or more machines fitted with speed sensing governors to share proportionally a variable total load, the following requirements must be met:

(1) Equal drop in speed (regulation) from no-load to full-load KW.
(2) Equal fall of voltage from no-load to full-load KVAR.

Failure to meet these requirements will result in unbalanced load-sharing though slight inaccuracies are normally acceptable.

Once the units are operating in parallel and carrying load the only means of varying load division between the units is to

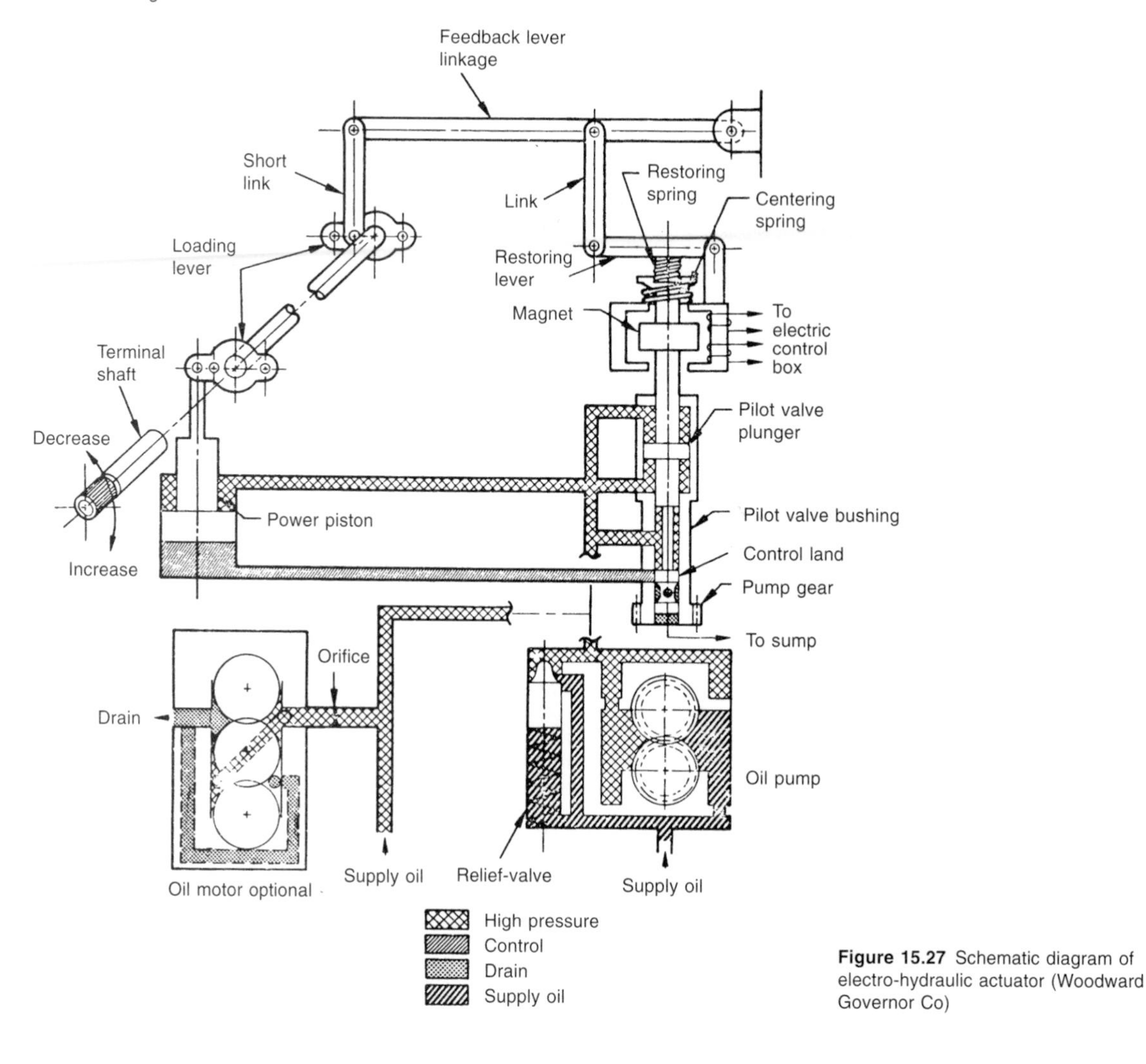

Figure 15.27 Schematic diagram of electro-hydraulic actuator (Woodward Governor Co)

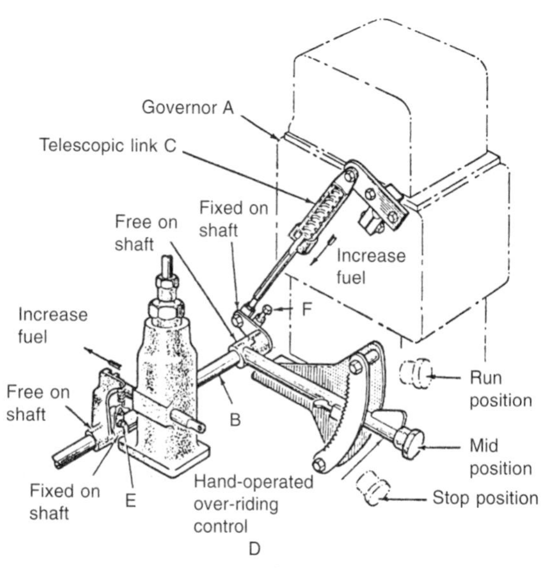

Figure 15.28 Typical linkage and hand controls

adjust the speed setting of the governors. Increasing the speed setting on a particular unit will result in an increase of load on that unit; decreasing the speed setting will result in decreasing the load on the unit.

An alternative method of operation when using speed-sensing governors is to have one engine (the master) set for isochronous operation and the other engines (slaves) set for droop operation. The advantage of using this method is that constant frequency is maintained but it should be noted that the isochronous unit must be capable of absorbing the load changes.

The main feature of electric governors with load sensing is evident when isochronous operation of paralleled generator sets is desirable. Electrical load on each generator is measured and signals sent directly into the electronic governor circuit to balance the load between sets. Other electric governor advantages include improved transient response and more flexible mounting arrangements. When a governor drive is not available on the engine an all-electric actuator is an obvious selection.

15.1.6.2 Marine propulsion

For these applications governors in use range from simple mechanical governors to the latest integrated electronic control systems. As marine installations have become more complex and the requirements more stringent the governing systems can become very complicated to meet the demands.

Prior to the controllable pitch propeller the speed of fixed-pitch propellers driven by diesel engines were in many instances controlled by simple throttle arrangements by which the quantity of fuel delivered to the engine was fixed. This type of control fixed the torque output of the engine and determined its operating speed, so, therefore, load changes on the propeller due to wave action, ship loading, or changes of ship's speed resulted in changes of speed (rev/min) of the prime mover. Normally, this control was backed up by an overspeed governor, which could override the throttle and control fuel to suit conditions encountered during heavy sea operation. The advent of the controllable-pitch propellers together with advanced engine designs brought about the need for more sophisticated governors.

Governors for marine propulsion engines are now available with a wide range of auxiliary features and devices. The most commonly used include:

Solenoid operated shutdowns.
Overspeed–trip test devices.
Load indicating (tail-rod) switches.
Speed setting fuel limiters.
Load (pitch) control and manifold pressure fuel limiters.
Load balance systems.

15.1.6.3 Rail traction control systems

Most governor applications require that the governor should control the fuel flow to the engine to maintain a set speed under varying load conditions. While the primary function of a diesel-electric rail governor is still to maintain a desired engine speed there is a secondary function, namely maintaining a constant engine horse power output for each specific speed setting, by means of a load control system. The load control system positions a variable resistance in the locomotive excitation system to change the electrical load on the engine and maintain a pre-determined fixed fuel setting.

Additional auxiliary devices which can be built into locomotive governors include:

Automatic safety shutdowns and alarms.
Load control override devices.
Starting and manifold air pressure fuel limiters.
Altitude compensation.

15.1.7 Conclusion

Finally it should be appreciated that the governor is only one part of the complete system which consists of the diesel engine, load and other controls. Each of these components contributes to the behaviour of the total system and the dynamic characteristics of each item of the system must be tuned to the other parts in order to provide optimum performance of the complete unit.

The author would like to acknowledge the help given by Woodward Governor Gmbh for the use of the photographs of their equipment illustrated in this section.

15.2 Starting gear and starting aids

15.2.1 Introduction

The diesel engine must be capable of starting reliably under extremes of climatic conditions ranging from tropical to arctic, and for a wide variety of applications. The form of starting gear adopted will be influenced significantly by the type and size of the engine, the purpose for which it is to be used and the climatic conditions in which it will be required to operate.

The combustion process in a diesel engine is initiated solely by raising the temperature of the air in a cylinder by compressing it during the compression stroke, so that it is above the self-ignition temperature of a spray of atomized fuel which is injected into it, towards the end of the compression stroke. It follows that if the air in the cylinder is not raised above the self-ignition temperature of the fuel employed, combustion will not occur and the engine will not start.

Chapter 4: Section (e), 'Ability to start', shows that, while under normal running conditions the air temperature in the cylinder at the end of the compression stroke is considerably above the self-ignition temperature of the fuel, the situation when starting a cold engine in a low-temperature environment is far less satisfactory. Indeed, under these conditions, the air in the cylinder may not attain the temperature required to initiate combustion by the end of compression stroke, and it then becomes necessary to provide a starting aid.

The above considerations apply to all diesel engines. There will be, for each engine, a minimum ambient air temperature below which it will not start solely with the heat generated during the compression stroke. Therefore as a preliminary to considering starting gear and starting aids, it is appropriate to review briefly the factors which influence the cold starting ability of an engine[11,12].

15.2.2 Unaided cold starting ability

The air temperature in the engine cylinder at the end of the compression stroke will be the air temperature in the cylinder at the beginning of the compression stroke, plus the rise in air temperature during the compression stroke. This rise will depend on the value of the exponent of the compression process and on the compression ratio of the engine.

15.2.2.1 The exponent of the compression process

The maximum rise in air temperature which can theoretically be achieved during compression will occur when the process is truly adiabatic. The exponent of the compression process will then have a value of 1.41, this is the ratio of the specific heat of the air at constant pressure to the specific heat of air at constant volume. To achieve this theoretical maximum, the mass of air contained in the cylinder must not change during the compression process and there must be no interchange of heat between the air and the enclosing surfaces of the combustion chamber, the piston and the upper portion of the cylinder bore.

These idealized conditions cannot occur in a conventional engine because there is inevitably some loss of air due to leakage past the piston rings and possibly the inlet and exhaust valves. There is also some heat loss from the air to the surfaces of the combustion chamber, the piston and the cylinder bore. The greater these losses, the greater the reduction in the exponent of the compression process and the smaller the rise in temperature of the air in the cylinder, during the compression stroke. The losses will be a maximum when starting a cold engine and their magnitudes depend on the following features of the engine:

(i) The stroke/bore ratio

For a cylinder of given swept volume, the lower the stroke/bore ratio, the greater the length of the circumference of the piston ring and therefore the greater the potential leakage of air during the compression stroke. Also, the surface area to volume ratio becomes greater over the upper portion of the compression stroke, where most of the heat loss from the air occurs.

(ii) The capacity of the cylinder

The surface-area/volume ratio increases with decrease in cylinder swept volume and therefore the loss of heat from the air will be proportionately greater as the cylinder capacity decreases.

(iii) The proportions of the combustion chamber

For a given volume, a sphere has the minimum surface area; the surface area of the combustion chamber will increase very rapidly as its proportions tend towards those of a shallow disc, and the loss of heat from the air will increase accordingly.

(iv) The type of combustion system

A quiescent, direct-injection system has a lower rate of heat loss than a high-turbulence system because the air movement in the latter increases the heat transfer coefficient between the air and the cold surfaces of the combustion chamber, the piston and the cylinder bore. Similarly an indirect or pre-combustion chamber system, which generally has a large surface area and induces high air velocities during the compression stroke, will also have a higher rate of heat loss from the air than a direct-injection system.

15.2.2.2 The compression ratio

An increase in compression ratio improves the unaided cold starting ability of the engine by increasing the rise in air temperature during the compression stroke. This compensates in some measure for the reduced value of the exponent of the compression process, due to air leakage and heat loss to the cold surfaces. It must be noted, however, that the combustion process is a 'mixed cycle' in which combustion occurs partly at constant volume and partly at constant pressure. There is therefore a significant increase in the cylinder pressure during combustion, and, unless engine performance and thermal efficiency are prejudiced, the higher the compression pressure, the greater will be the maximum cylinder pressure at a given brake mean effective pressure.

The maximum cylinder pressure which can be employed without detriment to the reliability and durability of the engine depends on the design of its structure and its components. The compression ratio adopted is therefore a compromise between the unaided cold starting requirements and the attainment of satisfactory conditions under operation at maximum power. In fact, in turbocharged engines as compressor pressure ratios increase, the need to limit the maximum cylinder pressure to an acceptable level sometimes results in the use of a compression ratio which is low for unaided cold starting.

The provision of a variable compression ratio which will enable a high ratio to be used for cold starting and a lower ratio for normal running on load is technically attractive; but except with the opposed-piston, two-stroke 'bell crank' engine[22], the added complexity of this provision is considerable. For example, it is possible to use a variable compression ratio piston[23] to provide a high ratio for starting and a progressively lower ratio as the brake mean effective pressure increases, but in general, alternative means of ensuring a cold starting ability are adopted.

15.2.2.3 Ambient air temperature

The most arduous cold starting condition which can exist is when the whole engine structure, the coolant and the lubricating oil have assumed the prevailing ambient air temperature—a condition which is common in small engines used for a variety of applications, including road vehicles, off-highway vehicles, contractors' plant and small marine craft. In a temperate climate, a minimum air temperature of −10°C can be expected, while in arctic regions the temperature may fall to below −30°C.

Under normal running conditions, the air for combustion picks up heat on its passage through the inlet manifolding; from the turbocharger if fitted; and from the surfaces of the cylinder head, combustion chamber, piston and cylinder bore, during its entry into the cylinder. Consequently, the temperature of this air at the commencement of the compression stroke will be significantly above the prevailing ambient air temperature.

In contrast, when starting a completely cold engine in a low-temperature environment, the whole engine will be at the prevailing ambient air temperature and therefore the air in the cylinder at the commencement of the compression stroke will also be at this low temperature. In addition, it has been shown in Section 15.2.2.1, that the value of the exponent of the compression process will be much reduced under these conditions, owing to air leakage and loss of heat. It follows that the unaided cold starting ability of an engine can be improved by creating starting conditions which help to increase the exponent of the compression process. This approach can be applied to all diesel engines but the methods of obtaining the increase vary with the type of engine and the constraints imposed by a particular application.

15.2.3 Improving the unaided cold starting ability

To achieve unaided cold starting of an engine, the temperature of the air in the cylinder must be raised to above the self-ignition temperature of the fuel employed, during the compression stroke. It has been shown that as the ambient air temperature decreases, a limiting figure will be reached, below which the temperature of the air in the cylinder will not be raised sufficiently to promote combustion and it will then be necessary to employ a starting aid. This limiting ambient air temperature will depend on factors which can themselves be varied in order to improve the unaided cold starting ability of the engine, and examples of these are described in the following paragraphs.

15.2.3.1 Air leakage during the compression stroke

The leakage of air past the piston rings and possibly past the inlet and exhaust valve seats, can fairly be regarded as leakage through a fixed orifice. Therefore, assuming that the pattern of cyclic variation in pressure difference across the orifice remains unchanged, the mass of air lost per compression stroke will be directly proportional to time. The time can be reduced by increasing the rate of compression; the mass of air lost will then be reduced and the value of the exponent of the compression process will be increased. Therefore for a given initial air temperature in the cylinder at the start of the compression stroke, the temperature rise during compression and the temperature of the air at the end of the compression stroke will be increased by increasing the cranking speed of the engine.

15.2.3.2 Heat loss during the compression stroke

The rate of heat loss from the air in the cylinder to the cold surfaces of the combustion chamber, the piston and the upper part of the cylinder bore will be proportional to the mean temperature difference between these surfaces and the air, while the amount of heat lost by the air will be proportional to time.

The duration of the heat flow can be reduced by increasing the rate of compression and this will achieve an increase in the value of the exponent of the compression process. Again, therefore, at a given initial air temperature in the cylinder at the start of the compression stroke, the temperature of the air at the end of the compression stroke will be increased by increasing the cranking speed.

15.2.3.3 Ignition quality of the fuel

The temperature to which the air in the cylinder must be raised during the compression stroke, in order to initiate combustion, will depend on the ignition quality of the fuel employed. This is defined by the cetane number; the higher the cetane number, the lower the self-ignition temperature. Therefore, the use of a fuel having a high cetane number will assist the unaided cold starting ability of an engine[13,14].

15.2.3.4 Prolonged cranking

During cranking, the full load fuel quantity or a significant proportion of it will be injected into the combustion chamber at each cycle. As cranking continues, this fuel will accumulate in the cylinder and reduce both the leakage of air past the piston rings and the effective clearance volume in the cylinders, thereby increasing the value of the exponent of the compression process and temporarily increasing the effective compression ratio.

Although combustion may not occur on the first compression stroke, there will be a progressive cycle-to-cycle increase in the temperature of the air in the cylinder at the end of the compression stroke, owing to partial oxidation of the fuel in the combustion chamber. If cranking is continued, the combustion process will become established and the engine will operate normally[15]. Therefore, the unaided cold starting ability of an engine can be improved by prolonged cranking and this may lower the minimum ambient temperature at which an unaided cold start can be achieved, by between 5°C and 10°C.

15.2.3.5 High-speed cranking

It has been shown in Sections 15.2.3.1 and 15.2.3.2 that an increase in the cranking speed of an engine will reduce both the leakage of air and the loss of heat from the air, during the compression stroke, thereby improving the value of the exponent of the compression process. It follows that the unaided cold starting ability of an engine can, in most cases, be improved significantly by employing a starting gear which will provide an increased cranking speed.

15.2.3.6 Inlet valve closing timing

The inlet valves close significantly after bottom dead centre (BDC) at the end of the induction stroke, so that, at normal running speeds the inertia effect of the column of air entering the cylinder will help to fill it in the very short time available and so increase the mass of air trapped in the cylinder. At the comparatively low cranking speed, the inertia effect of the air is negligible and the late closing of the inlet valves will result in a reduction of compression ratio to that represented by the closing point of the inlet valves, accompanied by a corresponding reduction in air temperature at the end of the compression stroke.

As a general guide, a timing later than 30° after BDC will result in some loss, and so in normally aspirated vehicle engines and the smaller hand-started engines a compromise may be necessary between the valve closing timing to give the best volumetric efficiency at full speed and that to give satisfactory starting. Some early engines overcame this difficulty by arranging a lever outside the rocker cover which rotated an eccentric rocker shaft to increase the tappet clearance of the inlet valves at starting. However most engines today avoid such extra complications and cost and achieve adequate cold starting by alternative means. Turbocharged vehicle engines can usually accept a compromise timing, perhaps 35°–40° after BDC. A long closing ramp on the inlet valve cam may also cause late valve closing and this can have a noticeable effect in small hand-started engines; especially directly-air-cooled engines where the increase in tappet clearance between cold and hot may be marked, so that the full length of the ramp may be in use when cold because a very small tappet clearance will then be present.

15.2.4 Engine cranking requirements

The starting gear must be capable of accelerating a completely 'cold' engine from rest to the required cranking speed and sustaining this cranking speed for a sufficient period to enable stable combustion conditions to become established, so that the engine will accelerate to its normal operating speed and develop power[16]. The demands on the starting gear will be greatest at the minimum ambient air temperature at which the engine will be required to operate and the relevant considerations are summarized in the following paragraphs.

15.2.4.1 'Break-away' torque

When an engine has been inoperative for a sufficient period for the whole unit to assume the prevailing ambient air temperature, the lubricating oil will have drained from the bearing surfaces and only boundary lubricating conditions will exist. The starting gear must therefore overcome the resulting 'break-away' torque but this demand will be of only short duration because relative movement between the bearing surfaces will rapidly establish a hydrodynamic regime which will be accompanied by a reduction in the frictional torque. However, as the rotational speed of the crankshaft increases, the frictional torque may become a significant consideration.

15.2.4.2 The initial compression

A major torque demand arises almost simultaneously with the beginning of rotation of the crankshaft of a multi-cylinder engine, because one cylinder will be at or near the start of its compression stroke. Consequently, the work of compression must be provided solely by the starting gear. Upon the completion of this compression stroke, the energy provided by the starting gear will be largely returned to the crankshaft during the expansion stroke of this cylinder and it will then contribute to the work of compression in another cylinder.

As the rotational speed of the crankshaft increases, the cyclic variation in torque due to the compression strokes will be absorbed increasingly by the inertia of the flywheel and the attached rotating and reciprocating masses of the engine. The surplus torque provided by the starting gear will then be dissipated by accelerating the crankshaft towards the equilibrium cranking speed.

In a single-cylinder engine the same principles apply. The work provided by the starting gear for the initial compression will be returned to the crankshaft during the ensuing expansion stroke. It will increase temporarily the kinetic energy in the shaft system and it will be available to provide the work required for the next compression stroke. The inertia of the rotating and reciprocating system, of which the flywheel is a significant part, is therefore of major importance because it affects the cyclic speed variation of the system during the compression stroke.

15.2.4.3 The sustained cranking speed

The sustained cranking speed is the equilibrium speed at which the torque provided by the starting gear equals the mean opposing torque due to the motoring losses in the engine. The required motoring torque increases with increase in crankshaft speed and its magnitude is related directly to the prevailing viscosity of the lubricating oil. The motoring torque therefore affects both the sustained cranking speed and the ability of the engine to become self-supporting and accelerate to its normal operating speed.

15.2.4.4 Importance of lubricating oil viscosity

It will be seen from the previous paragraph that the shear strength of the lubricating oil film at the minimum air temperature experienced will have a significant effect on the torque required from the starting gear to attain the specified cranking speed. The temperature/viscosity characteristic of the lubricating oil is therefore extremely important and it is noteworthy that the SAE have introduced a 0W grade of oil to provide an even thinner oil than the previous 5W grade for arctic climates: see Chapter 16. It must be observed however, that different lubricating oils which comply with a given specification may show wide variations in the torque required to overcome the shear strength of the oil film under low-temperature conditions and it is therefore highly desirable that this feature should be checked experi-mentally, for each recommended lubricant [15,17].

15.2.5 Methods of starting

The starting gear consists essentially of a means of applying to the crankshaft of the engine a torque of sufficient magnitude to rotate it at a speed above the specified minimum cranking speed, and the provision of a sufficient source of energy to ensure this. Some typical methods of starting diesel engines are described in the following paragraphs.

15.2.5.1 Manual starting

Manual starting is attractive because it is both inexpensive and simple. It does not require an external source of energy which must be stored and replenished periodically, but unfortunately it can only be applied to relatively small engines. Even then, unless adequate inertia can be provided by employing a large flywheel, which may be an embarrassment from considerations of installation and cost, it is necessary to incorporate a decom-pressing device in the design of the engine, and this should preferably be automatic in operation.

The starting sequence is first to accelerate the crankshaft to the required cranking speed with the decompressing device in operation, so that kinetic energy is stored in the flywheel and attached rotating and reciprocating masses of the engine, and then to release the decompressing device and use this kinetic energy, supplemented by continued cranking effort, to maintain the mean cranking speed at a satisfactorily high level.

Rope and pulley system

The smallest engines require a cranking speed of about 7 rev/s in an ambient air temperature of 15°C and a higher speed at lower ambient temperatures, unless a starting aid is employed. An economic and effective means of cranking the engine at these speeds is by a pulley attached to the crankshaft. To crank the engine, a rope is wound several times around the pulley and the operator then applies a sustained pull to it, which accelerates the crankshaft to the required cranking speed. A considerable length of rope is necessary to sustain the torque, and in some applications this cannot be accommodated in the space available. It is then necessary to adopt the more usual form of hand cranking[18].

Hand cranking system

The cranking speed required varies inversely with the swept volume of the cylinder. The smallest engines may employ a cranking speed of about 7 rev/s, while engines having a cylinder swept volume of the order of 0.70 litre may require a cranking speed of approximately 2 rev/s. In all cases, the total inertia of the rotating and reciprocating masses of the engine is most important and in some engines, the inertia of the flywheel employed is dictated largely by starting considerations.

The cranking speeds required by the smaller engines are higher than can be achieved directly with a conventional crank handle, even if this has a small throw. It is therefore necessary to interpose gearing between the crank handle and the crankshaft. On fourstroke engines, the 2:1 gear ratio between the crankshaft and the camshaft is utilized, so that the crankshaft rotates at twice the speed of the crank handle which is attached to the camshaft. On smaller engines, an even higher cranking speed may necessitate the use of speed increase ratios of 3:1 or 4:1.

15.2.5.2 Inertia starting system

Although not widely used at the present time. the inertia starting system should be mentioned because it is a variant of the manual starting system. It consists of a unit incorporating a small flywheel, which is connected to a conventional crank handle through speed increase gearing having a ratio of 100:1. The crank handle is rotated manually and accelerated to a speed of approximately 2 rev/s, so that the flywheel is rotating at a speed of some 200 rev/s. The kinetic energy in this flywheel is then transmitted through a clutch, to a pinion which engages with the flywheel of the engine, so that it provides the torque required to start the engine.

15.2.5.3 Electric starting system

The starting system consists essentially of a direct current starter motor, a battery to provide the electrical energy and a means of charging the battery. This type of system is widely used on high-speed and medium-speed diesel engines having cylinder diameters of up to approximately 200 mm[19].

The starter motor is mounted adjacent to the engine flywheel which has a gear ring attached to its periphery. The pinion of the starter motor engages with this by means of a Bendix-type drive and it is usually arranged so that it engages at a low rotational speed in order to minimize damage and wear of the gear teeth. In addition, provision is made to ensure that the pinion will remain in contact with the gear ring even if temporary torque reversals occur during the engine starting process. There is also provision to safeguard the starter motor from damage due to serious overspeeding, when the engine accelerates rapidly to its normal running speed.

Manufacturers of electric starting equipment include Robert Bosch GmbH in Germany, Delco Remy in the United States, Ducellier in France and Lucas CAV in Great Britain. The features described in the previous paragraph are usually incorporated in heavy-duty machines but there are variations in the method of providing them. The Lucas CAV type S 130 L starter motor is therefore taken as typical.

The starter motor

A sectional illustration of the Lucas CAV type S 130 L starter motor is reproduced in *Figure 15.29*, while *Figure 15.30* shows diagrammatically the operating sequence and the internal wiring.

In the At Rest Position, the pinion of the starter motor is clear of the gear ring on the engine flywheel (*Figure 15.30(a)*). When the Start button is pressed, the solenoid 15 in *Figure 15.29* is energized and its hollow plunger 3 moves axially, carrying four spring-loaded steel segments which bear against the shoulder of the pinion sleeve 2 and move the pinion 16 towards the gear ring on the engine flywheel.

At the same time, the plunger of the solenoid closes the first stage contacts 13 and current is supplied to the field coils and armature windings, through the resistance R (*Figure 15.30(b)*).

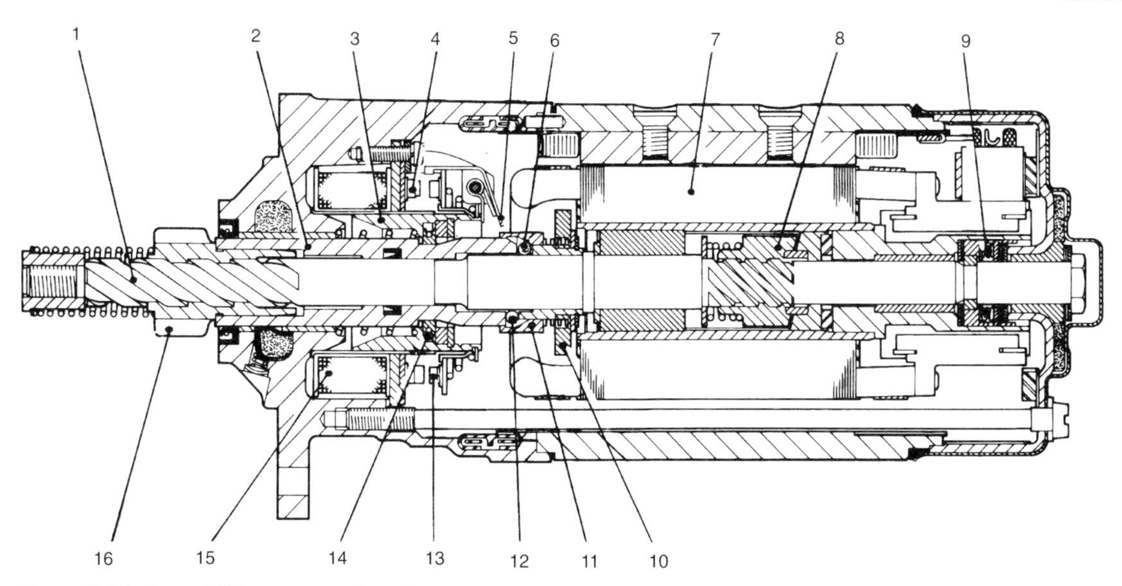

Figure 15.29 Lucas CAV starter motor Type S 130 L

1. Armature shaft helix
2. Pinion sleeve
3. Solenoid plunger
4. 2nd stage contacts
5. Switch trigger
6. Locking ball
7. Armature
8. Ratchet assembly
9. Recoil unit
10. Trip plate
11. Locking collar
12. Overspeed ball
13. 1st stage contacts
14. Spring loaded segments
15. Solenoid coil
16. Pinion

This limits the current so that the armature 7 in *Figure 15.29* rotates slowly and moves the pinion into mesh with the gear ring, by means of the armature shaft helix 1. Just before the pinion reaches the end of its axial travel the trip plate 10, on the pinion sleeve 2, operates the switch trigger 5. This closes the second stage contacts 4, which by-pass the resistance R, so that the full battery voltage is applied to the field coils and armature windings, and the starter motor then develops its full torque (*Figure 15.30*(*c*)).

When the pinion is in full engagement with the gear ring on the engine flywheel, it is locked in position by four steel balls 6 (*Figure 15.30*), which are located in holes in the pinion sleeve 2. These balls drop in front of the radiused shoulder on the shaft, when the pinion has moved axially to the limit of its travel, and the spring-loaded collar 11 slides over the balls to hold them in position (*Figure 15.30*(c)). Therefore the pinion is held permanently in mesh with the gear ring on the engine flywheel until either the Start button is released or the overspeed device operates. If the pinion is prevented from engaging with the gear ring due to tooth-to-tooth abutment, the action of the armature shaft helix 1 in *Figure 15.29*, will force the slowly rotating armature shaft against the Bellville washers in the armature recoil unit 9 and compress them. As soon as the Start button is released, the compressed Bellville washers will thrust the armature towards the pinion and so ensure that it takes up a different angular position for the next engagement with the gear ring.

When the engine starts, the pinion is accelerated rapidly by the engine flywheel, and the ratchet device 8 within the armature assembly then operates to allow the pinion to accelerate at a higher rate than the heavier armature, until the speeds are again equalized.

In the event of the flywheel driving the armature at an unacceptably high speed, the overspeed device operates. This consists of four steel balls 12 in *Figure 15.29*, housed in the pinion sleeve 2, which at a speed of approximately 200 rev/s, exert sufficient centrifugal force on the inside ramp of the locking collar 11, to move it axially against a spring; so that it releases the locking balls 6 and allows the pinion to move axially, out of engagement with the gear ring on the flywheel (*Figure 15.30*(*d*)).

The battery

In many application, the 12-volt or 24-volt engine starting system is integrated with other electrical systems: a typical example is a road vehicle in which the battery provides energy for lighting, for air conditioning, for the radiator fan and possibly the operation of ancillary control systems. A lead-acid battery is usually employed and while this is an economic and satisfactory source of electrical energy in temperate climatic conditions, special consideration is necessary if the engine is required to operate in very low ambient temperatures. This is particularly so because engine starting aids which require electrical energy may then increase the total demand on the battery[12].

The effect of low temperature on the performance of a lead-acid battery is shown in the graph of *Figure 15.31*. This is the performance of a fully charged battery when discharged at a rate of 340 amperes at various battery temperatures—a condition simulating a piece of diesel-driven equipment which has been exposed for an extended period to a low-temperature environment and must then be started. A problem with the lead-acid battery in an arctic environment is that as the state of charge of the battery falls, the density of the electrolyte falls and its freezing point rises. It will be seen from the graph of *Figure 15.32*, that in a low-temperature environment there is an increasing risk of the electrolyte freezing as the battery discharges. If this occurs, the battery output is reduced to zero and, of course, the freezing may damage the container and the plates.

Nickel-cadmium alkaline batteries are also used for engine starting but they are larger and significantly more costly than the equivalent lead-acid battery; consequently their use is generally confined to specialized applications. Their advantages are long life, very low self-discharging rate, robust construction,

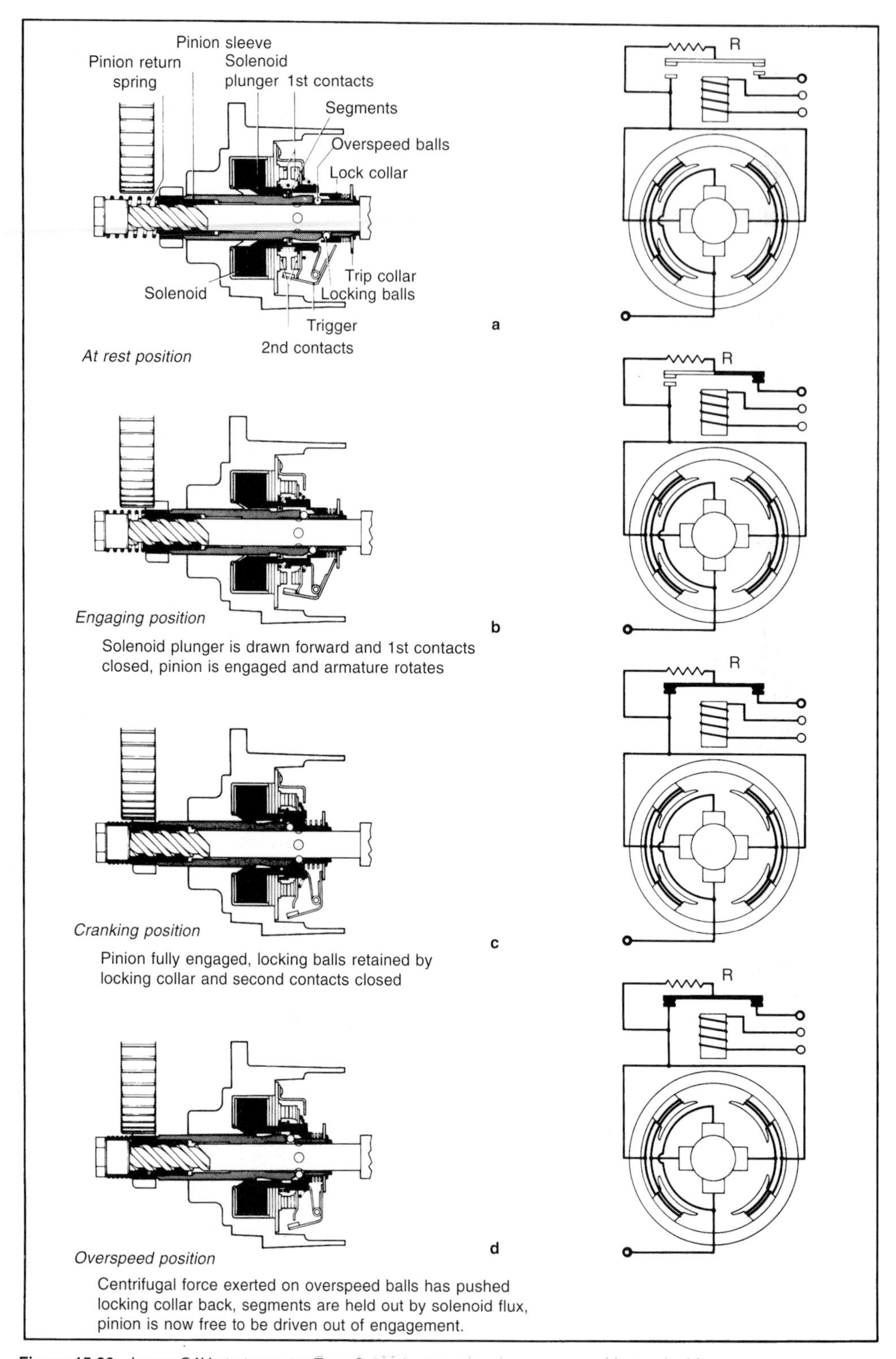

Figure 15.30 Lucas CAV starter motor Type S 130 L: operational sequence and internal wiring

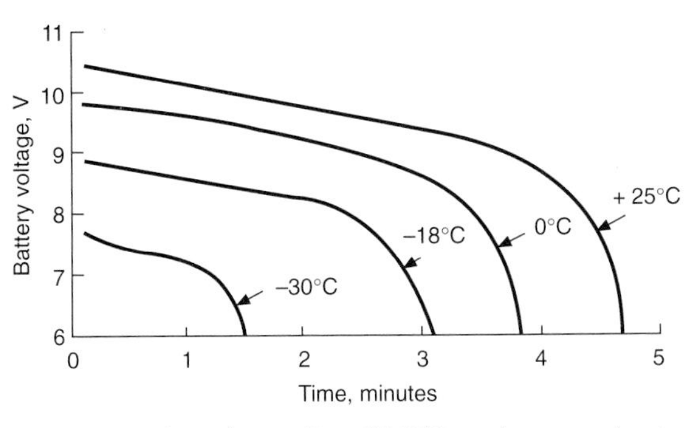

Figure 15.31 Lucas battery Type CP 13/11: performance at various temperatures

and an ability to stand for long periods at any stage of charge without deterioration. The nickel-cadmium alkaline battery is also temperature-sensitive and it has performance characteristics similar to those of the lead-acid battery; but it has the advantage that the density of the electrolyte does not vary with the state of charge. The normal specific gravity is 1.190, which corresponds to a freezing point of about −25°C, but for very low temperature applications a stronger electrolyte can be used which has a freezing point of about −42°C.

It will be seen from the above, that if equipment is required to operate in a very low temperature environment, there is a good case for thermally insulating the battery and providing some form of heating, possibly from the battery itself.

Battery charging

A battery which is in good condition and properly maintained will hold its charge for a considerable period but the energy taken from it to start the engine must be replaced to ensure future starts, and therefore a means of recharging the battery is provided. The most common means of recharging the battery is by the use of a proprietary direct current generator or an alternator with a rectifier, which is driven by vee belt from the diesel engine and supplies the battery through a voltage regulator and a cut-out.

On some small high-speed engines, a simple alternator is built into the engine itself by attaching a moulded stator winding to the end wall of the crankcase so that it is in close proximity to two permanent magnets which are attached to the flywheel.

If the diesel engine is the power source for a generating set, the battery can conveniently be charged through a transformer and rectifier, from the electrical energy produced; while in the case of stand-by, emergency or no-break sets, the battery can be charged by the primary power source which will probably be the public supply. Skid-mounted and trailer-mounted package generating sets are sometimes equipped with a small independent battery charging generator powered by a diesel or gasoline engine, so that when it is not possible to run the main engine, the battery can be charged.

15.2.5.4 Pneumatic (air motor) starting system

The starting system consists basically of an air-operated motor, an air reservoir in which to store the compressed air and a means of compressing air to recharge the reservoir. This type of system is widely used to start high-speed and medium-speed diesel engines which require more power than can conveniently be provided by electric starting equipment. As in the case of the electric starting system, the compressed-air-operated starter motor is mounted adjacent to the engine flywheel and the pinion of the starter motor engages with a gear ring attached to its periphery.

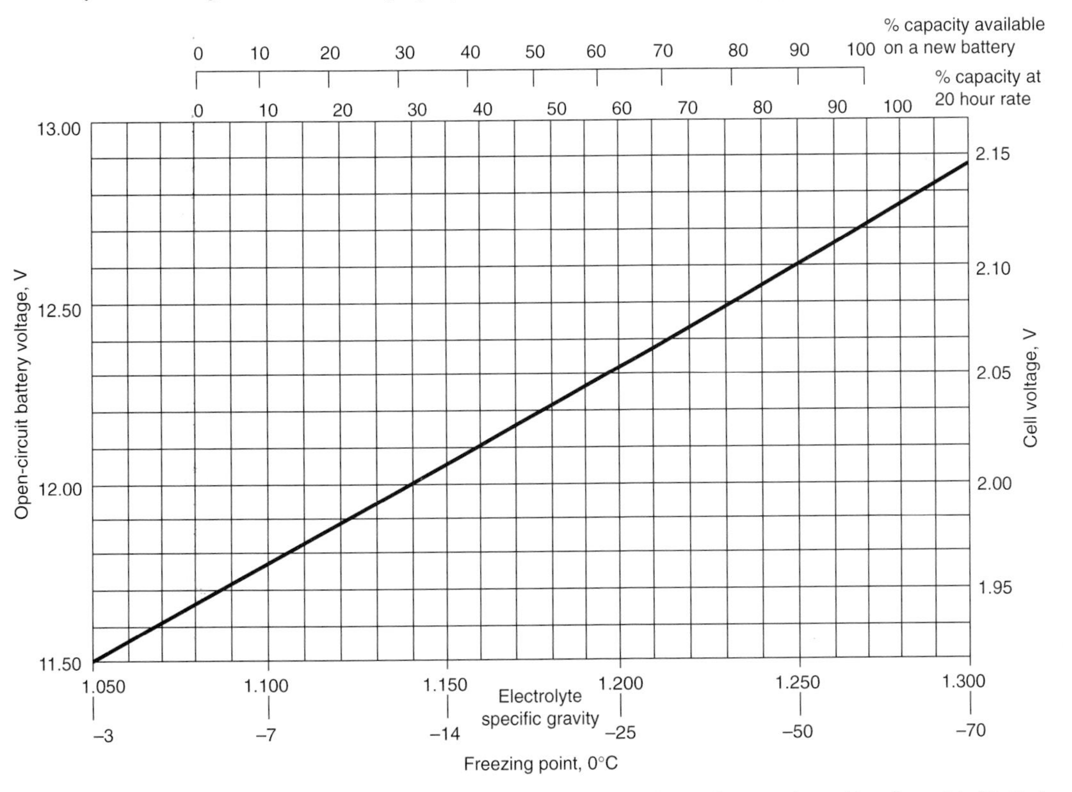

Figure 15.32 Characteristics of lead-acid battery. Open circuit voltage specific gravity, capacity and freezing point of batteries using normal temperature electrolytes. Curve relates to 25°C

The pinion is usually driven through a Bendix coupling and speed reduction gearing by a positive displacement rotary air motor but some designs employ a simple turbine.

Figure 15.33 shows a sectional arrangement of a vane-type starter motor produced by Pow. R. Quik Limited, in the United States; while *Figure 15.34* shows a gear-type starter motor which is manufactured by Tecnic Gali sa, in Spain. The characteristics of the A45 starter motor are shown in *Figure 15.35*.

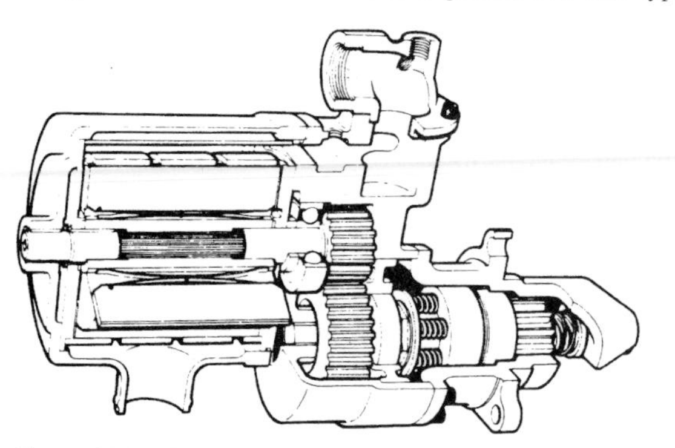

Figure 15.33 Pow. R. Quik air starter motor

The starter motor

Like its electrical counterpart, the air-operated starter motor must incorporate a means of engaging the pinion with the stationary gear ring on the engine flywheel, at a low speed and torque, before the full cranking torque is applied. Similarly, provision must be made to safeguard the starter against damage due to overspeeding. The Tecnic Gali starter has therefore been taken as an example and its method of operation is described in the following paragraphs.

Referring to *Figure 15.34(a)* the compressed air supply enters the starter motor at 1 and is conveyed through the drilling 2 and the pipe 3 to the valve 4 in the Start button 5. Depressing the Start button opens the valve 4 and closes the vent to atmosphere 7. The compressed air then flows through the pipe 8 and depresses the piston 9, which is vented to atmosphere on its lower side by

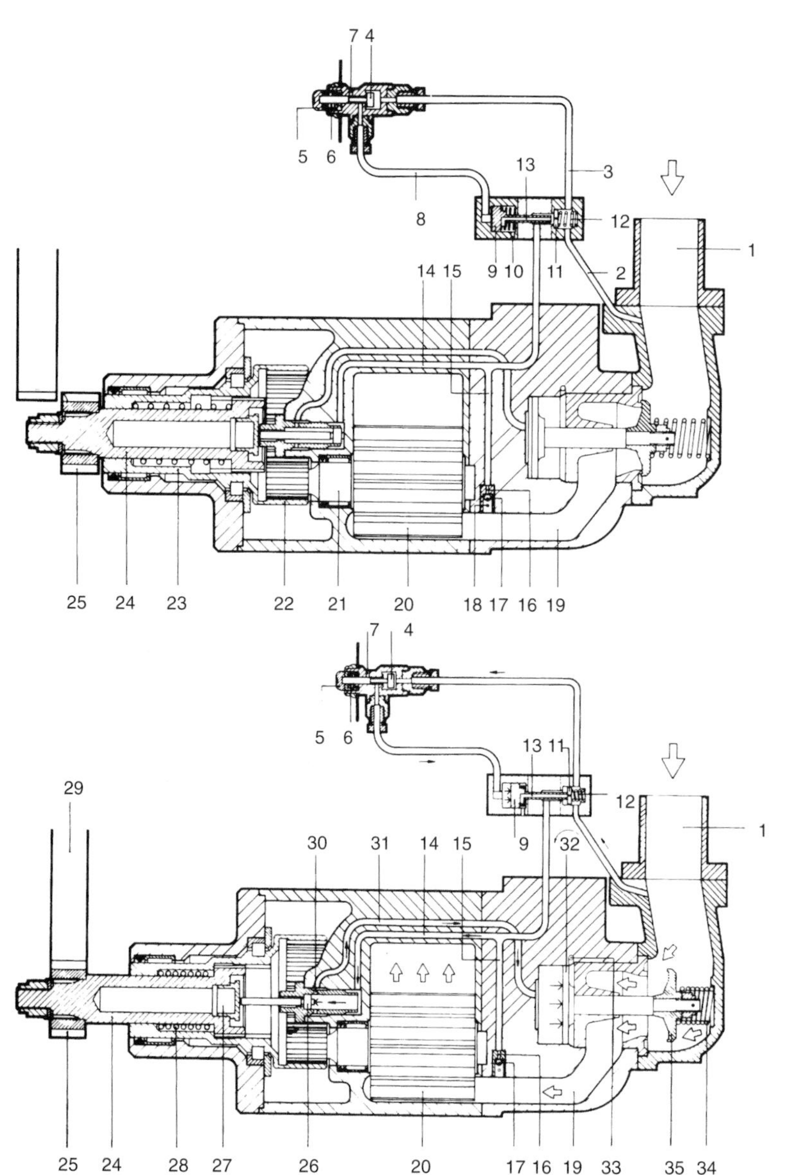

Figure 15.34(a) Tecnic Gali air starter motor

Figure 15.34(b) Showing operation of the air starter motor

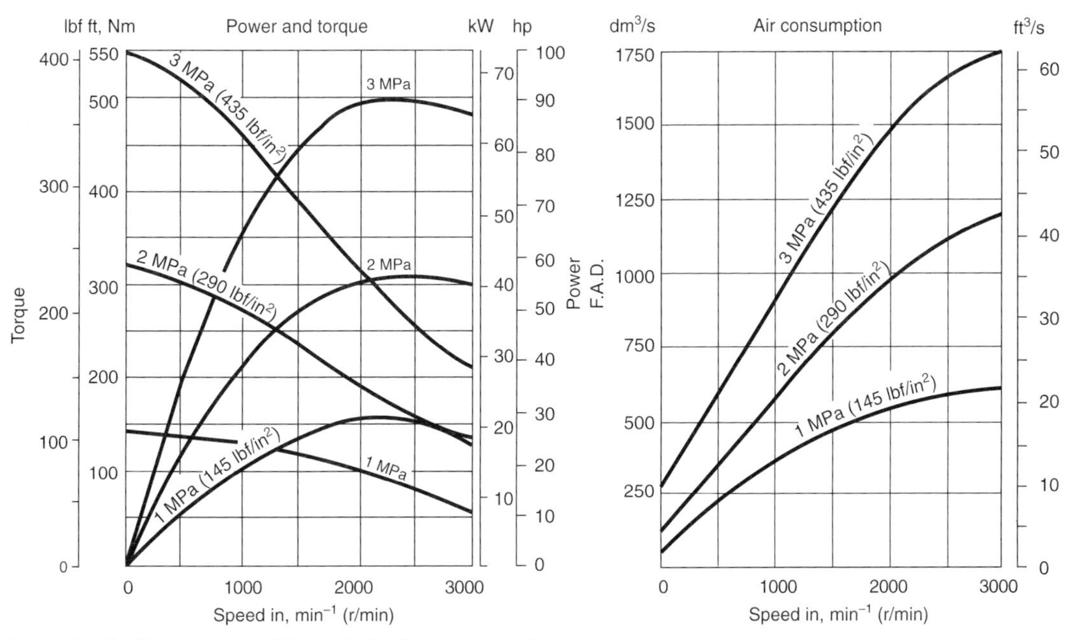

Figure 15.35 Characteristics of Tecnic Gali A45 air starter motor

TECHNICAL DATA

Max working pressure	3 MPa (435 lbf/in^2)
Maz power	66 kW (90 hp)
Speed at max power	2300 min^{-1} (r/min)
Max torque	540 Nm (398 lbf ft)
F.A.D.	250–1750 dm^3/s (8.83–61.8 ft^3/s)
Minimum diameter of air feed pipework	45 mm (1$^3/_4$ in)
Net weight	255 N (57.3 lbf)

the drilling 10. In doing so, it opens the valve 11 and closes the vent to atmosphere 13; thereby permitting compressed air from the duct 1 to pass through the drilling 2 and into the pipe 14, which communicates with the pipe 15. This compressed air passes through the orifice 16 and the non-return valve 17, to the duct 19. It provides a restricted flow of air to the rotary gear-type motor 20, which rotates slowly, driving the pinion 25 through the speed reduction gearing 21 and 22 and the helical splines formed on the inside of the sleeve 23 and on the shaft 24.

At the same time air pressure in the pipe 14 (see *Figure 15.34(b)*) acts on the piston 26 and exerts a thrust through the cap 27 which compresses the spring 28 and moves the slowly rotating pinion towards the gear ring on the engine flywheel. As the pinion 25 engages fully with the gear ring 29, the piston 26 opens the orifice 30 and permits compressed air to flow into pipe 31 and so apply pressure to the piston 32, which is vented to atmosphere on its opposite side, through the drilling 33. The area of the piston 32 is greater than that of the valve 35 and therefore, the piston opens the valve against the air pressure in the duct 1 and the force of the spring 34, and so connects the main air supply to duct 19. Consequently, the rotary gear-type motor 20 accelerates rapidly under the full torque available and rotates the engine flywheel through the pinion 25 and the gear ring 29; so that the engine starts and accelerates to its normal running speed.

Releasing the Start button closes the valve 4 and vents the pipe 8 to atmosphere through the orifice 7. The spring 12 then closes the valve 11 and in so doing, vents the pipe 14 to atmosphere through the drilling 13. The air pressure in the duct 19 then closes the non-return valve 17 and isolates the pipes 14 and 15 which have already been vented to atmosphere. The air pressure acting on the piston 33 will also be vented to atmosphere through pipes 31 and 14, so that the valve 35 will close under the action of the spring 34. The rotary motor 20, which is running at high speed, will then create a considerable depression in the duct 19 and this will reopen the non-return valve 17. The depression will cause a rapid deceleration of the motor and this, aided by the force from the spring 28, by the depression acting on the piston 26 and by the action of the helical splines, will withdraw the pinion 25 from engagement with the gear ring 29.

If the Start button in not released when the engine starts, the pinion 25 will be driven by the gear ring 29, and, owing to this torque reversal, the pinion will move axially under the action of the helical splines in the sleeve 23 and on the shaft 24, and the force exerted by the spring 28, out of engagement with the gear ring on the engine flywheel. In doing so the piston 26 is moved against the air pressure in the pipe 14 so that it uncovers the orifices 30 and 36 which vent the pipe 31 to atmoshphere. This releases the air pressure on the piston 32, so that the spring 34 closes the valve 35, which cuts off the main supply of compressed air to the rotary gear-type motor 20; although it will continue to rotate slowly as a result of restricted air supply through the orifice 16, until the Start button is released. When the starter motor is used on a remotely controlled or automatic starting installation, an overspeed device is employed which interrupts the air supply to the Start button if the starter motor overspeeds.

The air reservoir

The operating pressure will depend on the type of air starter motor employed. The air pressures range from approximately 30 bar for vane-type or gear-type air motors, to about 3 bar for turbine-type motors. In many applications, compressed air is used for a variety of purposes and the air starting system is often integrated with other services such as, steering, braking and gear changing on heavy vehicles, and control and monitoring systems on shipboard or in stationary plant.

The air reservoir is a pressure vessel and it may therefore be required to comply with statutory regulations regarding its design, its installation and its regular pressure testing and certification. It should be fitted with a safety valve, an isolating valve, a pressure gauge and an efficient drain valve. In some installations more than one air reservoir is employed, and one can then be isolated from the main system and held in reserve to provide an immediate source of compressed air in an emergency. An alternative is to use a bottle of nitrogen as a reserve supply.

The compressed-air starting system is unaffected by variations in climatic conditions. It will operate satisfactorily in the lowest ambient air temperatures which occur, providing it is designed to ensure that water from the condensation of humidity in the atmosphere does not collect and freeze in the pipework. To minimize this risk, the runs of all pipework should be arranged to drain naturally into the air reservoir or bottle; and where that is not possible drain traps should be provided at the lowest points in the pipe runs, and these should preferably be of the automatic self-emptying type.

The air compressor

Particularly with high-speed engines, which are widely used for vehicles, marine and portable applications, the air compressor is air cooled and it is often driven mechanically from the engine. It is controlled automatically by means of an unloading valve so that the air pressure in the reservoir or air bottle is maintained at the designed value. Depending on the air pressure required, the machine is either single-stage or, for the higher pressures, two stage. This may be of the single-cylinder stepped-piston type or arranged with separate cylinders for the low-pressure and high-pressure stages, having a simple copper coil intercooler between them.

In stationary and marine installations and in some package generating sets, an independent air compressor may be employed. This may be a single compressor which can be driven by an electric motor or a small internal combustion engine, or two independent compressors. In some small installations, a manually operated compressor is provided for emergency use.

15.2.5.5 Pneumatic (air-in-cylinder) starting system

The timed admission of compressed air into some or all of the cylinders is the traditional method of starting medium- and low-speed diesel engines. The system can also be applied to smaller bore high-speed engines but it tends to be more costly than the motor-type starting systems.

The air-in-cylinder system admits compressed air to a cylinder when the piston is a few degrees past top dead centre at the end of the compression stroke, and cuts off the air supply at about half-stroke, in the case of four-stroke engines. Thereafter, the air expands and continues to do work on the piston until the exhaust valves open and vent the cylinder to atmosphere in the usual manner. In the case of two-stroke engines, the starting air supply must be cut off before the piston uncovers the ports in the cylinder liner or the exhaust valves open.

Four-stroke engines having six cylinders, and two-stroke engines having four cylinders, will start from rest with the flywheel in any angular position when air starting is provided on all cylinders. With fewer cylinders, there will be blind spots from which the crankshaft will not commence to rotate until the flywheel has been turned by means of the barring gear, to a position at which air will be admitted to a cylinder having its piston at the commencement of the combustion stroke.

Air-in-cylinder systems usually operate at pressures between 20 bar and 30 bar and the requirements in regard to air reservoirs and air compressors, are similar to those described in later sections. The systems employed on the engines themselves are often designed and produced by the engine manufacturers but they fall into two broad categories: the simple directly controlled system and the more complicated indirectly controlled system, which has merits for larger engines. Therefore, both systems are described.

Directly controlled system

Compressed air is admitted from the air reservoir, through a stop valve, to a starting air control valve which is usually mounted on the engine itself. This may be operated manually or automatically as part of a programmed starting sequence. It admits air to a starting air distributor and this directs the air to an appropriate cylinder, in which the piston is at the beginning of the combustion stroke. The distributor is driven at half the crankshaft speed on a four-stroke engine, usually from the camshaft, and at the crankshaft speed on a two-stroke engine.

A disc-type air distributor is shown in *Figure 15.36*. The shaft 1 has a disc 2 attached to it which has a port 3 cut through it. When the shaft rotates, this port uncovers in turn, drillings such as 4, which are spaced on a common pitch circle diameter and each communicates with a pipe union screwed into the body 5. The camshaft 6 drives the shaft 1 through a peg 7, so that the shaft of the distributor can move axially in relation to it. Until air pressure is applied, the shaft and disc assembly are held against the stop 8 formed in the cover 9, by the spring 10; so that the disc is not in contact with the machined face 11. This is the normal engine running position.

When the engine is to be started compressed air from the starting air control valve enters through the port 12. The air pressure causes the shaft and disc assembly to move axially, compressing the spring 10 until the disc 2 contacts and seals against the machined face 11. The starting air then flows through the port 3 in the disc, to only one drilling such as 4 and its associated pipe union. each pipe union is connected, in the correct sequence, to a starting valve situated in a cylinder head; the correct sequence is dictated by the firing order of the engine. The starting valve shown in *Figure 15.37* is a simple non-return

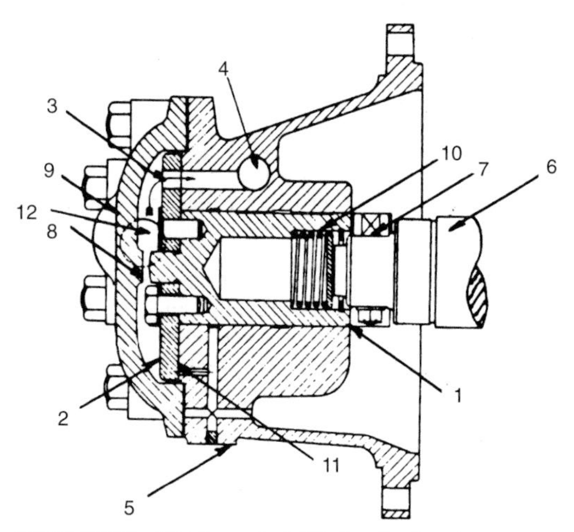

Figure 15.36 Disc-type air distributor-direct system

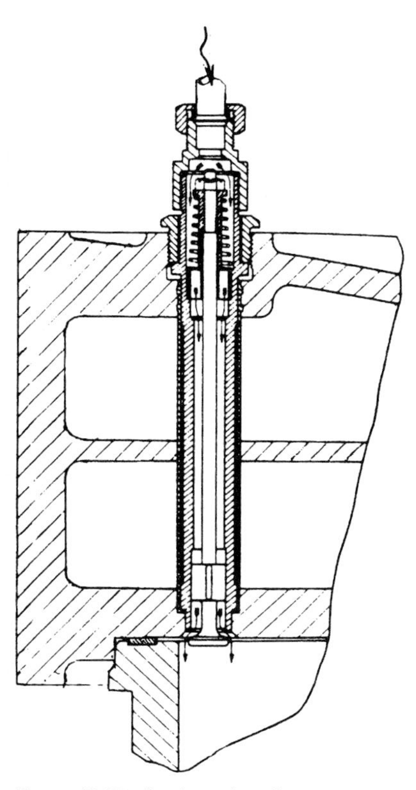

Figure 15.37 Starting valve: direct system

valve which communicates with the combustion chamber. It is opened by the pressure of the starting air and it closes automatically when the pressure in the cylinder exceeds the pressure in the pipe from the distributor—a situation which exists every time the distributor cuts off the air supply or when combustion occurs in the cylinder.

When the engine starts and begins to accelerate, the starting air control valve is closed cutting off the air supply to the distributor. The shaft and disc (1 and 2 in *Figure 15.36*) then move axially under the action of the spring 10 until the end of the shaft contacts the stop 8, so that the disc is not in contact with the machined face 11.

Indirectly controlled system

Compressed air is delivered from the air reservoir through a stop valve to the starting air control valve, as in the case of the directly controlled system described in the previous section, but here the similarity ends.

When the starting air control valve is opened, compressed air is conveyed to the starting valves situated in the cylinder head, through a large-diameter bus pipe and to a small air distributor which actuates the starting valves in the sequence dictated by the firing order of the engine, by means of a pneumatic servo system. The arrangement of a starting valve in the cylinder head is shown in *Figure 15.38*. The starting air supply from the large-diameter bus pipe enters through a non-return valve and the drilling 21, but the valve is of the balanced type. The force due to air pressure acting on the head of the valve 22 is balanced by the air pressure acting on the large-diameter land 23, which is formed on the valve stem. Consequently the valve does not open and it is held firmly on its seat by the spring 24, until air pressure is applied to the servo piston 25 through the distributor.

The air distributor is driven at half crankshaft speed in a four-stroke engine and at crankshaft speed on a two-stroke engine.

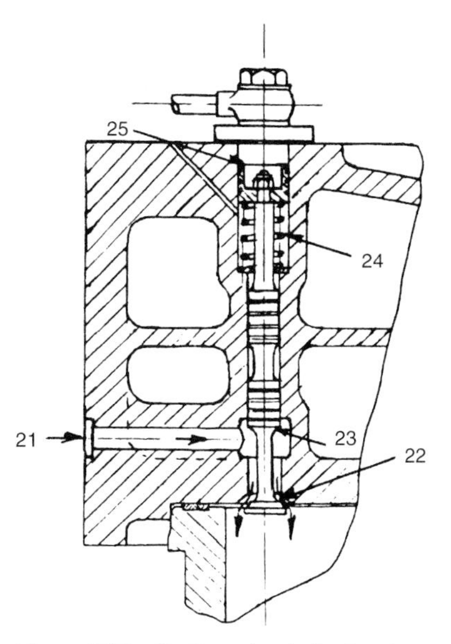

Figure 15.38 Starting valve: indirect system

The distributor shown in *Figure 15.39* is similar in general construction to that shown in *Figure 15.36*, but it is smaller because it passes only the small rate of air flow required to operate the servo pistons of the starting valves. To facilitate a comparison of the essential differences between the starting air distributor employed on the directly controlled system with the distributor shown in *Figure 15.39*, the numerical identities of the components have been made the same in both illustrations.

The compressed air enters the distributor through the port 12 (*Figure 15.39*), and the air pressure causes the shaft and disc assembly (1 & 2) to move axially, compressing the spring 10; until the disc 2 contacts and seals against the machined face 11. The disc 2 has one port 3 cut through it and apart from a sealing land at either end of this; a continuous circumferential groove 13 is machined in the face of the disc which seals with the machined face 11. The small-bore drillings such as 4 are spaced on a common-pitch circle diameter and each communicates with a pipe union which is screwed into the body 5.

Each outlet union on the distributor is connected, in correct sequence, to a starting valve through a small-bore pipe. When the port 3 in the disc 2 of the distributor uncovers a drilling such

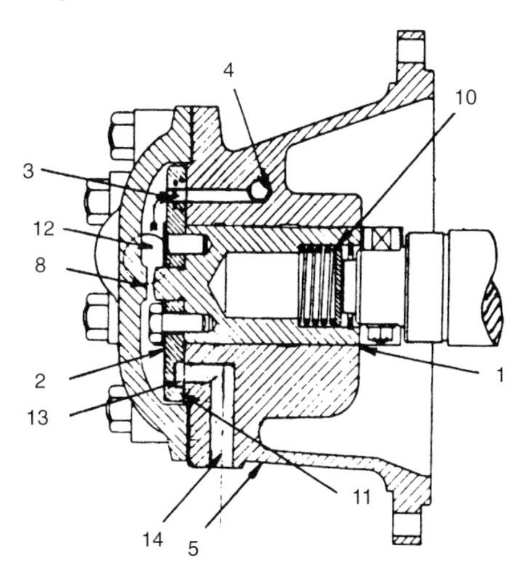

Figure 15.39 Disc-type air distributor-indirect system

as 4, compressed air is conveyed to the piston 25 of the starting valve (*Figure 15.38*). This opens the valve 22 and admits starting air from the large-diameter bus pipe into the engine cylinder, and so rotates the crankshaft. When the trailing end of the port 3 in the disc 2 (*Figure 15.39*) passes over a drilling such as 4, it cuts off the air supply to the related servo piston. Further rotation of the distributor disc 2 releases this air to atmoshpere through the circumferential groove 13 and the drilling 14 in the body of the distributor. This allows the valve 22 (*Figure 15.38*) to close under the action of the spring 24 and so cut off the supply of starting air to the cylinder. If combustion occurs in the cylinder while the starting valve is open, the flow of high-pressure combustion gases into the large-diameter bus pipe is prevented by a non-return valve, which is often accommodated in an elbow connecting the bus pipe to the cylinder head.

When combustion becomes established and the engine begins to increase in speed, the starting air control valve is closed and this cuts off the supply of compressed air to both the large-diameter bus pipe and the distributor. The shaft and disc assembly (1 & 2 in *Figure 15.39*) then moves axially under the action of the spring 10 until it is constrained by the stop 8, which is the normal engine running position.

Proprietary air starting systems

The manufacturers of medium- and low-speed engines often design and produce the starting equipment as part of the engine development programme but there are some companies which specialize in this field. The combined air starting and cylinder drainage system produced by Nova Werke AG, in Switzerland, is an interesting example and it is described briefly in the following paragraphs.

The damage which can result from attempting to start an engine when an incompressible fluid such as water has collected in a cylinder can be extremely dangerous and costly and it may well put the engine out of service for a considerable period. Advances in design techniques have reduced the risk of such an occurrence but with the increasing use of unattended diesel plant, which is often remotely controlled, the provision of a safeguard merits consideration.

Figure 15.40 shows a schematic layout of the NovaSwiss indirectly controlled air starting and drainage system: the related electrical control circuit is also shown. Depressing the button d1 on the solenoid valve 2 admits low-pressure air to the operating piston of the cylinder drainage valve 9 and to the distributor 4, which controls the cylinder drainage valves 10, so that they open only during the compression stroke of the cylinder to which they are fitted. The solenoid valve 6 then opens to provide a restricted high-presure air supply through the throttle valve 8, to the starting valves 12. These are controlled by the distributor 4, to open only during the combustion stroke of the cylinder to which they are fitted; and so the engine crankshaft rotates slowly throughout the cylinder drainage phase. At the conclusion of this, the control pistons of the drainage valves are vented to atmosphere through the distributor 4 and they are then are then inoperative. Depressing the button d2 opens the solenoid valves 6 and 7, which provide an unrestricted supply of high-pressure air to the starting valves 12, through the distributor 4; so that the crankshaft rotates at an increased speed. When combustion occurs and the engine accelerates, the speed-sensing switch (e) breaks the electrical circuit. The solenoid valves then close and the operating pistons of the starting valves are vented to atmosphere through the distributor 4. These cease to operate and so the engine continues to run normally.

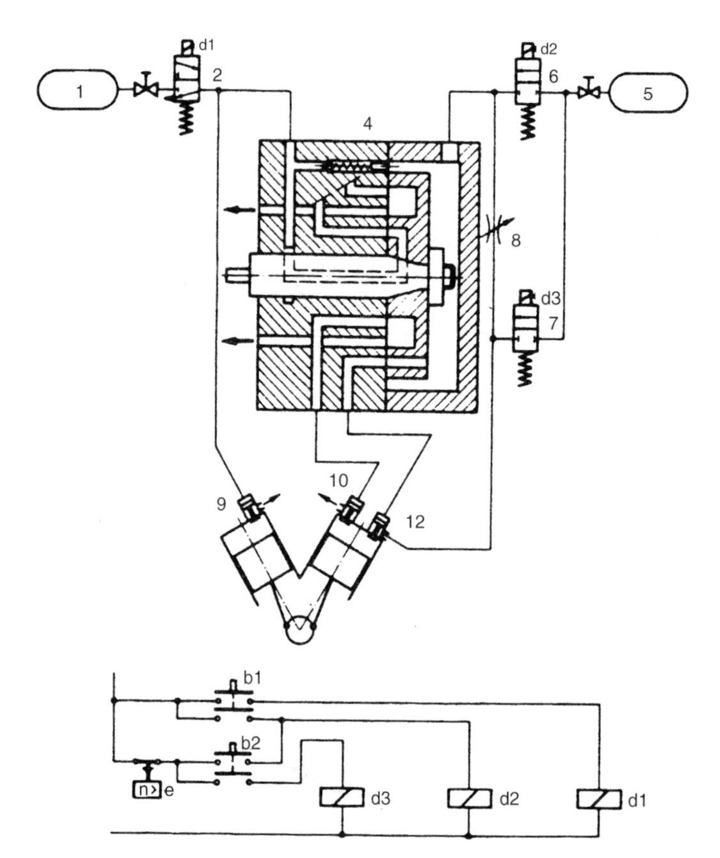

Figure 15.40 NovaSwiss indirectly controlled air starting and drainage system

1. Low-pressure air reservoir
2. Solenoid valve–drainage air control (d1)
4. Air distributor
5. High-pressure air reservoir
6. Solenoid valve-restricted starting air control (d2)
7. Solenoid valve-main starting air control (d3)
8. Throttle valve-restricted starting air supply
9. Drainage valve
10. Drainage valve
12. Starting valve

b1. Drainage solenoid control button
b2. Starting solenoid control button

15.2.5.6 Hydraulic starting systems

Hydraulic starting systems are unaffected by extremes in climatic conditions and they are independent of all ancillary services because the energy required can, if necessary, be produced manually. They are therefore well suited to the operating environment of engines employed for contractors' plant, agricultural equipment, portable generating sets and marine service, including the propulsion of workboats and lifeboats.

Two well-established desings are the Hydrotor Starter which is produced in the United States by American Bosch, and the Handraulic Starter which is manufactured by Lucas Bryce of Great Britain. The Hydrotor Starter employs a swash-plate-type hydraulic motor which rotates the flywheel of the diesel engine through gearing and a Bendix-type drive, in the conventional manner; while the Handraulic Starter applies an impulse to the engine crankshaft by hydraulic rams, and this system is described below.

The power operated starting systems described so far, are designed to accelerate the reciprocating and rotating masses of the engine and any permanently attached driven machine; from rest to a specified cranking speed, which must be above the minimum starting speed of the engine; the horizontal line S-S on *Figure 15.41*. At this cranking speed, the system will be in a state of equilibrium and the energy input to the starter will be equal to the total motoring resistance of the engine and any permanently connected driven machine. This is represented by the curve 0-E (dashed line) on *Figure 15.41*. The starting system will maintain this cranking speed until the engine starts or the supply of energy is either cut off or becomes exhausted.

In contrast the Handraulic Starter applies a very high torque to the crankshaft for approximately one revolution, during which the reciprocating and rotating masses of the engine and any permanently attached driven machine are accelerated to a speed well above the minimum starting speed of the engine, curve 0-M-A in *Figure 15.41*. The torque ceases at point A, but the momentum in the system continues to rotate the crankshaft, along the curve A-N-R. The effective cranking period is represented by M-N and the shaded area M-A-N in *Figure 15.41* shows that, during the whole of this interval, the cranking speed is above the specified minimum starting speed of the engine.

The Lucas Bryce Handraulic starting system is shown schematically in *Figure 15.42*. It is operated by energy stored in a piston-type hydraulic accumulator C, which contains nitrogen in the sealed volume above the piston and hydraulic fluid below it. When the hand pump B is operated, hydraulic fluid is transferred from the supply tank A to the lower portion of the accumulator C. This raises the piston and compresses the nitrogen. The pressure is limited to 340 bar by a relief valve and it is shown on the pressure indicator E. The system is controlled by a two-stage relay valve attached to the base of the accumulator. This can be operated by a lever, a pushbutton or a solenoid.

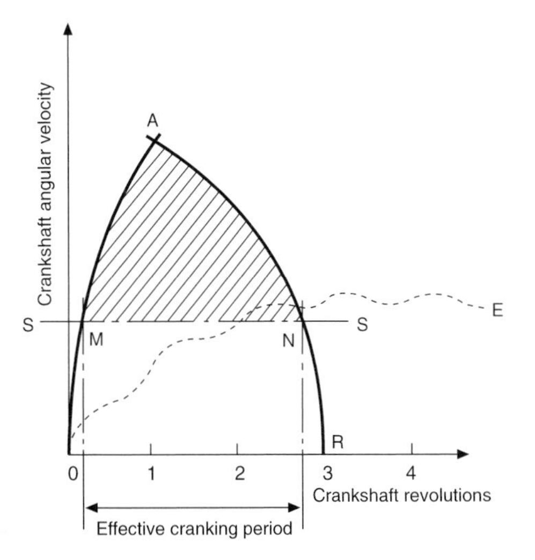

Figure 15.41 Comparison of cranking characteristics of Lucas Bryce Handraulic Starter with typical electric starter

The two-stage relay valve D is connected to the starter unit G, which is shown partly sectioned in *Figure 15.42*. This contains two horizontally opposed hydraulic rams, each attached to a toothed rack. These racks engage with a helically toothed pinion, having one part of a serrated, face-type coupling formed integrally with it: The mating part of this coupling is attached to the end of the engine crankshaft. Initial movement of the operating lever F on the relay valve D permits slow movement of the rams in the starter unit G. These apply a pure torque to the pinion H and owing to the helical teeth, cause it to move axially and engage with the crankshaft. Further movement of the control lever F admits the full hydraulic pressure to the rams of the starter unit G, which applies an impulse to the crankshaft, causing it to accelerate very rapidly to an angular velocity which is significantly above the minimum cranking speed required, thereby achieving the advantage of high-speed cranking (see Section 15.2.3). Release of the operating lever F on the relay valve D isolates the starter unit from the hydraulic accumulator and vents the hydraulic rams to the feed tank A, so that they return to their original positions. This starting cycle can be repeated as often as required and the system can be arranged to provide several starts without the need to recharge the hydraulic accumulator. This can be accomplished by means of the hand pump B or automatically, by employing the auto-charging pump K, which is usually driven from the engine.

15.2.6 Starting aids

As a generalization, the smallest high-speed diesel engines will start without the assistance of a starting aid, in ambient air temperatures down to about 15°C, while commercial-vehicle type engines will start in air temperatures down to about 0°C, but these temperatures will be influenced significantly by the general condition of the engine and the cranking speed attained. High-speed diesel engines having cylinders of larger swept volume will usually start without the assistance of a starting aid, at somewhat lower ambient air temperatures. In arctic areas, diesel engines must be capable of starting and operating satisfactorily in ambient air temperatures below –30°C. Appropriate starting aids have therefore been developed and these can be grouped broadly within the following categories[20].

15.2.6.1 Oil priming

It is a common practice with small hand-started diesel engines, to inject a measured quantity of lubricating oil or diesel fuel into the air manifold before commencing to crank the engine, so that it will be carried into the cylinder when the inlet valves open. This oil both improves the sealing of the piston rings and reduces the effective clearance volume in the cylinder; thereby temporarily increasing the compression ratio and these effects together, can lower the minimum starting air temperature by between 5°C and 10°C[18].

15.2.6.2 Excess fuel

In Section 15.2.3.4 reference is made to the fact that, by prolonged cranking combined with the injection of the full load fuel quantity per cycle, the minimum unaided cold starting temperature of an engine can be lowered by between 5°C and 10°C. This is not

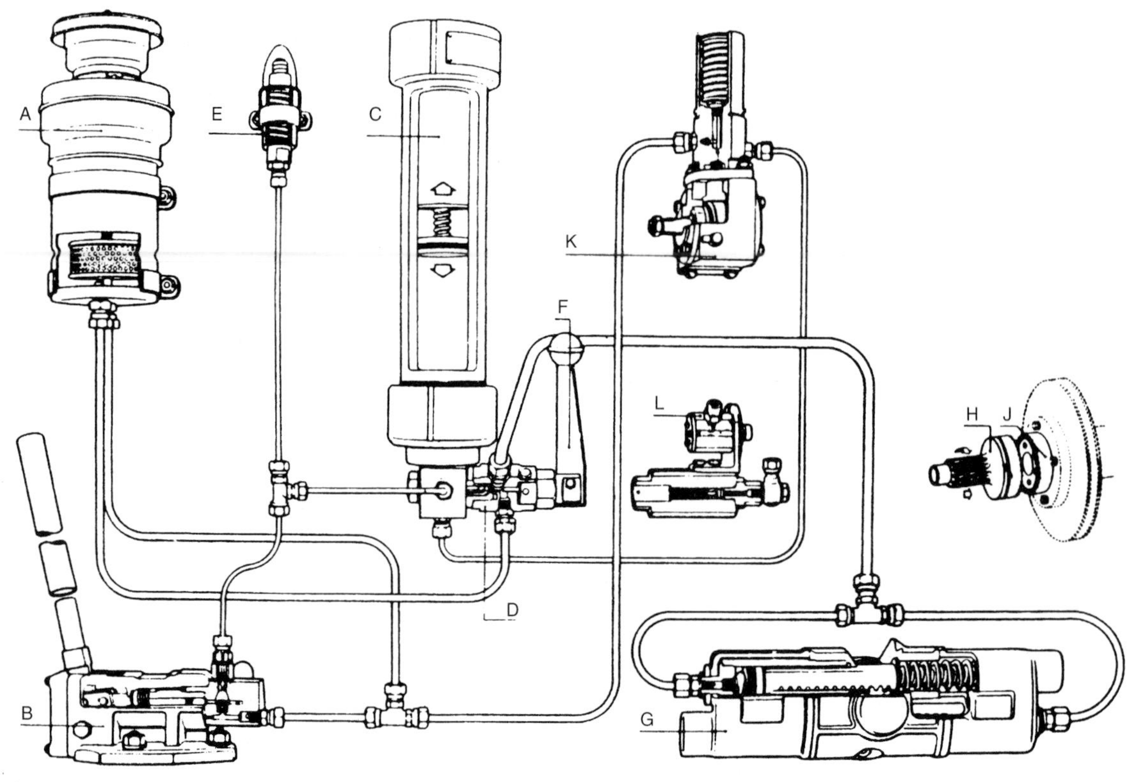

Figure 15.42 Lucas Bryce Handraulic starting system

A. Feed tank
B. Hand pump
C. Hydraulic accumulator
D. Relay valve assembly
E. Pressure indicator
F. Operating lever
G. Starter unit
H. Pinion
J. Engine clog on crankshaft pulley adaptor
K. Auto-charging pumps
L. Push-button valve and actuator

strictly a starting aid but it comes near to being one, because by over-fuelling to the extent of about 50 per cent, many direct injection engines can be started in air temperatures down to –15°C.

The excess fuel can be provided by means of a manual override on the maximum fuel stop. This practice has the disadvantage that the fuel which collects in the cylinders during cranking and the products of incomplete combustion during the initial firing strokes produce exhaust smoke which, in some applications, may not be acceptable on grounds of environmental pollution[15].

15.2.6.3 Heater plugs

The heater plug is an electrically heated plug which is somewhat similar to the familiar sparking plug used on gasolene engines. It commonly has an energy consumption of 50–60 W and when supplied with current from the starter battery, the heating element attains a temperature of about 1000°C in a period of about 30 seconds. The heater plug is fitted in the combustion chamber, where it provides a hot-spot, and consequently it can start combustion of the fuel only in its immediate vicinity or actually in contact with it. The heater plug is therefore most effective as a starting aid in engines which induce rapid air movement in the combustion chamber, at the relatively low speeds of cranking. This is a characteristic of the indirect-injection or pre-chamber combustion system and of some direct-injection systems used on small engines. In these engines, the use of heater plugs can reduce by some 25°C, the minimum ambient air temperature in which the engine will start[21].

Heater plugs are manufactured by Robert Bosch in Germany, Delco Remy in the United States and the Lucas organization in Great Britain. The usual starting procedure is first to energize the heater plugs, which are connected in parallel, for a period of about 30 seconds before beginning to crank the engine. Under the most arduous starting conditions, the energy available from the battery may be insufficient both to maintain the heater plugs at their full operating temperature and to provide sufficient current to enable the starter to crank the engine at the designed speed. An effective practice in such cases is first to apply all the electrical energy to the heater plugs and then transfer the total battery capacity to the starter motor, so that the maximum available cranking speed is attained.

The interval of 30 seconds required for the heater plugs to attain their full temperature is unacceptable in some applications and therefore, rapid warm-up systems have been developed. These use conventional sheathed element heater plugs which are temporarily overloaded electrically, so that they warm up rapidly under the control of a thermostat which reduces the rate of energy input when the full temperature is attained. By this means, the warm-up period is reduced to between 5 and 10 seconds without damage to the heater plugs. The Lucas CAV 'Micronova' Fast Start Aid System is a further development of this philosophy. It empolys an advanced design of heater plug and an electronic controller which virtually eliminate the warm-up period before cranking is commenced. In addition, the heater plug continues to be energized for a short period after combustion has commenced, in order to minimize the possibility of the engine stalling.

15.2.6.4 Manifold air heaters

A manifold air heater raises the temperature of the combustion air during its passage through the inlet manifold of the engine by means of an electrically heated element or a combustion burner using a liquid or gaseous fuel. Thermally the heater is inherently very inefficient because a large proportion of the heat added to the combustion air is lost by radiation and convection from the surfaces of the inlet manifold and to the cold surfaces of the inlet ports, valves, pistons, combustion chambers and cylinder walls. Even so, providing the required amount of energy can be made available, the manifold air heater is an effective starting aid which can be applied to both indirect-injection and direct-injection engines.

Electrically heated manifold air heater

The electrically heated manifold air heater has the merit that it does not consume any of the oxygen contained in the engine combustion air but its field of application is restricted by the high-energy requirement. The electrical energy required is of the order of 400 W/l of engine cylinder swept volume and this demand occurs concurrently with a high demand from the starter. It is therefore essential that batteries of very generous capacity should be provided to ensure that the cranking speed of the engine is not prejudiced by the additional electrical load. A situation could arise in which the beneficial effect of heating the combustion air is more than offset by a reduction in cranking speed, which is a very significant factor in regard to the cold starting ability of a diesel engine.

An electrically heated manifold air heater is marketed by Robert Bosch in Germany, for use with direct injection engines of up to 2 litres total cylinder capacity. It has an energy consumption of 600 W and it is capable of lowering the minimum cold starting temperature of the engine by about 10°C.

Combustion-type manifold air heaters

The combustion-type manifold air heater is a small burner which is fitted in the intake manifold of the engine to burn a limited quantity of liquid or gaseous fuel in order to raise the temperature of the engine combustion air as it passes through the inlet manifold. The oxygen for the combustion of the fuel used in the burner is taken from the engine air supply and therefore the amount of fuel burnt must be kept at a minimum to avoid significantly depleting the oxygen available for combustion in the engine cylinders. The fuel quantity must also be limited to avoid the risk of the engine overspeeding due to the carry-over of unburnt fuel into the engine itself. These safeguards are inherent in the established designs which are manufactured by Robert Bosch in Germany and by Lucas CAV in Great Britain.

The Lucas CAV Thermostart consists essentially of a gravity fed burner which uses diesel fuel from the engine system. This is supplied to the burner through a valve which is thermostatically controlled. It is opened by the expansion of a sleeve which is heated by a coil connected in series with the electrically heated igniter. Therefore, fuel cannot be admitted to the burner or the inlet manifold until the igniter has attained a satisfactorily high temperature which will ensure the immediate combustion of the fuel.

The starting procedure is first to energize the igniter, which is raised to its operating temperature in about 30 seconds. The current also heats the coil which admits fuel to the burner. Next the starter button is pressed to begin cranking the engine and this provides the oxygen required for the continued combustion of the fuel supplied to the burner. The Thermostart can be kept in operation after the engine has begun to fire, until the conditions have stabilized and the engine is running normally. The heater is then shut down by cutting off the electrical supply to the heater coil and igniter, which will automatically cut off the fuel supply. The Thermostart usually enables satisfactory starting to be achieved in ambient air temperatures down to about –18°C.

15.2.6.5 Special starting fuels

The starting aids described so far are all means of increasing the temperature of the air in the cylinder of a diesel engine at the end of the compression stroke, so that it will be satisfactorily above the self-ignition temperature of the diesel fuel employed, at the lowest ambient air temperature at which the engine will be required to operate. The alternative approach is to employ a fuel for cold-starting the diesel engine, which has a self-ignition temperature below the temperature of the air in the cylinder at the end of the compression stroke, when starting a completely 'cold' engine at the minimum ambient air temperature at which it will be required to start.

The starting fuels employed are usually ether-based but because ether itself produces an undesirably high rate of pressure rise during combustion in the engine cylinder and has virtually no inherent lubricating properties, appropriate additives are included in the formulation of the fuel to protect the engine against the effects of these undesirable characteristics. This form of starting aid is now widely used and in its simplest form is an aerosol container marketed under such trade names as Aerostart, Gasomatic and Quickstart. Some fluid from the aerosol is sprayed into the air filter or air intake while the engine is being cranked. This is effective as a means of obtaining a cold start under occasionally extreme conditions but, for general use, a more precise control to the process is highly desirable.

A comprehensive range of starting aids which employ ether-based fuels is produced by Start Pilot Limited. These aids include Gasomatic aerosols; hand-pump-operated fuelling units which spray the starting fuel into the engine air manifold; and completely automated equipment which not only facilitates the initiation of combustion in a cold engine, but continues to support the combustion so that it rapidly becomes stabilized. This increases the rate of load acceptance of the engine and is particularly useful in the case of emergency, stand-by and no-break installations. The illustration reproduced in *Figure 15.43* shows a Series 450 Start Pilot fully automatic equipment installed on a 2000 kW Pielstick 18 PA 4 high-speed diesel engine, which is the power source for a no-break generating set. Ether-based starting fuels are available in several grades and these enable engines to be started in ambient air temperatures as low as –50°C providing the engine can be cranked satisfactorily. It must be observed however, that in order to attain an acceptable cranking speed, it is usually necessary to preheat both the coolant and the lubricating oil.

15.2.6.6 Engine heaters

The starting aids which have been described in this chapter greatly extend the field of application of the diesel engine by promoting combustion in a completely 'cold' engine exposed to a low-temperature environment, but the most effective starting aid for these adverse conditions is the provision of an artificial environment for the engine and its equipment, which will be independent of climatic conditions. In addition to providing temperature conditions which are favourable to the initiation of combustion, this enables the lubricating oil viscosity, engine coolant temperature and starter battery temperature to be maintained at 'temperate' levels. This solution is not always

Figure 15.43 Start Pilot Model 450-TZ fitted to Pielstick 18PA4 18-cylinder engine, 2000 kW, installed in a no-break generating set

possible owing to constraints imposed by the particular duty but it is virtually inherent in such applications as base-load power stations and ship propulsion.

In a base-load power station, some generating plant is always running on load; the air temperature in the station is usually within the 'comfort zone' irrespective of the climatic conditions and the ventilating system can be arranged to ensure this. The cooling systems of the engines can be interconnected, so that heat from engines which are on load can be utilized to preheat engines not in service: heat radiated from the surfaces of these engines also contributes to the ambient air temperature in the station. In addition, warm water from the same source can be passed through the oil cooler and the lubricating oil circulated around an engine by the electrically driven priming pump, to reduce its viscosity, before the engine is started.

Ships are usually equipped with dual-fired waste-heat boilers which utilize exhaust heat from the propulsion engines to provide shipboard services when the vessel is at sea. When the ship is in port and the propulsion engines are shut down, the boilers are oil-fired and therefore a source of heat is readily available to preheat the cooling systems of the propulsion engines and their related lubricating oil and fuel systems, prior to starting.

In countries which regularly experience arctic conditions, preheating is applied to a wide range of diesel-powered plant including contractors' equipment, mobile plant and locomotives, by providing thermally insulated enclosures and a source of energy to either maintain a satisfactory temperature overnight, or preheat the equipment before starting. Where electrical energy from a mains supply is available, this is the most convenient form of heating, but diesel fuel and bottled gas are alternative sources of energy. In one such equipment of Swedish origin, which employs bottled gas, a timer is incorporated which can be preset to automatically start the heating cycle of the coolant and lubricating oil at an appropriate time before the engine will be required to start.

For emergency generating sets and 'no-break' installations, the coolant and lubricating oil are usually maintained at a temperature of about 30°C, to enable the engine to start immediately and accept the prevailing load very rapidly, without detriment to the reliability of the unit. The energy required is often provided by the primary source of power, which may be the public supply, by means of thermostatically controlled immersion heaters in the engine coolant system and in the engine sump or lubricating oil tank.

15.3 Heat exchangers

15.3.1 Introduction

The application of heat exchangers to diesel engines is principally in the dissipation of heat to the environment, either to water or to atmospheric air, although in a few instances some is abstracted for use in providing heat elsewhere. The functions which they fulfil can be classified as:

(i) Engine cooling, to maintain within acceptable temperature limits such components as liners, pistons, and cylinder covers, etc. The pistons are cooled by a separate circuit in some larger engines. The materials used in the construction of engines are unable to withstand the thermal stresses that would be imposed upon them, for any appreciable length of time, without some form of forced cooling.
(ii) Oil cooling, to control temperature and hence the viscosity of lubricating oil within the limits required to provide effective lubrication. The lubricating oil is sometimes used also for cooling the pistons.
(iii) Charge air cooling in pressure-charged engines, to increase the density of the combustion air entering the cylinders as well as reduce its temperature, enabling engine power to be increased

by burning more fuel on each piston firing stroke, whilst maintaining acceptable exhaust value temperatures.

15.3.2 Operating conditions

To enable the size of a heat exchanger to be determined, it is necessary to know the rate at which heat must be dissipated, as well as the temperatures and flows of the two fluids involved in the heat exchange relationship.

The rate of heat rejection to engine coolant depends upon many factors, including the b.m.e.p., rev/min, and the design characteristics of the engine, particularly whether it has pre-combustion, turbulence chambers, or direct injection. The heat to lubricating oil in relation to the shaft power generated is much dependent on engine speed and to what extent the pistons are cooled by the oil. As far as the charge air is concerned, the heat rejection rate is primarily a function of the pressure ratio of the turbocharger.

Values of heat rejection rate and flow rate vary widely. Those given in *Table 15.1* for different classes of engine should be taken as a guide only.

For high speed engines, the trend over many years has been towards greater circulation rates. Where pistons are directly oil cooled, the circulation rate is normally two to three times that without oil cooled pistons. If the exhaust manifolds are water-cooled, 25 to 30% is added to the heat rejected from the jackets and covers.

The maintenance of the temperature of each hot fluid at or near design value contributes not only to the safe and efficient operation of the engine but to extension of its life. Excessively high temperatures can lead to breakdown of lubrication, whereas low temperature operation usually results in high wear rates of rings and liners. The charge air, however, can be cooled substantially below the maximum recommended operating value without detriment to the engine, although excessively low temperature can give rise to starting difficulty in some engines. Condensation of water vapour from the charge air may occur under conditions of high humidity and low charge air coolant temperature.

15.3.3 Water-cooled systems

Where there is a suitably cool supply of water in sufficient quantity, this usually provides the most convenient heat sink into which the unwanted heat from an engine can be discharged. This is the case with almost all marine installations and some land based stationary applications near to the coast or a substantial river. Cool water can also be provided from a cooling tower or, where space permits, a spray pond.

Open, straight-through cooling, for the removal of heat from the engine jacket, is very rarely used because of the risk of fouling by dirt or scale of the passages within the engine. Instead, closed systems are now almost universal, employing mostly tubular heat exchangers to transfer the heat from the engine jacket coolant to the raw cooling water. Scale formation in cylinder jackets, covers, etc., is virtually eliminated and evaporation and water make-up losses are negligible. It is possible, with such systems, to operate at the highest temperatures compatible with the engine requirements, using coolant circulation rates to give a small temperature rise through the engine. With a suitable temperature control system, the temperature of the engine coolant can be made independent of that of the raw water supply.

Figure 15.44 illustrates, in diagram (a), the closed parallel system, which is suitable for most engines. The pump draws cooled water from the heat exchanger and discharges it through the engine, back to the heat exchanger; a small header tank guards against casual leakage, being connected in parallel, by a small bore pipe, to a point near the pump suction. For temperature control, thermostats can be introduced; these are usually of the by-pass type, located between the engine and the heat exchanger, with a by-pass led to the pump suction. At all speeds the engine is under the highest pressure in the circuit.

In scheme (b) the fresh-water pump draws from a hotwell tank and discharges, at the highest pressure, through the heat exchanger to an overhead tank. From this tank the water flows through the engine, by gravity, returning to the hotwell tank. Such an arrangement is suitable for after-cooling, particularly when the pumps are engine driven.

In system (c), the heat exchanger is on the fresh-water pump

Table 15.1

Speed range (rev/min)	*Low speed 80–200 PC*	*Medium speed 300–1000 PC*	*High speed[d] 1000–2500 PC[c]*	*NA*
Direct injection engine	2-stroke	4-stroke	4-stroke	4-stroke
Heat rejection rates:% power output				
jackets and covers	15–30	25–40	20–40	50–70
pistons	8–9	—	—	—
lubrication oil	2–8	10—15[a]	10–15[a]	8
charge air[b]	25–35	22–30	15–30	—
b.m.e.p. (bar)	13–15	16–19	14–21	6–7
Circulation rates: 1/h kW				
fresh water to jackets and covers	10–20	30–35	50–100	
fresh water to pistons	4–6	—	—	
lubricating oil	10–30	15–25	15–30	
Temperatures: °C				
fresh water to jackets and covers	65–70	65–75	80–90	
fresh water to pistons	45–55	—	—	
lubricating oil	40–45	60–75	90–115	

PC Pressure charged.
NA Naturally aspirated.
[a]Oil cooled pistons
[b]Charge air cooling heat rejection rates rise with increase of pressure ratio and jacket losses fall, with the sum remaining much the same.
[c]High speed PC engines include engines for rail traction with b.m.e.p. up to 21 bar as well as engines for road vehicles with b.m.e.p. about 14. Hence the wider range of heat rejection rates given.
[d]High speed engines using prechamber combustion systems have losses to jackets and lubricating oil some 10% greater than those of direct injection engines.

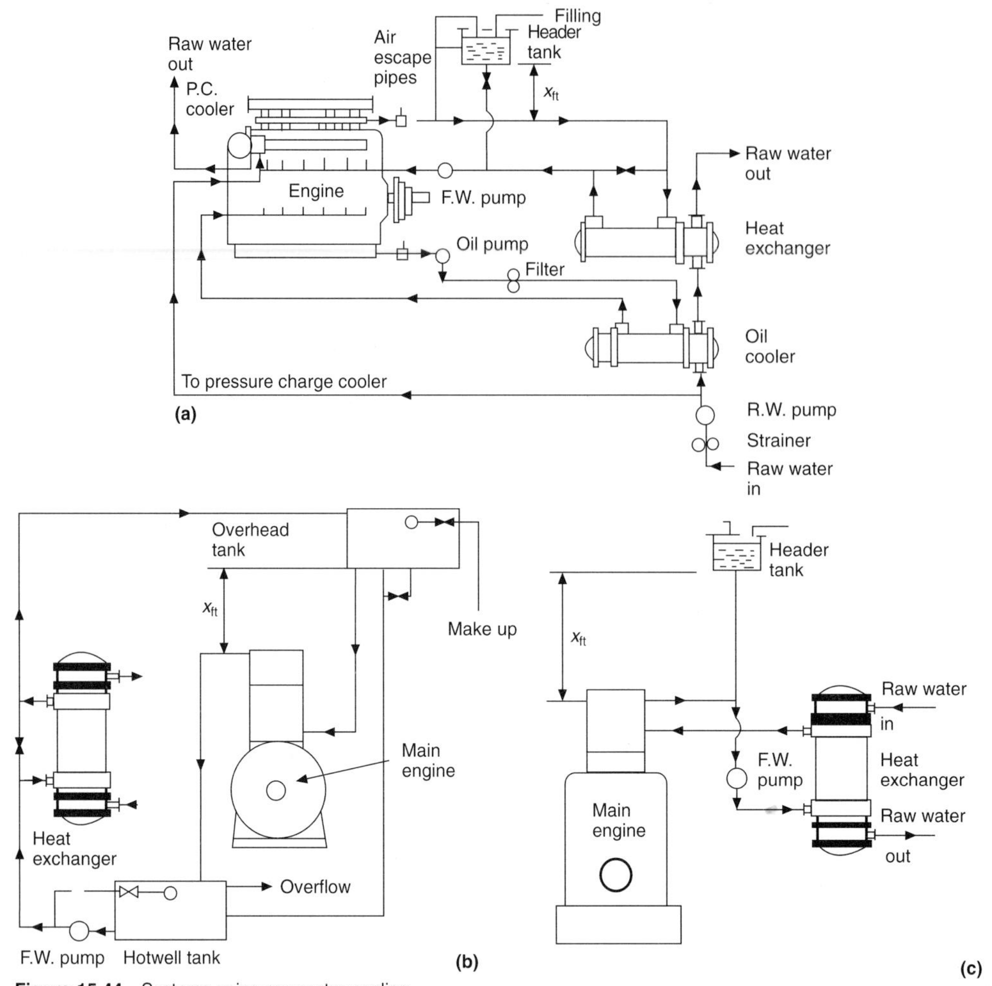

Figure 15.44 Systems using raw water cooling

discharge; the water, after passing through the engine, returns to the pump suction. The header tank is connected to the pump suction on the engine outlet. Under running conditions this system puts the highest pressure on the heat exchanger so that, in the event of tube failure, the engine water is likely to leak into the raw water. Although this is a safeguard against contamination, there is a risk of losing the cooling water, hence an alarm in the header tank is sometimes fitted.

The engine lubricating oil may sometimes be cooled by the engine coolant. This arrangement is described later under 'Aircooled systems'.

15.3.3.1 Economizing on cooling water flow

Although in many instances, such as in marine applications, the raw cooling water itself is 'free', the provision of its flow is not. The higher its circulation rate through the cooling system, the greater is the required sea water pumping capacity, necessitating larger or more numerous pumps as well as absorbing more power to drive them—raw water pipework, fittings and valves must also be correspondingly larger to cater for the higher flow. There is thus a first cost as well as a running cost related to the provision of flow and it is important to consider, at the system design stage, what may be done to optimize the cooling water flow for minimum system cost.

The way in which the size of a heat exchanger for a given thermal efficiency depends on raw water flow is shown in *Figure 15.45*. This is a qualitative representation only, the scales depending on the rate of heat exchange as well as some other factors. It does, however, serve to illustrate that, for a fixed value of thermal efficiency (which is related solely to the hot fluid inlet and outlet temperatures and the raw water inlet temperature) reduction in raw water flow much below the 'knee' of the corresponding curve will necessitate an excessively large (and therefore expensive) heat exchanger but heat exchanger size will be little reduced by increase in water flow beyond this area. The most economical value of raw water flow for an individual heat exchanger is therefore likely to lie within fairly narrow limits.

Where the required hot fluid temperatures in any one heat exchanger in a group is significantly higher than those in another, it is worth considering the possibility of passing the raw water through the heat exchangers in series, the higher temperature unit being downstream. Alternatively, the raw water from two heat exchangers operating in parallel may merge and then pass through a third. Such an arrangement is not uncommon in a ship's engine cooling system, the raw-water passing first through the lubricating oil and piston water coolers in parallel and then through the jacket water cooler. This approach can affect a reduction of cooling water flow for the system by as much as

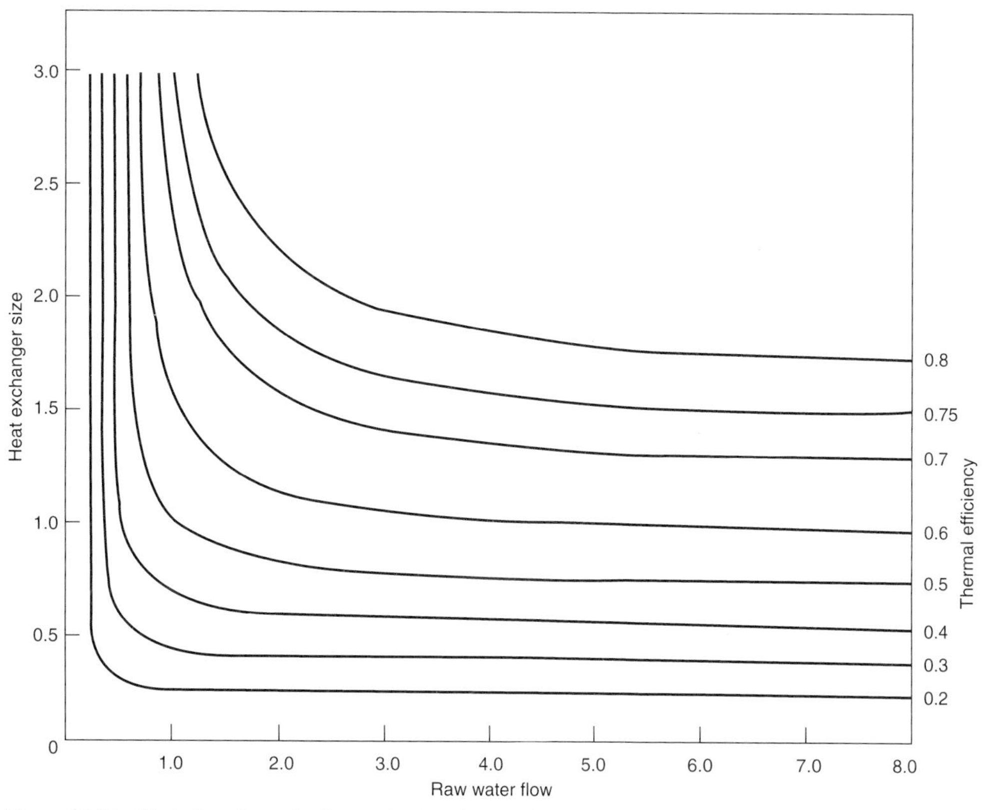

Figure 15.45 Effect of cooling water flow on heat exchanger size

half with little increase in the size of cooler required for the jacket water due to the slightly heated sea water entering it from the other two coolers.

Figure 15.46 shows a system in which this economy is carried further by passing the raw cooling water first through the charge air cooler, before using it for cooling the other engine fluids.

15.3.3.2 Intermediate water systems

In sea-water-cooled installations, there can be an operational advantage in using clean, fresh water as the cooling medium in all, or at least the majority, of the heat exchangers in a system. This requires an additional heat exchanger to cool the fresh water with sea water, the fresh water being circulated through it, as well as the other heat exchangers, in a closed circuit. This arrangement confines the problems associated with sea water, namely corrosion and fouling, to the one heat exchanger, which can then be sited conveniently for easy access for inspection and cleaning and, if desired, made from a highly corrosion-resistant material. The sea-water pipework can also be made much shorter than in the usual system where all heat exchangers are sea-water-cooled.

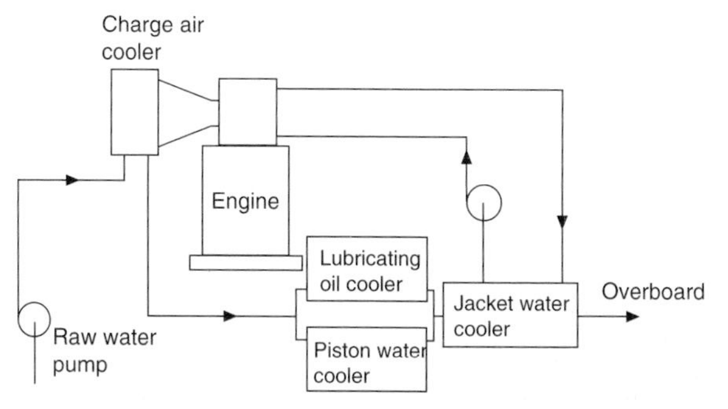

Figure 15.46 Cooling system giving cooling water flow economy

Figure 15.47 shows a line diagram of a circuit for a shipboard system, in which the fresh water leaving the intermediate heat exchanger passes first through the lubricating oil cooler and piston water cooler in parallel and then through the jacket water cooler before returning to the intermediate heat exchanger again. A spur provides cooling water for the charge air cooler. As illustrated, some of the fresh water being circulated can be employed for cooling auxiliary equipment.

The drawback to this approach to cooling with sea water is that the total cost of the system is very high compared with that of a conventional system. This is due mainly to the fact that the need to maintain the fresh water temperature near to that of the sea water requires a heat exchanger to cool the fresh water of very large size compared with the others in the system. This applies particularly in the case of a ship required to operate in tropical waters where the design value of temperature difference between lubricating oil and sea water may be as little as 15°C. In such a case, the charge air cooler may not be included in the fresh water cooling system but rather directly cooled by sea water, in order to obtain the minimum possible charge air temperature. The various engine coolers must also be enlarged to allow for the higher cooling water temperature.

The temperature of the circulating fresh water in these systems can, of course, be controlled by the usual means at the fresh water/sea water heat exchanger, so improving temperature control of the individual hot fluids.

The jacket water/intermediate fresh water heat exchanger may be eliminated from a system of this nature by using the

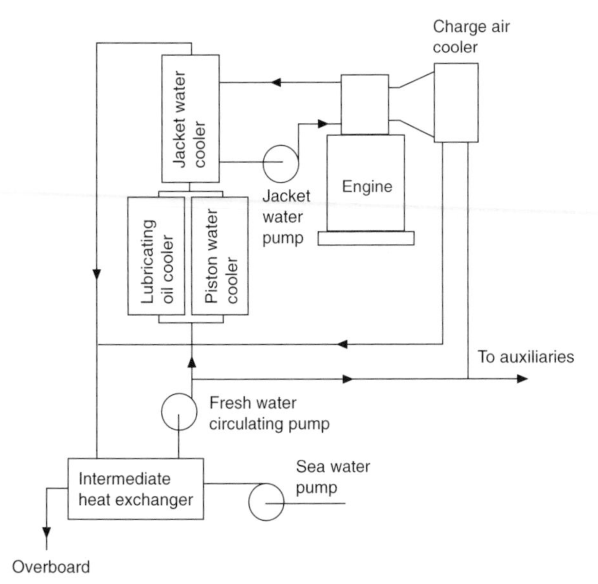

Figure 15.47 Intermediate cooling water systems

fresh water itself as the engine coolant. The higher temperature required by the engine is achieved by linking the engine circulating system into the intermediate water system from which the colder water is drawn as required, and to which it is discharged again after passing through the engine. A 3-way temperature control valve regulates the flow.

15.3.3.3 Heat recovery

At best, little more than 40% of the energy released in burning the fuel in a diesel engine is converted into mechanical output. The remainder, in the form of heat, is available in part for use elsewhere, for instance for space heating.

In most heat recovery systems, fresh water is used as the heat transmission medium, and is circulated in a closed circuit through the engine jacket water heat exchanger and then through an exhaust gas heater in series before passing to the heat exchangers which abstract the usable heat. Some 60% of the heat available in the engine exhaust as well as all of that from the engine jacket, sometimes with the small addition of heat from the lubricating oil, can be taken into the heat recovery system. The system may be pressurized to enable the water outlet temperature from the exhaust gas heated unit to exceed 100°C so increasing the utility of the heat removed.

In a few systems, an organic liquid is employed as the heat transmission medium, enabling these higher temperatures without pressurization. Alternatively, steam may be generated in the exhaust gas unit and either used for heating purposes or to generate power from a turbine in a condensed system. There is also some use of volatile organic fluids in such auxiliary power generation systems.

The overall efficiency of utilization of energy available from the fuel burnt can be 70% or more whilst the heat derived from the engine is being used for heating purposes in addition to the power generated by the engine. Such systems are generally economically viable only when the engine is in substantially continuous operation and the heat generated used usefully most of the time. Overall energy utilization efficiency of an engine with a system producing power from an auxiliary turbine is naturally somewhat lower at 50%–60%.

Use of waste heat from an engine for fresh water generation, either using the engine jacket coolant or steam from an exhaust gas boiler as the heating medium, is a long established practice in shipboard installations.

15.3.4 Evaporative systems

The effect of cooling by evaporation of water can be made use of in diesel engine applications in a number of different ways, the more common involving evaporation outside the engine to provide a source of cooling water.

Such evaporative systems include the use of cooling ponds, in which heat is lost to the atmosphere by evaporation from the surface. In temperate climates it is usual to provide a pond surface area in the region of 0.25 m^2/kW. of engine power and a volume of about 0.25 m^3/kW. Both figures should be doubled for tropical climates. The surface area of pond required can be reduced to about 0.1 m^2/kW by adding a water spray.

The use of a cooling tower such as that shown in *Figure 15.48* can be particularly beneficial in hot climates with low humidity for providing cooling water for charge air cooling, as the water temperature can be reduced to below the dry bulb temperature of the air. This may, in some instances, avoid derating an engine because of high charge air temperatures.

15.3.5 Temperature control

Control of temperature is important for engine jacket water and lubricating oil, to prevent over-cooling when the raw water is much below the design value for the system. This condition, for a ship designed for worldwide operation, may apply most of its time at sea. There are two basic ways in which the temperature of these hot fluids may be controlled.

First, by employing a three-way thermostatic valve, such as that illustrated in *Figure 15.49* controlling the proportion of hot fluid which passes through to that which bypasses the cooler. The motion of the valve between its two extreme positions is

Figure 15.48 Forced draught cooling tower

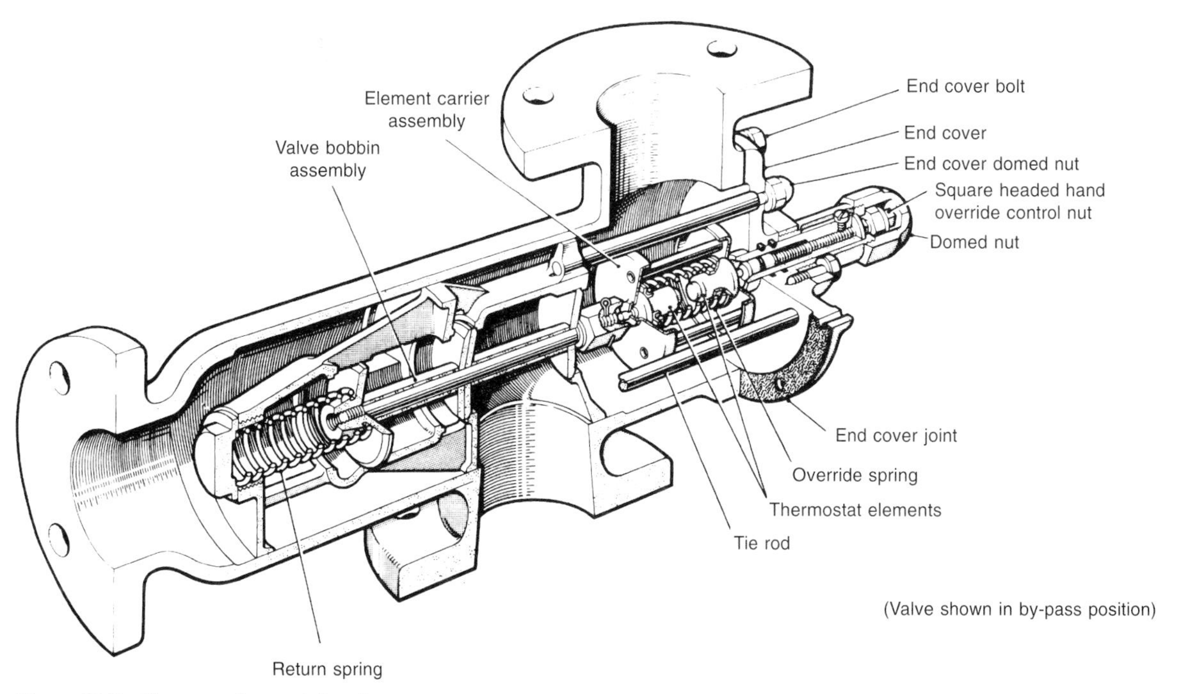

Figure 15.49 Three-way thermostatic valve

controlled by a temperature-sensing actuator operating over a preset temperature range, so that the flow through the cooler is stopped when the hot fluid is below the minimum desired temperature, whilst the by-passs is closed and the total flow passes through the cooler when the maximum value is reached. In a correctly designed system, the valve will adopt some intermediate position, depending upon operating conditions, whilst the engine is in operation. Such thermostatic valves can be fitted so as to control the inlet or the outlet temperature of the hot fluid.

Second, the flow of raw water through the cooler may be controlled, increasing the flow to lower the hot fluid temperature and restricting it to reduce its temperature. This is usually done by fitting a pneumatically operated throttling valve in the raw water pipe at discharge from the cooler, with the temperature sensor fitted in the hot fluid line.

There are some systems in which the temperature of the raw water is held substantially constant by partial recirculation of the raw water, using a three-way temperature controlled valve by passing a controlled proportion into the system discharge. The temperature maintained is normally that of the warmest raw water to be encountered. Whilst additional thermostatic control valves are required to supervise some individual coolers, accuracy of temperature control can be much improved through eliminating the variant of raw water temperature.

15.3.6 Air-cooled systems

In mobile systems, or where water is scarce, fan-cooled radiators (air blast coolers) are used in closed systems. They need a negligible amount of make-up water and are not normally subject to corrosion.

The heat storage capacity of the radiator is inherently small and so, provided the radiator can be placed near to the engine, the system can reach its design operating temperature quickly. When an engine is placed within a building, ducts may be needed to lead air to or from the radiator.

In many high speed engine cooling installations, the engine jacket water circuit is pressurized to enable a temperature above the boiling of water at atmospheric pressure to be reached. This is acceptable provided that the engine itself is suitably designed and has the benefit that the difference in temperature between the water and the cooling air is increased, so that the radiator may be smaller. This is of particular importance in installations operating in tropical climates. A pressure between 0.5 and 1.5 bar above atmosphere is usual.

For this necessary pressure to be generated, the whole water circuit must be sealed and a pressure relief valve provided to avoid over-pressurization. In smaller systems, this is often combined with the filler cap fitted to the upper loader of the radiator. *Figure 15.50* shows a typical such cap, which also incorporates a vacuum-break valve, so that the system pressure will not become sub-atmospheric when the water cools after engine shut-down.

As shown in *Figure 15.51* the lubricating oil may be cooled by the engine fresh water coolant provided that an oil temperature above the temperature of the water returning from the radiator can be tolerated. A temperature difference of between 5°C and 10°C is not unusual. Whilst this arrangement is sometimes used in auxiliary high speed diesel engines for marine and other applications using raw water as the ultimate heat sink, it is an arrangement more often employed in air cooled systems.

One particular advantage of this arrangement is that the engine coolant temperature, brought to its operating level quickly under

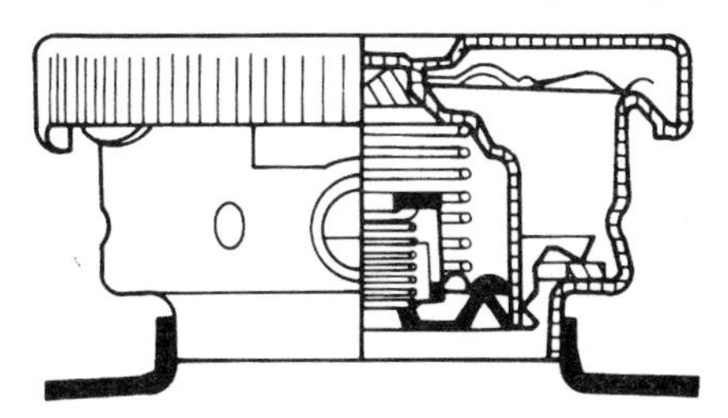

Figure 15.50 Combined filler cap and pressure relief valve

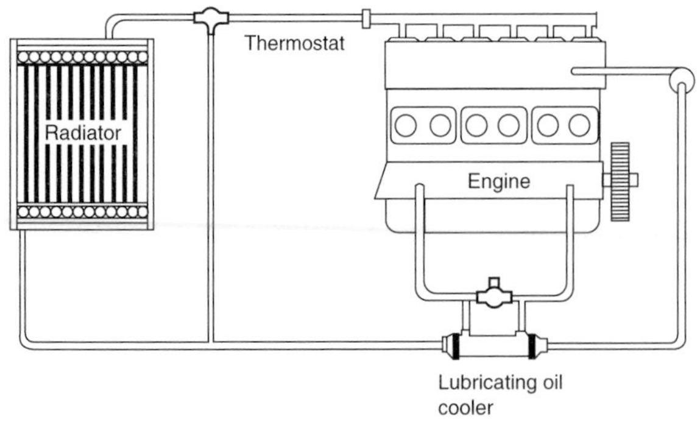

Figure 15.51 Circuit with jacket-water-cooled oil cooler

the action of the thermostatic control valve fitted in that system, acts as a heating medium for the lubricating oil for a period after start up from cold. *Figure 15.52* shows the relative speed with which the lubricating oil reaches its operating temperature with this system compared with direct air cooling of the lubricating oil. With direct air cooling of the lubricating oil, unless a thermostatic valve is employed in the lubricating oil circuit, the oil temperature will remain below its design operating value when the air is colder than the maximum for which the oil cooler was designed.

Where charge air cooling is required in an air-cooled system, the charge air may be cooled by water circulating through the engine jacket, as in the circuit shown in *Figure 15.53*, which also incorporates the lubricating oil cooler. This imposes a lower limit of temperature on the cooling of the charge air, since the temperature of the water entering the engine jacket must be maintained at a value high enough for the efficient operation of

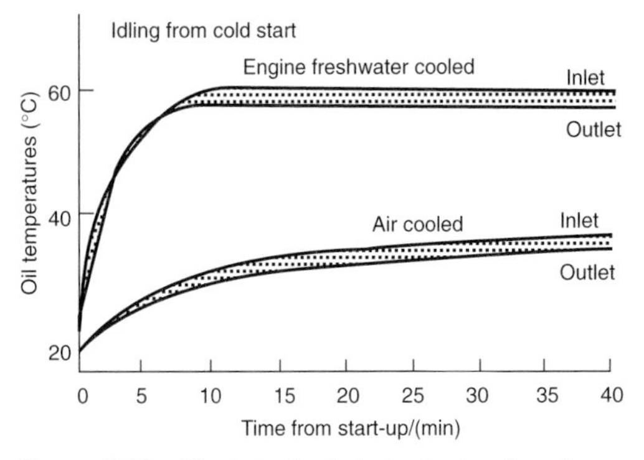

Figure 15.52 Effect of using jacket water for oil cooling

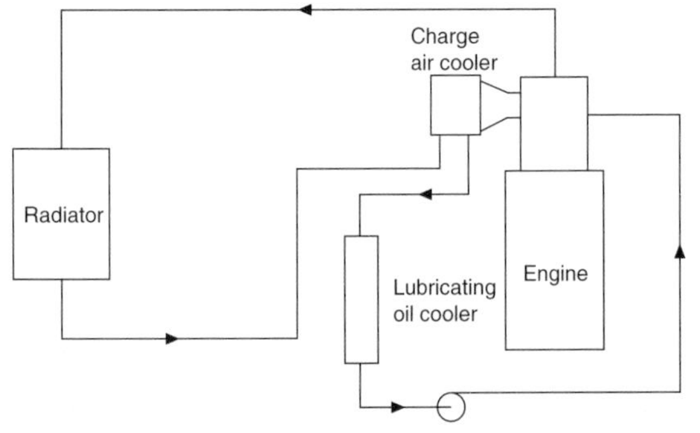

Figure 15.53 Circuit using jacket water for cooling oil and charge air

the engine. A charge air temperature lower than about 90°C cannot be achieved with this system operating under normal conditions.

In the system of *Figure 15.54* the charge air and lubricating oil are cooled by water in a separate air-cooled closed circuit. With this method, the charge air can be cooled economically to some 20°C above the ambient air temperature but there is the added complication of an additional pipework system and pump.

A third method is to cool the charge air directly by atmospheric air, usually in a separate section of the radiator cooling the jacket water and possibly the lubricating oil as well. This enables a charge air temperature down to within 5°C above ambient air temperature to be achieved. It is therefore of particular advantage in hot climates. Its disadvantage is that two charge air ducts of adequate cross section must be provided between the engine and the radiator section cooling the charge air and these will not only be large in diameter for a high power engine but they may also extend a considerable distance if, for instance, the radiator is placed outside an engine house.

15.3.7 Heat transfer

Before the proportions of any heat exchanger can be determined, some knowledge of the laws governing heat transfer is necessary. Only an outline of the subject can be given here. For further information, reference should be made to specialist works and papers.

Heat may be transferred by convection, by conduction and by radiation. The first of these processes is the one which is the most important in connection with heat exchangers for diesel engines.

When a metal wall separates two fluids flowing over the wall surfaces, the fluids being at different temperatures, heat is transferred from one to the other. The diagram shown in *Figure 15.54* shows, in simplified form, the temperature gradients encountered.

The bulk of the fluid on the left-hand side of the diagram is at a temperature T but, on approaching the wall, the temperature

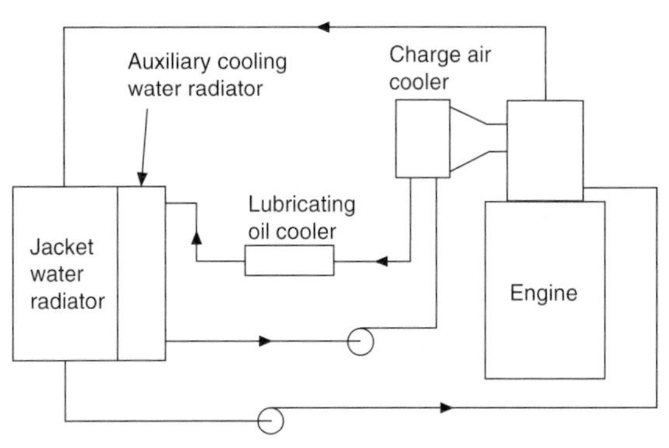

Figure 15.54 Closed circuit charge air cooling system

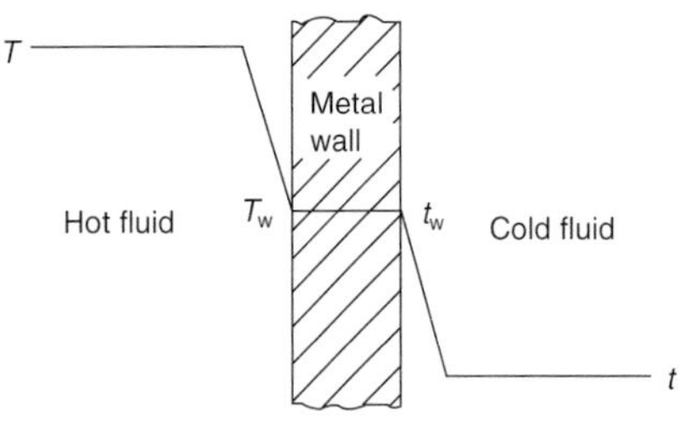

Figure 15.55 Heat transfer from hot to cold fluids

falls rapidly through the fluid boundary there, reaching a temperature of T_w at the fluid-to-wall interface. There is then a further fall in temperature, usually very slight, from T_w to t_w across the highly conductive metal wall, towards the right-hand side. Moving further to the right, the temperature falls rapidly from t_w to t, the bulk temperature of the second fluid, through its boundary layer.

The magnitude of each of these steps, occasioned by thermal resistances, depends upon a number of factors but the separate resistances can be combined into a single overall resistance. This is usually expressed in the reciprocal form of an overall heat transfer coefficient U, defined as the rate of heat flow from the one fluid to the other, per unit surface area, per unit temperature difference. In SI units, this heat transfer coefficient is expressed in Watts per square metre of wall surface, per degree K of temperature difference between the two fluids thus:

$$E = U\theta S \qquad (15.1)$$

where

E = rate of heat flow from hot to cold fluid (W);
U = overall heat transfer coefficient (W/m^2 K);
θ = temperature difference between hot fluid T and cold fluid t(K);
S = area of separating surface (m^2).

15.3.7.1 Mean temperature difference

Equation (15.1) assumes taht each fluid is at uniform temperature, i.e. that all the hot fluid is at temperature T and all the cold fluid at temperature t. In heat exchangers associated with diesel engines, however, this is not so, as the temperature of the hot fluid is reduced and of the colder fluid increased during the passage of both through the heat exchanger.

The temperature drop of the hot fluid is obtained from the expression

$$T_1 = T_2 = \frac{E}{w_1 s_1} \qquad (15.2)$$

where

T_1 and T_2 are respectively the inlet and outlet temperature in °C, of the hot fluid.
w_1 = flow rate of the hot fluid (kg/s);
s_1 = specific heat of the hot fluid (J/kg K).

Similarly, the temperature rise of the cold fluid is obtained from the expression

$$t_2 = t_1 = \frac{E}{w_2 s_2} \qquad (15.3)$$

where

t_1 and t_2 are respectively the inlet and outlet temperatures in °C, of the cold fluid;
w_2 = flow rate of the cold fluid (kg/s);
s_2 = specific heat of the cold fluid (J/kg K).

The temperatures are not constant at all points in the heat exchanger. If the hot and cold fluids flow in opposite directions—the so-called 'counterflow' arrangement—the temperature distribution takes the form shown in *Figure 15.56a* and the mean temperature difference is given by the equation:

$$\theta = \frac{(T_2 - t_1) - (T_1 - t_2)}{\log_e \left[\dfrac{T_2 - t_1}{T_1 - t_2}\right]} \qquad (15.4)$$

If the fluids flow in the same direction through the heat exchanger—the 'parallel flow' arrangement, as shown in *Figure 15.13b*, the corresponding expression is:

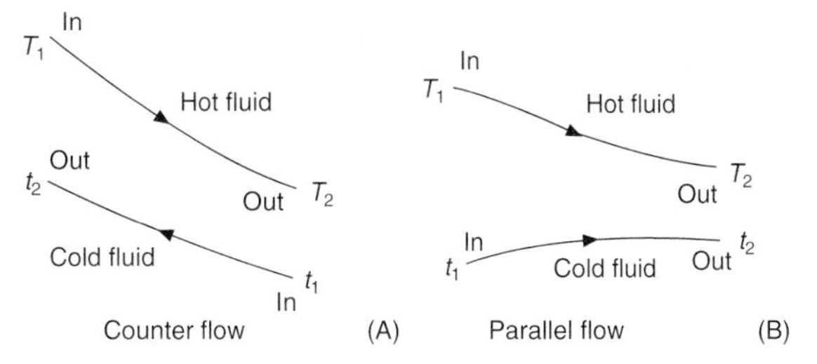

Figure 15.56 Temperature gradients through heat exchangers

$$\theta = \frac{(T_2 - t_2) - (T_1 - t_1)}{\log_e \left(\dfrac{T_2 - t_2}{T_1 - t_1}\right)} \qquad (15.5)$$

When the two fluids flow substantially at right angles through the heat exchanger the formulae become much more complicated, but the result is always intermediate between the mean temperature difference given by eqn (15.4) and that given by eqn (15.5). Usually, it is sufficiently accurate to take the mean of the two values. *Figure 15.57* is a nomogram which enables results to be calculated readily in the cases covered by eqns (15.4) and (15.5).

15.3.7.2 Components of heat transfer coefficient

The overall heat transfer coefficient U, itself the reciprocal of the overall thermal resistance, is the combination of three separate thermal resistances, linked by the reciprocal relationship:

$$\frac{1}{U} = \frac{1}{h_1} + \frac{x}{k_w} + \frac{1}{h_2} \qquad (15.6)$$

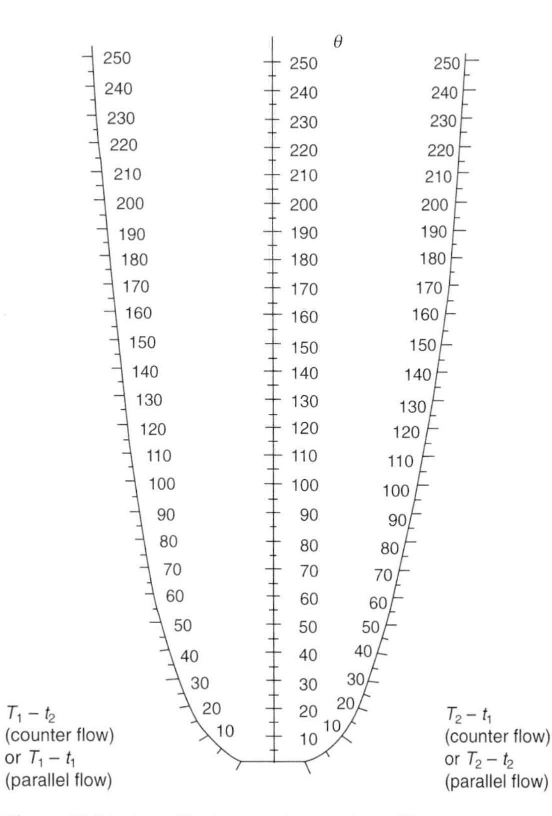

Figure 15.57 Logarithmic mean temperature difference

where
- U = overall heat transfer coefficient (W/m^2 K);
- h_1 = the partial heat transfer coefficient on the hot fluid side (W/m^2 K);
- h_2 = partial heat transfer coefficient on the cold fluid side (W/m^2 K);
- x = thickness of wall (m);
- k_w = thermal conductivity of wall (W/mK).

With practical heat exchangers used in diesel engine cooling applications, the resistance of the metal wall may normally be neglected, because of the high conductivity of the metals used for the heat transfer surface. When a film of dirt is expected to build up on the metal surface, the overall conductivity of the wall must be taken into consideration, which may be done by substituting a fouling factor for the term x/k_w.

Sometimes additional, or secondary, heat transfer surface is applied to either the hot or the cold fluid side of the heat exchanger, as shown diagrammatically in *Figure 15.58*. It is often economical to provide secondary surface in this way, particularly when the partial heat transfer coefficient for one fluid is very much lower than that for the other fluid. Secondary surface is arranged on the side having the lower heat transfer coefficient, thus enabling the overall size and cost of the equipment to be reduced. Then, instead of eqn (15.6), the following expression is used:

$$\frac{1}{U} = \frac{n}{h_1} + \frac{nx}{k_w} + \frac{1}{h_2} \qquad (15.7)$$

where
- U = overall heat transfer coefficient (W/m^2 K);
- n = ratio of surface area on side provided with secondary surface to that on plain side;
- h_1 = partial heat transfer coefficient on side having no secondary surface (W/m^2 K);
- x = effective thickness of metal wall (m);
- k_w = conductivity of material of which metal wall is made (W/m K);
- h_2 = partial heat transfer coefficient on side provided with secondary surface (W/m^2 K).

It should be noted that in eqn (15.7), the overall heat transfer coefficient is referred to the total surface area on the side provided with secondary surface. In using this value in eqn (15.1) it is that surface area that must be employed.

If the secondary surface is made of very thin metal, as is usually the case, there will be a drop in temperature along each fin. This reduces the effectiveness of the secondary surface and account must be taken of this. The efficiency of the fin depends, in a complicated manner, on the exact form of the secondary surface and on the value of the heat transfer coefficient between it and the fluid flowing over it.

15.3.7.3 Partial heat transfer coefficients

The partial heat transfer coefficients h_1 and h_2 can each be

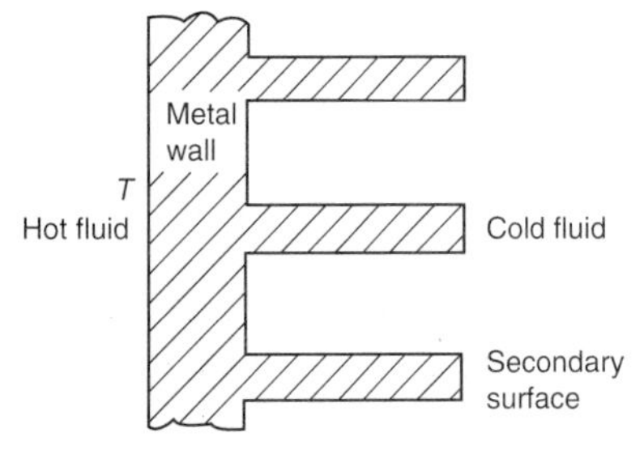

Figure 15.58 Conducting wall with secondary surface

considered dependent upon the existence of a stagnant film of fluid near the metal wall. The thickness of this film is, in turn, dependent upon the value of the rate of fluid flow past the surface, and upon the exact shape of the surface. Normally, forced convection of the two fluids occurs; that is, the fluids are pumped or blown past the separating surfaces, in contradistinction to natural convection where the fluids are allowed to flow under the natural influence of their temperature differences. It is usual to study the efects by the use of certain dimensionless numbers, or ratios, relating to the quantities involved. Of these the most important is the Reynolds number, given by:

$$R = \frac{D_e G}{\mu} \qquad (15.8)$$

where
- D_e = equivalent diameter of the passage through which fluid is flowing (m); (see also eqn (15.12));
- G = mass velocity, i.e. mass rate of flow divided by cross-sectional area of flow passage (kg/sm^2);
- μ = dynamic viscosity of fluid (daP) (1 daP = 10P = 1 Pa s).

This number represents the nature of the fluid flow. For undisturbed flow through a straight duct, a value of Reynolds number below about 2000 the flow is laminar. Above this value the flow is turbulent, although it may not be fully turbulent until the Reynolds number rises above 10 000. The range between R = 2000 and R = 10 000 is sometimes known as the transition region.

The second dimensionless number is the Prandtl number, given by:

$$Pr = \frac{s\mu}{k} \qquad (15.9)$$

where
- s = specific heat of fluid (J/kg K);
- μ = dynamic viscosity of fluid (daP);
- k = thermal conductivity of fluid (W/mK)

This number represents the thermal nature of the fluid.

The third dimensionless ratio is the Nusselt number, given by:

$$N = \frac{hD_e}{k} \qquad (15.10)$$

where
- h = partial heat transfer coefficient (W/m^{2}K);
- D_e = equivalent diameter of flow passage (m); (see also eqn (15.12));
- k = thermal conductivity of fluid (W/mK).

This number represents the generalized heat transfer coefficient. For any particular case, an expression of the form $N = f(RPr)$ exists, giving the generalized heat transfer coefficient, or Nusselt number, in terms of the Reynolds number and the Prandtl number.

Different expressions are usually necessary for the laminar and turbulent regions. In the laminar region, with viscous fluids, other dimensionless ratios may have an effect, such as the length/diameter ratio of the flow passages and the ratio of viscosity at the bulk fluid temperature to that at the wall temperature.

For fluids passing through straight tubes in turbulent flow, the recommended equation is:

$$N = 0.023R^{0.6}Pr^{0.4} \qquad (15.11)$$

For round tubes, the equivalent diameter used in eqns (15.8) and (15.10) is the tube bore, but for other shapes, the value can be obtained from the equation:

$$D_e = \frac{4a}{p} \qquad (15.12)$$

where
a = cross-sectional area of tube (m^2);
p = perimeter (m).

15.3.7.4 Practical heat transfer coefficients

For water flowing through straight round tubes *Figure 15.59* gives the relationship between partial heat transfer coefficient, velocity and diameter.

For liquids flowing outside tubes in plain tube heat exchangers, heat transfer coefficients can be derived from *Figure 15.60*, where the physical properties of the fluid are known. In this diagram:

h = partial heat transfer coefficient outside tubes (W/m^2K);
D_o = outside diameter of tubes (m);
k_f = thermal conductivity of fluid, measured at a temperature midway between that of metal wall and bulk temperature of fluid (W/m K);
s = specific heat of fluid, measured at bulk temperature (J/kg K);
μ = dynamic viscosity of fluid measured at a temperature midway between that of metal wall and bulk temperature (da P). (1 da P = 10P = 1 Pa s);
G_{max} = mass velocity of fluid at most restricted part of flow passage between adjacent tubes (kg/s m^2).

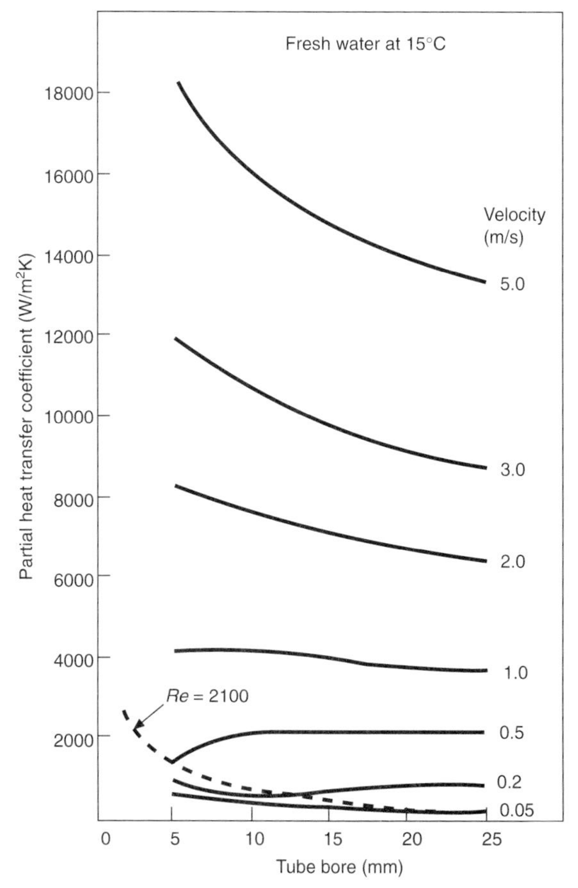

Figure 15.59 Heat transfer coefficient for water inside tubes

For water flowing over tubes a typical value of partial heat transfer coefficient is 7000 W/m^2 K and for lubricating oil 700 W/m^2K.

When air is flowing across secondary surface tubes in a finned tube radiator matrix, the value of the partial heat transfer coefficient will depend very much upon the exact arrangement of the surface and especially upon the nature of any special turbulence-promoting ripples, dimples or louvres, which may have been included to improve the efficiency. It is therefore not practicable to give precise values but typically a heat transfer coefficient will lie between 80 and 130 W/m^2 K.

15.3.8 Construction and design

15.3.8.1 Tubular heat exchangers

This form of construction, in which the hot fluid normally passes over the tubes, is the most convenient and the cheapest when a supply of raw water is available, which may be discharged to waste or recirculated and cooled. The cylindrical shape of such heat exchangers enables them to withstand the pressures involved and this general form also permits flexibility in design. They are reliable in operation and are easy to service.

A typical unit, shown in *Figure 15.61*, comprises a cylinder, with adequate inlet and outlet distribution belts. It is bored to receive a tube-stack with baffles turned to a predetermined diameter to register with the bore. Under temperature changes the tube-stack expands in the cylinder as a unit, because the tubes are mechanically secured to the tube-plates. One plate has a fixed end-joint and the other is allowed to expand freely within an expansion joint. Water boxes, with removable covers, or doors, complete the assembly. Usual test pressures range from 7 to 15 bar.

Figure 15.62 shows two types of baffle arrangements commonly used in tubular heat exchangers: (A) the radial flow or disc-and-doughnut arrangement, where the fluid in contact with the outside of the tubes flows radially, alternately inwards and outwards, at right angles to the tubes: (B) the segmental form, widely used in American designs.

In most tubular heat exchangers, provision is made for differential thermal expansion of the tube stack relative to the cylinder. This may be made within the cylinder itself, or, more usually, by allowing the tube-stack to expand and maintaining the tubes in tension where possible. *Figure 15.63* shows four examples of the arrangements for expansion, and in each example the tube loading with pressure is indicated. Very careful consideration should be given to these stresses at the design stage. Flat joints are used at the fixed end and moulded, synthetic rubber is used at expansion end.

Roller expanding is the usual method of obtaining a leak-tight joint between the tubes and the tube plates. Using this method, the hardness of the tube ends must be closely controlled, if necessary by annealing, to compensate for the effect of work-hardening during the roller expanding process. The amount of expansion must also be closely controlled, often by an electronic control of the torque applied to the roller-expanding tool.

In the design of a heat exchanger there must be maintained a balance between heat transfer and pressure loss, because the former, for turbulent flow, varies approximately as $G^{0.6}$ for outside tubes and $G^{0.8}$ for inside tubes, while the latter varies as G^2 and the pumping power as G^3. The higher the velocity the more expensive the heat transfer in terms of pressure loss, but a higher velocity means a smaller total surface and a more compact unit. Pressure loss should not normally exceed 0.4 to 0.7 bar on the engine water side or 0.2 to 0.3 bar on the cooling water side in an engine jacket water cooler.

The size of a heat exchanger depends upon the amount of cooling surface and upon the arrangement and distribution of

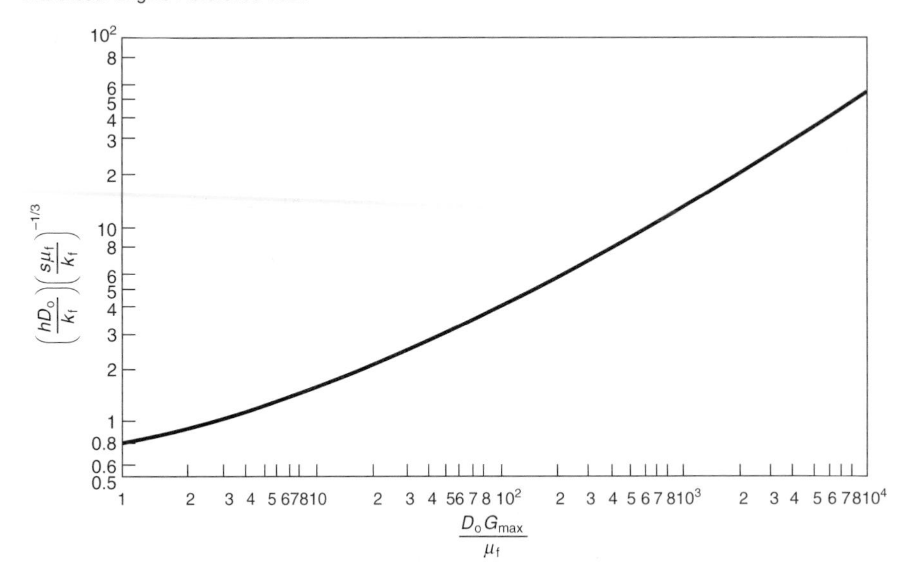

Figure 15.60 Heat transfer coefficient outside plain round tubes

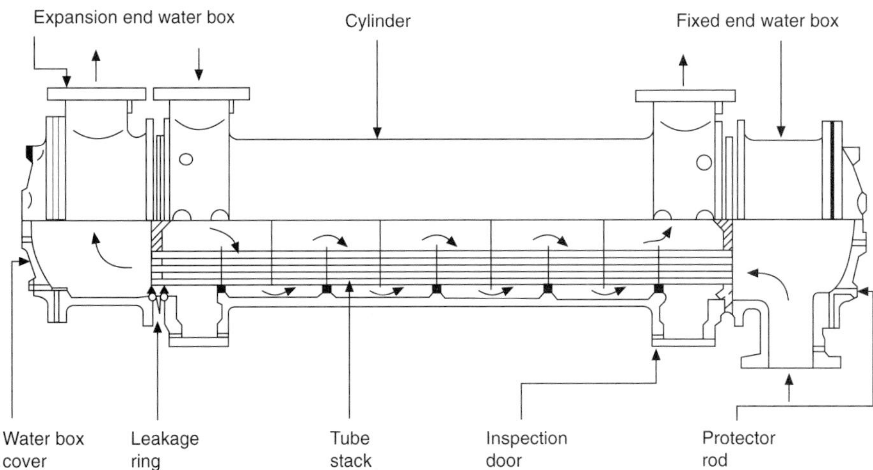

Figure 15.61 Tubular heat exchanger

the fluid paths. Accordingly, small tubes of close pitch are attractive but, against this, practical considerations of maintenance, installation, accessibility and tube life are important factors, Small diameter tubes should be used only where the fluids are exceptionally clean or where high performance and low weight are the chief factors, as in the engine installations of high-speed motor-boats, tractors and other vehicles. Large tubes are essential for river installations, where the water is subjected to seasonal deposits or to pollution.

The effect of tube size on relative surface area and relative heat transfer for a triangular pitch with pitch/diameter ratio 1.25:1 is given in *Table 15.2*.

The relative surface is established by using the equation:

$$\left[\begin{array}{c}\text{Number of tubes per square foot}\\ \text{of tube-plate occupied by tubes}\end{array}\right] \times \left[\begin{array}{c}\text{pitch of tubes}\\ \text{in inches}\end{array}\right]^2 = 166 \qquad (15.13)$$

This, for a constant pitch-to-diameter ratio, means that the relative surface area is proportional to 1/(tube diameter). The relative heat transfer, in this comparison, is proportional to the product of surface area and the partial heat transfer coefficient, assuming the temperature conditions and the mass velocity are the same. Thus, the partial heat transfer coefficient has been taken as proportional to:

$$\text{(a)}\ \frac{1}{D^{0.2}};\quad \text{(b)}\ \frac{1}{D^{0.4}};\quad \text{(c)}\ \frac{1}{D^{0.5}};$$

for (a) water through; (b) water over; and (c) oil over tubes respectively; D being the tube diameter.

Normally the heat transfer coefficient can be determined for flow through the tubes, because the speed of the cooling medium can be controlled, but the coefficient for flow outside the tubes is affected by the baffle arrangement and by the permissible clearances in the tube-stack. Fluid which deviates from the predetermined flow-path will have an adverse effect upon performance.

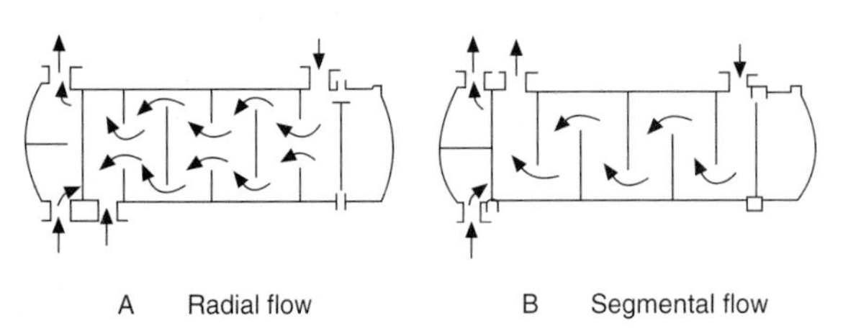

Figure 15.62 Common baffle arrangements

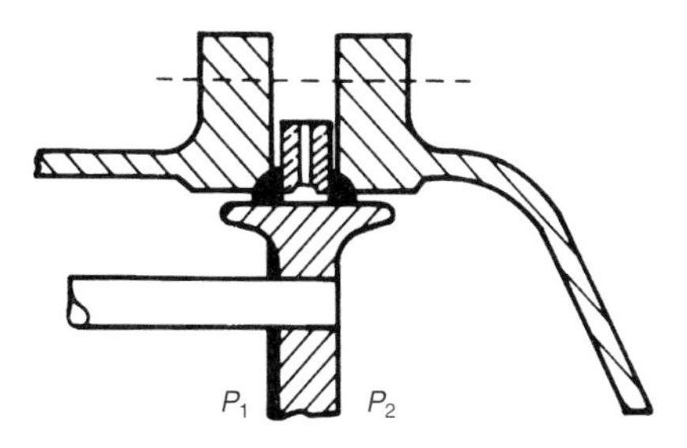

Tube in tension due to
$P_1 - P_2$

Safety leakage ring

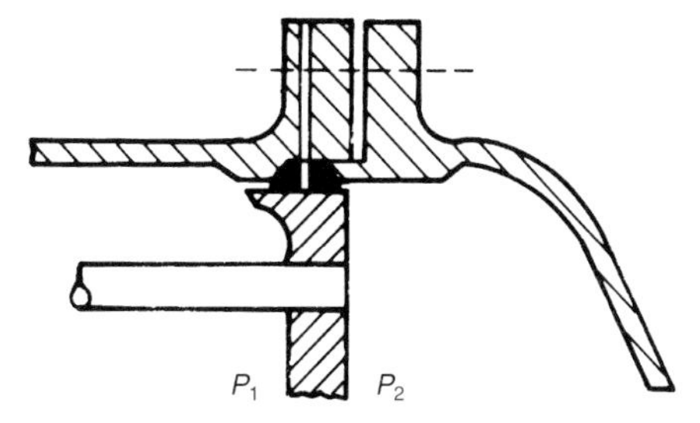

Tube in tension due to
$P_1 - P_2$

Packed gland

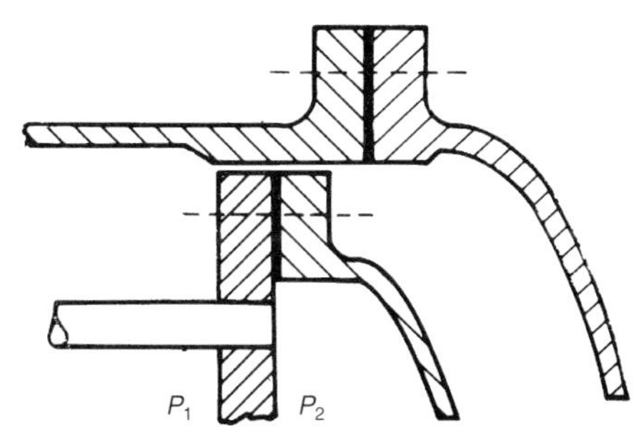

Tube in tension due to
$P_2 - P_1$

Internal floating box

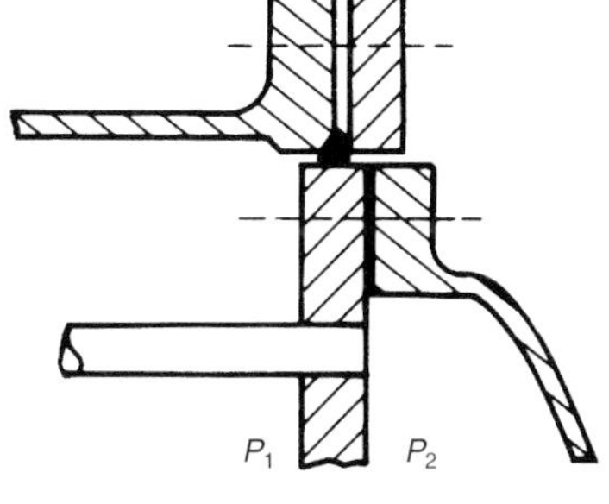

Tube in tension due to
$P_1 + P_2$

External floating box

Tube loading conditions

	Safety leakage type	Packed gland type	Internal floating box type	External floating box type
Shell side pressure-high Tube side pressure-low Load on shell bolts	Tubes in tension Low	Tubes in tension Low	Tubes in compression High	Tubes in tension Low
Shell side pressure-low Tube side pressure-high Load on shell bolts	Tubes in compression High	Tubes in compression High	Tubes in tension Low	Tubes in tension Low

Figure 15.63 Types of expansion joints

Table 15.2

Diameter of tube (mm)	*Relative surface area*	*Relative heat transfer*		
		Water through tubes	*Water over tubes*	*Oil over tubes*
25	1.0	1.0	1.0	1.0
20	1.25	1.31	1.36	1.40
15	1.67	1.85	2.05	2.16
10	2.5	3.00	3.61	3.95
5	5.0	6.90	9.52	11.18

The size of a fresh-water heat exchanger is determined by the amount of cooling surface required. This is obtained from the equation:

$$S = \frac{E}{\theta U} \qquad (15.14)$$

where
S = heat transfer surface area (m^2);
E = heat to be transferred by the cooler (W);
θ = desired mean temperature difference (K);
U = overall heat transfer coefficient (W/m^2 K).

The value of E is normally given by the engine builder. U is determined from practical experience: for normal conditions it varies between 1500 and 3000 W/m^2 K.

By way of example: if 150 kW are to be transferred with a mean temperature difference of say 15 K and a heat transfer coefficient of 2000 (referred to the outside surface), the surface required is 5 m^2. It should be noted that, as surface is inversely proportional to the mean temperature difference, it is as important to establish proper temperatures as it is to obtain suitable heat transfer rates.

Tubular heat exchangers are used as oil coolers as well as for water cooling applications but their performance is more affected by variations of tube pitch, baffle, and tube clearance, etc. The materials used are similar but it is essential that all surfaces in contact with the oil are free from foundry sand or mill scale. The size of an oil cooler is calculated in a similar way to that used for heat exchangers for water, but it is the partial heat transfer coefficient on the oil side which has the determining effect on size.

There can be benefit, in size and cost reduction, by adding secondary surface to the oil side of the heat exchanger. This is because the heat transfer coefficient between the oil and the heat transfer surfaces is much lower, by a factor of as much as 10 or 15, than the coefficient on the cooling water side.

This may be achieved by rolling a thread form into the surface of the tubes used, so creating helical fins of low height. A more usual construction for smaller coolers uses copper fins, perforated with a pattern of holes to receive the tubes, spaced evenly between the baffles and solder-bonded to the tubes. The unit illustrated in *Figure 15.64* has segmental baffles and the fins are correspondingly cut away.

Figure 15.65 shows the general effect of water velocity on the heat transfer coefficient. Cooling water velocity in excess of 1 m/s usually has little influence on oil cooler size. Excessive water speed only encourages corrosion. The viscosity of the oil at operating temperatures has an effect on the partial heat transfer coefficient; the value of the latter decreases with increase in oil viscosity.

Pressure losses of 1.0 to 1.5 bar on the oil side and 0.2 to 0.3 bar on the cooling water side are normal. Practical values of the overall coefficient U may vary between 300 to 700 W/m^2 K.

For piston and lubricating oil coolers, or where there is a risk of excessive fouling, consideration must be given to the addition of allowable margins.

Figure 15.66 shows the progressive reduction in performance of a piston oil cooler caused by carbonized sludge.

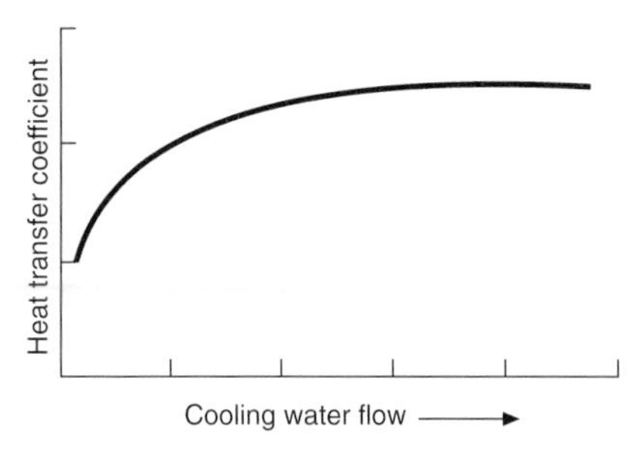

Figure 15.65 Effect of cooling water flow on overall heat transfer coefficient in an oil cooler

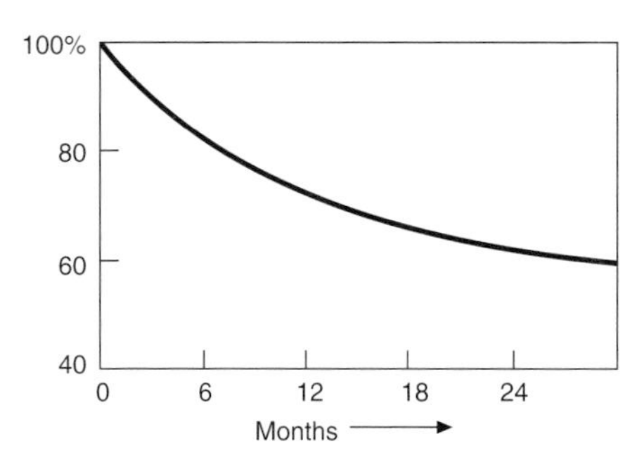

Figure 15.66 Reduction in oil cooler performance due to fouling

15.3.8.2 Plate heat exchangers

Plate heat exchangers were originally developed for applications such as milk cooling. They are now employed for engine cooling.

The element of heat transfer surface is a plate pressing containing four ports, one at each corner of the plate. A gasket, housed in a continuous groove formed in the pressing, follows the periphery of the plate and also surrounds two of the ports, usually on the same side of the plate. The pressing usually incorporates corrugations, in any one of a variety of more-or-less complex patterns, in the area of heat exchange, the function of which is to promote additional turbulence, thereby enhancing heat transfer, and to provide frequent contacts between adjacent plates for support against the load imposed by difference in pressure between the two fluids on opposite sides of the plate. A typical plate is shown in *Figure 15.67*.

A further feature of the pressing is a portion removed between the ports at each end, so shaped that it will form a means of hanging the plate on a suitably shaped rail at the top and locating it on a guide at the bottom.

The mode of operation of this type of heat exchanger is illustrated in *Figure 15.68*. Several heat exchanger plates are fitted into a frame comprising an end plate, a stanchion and top and bottom rails. The plates are hung from the upper and also register on the lower rail so that they are aligned to ensure correct engagement of each seal against the neighbouring plate.

Figure 15.64 Secondary surface water-cooled oil cooler

Figure 15.67 Heat exchanger plate

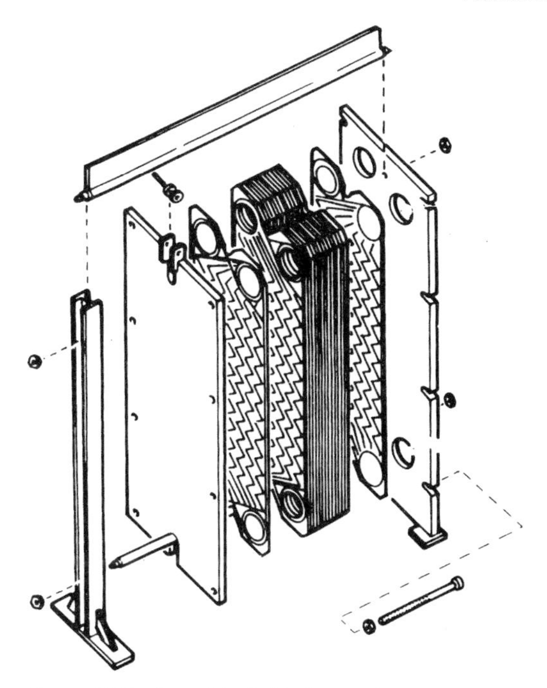

Figure 15.68 Assembly of plate heat exchanger

Alternate plates are inverted so that the fluid flowing through one port can enter the space between two adjacent plates, when the plates are in contact, and leave by the port below it, whilst the other two ports are isolated from that space by the portion of the seal surrounding those ports. The second fluid enters and leaves the intervening spaces between plates by menas of those other ports.

The seal is so designed that there is a double area of contact with the adjoining plate in those portions isolating the two ports. With the channel between these two areas open to atmosphere, there can be no leakage past the seal between the two fluids. A substantial compression plate, also hung from the upper rail, serves to compress the stack of plates against the end plate, bringing the seal on each plate into intimate contact with the adjacent plate. Screwed tie bars effect the compression. The end plate incorporates branches for connection to the system pipework. The plates are usually of a highly corrosion resistant material such as titanium.

15.3.8.3 Air cooled secondary surface units

The addition of fins on the cooling air side of heat exchangers used in the cooling of the hot liquids associated with the diesel engine provides considerable benefit by way of reduction in size and cost, compared with plain tube units. The hot liquid passes through tubes of circular, oval, or flat-sided section and the cooling air between the fins attached to the outside of the tubes.

There are several forms of this type of heat exchanger. Four typical examples are shown in *Figure 15.69*. The pack-construction, or tube-and-centre, type (a) is largely confined to use on motor cars and light vehicles where vibration and shock are at a low level. This type is constructed using flat-sided tubes interlaced with lengths of corrugated copper fin soldered to the tubes. For cooling engines in the low and medium power range, in other mobile and for stationary applications, the single block fin-and-tube radiator (b) is mostly used. In this, flat-sided tubes are laced through and soldered to continuous fins.

The much larger unit (c) is constructed from separate, narrow sections of fin-and-tube type, each with its own headers at each end. The header branches are inserted and sealed into manifolds at top and bottom, so enabling the hot fluid to flow in parallel through the tubes of several sections together. The sections are removable for ease of cleaning and replacement. This type is relatively easy to design so that different parts of the same unit are devoted to the cooling of the various engine hot fluids.

Ribbon tube coolers (d) are used in large stationary installations where weight and compactness are less important but where low running cost is sought.

A typical value of overall heat transfer coefficient, referred to the air-side (finned) surface, for an air cooled water radiator such as might be used for cooling engine jacket water is 100 W/m^2 K. If the mean temperature difference between water and air is 60 K and the rate of heat dissipation required of the radiator is 600 kW, the total heat transfer surface in contact with the air must be 100 m^2, taking the typical value for overall heat transfer coefficient above.

The rates of heat transfer surface area to the projected facial area of a radiator varies, of course, with both pitch of fin and with depth in the direction of air flow but with a pitch of 2.5 mm and depth of 100 mm is about 40. This gives an approximate indication of the facial area required.

The heat transfer properties of oil are considerably poorer than those of water and consequently the lower heat transfer coefficient between the tubes and the lubricating oil flowing through them leads to a lower overall heat transfer coefficient. Since, however, the heat rejected to the lubricating oil is normally substantially lower than from the jacket water, the oil radiator is normally considerably smaller.

(a)

(b)

(c)

(d)

Figure 15.69 a, b, c and d Types of secondary surface air-cooled units

The power required to drive the fan passing the cooling air through a radiator varies considerably with the application but 2% of engine power is a reasonable figure for a large installation where compactness of the cooler is not of great importance. For compact, low-weight radiators, where space saving is essential, power consumed may exceed 5% of engine power. Normally, high efficiency axial flow fans, with aerofoil blades are used, giving a static efficiency in the region of 70%.

For locomotive engine cooling applications, it is not unusual to build all the cooling equipment, with its accessories, into a cooling group. A modern example of this is shown in *Figure 15.70*. In this particular design, the whole structure is made of aluminium, including the heat exchange surfaces, in order to minimize weight.

A hydrostatic drive is used to transmit power from the engine to the fan and the speed of the fan is controlled automatically according to the engine coolant temperature. The basic circuit is shown in *Figure 15.71*. A hydraulic pump is driven from the

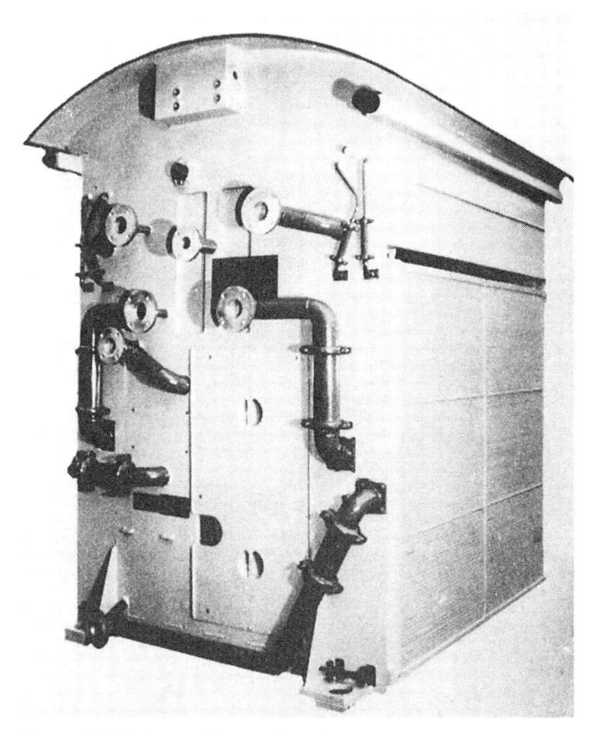

Figure 15.70 Locomotive cooling group

engine and, when the engine is cold, the flow of hydraulic fluid is diverted, through the temperature controlled valve, back to the pump suction. As the engine warms up, the control valve begins to close, generating a pressure difference across the hydraulic motor, which then commences to rotate, driving the fan. Further increase in engine coolant temperature increases fan speed.

The system therefore provides a means of thermostatic control of the engine coolant temperature by varying fan speed, the fan rotating only at the speed necessary to provide the cooling requirements of the moment in time. Since the cooling air flow required will vary not only with engine load but with ambient air temperature also, fan speed can vary over a very wide range, particularly in such applications as locomotive engine cooling. In practice, over a year's running in a temperate climate, the power absorbed on average throughout the year is typically only 20% of what would be required if the fan were driven at maximum speed continuously. No other thermostatic control device is required in the system.

In such systems, positive displacement hydraulic pumps and motors are used, such as the one illustrated in *Figure 15.72*. Units of this type operate at very high efficiency and enable 80% of the power taken from the engine to be transmitted to the fan.

15.3.8.4 Pressure-charge air coolers

In cooling engine charge air by means of water, greater compactness can be achieved by the addition of fins to the tubes carrying the water. *Figure 15.73* illustrates three forms of finned surface commonly used for this application. Two of these are of the fin-and-tube variety, the one with round and the other with oval tubes, whilst the third is in the form of a single tube, ribbon wound.

The choice of construction depends in part on the application of the charge air cooler, the round tube fin-and-tube matrix being used more extensively for smaller coolers, whereas the large oval tube construction is employed for larger, medium and low-speed engines. The oval tube is also capable of passing somewhat larger flakes of material that might be entrained in the cooling water than are the round tubes of bore similar to the minimum dimension of the oval tubes. Where particularly dirty cooling water is expected, the use of large diameter (as much as 25 mm) ribbon tubes are sometimes employed.

Figure 15.74 shows two forms of construction of water-cooled charge air cooler, both of them suitable for use with sea or other raw water as the cooling medium. Side-plates are robust to withstand the pressure of the charge air which may be as high as $2\frac{1}{2}$ bar above atmosphere. The cooler is engine-mounted and must therefore be of generally robust construction in order to withstand the vibration transmitted to it. The cooler must be sealed into the air trunking conducting the compressed air between the turboblower and the engine inlet manifold and the cooler flanges which mate with the air trunking must be made flush.

The most frequent cause of reduction in performance of a charge air cooler is the fouling of the leading edges of the fins due to the adhesion of dust ingested into the induction system, some of it passing through the air filters, enhanced by some wetting of the surfaces by traces of oil from the turboblower. It is therefore necessary to be able to clean at least the inlet face of the charge air cooler periodically. To this end, some installations provide for removal of the charge air cooler 'stack' from the air trunking whilst in other instances, a side plate may be removed giving access to cleaning lanes between the sections of the cooling surface.

Although charge air cooling, as a means of raising engine power output without increasing thermal loading has been cost effective since the early days of introduction of pressure charging, progressive rise in charge air pressure achieved by development of turboblowers has made charge air cooling not only more economically worth while but, in some cases, essential for the operation of the engine.

The criterion of heat transfer performance in a water cooled charge air cooler is often taken as the 'thermal ratio' defined as:

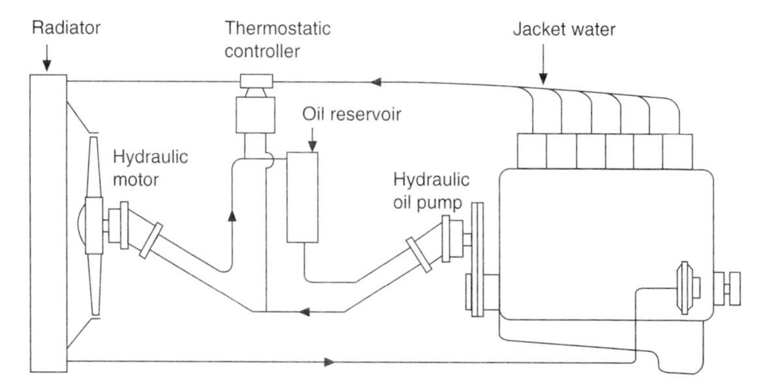

Figure 15.71 Hydrostatic fan drive circuit

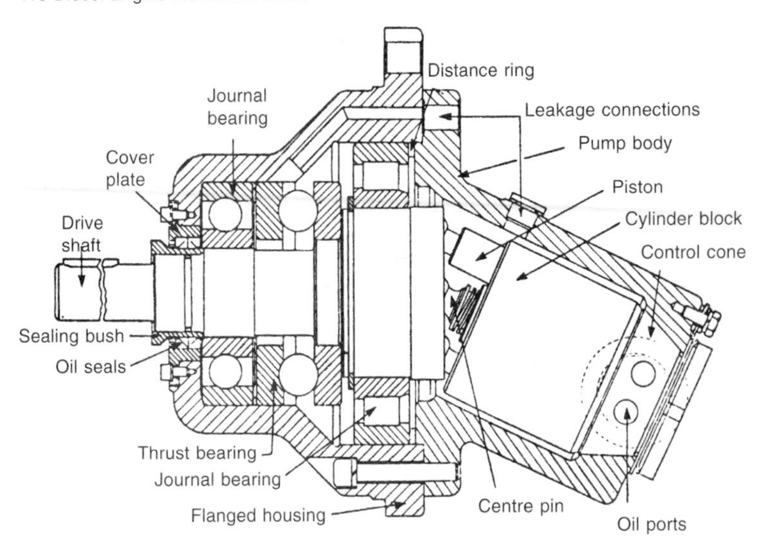

Figure 15.72 Axial piston pump or motor

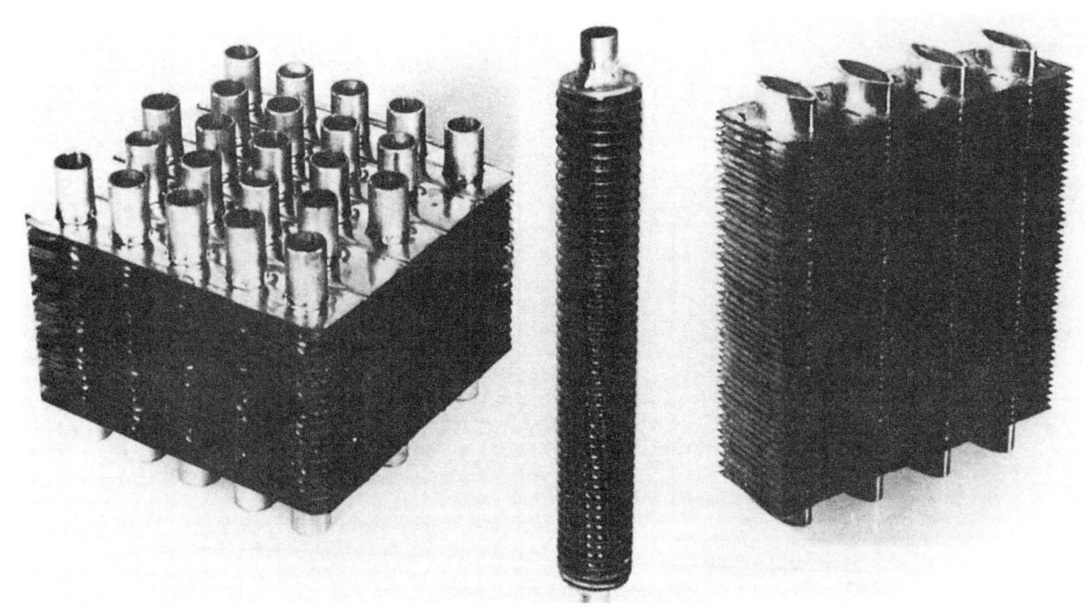

Figure 15.73 Forms of secondary surface for charge air coolers

$$\eta = \frac{\text{(air temperature in)} - \text{(air temperature out)}}{\text{(air temperature in)} - \text{(water temperature in)}}$$

The maximum value of η obtainable, assuming that the unit can be fitted with any desired value of cooling surface area, is dependent on the arrangement of the water flows. Curves of η against specific cooling surface, defined thus:

$$\frac{\text{(cooling surface area)} \times \text{(overall heat transfer coefficient)}}{\text{(mass air flow rate)} \times \text{(specific heat of air)}}$$

are shown in *Figure 15.75* for various flow arrangements and for a ratio of air temperature drop to water temperature rise of 5:1, which may be taken as typical.

The curves for counterflow (the case where the two fluids are flowing in substantially the opposite direction) and parallel flow (where they are flowing in the same direction) are shown in the graph to illustrate the limits that can be achieved. In fact, in normal charge air coolers the cooling water flows at right angles to the air flow and provides what is known as the crossflow arrangement. This may be modified by passing the cooling water first through tubes in the rear half of the cooler and then, using a suitable header arrangement, returning it through the front half, thus giving a two pass arrangement.

In yet other coolers, four or more passes may be employed, the greater number of passes, the nearer the performance coming to the countrerflow case. It can be seen, from *Figure 15.75* that, with this ratio of air temperature drop to a water temperature rise, the cooling surface area required in the cooler must be approximately doubled to achieve a thermal ratio of 0.9 as against 0.8, assuming that the overall heat transfer coefficient remains constant. Most charge air coolers are designed to operate with a thermal ratio of around 0.8, although there has been a tendency to design for a somewhat higher value in engine installations using a turbocharger with a pressure ratio exceeding 3.

If the charge air enters the charge air cooler at a temperature

(a)

(b)

Figure 15.74 a and b Water cooled charge air coolers

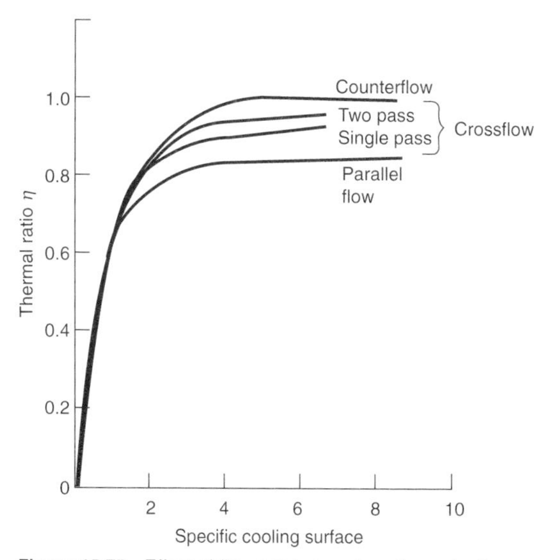

Figure 15.75 Effect of flow arrangement on thermal ratio

of 150°C and the cooling water is available at 30°C, this gives an extreme temperature difference between the two of 120 K. If the charge air cooler is designed to operate at a thermal ratio of 0.8, the charge air will emerge from the cooler at 54°C. With a charge air cooler giving a 0.9 thermal ratio, the air outlet temperature will be 42, giving a 12 K improvement. If the air inlet temperature, representing the case with a turbocharger of higher pressure ratio, were 200°C, the improvement in temperature drop afforded by the higher performance cooler would be increased by 50%.

As the density of the charge air is increased by cooling, it is lowered by reduction in pressure occasioned by the air passing through the cooler. A reduction in density due to this effect is clearly in direct proportion to the ratio of the pressure at outlet from and inlet to the cooler. For this reason, it is possible to tolerate a higher pressure drop through a cooler operating with a turbocharger giving a higher rather than lower pressure ratio Whilst for a system operating at a pressure of 1 bar above atmosphere, a cooler may be designed to operate with pressure loss not exceeding 0.02 bar, in a system operating at 2.5 bar above atmospheric pressure, a pressure drop of in excess of 0.05 bar may be acceptable. Since the air pressure drop through a given charge air cooler with a fixed mass flow of air pumped through it is inversely proportional to the mean density of the charge air, the task of designing a cooler to meet the required pressure loss within a given space envelope is easier in high pressure than in low pressure systems.

Figure 15.76 illustrates a typical performance for a watercooled charge air cooler, showing how the mass water flow and mass air flow rates influence the thermal ratio. The curve of air pressure loss is related to standard density at 1 bar and 15°C.

For large engine installations operating in tropical climates, the charge air is not infrequently cooled directly in a radiator-like air blast cooler, such as that illustrated in *Figure 15.77.* Substantial air ducts are required between the engine and the cooler and the manifolds employed in distributing the charge air throughout the many tubes in the cooler must be of ample proportions to ensure good distribution. Such air cooled charge air coolers are often designed to operate with a thermal ratio exceeding 0.9 and sometimes even above 0.95. This approaches

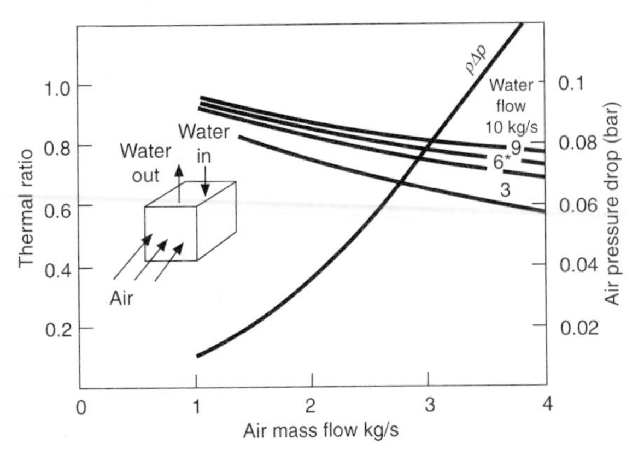

Figure 15.76 Typical performance of charge air cooler

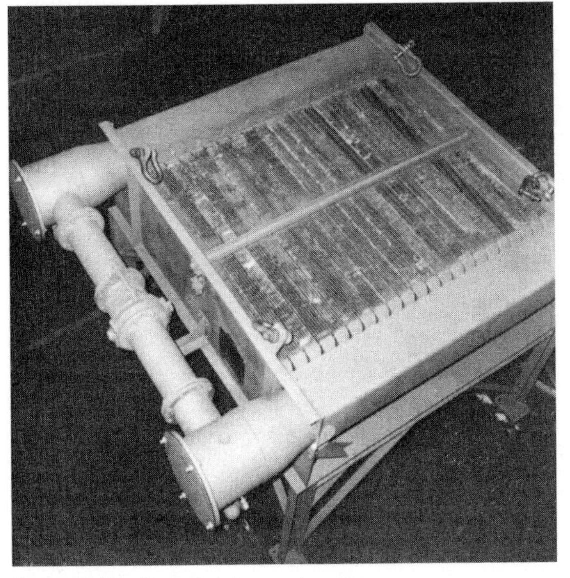

Figure 15.77 Air cooled charge air cooler

the maximum possible cooling (apart from resorting to an evaporative system, which is unacceptable in many hot climates, because of the lack of availability of make-up water).

Direct air cooling is often of particular value in reducing the degree of derating of engine power output in tropical climates.

15.3.9 Materials

15.3.9.1 Tubular heat exchangers

Cylinders are usually made of cast iron or welded steel but may be of cast aluminium for oil cooler applications. Light alloys are also used where non-magnetic properties are desired. Water boxes are mostly made of cast iron, sometimes protected by a coating of synthetic rubber, nylon or a synthetic resin, with gun metal as an alternative for use in more corrsive waters.

Tube plates are usually of naval brass or Munz metal. The most usual material for the tubes is aluminium brass but the more expensive 70/30 copper nickel alloy sometimes being necessary for use in polluted waters, such as are encountered in some estuaries as well as a few marine applications. The general specifications for these materials are given in *Table 15.3*.

For those heat exchangers which employ the engine jacket water as the cooling medium, such as oil coolers, the tubes may be made of copper and the tube plates and headers of 70/30 brass.

In tubular heat exchangers with secondary surface, the fins are normally of copper, soldered to the tubes, but may also, in some constructions, be of aluminium, intimate contact with the tubes being achieved by expansion of the tubes themselves.

15.3.9.2 Plate heat exchangers

For diesel engine cooling, using raw water as the cooling medium, the plates of plate heat exchangers are now almost invariably of titanium, giving complete assurance against corrosion under all normal operating conditions, even with heavily polluted water.

The frame, end plate, and compression plates of the heat exchanger may be of carbon steel but the end plate usually contains branch inserts of titanium, conducting the raw water to and from the ports in the plates. Seals fitted to the plates are of synthetic rubber.

15.3.9.3 Secondary surface radiators

Tubes and tube plates are usually of brass, either the 70/30 or 63/37 alloy, whereas the fins are invariably of high conuctivity copper containing a minor element of either cadmium or tin to enhance the stiffness and reduce softening during the soldering process of bonding the tubes to the fins.

In small radiators, brass headers are nrmally used, soldered to the tube plates,whereas for somewhat larger radiators headers bolted to the tube plates may be of cast iron or fabricated from carbon steel. Side plates are usually fabricated from carbon steel but may be lead-coated.

Carbon steel is also used in the fabrication of manifolds for still larger radiators based on the use of sectional construction or with ribbon tubes.

Table 15.3

(a) Aluminium brass

Specification	*BS 2871 Pt. 3*	*ASTM B111/79*
	%	%
Copper	76–78	76–79
Aluminium	1.8–2.3	1.8–2.5
Arsenic	0.02–0.06	0.02–0.10
Total impurities	0.30 max.	*0.13 max.
Zinc	remainder	remainder

* Of which lead must not exceed 0.07%.

(b) 70/30 Copper nickel

Specification	*BS 2871 Pt. 3*	*ASTM B111/79*
	%	%
Nickel	30–32	29–33
Iron	0.4–1.0	0.4–1.0
Manganese	0.5–1.5	1.0 max.
Zinc	—	1.0 max.
Lead	—	0.05 max.
Total impurities	0.30 max.	—
including sulphur	0.08 max.	—
Copper	remainder	65 min.

15.3.9.4 Charge air coolers

For those parts in contact with the sea water, the materials used are the same as for tubular heat exchangers.

Fins are of copper, except in the case of ribbon tubes where the ribbon itself may be aluminium, tension wound on to the tube, to provide the intimate contact between the tube and the ribbon.

15.3.10 Corrosion

Tube corrosion may be due to deposit attack, impingement, pitting or de-zincification.

15.3.10.1 Deposit attack

Deposit attack is caused by a layer of non-adherent matter, which usually lies on the lower half of the tubes, where these are in a horizontal position. This is a usual hazard in inshore installations, such as estuarine power stations, in coastal vessels and, indeed, in any installation which is liable to shut-down periods of considerable length. In all such instances there is likelihood of mud or other solid matter being deposited on the tubes. As a result there is danger of corrosion taking place beneath this layer, particularly at temperatures much above normal.

Methods of prevention include the use of vertical heat exchangers, so that deposits cannot lie in the tubes. Reasonably high water speeds also help, by sweeping away any deposits that are likely to form. Mud doors should be provided on heat exchangers for examination purposes. If a system is being shut down for any length of time, it is advisable to drain and wash out the tubes.

15.3.10.2 Impingement attack

This type of corrosion is caused by the mechanical effect of a stream of water at high velocity. This removes the protective film on which the corrosion resistance of tube material depends. The effect is aggravated by the presence of bubbles of entrained air in the water. Until the advent of the newer corrosion-resistant alloys, this was probably the most common and certainly the most serious form of corrosion. Today it is less frequently experienced.

Methods of prevention of this type of attack include:

(a) Endeavouring to obtain an even velocity of flow through all tubes. In certain circumstances it is possible to have a water speed through one particular tube, or group of tubes, or cooler in the case of a number of coolers in parallel, many times higher than the nominal, or general, velocity.
(b) Avoiding the cutting down of the size of water boxes because, if these are too small, uneven water distribution and the creation of eddies will be encouraged; small boxes also tend to induce concentration of entrained air in one part of the heat exchanger. Entrained air encourages corrosion by impingement attack, hence provision should be made for the release of air.

In multi-flow heat exchangers there should always be a sound joint between the water box and the tube-plate; otherwise there may be very high velocities at this point, with risk of corrosion. If there is no iron or steel in the water system steel protector plates, or rods, should be provided. Steel has a beneficial effect on the resistance of modern tube alloys. In gunmetal water boxes the iron can be provided by steel-spraying the surface.

Heat exchangers of widely differing sizes should not be circulated in series by the cooling water as this can cause unduly high water velocities in the smallest unit. The minimum amount of cooling water necessary is a characteristic feature of a good design. It is always desirable to avoid partial obstructions near or in the tubes. All connecting pipework should be thoroughly cleaned before installation, and weed boxes should be fitted where there is a risk of mussels, shells, etc, obtaining ingress to the tubes. The latter cause local increase in water velocity and the products of decomposition are extremely corrosive. This attack is also influenced by pressure and can be greatly accentuated by excessive vacuum.

15.3.10.3 Pitting

Pitting may be due to one or more of several causes. The deciding factor often lies in the operation circumstances at the outset of the life of the heat exchanger. Pitting is, in actuality caused by a breakdown—usually quite local—in the protective film of the tube.

The essential factors are often most difficult to determine; accordingly it is not easy to enumerate definite rules for prevention, but the following points should be emphasized:

(a) The use of steel protector rods, or plates, assists greatly in the early formation of a good protective film;
(b) It is desirable to avoid anything which may destroy the film;
(c) Endeavours should be made to ensure good conditions during the initial stages of operation, thereby assisting film formation on the inside of the tubes.

15.3.10.4 De-zincification

This is a very distinctive form of attack, where large or small areas are corroded, copper remaining or being redeposited in a spongy condition. Copper nickel, containing no zinc, is by definition free from this type of corrosion. In aluminium brass the inclusion in the alloy of small amounts of arsenic makes it extremely resistant.

The usual cause of de-zincification is high temperature on the sea water side. This is liable to happen during a shut-down, because, in addition to the likelihood of high water temperature, there is stagnation—which is an additional predisposing cause.

Methods of prevention, particularly with brass tubes, include the use of alloys containing arsenic. When in operation, it is advisable to avoid very high temperatures at the sea water outlet. When shutting down the system, it is desirable to drain the raw water from the heat exchanger and so keep the internal surfaces free from deposits.

15.3.11 Maintenance

The maintenance of charge air coolers is particularly important, since build up of dirt on the air inlet face can cause excessive pressure loss, which can eventually lead to malfunction of the engine. There are many methods of cleaning the air side heat transfer surface of the cooler, depending upon the particular installation. Before undertaking the task the engine-maker's manual should be consulted. The heat transfer performance of a cooler may also be substantially diminished by partial blockage of the tubes and the resulting high charge air temperature leaving the cooler may ultimately result in damage to the engine. Monitoring of charge air temperature to pre-empt this is therefore highly desirable, so that action can be taken to clean the water side of the cooler before damage occurs.

Closed system heat exchangers and radiators seldom require attention but surfaces should be examined from time to time and, if necessary, should be cleaned by the method recommended by manufacturers.

When an installation is to be laid-up for long periods the raw water side of tubular units should be washed out, thoroughly dried, and washed out again with clean fresh water. On the engine water side the procedure recommended by the engine manufacturer should be followed. In replacing a tube it is very important to ensure that the new tube is of the same material.

The outside surfaces of secondary surface air cooled units should be brushed clean, washed with a degreasant or blown out with a high pressure air line at intervals necessary to maintain performance. A caustic degreasant must not be used with aluminium fins.

15.3.12 Water treatment

For the initial filling of closed systems, the water used should be either distilled or very soft fresh water wherever possible: similar precautions should be taken in adding make-up quantities. Impure water containing mud, vegetable matter or chemicals in solution should not be used in a closed system. Where corrosion inhibitors are added to a system, these should be entirely according to the engine maker's specification, since use of an inhibitor not suited to the various metals in the system could result in accelerated corrosion of some components.

Where anti-freeze is required in the water circuit of an air cooled system, again the corrosion inhibitors contained in the anti-freeze solution must be compatible with the materials in the system and the maker's manual should be consulted on this point.

Dosing of raw water using intermittent chlorination has proved effective in reducing the growth of marine organisms, particularly in shipboard and estuarine installations. The addition of ferrous sulphate is thought to have a beneficial effect in the promotion of a stable, corrosion-resistant film on the surface of aluminium brass and cupronickel tubes.

References

1 FRENKEL, Dr. M. S., 'A new speed responsive device for governors', *Engineering and Boiler House Review*, 294–300, October (1951)

2 HUTAREW, G., SCHMID, A. and WUHRER, M., 'Static and dynamic tests of speed governors for diesel engines', *M.T.Z.*, **26,** No. 1, October (1965) (in German) or ASME Paper 67-OGP-3 (in English)

3 WELBOURN, D. B., ROBERTS, D. K. and FULLER, R. A., 'Governing of compression ignition oil engines', *Proc. I MechE*, **173**, No. 22, 575 (1959)

4 BUJAK, J. Z., 'The variable speed hydraulic governor', *Proc. IMechE* **153**, 193 (1945)

5 'Critical considerations in the assessment of different construction methods for centrifugal governors', *M.T.Z.*, February (1954) (in German)

6 CHUTE, R., HEISE, C. J., JOKL, A. L. and GOSS, M. L., 'Engine governors analysis', SAE Paper 784 B (1964)

7 MASSEY, A. G. and OLDENBURGER, L. R., 'Scientific design of a diesel governor', ASME Paper 58-OGP-11 (1958)

8 WELBOURN, D. B., *Essentials of Control Theory for Mechanical Engineers*, Edward Arnold (1963) (particularly Chapters 5–8)

9 'Speed governing systems for internal combustion engine generator units', ASME, Power Test Codes—PTC 26 (1962)

10 CHRISTENSEN, T., 'Future aspects of engine speed controllers', *Motorship*, 93, 94, March (1980)

11 AUSTEN, A. E. W. and LYN, W. T., 'Some investigations on cold starting phenomena in diesel engines', *Proc. I MechE*, Automobile Division, No. 5 (1959–60)

12 BRUNNER, M. and RUF, H., 'Conrtibution to the problem of starting and operating diesel vehicles at low temperatures', *Proc. IMechE*, Automobile Division, No. 5 (1959–60)

13 BURK, F. C., CLOUD,G. H. and AUG, W. E., 'Fuel requirements of automotive diesel engines', *SAE Journal (Transactions)*, **53**, No.1

14 DERRY, L. D. and EVANS, E. B., 'Cold starting performance of high speed diesel engines and their fuels', *Institute of Petroleum Journal*, **36**, No. 319 (1950)

15 BIDDULPH, T. W. and LYN, W. T., 'Unaided starting of diesel engines', *Proc. IMechE*, Automobile Division, **181**, Pt 2A (1966-67)

16 KOMIYAMA, K., OKAZAKI, T. and YAMAZAKI, Y., 'Investigation of cold starting mechanism of diesel engines, Paper D71, CIMAC Conference, Vienna (1979)

17 HOLLINGHURST, R. and NYSTRÖM, C. G., 'European low temperature viscosity requirements for engine oils, and their impact on SAE Classification utilisation', SAE Paper 740975

18 SMITH, J., 'Air-cooled diesel engines', DEUA Publication 326 (1969)

19 HUTCHEON, K. F. and MARKS, R. L., 'Developments in starter motor application to diesel engines', *Proc. IMechE*, **184**, Pt 3A (1969–70)

20 MAYER, W. E., DECAROLIS, J. J. and ESPENSCHADE, P. W., 'The starting of diesel engines at very low temperatures', Paper A19, CIMAC Conference, Wiesbaden (1959)

21 OSHIKA, S. and KAWAHARA, K., 'Study on the starting of diesel engines for passenger cars', *Japan SAE Journal*, September (1979) (Translated by J. Everest, MIRA, Translation 20/82)

22 TIMONEY, S. G., 'Compact long-life diesel engine', *Proc. IMechE*, **183**, Pt 3B, Paper 5 (1968–69)

23 MANSFIELD, W. P., 'Transport engines of exceptionally high specific output', *Proc. IMechE*, **183**, Pt 3B (1968–69)

24 McADAMS, W. H., *Heat Transmission*, 3rd ed., McGraw-Hill, New York (1954).

25 RUDGE, T. H. and FORBES, M. K., 'Cooling Equipment for Diesel Locomotives', *Inst. Loco. Eng.*, **51**, Pt. 2, 202–255 (1961/62).

26 FORBES, M. K. and HAMPSON R. J., *Charge Air Cooling in C.I. Engines*, Diesel Engine Users' Association, Pub. S296 (1964).

27 MONTGOMERIE, G. A. and FORBES, M. K., 'Cooling Equipment for Internal Combustion Engines', *Proc. I. Mech. E.*, **181**, Pt. 1, No. 6 (1966/67).

28 MONTGOMERIE, G.A., FORBES, M.K. and JONES, T.D., *Diesel Engine Cooling and Coolers*, Diesel Engine Users' Association, Pub. 319 (1968).

29 FORBES, M. K. and HAMPSON, R. J., *Some Present-day Aspects of Charge Air Cooling*, Diesel Engine Users' Association, Pub. 347 (1972).

30 MONTGOMERIE, G. A., FORBES, M. K. and GILBERTSON, J. H., 'Optimum Design of Cooling Systems for Today's Environment', *Trans. I. Mar. E.*, **86**, 65–76 (1974).

31 SIMONSON, J. R., *Engineering Heat Transfer*, Macmillan Press Ltd, London (1975).

32 *Standards of Tubular Exchanger Manufacturers' Association*, 6th ed., TEMA Inc., New York (1978).

16 Aircooled engines

Contents

16.1 Introduction

In principle, the ultimate cooling medium for all internal combustion engines is air. However, in 'watercooled' engines the heat from critical components is being extracted by means of the intermediate cooling medium water, and then transmitted to the cooling air in a radiator. The term 'aircooled' is used for engines where the heat of the component is directly transmitted to the cooling air.

Early internal combustion engines used water evaporation for cooling. De Bisschop built an atmospheric engine in 1873 with cooling fins. The first Otto cycle engine with air cooling was probably designed by Forest in 1881. Up to about 1930, however, air cooling was restricted to small engines of low output and for airplane propulsion. The first aircooled high speed diesel engine was built by Austro Daimler in 1927 and produced 15 kW (*Figure 16.1*)[33]. In 1933 Krupp exhibited an aircooled diesel engine of 40 kW output. Little research was done on aircooled engines before 1931 when Marks and Dorman reported on intensive tests[1].

Prior to and during the 1939–45 World War, Army requirements initiated the development of a number of aircooled engines in Europe and in the USA not only for aircraft and vehicular use, but also for small independent power plants. Examples are an 8-cylinder aero-engine of 1937 (*Figure 16.2*) and the truck engine series FL 514 (*Figure 16.3*) of Deutz, developed in 1942[59] which went into production in 1944. This engine shows all the typical features of modern aircooled engines. After the war, a few companies intensified the development for civil applications where the advantages of air cooling summarized below could be best utilized:

Reliability and minimum maintenance requirements
The absence of cooling water avoids problems and damage associated with freezing, boiling, corrosion and cavitation. There is no necessity for a watercooler, cooling water pump, waterhoses, joints and sealing elements.

Indifference to climate
With aircooled engines, the difference between the ambient air temperature and the average temperature of the cylinders remains constant, and the amount of heat extracted is almost independent of the ambient temperature. In watercooled engines the temperature difference between the cooling medium and ambient air and hence the amount of heat extracted decreases with increasing ambient temperature, as water has a definite boiling point. This has to be compensated for by overdimensioned water radiators of large volume and high cost.

Figure 16.1 First aircooled high speed diesel engine by Austro Daimler (1927)

Figure 16.2 Aircooled aircraft diesel engine by Deutz (1937)

Figure 16.3 Aircooled diesel engine FL 514 by Deutz (1944)

Operational temperature is reached very quickly[10]
After starting, the condensation temperature is reached much more quickly in aircooled than in watercooled engines (*Figure 16.4*). As a consequence corrosion resulting from operation below the dew point is minimized, and exhaust hydrocarbon emissions are reduced.

Optimum dimensions and small weight
The cooling system is integrated in the engine design which results in an optimum package volume and reduction of weight. When comparing the weights of watercooled and aircooled engines, it is necessary to ensure that the weight of the watercooled engine includes the cooling system.

Good prerequisites for incapsulation and noise reduction
Investigations have shown that the cooling method of an engine does not influence its noise. The aircooling principle, however, favours further noise reduction. The cooling fan is integrated in the engine design and the amount of cooling air is smaller than for watercooled engines. Therefore it is possible to encapsulate the whole engine quite easily and to use noise-reducing insulation at the duct for the cooling air.

Modular system
The application of separated single cylinders and single-cylinder

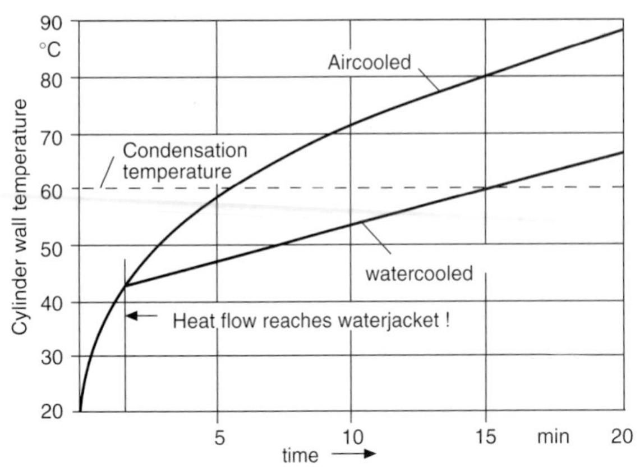

Figure 16.4 Rise of cylinder wall temperature after no-load start, measured in the cylinder wall 1 mm from the inner surface

heads in multi-cylinder engines lends itself to a modular system. This offers the opportunity of using a maximum of common parts for an engine family. In the case illustrated in *Figure 16.5*, 85% of the parts are common to all engines of the family[53,57]. The modular system also allows easy adaption for special applications by changing only a few parts.

Good maintenance (Figure 16.6)

For the same reasons, all maintenance points can be made readily accessible and the exchange and repair of parts is very easy. Within an engine family the parts are interchangeable. This reduces the number of shelf parts and allows for simpler spare part planning.

These advantages have led to the fact that aircooled engines in smaller units, up to about 20 kW, are in widespread use. In the medium power range, up to about 120 kW, aircooled engines are preferred in some areas, for example in construction equipment and in the mining industry, and are successful worldwide also in truck applications.

The disadvantages of aircooled engines can be attributed to the higher component temperatures and to the difficulties in removing heat from some critical areas where sufficient space for the cooling air can not easily be provided (see Section 16.2).

It also must be understood that it becomes less easy to aircool an engine as the cylinder diameter and the length of the flow path of the cooling air increase. This is because the ratio of volume to surface area is increasing, even though the total heat supplied does not increase as fast, especially when piston oil cooling is applied. An increase of the cooling surface by using

Figure 16.6 Accessibility of the cylinder unit

higher fins decreases fin efficiency and is, therefore, not very meaningful. An increase of the cooling air velocity in turn leads to unacceptably high pressure losses. In practice, therefore, bore sizes over 150 mm today are usually watercooled.

These facts require a somewhat more sophisticated technology and intensified development work for aircooled engines, especially for higher power classes, and lead to the introduction of new technologies (Section 16.2).

It is a tribute to such development work that today aircooled engines up to 400 kW are commercially available (Example: *Figure 16.7*). Aircooled diesel engines of higher output, such as the Teledyne Continental Tank Engine[14,20,56] show the potential of air cooled engines, but are limited to special, for example military applications.

Figure 16.7 Deutz BF 12L 413 FC, 400 kW, 2500 l/min (with aftercooler and hydraulic oil cooler)

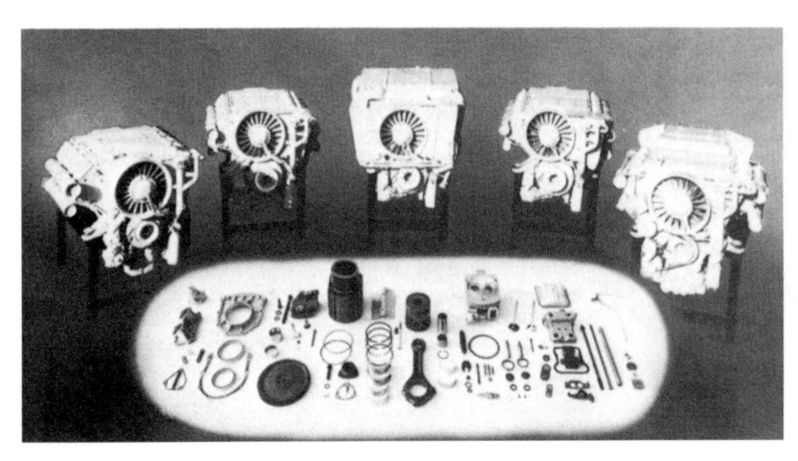

Figure 16.5 Common parts of the engine family FL 413 (Deutz)

The following sections deal only with the features and design aspects of aircooled engines, that are different from their watercooled counterparts. Development efforts in many companies have led to a wide variety of solutions to the problems of aircooled engines, which cannot be covered completely within the scope of this chapter. The designs mentioned here have been used for illustration purpose only, and show mainly the solutions developed over many years by one such company.

16.2 Design features and functional aspects

The direct cooling system makes it necessary to enlarge the cooling surfaces of all the parts to be cooled. This is usually achieved by ribs or fins and is very important as it is ultimately the reason for the individual arrangement of liners and cylinder heads. The design of aircooled engines is mainly determined by this 'unit cylinder' construction and the rest of the design follows on from it—i.e. the low height of the crankcase, the layout and the drive line of the cooling fan, the proper arrangement for the heat exchangers (for engine oil, transmission oil, hydraulic oil, after-cooling), the guidance for the cooling air flow and the optimized air distribution.

To enable the reader to understand these points, *Figures 16.8 and 16.9* show examples of in-line engines. *Figure 16.10* shows a cross-section of a V-engine.

16.2.1 Crankcase

Because of the low overall height the flexural and torsional rigidity of the crankcase is reduced. Therefore the structural strength of the crankcase must be increased.

Whereas the tunnel-bored crankcase is very often adopted for single- and twin-cylinder engines, there are basically important differences in the design of watercooled and air-cooled multi-cylinder engines. These are apparent from *Figure 16.11* where the configurations determining torsional and flexural rigidity are compared.

Figure 16.12 gives some design details. For vertical anchoring of each main bearing cap two (sometimes four) special securing bolts are used: also a good horizontal fit of the cap is most important. The flexural and—more particularly—the torsional rigidity of the crankcase are significantly improved by crossbolts. The hatched areas give an idea at which points the cross-section of the crankcase should be strengthened for optimal flexural rigidity[35].

Figure 16.8 Cut-away view of four-cylinder in-line engine

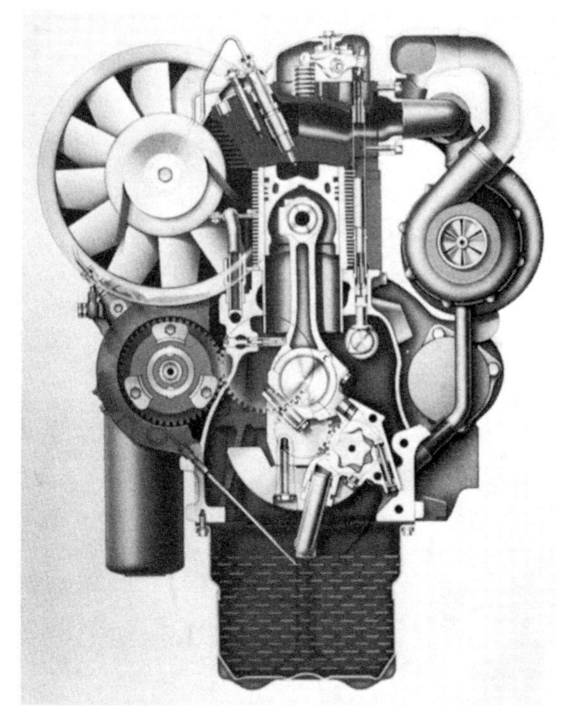

Figure 16.9 Cross-section of a turbocharged in-line engine

16.2.2 Cylinder unit

The design of cylinder head, liner, piston and bolts which connect the cylinder unit to the crankcase are most important for the behaviour of an aircooled diesel engine. Development work therefore is mainly concentrated on these parts, especially with supercharged engines.

Progress in the properties of material, optimized cooling airflow and heat extraction, together with minimizing the mechanical and thermal loading of these parts, make aircooled engines fully competitive with heavy duty watercooled engines.

16.2.2.1 Cylinder head

Figure 16.13 shows sections through two modern heavy duty U-flow heads with the following features[45].

The flanges of the inlet and exhaust ducts are located on the exit side of the cooling air. Apart from advantages with respect to installation, this layout also provides very good access to servicing points especially for V-type engines.

A large finning area is provided on the cooling air inlet side for improved cooling of the fuel injector. The layout of the fins is the result of elaborate aerodynamic studies and the determination of optimum heat flow areas. The inclined position of the valves provides maximum finning of the metal land between them.

A parallel position of valves has some advantages on its side; easier manufacture and less dead volume in the combustion chamber. Smaller engines therefore favour this more simple valve arrangement.

The valve seat inserts of special centrifugally cast iron are shrinkfitted at a temperature difference of 250°C.

Cross flow cylinder heads—with the inlet duct on the cooling air entrance side—are also used. Special care has to be taken to provide sufficient cross-section for the cooling air and for the combustion air especially with high swirl DI-combustion systems.

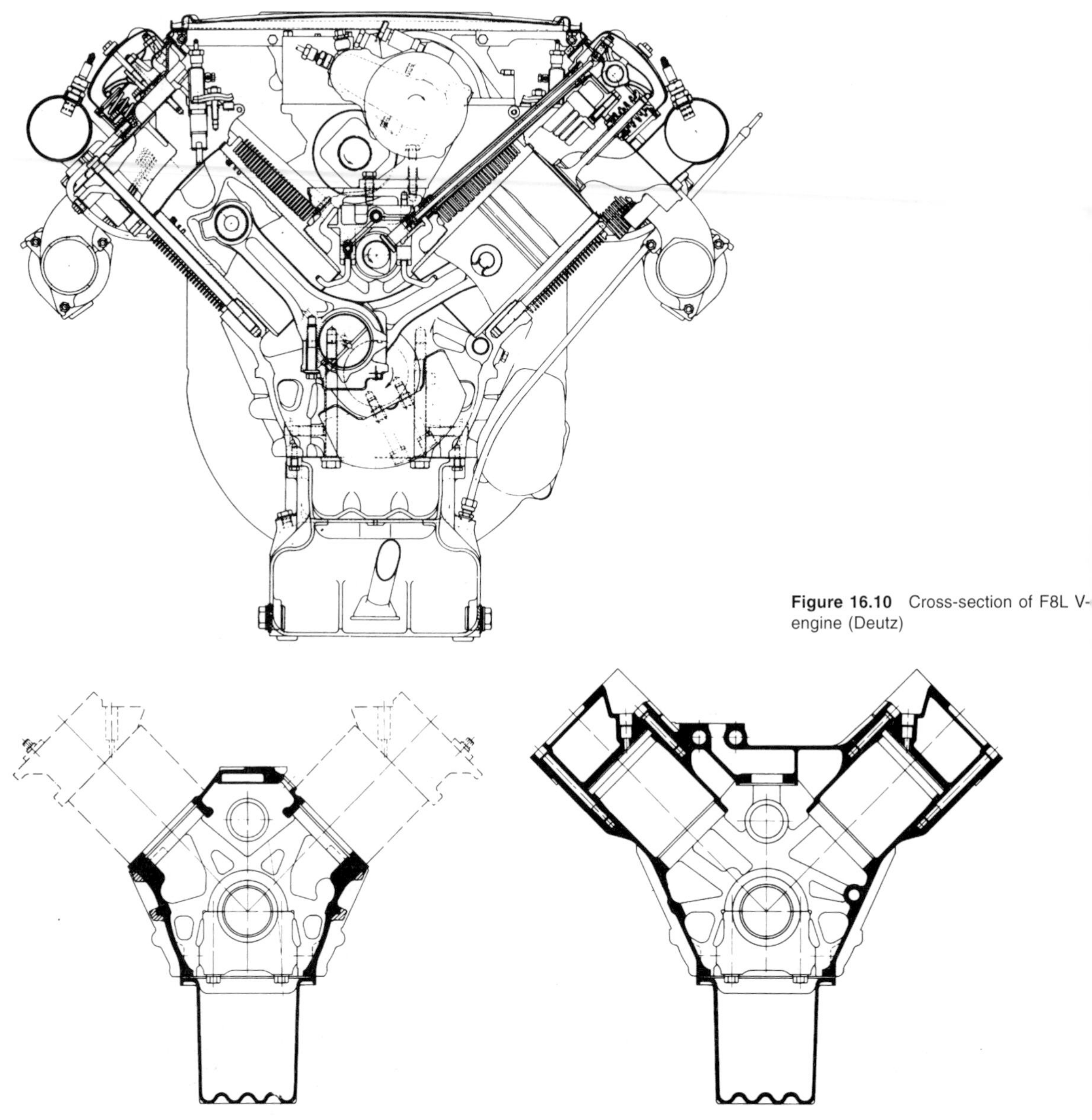

Figure 16.10 Cross-section of F8L V-engine (Deutz)

Figure 16.11 Configurations determining rigidity; comparison of an aircooled engine and a watercooled engine

Provision of integrally cast sheet-metal valve bridge inserts keep the high thermal loads on the ridge away from the actual head while not affecting heat flow in the direction of the bridge fins. So even extreme changes of load will not give rise to thermal stresses in this critical zone of the head (*Figure 16.14*).

The combustion forces are introduced into the head in an optimum way by supporting the holding-down studs or bolts at the upper head surface. Even distribution of the forces on head and joint thus ensures maximum resistance to distortion, and also enhances the head/liner sealing effect at heavy load changes. This makes it possible to manage with only three bolts, but more often four and in some cases five or six bolts are used.

The head is made of material of high temperature fatigue strength (*Figure 16.15*). Significant progress has been achieved over the past few years by intensive metallurgical investigations and improved casting techniques in conjunction with specially developed testing procedures. These procedures also permit better quality control so as to ensure consistent head quality.

By improved heat treatment processes for aluminium alloys, material hardness is considerably increased without affecting ductility (*Figure 16.16*). This feature opens up further perspectives for the use of light alloy cylinder heads, whereas grey cast iron and its alloys seem to be suitable for engines with limited heat rejection only. New developments in cylinder head design include efforts to reduce the heat rejection from the exhaust duct to the head. *Figure 16.17* shows a cast-in ceramic port liner.

Some development stages improving the head/liner joint are indicated in *Figure 16.18*. Stage 1 uses a composite steel gasket in connection with an internal locating recess. This, however, resulted in sticking of the cylinder head to the liner by coking. The problem was solved by adopting an externally located recess in connection with a compressive sealing land which dispensed

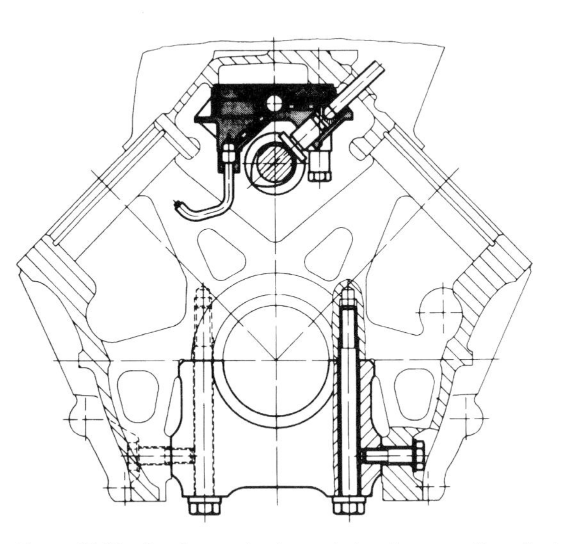

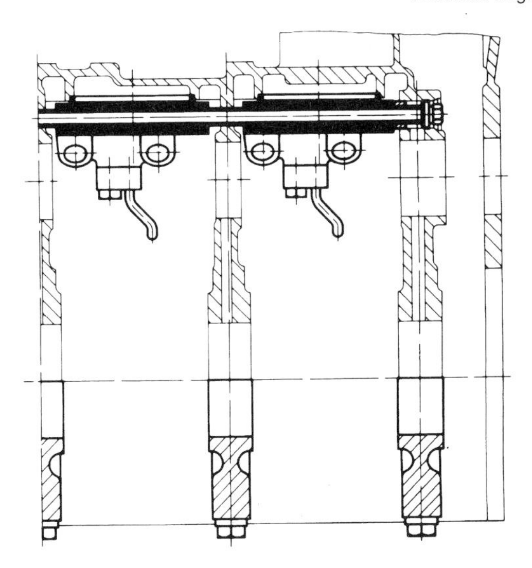

Figure 16.12 Crankcase showing main bearing cap with vertical bolts, cross bolts and tappet carrier

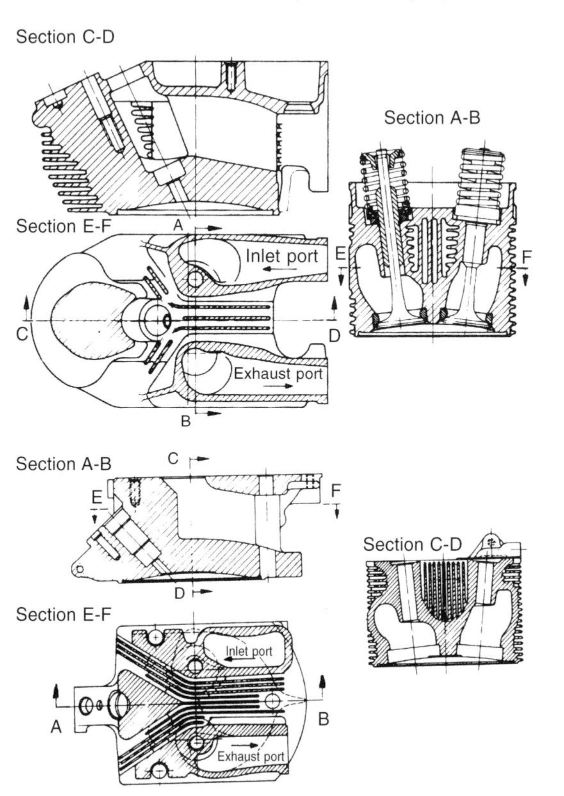

Figure 16.13 Aircooled cylinder heads

with the need for a steel gasket (stage 2). A sturdy design of the cylinder head together with higher hardness of the cylinder head material allowed elimination of the sealing land (stage 3).

Stage 4 introduces a special steel ring which also allows shimming for minimum piston clearance. This is of special importance for modern low swirl DI-combustion systems. Their low thermal load factor is a prerequisite for a higher degree of supercharging. Temperatures in the bottom deck of a light alloy cylinder head with a DI-and IDI-combustion system at full load are compared in Figure *16.19*.

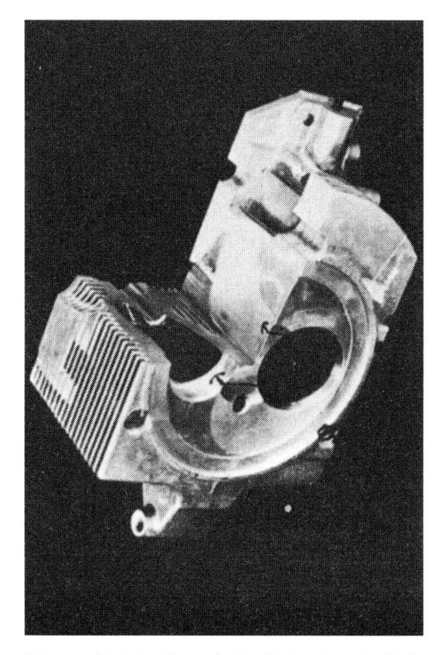

Figure 16.14 Aircooled cylinder head with inserts in the valve land

16.2.2.2 Liner

Intensive development work has been carried out and still continues, to transfer the heat rejection to the cooling air at acceptable cylinder-wall temperatures.

Many publications are available dealing with the heat dissipation of the cylinder liner, the design of cooling fins and ducting of the cooling air, especially baffle plates[12,13,15,16,21,24,26]. Therefore, only the main parameters are mentioned here. The heat to be transferred to the cooling air from the cylinder-wall amounts to

$$q = \Delta T \times \alpha \times \frac{2h + \tau}{\tau} \times \eta_R$$

q = specific heat rejection (W/m^2);

ΔT[K] = difference between wall temperature in the fin base and cooling air temperature (K);

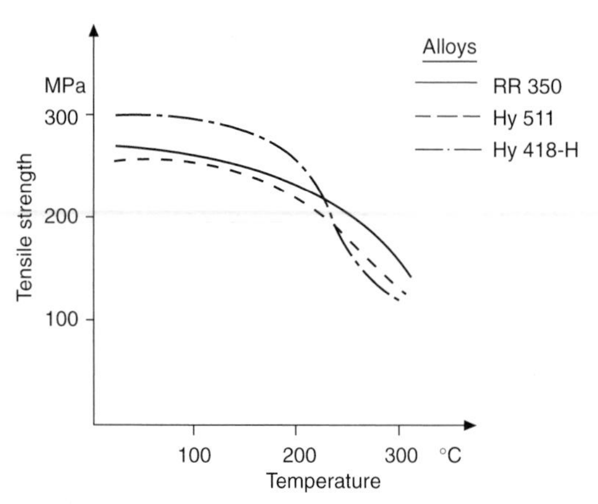

Figure 16.15 Tensile strength of die casting aluminium alloys at high temperatures

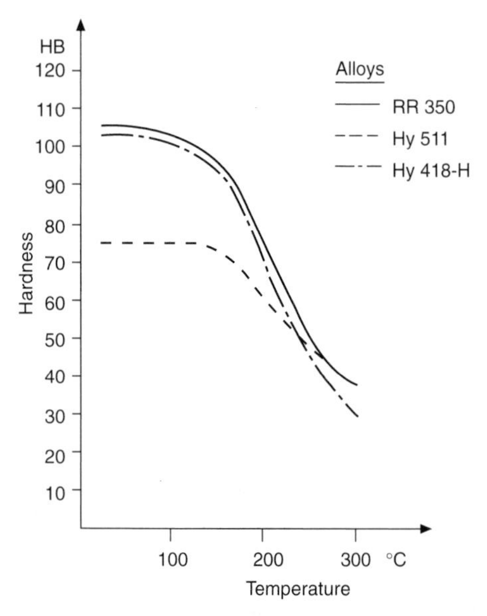

Figure 16.16 Hardness of die casting aluminium alloys at high temperatures

α = heat transfer coefficient (W/m^2.K);
h = fin height (m)
τ = fin spacing (m)
η_R = fin reduction factor (fin efficiency).

The term $\dfrac{2h + \tau}{\tau}$ indicates the increase of the cooling surface.

The fin efficiency η_R takes into consideration the decrease of fin surface temperature with increasing fin height.

$$\eta_R = \frac{\tan h H}{H}, \text{ where } H = h \times \sqrt{\frac{2\alpha}{\lambda \cdot S_m}}$$

with S_m = mean fin thickness (m);
λ = thermal conductivity (W/m.K).

Figure 16.20 plots the fin reduction factor η_R against H. The effectiveness of a finning decreases with

increasing height h;
increasing heat transfer α;
decreasing thermal conductivity λ;

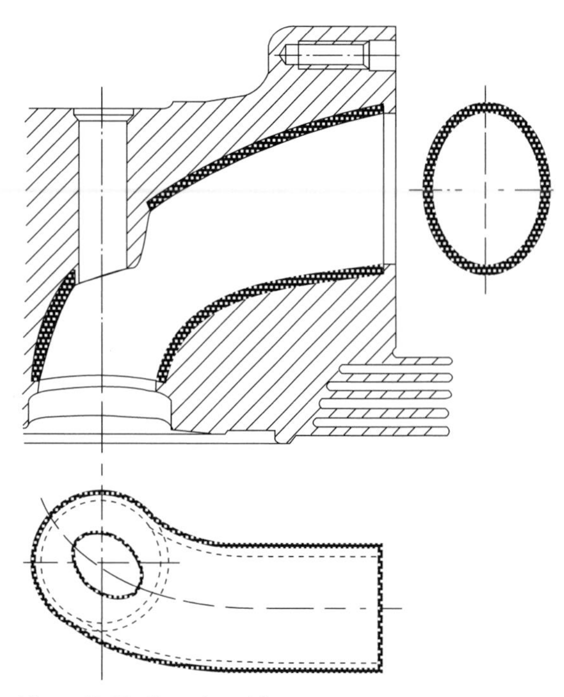

Figure 16.17 Ceramic port liner

decreasing fin thickness s_m;
increasing outer cylinder diameter $d_b = 2r_b$.

For cast iron cylinders of common design the following data may serve as examples:

fin height h = 20–25 mm
fin thickness s_m = 1.5–2.5 mm
fin spacing τ = 4–8 mm
thermal conductivity λ = 49 W/m.K
heat transfer coefficient $\alpha \leq$ 120 W/m^2.K

These figures apply for a fin design similar to *Figure 16.20*, a cooling air velocity of 15–20 m/s and a static pressure drop of the cooling air of $\Delta p_{stat} \leq$ 20 mbar. They lead to

$$\frac{2h + \tau}{\tau} \cong 6\text{–}14$$

$$H \cong 0.8\text{–}1.2$$

$$\eta_R \cong 0.6\text{–}0.8$$

$$\frac{r_b + h}{r_b} = 1.2\text{–}1.6$$

With ΔT = 170°C at the cylinder entrance, the specific heat rejection results in $q \leq$ 180 kW/m^2 which is sufficient for naturally aspirated, or slightly turbocharged engines.

As the cooling air is heated up passing the fins the full value for the liner heat rejection must be calculated step by step. Under similar conditions a grey-iron cylinder with cast-around light alloy fins (λ = 155 W/m.K) would roughly double the cooling capacity, if the fin height is increased accordingly. With machined fins, fin spacing and/or fin thickness could be decreased to further enhance the cooling capacity.

In multicylinder engines, however, the fin height is limited, as it leads to an increase in cylinder centre distance which normally cannot be accepted. To avoid this cylinders are flattened to keep the centre distance as small as possible. Modern aircooled engines

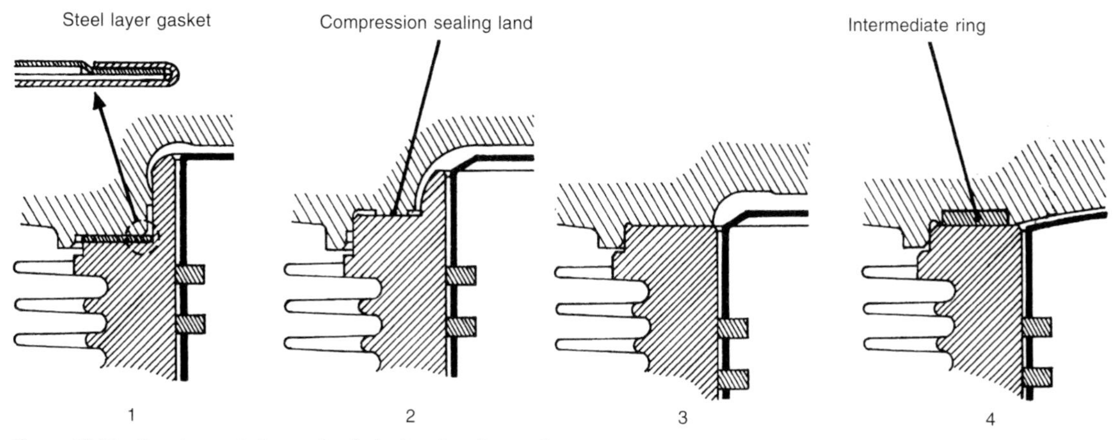

Figure 16.18 Development stages of cylinder head sealing surfaces

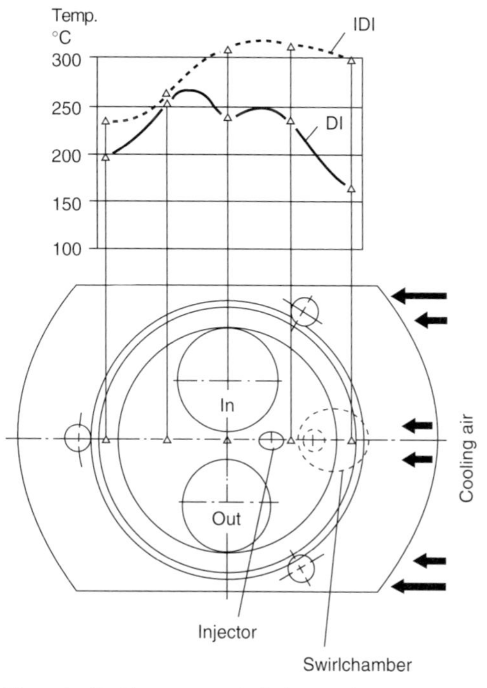

Figure 16.19 Temperatures in the bottom deck of a light alloy cylinder head with a DI and IDI-combustion system

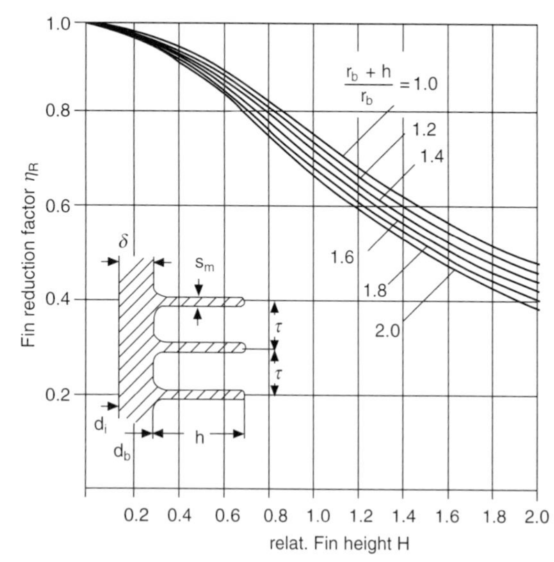

Figure 16.20 Fin reduction factor

reach a centre distance to bore ratio of 1.27 which is only slightly higher than that for well-designed watercooled engines.

As well as guiding the piston and providing cooling, the liners of most of the aircooled engines are designed to transmit the gas and sealing forces between crankcase and head with minimum distortion. These forces are active both in the axial and radial direction, and induce cylinder stresses which are superimposed by thermal stresses caused by asymmetrical temperature distribution. For the purpose of transferring the holding-down bolt/stud forces in the axial direction, the liner wall is thickened toward the liner seat on the crankcase. At the top, the cooling fins are not interrupted circumferentially, in order to ensure sufficient stiffness against the gas pressure, and to reduce thermal expansion of the liner in this zone.

Fins in the lower portion are slotted alternately to permit balancing of thermal stresses and thus maintain optimum roundness. The thermal compensation effect is also facilitated by alternately embracing and straddling the holding-down bolts or studs.

A high standard of alloy engineering has been achieved by systematic material testing. In fact, current alloy iron castings are used without separate liner sleeves in the design of highly turbocharged engines. The liner bore is specially honed. In addition phosphating or nitriding may improve the running performance of heavily loaded engines.

Figure 16.21 gives an idea of such a modern grey cast iron cylinder as it is used in most commercial aircooled engines[57]. Finally the importance of an optimized guiding of the cooling air around the liner should be emphasized. Circumferential differences of the wall temperature must be minimized as well as the pressure loss of the cooling air.

16.2.2.3 Piston

In the past (especially with the IDI version) pistons of aircooled engines generally had to cope with slightly higher temperatures in the top land and in the first ring groove. The clearances between liner and piston had to be adapted accordingly. Meanwhile, piston cooling by using an oil jet splashed against the piston bottom has become a common feature in high-speed

Figure 16.21 Cylinder liners, grey cast iron

heavy-duty engines. The piston with a cooling duct (oil gallery) is a further refinement that finds growing acceptance in turbocharged versions of both air- and watercooled engines (see Chapter 14).

Figure 16.22 gives a section through a piston with an oil gallery. Jets in the crankcase provide for forced coolant circulation. The piston coolant passes through the existing engine oil cooler.

Lower piston temperature also reduces liner temperature, an effect that permits decreasing the finned surface area of aircooled engines. This makes it possible to provide the extremely small cylinder centre line distances mentioned previously.

In *Figure 16.23*, the piston and cylinder temperatures of pistons using jet cooling, both with and without gallery are plotted against the load. The effect of these two cooling systems on piston and liner temperatures is clearly distinguishable. In the case of the gallery piston the heat flow in the upper part of the liner is even reversed.

16.2.2.4 Cylinder head bolts

The bolts securing each cylinder head to the crankcase usually pass through the cylinder fins. The lowest possible forces should be applied to keep head and liner as free as possible from distortion, even in quickly alternating load conditions and extreme ambient temperatures (tropic or arctic).

Figure 16.22 Section of a piston with oil gallery

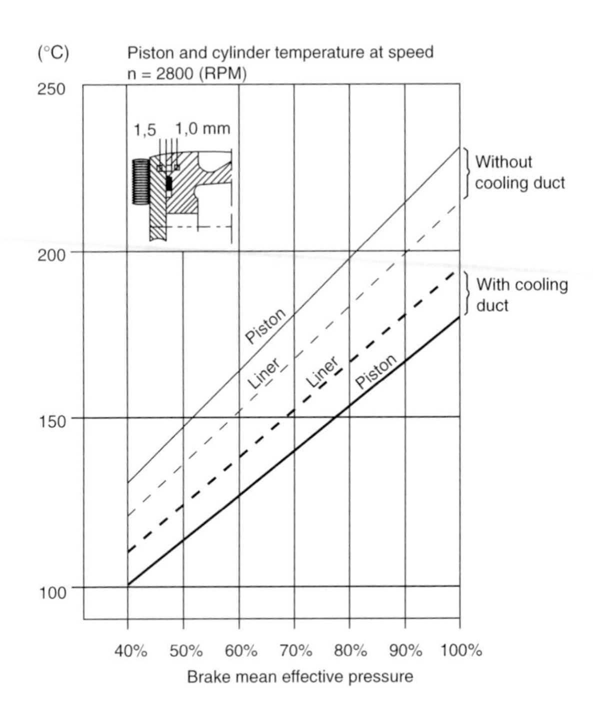

Figure 16.23 Piston and liner temperatures with pistons of different oil cooling methods

Modern high output engines therefore use extra slim antifatigue bolts of high tensile strength (about 1250 MPa). Plastic deformation should be avoided under all conditions to ensure positive tightness even after an unlimited number of hot/cold cycles. Hence sufficient margin from the yield point has to be provided[35].

By employing bolts with a shank diameter of about 0.8 of the outer thread diameter, resilience is enhanced and tension buildup with increasing engine thermal load is minimized, so are cylinder head and liner deformation. The effectiveness of this practice is revealed by *Figure 16.24*, which compares the deformation caused by the force of a full-shank bolt and an antifatigue bolt of equal thread diameter.

The thermal expansion caused by heating the cylinder and cylinder head or by heating the bolts is the same for antifatigue and full bolts. Elongation of the cylinder and cylinder head is indicated by Δl_T; that of the bolts by Δl_S. The value

$$\Delta l_{(therm)} = \Delta l_T - \Delta l_S$$

constitutes the degree of variation in the preloading forces. When the full bolts are stressed under warm-up conditions, the steeper angle of the spring force line leads to slightly less additional elongation of the bolts ($\Delta l_{SZ'}$) but to greater compression of the bolted components $\Delta l_{TZ'}$, as compared with antifatigue bolts; and thus to a bigger increase in preload force $\Delta P'$.

To ensure a given minimum cold tightening force P_{TIGHT} for conditions giving an operating force range P_{OP}, a full-shank bolt requires a slightly smaller preload force $P_{1'}$, than an antifatigue bolt $P_{1'}$. This fact is taken into account, No account was taken of the change in the spring force lines brought about by shifting the point of force initiation as firing pressure became effective.

The cross-section in *Figure 16.24* shows the thread deeply immersed in the crankcase. This is important and has two advantages, namely longer bolts and better distribution of the bolt forces in the crankcase. So the deformation of the crankcase on the surface is limited.

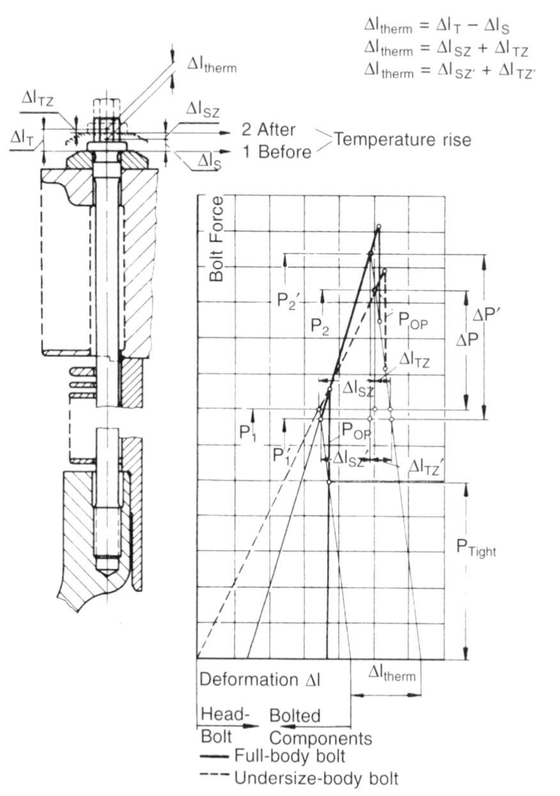

Figure 16.24 Force/deformation diagram of a full-shank and an antifatigue bolt of equal thread diameter

16.2.3 Heat exchangers

16.2.3.1 Oil cooler

In smaller engines or in engines with low thermal loading, finned tubes are favoured. They are inexpensive and can be installed in the plenum chamber for the cooling air of the engine (*Figure 16.25*). With this arrangement the oil cooler does not require separate space, but the heat rejection of the oil increases the cooling air temperature before it reaches the cylinder. The cooler units of larger engines mostly require additional air from the fan.

Due to the consistent uprating of diesel engines in the past years, space requirements for the engine oil cooler have grown accordingly. Moreover, there is an increasing trend towards integrating further units such as converter or hydraulic oil coolers into the cooling system of aircooled engines. Such integration reduces the dimensions of the entire equipment concerned. Nevertheless efforts are made to keep weight and size of the cooler small. Today most engine oil coolers are fabricated from aluminium and with a special type of cooling fin both on the air and oil side, with particular attention to the prevention of clogging of the air side fins by dust or dirt.

Figure 16.26 illustrates such a cooler of 'shell-type' construction. Turbulators for increasing surface areas on the oil side as well as suitable fins on the air side, give rates of heat transfer which approximate those of the conventional radiators of watercooled engines.

In hydraulic oil coolers the configuration of turbulators on the oil side and finning on the air side is similar to that of motor oil coolers. For high stability and easy installation on the engine, the plate-type construction has proved favourable.

Both the engine oil cooler and the hydraulic oil cooler are integrated into the engine in such a way as to ensure optimum air supply to these units. The layout of the fan and distribution of the air are specially matched. As an example the diagrams in *Figure 16.27* allow the proper adaptation of a given oil cooler to the cooling system together with the performance map of the fan (*Figure 16.36*). The layout is made for a temperature difference between oil entrance and cooling air entrance ETD = 80 K and considers the different flow volume and pressure loss of the oil and the cooling air.

The cooling air distribution in a commercial truck engine is given in *Figure 16.28*. This diagram gives a rough idea of the amount of heat rejected in the cylinder head, liner and engine oil cooler.

16.2.3.2 Aftercooler

Aftercooling with highly supercharged engines is in common

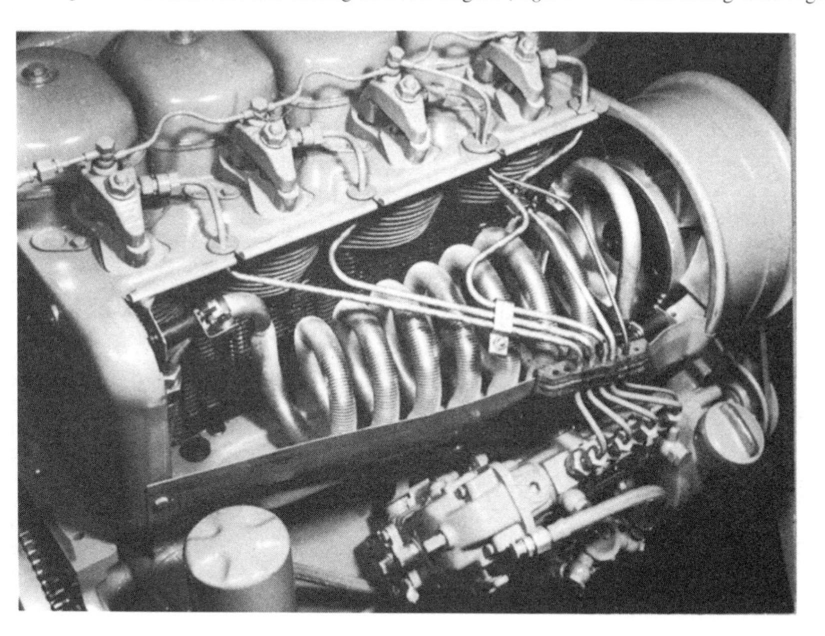

Figure 16.25 Finned tube oil cooler in the cooling air plenum chamber of a four-cylinder engine

Figure 16.26 Oil cooler of 'shell-type' construction

use. In the case of aircooled engines the aftercooler can be installed in the main cooling air flow in front of the fan, similarly to modern watercooled engines where it is installed in front of the radiator. It can also be incorporated with comparatively little additional effort into the engine's cooling system—like the hydraulic oil cooler.

Figure 16.7 gives an example of the installation of the aftercooler and transmission oil cooler in a 12-cylinder truck engine. Air-to-air coolers basically are of the same shell-type system as oil coolers. They are of the crossflow type and the fins are matched to the cooling air and combustion air flow.

Aftercooling, of course, also appreciably lowers component temperatures which is especially important in the case of aircooled engines. *Figure 16.29* demonstrates this effect by showing the temperature in the valve land. This temperature is plotted against performance, both with and without aftercooling. The graph additionally indicates the charge air temperature in the manifold at the inlet to the head and shows the effectiveness of air-to-air coolers.

16.2.4 Fan control

The power needed for cooling is about equal in aircooled and watercooled engines. A variable fan speed is used to reduce this power loss at part load. In the case of aircooled engines this is usually obtained through a fluid coupling, operated with oil from the engine's lubrication system. An example of a fan with a hydraulic coupling is given in *Figure 16.30*.

Oil flow rate and thus the degree of coupling slip is varied by a valve, which—for instance—could be controlled by the temperature of the cooling air leaving the engine or by the exhaust temperature. The operative range is adjustable at the temperature controlled valve. Coupling slip is, to a minor extent, also influenced by the input speed. An example for a modern electronic fan speed control is given in *Figure 16.31*, where the

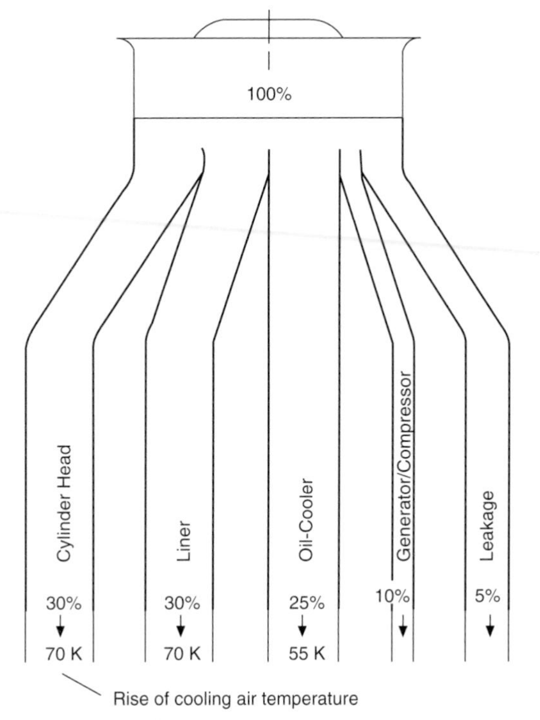

Figure 16.28 Cooling air distribution in a truck engine

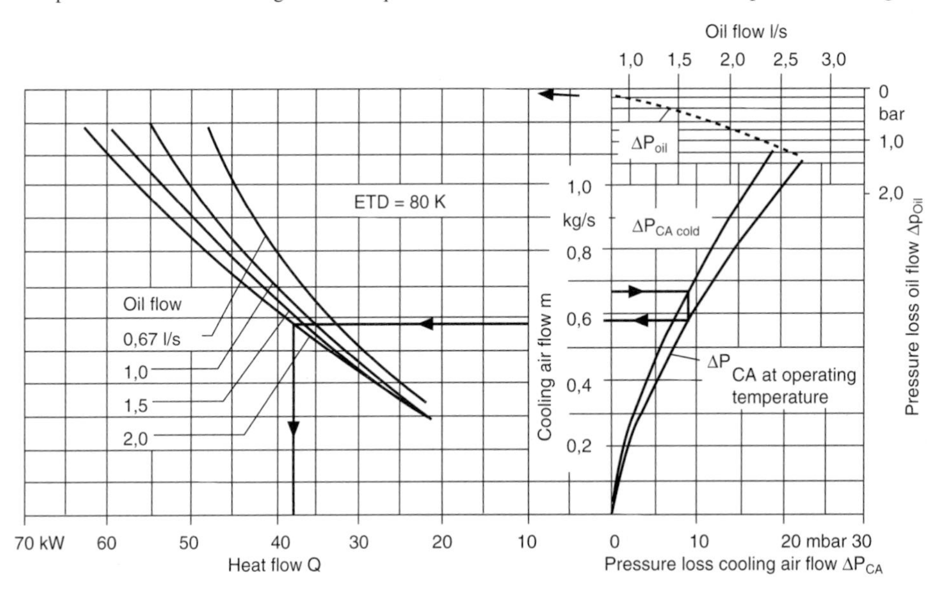

Figure 16.27 Heat flow, cooling air flow and pressure losses in an oil cooler

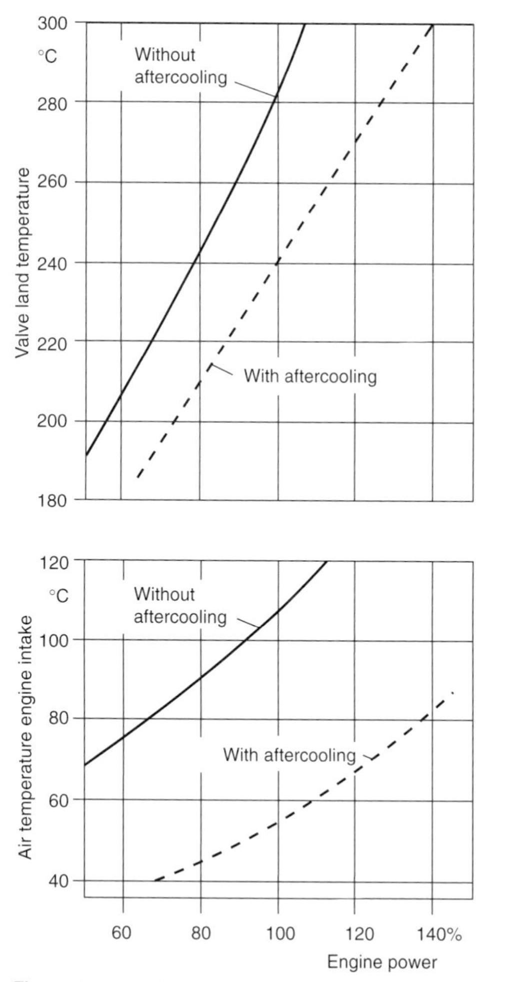

Figure 16.29 Influence of aftercooling on valve land temperature

oil flow is regulated by a solenoid valve, which again is influenced by one or more component—and/or oil-temperature sensors. In *Figure 16.32* the temperature of one component is plotted against the engine load at different engine speeds.

Besides the uniformity of the component temperature, a rather high temperature level is equally important as can be seen from *Figure 16.33*. The relative friction loss of piston and piston rings is plotted against the cylinder wall temperature. With 25°C higher wall temperature the friction loss decreases about 3%.

The design, the cooling requirements and the functional aspects of aircooled engines will be influenced in the future by the use of ceramic components. An outlook to these developments is given in the *Figures 16.17 and 16.34*[84].

16.3 Cooling fan

16.3.1 General aspects

For watercooling, the fan and radiator are designed apart from the engine as a separate unit. In the case of aircooled engines, however, the fan and associated ducting that guides the cooling air form an integral part of the engine unit. The space allowance for the fan is therefore determined by engine dimensions and is often very stringent. The same is true for the flow area between cooling fins.

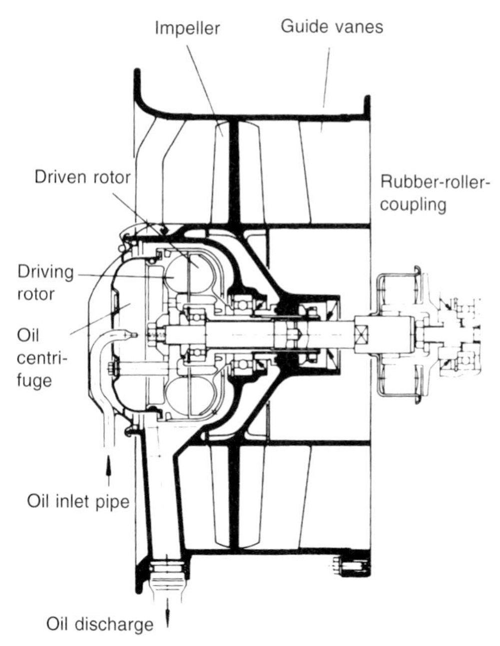

Figure 16.30 Exit guide vane fan with a hydraulic coupling

Fans for aircooled engines therefore normally have to supply a higher pressure rise than fans for watercooled engines. The rise in the cooling air temperature across the cooling-fins and surfaces, however, is not restricted by the temperature of any cooling water and can be much higher. Consequently, the airflow rate for transporting the same amount of heat can be smaller. The power required to drive the fan, which results from pressure and airflow rate, is therefore about the same for both types of engines. Fan power reduces engine output and increases the specific fuel consumption. It is essential therefore to achieve maximum fan efficiency, and this includes the reduction of fan noise, which is directly related to it.

For these reasons, all aircooled engines with more than two cylinders use axial-type fans. For a given airflow they are of higher efficiency and are smaller than radial fans. This permits simpler design of the air-ducts. Radial type fans, sometimes integrated in the flywheel, are still used in one- and two-cylinder engines for the sake of simplicity and because of the relatively low fan power required. Other types of fans, such as the drum or cross-flow types, have much lower efficiency and are normally not used for engine cooling.

16.3.2 Layout and design of axial fans

Steady and intensive development work influenced by know-how from the field of aircraft engine design has improved the aerodynamic and acoustic properties of axial fans in recent years.

Engine power and design determine the required air flow volume $V(m^3/s)$. The cooling fan must be capable of maintaining the necessary flow of air at all speeds and engine loads. Generally, the fan supplies air to a plenum chamber, from which the total air flow has to be distributed in correct proportions to the various components, (see *Figures 16.3, 16.10, 16.25*). The air pressure required in the plenum chamber is determined by the pressure drop in the components to be cooled, and by the additional flow resistance due to the installation of the engine. In most cases, the total flow resistance has to be determined by experiments. It varies between 15 and 25 mbar and reaches in special cases

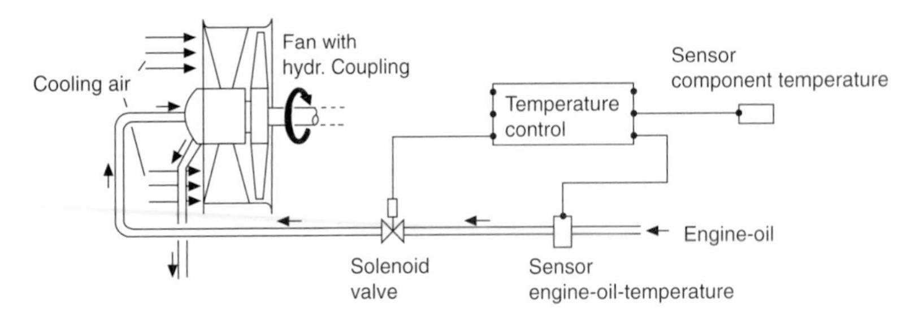

Figure 16.31 Electronic fan speed control

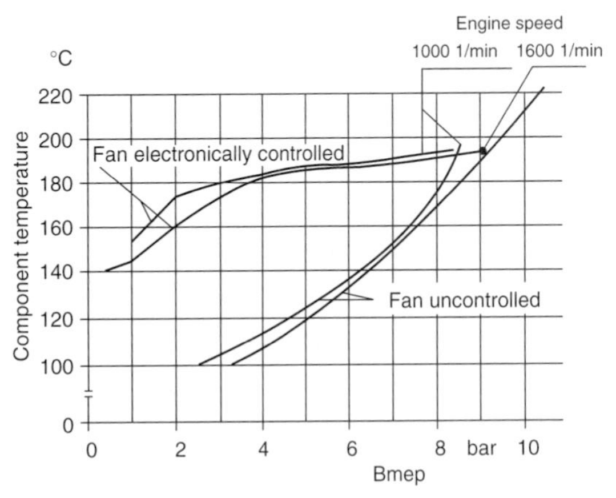

Figure 16.32 Component temperature with electronically controlled fan

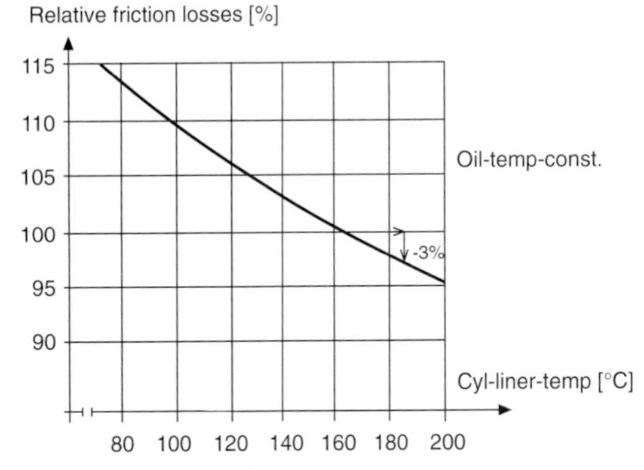

Figure 16.33 Friction losses versus cylinder liner temperature

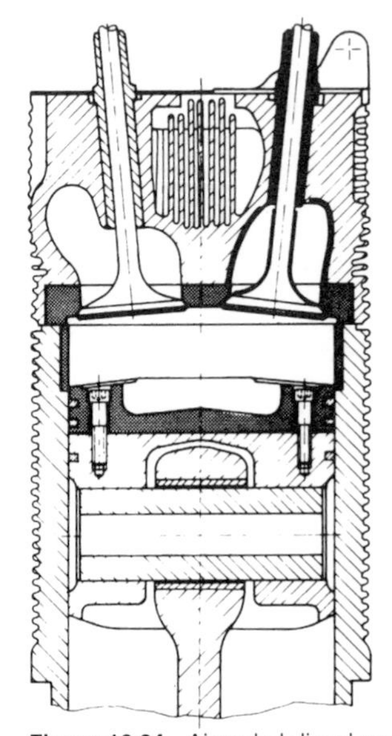

Figure 16.34 Aircooled diesel engines (with ceramic components)

40 mbar[26]. It defines the static pressure rise Δp_{stat} to be built up by the fan.

Besides this static part, the total energy transmitted to the air by the cooling fan (represented by the total pressure Δp_{tot}) also includes a dynamic part which consists of the energy contained in the velocity '*c*' of the air entering the plenum chamber.

The velocity '*c*' has an axial (meridional) component c_m in the direction of the fan axis and a circumferential component c_u perpendicular to c_m.

$$\frac{\Delta p_{tot}}{\rho} = \frac{\Delta p_{stat}}{\rho} + \frac{c^2}{2} = \frac{\Delta p_{stat}}{\rho} + \frac{c_m^2}{2} + \frac{c_u^2}{2} \qquad (16.1)$$

In this equation, ρ is the density of the air.

The energy of the circumferential velocity c_u is the vortex energy behind the fan. Most of this can be converted to static pressure by means of a stator cascade that changes the flow direction, thus the remaining term $\frac{c_u^2}{2}$ is normally very small. The energy contained in the axial velocity c_m can only be recovered by means of a diffuser, and this is normally not possible. This dynamic energy can therefore not be converted to pressure, but is dissipated in the plenum chamber and lost.

With flow volume V, fan tip diameter D, fan hub diameter d and the ratio of hub to tip $v = \frac{d}{D}$, c_m becomes

$$c_m = \frac{V}{(D^2 - d^2)\frac{\pi}{4}} = \frac{4}{\pi}\frac{V}{(1 - v^2)} \times \frac{1}{D^2} \qquad (16.2)$$

and equation (16.1) can be transformed to

$$\frac{\Delta p_{tot}}{\rho} = \frac{\Delta p_{stat}}{\rho} + \frac{8}{\pi^2} \times \frac{V}{(1 - v^2)^2} \times V^2 \times \frac{1}{D^4} \qquad (16.3)$$

The dynamic part of this equation is the same order of magnitude as $\frac{\Delta p_{stat}}{\rho}$. As v is determined by fan layout criteria and V by the cooling requirements, only the fan tip diameter D remains as a variable to influence this flow loss. It appears to the fourth power, therefore increasing the tip diameter reduces drastically the power necessary to drive a fan for a given mass flow rate and a given pressure in the plenum chamber.

For example, it was found that by changing the fan diameter from 389 mm to 450 mm, the power absorbed by the fan for a large production engine could actually be reduced from 26.8

kW to 20.2 kW, at the same time lowering the noise level. In combination with blade optimization a noise reduction of 4 dB was achieved.

For the determination of the main fan dimensions, diagrams have been developed by Esche which are recommended for use[61,69].

An axial fan (with stators) can be seen in *Figure 16.35*, a typical performance map in *Figure 16.36*. A wide operating range is advantageous as several applications can be covered by one fan. In the example given, the operating lines of engines equipped with hydraulic oil coolers of different cooling rates are shown. These coolers take additional air from the fan.

On the other hand, a steep slope of the curves n = constant can be of advantage as it provides some self-regulation for constant flow volume[23]. Through fan layout and blade design such features can be optimized.

The lines

$$\frac{\Delta p_{\text{stat}}}{p_0} = \text{constant } (p_0 \text{ is ambient pressure})$$

clearly show that the static pressure diminishes with increased mass flow as the dynamic part of equation (3) increases. Along the operating line 'with hydraulic oil cooler 50%' the static part Δp_{stat} approximately equals the dynamic part.

$$\Delta p_{\text{stat}} \approx \frac{\Delta p_{\text{tot}}}{2}.$$

Operating the fan beyond this line with higher mass flows leads to small static efficiencies.

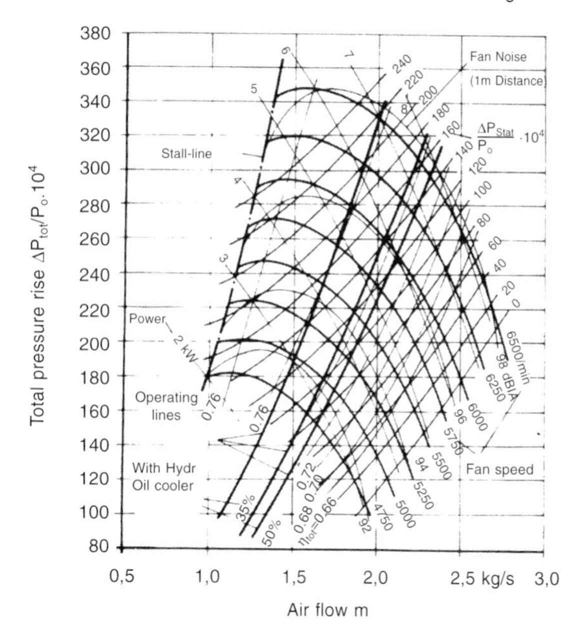

Figure 16.36 Performance map of an axial cooling fan with inlet guide vanes.
Fan diameter D = 276 mm;
Reference state: T_0 = 288 K; p_0 = 1.013 bar

16.3.3 Stators

As already explained, the energy of the circumferential velocity c_u (vortex energy behind the rotor) is also lost unless it is converted into static pressure by means of a stator cascade changing the flow direction. However, a stator has the disadvantage that it creates other flow losses. The relative magnitudes of the vortex energy and the stator losses induced when recovering this energy are therefore the determining factors as to whether or not a stator would be worthwhile in any particular application.

Euler's equation

$$\frac{\Delta p_{\text{tot}}}{\rho} = u \times c_u \times \eta_{\text{tot}} \tag{16.4}$$

Figure 16.35 Axial fan with inlet guide vanes

shows that c_u is small if either Δp_{tot} is small or the circumferential velocity of the rotor tip 'u' is large. As long as the pressure coefficient

$$\psi = \frac{\Delta p_{\text{tot}}}{\rho} \times \frac{2}{u^2} \tag{16.5}$$

is smaller than 0.25, a stator should be omitted. This is normally the case for the cooling system of watercooled engines.

In aircooled engines axial fans without stators can only be used if the speed of the fan can be made high enough to produce a pressure coefficient less than ψ = 0.25. For the design and layout of such fans, the Marcinowsky method[19] is recommended. This method makes it possible to establish the diameter, speed and power requirements of the fan by means of diagrams and tables.

Usually, diameter constraints due to the space restrictions, and the relatively small air flow combined with high pressure resistance normally lead to pressure coefficients ψ of 0.25 to 0.6 for aircooled engines. This makes it necessary to use a stator, which can be located either in front of or behind the rotor (*Figure 16.37*).

Fan efficiency is usually higher with exit guide vanes (stator behind the rotor) than with inlet guide vanes. This is due to the prevailing aerodynamic conditions which are illustrated in *Figure 16.37*[43,48].

All other conditions being equal, the relative velocity of the air within the rotor cascade is considerably larger with inlet guide vanes and therefore creates higher losses, as they increase with the square of the velocity. It is also necessary to increase the velocity at the rotor inlet and change its direction when inlet guide vanes are used. This results in a pressure decrease in front of the rotor which must be compensated for by an increase in the pressure difference created by the rotor. In the case of the exit guide vanes, both rotor and stator contribute to the pressure rise.

Fan noise is closely related to fan efficiency. It is therefore usually easier to reach low noise levels with exit guide vanes.

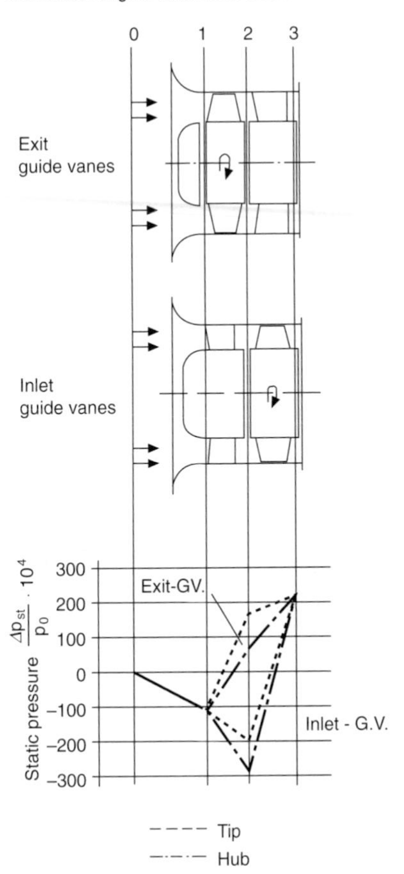

Figure 16.37 Pressure diagrams for fans with inlet and with exit guide vanes

However, in most applications the fan is located in front of the engine and inlet guide vanes can serve as a safeguard against the running rotor blades. In the case of exit guide vanes a safety mesh must be used and this reduces the efficiency advantage. Also, as the air velocity is retarded in the stator vanes when exit guide vanes are used, this type of arrangement is more prone to the deposition of dust and dirt. The design of the bearing arrangement and the direction of the air flow also influence this choice. For these reasons, it is not possible to give a general recommendation as to which type is preferable.

In axial fans with guide vanes, the number of rotor blades is greater than in fans without stators, so cascade theory rather than air foil theory should be applied here. Some principles of blade design are given in References 7 and 27.

16.3.4 Fan noise and its reduction

When dealing with fan noise, one has to differentiate between two different types of noise. On the one hand, there is broad band noise, spread over a wide range of frequencies and caused by vortices around the blading and in the inlet and exit ducts. On the other hand, there is tonal sound, occurring at discrete frequencies and caused by the interaction of the rotor blades and the stators, struts and other obstacles in the flow path. It can always be traced back to the basic frequency defined by the rotor speed and by the number of blades and its harmonics.

Broad band noise arises from the vortices behind each blade caused by changes in the circulation and therefore in the pressure distribution at the aerofoil. This in turn results in sound waves. The acoustic frequencies are identical with the frequencies of the vortex origin, and these depend on the air velocity and the thickness of the profile. They therefore change across the height of the blade, resulting in a noise spread over a wide range of frequencies.

Judin[18] has developed a method for calculating this broad band noise. Extensive testing has shown that his formula makes very accurate predictions of vortex noise possible.

$$S = \frac{1.78 \times 10^{-2}}{\rho \times a^2} \times \frac{\sqrt{1+v^2}}{(1-v)(1+v)^2} \times \frac{t}{l} \times \frac{(\Delta p \cdot V)(1-\eta)^2}{D^2} \tag{16.6}$$

S is the noise power, a the velocity of sound; t blade spacing; l, blade length; V, the air flow volume.

The noise power is proportional to the square of the fan power (ΔpV) and fan power changes approximately with the third power of speed. Consequently, the noise of an axial fan is approximately proportional to the sixth power of the speed. This is borne out very well by test results—for example, doubling the speed results in an increase of fan noise level of 10 dB.

For optimum fan design, the ratio v of hub to tip diameter and the ratio $t{:}l$ of blade spacing to blade length must be kept within certain limits. For practical purposes therefore, most of the terms shown in the formula can be assumed to remain constant. The most important influences remaining are—fan power, fan diameter and fan efficiency.

With the noise level L defined as

$$L = 10 \times \log S \tag{16.7}$$

measured in decibels (dB), the difference in the noise levels of two fans designated by the subscripts 1 and 2 can be expressed in logarithmic terms. This relation is illustrated in *Figure 16.38*. A 20% reduction of cooling power, about 5% improvement of efficiency, or a 25% increase of fan diameter could lower the fan noise by 2 dB.

The other element in fan noise, i.e. discrete frequency tonal sound, is usually much more annoying. Each rotating flow cascade creates a rotating pressure field, the periodic pressure changes of which are perceived as noise. The frequencies of these tones are determined by the number of blades, multiplied by the speed and multiples thereof. *Figure 16.39* shows the noise spectrum of an axial fan with discrete frequency noise.

The discrete frequency noise can be eliminated almost entirely by modulation of the tones. This is done by irregular distribution

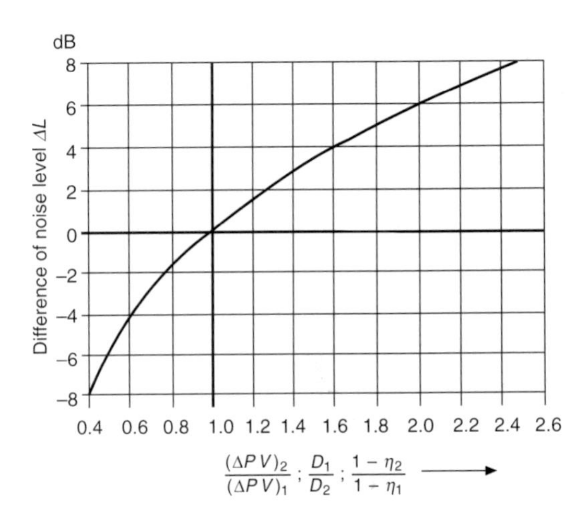

Figure 16.38 Influence of the cooling requirements ΔpV, fan diameter D and the fan efficiency η on fan noise level L

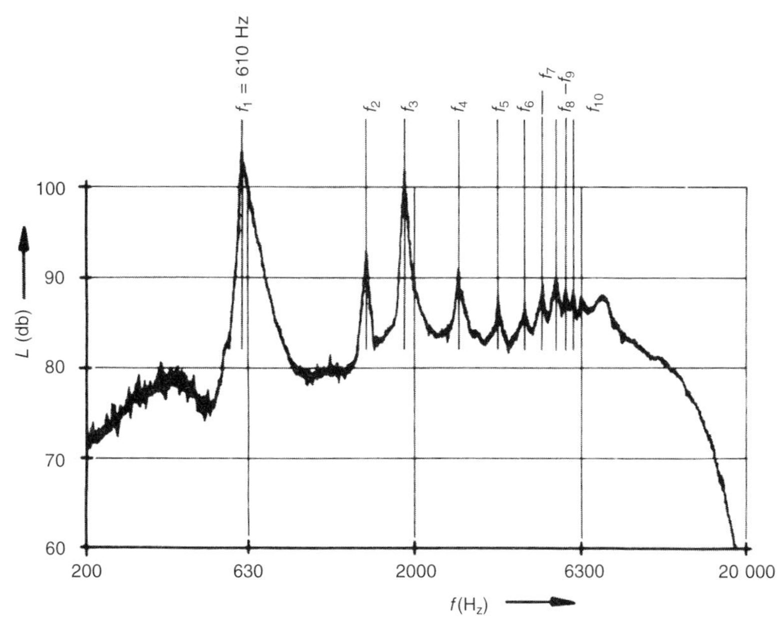

Figure 16.39 Noise spectrum of an axial fan with uniform pitch

of blades around the circumference of the rotor. The flow within the cascade is thus somewhat distorted and higher flow losses are the result. The loss of efficiency can, however, be kept to a minimum by using the smallest effective differences in pitch. When determining the distribution of blades, its effect on rotor imbalance also has to be taken into consideration. Mellin and Sovran[36] have published a method to determine the spectrum of the noise from the spectrum of excitation. The application of the blade spacing as recommended by them can be taken also from Reference[69]. *Figure 16.40* shows the spectrum of an axial fan with the tonal spectrum completely eliminated and *Figure 16.41* illustrates the fan rotor.

16.3.5 Other design considerations

Besides the selection of fan dimensions for best efficiency and low noise, it is necessary not only to optimize the design of the blades and vanes of the fan but also to look carefully at the flow path close to the fan inlet and exit.

The gap between the outer tip and the shroud should be kept to a minimum. When using exit guide vanes, the shroud should extend in front of the rotor and show a smoothly rounded edge to the inlet (*Figure 16.35*). If a grid is to be included for safety reasons, as, for example, in *Figure 16.54*, it should be designed for minimum flow resistance. Dependent upon space requirements, the grid should be as far away from the blades as possible and be located in an area of smaller air velocities.

The same applies to all obstacles in the flow path. Hoses, tubes or bars leading through the space in front of the fan are always detrimental to fan efficiency and increase noise, thus should be avoided if at all possible. Any such obstacles should at least be placed as far away from the fan inlet as engine compartment space permits. The fan drive, either by gear train or by means of V-belts, also forms an obstacle in the air flow. *Figure 16.42* shows the effect of such obstacles. The total noise level difference of the two designs shown is 3 dB.

Fans with stators often fail to meet the requirements for axial core removal if the blades are designed for best efficiency. In such cases, the cores have to be withdrawn radially. Even then, the fan design has to be such as to ensure that the blade shape across the radius does not show undercuts due to extreme changes of curvature. Layout parameters which allow axial cores are indicated in the diagrams in References 61 and 69.

As a material, aluminium alloy is used almost exclusively.

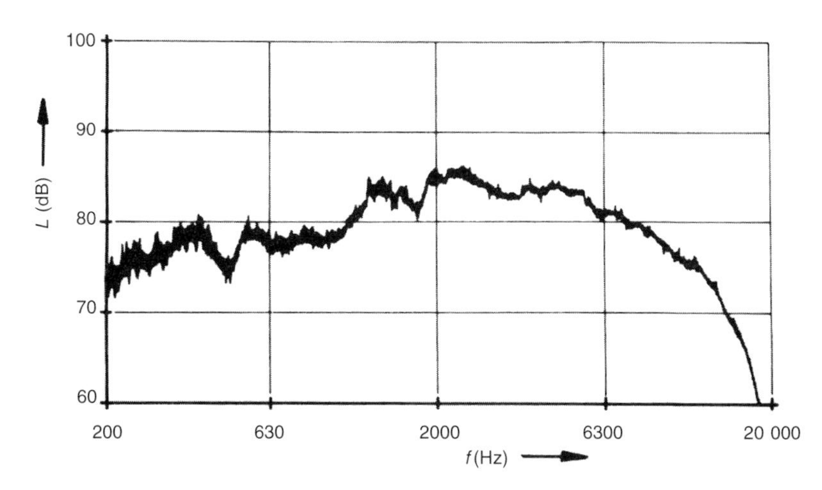

Figure 16.40 Noise spectrum of an axial fan with irregular pitch

$\alpha_{1,2}$ = 26,9° $\alpha_{7,8}$ = 33,4°
$\alpha_{2,3}$ = 27,6° $\alpha_{8,9}$ = 32,4°
$\alpha_{3,4}$ = 29,0° $\alpha_{9,10}$ = 30,7°
$\alpha_{4,5}$ = 30,7° $\alpha_{10,11}$ = 29,0°
$\alpha_{5,6}$ = 32,4° $\alpha_{11,12}$ = 27,6°
$\alpha_{6,7}$ = 33,4° $\alpha_{12,1}$ = 26,9°

Figure 16.41 Fan rotor with irregular pitch

Plastics, which are increasingly used in watercooled engines, have not yet found their application for fans in aircooled engines because of the higher fan speed and the more extreme temperature conditions after engine shutdown. Riveted fan blades such as those used in watercooled engines cannot be applied in aircooled engines for reasons of efficiency and noise.

16.3.6 Manufacturing considerations

Fans are normally produced by pressure die casting because of the high production rate possible in this way. The investment costs for the dies are dependent upon the way the cores can be drawn. Casting requirements therefore have to be taken into consideration in the blade design. In the case of rotors without guide vanes, the shape of the blades permits cores to be formed so that they can be withdrawn in the axial direction, which makes the casting process more economical.

16.4 Environmental aspects

16.4.1 Exhaust emissions

Particulate (smoke) and gaseous exhaust emissions are almost entirely determined by the combustion system and its characteristics and are, therefore, independent of the cooling system. One exception to this is during the start and warm-up periods when the emission of unburnt hydrocarbons is considerably higher than after the engine has reached its operating temperature.

Due to the absence of a liquid coolant which has to be first heated, aircooled engines reach their operating temperature much quicker than their watercooled counterparts (*Figure 16.4*). By controlling the fan speed, the warm-up time can be further reduced. As a consequence the duration of unfavourable HC-emissions is minimized and the total emissions over a driving cycle are reduced.

16.4.2 Engine noise

It was originally thought that aircooled engines generally tend to be noisier than watercooled engines. In 1970, CIMAC carried out extensive investigations to determine the noise level of diesel engines[38,39]. One of the major results of this study was the conclusion that the cooling system has no significant effect on engine noise level. *Figure 16.43 and 16.44*, which show the noise level of several engines plotted over speed, provide support for this statement. The hatched area in *Figure 16.43* represents the tolerance range for the noise level of ten watercooled engines of 150–260 kW, the same range for engines below 150 kW is shown in *Figure 16.44*[54].

It can be seen that the noise levels of aircooled engines currently in production are at the lower end or even below a number of watercooled ones. *Figure 16.44* also shows the noise difference of the same aircooled engine equipped with direct injection (912) and swirl chamber (912W). It should be noted that the noise levels of the aircooled engines include fan noise while the watercooled engines have been measured without fan.

If little or no emphasis is placed on the need to design for low noise, aircooling is at a disadvantage. This was the case before noise became a major issue and as a consequence, early designs have established the image of the aircooled engines as being noisy. Such prejudices are not easy to change.

16.4.3 Noise characteristics of aircooled engines

To describe the features that are peculiar to aircooled diesel

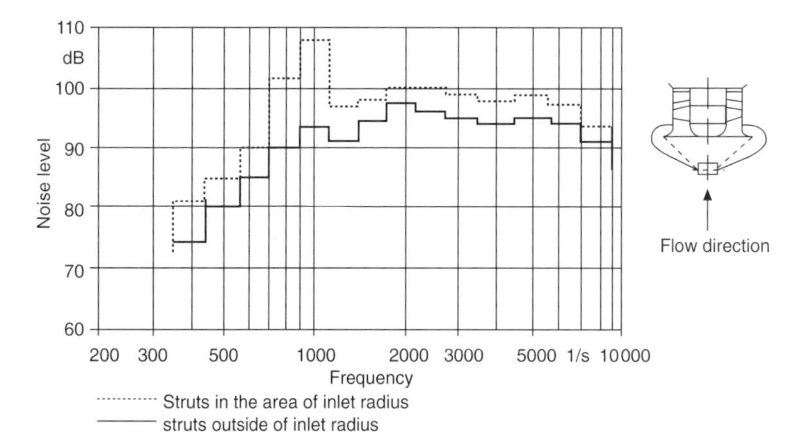

Figure 16.42 Influence of different struts in the inlet area on the fan noise

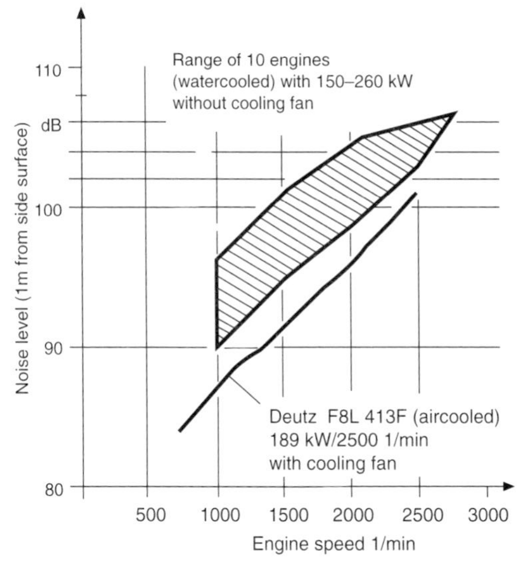

Figure 16.43 Noise of direct injection diesel engines of 150–260 kW at full load

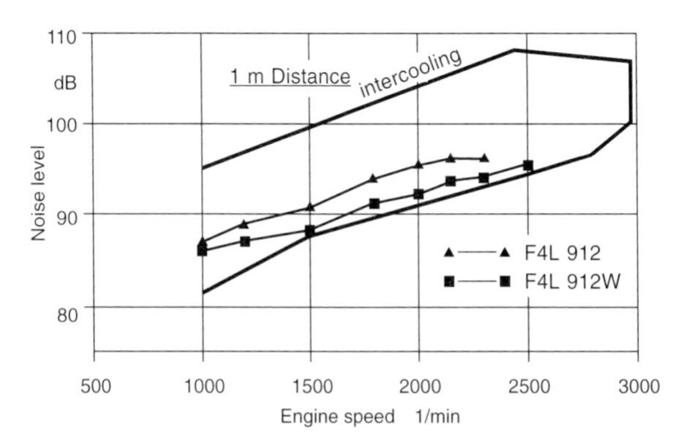

Figure 16.44 Noise of diesel engines up to 150 kW. A range of watercooled engines is compared with aircooled direct injection (F4L912) and swirl chamber engine (FL912W)

engine noise, it is first necessary to identify the origins or sources of noise generation. Vibrations within the engine system are caused by flow oscillations, combustion and mechanical excitation (*Figure 16.45*). These vibrations are transformed directly or indirectly into airborne noise. The relative contributions of the various sources to total engine noise are of similar magnitude (*Figure 16.46*).

Flow oscillations create noise directly. The contribution to total engine noise of the induction and exhaust processes are the same in aircooled and in watercooled engines, provided the same effort has been taken for attenuation. The noise radiated by the cooling system, especially the cooling fan, represents one of the main differences.

As demonstrated in section 16.3, aircooled engines have more stringent space restrictions dictating the use of small air passages and fan diameters which in turn lead to high air velocities and fan speeds. In combination with difficult or careless installations, this often results in increased noise levels in the high frequency part of the spectrum which are particularly obtrusive. In addition, poor fan design may result in discrete frequency peaks. In the early days of aircooled engines, the importance of these factors had not been fully recognized. The high frequency noise originating from the fan and from the cooling air flow is the main reason for the high noise levels associated with early aircooled engines.

Fan noise and methods for its reduction have been discussed in section 16.3.4. Fan noise emission can be further reduced by ducting the cooling air to the fan inlet in such a way that a shield prevents direct radiation from the fan cascade to the outside. The same principles are valid for the cooling air exit. Noise attenuation can be provided by a Z-shaped channel or by a plate that shields the area of fan inlet or air exit (see *Figures 16.47, 16.49 and 16.50*). Such measures also prevent any direct noise radiation of other adjacent engine components. By providing these ducts with layers of sound absorbing material it is possible to substantially reduce the high frequency content of fan and engine noise.

The crankcase is one of the dominating engine components for noise emission. Combustion induced noise is transmitted to the crankcase via two different paths: firstly through the piston—connecting rod—crankshaft, and secondly through the cylinder head and cylinder liner.

Research results[40,60] proved that the path across the crankshaft is more important and, as expected, is not influenced by the cooling system. In the case of aircooling, the sound path via the cylinder head influences the noise level of the crankcase only at very high frequencies; whilst for watercooling, a noticeable influence was also recorded in the lower frequency range (up to 1000 Hz).

The fundamental difference between air- and watercooled engines in crankcase design has been explained in section 16.2.1. The individual cylinders of aircooled engines provide two changes of the acoustic impedance along the noise paths, i.e., between the cylinder head and liner and the liner and crankcase, thus reducing the structure-borne noise.

Finally, the noise of the auxiliary units is radiated directly from their surfaces as well as transmitted via their flanges and brackets to the structure of the engine and radiated from the engine surface. Their direct contribution to total engine noise should be independent of the cooling system.

Therefore, there are three inherent differences: in the noise excitation by the cooling system, in the noise transformation by the engine structure and in the noise radiation of the engine surface.

Summarizing, the total noise level of an engine is independent of the cooling system. The character of the noise, however, can be different: aircooled engines have a tendency towards higher frequencies while the lower frequency range dominates in watercooled engines. High frequency noise can be shielded more effectively.

16.4.4 Noise attenuation by secondary measures

It is important to stress that careless installation of an engine can ruin the noise reduction achieved through careful engine design. On the other hand, close cooperation between engine and equipment manufacturers can lead to remarkably low noise levels of the end product.

As an example, two different installations of a screw compressor are shown in *Figures 16.48* and *16.49*. While simple installation of the engine without any special soundproofing measures (*Figure 16.48*) will reduce the bare engine noise by 5 dB, a further reduction of 13 dB is possible through redesign of the hood (*Figure 16.49*). The hood was enlarged to bring the muffler into the enclosure (1), the duct for the cooling air exit was redesigned (2), the hood was foam coated (3) and the engine-compressor unit was encased by complete walls (4), shielding the direct radiation of air inlet and exit by special plates. Similar results are possible in truck installations[76,77,78].

In many cases such a joint effort between engine producer and equipment manufacturer is the most efficient way to achieve low overall noise level, but sometimes it is not practicable for

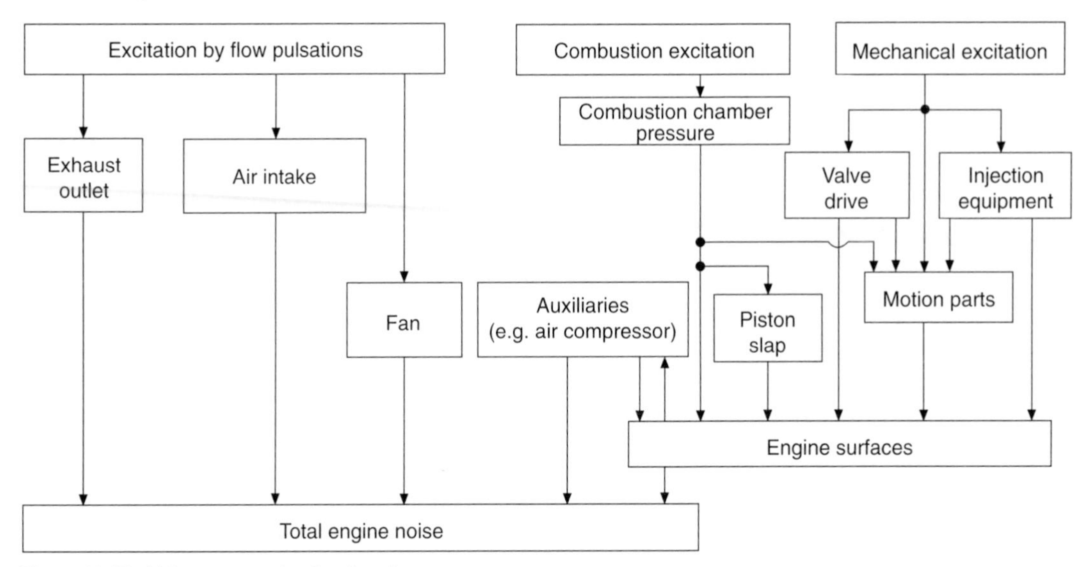

Figure 16.45 Noise sources of a diesel engine

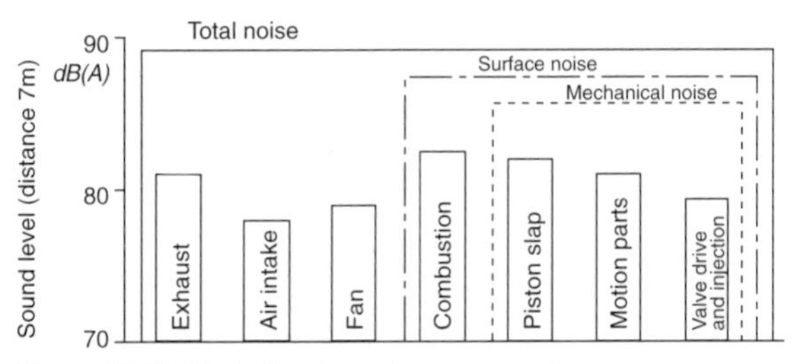

Figure 16.46 Typical engine noise components

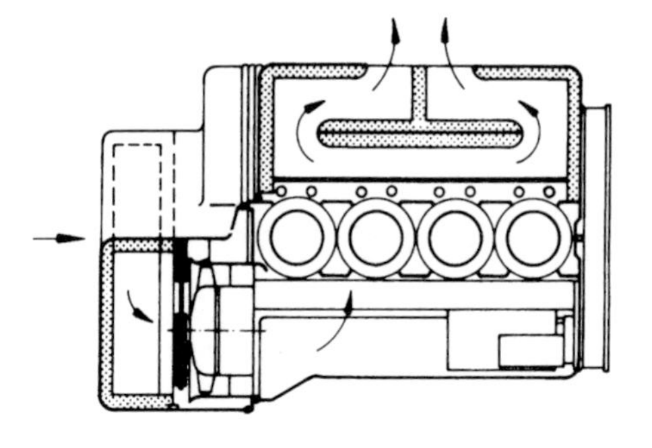

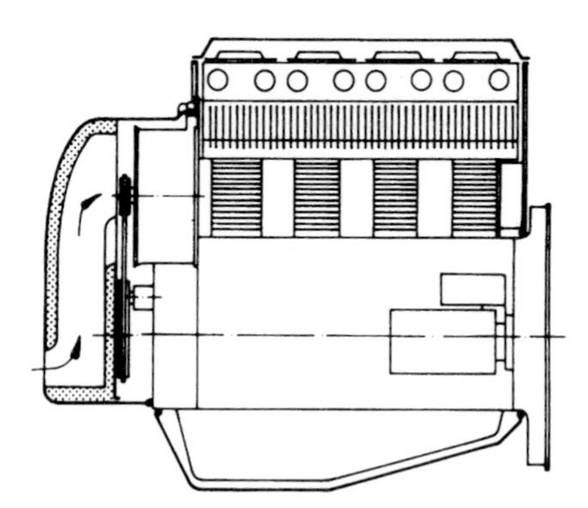

Figure 16.47 Aircooled engine with noise attenuation to cooling air flow

an engine company selling to many different customers. In this case the answer is to market the engine totally enclosed by an engine mounted capsule. The design principles for such an enclosure were established in the early seventies at AVL in Graz, Austria in a joint effort of the German research association (FVV)[51,53].

Figure 16.50 shows an experimental enclosure for a 4-cylinder aircooled engine, achieving a 19 dB reduction of engine noise. When designing the encapsulation, care has to be taken to ensure reliability of the enclosure parts and easy access for maintenance of the engine. Other important points are the attachment of the enclosure to the engine (which must interrupt the noise path in the structure); holes in the enclosure for passing hoses, links and mounting pads; and special maintenance covers which must be easily removable.

Total engine enclosures provide the maximum possible noise reduction. They are especially effective at high frequencies; and therefore in combination with aircooled engines. This approach has therefore already been adopted by some manufacturers[63,83]. An example is shown in the *Figures 16.51* and *16.52*.

16.5 Applications

Because of their various advantages aircooled engines are used in almost any field of application. We find them in:

Commercial and special vehicles (vans, trucks and buses);
Agricultural machinery;
Construction machinery;
Compressors;
Generating and pumping sets;
Underground mining equipment;
Vessels and ships;
Railborne vehicles;
Snow removers, etc.

Some special principles have to be observed when installing aircooled engines, in order to ensure long life operation without problems. Most engine manufacturers therefore provide specifications and/or have a group of specialists available, which and who advise the customer how to obtain maximum benefit from

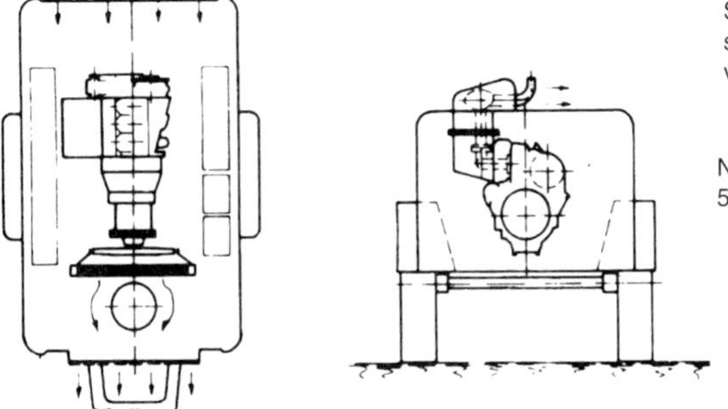

Figure 16.48 Screw compressor without soundproofing measures

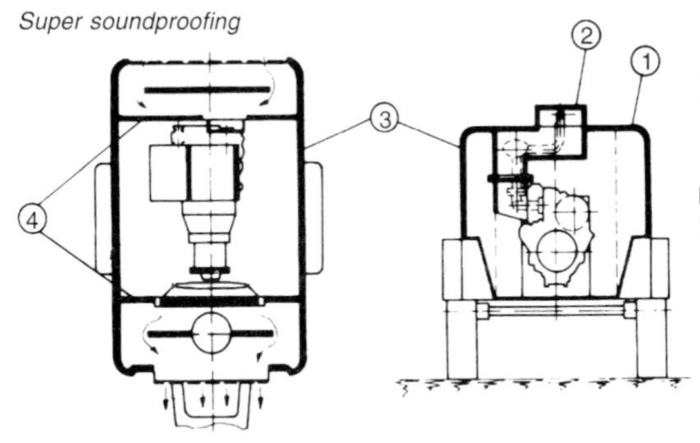

Figure 16.49 Screw compressor with special soundproofing

Figure 16.50 Experimental enclosure for an aircooled engine

Figure 16.51 HATZ engine 4L30S

Figure 16.52 HATZ engine 4L30C (completely enclosed)

their aircooled engines. For instance, it is most important to keep the discharging, heated cooling air away from mixing with the intake cooling and combustion air. Usually this can be done rather simply.

The importance of optimum air inlet condition to the cooling fan has already been stressed. No obstacles such as hoses or sharp edges are allowed at the fan inlet radius. Closed ductings for the cooling air need a large area especially around the fan entrance to keep the parasitic pressure losses as low as possible.

When the guidelines for installation are properly observed, aircooled engines work satisfactorily in all applications and in all climates, tropic and arctic, in the desert as well as in the mountains. Their versatility is unequalled. Some examples of aircooled engines and their installations are given in the *Figures 16.53 to 16.61*.

Figure 16.53 F8L610 engine

No. of cylinders	8
Bore/stroke	102/100 mm
Displacement	6.54 litre
Net power (ISO 1585)	119 kW
Speed	3200 min^{1}
Weight (dry)	348 kg

Figure 16.54 BF6L913 engine

No. of cylinders	6
Bore/stroke	102/125 mm
Displacement	6.13 litre
Net power (ISO 1585) (turbocharged)	118 kW
Speed	2500/2650 min^{1}
Weight (dry)	485 kg

A list of aircooled diesel engine manufacturers is given at the end of this book.

Figure 16.55 F2L511 engine

No. of cylinders	2
Bore/stroke	100/105 mm
Displacement	1.65 litre
Net power (ISO 1585)	25.6 kW
Speed	3000 min^{1}
Weight (dry)	164 kg

Figure 16.56 F1L210D engine

No. of cylinders	1
Bore/stroke	95/95 mm
Displacement	0.673 litre
Power	11 kW
Speed	3000 min^{1}
Weight (dry)	90 kg

Figure 16.57 Truck installation, tilted cab

Figure 16.58 Loader in underground mining operation

Figure 16.59 Floating crane with Schottel drive

Figure 16.60 Pumping station

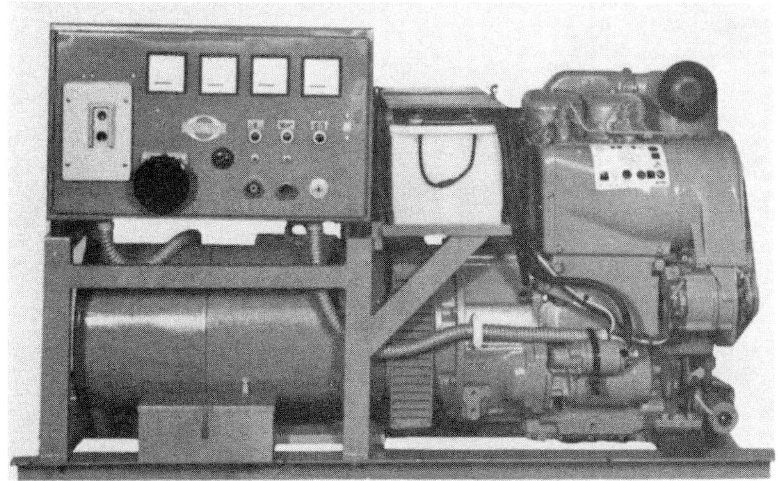

Figure 16.61 Generating set

References and bibliography

This list gives a chronological survey over more recent literature pertaining to aircooled engines and related subjects but by no means should be considered complete. The references given in the books and papers cited will guide to further and more specialized publications.

1 MARKS, E. S. and DORMAN, C. T.; 'Development of the Franklin Direct Air-Cooled Engine', *SAE Journal* (February 1931)

2 LÖHNER, Kurt, 'Die Grundlagen der Luftkühlung der Brennkraftmaschinen', *MTZ Motortechnische Zeitschrift* **12**, 3, pp. 53–62 (1951)

3 STEGEMANN, W., 'Einige Erörterungen über das Kühlungsproblem von luftgekühlten Motoren', *MTZ Motortechnische Zeitschrift* **12**, 3, pp. 63–69 (1951)

4 KLOSS, Richard, 'Der luftgekühlte Fahrzeug-Dieselmotor', *MTZ Motortechnische Zeitschrift* **12**, 3, pp. 69–76 (1951)

5 HAUSFELDER, C. L., 'Luftgekühlte englische Dieselmotoren kleinerer Leistung', *MTZ Motortechnische Zeitschrift* **12**, 3, pp. 82–85 (1951)

6 POLLMANN, E., 'Die Berechnung der Gebläse für luftgekühlte Motoren', *MTZ Motortechnische Zeitschrift* **12**, 3, pp. 77–82 and 5, pp. 140–144 (1951)

7 ECKERT, B. and SCHNELL, Erwin, *Axialkompressoren und Radialkompressoren, Answendung/Theorie/Berechnung,* Springer-Verlag Berlin/Göttingen/Heidelberg (1953), 2nd edition 1961

8 BERNDORFER, H., 'Vorausberechnung des Kühlungsaufwandes von Motoren mit luftgekühlten Zylindern bei gegebener Verrippung and Luftführung', *MTZ Motortechnische Zeitschrift* **14**, 3, pp. 57–59 (1953)

9 BRÜGGER, F., 'Neuentwicklungen auf dem Gebiet der Herstellung von Grauguß-Rippenzylindern für luftgekühlte Motoren', (Auszug aus einem Vortrag, gehalten anläßlich der VDI-Tagung 'Luftkühlung bei Verbrennungsmotoren' in Hamburg am 12.11.1952) *MTZ Motortechnische Zeitschrift* **14**, 3, pp. 64–66 (1953)

10 KLOSS, Richard, 'Betriebserfahrungen mit luftgekühlten Dieselmotoren', *MTZ Motortechnische Zeitschrift* **14**, 3, pp. 66–71 (1953)

11 GÜRTLER, G. and THIELE, W., 'Messungen an Eisen-Aluminium-Verbundguß-Zylindern für luftgekühlte Motoren. (Mitteilung aus dem Metall-Laboratorium der Metallgesellschaft A.-G.), *MTZ Motortechnische Zeitschrift* **14**, 3, pp. 71–76 (1953)

12 CORDIER, Otto, 'Ähnlichkeitsbedingungen für Strömungsmaschinen', *Brennstoff-Wärme-Kraft* **5**, 10, pp. 337–340 (1953)

13 POUNDER, C. C., *Diesel Engine Principles and Practice,* Chapter XIX: aircooled diesel engines, pp. 19.1–19.20. London, George Newnes Limited (1955)

14 HAAS, Herbert H. and KLINGE, Earl R., 'The Continental 750-Horsepower Aircooled Diesel Engine', (a) *Automotive Industries* pp. 70–72 (February 1957) (b) SAE 47, Detroit Michigan, (Jan. 14–18).

15 LḦNER, Kurt and CHONÉ, G., 'Wärmeübergang an Rippen und Rippenzylindern', *MTZ Motortechnische Zeitschrift* **18**, 12, pp. 373–377 (1957)

16 ZEYNS, Johannes, 'Probleme der Kühlung im Verbrennungsmotorenbau', *MTZ Motortechnische Zeitschrift* **18**, 12, pp. 378–383 (1957)

17 BACHLE, C. F., 'Air Cooled Diesel Engine Appraisal', *SAE Transactions* **66**, pp. 304–311 (1958)

18 JUDIN, E. J., 'Untersuchung des Lärmes von Lüfteranlagen und die Methode zu seiner Bekämpfung', (in russischer Sprache), *ZAGI Report Nr. 713,* Moskau (1958)

19 MARCINOWSKI, Heinz, 'Die Auslegung von Ventilatoren für Kraftfahrzeug-Kühlanlagen', *MTZ Motortechnische Zeitschrift* **19**, 9, pp. 304–311 (1958)
20 Large Special Purpose Air-Cooled Engine. The Continental AVDS-1790', *The Oil Engine and Gas Turbine* pp. 374–375 (April 1960)
21 HAAS, Herbert H., 'Air-Cooling versus Water Cooling in Agricultural and Construction Equipment', SAE Paper 222A, *SAE National Farm, Construction and Industrial Machinery Meeting*, Milwaukee, (September 12–15, 1960)
22 MACKERLE, Julius, (a) *Air-Cooled Motor Engines*, Cleaver-Hume Press Ltd, London (1961) (English Translation). (b) *Luftgekühlte Fahrzeugmotoren Franckh'sche Verlagshandlung*, Stuttgart (1963) (Deutsche Übersetzung). (c) *Vzduchem chlazené vozidlové motory* Prague (1960) (Tschechische Originalausgabe), SNTL, Publisher of Technical Literature
23 ECK, Bruno, *Ventilatoren. Entwurf und Betrieb der Radial-, Axial-und Querstromventilatoren.* Springer-Verlag; Berlin/Göttingen/Heidelberg, 4th edition (1962). English edn. FANS Pergamon (1973)
24 SCHMIDT, Ernst, *Einführung in die Technische Thermodynamik* 10th edn, Springer Verlag, Berlin/Göttingen/Heidelberg (1963)
25 WITTEK, Hans, SCHEITERLEIN, Andreas and LEBER, Friedrich, 'A Family of Lightweight Air-Cooled Industrial Diesel Engines', *SAE Paper 761 A*, National Power Plant Meeting, Chicago, Illinois, (October 14–17, 1963)
26 MÜLLER, Reinhard, *Luftkühlung Kapitel 2.35* in *Bussien: Automobiltechnisches Handbuch* 18th, pp. 550–645 (1965) Technischer Verlag Herbert Cram, Berlin (1965)
27 SCHOLZ, N., *Aerodynamik der Schaufelgitter, Vol. I*, Verlag G. Braun, Karlsruhe (1965)
28 JUDGE, Arther W., 'Modern Smaller Diesel Engines in Theory, Construction, Operation and Maintenance', *Motor Manuals Vol. Seven*, Chapter 7: Air-Cooled Engines, pp. 134–181. Chapman and Hall Ltd., London (1965)
29 BASILETTE, J. C. and BLACKBURNE, E. F., 'Recent Developments in Variable Compression Ratio Engines', *SAE Paper 660344* (June 1966)
30 LUDECKE, Otto A., 'Blowers for Air Cooled Engines', *SAE 660351*, Mid-Year Meeting, Detroit, Michigan (June 6–10 1966)
31 HOWE, Hans-Ulrich and PISCHINGER, Franz, 'Der luftgekühlte Deutz-Dieselmotor FL 912', *MTZ Motortechnische Zeitschrift* **29**, 4, pp. 132–138 (1968)
32 CORDIER, Otto and REYL, Gaston, 'The Noise Problem of Air Cooled Diesel Engines—Measures toward its Reduction with General Observations and Specific Results', *SAE 680405*, Mid-Year Meeting, Detroit, Michigan, (May 20–24, 1968)
33 KLOSS, Richard, '25 Jahre luftgekühlter Deutz-Dieselmotor. Entwicklungsfragen und Betriebserfahrungen', *Automobil Industrie* **14**, 3 (9.9.1969)
34 BROOKS, Frank, 'A Range of Multicylinder, Aircooled Diesel Engines', *SAE 700027*, Automotive Engineering Congress, Detroit, Michigan (Jan. 12–16, 1970)
35 HOWE, Hans-Ulrich, 'The FL 413—A new series of Deutz Air-Cooled, V-Type Diesel Engines', *SAE 700028*, Automotive Engineering Congress, Detroit, Michigan, January 12–16 (1970)
36 MELLIN, R. C. and SOVRAN, G., 'Controlling the Tonal Characteristics of the Aerodynamic Noise Generated by Fan Rotors', *Trans ASME*, Journal of Basic Engineering, pp. 143–154 (March 1970)
37 LÖHNER, Kurt and KRUGGEL, Otto, 'Verformungsmessungen am Zylinder eines laufenden luftgekühlten Viertakt-Diesel-motors', *MTZ Motortechnische Zeitschrift* **31**, 3, pp. 93–101 (1970)
38 HEMPEL, W., 'Statistical Investigation into Diesel Engine Noise, conducted by the CIMAC Working group "Noise", *Trans I.M.E.* **82**, No. 12 (1970)
39 HEMPEL, W. and SEIDL, T., 'Statistische Erhebung über Dieselmotorengeräusche', *CIMAC-Arbeitsgruppe 'Geräusch', MTZ Motortechnische Zeitschrift* **31**, 4, pp. 153–156 (1970)
40 THIEN, Gerhard and NOVOTNY, Bernd, 'Untersuchungen über den Einfluß von Körperschallvorgängen auf das Geräusch von Dieselmotoren', *MTZ Motortechnische Zeitschrift* **32**, 6, pp. 185–193 (1971)
41 KUNBERGER, Klaus, 'Progress with air-cooled Diesels', *Diesel & Gas Turbine Progress, Worldwide Edition* (July/August 1971)
42 HOWE, Hans-Ulrich, 'Die Baureihe luftgekühlter Deutz-Dieselmotoren FL 413' (Vortrag auf der VDI-Jubiläumstagung "75 Jahre Dieselmotor" in Augsburg, 15–17.3.1972), *ATZ Automobiltechnische Zeitschrift* **74**, 3, pp. 102–111 (1972)
43 KILLMANN, Irolt, 'Progress in Fan Noise Reduction', *Diesel and Gas Turbine Progress, Worldwide Edition*, pp. 30–31 (Nov/Dec 1972)
44 HENSHALL, *Medium and High Speed Diesel Engines for Marine Use*, Chapter 18: Air Cooled Engines, pp. 374–385 (1972)
45 THOLEN, Paul and STREICHER, Kurt, 'New Experiences in the Development of Air-Cooled Diesel Engines with Particular Regard to the Human Environment', (a) *10th International Congress on Combustion Engines* (CIMAC) *Washington D.C.*, Paper 21, pp. 511–536 (April 5–9, 1973. (b) 'Neue Erkenntnisse bei der Entwicklung luftgekühlter Dieselmotoren unter besonderer Berücksichtigung der Umweltbelastung', *MTZ Motortechnische Zeitschrift* **34**, 8, pp. 263–264 (1973)
46 METTIG, Hermann, 'Die Konstruktion Schnellaufender Verbrenungsmotoren, Chapter 3.5.2; *Luftkühlung*, pp. 328–350, Walter de Gruyter-Berlin-New York (1973)
47 ALLSOPP, L. E. and JONES, D. L., 'Turbocharging Air-Cooled Diesel Engines', *GEC Journal of Science and Technology* **45**, 2, pp. 56–62.
48 KILLMANN, Irolt, *Fan Development for Air Cooled Engines with Special Emphasis on Noise Reduction*, Science and Motor Vehicles 73, Compendium, Beograd (April 1973), pp. 43–57. Published by Jugoslovensko Drustvo za Motore i vozila, 27. Marta 80, Beograd
49 WEIDENMÜLLER, Manfred, 'Zum Entwicklungsstand luftgekühlter Dieselmotoren', *MTZ Motortechnische Zeitschrift* **34**, 4, pp. 115–121 (1973)
50 ROGGENDORF, M. and STREICHER, Kurt, 'Die luftgekühlten Deutz-Fahrzeug-Dieselmotoren F 4/6L 913 und BF 6L 913', *ATZ Automobilechnische Zeitschrift* **76**, 1, pp. 1–5 (1974)
51 THIEN, Gerhard and FACHBACH, Heinz, 'Geräuscharme Dieselmotoren in neuartiger Bauweise', *MTZ Motortechnische Zeitschrift* **35**, 8, pp. 237–246 (1974)
52 GARTHE, Hellmut, 'Oil Bath—Dry Type Air Cleaner Combination for Heavy-Duty Operation', *Diesel & Gas Turbine Progress, Worldwide Edition* pp. 23–24 (April 1975)
53 GARTHE, Hellmut, 'Air-Cooling Systems of Deutz Diesel Installations', *VDBUM Seminar at Braunlage 1975*, VDBUM 4/75, Published by A.Krieghoff, 2805 Stuhr 1, FRG
54 PRIEDE, T., 'The Problems of Noise of Engines in Different Vehicle Groups', *SAE 750 795 (SP-397)*, Diesel Engine Noise Conference (August 1975)
55 FACHBACK, Heinz and THIEN, Gerhard, 'Musterausführungen geräuscharmer Dieselmotoren', *MTZ Motortechnische Zeitschrift* **36**, 10, pp. 261–266 (1975)
56 GRUNDY, J. R., KILEY, L. R. and BREVICK, E. A., 'AVCR 1360-2 High Specific Output-Variable Compression Ratio Diesel Engine', *SAE 760 051, Automotive Engineering Congress, Detroit, Michigan*, (February 23–27, 1976)
57 WÖPKEMEIER, H.-F. and HONRATH, K., 'Besonderheiten der Fertigung luftgekühlter Dieselmotoren', *ATZ Automobiltechnische Zeitschrift* **78**, 3, pp. 87–92 (1976)
58 GROSS, Hans-Jürgen, 'Einbluß der Verbrennung auf das Motorgeräusch' (Vortrag bei der Tagung der VDI-Gesellschaft Fahrzeugtechnik '100 Jahre Viertakt-Otto-Motor' 26–28.4.1976 in Köln), *ATZ Automobiltechnische Zeitschrift* **79**, 4, pp. 133–136 (1977)
59 HERSCHMANN, Otto, H, 40 Jahre luftgekühlte Deutz-Dieselmotoren—eine technisch-wirtschaftliche Bilanz, *MTZ Motortechnische Zeitschrift* **37**, 4, pp. 117–123 (1976)
60 FACHBACH, Heinz and THIEN, Gerhard, 'Körperschallausbreitung bei Dieselmotoren', *MTZ Motortechnische Zeitschrift* **37**, 7/8, pp. 269–274 (1976)
61 ESCHE, Dieter, 'Beitrag zur Entwicklung von Kühlgebläsen für Vebrennungsmotoren', *MTZ Motortechnische Zeitschrift* **37**, 10, pp. 399–403 (1976)
62 BARTSCH, Christian, 'Motorlärm ist vermeidbar. Mit konstruktiven Maßnahmen auf dem Weg zum leiseren Einbaudiesel', *VDI-Nachrichten* 14 (April 8, 1977)
63 KUNBERGER, Klaus, 'Progress with Quiet Small Diesels', *Diesel & Gas Turbine Progress Worldwide* pp. 56–57 (May 1977)
64 THOLEN, Paul and KILLMANN, Irolt, 'Investigations on Highly Turbocharged Air-Cooled Diesel Engines', (a) 12th International Congress on Combustion Engines (CIMAC), Tokyo (May 23–26, 1977). Paper A 11, pp. 505–543. (b) ASME Paper 77-DGP-11, Energy Technology Conference and Exhibit, Houston, Texas, USA (September 18–22 1977)
65 LICHTNER, Emil and WAHNSCHAFFE, Jürgen, 'Advancement of the Air-Cooled Deutz FL 413 Diesel Series', *Diesel & Gas Turbine Progress Worldwide*, pp. 33–36 (June 1977)

66 GARTHE, Hellmut, 'The Development of Deutz Air-Cooled Diesel Engines', *VDBUM Information 5*, 3/77, Published by A. Krieghoff 2805 Stuhr 1, FRG (August 1977)
67 ALLSOPP, L. E. and JONES, D. L., 'The Design and Development of Turbocharged 1 Litre per Cylinder 4 and 6 Cylinder Air Cooled Diesel Engines', *ASME Paper 77-DGP-9*, Energy Technology Conference and Exhibit, Houston, Texas, USA (September 18–22 1977)
68 GROSS, Hans-Jürgen and HARTGES, Hans, 'Noise Control in Air-Cooled Diesel Engines, University of Wisconsin, Department of Engineering, Madison, Wisconsin', *Tenth Annual Noise Control in Internal Combustion Engines* (April 27–28 1978)
69 ESCHE, Dieter, Kühlung. BUSSIEN: *Automobiltechnisches Handbuch Herausgeber*: Gustav Goldbeck Ergänzungsband zur 18. Auflage, Kapitel 1.7, pp. 267–281, Walter de Gruyter, Berlin, New York (1979)
70 GROSS, Hans-Jürgen, Lärmschutz, *Bussien: Automobiltechnisches Handbuch Herausgeber*: Gustav Goldbeck Ergänzungsband zur 18. Auflage, Kapitel 6.2, pp. 1349–1385, Walter de Gruyter, Berlin, New York (1979)
71 KOCHANOWSKI, H. A., KAISER, W. and ESCHE, Dieter, 'Noise Emission of Air-Cooled Automotive Diesel Engines and Trucks', *SAE-Paper 790451*, p. 79
72 KOCHANOWSKI, H. A., 'Geräuschdämmung an luftgekühlten Dieselmotoren and Lastkraftwagen', *Umwelt* 4/79, pp. 323–327
73 *Magirus in Sibirien*, Lastauto—Omnibus, No. 2–5 (1979)
74 'New Air Cooled Diesels: Deutz adds models', *Diesel & Gas Turbine Progress Worldwide* (April 1979)
75 THOLEN, Paul, 'Aufgeladene Nutzfahrzeug-Dieselmotoren, Probleme und Entwicklungstendenzen', *Automobil Industrie* 4, pp. 23–34 (1980)
76 ESSERS, U., LIEDL, W., DENKER, D. and GERNGROSS, H. G., 'Untersuchung von versuchmäßig dargestellten Lösungen zur Geräuschminderung eines Serien-Lastkraftwagens', *VDI Verein Deutscher Ingenieure, VDI-Gesellschaft-Fahrzeugtechnik*, XVIII. Internationaler Congress FISITA Hamburg (May 5–8, 1980), VDI-Berichte Nr. 367 (1980, pp. 331–341)
77 FISCHER, J., MÜHE, P. and STANGL, G., 'Darstellung von in Serie realisierbaren Lösungen zur Geräuschminderung an einem Lastkraftwagen für den Verteilerverkehr. 'Verein Deutscher Ingenieure', *VDI-Gesellschaft-Fahrzeugtechnik, XVIII.* Internationaler Congress FISITA Hamburg. VDI-Berichte No. 367 (1980), pp. 259–266 (May 5–8, 1980)
78 KOCHANOWSKI, H. A. and HALLER, H., 'Geräuschemission luftgekühlter Fahrzeug-Dieselmotoren', Verein Deutscher Ingenieure, *VDI-Gesellschaft-Fahrzeugtechnik,* XVIII. Internationaler Congress FISITA Hamburg. VDI-Berichte Nr. 367 (1980), pp. 267–274 (May 5–8, 1980)
79 GARTHE, Hellmut, 'Leistungskennfelder und Charakteristiken luftgekühlter Dieselmotoren für den Güterfern- und Verteilerverkehr, *Internationales Verkehrswesen* **32**, 4, pp. 278–281 (July/August 1980)
80 GARTHE, Hellmut, 'Air-Cooled Deutz Diesel Engines for Underground Mining Applications', *SAE 801000, International Off-Highway Meeting and Exposition, Mecca, Milwaukee,* (September 8–11, 1980)
81 THOLEN, Paul, 'Entwicklungstendenzen auf dem Gebiet der Motorentechnik im Hinblick auf eine wehrtechnische Anwendung', *Forum 'Mobile Stromversorgung im Felde' der Deutschen Gesellschaft für Wehrtechnik e.V. Bonn* (Nov 20–21, 1980) in Aachen
82 SLEZAK, Paul J. and VOSSMEYER, Wilhelm, 'New Deutz High Performance Diesel Engine', *SAE 810905, International Off-Highway Meeting and Exposition, Milwaukee, Wisconsin* (September 14–17, 1981)
83 KUNBERGER, Klaus, 'Four years of Hatz Silent Power', *High Speed Diesel Report* **I**, 3, pp. 12–15 (May/June 1982)
84 WÖPKEMEIER, H.-Friedrich, 'Möglichkeiten und Grenzen der Entwicklung verbrauchsgünstiger und schadstoffarmer Dieselmotoren', Vortrag beim TÜV, Rheinland in Köln, (April 27, 1982)

17 Crankcase explosions

Contents

17.1 Introduction

Crankcase explosions happily are not an everyday occurrence but when they do occur can lead to extensive damage, fire and possible loss of life. Experience indicates that such explosions can be most dangerous in the large slow speed engines, but crankcase explosions in smaller engines for vehicles, locomotives and stationary plant are not unknown, although here the effects have, in general, been contained within the crankcase. The phenomena are not restricted to oil engines as cases have also been reported in gas compressors, steam engines and other machinery with enclosed crankcases.

The consequences have been more pronounced in the large engines because the design must be such that the crankcase doors are light enough to handle manually and, at the same time, large enough to permit access; also the doors are not normally constructed to withstand excessive pressure effects or explosion. This chapter will therefore concentrate largely on engines used in motorships and in power stations and will range from the large slow speed engine to the medium and high speed types.

There is a continuing uprating of all engine designs with higher bearing pressures, increased thermal stresses and temperatures, thus making increased demands on cooling and lubricating systems. It is therefore all the more vital to ensure that when a failure does occur and may produce conditions favourable to an explosion, firstly the explosion should be averted or, in the extreme case, the effects be controlled. This leads us to a consideration of the cause of crankcase explosions and the methods used both to control their effects and to monitor conditions in a crankcase.

17.2 Oil mist in crankcases

Following a disastrous crankcase explosion in 1947 extensive investigations were carried out during the following ten years and these provided information which still remains valid, concerning the properties and generation of oil/air mixtures in enclosed spaces[1, 2, 3].

Figure 17.1, from Smith and Thomas[4], gives the relationship between the air/oil ratio and the temperature for a typical crankcase oil. This shows the upper and lower limits of inflammability and indicates that spontaneous combustion can occur at a temperature of 270°C with a mixture having an oil content of 13% by weight. This figure also shows the two separate regions in which ignition can take place with a band of temperature between where no ignition is possible.

The above work also established that overheating of mechanical parts and the creation of what has commonly been called a 'hot spot' can result in the formation of an explosive mixture which, in turn, is ignited by the hot spot itself. The overheated area thus becomes both source and ignitor of the explosive medium. It is important to note that it is the condition of the air/oil mixture in the vicinity of the overheating which is the controlling factor in determining whether an explosion will take place and not the general condition in the crankcase. In a trunk piston engine entry of flame past the piston into the crankcase could also act as ignitor.

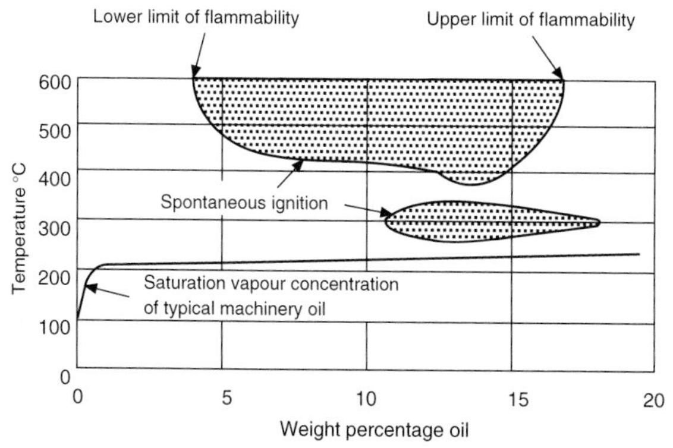

Figure 17.1 Spontaneous ignition limits for mineral oil vapour (mist) mixtures with air

Under normal conditions a crankcase contains a mixture of mechanically generated spray and a small amount of condensed mist resulting from the vaporizing of oil from bearings and other parts at working temperature. The larger droplets of oil do not constitute a hazard in themselves and normal crankcase breathing and circulation of the air maintain a stable condition whereby any vaporized oil is condensed in the cooler parts of the space but with the mixture remaining outside the limits of flammability.

In the normal course of events, with a typical crankcase oil, the mixture becomes overrich and above the limit of flammability before the minimum ignition temperature is reached. Provided the stable conditions in the crankcase persist the presence of a hot spot will not inevitably lead to an explosion.

The dangerous situation arises when the degree of overheating leads to accelerated vaporization of the oil which may be accompanied by cracking and oxidation to form gases and vapours even more hazardous than the oil mist[3]. An explosive mixture can then result which may be ignited by the overheated part.

Any changes in the stable conditions in the crankcase can lead to hazard. These can arise from dilution of the overrich mixture caused, for instance, by removal of an inspection door and changes in the air circulation patterns induced by the moving parts, including stopping and starting of the machine. Each may have the effect of bringing an explosive mixture to the hot spot area with consequent risk of ignition.

An alternative explanation of delayed ignition has been put forward by Freeston, Roberts and Thomas[1] who have pointed out that it is likely that the machinery part heats up through the low temprature ignition region without producing flame, because of the length of the ignition delay at low temperatures. When the temperature reaches the non-ignition band between 350°C and 400°C, shown in *Figure 17.1*, there will be vaporization of the oil without ignition, which can promote formation of an explosive mixture. If now the engine is stopped, or for any other reason cooling is introduced, the temperature of the heated part will fall into the region where an explosion can occur. The rate of cooling will determine when the explosion will take place and can be some minutes after stopping the machine.

17.3 Explosion effects

The magnitude of the explosion may vary between a mild 'puff', which produces a small amount of white smoke, and detonation, which causes extensive damage and the emission of flame. Research has indicated that ignition of the hydrocarbon/air mixture within an enclosure, initially at atmospheric pressure, does not create a pressure greater than 7 bar. The violence of an explosion will, however, depend on the composition and quantity of the explosive mist, the speed of propagation of flame and of pressure rise within the crankcase.

The pressure wave of the primary explosion is followed by a negative pressure, and if air is permitted to enter the crankcase, a further mixing of oil and air occurs which may result in a secondary explosion. *Figure 17.2* from Pounder[5] shows that the instantaneous pressure wave is extremely short and is followed by a relatively long negative pressure during which the incoming

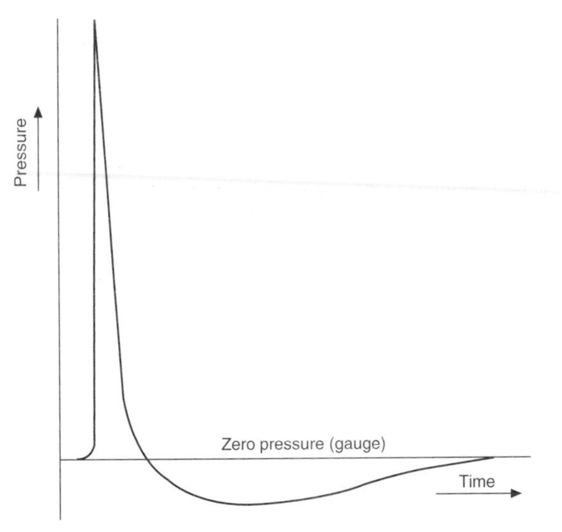

Figure 17.2 Pressure/time explosion curve showing long duration of suction phase (no attempt at scales has been made)

air mixes thoroughly with the oil mist, with the result that the secondary explosion can be more violent than the first.

17.4 Incidence of crankcase explosions

During the 1950s, developments were in progress to reduce the serious consequences of crankcase explosions. Some time elapsed before satisfactory solutions were found, and were incorporated in new and existing engines. The records of Lloyd's Register of Shipping show that, between 1957 and 1961, eight explosions occurred in a total of 1834 four-stroke oil engines, and twelve in 2926 two-stroke engines. These resulted in the loss of four lives and nine injuries to personnel, and there was serious damage to a number of the affected engines. In the years 1974 to 1979, only four explosions were reported in a total of 4757 two-stroke oil engines. There were neither casualties to personnel nor undue damage; however during the same period thirty-seven explosions occurred in a total of 6239 four-stroke engines, and four persons were injured as a consequence.

Reports from other sources appear to be less complete; however at least ten cases were reported in the years 1957 to 1961 which led to deaths or injuries, the totals being eight killed and twently-eight injured. So far as can be ascertained actions subsequently taken to safeguard such engines have also been effective and reflect the value of modern preventative measures.

In the assessments of all the reported incidences it is apparent that the chief causes of explosions were associated with the overheating or seizure of pistons and bearings.

17.5 Prevention of explosions

It is clear that elimination of either an explosive mixture or a source of ignition will completely remove the possibility of an explosion. For the latter, overheating of parts is an ever present possibility, while the former can only be achieved by flooding the crankcase with an inert gas to reduce the oxygen level below 10%. This would necessitate a 'tight' crankcase, monitoring of the oxygen level and a supply of inert gas.

Inerting has not been adopted in crankcases because of the practical difficulties and associated economic aspects, and accordingly the trend has been towards improvement in design, with the introduction of good safety features and the provision of means to monitor conditions in the crankcase.

17.6 Design aspects

In a small engine it is possible to design the crankcase to withstand the forces likely to be produced by an explosion; however this would be impractical in the larger engine. In these engines the maximum realizable pressure resulting from an explosion can be reduced by sub-divisions since the maximum pressure is proportional to the distance of flame travel and the crankcase volume. In practice, however, only partial sub-division is feasible as access is required for assembly and maintenance, and the solution has been to incorporate other means of limiting the maximum pressure.

Designers have done much to reduce hazards by minimizing the risk of overheating and seizure, and considerable attention has been given to the design and securing of crankcase doors. Such doors on larger engines are capable of withstanding reasonably high pressure while retaining a light weight. Crankcase doors for some small and medium size engines are, in fact, made from reinforced plastic materials. Construction from such materials allows easy forming of a dished shape to obtain the benefits associated with hoop stress, and approval of such designs has been given by Classification Societies after rigorous tests. Similar attention is given to securing arrangements since the spacing of bolts and clamps on the door must be such as to ensure that the door is not blown off by the initial explosion or sucked in during the negative pressure period.

Crankcase breather pipes are fitted to relieve internal crankcase pressure due to expansion and these should be as small as practicable. For marine engines, classification societies stipulate that these pipes should be led to a safe space on deck. Preferably they should be fitted with flame arrestors and, where there is more than one engine, the pipes should not be inter-connected.

17.7 Explosion relief valves

It has already been pointed out that it is possible to design the crankcase of the smaller engine to withstand the anticipated maximum internal pressure of 7 bar but in the case of larger engines protection is normally by means of relief valves. These must be of sufficient size, in sufficient number and be placed in positions most likely to limit the pressure. They should also have the capacity to close quickly after lifting in order to prevent ingress of air after the initial explosion. The valves are normally fitted with flame traps to prevent emission of flame.

The British Department of Trade and Classification Societies have similar Rules which state that valves can only be omitted in engines having cylinder bores not exceeding 200 mm and a gross crankcase volume not exceeding 0.6 m^3[6,7]. Above the 200 mm bore size there is a progressive increase in the number of valves to be fitted until, in engines with cylinders above 300 mm bore, at least one relief valve is to be fitted in way of each crank throw, plus a separate valve for each separate space, e.g. chaincase, gear case, etc. The requirements further state that the free area of each relief valve should not be less than 45 cm^2 and the combined free area of the relief valves fitted to an engine should not be less than 115 cm^2/m^3 based on the volume of the crankcase. The quick-acting relief valves should be made as light as possible to reduce inertia effects and should be designed to open at a pressure not greater than 0.02 N/mm^2. The material of the valves must be capable of withstanding the shock of contact with stoppers at the full open position.

As mentioned earlier, a crankcase explosion can generate flame and there have been many cases where the discharge of burning gases has caused injury to personnel. For this reason flame guards or flame traps are fitted to minimize the danger arising from the emission of flame. Research has established that an oil wetted gauze greatly increases the effectiveness of a flame trap, wetting of the gauze being achieved by arranging the gauze assembly before the valve and in the path of lubricating oil spray within the crankcase. As will be noted from the descriptions which follow later, oil wetted gauze is not, however, fitted to relief valves in all cases.

Swing automatic self-closing relief doors of the type described by Pounder[5] were among the early designs of crankcase explosion relief devices. The swing door type allowed a direct discharge of the gas but it was difficult to provide complete protection by shielding in all cases. Over the years further designs have been developed, all having the same objectives, namely relief of the pressure ond protection of personnel from issuing flame and hot gases. The following two are well-known examples.

Figures 17.3 and 17.4 illustrate an explosion relief valve devised by the British Internal Combustion Engine Research Institute where the flame trap is arranged to form a single unit with the valve, the crankcase door being sandwiched between the two components. The internal element supplies the frame for the mild steel gauze layers and for the spider which supports the spindle on which the relief valve rides. The external part of the assembly is a combined valve cover and deflector and is of aluminium. The deflector opening extends 120 degrees and can be turned to the direction which is safest for the issuing hot gases. The valve itself is an aluminium alloy casting having synthetic rubber, oil and heat resisting seals. A spring, which presses against a stop in the spider, is used to retain the valve lid in the closed position. The licensee for the manufacture of BICERI valves is The Pyropress Engineering Co. Ltd of Plymouth. The patented Hoerbiger crankcase relief valve is shown in *Figure 17.5*. This type is supplied with an external dry flame trap.

The valve has a valve lid of deep drawn steel which seats against an oil resistant rubber ring embedded in the machined seat. The valve movement is guided by a collapsing coil spring, the end coils being retained by the shape of the cover and the valve lid respectively. The cover, which is also the flame deflector allows the escaping gases to exit through 360 degrees.

The Hoerbiger valve responds very quickly to increased internal pressure—possibly faster than the internal flame trap design due to the internal flame trap acting as a slight flow restrictor.

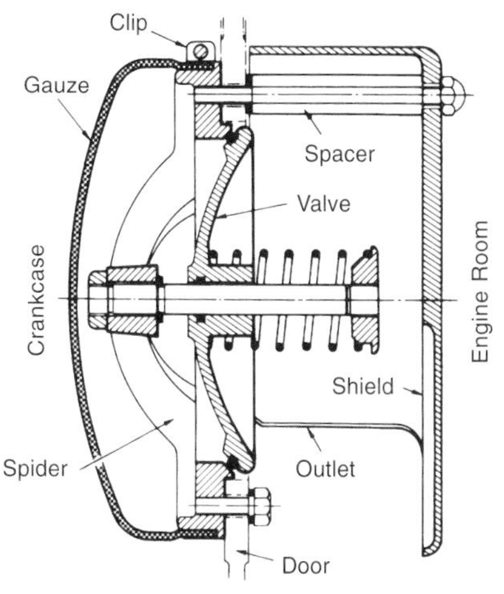

Figure 17.3 Sectional arrangement of BICERI explosion relief valve

Figure 17.4 External view of BICERI relief valve

17.8 Crankcase monitoring systems

There are a number of methods by which conditions in a crankcase may be monitored and these include temperature sensing of parts likely to overheat, checking the temperature of the lubricating oil itself and monitoring the metallic debris in the lubricating oil returns. Temperature sensing would require a large number of sensors to be fitted to efficiently monitor a crankcase besides presenting practical difficulties in fitting detectors to moving parts. However such sensing of main journal bearings and cylinder liners where they enter the crankcase is common practice. Monitoring of the temperature and of the metallic debris in lubricating oil has the disadvantage that there is a delay before the effects of abnormality are detected. One method which has, however, found favour is that of oil mist detection.

The formation of oil mist or condensed oil vapour is largely dependent on the relationship of oil vapour pressure and temperature, the oil mist increasing in quantity as the temperature rises. See *Table 17.1*. Oil mist detection is based on the principles that there is a non-linear relationship between oil mist density and its opacity to light (see *Figure 17.6*). The optical density of oil mists is fairly high and, with rising concentration, the intensity of a light beam directed towards a photocell is reduced.

17.9 Oil mist detectors

17.9.1 Graviner systems

With the Graviner systems there are two basic methods of measuring the oil mist density (see *Figure 17.7*). The first is the

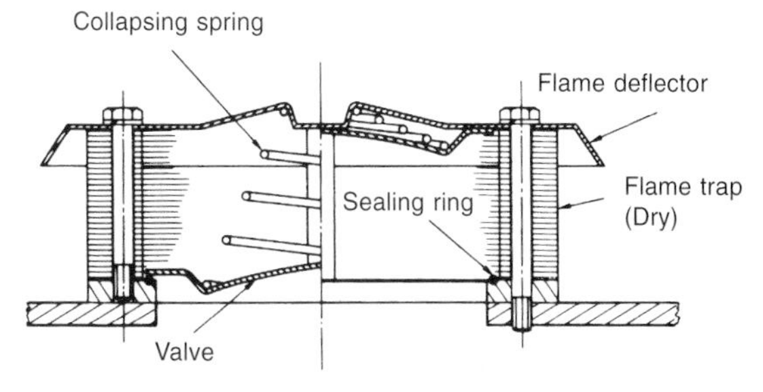

Figure 17.5 Patented Hoerbiger crankcase relief valve with external dry flame trap

Table 17.1

Temperature °C	*Oil vapour pressure (mm of Hg)*
80	0.0002
100	0.002
150	0.06
200	1.0
250	10.0

Relation between temperature and oil vapour pressure

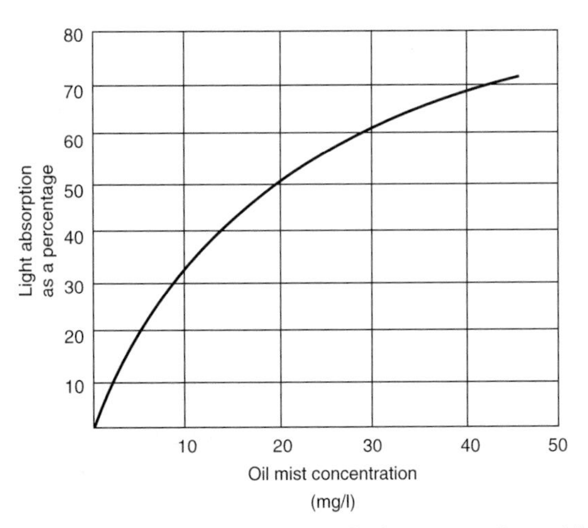

Figure 17.6 Relationship between oil mist concentration and light absorption

Comparator type where a sample taken from each crankchamber is compared in turn with the combined mist from the remaining chambers.

The second method, called the Level type, involves comparing a sample from each crankchamber in turn with clear air. The first method is particularly applicable to crosshead type two-stroke engines where there is a low background of oil mist, a sub-divided crankcase and relatively slow speeds. The second is used for engines where there is a high background of oil mist and litle or no sub-division of the crankcase, i.e. high and medium speed trunk piston engines.

17.9.1.1 Comparator type

The operation follows a procedure whereby oil mist is continually drawn by a fan in the detector from sampling points on the engine. A rotary valve passes each sample in turn to the Measuring Tube while the remaining samples (representing the average oil mist density) are simultaneously passed to the Reference Tube. The photocells in the Measuring and Reference Tubes combine to give an output proportional to the difference between the density of the oil mist in the one crankchamber as against the average.

Once per cycle the rotary valve passes clear air into the Measuring Tube, the photocells then giving an output proportional to the average oil mist density. Further, as a manual daily operation, air is admitted to both tubes as a check for meter zero. If an out-of-balance is produced the unit can easily be adjusted to the zero position.

In the case of increased oil mist in an individual crankchamber, the system is so arranged that an alarm sounds and simultaneously the rotary valve stops at the position which caused the alarm. A scale enables the area with excessive oil mist to be identified.

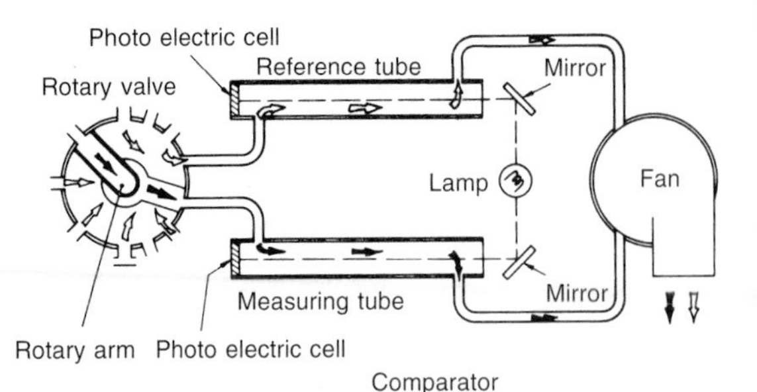

Rotary valve
Reference tube
Mirror
Photo electric cell
Lamp
Fan
Measuring tube
Mirror
Rotary arm
Photo electric cell
Level

Figure 17.7 Graviner oil mist detector

An alarm condition also operates in the event of the overall oil mist density being excessive when measured against clear air.

17.9.1.2 Level type

This is similar in many respects to that of the Comparator except that the Reference Tube, which is sealed, contains clear air against which the oil mist density from the crankcase is measured. An offset control is also incorporated in this type to allow for the high density of the normal crankcase oil mist in trunk piston engines. The sensitivity to an increase in oil mist is the same for all settings of this control; however with the offset applied, greater density of the oil mist is required before the meter starts to read on the scale. The control is calibrated to suit the individual engine to which the detector is fitted, but even with maximum offset applied, warning is given before the concentration reaches 5% of the lower explosion limit.

The Level type is used either as a sequential scanning unit or by combining pairs of sampling points. The first enables samples from each crankchamber to be checked individually which gives good fault location. The second method gives less accurate fault location but is applicable where the scanning cycle time, and hence response time, are important, as for instance in four-stroke engines where crankcase sepration is minimal.

17.9.2 Schaller Visatron systems

In the Schaller Visatron systems the samples drawn from the crankcase are passed through a cyclone separator which removes larger oil droplets, these being returned to the crankcase. The remaining air/oil mixture passes through an optical-electrical measuring system where, in the case of rising opacity, the reduction in the intensity of light falling on the photo-diode is reduced and this leads to actuation of the alarm.

This range of oil mist detectors operate on the manifold system of sampling, i.e. crankchamber samples are led to a common header. As with other types which combine samples, a high concentration of oil mist in any one crankchamber will be diluted in the manifold by the remaining samples. The means used in this case to overcome this disadvantage is a high sensitivity alarm system incorporating a ‘floating threshold’ which is able to follow slow variations of the opacity, thus permitting high sensitivity but without the risk of false alarm caused by low settings. If a hot spot develops the opacity increases quicker than the 'floating threshold' rate, thus causing the detector to react.

The capability of the Schaller VN 115 detector to indicate hazard is illustrated in *Figure 17.8*. These curves were obtained during tests on a six-cylinder medium speed engine developing 2200 kW where a piston seizure was induced intentionally. It will be noted that the rapid increase in opacity was detected after 4s and an alarm was given shortly after the commencement of loss of speed due to the impending seizure of the piston.

For higher sensitivity the VN116 unit is used, which takes account of slow increases in oil mist concentration. To achieve this without lowering the alarm setting, with the possibility of false alarms, a logic circuit is introduced to provide a 'damage check' i.e. to determine whether the oil mist concentration is distributed in a non-uniform way throughout the whole crankcase—this being a sure sign of a hot spot. As a refinement of this system the VN215 gives specific crankchamber location at the hot spot. The system sets off the alarm in the same way as the VN116 but continues to perform independent of alarm indication, to detect the hot spot crankchamber—this crankchamber is then indicated on the instrument, with the possibility of remote indication.

17.9.3 Location of sampling points

The position of the sampling points on the crankcase is obviously of importance and on large slow speed engines the samples are generally extracted from the upper part of the single compartments where partitioning is complete. It is in this area that the mist generated from moving parts tends to accumulate.

In medium and high speed engines the sampling points are best located between the bottom of the liners and the main bearings, preferably at positions where oil mist will accumulate. Experience has shown that pipe runs from the sampling points should be as short as practicable, avoiding sharp bends where oil could accumulate and there should be adequate fall of pipes in order to provide drainage of condensed oil. Sampling speeds require to be adjusted to suit particular engine crankcase volumes and anticipated oil mist levels, it being the case the at excessively high sampling speeds can give rise to errors in measurement.

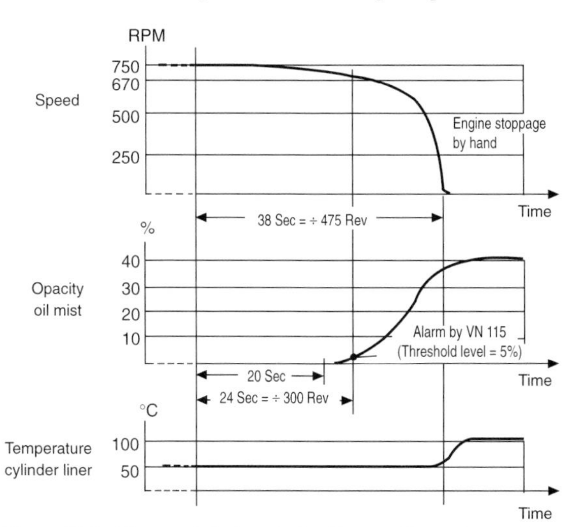

Figure 17.8 Characteristic data obtained by provoking a piston seizure on a diesel engine fitted with a Schaller oil mist detector

17.10 Practical aspects

The risk of hot spots is always present but is more likely when the moving parts may not have fully bedded in, i.e. after overhaul or on new engines. If overheating is suspected or smoke is seen issuing from the crankcase, the engine should be stopped, but the crankcase doors, or other closing covers likely to allow ingress of air, should not be removed until sufficient time has been allowed for the parts to cool. It is equally important that an engine should not be re-started before a fault has been rectified since this has, on occasion, also led to crankcase explosions.

Although there is no indication that contamination of lubricating oil by fuel oils or other media has led to crankcase explosions, it is essential to carry out routine checks since contamination can increase the chance of overheating due to lubrication break down. By routine tests on used lubricating oils to determine viscosity and flashpoint, it is possible to assess accurately the degree of dilution.

Certain manufacturers of larger slow speed engines supply inert gas systems which can be used to flood crankcases in the event of suspected danger, but these are not invariably fitted to all engines. Classification societies do, however, require that fire extinguishing arrangements be fitted to scavenge spaces of crosshead engines where the scavenge spaces are in open connection with the cylinders. Such spaces should also be fitted with explosion relief valves.

Crankcase explosions can be avoided but to achieve this it is necessary to ensure that mechanical parts are well designed, constructed and fitted and that overhauls are carried out with due diligence. It is also important that operation of the engine be such as to keep temperatures and pressures and loading within the designed parameters, and that oil mist detectors, explosion relief valves and all other safety devices be maintained in good order.

References

1 FREESTON, H.G., ROBERTS, J.D. and THOMAS, A., 'Crankcase Explosions: An Investigation Into Some Factors Governing the Selection of Protective Devices', *Proc. I. Mech. E.* (1956)

2 BURGOYNE, J.H., NEWITT, D.M. and THOMAS, A., 'Explosion Characteristics of a Lubricating Oil Mist', *The Engineer* (1954)

3 BURGOYNE, J.H. and NEWITT, D.M., 'Crankcase Explosions in Marine Engines', *Trans. Inst. Marine Eng.* (1955)

4 SMITH, A.C. and THOMAS, A., 'Fire and Explosion Hazards in Compressed Air Systems', *Proc. Prod. E. Compressed Air Conference* (1958)

5 POUNDER, C.C., 'Diesel Engine Principles and Practice'. 2nd Ed. (1962) and 'Marine Diesel Engines', 3rd Imp. of 5th Ed. (1976), Butterworths

6 *Survey of Passenger Ships*, Vol. 1, Department of Trade

7 *Rules and Regulations for the Classification of Ships*, Lloyd's Register of Shipping

Part 4

Environmental aspects

18 Exhaust smoke, measurement and regulation

Contents

18.1 General considerations

It has long been appreciated by engineers that the presence of smoke in the diesel engine exhaust is an indication of poor combustion resulting from some malfunction or maladjustment. Nevertheless, with increasing concern for the effect of air pollution on the environment, particularly in the field of road transport, vehicle exhaust emissions have, in recent years, been subjected to increasingly stringent regulations.

Most industrialized countries have therefore introduced regulations of varying degrees of complexity to control smoke emission from road vehicles. The regulations have been in addition to the relatively simple existing regulations covering industrial plant and have involved much development both of test methods and instrumentation.

Smoke may be defined as particles, either solid or liquid (aerosols), suspended in the exhaust gases, which obstruct, reflect, or refract light. Diesel engine exhaust smoke can be categorized under two headings:

1. Blue/white in appearance under direct illumination, and consisting of a mixture of fuel and lubricating oil particles in an unburnt, partly burnt, or cracked state.
2. Grey/black in appearance, and consisting of solid particles of carbon from otherwise complete combustion of fuel.

The blue component derives mainly from an excess of lubricating oil in the combustion chamber, resulting from deterioration of piston ring sealing, or value guide wear, and is thus an indication of a need for mechanical overhaul. However unburnt fuel can also appear as blue smoke if the droplet size is circa 0.5 μm.

The white component, on the other hand, is mainly a result of too low a temperature in the combustion chamber during the fuel injection period. It has a droplet size of circa 1.3 μm. This can occur as a transient condition during the starting period, in low ambient temperatures or at high altitude, disappearing as the engine warms up. On the other hand, it can result from too late fuel injection or may even be an indication of a design fault, in the sense that the compression ratio is too low, or has been optimized for an inappropriate combination of operating conditions.

Grey/black smoke is produced at or near full load if fuel in excess of the maximum designed value is injected, or if the air intake is restricted. In normal operation its onset is associated with reduced thermal efficiency and sets a limit to power output before any serious proportion of toxic component such as carbon monoxide is discharged. The main causes of excessive black smoke emission in service are either poor maintenance of air filters and/or fuel injectors, or incorrect setting of the fuel injection pump.

Such smoke consists essentially of carbon particles or coagulates of a wide range of sizes, ranging from 0.02 μm upwards to over 0.12 μm mean diameter. This size distribution depends to some extent on the type of combustion system, which also affects the onset of smoke emission as fuel input quantity is increased. Thus, in general, open chamber (direct injection) systems show a rather gradual increase in exhaust visibility with increasing fuelling, whereas swirl chamber (indirect injection) systems tend to have a critical fuelling level above which smoke emission increases very rapidly. It should be appreciated however, that there are some carbon particles present in the diesel exhaust under any operating conditions, so that zero smoke emissions is impossible.

Fuel properties are also capable of influencing smoke emissions. Thus, increasing the cetane number will reduce the tendency to produce white smoke, as also will increased volatility, usually indicated by reduction in mid-boiling point. On the other hand, chemical composition, cetane number and volatility all affect black smoke in a complex way, while increasing relative density will increase black smoke, for the same fuel pump setting, merely as a result of the increased mass of fuel injected. Compounds of the alkaline earths, typically barium, are effective as fuel additives, in small quantities, in reducing black smoke, but their use poses problems in administering regulations, as well as in fuel distribution, and they remain only of technical interest[1,2].

In seeking to control 'excessive' emission of smoke by regulation and inspection it is clearly not satisfactory to rely on subjective impression. Instrumentation is necessary to quantify smoke objectively as to its degree of visibility. It is also necessary to define a test method to relate the objective measurement to the subjective impression in a meaningful way.

Much early work on the optical properties of smoke plumes was carried out by the United States Public Health Service (USPHS) of the Department of Health, Education and Welfare[3]. This demonstrated, in viewing a stationary smoke stack in various conditions of lighting, and at different times of day, that the subjective visual assessment of identical plumes varied widely, particularly in the case of white smoke. The conclusion was reached that the optical property easiest to measure is the light transmittance of the plume.

18.2 Instrumentation

It would be quite impossible in the space available to describe the multitude of smoke meters and indicators that have been devised over the years. Only those which have achieved some common acceptance, or have been specified in connection with Standards or Regulations will therefore be considered in brief detail, as representing a generic type. These types fall into fairly well-defined classes.

18.2.1 Comparators

Typical of this class, in which comparison of the visual appearance of a smoke plume is compared directly with a standard scale of grey, is the Ringelmann Chart[4]. In this, a white card, on which has been printed a series of black grids obscuring, respectively, 20%, 40%, 60% and 80% of the surface, is viewed in optical proximity to the plume. The degrees of obscuration listed are arbitrarily numbered 1, 2, 3 and 4 Ringelmann. On this scale the white card is numbered 0 and a totally black card 5 Ringelmann.

Although there are obvious objections to comparing the appearance of smoke transmitting light from behind the plume with that of a chart reflecting light from a quite different part of the sky, the Ringelmann Chart has been in common use for external surveillance of industrial plant.

A preferable comparator would appear to be a photographic grey scale, having varying shades of blackness on a transparent base. With its use, the smoke can be compared against a similar background and with similar transmitted light. The grey scale can be accurately calibrated in terms of per cent transmittance and thus fulfils the recommendations of the USPHS referred to earlier. Even so, if the smoke is not black, errors can clearly arise. Various forms of this type of comparator are available, including the 'USPHS Film Strip'.

It will be obvious that such comparators are of very limited use in assessing smoke emission from a moving vehicle, with the constantly varying lighting, background and viewing angle, to say nothing of the varying emission from the vehicle itself, as speed and load change.

We wish to acknowledge the help of Ricardo Consulting Engineers in the updating of this chapter.

18.2.2 Filter-soiling 'spot' meters

If exhaust gas is passed through a white filter paper, the carbon particles are deposited, and the darkening of the paper can be taken as a measure of the smoke density. For consistency of measurement it is essential that a fixed volume of gas is passed through a defined area of filter paper, and the paper itself needs to be closely specified. The gas sample should be passed through the paper at a constant rate, and excessive pressure fluctuations at the point in the exhaust system from which the gas sample is extracted will produce erroneous results, as will condensation of moisture on the filter paper. Furthermore, it has been shown[6] that a high proportion of aerosols in the exhaust gives a reduced value of smoke density, since the paper is rendered transparent, to some extent. Such smokemeters are therefore of no use in cases where blue/white smoke is present.

The simplest usable form of this type of instrument is represented by the Bacharach Type RCC-B. This is a hand operated suction pump drawing an exhaust sample through a $\frac{1}{4}$ in sampling probe inserted $2\frac{1}{2}$ in into the stack. The stroke volume to filter area is 225 cu in per sq in (363 cc per sq. cm). The soiled filter paper, backed by a white plastic, is compared visually with a 10-step series of grey shades, ranging from 0 (white) to black (9).

The matching of the darkened filter to the steps of grey involves some degree of subjective judgement, while the manual operation of the pump is likely to lead to some error as a result of uneven rates of withdrawal of the sample. This instrument is really only of use in monitoring the smoke from boiler plants, and is not suitable for assessing diesel engine exhaust.

Probably the ultimate development of the 'spot' type of smokemeter is that developed in Germany by Bosch[7]. This Sampling Pump. Type EFAW/65, overcomes many of the objections raised above, and avoids the need for an external power source (*see Figure 18.1*). Before taking the sample, the pump piston is set manually in the inner (minimum volume) position and is held there by a spring-loaded ball detent. To take the sample this detent is remotely released pneumatically, permitting the piston to be returned by a spring to the outer (maximum volume) position, the movement being accomplished in 1 to $1\frac{1}{2}$ seconds. The volume displaced is 0.33 litres and the gas is drawn through a circular filter area of 8 sq. cm. It is mandatory that the sampling probe is that specified by the makers, as this is designed to prevent dynamic pressures at the sampling point being transmitted into the sample line and so affecting the pump piston motion. Static pressure at the sampling point should not exceed 15 in (380 mm) of water.

A detailed procedure has been laid down for the sampling operation[8], and this should be followed precisely if the results are to be consistent between different operators. At the same time, maintenance procedures on the sampling pump, and checks for leakage must also be carried out.

The darkening of the filter paper is assessed by means of an evaluating unit. Type EFAW/68 (*see Figure 18.2*). This is, in effect, a reflectometer, the light from a filament lamp being reflected from the soiled filter disc onto an annular photocell. Lamp, photocell and filter disc are arranged coaxially, the disc resting on a stack of at least twelve clean filters. The output from the photocell is fed to a microammeter, scaled arbitrarily from 0 to 10 Bosch Numbers. Again, a detailed procedure has been devised to ensure accuracy and consistency[8,19], including periodic checks for zero and linearity.

Use of the Bosch smokemeter is largely confined to test bed operation under steady state engine operating conditions, and it is clearly outclassed by more versatile instruments. Nevertheless, the sampling unit possesses the practical merits of robustness and mechanical simplicity, which enable it to endure the rigours of test bed use and operation by relatively unskilled personnel.

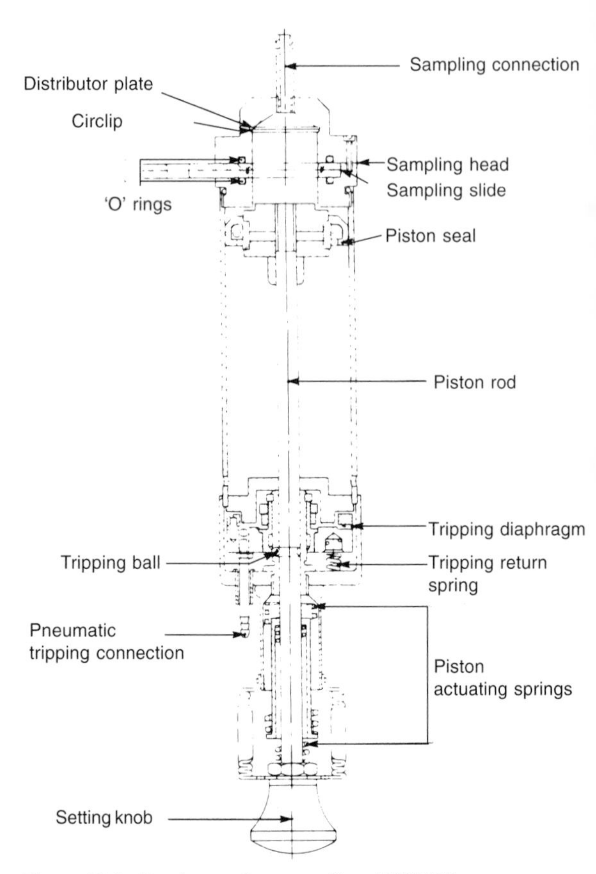

Figure 18.1 Bosch sampling pump-Type EFAW/65

The more delicate evaluating unit can be kept remote from the test area, can deal with the output from several sampling units, and provides a semi-permanent record. With increasing stringency of smoke legislation the Bosch is somewhat lacking in sensitivity to the lower levels of smoke density demanded. In a more elaborate and sophisticated form, with remote control, the various AVL meters—including fully automatic units—(designed and made by AVL Professor List, Graz. Austria) overcome many of the shortcomings of simpler units, giving high resolution and good repeatability. The AVL 415 smoke meter looks likely to become the new industry standard for engine development work. This unit is automated and is particularly sensitive at very low smoke numbers.

A variant of the sampling pump to meet an increasing requirement to assess smoke during a short period of full-load engine acceleration has also been devised. Named as an 'integrating' smokemeter the movement of the pump piston has been pneumatically damped so that it is extended to some 7 s, so that the carbon deposit on the filter is an average representation of the smoke emission over the acceleration period. The filter area is reduced to 1.1 sq cm and the smoke level is evaluated visually by use of the Bacharach Grey Scale referred to earlier. This form of sampling pump is designated Type EFAW/65B.

As mentioned earlier, and in common with all filter-soiling smokemeters, the Bosch meters cannot give useful or accurate results if there is appreciable blue/white smoke present in the exhaust.

18.2.3 Opacimeters

The visibility of smoke is by definition an optical phenomenon, and its density most easily measured in terms of light absorption

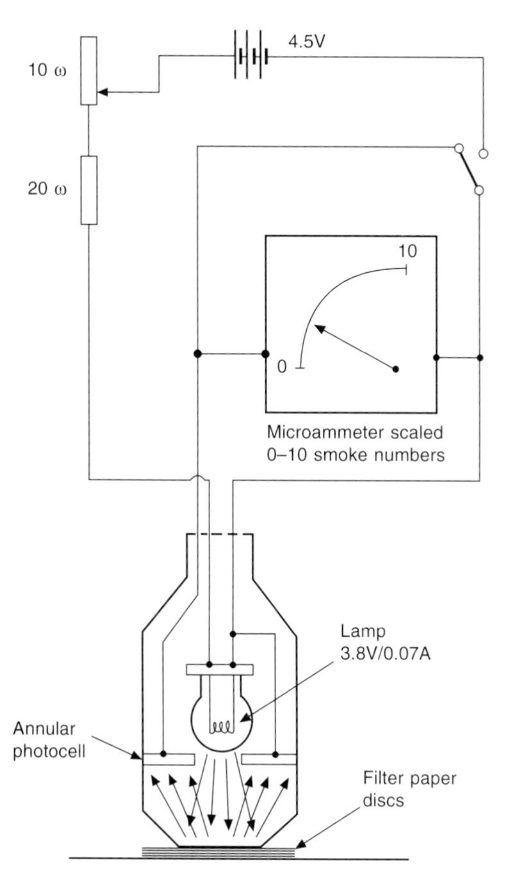

Figure 18.2 Bosch evaluating unit reflectance meter)—Type EFAW/68

either across the width of the flue through which the smoke is passing, or through a chamber into which a sample of the gas is diverted. The essential elements are therefore a light source, a defined length of light path filled with smoke, and a device (photocell) placed at the opposite extremity of that light path from the light source to convert the transmitted light into electrical current.

Photocell output is related linearly to the reduction in light intensity (opacity) resulting from the presence of smoke, and opacity is usually expressed as a percentage:

$$N = 100\left(1 - \frac{I}{I_0}\right) \text{per cent opacity} \qquad (18.1)$$

where

I is the light intensity at the photocell with smoke present in the light path;

I_0 is the light intensity at the photocell with only clean air present in the light path.

The reduction in light intensity can be expressed in accordance with the Beer-Lambert Law as:

$$I/I_0 = e^{-naQL} \qquad (18.2)$$

where

n is the concentration of smoke particles (for black smoke gm/cu m carbon);

a is the average particle projected area;

Q is the average particle extinction coefficient;

L is the effective light path length within the smoke (in meters).

Smoke density is defined by $naQ = k$, the parameter k being referred to as either the 'extinction coefficient', or the 'coefficient of light absorption'. This is related to the opacity and effective length of light path by the equation:

$$k = \frac{-1}{L} \log_e\left(1 - \frac{N}{100}\right) \qquad (18.3)$$

k is expressed in units of metres^{-1} and is thus dimensionally similar to the ratio of filter area to gas volume of spot meters referred to previously. Also if, as seems likely, a and Q are similar for the carbon particles produced under most engine operating conditions, k is linearly related to the gravimetric concentration of carbon in the exhaust.

k thus represents a smoke density parameter independent of the particular design configuration of the opacimeter. It should also be realized that the efffective length L is not necessarily identical to the geometric distance between the light source and the photocell.

Opacimeters may be classified as either sampling, or full-flow, the latter being further subdivided into in-line and end-of-line types. Sampling opacimeters differ from spot meters, of course, in that they can operate more or less continuously, and may thus be used to investigate varying operating conditions. Full-flow meters measure, by definition, the smoke density of the whole of the exhaust emission. In the case of in-line meters the instrument forms a permanent part of the exhaust system either of a test bed or an industrial installation, while end-of-line instruments are of course fitted to the outlet of the system, either permanently, or for an individual test.

The end-of-line meter may be arranged to pass the whole of the exhaust gas through a chamber whose dimensions define the light path, or the light beam may be arranged to pass through the freely emergent smoke plume to the photocell, when the exhaust pipe dimensions determine the value of L.

18.2.3.1 Sampling opacimeters

In its simplest classical form, the exhaust gas sample is extracted from the system by a probe, and passed through a tube having a photocell at one end and a filament bulb at the other. Zero is checked by passing scavenging air through the tube. Not only is this scavenging uncertain in its efficiency, but zero errors occur from soiling of the light source and the photocell. Diffusion of light from both smoke particles and condensation droplets also forms a source of error.

Innovations which provide an acceptable instrument of the sampling type were carried out in the design and development of the Hartridge smokemeter (*Figure 18.3*). Two identical measuring tubes 18 in (456 mm) long are provided, one carrying exhaust gas only, supplied from a sampling probe with pressure control by a relief valve, while the other is continuously scavenged with clean air from a motor-driven fan. The filament lamp and photocell are carried on pivoted arms and can be swung simultaneously from one tube to the other, so that they are only exposed to smoke while a measurement is being made. Even in the measuring position the ventilating air tends to deflect smoke from them. The black interior surface of the tubes, and circumferential fins, minimize the effects of diffusion and reflection.

The instrument has, however, been shown to be adversely affected by pressure pulsations in the sampling line, and if the sample is taken from a point in the exhaust system upstream of the silencer, some damping volume must be introduced into the line[6]. Care must also be exercised if continuous operation is required, to avoid high temperatures near the photocell. It may be necessary to use a cooler in the sample line[9].

Because of the introduction of a variety of volumes in the sampling system, and because the milliammeter has a relatively slow response, rapid changes in smoke level cannot be accurately

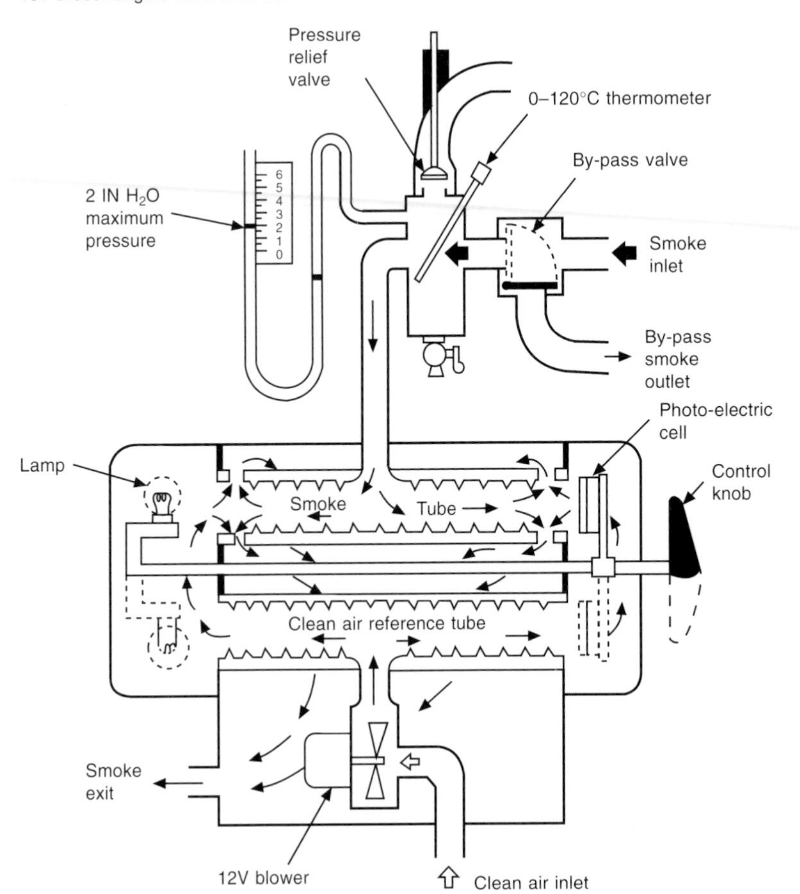

Figure 18.3 Hartridge Smokemeter (diagrammatic)

indicated. Nevertheless, the instrument is still used in Europe for certification testing.

18.2.3.2 Full-flow opacimeters

The full-flow end-of-line opacimeter designed by USPHS for the measurement of smoke emitted by heavy-duty vehicle engines is based logically on the premise that the appearance of the smoke plume discharged from the tail pipe is the essential quality to be assessed. The sensor, as shown in *Figure 18.4,* consisting of the light source and photocell, is carried on a rigid ring which is mounted close above the vertical exhaust pipe, so that the collimated light beam is transmitted diametrically through the plume. By connecting the photocell to suitable indicating or recording instruments rapid response to changing engine conditions can be achieved. A supply of clean air under pressure to the optical system is required both to keep the system cool and avoid soiling by smoke.

The instrument is thus suitable for investigating both steady-state and transient conditions, but suffers from lack of sensitivity on account of the small diameter of smoke plume in which light is absorbed. Also, with smoke other than completely black, changes in ambient light may influence readings. The opacimeter is essentially an instrument for use on engine test bed or chassis dynamometer and an exhaust extraction system is needed; this should not cause distortion of the plume. This distortion can occur with small rates of exhaust discharge, while in the case of large diameter tail pipes turbulence modifies the plume profile, so that there is a limit to the diameter of tail pipe which can be accommodated. Some interference with the plume occurs at 127 mm (5 in) dia., and the instrument is probably not usable with larger pipes[11]. If used in the open air it is clear that the plume must be shielded from wind.

The Celesco Model 107 Smokemeter (*Figure 18.5a*) has been developed as a result of experience with earlier models of generally similar design. Intended for installation in test bed exhaust systems, the detector unit is carried on a length of 150 mm dia. pipe inserted into a vertical part of the system. Protected by a stainless steel radiation shield concentric with the pipe, the light source, detector and collimating devices are carried on a rigid steel ring and are water cooled to prevent thermal drift. Air under pressure is used to ventilate the optical system and to prevent soot deposition. The light source is a light emitting diode (l.e.d.) giving green light peaking at 565 nm and the emission is pulsed at a frequency of 600 Hz.

The photo diode detector is incorporated in a circuit which is tuned and gated to the light source pulse frequency, and is thus rendered insensitive to changes in ambient lighting. The output amplifier (*Figure 18.5b*) also provides digital display of either percentage opacity, or coefficient of light absorption and can simultaneously operate a recorder. Linear correlation with both USPHS and Hartridge instruments has been demonstrated under specific operating conditions.

Such correlation only applies if the instruments are operating with exhaust gas at the same temperature. The effect is not entirely attributable to the functioning of the normal gas laws. At the time of writing, several new models had recently appeared. These are designed to meet the new SAE J1007 (and expected ISO 8178-9) standard, with modern data processing digital electronics. Details on these standards follow below.

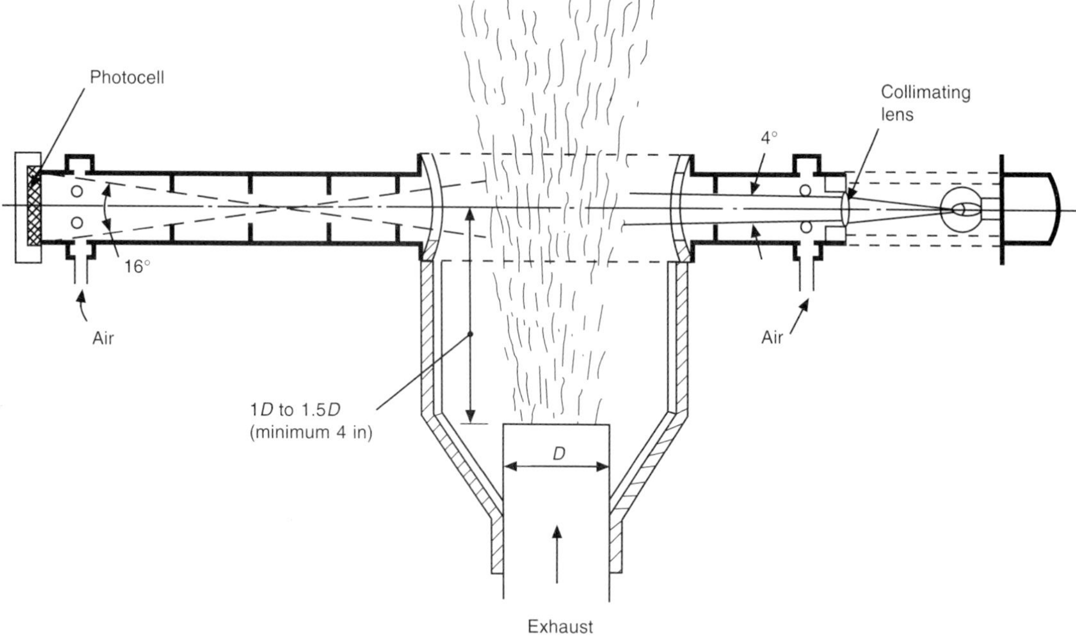

Figure 18.4 USPHS Smokemeter (end-of-line sensor–diagrammatic)

18.3 Calibration and correlation of smokemeters

The linear scale of photocell output of an opacimeter can be checked by inserting neutral density filters of known opacity in the light path. This is subject to a note of caution as regards the spectral distribution of the light source and the spectral response of the detector, as discussed later. Analogously the Bosch evaluating unit, Type EFAW/68, is checked at the 50% opacity point by placing on a stack of clean filter discs a matt black disc pierced with holes reducing its area by half.

The crucial importance of the effective length of the opacimeter light path can be inferred from the curves of *Figure 18.6*. This compares the opacity, calculated in accordance with the Beer-Lambert Law, for meters of different effective length with measurements of the same smoke made on the Hartridge meter (effective length 0.43 m). It will be seen that, at small values of L, a small change in effective length results in a large change in opacity reading, but the instrument is insensitive to smoke density variation. On the other hand, a large effective length gives a high degree of sensitivity to smoke density changes and is insensitive to changes in effective length. The most desirable compromise appears to lie with an opacimeter of about 0.5 m effective length.

Unfortunately, the effective length of the smoke-filled light path of an opacimeter is not subject to determination by absolute standards. Particularly where, as is usual, ventilating air is blown across the faces of the photocell and light source, there is no clearcut termination of the smoke column, some dilution of the smoke in these regions being inevitable. The determination of effective length can therefore be made by passing smoke from an engine simultaneously through the meter and through an opacimeter of known effective length. From the readings of opacity on both instruments the unknown effective length can be calculated using the equation:

$$L = L_o \times \frac{t + 273}{t_o + 273} \times \frac{\log\left(1 - \dfrac{N}{100}\right)}{\log\left(1 - \dfrac{N_o}{100}\right)} \tag{18.4}$$

where

t is the exhaust gas temperature in °C in the smoke measuring zone;

L, N and t refer to the opacimeter under test; L_o, N_o and t_o refer to the known opacimeter.

Alternatively, the value of L may be determined by passing smoke through the opacimeter normally, and then with the ventilating air temporarily cut off, the two values on N recorded again enabling L to be calculated again using eqn (18.4), but where L, N and t refer to the unmodified condition, and L_o, N_o and t_o refer to conditions with the ventilating air cut off.

In either case, the result is subject to the variability of smoke emission from the engine, this introducing an uncontrollable element into the procedure.

When L is known comparability of reading between different opacimeters is secured by scaling the indicating instrument in terms of k in addition to the linear opacity scale. Correlation is only secured, however, by simultaneous operation of the meters on the smoke emission from the same engine. This presupposes also that the gas temperature at the two opacimeters is the same, and this will not be so when an in-line instrument is being compared with an end-of-line one. In fact, the effective path lengths will be inversely proportional to the absolute temperatures.

Alternatively, on the assumption that the smoke is only due to the presence of carbon particles, a determination can be made, under constant engine conditions, of the carbon concentration in gm per cubic metre. A measured volume of gas is passed through an 'absolute' filter which will retain all the carbon suspended in the gas sample. The weight of carbon collected is determined by measuring the increase in weight of the filter. In the case of the Hartridge and Bosch smokemeters the correlation

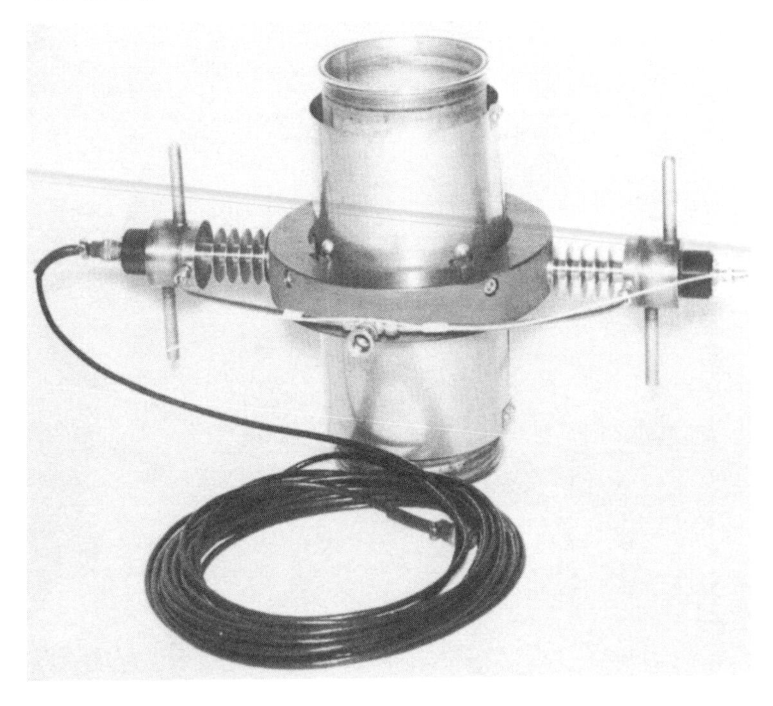

Figure 18.5a Celesco Model 107 Smokemeter (in-line sensor)

Figure 18.5b Celesco Model 107 Smokemeter output amplifier

has been made in both ways[6] and the results are given in *Table 18.1*. The errors incurred in the carbon concentration measurement are not inconsiderable, and amount to about ± 7 Hartridge Smoke Units (H.Sm.U.).

18.4 Optical system—spectral response

Since a smokemeter is concerned with visibility (or opacity) as appreciated by the human eye, it is clearly desirable that the optical system comprising photocell, detector, and the light source, should have a similar sensitivity to the spectral distribution. This implies maximum sensitivity between 550 and 570 nm, with much reduced response below 430 nm and above 680 nm.

When a tungsten filament light source is used with a wide band detector, the colour temperature must be controlled and a selective filter combined with the photocell. Alternatively, a similar result can be obtained by the use of a light source whose emission characteristics meet the required spectral distribution. There is some evidence that diesel smoke attenuates light of long wavelength (red light) less than light of short wavelength (blue light)[11]. It has also been inferred that as gas temperature falls below about 300°C and coagulation of carbon particles increases, the absorption characteristic of the smoke is shifted reducing the apparent absorption[13].

18.5 Opacimeter specifications

Opacimeter specifications can likewise be classified by their jurisdiction and intended use.

For in-service automotive applications the most important equipment specification is the SAE J1667, published by the US Society of Automotive Engineers. This recommended practice specifies the procedures for a smoke test, and the method of analysis of the results. As a snap acceleration test, J1667 is intended for in-service field use, and is designed for high smoke emitters, not marginal cases. The details of the smokemeter are not specified: it may be full-flow or sampling, with digital or analogue data processing. It is expected that the upcoming ISO standard 8178-9 for off-road vehicular smoke emissions will adopt many of the SAE J1667 provisions.

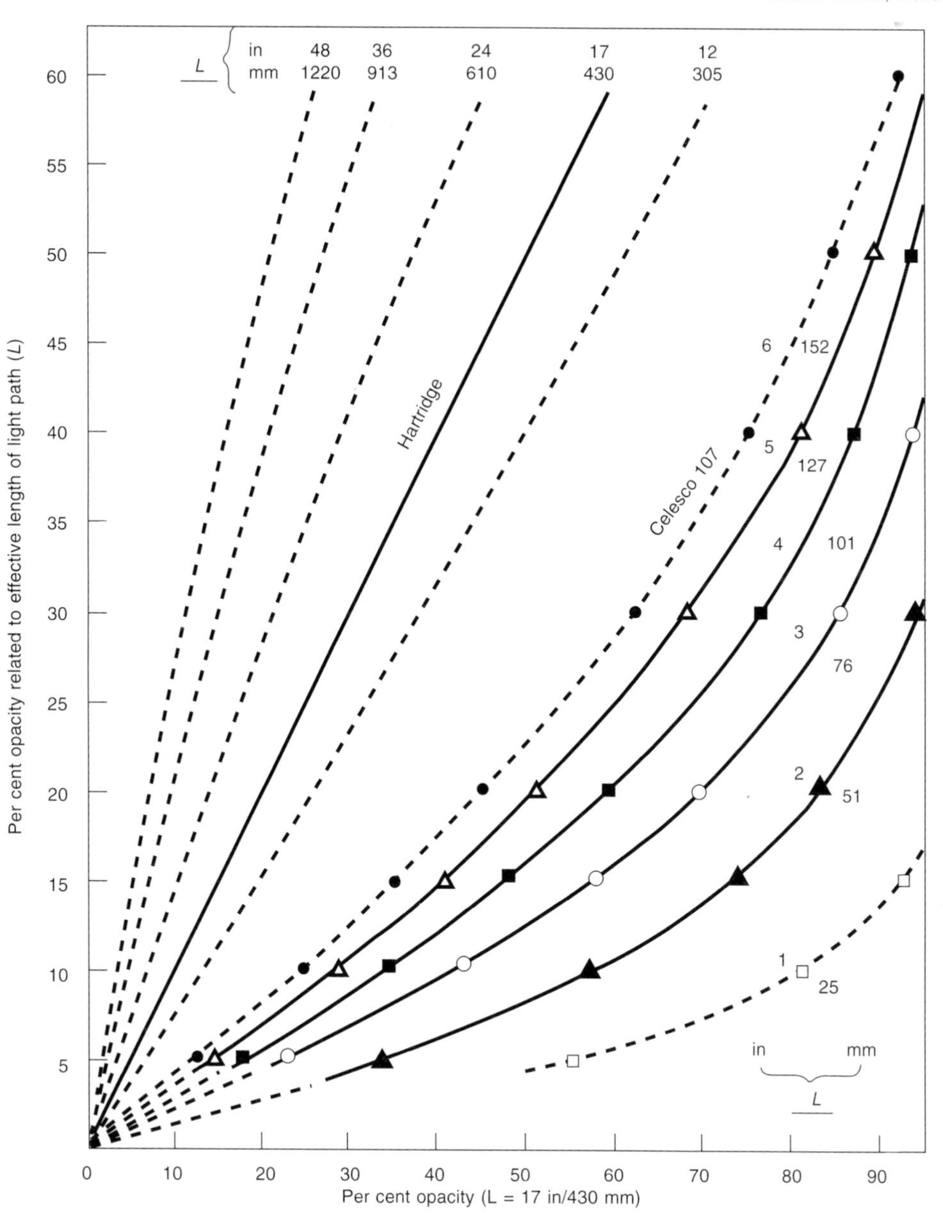

Figure 18.6 Influence of effective length of light path (*L*) on opacity reading

Table 18.1 Approximate correlation of Bosch and Hartridge smokemeters

Hartridge Number	*Carbon (g/m^3)*	*Bosch Number*	*Coeff. of light abs $-k$ (m^{-1})*
10	0.038	1.1	0.26
20	0.100	2.0	0.53
30	0.142	2.8	0.84
40	0.197	3.4	1.19
50	0.265	3.9	1.62
60	0.350	4.4	2.11
70	0.460	4.9	2.81
80	0.620	5.5	3.75
90	0.835	6.2	—
100	—	—	∞

Certification tests typically specify the type of smokemeter to be used. For certification testing in the US the USPHS smokemeter described above is required. In Europe measurements are by an opacimeter such as the Hartridge described above. Both opacimeters and filtering smokemeters are acceptable for certification testing in Japan.

Figure 18.7 illustrates the differences between the typical European sampling opacimeter (Hartridge) and the American full-flow, end-of-line instrument required by Federal Regulations (USPHS opacimeter). *Figure 18.8* shows the set up for a typical use of the Hartridge.

18.6 Visibility criterion—public objection

In the case of road vehicles, the basis of public objection to the diesel engine is the degree of visibility of the exhaust smoke.

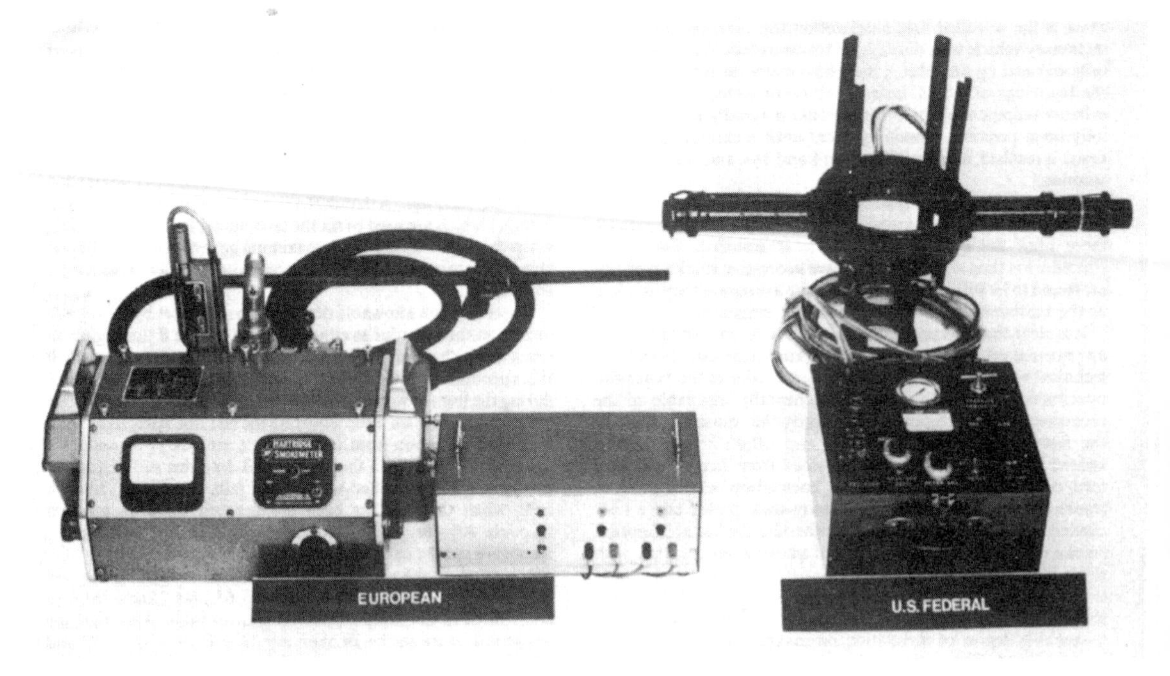

Figure 18.7 Typical European (Hartridge) and US Federal (USPHS) opacimeters

To meet the need for a quantitative (numerical) criterion of such objection, the Motor Industry Research Association (MIRA) and the Warren Spring Laboratory of the UK Department of Trade and Industry carried out tests in which a wide variety of vehicles were driven under full-load constant-speed conditions past neutral juries of people who were required to register whether they considered the smoke emitted as 'acceptable' or 'not acceptable'[18].

The smoke density (opacity) was measured at the same time with Hartridge smokemeters carried in the vehicles. A simple relationship between smoke density, size and speed of the engine, and acceptability, emerged. This is given by the formula:

$$C\sqrt{G} = K \quad (18.5)$$

where

C is the carbon concentration (gm/m^3);

G is the nominal rate of gas emission (engine displacement rate in litres/sec), and

K is a constant whose value depends on the degree of visual acceptance.

For $K = 3$, 75% of viewers found the smoke unacceptable, for $K = 2$, 50%, and for $K = 1.5$, 25% of viewers found the smoke unacceptable.

A similar series of tests was carried out by a committee of the British Standards Institution in drawing up an automobile smoke standard[19] with similar results. Furthermore, MIRA and the Warren Spring Laboratory repeated their tests more recently with more up-to-date vehicles, with the same type of relationship resulting, although the index for the 'G' term differed slightly[20].

18.7 Test methods and procedures

In the case of stationary plant it is sufficient to measure smoke emission at the rated speed at both full rated load and at any permitted overload. This could be done either on the manufacturer's test bed before delivery, or on site as finally installed, and, subject to atmospheric conditions, identical results should be obtained.

For vehicle engines the problem is more complex. Manufacturers need to establish a procedure ensuring repeatable and consistent results while covering the normal performance test. This is most easily accomplished on an engine dynamometer test by measuring steady-state full-throttle behaviour (torque, power, fuel consumption and exhaust smoke opacity) at a sufficient number of individual speeds covering the operational range, i.e. from maximum governed speed to a speed below that giving maximum torque. In order to confirm that vehicles in service conform to environmental requirements such a procedure is clearly impracticable, even if such expensive equipment as chassis dynamometers could be made available. Legislators require a much quicker and simpler surveillance test.

Such a simple test, originally devised by the Belgian authorities, is the so-called 'free acceleration test' 'snap acceleration test' or 'Snap idle test', carried out on the stationary vehicle with disengaged transmission. With the engine fully warmed up and idling, the smokemeter (in Europe usually the Hartridge or UTAC instrument) is connected to the vehicle exhaust tailpipe and the throttle pedal is rapidly moved to the fully open position, remaining there until maximum governed speed is reached, usually in between 1 and $1\frac{1}{2}$ s and held for a few seconds.

The maximum smoke reading reached during the operation is noted and the engine returned to the idling condition, remaining there until the original idling state is restored. The whole procedure is then repeated until three successive smoke readings are found to lie within ± 2% opacity. The average of these is taken as the representative value of the smoke emission.

It is clear that this procedure bears little or no resemblance to any normal vehicle engine operation, and many doubts as to its technical value have been expressed, especially as the European opacimeters used with it are fundamentally incapable of the necessary speed of response required by the transient nature of the test. Investigations by MIRA and others[10,21,22] demonstrated that the smoke values obtained from free acceleration tests on various engines gave no correlation with maximum smoke readings obtained from steady-state power curve tests carried out on the same engines. Neither did the free acceleration smoke

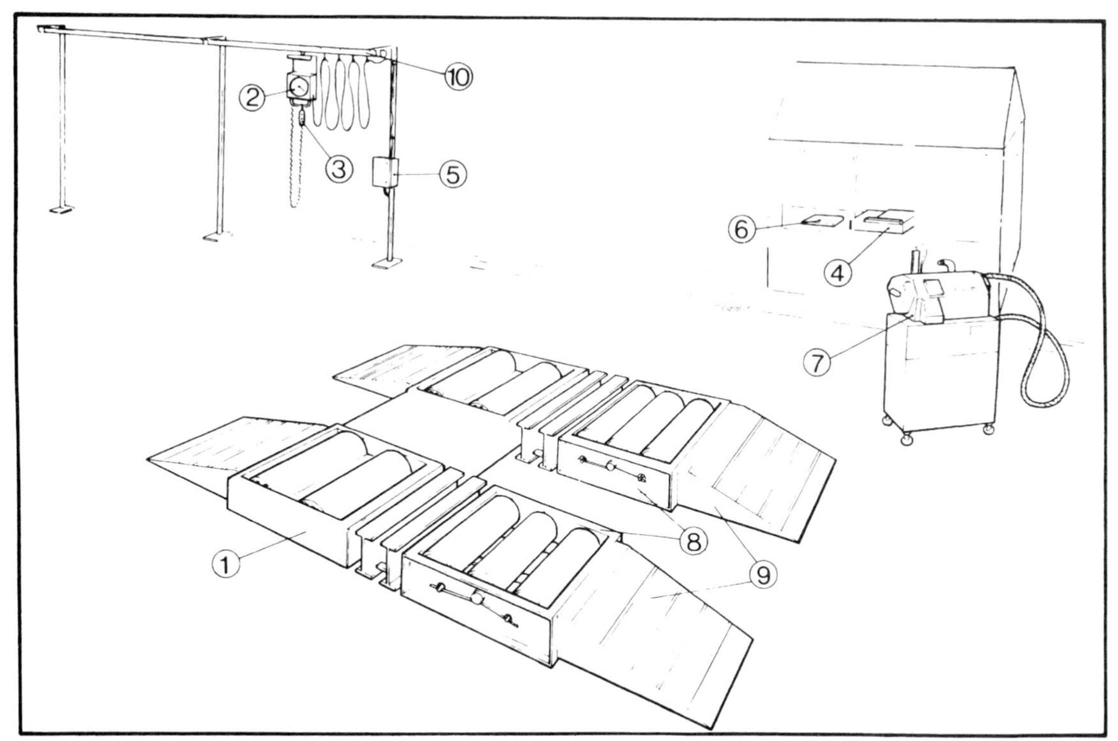

Figure 18.8 Equipment layout for British 'Lug-Down' test

1 Lug-down rolls
2 Driver's aid tachometer
3 Handset
4 Plotter
5 Switch
6 Test record pad
7 Hartridge Mk III smokemeter and power pack
8 Bogie rolls
9 Ramps
10 Stand

readings made on one type of smokemeter correlate with those obtained on another type. However, a snap acceleration test has been accepted by the SAE, and is expected to be adopted by ISO (details of these appear above). These tests are designed to catch gross emitters in need of maintenance, and are fast, cheap, and easy to perform.

For certification testing in the US, the USPHS smokemeter described above is required. The test procedure is the US Federal Smoke Test, which consists of a six step cycle that is designed to produce the most severe smoking conditions. The test is performed on an engine test stand, and is repeated three times.

The ECE R24 cycle is used for certification testing in Europe. The engine is run on a test stand at steady state, at full load at six defined speeds between 45% and 100% of its rated speed. Smoke measurements are by an opacimeter such as the Hartridge described above.

The smoke certification for Japan and Korea is similar to the European R24 test, but consists of only three speed steps (40%, 60% and 100% of rated speed). Opacimeters or filtering smokemeters are acceptable.
Reference: *Diesel Fuel Injection,* Robert Bosch GmbH, 1994.

18.8 Typical smoke regulations

18.8.1 Road vehicle applications

The regulated level of smoke limits has not moved lower in recent years with either the severity or frequency of the limits for gaseous and particulate emissions. This may reflect the fact that as particulate limits continue to be reduced, smoke emissions are simultaneously curbed too. A vehicle cannot hope to meet particulates limits if it suffers from high smoke emissions.

For on-highway vehicles in the US Federal Smoke Test, opacity limits in the acceleration and lugging phases are 20% and 15% respectively. The peak opacity may not exceed 50% (this limit will fall to 35% after 2001). In Europe the maximum permissible absorption coefficient is defined by a curve which is a function of the nominal exhaust gas flow rate; higher flow rates must achieve lower smoke levels. The maximum smoke level over the Japanese 3-step smoke test is 40% opacity, lowering to 25% opacity around 2005.
(Reference: *Diesel Fuel Injection,* Robert Bosch GmbH, 1994).

18.8.2 Regulations other than for road vehicles

In the USA, off-highway vehicles are subject to the same standards as the on-highway ones. Europe and Japan might follow ISO 8178-9 on smoke when that standard is finalized, as these countries have already adopted other ISO 8178 provisions for gaseous and particulate emissions.

18.9 Conclusions—future legislation

Smokemeter technology continues to advance, with consequent improvements in the resolution and repeatability of the measurement. Smoke regulations are among the most well-harmonized: between nations around the world, and between on-road and off-road applications. However, as particulate emission limits continue to be reduced, smoke limits are becoming somewhat—though not completely—academic. The USA and Japan are both proposing lower smoke limits in the future.

References

1 GOLOTHAN, D. W., 'Diesel Engine Exhaust Smoke: the Influence of Fuel Properties and the Effects of Using Barium-Containing Fuel Additives,' *SAE Paper No. 670092* (January 1967)
2 BURT, R. and TROTH, K. A., 'The Influence of Fuel Properties on Diesel Exhaust Emissions', *Paper No. 11, Inst. Mech. Eng. Symposium on Motor Vehicle Air Pollution Control* (November 1968)
3 CONNOR, W. D. and HODKINSON, J. R., 'Optical Properties and Visual Effects of Smoke Stack Plumes, *USPHS Publication No. 999-Ap-30,* Washington, DC, US Government Printing Office (1967)
4 RINGELMANN, M., 'Method of Estimating Smoke Produced by Industrial Installations', *Rev. Technique,* 268 (June 1968)
5 DODD, A. E. and HOLUBECKI, Z., 'The Measurement of Diesel Exhaust Smoke', *MIRA Report No. 1965/10* (April 1965)
6 STOLL, H. and BAUER, H., 'Rauchmessung bei Dieselmotoren', *MTZ, 18/5, pp.* 127–131 (May 1957)
7 VULLIAMY, M. and SPIERS, J., 'Diesel Engine Exhaust Smoke, its Measurement, Regulation and Control, *SAE Paper No. 670090* (January 1967)
8 SOCIETY OF AUTOMOTIVE ENGINEERS, 'Diesel Engine Smoke Measurement (Steady State)', *SAE information Report J. 255,* SAE Handbook (1972)
9 DODD, A. E. and GARRATT, G., 'The Measurement of Diesel Exhaust Smoke—The UTAC Smokemeter', *MIRA Report No. 1968/8* (May 1968)
10 BASCOM, R. C., CHIU, W. S. and PADD, P. J., 'Measurement and Evaluation of Diesel Smoke', *SAE Paper No. 730212* (January 1973)
11 BS 2811:1969, *Smoke Density Indicators and Recorders,* with Amendment No. 1, 1972, British Standards Institution
12 WALLACE, D. A., *Results of Measurements made with a Celesco/Berkeley Model 107 Smokemeter,* Telonic/Berkeley UK Publication (1977)
13 INTERNATIONAL STANDARDS ORGANISATION, 'Apparatus for Measurement of the Opacity of Exhaust Gas from Diesel Engines Operating under Steady-State Conditions', *ISO/DIS 3173 (Draft)* (1975)
14 SOCIETY OF AUTOMOTIVE ENGINEERS, 'Measurement Procedure for Evaluation of Full-Flow, Light-Extinction Smokemeter Performance', *SAE Recommended Practice J1157,* SAE Handbook (1978)
15 COORDINATING RESEARCH COUNCIL, INC., 'Evaluation of Research Techniques for Evaluating Full-Flow Light-Extinction Smokemeters', *CRC Report No. 453,* New York (January 1973)
16 JAPANESE STANDARDS ASSOCIATION, 'Reflection-Type Smokemeters for Measuring Carbon Concentration of Exhaust Smoke for Diesel Automobiles', *JIS D 8004* (1971)
17 DODD, A. E. and REED, L. E., 'The Relationship between Subjective Assessment and Objective Measurement of Exhaust Smoke from Diesel-Engined Road Vehicles', *MIRA Report No. 1964/12* (May 1964)
18 BS AU: 141a: 1971, *The Performance of Diesel Engines for Road Vehicles,* British Standards Institution
19 DODD, A. E. and WALLIN, S. C., 'The Subjective Assessment of Exhaust Smoke from Diesel-Engined Road Vehicles's, *MIRA Report No. 1971/10* (November 1971)
20 DODD, A. E. and GARRATT, G., 'A Comparison of Constant Speed and Acceleration Tests for the Measurement of Smoke from Diesel-Engined Vehicles', *MIRA Report No. 1968/6* (March 1968)
21 PINOLINI, F. and SPIERS, J., 'Diesel Smoke—a Comparison of Test Methods and Smokemeters on Engine Test Bed and Vehicle', *SAE Paper No. 690491* (May 1969)
22 SOCIETY OF AUTOMOTIVE ENGINEERS, 'Diesel Smoke Measurement Procedure', *SAE Recommended Pratice J 35,* SAE Handbook (1974)
23 UNITED NATIONS ECONOMIC COMMISSION FOR EUROPE, 'Uniform Provisions Concerning the Approval of Vehicles Equipped with Diesel Engines with regard to the Emission of Pollutants by the Engine', *ECE Regulation 24* (E/ECE/324), Geneva (March 1974)
24 EUROPEAN ECONOMIC COMMUNITY, 'On the Approximation of the Laws of the Member States Relating to the Measures to be taken Against the Emission of Pollutants from Diesel Engines for Use in Vehicles' Council Directive 73/306 (Brussels 1973)
25 EUROPEAN ECONOMIC COMMUNITY, 'On the Approximation of the Laws of the Member States relating to the Measures to be taken Against the Emission of Pollutants from Diesel Engines for Use in Wheeled Agricultural and Forestry Tractors', Council Directive 77/537/EEC (Brussels 1977)
26 US ENVIRONMENTAL PROTECTION AGENCY, 'New Motor Vehicles and New Motor Vehicle Engines—Control of Air Pollution', *US Federal Register,* **40,** No. 126 (June 30, 1975), Washington, DC, US Government Printing Office
27 JAPANESE STANDARDS ASSOCIATION, *The Measurement of Exhaust Smoke Concentration from Diesel Automobiles*, JIS D 1101 (1971)
28 UNION INTERNATIONALE DES CHEMINS DE FER (ORE), 'Limits for Pollutants in Diesel Engine Exhaust', *Report B13/RP 22* (Utrecht)

19 Exhaust emissions

Contents

19.1 Introduction

Concern over global warming has in recent years led to international government and industry efforts being focused on the reduction of emissions of greenhouse gases. In the area of automotive transport, the interest in more fuel-efficient technologies to achieve better fuel economy and reduce CO_2 emission has been fundamental to the increasing popularity of the diesel passenger car, particularly in Europe.

As emissions legislation worldwide becomes ever tighter, it has been recognized that the capabilities of the gasoline and diesel engine are different. Hence the light-duty diesel car has, in most countries, a separate set of emissions limits to the gasoline car, including a requirement for particulate emissions which is not currently legislated for gasoline. Future legislation continues to evolve as new knowledge becomes available. The so-called 'Auto/Oil' programmes which have been undertaken in the United States, and continue to run in Europe and Japan, represent an unprecedented cooperation between the automobile manufacturers and the oil industries, in an effort to identify which vehicle and fuel technologies will represent cost-effective solutions to achieve future air quality objectives. The aim is to establish emissions limits which are technically feasible and also achievable within a defined timescale.

In the diesel engine, a flammable fuel–air mixture is obtained by injecting the fuel at high pressure into high temperature compressed air in the combustion chamber. The fuel self-ignites and the piston is forced downwards producing the work which is taken from the crankshaft. The use of high compression ratios—typically in the range of 15:1 to 23:1 in automotive diesel engines—ensures that sufficiently high temperatures are reached for autoignition to occur.

If the diesel fuel, or indeed any hydrocarbon fuel, were to completely combust in the engine, then just water (H_2O) and carbon dioxide (CO_2) would be formed in the exhaust. However, in reality combustion is always incomplete, and a whole cocktail of species are present in the exhaust including unburned hydrocarbons from the fuel and partially oxidized products such as aldehydes and carbon monoxide. Oxides of nitrogen, sulphur compounds and particulates are other important components of diesel exhaust.

Presently, oxides of nitrogen (NOx), carbon monoxide (CO), total hydrocarbons (HC), particulates and visible smoke are the emissions which are legislated for diesel engines. Attention on the visible sooty exhaust emissions, and their associated health risks, typically associated with diesel engines, led to the introduction of controls on smoke and particulates.

Other emissions, such as formaldehyde for example, can be an important component of diesel exhaust but are not currently regulated in many countries. Such components remain of interest, however, due to environmental concerns such as greenhouse effects, acid rain, and low-level ozone formation, or potential health effects.

19.2 Legislation

In principle, carbon monoxide, unburned hydrocarbons and nitrogen oxides, together with the unregulated emissions can be measured in the raw undiluted exhaust. However, the regulations relating to specific classes of engine in certain countries demand that the exhaust components are not measured in the raw exhaust but that other methods are employed. These methods are not necessarily dictated by technical considerations but rather by the need to express the result in a specific manner. These procedures are current at the present time, but are likely to change in the future.

19.2.1 USA

The legislation of emissions has been led by the USA. This has largely been a result of the Clean Air Act which was formulated to produce and maintain an acceptable quality of ambient air. Clearly passenger cars, light and heavy trucks, railway locomotives, ships and stationary plants using diesel engines all combine to pollute the atmosphere.

However, the relative contributions vary and consequently the regulations are different for the different categories. Details of the regulations and measurements are published in the Federal Register.

19.2.1.1 Light duty vehicles

Diesel engines fitted to pasenger cars and delivery vans have to be tested installed in each vehicle to which they are fitted. The vehicle is driven on a chassis dynamometer according to a prescribed drive cycle which simulates urban driving conditions. In this case the gaseous emissions, carbon monoxide, carbon dioxide hydrocarbons and the nitrogen oxides are determined using a device called a constant volume sampler.

A schematic is shown in *Figure 19.1*. The main pump draws air through the dilution tunnel at a constant rate. Into the air stream is introduced the whole of the exhaust gas from the vehicle, the diluting air making up the difference between the exhaust flow and the pump flow. A sample of the diluted gas, typically of the order of 1% is removed from the gas stream and passed into sample bags. At the end of the prescribed driving distance the contents of the bag are analysed using the relevant analysers.

The hydrocarbons found in diesel exhaust have relatively high boiling points and would, if collected in bags, tend to condense out on the walls giving a spuriously low emission rate. To avoid this, a heated probe continuously removes a sample directly from the dilution tunnel. The sample lines to the analyser are maintained at about 191°C, the temperature which yields maximum analyser response. The analyser itself, a flame ionization detector is also maintained at 191°C. It is operated continuously and the readings integrated over the cycle.

Particulate emissions are measured at the same time using the same equipment. A second probe is positioned in the dilution tunnel. To this probe is attached a filter, which is of fluorocarbon coated glass fibre, through which a sample is drawn to collect the filterable particulate matter. This filter is subsequently weighed to determine the mass of particulate collected. Knowing the volume which has passed through the dilution tunnel and the concentration of the gaseous emissions the total vehicle emissions can be calculated. The results are expressed as mass per unit distance.

19.2.1.2 Heavy duty engines

Heavy duty engines tend not to be sold as a package in a vehicle as are light duty engines. To avoid the problem of having to test every possible engine, transmission, chassis, and body combination the engines themselves are tested on a test bench. Regulations at present only apply to on-highway automotive engines.

The engine is operated over a transient test cycle involving steady states, accelerations, decelerations and overrun conditions. The cycle lasts almost twenty minutes. This requires a computer-controlled test bed installation.

The analysis equipment consists of a constant volume sampling system very similar to that used for the testing of diesel engined passenger cars and light duty trucks. The gas flows involved are however much larger to maintain adequately low and stable temperatures. The temperature limits are defined in the Federal

We wish to acknowledge the help of Ricardo Consulting Engineers in the updating of this chapter.

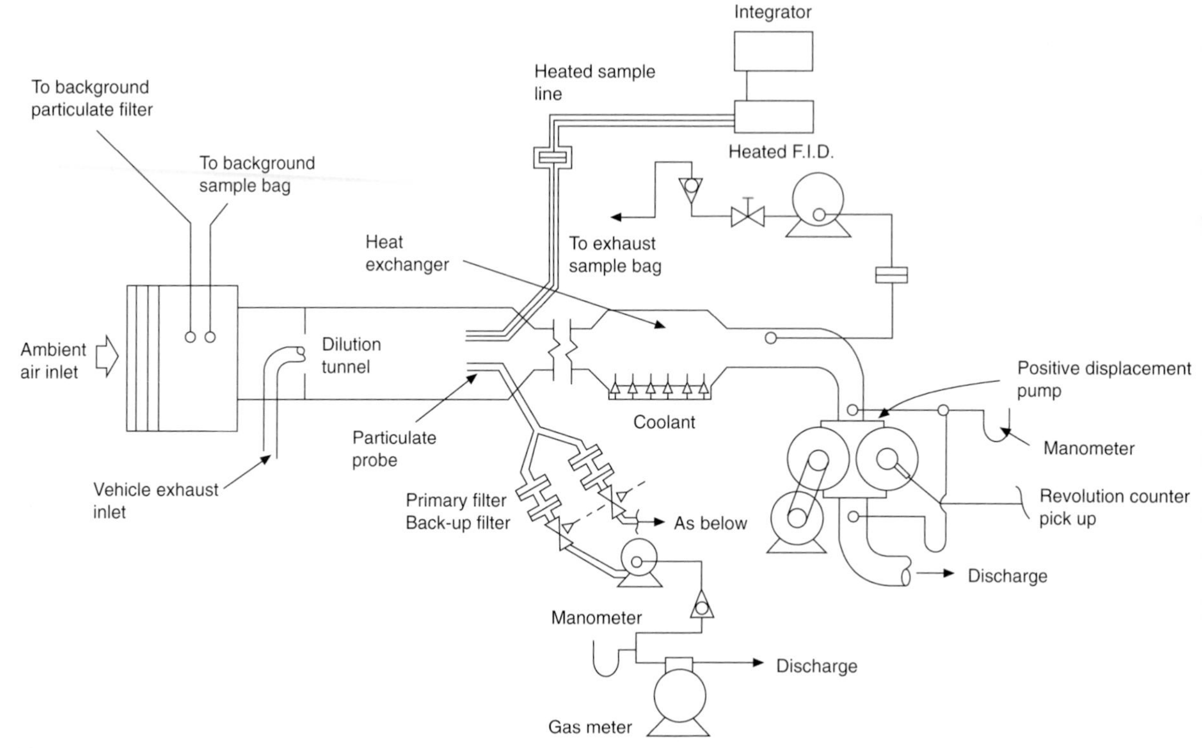

Figure 19.1 Schematic of constant volume sampling apparatus for measuring gaseous and particulate emissions from light duty vehicles

Register. The carbon monoxide, carbon dioxide and nitrogen oxide emissions can be measured either continuously or collected in a bag and analysed at the end of the test. Hydrocarbons are measured continuously from a sample point in the dilution tunnel.

Particulates are also legislated for and measured in a similar manner to those emitted by light duty vehicles. The maximum temperature allowed at the sample point in the dilute exhaust gas stream is 52°C. This value was chosen because at temperatures higher than 52°C the hydrocarbon element of the particulate material can be lost. If this occurs a low particulate mass emission is recorded.

19.2.1.3 Off-highway heavy duty engines

Heavy duty

Legislation covering off-highway engines is now in place and being phased in from 1996. It applies initially to engines rated between 175 hp and 750 hp but standards for all engines above 50 hp will be in force by 2000. The steady state test cycle employed is an 8-mode version of ISO 8178. California has similar legislation for the 175–750 hp range and has published a second level of standards for these for 2001. The first level will be applied to engines over 750 hp from 2000.

A proposed federal rule for locomotive emissions was published early in 1997 and, at time of going to press, a final ruling is expected very soon. Three different sets of standards have been proposed with the applicability of each being dependent upon the date of first manufacture of the locomotive (pre-1999, 2000–2004, post-2005). Three different types of locomotive operation have been identified; switch, passenger and line-haul, and there are some weighting differences in the proposed standards to reflect this. The EPA is proposing an alternative set of CO and PM standards intended primarily to apply to locomotives operating on alternative fuels such as natural gas. Such locomotives are expected to have higher, and more difficult to control, CO emissions but lower PM emissions.

Individual states may set standards for in-use locomotives but not for new ones.

Light duty

California has emissions regulations covering small utility engines as used in lawn, garden equipment and chain saws, etc. The first level of standards was implemented in 1995 and a second, more severe level will take effect from 1999. The test procedure is that of SAE J1088 in different versions according to the type of equipment being tested. There is now similar US federal legislation which came into effect 1 September 1996 but, thus far, applies to spark ignition engines only.

California has also introduced emissions legislation for light duty off-highway recreational equipment and vehicles, including go-karts and speciality vehicles. The regulations take effect progressively from 1997 and the test procedure is similar to that for small utility engines.

19.2.1.4 Stationary engines

The gaseous emissions from stationary engines are usually determined by sampling the raw exhaust. Legislation in the USA for diesel engines demands a maximum nitrogen oxides concentration of 600 ppm with the exhaust gas volume corrected to standard conditions and 15% oxygen content.

The rationale behind specifying nitrogen oxides is that economy of operation is of prime importance in operating stationary plants. Hence other emissions will, or can be, controlled without too much difficulty, but nitrogen oxides are more difficult to control and can contribute to smog formation.

19.2.1.5 Marine engines

In 1994, the EPA issued proposals for the emissions regulation of all marine engines of all types and power rating. The Final Rule was published in October 1996. A significant feature of this is that only Outboard and Personal Watercraft engines are now being regulated at this time.

Diesels would be required to meet the existing non-road (off highway) standards.

19.2.2 Europe

Originally, Europe emissions regulations were primarily formulated by the United Nations Economic Commission for Europe (UN-ECE) via its technical body GRPE. Most European countries support the ECE but the adoption, or not, of its standards by individual states is entirely voluntary. The European Community tended to issue regulations which were technically identical to those of the ECE. This situation has gradually changed in that the European Community (now the EU) has come to play the leading role in the development of new regulations, which are published as EU Directives. The application of these is mandatory for the member nations of the EU. Similar ECE regulations generally follow.

9.2.2.1 Light duty vehicles

Light duty vehicles are generally taken as those not exceeding 3500 kg gross vehicle weight. The test procedure is now that specified in the 04 amendment of ECE Regulation 15, with the addition of the new Extra Urban Driving Cycle (EUDC). In the basic cycle the vehicle is driven four times without interruption over the prescribed 15 modes to give a total distance of 4 km. A constant volume sampling system is used, similar to that defined in the Federal Register for US Regulations. Carbon Monoxide, cabon dioxide, hydrocarbons and nitrogen oxides are measured and also emissions of particulate material (PM).

The EUDC was added because the ECE 15 cycle with its maximum speed of 50 km/h was considered to be insufficiently representative of many driving modes and to give unrealistically low levels of NOx emissions. This additional cycle has a maximum speed of 120 km/h and is carried out after the standard R 15 test.

As from January 1997, type approval of new models of Class I have been subjected to the limits of Directive 96/69/EC, issued in 1996. All production Class I vehicles must comply from 1 October 1997. For Class I and II, new models must comply with this Directive from 1 January 1998 and all production by 1 October 1998. Prior to these dates, Directive 93/59/EC is relevant.

Proposal for Directive COM(97) 61 Final includes requirements for the mandatory introduction of on board diagnostic (OBD) systems on light duty trucks and passenger cars. The OBD system will be required to indicate any failure that may lead to an increase in emissions levels above regulated limits. It is likely that compliance by diesels will remain optional until 2005.

19.2.2.2 Heavy duty engines

Emissions legislation for heavy duty engines is mandated in the EU by Directive 91/542/EEC. The test procedure is known as ECE 49 and the engines are run on a test bench with emissions measured at a series of 13 steady state conditions. The test procedure is similar to that of the earlier, now obsolete, US 13-mode procedure although weighting factors have been changed to represent European driving conditions. The weighted emissions of CO, HC, NOx and PM are expressed as g/kWh. Modifications to the test procedure are anticipated for the next stage of legislation, expected around 2000–2001.

19.2.2.3 Medium speed engines

Legislation exists for engines over 200 hp for railway use. The testing is carried out on a test bench and the engine operated at four defined points. Otherwise there are no specific regulations relating to diesel engines. There are however environmental standards defining maximum permissible pollutant concentrations in terms of mass per unit volume of air. Engines have to be operated in a manner such that these regulations are not infringed.

19.2.3 Japan

19.2.3.1 Light duty vehicles

As compared with Europe, there are still relatively few diesel-engined passenger cars in Japan. The emissions test procedure for these is driven on a dynamometer and is known as the 10.15. This proedure superseded the earlier 10-mode test in 1991. The test distance covered is 4.16 km and the maximum speed is 22.7 km/h. CO, HC, NOx and PM are measured and the test is also used to determine fuel economy. Direct comparison with European and US standards cannot be made due to the differences in test procedures and weightings.

The 10.15 procedure is also used for diesel powered light trucks and buses of up to 2.5 tonnes GVW.

19.2.3.2 Heavy duty engines

All diesel engines powering heavy duty (above 2.5 t GVW) vehicles are now emissions tested over the Japanese 13-mode diesel cycle, which replaced the earlier 6-mode test in 1994.

Special NOx emissions standards for six specified areas have been implemented since 1 December 1993. These include the 23 wards which make up Tokyo, Osaka and 173 towns.

19.2.3.3 Stationary engines

Legislation falls into two categories, national general limits and various local limits which govern specific cities and their surrounding countryside. There is a general environmental standard covering the concentrations of sulphur dioxide, carbon monoxide, nitrogen oxides and particulate material. This inherently affects all on land stationary sources, automobiles and marine sources.

19.2.4 Concluding remarks

The need to measure exhaust gas components has been driven by legislation. The bulk of the legislation is derived from the need to meet overall ambient air quality standards.

Since the legislative scene relating to emissions is constantly being reviewed and legislative limits changing, readers are recommended to consult the relevant legislative documentation for details of definitions of categories, latest test procedures and permitted emissions levels.

For the USA the Federal Register and the California code of Regulations should be examined, while for Europe the appropriate EU directives and ECE Regulations should be studied.

19.3 Analysers

Analysers used for measuring diesel exhaust gases must be sensitive enough to detect the sometimes low levels of gases in the exhaust, especially in diluted exhaust streams, and be devoid of any significant interference from other gases which might be present.

For the emissions which are covered by legislation, carbon monoxide, hydrocarbons and nitrogen oxides, together with carbon dioxide there are specific analysers whose types have been approved by the relevant regulatory bodies. However for the non-regulated emissions there are often a number of procedures which can be employed. The choice of method is often controlled by the equipment available in a specific laboratory.

In the following sections, the principles of operation of a number of analysers or analysis systems currently in common use are described. Clearly individual features differ from manufacturer to manufacturer and from model to model. The information presented here should therefore be taken as an indication of the principles employed. The analysers are listed by the gas to be analysed.

19.3.1 Carbon dioxide

Although not legislated for, the analysis system for carbon dioxide is defined by a number of regulatory bodies. Carbon dioxide is almost invariably measured using a non-dispersive infra-red (NDIR) analyser. This is possible because carbon dioxide absorbs radiation in the infra-red region. Thus if a beam of infra-red radiation is passed through carbon dioxide the intensity of the beam is attenuated. The degree of attenuation depends on the amount of carbon dioxide present in the path of the beam; the more carbon dioxide the greater the attenuation.

A schematic diagram is shown in *Figure 19.2*. As can be seen, the analyser consists of four elements, a source of infra-red radiation, a reference cell, a sample cell and a detector with its associated electronics.

The infra-red source is often a heated wire. The radiation is divided, part passing through the reference cell and part through the sample cell. Before the radiation reaches the cells proper it passes through a filter. This can be either a thin film interference filter or a cell filled with an appropriate non-absorbing gas, typically carbon dioxide free air or nitrogen. The reference cell is filled with carbon dioxide free dry air. The sample cell contains the sample gas. The gas can either be sampled discretely or, as is more common for diesel engine analysis work, continuously. Some analysers have several output ranges available. This is achieved partly by electronics and partly by using a sample cell consisting of two elements of different length, the shorter being used for the more concentrated ranges and the longer for more dilute gas mixtures.

There are a number of different detectors available. Probably the most common is the Luft type detector. This consists of two sealed chambers separated by a thin membrane which can move as the pressure changes in either of the chambers. The movement of the membrane is detected capacitatively.

The pressure changes occur as follows. Both chambers of the detector are filled with the gas to be analysed, in this case carbon dioxide. The radiation arriving at each side of the detector is absorbed by the carbon dioxide and increases its temperature. Since it is in a fixed volume the pressure increases. The reference side receives the full amount of the radiation from the source whereas the sample side receives less, the difference being due to that absorbed by the carbon dioxide present in the sample gas. The resultant pressure difference causes the membrane to move towards the sample side. This is the movement which is detected.

An alternative type of detector consists of an absorption chamber and a compensation chamber which are linked to each other by a gas path. Both chambers are filled with the gas to be analysed, namely carbon dioxide. The received radiation causes the gas in the absorption cell to heat up, expand and hence flow into the compensation chamber. The flow between the chambers is determined using an appropriate detector. The use of a chopper

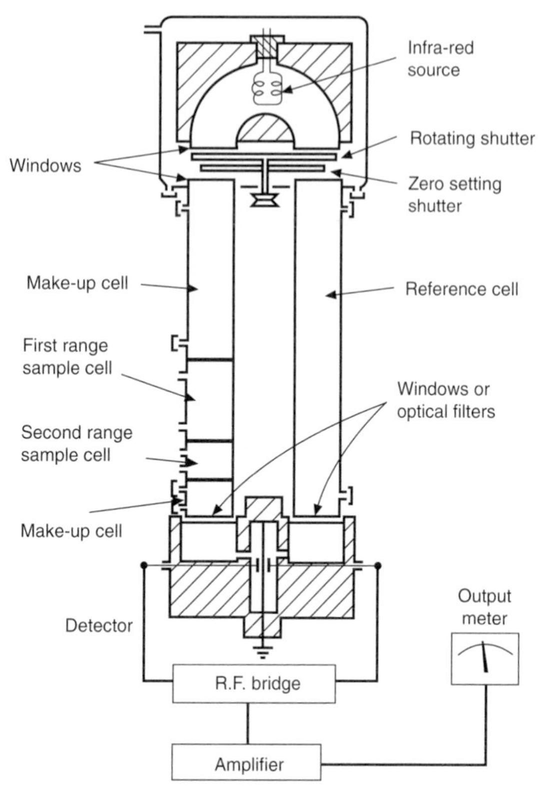

Figure 19.2 Schematic of non-dispersive infra-red (NDIR) analyser for measuring carbon dioxide, carbon monoxide and nitric oxide

produces an alternating signal as the gas oscillates between the two chambers, which is easier to handle electronically and which also reduces interferences from, for example, thermal effects.

Clearly there is some art in the design of analysers, the specification of the filling gases and the construction of the detector.

19.3.2 Carbon monoxide

Carbon monoxide is also measured using a non-dispersive infra-red detector. This would be identical in principle to that described in the previous section for carbon dioxide except that, if a thin film interference filter were not used, the filter cells would be filled with pure carbon dioxide to avoid carbon dioxide interference, and the detector would contain carbon monoxide.

19.3.3 Nitric oxide

Nitric oxide can be measured using the non-dispersive infra-red principle. In this instance, if thin film interference filters were not used the filter cells would be filled with a mixture of carbon dioxide and carbon monoxide to avoid their interfering with the nitric oxide measurement. The detector would of course contain nitric oxide.

Water vapour absorbs infra-red radiation and since diesel exhaust, whether it be raw or diluted with air, contains water vapour the sample has to be dried before it passes through the sample cell. This can be achieved using any one of a number of drying agents, such as self-indicating silica gel or magnesium perchlorate. Such a drying column is incorporated in the analyser.

Rather displacing the non-dispersive infra-red detector is the chemiluminescence analyser. This has the advantage that it can be used to detect not just nitric oxide but also nitrogen dioxide (or dinitrogen tetroxide). This is particularly important for the measurement of the exhaust from the larger medium speed engines. Nitrogen oxide emissions from high speed diesel engines tend to be mostly as nitric oxide, although up to 30% dioxide can be detected under certain operating conditions such as at low speed and high air–fuel ratios. However the larger engines tend to emit proportionally more NO_2 and hence the measurement of nitrogen dioxide becomes increasingly important.

A schematic arrangement of a chemiluminescence analyser is shown in *Figure 19.3*. To measure nitric oxide the exhaust gas is passed into a reaction chamber where it reacts with ozone generated within the instrument. This produces nitrogen dioxide in an electronically excited state. The molecules 'relax' back to their normal ground state by emitting photons of radiation. These are detected and amplified by a filter and photomultiplier tube housed within the reaction chamber. The output of the photomultiplier is proportional to the nitric oxide concentration.

If the instrument is to be used for the determination of nitrogen dioxide, instead of being passed directly into the chamber where it reacts with ozone, the exhaust gas sample is diverted through a converter. This thermally or catalytically reduces nitrogen dioxide to nitric oxide. The reduced gas is then passed into the ozone reaction chamber where the total intric oxide content is measured. The total nitric oxide content is the sum of the nitric oxide produced from the introgen dioxide and the nitric oxide originally in the exhaust sample. This sum is usually referred to as nitrogen oxides (NOx). The nitrogen dioxide level is thus calculated by subtracting the nitric oxide reading from the NOx reading.

Since most regulations require that the nitrogen oxides be expressed as NOx the chemiluminescence analyser has the advantage of giving a direct reading.

19.3.4 Hydrocarbons

Hydrocarbons in diesel exhaust are universally measured using a heated flame ionization detector (HFID). This comprises, as is shown in *Figure 19.4*, a flame ionization detector cell, such as that used on gas chromatographs, together with the necessary electronic signal processing and readout equipment. For diesel exhaust measurement where the hydrocarbons are of fairly high molecular weight and consequently of higher boiling points, it is essential to avoid losses due to condensation on any surfaces in contact with the gas sample. This is achieved by maintaining all sample lines and all parts within the analyser which come into contact with the sample at about 191°C. This temperature has been selected as that which gives the maximum detector response for diesel hydrocarbons.

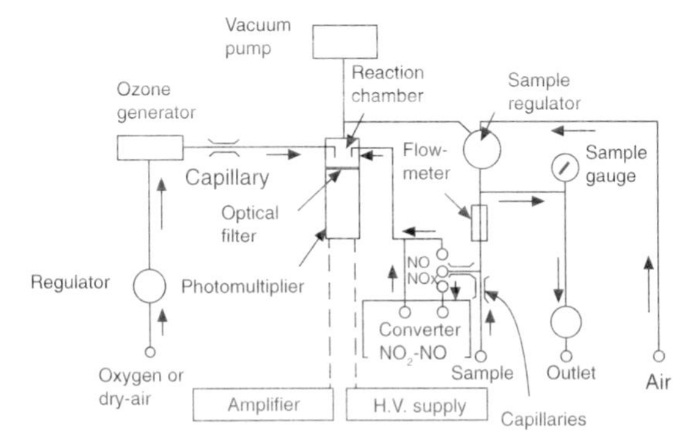

Figure 19.3 Schematic of chemiluminescence nitrogen oxides analyser

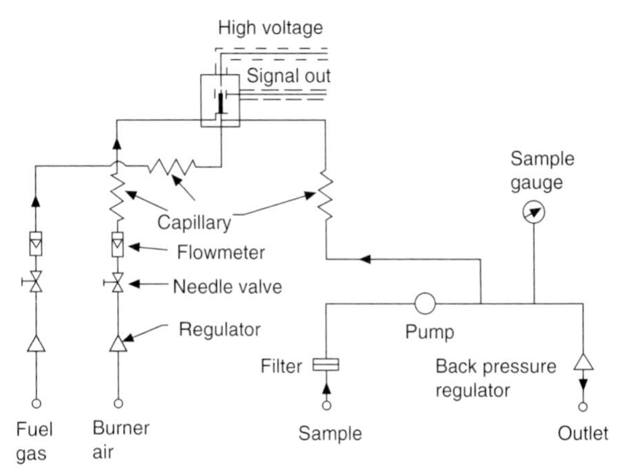

Figure 19.4 Schematic of flame ionization detector (FID) for measuring unburnt hydrocarbons

When a hydrocarbon is burnt in a flame, ions are produced. In the flame ionization detector the exhaust gas sample is passed through a capillary, to control the flow rate, and into a flame obtained by burning either hydrogen or a mixture of hydrogen and helium or hydrogen and nitrogen in air. The latter mixtures suffer less from interference from the different levels of oxygen found in the exhaust sample and are therefore to be preferred.

A potential difference is applied across the flame and as the hydrocarbons burn the ions produced move towards one of the electrodes applying the field. This produces an electric current which is amplified and measured, the magnitude of the current being proportional to the number of carbon atoms passing into the flame. The analyser has to be calibrated with a known hydrocarbon gas, usually propane. The results are often specified as parts per million carbon which gives a true measure of the amount of hydrocarbon present without the need to specify the individual hydrocarbons.

19.3.5 Oxygen

Not surprisingly, oxygen is not legislated against. However it is very useful to be able to measure its concentration in the raw exhaust from a diesel engine since this provides a check on other engine parameters. There are two types of oxygen analyser. One uses an electrochemical cell, and the other is based on the fact that oxygen is paramagnetic.

Electrochemical cell analysers typically consist of two electrodes separated by an electrically conducting liquid or gel. The cell is mounted behind a polytetrafluorethene membrane through which oxygen can diffuse. The device therefore measures oxygen partial pressure.

If the device were to have a gold cathode and silver anode, not an uncommon combination, with potassium chloride separating them, as oxygen diffused through the membrane the following reaction would occur at the cathode:

$$O_2 + 2H_2O + 4e^- \rightarrow 4OH^-$$

The reaction at the anode would be:

$$4Ag + 4Cl^- \rightarrow 4AgCl + 4e^-$$

If a polarizing voltage is applied between the electrodes the resultant current is proportional to the oxygen partial pressure.

Clearly electrochemical cells are consumed. This is not

necessarily a disadvantage since the cells themselves can be made as readily replaceable cartridges.

A paramagnetic substance is one which aligns itself with the lines of force in a magnetic field. This was demonstrated by Faraday in 1851. He demonstrated that a glass sphere filled with oxygen was attracted by a magnet. A modern oxygen analyser consists of two spheres, one at each end of a bar, forming a dumbell with oxygen sealed in each sphere. The whole is suspended in the test gas in a symmetrical non-uniform magnetic field. The dumbell aligns itself away from the most intense part of the magnetic field.

When oxygen is present in the test gas it affects the magnetic field and causes the spheres to be pushed further out of the field. The torque acting on the dumbell is thus proportional to the paramagnetism of the sample gas. This torque is measured, either directly, or via a feedback system often including a coil around the dumbell, the current required to keep the dumbell stationary being measured.

Clearly any other gases which are paramagnetic will interfere. Nitric oxide and nitrogen dioxide both interfere. An atmosphere of nitric oxide gives about half of the oxygen reading, whilst nitrogen dioxide gives about one-third the response. Fortunately these gases, although present in diesel exhaust, only occur at low levels, therefore for practical purposes their effect can normally be ignored.

The paramagnetic and electrochemical cell oxygen analysers find widespread use in the engine industry for measuring oxygen at any concentration in engine exhaust gases. An electrochemical device based on zirconium oxide is used extensively for the measurement of oxygen or more strictly carbon monoxide. In the exhaust of gasoline vehicles fitted with three way catalysts which require the air–fuel ratio to be oscillated at about stoichiometric.

Zirconia sensors work because a potential difference is generated across a ziroconium oxide wall when the concentration of oxygen on each side differs. The potential difference is detected by the paltinum electrodes deposited on each side of the wall. The potential difference is often amplified by an FET which takes a very small current and hence does not significantly affect the zirconia/oxygen cell.

A problem with this type of analyser is that the oxygen concentration in the reference gas always has to be considerably higher than that in the sample. This means that for diesel exhaust studies, several reference gases are required to cover the normal range of operation. Similarly at low oxygen concentrations carbon monoxide in the exhaust reacts with the oxygen on the probe and hence reduces the concentration of oxygen on the surface; hence the analyser gives a spuriously low reading. It is this characteristic which is employed in the exhaust sensors used in conjunction with three-way catalysts since near the stoichiometric air–fuel ratio this reaction dominates and a rapid rate of change of the signal from the zirconia cell occurs.

The zirconia cell oxygen analyser is not usually used in exhaust gas analysis sets for research purposes. Neither are the sampling methods which rely on removing oxygen from a known volume of gas and measuring the volume decrease. These devices which include Orsat, Haldane, Lloyd Haldane. Bone and Wheeler and Scholander, depend on the skill of the operator and do not give any output signal which could be used for recording or control.

19.3.6 Particulates

Particulates are defined as any material other than water, in the exhaust of an internal combustion engine which can be filtered after dilution with ambient air. The measurement apparatus therefore consists of a tunnel, in which the engine's exhaust is mixed with filtered ambient air, a sample probe, to remove a portion of the diluted exhaust and a filter through which the sample is passed. This is shown schematically in *Figure 19.1* where it comprises part of the CVS system.

The proportions of the tunnel are specified in the relevant regulations. Basically the exhaust gas and diluting air should be well mixed before a sample is taken.

The sample probe does not have to sample the diluted exhaust gas stream isokinetically since the individual masses of the particles are so small that they effectiely behave as a gas. However, according to the US regulation, the temperature at the probe should not exceed 52°C. There are good technical reasons for defining a temperature at least for the filter. If the temperature of the filter is maintained at, say, 50°C then the materials collected would be those which are solids or liquids below that temperature. Any other material would be gaseous and hence would pass through the filter. However if the temperature were to be increased to, say, 150°C then material boiling between 50 and 150°C would be driven off. A lower particulate mass would then be recorded. It is clear therefore that for comparative purposes a filter temperature should be specified. The amount of material collected on the filter is determined gravimetrically.

The particulate material has been shown to consist of substances which might be mutagenic or carcinogenic and thus there is some interest in the composition of the particulate matter.

It should be pointed out that it is not possible to take a smoke reading and relate it to a particulate emission rate. This is because smoke meters, whether they be of the opacimeter type, such as the Hartridge meter, or of the filter soiling type such as the Bosch Smokemeter, only measure optional properties of the emission.

The composition of particulate can vary considerably from 5% hydrocarbons and 95% carbon to 20% carbon and 80% hydrocarbons depending on the engine operating conditions. Carbon will give a high smoke reading since it is opaque and essentially non-reflective, whereas a stream of hydrocarbon droplets would show a low smoke reading because they would have a relatively high optical transmission and reflectivity. The issue is further complicated by the fact that different hydrocarbons have very different optical properties; in fact refractive index is used in the characterisation of liquid hydrocarbons.

There are, however, optical, photoacoustic and oscillatory methods being developed to give a continuous output of particulate emission. At the time of writing none of these methods are included in any regulatory procedures or are in widespread use.

19.4 Formation and control

19.4.1 Carbon dioxide

Carbon dioxide (CO_2) occurs naturally in the atmosphere and is a normal product of combustion. Ideally, combustion of a hydrocarbon fuel should produce only carbon dioxide and water (H_2O). The relative proportion of these two depends on the carbon-to-hydrogen ratio in the fuel, about 1 : 1.75 for ordinary diesel fuel. Thus, an engine's CO_2 emissions can be reduced by reducing the fuel's carbon content per unit energy, or by improving the fuel efficiency of the engine. The high fuel efficiency of diesel engines gives them an environmental advantage over some fossil fuels, though the processing of crude oil into diesel fuel has fairly high CO_2 emissions.

Long regarded as benign, the emission of CO_2 from the combustion of fossil fuels has recently attracted considerable attention. Water vapour and CO_2 (along with other gases) allow solar energy to reach the earth, but trap some of the thermal radiation then given off by the earth. This 'greenhouse effect' makes the earth much warmer than it would otherwise be. It is vital to the our climate; we need the greenhouse effect to survive. However, atmospheric levels of CO_2 have been rising since the beginning of the Industrial Revolution, apparently because of the widespread and still growing combustion of fossil fuels for power. There is evidence that the global climate is being affected by this change—some measures of world average temperature have been rising steadily, the phenomenon known as 'global warming'. But the data is contentious: other measures (such as satellite data) show no change in world temperatures. Moreover, identifying CO_2 emissions as the cause of the possible global warming is difficult. Many other world climatic factors, such as oceans, cloud cover, polar ice, volcanic eruptions, other atmospheric constituents, and even sunspot cycles, are thought to be important. Still, in 1996 the Intergovernmental Panel on Climate Change (IPCC), which advises the United Nations on climatic matters, opined that 'the balance of evidence suggests a discernible human influence on climate'. However, legislation restricting specifically CO_2 emissions is virtually non-existent. Owing to their close connection with CO_2 emissions, fuel consumption standards (such as the CAFE standards in the USA) may be viewed as an indirect CO_2 regulation.

As an inevitable end-point of combustion, CO_2 cannot practically be decreased by after-treatment (such as by catalytic converters). Indeed, the catalytic oxidation of CO and HCs will increase CO_2 emissions very slightly.

19.4.2 Carbon monoxide

Carbon monoxide (CO) is toxic. It is an intermediate product in the combustion of a hydrocarbon fuel, so its emission results from incomplete combustion. Emissions of CO are therefore greatly dependent on the air–fuel ratio relative to the stoichiometric proportions. Fuel-rich combustion invariably produces CO, and emissions increase nearly linearly with the deviation from stoichiometric. As diesel engines operate with an overall lean mixture, their CO emissions are normally well below legislated limits and not of much concern. Any CO from a diesel engine is due to incomplete mixing: combustion taking place in locally rich conditions. An oxidation catalyst in the exhaust can further decrease CO and unburnt hydrocarbon emissions. This process is aided by the excess air in the exhaust gas.

19.4.3 Unburnt hydrocarbons

Unburnt hydrocarbon (HC) emissions consist of fuel that is completely unburned or only partially burned. The term HCs means organic compounds in the gaseous state; solid hydrocarbons are part of the particulate matter. The mechanisms leading to HC emissions from diesel engines are completely different from those leading to HC emissions from spark ignition (SI) engines. In the latter, a nearly homogeneous fuel–air mixture is compressed, and then a flame passes across the combustion chamber. Hydrocarbon emissions arise in SI engines when fuel–air mixture hides out in places inaccessible to this flame. The most significant of these places is the volume between the piston and the cylinder wall, above the top piston ring. In a diesel engine only air is compressed into this volume. In effect, this reduces the air available for combustion, but it does not allow a significant amount of fuel to escape combustion. In a diesel engine, with its non-homogeneous combustion, HC emissions result from problems of fuel and air mixing, and are largely unaffected by the overall air–fuel ratio. There are two primary mechanisms by which fuel escapes the main combustion in a diesel; over-mixed, over-lean regions formed before ignition, and under-mixed fuel injected at low velocity near the end of combustion.

The first of these mechanisms is over-mixed over-lean regions formed before ignition. Diesel fuel is injected into the hot compressed air as the piston nears topdead-centre. After a short delay period–during wihich fuel and air mixing occurs, the fuel is heated by the air, and chemical reactions begin–spontaneous ignition of the fuel takes place. Only mixtures within combustible limits will burn; freshly-injected fuel that is too rich must wait until it mixes with enough air to support its combustion. Some of the fuel injected before ignition, however, has already mixed with too much air, and so is already too lean to burn. Further mixing is unlikely to reverse this condition. Slower thermal-oxidation reactions can occur, but these are too slow to consume this fuel in the time available. Fuel injected after ignition cannot over-mix with air, because as it comes within combustible limits, it burns. Thus, over-mixed over-lean fuel injected during the ignition delay period escapes the main combustion, and is a significant source of unburnt fuel. Factors that lengthen the ignition delay will increase these HC emissions.

The second major source of HC emissions from diesel engines is under-mixed fuel injected near the end of combustion. As fuel is injected, its mixing rate with air depends on their vigorous relative motion. After injection ends, secondary injections can occur, or fuel left in the small volume at the nozzle tip (the 'sac' volume) may enter the combustion chamber. With the latter, there is some delay while this fuel is evaporated. Either way, some fuel (either as vapour or liquid) enters at low velocity into a rapidly cooling combustion chamber. This fuel does not mix effectively with the air, and some of it leaves the cylinder unburnt or only partially burnt. This HC source is controlled by designing fuel injection systems which have rapid and clean ends to the injection (the so-called 'spill rate'). Injectors with minimized sac volumes are now standard. In a valve-covers-orifice (VCO) type injector, the injector needle closes over the injector holes, eliminating most of this source (the small volume of the holes remains). This design can create problems in maintaining uniform injection spray patterns and end-of-injection timing between the holes, however.

Additional, though less important, sources of HCs from diesel engines are wall quench and misfire. Misfire is rare in diesel engines. Liquid fuel which impinges on the cylinder walls is neither hot enough nor sufficiently well mixed with air to burn rapidly. A properly set-up fuel injection system should eliminate droplet wall impingement, though the M.A.M. M system deliberately wets the bowl wall.

Catalytic oxidation of diesel HCs is possible. Aiding this is the plentiful excess oxygen in the exhaust of a diesel engine, though the relatively low exhaust temperatures slow the conversion. Engine changes that reduce particulate emissions and fuel consumption also tend to reduce HC emissions: all derive from combustion inefficiency.

19.4.4 Nitrogen oxides

Nitrogen oxides (NO_x) are comprised of nitric oxide (NO) and nitrogen dioxide (NO_2), with the former making up 70–90% of the total NO_x from diesel engines. Unlike the other pollutants described here, NO_x is a side-effect of combustion, not an incomplete step in it. Burning HC fuel with oxygen powers the engine; atmospheric nitrogen (whih supplies virtually all of the nitrogen in NO_x, fuel-borne nitrogen being negligibly small) is just caught in the reaction process.

The formation of NO is well understood. It is accepted that nitric oxide (NO) is formed by the extended Zeldovich mechanism:

$$O + N_2 = NO + N$$
$$N + O_2 = NO + O$$
$$N + OH = NO + H$$

Nitrogen dioxide (NO_2) forms from NO; quenching by excess air in the cylinder can freeze NO_2 levels at well above equilibrium concentrations.

The formation of NO depends on plentiful oxygen and high temperatures. Gas that burns before the time of peak cylinder pressure is particularly important. After it has burned, it is compressed to a higher pressure and temperature, and so reaches the highest temperature of any portion of the cylinder charge. Thus, the early part of combustion is important for NO_x, almost all NO_x is formed in the first 20° of crank angle after the start of combustion. Techniques to control NO_x therefore focus on this stage of combustion. However, most of these techniques reduce combustion temperatures, and so extract penalties in hydrocarbon emissions, particulate emissions, and fuel consumption. It is common to refer to the 'trade-offs' between NO_x emissions and particulates and fuel consumption for diesel engines.

Most factors affecting the initial rate of heat release will affect the NO_x formation rate in the same direction. The amount of fuel burned in the pre-mixed burning phase can be reduced by rate shaping (a lower injection rate early in the injection period) and pilot injection (a separate small injection to initiate combustion with a minimum of fuel). Rate shaping is already used. While the pilot injection requires extra equipment, it holds some promise for the future. Shortening the delay period can also reduce the amount of fuel burned in the pre-mixed burning stage: higher cetane-number fuels produce somewhat lower NO_x emissions. Higher compressed air temperatures (via higher intake temperatures, higher compression ratio, or turbocharging) may reduce NO_x emissions at light loads for this reason, but at higher loads the higher peak combustion temperatures lead to higher NO_x emissions. Any factors which promote air–fuel mixing in the delay period—air swirl, the size and number of injector nozzle holes, combustion chamber design, and the fuel injection pressure—also increase the initial rate of heat release, and therefore NO_x emissions. Ignition timing can easily be retarded to reduce not only the peak gas temperature but also the time spent at that temperature, but fuel consumption and particulate penalties can become severe. Aftercooling also reduces the air temperature, but with a superior NO_x-fuel consumption trade-off and higher engine output. Diluents such as EGR or water injection reduce the peak local temperatures, and are among the most effective in-cylinder NO_x control strategies. However, to date their practicable application has been limited to a few specific engine types (small-to-medium duty automotive engines for EGR, large stationary engines for water injection). All of these factors are design or operational parameters that change the NO_x-particulate and NO_x-fuel consumption trade-offs.

The other principal factor in NOx productin is the local air–fuel ratio. Indirect injection (IDI) diesel engines produce low NO_x emissions not only because of their higher heat losses to the swirl chamber walls and throat, but also because their pre-mixed burning takes place under locally rich (oxygen deficient) conditions.

The technology for the catalytic after-treatment of NO_x emissions is still developing. The exhaust of a diesel engine has excess oxygen and is relatively cool, so the catalytic reduction of NO_x is difficult. The two most effective technologies use additives in the exhaust to act as reducing agents. Selective catalytic reduction (SCR) uses ammonia or urea injected upstream of a precious metal catalyst. Reduction efficiencies are high (up to c. 90%). but cost, practical, and controls issues have so far limited SCR to large stationary engines. Active de-NO_x catalysts use a hydrocarbon, often the fuel itself, as the reducing agent. Their conversion efficiency is low (up to c. 20%) and their application has been very limited. The passive de-NO_x catalyst uses no reducing reagent. While this solves many of the problems of active de-NO_x catalysts, the reduction efficiency of the passive de-NO_x catalyst is very low.

19.4.5 Odour

Diesel exhaust odour results from a combination of aromatic hydrocarbons and oxygenated orgainc molecules such as aldehydes. The latter tend to be found at the edges of the fuel sprays and during starting. Although aldehydes do contribute there appears to be no single group of compounds which account for a diesel engine's odour.

Since odour is not very clearly defined, methods for reducing it are also not well defined. However for a given engine, everything which is done to reduce unburnt hydrocarbon emissions also tends to reduce odour.

19.4.6 Particulates

Particulate matter (Pm) is the other diesel emission—along with NO_x—of most concern. It is composed of *soot* (carbonaceous solid matter similar to carbon black), an *extractable fraction* (hydrocarbons extractable with a strong solvent) adsorbed onto the soot, and other contained inorganic compounds (largely sulphates, water and ash). Particulate concentrations are measured by drawing exhaust gas through a filter maintained at 52 °C, and computing the change in filter weight. Because of the methods, there is some overlap between this measurement and both the hydrocarbon measurement and the various smoke measurements. The soot component of the Pm corresponds to the smoke measurement, while the extractable fraction corresponds to a portion (ranging from about 25–50%) of the gaseous HC emissions. The exact fraction depends on the engine type and operating conditions, as these affect the distribution of the boiling range of the gaseous HCs emissions.

Soot *particles* are made up of roughly spherical *spherules* arranged in irregularly-shaped *clusters* or *chains*. Soot is formed in the cylinder, from heavy hydrocarbons in the gas phase which condense and coalesce in the oxygen deficient regions in the very rich core of the fuel sprays. Soot forms by pyrolysis reactions: these need rich conditions and high temperatures. Adjusting the balance between the mixing rate and the temperature rise can change the root formation rate. Faster mixing before temperatures rise will reduce soot formation. Later, as soot mixes with air, it encounters oxygen and burns if the temperature is high enough. Peak in-cylinder soot concentrations are much higher than exhaust levels; about 90% of the soot formed is oxidized within the cylinder. However, there is evidence that it is the soot *formation* that is adjusted by changes in engine design and operating parameters. The main strategies to reduce soot emissions aim to increase the mixing rate, and so require more developed fuel injection equipment (FIE) and optimized combustion chambers. Small fuel droplets have more surface area for their volume, and so evaporate and mix with air faster. To make small droplets, smaller injector holes and higher injection velocities are required—both of these require higher injection pressures for a given fuelling rate. More symmetrical fuel sprays (implying centrally-located and vertical injectors on DI diesel engines) will produce a wider and more uniform dispersion of droplets. Turbocharging and aftercooling supply more air for a given fuelling rate (i.e. load), and so raise the overall air–fuel ratio. Raising the combustion temperature (by using a higher compression ratio, for example) can increase soot formation and emissions at high loads, but may reduce these at low loads because of the shortening of the ignition delay, and consequent reduction in overmixed HCs. Turbocharging without aftercooling also raises the gas temperature, but the increased oxygen availability offsets this, so soot emissions are reduced overall.

The extractable fraction of particulate matter has traditionally been removed from the soot by a strong solvent, though more sensitive chromatographic methods are now available. The extractable fraction is composed of heavy hydrocarbons that condense as they cool in the exhaust system and atmosphere, and are absorbed onto the soild soot particles. About 15–45% by mass of the particulate matter is extractable, though this fraction varies with engine type, size, and operating conditions. While fuel is the main source for these HCs, the lubricating oil is also a significant source. It is the extractable fraction of particulate matter causes concern about the health risk of particulates. Many of these compounds are harmful to human health and to the natural environment. Retarded injection timing, while it reduces the soot forming combustion temperature, increases Pm emissions because of the lengthening of the delay period and the consequent increase in over-mixed HCs. There is some evidence that complex polyaromatic compounds in the fuel contribute to Pm, perhaps in the extractable fraction.

Fuel sulphur is the source for most of the inorganic remainder. Sulphates concentrations in the particulate matter are linearly related to the sulphur concentration in the fuel, with about 1–2% of fuel sulphur being converted and subsequently deposited in the Pm. Reducing the sulphur content of fuel can therefore reduce the inorganic fraction in the particulates. However, the scope for improvement is limited, given the already low sulphur content of fuel, and the small proportion of Pm composed of sulphates (10–15%).

The particle size distribution can vary with engine speed, load, and probably with engine and fuel type too. Recently attention has focused on the Pm10 component of particulates, those particulates below 10 μm. These are a health concern since their small size allows them to penetrate further into the human lung. The exact health effects of Pm10 still need to be clearly defined, however.

The after-treatment of particulate matter is possible. Catalytic converters are effective at controlling the HC (*i.e.* extractable) component of Pm, but only if the exhaust temperature is high enough. Simple throw away or cleanable filters do not seem practical for automotive applications. Oxidizing filters or traps are regenerated: Pm is periodically oxidized in-situ to clean the filter, but these technologies normally require additional fuel injection at the filter to start the oxidation. The continuously regenerating trap (CRT) makes this process continuous, and does not need additional heating.

19.5 Unregulated emissions

Unregulated emissions are those which are not currently covered by legislation, i.e. all other compounds emitted from the exhaust except: carbon monoxide, total hydrocarbons (HC), oxides of nitrogen (NOx) and particulate matter (PM). However, the unregulated portion of the exhaust is still of interest for various reasons. For example, in Europe data is being collected on CO_2 emissions from vehicles with a view to the possibility of these being legislated in the future. In addition, much attention has been given over recent years to the potential carcinogenicity/mutagenicity of diesel prticulate. Even more recently, the question of particle size and related health effects has been the focus of much public debate.

By contrast, some emissions have become less of a concern. Formation of sulphur oxides across diesel oxidation catalysts is directly related to the sulphur content of the fuel and can account for a significant proportion of total particulate matter. The introduction of low sulphur (0.05% wt.) diesel fuel, firstly in the US in 1993, then Europe from October 1996, to be followed by Japan in Oct 1997; and of very low sulphur (< 0.005% wt.) diesel fuel for certain urban applications, has meant that the sulphation issue has largely been resolved. Low sulphur fuels will also help to maintain catalyst performance over the lifetime of the vehicle, since sulphur is a catalytic poison.

The following sections describe those unregulated emissions which are particularly pertinent to diesel exhaust and have come under recent scrutiny.

19.5.1 Aldehydes

Aldehydes are oxygenates with the general formula RCHO where R is an organic side chain. Aldehydes are formed when combustion is incomplete; their formation is favoured when the environment is oxygen-rich. These compounds can be irritants or odorous; they may also be toxic and can contribute towards photochemical smog. Formaldehyde, for example, is both toxic and a highly active ozone-precursor.

Aldehyde levels can be significant in diesel exhaust, and may become of more concern in the future as oxygenate additives are increasingly used in diesel fuels. Formaldehyde and acetaldehyde are typically the dominant compounds and can be present at levels equivalent to or in excess of those seen from gasoline exhaust. The characteristic odour which has been associated with diesel engines in the past can be partly attributed to the presence of aldehydes. However, these compounds can be effectively controlled by oxidation catalysts, where typical average conversion levels would be around 70%.

Aldehydes are usually analysed as their hydrazones via High Performance Liquid Chromatography (HPLC). The method is highly sensitive; compounds are measured by Ultra-Violet (UV) detection at optimum wavelength.

19.5.2 Polycyclic aromatic hydrocarbons

Polycyclic aromatic compounds (PAHs), are 3-, 4-, 5- or 6-ring compounds, and are of concern because some of them are carcinogenic. The smaller PAHs, such as naphthalene and phenanthrene, are present in the vapour phase as well as the particulate phase. The most important PAHs, with regard to health effects, are present in particulate (i.e. those having 4 or more fused rings.) Benzo(a) pyrene, for example, a 5-ring compound, is a potent carcinogen.

PAHs can be emitted at significant levels from diesel engines and are responsible for much of the concern about the carcinogenicity of diesel particulate. They can be controlled to some extent by an oxidation catalyst, however the degree of conversion over the catalyst is highly dependent on the individual compound. The lighter compounds, for example, would be effectively controlled (about 80%+), whereas in contrast, the heavier compounds show poor (about 30%), or in some cases, very little conversion.

Particle-bound PAHs are usually collected on a filter paper. They are extracted into a suitable solvent, the sample is then cleaned and analysed via HPLC with fluorescence detection. The analysis of vapour-phase PAHs is described in the section on vapour-phase hydrocarbons, below.

19.5.3 Nitrated polycyclic aromatic hydrocarbons

These compounds are PAHs with a nitro group attached to one or more of the rings. Nitrated polycyclic aromatic hydrocarbons (NPAHs) are formed by the reaction of PAHs with nitrating agents in exhaust gases, such as nitric acid. They may also sometimes be artefacts of sampling procedures.

Some NPAHs are known to be the potent carcinogens, and 1,8-dinitropyrene, which is commonly detected in diesel exhaust, is reported to be the most mutagenic compound known. NPAHs are responsible for a substantial proportion of the direct-acting Ames activity of diesel particulate.

NPAHs are quantified using HPLC with fluorescence detection.

19.5.4 Vapour-phase hydrocarbons

The lighter hydrocarbons (C1–C12) which are present in the gaseous phase, are important in gasoline exhaust but tend to be

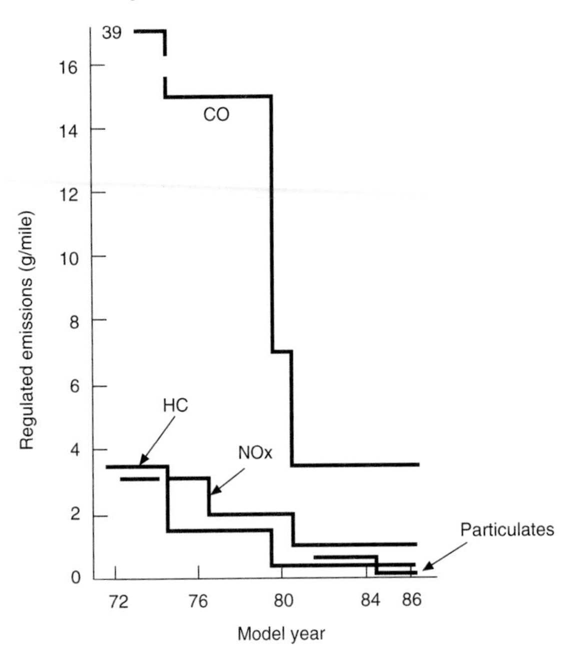

Figure 19.5 US Federal exhaust emission regulations for light duty passenger vehicles

Model Year	1981	1982	1983	1984	1985
HC g/mile	1.7 →			0.8	
CO g/mile	18 →			10	
NOx g/mile	2.3 →				1.2*
Particulates	–	0.6 →			0.26**

*NOx averaging proposed
** particulate averaging proposed

Figure 19.6 US Federal exhaust emissions regulations for light duty trucks (i.e. up to 8500 lb GVW)

present in insignificant quantities in diesel exhaust. However, the heavier hydrocarbons (C14+), which mainly exist in the vapour phase, are more commonly found in diesel exhaust. Typically occurring chemical types are: some PAHs, substituted naphthalenes, heavier aliphatics, alkylbenzenes, and some sulphur- or *updated* oxygen-containing compounds such as naphthaldehydes.

Vapour-phase hydrocarbons are sampled via tubes containing an adsorbent material. They can then be analysed via gas chromatography/mass spectrometry (GC/MS). A cryogenic system is usually required.

19.5.5 Particle size

Considerable information has been presented recently which suggests empirical links between gasoline and diesel engine exhaust particulate size and health effects. Particles with an aerodynamic diameter of less than 10 μm, so-called PM_{10}, have been highlighted as being of special concern since these can be inhaled deep into the lungs and may remain there to do physiological damage.

At the time of writing, very little is known about the effects of engine design and operation on the formation of such particles. In addition, the methodology for sampling and measuring exhaust particulate size is not well-defined, although several instruments are available.

Several countries have in place, or are in the process of defining, an ambient air quality standard for PM_{10}. It is likely, therefore, that particle size will remain of concern to the automotive industry and will be the focus of future research.

19.6 Conclusions

The emissions carbon monoxide, hydrocarbons, nitrogen oxides and particulates are legislated in many areas, mainly in the developed regions, of the world. Regulated emissions are required to be measured over defined drive cycles or steady-state test procedures, the requirements of which can vary depending on the country.

Recently, attention has focused on certain emissions which are not currently regulated, such as the polycyclic aromatic hydrocarbons or the very small (micron and sub-micron) particulates. Reports issued into the public domain highlighting potential health-related and/or environmental risks have stimulated the interest in these emissions. Research continues into establishing good methodologies for their measurement from mobile sources.

Low emissions, combined with good fuel economy, have been the main drivers for technological developments in both gasoline and diesel for several years, and this continues to be the case.

Model year	*1981*	*1982*	*1983*	*1984*	*1985*	*1987**
Test procedure	Diesel : 13 mode →			13 mode or transient cycle	Transient cycle	
	Gasoline : 9 mode			Transient cycle	→	
HC g/bhp.h	1.5 →			0.5 (13 mode) or 1.3 (transient)	1.3 →	
NOx g/bhp.h	–	–	–	9.0 (13 mode) or 10.7 (transient)	→	
HC + NOx g/bhp.h	10 →			–	–	–
CO g/bhp.h	25 →			35* (13 mode transient)	35*	15.5

*proposed

Figure 19.7 US Federal exhaust emissions regulations for on-highway heavy duty engines (diesel or gasoline)

Appendix (See Figures 19.5–19.12)

This Appendix contains a summary of the emissions regulations relating to the USA Federal California, Europe and Japan as perceived in 1982. For up to date information the US Federal Register, California Air Resources Board publications, ECE Regulations Number R15, R24 and R29 and the Japanese Auto Pollution Board publications should be consulted.

	Non-methane HC (g/mile)	Carbon monoxide (g/mile)	NOx (g/mile)
1981			
PC	0.39	3.4/7.0	1.0/0.7(1.5)*
LDT (4000 lb max)	0.39	9.0	1.0(1.5)*
LDT (6000 lb max)	0.50	9.0	1.5(2.0)*
1982			
PC	0.39	7.0	0.4/0.7(1.5)*
LDT (4000 lb max)	0.39	9.0	1.0(1.5)*
LDT (6000 lb max)	0.50	9.0	1.5(2.0)*
From 1983			
PC	0.39	7.0	0.4/0.7(1.0)*
LDT (4000 lb max)	0.39	9.0	0.4/1.0(1.0)*
LDT (6000 lb max)	0.50	9.0	1.0(1.5)*

* 100 000 mile Emission Standards.
PC: Passenger Cars.
LDT : Light Duty Trucks.

Figure 19.8 Californian exhaust emissions regulations for passenger cares and light duty trucks

13 MODE CYCLE (g/bhp h)

Model year		HC	CO	HC + NOx
1981–1983		1.0	25	6.0
	OR	–	25	5.0
1984		0.5	25	4.5

TRANSIENT CYCLE (g/bhp h) (OPTIONAL)

Model year	HC	CO	NOx
1984	1.3	15.5	5.1

Figure 19.9 Californian exhaust emissions regulations for heavy duty engines

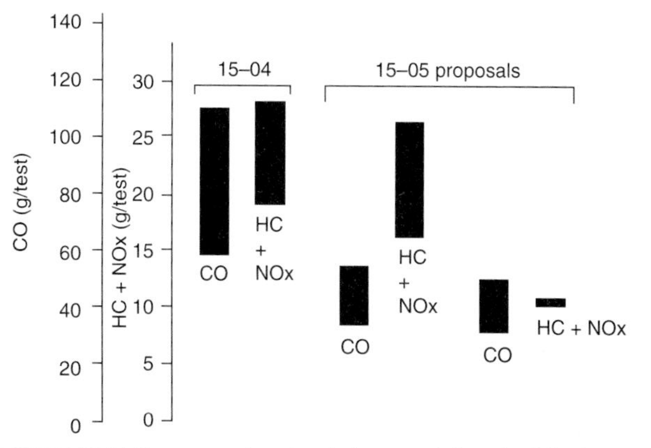

Figure 19.10 European exhaust emissions regulations for light duty vehicles (up to 3500 kg GVW). Exact emission value depends on the weight of the vehicle under test

Model year	HC	CO	NOx
1984	3.5(2.6)	14(10)	18(13.4)

13 mode cycle g/kW h (g/bhp h).

Figure 19.11 European exhaust emissions regulations for heavy duty diesel engines. Engine is tested over 13 mode cycle

Diesel engines over 6 mode mean figures (ppm)

Model year		1981	1982	1983	1984
HC	DI	510	→	→	→
	IDI	510			
CO	DI	790	→	→	→
	IDI	790			
NO_x	DI	540	540	→	470
	IDI	340	290	→	290

Figure 19.12 Japanese exhaust emissions regulations for diesel engines. Engine is tested over 6 mode cycle

Bibliography

1 BRADDOCK, J. N. and BRADOW, R. L., 'Emissions Patterns of Diesel-Powered Passenger Cars', *SAE 750682*

2 BROOME, D. and KHAN, I. M., 'The Mechanisms of Soot Release from Combustion of Hydrocarbon Fuels with Particular Reference to the Diesel Engine,' *I. Mech. E.* C140/71 (1971)

3 GREEVES, G., KHAN, I. M., WANG, C. H. T. and FENNE, I., 'Origins of Hydrocarbon Emissions from Diesel Engines', *SAE 770259*

4 SPRINGER, G.S. and PATTERSON, D. J. (ed.), *Engine Emissions—Pollutant Formation and Measurement*, Plenum Press, New York, London (1973)

5 YU, R. C., WONG, V. W. and SHAHED, S. M., 'Sources of Hydrocarbon Emissions from Direct Injection Diesel Engines', *SAE 800048*

6 ECONOMIC COMMISSION FOR EUROPE, 'Uniform Provisions Concerning the Approval of Vehicles Equipped with a Positive-Ignition Engine with Regard to the Emission of Gaseous Pollutants by the Engine', ECE 1503 Series of Amendments, Trans/SCI/WP29/R107 (18 February 1977) and ECE 1504 Series of Amendments (second draft) Trans/SCI/WP29/204 (28 March 1980)

7 JAPAN MINISTRY OF TRANSPORT, 'Prevention of Emission of Smoke, Bad-smelling and Harmful Gases', *Article 31* (1978 Emission Standards)

8 US EPA, 'Control of Air Pollution from New Motor Vehicles and Motor Vehicle Engines; Gaseous Emission Regulations for 1984 and Later Model Year Light-Duty Trucks', Red. Reg. Vol. 45 No. 188 pp. 63734–63784 (25 September 1981)

9 US EPA, 'Averaging of Particulate Emissions from 1985 and Later Model Year Diesel-Fueled Light-Duty Vehicles and Light-Duty Trucks; Proposed Rulemaking', Fed. Reg. Vol. 46 No. 247 pp. 62608–62625 (24 December 1980)

10 US EPA, 'Control of Air Pollution from New Motor Vehicles and Motor Vehicle Engines; Revised Gaseous Emission Regulations for 1984 and Later Model Year Light-Duty Trucks and Heavy-Duty Engines', Fed. Reg. Vol. 47 No. 8 pp. 1642–1651 (13 January 1982)

11 EPA, 'Stationary Internal Combustion Engines—Proposed NOx Limits', Fed. Regulations Title 40, Part 60/FF

12 BICEMA, 'Guide Lines for Measuring Exhaust Emissions from Diesel and Dual Fuel Engines Excluding Automotive Engines', British Internal Combustion Engine Manufacturers Association

13 DEMA, 'Exhaust Emission Measurement Procedure for Low and Medium Speed Internal Combustion Engines', USA—Diesel Engine Manufacturers Association

14 ECE, 'Draft Regulation 49—Uniform Provision Concerning the Approval of Diesel Engines with Regard to the Emission of Gaseous Pollutants (13 Mode Diesel Test), TRANS/SCI/WP29 R195

15 BURT, R. and TROTH, K.A., 'Influence of Fuel Properties on Diesel Exhaust Emissions', *I. Mech. E. Motor Vehicle Air Pollution Control Proceedings 1968–69*, Vol. 183 Part 3E Paper 11, pp. 171–178

16 COLLINS, D., CUTHBERTSON, R. D., GAWEN, R. W. and WHEELER, R. W., 'The Use of Constant Volume Sampler and Dilution Tunnel to

Compare the Total Particulates from a Range of Automotive Engines', *Automobile Engineering Meeting, Detroit, Michigan* (October 13–17 1975)
17 GRIGG, H. C., 'The Role of Fuel Injection Equipment in Reducing 4-Stroke Diesel Engine Emissions', Presented at the Automotive Engineering Congress and Exposition Detroit, Michigan (February 23–27 1976)
18 HOWITT, J. S. and MONTIERTH, M. R., 'Cellular Ceramic Diesel Particulate Filter', *International Congress and Exposition, Detroit, Michigan* (February 23–27 1981)
19 KHAN, I. M. and GREEN, A.C., Scheduling Injection Timing for Reduction of Diesel Emissions', *Automotive Engineering Congress and Exposition, Detroit, Michigan* (February 24–28 1975)
20 STUMPP, G. and BANZHAF, W., 'An Exhaust Gas Recirculation System for Diesel Engines', *Congress and Exposition Cobo Hall, Detroit* (February 27–March 3 1978)
21 WALDER, C. J., 'Reduction of Emissions from Diesel Engines', *International Automotive Engineering Congress, Detroit, Michigan* (January 8–12 1973)
22 GROSS, G. P. and MURPHY, K. E., 'The Effects of Diesel Fuel Properties on Performance, Smoke, and Emissions', ASME 78–DGP-26 (1978)
23 CAMPBELL, J., SCHOLL, J., HIBBLER, F., BAGLEY, S., LEDDY, D., ABATA, D. and JOHNSON, J., 'The Effect of Fuel Injection Rate and Timing on the Physical, Chemical and Biological Character of Particulate Emissions from a Direct Injection Engine', *SAE 810996*
24 DANIELSON, E., 'Draft Recommended Practice for Measurement of Gaseous and Particulate Emissions from Heavy-Duty Diesel Engines Under Transient Conditions', EPA SDSB-79–18, EPA, 83 pp. (April 1979)
25 PIERSON, W. R., BRACHACZEK, W. E., HAMMERLE, R. H., McKEE, D. E. and BUTLER, J. W.,'Sulphate Emissions from Vehicles on the Road', *J. Air Pol. Control Assoc.*, pp. 123–132 (February 1978)
26 ULLMAN, T. L., SPRINGER, K. J. and BAINES, T. M., 'Effects of Six Variables on Diesel Exhaust Particulate', *ASME 80–DGP-42* (1980)
27 BRICKLEMEYER, B. A. and SPINDT, R. S., 'Measurement of Polynuclear Aromatic Hydrocarbons in Diesel Exhaust Gases', *SAE 780115*
28 CADLE, S. H., NEBEL, G. J. and WILLIAMS, R. L., 'Measurements of Unregulated Emissions from General Motors'Light-Duty Vehicles', *SAE 790694*
29 CANDELI, A., MASTRANDREA, V., MOROZZI, G. and TOCCACELI, S., 'Carcinogenic Air Pollutants in the Exhaust from A European Car Operating on Various Fuels', *Atmos. Environ.,* 8, 693–705 (1974)
30 DIETZMANN, H. E. and BLACK, F. M., 'Unregulated Emissions Measurement Methodology', *SAE 790816*
31 CUTHBERTSON, R. D., STINTON, H. C. and WHEELER, R. W., 'The Use of a Thermogravimetric Analyser for the Investigation of Particulates and Hydrocarbons in Diesel Engine Exhaust', *SAE 790814*
32 FUNKENBUSCH, E. F., LEDDY, D. G. and JOHNSON, J. H., 'The Characterization of the Soluble Organic Fraction of Diesel Particulate Matter', *SAE 790418*
33 GRIMMER, G., HILDEBRANDT, A. and BÖHNKE, H., 'Sampling and Analysis of Polycyclic Aromatic Hydrocarbons in Automotive Vehicular Exhaust Gases. Part 1. Optimization of the Collecting Device—Enrichment of the Polycyclic Aromatic Hydrocarbons', Erdöl U. Kohle 25, 442–447 (1972)
34 HANGEBRAUCK, R. P., LAUCH, R. P. and MEEKER, J. E., 'Emissions of Polynuclear Hydrocarbons from Automobiles and Trucks', *Amer. Ind. Hyg. Assoc. J.*, **27**, 47–56 (1966)
35 JÄGER, J., 'Detection and Characterisation of Nitro Derivatives of some Polycyclic Aromatic Hydrocarbons by Fluorescence Quenching after thin-Layer Chromatography: Application to Air Pollution Analysis', *J. Chromatogr.,* **152**, 575–578 (1978)
36 KHATRI, N. J., JOHNSON, J. H. and LEDDY, D. G., 'The Characterization of the Hydrocarbon and Sulphate Fractions of Diesel Particulate Matter', *SAE 780111*
37 POSTULKA, A. and LIES, K.-H., 'Chemical Characterisation of Particulates from Diesel-Powered Passenger Cars', *SAE 810083*
38 SPRINGER, K. J. and BAINES, T. M., 'Emissions from Diesel Versions of Production Passenger Cars', *SAE 770818*
39 LAITY, J. L., MALBIN, M. D., HASKELL. W. W. and DOTY, W. I., 'Mechanisms of Polynuclear Aromatic Hydrocarbon Emissions from Automotive Engines', *SAE 730835*
40 MOHR, U., SCHMÄHL, D. and TOMATIS, L. (ed.), 'Air Pollution and Cancer in Man', Proceedings of the Second Hanover International Carcinogenesis Meeting, Hanover, 22–24 October 1975, *IARC Scientific Publication No. 16*
41 'Impacts of Diesel-Powered Light-Duty Vehicles: Health Effects of Exposure to Diesel Exhaust', *Report of the Health Effects Panel of the Diesel Impacts Study Committee, National Research Council,* National Academy Press, Washington, DC (1981)
42 WEI, E. T., WANG, Y. Y. and RAPPAPORT, S. M., 'Diesel Emissions and the Ames Test: A Commentary', *J. Air Poll. Control Assoc.,* **30**, 267–271 (1980)
43 YU, M.-L. and HITES, R. A., 'Identification of Organic Compounds on Diesel Engine Soot', *Anal. Chem.,* **53**, 951–954 (1981)
44 HEYWOOD, J.B., *Internal Combustion Engine Fundamentals*, McGraw-Hill (1988).
45 LILLY L.R.C. (ed.) *Diesel Engine Reference Book*, Butterworths (1984).
46 Stormy Weather Ahead, *The Economist,* March 23rd, 1996.
47 GREEVES, G. and WANG, C.H.T., Origins of Diesel Paticulate Mass Emission, Society of Automotive Engineers Technical Paper Series, *SAE paper 810260* (1981).
48 COWLEY, L. T., STRADLING, R. J. and DOYON, J., The Influence of Composition and Properties of Diesel Fuel on Particulate Emissions from Heavy-Duty Engines'. Society of Automotive Engineers Technical Paper Series, *SAE paper 932732* (1993).
49 GREEVES, G. and TULLIS, S., 'Contribution of EUI-200 and Quiescent Combustion System Towards US94 Emissions'. Society of Automotive Engineers Technical Paper Series, *SAE paper 930274* (1993).
50 GREEVES, G. and TULLIS, S. 'Improving NOx Versus BSFC with EUI 200 Using EGR and Pilot Injection for Heavy-Duty Diesel Engines'. Society of Automotive Engineers Technical Paper Series, *SAE paper 960843* (1996).
51 Particulates—a major health hazard *Automotive Engineer,* October/November 1995.
52 NIGHTINGALE, D. R., 'A Fundamental Investigation into the Problem of NO Formation in Diesel Engines'. Society of Automotive Engineers Technical Paper Series, *SAE paper 750848* (1975).
53 SUCH, C., NEEDHAM, J., EDWARDS, S., and FREEMAN, H. 'Development of Heavy-Duty Diesel Engines for Low Emissions Using Exhaust Gas Recirculation (EGR).' *Proc. 8th Int. Pacific Conference on Automotive Eng.,* Soc. Auto Eng. Japan, Nov. 4–9 (1995).

20 Engine noise

Contents

20.1 Introduction

This chapter considers the noise of diesel engines used in vehicles and machines which move on wheels or caterpillar tracks. These include passenger cars, heavy trucks, other commercial vehicles, industrial (material handling) trucks, construction equipment, agricultural and forestry machinery, and other similar applications.

Noise is simply unwanted sound. This, however, means that what constitutes noise may be very subjective; the sound of 'discothèque' music is enjoyed by many people but is often regarded as a noise nuisance by neighbours who are not participating. Noise is one of the major pollutants which affect the 'quality of life', particularly in the urban environment. Exposure to excessive sound (wanted or unwanted) will cause some hearing loss. Short-term exposure to high sound levels usually causes a temporary shift in hearing sensitivity but if hearing damage is sustained, as a result of prolonged exposure, it is permanent.

Where diesel engine applications are concerned, the radiated noise affects many people, ranging from the operator and passengers to external observers who may be in the local vicinity or some distance (kilometres) away. Most governments have introduced legislation, with strict limits on the maximum level of noise which can be radiated from the whole vehicle/machine during designated test procedures. There is little noise legislation which relates to the engine on its own.

20.2 Theory and definitions

In the context of this chapter, sound is the sensation which results when the human ear detects a series of small, repidly varying, pressure fluctuations (compressions and rarefactions) that propagate through the air; in most sounds the pressure fluctuations are irregular. The human ear is capable of sensing these fluctuations over an extremely wide range of both amplitude and frequency.

The practical measurement of sound has developed over the last 70 years into a specialization of its own. The essentials are reviewed here, but for detailed treatments the reader is referred to the many available books on the subject; details of some are included in the References.

20.2.1 Amplitude

The pressure fluctuations which can be sensed by the healthy ear are in the range from slightly below 20 μPa, the nominal threshold of hearing, to 200 000 000 μPa, the threshold of pain, and above. In this wide range the human response is non-linear: typically a trebling of actual pressure fluctuations is perceived as a doubling of the loudness of the sound. This non-linearity and the very wide pressure range has led to the use of a logarithmic scale to make objective measurements more manageable. The quantitative value is called 'sound pressure level' and the measurement units are decibels (dB). It is obtained from the following equation:

$$\text{Sound pressure level, dB} = 10 \cdot \log_{10} [(p_{rms}/p_{ref})^2]$$

where

p_{rms} = root mean square (rms) value of the measured pressure fluctuations, in pascals, Pa,
p_{ref} = the standard reference pressure of 20×10^{-6} Pa (20 μPa is the hearing threshold).

Note that the sound pressure level is based on the square of the pressure ratio (p_{rms}/p_{ref}) to give an energy related unit. The equation is commonly written as:

$$\text{Sound pressure level, dB} = 20 \cdot \log_{10} (p/p_{ref})$$

where
$p = p_{rms}$

The range of commonly experienced values is illustrated in *Figure 20.1*.

20.2.2 Effect of distance on sound pressure level

It is important to understand that the sound pressure level relates to the observer (i.e. where the ear or microphone is positioned); it does not directly indicate the sound power being radiated by a sound source: see 20.2.4 below. In free space the amplitude of the pressure fluctuations decrease in inverse proportion to the distance from the sound source. Consequently, the sound pressure level measurements will reduce according to the inverse square law. If a source produces a sound pressure level of 100 dB at 1 m, then the levels measured at increasing distances over free ground or in free space will decrease by 6 dB for each doubling of distance: 94 dB at 2 m, 88 dB at 4 m, 82 dB at 8 m, etc. The equation for the sound pressure level at point 2, knowing the value at point 1 is:

$$L_{p2} = L_{p1} - 10 \cdot \log_{10} [(r_2/r_1)^2]$$

where
L_p = sound pressure level,
r = distance from the sound source.

This relationship between sound pressure and distance does not apply when close to the source, in the so-called 'near field'. It is also more complicated when reflections from walls, buildings, natural obstructions, etc. are encountered and no simple rules can be used in such cases.

20.2.3 Frequency and wavelength

The number of times an event is repeated per unit time is called its frequency. In the measurement of sound, frequency is expressed in cycles per second; the units are called hertz (Hz). The frequency of pressure fluctuations which can be sensed by the ear is in the range from 20 Hz to 20 000 Hz. This range reduces with individual ageing, but can be severely reduced by damage due to exposure to excessive sound.

The pressure fluctuations propagate through the air at around 344 m/s; this varies a little with temperature and pressure. The

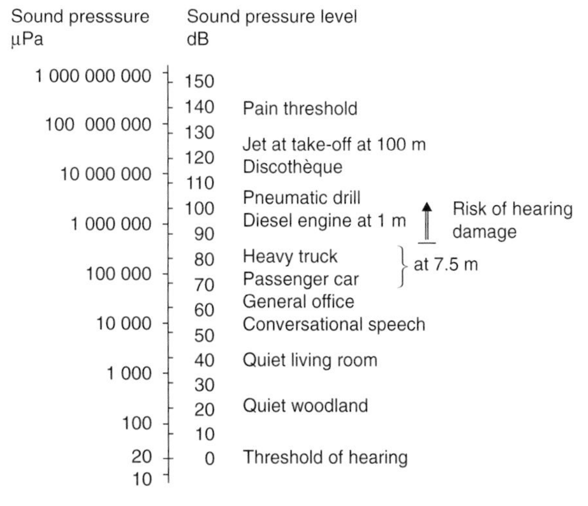

Figure 20.1 Typical sound pressure levels

distance between successive peaks of a pressure fluctuation produced at a single frequency is called the wavelength (λ); the units are metres. It is related to the frequency (f) and the speed of propagation of the sound wave (c) by the equation

$$\lambda = c/f$$

In air at 20°C and at normal atmospheric pressure, a sound at a frequency of 1000 Hz has a wavelength of 0.344 m, and at 100 Hz it has a wavelength of 3.44 m.

20.2.4 Sound power level

Sound power level is an absolute value related to the energy being radiated from a sound source. It is the only parameter which allows direct comparisons between the noise levels of different engines or machines. Sound pressure measurements cannot be compared without a detailed knowledge of the microphone positions: so-called 1 m microphone positions vary between different test procedures. Unfortunately there is no direct measure of sound power. Most techniques use microphones with measurements taken on an imaginary surface which surrounds the sound source. Sound power is then calculated from either intensity values or pressure levels. The theory is as follows:

Defining equation:

$$\text{Sound Power, watts} = \int_s I_s \cdot dS$$

where
I_s = intensity through each incremental area dS, of the imaginary surface surrounding the sound, W/m^2,
S = area of the enclosing imaginary surface, m^2

For any free, progressive sound wave there is a unique and simple relationship between the mean square sound pressure and the intensity. This relationship, at a particular point and in the direction the wave is travelling, is

$$I = (p_{rms})^2/\rho c \qquad (20.1)$$

where
I = intensity, W/m^2
p_{rms} = root mean square of the measured pressure fluctuations, Pa, see 20.2.1 above,
ρc = characteristic impedance of the medium; for air at 20°C, ρc = 413 kg/m^{2}s (density of air, ρ = 1.20 kg/m^3; acoustic velocity, c = 344 m/s).

In practice, the above equations are simplified as follows:

$$\text{Sound power} = \Sigma I_n \cdot S_n$$

where
I_n = mean intensity through an element (small part) of the imaginary surface surrounding the sound, W/m^2
S_n = area of the element of the imaginary surface, m^2

It is common to express sound power and intensity on logarithmic scales, in decibels (dB), as with sound pressure. The equations are

$$\text{Sound power level, dB} = L_W = 10 \cdot \log_{10} (W_n/W_{ref})$$
$$\text{Sound intensity level, dB} = L_I = 10 \cdot \log_{10} (I_n/I_{ref})$$

therefore, the relationship for practical measurements is

$$10 \cdot \log_{10} (W_n/W_{ref}) = 10 \cdot \log_{10} (I_n/I_{ref}) + 10 \cdot \log_{10} (S_n/S_{ref})$$

where
W_n = the sound power for the element of the surface, watts,
W_{ref} = the standard reference power of 1×10^{-12} watts (1 pW),
I_n = the mean sound intensity through the element, W/m^2,
I_{ref} = the standard reference intensity of 1×10^{-12} W/m^2 (1 pW/m^2),
S_{ref} = the standard reference surface area of 1 m^2.

The general addition of decibel levels is considered in 20.2.5, below. However, it can be shown that when all the elemental surfaces have the same area (i.e. $S_1 = S_2 = S_3 = S_4$ . . . etc.):

$$\text{Sound power level, dB} = 10.\log_{10} (I_{average}/I_{ref}) + 10.\log_{10} (S/S_{ref})$$

where
$I_{average}$ = average intensity through the whole surface enclosing the sound (averaging of decibel levels is considered in 20.2.6, below),
S = area of the enclosing surface, m^2.

The accurate measurement of sound intensity is not easy and requires more sophisticated equipment than that used for sound pressure measurements. For an alternative approach, consider the sound pressure in air at 20°C at the reference intensity level of 1 pW/m^2:

$$p_{rms} = (\text{I} \cdot \rho c)^{0.5} \qquad \text{(see eqn. 20.1, above)}$$
$$= (10^{-12} \times 413)^{0.5}$$
$$= 20.3\ \mu\text{Pa}$$

This is close to the standard reference sound pressure of 20 μPa, and if it were to be used as the reference level, sound pressure levels would be reduced by 0.41 dB. Knowing this relationship, a more commonly used method for calculating the approximate under power of whole machines or engines involves obtaining average sound pressure levels from a number of microphones evenly distributed on an imaginary surface surrounding the source. The sound power level is then approximated by the equation:

$$\text{Sound power level, dB} = 10.\log_{10} [(p_{average}/p_{ref})^2] + 10.\log_{10} (S/S_{ref})$$

where
$p_{average}$ = average sound pressure level on the surface enclosing the sound.

The 0.14 dB error, from the substitution of using pressure instead of intensity, is small when compared with the other potential errors in the practical test procedures which are used and, since it leads to a small overestimation of the sound power, it is normally ignored.

20.2.5 Addition and subtraction of sound sources

The additionl of individual sound sources must be carried out using energy-related terms. Except for the special case where two sources are producing dominant levels at exactly the same frequency (a condition not normally encountered in diesel engine applications), the summation is carried out as follows:

$$x = L_n/10$$
$$e_n = 10^x$$
$$e_{total} = \Sigma e_n$$
$$L_{total} = 10 \cdot \log_{10} (e_{total})$$

where
L_n = the level, in decibels, of an individual source,
e_n = the energy related term for the individual source,
e_{total} = total energy related term for the sum of all the sources,
L_{total} = the level, in decibels, for the total of all the sources summed together.

In a similar manner, sound sources must be subtracted using energy related terms:

$$e_{remain} = e_{total} - e_1 - e_2 - e_3 - e_4 \ldots \text{etc.}$$
$$L_{remain} = 10 \cdot \log_{10} (e_{remain})$$

where

e_{remain} = remaining energy related term after sources have been subtracted,
L_{remain} = the level, in decibels, for the remaining sound source(s).

The above equations can be used to add or subtract sound pressure levels, sound intensities or sound powers, but these quantities must not be mixed together; although they are all ratios in decibels, they do not relate to the same reference values. In any addition or subtraction, all the L_n terms must be of the same type: either all pressure levels, or all intensities, or all power values, and all must be based on the same reference level.

The addition and subtraction of two sources are graphically represented in *Figures 20.2a* and *20.2b*, respectively. To determine the sum, or difference, of the two sound sources, L_2 and L_1 (where $L_2 > L_1$), the arithmetic difference between them must first be obtained using

$$\Delta L = L_2 - L_1$$

For addition or subtraction the value of ΔL should be found on the x-axis of the appropriate graph. By moving vertically up to the curve and then horizontally across to the y-axis, the corresponding value of L_{add} or L_{sub} should be found. Then simply:

the sum of the two sources = $L_2 + L_{add}$

and

the difference of two sources = $L_2 - L_{sub}$

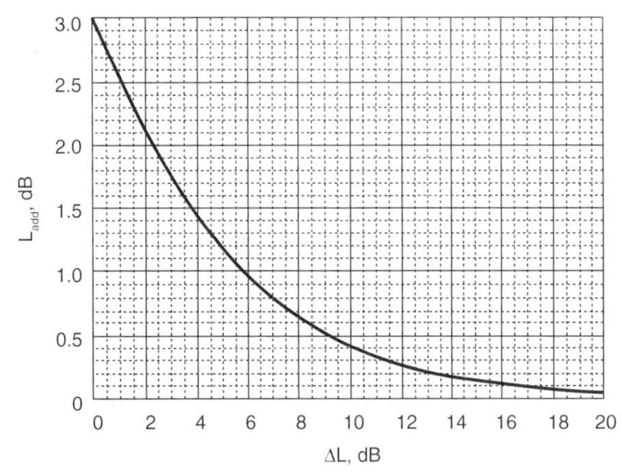

Figure 20.2a Curve for decibel addition

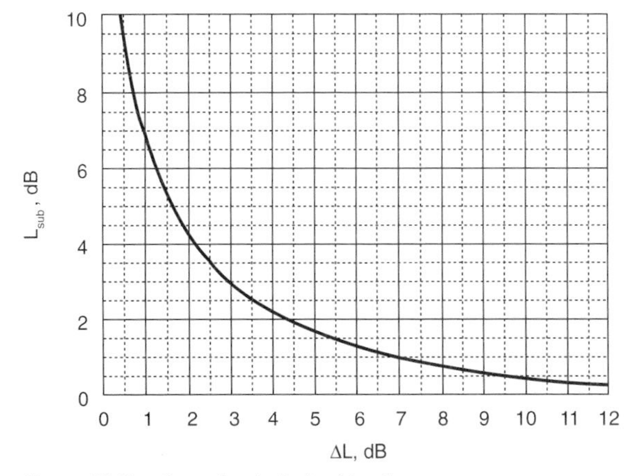

Figure 20.2b Curve for decibel subtraction

20.2.6 Averaging decibel levels

There are two averaging methods commonly used when dealing with decibel levels:

(a) Arithmetic averaging—addition of all the values as simple numbers and division by the number of values in the sample. This is only used when a **typical** value is required, e.g. the average of several measurements taken at the same position for nominally the same operating conditions on one machine, or the average level for several, nominally identical, machines tested to the same procedure.

(b) Averaging of energy related terms, as used in addition and substraction; to distinguish it from arithmetic averaging, it is sometimes (incorrectly) referred to as 'logarithmic averaging':

$$e_{total} = \Sigma\, e_n$$
$$e_{average} = e_{total}/n$$
$$L_{average} = 10 \cdot \log_{10} (e_{average})$$

where

$e_{average}$ = the average energy related term,
$L_{average}$ = the average level, in decibels.

This averaging method is used to obtain a true decibel average. It should be used for the calculation of average sound pressure levels or intensities, when calculating sound power levels.

20.2.7 Calculating relative levels

Since decibel levels are a logarithmic scale, the relative contribution (or importance) of each noise source to the overall level, is not readily apparent. It is a useful practice to express the contribution from each source as a percentage of the total sound and, again, this is done by using the energy-related terms:

$$C_n = (e_n/e_{total}) \times 100$$

where

C_n = the percentage contribution from an individual source relative to the total level.

20.2.8 Weighting curves

The apparent loudness of a sound (i.e. the subjective evaluation made by the human ear) varies with frequency as well as with sound pressure. Compared with the response to sounds in the frequency range 250 to 8000 Hz, the ear is less sensitive at other frequencies, particularly below 100 Hz. Progressively lower loudness ratings will be given to the same sound pressure level as the frequency moves away from this preferred range. The amount of variation of loudness with frequency also depends to some extent on the sound pressure level itself. At high sound pressures the perceived variation in loudness with frequency is smaller than it is at low pressures. Loudness rating curves are available in the many books on the subject.

To adjust the frequency response of measuring systems to be similar to that of the human ear, several weighting curves were proposed; see *Figure 20.3*. Originally, the 'A' weighted curve was designed to approximate to the human response at low sound pressure levels (< 55 dB). Similarly, the 'B' and 'C'

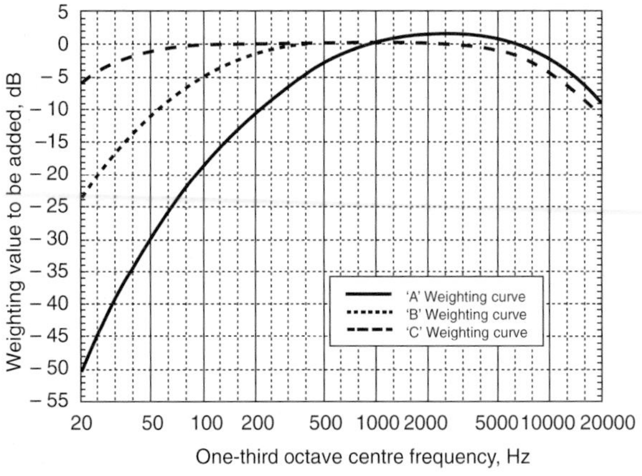

Figure 20.3 Measurement weighting curves

weightings were intended for use at sound pressure levels of 55 to 85 dB and above 85 dB, respectively. However, the trend is towards the exclusive use of the 'A' weighting curve for noise measurements associated with diesel engine application. When a more detailed understanding of the human response is required, loudness curves or 'sound quality' analysis systems are employed.

20.2.9 Noise dose level

The hearing damage potential of sound depends not only on its level but also on the exposure time. To assess the potential damage in environments where noise levels are varying, as in most diesel engine applications, the noise exposure is expressed as being equivalent to a constant A-weighted level over the same time period, or L_{Aeq} value. In the simplest of cases, where the noise levels change in a stepwise manner and are fairly constant, during each step, the L_{Aeq} value can be calculated from simple time and sound pressure measurements. However, in most cases operators are working with continuously varying noise levels and so it is normal practice to assess their exposure using a dose meter.

20.3 Legislation

Noise in the environment has been identified as one of the main problems which reduces the 'quality of life', and it is the subject of an increasing number of complaints from the public. It has been estimated that around 20% of the inhabitants in western Europe (about 65 million people) suffer noise levels that scientists and health experts consider to be unacceptable. Problems associated with noise include annoyance, sleep disturbance and other auditory and non-auditory physiological effects. It is reasonable to assume that similar problems exist throughout the urbanized world.

Noise from transportation has been the major contributor, but the contribution from other machines used outdoors is growing. Consequently, governments are under increasing pressure to introduce legislation to restrict noise emissions from vehicles and machines and to steadily reduce the permissible limits. These reductions are required not only to reduce the noise in the environment, but also to keep pace with the increasing number of noise sources. If the number of vehicles and machines being used is doubled, then ideally the noise emitted by each of them should be halved (i.e. reduced by 3 dB) just to keep the general environmental noise constant. Since older machines still in service are not recalled for noise reduction treatments, it is necessary to produce new machines with substantially lower noise emissions if long-term environmental improvements are to be achieved.

20.3.1 On-highway vehicles

In western Europe since the mid-1970s, the legislative 'drive-by' limit for the exterior noise emitted by passenger cars has been reduced by 8 dB(A); this represents an 84% reduction. During the same period, the limit for heavy trucks has been reduced by 11dB(A); a 92% reduction. These changes are illustrated in *Figure 20.4;* the change of test procedure, which occurred in October 1984, was also effectively a noise reduction point as some noise reduction was required to achieved the same level measured to the new procedure. The exterior noise limits which were applied to other on-highway vehicles lie between the two curves shown. Details of the current test procedure and the appropriate limits are in the European Communities Council Directive 92/97/EEC. Similar legislative changes have taken place in some other countries and, where applicable, the appropriate legislation documents should be consulted.

There is still a desire to further reduce the exterior noise of on-highway vehicles, but at present the test procedures are under review. Following the substantial reductions achieved in vehicle noise, noted above, the measured levels are now tending to be dominated by tyre and road surface noise. There is some concern that the current procedures no longer adequately represent urban driving conditions and are not, therefore, appropriate for achieving lower environmental noise levels.

The vehicle refinements introduced to achieve the exterior noise reductions, discussed above, have contributed to substantial reductions of interior noise levels. The levels now experienced by drivers and passengers are normally well below any legislative limits set on the allowable exposure to noise. Interior noise signatures are influenced by 'market forces' and the manufacturers' desire to have some 'saleable' product discrimination. The refinement of vehicle interior noise now involves the use of 'Sound Quality' analysis techniques; see section 20.4.5, below.

20.3.2 Off-highway machines

Legislation relating to the exterior noise levels of off-highway machines has tended to lag behind that which has been applied to on-highway vehicles. Consequently, the sound power emitted by machines is typically much higher. As an example: in western Europe in 1997, the limit for the allowable 'drive-by' sound pressure level of a heavy truck fitted with a 150 kW engine,

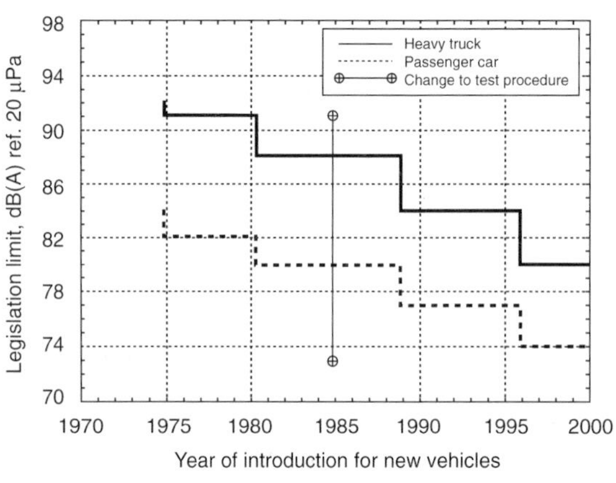

Figure 20.4 EEC legislative on-highway vehicle 'drive-by' noise limits

measured at a distance of 7.5 m, was 78 dB(A), whereas the sound pressure level of a typical earth-moving machine (an excavator/loader), with the same engine power and at the same measurement distance, was around 84 dB(A). This 6 dB higher level represents a four fold increase in sound power.

Since the late 1980s, only some types of off-highway machines have been the subject of noise legislation and the reductions in the allowable limits have been modest. However, this trend may be changing due to concerns with urban construction sites, the latest 'noise at work' regulations, etc. To illustrate this, the past and future limits for the sound power emitted by some earth moving machines are shown in *Figure 20.5*. The change to the test procedure, in January 1997, was the introduction of a dynamic test cycle which better represents the likely operating life of the machine. Although the values of the noise limits, before and after the change to a dynamic test cycle, are similar and in some instances higher, most machines had to be 'noise reduced' to meet the requirements. The 3 dB(A) reduction required in January 2002 is a halving of machine noise. Details of the test procedures and the appropriate limits for earth moving machines are in the European Communities Council Directive 95/27/EC. Other EC legislation applies to the majority of diesel engine applications.

To consolidate and extend the legislation relating to machines there is a 'Proposal for a European Parliament and Council Directive relating to the noise emissions by equipment used outdoors'. Similar legislative changes are taking place in some other countries and the appropriate documents should be consulted.

The allowable operator's exposure, or interior noise levels, is specified by legislation, such as the Noise at Work Directive 86/188/EEC; see also the Machinery Directive 89/392/EEC.

20.4 Measurement and analysis of noise

20.4.1 Measurement environments

In order to measure engine or machine noise with reasonable accuracy the acoustic characteristics of the measurement environment must be known. On an acoustic basis, the ideal environment is a space with no reflecting surfaces and no background noise. In practical terms the 'best' environment, for diesel engine applications, is an open-air site with one hard reflecting surface (the ground), no other obstructions for at least 50 m from the noise source and all the microphone positions, and background noise levels which are at least 10 dB (and preferably 20 dB) below those to be measured. In most other measurement environments it is necessary to apply correction

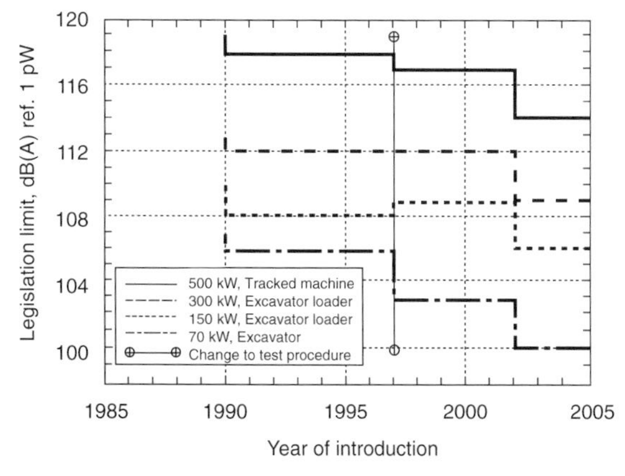

Figure 20.5 EC legislative noise limits for some off-highway machines

factors to the measured results in order to simulate those which would have been measured in this 'best' environment. It should also be clear that to obtain the correct results, the noise levels of ancillary equipment, such as a dynamometer, and any reflected noise will also have to be subtracted.

To avoid the effect of weather, the open-air environment can be simulated in a test cell which has acoustic absorption materials on the internal surfaces, usually in the form of wedges. When built with a hard reflecting floor this type of test environment is called 'semi-anechoic'; see *Figure 20.6*. The absorption must be effective down to frequencies below those which contribute significantly to the overall noise level. The lowest frequency which can be measured accurately is called the 'cut-off' frequency for the test cell. It is defined as the frequency at which the absorption coefficient for the test environment is 0.9 and this is a function of the cell dimensions and the properties of the sound absorbing materials. In practical terms, at this frequency and above, at least 90% of the sound energy is absorbed on the first contact with the cell walls. For diesel engine applications, a test cell with a 'cut-off' frequency of 200 Hz is usually adequate, although it may be desirable to measure down to 100 Hz in some cases. The minimum size of the test cell will be determined by the type of applications to be tested and the relative microphone positions, allowing for sufficient clearance from the walls. It is normal practice for all ancillary equipment to be heavily shielded to reduce their noise contributions or positioned outside the test environment.

Alternatively, engine noise measurements could be made in 'non-anechoic' test cells. However, this approach is generally non-preferred since it requires the use of correction factors and/or test procedures which are cumbersome to use for detailed assessment and development work involving measurements at many operating and/or build conditions. To summarize there are two types of 'non-anechoic' test cells:

(a) A specialized test cell with no absorption; this is known as a reverberant room. Here the noise builds up to a high level until the losses through the walls matches the acoustic output from the engine. Measurements made under these conditions require correction to correlate with open-air results. The correction is obtained by measuring the reverberation time of the room: that is the time taken for the noise level to fall by 60 dB when an acoustic source is suddenly (instantaneously) stopped. More details can be found in ISO 354, 3741, 3742 and 3743.

(b) Standard test cells which have some acoustic absorption; these fall between 'semi-anechoic' and reverberant. For tests involving sound pressure measurements, these cells have to be calibrated on each occasion either by the measurement of the reverberation time, or by the use of a calibrated sound power source; see ISO 354, 3741, 3742, 3743 and 3747. Alternatively, sound power levels can be obtained using acoustic intensity analysis equipment; see sections 20.4.2 and 20.5.4, below.

20.4.2 Equipement

The type of equipment available for the measurement and analysis of sound is diverse and continues to be developed at a pace. This section provides a brief overview, but no attempt is made to give detailed descriptions. It is, however, important that the equipment is appropriate for the job and that the user understands its accuracy and limitations; detailed specifications and user instructions should be obtained from the manufacturer. It is also important that calibration checks are done regularly; i.e. at least daily, and before and after each test sequence.

With the exception of surface vibration techniques, all noise measurements start with the microphone. The microphone contains a diaphragm, or sensor, which responds to pressure

Figure 20.6 A typical semi-anechoic test cell

fluctuations in the air and, together with power supplies and amplifiers, produces an equivalent electrical signal. This signal can then be recorded and/or analysed using a range of equipment including:

(a) The precision sound level meter. This is the most widely used noise measurement instrument and in its basic form it should be able to display linear (un-weighted) and weighted overall sound pressure levels in dB(lin) and dB(A). Some instruments will also provide readings of dB(B) and dB(C). Many sound level meters will also analyse the noise signature into one-third octave bands and some will store the narrow band spectrum; see section 20.4.3, below.
(b) The dose meter. This is a sound level meter which can integrate the noise signature in real time and display an equivalent continuous level. It is used to assess the human exposure level; see section 20.2.9, above.
(c) Signal analysers. There is an almost unlimited variation in the capabilities of the available signal analysers. Most are able to perform all the types of signal analysis discussed in sections 20.4.3 and 20.4.4 below, and often much more.
(d) Acoustic intensity systems. These employ an accurately matched pair of microphones and analysis channels to calculate sound intensity. They can be used to establish the noise levels of particular machines, or individual components, in non-ideal environments. They can be used to assess the radiated noise field and display the results as a contour map. An in-depth study of acoustic intensity and its uses is given in Reference 1.
(e) Tape recorders. FM and DAT recorders with a dynamic range of at least 60 dB and a frequency range of at least 10 kHz can be used to store noise signatures for subsequent analysis. If 'sound quality' characteristics are to be assessed, the recorder's dynamic range should be at least 80 dB and the frequency range should be up to 16 kHz or 20 kHz.

20.4.3 Frequency analysis

Simple overall sound pressure level measurements provide sufficient detail to assess a vehicle's or machine's performance against any legislation requirements, or to carry out basic noise exposure surveys. However, before noise reduction work can be undertaken, a more detailed understanding of the frequency content of any noise signatures is normally required.

For comparative test work, the most commonly used frequency information comes from one-third octave band analysis; this is considered to provide a good trade-off between detail and the quantity of data to be stored. Each one-third octave frequency band is known (labelled) by its centre frequency, or by a band number (an integer) which is ten times the logarithmic (base 10) value of the centre frequency: band 20 is at 100 Hz, band 30 is at 1000 Hz, band 33 is at 2000 Hz, etc. The centre frequency of each band is related to the centre frequencies of adjacent bands by a factor approximately equal to the cube root of two (1.26). Numerically, the standard centre frequency sequence is 1.25, 1.6, 2.0, 2.5, 3.15, 4.0, 5.0, 6.3, 8.0, and 10.0, and this sequence is repeated by multiplying the terms by powers of ten. It is normal to display one-third octave data as bands of equal width by having a logarithmic frequency scale. However, it should be remembered that the 100 Hz band includes all energy in the frequency range 89.1 to 112.2 Hz, while the 10 kHz band spans the range of 8913 to 11220 Hz, etc.

With the ever increasing analysis and storage capacity of signal analysers the use of 12th and 24th octave band displays is becoming more common. These simply have four or eight times the resolution of the equivalent one-third octave band analysis, and again with a logarithmic frequency scale they are displayed as being of equal width. Adjacent bands are related by factors equal to the 12th or 24th root of two.

To relate particular frequency characteristics to individual engine or machine components, more detail is required. This is obtained from 'narrow band' analysis, where the signal is analysed into a large number of bands all of which have the same frequency span. A basic analyser will typically have 400 or 800 line resolution. For example, a 400 line analyser, set to a frequency range of 8000 Hz, will display 400 frequency bands each with a resolution of 20 Hz; the 60 Hz band includes all energy in the frequency range 50 to 70 Hz, while the 7840 Hz band range is from 7830 to 7850 Hz. A comparison between a one-third octave and a narrow band frequency analysis of the same signal is shown on *Figure 20.7*. Powerful analysers provide higher resolution using more lines and/or 'zoom' facilities.

20.4.4 Tracking analysis

Tracking analysis is used to determine the nature of frequency peaks in a noise signature by observing how they change throughout the engine's operating speed range. It readily identifies the frequency peaks which change with engine speed; these are most likely to be generated by force mechanisms within the engine, or synchronized equipment such as transmissions, engine driven pumps, etc. It also highlights frequencies which do not change with speed, although amplitudes may rise and fall as speed-related forces pass through the frequency; these are likely to be associated with system resonances.

The data is commonly presented as 'waterfall' plots or 'Campbell' plots. 'Waterfall' plots are simply frequency spectra taken at small speed increments and stacked behind one another at spacings equal to the speed changes. By altering the orientation of the 3-D image that is generated, the constant frequencies show up as lines of peaks running parallel to the speed axis, while the speed-related frequencies appear as lines of peaks radiating out from the origin of the speed and frequency axes. 'Campbell' plots are simply a plan view of 'waterfall' plots where the frequency and speed axes are shown, and the amplitudes of the peaks in the original frequency spectra are represented by circles (or squares) of varying diameters (and usually different colours); only the higher peaks are shown. The lines of major peaks, discussed above, show up as distinctive lines of circles; an example is shown in *Figure 20.8*. Here many orders, including $2e$ (e = engine order) the firing frequency, and the oil pumping frequency and its 2nd harmonic ($8.85e$ and $17.7e$), can be seen together with a component resonance at about 325 Hz.

An alternative approach is to use 'order tracking' whereby the influence of a particular speed related frequency can be observed. Essentially, the analyser applies a narrow band filter, with frequency characteristics that change with engine speed, to the noise signature; the levels that are observed are due only to the speed-related frequency. In four-stroke engine applications, it is common to track the frequency of half engine speed (or $0.5e$) and many of its harmonics, particularly those of firing frequency and its harmonics. There may, however, be other speed-related frequency components which are not harmonics of half engine order, and tracking these will depend on the capabilities of the analyser and the source of the speed signal.

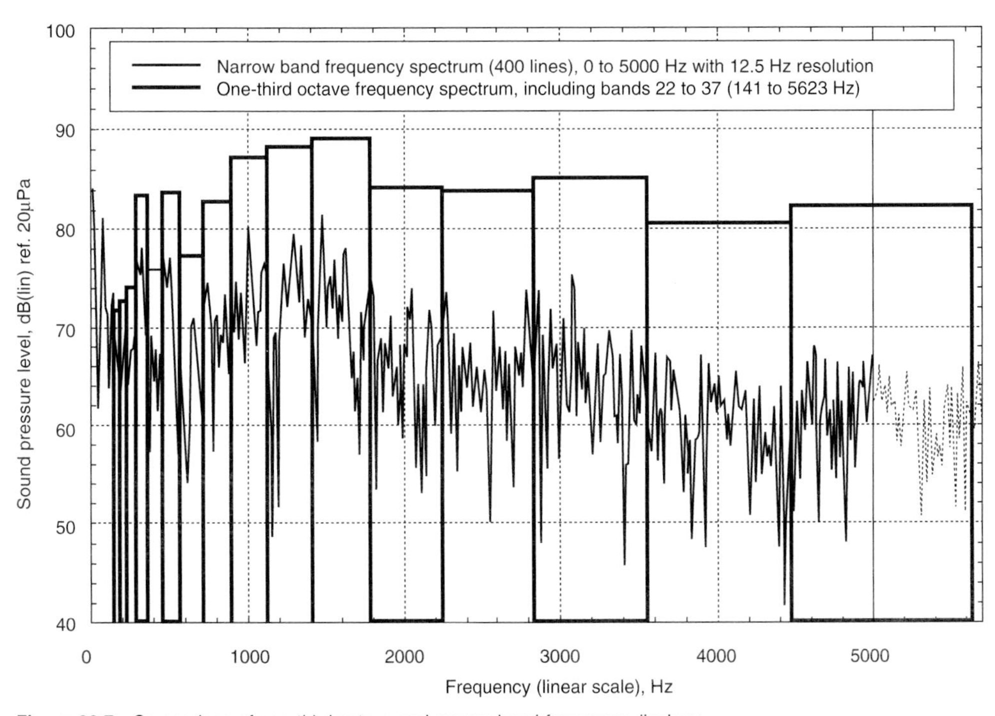

Figure 20.7 Comparison of one-third octave and narrow band frequency displays

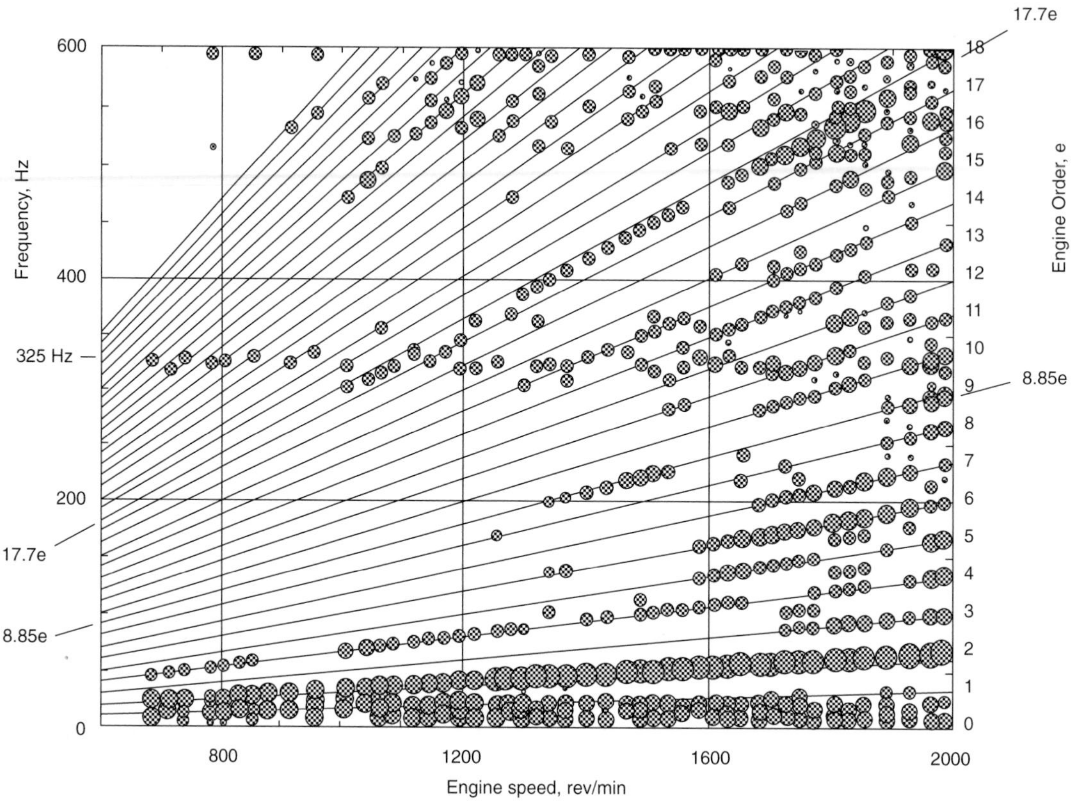

Figure 20.8 An example of a 'Campbell' plot

20.4.5 Sound quality analysis

Sound quality analysis includes relatively new techniques now available to engineers, made possible by the ever increasing power of computers at affordable prices. A detailed understanding of the component frequencies that influence the sound quality of an engine or installation facilitates better target setting. Sounds can be recorded on to the computer, modified in the frequency and/or time domains, and replayed as new sounds. By manipulation of the software and subjective evaluation of the modified sounds, problem frequencies or frequency combinations can be identified, although their source(s) may have to be determined by other techniques. Replaying computer modified sounds can also demonstrate the potential improvements before costly modification and verification work is undertaken. Apart from the subjective assessment of sound, there are several indicators which give an objective value to sound quality, which include: 'loudness', 'roughness' 'Kurtosis', 'speech interference' and 'intelligibility'. For further details of these measurement techniques the reader is referred to the available books on the subject of sound quality and signal analysis.

20.5 Noise characteristics of diesel engines

With respect to noise, the diesel engine is a very complex system of interacting dynamic forces acting on an equally complex structures of varying stiffness, damping and response characteristics. Similarly, the vehicles and machines in which diesel engines are installed are also complex structures and include many other noise sources. A simple model showing the major machine noise sources and their transmission paths to the environment is given in *Figure 20.9;* for on-highway vehicles other sources such as tyre and wind noise can be added to the model. It is intended that most of this section will be devoted to the noise radiated by the diesel engine, but for completeness there is a discussion on the other machine, or vehicle, sources in section 20.5.5, below. During the engine evaluation, it is recommended that only the base engine, i.e. the 'combustion' and 'engine mechanical systems', be considered; all other noise sources should be removed or silenced.

The most significant features which inherently make a diesel engine quiet should be included in the initial design. Once the engine's basic structure and layout have been fixed, the potential for subsequent cost effective noise reduction, through component modification and other palliative treatments, is limited. There are now available advanced computer software packages which can predict engine performance characteristics, dynamic loads, and component and structural responses. These should be used to develop and optimise the initial engine design. The same tools can also be used to improve existing engines and a description of them is included in section 20.6, below.

For an existing engine, the noise assessment programme discussed below is designed to establish the general noise characteristics with speed and load, the noise-generating mechanisms within the engine, and the dominant components which radiate the sound energy to the environment. The data that is acquired from the programme provides a good understanding of engine noise and the development team can then focus their efforts on the most significant issues. Engine noise measurements are normally carried out in a semi-anechoic test cell with an array of microphones from which sound power levels can be calculated; see ISO 3744.

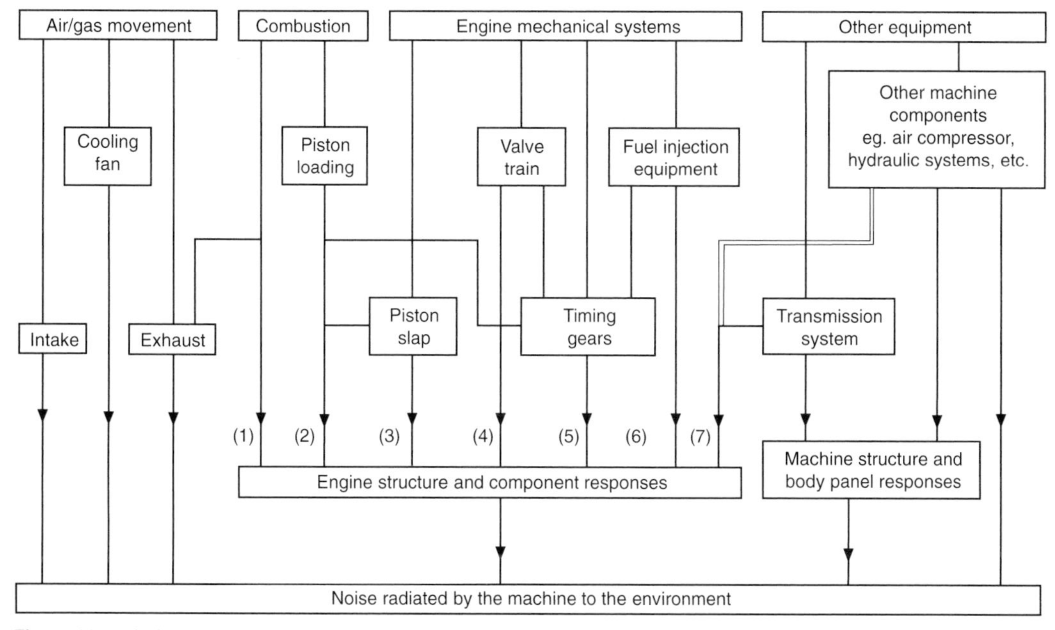

Figure 20.9 A simple model of machine radiated noise

20.5.1 Engine overall noise levels

The steady state noise levels of the engine, throughout its speed and load envelope, should be established at the start of any development programme; a grid matrix of up to ten speeds at five load conditions is normally adequate. All engine characteristics, particularly temperatures, and components such as crankshaft dampers, that may influence the noise levels, should be stable at each condition. One approach, to reduce the effects of any thermal lags, is to average noise levels measured during speed increasing steps and speed decreasing steps. It should also be confirmed, or otherwise, that the steady state test points are representative of all the possible operating conditions by recording the levels during slow transient sweeps through the speed range. The rate of speed change should not be more than 15 rev/min per second; i.e. a speed sweep from 1000 to 2500 rev/min should take at least 100 seconds. *Figure 20.10* shows a plot of steady state and transient noise levels. In the example the transient data shows the presence of a significant problem not identified by the initial steady state noise levels, and an additional test point at this speed/load condition should be included for all subsequent work.

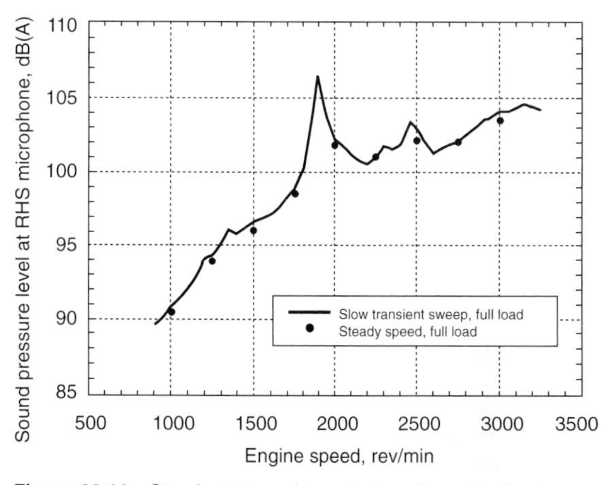

Figure 20.10 Steady state and transient engine noise levels

The overall engine noise levels should then be compared with any noise targets developed from installation appraisals (see 20.5.5 below), market requirements, and/or competitor engine data, to establish the desired noise reductions. Fundamentally, the best approach to engine noise improvement is to reduce the levels 'at source' and this is normally required for larger noise reductions (i.e. of more than 2 or 3 dB). It will involve development of the combustion system and/or the mechanical systems associated with pistons, timing gears, etc. An alternative approach where smaller reductions are required, particularly for mature products, involves increasing the effective structure attenuation with minor block modifications, isolation of external components, the addition of shields, etc. These simply reduce the amount of sound energy that is radiated to the environment; the internal generators are unchanged.

20.5.2 Assessment of combustion noise

The proposed approach for understanding the mechanisms of noise generation within the engine is to first establish the noise associated with combustion (i.e. the in-cylinder 'bang', or pressure change characteristics); see the engine structure input (1) in *Figure 20.9*. Initially, the combustion noise levels of an engine can be estimated using a 'combustion noise' meter. The meter incorporates a filter system that represents the attenuation due to the structure of a 'typical' engine, and gain settings to give the same average noise levels as would be measured with a 'standard' microphone array. It modifies the in-cylinder noise signature to indicate the exterior noise level associated with combustion at an average microphone position. If the indicated levels are particularly small when compared with the overall engine noise (i.e. at least 12 dB lower), further evaluation work may not be required. However, for most development programmes an understanding of the engine's actual structure attenuation and combustion noise signatures is very useful. The technique offered in this text to establish these parameters aims to make changes to the in-cylinder noise, while having only minimal effect on the mechanical noise sources, including the effects of peak cylinder pressure (also see Reference 2). In-cylinder noise can be changed by changing the induction air temperature, by using fuels with a range of cetane number, and/or by changing the fuel injection timing. Timing changes should be limited to ± 5 degrees to avoid having a significant effect on mechanical noise. The two assumptions made are first, that the mechanical noise remains constant and secondly, that the structural response to in-cylinder noise changes is linear. When these two assumptions hold, the following equation can be written:

$$SP = m \cdot CP + a$$

where

SP = there energy related value of the sound pressure level; either for a particular microphone or, more commonly, for the average level from around the engine,

CP = the energy related value of the in-cylinder pressure level; it is usual to monitor only one cylinder, but several may be averaged together,

a = the y-axis intercept and is the energy related value of the mechanical noise (i.e. when there is no combustion noise).

m = the slope of the straight line; the inverse of 'm' represents the engine's structure attenuation.

The in-cylinder pressure signature is measured in decibels, with a standard reference of 20 μPa as used for sound pressure level, using a high temperature pressure sensor designed for use in internal combustion engines. It is normal practice to make at least five changes to the engine to alter the in-cylinder noise levels, and to test at several engine operating conditions; e.g. at peak power, peak torque or any other operating condition of interest. The external engine sound pressure levels and the cylinder pressure signatures are analysed into one-third octave frequency spectra. For each one-third octave band the sound pressure and cylinder pressure levels are converted to energy related terms using

$$x = L_n/10$$
$$e_n = 10^x$$

where

L_n = the measured sound pressure level (spl) or cylinder pressure level (cpl),

e_n = the energy related term.

Hence:

$$SP = [(p_{rms}/p_{ref})^2]_{spl}$$
$$CP = [(p_{rms}/p_{ref})^2]_{cpl}$$

For each one-third octave frequency band the data from the above terms is plotted, on an x–y graph, and the values of 'a' and 'm' can be established; a sample plot is shown on *Figure 20.11*.

The total mechanical noise, the engine's structure attenuation and the combustion noise heard outside the engine can then be calculated for each one-third octave frequency band, as follow:

$$\text{Mechanical noise, dB} = 10 \cdot \log_{10}(a)$$
$$\text{Structure attenuation, dB} = 10 \cdot \log_{10}(1/m)$$
$$\text{Combustion noise, dB} = 10 \cdot \log_{10}(b - a)$$

or

$$\text{Combustion noise, dB} = \text{In-cylinder pressure level (dB)} - \text{Structure attenuation (dB)}$$

where

b = the value of the SP term at the normal operating conditions.

Sample data for the four calculations described above is also shown of *Figure 20.11*.

The structure attenuation for a fairly typical engine is shown on *Figure 20.12*. However, the attenuation curves for other engines can be significantly different; ± 10 dB in any frequency band is not unusual. Structure attenuation is a function of the cylinder

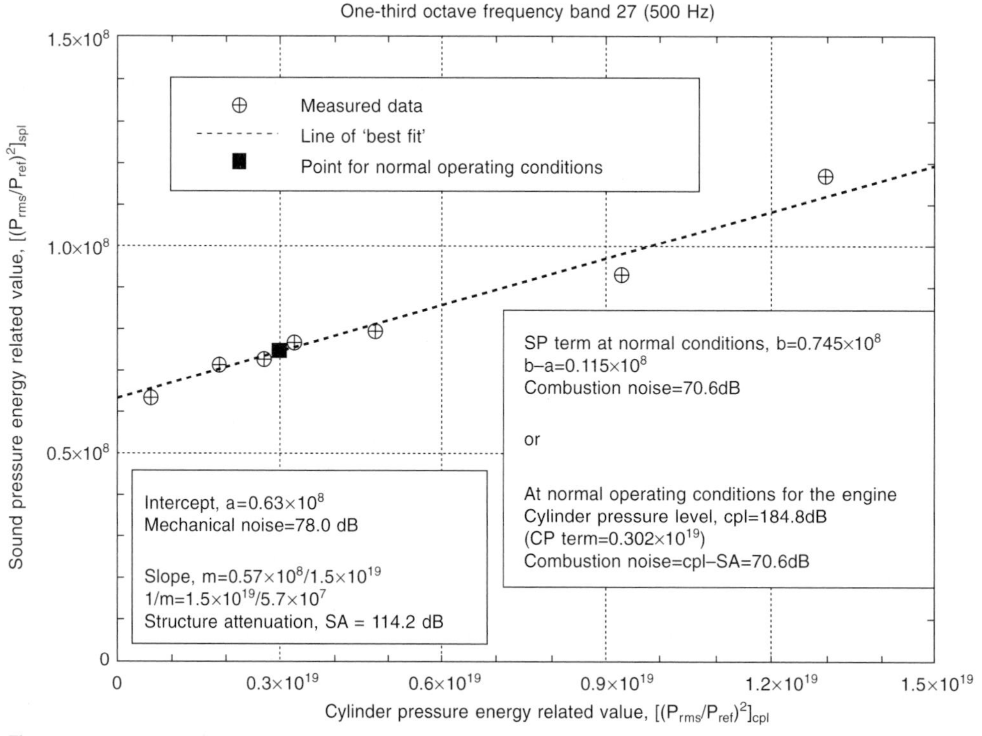

Figure 20.11 An Example of a plot of overall sound versus cylinder pressure data

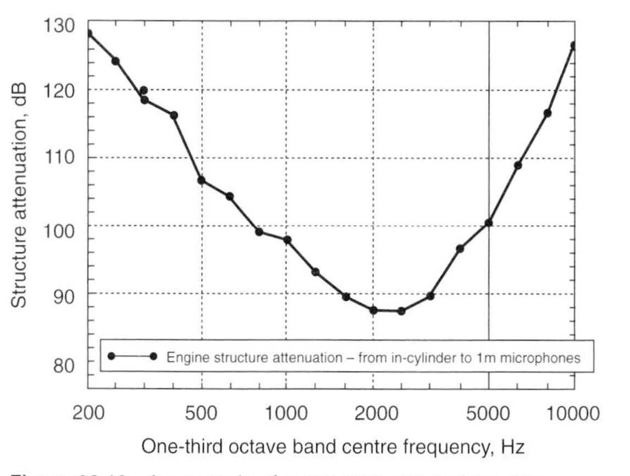

Figure 20.12 An example of a structure attenuation curve

block design and the response characteristics of the major external components and it can differ widely from engine to engine, particularly where some noise reduction treatments are already fitted.

20.5.3 Assessment of mechanical noise

There are no simple approaches to establishing the contribution from each of the major mechanical systems of an engine; see the engine structure inputs (2) to (6) and possibly (7) on *Figure 20.9*. This clearly shows the interactions involved and any changes to one will, to some extent, affect the others. At best, any test work can only produce an understanding of the relative importance of the contributions from each of these systems by skilful interpretation of the noise characteristics of any changes. The following test techniques have proved useful in assessing trends and identifying areas for concern:

(1) Component changes, to modify timing gear backlashes, pistons and tappets clearances, etc. Changing the engine build to introduce components which have been selected or modified to achieve maximum, mean and minimum clearances, can show the sensitivity of engine noise to particular parameters. It is important, but sometimes difficult to ensure, that all other noise sources remain unchanged. Often the biggest sources of errors come from assembly variations of other components which have to be disturbed to make the desired change. It is also important to monitor combustion noise to ensure that either any mechanical changes have not affected the in-cylinder characteristics or that, if they have, the differences can be removed from any assessment of the component changes.

(2) 'Motoring' of the engine–that is, driving the engine at the appropriate speed without any fuel delivery or combustion. The measured noise signature is the inherent 'rotational noise' characteristic of the relatively **cold** engine and on its own is of little value. The valve train noise should essentially be the same as it would be on the normal running engine. The piston loading is lower without forces from combustion. However, the piston/bore temperatures are also much lower and their clearances will probably be higher, therefore piston slap noise is still likely to be significant and, in some cases can be higher than in the running engine. The loads on the fuel injection equipment are small and the noise contribution should be low. The contribution from the timing gears is the most variable factor. Without the high torque fluctuations from the crankshaft and fuel system of the loaded engine, gear noise is expected to be substantially lower but this may not always be the case.

(3) 'Motoring' of the engine and loading of the fuel system–that is, driving the engine without combustion but with the fuel system delivering fuel through a set of slave injectors, fitted in a fuel collection rail. It is important that the fuel system sees the correct loads. Therefore, fuel quantity, injector nozzle characteristics and pipe lengths must, where possible, be the same. The difference between the measured noise signature during this test and the engine 'rotational noise' from the 'motoring' test above, gives a good indication of the noise associated with the fuel system. For the two tests, the piston dynamics and valve train noise should not have altered significantly, but the lack of torsional activity from the crankshaft gear may affect the absolute levels from the fuel system.

(4) An evaluation of noise and vibration frequency spectra from the engine, at various load and speed conditions, can indicate the likely noise generating mechanisms. However, this requires considerable experience and the reader is referred to other texts, such as chapter 19 of Reference 7, which consider this difficult topic more rigorously.

Figure 20.13 shows an example of the type of results which can be obtained from the above techniques. It clearly identified that the backlash in the timing gear meshes was an important factor, particularly as engine speed was reduced, at the part load conditions. Other results from the assessment of fuel system noise (item (3) above) inferred that gear noise was a major factor at the higher speeds and loads, for reasons other than backlash. Clearly, in this example, refinement of the timing gear system, i.e. the load characteristics, the geometry and the response of the supporting structure and covers, is essential if substantial reductions in overall engine noise are to be achieved.

20.5.4 Engine radiated noise

The combustion and mechanical systems, discussed above, are the internal generators of structural vibrations which can be transmitted to the surrounding environment as noise, but the external components and covers are of equal importance in determining the characteristics and levels of the actual engine radiated noise signature. Therefore, a detailed understanding of the noise levels radiated by the components is essential before cost effective noise reduction treatments can be developed. Generally, there are three techniques used for determining component noise levels: direct measurement using acoustic intensity analysis; noise source elimination by heavy shielding or removal of the component; and calculation from surface vibration measurements. All three techniques have advantages and disadvantages:

(1) Acoustic intensity survey methods are relatively quick but to monitor the frequency range of 50 Hz to 10 KHz, two or three measurements with different microphone spacings and sizes may be required. The microphone carrier is relatively large and can be difficult to manoeuvre between components to measure individual sources. Alternatively, noise 'contour' plots of the whole engine can be obtained, but it may be difficult to discriminate between adjacent sources and identify each component's noise level. Acoustic intensity measurements, however, are the best method for assessing moving components, such as pulleys, etc.

(2) The heavy shielding technique for whole engine noise evaluation has generally been replaced by the other methods. Wrapping the engine in lead sheet plus absorption material, to reduce the overall noise by at least 10 dB, and then to successively expose small areas to establish the noise level of each component did give reliable results. However, when compared with the other techniques, it is very tedious, time consuming and costly. The shielding or removal of individual

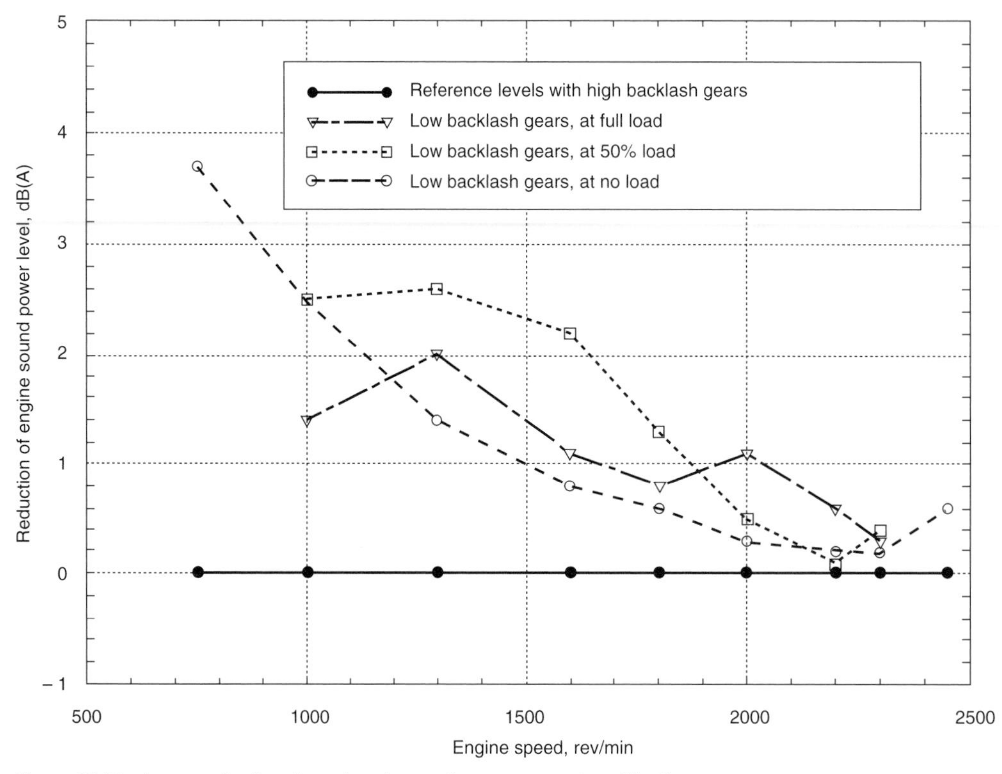

Figure 20.13 An example of engine noise changes from component modifications

components can, however, be used to confirm, or otherwise, unexpectedly high results. Absorption and barrier materials can also be used to eliminate cavity resonances which sometimes occur between parallel surfaces: e.g. between the rear of a crankshaft damper and the front of the timing case.

(3) Noise levels calculated from surface vibration measurements are discrete to the panel area being monitored, and are not normally affected by other noise sources. Generally, five or six measurements distributed at random, but not too close to the edges of the radiating surface, are sufficient. However, for a full engine survey the technique requires all the components to be partitioned into about 60 or 70 individual panels and this requires the measurement of surface vibration at 300 to 400 positions.

This technique uses the equation below, which contains a radiation efficiency term σ_{rad}. This is a measure of the efficiency of noise radiation from the component surface to the air and is sometimes called the radiation ratio, since it can be larger than unity. Where possible the appropriate radiation efficiency curves should be used, but for some irregular shaped engine panels accurate values may not be readily available. In such cases the curve shown on *Figure 20.14* is normally adequate. See References 3, 4 and 5.

$$\text{Sound Power, watts} = \rho c \cdot A \cdot \sigma_{rad} \cdot \langle v^2 \rangle_T$$
$$\text{Sound power level, dB} = 10 \cdot \log_{10}(\rho c/W_{ref}) + 10 \cdot \log_{10}(A) + 10 \cdot \log_{10}(\sigma_{rad}) + 10 \cdot \log_{10}(\langle v^2 \rangle_T)$$

where

ρc = characteristic impedance of the air around the engine, kg/m^{2}s
(ρ = density of air, c = acoustic velocity)
(at 20°C, ρc = 413 kg/m^{2}s; at 60°C, ρc = 388 kg/m^{2}s).
A = surface area of engine panel, m^2
σ_{rad} = radiation efficiency
$\langle v^2 \rangle_T$ = space and time averaged squared surface velocity, (m/s)^2
W_{ref} = the standard reference power of 1×10^{-12} watts (1 pW).

There may also be a requirement to correct the measured vibration signatures, to account for the effect of the mass of the transducers. A correction factor can be obtained from equations relating to plate vibration; see References 6 and 7. Since the average impedance of many vibration modes approaches a pure resistance for wideband excitation, to a first approximation the correction factor can be taken as:

$$\text{Correction Factor} = 1 + \{[3\pi^2 \cdot f^2 \cdot m^2 \cdot (1 - v^2)]/[4 \cdot \rho \cdot E \cdot t_m^4]\}$$

where

f = one-third octave centre frequency, Hz
m = mass of vibration transducers, kg
v = Poisson's ratio of the component's material
ρ = density of the component's material, kg/m^3
E = Young's modulus of the component's material, GN/m^2
t_m = mean thickness of the panel, m

When all of these techniques are used, as appropriate, a comprehensive understanding of the engine external component noise levels can be achieved. It is strongly recommended that the total engine noise levels obtained by these techniques are compared with overall engine sound power levels obtained by another method (e.g. from standard microphone tests; see ISO3744), so that any errors tend to be highlighted rather than masked. A good correlation between the overall engine sound power level and the total sound power of all the components would give results differing by less than 1 dB (A) overall and by less than 3 dB in any one-third octave band; see *Figure 20.15a*. If this level of correlation is not achieved, there are

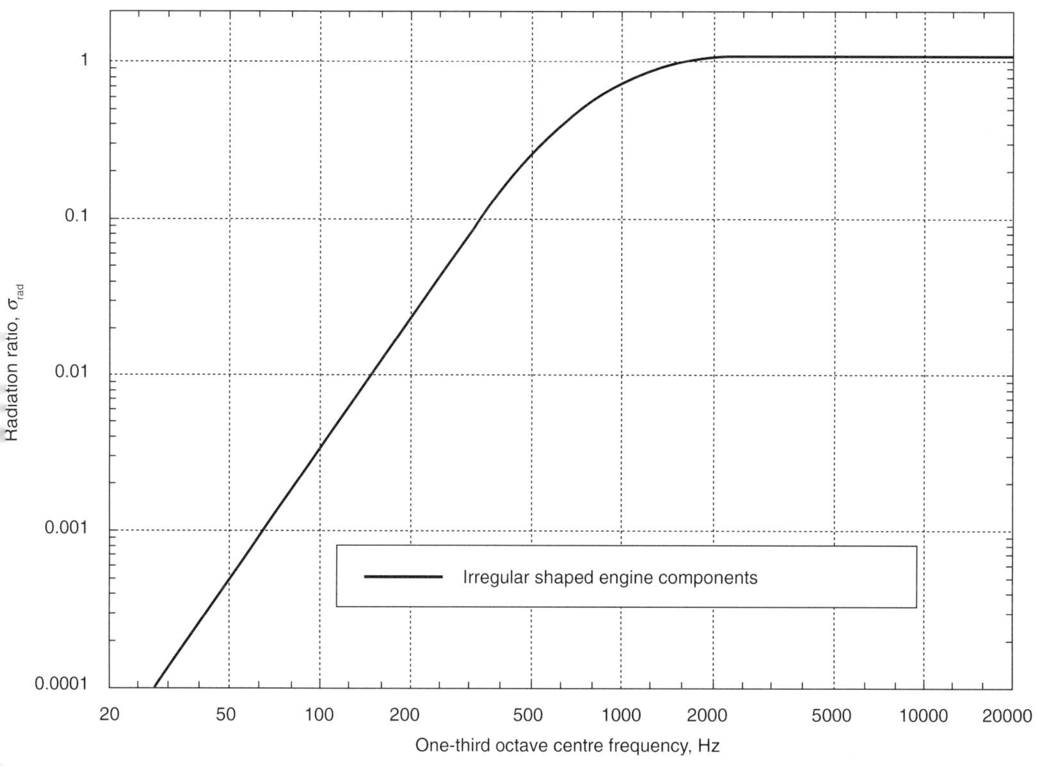

Figure 20.14 Radiation efficiency curve for irregular engine components

either significant errors in the data or some source(s) has not been included. If the overall correlation is not achieved, the following points should be reconsidered: the positions and calibration of the vibration transducers; the material data and dimensions of each panel; the radiation efficiency curves used; the possibility of other significant noise sources. If the correlation is poor in only one or two frequency bands, the results may be indicating the presence of a cavity resonance. A sample of typical results is shown in *Figure 20.15a* and *20.15b*.

20.5.5 Vehicle and machine noise assessment

The induction, exhaust and cooling systems are usually the most significant noise generators associated with air or gas movement in vehicles and machines, but other components such as the alternator fan should also be considered; see *Figure 20.9*. Gas-borne noise in the intake and exhaust systems' tends to be dominated by the lower harmonics of firing frequency, caused by pulses generated each time a valve opens or closes. Although the gas-borne noise is usually the largest contributor from these systems, their structural responses can also be significant in many cases. The noise refinement of intake and exhaust systems is usually undertaken by the suppliers using specific predictive software. As changes to these systems can alter engine performance, their effects should be monitored, so that the system parameters remain within the ranges set by the engine manufacturer to ensure that gases emissions limits are not exceeded.

Refinement of noise from the cooling system fan is also a specialized subject and will not be dealt with in detail here. Basically, fan noise is a function of speed (approximately to the power five), diameter (approximately to the power seven), the number and shape of the blades, and the cowl and radiator design. Disturbances (obstructions: brackets, badges, other components, etc.), either upstream or downstream, can have a great influence on the fan noise characteristics.

Other equipment such as the transmission system, air compressor, hydraulic pump and pipes, body panels and tyres, can make significant contributions to the overall noise level, due to their own noise signature, and the response characteristics of the vehicle or machine structures to which they are attached.

Vehicle or machine noise levels may have to be reduced for several reasons, including operator/passenger noise exposure levels, legislation limitations, and/or general marketing (image) requirements. Ideally, assessment of the noise levels should be carried out at the critical operating conditions, with respect to one or more of these. However, this may be very difficult if the vehicle or machine is moving at the time, and alternative stationary conditions, which are reasonably representative, may need to be identified for at least some of the test programme outlined below. Probably the most important parameter which should not be changed is engine speed, because it controls the frequency content of the noise signatures from almost all vehicle and machine components.

The noise assessment programme, outlined below, is designed to establish the general airborne noise contributions from the major components by using relatively simple noise source reduction and elimination techniques. For the required operating condition(s), the change in the noise levels at each observation point should be recorded as each of the treatments given below is applied to the components consecutively, on a cumulative basis. It is essential to monitor one change at a time.

(1) The air intake and exhaust systems should be piped to additional remote silencers, either vehicle or machine mounted, or preferably positioned several metres away. Care has to be taken to ensure that the induction depression and exhaust back pressures remain similar, since large

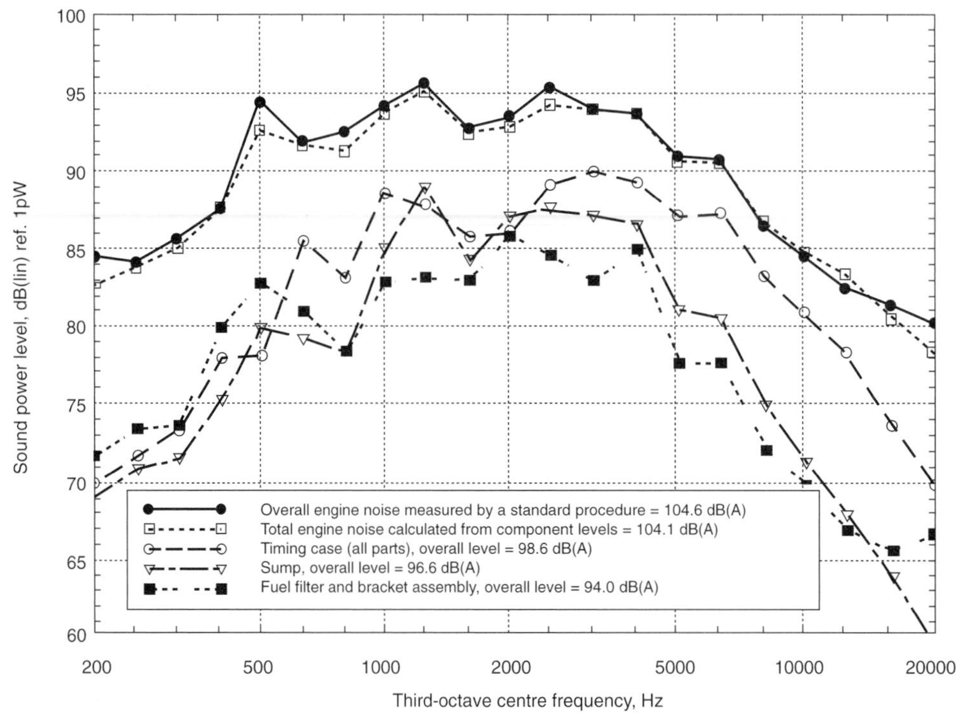

Figure 20.15a Correlation of overall noise levels and component spectra

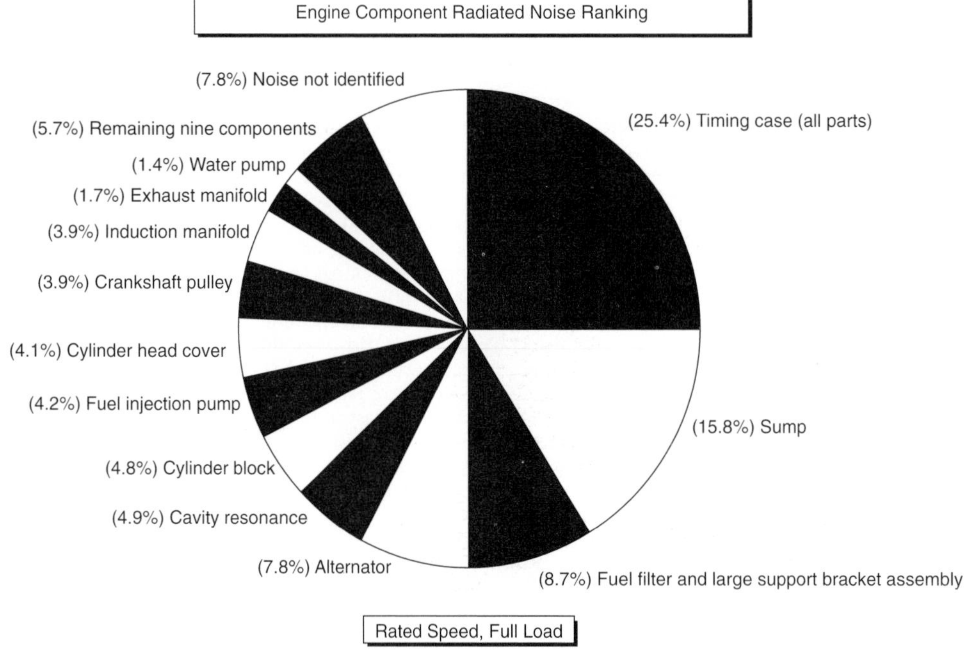

Figure 20.15b Relative noise contributions from the external engine components

changes will affect engine noise, by changing engine performance. The additional silencers and connecting pipework should also be wrapped (see (2) below) so as not to add to the noise signature.

(2) The pipework and silencer(s) of the existing exhaust system should be wrapped with glass fibre bandage or a similar, non-flammable, sound- and vibration-absorbing material. It is normal to include all the system except for the engine-mounted manifold.

(3) Components like the engine (or parts of it), the transmission housing, air intake filter, hydraulic pump, etc., should be wrapped with sound-absorbing materials and a heavy barrier layer; typically, foam and lead sheet are used.

(4) Body, panels should be shielded, isolated from the machine

structure, or fitted with vibration damping materials. They should not be removed as they are likely to be providing some shielding affects for other noise sources.

(5) Some component noise levels are best eliminated by removing them from the vehicle or machine. This is particularly true of the cooling fan, but precautions must be taken to ensure that engine temperatures do not exceed permitted limits.

(6) In the case of vehicles moving at relatively high speed, the approximate levels associated with tyre and wind noise can be othained during a 'coasting' test. It is necessary to complete the 'coasting' test over the same speed range, albeit with the speed decreasing rather than increasing. However, some aspects of tyre noise will not be replicated by this method, notably those associated with tyre excitation due to the forces encountered during vehicle acceleration; see Reference 8.

When the above treatments have been applied to all the major vehicle or machine airborne noise sources the overall levels should be reduced by 8 to 10 dB(A). If this is not achieved there are other significant sources and, in the case of interior noise, these could well be structure borne. It is important, however, to be sure that none of the noise reduction treatments already applied are causing significant noise levels. By calculating the difference in noise levels, before and after each component treatment, the airborne contributions can be established. A typical set of results from a machine are shown in *Figure 20.16*; here the top, sides and rear of the engine were enclosed by body panels. Normally, similar pie charts would be generated for measurements at other observation points (e.g. the operator's ear) and/or other operating conditions. To achieve a 3 dB(A) reduction in the overall noise, the area of the pie chart has to be halved, and the optimum combination of component refinements will be dependent on the relative contributions from each set of measurements. If structure borne noise is considered to be a major contributor, other analysis techniques including 'route tracking' need to be used to identify the energy inputs from each contact point between the vehicle body and the noise generating systems. Consideration needs to be given to the powertrain (engine and gearbox assembly), intake and exhaust system, gear linkages, pipework, etc.

An alternative approach for investigating the external noise characteristics of a stationary vehicle or machine is to use an acoustic intensity measuring system to generate 'contour' plots of the radiated noise. When measured close to the installation, these plots can clearly show the areas which are radiating the highest noise levels, and consequently the likely source(s) of the sound energy. The technique is relatively quick but is normally only used for stationary operating modes; the investigation of tyre noise in Reference 8 is a notable exception.

20.6 Methods for control of diesel engine noise

The techniques for assessing the noise characteristics of the diesel engine, discussed in 20.5 above, should identify the areas which require improvement. As with overall vehicle or machine noise, to achieve a 3 dB(A) reduction in overall engine noise the area of the pie chart, shown in *Figure 20.15b*, has to be halved. This can be done either by reducing its diameter (i.e. reducing the inputs from the combustion and mechanical noise generators) or by reducing the size of the major slices (i.e. reducing the noise radiated by the external components). The approaches which are available are outlined below, but the execution will depend on many factors including, the required improvements, legislation limits, cost and time scales. Generally, the best approach is to reduce the noise generating mechanisms and this principle should be applied to all new engine designs, but it may be too costly for small improvements on mature products.

20.6.1 Combustion noise

With the need to meet ever more stringent gaseous emissions legislation, it is not realistic to change combustion noise without a substantial amount of performance development work. On

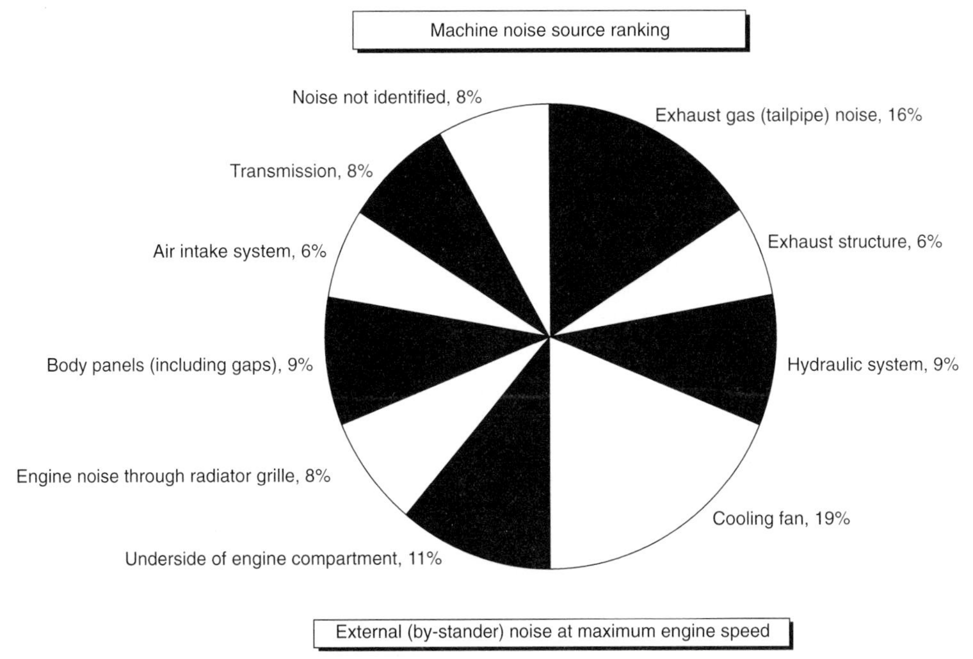

Figure 20.16 Relative noise contributions from machine components

modern diesel engines combustion noise is unlikely to be a major contributor to the overall noise level, since the emissions legislation has generally produced substantial noise reductions. It has encouraged the development of combustion systems that have relatively short ignition delay periods, and hence has reduced the 'maximum rate of pressure rise' which controls in-cylinder noise levels. Older designs of the naturally aspirated direct injection (D.I.) diesel engine were dominated by combustion noise which typically contributed around 70% of engine noise; today, this does not need to be the case.

The problem of combustion noise has not disappeared completely. The relatively low in-cylinder noise levels of the modern diesel engine can easily be increased 'accidentally'. *Figure 20.17a* is one example showing the effect of simply increasing the fuel injector nozzle opening pressure by 5%; engine noise increased at all operating conditions but the change was most dramatic during no load. With the advent of electronically controlled fuel injection systems it is possible to optimize the fuel delivery characteristics for every engine operating condition. The electronically controlled fuel system can reduce noise levels but, if it is optimized for performance (power and economy) and gaseous emissions without due attention to in-cylinder noise, combustion noise problems can recur. *Figure 20.17b* compares the noise levels of two engines, which were of the same base design with many common components but fitted with alternative fuel systems. In this example, the new fuel system reduced noise levels at full load above 1500 rev/min but increased engine noise at most other operating conditions by up to 6 dB(A). In both of the above examples, in-cylinder noise changes accounted for more than half of the noise increases; the remainder were mainly due to increased loads and torsional activity on the timing gears. It is, therefore, strongly recommended that during any combustion system development programme in-cylinder noise is one of the main parameters that is monitored and optimized for **all** operating conditions.

It is now possible to carry out a substantial part of the combustion system and in-cylinder noise optimization using computer software. With regard to combustion noise (i.e. the in-cylinder 'bang'), the important feature is the frequency content of the in-cylinder pressure signature and, in particular, the higher frequency content above 500 Hz; see References 9 and 10. This is a function of the rapid initial combustion which can produce dramatic increases in the in-cylinder pressure. Simulation of engine breathing and combustion is an important part of the development process used by engine manufacturers in achieving legislative requirements relating to gaseous emissions. The application of the technology to noise centres around understanding the influence of the parameters which affect combustion and the subsequent in-cylinder pressure signature. In diesel engines the prediction of port and in-cylinder gas flows using computational fluid dynamics (CFD) provides an understanding of air management and combustion chamber shape. The fuel injection, including injector nozzle configurations and injection rate, can be investigated using specialist adaptations of the CFD calculations to assess the effect of the spray atomization and interaction with the in-cylinder gas flows. Finally, simulation techniques are available (see Reference 13) to predict the compression ignition of the fuel and heat release allowing the calculation of the in-cylinder pressure signature. The frequency content of this signature can then be related to the temporal events within the combustion event. Combustion and its characteristics will fundamentally be controlled by the demand for lower emissions but simultaneous consideration of the effect on the in-cylinder pressure frequency content will identify beneficial developments for noise reduction.

20.6.2 Mechanical noise

The 'mechnical noise' of an engine has many interrelated sources as illustrated in *Figure 20.9*, above. The main sub-systems which need to be considered for noise reduction include:

(1) Those parts involved in the primary function of the engine, namely the production of usable power. This system starts at the top of the piston, where gas pressures generate the forces that move the piston down the cylinder bore and it finishes beyond the vehicle's or machine's transmission system. It involves the dynamic characteristics of the pistons, connecting rods, crankshaft, damper (if fitted), flywheel, components of the transmission, etc. and is influenced by the reactions of bearings and the whole engine supporting structure.
(2) Aspects of piston motion and bore distortion which arise from the geometry of the engine and the need to have finite clearances between moving surfaces. These control the piston to liner impacts known as 'piston slap'.
(3) The timing gears, chains or belts. Their primary function is to synchronize the events which take place within the engine to produce output power but they are often used as a secondary power take-off for internal oil and water pumps, and for driving external equipment. The clearances necessary for manufacturing tolerances and to ensure long-term service, the response characteristics of the supporting structures, and the varying loads from the crankshaft, camshaft and fuel pump are the major factors in noise generation.
(4) The valve operating system from the cam lobe and follower to the valve and seat impacts.
(5) Any engine mounted (internal or external) ancillary equipment: oil pump, turbo-charger, alternator, compressor, etc.

The development of these systems, in a cost effective manner while meeting the need for ever shorter time scales, is a complex process which needs to integrate practical experience, test data (if available) and the latest predictive software. It is not yet possible to accurately predict the absolute noise level of an engine. The size of an engine model which includes all components, defined with meshes that are sufficiently fine to allow the calculation of model fequencies up to 10 kHz, is prohibitively large. In addition, the damping characteristics of many parts of the model, particularly relating to joints, is not easily represented. However, predictive techniques are now an essential part of any development programme and should be used to optimize a design with respect to performance, strength, weight, noise, cost, etc.

To develop a quiet engine, predictive analysis can be used to minimize the input forces to both the main cylinder block structure and to the attached components. By reducing the internal metal-to-metal impacts, associated with piston, gears, etc., the higher frequency (above 500 Hz) energy which dominates engine noise will be reduced. In addition, it is necessary to optimize the cylinder block structure to reduce the response characteristics of both its external surfaces and the attachment interfaces with the larger components such as the sump, timing case, manifolds and covers. Finally, the structural response characteristics of the major components can also be investigated and improved using predictive techniques. Although absolute values of radiated noise cannot be calculated, the outlined approach can be used to optimize trends that will lead to low noise designs, by comparative assessments.

20.6.3 Predictive analysis

Some of the available predictive techniques for engine development and their noise refinement capabilities are outlined below.

(1) The development of the combustion system using predictive techniques to study gas flows, fuel atomization and the combustion event, and to reduce the in-cylinder noise levels has been outlined in 20.6.1, above.
(2) Combustion loads acting on the flame face excite an effect-

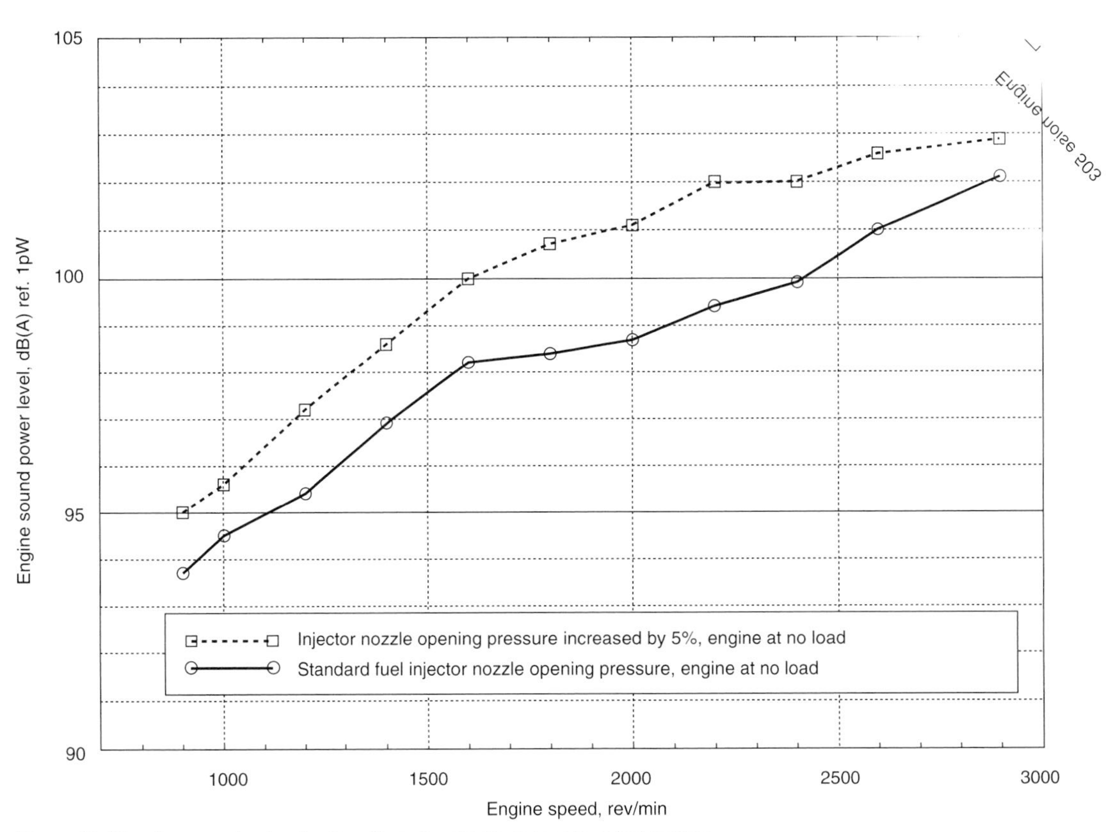

Figure 20.17a An example showing the effect of a simple change to a fuel system

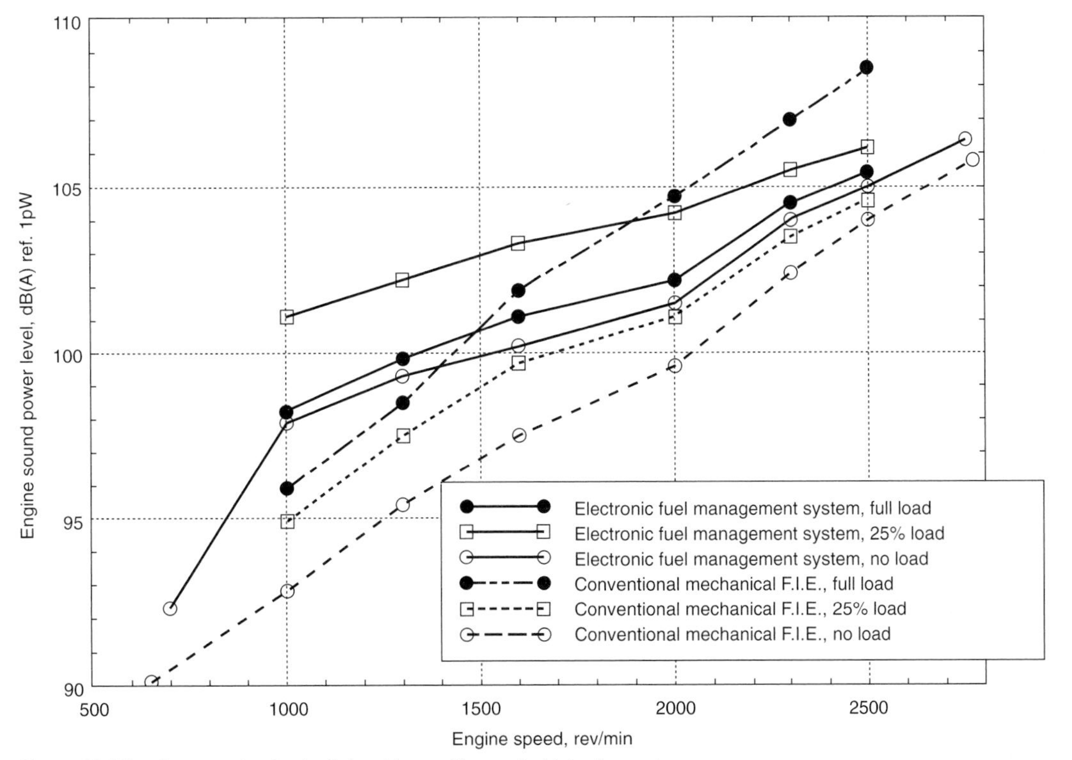

Figure 20.17b An example of potential problems with new fuel injection systems

ively inert region of the engine's cylinder head. However, the same loads transmitted into the crankcase via the piston and connecting rod have a significant effect due to the relatively flexible and highly resonant nature of the crankshaft and crankcase system. The noise generated by an engine is therefore significantly affected by the design of the crankcase and how the structure transfers vibration to other attached engine components. To provide an accurate prediction of the main bearing loads the simulation method must include all of the important characteristics of the crankshaft system. The crankshaft assembly should include an accurate representation of the crankshaft geometry and the dynamics of the vibration damper and flywheel. The main bearing hydrodynamic behaviour is very complex and, at small oil film thicknesses, is influenced by the local flexibility of the main bearing housing. This is particularly true where the tilting of the crankshaft journal within the bearing leads to edge loading, which is a function of the crankcase structural design and its stiffness in relation to the crankshaft. The most accurate representation of the system is currently made using specific application software which couples the structural components via an elasto-hydrodynamic (EHD) bearing in a transient solution over the engine cycle. The degree of sophistication required is dependent upon the problem under investigation. Evaluation of component and sub-system dynamics has been successful in understanding and resolving specific noise issues; see Reference 11. However, to ensure that the system response is correctly characterized requires that the full system is evaluated; see Reference 12.

(3) The motion of the piston is very complex due to the requirement for a finite clearance between the piston and cylinder walls. The piston adopts an operating shape as a result of its machined profiles and the distortion caused by the temperature gradient from the transmission of the heat energy from combustion. The cylinder barrel also experiences distortions due to temperature gradients and is further distorted due to cylinder head bolt loads transmitted through the highly non-linear gasket. Piston motion and impacts, due to rotation about the gudgeon pin and non-thrust to thrust side movement across the clearances, are further complicated by the oil film behaviour and the surface finish from the bore honing process. Prediction of the distortion of engine components under assembly and thermal operating loads is well documented. Simulation of the kinematic motion of the deformed piston within the distorted bore allows the development of a cold piston shape which will maintain the necessary minimum operating clearances over the whole engine cycle; see Reference 13. Assessment of the stability of the piston, with the predictive software, allows the optimization of, for example, the gudgeon pin offset. Development of computer code which includes the oil film behaviour and component flexibility allows the prediction of the impact forces and thus provides a direct indication of the loading that the engine structure experiences. The piston structural design can then be optimized to reduce overall engine noise by reducing the piston impacts and minimizing the engine's structural responses.

(4) Gear noise can be subdivided into rattle and whine. Gear rattle is a function of the clearances within the tooth mesh. If the torsional vibration within the system exceeds the mean torque being transmitted the result will be tooth separation and, consequently, tooth impacts, Gear rattle occurs in engine timing drives and gearboxes. An analysis for the onset of gear rattle can be made by assuming that the system is linear and by ignoring clearances. When cyclicly varying loads in the reverse direction exceed the mean transmitted torque, tooth separation will have occurred and rattle will result. This is not a satisfactory approach in most rattle conditions where continuous tooth mating is not easily achieved and the degree of rattle causes the unacceptable nature of the resulting noise. Methodologies are under development to perform non-linear simulations including the clearance effects. However, the impulsive nature of the rattle event means that these simulations may be unstable. References 14 to 19 give some recently published work on this subject.

Gear whine is a displacement-controlled noise equated to the ability, or not, of the gear mesh to transfer perfect rotation. The whine can be at tooth passing frequency due to imperfection in the tooth profile or at shaft rotational speed due to imperfect tooth spacing. Misalignment of the shafts in both proximity and orientation also leads to whine. The complexity of the problem is demonstrated by the significant variability which is found in nominally identical production units. Analysis methods can be used to improve the stiffness of the system to control the mesh alignment under load and to understand the dynamics of the system to keep modal activity outside the maximum tooth passing frequency. Fundamentally, reduction of gear whine is associated with manufacturing accuracy.

(5) To optimize the engine structure for minimum radiated noise, requires the calculation of the surface vibration levels and the use of vibro-acoustic software to calculate the far field pressures. Various methods exist for predicting radiated noise. The simplest approach is to use the equations given in section 20.5.4 for the calculation of radiated noise from surface vibration measurements; this provides a direct comparison between predictive and experimental results. Alternatively, the application of the Rayleigh Piston theory is ideally suited to Finite Element models in that it treats each element as a rigid, infinitely baffled piston and converts the nodally averaged vibration into pressure at specified far field locations. The total sound pressure is calculated based upon the accumulated pressure from all the elements. The benefit of this approach is that it provides a single spectrum of noise from a large database of vibration data in a quick calculation. The limitations are that near field effects are ignored and that low frequency oscillation is not well represented due to the infinite baffle assumption. Boundary element methods provide the ability for improved vibro-acoustic calculation but with a significant increase in computing time.

Ignoring the vibro-acoustic calculation, the accuracy of any noise prediction is a function of the quality of the surface vibration prediction. This in turn is a function of the complexity of the structural model and the accurate simulation of any load characteristics applied to it. Only by the use of fully validated, comprehensive software packages can reliable predictions be made. At present it is not possible to make accurate predictions of engine noise level. Many of the design changes which improve low frequency dynamics will produce improvements at higher frequencies but considerable experience is required to identify those occasions where this is not likely to be the case. Generally, if the broad requency-band force inputs associated with impacts, combustion, etc., are reduced at the lower frequencies, then they are usually reduced at the high frequencies. However, structural changes which make low frequency improvements by increasing stiffness or reducing the effective mass of components can increase noise at higher frequencies.

(6) In addition to the reduction of engine noise, noted above, predictive analysis of the whole powertrain can play an important role in achieving good in-vehicle noise characteristics. The low frequency (below 500 Hz) bending and torsional vibration modes of the powertrain and the potentially large force inputs at some of the lower harmonics

of half engine speed (0.5*e*) are the major parameters that determine the required characteristics of any mounting system and the vibration inputs to a vehicle's structure. Reduction of the internal combustion and inertia loads coupled with improvements in the resonant activity of the powertrain will lead to reductions in the noise levels experienced by the vehicle's passengers.

20.6.4 Palliative treatments and enclosures

The vibration of the external surfaces of the engine cause pressure fluctuations in the surrounding air which then propagate to the environment as radiated noise. Engine noise levels can be reduced by reducing the noise generators and/or the structural responses, as described in sections 20.6.1 to 20.6.3. Alternatively, engine radiated noise can be suppressed by the use of engine-mounted palliative treatments or machine-mounted noise absorption panels and engine enclosures. The palliative treatments commonly used on engines fall in to two groups: shielding and increased damping to absorb energy from the major noise radiating areas; and isolation or stiffness changes to substantially reduce energy transfer and/or structural response of the major components.

(1) The shielding and increased damping techniques usually produce around a 6 dB(A) (or fourfold) reduction in the sound power radiated by the surface area that they affect. Two simple examples can be made by considering the noisiest components on *Figure 20.15b*, the timing case and sump, as follows:

   (a) Shielding of 75% of the surface of the timing case should reduce its noise level by about a factor of three (i.e. a fourfold reduction applied to 75% of area, assuming fairly uniform noise distribution from the surface). This removes two-thirds of the noise energy radiated by the timing case and reduces overall engine noise by around 17% (i.e. 0.67 × 25.4%) or 0.8 dB(A).

   (b) Manufacturing the whole sump from an internally damped steel, often called 'MPM' or metal-plastic-metal steel, should reduce sump noise by a factor of four. This removes three-quarters of the noise energy radiated by the sump and reduces overall engine noise by around 12% (i.e. 0.75 × 15.8%) or 0.5 dB(A).

(2) Isolation of components usually produces a 10 to 12 dB(A) (or ten to sixteen fold) reduction in the radiated sound power. As an example, reconsider the sump data from *Figure 20.15b*. If the sump were isolated from the main cylinder block the overall noise level of the engine should be reduced by around 14.4% (i.e. 0.91 × 15.8%) or 0.7 dB(A). Isolation requires the use of relatively thick soft joints with no metal-to-metal contact. The aim is to achieve the lowest possible natural frequency for the mass-elastic system of the component and joint assembly. However, the natural frequency must be significantly (at least 40%) higher than the major forcing frequency of the engine, namely firing frequency, to avoid resonance and durability problems.

(3) Modification of engine components including changes to panel dimensions, stiffness characteristics and/or effective mass can also produce noise reductions but these type of changes usually require some input from predictive analysis; see section 20.6.3. A 'trial and error' approach usually proves to be less effective and more costly.

The combined effect of the sump isolation and the timing case shield, discussed above in items 2 and 1a, respectively, is

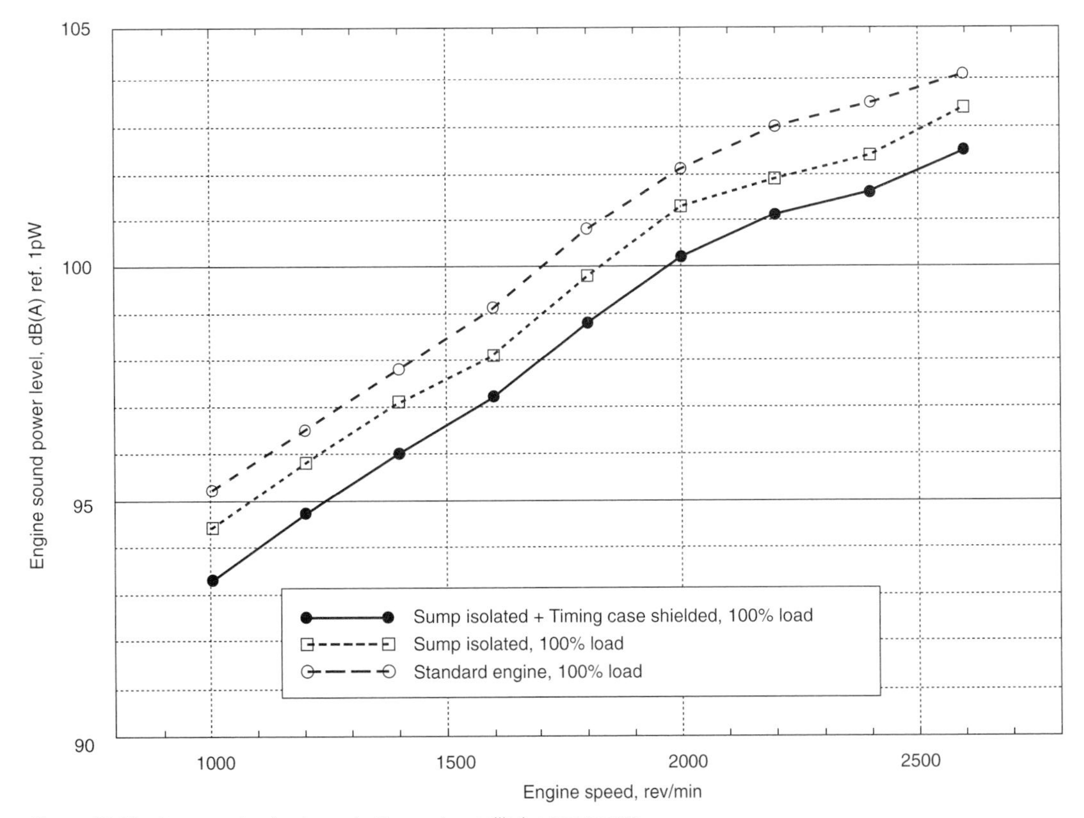

Figure 20.18 An example of noise reductions using palliative treatments

shown on *Figure 20.18*. The plot shows the limitation of palliative treatments. After applying good practical improvement techniques to the two noisiest components, the overall engine noise reduction was less than 2 dB(A) at all operating conditions. On some engines it is practical to reduce noise levels by up to 3 dB(A) using palliative treatments alone but it is rare for more than this to be achieved.

Machine or vehicle mounted noise absorption materials can reduce external noise levels; this type of technology now appears in the form of absorption panels in most diesel engine installations. However, complete engine enclosures are not so common because if they are effective for noise control they usually restrict access for cooling air and engine maintenance.

20.6.5 Vehicle and machine refinement

Reduction of the airborne noise levels radiated from vehicles and machines requires the refinement of the components identified by the techniques outlined in section 20.5.5. The general approach should be to reduce the largest noise sources first. Simple calculations will show that if the largest sources are not reduced other refinement activities will be less effective.

Noise within the passenger car is usually dominated by structurally transmitted noise. The powertrain vibration is transferred through the mounts into the body structure which drives the passenger compartment acoustic cavity. Control of in-vehicle noise is achieved by optimizing the powertrain vibration at the mounting points, the configuration and 'elastomer' characteristics of the mounts themselves, and by reducing the sensitivity of the body structure to excitation. Consideration must also be given to the inputs from the powertrain via other routes such as exhaust hangers, gear linkages, pipes, cables, etc. Powertrain vibrations are a function of its own vibration modes and the vibration characteristics of engine-mounted ancillary equipment. All components attached to the powertrain should be considered in any optimization analysis.

In a similar manner in-cab noise levels of machines will be related to the response of the cab to structural vibrations through engine and/or cab mounts, and other contact points. In addition, some airborne sources such as the engine air intake system or cooling fan can excite cab panels into resonance creating high in-cab noise levels.

20.7 Conclusion

As will have been evident from the above, noise refinement of the diesel engine, and its installations, is a series of complex processes. The most efficient application of noise reductions techniques are made in the early design stages and during prototype development, where trade-offs against other aspects of the design can be fully investigated. The *ad hoc* application of so-called noise reduction treatments and materials is fraught with difficulties and, in general, it is advisable to seek the assistance of noise reduction experts. There are many examples of successful noise reduction programmes listed in the Bibliography.

References

1. FAHY, F. J., *Sound Intensity*, Elsevier Science (1989)
2. CHAN, C. M. P., MONCRIEFF, I. D. and PETTITT, R. A., 'Diesel Engine Combustion Noise with Alternative Fuels', *SAE Paper 820236* (1982)
3. BIES, D. A. and HANSEN, C. H., *Engineering Noise Control*, Unwin Hyman (1988)
4. YORKE, P. J., 'The Application of Idealization and Response Analysis to Diesel Engine Noise Assessment', *SAE Paper 750836* (1975)
5. WHITE, R. G. and WALKER, J. G., *Noise and Vibration*, Ellis Horwood (1982)
6. BERANEK, L. L. *Noise and Vibration Control*, McGraw-Hill (1971)
7. BERANEK, L. L. and VÉR, I. L., *Noise and Vibration Control Engineering*, John Wiley & Sons (1992)
8. BOLTON, J. S., HALL, H. R., SCHUMACHER, R. F. and STOTT, J., 'Correlation of Tire Intensity Levels and Passby Sound Pressure Levels', *SAE Paper 951355* (1995)
9. RUSSELL, M. F., 'Diesel Engine Noise; Control at Source', *SAE Paper 820238* (1982)
10. RUSSELL, M. F. and HAWORTH, R., 'Combustion Noise from High Speed Direct Injection Diesel Engines' *SAE Paper 850973* (1985)
11. KATANO, H., IWAMOTO, A. and SAITOH, T., 'Dynamic Behaviour Analysis of Internal Combustion Engine Crankshafts Under Operating Conditions', *IMechE Paper C430/049* (1991)
12. RICHARDSON, S. H. and RIDING, D. H., 'Predictive Design Support in the Achievement of Refined Power for the Jaguar XK8', *SAE Paper 972041* (1997)
13. HANKS, P., 'Engine Design Analysis–An Integrated Methodology', *Fourth Symposium on Indian Automotive Technology*, 7-10/12/94–*SIAT-94* pp. 251–260.
14. WILHELM, M., LAURIN, S., SCHMILLEN, K. and SPESSERT, B., 'Structure Vibration Excitation by Timing Gear Impacts', *SAE Paper 900011* (1990)
15. SPESSERT, B. and PONSA, R., 'Investigation in the Noise from Main Running Gear, Timing Gears and Injection Pump of DI Diesel Engines', *SAE Paper 900012* (1990)
16. CROKER, M. D., AMPHLETT, S. A. and BARNARD, A. I., 'Heavy Duty Diesel Engine Gear Train Modelling to Reduce Radiated Noise', *SAE Paper 951315* (1995)
17. MEISNER, S. and CAMPBELL, B., 'Development of Gear Rattle Analytical Simulation Methodology', *SAE paper 951317* (1995)
18. MIURA, Y. and NAKAMURA, S., 'Gear Rattling Noise Analysis for a Diesel Engine', *IMechE Paper C521/001/98* (1998)
19. WANG, Y.M., 'Transmission Modelling for Gear Rattle Analysis', *IMechE Paper C521/004/98* (1998)

Bibliography

1. ANDERTON, D., GROVER, E. C., LALOR, N. and PRIEDE, T., 'Origins of Reciprocating Engine Noise—Its Characteristics, Prediction and Control', ASME Publication 70-WA/DGP-3 (1970)
2. RUSSELL, M. F., 'Automotive Diesel Engine Noise and Its Control', *SAE Paper 730243* (1973)
3. ANDERTON, D., 'Relationship Between Combustion System and Engine Noise', *SAE Paper 790270* (1979)
4. TUNG, V. T. C. and CROCKER, M. J., 'Diesel Engine Noise and Relationship to Cylinder Pressure', *SAE Paper 820237* (1982)
5. TURNER, G. L., MILSTED, M. G. and HANKS, P., 'Vibration Characteristics of an In-Line Engine Structure', *IMechE Paper C138/84* (1984)
6. SAGDEO, P. M., 'Frequency Spectra of Diesel Engine Heat Release Rate, Pressure Development and Engine Noise: An Experimental Investigation of Cause and Effect Relationship', *SAE Paper 871620* (1987)
7. BRANDL, F. K., AFFENZELLER, J. and THIEN, G. E., 'Some Strategies to Meet Future Noise Regulations for Truck Engines', *SAE Paper 870950* (1987)
8. PETTITT, R. A. and TOWCH, B. W., 'Noise Reduction of a Four Litre Direct Injection Diesel Engine', *IMechE Paper C22/88* (1988)
9. FARNELL, R. A., PETTITT, R. A. and TOWCH, B. W., 'The Installation of a Low Noise Diesel Engine into a 5.6 tonne Truck', *IMechE Paper C382/073* (1989)
10. BRANDL, F. K., WUNSCHE, P. and GSCHWEITL, E., 'Design Strategies for Low Noise Engine Concepts', *SAE Paper 911070* (1991)
11. VORA, K. C. and GHOSH, B., 'Vibration Due to Piston Slap and Combustion in Gasoline and Diesel Engines', *SAE Paper 911060* (1991)
12. MUTOH, H., NIIKURA, T., YAMAMOTO, A., ARAKAWA, M. and KIKUCHI, K., 'The Development of Low Noise Engine on Hino New "H" Series', *SAE Paper 900347* (1990)

13. HOWER, M. J., MUELLERE, R. A., OEHLERKING, D. A. and ZIELKE, M. R., 'The New Navistar T444E Direct-Injection Turbocharged Diesel Engine', *SAE Paper 930269* (1993)
14. BALEK, S. J. and HEITZMAN, R. C., 'The Caterpillar 3406E Heavy Duty Diesel Engine', ASME 1994, ICE-Vol. 22, *Heavy Duty Engines: A Look at the Future*
15. KNOWLAND, C. G., CHALLEN, B. J. and FARNELL, R. A., 'Practical and Analytical Studies in Powertrain and Vehicle Refinement', *SAE Paper 951295* (1995)
16. TOGASHI, C. and NAKADA, T., 'A Study on the Noise Generating Mechanism of a Fuel Injection Pump', *SAE Paper 951345* (1995)
17. SOUTHALL, R. and TRIMM, M. A., 'Noise and Vibration Technology for the Perkins V6 HSDI Demonstration Engine', *SAE Paper 972044* (1997)
18. SAE International, *'Proceedings of the 1995 Noise and Vibration Conference'*, SAE P-291
19. SAE International, *'Proceedings of the 1997 Noise and Vibration Conference'*, SAE P-309
20. ImechE, 'European Conference on Vehicle Noise and Vibration', *ImechE Conference Transaction 1998-5*

21 Larger engine noise and vibration control

Contents

21.1 Introduction

The advantages offered by large diesel engines, i.e. low fuel consumption, durability, etc., are partially offset by the higher levels of noise and vibration typically encountered with their use. However, with proper application of noise control and vibration isolation technology, these higher levels can be significantly reduced to acceptable levels for most applications.

The principal sources of noise and vibration from large diesel engines include:

exhaust noise
inlet noise
fan and accessory noise and vibration
radiation from engine casing

due to combustion
due to reciprocating components (valves, crankshaft, etc.)

All of the above sources produce disturbances that can be arbitrarily grouped into two general categories and frequency bands:

Noise 50–20 kHz
Mechanical vibration 0–50 Hz

21.2 Noise

The category of Noise can be subdivided into airborne noise and structureborne noise. Structureborne noise is transmitted along solid structures and is usually generated by a lower frequency vibration source.

Testing of airborne noise levels produced by diesel engines has shown that at frequencies below 800 Hz, exhaust, inlet, and accessory equipment are the predominant noise sources. Engine casing radiation is the dominant source at higher frequencies, with maximum values usually occurring in the 1000–1600 Hz region. For vehicular applications, the Society of Automotive Engineers (SAE) Standard J366b is commonly used in the USA for measuring the noise emission levels of diesel trucks. Combining the results of many tests[1], an approximate relationship for the mean A-weighted sound level attributable to the truck engine noise sources only, as a function of truck speed, is given by:

$$L_A = 82.4 \text{ dB(A)} \quad v < 56 \text{ km/h (35 mph)} \qquad (21.1a)$$

$$L_A = 83.6 + 6 \log_{10} (v /88) \text{ dB(A)} \quad v > 56 \text{ km/h (35 mph)} \qquad (21.1b)$$

Exhaust noise is produced by the individual gas pressure pulses created as each cylinder fires, and therefore the maximum noise level normally occurs at the firing frequency of the engine. For an 8 cylinder diesel engine running at 2400 rpm, the firing frequency is calculated by multiplying the rmp by the number of cylinders/2 (since each cylinder fires on every 2nd revolution in a 4 cycle engine), and then dividing this result by 60, to convert to hertz:

$$\text{Firing frequency} = (2400 \text{ rpm}) (8 \text{ cylinders}/2) / 60 = 160 \text{ Hz}$$

For an engine with a straight stack (unmufflered) exhaust, sound level values in an octave band centred at the firing frequency typically range from 100–120 dB, 3 feet from the stack outlet[2]. The use of a well-designed muffler can be expected to reduce this level by approximately 15–25 dB. There are very specific requirements for each engine application to ensure the level of sound reduction is optimized without increasing exhaust back pressure to unacceptable levels, thereby limiting power output. For this reason, it is recommended that the engine manufacturer be consulted to determine the proper exhaust system and muffler design.

New technology in the area of exhaust noise reduction has employed the principle of active noise cancellation. In this approach, sensors at the exhaust outlet sense the frequency and amplitude of the sound produced and, using the proper control algorithms, create a drive signal to a loudspeaker that generates an 'antinoise' signature that, when added to the exhaust noise, cancels out a significant portion of it. This technology works well on exhaust noise because it is quite periodic, with frequencies proportional to the engine rpm. It is less applicable to noise that is more random in nature.

Inlet noise is primarily caused by the motion of air through the air cleaner and inlet manifold, and generally is proportional to engine speed. A common method of reducing inlet-related noise is to use a material that is inherently well damped in place of the usual metal casting alloys. A number of engineered polymers are now available that have suitable temperature resistance and structural strength along with the inherent material damping required to reduce structureborne noise transmission, and consequently, noise radiation. Finally, surrounding the inlet (or entire engine) with a sound absorbing barrier can be effective in reducing noise once it has become airborne. Both of these approaches will be discussed later in the chapter.

Fan noise, which usually is the largest single noise contributor in the accessory category, is caused by disturbance of air by the fan blade passage. It is highly influenced by fan rotation speed, number of blades, and fan blade tip configuration. Fan noise typically peaks at the blade pass frequency [(number of blades) (fan rpm/60)]. Whenever possible, fan or shroud support struts should be kept far from the moving fan blades, to prevent blade pass noise and minimize resistance to the cooling air flow. Other accessories often responsible for engine noise include water pumps, hydraulic pumps, fuel injectors, and drive belts.

Noise radiated from the engine casing is created by both the act of fuel combustion and from the reaction forces caused by rotating or reciprocating masses within the engine. The combustion-related noise is created by gas pressure pulses impinging on the cylinder walls and inducing high frequency vibration (structureborne noise) in the engine structure. The structureborne noise becomes audible airborne noise when the engine surfaces interact with the layer of air surrounding the engine, creating the variations in air pressure density that are perceived as noise to the human ear. The vibration produced by moving masses, such as valves, crankshaft, camshafts, etc. is normally lower in frequency, and will be examined later in Section 21.3, but it also can create noise by inducing higher frequency harmonics of the lower frequency vibration in the engine and component structures.

Reduction of engine casing radiation is accomplished in two ways. The first is to add structural damping to the engine components, similar to the approach for inlet noise. This has been most successful with relatively thin components, such as oil pans, valve covers and heat shields, and a number of metallic and polymer materials have been developed for these applications. This approach controls the noise at its source, which is always preferable.

The other approach is to surround the entire engine with a barrier of sound-absorbing material, to prevent the escape of noise into the environment. This obviously is more practical in stationary engine applications, such as generators or pumps, than it is for mobile applications.

There are usually trade-offs that must be made when attempting to completely surround a large stationary diesel for noise control, primarily in the areas of heat dissipation, inlet air supply and exhaust, but it has been demonstrated that if as little as 10% of the surrounding space is open, then it will be impossible to reduce the noise level transmitted through the barrier by more than 10 dB. If openings in the barrier are necessary, they should not be straight, line of sight holes, but instead should be as convoluted as is practical[3].

A typical diesel engine noise absorbing barrier is shown in *Figure 21.1*. There is an optional inner Mylar™ film layer to reflect heat that might deteriorate the adhesives used in the barrier material construction, followed by a relatively thick layer (usually 5–15 cm/2–6 in) of open cell acoustical foam. This material absorbs the noise by effectively changing the acoustic vibrational motion into heat energy. The foam is bonded to a thinner septum layer of dense, inherently damped material. This layer was typically made of lead, but environmental considerations have led to the development of polymer composite replacement materials. The septum layer reflects the acoustical energy back into the enclosed space, primarily due to the dissimilar value of the speed of sound transmission between the material and air. A second, typically thinner layer of foam serves to prevent contact, and therefore noise transmission, between the septum layer and the structural outer material.

With the global proliferation of laws limiting the allowable levels of noise that can be generated for both vehicular and stationary applications, noise reduction has become a primary consideration for both diesel engine manufacturers and users.

21.3 Vibration

There are two primary mechanisms by which any internal combustion engine produces vibration. The first is due to unbalanced forces and coupled moments created by rotation or linear reciprocating motion of the crankshaft and related components (see *Figure 21.2*). These disturbances can be balanced in engines with six or more cylinders, but in engines of less than six cylinders, there is always at least one inertia force or couple that is impossible to cancel by balancing. The primary vibration created by this imbalance occurs at the same frequency as the engine rpm, and therefore is called first-order vibration. (harmonics of this vibration occurring at two or three times the engine rpm would be classified as second order, third order, etc.). Depending on the number of cylinders and the crank arrangement, the inertial forces combine or cancel in varying degrees. As the number of cylinders increases, these forces are usually more balanced, and the vibration created is reduced in magnitude.

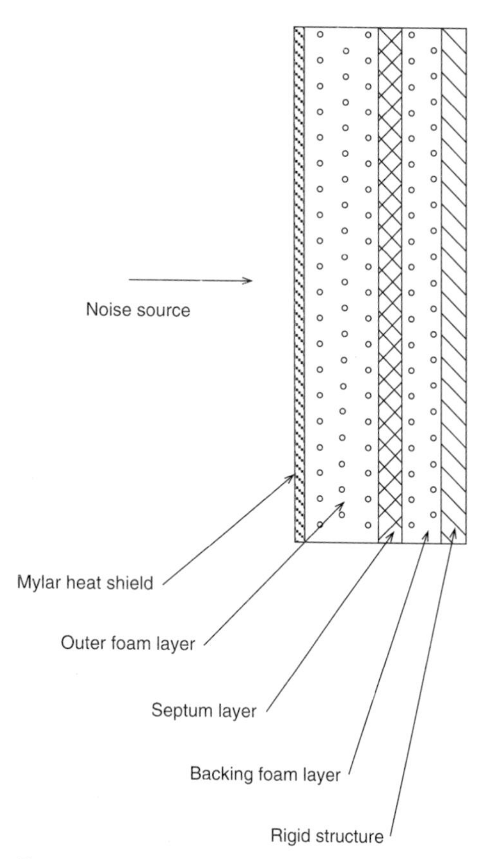

Figure 21.1 Engine noise absorbing barrier schematic

The second source of diesel engine vibration is caused by the torque pulses created at the crankshaft by the increase in cylinder gas pressure as each spark plug fires (see *Figure 21.2*) These torque pulses by their very nature create the rotational output power that the engine is designed for, so reducing their magnitude is not an option. When these torque pulses are produced, however, the non-reciprocating mass of the engine is subjected to a reactional rotational force of equal and opposite magnitude, which tries to rotate the engine in the opposite direction from the crankshaft rotation. The resulting torsional vibration is almost always the major source of vibration in large diesel engines[3]. This vibration occurs at the firing frequency of the engine, as discussed earlier. Since a power stroke occurs on every second revolution of a 4 cycle diesel engine. The order of the firing frequency is half the number of cylinders, i.e. a 5 cylinder engine would create $2\frac{1}{2}$ order vibration as its primary disturbance.

The firing frequency vibration, much like noise, becomes a problem when it is allowed to be transmitted from the engine to the surrounding environment. This occurs principally at the mounting locations where the engine is attached to the chassis (for vehicles) or the supporting floor (for stationary applications). If this transmission is not prevented, then the vibration will be perceived and possibly found objectionable by drivers or workers in the surrounding area. The most common method of reducing this transmitted vibration is with the use of resilient isolators.

Resilient isolators reduce vibration by temporarily storing the incoming vibrational motion, and then releasing it to the support structure over a time interval that affords a reduction of the magnitude of motion. This mechanism is shown in *Figure 21.3* which shows a transmissibility curve for a single degree of freedom (vertical motion only) spring/mass system with no damping. Transmissibility is a measure of the amount of vibration from the engine that passes through the resilient mount into the support structure. It is simply by the displacement motion out of the isolator divided by the input motion. Equation (21.2) shows the formula used to calculate the curve given in *Figure 21.3*.

$$\text{Transmissibility} = 1/[1 - (f/f_0)^2] \qquad (21.2)$$

At very low frequencies the transmissibility is approximately 1; whatever displacement is input to the isolator passes directly through it unchanged, and the isolator acts like a rigid mass. As the frequency is increased, it begins to approach the natural resonant frequency of the spring mass system, and the output across the isolator is greatly increased. This is obviously the **worst** possible frequency for an isolator to operate at, since its purpose is to reduce the vibration transmitted across it. As the frequency increases above the natural resonant frequency of the spring mass system, the transmissibility decreases, recrossing the value of 1 at $\sqrt{2}$ times the resonant frequency. At all higher frequencies the output across the isolator is less than the input, so the vibration is reduced. Since it is the firing frequency of the engine that must be reduced, then it follows that the lower in

Type of Engine	Crank Arrangement	Inertial forces		Inertial moments	
		Primary (RPM)	Secondary (2 x RPM)	Primary (RPM)	Secondary (2 x RPM)
One cylinder	1	Vertical Horizontal	Vertical	None	None
Two cylinder	1 2		Vertical	Pitch Yaw	
Two cylinder opposed	1 2			Yaw	Yaw
Two cylinder horizontal	1 2	Vertical			
Two cylinder 90°V	1 2		Horizontal		
Two cylinder 60°V	1 2	Vertical			
Three cylinder vertical	1-2-3			Pitch yaw	Yaw
Four cylinder vertical	1-2-3-4		Vertical		
Four cylinder vertical	1-2-3-4			Pitch yaw	
Four cylinder horizontal	1 3 2 4			Yaw	
Four cylinder horizontal	1 2 4 3				Yaw
Four cylinder 60°V	1 2 3 4			Yaw	Yaw
Six cylinder vertical	1-2-3-4-5-6				

Dynamic Disturbances

← Unbalanced inertia forces and moments

→ Torsional moment created by crank mechanism (at firing frequency)

ωt

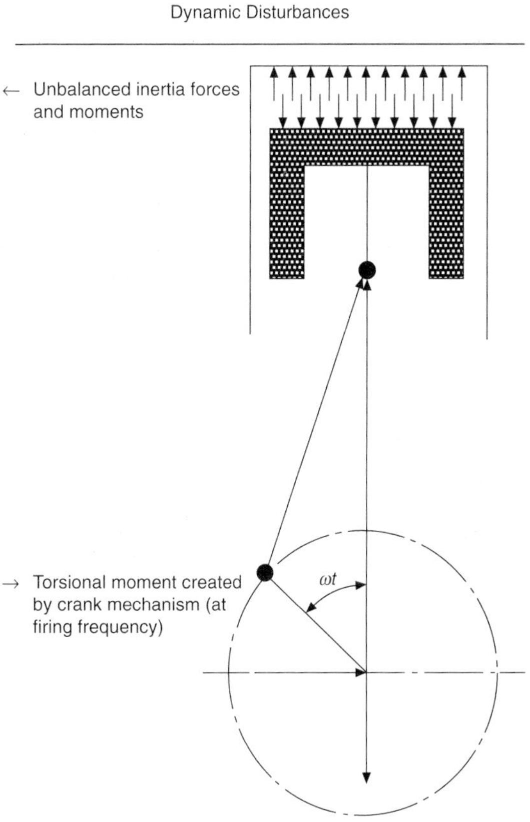

Figure 21.2 Reciprocating engine mechanism

natural frequency (i.e. 'softer') the isolator is, as compared to the firing frequency, the more reduction in the firing frequency vibration will be accomplished.

The calculation of isolator natural frequency is shown in eqns (21.3a, b, or c).

K = stiffness of the isolator (lbs/in)
W = weight of isolator (lbs)
g = 383 in/s^2
Δ_{st} = static deflection of isolator due to supported weight W (in)

$$f_n = (1 - 2\pi)(\sqrt{[Kg/W]}) \quad (21.3a)$$

$$f_n = (3.13)(\sqrt{[K/W]}) \quad (21.3b)$$

$$f_n = (3.13)(\sqrt{[1/\Delta_{st}]}) \quad (21.3c)$$

The preceding discussion has addressed vibration in a single direction only. To effectively isolate engine vibration in all directions, it is necessary to take into account all six degrees of freedom that an engine can move in 3-dimensional space, as shown in *Figure 21.4*. This is done by first determining the three principal axes of inertia of the engine. These axes by definition are the axes with the lowest moments of inertia in the X, Y, and Z planes, and because of this, the engine rotates around or moves along these axes with the least expenditure of energy. Thus, an engine on resilient mounts will always try to move or rotate along the principal axes when driven by the firing frequency or any other vibration, since it takes the least relative expenditure of energy to do so. *Figure 21.5* shows the principal axes of typical diesel engine. By convention, the translational modes are called longitudinal (X), lateral (Y), and vertical (Z). The rotational modes are roll (rotation around X), pitch (rotation around Y), and yaw (rotation around Z). It is important to note that the X (roll) axis does not normally coincide with the crankshaft

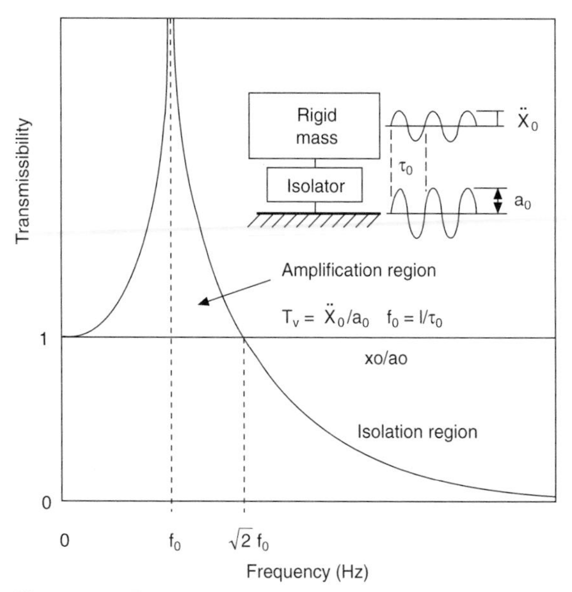

Figure 21.3 Transmissibility of a simple degree of freedom mass-spring system

centreline, but is instead usually inclined 8–20 degrees from it (Figure 21.4).

Since the torque reaction of firing frequency vibration tends to create roll rotation as the primary engine vibration, the most important natural frequency of the isolator/engine system is the roll natural frequency. Designing an isolation system with a roll frequency sufficiently *lower* than the engine firing frequency is necessary for proper isolation. Therefore, effective engine isolation depends on the placement of isolators with respect to the principal (X) roll axis of inertia, *not* the crankshaft centreline, as well as selection of proper isolator stiffness in the X, Y, and Z axes.

To determine the six degree of freedom natural frequencies of an engine mounted on isolators, it is necessary to solve the six differential equations of motion that describe the freebody motion simultaneously. This is best done with one of the many computer modal analysis programs available today. A typical program input and predicted modal output is shown in *Figure 21.6*[4].

The six natural frequencies and mode shapes shown as output represent the responses that the engine/ mount system will display when it is subjected to any dynamic disturbance. As previously shown, the roll natural frequency is the most important to consider, since it will be responsible for the attenuation of the firing frequency induced vibration. In the example given in *Figure 21.6*, the firing frequency at idle is 30 Hz, [(500 rpm) (6 cyl./ 2) (60)] and the engine/mount calculated roll natural frequency is 14.224 Hz. From eqn. (21.2) the transmissibility will equal 0.29, therefore 71% of the roll vibration will be eliminated by the mounts at idle. If the normal operating speed of the engine is 2800 rpm, the firing frequency will increase to 168 HZ, and the transmissibility will be improved to 0.007, or 99.3% isolation.

As shown in the preceding example, a general rule of thumb to ensure excellent vibration isolation for large diesel engines is to design the mounting system to have a roll frequency of 15 Hz or lower. If the isolators are too soft, however, the excessive deflections they see may lead to premature fatigue failure. To guard against this, it is usually good practice to keep the system vertical frequency above 10 Hz for mobile applications, and above 7 Hz for stationary applications.

In addition to the natural frequencies predicted by the computer program, the mode shapes associated with each frequency are given. For each mode, the relative magnitude of the displacement in each translational and rotational axis are listed, so it is possible to examine the listed values and determine the primary direction of motion each represents. In the example given, examination of mode 6 shows that it exhibits 1 cm of motion along the Z axis, 0.506289 cm along the X axis, and –0.001607 radians of rotation around the Y axis (the motions given by the computer program for mode shapes are all expressed in relation to a largest

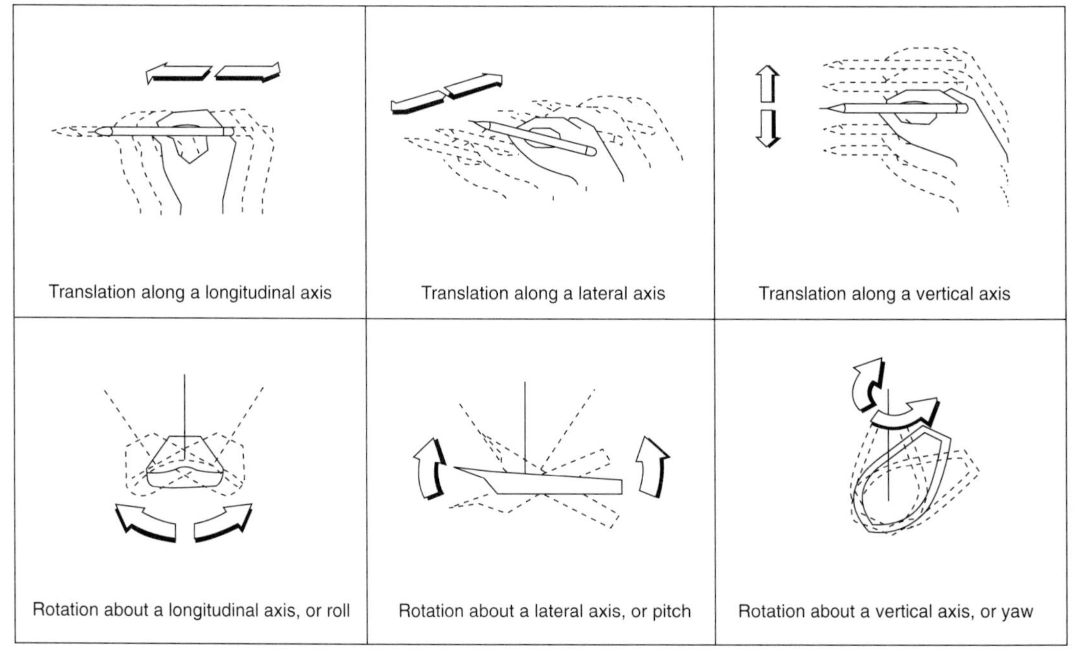

Figure 21.4 Degrees of freedom for the motion of a body in three-dimensional space

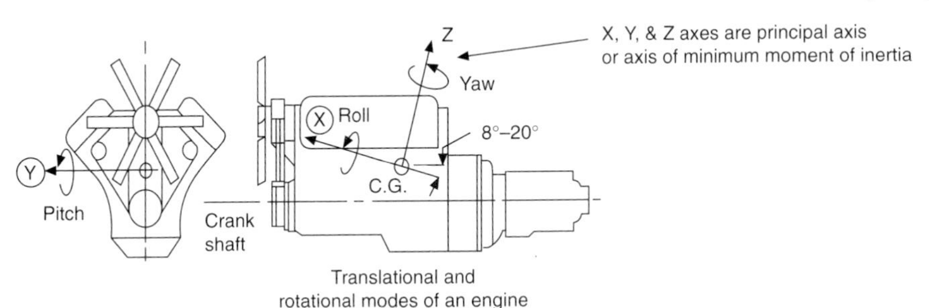

Figure 21.5 Principal axes of a typical diesel engine and rotational modes—roll, pitch and yaw

Date: 1 Jun 1998 Time 09:27 AM

Title: Large Diesel Engine on Barry Controls Mounts

		X	Y	Z
Weight (kg) = 1445.000				
CG Location (mm)	=	– 376.100	0.000	56.800
Moment of Inertia (Kg-m* *2)	=	62.319	492.000	449.358
Euler Angles of Mass (Deg)		Phi	Theta	Psi
		0.000	0.000	0.000

Output in Mass Coordinates

Isolators

Isol No.	Type No.	Euler Ang. No.	Coordinates (mm) X	Y	Z
1	1	1	– 1152.2000	– 50.000	– 360.0000
2	1	1	– 1152.2000	50.0000	– 360.0000
3	2	1	38.1000	353.6000	97.0000
4	2	1	38.1000	– 353.6000	97.0000

Isolator Stiffness Types

Type	Stiffness (N/mm) X	Y	Z
1	193.000	193.000	1313.000
2	2452.000	814.000	1646.000

Isolator Orientation angles

Number	Phi	Theeta	Psi (Degrees)
1	0.00000	0.00000	0.00000

Natural Frequencies

Mode	1 Fore-Aft	2 Roll	3 Pitch	4 Vertical	5 Lateral	6 Yaw
Freq (HZ)	9.634	14.224	11.403	9.382	5.435	8.075
(RPM)	578.015	853.461	684.170	562.892	326.078	484.490
Mode Shape	X, Y, Z	(mm)	Rot X,	Rot Y,	Rot Z (Radians)	
X	1.000000	0.000000	0.037885	0.179166	0.000000	0.000000
Y	0.000000	1.000000	0.000000	0.000000	1.000000	1.000000
Z	– 0.126277	0.000000	1.000000	1.000000	0.000000	0.000000
Rot X	0.000000	0.130962	0.000000	0.000000	– 0.000504	0.001628
Rot Y	0.000117	0.000000	0.002223	– 0.001330	0.000000	0.000000
Rot Z	0.000000	– 0.007927	0.000000	0.000000	– 0.000750	0.004137

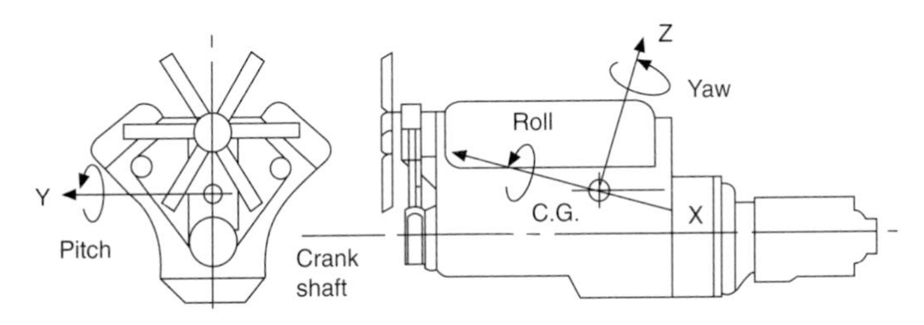

Figure 21.6 Program omega natural frequencies/mode shape

arbitrary deflection value of 1 cm for the axis with the maximum deflection). Since the primary motion in this mode is in the vertical Z axis, this mode is labelled as the vertical mode. Note that mode 3 also has 1 cm of vertical motion, but since it has almost twice as much rotation around the Y axis as mode 6, it is more properly considered as the pitch mode.

If the mount locations and stiffnesses are chosen to perfectly balance the system so all six modes show motion in only one translational or rotational direction, the system is said to be perfectly 'decoupled'. This means that an input disturbance in one axis will result in an output in that axis and no other. If the system is not perfectly decoupled, a single axis input will result in output motion on two or more axes. This 'coupled' type of system is shown in *Figure 21. 7.*

A force applied in the horizontal direction will cause the mass to both translate horizontally and 'rock' or rotate about the mounts located at the bottom of the mass. This system could be decoupled if the mounts were moved up to be in the same horizontal plane as the centre of gravity, or, as will be shown later, if the mounts are 'focused' by inclining them at the proper angle. In the real world, it is impractical to perfectly decouple an isolation system, but the more decoupled a given system is, the better relative isolation performance it will exhibit. This occurs because a coupled system produces a transmissibility curve with two resonance peaks, as shown in *Figure 21.8.*

The first natural frequency peak is mostly translational motion, and is referred to as the lower rocking frequency. The second peak is mostly rotation of the mass around the centre of gravity, and is called the upper rocking frequency. The reason coupling is undesirable is that it 'spreads' the two frequency peaks, requiring softer mounts to obtain the same degree of isolation, as compared to an uncoupled system with only one peak. For example, in *Figure 21.8,* at 50 Hz the coupled system has a transmissibility of 0.75, while the decoupled system has a transmissibility of 0.25, and therefore isolates three times as well at that frequency.

As mentioned previously, another means of decoupling a mounting system is by inclining or 'focusing' the mounts so their effective elastic centre is raised into the plane of the centre of gravity, as shown in the decoupled system of *Figure 21.7.* Two mounts inclined at an arbitrary angle are far from necessarily focused. To focus isolators properly, a certain stiffness ratio of horizontal to vertical stiffness must be provided. The required stiffness ratios and inclination angles are shown in *Figure 21.9,* where ($L = K_v/K_h$), and K_v = vertical isolator stiffness, K_h = horizontal isolator stiffness.

Figure 21.10 illustrates a typical vehicular application of focused mounts on both the front and rear of an engine, and represents, perhaps, an ideal mounting arrangement from an isolation point of view. In addition to focusing the mounts, they are located as close to the principal roll axis as possible, providing the lowest possible stiffness around the roll axis, and consequently the lowest roll natural frequency and maximum firing frequency vibration isolation. Also, the mounts are located longitudinally along the engine at its fundamental bending frequency nodal points. Since a nodal point is by definition a point of no motion during engine bending, then location of the mounts at these points will ensure that no vibration can be transmitted through the mount by engine flexure.

Another consideration in engine mounting that must be addressed is the stiffness of the support structure. If a mount is to isolate vibration, it must be able to deflect. And the more it moves or deflects, generally, the better the isolation provided. But if the supporting structure between the engine and the mount or the mount and the support frame is too soft, the relative motion across the isolator may be inadequate. This occurs because the support structure works in series with the isolators, and will deflect proportionately to their respective spring rates. A general rule is that the support structure spring rate should be a minimum of *ten* times stiffer than that of the isolators. The total stiffness of the mount and structure, when the structure stiffness is less than the required ten times stiffer, can be calculated by eqn. (21.4)

$$K_{total} = (K_{support} \times K_{mounts})/(K_{support} + K_{mounts}) \quad (21.4)$$

This is shown graphically in *Figure 21.11.*

Finally, when selecting mounts, environmental exposure and expected fatigue life must be considered to ensure a long lasting solution. Elastomeric resilient mounts are most commonly used for diesel engine mounting applications due to their low cost, small size, and ease of design to achieve required stiffnesses. The most commonly used elastomers are natural rubber, which usually has the best fatigue life, and neoprene™, which has the best overall environmental compatibility. Only when especially

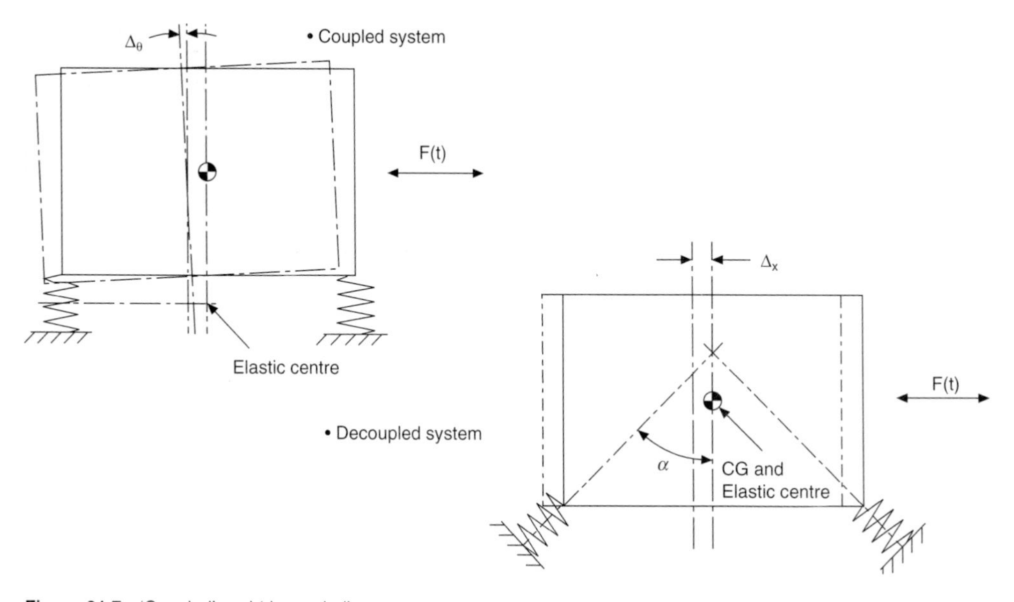

Figure 21.7 'Coupled' and 'decoupled' mount systems

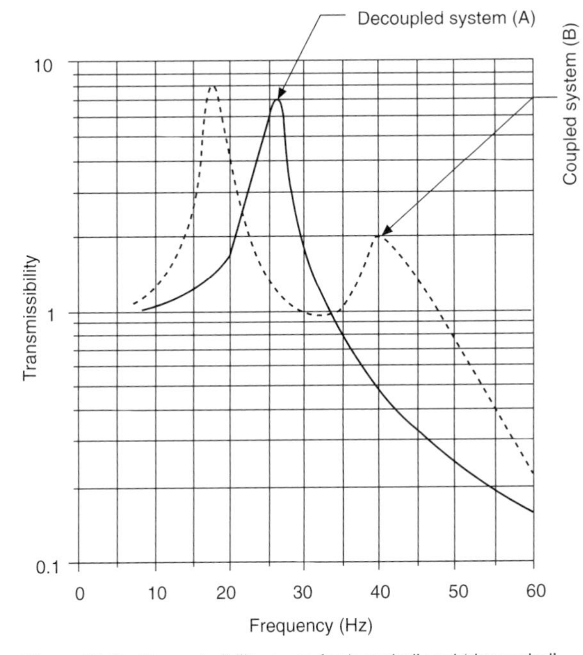

Figure 21.8 Transmissibility cures for 'coupled' and 'decoupled' systems

hostile environments are expected, such as operating temperatures in excess of 107°C (225°F), or severe oil exposure, should consideration be given to other elastomers, such as nitrile, EPDM, silicone, etc.

Regardless of their geometry, elastomers, like metals, exhibit fatigue failure properties from repeated cyclic loadings[5]. The predicted fatigue life of elastomers is given by eqn. (21.5), where

N = number of cycles to failure
ε = maximum shear strain resulting from cyclic load
K, b = empirical constants for each type of elastomer

$$N = (K / \varepsilon)^b \tag{21.5}$$

The empirical constant K can be as high as 1250 for natural rubber and 1000 for neoprene. For both, $b = 5$. From this equation, it is obvious how important shear strain is, in determining the fatigue life of an elastomeric part. If the shear strain can be halved, for example, the fatigue life of the mount will be increased by a factor of 32.

Examples of typical isolation mounts that have been used successfully to isolate large diesel engines are shown in *Figure 21.12*[6] on p. 519.

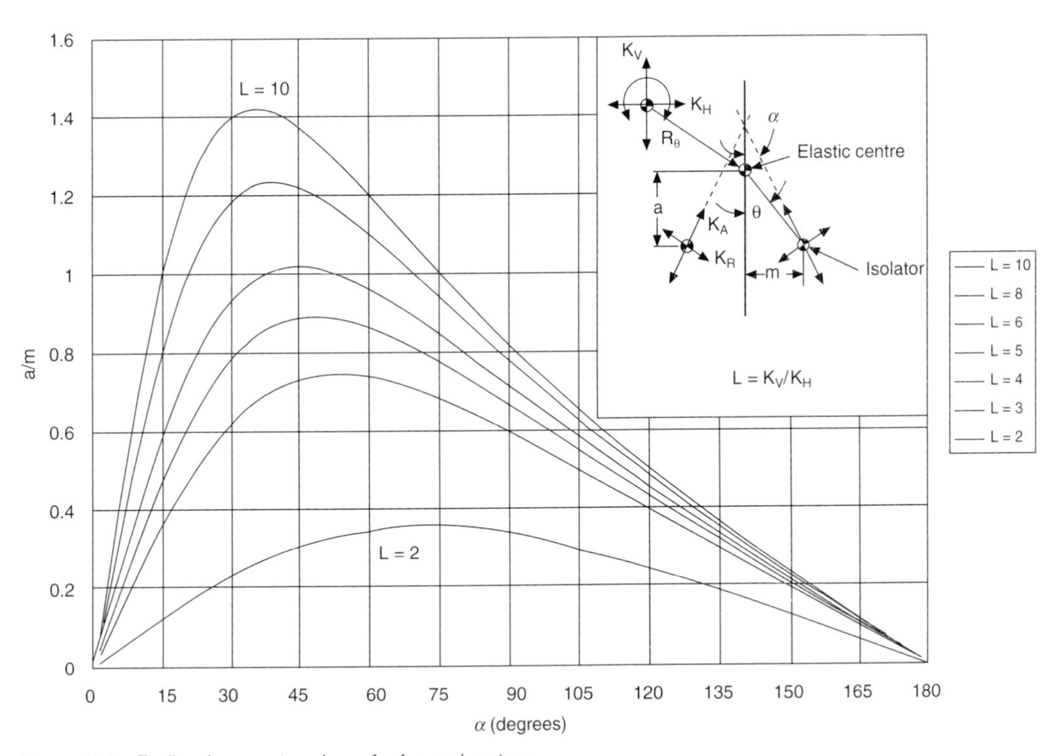

Figure 21.9 Finding the mount angle, α, for focused system

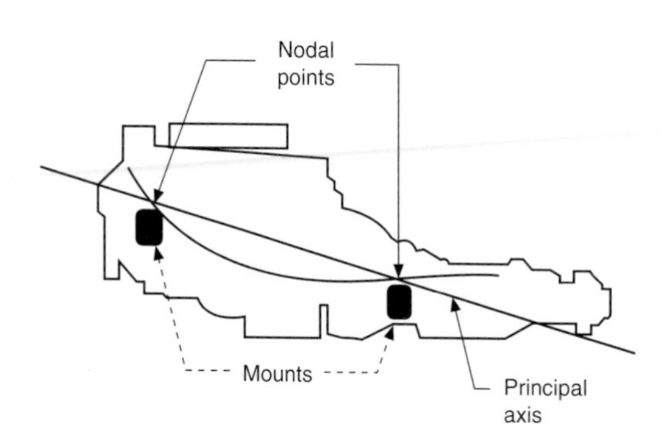

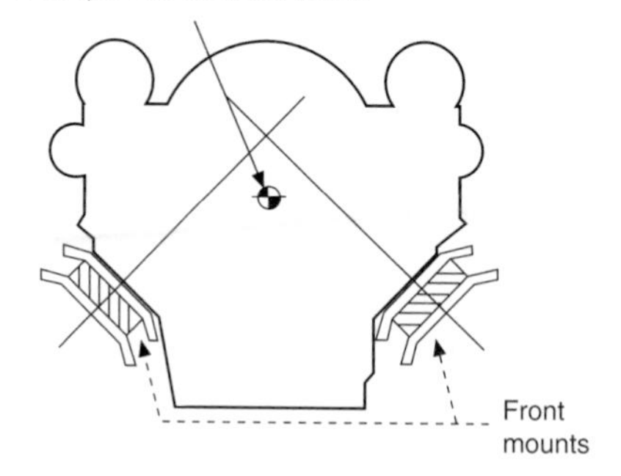

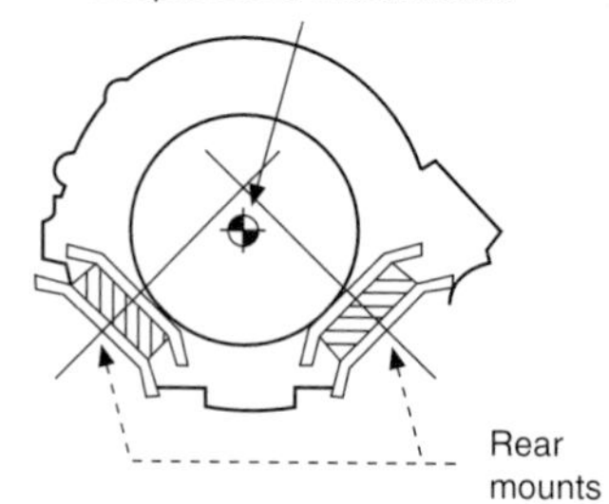

Figure 21.10 Mounting arrangement of an engine on vehicular application; focussed mounts on front and rear

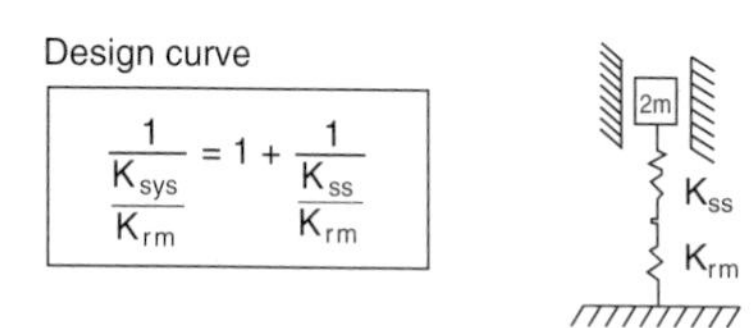

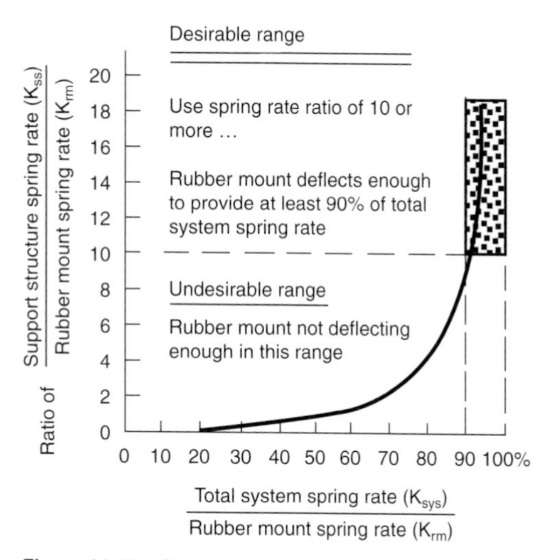

Figure 21.11 'Desirable' and 'undesirable' support stiffness for vibration isolation systems

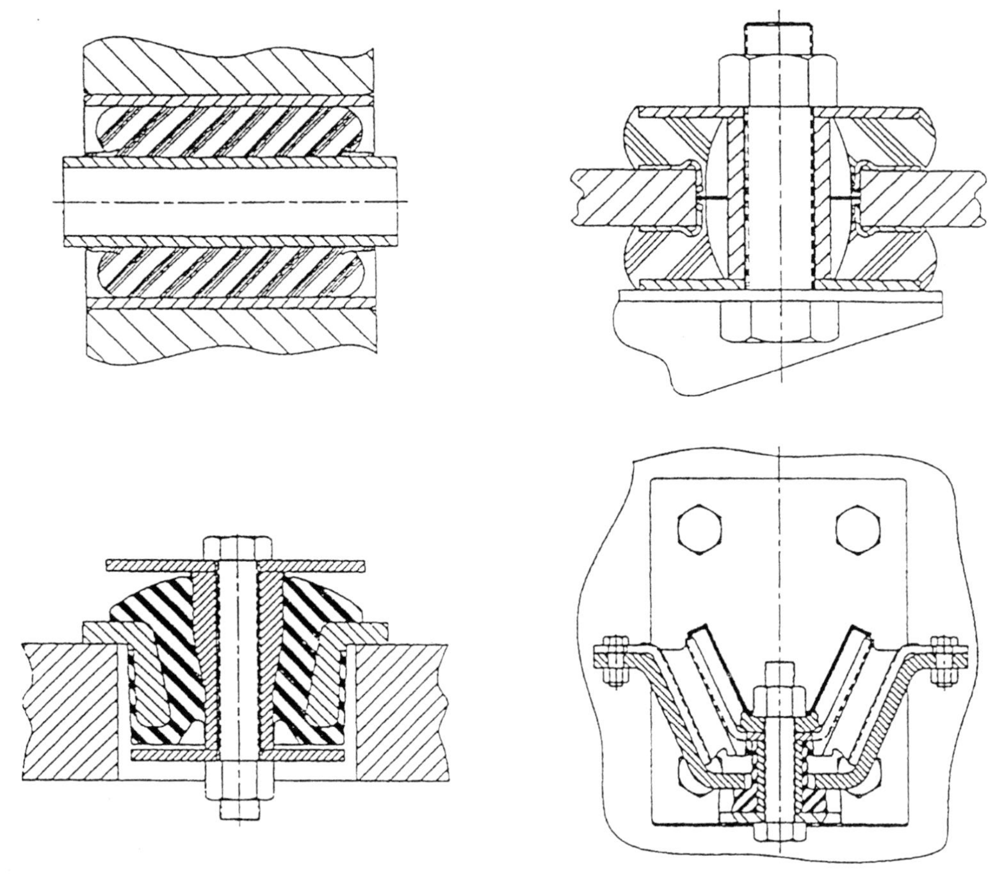

Figure 21.12 Several examples of isolation mounts for large diesel engines

References

1. HARRIS C. M. (ed.), *Handbook of Acoustical Measurements and Noise Control,* third edition, McGraw-Hill p. 48.8 (1991)
2. LYON, R. H.,'*Lectures in Transportation Noise,* Grozier Publishing, Cambridge, MA, p. 141
3. *Noise Control,* BRUEL and KJAER, p. 121 (1982)
4. RACCA, R. and WALKER, S. 'Effective Reduction of Noise and Vibration of Internal Combustion Engines in Marine Installations', *Society of Automotive Engineers Technical Paper no. 941695,* p. 3 (1994)
5. RACCA, R., 'How to Select Power-Train Isolators for Good Performance and Long Service Life', *Society of Automotive Engineers Technical Paper no. 821095,* p. 9 (1982)
6. *Automotive and Bus Engine Mounting Installation Recommendations,* Cummins Engine Co. Bulletin 3382382, Columbus, Indiana, pp. 11–12 (1998)

Part 5

Applications

22

Passenger car engines

Contents

22.1 Introduction

Although introduced more than 60 years ago, the performance characteristics of passenger car diesel engines did not show much progress during the first 40 years of their history. This can be seen in *Figure 22.1* which shows the 'History of Specific Power' for passenger car diesel engines. Those first passenger car diesels were rather heavy engines which were typically installed in larger vehicles. This engine/vehicle combination offered poor vehicle performance and was limited in market penetration to special applications such as taxi cabs.

The development of the Volkswagen 1.5L IDI diesel engine in the middle of the 1970s, was definitely an important milestone for the diesel engine in passenger car applications. This engine was derived from a gasoline engine, possessed an aluminum cylinder head and exhibited almost the same weight as its Otto-cycle cousin. A tooth belt was used to drive not only the overhead camshaft but also the distributor fuel injection pump. Moreover, the engine could be manufactured with almost the same equipment as the gasoline engine. This engine achieved a specific power output of about 25 kW/l, while most competitive diesel engines showed values of only 20 kW/l at rated conditions.

When installed in a light vehicle (VW Golf), this engine offered relatively good vehicle performance and excellent fuel consumption characteristics at the time. Moreover, a VW Golf diesel, compared to a Mercedes-Benz diesel for example, was affordable for many customers.

The next very important step was the application of turbochargers to passenger car diesel engines. Mercedes-Benz, Peugeot and Volkswagen pioneered this approach. Here, again, it was Volkswagen which introduced the most successful package with the Golf application, while the other manufacturers used the turbocharger only on diesel engines for their high-end vehicles.

The most rapid development in passenger car diesel engine technology, however, occurred over the last 10 years, when almost every manufacturer of passenger cars introduced vehicles with diesel engines. Engine families which include both gasoline and diesel variants became the standard approach allowing both types of engines, to be produced with the same manufacturing equipment and to be easily adapted to changing customer demands for one engine type or the other.

In addition to turbocharging, many new features, such as intercooling, electronic injection system control, exhaust gas recirculation (EGR), diesel oxidation catalysts and multi-valve cylinder heads with three or four valves per cylinder were developed. Also, direct fuel injection (DI) technology was introduced during this period. Compared to IDI engines with swirl or prechambers, DI engines offer another 15–20% reduction in fuel consumption. The most recent combustion system development steps have been the development of 4-valve DI diesel engines and the application of high pressure common rail fuel injection systems.

A very important boundary condition for the introduction of these new features was significant progress in the development of fuel injection systems, electronic control systems, turbochargers, and catalysts. This was supported through the development of base engine components such as pistons, crankshafts or bearing shells which could resist the high thermal and mechanical loads of highly boosted high speed diesel engines. Advances in diesel fuel quality ('winter diesel', reduced sulphur content) helped to improve the critical operation of diesel engines at low ambient temperatures and reduce the exhaust emissions.

Today, the diesel powered cars are readily accepted by European customers, since most of the drawbacks which historically came along with diesels could be eliminated as poor vehicle performance, visible smoke, strong exhaust smell, objectionable noise, vibration and harshness (NVH) characteristics (especially after cold starts and during engine warm up) and poor cold startability. Today's turbocharged passenger car diesel engines achieve specific power values of more than 50 kW/l which is similar to good gasoline engines (see *Figure 22.1*). They also show specific torque values of up to 144 Nm/l, even at 1500–2000 rpm (see *Table 22.1*), while state-of-the-art naturally aspirated gasoline engines do not exceed 90–95 Nm/l at much higher speeds. The combination of very high low end torque and gasoline-engine-like power output offers superior vehicle performance and driveability.

Together with these achievements came significant improvements in fuel consumption and NVH characteristics, which make it very difficult under most driving conditions to determine that a diesel engine is under the hood when driving the vehicle. These characteristics and the European fuel and vehicle tax legislation are the reasons why diesel engines power some 50% of all new vehicles in some European countries today. The diesel engine penetration in most light-duty truck applications is even

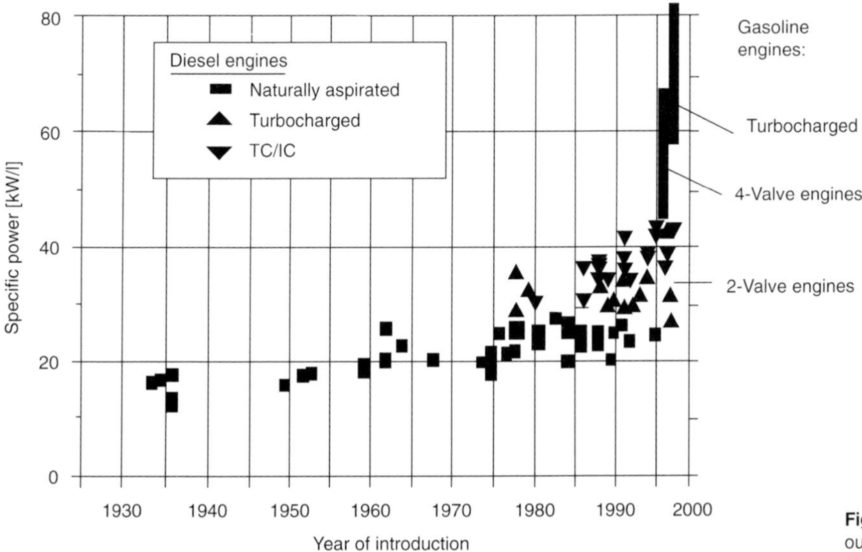

Figure 22.1 History of the specific power output of passenger car diesel engines

Table 22.1 Design and performance data passenger car diesel engines

State-of-the-art passenger car diesel engines (I)

Manufacturer		*Audi*	*BMW*	*BMW*	*BMW*	*BMW*	*Citroen*	*Citroen*	*Citroen*	*Citroen*	*Citroen*
Combustion system		DI	DI	IDI/SC	IDI/SC	IDI/SC	IDI/SC	IDI/SC	IDI/SC	IDI/SC	IDI/SC
Fuel injection system		VP 44	VP 44								
Valves per cylinder		4	4	2	2	2	2	2	2	3	3
Boosting system		TC/IC/ VGT	TC/IC	TC/IC	TC	TC/IC	NA	NA	TC/IC	TC/IC	TC/IC
Bore	mm	78.3	84	80	80	80	77	83	83	85	92
Stroke	mm	86.4	88	82.8	82.8	82.8	82	88	88	92	92
Cylinder size	cc	416	488	416	416	416	382	476	476	522	612
Cylinder arrangement		V6	I4	I4	I6	I6	I4	I4	I4	I4	I4
Displacement	cc	2496	1951	1665	2497	2497	1527	1905	1905	2088	2446
Stroke/bore ratio		1.10	1.05	1.04	1.04	1.04	1.06	1.06	1.06	1.08	1.00
Compression ratio		19.5	19	22	22	22	23	23	21.8	21.5	22
Cylinder spacing	mm	88	91	91	91	91					
Conrod length	mm	158	135	130	130	130					
Main bearing dia	mm	65	60	60	60	60					
Conrod bearing dia.	mm	58	45	45	45	45					
Piston pin dia.	mm		30	27	27	27					
Rated power	kW	110	100	66	85	105	40	51	66	80	95
Rated speed	rpm	4000	4000	4400	4800	4800	5000	4600	4000	4300	4300
Mean piston speed	m/s	11.5	11.7	12.1	13.2	13.2	13.7	13.5	11.7	13.2	13.2
Specific power	kW/l	44.1	51.3	39.6	34.0	42.0	26.2	26.8	34.7	38.3	38.8
BMEP at rated power	bar	13.2	15.4	10.8	8.5	10.5	6.3	7.0	10.4	10.7	10.8
Max. torque	Nm	310	280	190	222	260	95	120	196	250	285
Max. torque speed	rpm	1500	1750	2000	2200	2200	2250	2000	2250	2000	2000
Specific torque	Nm/l	124	144	114	89	104	62	63	103	120	117
Max. BMEP	bar	15.6	18.0	14.3	11.2	13.1	7.8	7.9	12.9	15.0	14.6
Max. cyl. pressure	bar	140	160								

State-of-the-art passenger car diesel engines (II)

Manufacturer		*Fiat*	*Fiat*	*Fiat*	*Fiat*	*Fiat*	*Fiat*	*Fiat*	*Fiat Alfa Romeo*	*Fiat Alfa Romeo*
Combustion system		IDI	IDI	IDI	IDI	IDI	IDI	IDI	DI	DI
Fuel injection system									Com. Rail	Com. Rail
Valves per cylinder		2	2	2	2	2	2	3	2	2
Boosting system		TC/IC	TC/IC	TC	TC/IC	TC/IC	TC/IC	TC/IC	TC/IC	TC/IC/VGT
Bore	mm	82.6	82.6	82	82	82	83	85	82	82
Stroke	mm	79.2	79.2	90.4	90.4	90.4	88	92	90.4	90.4
Cylinder size	cc	424	424	477	477	477	476	522	477	477
Cylinder arrangement		I4	I4	I4	I4	I5	I4	I4	I4	I5
Displacement	cc	1698	1698	1910	1910	2387	1905	2088	1910	2387
Stroke/bore ratio		0.96	0.96	1.10	1.10	1.10	1.06	1.08	1.10	1.10
Compression ratio		19	19	20.7	20.7	20.7	19.2	21.5	18.5	18.5
Cylinder spacing	mm									
Conrod length	mm									
Main bearing dia.	mm									
Conrod bearing dia.	mm									
Piston pin dia.	mm									
Rated power	kW	46	51	55	74	91	66	80	77	100
Rated speed	rpm	4500	4500	4200	4200	4000	4000	4300	4000	4200
Mean piston speed	m/s	11.9	11.9	12.7	12.7	12.1	11.7	13.2	12.1	12.7
Specific power	kW/l	27.1	30.0	28.8	38.8	38.1	34.7	38.3	40.3	41.9
BMEP at rated power	bar	7.2	8.0	8.2	11.1	11.4	10.4	10.7	12.1	12.0
Max. torque	Nm	118	134	147	200	265	196	250	255	304
Max. torque speed	rpm	2500	2500	2750	2250	2000	2250	2000	2000	2000
Specific torque	Nm/l	70	79	77	105	111	103	120	134	127
Max. BMEP	bar	8.7	9.9	9.7	13.2	14.0	12.9	15.0	16.8	16.0
Max. cyl. pressure	bar									

(Contd.)

Table 22.1 Contd

State-of-the-art passenger car diesel engines (III)

Manufacturer		*Ford*	*Ford*	*Ford*
Combustion system		IDI/SC	IDI/SC	IDI/SC
Fuel injection system				
Valves per cylinder		2	2	2
Boosting system		NA	TC	TC/IC
Bore	mm	82.5	82.5	82.5
Stroke	mm	82	82	82
Cylinder size	cc	438	438	438
Cylinder arrangement		I4	I4	I4
Displacement	cc	1753	1753	1753
Stroke/bore ratio		0.99	0.99	0.99
Compression ratio		21.5	21.5	21.5
Cylinder spacing	mm			
Conrod length	mm			
Main bearing dia.	mm			
Conrod bearing dia.	mm			
Piston pin dia.	mm			
Rated power	kW	44	51	66
Rated speed	rpm	4800	4500	4500
Mean piston speed	m/s	13.1	12.3	12.3
Specific power	kW/l	25.1	29.1	37.6
BMEP at rated power	bar	6.3	7.8	10.0
Max. torque	Nm	105	135	180
Max. torque speed	rpm	2500	2500	2000
Specific torque	Nm/l	60	77	103
Max. BMEP	bar	7.5	9.7	12.9
Max. cyl. pressure	bar			

State-of-the-art passenger car diesel engines (IV)

Manufacturer		*Mercedes*	*Mercedes*	*Mercedes*	*Mercedes*	*Mercedes*	*Mercedes*	*Mercedes*	*Mercedes*	*Mercedes*
Combustion system		DI	DI	DI	DI	IDI/PC	IDI/PC	IDI/PC	IDI/PC	IDI/PC
Fuel injection system		Com. Rail	Com. Rail	Com. Rail	VP 37		Dist. Pump	n-Line Pum	n-Line Pum	n-Line Pump
Valves per cylinder		4	4	4	2	2	4	4	4	4
Boosting system		TC	TC/IC	TC/IC/VGT	TC/IC	TC??	NA	TC/IC	NA	TC/IC
Bore	mm	80	80	88	89	89	89	87	87	87
Stroke	mm	84	84	88.4	92.4	92.4	86.6	84	84	84
Cylinder size	cc	422	422	538	575	575	539	499	499	499
Cylinder arrangement		I4	I4	I4	I5	I4	I4	I5	I6	I6
Displacement	cc	1689	1689	2151	2874	2299	2155	2497	2996	2996
Stroke/bore ratio		1.05	1.05	1.00	1.04	1.04	0.97	0.97	0.97	0.97
Compression ratio		19	19	19	19.5	22	22	22	22	22
Cylinder spacing	mm	90	90	97	97	97	97	97	97	97
Conrod length	mm	140	140	149	145		149	149	149	149
Main bearing dia.	mm			58	61		58	58	58	58
Conrod bearing dia.	mm			48	53		48	48	48	48
Piston pin dia.	mm			28	30					
Rated power	kW	44	66	92	95	72	70	110	100	130
Rated speed	rpm	3600	4200	4200	4000	3800	5000	4400	5000	4400
Mean piston speed	m/s	10.1	11.8	12.4	12.3	11.7	14.4	12.3	14.0	12.3
Specific power	kW/l	26.1	39.1	42.8	33.1	31.3	32.5	44.1	33.4	43.4
BMEP at rated power	bar	8.7	11.2	12.2	9.9	9.9	7.8	12.0	8.0	11.8
Max. torque	Nm	160	180	300	300	230	150	280	210	330
Max. torque speed	rpm	1500	1600	1800	1800	1700	3100	1800	2200	1600
Specific torque	Nm/l	95	107	139	104	100	70	112	70	110
Max. BMEP	bar	11.9	13.4	17.5	13.1	12.6	8.7	14.1	8.8	13.8
Max. cyl. pressure	bar			140			70			

(Contd)

Table 22.1 Contd

State-of-the-art passenger car diesel engines (VI)

Manufacturer		*Opel*	*Opel*	*Opel*	*Opel*	*Opel*	*Opel*	*Opel*	*Opel/Isuzu*
Combustion system		IDI/SC	IDI/SC	IDI/SC	IDI/SC	DI	DI	DI	IDI
Fuel injection system						VP 44	VP 44	VP 44	
Valves per cylinder		2	2	2	2	4	4	4	2
Boosting system			TC	TC	TC	TC	TC/IC	TC/IC	TC
Bore	mm	76	79	79	79	84	84	84	95.4
Stroke	mm	82	86	86	86	90	90	98	107
Cylinder size	cc	372	422	422	422	499	499	543	765
Cylinder arrangement		I4	I4	I4	I4	I4	I4	I4	I4
Displacement	cc	1488	1686	1686	1686	1995	1995	2172	3059
Stroke/bore ratio		1.08	1.09	1.09	1.09	1.07	1.07	1.17	1.12
Compression ratio		22	22	22	22	18.5	18.5	18.5	20
Cylinder spacing	mm					93	93	93	
Conrod length	mm					152	152	148	
Main bearing dia.	mm					68	68	68	
Conrod bearing dia.	mm					48	48	48	
Piston pin dia.	mm					31	31	31	
Rated power	kW	49	44	50	60	60	74	85	84
Rated speed	rpm	4600	4400	4500	4400	4300	4300	4300	3600
Mean piston speed	m/s	12.6	12.6	12.9	12.6	12.9	12.9	14.0	12.8
Specific power	kW/l	32.9	26.1	29.7	35.6	30.1	37.1	39.1	27.5
BMEP at rated power	bar	8.6	7.1	7.9	9.7	8.4	10.4	10.9	9.2
Max. torque	Nm	132	112	132	168	185	205	260	260
Max. torque speed	rpm	2600	2650	2400	2400	1800	1600	1900	2000
Specific torque	Nm/l	89	66	78	100	93	103	120	85
Max. BMEP	bar	11.1	8.3	9.8	12.5	11.7	12.9	15.0	10.7
Max. cyl. pressure	bar						150	150	

State-of-the-art passenger car diesel engines (VII)

Manufacturer		*PSA*	*PSA*	*PSA*	*PSA*	*PSA*	*Renault*	*Renault*	*Renault*	*Renault*	*Renault*
Combustion system		IDI/SC	IDI/SC	IDI/SC	IDI/SC	IDI/SC	IDI/SC	IDI/SC	IDI/SC	IDI/SC	DI
Fuel injection system											VP 37
Valves per cylinder		2	2	2	3	3	2	2	3	3	2
Boosting system		NA	NA	TC	TC/IC	TC/IC	NA	TC/IC		TC	TC/IC
Bore	mm	77	83	83	85	92	80	80	87	87	80
Stroke	mm	82	88	88	92	92	93	93	92	92	93
Cylinder size	cc	382	476	476	522	612	467	467	547	547	467
Cylinder arrangement		I4	I4	I4	I4	I4	I4	I4	I4	I4	I4
Displacement	cc	1527	1905	1905	2088	2446	1870	1870	2188	2188	1870
Stroke/bore ratio		1.06	1.06	1.06	1.08	1.00					1.16
Compression ratio		23	23	21.8	21.5	22	21.5	20.5	23		18.3
Cylinder spacing	mm				93						89
Conrod length	mm										139
Main bearing dia.	mm										53
Conrod bearing dia.	mm										48.3
Piston pin dia.	mm										28
Rated power	kW	40	50	66	80	95	47	66	61	83	72
Rated speed	rpm	5000	4600	4000	4300	4300	4500	4250	4500	4300	4000
Mean piston speed	m/s	13.7	13.5	11.7	13.2	13.2	13.9	13.2	13.8	13.2	12.4
Specific power	kW/l	26.2	26.3	34.7	38.3	38.8	25.1	35.3	27.9	37.9	38.5
BMEP at rated power	bar	6.3	6.8	10.4	10.7	10.8	6.7	10.0	7.4	10.6	11.6
Max. torque	Nm	95	120	196	250	285	118	176	142	234	200
Max. torque Speed	rpm	2250	2000	2250	2000	2000	2250	2000	2250	2000	2000
Specific torque	Nm/l	62	63	103	120	117	63	94	65	107	107
Max. BMEP	bar	7.8	7.9	12.9	15.0	14.6	7.9	11.8	8.2	13.4	13.4
Max. cyl. pressure	bar										

(Contd)

Table 22.1 Contd

State-of-the-art passenger car diesel engines (VIII)

Manufacturer		*Rover*	*Rover*	*Toyota*	*Toyota*	*Toyota*	*VM Motori DDC*	*VM Motori DDC*	*VM Motori DDC*	*VM Motori DDC*	*VM Motori DDC*
Combustion system		DI	DI	IDI	IDI	IDI	IDI/SC	IDI/SC	IDI/SC	DI	DI
Fuel injection system		VP 37	VP 37								
Valves per cylinder		2	2	2	4	2	2	2	2	2	4
Boosting system		TC	TC/IC	NA	TC	TC/IC	TC/IC	TC/IC	TC/IC	TC/IC	TC/IC
Bore	mm	84.5	84.5	86	86	86	92	92	92	92	92
Stroke	mm	88.9	88.9	85	85	94	94	94	94	94	94
Cylinder size	cc	499	499	494	494	546	625	625	625	625	625
Cylinder arangement		I4	I4	I4	I4	I4	I4	I5	I6	I4	I4
Displacement	cc	1994	1994	1975	1975	2184	2500	3124	3749	2500	2500
Stroke/bore ratio				0.99	0.99	1.09	1.02	1.02	1.02	1.02	1.02
Compression ratio		19.5	19.5	23	23	22	21.5	21.5	21.5		
Cylinder spacing	mm										
Conrod length	mm										
Main bearing dia.	mm	57	57								
Conrod bearing dia.	mm	54	54								
Piston pin dia.	mm										
Rated power	kW	63	77	53	61	74	92	110	118	100	114
Rated speed	rpm	4500	4200	4600	4000	4200	4200	4200	3500		
Mean piston speed	m/s	13.3	12.4	13.0	11.3	13.2	13.2	13.2	11.0		
Specific power	kW/l	31.6	38.6	26.8	30.9	33.9	36.8	35.2	31.5	40.0	45.6
BMEP at rated power	bar	8.4	11.0	7.0	9.3	9.7	10.5	10.1	10.8		
Max. torque	Nm	170	210	131	174	216	300	380	400	320	340
Max. torque speed	rpm	2000	2000	2500	2000	2600	2000	2000	2000		
Specific torque	Nm/l	85	105	66	88	99	120	122	107	128	136
Max. BMEP	bar	10.7	13.2	8.3	11.1	12.4	15.1	15.3	13.4	16.1	17.1
Max. cyl. pressure	bar										

State-of-the-art passenger car diesel engines (IX)

Manufacturer		*VW*	*VW*	*VW*	*VW*	*VW*	*VW*
Combustion system		IDI/SC	DI	DI	DI	IDI/SC	DI
Fuel injection system			VP 37	VP 37	VP 37		VP 37
Valves per cylinder		2	2	2	2	2	2
Boosting system		NA	NA	TC/IC	TC/IC/VGT	NA	TC
Bore	mm	79.5	79.5	79.5	79.5	79.5	81
Stroke	mm	95.5	95.5	95.5	95.5	95.5	95.5
Cylinder size	cc	474	474	474	474	474	492
Cylinder arrangement		I4	I4	I4	I4	I5	I5
Displacement	cc	1896	1896	1896	1896	2370	2461
Stroke/bore ratio		1.20	1.20	1.20	1.20	1.20	1.179
Compression ratio		22.5	19.5	19.5	19.5	22.5	19.5
Cylinder spacing	mm	88	88	88	88	88	88
Conrod length	mm	144	144	144	144	144	144
Main bearing dia.	mm	54	54	54	54	54	54
Conrod bearing dia.	mm	47.8	47.8	47.8	47.8	47.8	47.8
Piston pin dia.	mm		26	26	26		
Rated power	kW	47	47	66	81	57	75
Rated speed	rpm	4400	4200	4000	4150	3700	3500
Mean piston speed	m/s	14.0	13.4	12.7	13.2	11.8	11.1
Specific power	kW/l	24.8	24.8	34.8	42.7	24.0	30.5
BMEP at rated power	bar	6.8	7.1	10.4	12.4	7.8	10.5
Max. torque	Nm	124	124	210	235	164	250
Max. torque speed	rpm	2200	2200	1900	1900	1800	1900
Specific torque	Nm/l	65	65	111	124	69	102
Max. BMEP	bar	8.2	8.2	13.9	15.6	8.7	12.8
Max. cyl. pressure	bar			135	155		

higher in Europe. Many experts forecast a bright future for the diesel engine in passenger car and light duty truck applications and even higher market shares.

Currently, several development programs intended to bring diesel engines into pick-up trucks and sport utility vehicles (SUV) in the United States have been launched. However, many challenges appear on the horizon for diesel engines. In particular exhaust emissions (especially NOx and particulates) must be drastically reduced, to cope with future exhaust emission legislation which may no longer differentiate between diesel and gasoline engine emission standards. While at the same time reducing the exhaust emissions, fuel consumption must be maintained or even further reduced.

Another major challenge is production cost. Compared to

gasoline engines, diesel engines have always been more costly to produce. With the need for application of electronically controlled high pressure fuel injection systems, in combination with very efficient exhaust gas aftertreatment systems, the cost penalty for the diesel engine may become even larger in the future. Accordingly alternatives to the diesel engine, such as gasoline engines with direct fuel injection, will become major competitors.

Therefore, to increase marketshare it will be necessary to develop future diesel engines for passenger car and light-duty applications which offer additional advantages for the customer, when evaluating such factors as initial investment cost, operational cost, vehicle driveability, engine durability or vehicle resale value. Against this backdrop, the following sections discuss current engine technology and performance characteristics as well as future development potential.

22.2 Vehicle specific requirements

Customers buy diesel passenger cars not engines. Of course, they want the traditional diesel engine advantages, such as low fuel consumption and high durability, but they also want to get good vehicle performance, driveability and driving comfort. To achieve these goals, the development process of the engine and the vehicle must be interrelated more than ever before. The diesel engine is just one element in the total vehicle system which consists of the major elements such as the body, chassis, axle and suspension elements, the powertrain and the electronic control systems for the engine and the vehicle. To fulfil the customer requirements, this entire system must be optimized for best vehicle performance/driveability, fuel consumption, safety and comfort. In addition, environmental aspects such as exhaust gas and noise emissions, energy and material consumption during manufacturing process and the ability for recycling are becoming increasingly important. Production cost and quality are always focal points.

To achieve superior *vehicle performance, driveability* and *fuel consumption*, the number of gears, the gear ratios and the final drive ratio of the transmission must be optimized together with the torque curve of the engine. In the past, when exhaust emissions were not such a concern, it was possible, to optimize these parameters without considering their influence on the emissions behaviour in the various emission certification test cycles. Today and even more in the future, this approach is no longer possible. The implications of the transmission and complete powertrain parameters on the emission test results must be taken into account. Therefore, different torque curves and transmission settings may result even if the same engine is used for different vehicle applications.

Of special importance is the final drive ratio of the transmission. Lower final drive ratios ('longer axles') result in lower speeds and higher engine loads at the same power output. This, in turn, results in engine operating points closer to the minimum fuel consumption 'island' in the BSFC map. Similar effects can be achieved with reduced engine displacements ('engine downsizing'). Smaller engine displacements lead to reduced exhaust gas mass flows which tends to improve exhaust emissions in the certification test cycles. However, vehicle driveability and launch characteristics set limitations regarding final drive ratio and engine displacement reduction.

Another important aspect is the speed range of the diesel engine compared to gasoline engines of similar specific power. High speed diesel engines are limited regarding their maximum speeds. As in heavy-duty truck applications, smaller speed regimes necessitate for a higher number of gears. This issue becomes more critical with increasing top speed of the vehicle. Consequently, Audi introduced a manual six-speed transmission already with their first high speed DI (HSDI) diesel engine.

Safety and *comfort* are other important aspects for the customer. The electronic engine controller must detect irregularities in the engine operation and provide the driver with appropriate feedback. It must also assure a 'limp-home' mode which allows limited engine operation, even, if components of the engine or its control systems malfunction. In addition, functions which indicate maintenance requirements may also be included in the electronic control system.

Passenger comfort refers to issues such as good NVH, favourable heating and air conditioning performance as well as ease of vehicle operation. To achieve favourable NVH characteristics, it is not only necessary to reduce the airborne noise but also to address the vibration excitation of the vehicle body through the engine. Vibration of steering wheel columns, dashboards, pedals, seats or rear mirrors which were historically typical for diesel-powered cars, especially during idling, are no longer acceptable. Optimization of the engine mounting system in the vehicle together with the structure of the vehicle body is required to achieve this goal.

Production costs must also be optimized for the entire engine/vehicle system. This refers to issues such as ease of installation of the engine and its subsystems in the vehicle, and even the engine testing itself. Typically, passenger car diesel engines undergo a hot test (fired engine operation) at the end of the engine production line, while cold testing (engine motoring without firing) has increasingly become the standard for gasoline engine production.

22.3 Current engine technology

When the previous edition of this book was published, a somewhat brief discussion of the state-of-the-art in passenger car diesel engine technology was possible. At that time, all passenger car diesel engines employed indirect fuel injection (IDI) with prechambers (Mercedes-Benz) or swirl chamber (all the others). The typical swirl chamber designs were similar to the Ricardo Comet V type and only two valves per cylinder were utilized. The standard fuel injection pump was a mechanically controlled distributor pump (only Mercedes-Benz used in-line pumps) and no exhaust gas aftertreatment was applied to production engines. All European passenger car and light-duty truck diesel engines had an in-line cylinder arrangement with 4 to 6 cylinders. Light-duty diesel engines, with V-cylinder arrangements, were built only in the US by General Motors and Navistar International. Most European and Japanese passenger car diesel engines used cast iron cylinder blocks in combination with slant aluminum cylinder heads.

Current passenger car and light-duty truck diesel engines exhibit a much wider variety of design and combustion system features. There are distinctly different fuel injection systems being used on the current engines and exhaust gas aftertreatment with oxidation catalysts has become a standard for all markets in which stringent exhaust emission regulations are enforced.

22.3.1 Combustion systems

IDI combustion systems with two separated combustion chambers (volume ratio about 1 : 1) which are connected through one 'throat' (swirl chamber) or several 'blow holes' (prechamber) were the standard until the late 1980s. These concepts are still utilized in the majority of current passenger car diesel engines.

During the compression stroke in an IDI diesel engine, air is forced through the passages from the main combustion chamber into the swirl or prechamber. Thereby, highly turbulent air flow is generated in which the fuel is injected at the end of the compression stroke. The compression itself and the intensive heat transfer from the hot walls of the swirl or prechamber, result in a very high temperature level at the start of injection and thus a short ignition delay and low combustion noise excitation. After the onset of combustion, a rapid pressure increase occurs in the air–fuel mixture which is rich at high engine loads. The unburned or partially burned gases are subsequently blown through the passages into the main combustion chamber, where the second stage of the combustion event takes place. During this second combustion stage, further mixing of the relatively rich gases from; the swirl or prechamber with the pure air in the main chamber is essential. This second stage mixture formation is accomplished through a suitable design of the throat or the blow holes between the swirl or prechamber and the main combustion chamber together with a carefully optimized design of the piston bowls which are typically very shallow.

Over the years, these well-proven IDI combustion systems have been continuously improved. *Figures 22.2* and *22.3* show two IDI system examples. BMW concentrated on the optimization of the Ricardo Comet V combustion chamber layout and introduced a modified shape of the piston bowl (V-bowl) to improve the second stage of the combustion process[1]. Through this design change, it was possible to improve the air utilization and to increase the full load performance at constant smoke emission levels.

Figure 22.3 shows another modification of the classic Ricardo Comet V swirl chamber design with a glow plug installed in a vertical position[2]. Relative to the air motion in the swirl chamber, this is downstream of the injection nozzle, while the traditional horizontal position is upstream of the injector. With this modified glow plug position, the impingement of the fuel spray on the glow plug is reduced and thus the mixture formation is improved with the result of a reduced primary soot formation.

Mercedes-Benz also improved its prechamber combustion system by modifying the general shape of the prechamber cross-section to a more round design and by inclining the injector relative to the combustion chamber axes (in the traditional prechamber design, the prechamber axis and the injector axis were identical). Together with a carefully optimized design of the so-called impingement pin, which exhibits an inclined flat surface on one side, a more controlled tumble motion, rather than an 'uncontrolled' highly turbulent air flow pattern, is created in the prechamber. Similar to the swirl chamber combustion system, the fuel is injected in the direction of the tumble motion. These design features and a moderate increase of the injection pressure helped to significantly reduce the particulate emission.

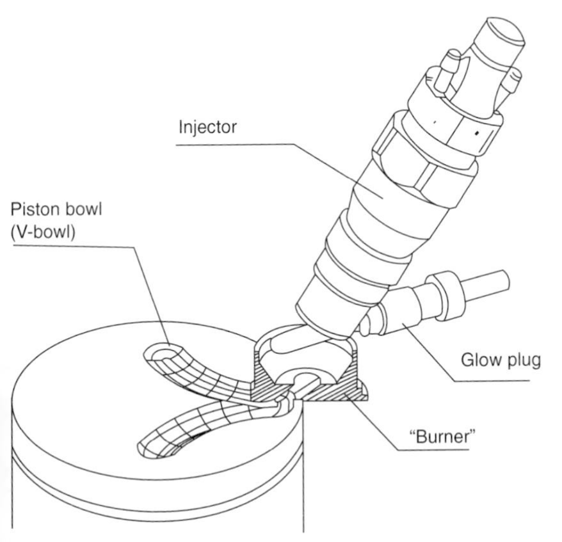

Figure 22.2 Combustion chamber layout BMW swirl chamber engines[1]

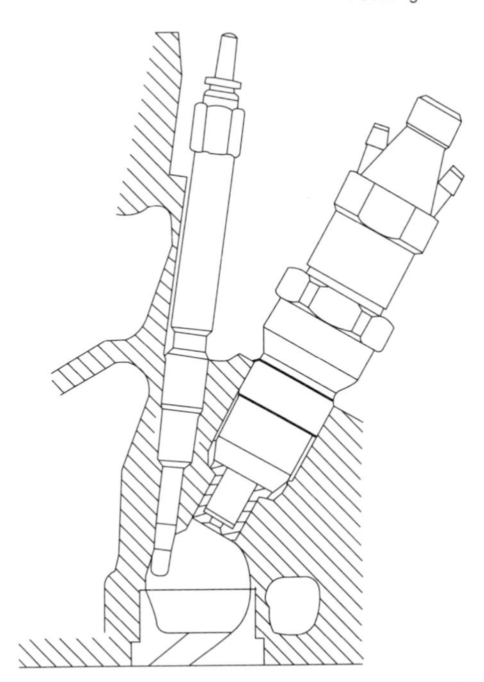

Figure 22.3 Swirl chamber with glow plug in vertical position[2]

The modified prechamber design was first introduced in 2-valve engines, but then carried over to 4-valve cylinder heads (see *Figure 22.5*).

In addition to the 'fine tuning' of the IDI combustion systems, the following major design features have been introduced over the last ten years:

- Turbocharging in combination with charge air cooling
- Multivalve cylinder heads for IDI engines
- Direct fuel injection (DI) together with electronically controlled high pressure fuel injection systems
- Turbochargers with variable turbine geometry
- Four-valve cylinder heads for DI engines
- Common rail fuel injection systems.

For many years, *charge air cooling* (intercooling) has been a standard approach to increase the power output and to reduce NOx emissions for truck and large marine diesel engines. Through this measure, the air mass which is trapped in the cylinder at the end of the gas exchange cycle can be significantly increased due to higher density. The engine can burn more fuel and deliver more power at the same boost pressure. The lower charge temperature at the beginning of the compression stroke also helps to reduce the thermal load of the cylinder head and the piston which are critical components, especially in highly boosted IDI diesel engines. In addition, lower charge air temperatures result in reduced NOx formation during the combustion process.

For vehicle applications, this is especially true if air-to-air intercooling is applied. Cost and packaging of the intercooler in the vehicle have been and remain major issues regarding the introduction of intercooling. Manufacturers have observed significant benefits due to charge air cooling, even if the size and/or the placement of the turbocharger in the vehicle were less than optimal.

Three primary companies developed *multivalve* car diesel engines with more than two valves per cylinder; PSA and Renault introduced swirl chamber engines with three valves per cylinder (see *Figure 22.4*)[3] and Mercedes-Benz developed prechamber engines with 4-valve cylinder heads[4].

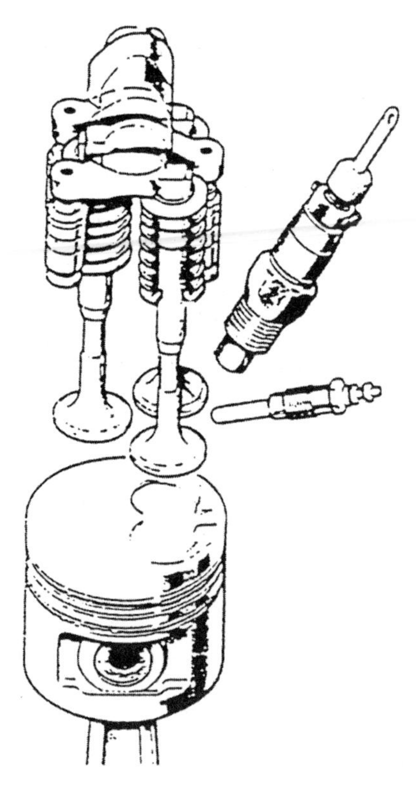

Figure 22.4 Injector, glow plug and valve arrangement in a PSA 3-valve IDI engine[3]

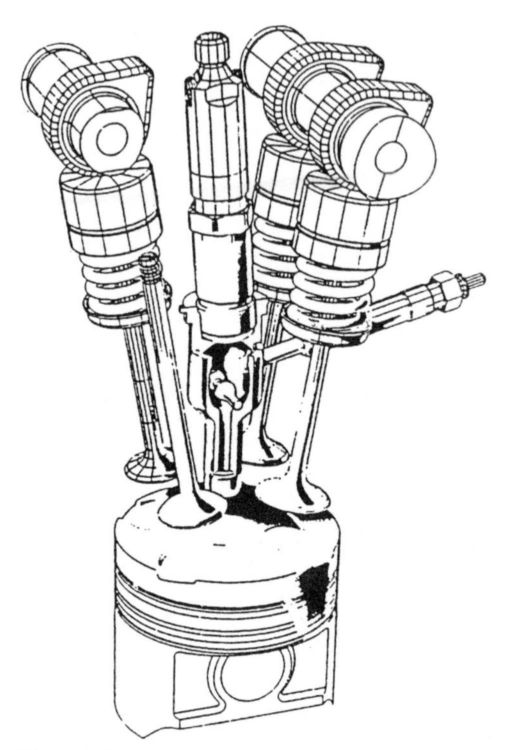

Figure 22.5 Prechamber and valve arrangement in Mercedes-Benz 4-valve IDI engines[4]

The 3-valve arrangement with two intake valves and one exhaust valve is the logical approach, if the traditional swirl chamber design is used. The typical shape of the swirl chamber as well as its eccentric position relative to the cylinder axis precludes space for a second exhaust valve. The major advantage of this arrangement is better volumetric efficiency, while the parameters for mixture formation and combustion remain similar to the swirl chamber arrangement in a 2-valve cylinder head. Opposed to Renault, PSA uses 3-valve cylinder heads only for turbocharged engines to increase the specific power output (see *Table 22.1*).

Mercedes-Benz applies its 4-valve prechamber technology to naturally aspirated and turbocharged engines (see *Figure 22.5*). Compared to the 3-valve swirl chamber arrangement, the 4-valve prechamber arrangement offers additional advantages which result from the central location of the prechamber in the cylinder axis.

This arrangement provides very good boundary conditions for even mixture distribution and combustion in the main chamber. Through optimization of combustion chamber design parameters, both good air utilization and low particulate emissions under part load operating conditions could be achieved. The 4-valve arrangement also reduces the pumping losses, especially at high speeds. Together with the favourable mixture preparation characteristics of IDI combustion systems at high speeds, it was possible to increase the rated speed to 5000 rpm and achieve a specific power output of 33 kW/l. This is the highest value of all current naturally aspirated diesel engines (see also *Table 22.2*). Another advantage of the 4-valve arrangement is more even thermal loading of the piston which is especially important for highly boosted engine versions.

Other manufacturers of car diesel engines chose not to follow the multi-valve IDI engine approach. They were either able to achieve almost the same specific power output with their 2-valve engines (e.g. BMW) or they had already concentrated on direct fuel injection (DI).

The introduction of *high speed direct injected* (*HSDI*) diesel engines is perhaps the most important recent achievement in light-duty diesel engine technology. Because of the fuel consumption advantages over IDI diesel engines, direct fuel injection had long ago become the standard for larger diesel engines used in truck, off-road or marine applications. Major technical hurdles had to be overcome before the DI technology could be applied to small high speed diesel engines. One major challenge was the tuning of the mixture formation and combustion parameters for the wider engine speed range which is typical for all light-duty applications. Experiments with derivatives from the MAN-M combustion system with deep piston bowls, high swirl levels and peripheral single spray fuel injection at relatively low injection pressures were not very successful[5]. Therefore, significant effort concentrated on the classical DI combustion system approach with 'Mexican hat' type combustion chambers and multi–hole injection nozzles. To date, the wide speed range of HSDI engines and limited pressure capabilities of the fuel injection equipment do not allow quiescent type combustion systems to be used which are increasingly becoming the standard for many larger DI diesel engines. Therefore, careful tuning of the intake swirl is required to achieve good combustion characteristics.

First generation HSDI diesel engines were introduced by Ford and Fiat (Iveco) for light-duty truck applications[6]. These engines were initially naturally aspirated and exhibited many disadvantages that had prevented earlier introduction of the DI technology for light-duty applications, such as unfavourable NVH characteristics, poor air utilization and high NOx and HC emissions. However, these first generation HSDI diesel engines achieved a large reduction in fuel consumption which historically has always been of primary concern in commercial applications.

The major breakthrough for the HSDI technology was the introduction of the so-called TDI (**t**urbocharged **d**irect **i**njection) five and four cylinder engines from Audi[7] and Volkswagen[8]. *Figure 22.6* shows the combustion system arrangement of these engines. A similar geometry can also be found on most 2-valve HSDI diesel engines from other manufacturers. However, in some cases 'crossflow', opposed to 'sideflow', cylinder heads have been realized[9].

A helical intake port is used to create sufficient swirl during the intake stroke. At the end of the compression stroke, the rotating air is forced into the re-entrant piston bowl which has a significantly smaller diameter than the cylinder, thereby, further increasing the rotational speed of the air. Squish effects and a careful design of the piston bowl rim further increase the turbulence level of the air in the bowl. However, through these measures, the air motion is still about one order of magnitude lower than is the case in a swirl or prechamber of an IDI diesel engine.

The use of multi-hole injection nozzles (with five spray holes in the case of Audi/Volkswagen engines) and much higher injection pressure than in IDI engines are necessary to achieve good air–fuel mixing in the combustion bowl. Injection pressures of up to 900 bar at the nozzle have been realized in the first generation TDI engines.

The shape of the piston bowl very much defines the combustion chamber of the DI diesel engine. It is essential to concentrate as much air as possible in the piston bowl at the end of the compression stroke to achieve good utilization of the air which has been trapped in the cylinder. In the TDC position of the piston, the parasitic volumes outside the piston bowl must be minimized. A long stroke design favours this requirement. A 2-valve cylinder head configuration does not allow the injection nozzle, to be installed in the middle of the cylinder, if valves with a sufficient diameter are used. Therefore, it is necessary to off-set the piston bowl relative to the piston axis. Also, it is typically necessary to incline the injector.

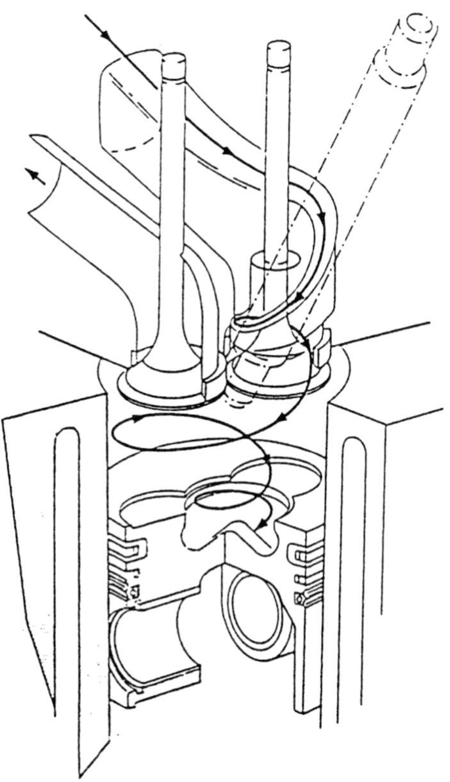

Figure 22.6 Combustion system of the Volkswagen/Audi 'TDI engines'[8]

The TDI engines were turbocharged and intercooled and incorporated an electronically-controlled, high pressure distributor fuel injection pump. Together with the application of so-called VCO (valve covers orifice) injector tips, exhaust gas recirculation and oxidation catalysts, it was possible to achieve a good compromise between full load performance and exhaust gas emissions. Two-stage injection, which was realized with a two-spring injector, limited the combustion noise excitation to an acceptable level even for use in up-scale vehicles.

The development of such a DI combustion system always necessities a compromise between several major parameters including: swirl level, piston bowl diameter, compression ratio, spray hole number, spray hole size, injection pressure and boost pressure. These parameters must be carefully tuned to achieve the desired power output and emissions characteristics at acceptable cylinder pressures and thermal loads of the power cylinder components, while maintaining good cold startability.

Compared to IDI combustion systems, the major advantages of DI combustion systems can be summarized as follows:

- 15–20% lower fuel consumption, through reduced heat losses during the combustion cycle
- Better cold startability (allowing the compression ratio to be reduced)
- Better preconditions for boosting (higher power output potential) because of the lower compression ratio and the lower thermal loading of the piston and the cylinder head
- Lower degradation of the lube oil through combustion residues (allows oil change intervals to be extended).

However, some important disadvantages of the DI combustion system, compared to the IDI combustion system still remain:

- Significantly higher fuel injection pressure requirements which result in higher costs for the injection system
- Higher NOx raw emissions require very high EGR rates and even EGR cooling to meet future emission standards.

Other disadvantages of the DI-combustion system which were historically cited, are no longer valid. Advanced 2-valve DI engines with high pressure fuel injection systems achieve the same air utilization as 2-valve IDI engines. One example is the naturally aspirated (NA) passenger car diesel engine from Volkswagen[10]. Also, the combustion noise excitation of a NA DI diesel engine can be limited to a level similar to that of swirl chamber diesel engines if staged injection with a two spring nozzle holder is applied. However, prechamber IDI diesel engines still exhibit the lowest combustion noise excitation of the three concepts.

Due to their unique combination of favourable vehicle performance and fuel combustion characteristics, the Audi/Volkswagen TDI 'diesel engines have been commercially very successful, gaining customers who had not previously considered buying a diesel. Therefore, a logical step was to further increase the power output to gain even better vehicle performance. Volkswagen realized this through application of a *variable geometry turbocharger* (*VGT*) to its 1.9L engine[11]. The VGT principle provides higher boost and thus higher engine torque at low speeds. However, it also delivers more air at rated conditions with the same exhaust back pressure and exhaust gas temperature level. Therefore, the power of the engine can be increased without an increase in thermal loading. Additional HSDI diesel engines with VGT turbochargers have also been introduced[12,13].

The next very important step to further improve HSDI diesel combustion was the application of *4-valve cylinder heads*. Compared to 2-valve engines, the 4-valve valve design offers the following advantages:

- Reduced pumping losses and improved volumetric efficiency, especially at higher speeds (higher power output potential and improved fuel economy).
- Vertical and central position of the injection nozzle and the piston bowl results in a more even distribution of the injected fuel into the combustion air, reduced requirements for air flow assistance in mixture formation (reduced swirl demand) and more symmetrical thermal loading of the piston.
- Improved mixture formation parameters in combination with moderate swirl levels, increased injection pressures and a higher number of spray holes (more quiescent type combustion system) improve the particulate-NOx trade-off significantly, especially at retarded injection timing.
- Two separated intake ports offer favourable preconditions for variable swirl concepts through deactivation of one intake port.

These advantages were well known from recent heavy-duty truck engine developments. Also, private engine development companies have demonstrated the improvement potential of the 4-valve technology for HSDI engines several years before their introduction[14,15].

Adam Opel AG/General Motors finally was the first manufacturer that introduced a HSDI engine with a 4-valve cylinder head[16]. *Figure 22.7* illustrates the cylinderhead design and the port deactivation (variable swirl) system of the Opel ECOTEC diesel engine. Shortly after the launch of the Opel 4-valve HSDI diesel engine, Audi[12] and Mercedes-Benz[13,17] also introduced new engines with 4-valve cylinder heads.

The most recent step in HSDI diesel development, is the application of *common rail fuel injection systems*[18] to HSDI diesel engines[13,17]. Compared to conventional pump-line-nozzle injection systems, electronically controlled common rail fuel injection systems offer major advantages through a free mapping of the injection timing, fuel quantity, and injection pressure (see also Section 22.3.3). Common rail systems also allow multiple injections (pre- and post-injection). In particular the possibility to achieve high injection pressures at low loads and speeds helps to overcome one of the classic conflicts in DI diesel combustion system tuning. With conventional fuel injection systems, the hydraulic nozzle flow and the swirl level typically have been optimized for the rated power operating point. This approach leads to nozzle spray hole diameters which are too large at lower speeds and loads, which results in relatively low injection pressures and unfavourable mixture formation properties. The other major advantage of common rail systems is

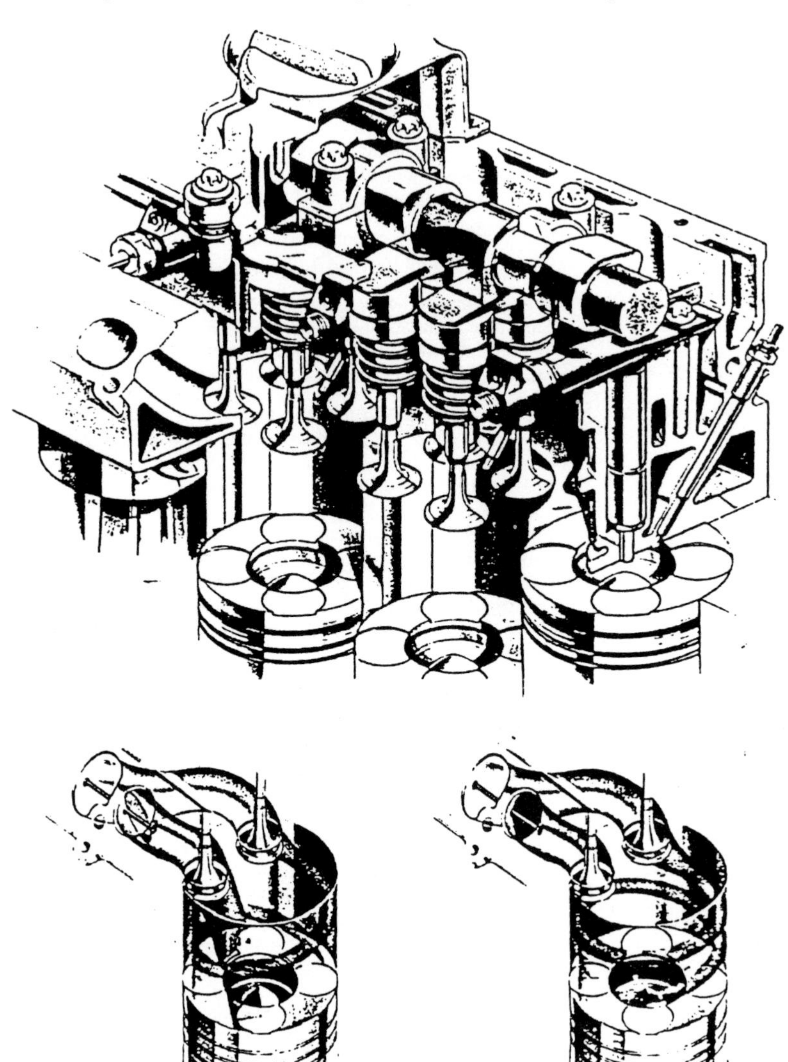

Figure 22.7 Cutaway of the Opel ECOTEC 4-valve cylinder head and port throttling arrangement[16]

the possibility to realize pre-injection which significantly reduces the combustion noise excitation. Post-injections, after the main injection event, will become more important as active lean NOx catalysts are developed to further reduce the NOx emission (see Section 22.5.3)

It is most likely that multi-valve IDI engines will disappear from production. They will be replaced by multi-valve DI engines. The only IDI engines which may remain in the future will be low cost (naturally aspirated) engines for smaller inexpensive vehicles.

22.3.2 Design features

Advanced high speed diesel engines for passenger car and light-duty applications often are members of larger engine families which are manufactured with the same equipment. They share many design features with the gasoline members of the engine families, such as cast iron cylinder blocks with integral liners and aluminum cylinder heads with overhead camshafts. Camshaft timing drives are typically accomplished with tooth belts or chains and the injection pump drive is often integrated into the camshaft timing drive.

The major differences in the design approaches for diesel engines with the different combustion systems, as previously described, involve the cylinder heads. No major differences in the block design for IDI and DI engines are, in principle, necessary. However, since the combustion chamber is in the piston of a DI engine, it may be necessary to increase the block height to a certain extent.

The following illustrations provide a short overview of current passenger car diesel engine design features. *Figure 22.8* shows a longitudinal and a cross section of the turbocharged and intercooled 1.7L 4-cylinder 2-valve swirl chamber diesel engine from BMW[1].

This engine achieves a relatively high specific power output of almost 40 kW/l (see also Table 22.1) while requiring gallery oil cooling for the pistons. It employs a cross-flow cylinder head with one camshaft which acts on the valves via bucket tappets with integrated hydraulic lash adjusters. The electronically controlled distributor fuel injection pump and the camshaft are chain driven. A single poly-V belt drive with automatic tensioner is used to drive the accessories. The disposable-cartridge oil filter and the oil cooler are integral and generously sized to allow long oil change intervals and to provide efficient oil cooling. The plastic intake system helps to limit engine weight.

Figure 22.9 shows another IDI engine example, the XUD 11A TE 2.1L 3-valve engine from PSA (Peugeot)[3]. The gray cast iron block for this engine was derived from a gasoline engine block. Like the BMW engines, this PSA diesel engine is turbocharged and intercooled. At a specific power output of 38 kW/l, it also uses gallery piston oil cooling. The two intake ports and the exhaust port for each cylinder are located on the same side of the aluminum cylinder head. The lower portion of the two-piece cylinder head assembly houses the gas exchange ports, valves, swirl chambers with injectors, glow plugs and the water jacket.

Sixteen (16) bolts are used to attach the upper portion of the cylinder head with the camshaft bearings to the lower portion. The valve train uses finger followers, hydraulic lash adjusters and one overhead camshaft with three (3) cam lobes for each cylinder. Both the camshaft and distributor injection pump are driven by a single tooth belt. A tuned intake system with long individual runners is used to improve the volumetric efficiency in the lower speed range (max. specific torque: 120 Nm/l @ 2000 rpm).

Figure 22.10 shows the cross-section of the Mercedes-Benz 4-valve prechamber 5-cylinder engine[4]. This engine belongs to a family of 4-, 5-, and 6-cylinder in-line engines which share the cylinder spacing (97 mm) with a gasoline engine family, allowing combined manufacturing. Naturally aspirated and turbocharged versions of the 5- and 6-cylinder engines are offered. The naturally aspirated 4-cylinder engine has a larger bore and stroke than the other engines (89 × 86.6 mm vs. 87 × 84 mm).

The engine block material is gray cast iron, whereas nodular cast iron is used for the crankshafts of the naturally aspirated 4-cylinder engines. However, steel crankshafts are employed in the 5- and 6-cylinder engines. The aluminum crossflow cylinder head is designed with a separate camshaft carrier. The valve train with two overhead camshafts and hydraulic bucket tappets activates the inclined intake and exhaust valves. Inclination of the valves in the aluminum cylinder head (valve angle 10°) is necessary to install the prechamber in the desired central location.

A duplex chain drives the exhaust camshaft and the fuel injection pump. Gears are used to synchronize the camshafts

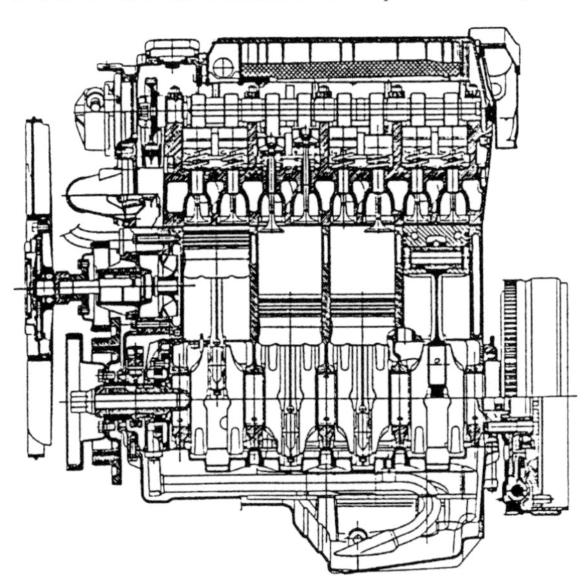

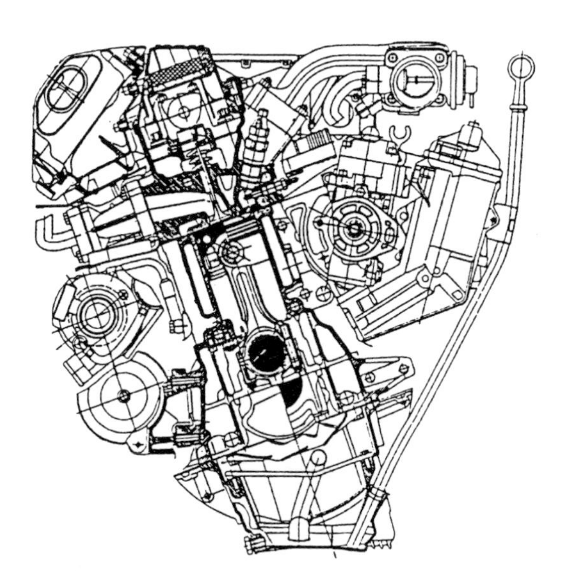

Figure 22.8 1.7L BMW turbocharged swirl chamber engine[1]

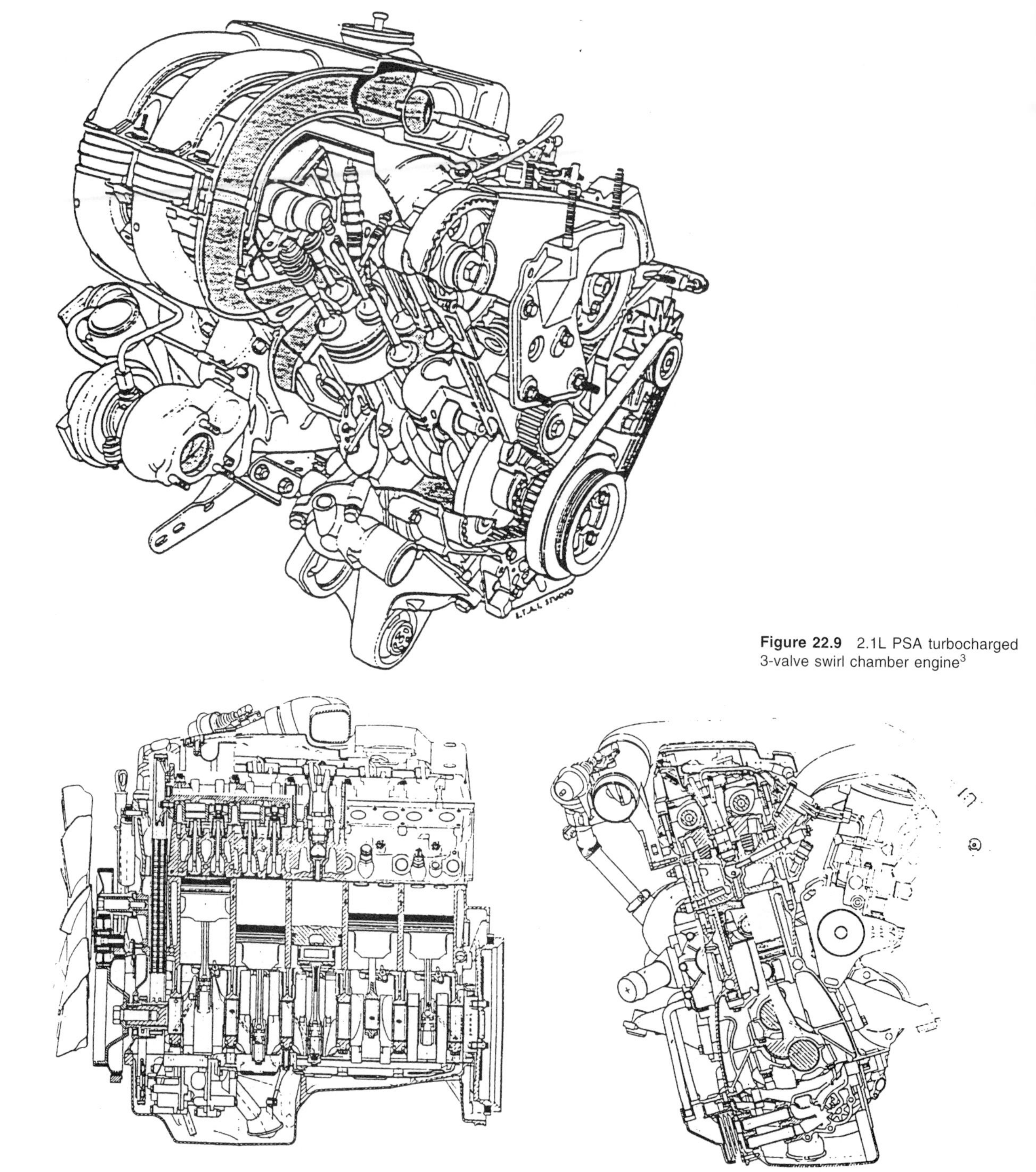

Figure 22.9 2.1L PSA turbocharged 3-valve swirl chamber engine[3]

Figure 22.10 2.5L Mercedes-Benz 4-valve prechamber engine[4]

and a second chain drives the oil pump. The 4-cylinder engine is equipped with a distributor fuel injection pump while the 5- and 6-cylinder engines use in-line injection pumps. All injection pumps are electronically controlled.

A single serpentine belt drives the accessories and the intake manifolds are made from plastic material. With a rated speed of 5000 rpm, the naturally aspirated version of the engine it is among the highest of all current passenger car diesel engines. This results in naturally aspirated specific power values of 33 kW/l which historically have been typical for turbocharged car diesel engines.

The next group of figures shows five examples of DI diesel engines. All engines are turbocharged and employ electronically controlled fuel injection systems. Although they have been introduced over a seven year period, each engine represents an evolution, each with major new design features.

Figure 22.4 shows the Volkswagen 1.9L TDI diesel engine[8,10,11]. This engine is a member of a large family of 4-, 5- and 6-cylinder in-line gasoline and diesel engines (IDI and DI) which are manufactured by Volkswagen and Audi. Even the Audi V6 and V8[12] engines share the same cylinder spacing of 88 mm with these in-line engines. Compared to the IDI engines presented previously, the basic design is relatively simple, exhibiting the following design features: cast iron engine block, one-piece side-flow 2-valve aluminum cylinder head, single overhead camshaft, hydraulic bucket tappets, tooth-belt driven camshaft and electronically controlled distributor injection pump (Bosch VP 37), and a very compact intake manifold without runners. Although this design appears unsophisticated (many design features are carried over from the original gasoline engine which was developed in the early 1970s), all engine components have been continuously optimized to cope with the high mechanical and thermal loads of a highly boosted DI diesel engine. This includes not only to the pistons and the bearings, but also the crankshaft, cylinder head and crankcase which, today feature minimum wall thicknesses of only 2.5 mm.

Due to these upgrades, it is possible to withstand operational cylinder pressures of up to 155 bar and thereby achieve the highest specific power output of all current 2-valve IDI and DI diesel engines (almost 43 kW/l) when a variable turbine geometry turbocharger is applied. This Volkswagen 4-cylinder DI diesel engine family, which is currently produced in three different power versions (NA, turbocharged and intercooled), is to date the most successful of all passenger car DI diesel engines, worldwide.

The Opel DI diesel engine shown in *Figure 22.12*, was the first 4-valve HSDI engine in production[16]. Two different displacement versions of this 4-cylinder eigine have been developed. Both versions share a bore of 84 mm and the same cylinder head. However, the 2.2L engine has a longer stroke (98 mm) than the 2.0L engine (90 mm). It also uses a second order mass force balancing system.

All engines are equipped with wastegate turbochargers and the 2.0L engine is available with and without intercooling, while the 2.2L engine is intercooled only. The gray cast iron engine block carries a one piece aluminum cross-flow cylinder head with an unique valve train concept. Valve actuation is accomplished with one overhead camshaft in combination with valve bridges for each of the two intake or exhaust valves and hydraulic bucket tappets. The axis of the intake and exhaust valves are parallel to the cylinder axis. A port deactivation device with throttle flaps is installed between the intake port flange and the intake manifold. This manifold is essentially just a 'hood' above the port flange (see *Figure 22.13*). Exhaust gas recirculation from the exhaust to the intake side is realized via a cast-in duct in the cylinder head. A staggered chain drive is used to drive the camshaft and fuel injection pump, an electronically controlled distributor pump with radial piston arrangement (Bosch VP 44). Other important design features are steel cracked conrods with trapezoidal small eyes and an integrated oil filter/cooler assembly.

The first production HSDI diesel engine with V-cylinder arrangement was introduced by Audi (see *Figure 22.13*)[12]. This 2.5L 4-valve engine was developed for installation in high-end vehicles and features several sophisticated design approaches. The cast iron block (with the typical 88 mm cylinder spacing of most Volkswagen/Audi engines) can be machined on the same line as the Audi gasoline V6-engines. A high strength steel split pin crankshaft (30° pin offset) without intermediate webs is used to obtain an even firing order. Compared to the Audi V6 gasoline engine, a larger crankpin diameter was introduced to react the cylinder pressures of up to 140 bar and to limit the conrod bearing loads which are always critical in highly boosted V-engines, compared to in-line engines. The big eye of the steel connecting rod is cracked under an angle of about 45°.

Gallery oil-cooled pistons are used to obtain sufficient piston durability of this high specific power engine (44 kW/l). The one-piece cross-flow aluminum cylinder heads each carry two camshafts which actuate the 24 valves (valve stems parallel to the cylinder axis) via finger follower/hydraulic lash adjuster

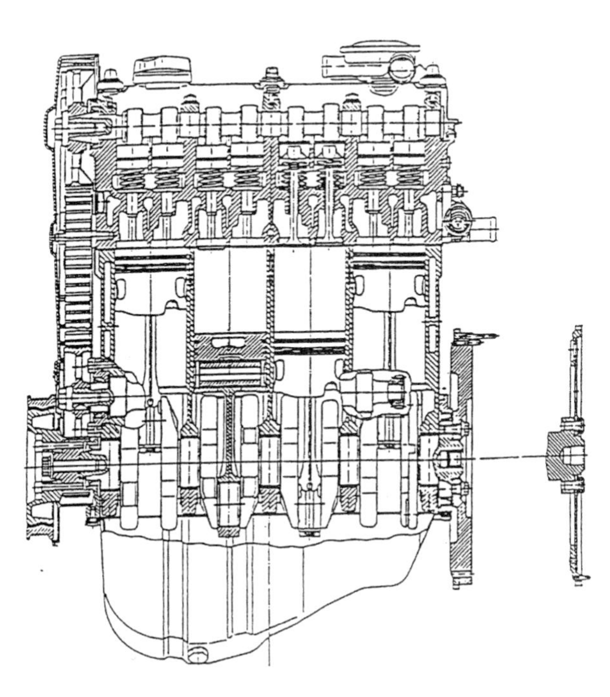

Figure 22.11 1.9L Volkswagen TDI diesel engine[11]

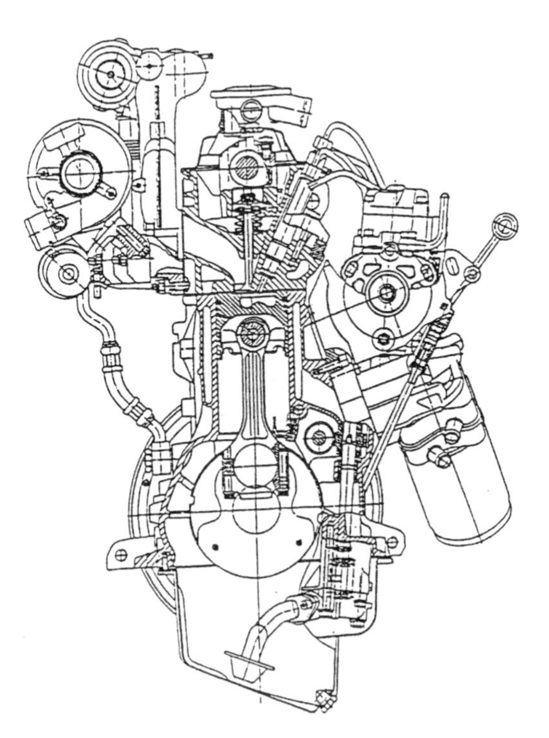

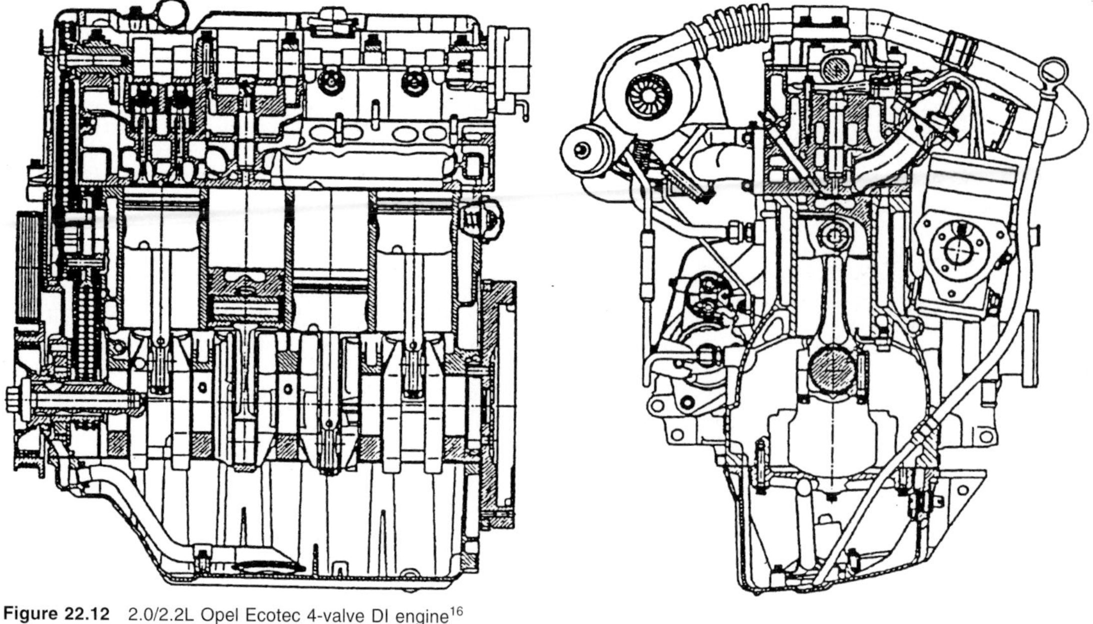

Figure 22.12 2.0/2.2L Opel Ecotec 4-valve DI engine[16]

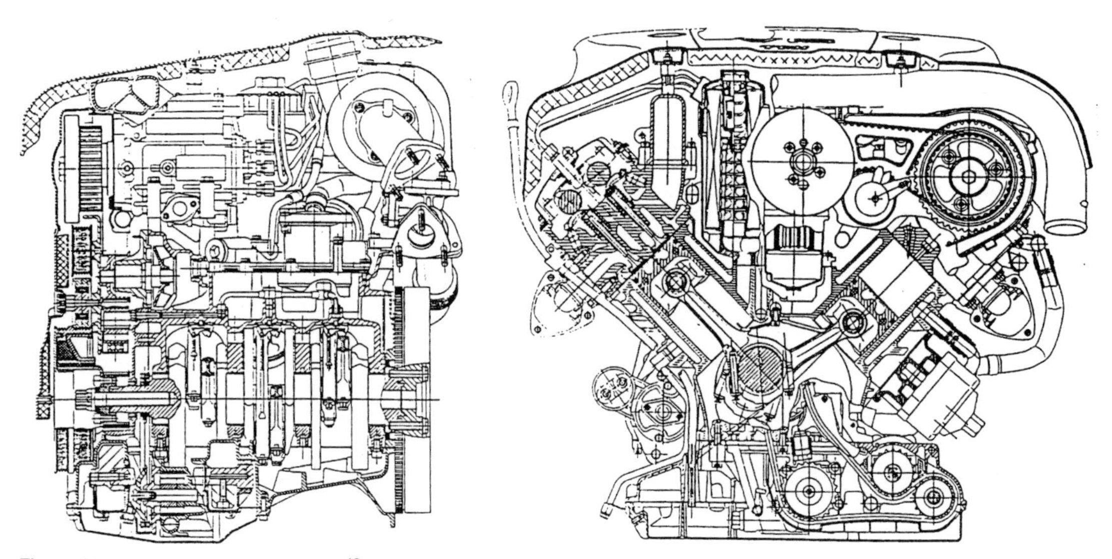

Figure 22.13 2.5L Audi V6 TDI engine[12]

assemblies. Tooth belts drive the Bosch VP 44 fuel injection pump and the camshafts of each cylinder head.

The camshafts of each cylinder head are synchronized by gears. An additional chain drives the oil pump and a mass balance shaft. To counteract the free first order mass couples of the 90°V6-engine, the balance shaft rotates in the opposite direction from the crankshaft. A VGT turbocharger, in combination with intercooling, is used for boosting.

In addition to the installation of the mass balance system, several other measures were taken to achieve favorable NVH characteristics. These include; application of a two-mass flywheel and a belt pulley damper, careful optimization of the complete powertrain structure (engine/transmission assembly) and other components such as oil pan, manifolds and valve covers to achieve maximum stiffness without weight penalty, application of switchable hydraulic engine mounts and application of several engine acoustic cover elements.

The Mercedes-Benz 4-valve DI diesel engine[13] shown in *Figure 22.14*, shares many basic design features with the Mercedes 4-valve prechamber diesel engine (see *Figure 22.10*). However, the crossflow cylinder head has been modified for the DI combustion system. The inclination of the valves which was introduced in the IDI engines to allow the installation of the prechamber, has been carried over for the DI engine cylinder head.

A very interesting design feature of this engine is the common rail fuel injection system (see also Section 22.3.3) which consists of a chain-driven fuel pump with three plungers, a high pressure fuel rail and solenoid-controlled injectors which are hydraulically linked with the rail through short individual fuel lines. In addition to the favourable potential of common rail fuel injection systems concerning performance, emissions and NVH, there is a major advantage with respect to the high pressure pump drive. The torque fluctuations in a common rail pump drive are significantly

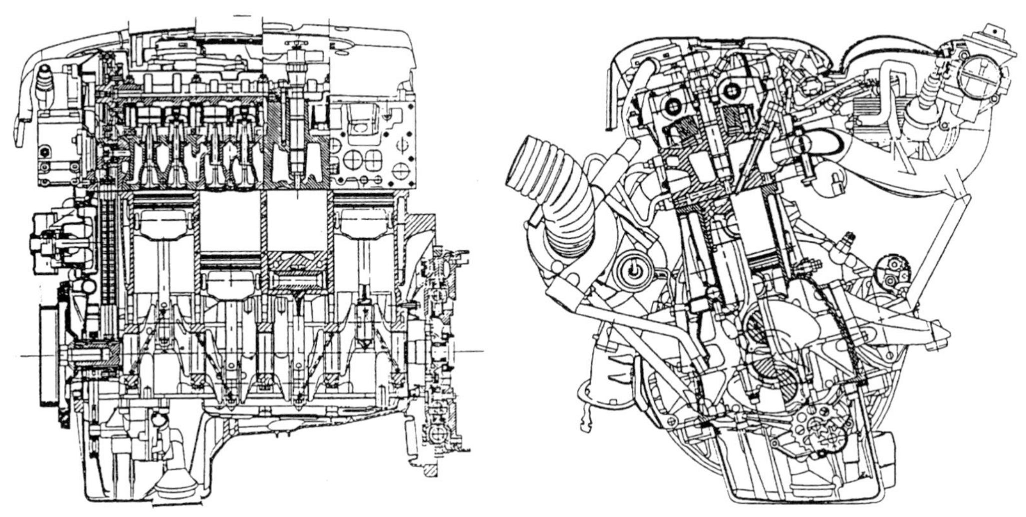

Figure 22.14 2.2L Mercedes-Benz 4-valve DI diesel engine OM 611 with Common Rail fuel injection system[13]

lower than in a pump drive of a conventional high pressure fuel injection pump. Therefore, it is possible to achieve very high injection pressures without the need for an extremely stiff pump drive.

To react the high cylinder pressures, sputter bearings are used not only for the upper bearing shells of the cracked steel conrods (which has become a standard on most turbocharged DI diesel engines) but also for the lower main bearing shells. The pistons are gallery oil cooled. The specific power output with a wastegate turbocharger is almost 43 kW/l and the specific torque reaches almost 140 Nm/l. A tuned intake system with relatively long runners is employed as is a port deactivation system.

A second Mercedes-Benz HSDI engine is shown in *Figure 22.15*[17]. This engine shares many combustion related design features with the 2.2L engine described above. Accordingly, a 4-valve crossflow cylinderhead with inclined valves and gallery-oil cooled pistons are used. In addition, a Bosch common rail fuel injection system is applied. However, completely different design approaches were taken for many other components.

The engine was designed for extreme forward-inclined (59°) transverse installation in a compact vehicle. Gasoline versions of this engine share the same basic design. Since weight reduction was a major concern, aluminum was chosen as material not only for the cylinder head but also for the open deck die cast engine block with cast-in cast iron cylinder liners. In addition, most components of the air handling system are manufactured from plastic. The valve train employs roller finger followers, opposed to bucket tappets, to minimize friction.

Table 22.1 summarizes design and performance data of several important passenger car diesel engines, which are currently available on the European market from European and Japanese manufacturers.

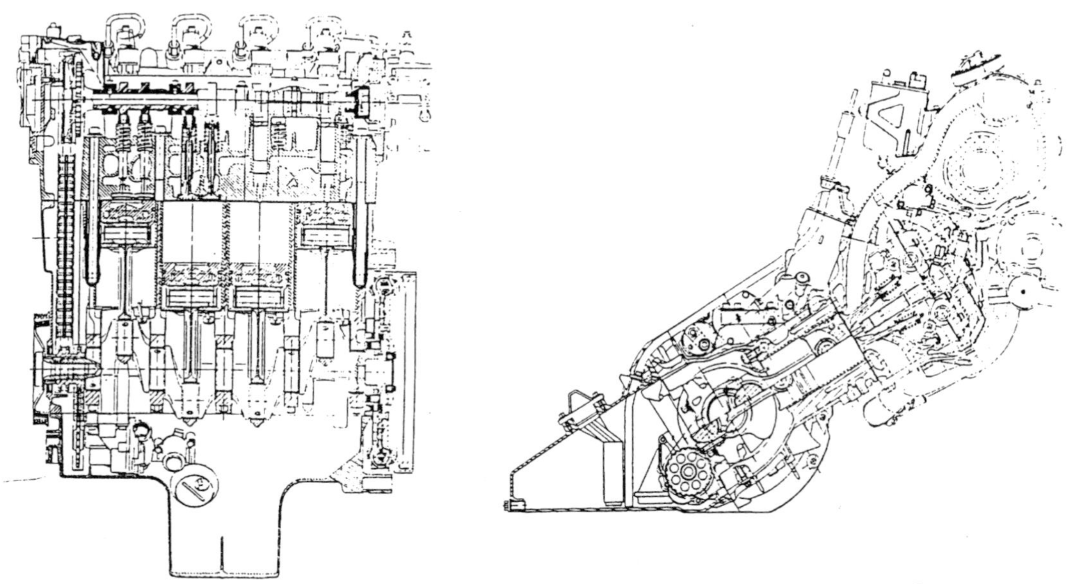

Figure 22.15 1.7L Mercedes-Benz 4-valve DI diesel engine OM 668 with Common Rail fuel injection system[17]

22.3.3 Fuel injection equipment

The state-of-the-art in fuel systems for passenger car IDI diesel engines is a pump-line-nozzle configuration with an axial or radial piston distributor pump. Mercedes-Benz, however, still applies in-line fuel injection pumps to some of their engines. Electronic control of the pump (fuel quantity, injection timing, etc.), in addition to control of other engine and vehicle systems, has become the standard (see also Section 22.3.5). These electronically-controlled injection pumps have been derived from mechanically-controlled pumps and still employ geometric fuel metering. However, the mechanical control mechanisms have been replaced by solenoid actuators.

All second generation passenger car HSDI diesel engines have been equipped with such electronically-controlled axial piston distributor fuel pumps (Bosch VP 37)[8,19,20]. The pressure capabilities of the existing distributor pumps for IDI engines have been upgraded significantly for these applications (800 bar at the pump and 1000 bar at the nozzle).

Further improvement in the fuel injection equipment is a key factor for increased power output of HSDI engines and the ability to meet more stringent exhaust emission standards while maintaining the current low fuel consumption levels. Increased pressure capabilities and injection rate shaping as well as a more flexible and precise control of injection quantity and timing are desirable. Also, the full potential of 4-valve configurations with a more quiescent type mixture formation can only be realized with sufficiently high injection pressure. There are different approaches to meet these requirements: application of high pressure in-line pumps or unit pumps as they are used for many European truck engines, further improvement in the distributor pump principle, application of unit injectors which have become a standard for most US heavy-duty-truck diesel engines or usage of high pressure common rail fuel injection systems.

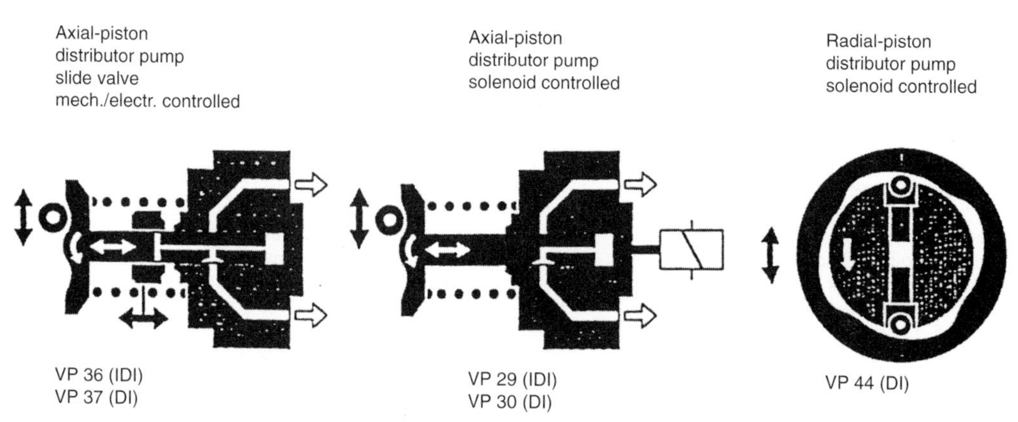

Figure 22.16 Distributor pump principles (Bosch)[21]

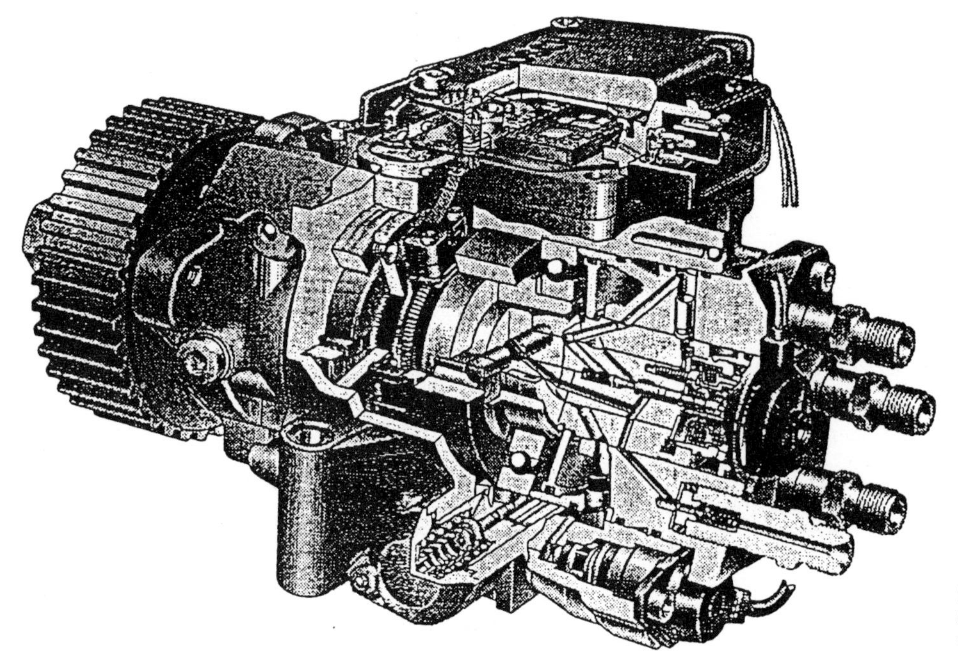

Figure 22.17 Solenoid controlled radial piston distributor pump (Bosch)[21]

The relatively high cost of in-line or unit pump configurations, as well as the desire to keep the engine block design common for diesel and gasoline engine versions (unit pumps are typically integrated into the engine block), are the major reasons to question using these systems.

Further development of the distributor pump principle appeared to be the most economic approach for the next generation of HSDI diesel engines with 4-valve cylinder heads. Bosch, the dominating fuel system supplier for passenger car diesel engines, has developed new distributor pumps, both with axial and radial piston configurations, with time-controlled fuel metering[21] (see Figure 22.16). A single solenoid valve controls both injection timing and quantity for all cylinders. The solenoid controlled radial piston pump from Bosch (VP 44) (see Figure 22.17) has higher pressure (1000+ bar at the pump) and quantity capabilities than the comparable solenoid controlled axial piston pump for DI engines (VP 30).

The VP 44 injection pump was originally developed for truck engines with cylinder sizes of about 1 l, such as the Cummins B engine family. However, it was also applied to 4-valve HSDI diesel engines from Opel and Audi[12,16]. It is interesting to note that Audi achieves up to 1500 bar injection pressure at the nozzle with a carefully optimized tooth belt drive for the pump which incroporates a flywheel on the pump to reduce the torque drive fluctuations. Many have considered a tooth belt drive insufficiently stiff to achieve such high injection pressures.

The major advantages of unit injectors, which omit the fuel line between injection pump and nozzle, are their high hydraulic stiffness which results in the highest pressure capabilities of all current injection systems and the possibility to achieve 'rapid spill' for a fast termination of the injection event. Major issues are the size of the injector which makes packaging difficult, especially on HSDI engines with 4-valve cylinder heads, and the necessity to provide a very stiff pump drive to utilize the full potential of the system. Also, system cost for the hydraulic components and the pump drive are major concerns for passenger car applications.

Nevertheless, Volkswagen announced its intention to apply electronically controlled unit pumps to a next generation 3-cylinder HSDI engine with a 2-valve cylinder head configuration[30]. *Figure 22.18* shows the impressive injection pressure map of the unit pump fuel injection system applied to that engine. At rated conditions, an injection pressure of 2000 bar is achieved. At maximum torque, 1500 bar is still achieved.

Compared to the other diesel fuel injection systems identified thus far, common rail fuel injection systems offer perhaps the highest flexibility and potential to meet many future demands. Consequently, the first engines with common rail systems have been developed by Mercedes-Benz and Fiat/Alfa-Romeo.[13,17] *Figure 22.19* illustrates the principle of a common rail system[18]. A fuel supply pump feeds diesel fuel to a multiple piston high pressure pump which is driven by the engine. This high pressure pump delivers fuel to the rail which is connected to the individual injectors via fuel lines. An electronic control unit (ECU) controls the injection timing and duration as well as the rail pressure.

Compared to conventional pump-line-nozzle injection systems and unit injectors, common rail systems offer several advantages: the place of generating the injection pressure (up to about 1350 bar in the current systems) is separated from the injection event control. This allows injection pressure and timing to be controlled essentially independently from engine speed and load. It also offers the potential for multiple injections (pre- and post injection). Very high injection pressures and therefore favourable spray characteristics can be achieved over the entire load and speed range (see *Figure 22.20*). This improves the air utilization and therefore the full load performance at lower engine speeds. It also significantly improves the trade-off between NOx and particulate emission in the engine operating points which are relevant for the emission test cycles.

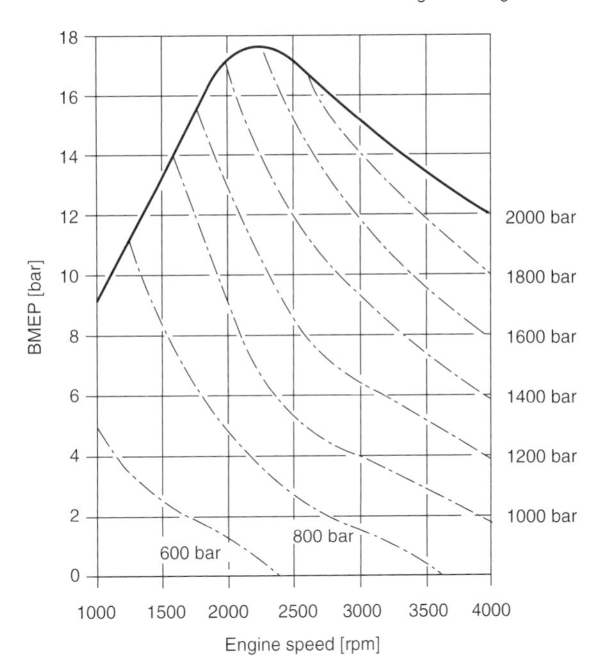

Figure 22.18 Injection pressure map for a unit pump fuel injection system[30]

Pre-injection reduces the combustion noise excitation of DI diesel engines[22], and it improves cold-start behaviour. Post injection during the expansion stroke can provide hydrocarbons to support active lean NOx catalysts. Another very important advantage of common rail systems is the relatively low torque fluctuation in the high pressure pump drive. The requirements regarding the stiffness for the drive system are significantly lower than for conventional fuel injection pump drives. This becomes even more important as injection pressure requirements increase. The pump drive geometry also limits the differences between diesel and gasoline engine versions of an engine family.

It can be expected that advanced distributor injection pumps, common rail fuel injection systems, and even unit injectors will all be used on HSDI engines over the next 10 years. However, unit injectors will probably be limited to very few applications. If the common rail systems can meet all the expectations and if the production costs can further be reduced with increased production quantities, it is most likely that the other injection system principles will slowly disappear from production.

22.3.4 Exhaust gas aftertreatment

Exhaust gas aftertreatment using oxidation catalysts has become a standard approach to meet the exhaust emission targets for passenger car diesel engines in Europe and the US. Diesel catalysts function in an exhaust gas environment which is significantly different from gasoline engines. The air–fuel ratio is always lean (generally higher than 20 : 1) and the exhaust gas temperatures are much lower (about 100–500°C for a typical turbocharger diesel engine downstream the turbine of the turbocharger). The composition of the raw exhaust gases of a diesel engine are considerably different from those of a gasoline engine. HC, CO and also NOx raw emissions in the diesel exhaust are much lower, however, particulate emissions are significantly higher. These carbonaceous particles are difficult to oxidize with a catalyst, since the ignition temperature of over 500°C is well above the typical diesel exhaust temperatures under normal conditions. However, it is possible to oxidize the condensible

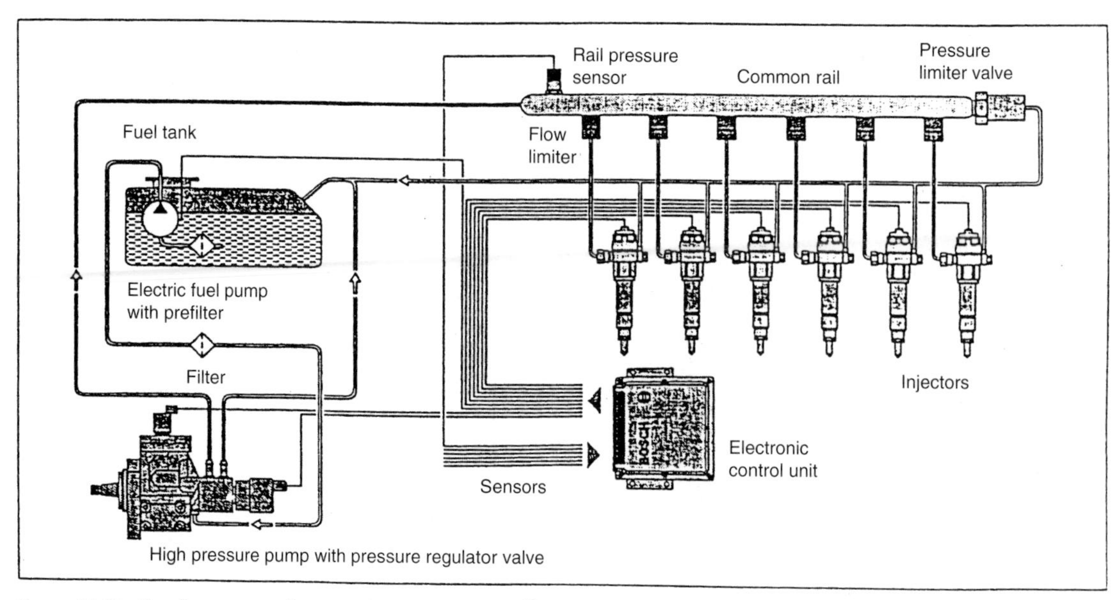

Figure 22.19 Bosch common rail system for passenger cars[18]

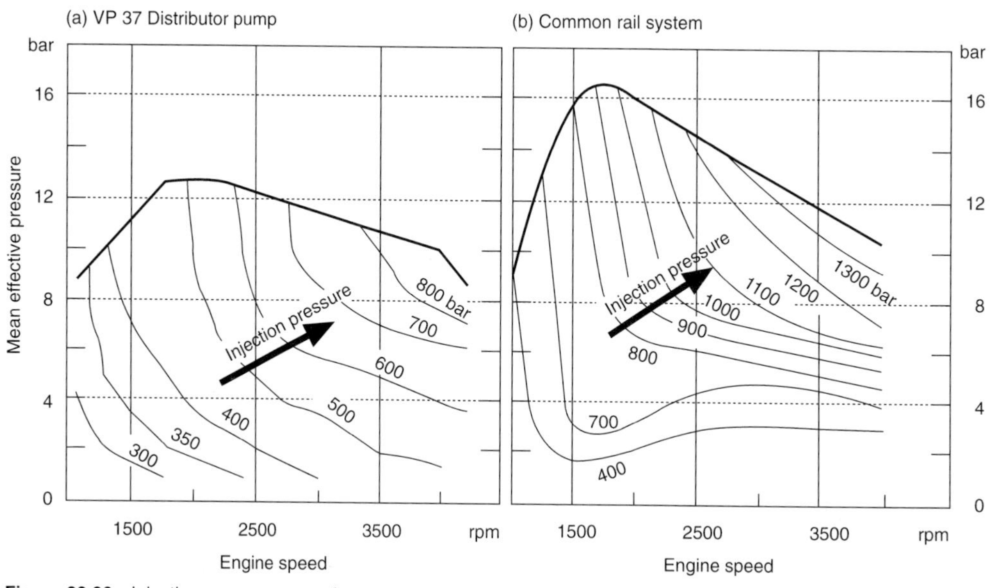

Figure 22.20 Injection pressure maps for a pump-line-nozzle and a common rail fuel injection system[18]

hydrocarbons which can amount to more than 50% of the total particulate mass in the exhaust of a HSDI engine.

The hydrocarbon portion (unburned fuel and lubrication oil) of the total particulate mass of IDI engine exhaust gas is typically lower (dry soot) than that of DI engine exhaust gases ('wet soot'). Since all diesel fuel contains some level of sulphur, the raw diesel exhaust gas contains sulphates which contribute to the particulate mass. Additional sulphates are produced through conversion of the SO_2 to SO_3 in the catalyst, especially at higher exhaust gas temperatures (higher loads). Reduction of the sulphur content in the diesel fuel offers the possibility to further reduce the particulate emission and use the full potential of oxidation catalysts. Therefore, in most countries with stringent emission regulations, the sulphur content in the diesel fuel has been limited to 0.05 weight %. An even further reduction of the sulphur content would result in an additional particulate reduction potential. On the other hand, reduction of the NOx emission in the oxygen-rich environment of diesel exhaust gas is very difficult with catalysts.

The level of raw emission reduction that currently can be achieved in the European emission test cycle through application of catalysts is illustrated in *Figure 22.21*. These results have been published by Mercedes-Benz for a turbocharged DI diesel engine with common rail fuel injection[13]. The catalyst system for this 2.2L engine (rating: 92 kW) consists out of two platinum catalysts. The first catalyst is close-coupled and has a volume of 2.1 dm^3. The second catalyst with a volume of 1.8 dm^3 is installed under the vehicle floor. There is an integrated bypass in the first catalyst which feeds some 'HC-rich' raw exhaust gas to the second catalyst, thereby achieving some NOx reduction effects in the second catalyst.

To further reduce the NOx emission in the US FTP test cycle,

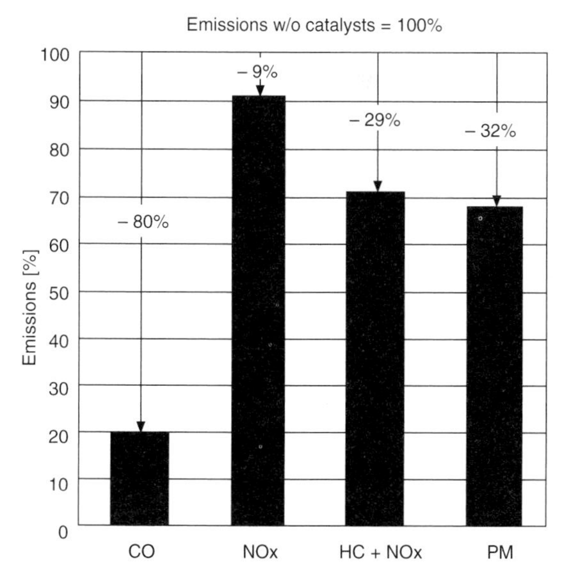

Figure 22.21 Emission reduction with catalytic converters in the European emission test cycle[13]

Volkswagen introduced a '3-way' catalyst system for their TDI engine applications on the US market. The hydrocarbons which are required for an active lean NOx process were provided through injection of small diesel fuel quantities into the exhaust system downstream of the turbocharger[23]. However, Volkswagen subsequently dropped this system when the emission standards were met with fuel injection and in-cylinder combustion system improvements.

There have been numerous research and development projects regarding particulate trap (filter) applications for passenger cars with diesel engines. Ceramic wall-flow filters, glass ceramic filters, silicon carbide traps, glass or metal fibre coil traps, sintered metal or wire mesh traps, etc. have been developed. With these filters, tailpipe particulate emission reductions between 70 and 90% have been achieved[24,25]. However, major issues with the usage of particulate traps are the increased exhaust back pressure which results in increased fuel consumption and the regeneration of the particulate trap. Regeneration support through active heating of the loaded trap, through palladium catalyzed coatings or through fuel additives as well as additive injection into the exhaust have been tested.

Mercedes-Benz introduced a particulate trap system in the US which was designed without any active filter generation support[24]. A wall-flow ceramic particulate trap was installed between the exhaust manifold outlet and the turbine inlet of a turbocharged 3L IDI engine. Thereby, a maximum exhaust gas temperature level for self-generation of the trap should occur. Additional measures to support the self generation were catalytic coating of the filter surfaces, insulation of the exhaust system and adaptation of a transmission which forced the engine to operate at relative high loads. However, problems such as insufficient durability of the ceramic traps in this specific installation upstream of the turbocharger turbine and no self generation at low ambient temperatures or continuous operation at very low loads, forced Mercedes-Benz to stop the production of this system shortly after its introduction.

No further attempts to introduce particulate filters in passenger car applications have since been made. All current particulate emission standards were met through advances in IDI and DI combustion system development. These advances have been supported through improved injection systems and electronic control strategies as well as through the introduction of low sulphur fuel and more efficient oxidation catalysts. However, future emission standards which are currently being discussed especially in the US, may necessitate the use of trap systems.

22.3.5 Electronic control systems

Usage of electronic control systems has become the standard on most passenger car diesel engine applications in countries with stringent emission legislation. In these countries, only a few IDI engines in smaller vehicles are left that still use mechanically controlled injection systems. However, these engines also will most likely need electronic control systems to meet the next level of emission standards.

The development of electronic diesel control systems was one decisive precondition for the successful introduction of the second generation passenger car HSDI engines by Volkswagen and Audi. DI diesel engines which are more critical regarding NOx emissions than IDI engines, require a very precise control of the injection timing and of the quantity of the exhaust gases which are recirculated to reduce the NOx emission.

Today's advanced electronic control systems for passenger car diesel engines include not only control functions for the fuel injection system itself, such as injection timing or injection quantity control, but they also include other engine related control functions for fuel feed pumps or exhaust gas recirculation (EGR), boost pressure and cold start aid (glow plug) systems. Various diagnostic functions, which will become even more important when OBD (on-board-diagnostics) becomes a mandatory part of the emission legislation for diesel engines, have become standard features. Electronic diesel control systems are becoming more an integral part of the vehicle control system as control of radiator fans, air conditioning systems or alternators are added. Communication with other vehicle controllers for transmissions, braking systems (ABS) or traction control systems is typically accomplished via CAN (controller area network) busses.

The functionality of electronic control systems is very similar among fuel systems which still use geometric fuel metering and the newer fuel systems which use time-controlled fuel metering (see Section 22.3). *Figure 22.22* shows an example of an electronic control system for an engine with an advanced common rail fuel injection system[18].

22.4 Performance and emissions characteristics

22.4.1 Power and torque

The specific power output of the current technology passenger car diesel engines can be inferred from Figure 22.1 and from Table 22.1. There is a wide range due to the fact that naturally aspirated (NA), turbocharged (TC) and turbocharged/intercooled (TC/IC) engines are listed. Rated speeds of DI engines presently do not exceed 4500 rpm, while IDI engines are operated with up to 5000 rpm at rated conditions. Specific power values of naturally aspirated engines range from about 25 kW/l (2-valve engines) to about 33 kW/l (4-valve Mercedes-Benz engines) with no major differences between IDI and DI engines. The latest turbocharged and intercooled DI diesel engines from Audi and Mercedes-Benz achieve more than 44 kW/l. Most recently, BMW introduced an all new 2.0L 4-valve DI engine which achieves even more then 51 kW/l. This engine is equipped with a Bosch VP 44 fuel injection pump and a variable geometry turbine (VGT) turbocharger. Also, the highest rated turbocharged and intercooled IDI diesel engines are build by BMW which achieve 42 kW/l.

On the lower end of the range are so-called 'low-blow' turbo IDI engines with specific power output values of about

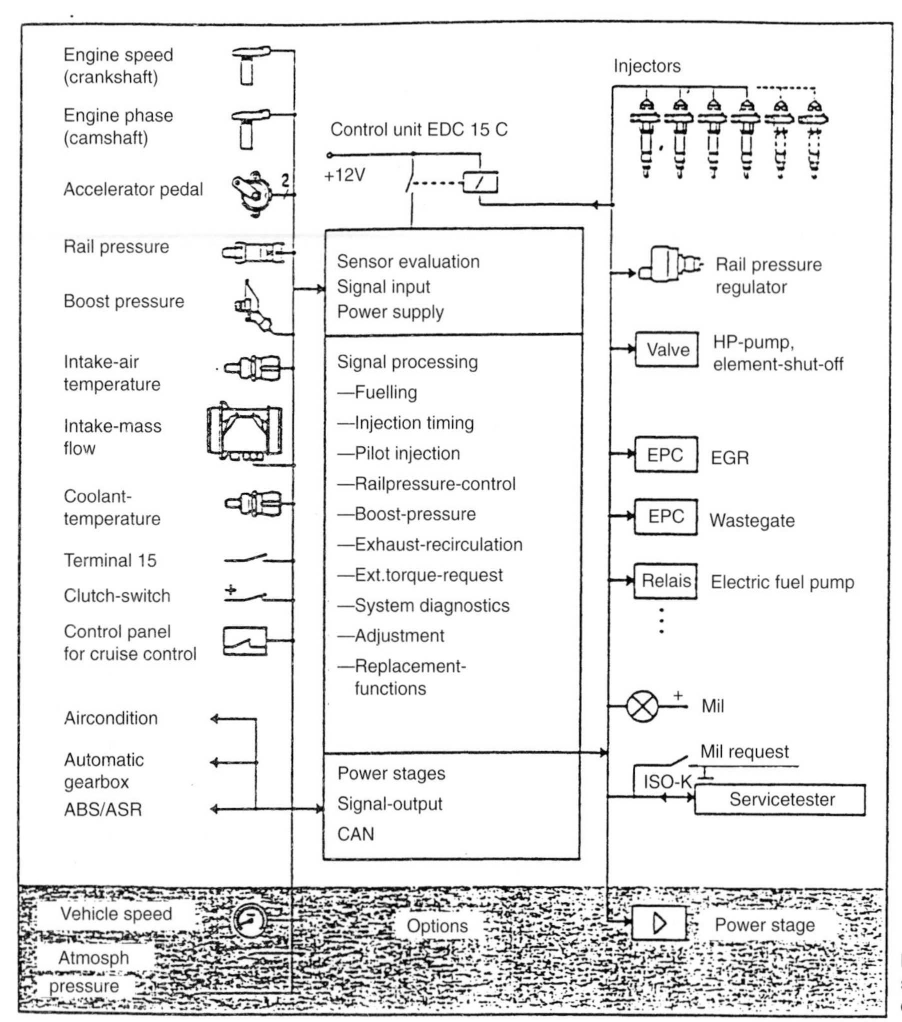

Figure 22.22 Electronic control system for an engine with common rail injection system[18]

27–30 kW/l. In these engines, turbocharging is used to provide additional intake air not so much for power increase, but for the reduction of emission, especially particulates.

Specific torque values of 2-valve NA engines (both IDI and DI) are about 65 Nm/l (BMEP = 8.2 bar) at 2000 to 2500 rpm (see Table 22.1). The torque curves of Mercedes-Benz 4-valve IDI engines peak at 70 Nm/l (BMEP = 8.8 bar). The highest specific torque values for turbocharged engines with intercooling so far have been realized by BMW with its 2.0L 4-valve DI engine (144 Nm/l, BMEP = 18 bar at 1750 rpm) and by Mercedes-Benz with its 2.2L 4-valve DI engine with Common Rail fuel injection system (139 Nm/l, BMEP = 17.5 bar at 1800 rpm).

Advanced 4-valve DI engines can easily reach even higher maximum BMEP values, especially when they are equipped with common rail fuel injection systems. However, the availability of transmissions which allow further torque increase has become a major hurdle. Similar to the torque characteristics of many heavy-duty truck diesel engines, the torque curves of some modern highly boosted passenger car diesel engines are flat over a wide speed range. The 2.5L V6 TDI engine from Audi has the same maximum torque of 310 Nm (124 Nm/l, BMEP = 15.6 bar) between 1500 and 3500 rpm[12]. A VGT turbocharger provides maximum boost pressure already at 1500 rpm. With such a torque characteristic, it is possible to always operate the vehicle in high gears at very low engine speeds without feeling a lacking power for sudden accelerations. This helps to significantly reduce the fuel consumption.

22.4.2 Fuel consumption

The significantly better fuel efficiency of the diesel engine compared to the gasoline engine has always been one of the most important attractions for its application. In *Figure 22.23*, the in-vehicle fuel consumption of passenger cars with gasoline engines and those with IDI and DI diesel engines in the MVEG emission test cycle is plotted versus vehicle curb weight. The fuel consumption of vehicles with IDI engines is about 25–30% lower than comparable gasoline vehicles. There are large differences in the fuel consumption of gasoline engine-powered vehicles with the same curb weight, which result out of the different engine technology levels. However, it is obvious that the best in-class gasoline engine-powered vehicles (which typically feature high compression ratio engines with multi valve cylinder heads) are almost as low in fuel consumption as some of the vehicles with IDI diesel engines. The DI diesel powered-vehicles, however, show another 15–20% improvement in fuel efficiency when compared to the vehicles with IDI diesel engines.

The fuel consumption of the diesel passenger cars in the MVEG emission test cycle, which combines city and highway driving, reflects relatively well the real world fuel efficiency as

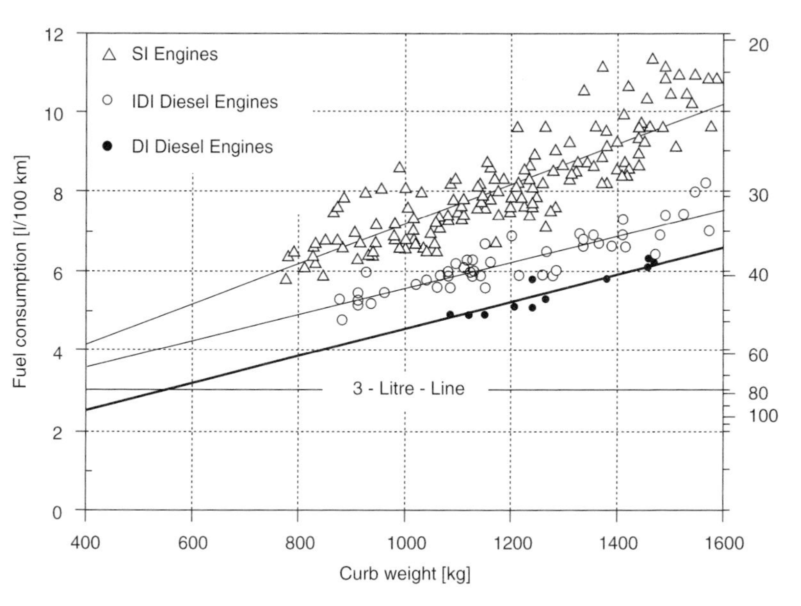

Figure 22.23 Fuel consumption of European passenger cars in the MVEG emission test cycle

it is experienced by an average driver. However, the gasoline values shown are typically exceeded by most vehicle owners by 10% or more. This is especially true when gasoline engine-powered vehicles are frequently operated over short distances, in slow urban traffic or at low ambient temperatures, where fuel enrichment is required to achieve acceptable driveability.

In the future, gasoline engines with direct fuel injection (GDI or DISI engines) compete to achieve fuel efficiencies which approach IDI diesel engines. DI diesel engines would still maintain a major advantage in volumetric fuel consumption, should this become reality. However, the CO_2 emission advantages of the DI diesel engine over such a very efficient DI gasoline engine would be reduced significantly, considering that the combustion of one litre diesel fuel produces about 2.6 kg CO_2 while only about 2.4 kg CO_2 are emitted if one litre of gasoline is burned.

Figure 22.24 depicts a map for the brake specific fuel consumption (BSFC) of a state-of-the-art 4-valve HSDI engine[16]. The minimum BSFC level of this engine is 195 g/kWh at a point close to maximum torque operation. This corresponds to a thermal efficiency of about 43%. Historically, such efficiency values could only be achieved with much larger DI diesel engines.

22.4.3 Exhaust emissions

To date, there is no system available for diesel engine emission reduction that approaches the very effective three-way catalyst for gasoline engines. Therefore, reducing the raw exhaust emissions through optimization of the in-cylinder combustion process and the air handling/exhaust gas recirculation systems, is still a major, if not the most important, task in diesel engine development. Concept decisions concerning new engines or

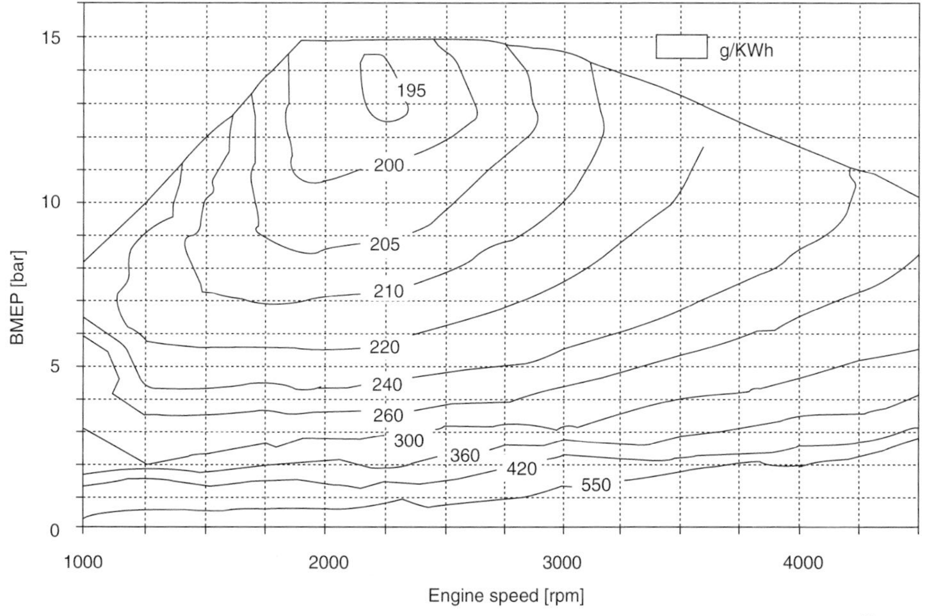

Figure 22.24 BSFC map of the turbocharged and intercooled 4-valve Opel ECOTEC DI diesel engine[16]

engine modifications are often driven by emission legislation. Particulate and NOx emissions are the major concern.

Figures 22.25 and *22.26* illustrate how much these pollutants have been reduced so far in diesel-powered vehicles for the European and the US market. Most of the particulate emission reduction is the result of improvements in mixture formation and combustion as well as in air handling (boosting). Decisive for progress in in-cylinder particulate emission reduction was a better understanding of the mixture formation and combustion phenomena and significant progress in fuel injection system hydraulics and control technology. Additional tailpipe particulate reduction has been achieved through application of oxidation catalysts which reduce the soluble organic particulate fraction (see also Section 22.4).

Most of the NOx reduction that has been thus far achieved can be attributed to the application of exhaust gas recirculation (EGR) which has become a standard feature for all passenger car diesel engines. Exhaust gas recirculation increases the amount of inert gas in the combustion chamber. These gases do not participate in the chemical reactions, but absorb some of the energy that is released during the combustion process. This results in a temperature decrease which inhibits the reactions between nitrogen and the remaining oxygen. DI diesel engines in particular require high EGR rates to meet the current emission standards (*Figure 22.27*).

Further NOx emission reduction can been achieved through EGR cooling and, of course, through even higher EGR rates. While EGR-cooling has already been introduced into production[26], it is difficult to further increase the EGR rates without other disadvantages. One problem is the increase in

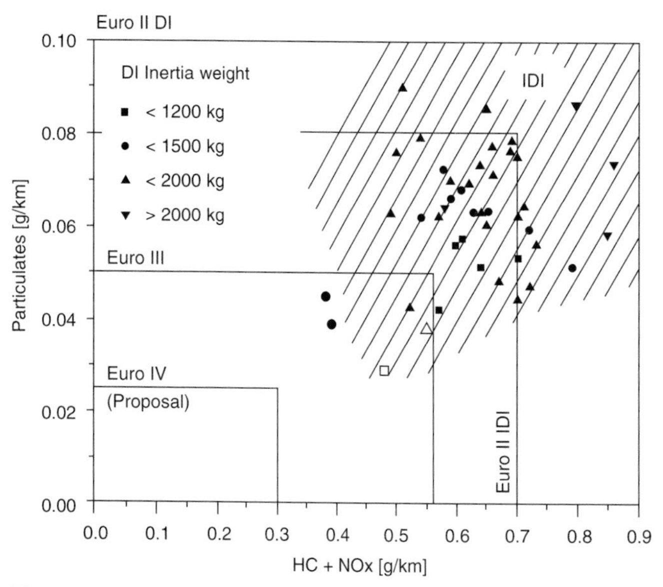

Figure 22.25 Emission certification test results (European driving cycle)

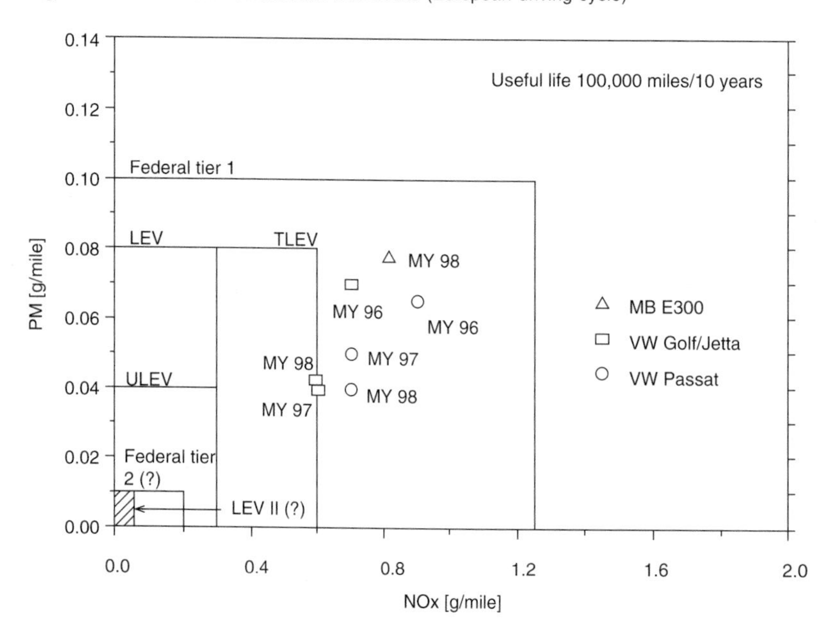

Figure 22.26 Emission certification test results (US FTP 75 test cycle)

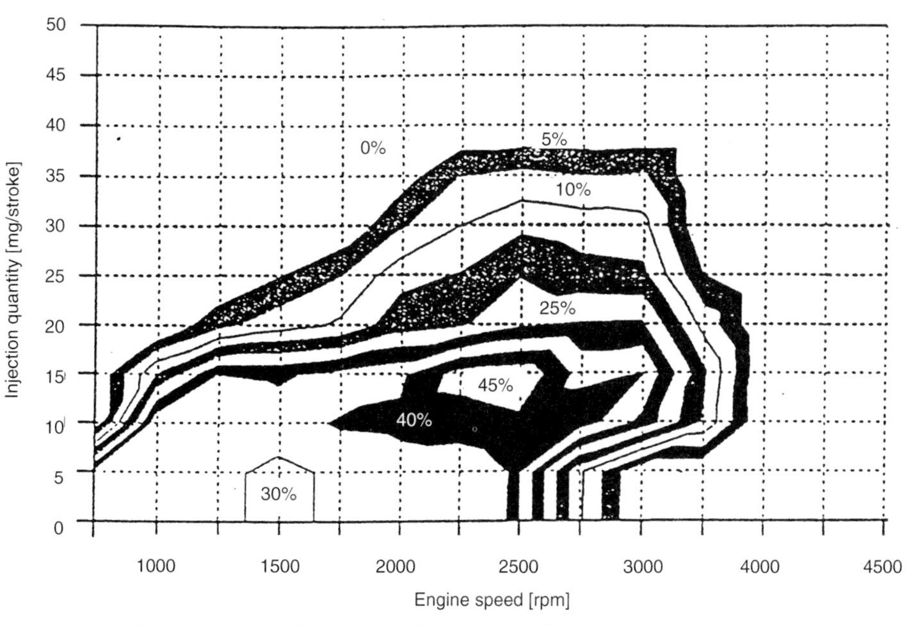

Figure 22.27 EGR map Opel ECOTEC 4-valve DI diesel engine[16]

particulate emissions which is associated with higher EGR rates. Another issue is the difficulty in providing a sufficient level of pressure difference between the intake and exhaust manifold which is required to force high amounts of exhaust gas from the exhaust manifold into the intake manifold. Also, precise control of the EGR rates under transient operating conditions becomes more challenging with increased EGR rates.

22.4.4 Noise, vibration and harshness (NVH)

NVH has always been a major concern for the application of diesel engines in passenger cars. As such, the future success of the diesel engine in the passenger car market strongly depends on favourable NVH characteristics. The term 'gasoline engine NVH transparency' very well describes the NVH development targets: passengers as well as by-standers should not be able to distinguish between a gasoline or diesel fuelled vehicle. To achieve this goal, reduction of the combustion noise excitation as well as reduced transmission of noise and vibration to the engine surfaces, into the engine mounts and brackets and into the transmission are required.

The ignition delay in a DI diesel engine is typically longer than in an IDI diesel engine, where fuel evaporation in the swirl or prechamber is supported through very high turbulence and temperature levels. The fuel which is direct injected during the ignition delay starts to burn rapidly, resulting in a high pressure increase. With higher injection rates, which are required to improve the trade-off between particulate, NOx emission and fuel consumption, this problem becomes exacerbated. In addition, a combustion noise disadvantage exists for the DI diesel engine during rapid load increases where the ignition delay is increased during the first combustion cycles, due to lower combustion chamber temperatures which result from the previous low load operation.

Figure 22.28 compares the sound pressure levels (1 m distance) of IDI and DI diesel engines with those of gasoline engines. The fast pressure rise after onset of the combustion in diesel engines results in high sound pressure levels, especially in the lower speed range. At higher speeds, where the mechanical noise becomes more dominant, the differences between diesel and gasoline engines diminish. Therefore, diesel combustion noise reduction measures must focus on the low speed range. Application of fuel injection systems with pilot injection capabilities (e.g. common rail systems) is one such measure. Well timed and quantity-controlled pilot injection significantly reduces the ignition delay of the main combustion without major exhaust emission or fuel consumption disadvantages.

Reduction of the combustion noise excitation through optimization of the in-cylinder process is not the only approach to improve the NVH characteristics of diesel engines for passenger car and other light-duty applications. The engine structure should also be as stiff as possible to limit the noise radiation from the surfaces of the engine and attached sub-systems. Optimization of the engine structure during the very early stages of the powertrain (engine + transmission) development has become a common practice. Dynamic finite element analysis (FEA) is a very suitable tool to identify weak areas (areas that show a high surface mobility) in the structure and to stiffen these weak areas with minimum weight increase. FEA is also used to optimize the bending stiffness of the complete powertrain, which has many times, historically, been too low.

Another NVH-critical aspect is the global vibration of the powertrain caused by the cyclic gas forces. Due to the high compression ratio and unthrottled operation, it is not possible to reduce the vibration level of diesel engines down to the values which are typical for gasoline engines. Therefore, additional measures are required to achieve favourable NVH characteristics with diesel engine installations in passenger cars.

Usage of two-mass flywheels reduces the excitation of the transmission gears and, thus, gear rattle which is often a problem with manual transmissions. Automatic transmissions typically do not require the application of two-mass flywheels.

Switchable engine mounts which allow the damping characteristics to be adjusted to engine speed are applied in some cases to significantly limit the vibration level at the engine mounts[27]. At low speeds, the damping characteristic of these mounts is switched to the 'soft' position, and then are switched to 'hard' at higher engine speeds.

Another efficient feature for overall NVH improvement is the use of massbalancing shafts. Audi introduced a balance shaft which rotates with engine speed to compensate the first order mass couple of its new 90° V6 TDI engine[27]. Opel uses

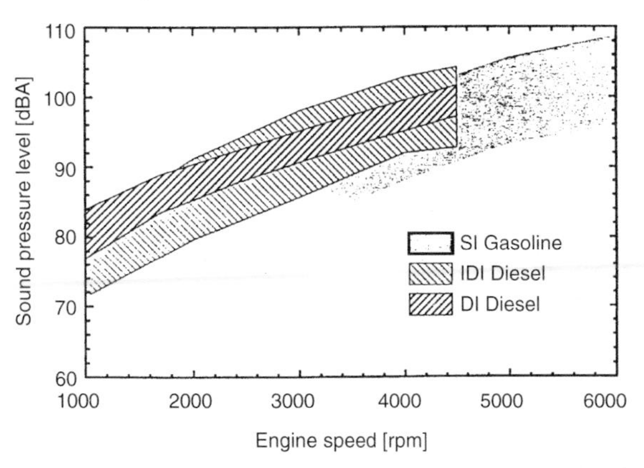

Figure 22.28 Full load sound pressure (1 m distance) of state-of-the-art passenger car engines

a balance shaft system (two shafts rotating at twice the engine speed) to compensate the second order mass forces of the 2.2L member of its 4-cylinder in-line (I4) HSDI diesel engine family.[16]

Second order mass forces increase with the square of the engine speed. Rated speeds of diesel engines are significantly lower than rated speeds of gasoline engines. Consequently, the application of second order mass balance systems to I-4 diesel has typically not been chosen unless the displacement exceeds 2.0L. However, highly boosted DI diesel engines have much heavier pistons than gasoline engines and thus higher second order mass forces at the same engine speed. Therefore, it is possible to significantly improve the vibration level of I-4 diesel engines, even at medium and low speeds, through application of second order mass balance systems.

Because of the high importance and 'visibility' of the NVH characteristics to the customer, it can be foreseen that second order mass balance systems will increasingly become a standard feature, even on smaller I-4 diesel engines.

22.5 Future developments

During the last decade, passenger car diesel engine technology has increasingly advanced. Among the new features were inter-cooling, electronic control systems, exhaust gas recirculation, exhaust gas aftertreatment with oxidation catalysts, multi-valve cylinder heads, direct fuel injection, high pressure fuel injection systems turbochargers with variable turbine geometry and, most recently, common rail injection systems and aluminum engine blocks. What can be anticipated for further developments?

Regarding the base engine concept, it can be expected that, both, larger and smaller engines than at present (with dis-placements between 1.8 and 2.5L in most of the cases) will be developed. Larger engines with eight (8), and may be in the future even more cylinders, will be used for European luxury car and eventually US light truck applications. Examples of such engines have already been presented by BWM and Mercedes-Benz (*Figures 22.29* and *22.30*).

Common features of these V8 engines are: direct fuel injection, 4-valve DOHC cylinder heads, common rail fuel injection systems, and twin turbochargers. One major difference is to the engine block material. Mercedes-Benz uses aluminum, while BWM's prototype still exhibits cast iron. The progress in the common rail fuel injection system technology was a major

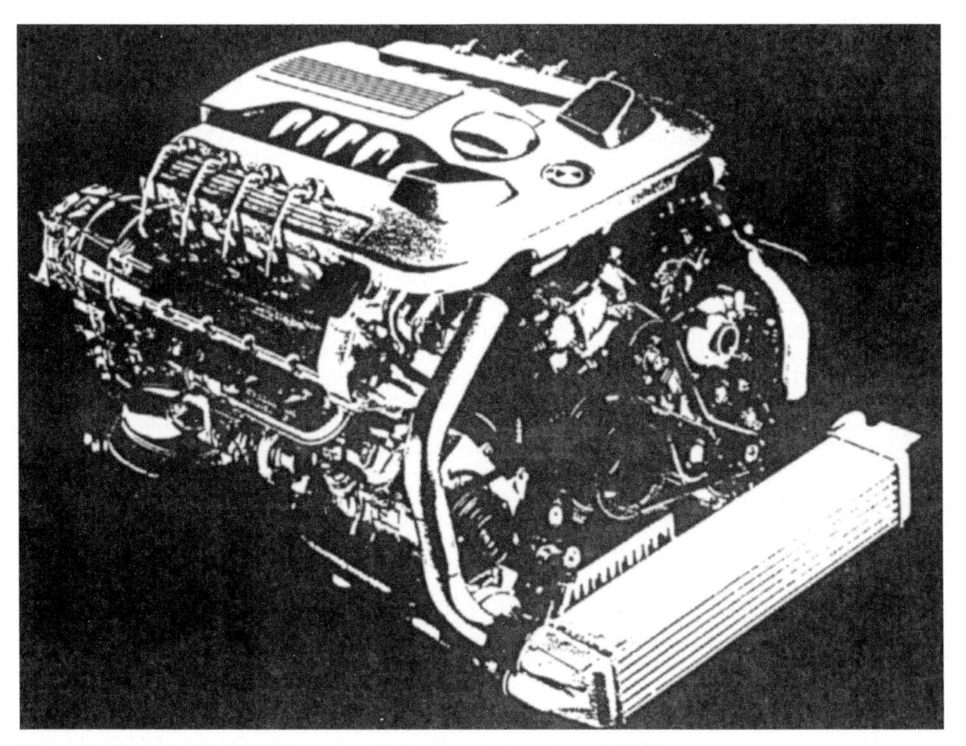

Figure 22.29 3.9L 170 kW BMW 4-valve DI diesel engine (courtesy of BMW)

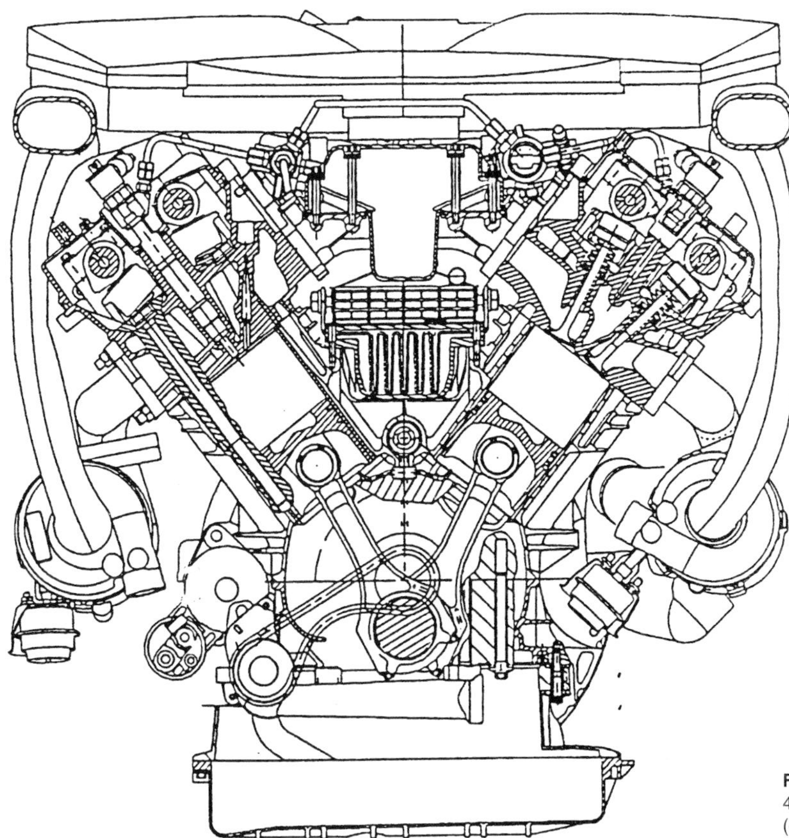

Figure 22.30 4.0L, 175 kW Mercedes-Benz 4-valve DI diesel engine (courtesy of Mercedes-Benz)

precondition for the development of these engines. Previously, no high pressure fuel injection system for HSDI diesel engines with more than six cylinders existed.

On the other end of the spectrum there will be engines with significantly smaller displacements than the current 4-cylinder engines which typically feature displacements of about 2 litres. These engines will most probably, also use direct fuel injection. The smaller displacements will be achieved through reduced cylinder numbers (e.g. three cylinders instead of four) and/or through smaller cylinder sizes. An extreme example has been announced by Mercedes-Benz. For their 'Smart' micro-car they are developing a 3-cylinder HSDI diesel engine with a displacement of only 758 cc (bore/stroke: 65.5/75 mm, rating: 37 kW)[28].

Ford is developing a 55 kW 1.2L 4-cylinder engine for a hybrid electric vehicle (HEV) application[29]. Volkswagen announced production of a 3-cylinder 1.42L HSDI diesel engine for a sub-compact car[30]. The design of the engine is based on Volkswagen's 1.9L TDI engine and it features the same bore and stroke (79.5/95.5 mm). However, it uses a completely new fuel injection system with electronically controlled unit injectors which are installed in an inclined position in the 2-valve cylinder head. The unit injectors are actuated via roller rocker arms by the same belt driven overhead camshaft which actuates the valves (*Figure 22.31*).

Both, the Mercedes-Benz 'Smart' diesel engine and Ford's 'DIATA' (**D**irect **I**njection **A**luminum **T**hrough-bolt **A**ssembly) feature aluminum engine blocks. While the usage of aluminum offers significant weight savings for larger engines, the weight reduction potential through the usage of aluminum as a block material reduces significantly with the size of the engine. Therefore, future developers must assess the usage of aluminum for the blocks of smaller HSDI engines.

Also, with the focus on small car applications, investigations into the feasibility of the 2-stroke principle for HSDI diesel engines are being conducted[31,32]. Favourable power-to-weight and torque-to-displacement ratios are significant incentives. However, only the uniflow scavenging principle with exhaust valves in the cylinder head and intake ports in the cylinder liners has thus offered sufficient gas exchange characteristics. Also, in addition to a turbocharger, a mechanically driven auxiliary blower is needed for engine start and low load operation, which adds cost and complexity to the system. Cost and emission concerns were probably the major reasons for Detroit Diesel Corporation, the formerly largest manufacturer of 2-stroke diesel engines in the world, to switch from 2-stroke to 4-stroke operation for its modern low emission heavy-duty diesel engines.

The future application of diesel engines heavily depends on exhaust emission legislation. However, thus far neither Europe nor the US, have the standards for the next level of emission reduction clearly defined. In particular, the particulate emission of the diesel engine is a major concern. Discussions are ongoing whether to limit not only the total particulate mass but also to relate the particulate mass to particulate size or even the total number of particles that are emitted in a test cycle. The current emission standards for diesel-powered passenger cars in Europe and the US (Federal and California) as well as some proposals for the future standards are summarized in *Table 22.2*.

The expected European EURO IV legislation, still appears within reach for advanced DI diesel engines, if the fuel quality is further improved (reduced sulphur content) and more efficient exhaust gas aftertreatment systems become available. However, today it is questionable whether the NOx and particulate values which have recently been proposed by the US environmental agencies (EPA Tier 2 and CARB LEV II) can be met by diesel-powered vehicles.

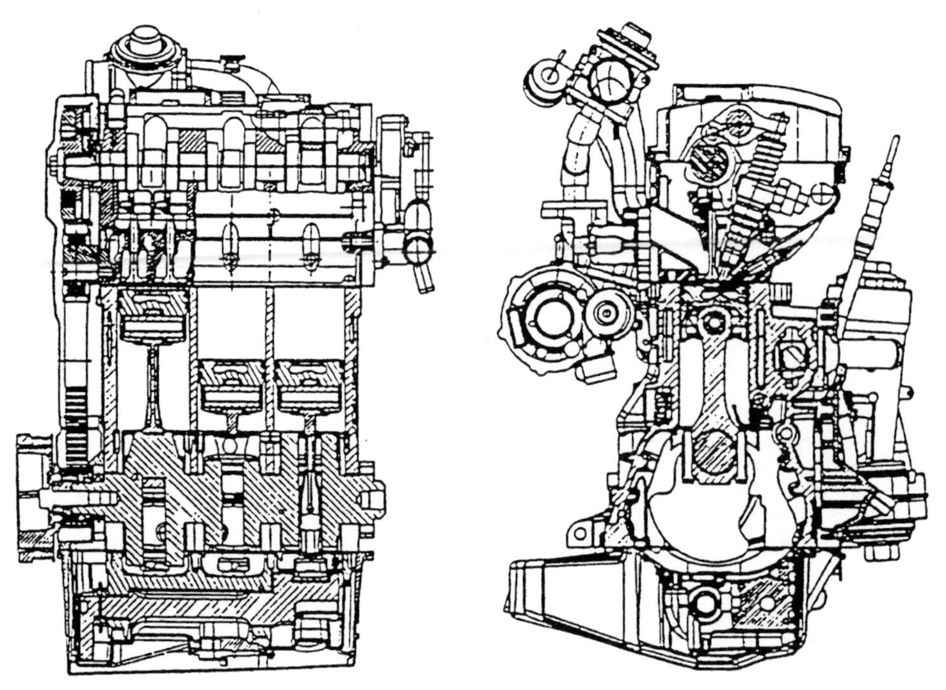

Figure 22.31 1.4L 3-cylinder 55 kW Volkswagen 2-valve DI diesel engine with unit injector fuel injection system[30]

Table 22.2 Current and possible future emission standards for passenger cars with diesel engines (100,000 miles for US standards)

	Euro II *1996+* g/km	*Euro III* *2000+* g/km	*Euro IV* *2005+* g/km	*Usa-Epa* *1994–2002* *Tier 1* g/mile	*USA-Epa* *2003+* *Tier 2* g/mile	*USA-Carb* *1994–1999* *Tier 1* g/mile	*USA-Carb* *1997–2003+* *ULEV* g/mile	*USA-Carb* *2004–2010* *LEV II* g/mile
Nmhc				0.31	0.125	0.310	0.053	
Nmog						0.319	0.055	
NOx+HC	0.70	0.56	0.30 ?					
NOx		0.50	0.25 ?	1.25	0.20	0.6	0.3	0.05 ?
CO	1.00	0.64	0.50 ?	4.20	1.70	4.2 (cold)	2.1 (cold)	
PM	0.08	0.05	0.025 ?	0.10	0.01 ?	0.10	0.04	0.01 ?

Significantly improved exhaust gas aftertreatment systems with conversion efficiencies close to those of current three-way catalysts for gasoline engines and/or the application of alternative fuels will be required to meet such extremely low emission standards. Major focus will be placed on particulate and NOx reduction devices. However, for particulate trap applications, significant improvements, especially in the regeneration technology, are required.

NOx reduction in the diesel exhaust must be accomplished in an oxygen rich atmosphere. Therefore, reductants such as hydrocarbons or urea must be fed into the exhaust gas upstream of the catalyst to accomplish sufficient NOx reduction[33,34,35,36]. Normally, engine-out hydrocarbon emission of modern diesel engines is not sufficient to achieve more than about 10% NOx reduction (passive lean NOx approach). Higher reduction rates of about 30% are currently possible if a small quantity of diesel fuel is injected late in the combustion cycle (post-injection) to increase the amount of HC in the exhaust gases.This HC enrichment can easily be accomplished with common rail fuel injection systems which allow multiple injections at any time during the cycle[33].

Very high NOx conversion rates can be accomplished with selective catalytic reaction devices using urea as a reductant. This technology has been successfully applied to stationary engines for several years, where conversion rates of more than 90% are typical. It is also being considered for European heavy-duty truck applications. In experimental light duty applications, efficiencies of about 70% have already been demonstrated. However, there is some concern that such a system is too complex, bulky (additional tank required) and expensive for passenger car applications. Compared to heavy-duty truck applications, where the emission test includes a significant portion of high engine load operation, light-duty applications result in the lower exhaust gas temperatures which reduce conversion rates.

Another alternative may be NOx adsorber catalysts which store NOx under lean operating conditions. However, the storage capacity is limited and the stored NOx must be converted into N_2 through frequent enrichment of the air–fuel ratio. This concept has already been introduced for direct injected gasoline engines where short fuel enrichment periods can be realized without other major implications on the emissions. In a diesel engine, it is more difficult to achieve engine operation at relative air–fuel ratios below $\lambda = 1$ without major increases in particulate emissions. However, first results are very promising with conversion rates of about 50% in the European emission test cycle[35].

All of these NOx reduction technologies require that one important boundary condition be fulfilled: the sulphur content in the diesel fuel must be reduced further to avoid catalyst poisoning through sulphate storage. In Europe as well as in the US, the sulphur content is currently limited to 500 ppm. Together with the implementation of the EURO IV emission standards, the sulphur content will probably be regulated to 50 ppm maximum. The engine and vehicle manufacturers are requesting a maximum limit of only 30 ppm.

Further modifications in the diesel fuel composition may also help to reduce the critical diesel exhaust emissions. Synthetic diesel fuels such as 'Fischer-Tropsch Diesel' are under discussion with very high cetane numbers and no aromatics or sulphur. Naphtha-water-emulsions or methanol blended diesel fuels are also being considered.

Additional improvements in the in-cylinder diesel process can be expected. Here, the fuel injection system will play a dominant role. Second generation common rail systems will provide higher injection pressures, smaller injection quantity tolerances (for the main and the pilot injection), smaller pilot quantities and reduced power requirements for the high pressure pump drive through application of on-demand pumps[37,38].

The combination of all these technological advancements will determine the future of the HSDI engine in the face of increasing stringent emission standards, engine/vehicle cost and customer acceptance.

References

1 HEUHAUSER, W., STASTNY. J. and STEINPARZER, F., 'The new BMW four cylinder diesel engine', *MTZ 55* 7/8 (1994)

2 CARNOCHAN, W. A., and HORROCKS, R. W., 'The Ford 1.8 TCI Diesel in the Mondeo and future developments', *4th Aachen Colloquium Automobile and Engine Technology '93*

3 NORBYE, J. P., '12 Valve, four cylinder diesel for the Citroen XM', *High Speed Diesel Engines and Drives,* May (1990)

4 CONRAD, U., FEUCHT, H. J., KLINGMANN, R. and KRAUSE, R., 'The new four valve diesel engines from Mercedes-Benz, *MTZ 54* 7/8 (1993)

5 NEITZ, A. and D'ALFONSO, N., 'The MAN combustion system with controlled direct injection for passenger car diesel engines' *SAE paper 810479* (1981)

6 BIRD, G.L., 'The Ford 2.5 litre high speed direct injection diesel engine—its performance and possibilities', *SAE paper 850262* (1985)

7 VAN BASSHUYSEN, R., STOCK, D. and BAUDER, R., 'Audi turbo diesel engine with direct injection', *MTZ 51,* 1 (1990)

8 RHODE, W., GOKESME, S., LIANG, J. R. and SCHMITT, J. L., 'The new direct injected 1.9l diesel engine from Volkswagen', *3rd Aachen Colloquium Automobile and Engine Technology' 91*

9 KRAUSE, R. and SÄLTZER, D., 'A new turbo diesel engine with direct injection for the E-class from Mercedes-Benz', *MTZ 56,* 6 (1995)

10 NEYER, D. and DORENKAMP, R., 'The new low emission direct injection 4-cylinder naturally aspirated diesel from Volkswagen', *5th Aachen Colloquium Automobile and Engine Technology' 95*

11 WILLMANN, M., JELDEN, H., POHLE, J., ROOST, G. and KACKE, A., 'The new 81 kW-TDI engine from Volkswagen', *MTZ 56 12 (1995)*

12 BAUDER, R., DORSCH, W., MIKULIC, L., PÖLZL, H.-W. and REUSS, T., 'The new V6-TDI engine from Audi', *MTZ 58,* 10 (1997)

13 PETERS, A. and PÜTZ, W., 'The new 4-cylinder diesel engine OM 611 with Common Rail injection ', *MTZ 58* 12 (1997)

14 HERRMANN, H.-O. and DÜRNHOLZ, M., 'Development of a DI-diesel engine with four valves for passenger cars', *SAE paper 950808* (1995)

15 WÜNSCHE, P. and KÖNIG, F., 'AVL LEADER—The development of a new generation of passenger car diesel engines', *MTZ 55* 7/8 (1994)

16 KRIEGER, R. B., SIEWERT, R. M., PINSON, J. A., GALLOPOULOS, N. E. HILDEN, D. L., MONROE, D. R., RASK, R. B., SOLOMON, A., S., P. and ZIMA, P., 'Diesel engines: one option to power future personal transportation vehicles', *SAE paper 972683* (1997)

17 BRÜGGEMANN, H. and WAMSER, M., 'The new OM 668 diesel engine with Common Rail direct injection for the Mercedes-Benz A-class', Special edition of ATZ and MTZ 1997

18 STUMPP, G. and RICCO, M., 'Common Rail—an attractive fuel injection system for passenger car DI diesel engines', *SAE paper 960870* (1996)

19 PETERS, A. and PÜTZ, W., 'A new turbo diesel engine with direct fuel injection for the Mercedes-Benz E-class', *MTZ 56* 7/8 (1995)

20 ADKIN, P., CHAPMAN, J. and PEPPERELL, J., 'The new Rover L-Series diesel engine', *Automotive Technology International '95*

21 WALZ, L., 'Modern diesel fuel injection systems', *5th Aachen Colloquium Automobile and Engine Technology '95*

22 DÜRNHOLZ, M., ENDRES, H. and FRISSE, P., 'Preinjection—a measure to optimize the emission behaviour of DI diesel engines', *SAE paper 940674* (1994)

23 United States Environmental Protection Agency, 'Technical Description of Volkswagen Passat Diesel Turbocharged Engine, EPA Engine Family TVW1.9V6DF2E-Certification Application Document'

24 ABTHOFF, J., SCHUSTER, H. D., LANGER, H. J. and LOOSE, G., 'The regenerable trap oxidizer—an emission control technique for diesel engines', *SAE paper 850015* (1985)

25 RAO, V. D. N., CIKANEK, H. A. and HORROCKS, R. W., 'Diesel particulate control system for Ford 1.8l Sierra turbo-diesel to meet 1997–2003 particulate standards', *SAE paper 940458* (1994)

26 GEORGI, B., HUNKERT, S., LIANG, J. and WILLMANN, M., 'Realizing future trends in diesel engine development', *SAE paper 972686* (1997).

27 HOFER, K., SCHEDL, H. and VAN DEN BOOM, J., 'The acoustics of the new V6 TDI engine in the Audi A8', *ATZ 99* 7/8 (1997)

28 ABTHOFF, J., DUVINAGE, F., PISCHINGER, S. and WEBER, S., 'The potential of small DI-diesel engines with 250 cm^3/cylinder for passenger car drive trains', *SAE paper 970838* (1997)

29 GOMES, E., 'Developments in small diesel technology', *SAE TOPTEC Workshop "Diesel technology for the new Millennium",* Chicago, April 21–22 (1998)

30 NEUMANN, K.-H., NEYER, D. and STEHR, H., 'The new 3 cylinder diesel engine with high pressure injection from Volkswagen', *19th International Vienna Engine Symposium,* May 7–8 (1998)

31 KNOLL, R., 'AVL two-stroke diesel engine', *SAE paper 981038* (1998)

32 ABTHOFF, J., DUVINAGE, F., HARDT, T. KRÄMER, M., and PAULE, M. 'The 2-stroke DI-diesel engine with Common-Rail-Injection for passenger car application', *SAE paper 981032* (1998)

33 PETERS, A., LANGER, H.-J., JOKL, B., MÜLLER, W., KLEIN, H. and OSTGATHE, K., 'Catalytic NOx reduction on a passenger car diesel Common Rail engine', *SAE paper 980191* (1998)

34 CARTUS, T., HOLY, G., and HERZOG, L., 'Integration of NOx-adsorber technology in future gasoline and diesel engine concepts', *19th International Vienna Engine Symposium,* May 7–8 (1998)

35 KRÄMER, M., ABTHOFF, J., DUVINAGE, F., KRUTZSCH, B., and LIEBSCHER, T., 'Possible exhaust gas aftertreatment concepts for passenger car diesel engines with sulphur-free fuel', *19th International Vienna Engine Symposium,* May 7–8 (1998)

36 AUST, M., TOST, R., WISSLER, G., FISCHER, S., and ZÜRBIG, J., 'Exhaust gas aftertreatment system to meet EURO IV emission standards for diesel passenger cars', *19th International Vienna Engine Symposium,* May 7–8 (1998)

37 KRIEGER, K. and HUMMEL, K., 'Bosch-Common Rail: New experiences and future challenges', *19th International Vienna Engine Symposium,* May 7–8 (1998)

38 EGGER, K., and SCHÖPPE, D., 'Diesel Common Rail II-injection technology for the challenges of the future', *19th International Vienna Engine Symposium,* May 7–8 (1998)

23 Trucks and buses

Contents

23.1 Market demands

Market demands for heavy-duty diesel engines used in trucks and buses include: size, weight, cost, durability, reliability, performance and fuel economy, gaseous and noise emissions, and amenities such as electronic controls and features.

23.1.1 Size and physical constraints

The engine must fit into the truck, bus, or specialty vehicle. With the onset of smaller, more aerodynamic vehicles, engine package and size become more important. Fitting the bottom of the engine between the frame rails is becoming more of a challenge. Physical constraints such as: location of the starter and the intake manifold on the left and location of the turbocharger on the right must be accommodated. Typically, in-line six-cylinder engine configurations are used; however, some in-line four-cylinders and vee-eight configurations can be found in North American applications.

23.1.2 Weight

The weight of the engine is very important. A balance between durability and hauling capacity must be achieved. In today's trucking market, a 3000 lb (1360 kg) engine is considered too heavy, while a 2000 lb (910 kg) engine is too light.

23.1.3 Cost

Overall cost objectives need to be set early in the design process so 'creeping elegance' does not set in. In addition to controlling material cost, Design for Manufacturing and Design for Assembly techniques need to be utilized to control manufacturing and assembly costs. Durability and reliability goals must be set with warranty periods, cost, and customer satisfaction in mind.

23.1.4 Durability and reliability

The durability (life) of a heavy-duty diesel engine is defined to be the mileage to first overhaul or as the mileage to an in-frame overhaul. From 1977 to 1983, the on-highway market expected the 250 000 mile life (402 000 km) to increase to 350 000 miles (563 000 km). By 1986, the market expected life to be increased to 500 000 miles (804 500 km). In 1994, the expectation grew to 750 000 miles (1 207 000 km). The goal now is 1 000 000 miles (1 609 000 km), four times more than two decades prior.

Failure modes and effects analysis (FMEA) is a tool commonly used by a project team to meet life goals and to eliminate failures occurring in the early portion of product life (due to manufacturing and quality problems), in the useful life portion (due to random occurrences of operating and environmental stresses exceeding the produces strength), and in the wear out portion (due to an accumulation of stresses).

23.1.5 Performance

Since the 1960s, the power requirement has continued to increase every year. By 1987, the requirement was 400 hp (300 kW). Less than ten years later, the requirement grew to 500 hp (373 kW). This increasing trend is likely to continue. Along with an increased need for power has come the need for more bottom end (low speed) torque. The current range is between 1150 and 1850 lb. ft (1560 to 2510 N.m), with 1450 lb.ft (1966 N.m) being most popular. Even though fuel economy is a trade-off with power, the market has demanded better fuel economy, while demanding increased power.

A maximum power of 400 hp (298 kW) at 1800 r/min with a peak torque of 1450 lb.ft (1966 N.m) at 1200 r/min was very popular in 1994. As a point of reference, the full load fuel consumption is typically plotted with the maximum power and torque curves. See *Figure 23.1*.

23.1.6 Fuel economy

While many people talk about full load fuel consumption, it is really the fuel consumption under part load and full load that sets the fuel economy. Fuel maps are typically used to show the full and part load fuel consumption. See *Figure 23.2*.

In 1994, the heavy-duty on-highway market demanded between 7 and 8 miles per gallon (42 to 49 km/l), compared to 5 and 6 miles per gallon (30 to 37 km/l) just five years prior. The engine manufacturers have met these challenges, mainly through the use of electronic controls, advanced materials, and innovative design approaches, coupled with more aerodynamic vehicles and matching the entire driveline for a given set of operating requirements. Drivers are beginning to understand the once desired top vehicle speed capabilities of 70 and 80 miles per hour (115 to 130 km/h) unnecessarily sacrifices fuel economy.

Engine, vehicle, tyre, and driveline component manufacturers have also begun their next round of improvements to meet the 10 miles per gallon (61 km/l) challenge. Engine manufacturers are looking for ways to desensitize the effects of driver and environment variability by making the engine perform the same, independent of load, over the entire operating speed range. Vehicle manufacturers are adding more aerodynamic features, while tyre and driveline manufacturers are designing new products for the newest expectations.

In 1995, Detroit Diesel Corporation marketed an improved version of its popular Series 60® engine that broke the 0.300 lb/hp.h (182 g/kW.h) brake specific fuel consumption (BSFC) level. At 0.296 lb/hp.h (180 g/kW.h), it still is the most fuel efficient engine available in the heavy-duty truck and bus markets. So,

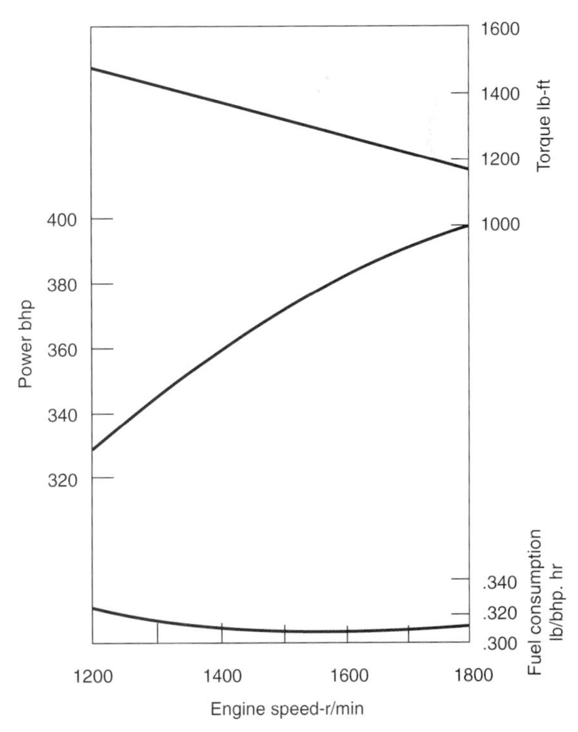

Figure 23.1 Typical performance curve (c.1994)

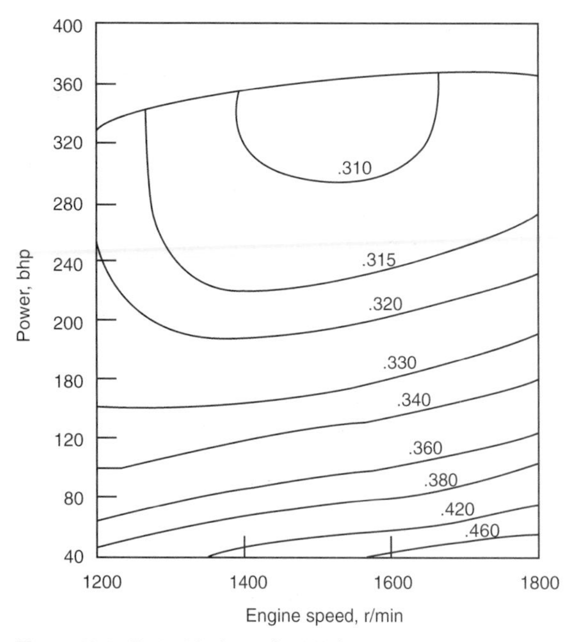

Figure 23.2 Typical fuel map (c. 1994)

while engine designers tried to develop engines in 1983 with a 0.315 lb/hp.h (192 g/kW.h) BSFC level, they must now design for 0.296 lb/hp.h (180 g/kW.h), and look for ways to meet a 0.250 lb/hp.h (152 g/kW.h) level.

23.1.7 Gaseous and noise emissions

Since the inception of the Clean Air Act in 1970, the reduction in gaseous emissions without a trade-off in fuel consumption has been a paramount effort by engine manufacturers. For instance in 1998, the NOx and PM standards (compared to uncontrolled levels) represent 65 and 95 per cent reductions, respectively.

Under the Federal Certification Program, engine manufacturers must demonstrate the standards are met when the engines are new and throughout the useful life, defined by the EPA to be 290 000 miles (466 610 km). Engineers generally set emission targets below the standards to be sure all requirements are met. This margin is typically 80 to 90 per cent of the standard limit.

Airborne sound sources include: mechanical impact, combustion pressure, gas expulsion, and air intake. Therefore, airborne sound source considerations must be a part of the initial design. Incorporating new materials and novel design concepts can eliminate the need and expense of shields and covers. SAE J366 details the N.A. vehicle passby test procedures for measuring sound pressure.

23.1.8 Electronics

Since the first introduction of fully controlled electronic engines in 1983, the changes in engine control and vehicle amenities have been revolutionary. Electronic features such as cruise control and engine protection, along with optimization of fuel economy, smoke, emissions, diagnostics, and performance tracking have become standard in the heavy-duty truck and bus markets. Optimizing idle operation, varying top vehicle speed, transmission-to-engine communication, remote data interchange, sophisticated diagnostic tools, and forward thinking (driver anticipation) electronics and software integration are new for 1996.

23.1.9 Product support

Each engine and vehicle manufacturer are also providing technical, service, and parts information electronically to aid consumers. Although the approaches vary, it is clear support must continue in this area.

23.2 Starting point

Typical configurations and operating levels of heavy-duty on-highway engines used in North America are very similar and are shown in *Table 23.1*. As technology advances and market demands change, the typical values will vary. Nevertheless, these values are used to determine the displacement and then the bore and stroke sizes.

Table 23.1 Typical configurations and operating values for the North American Market of the 1990s

Description	*Typical values*
Configuration	In-line six-cylinder, turbocharged, and air-to-air charge cooling
Controls	Electronic fuel injection and engine control
Cylinder pressure	>2000 lb/in^2 (13 800 kPa)
Brake mean effective pressure (BMEP)	245 lb/in^2 (1690 kPa)
Piston speed	2,200 ft/min (670 m/min)
Stroke-to-bore ratio	1.25:1
Rated power and peak torque	500 hp (373 kW) 1550 lb.ft (2101 N.m)
Full load speed (FLS)	1800 to 2100 r/min

To determine the displacement, use the following equation:

$$\text{Displacement} = \frac{(\text{Power @ FLS} \times 792\,000)}{(\text{BMEP} \times \text{FLS})}$$

By using the values from *Table 23.1*, the typical displacement can be determined.

$$\textbf{Typical N.A. Displacement} = \frac{500 \times 792\,000}{(\text{BMEP} \times 2100)} = 770 \text{ in.}^3 \text{ (12.7 l)}$$

For a stroke of 6.3 inches (160 mm), the resulting bore should be 5.2 inches (130 mm). With these parameters, the component design can start.

23.2.1 Cylinder block and head

The cylinder block and head of modern heavy-duty diesel engines are generally the largest single components in the engine. They are generally green sand castings made from gray or alloyed cast iron and serve as the foundation for the engine. In many cases, the cylinder block and head are used to connect the engine to the vehicle.

The design of a cylinder block and head is driven by:

- the mechanical and thermal performance requirements
- the fluid circuits
- the intake air and exhaust flow and port configurations
- the reciprocating and rotating components and bearing sizes
- the camshaft and accessory drive requirements
- starter motor positioning
- electronic or mechanical control systems
- noise suppression features
- casting technology limitations

- reliability and durability expectations
- assembly and serviceability requirements.

These requirements can be met in different ways. Some engines have the camshaft in the cylinder block, while most others use overhead camshafts. Some engines have multiple-piece cylinder heads, while most others use a single-piece cylinder head.

To maximize volumetric efficiency, the port dimensions must be carefully selected. Optimal port design is characterized by two dimensionless parameters: the port length-to-cylinder bore ratio and the port diameter-to-port length ratio. Engine speed is another key factor, since the piston speed affects air flow dynamics within the ports, valve openings, and combustion chamber.

Performance requirements for the engine dictates the use of a four-valve, centrally-located unit injector with a cross-flow air system. This usually leads to parallel flow intake and exhaust ports that merge together at the sides of the head. The length of the exhaust port should be minimized to reduce the heat passed to the coolant, reducing the cooling load and increasing the available energy to the turbocharger.

The water manifold on the side of the cylinder block must be designed to distribute coolant to the cylinders. Ribbing and serpentined shaping of the cylinder block results in reduced noise emissions. Machined stress relief slots between the cylinders in the cylinder head, positioning of the exhaust and intake ports, and efficient water flow through the cylinder head result in a design tolerant of mechanical and thermal stresses.

Top liner cooling concepts and free hanging liners reduce top ring reversal temperatures, promoting longer life of the cylinder kit.

Reliability expectations generally drive designs toward the fewest number of components. Recently, these requirements have driven designers to direct-acting overhead camshaft and valve operating systems, eliminating numerous components and wear surfaces. This in turn,

- eliminates the need to use the block structure as the camshaft support
- simplifies machining by eliminating guide holes for the push rods
- offers assembly and service advantages
- allows cylinder head bolt positioning such that very high, yet uniform clamp loading and sealing between the cylinder head and block occurs. (Note: with 8 bolts per cylinder, over a million pounds of force can be achieved.)

Assembly and serviceability concerns force designs to be directly accessible, avoiding complicated assembly techniques or removal of extraneous components during regular maintenance or in the event of failure.

Manufacturing and serviceability concerns can be met in a number of ways. For example, the cylinder block should have an extra amount of stock at the top to allow repair in the event of damage to this surface.

23.3 Cylinder kit components

The cylinder kit components and connecting rods provide the means to transfer the combustion gas forces to the crankshaft. The basic operational principle is described by the slider-crank mechanism, where linear motion of the piston is converted into rotational motion about a pivot (the crankshaft) through the connecting rod. The components of the cylinder kit are a piston, a set of piston rings, and a cylinder liner. Design considerations should focus on obtaining good combustion efficiency, transferring force through the linkages, and minimizing emissions, friction, wear oil consumption and heat transfer.

23.3.1 Pistons

The piston designer must consider the need to guide the connecting rod in the cylinder, provide a bearing surface for the normal forces being transferred to the cylinder walls, provide a means to seal the combustion gases within the combustion chamber, and withstand high mechanical and thermal stresses. Other important areas of piston design are the combustion bowl shape and the ability to dissipate heat energy.

Pistons used in heavy-duty diesel engines are either trunk style or articulated. Trunk style pistons combine the crosshead (or piston crown) with the side loaded bearing surface (or skirt) in one unit. Articulated pistons separate these two parts, allowing relative motion when the skirt and crosshead rotate about the piston pin.

The crosshead design is influenced by the overall combustion chamber design, the peak combustion pressure forces it must transfer to the connecting rod, the combustion gas seals (piston rings) and its operating temperature, all while minimizing weight.

The piston bowl design is an integral part of the combustion design process. During the engine cycle, the cylinder must receive a fresh air charge, compress the air charge, inject and mix the fuel into the compressed air charge, burn the fuel, expand the hot combustion gases, and then exhaust the expanded gases. At the top of the piston stroke, sufficient clearance is required to open the intake and exhaust valves. This may be done by adding cut-outs for the valves in the top of the piston or by carefully specifying heights.

The volume and shape of the bowl must provide the correct compression ratio (a factor especially important in start-up and cold smoke control) and assist in the mixing of air and fuel. In low swirl engines, the bowl is shaped to minimize contact between the injected fuel plume and the piston dome. In medium and high swirl engines, the bowl is shaped to help promote swirl and take advantage of the atomization effects of the fuel plume. This results in a tendency towards smaller, deeper, and reentrant bowl designs, to the point where the highest swirl engines have spherical bowls.

The volume of dead (or unused) air between the piston and the cylinder head should be minimized, as should the volume between the top compression ring and the top edge of the piston. The bowl volume should be maximized. These requirements typically result in piston designs with high top compression rings or tight piston-to-liner clearances.

Structurally, pistons must handle high thermal and mechanical stresses. Since the materials used for pistons, cast iron and aluminum, have different mechanical and thermal properties, their structural designs differ particularly in regard to cooling and structural load capability.

In aluminum piston designs, the use of a top ring groove insert is required. Occasionally a second ring groove insert is required. These ring groove inserts require additional structural support. Improvements in aluminum metallurgy have improved material properties through squeeze cast fibre reinforcement. In cast iron pistons, the high temperature strength of cast iron does not require the use of inserts. The lower thermal conductivity of cast iron relative to aluminum requires additional cooling consideration. So, maximum bowl volume is compromised by the need to providing adequate ring groove support and cooling. Cooling strategies, in iron and aluminum pistons, include spray jets and cocktail shakers.

A spray jet squirts engine oil through a drilled hole in the connecting rod, through a fixed pipe under the piston crown, or into a piston cooling channel. Cocktail shakers consist of a chamber under the piston crown, which is fed with oil to a controlled level. The oscillating motion of the piston shakes the oil, providing efficient heat transfer.

Modelling can predict piston strut cracking due to gas pressure loads and thermal hoop stress. In general, the combination of thermal and mechanical stresses require an iterative finite elemental analysis (FEA) approach to ensure reasonable stress levels.

The piston skirt is the main bearing surface for the normal forces transferred to the cylinder walls from the piston. Like most bearings, the skirt is frequently coated with a layer of tin or babbit to reduce the potential of skirt scuffing. The skirt design must enclose the piston pin and rod bearing, and if necessary, provide cooling to the lower half of the piston.

Lighter trunk style pistons are used for lower power density applications and are cheaper to manufacture. In both trunk and articulated pistons, the most effective bearing surface of the skirt under thermal and mechanical loads is frequently provided by a cam ground elliptical profile.

In higher power density engine designs, articulated pistons provide better load transfer to the cylinder wall by separating the normal and axial loads, with the crosshead carrying the axial load and the skirt carrying the normal load. This separation reduces the heat transfer from the piston into the skirt, resulting in less thermal and mechanical distortion of the skirt.

23.3.2 Piston rings

Piston ring design is very complex because the rings must seal the combustion gases, assist in heat transfer from the piston to the cylinder wall, and control lubrication in the cylinder kit area. Piston ring design should be a part of the dome design process to minimize gas leakage (blowby), minimize ring sticking, and optimize ring dynamic motion. Typical cylinder kit designs incorporate three or more rings:

- a top compression or fire ring
- one or more secondary compression rings
- one or more oil control rings

Typically, uniform contact pressure is desired for all engine operating conditions. However, due to thermal distortion of the rings and the cylinder liner, or special considerations such as liner ports in two-cycle engines, variable contact pressures are specified at room temperature. In all cases, rings should not have an opening between the ring face and the cylinder wall when installed in the cylinder. This specification is known as 'light tight'.

There are five common compression ring cross-sections:

- rectangular
- taper faced
- keystone
- internally bevelled
- L-shaped

The rectangular ring is geometrically simple, thus it is the least expensive. It provides sufficient sealing for most applications. Taper faced rings provide line contact, which scrape oil, provide oil control, and shorten run-in through their higher unit pressures. Keystone rings have tapered sides that, through movement in and out of the ring groove, minimize the build up of combustion deposits that often causes ring sticking. Internal bevelling of a ring causes the asymmetrical cross-section to twist when installed, resulting in bottom edge contact of the ring face. This is beneficial toward keeping piston face contact centred when the ring distorts under thermal loads. L-section rings are used in very high piston groove applications to minimize dead space. They are very sensitive to gas activation, requiring very little ring tension. They are also resistant to ring flutter.

There are two basic oil control ring types: with and without expander springs. All oil control rings are designed to be flexible, with high unit loading and scraping action. By making oil control rings flexible, their inherent spring action is reduced, making expander or coil springs necessary.

Oil control rings without expander springs are frequently slotted with bevelled edges to provide high unit loads and directional oil scraping. Spring loaded oil control rings also have bevelled edges. With this design, manufacturing flexibility results from the multiple-piece approach.

Ring design must include detailed attention to the size and type of the ring joint (gap). Practically, a ring must have a gap large enough to install onto a piston and to allow for the thermal expansion during engine operation. This is necessary to prevent ring end butting, which could result in ring face scuffing and failure. Since this gap produces a path for combustion gases to leak into the crankcase, a second ring is usually required. Two-stroke engines require pinning of the ring or camber to prevent the ring tips from protruding into the liner ports, where clipping could cause ring damage.

Since the ring face is a bearing surface that experiences high temperature and pressure, excessive ring wear, scuffing and failure can occur. This should be addressed by the appropriate use of coatings or surface treatments, surface finishes, and ring face profiles. Generally, the closer the ring is to the combustion chamber, the more likely a protective coating is needed. Typical coating materials include chromium (applied by plating) and molybdenum, cermet, and ceramics (applied by plasma spray processes). Surface treatments typically include phosphating, ferroxiding, and nitriding. Ring faces are typically lapped to a very fine surface finish and have a characteristic profile, promoting good oil film lubrication over the entire piston stoke and minimizing film break-down from high unit loading when the piston tilts.

Asymmetric barrel faced rings are used to provide thicker oil film formation on the upward stroke. While face coating, profile, and ring gap make up a large proportion of the factors related to friction and wear, corrosive wear from combustion by-products, adhesive wear between similar materials in the cylinder kit, and abrasive wear due to dust, dirt, and wear particles in the lubricant, need to be considered in the ring design process.

Ring groove design is very important for optimal piston ring function. Since piston rings seal against the groove, groove wear occasionally needs to be considered. The width and depth of the groove must be sufficient for thermal expansion of the rings and provide clearance for deposit packing.

Keystone groove geometry and cooling of the ring pack area provide resistance to ring sticking caused by deposits. Sufficient volume must be provided behind the rings to allow combustion pressure pulses to activate the rings. Since the rings shuttle between the top and bottom of the ring, inertial, gas pressure, and hydraulic forces must be controlled to provide proper ring seating over the engine speed and load range. Oil consumption should be minimized to a level necessary for proper lubrication without impacting emissions. Computational fluid dynamic modelling coupled with extensive testing of the ring pack can provide insight into these factors.

23.3.3 Cylinder liner

The cylinder bore in heavy-duty diesel engines is typically replaceable. Dry cylinder liners are typically stiff, have poor heat transfer, and are noisier than wet liners. They typically do not need engine coolant seals. Wet cylinder liners have good heat transfer characteristics, but tend to suffer from coolant cavitation and experience liner-to-head joint motion. If cylinder liners are not supported properly, thermal distortion can result in out-of-round conditions and blowby.

The cylinder liner bore is usually honed to provide a good lubrication surface finish and to minimize ring wear, while minimizing exhaust emissions. Plateau honing is one method used to increase the bearing area.

23.4 Connecting rod assembly

The connecting rod assembly is the linkage between the linear motion of the piston and the rotational motion of the crankshaft. The connecting rod is connected to the piston pin with a bolt, an interference fit, or is retained within the skirt by clips. The crankshaft end of the connecting rod has a separable piece, called a bearing cap, that secures the connecting rod to the crankshaft and holds the connecting rod-to-crankshaft bearing.

23.4.1 Connecting rods and bearing caps

The connecting rods in four-stroke heavy-duty diesel engines carry very high tensile, compressive, and bending loads, while two-stroke engines have much lower tensile loads. These loads also fluctuate, making fatigue strength a major design criterion.

For the piston pin end, two types of connecting rod designs have been developed: square and trapezoidal sections. The trapezoidal design provides increased pin bearing area on the connecting rod side, while reducing the bending stresses in the piston pin. With square rod ends, these same features are typically incorporated into the piston.

Two types of connecting rod bearing cap designs have been developed: normal and angle cuts. Normal cut bearing caps are the most common. In this design, the split of the rod is normal to the axis of the rod, making it the easiest to manufacture, but has the disadvantage of placing the bolts directly in line with rod loads. Angle cut connecting rods, reduce this loading concern, but at the expense of manufacturing. Angle cut connecting rod designs are sometimes selected to avoid interference and to make assembly and maintenance easier. For both types of bearing caps, serrated (or stepped) surfaces are frequently incorporated to improve alignment and to reduce clamp load distortion.

Finally, the connecting rod must be the proper length to provide the intended piston stroke, operate within the constraints of the cylinder bore, and be of minimal weight.

23.4.2 Piston pin bearings and connecting rod-to-crankshaft bearings

Piston pin bearings and the connecting rod-to-crankshaft bearings in heavy-duty diesel engine applications are hydrodynamic journal bearings. Many standard materials and designs exist, although modern engine requirements often require new material developments, innovative manufacturing processes, and new component configurations.

The piston pin bearing carries high loads and oscillates with low average surface speeds. This makes wear and heat generation of less concern and high endurance strength of more concern.

The piston pin frequently contacts the bearing, so it should be hard and highly polished. Since the highest loads are from cylinder pressure, the majority of the bearing area should be on the piston dome side.

The connecting rod-to-crankshaft bearings are some of the most critical bearings in the engine, so additional attention should be taken in their design. Since the loading of this bearing rotates with the crankshaft and the oil passes through it to the connecting rod, care should be taken in the location of lubrication grooves and supply holes. Since most bearings have pressure fed lubrication, manufacturing considerations for the drilled hole through the centre of the rod impacts connecting rod design.

Finally, for all journal bearings, specification of sufficient clearance is mandatory, since inadequate, rather than excessive clearance is more frequently the cause of bearing failures.

23.5 Crankshaft assembly

The function of the crankshaft assembly is to transmit the energy in the combustion chamber to the customer's application by transforming the reciprocating energy of the piston to rotational energy. The crankshaft assembly consists of the crankshaft, crankshaft oil seals, crankshaft main bearings, crankshaft pulley, and crankshaft vibration damper.

23.5.1 Crankshaft

Crankshafts are typically forged from a carbon steel; typical materials are SAE 1046, 4140 and 4340. Less typical materials include cast steel SAE J435a, iron J433 and J434; these materials are chosen for their high strength-to-weight ratio and castability. A modified version of SAE 1548 is also used in forged crankshafts. Alloying elements can be added to improve strength.

The crankshaft must be able to withstand forces from the combustion chamber gas pressures, rotating and reciprocating inertias, and vibratory forces. The crankshaft should be sized for the power rating of the engine, plus a safety factor, and future growth expectations. Its physical size is governed by the engine design envelope. Oil passages transferring lubricating oil from the oil pump to the main and connecting rod bearings frequently penetrate the length of the crankshaft. Location and surface blending of the oil passages requires considerable engineering effort.

The crankshaft butt provides the attachment point for the flywheel and is a sealing surface for a rear oil seal.

Attachment of the flywheel is provided by a bolt circle. A calculation of flywheel to crankshaft joint capacity will determine the number of bolts, bolt size, bolt torque, and bolt circle diameter.

The journals are hardened surfaces. A hard surface on the main and connecting rod journals will provide resistance to material failure, such as galling and provide resistance to fillets and radii fractures. The main journal size is based on the allowable space, bearing load, and maximum engine speed. The surface finish affects crankshaft and bearing durability. Hardening affects the strength of the crankshaft.

When a cylinder fires, a power impulse is sent through the crankshaft that causes torque oscillations. Oscillations of torque within the crankshaft need to be considered due to the extremes of the torque peak and the fatigue strength of the crankshaft material. An equivalent torsional stiffness of the main journal diameters and lengths and cheek thickness along with the inertia of each crankshaft section must be calculated. Additionally, the inertias of the connecting components (e.g. gear train, pulley, and flywheel) and the rotating and reciprocating inertias of the cylinder kits must be accounted for. With an equivalent stiffness and inertia, an estimate of the natural frequencies of vibration can be made. If a natural frequency of torsional vibration lies within the operating range, the crankshaft vibration could lead to fatigue failure. Every effort needs to be made to ensure that this does not happen. Therefore the fundamental frequency must be located above the operating speed of the engine.

A mode of torsional vibration is the natural frequency of vibration. The first mode equates to the fundamental frequency, the second mode equates to the second natural frequency and so on. The modes are determined by the stiffness of the crankshaft and the inertias within and attached. The fundamental mode is typically termed 'the crankshaft mode' as it is directly related to the crankshaft; remaining modes are determined by the characteristics of the power train.

Orders are fractional parts of the frequencies of vibration. For example, if the engine has a second mode critical speed of 9242 r/min, the sixth order critical speed is 2054 r/min. The dominant orders are dependent upon the number of cycles (two vs. four), the engine configuration (in-line vs. vee) and the number of cylinders. An in-line, four-cycle, six-cylinder engine will have the third, sixth, and ninth of the dominant orders of torsional vibration. These orders will exhibit higher amplitudes of torsional vibration than the other orders.

Nodes are locations within the power train where the torsional oscillations are directly out of phase with each other; the result is zero relative movement between adjacent components and a loci of highest torsional stresses. Nodes will occur within the crankshaft and power train; their number is equal to the mode number. For example, a second mode frequency will have two nodes where the stress of torsional vibration will be higher.

The centrifugal forces should be balanced by the crankshaft counterweights, so only the inertial forces remain unbalanced. The unbalanced forces can be analysed using the imaginary-mass approach.

In practical terms, the main contributors to the unbalanced force for a six-cylinder, four-cycle engine are the first harmonic, occurring at the same frequency as the engine speed, the second harmonic, occurring at twice the engine speed, and the fourth harmonic, occurring at four times engine speed. The contributions by the higher harmonics to the unbalanced force are negligible by comparison.

23.5.2 Crankshaft oil seals

The crankshaft oil seal performs a dual function; it prevents engine lubricating oil from leaving the engine and it prevents dust from entering. The seals are located at each end of the crankshaft where it projects through the cylinder block. Crankshaft seal design usually incorporates a steel form for rigidity and a synthetic rubber compound for sealing between the cylinder block and crankshaft seal and between the crankshaft and crankshaft seal.

The seal lip is usually made from TPFE or Teflon. Some seal designs incorporate a coil spring to provide a circumferential force on the lip. In this design, the entire seal remains stationary. In other designs, the majority of the seal is stationary with the exception of a press fit sleeve that rotates with the crankshaft.

The seal design must take into account expected temperature variations. The upper temperatures are determined by the heat from the engine and transmission. A range of 300 to 350°F (149–177°C) is considered normal. The lower temperature range depends upon geographical location. A typical lower temperature rating of – 25°F (– 32°C) is not uncommon in North America and Europe.

Crankshaft seal designs can be of a multiple-piece or unitized. Multiple-piece designs have advantages in ease of manufacture and installation. Multiple-piece seal designs are common. Unitized seals have the advantage of a protected seal lip. The seal lip rides on the sealing surface of the rotating crankshaft; any nicks or tears to the lip occurring during seal installation will affect seal durability.

On the oil side of a crankshaft seal, a method of returning oil back to the engine is required. This can be accomplished by the use of an oil slinger or by incorporating a helical return groove in the seal. The helical return groove forms a hydrodynamic dam and pumps the oil back toward oil pan. When the engine has a wet flywheel housing (typical with wet clutch installations and some torque converters), a helical return groove and a second seal lip will be used to prevent entrance of transmission fluid into the engine.

On the air or transmission side of the oil seal, a dust cover is usually incorporated to protect the lip seal from foreign object damage and particulate abrasion. In others, two dust covers are provided. The additional protection increases seal durability.

23.5.3 Crankshaft main bearings

The main crankshaft bearings are the components that transmit the gas forces, via the rotating crankshaft to the stationary structure of the cylinder block. The bearings are typically manufactured from a steel backed plate with a thin layer of bearing material overlay. Two-piece bearings are the typical in the heavy-duty diesel engine markets. The crankshaft main bearings require the following characteristics:

- adequate endurance strength
- sufficient deformability, the ability to deform plastically at points of very high pressure and to conform to the shape of the journal
- sufficient embeddability, the ability to allow particles of foreign matter to embed in the bearing material rather than to circulate in the oil film where they would score or wear the journal and bearing surfaces
- adequate resistance to scoring and seizure of the bearing material to the shaft
- high resistance to corrosion, due to elements in the lubricant
- adequate oil film and good thermal loading characteristics to support the journals are essential.

The four principal bearing materials are:

- tin based (80% Sn, 10%, Cu, 10% Sb)
- lead based (82–86% Pb, 9–11% Sb, 5–7% Sn)
- copper based (approx. 70% Cu, 30% Pb)
- aluminum based (100% Al).

Copper based bearings should be limited to 3000 to 4000 lb/in^2 (20 700–27 600 kPa) pressures. Lead based bearings should be limited to 1800 lb/in^2 (12 420 kPa) pressures.

Main bearings should be lubricated with pressurized oil to facilitate the formation of an oil film. In general, the oil holes should be located in areas of minimum loads. Exact location of the oil feed holes and the oil grooves in the bearings should be carefully considered to provide an adequate supply of oil to the entire bearing surface, especially to the highly loaded areas where maintenance of the oil film thickness is critical for good bearing durability.

Bearing pressure (P_B), a measure of the tendency to squeeze out the oil film, is predicted by the following equation;

$$P_B = \pi \times D^2 (p_c - p_L) / (4 \times d \times l)$$

where

P_B = Maximum bearing pressure, lb/in^2
P_C = Maximum combustion pressure, lb/in^2
p_L = Inertia pressure of piston area, lb/in^2
D = Cylinder bore diameter, in
d = Crankpin diameter, in
l = Length of crankshaft bearing, in.

The inertia pressure of piston area (p_L) is derived from the following equation:

$$p_L = 28.4 \times W \times R \times (N/1000)^2$$

where

W = Weight of piston assembly and connecting rod, expressed in lb/in^2 of piston area
R = Crank radius, in
N = Crankshaft speed, r/min.

23.5.4 Crankshaft pulley

The crankshaft pulley transmits crankshaft power to auxiliary devices such as cooling fans, alternators, water pumps, and hydraulic pumps, Pulleys are manufactured from cast iron or steel. Some pulleys will have a pilot diameter and bolt circle available for the attachment of additional grooves or for an axial drive.

Two styles of crankshaft pulleys are typically used: the vee and multi-ribbed (poly-vee). Vee belt systems are common throughout the diesel engine industry; poly-vee belt systems are used in higher auxiliary horsepower installations (more than 50 hp, 37 kW). The pulley diameter and width are determined by an analysis of the auxiliary systems. Power requirements, duty cycles and orientation of the auxiliary devices are examined and resolved to the recommendations of drive belt manufacturers. For example, cooling fans for Class 8 tractors will typically absorb 25–35 bhp (19–26 kW); meeting this demand requires two or three vee belts.

An analysis of the auxiliary systems will indicate if the required power is within the drive capability. Several designs of pulley drives include keyed shaft, press fit, and tapered shaft; each will have a limit for allowable drive torque. Consideration should be given to the resulting force vector from the belt loads. The analysis of the auxiliary systems will indicate if it imposes an objectionable bending moment upon the front of the crankshaft.

23.5.5 Crankshaft vibration damper

Reciprocating engines produce torsional oscillations in each rotation of the crankshaft. The oscillations or vibrations are present to some extent throughout the speed range of the engine, but will be concentrated about certain frequencies or modes. Sometimes the additional stress is sufficient to cause a fatigue failure of the crankshaft or attached components. The crankshaft vibration damper is a component used to control the torsional vibrations within the crankshaft.

Two types of dampers exist: tuned and untuned. A tuned damper is designed to a specific frequency or engine speed. A tuned damper will have a seismic mass coupled to the crankshaft by an elastic element (usually a synthetic rubber compound) with a known torsional stiffness. Because of the relatively soft torsional stiffness of the elastic element, the oscillation of the seismic mass will tend to negate or limit the amplitude of vibration; the lower amplitudes of vibration relate directly to lowered torsional stresses within the crankshaft.

An untuned damper has a seismic element encased within a tight fitting case filled with a viscous silicone fluid. As the seismic mass moves opposite of the torsional oscillations, the fluid shears, transforming rotational energy into heat. Since the seismic mass is not solidly coupled to the casing, it functions at more than one frequency. Physical size of the damper is predicated upon the torsional energy to be dissipated, the shear rate of the silicone fluid, and the size of the gap between the case and seismic mass.

23.6 Camshaft assembly

The camshaft assembly includes the camshaft, camshaft bearings, bearing caps, and drive gear. Its purpose is to displace the valve and injector mechanisms.

23.6.1 Camshaft

The camshaft location must be consistent with primary engine design objectives; cost, durability, size, weight, and engine performance goals. The camshaft lobe geometry is driven by; valve lift (or injector plunger travel) requirements, manufacturing limitations (mainly grinding wheel size), and contact stress considerations (component durability). The determination of lobe radius and rate of lift must be carefully balanced to achieve engine performance goals and to avoid excessive contact stress at the roller-to-camshaft lobe interfaces.

Two common camshaft locations exist: block-mounted camshaft; and overhead camshaft (OHC). The OHC design is either a single overhead camshaft (SOHC) or a dual overhead camshaft (DOHC). The SOHC is most commonly used in in-line engines. Block-mounted camshafts are typically used in concert with lower pressure unit pump injection systems. Since the camshaft is located near the side of the engine block, a complex push rod system must be employed between the camshaft and injector rocker assembly. This increases actuator inertia, friction, and compliance. An excessively compliant fuel injector drive develops unacceptable injection characteristics and adversely affects emissions and fuel economy. Despite this disadvantage, block-mounted camshaft engines are commercially produced with unit fuel injectors. However, these combinations typically occur when older pump-line-nozzle engines are upgraded with unit fuel injectors.

The most common approach for new engines is the single overhead camshaft (SOHC) design. This design has several key disadvantages, however. The most significant is the need for a more complex gear train. Since the camshaft is located above the cylinder head, crankshaft power must be transmitted to the camshaft through the gear train. This typically adds weight and friction to the engine.

In general, single overhead camshaft designs require fewer moving components and have fewer wear locations than block-mounted camshaft configurations. Since the camshaft directly drives the rocker arm, push rods and lifters are not required. This provides added flexibility in cylinder head design, since more space is available for: valves, ports, fuel injectors, fuel galleries, oil galleries, coolant passages, engine retarder mechanisms, and cylinder head bolts.

23.6.2 Camshaft bearings and caps

Many camshaft installations use bearing caps, for ease of installation, reduced engine height, and to lighten the cylinder head. The camshaft bearings should be hydrodynamically lubricated to meet durability goals. Bearing size and geometry should be determined by considering the effects of lubricating oil shear rate, operating speed range, and load on oil film load carrying capacity.

23.6.3 Camshaft drive gear

The camshaft drive gear must be designed to withstand gear train and camshaft-induced torsional excitations. The substantial load transmitted through the camshaft gear when driving high pressure unit injectors is particularly important when designing a robust gear tooth profile, selecting an appropriate material (usually high grade steel), and determining what type of heat treatment may be required. Some engine manufacturers use spring-pack dampers to minimize the torsional excitations.

23.7 Overhead components

Overhead components consist of the following major assemblies: valve train; rocker; rocker cover; and the engine retarder. They must perform the following functions: regulate cylinder breathing; supply and meter high pressure fuel; seal the combustion chamber; transfer power to valve and injector mechanisms; provide lubrication and wear reduction; provide structural rigidity; reduce engine acoustic emissions; and provide cooling.

In general, there are four types of overhead rocker arm configurations: direct-acting, end pivot, centre pivot, and centre pivot with lifters. With the exception of the direct-acting rocker design, the rocker arm is driven by a roller follower, acting in response to the camshaft lobe. The centre pivot rocker with lifter design has an additional lifter located between the camshaft lobe and the rocker arm. The driven end of the rocker actuates the valve or injector. The direct-acting rocker design simplifies the overhead with the fewest moving parts; however, it increases the height of in-line engines. The centre pivot design provides significant friction and wear benefits, with a high degree of design flexibility. It also has fewer moving parts than most other overhead systems.

Overhead components interact with several systems including the air, fuel coolant, lubrication, and combustion systems. Therefore, the overhead design frequently requires unique engineering considerations:

- fuel injector actuation requirements result in non-trivial space claims, limiting the overhead design flexibility
- combustion chamber design requires specific scavenging capabilities, which the valve train must satisfy
- overhead designs must be compact and robust, while satisfying engine performance objectives
- cost, packaging, and durability goals must also be met.

With an established engine design philosophy and knowledge of design constraints (created by other engine systems), a detailed overhead design may be pursued.

23.7.1 Valve train assembly

The valve train assembly includes exhaust and intake poppet valves and the components that guide and return the valves to their closed positions. These components include: valve guides; valve seats; valve stem oil seals; valve springs; valve rotator cap; oil seal protector cap; valve locks; and valve spring seats. See *Figure 23.3*

The primary design issues involve: achieving satisfactory gas flow (high volumetric efficiency); valve assembly, cooling and heat flow; structural strength; lubrication and wear; and service provisions. Numerous options exist and can be classified within three categories:

- poppet valves
- sleeve and piston valves
- rotary and slide valves.

Poppet valves are actuated by the valve train, which converts rotary camshaft motion into linear valve movement. Poppet valves are relatively inexpensive to manufacture, when compared with the other designs. With poppet valves, large valve flow areas, favourable flow coefficients, and low friction can be achieved. Since poppet valves are not subjected to high cylinder gas pressures, lubrication requirements are less severe and friction is reduced with this approach. The primary weakness with the poppet valve approach is exhaust valve cooling. For these reasons, poppet valves are common in heavy-duty diesel engines.

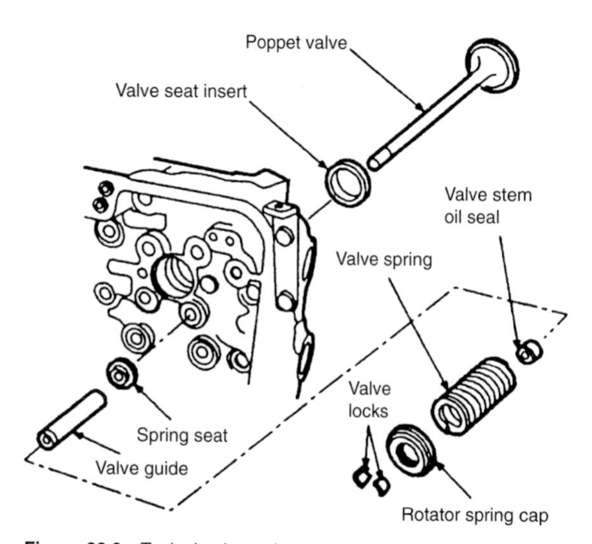

Figure 23.3 Typical valve train components

The valve orientation is considered to be a variable. Some manufacturers use a 90 degree rotation. While other manufacturers use a 45 degree rotation. The 45 degree rotation is common for push rod actuation; however, the 90 degree orientation offers several advantages over the 45 degree orientation. First, the 90 degree orientation reduces the length of the intake and exhaust ports. Therefore, flow restrictions are reduced and exhaust port tuning becomes less important. It also allows the intake and exhaust ports to be located on unique sides of the engine, simplifying the manifolding. Second, the 90 degree orientation decreases the exhaust port wall area, which improves heat transfer characteristics by minimizing heat and intake port heating.

With the camshaft location, injector design, combustion chamber geometry, and valve type selected, valve and port location and design may be defined. In engines with a flat firedeck, maximum valve areas are limited by the number of valves, valve bridge stress limitations, and the cylinder bore. For centrally-mounted fuel injector designs, a four-valve configuration is used to optimize gas flow.

Valve areas should be maximized. The intake valve capacities should exceed exhaust valve capacities. The ratio of exhaust-to-intake flow capacity usually is between 0.70 and 0.75 and the exhaust-to-intake flow area ratio is usually between 0.83 and 0.87. So, for 40 mm diameter exhaust valves, the intake valve diameters would be sized at about 44 mm, resulting in a 0.83 exhaust-to-intake valve flow area ratio.

Once the valve mechanism, ports, and combustion chamber are designed, valve timing must be determined. The amount of valve overlap and the exhaust valve closing time are key characteristics in this process. Four stroke engines require significant amounts of valve overlap to properly scavenge the combustion chamber. Computerized cycle simulation and computational fluid dynamics (CFD) codes like KIVA are valuable tools for analysing scavenging efficiencies and valve timing strategies. Scavenging efficiency (EFFs) is the ability of an engine to exchange fresh charge air for exhaust byproducts, as described by the following equation:

$$\text{EFFs} = \frac{\text{Trapped Fresh Charge}}{\text{Fresh Trapped Charge + Residuals}}$$

Overlap is defined as the number of crank angle degrees between exhaust valve closing and intake valve opening. Overlap is necessary to effectively scavenge exhaust gases from the cylinder and account for inertia effects associated with the air column. During the exhaust stroke, combustion byproducts are expelled through the exhaust valve. The exhaust valve remains open until after the intake valves open. This permits the boosted charge air to enter the combustion chamber and displace additional exhaust gases that exit through the exhaust port. During this process, a portion of the charge air is lost with escaping exhaust gases. Trapping efficiency is used to quantify the efficiency of filling the combustion chamber with fresh charge. The trapping efficiency (EFF_{tr}) is defined by the the following equation:

$$\text{EFF}_{tr} = \frac{\text{Mass of Air Returned}}{\text{Mass of Air Supplied}}$$

Several basic engine parameters influence the charging process. The two most significant are intake manifold conditions and valve timing. With turbocharged engines, the intake manifold pressure is higher than the cylinder pressure, during the scavenging process. Therefore, more excess air is pumped through the combustion chamber and out the exhaust port than in naturally aspirated engines.

The valve timing is established as a function of the exhaust-to-intake manifold pressure ratio, port geometry, and combustion chamber design. Computerized cycle simulation and engine performance testing provide valuable inputs to this design process.

23.7.2 Rocker assemblies

Rocker assemblies transmit the cam force to the valve and fuel injector mechanisms. The rocker assembly includes: exhaust valve rocker arm: intake valve rocker arm; injector rocker arm; roller followers; bearings; rocker arm shaft; rocker arm end plug; valve and injector buttons, adjustment screws, lock nuts, and clips.

A typical design is a side-by-side rocker arm assembly arrangement. FEA should be used during the rocker arm design process. The Detroit Diesel Series 60® rocker assembly is unique in that it uses a bridged rocker arm design to facilitate preferred manifolding without sacrificing durability. Valve train simulation programs should be used to evaluate dynamic performance characteristics and overall system compliance.

Particular attention should be given to the cam roller design because the size and shape of the roller affects the following critical component functions:

- rocker arm ratio
- valve lift and injection device travel
- camshaft lobe-to-roller Hertzian load
- roller-to-roller pin bearing load.

Insufficiently sized rollers can result in excessive wear rates or contact fatigue. Changes in the roller size over the life of the engine can reduce valve lift, injector travel, change injection timing, or change valve timing. Alloyed steel can be used to extend the useful life of the roller and cam lobe, as well as valve and injector functionality. Load and volumetric constraints may make this solution insufficient. Other alternatives include, polishing, tumbling, axially contoured surfaces, or ceramic materials. In 1994, Detroit Diesel introduced the use of monolithic silicon nitride rollers into heavy-duty diesel engines as a way to cost-effectively offset the increased loading associated from higher injection pressure required to meet emission standards. Ceramic rollers reduced roller skidding and friction at the cam lobe and roller pin interfaces. They also eliminated the need for a roller bushing and forced lubrication. So, the higher material cost actually reduced the overall engine cost by eliminating the bushing and the machining of the drilled oil passages within the rocker arm. Since the ceramic material has roughly half the mass of the metallic roller, dynamics were also improved.

23.7.3 Rocker cover assembly

The rocker cover assembly must seal the overhead cavity from oil leakage; prevent foreign objects from falling into moving parts; protect overhead components from dirt and corrosive materials; and reduce acoustic emissions. The rocker cover assembly includes: the rocker cover; rocker cover rim seal (isolator gasket); a breather to vent crankcase pressure; rocker cover isolators; and the rocker cover fasteners.

Rocker covers are typically composed of a reinforced fibre glass compound to dampen vibrations, minimizing acoustic emissions. Rocker cover assemblies are usually separated from the cylinder head by a silicone oil seal.

Rocker covers are typically centrally-bolted or perimeter-bolted. Centrally-bolted rocker covers induce more demanding stress regimes within the cover compared to perimeter-bolted designs, but reduce engine noise and component quantities.

The fastening arrangement should provide an even pressure distribution over the silicone oil seal to prevent oil leaks. The use of silicone isolators limit vibration transmission through the retaining bolts. By minimizing the transmission of engine vibrations to the rocker cover, acoustic emissions are reduced. However, the isolator attachment strategy increases rocker cover stresses, requiring added strengthening precautions. Internal ribs can be developed to provide stress relief using FEA.

Spatial claims are an important consideration in rocker cover design. Engine height should be minimized while providing adequate noise cancellation and space for the overhead and valve train. For these reasons, manufacturers frequently offer several rocker cover options to meet a variety of customer expectations.

Typically, one-piece rocker cover assemblies are used. However, two-and three-piece rocker cover assemblies have been successfully utilized to accommodate various accessories and application requirements. When engine retarder mechanisms are required, a higher profile cover must be selected at the expense of noise suppression and a higher engine profile.

23.7.4 Engine retarders

Engine retarders play an important part in the heavy-duty trucks' slowing and/or stopping systems, they supplement or temporarily substitute for the service brakes. By retarding the vehicle through the power train, improved anti-skid characteristics result in lieu of service brakes. Vehicle retardation can be accomplished without locking the vehicle wheels by uniformly applying the differential action to all driving wheels.

The market demand for vehicle retarders with improved performance is increasing as heavy-duty diesel trucks become more aerodynamically designed, as the legal gross vehicle weight (GVW) increases, and as engines and equipment become more efficient.

There are two types of retarders: input retarders and output retarders. The group of output retarders includes: electric, hydraulic, and friction retarders. Input retarders, also known as engine retarders and engine brakes. Engine retarders include: exhaust brakes, bleeder brakes and compression release brakes.

Exhaust brakes restrict exhaust flow from the engine during each exhaust stroke. An air cylinder is commonly used to operate it. The degree of restriction should be chosen to ensure the exhaust valves are not held off their seats. The exhaust brake operation can easily be integrated with the service braking system, so that it is applied by the first movement of the brake pedal. The cost of installation is relatively low. It should be noted that because the operating member of the slide type valve is fully retracted when not in use, it is not the subject to the effects of the gas stream and it is self-cleaning of carbon deposits. Exhaust brakes have been in use for decades in the mountainous regions of Europe with diesel engines made by Volvo, among others, but is much less common throughout the rest of the world.

A bleeder brake consists of a decompression valve in the head, at each cylinder, that is integrated into a bypass to the exhaust valve. At high piston speeds, little air escapes through the transfer duct into the exhaust manifold, so required combustion compression work is performed. Near top dead centre of the piston's stroke, the piston slows and a large amount of combustion chamber air overflows into the exhaust manifold. This overflow reduces expansion work and combustion work adds to the engine's braking power. This system is used on the Mercedes-Benz Series 400 six- and eight-cylinder engines and on Detroit Diesel Series 55™ engines. See *Figure 23.4*.

The compression release brake is most effective with turbo-charged engines. More air mass introduced into the cylinder means more work to compress the charge, resulting in higher retarding power. In the case of engines equipped with unit injectors and overhead camshaft, the braking device can be operated off the injector camshaft lobe. Valve train operated retarders develop

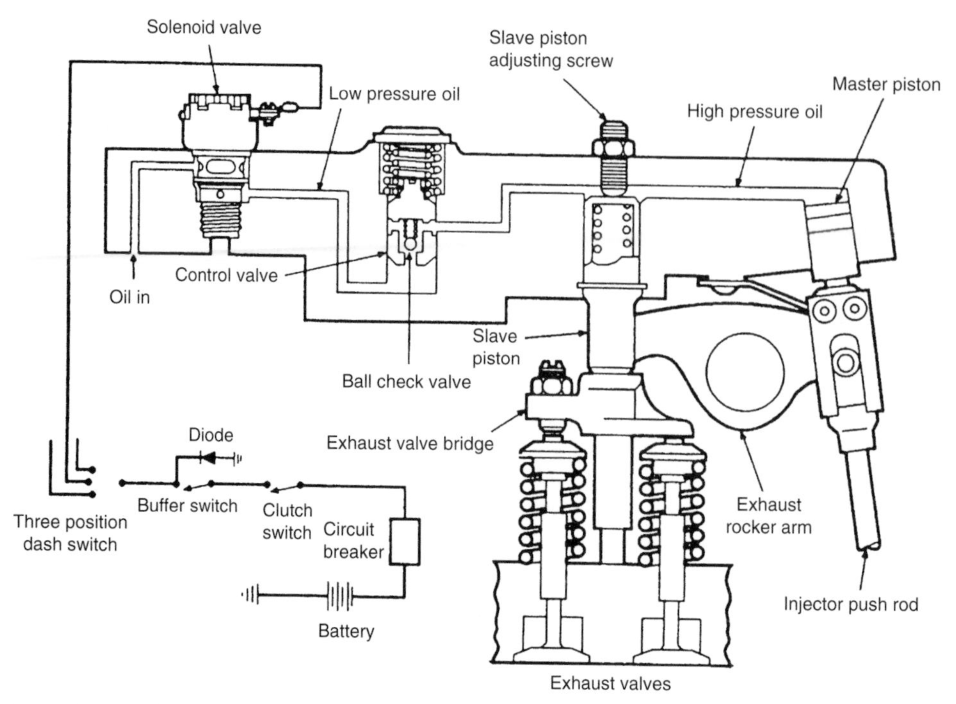

Figure 23.4 Braking schematic

less retarding power than an optimized injector train operated braking retarder.

By way of example, the requirements of engine retarders can be shown. For trucks weighing 80 000 lb (36 360 kg) and having a natural retarding power of approximately 190 hp (142 kW) an additional 240 hp (180 kW) of retarding capability would be required to maintain a controlled speed of 60 miles per hour (96 km/h) on a 4 per cent hill without employment of the service brake. The same amount of power would be required to decelerate the truck from 55 to 25 miles per hour (88 to 40 km/h) over a 1/4 mile (0.4 km) distance.

Typically, compression braking devices are located between the top of the rocker arms and the rocker cover, where one brake unit is used for each pair of cylinders. Usually, the units are bolted to the overhead and are operated electronically by a control module.

Operation of the compression brakes occurs by toggling a dash-mounted on /off switch. An intensity switch is used to select brake actuation of low, medium, or high braking power. These braking device systems usually have two additional switches, one activated by the position of the clutch pedal and the other by the position of the throttle. When the driver removes their foot from the throttle, the compression release brake converts a power producing diesel engine into a power absorbing air compressor. This is accomplished by motion transfer through a master-slave piston arrangement that opens the exhaust valves near the top of the normal power stroke releasing the compressed cylinder air charge to the exhaust system. The blowdown of compressed air to atmospheric pressure prevents the return of energy to the piston on the expansion stroke. The net effect is an energy loss. Different versions of the compression release brake are common in North American engines.

23.8 Flywheel

The flywheel is a rotating device attached to the crankshaft to reduce rotational speed fluctuations from friction, scavenging, compression, and combustion pulses.

A flywheel stores kinetic energy when angular velocity increases and releases kinetic energy when angular velocity decreases. The storing of energy reduces the rotational acceleration rate and the releasing of energy decreases and rotational deceleration rate providing a more constant rotational velocity. The flywheel cannot eliminate rotational speed fluctuation, therefore it is necessary to define an acceptable rotational speed fluctuation. An excessively large flywheel will have a poor transient response, while small flywheels will have large fluctuations in engine rotational speed.

The frequency of the speed fluctuation varies with engine speed and is most noticeable at idle. The frequency of speed fluctuation (F) is defined by:

$$F = N \times n/(C \times 0.5)$$

where,

F = Frequency of speed fluctuation, Hz
N = Crankshaft speed, r/min
n = Number of cylinders
C = Engine strokes per cycle (2 or 4).

Satisfactory idle speed fluctuations can usually be achieved with a 1/3 coefficient of speed fluctuation. The coefficient of speed fluctuation (C_S), is calculated by:

$$C_S = w_1 - w_2/w_o$$

where,

C_S = Coefficient of speed fluctuation
w_1 = Maximum speed in one revolution, r/min
w_2 = Minimum speed in one revolution, r/min
w_o = Mean crankshaft speed, r/min.

The inertia needed to maintain a desired instantaneous speed fluctuation can be calculated using the excess energy method. Using a torque vs. crankshaft angle diagram, the mean indicated torque can be calculated by summing the area under the torque curve. The excess energy must be equal to the energy stored as rotating inertia energy to limit speed fluctuations.

The mass polar moment of inertia, (J) involved in controlling speed fluctuations includes the inertia of all rotating masses: the crankshaft, driven machine, and flywheel, plus the mass polar moment of inertia of the connecting rods, etc. For rotating cylindrical discs, J is calculated by:

$$J = m \times d^2/8$$

where,
J = Mass polar moment inertia, lb. in^2
(to convert to lb in^2. s^2, divide by 386)
m = Mass, lb
d = Outside diameter, in.

The kinetic energy (U_K) for the rotating engine can now be calculated using:

$$U_K = (1/2)\ J \times w^2$$

where,
U_K = Kinetic energy, lb. in
w = Rotational velocity, rad/s

The change in kinetic energy, due to a speed fluctuation is then calculated by:

$$\Delta U_K = 1/2 \times J \times (w_1 - w_2)$$

The mass polar moment of inertia needed to achieve the desired speed fluctuation can now be determined by setting the excess energy equal to the kinetic energy change.

$$\Delta U_T = 1/2 \times\ J \times (w_1 - w_2)$$

The mass polar moment of inertia needed for the flywheel can now be determined by solving for J.

$$J = U_T \times 2/(w_1 - w_2)$$

The kinetic energy contained in a flywheel at high rotating speeds is capable of causing great damage to the engine and user if failure or release from the crankshaft should occur. Because of the potential for disaster, testing is necessary. Testing involves subjecting the flywheel to rotational velocities exceeding the velocity reached during normal operation. A flywheel must be able to withstand 2.5 times maximum engine speed before bursting. Alternatively, tensile specimens can be cut from flywheels and tested to failure, per SAE J416B.

The friction between the crankshaft and flywheel must be able to withstand all torque output of the engine. If the torque capacity (friction) is sufficient, the mounting bolts will not be subjected to bending torque or fatigue failure. The total torque is the sum of the mean and vibratory torques.

The friction capacity (Tc) of the crankshaft-to-flywheel joint is given by:

$$T_c = N_B \times F_B \times D_B \times C_F/2$$

where,
T_C = Torque capacity, lb.in
N_B = Number of bolts
F_B = Clamp force per bolt, lb
D_B = Diameter of bolt circle, in
C_F = Coefficient of friction between the faces

Typical values of C_F range from 0.1 to 0.15

23.9 Flywheel housing

The flywheel housing geometry is generally complex requiring sand casting. Because of the size, light weight materials should be considered; cast aluminum has become common. SAE J617C describes typical flywheel housing sizes.

Typically, the engine couples directly to the transmission at the flywheel housing. The engine output and transmission input must be accurately aligned. The need for critical static alignment demonstrates a need to resist service loads with minimal distortion.

Flywheel housings must also provide a rigid connection to resist engine torque loads and shock loads input through the system mounts. Shock loads are induced in the flywheel housing when the vehicle undergoes any shock loading. For instance, the housing will experience 7g loads during trailer engagement.

Natural bending frequencies above excitation frequencies of the engine and maximum output shaft speed are mandatory to achieve necessary rigidity. Bending, shock, shaft oscillation, and torque output loads must all be considered simultaneously. Ideally, the worst case installation should be modelled to determine the natural bending frequency of the flywheel housing. A freebody diagram using engine and transmission weights, centre of gravity, and mounting locations can be used to determine bending loads exerted on the flywheel housing. By adding shock loads to the freebody diagram, maximum shock loads can be computed. The complexity of a flywheel housing makes analysis by computer modelling a necessity.

If a single mount is used at the front centre of the engine with multiple mounts on the transmission, no torsional loads will be induced in the flywheel housing. However, if multiple mounts located away from the crankshaft's axis of rotation are used, torsional loads will be induced in the flywheel housing.

23.10 Geartrain

The geartrain provides power from the crankshaft to drive the camshaft and auxiliary components and provides timing between combustion and valve vents. A typical geartrain consists of a crankshaft gear, an oil pump drive gear, a bull gear, an adjustable idler gear, and a camshaft drive gear. Auxiliary components driven by the geartrain include a water pump, a fuel pump, an accessory drive for alternators, and a drive for an air compressor.

Three types of gears are used in heavy-duty diesel engines: spur, helical, and herring bone. Spur gears are economical and easy to manufacture. Helical and herring bone gear types offer the advantage of greater tooth loading capability and noise reduction. Herring bone gears neutralize gear thrust loads, but are expensive to manufacture.

In electronic engines, the timing of the combustion events in relation to the rotation of the crankshaft must by relayed to the electronic governing system. This is accomplished with a steel pin and a magnetic pickup. The steel pin is typically pressed into the bull gear. The electronic governing system also needs to know when each cylinder is approaching top dead centre, so fuelling events may be scheduled. This is accomplished with a timing wheel and a magnetic pickup.

All gear systems must have the ability to adjust the amount of lash in the system. Gear lash (back lash) is the difference between the width of a tooth space and the thickness of the engaging tooth, as measured on the pitch diameter. Some backlash is required to satisfy manufacturing tolerances, lubrication requirements, and thermal growth of the gears. Too much backlash leads to gear rattle and noise; severe cases reduce gear life, while too little backlash increases gear tooth side loading and may lead to gear binding and noise.

Gear teeth are also loaded by a mean driving torque, variations in engine torque, driveline resonance, geometric errors, and centre distance deviations. Individual teeth are subjected to combined stresses from bending due to tangential loads, side loads due to gear cut and thrust. The American Gear Manufacturers

Association (AGMA) and the American National Standards Institute (ANSI) are organizations that provide guidance in predicting tooth stresses for individual gear types.

An involute tooth form is invariably used in heavy-duty engines. The minimum number of teeth should be chosen to avoid undercutting, that is, a reduction of tooth thickness below the pitch line. In general, the face width of gears should be no more than one half the pitch radius; larger gear face widths tend to lead to poor load distribution.

Gear materials must have the ability to be case hardened and provide a good surface finish. Alloyed materials will, in combination with a core heat treatment and a surface hardening process, provide the strongest and most wear resistant gears. SAE 4140 with specific microstructure and case hardening is commonly used.

23.11 Gear case and cover

The function of the timing gear case is to provide:

- structural support and protection for the timing gear
- a mounting structure and access to engine power for auxiliary components, such as air compressors and cooling fans
- alignment between all gears regardless of engine load and speed.

The timing gear case and cover are typically cast iron components bolted about their periphery to form a sealed compartment. The case contains the crankshaft gear, the idler gear, bull gear, and camshaft gear.

On the exterior, the gear case and cover may be required to drive and provide support for auxiliary components such as an air compressor, fuel pump, hydraulic pump, cooling fan, and alternator. Special consideration of the effects of the exterior auxiliary components on the gear case and cover should be given. Heavy components may require additional mounting brackets to reduce the bending moment imposed upon the gear case.

The gear case and cover are the subject of finite element structural analysis for several reasons. First, the case must be rigid enough to maintain accurate locations for the gears within. Second, the gear case and cover could become a radiator of noise. Being a planer structure, resonance in a primary or secondary mode of vibration must be considered during the design phase.

23.12 Electronic control system

The application of electronics to heavy-duty diesel engines is motivated by the basic needs of competitive fuel economy and performance at legislated exhaust emission levels. As such, it must monitor all engine critical sensor inputs, calculate parameters for functional control requirements, and actuate output controls in a total time period that maintains optimal engine operation. The most challenging design issues of electronic control systems are the electronic control module (ECM) and the sensors.

The flexibility of electronic based systems can maximize customer benefits such as improved cold startability, engine protection and system diagnostics, idle shutdown, vehicle speed limiting, cruise control, etc., for little or no extra cost. In the planning stages of a diesel engine electronic control system it is important to work closely with original equipment manufacturers (OEMs) and end users to address the issues of customer acceptance, reliability, serviceability, and economics.

In the heavy-duty trucking and bus industry, standardized serial data communications, components, diagnostics, and service tools are considered a priority. Most major truck OEMs are currently involved in integrating electronic controls for engines, brake systems, traction control systems, transmissions, instrumentation, trailers, etc.

If each electronic system manufacturer required different proprietary serial communications messages, diagnostics and service tools, and serviceability, sharing between systems would become too difficult. Therefore, it is important for the diesel engine electronic control system to provide a communication interface that follows the standards created by the heavy-duty on-highway industry. The entire electrical system should be designed so it can function properly in its intended environment.

The Society of Automotive Engineers (SAE) and The Maintenance Council (TMC) of the American Trucking Association (ATA) have developed recommended practices to provide interface requirements for communications relevant to the industry. Independent suppliers offer industry compatible off-board service tools that can provide insight to the on-board engine diagnostic design.

The complexity of electronic systems and their ability to impact the overall performance of the engine makes packaging a critical issue for the system designer. To maintain a simple engine installation for the OEM, the engine electronic control system should be mounted on the engine. Designing the system with a single electronic control module is more cost-effective than systems with two.

23.12.1 ECM

The ECM typically consists of: a central processing unit, memory, software, integrated circuits, and peripheral devices. The system memory typically stores the engine control software and operating data, which can be uniquely addressed by the central processing unit. The peripheral devices typically perform input and output control functions. The engine sensors typically provide the ECM with absolute measurements and the engine wiring harness typically provides the interconnection between the ECM, the injector solenoids, the sensors, and the other vehicle systems.

Basic design principles of system architecture; circuit architecture; power analysis and distribution; environmental extremes; and safety, assembly, and service requirements must be applied to all phases of the ECM design. A properly designed ECM will improve engine starting, by accurately determining engine cranking speeds and temperatures to determine optimal fuel injection quantity and timing.

The following requirements need to be met to be competitive:

- The ECM must control exhaust smoke by measuring the intake manifold air pressure and engine speed to determine a fuel rate limit.
- The ECM should have multiple engine torque choices the user can select from.
- The ECM should contain balance cylinder-to-cylinder power to smooth engine operation and improve exhaust emissions.
- The ECM must provide flexibility to the user to set the amount of braking and vehicle speed at which the engine retarder will be actuated.
- The ECM needs to control the fan (on vs. off), based on sensed temperatures and braking power demand.
- The ECM should have a cruise control feature to maintain vehicle speed. The cruise mode should be interruptible by the activation of either a brake or clutch switch.
- The ECM should contain a reduced power to prevent engine damage under certain circumstances.
- The ECM should contain a shutdown feature that can be used to prevent engine damage or after extended idling periods to save fuel.

The designer should also consider limiting excessive vehicle speeds, because fuel consumption efficiency can be achieved and vehicle safety can be improved and progressive shifting logic to encourage the driver to upshift from a lower to a higher gear prior to reaching the engine's governed speed. Fuel economy is improved by limiting the rate of acceleration of the vehicle.

23.12.1.1 Central processing unit

Generally, the central processing unit (CPU) is referred to as a microprocessor. Stand-alone microprocessors of a decade ago have evolved from simple arithmetic logic units (ALUs) with limited memory space to 8-,16-, and 32-bit architecture combined with random access memory (RAM), read only memory (ROM), various input/output (I/O) facilities, programmable interrupt timers, serial port interfaces, etc. to a single microcontroller chip.

The central processing unit controls the operation of the system and performs a number of logical functions simultaneously to fulfill the specific diesel engine control requirements.

The microcontroller device, with its on-chip resources provides an integrated approach to real-time control, minimizes ECM chip count, and reduces power and heat dissipation. The microcontroller must provide enough processing power to perform vehicle functions, diagnostics, etc., and be able to execute new control programs for future growth. The microcontroller must have adequate clock speed and response time to execute instructions at all engine speeds. The quantity, frequency, and accuracy of input and output signals need to be understood to size a microcontroller properly.

Radio frequency interference (RFI) and electromagnetic interference (EMI) compatibility must be considered in the design microprocessor.

Measurements should be made at the actual mounting site of the electronic control system during all vehicle operating conditions, while the system is subjected to the maximum heat generated by adjacent equipment, and while they are at the maximum ambient conditions. Exposure to moisture, debris (from fine dust to surface gravel), and varying amounts of water, chemicals and oil must be tolerated.

23.12.1.2 Memory

The ECM's memory structure should consider three types of data retention: (1) permanent, (2) reprogrammable, and (3) dynamic. The engine management main program code should reside in permanent type memory. A permanent memory device retains its contents over an ignition cycle and when the ECM supply power is disconnected. The ability to reprogram the permanent memory device should be considered a key issue when the main program code is to be updated or if features are to be added to existing ECMs after the engine has left the factory.

There are currently three basic types of reprogrammable permanent memory devices: (1) erasable programmable ROM (EPROM), (2) electrically erasable programmable ROM (EEPROM), and (3) flash ROM. Of these three types of reprogrammable memory, EPROMs are typically the best choice for applications where data almost never needs changing. Otherwise, flash devices should be considered. Although the average EPROM costs about a third of what a flash device costs, the differential is wiped out by the expense of a single reprogramming. The in-ECM reprogramming of a flash ROM device is minimal as compared to removing, erasing, and reprogramming an EPROM, when vehicle downtime and labour are factored in.

Parameters, which can be set by OEMs and customers, provide the required system configuration flexibility. This flexibility necessitates the need to erase and program individual ECM memory bytes selectively. Therefore, EEPROMs should be considered as the storage device for engine performance data, customer-specified parameters, engine audit trail information, etc. Although currently about twice the cost of flash devices, EEPROM technology enables individual byte erasure and programming. Flash devices are typically erased in their entirety or large section by section.

Static RAM (or fast read/write memory) should be used to provide temporary storage of intermediate values for calculations, for parameters passed to subroutines, for program stacks, etc. A substantial reserve of memory space should be considered for future growth and control features.

23.12.1.3 Software

The software development process for the ECM should be modelled after the conventional engineering cycle that begins at the system requirements level and progresses through analysis, design, coding testing, and maintenance. In theory, this is a sequential process, and iterations are generally required. When the system requirements become clear, the analysis and design must be visited again to determine the impact on the software.

Since the software is only a part of the entire ECM, a system requirements specification should be generated which states what the ECM is supposed to do. From this specification, a subset of the system requirements should be allocated to software. Many times, this allocation of requirements to software will influence the microprocessor and memory decisions made in the hardware determination process.

A software requirements analysis must be done to understand the required functionality, interfaces, and performance of the software. This analysis should be done using some structured analysis and a Computer Aided Software Engineering (CASE) tool. The analysis should show the essential interfaces and intended functionality of the software without specifying implementation details.

The coding process typically consists of translating the design into a machine-readable format. Coding with a high-level language can provide the advantages of, sophisticated data structures, a large set of operators, and low-level, assembler-like, features for efficiency. The high-level language is easier to read than assembler code and more portable. Many times the coding phase can be a mechanical step if the design is very detailed. The coder should follow established coding standards and attempt to write code that is efficient, readable (well-structured and commented), and maintainable.

Most ECM memory programming is performed using an industry standard personal computer (PC). The standard serial interface on a PC is RS-232, which is not compatible with SAE J1708. The SAE J1683 defines a set of software functions that allows software designers to write applications independent of the interface design.

The engine calibration programmed in ECM memory uniquely defines the operational characteristics of the engine in the vehicle. There are two components to this calibration; basic engine performance and customer-specified parameters. The basic engine performance calibration is developed by a calibration engineer to ensure that the engine will produce the required torque curve, will meet customer performance expectations, and adhere to regulated smoke, emissions and noise criteria. The customer-specified parameters, such as cruise control operation, vehicle speed limiting, engine protection levels, etc., will vary depending on the type of vehicle and the environment. While many of these parameters may be customized by the user with the aid of a manufacturer proprietary hand-held tool, many users may choose to rely on the engine or vehicle manufacturer to perform their initial customization.

The testing phase should begin with code level or unit tests where the code statements are tested. Next, the code should be tested at a module level where the interfaces are tested and finally, at the system or functional level where testing is done using the ECM and an engine simulator or a test vehicle to ensure that the ECM system requirements have been met. Testing should consider using simulators, in-circuit emulators, source code browsers and debuggers to reduce the time spent finding and fixing software bugs.

The software will almost always be changed after it has been delivered to the customer because of feature enhancements or additions, etc. Because of this, the maintenance phase of software development is very important and consists of executing each of the preceding steps to modify the software. Changes to the software should never be performed by simply changing the code because the system requirements and design become obsolete and make future enhancements and understanding of the ECM software very difficult.

23.12.1.4 Integrated circuits

To attain cost effective packaging and to minimize size, surface mounted devices (SMD) should be used exclusively throughout the ECM. The benefits of designing with SMD technology are closely spaced component leads, thinner and more closely spaced circuit board conductor lines, elimination of drilled holes in circuit boards, an a circuit board assembly process that is more easily automated. An optimized circuit-board layout should make room for heat-dissipaters, if necessary.

Engine vibration occurs along all three axes and over a wide range of frequencies. Vibration can cause loss of wiring harness electrical connection due to improper design or assembly, and induce connector terminal fretting corrosion. Vibration levels may also vary during vehicle operation, like when operating on rough roads at high speeds. Electronic devices mounted on a circuit board can go into resonance, causing relative motion and possibly failure. The designer should therefore be alert to intermittent failures or faulty operation during applied vibration that may revert to normal operation after the vibration excitation is removed.

The effects of thermal shock and cycling should be understood to avoid the following problems: printed circuit board cracking, ceramic substrate cracking, solder joint fatigue, and component delamination.

23.12.1.5 Peripheral devices

The ECM hardware design will require power devices to interface with external high operating current devices. Power device applications typically include: system power supply, amplifiers, injector solenoid drivers, lamp drivers, relay drivers, etc.

Power metal oxide semiconductor field-effect transistors (MOSFETs) offer the designer a high speed switching device, an exceptionally wide range of voltages and currents, and low turn on resistance. Suppliers offer power MOSFETs with various characteristics and capabilities. Selection of the proper devices will depend on the external devices that are to be operated. The ECM circuit designer must determine the load characteristics of all devices to ensure the MOSFET will function within its designed safe operating area with regard to its power handling capability. Other key MOSFET design parameters are: switching speed, turn-on voltage and resistance, and gate drive requirements. Power devices along with control and protection circuitry, have been integrated on a single IC, resulting in a smart power device. The designer should consider this combination of power and logic to reduce circuit complexity, and improve the cost and reliability of the ECM.

The design and selection of interface devices, commonly referred to as input/output (I/O) devices are critical to system performance. The microcontroller interface to external data is accomplished through peripheral interface devices. These I/O devices provide data acquisition and perform tasks of conversion, storage or latching, buffering, timing, and control. Therefore, the performance and accuracy of the I/O devices must meet the data processing requirements of the engine control system. The designer must have knowledge of the external signals from engine sensors, vehicle switches, etc., as they can take on a variety of wave forms such as discrete ON/OFF, variable frequency, variable duty cycle pulse width modulated (PWM), and continuous analogue levels. Typical acquisition of these signals is achieved with peripheral I/O devices such as analogue-to-digital converters (ADC), parallel interfaces, and serial interfaces. Many integrated microcontrollers incorporate these functions, which can simplify the I/O design structure of the ECM.

The design and implementation of heavy-duty industry standardized serial data links (serial interface devices) can facilitate the diesel engine-to-vehicle integration. The SAE Vehicle Network for Multiplexing and Data Communications Sub-committee has defined three basic serial data link hardware interfaces:

- Class A: Low-speed data bit rate for remote device load control
- Class B: Mid-range data bit rate for parametric data exchange
- Class C: High-speed data bit rate for real-time control information transfer.

The serial data networks described in Classes B and C have been accepted and implemented by the heavy-duty truck and bus industries. The SAE J1708 document addresses the hardware to be used to design a Class B network. The Class C network is based around the controller area network (CAN) protocol defined by Robert Bosch GmbH. The first electronic devices implementing this protocol were stand-alone ICs requiring additional PC board space and logic devices to interface the IC to the microcontroller. Newer devices integrate the CAN protocol into the microcontroller to simplify system design.

The ECM power supply should be designed to accept 12 or 24 volt supply voltages. Designing reliable power supplies requires accounting for the effects of peak instantaneous transients, using high-reliability components, and using conservative derating practices, which result in operation at low device junction temperatures.

The designer should compare the operational aspects of the two typical power supply designs, linear mode, and switching mode. A high frequency switching mode design can minimize power dissipation compared to linear mode, although this switching may be an EMI noise source.

23.12.1.6 Other considerations

Voltage transients are typically the result of load dump, inductive-load switching, alternator field decay, or electrostatic discharge. Proper selection of suppression devices depends on the energy levels and time duration of the voltage transients.

Battery leads are sometimes connected in reverse during vehicle repair and engine installation, resulting in reverse polarity that could exceed the reverse bias rating of electronic components. The designer should determine the actual values of peak voltages, peak current, source impedance, repetition rate, and frequency of occurrence at the interface between the engine electronic control system and the vehicle electrical distribution system.

23.12.2 Sensors

Sensors provide the ECM information that is used to enhance engine performance during various operating conditions. Also, diagnostic limits can be programmed into the ECM to provide engine protection based on sensor information. It will be the decision of the system designer to identify the required types of engine sensors. Designing and manufacturing sensors is a formidable task. Therefore, working with a supplier who offers the desired components is typically the best approach for incorporating them into the engine control system design. Sensors are available from many suppliers who offer sensing elements of various technologies. When considering these technologies it is also important to thoroughly understand the sensor application environment.

Key design features of sensors are: electrical accuracy of the sensing element, strong output signal without the need for compensation, response time, supply voltage requirements, transfer function, and robust packaging for survivability. Sensors are typically mounted directly on the engine with the exception of the throttle position sensor, coolant level sensor, and vehicle speed sensor. The mounting locations of these sensors are typically in the vehicle cab, radiator, and wheel or transmission tail shaft respectively.

23.12.2.1 Crank position sensor

The crank position sensor is a key input to the ECM that is used to calculate engine speed and schedule fuel injection events. Crank position sensors are typically magnetic sensor devices that can detect either teeth or holes in a ferrous wheel rotating within a small air gap. The sensor along with the timing wheel that has known tooth or hole crank angle indexing is used to determine engine speed, cylinder top dead centre (TDC) positions, and injection timing and duration. Variable reluctance and Hall Effect sensors are commonly used as engine position sensors.

The system designer must determine the most effective type of sensor technology and also define the optimum tooth or hole crank angle index.

23.12.2.2 Throttle position sensor

The throttle position sensor is typically a rotary potentiometer and is an integral part of the electronic foot pedal assembly that is mounted in the vehicle's cab. An idle validation switch can be added to the pedal for an independent idle position indication to detect in-range electrical failures. To maintain a standardized hardware design approach, the system designer should consider selecting an electronic foot pedal that meets SAE J1843 'Accelerator Pedal Position Sensor for use with Electronic Controls in Medium-and Heavy-Duty Diesel On-Highway Vehicle Applications.'

23.12.2.3 Pressure sensors

Typical engine pressure sensors include: turbo boost, oil fuel, and barometer sensors. The turbo boost sensor is used to measure the intake manifold air pressure. The ECM can use this information to maintain optimum air-to-fuel ratio (to limit smoke). Oil and fuel pressure sensors are normally used as engine protection devices (to detect low conditions). Plugged engine fuel filters can be detected by a fuel pressure sensor. Barometric pressure information is useful to the ECM for fuel rate adjustment when operating at elevated altitudes.

23.12.2.4 Temperature sensors

The engine temperature sensors typically include; oil coolant, fuel, and air. Oil and coolant sensors are common engine protection devices to detect overtemperature conditions. Information from these sensors is also useful for cold engine starting, fuel timing optimization, and fast idle strategies. Continuous ECM monitoring of the engine's fuel temperature is essential for maintaining consistent fuel delivery. Air temperature can be used for a variety of engine controls.

23.12.2.5 Fluid level sensors

The ECM should allow for inputs from oil and coolant level sensors. Detecting out-of-range levels is an effective engine protection feature.

23.12.2.6 Vehicle speed sensor

The vehicle speed sensor (VSS) information is the essential data needed to perform vehicle speed limiting and vehicle cruise control. Vehicle speed sensors are of the same technology as engine position sensors. The VSS can be located at a vehicle's wheel or on a transmission tail shaft.

23.12.3 Interconnections and wiring

Electrical connectors might seem to be basic parts, but they are crucial components that must maintain integrity under all operating conditions.

Electrical connectors are available in a wide variety of configurations and styles. As the designer, one must select the type of connector that will best meet the requirements of all the anticipated applications, and stay within the limiting guidelines of industry standardization. Also, regardless of the connector type selected, existing equipment and systems will also be committed to various connector types, and the OEMs and customers must be familiar with them. In the initial design, coordination with the customer is important to establish a preferred connector system.

Choice of electrical connectors and associated hardware must consider the following: types of connectors, number of circuits required, electrical capabilities, mechanical capabilities, environmental capabilities, ease of contact replacement, and ease of mating. Connector mounting constraints and considerations are as follows: corrosion resistant bodies and accessories, minimum galvanic coupling of dissimilar metals, effect of cross-connecting cables, and adequate cable strain relief. The system wiring harness designer must consider wiring circuit number coding and a wire colour identification structure. The SAE is currently developing a recommended practice for a method of identifying electrical circuits in heavy-duty vehicles. The purpose of this document is to provide an industry-wide standard to enhance engine and vehicle serviceability. Circuit codes and colour identification can reduce the cost of maintenance and vehicle downtime.

23.12.4 Communications

Implementing serial data link interfaces in the ECM is an essential design consideration because of the benefits that can be gained from bi-directional electronic information transfers. These benefits can be found in diagnostics and maintenance through available system operating information, and with vehicle performance and safety improvements through power train control functions such as, anti-lock braking systems (ABS), automatic traction control (ATC), transmission control, etc. The ability to share information between vehicle electronic subsystems and off-board information systems requires a data link network topology that is supported by all connecting functions.

To achieve a high level of interconnect compatibility, the ECM design should support data link interfaces common to heavy-duty vehicles. The heavy-duty industry has developed recommended practices through the combined efforts of the SAE and TMC to describe the basic hardware interfaces and the format of messages and data being communicated between microprocessor-based systems used in heavy-duty vehicle applications.

These recommended practices are as follows:

1. **SAE J1708** 'Serial Data Communications Between Microcomputer Systems in Heavy duty Vehicle Applications.'
2. **SAE J1587** 'Joint SAE/TMC Electronic Data Interchange Between Microcomputers Systems in Heavy-duty Vehicle Applications.'
3. **SAE J1922** 'Power Train Control Interface for Electronic Controls Used in Medium and Heavy-duty Diesel On-Highway Vehicle Applications.'
4. **SAE J1939** 'Serial Control and Communications Network (Class C) for Truck and Bus Applications.'

The SAE J1708 document describes the electrical hardware requirements that the ECM must have in order to connect into and access the data link network to send or receive messages. The SAE J1587 document provides a specification for communicating diagnostic information from system devices, sensors and/or components, and failures on a SAE J1708 network. The SAE J1922 document is currently being used in some heavy-duty vehicles for control of drive train systems. SAE J1922 similar to SAE J1587, also utilizes a SAE J1708 specified network. SAE J1922 specifies a serial data message format for real-time control of ABS, ATC, and transmissions.

The ECM designer should consider a dedicated SAE J1922 link to achieve the benefits of driver safety and comfort from these vehicle control functions. A SAE J1939 (Class C) data link can provide the diesel engine electronic control system with a high-speed data transfer method for real-time control and communications. As this protocol becomes more and more accepted by OEMs, it is expected that this link will perform all of the information and control tasks required on heavy-duty vehicles.

The SAE J1587 document defines the format of messages and data transmitted on a SAE J1708 data link network. This protocol provides a technique for message source and destination addressing. This technique categorizes and numbers devices and systems on the vehicle's SAE J1708 link. These numbers are unique to the devices and systems and are defined as message identifiers (MIDs). Parameter identifiers (PIDs) are used to describe various system operating parameters common to heavy-duty vehicles. The PIDs define the number of data bytes used by the parameter, data range and resolution, link priority, and frequency of transmission.

The SAE J1587 document supports system diagnostics through the use of failure mode identifiers (FMIs) and subsystem identifications (SIDs). SIDs are assigned by the electronic system designer to identify field-repairable or replaceable subsystems for which failures can be detected and isolated. FMIs. describe the type of failure detected in the subsystem.

Major opportunities exist to help the vehicle owner manage the business through the use of more information made available from electronics on the vehicle. Productivity can be optimized by knowing the drivers' habits and the vehicle's condition and capability. Maintenance cost can be better controlled with better diagnostics, and with prognostic data. Improved information for the driver can enhance the efficiency and effectiveness of the vehicle operation. For example, knowledge of the miles per gallon at a particular time or for a certain prior segment, can allow the operator to optimize fuel economy. Trip recorders and navigational equipment can be used to convey maintenance information to the service manager who can then track vehicle failures and predict future problems instead of just diagnosing current ones.

23.13 Fuel injection system

The fuel injection system consists of: the injection devices and a fuel pump for unit injectors, fuel lines, fuel filter(s), a fuel cooler, and possibly a water separator. The function of the fuel injection system is to meter the correct quantity of fuel to each cylinder for all engine speeds and loads.

Unit fuel injection is not new, as it has been around since the 1930s, but the integration of electronic control and unit fuel injection did not occur until 1985.

The current fuel injection system design philosophy is to achieve: high injection pressures (to enable the entrained fuel to enter the combustion chamber in a more atomized state) and high injection rates (to allow fuel to travel more quickly and further within the cylinder, resulting in more complete combustion). High pressure unit fuel injectors work well with the SOHC design approach.

The injection rate can be increased by increasing cam velocity, cam lift, or both. Since increasing lift will increase the plunger stroke, the designer must be careful not to continue injecting beyond the constant velocity portion of the cam. Also, the rapid change in lobe contour coupled with the high injection load results in radical increases in cam lobe Hertzian stresses. Since prolonged exposure to high Hertzian stresses results in lobe pitting and flaking the designer must also consider this factor.

In the event that the deceleration portion of the lobe is too great or abrupt, the potential for plunger follower-to-cam lobe no follow may occur. This is highly undesirable from a durability standpoint, as well as from an increased injector train noise aspect.

Another key factor of fuel system design is the ability to precisely manage fuel injection with electronic control modules. With electronic control, fuel delivery can be optimized at all operating conditions, resulting in low emissions, good driveability, and maximized fuel economy.

23.13.1 Electronic fuel injection devices

Two types of electronic fuel injection devices are used; electronic unit injectors and electronic unit pumps.

Since fuel injection is controlled electronically and is not tied to the injector in a mechanical sense, fuel metering becomes a function of throttle position, engine speed, oil, water and air temperatures, turbocharger boost levels, and barometric conditions.

Traditional trouble areas of mechanically operated fuel injectors, such as: cold start-up and white smoke at light load operation is improved, since cylinder-to-cylinder fuelling differences are reduced with solenoid control. For the same reasons, engine-to-engine power variability is also reduced.

Through cylinder cut-out, an electronically controlled engine can isolate and identify individual injectors as being faulty. Therefore, diagnostic capability is another advantage, since fuelling, timing, and proper injector solenoid operation can be continuously monitored.

An additional benefit of electronic control is the reduced development time associated with generating new ratings and new torque curves, since the injector can cover a wider range of operation. By changing the injector tip size, engine operating ranges can be extended further. Governor controls are also enhanced, whereby fuelling is more accurately defined for maximum and minimum engine speeds and engine droop, resulting in more efficient engine operation.

23.13.1.1 Electronic unit injectors

The electronic unit injector (EUI) is a direct adaptation of its mechanical predecessor. The EUI utilizes an electric solenoid-activated poppet valve to meter fuel. Closure of the solenoid valve initiates pressurization and injection and opening of the valve causes injection pressure decay and end of injection See *Figure 23.5*

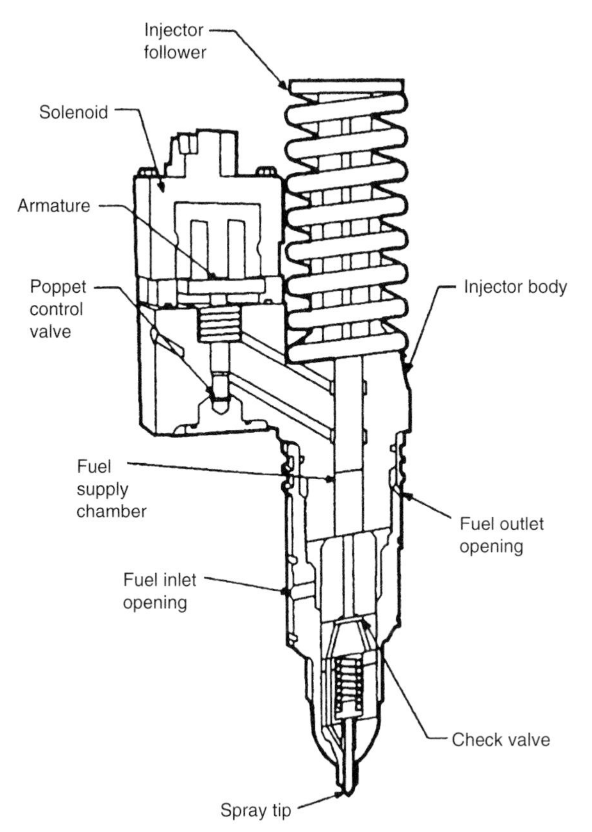

Figure 23.5 Electronic unit injector cross-section

With the absence of helices, racks, and linkages, fuel control is constant over the life of the engine, whereas mechanical systems were prone to component wear, in loss of fuel metering exactness, and the need for continued adjustments.

Proper sizing of the injector body is critical. Ideally, the injector should be sized to flow enough fuel into the combustion chamber for all operating conditions.

O-ring seals are used to keep the fuel from flowing into the combustion chamber or the overhead. They also prevent the fuel from choosing the path of least resistance and spilling into the return passages.

A filtering screen should cover the injector inlet to prevent unwanted material from entering. Fuel that enters the injector and is not used during the injection process is spilled to the return passages through spill holes in the injector body. Insufficient injector filling results in low power and loss of engine performance. The spill flow is also a means of cooling internal injector componentry subject to intense heat of combustion and working parts.

Drawbacks to the EUI include the need for a high pressure plug to close off the fuel passage and the high stress concentrations that exist at the corners of the high pressure fuel passage.

23.13.1.2 Electronic unit pumps

The electronic unit pump (EUP) uses electronically controlled nozzles to meter fuel at each cylinder. Unlike the unit injector, high pressure fuel is provided by separate camshaft driven pumps. The flexibility of this design permits a variety of camshaft and pump locations, while maintaining desirable unit injector characteristics.

The injection pump sends highly pressurized fuel via high pressure delivery lines or passages to the injection nozzles where fuel is then sprayed into the combustion chamber. All excessive fuel not needed for injection is returned to the fuel tank.

The nozzle body contains a needle valve, a needle spring and an injector tip. Like the unit injector, once fuel pressure overcomes the needle spring force, the needle valve unseats allowing fuel to pass through the injector tip.

Where the unit pump differs in design from the unit injector is that the nozzle assembly is separate from the remainder of the pump assembly. The injection pump and solenoid connect to the nozzle assembly through a high pressure line.

Depending on the engine design, the EUP can offer alternative degrees of freedom in contrast to the EUI. For example, overhead crowding can be minimized, since only the nozzle assembly is located above the firedeck rather than the entire injection unit. This adds flexibility to valve sizing and positioning. Also, overhead rocker arm assembly design is less complex with a EUP system, since the injector plunger is not located in the overhead.

A key benefit to the EUP body design is that the high pressure fuel passage is centrally located along the body's axis in the direction of flow and is uniformly surrounded by material. From a service standpoint, if a failure occurs, the entire pump assembly does not need to be replaced, only the component that failed.

EUP systems are not free from fault, though. The use of high pressure jumper lines provide potential leak sources. In addition, long passages are conducive to pressure oscillations, therefore, design considerations must be incorporated to eliminate pressure waves that can re-open the needle valve and cause cavitation damage. Typically the EUP nozzle assemblies include damping valves, restrictive fittings, or both to eliminate pressure oscillations.

23.13.1.3 Injector tip

The most important injector device parameter for today's on-highway engine manufacturers is clearly the injector tip. Conventional design practices include two injector tip configurations; the valve covered orifice (VCO) tip and the low-sac tip design configuration. With the VCO design the needle valve covers the tip holes when closed. In the low-sac tip design the needle valve does not cover the tip holes when seated. The VCO configuration is favoured because of its positive impact on particulates and smoke. Once an injector tip configuration is chosen, the injector tip hole characteristics must be optimized. See *Figure 23.6*

The number of holes and their diameter are based on the injector output and the pressure level requirements. It is not uncommon to have nine holes, nor is it uncommon to have injector tip hole diameters of 0.0060 in (0.152 mm). By changing the hole number, one can influence the emission characteristics of the engine. For example, by increasing hole number, while maintaining a constant injector output, one can expect fuel atomization to increase and spray plume penetration to decrease.

When specifying a diameter of this size, the manufacturer usually uses electrical discharge machining (EDM). After the initial hole is machined, a fluid containing abrasive particles is flowed through the holes. The abrasive nature of the fluid acts to smooth the holes and to obtain the final hole size. This flow also smoothes the entrance and exit areas of the orifices, producing an optimal spray plume and lowering particulate levels.

Although hole length is essentially equal to the wall thickness, this parameter can influence emission characteristics. A hole length-to-diameter ratio of approximately 8 is considered acceptable.

The injector hole spray angle relative to the firedeck is defined as the tip hole angle. Plume penetration and formation depend on the proper hole angle and injector spray tip protrusion into

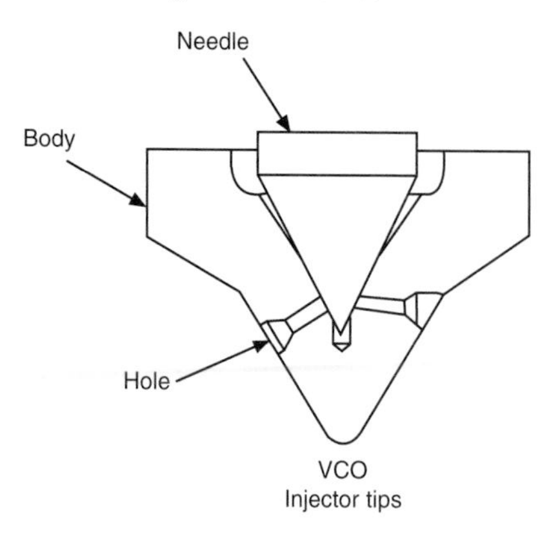

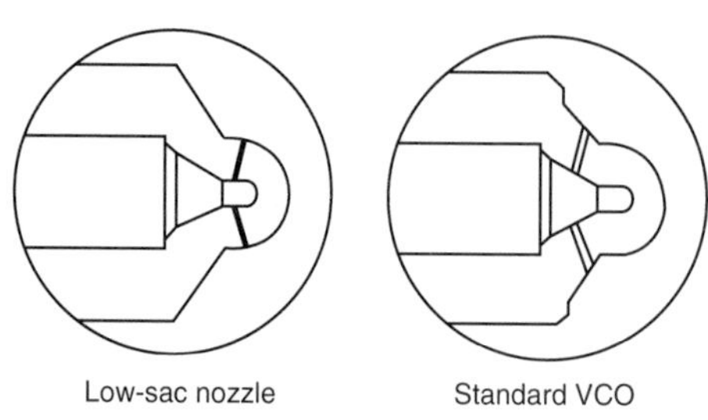

Figure 23.6 Common injector tip designs

the combustion chamber. Initially, based on piston bowl design, swirl characteristics, bore diameter, and compression ratio, the design engineer defines a tip hole angle. Considering engine durability, emissions and fuel consumption feedback, the optimum hole angle is chosen.

Inevitably, fuel will reach both the cylinder walls and the piston bowl. Ideally, some plume contact with the piston bowl is desirable because this allows the evaporating fuel spray to bounce back into the air volume above the bowl contributing to further fuel and air mixture. Careful consideration must be given to the final configuration, because spray plumes reaching the cylinder liner wall may lead to premature liner scuffing and too much fuel impinging the piston may lead to scorching and premature failure.

As a reference, the hole angle of the Detroit Diesel Series 60® 11.1 litre engine with its 16:1 compression ratio uses 160 degree hole angles, while the Detroit Diesel Series 60® 12.7 litre engine with its 15:1 compression ratio uses 155 degree hole angles.

23.13.2 Fuel (transfer) pump

Depending on the type of fuel injection system being utilized, the fuel pump may be referred to as the fuel transfer pump, In the case of the Detroit Diesel Series 60®, which utilizes unit injection, the fuel pump is needed to transfer fuel from the fuel tanks to the unit injectors. The Series 60® fuel pump in particular is a positive displacement gear-type pump, driven by the engine gear train. Some fuel pumps are mounted directly to the gear train and others by the air compressor, where the air compressor is driven off the engine gear train.

In operation, fuel enters the pump on the suction side and fills the space between the gear teeth that are exposed at that instant. The gear teeth then carry the fuel oil to the discharge side of the pump and as the gear teeth mesh in the centre of the pump, the fuel is forced out into an outlet cavity. Since this is a continuous cycle and fuel is continually being forced into the outlet cavity, the fuel flows from the outlet cavity into the fuel lines and through the engine fuel system are pressurized.

A pressure relief valves is used to discharge excess pressure, by bypassing the fuel from the outlet side of the pump to the inlet side. The pump's drive shaft rotates at about 1.2 times engine speed.

Regardless of the pump type employed several design considerations are imperative to ensure proper engine performance and fuel system longevity. First, the fuel pump must provide sufficient fuel flow to cover the wide range of horsepower ratings offered; second, it must provide adequate injector cooling; and third, it must overcome all engine and system restrictions.

Though not utilized on all engines, fuel heaters and/or coolers, add additional restriction to the fuel system. The design engineer must take this into account. End users may also incorporate a variety of line sizes, filters and/or check valves that may not meet the engine manufacturers' specifications. Excessive fuel system restriction will lead to a multitude of problems.

Although each manufacturer's fuel system restriction limits are different, they typically specify a maximum of 6 in Hg (20 kPa) on the suction side with clean filters and a maximum of 12 in Hg (40 kPa) for dirty filters. Prolonged operation of the engine under excessive fuel system restriction will cause fuel pump seal leaks and ultimately pump failure. An expensive secondary failure mode associated with fuel pump seal leaks is injector failure, since aerated fuel reduces the cooling ability of the fuel.

23.13.3 Fuel lines

The fuel passage design must allow excess fuel to bypass each injector and continue onward toward the next injector. Reliability requirements have driven fuel supply system designers to eliminate external fuel jumper lines in lieu of internal passages within the cylinder head.

23.13.4 Fuel filters

Fuel filters are typically found on the suction and on the supply sides. Micron rating and filter capacity are the two most important factors in filter design. The sizing of fuel filters should be based on their ability to clean unwanted debris from the fuel, but not restrict the fuel flow to the engine.

23.13.5 Fuel heaters and coolers

Although fuel heaters and coolers are not common, they can be used. The additional restriction must not be overlooked.

23.13.6 Fuel and water separators

A fuel and water separator, placed between the fuel tank and the inlet side of the engine fuel pump is sometimes recommended to avoid damage to the injector plunger and barrel assembly due to the injection of water. The centrifugal and gravity forces combine to separate the heavier water and dirt from the spinning fuel. The fuel then rises to the top, following the least resistive path, while the water and dirt collect at the bottom.

23.14 Air system

The air system is composed of the following components: turbocharger, charge cooler (or intercooler), intake manifold, and exhaust manifold. The purpose of the engine intake air system is to supply fresh air to the engine power cylinders in the proper quantity and under the proper conditions of pressure and temperature to meet the needs for cylinder scavenging and combustion. Similarly, the purpose of the engine exhaust system is to collect, transport, and discharge the exhaust gases from the engine cylinders to the vehicle exhaust system.

The engine designer should define the maximum intake air flow, the maximum exhaust temperature, and the maximum exhaust volume flow for the restrictions that can occur when the engine is installed in the vehicle. Typically, the maximum air intake restriction set by the individual OEM is 12 in H_2O (3 kPa) with new air cleaners and limited to 20 in H_2O (5 kPa) with dirty air cleaners. Where the total intake restriction is determined by the summation of the individual engine components and the rain shield, air cleaner, ducting and bending restrictions, etc. Typically, the maximum exhaust pressure at rated speed and load set by the individual OEM is 3 in Hg (10.1 kPa). This requirement is a fine balance between noise control using high back pressure mufflers, the effect back pressure has on the turbocharger performance, and the resulting engine efficiency.

Although naturally aspirated engines have the advantage of simplicity of air system design, their specific power output is limited by the quantity of air available for combustion. Increased power is typically accomplished by increasing the charge air density through the use of a pressure boost device. Although turbocharging is the primary means used to increase charge air density, it has the undesirable side effect of raising the charge air temperature. This temperature rise is offset, by installing a heat exchanger (charge air cooler) between the turbocharger and the intake manifold(s). Other considerations include operation at altitudes of up to 12 000 ft (3660 m).

23.14.1 Turbocharger

A turbocharger has three main sections: the turbine section, the centre section, and the compressor section. The most common type used in on-highway applications is a radial flow turbocharger.

With this design, the turbine housing includes an inlet that bolts to a mating flange on the exhaust manifold and is shaped such that the exhaust gas flows inward radially through a converging section or nozzle and then impinges on the turbine blades. The turbine blades turn the gas flow and cause it to exit axially from the turbine section. The change in angular momentum of the gas flow causes a reaction force on the turbine blades and imparts a torque on the shaft joining the turbine and compressor wheels. The nozzle portion of the turbine section must be designed to cause the exhaust gases to impinge on the turbine blades at the optimum angle and velocity.

The compressor section consists of a compressor wheel, a housing, and a diffuser. Its construction is similiar to the turbine section except that the intake air enters the compressor axially and exits radially via the diffuser. The diffuser must be designed to maximize the total pressure at the compressor outlet.

The centre section contains the bearings, which support the shaft joining the turbine and compressor wheels and provides mounting surfaces for the turbine and compressor housings. The centre section also includes passages that allow engine oil to be pressure fed to the shaft bearings and to drain back to the engine crankcase.

In designing the turbocharger lubrication system, the designer must take precautions to ensure that oil in the bearing area does not overheat and breakdown either during normal operation or during the soak period following the shutdown of a hot engine. Breakdown of the oil can result in the shaft sticking and/or damage to the bearings. The designer must also provide appropriate seals to prevent the oil from leaking into the compressor and turbine sections and contaminating intake and exhaust gas flow streams and to prevent exhaust gas from entering the engine crankcase via the turbocharger oil drain.

23.14.1.1 Operating conditions

Turbocharger performance and operating characteristics are normally communicated through the use of compressor and turbine maps. In a compressor map, lines of constant rotational speed and thermal efficiency 'islands' are plotted on axes representing the compressor pressure ratio and the air mass flow rate.

The compressor map also shows the compressor operating limits. The left edge of the map is the surge limit. Operation in the low flow region to the left of this boundary is unstable and characterized by flow reversals or surges through the compressor. Surging flow can destroy the compressor and represents a safety hazard to the engine operator and others in close proximity to the engine. The right edge of the map gives the maximum flow capability of the compressor and is defined by the point where the speed lines become vertical. The upper edge of the map is the rotational speed limit and defines the maximum pressure ratio capability of the compressor. Operation above the speed limit will result in centrifugal forces that are beyond the mechanical design limits of the compressor wheel. See *Figure 23.7*.

Compressor efficiency is a measure of how closely the actual compression process approaches the ideal isentropic process. Since any departure from isentropic compression results in reduced pressure rise and an excess temperature rise, the highest possible compressor efficiency is always sought. However, since the excess compressor temperature rise can be offset by charge cooling, the importance of compressor efficiency diminishes as charge cooling is employed and as the cooling effectiveness is increased.

The compressor flow characteristics and the size, shape, and position of the compressor efficiency islands can, to some degree, be influenced by compressor trim. Compressor trim is defined by the wheel inducer diameter, the wheel and shroud contours, and the diffuser width. To affect larger differences in the

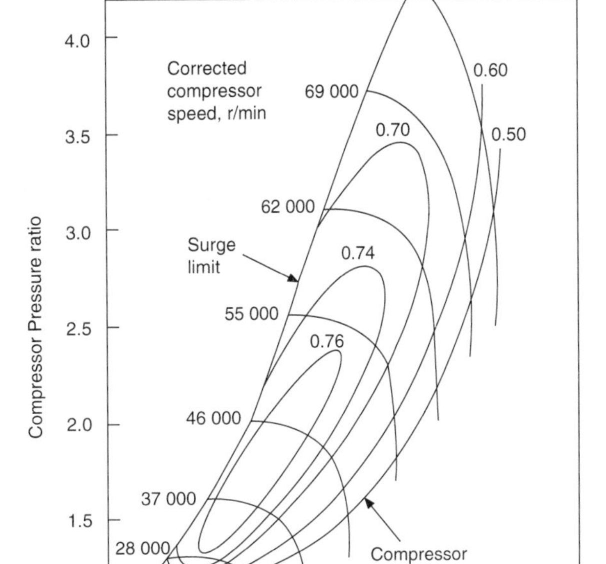

Figure 23.7 A typical compressor map

compressor flow range the fundamental frame size of the compressor unit must be modified.

Typically the efficiency shown in the turbine map is the product of the turbine's adiabatic and mechanical efficiencies. The mechanical efficiency accounts for the frictional losses of the shaft bearings. As the mechanical efficiency is increased, the shaft work required by the compressor can be achieved with less back pressure being imposed on the engine cylinders. The turbine flow range is fixed by the ratio of the turbine inlet flow area to the radial distance from the shaft centreline to the centroid of the inlet area. This is commonly known as the turbine A/R ratio. Decreasing the A/R ratio, increases the angular momentum and increases the energy available to the turbine wheel. Rotor shaft speed and compressor pressure ratio and flow are correspondingly increased. Turbine pressure ratio is also increased as a result of the increased exhaust back pressure imposed on the engine cylinders. Increasing the turbine A/R has the opposite effects.

For an engine designer, the challenge is to select the proper turbocharger for use on a particular engine. This is not an easy task and inevitably requires some trade-off. Typically, a four-step matching process is taken in selecting a turbocharger for use on an engine.

(1) Determine the engine air mass flow requirement. This is a speed dependent function that depends on the desired engine torque curve and the air utilization efficiency of the combustion system.
(2) Establish the air density-to-compressor pressure ratio that meets the air flow requirement and the engine cylinder displacement.
(3) Select a suitable compressor, by overlaying engine pressure ratio and air mass flow operating lines on candidate compressor maps and selecting the best overall match.
(4) The final step is to select a turbine design that produces the required shaft torque at each compressor match airflow condition.

The compressor selection process is complicated by the fact that it is not possible to simultaneously maximize the turbocharger efficiency at all engine operating conditions. The need to maintain adequate compressor surge margin may further restrict compressor selection. More importantly, however, there is a fundamental trade-off between the competing requirements to have adequate boost and air-to-fuel ratio at the engine peak torque speed while avoiding excessive boost that causes peak cylinder pressures to exceed engine structural limits at high speed.

Besides the steady-state design trade-off, the engine designer must also carefully consider the transient operation of the turbocharger. A major concern is known as turbocharger lag. Turbocharger lag refers to the inability of the turbocharger to instantaneously respond to changes in engine speed and load because of its rotating inertia. Turbocharger lag can be a problem under conditions of rapid engine acceleration because the compressor flow may not be sufficient to meet engine requirements during the lag period. This temporary shortage of air may cause excessive exhaust smoke and can limit engine responsiveness. To minimize the effects of turbocharger lag, it is desirable to reduce the turbine A/R ratio. This increases airflow and keeps the rotational speed of the turbocharger high, but as already noted, may result in excessive boost and cylinder pressure under high engine speed and load conditions.

A number of turbocharger technologies have evolved to assist the designer in minimizing the negative consequences of the design trade-off noted. One of these technologies is wastegating.

23.14.1.2 Wastegating

A wastegate is a passage in the turbine housing that allows exhaust gas to bypass the turbine and pass directly into the vehicle exhaust system. The wastegate is closed at low intake manifold pressure and is opened at manifold pressures above some preset trigger level. In essence, the wastegate is a boost limiting device. The valve opening pressure and the flow passage must be designed to ensure that engine boost and cylinder pressure limits are not exceeded. Use of a wastegate allows the designer to optimize the engine/turbocharger match for low speed and transient operation without being constrained by high speed boost limits. In most cases this results in the selection of a lower turbine A/R ratio than would otherwise be possible. The lower A/R ratio ensures sufficient boost and airflow to meet low speed engine demand, and because it increases turbine speed, it reduces turbocharger lag and improves turbocharger and engine response.

While wastegating is an effective way to allow the designer to maximize low speed and transient engine performance, the addition of the wastegate adds to the cost and complexity of the turbocharger and necessitates additional development to prove the reliability and durability of the wastegate valve and valve actuator. In addition, wasting of exhaust energy results in reduced overall engine efficiency at high speed. If this later trade-off is not acceptable, the designer may choose to go to the next level of turbocharger design sophistication known as variable geometry turbocharging.

23.14.1.3 Variable geometry

With variable geometry turbocharging, moveable components are incorporated in the turbine that allow the turbine flow area or the flow angle of the gas entering the turbine to be continuously varied. The variable geometry feature allows turbine performance to be optimized over a wide range of operating conditions.

Some variable geometry turbine housings include a plate, which can be moved during operation to vary the inlet flow area and the turbine A/R ratio. In other variable geometry turbines, the A/R ratio remains fixed, but efficiency is varied via adjustable nozzle vanes that control the angle of gas flow impact at the turbine wheel.

Variable vane turbine designs not only overcome the high and low speed boost and flow compromises which must be made with conventional turbocharging, but allow the engine exhaust energy to be utilized reasonably efficiently throughout the engine operating range. There are, however, some concerns with the use of variable geometry technologies that must be addressed. First, as with wastegating, the turbocharger's variable geometry mechanism complicates the turbocharger design and must be proven to be reliable and to meet engine durability requirements. Secondly, a control system must be provided. The control logic must be developed and optimized for the full range of engine operating conditions, and the control system sensing and actuating elements must, of course, be shown to be reliable and durable. Finally, the introduction of the variable geometry feature necessitates increased turbine clearances and flow disturbances that may negatively impact the aerodynamic efficiency of the turbine.

Finally, the introduction of the variable geometry feature necessitates increased turbine clearances and flow disturbances that may negatively impact the aerodynamic efficiency of the turbine. Compressor efficiencies may also need to be compromised to accommodate the flow and boost schedules achieved with the variable geometry turbine. Frequently, these practical considerations outweigh the theoretical advantages of variable geometry turbocharging in heavy-duty diesel engine applications.

23.14.1.4 Ceramic turbine wheels

Another turbocharger technology that is receiving increasing interest from heavy-duty engine designers is the use of ceramic turbine wheels. Use of these wheels has little effect on steady-state turbocharger performance, but because they have much less inertia than conventional metallic rotors, they can be used to significantly reduce turbocharger lag. Engine response and transient smoke characteristics can be improved without resorting to an A/R ratio that overboosts the engine at high speed and without applying wastegate or variable geometry turbocharging.

Ceramic turbine wheels are fairly commonplace on the smaller turbochargers used on passenger car engines, however, their usage on larger heavy-duty diesel turbochargers has been limited because of costs and difficulties in producing larger defect-free ceramic castings. As ceramic casting technology improves, wider utilization of ceramic turbine wheels on heavy-duty engines can be anticipated.

23.14.2 Charge cooler

Charge cooling has a number of important benefits. First, because it provides an additional charge air density increase, it allows the specific output of the engine to be increased. Also as a result of the increased charge density, the altitude capability of the engine is increased and the brake specific fuel consumption is improved. Another major benefit of charge cooling is that it reduces the operating temperature of cylinder and exhaust system components. As a result, component durability can be improved while avoiding the need to make these components from more expensive high temperature materials.

Perhaps the greatest benefit from charge cooling is its effect on exhaust emissions. Carbon monoxide and particulate emissions are reduced as a direct result of the increased charge density and air-to-fuel ratio. The effect on NOx emissions can be even more dramatic. The in-cylinder NOx formation reactions are temperature dependent and are controlled by the peak cylinder temperature. Depending on the compression ratio of the engine, a 100°F (38.5°C) reduction in intake manifold temperature lowers the peak cylinder temperature by 250 to 300°F (121 to 149°C) and can reduce NOx emissions by 30% or more.

Because charge cooling is one of the few ways to reduce NOx emissions without degrading fuel economy or increasing other emissions, it has become a preferred NOx control technology for turbocharged diesel engines.

23.14.2.1 Air-to-water charge air cooling

Air-to-water charge cooling uses the engine coolant as the cooling medium for cooling the charge air. In its simplest form, engine coolant returning from the radiator is passed through the charge cooler before it circulates through the engine water jacket. While this type of system is easily integrated into the engine design, the cooling potential is limited by the temperature of the engine coolant which is typically in the range of 170 to 190°F (77 to 88°C).

Lower temperature air-to-water charge cooling can be obtained by adding a separate radiator and coolant circuit for the charge cooler or by regulating the flow through the engine radiator. While these systems can increase the cooling potential by 40 to 60°F (4 to 16°C) relative to a standard air-to-water system, they add to the overall complexity of the cooling system design.

23.14.2.2 Air-to-air charge cooling

Air-to-air charge cooling uses ambient air as the cooling medium. Because ambient air is used to cool the charge air, the charge is considerably cooler than with any of the air-to-water cooling systems. With a well designed air-to-air system, charge temperatures within 30°F (17°C) of ambient can be achieved.

Besides producing lower charge temperatures, air-to-air systems are quite simple and can be installed without modifying the existing engine cooling system. One disadvantage of these systems is that the air-to-air heat exchanger and the interconnecting ducting represent a significant space claim in the vehicle engine compartment. Packaging concerns can be resolved, but close coordination between the engine and vehicle design teams is required to ensure an optimum system design.

It should also be noted that the charge air temperature achieved with an air-to-air cooling system varies with the ambient air temperature. As a result, the charge temperature is not as closely controlled as with an air-to-water charge cooling system where the cooling water temperature is kept within fairly narrow limits via thermostatic control.

Charge air refrigeration is the most powerful form of charge air cooling and can, in theory, achieve charge temperatures below ambient. This type of system uses freon or some other refrigerant and requires the addition of a complete refrigeration circuit, including a compressor and heat exchangers to cool the intake air and to condense the refrigerant. Use of a refrigeration system to accomplish the total charge air cooling requirements would require a large refrigeration unit that would be impractical for truck applications.

In order to achieve maximum charge cooling, the heat exchanger is normally placed just ahead of the engine radiator. If the vehicle does not have air conditioning, the heat exchanger and the radiator usually have the same frontal area. In vehicles with air conditioning, the heat exchanger may be mounted either in parallel or in series with the air conditioning condenser.

In parallel flow designs, the cooling air flow restrictions of the charge cooler and the air conditioning compressor should be matched so that the radiator receives an even airflow distribution. It is always desirable to minimize the cooling air flow restriction since this maximizes the opportunity to achieve engine cooling with ram air and minimizes the parasitic engine power losses associated with cooling fan operation. Series designs place a premium on minimizing the cooling air restriction. The heat exchanger design may need to be customized for each engine and vehicle combination.

The effectiveness of the heat exchanger (E) is the ratio of the actual charge air temperature drop across the cooler core to the temperature differential that is available for cooling. E is defined as:

$$E = (T_2 - T_3)/(T_2 - T_1)$$

where:

E = Heat exchanger effectiveness
T_1 = Temperature of the ambient cooling air
T_2 = Temperature of the charge air entering the heat exchanger (approximately equal to the compressor discharge temperature)
T_3 = Temperature of the charge air exiting the heat exchanger (approximately equal to the intake manifold temperature).

Since the overall objective of the charge cooling system is to maximize the intake manifold density, it is desirable to maximize the heat exchanger effectiveness and, at the same time, minimize the charge pressure drop across the cooler. Optimizing the trade-off between cooling effectiveness and flow restriction within the vehicle spatial and configurational constraints require careful design of the cooling core flow passages and fin density (fins per inch).

Similarly, the heat exchanger must be designed to minimize flow losses and to provide a uniform charge air flow distribution through the core. Because the air-to-air heat exchanger is located remotely from the engine, ducting is required to deliver the charge air from the turbocharger compressor discharge to the charge cooler and to return the cooled air to the intake manifold. The underhood configuration of the vehicle and serviceability requirements often limit options for routing these ducts. Within the available space, however, it is important to design the ducts to minimize pressure drop flow losses and to avoid any temperature rise particularly on the cold side return ducting.

To minimize pressure drop, it is desirable to use large diameter ducting and to minimize the number and sharpness of bends in the ducting. Overall length of the ducting should also be kept to a minimum. To minimize temperature rise, the ducting should be routed away from exhaust manifolds and other hot surfaces that radiate heat. Alternatively, insulation or heat shields can be used to insulate the ducting from nearby heat sources. The inlet and outlet connections on the heat exchanger must be customized for each vehicle in order to ensure optimal ducting design.

While it is beneficial from a pressure drop standpoint to have large heat exchanger flow passages and large diameter ducting, this may not be desirable from the standpoint of transient engine performance. The volume of the charge air system between the turbocharger compressor discharge and the engine intake valve acts as a pneumatic reservoir during transient engine operation. Air is supplied to the reservoir by the compressor and withdrawn by the engine cylinders. During an engine acceleration, it may be desirable to have a large system volume to buffer the cylinder charging pressure from turbocharger lag effects. If a more responsive turbocharger is used, there may be an advantage to reducing the volume that the turbocharger must pressurize to effect a charging pressure increase. To optimize the design, the designer must consider the complex interactions between the turbocharger and the charge cooling system during steady-state and transient operation.

23.14.3 Intake and exhaust manifolds

Although intake and exhaust manifolds are simple components, they play a major role in engine performance. The purpose of the intake manifold is to distribute charge air to the cylinder intake ports. Similarly, the purpose of the exhaust manifold is to collect the exhaust from the cylinder exhaust ports and deliver it to the turbocharger turbine or to the vehicle exhaust system. For both manifolds it is important that flow losses be kept to a minimum to ensure adequate cylinder scavenging and charging and to minimize cylinder pumping work. Typically, flow bench testing is required to optimize manifold flow characteristics.

To assist engine breathing the designer may wish to consider the use of 'tuned' manifolds. In a properly tuned manifold, the compression pulse is reflected and returned to the exhaust valve as a rarefaction prior to the valve closing. The low pressure from the rarefaction improves cylinder scavenging and reduces exhaust stroke pumping work. On the intake side, a rarefaction is generated at valve opening and is returned to the intake valve as a compression pulse just before valve closing. The pulse improves cylinder charging and reduces intake stroke pumping work. Because turning relies on the timed arrival of reflected pressure disturbances, benefits are obtained only at certain engine speeds. Typically, designs are optimized at the engine's peak torque speed because the greatest need for increased airflow occurs at this speed.

For turbocharged engines, exhaust manifold design is influenced by the constant pressure or pulse flow. Constant pressure turbocharging requires the use of a large manifold to absorb the exhaust blowdown pulses before they reach the turbine. Pulse turbocharging requires small manifold flow passages to preserve as much of the blowdown pulse energy as possible. To minimize pulse interactions and preserve pulse energy, it is customary to design pulse turbocharged manifolds to separate the exhaust flow from consecutive cylinders in the engine firing order.

The structural aspects of manifold design must not be ignored. This is especially critical for the exhaust manifold that frequently supports the weight of the turbocharger and other exhaust system components at elevated temperatures. The manifold sealing surfaces and fasteners must be designed to seal intake and exhaust gases and retain sealing integrity over thousands of thermal cycles.

23.15 Lubrication system

The lubrication system provides both lubrication and cooling of engine components. In some engines it is also used as an integral part of the fuel system, providing the high pressure hydraulic fluid used to pressurize fuel for injection into the combustion chamber.

Typically, a multi-grade lubricant (for example, 15W30 API classified CG-4 oil) is pumped through the engine from the oil pan through filters at a regulated level. If the oil pressure becomes too high, a portion is returned to the pan. From the filters, oil flows to the oil cooler and to the turbocharger. The oil that is delivered to the turbocharger is drained back to the oil pan, and the oil that is passing through the oil cooler is delivered through a regulator valve to the oil gallery.

Typically, a relief valve on the oil pump limits the oil pressure to 105 lb/in^2 (725 kPa) and the filters' bypass valve allows oil to bypass the filters if a pressure drop of 25 lb/in^2 (175 kPa) or more between the filter inlet and outlet occurs. The oil cooler usually has a bypass to maintain optimum flow rate of oil through the cooler. The regulator valve stabilizes the oil pressure at about 45 lb/in^2 (310 kPa) by returning excess oil to the oil pan.

Oil from the cooler is directed to the main oil gallery. This gallery distributes the oil, under pressure, to the crankshaft main bearings; the connecting rod and piston bearings; the camshaft bearings; the rocker shaft; the camshaft follower, roller pin and bushing and to any other components requiring lubrication, like the valve button and the fuel injector follower.

Obviously several components must have internal oil passages to get the oil from the main gallery to the bearing interface. External provisions, like squirters may also be required when internal passages are impractical. The rocker arm assemblies, the connecting rods, and crankshaft are examples of components that frequently have internal passages.

Lubrication of the gear train may be facilitated by holes in the bull gear recess area of the cylinder block that are connected to the gallery. This passage should provide oil from the oil gallery to the bull gear bearings, bull gear and camshaft idler gear, and hub. Oil from the bull gear lubricates the crankshaft timing gear and oil pump drive gear. Oil from the adjustable idler gear lubricates the accessory drive and water pump drive gears. The bearings and shafts of these two drive assemblies are frequently splash-fed through holes in their housings.

A flexible, external oil line from the cylinder block can be used to lubricate the air compressor assembly. This oil should be routed back into the gear case, though.

The oil passages must be designed to provide satisfactory oil flow rates at cold and warm engine conditions. Analysis using incompressible fluid mechanics should be used to determine sizes of orifices and passages. In addition to the design analysis, extensive performance and durability testing for cold, hot, used oil, and new oil conditions must be conducted to ensure that this critical system performs properly under all conditions.

23.15.1 Oil pump

The oil pump should be designed to provide the proper volume of oil needed (not an excess amount) to avoid unnecessary power use. A 10% margin should be added to cover the increased volume requirement of older, worn components. The oil pump must be designed for extremely stringent reliability and durability. Since most oil pump failures occur in the bearings, good design and lubrication of the bearings should be emphasized.

An important consideration for a gear type oil pump is the maximum speed and the ability to fill the pumping cavities. For this reason, pump inlet restriction must be kept to a minimum. The inlet tube should have a large cross-sectional area and bends with generous radii. The oil pick-up screen should have a total flow area at least 1.5 times that of the inlet tube and be as low in the oil pan as possible to prevent admission of air into the pump, but it can't be so close to the bottom of the oil pan that flow is restricted. The oil pump also needs to be self-priming, such that the dry pump (no oil) suction pressure is high enough to prime at engine cranking speed, even for cold oil conditions.

23.15.2 Regulator

The oil pressure regulator valve must stabilize lubricating oil pressure, regardless of temperature. The regulator should be designed to minimize energy loss when oil is being bypassed. Minimizing the amount of oil bypassed and the pressure drop are two ways to reduce wasted energy that occurs when bypassing oil.

23.15.3 Relief valve

The relief valve is usually located on the outlet of the oil pump. It should be designed to bypass oil at as low a pressure as the above considerations will allow, in order to minimize wasted energy. Some manufacturers use a common body for the relief valve and the regulator. With this approach, pin location established the pressure limit.

23.15.4 Filters

The filters should be designed to remove particles of a minimum size and to keep functioning to at least the duration of the oil change interval, without exceeding the maximum pressure drop allowed across the filters. Some manufacturers use a bypass valve in the filter adapter to bypass the filters, if the pressure drop becomes excessive. This is an undesirable, but necessary, condition in order to maintain satisfactory flow of oil to the engine if the filters become excessively plugged. Two filters, rated to filter particles down to 28 micron diameter at 98% efficiency, provide satisfactory filtration.

23.15.5 Oil cooler

The oil cooler maintains the oil temperature within its optimum temperature range. If the oil is too cold, it will not flow freely, and will require more power from the engine for proper circulation. If the oil is too warm, the oil film will not be thick enough to support bearing loads, and it cannot carry enough heat away from components such as the piston dome. Therefore, the oil cooler must be sized to provide the right amount of cooling to maintain the optimum oil temperature.

An oil bypass with a small orifice should be provided around the oil cooler to maintain optimal oil flow through the cooler and to the engine, for varying conditions such as cold oil or a plugged (restrictive) cooler.

The type and design of the cooler should be based on heat transfer analyses.

23.15.6 Dipstick

The dipstick is inserted into the oil pan in order to check the amount of oil in the engine oil pan. Marks on the dipstick indicate 'low' and 'full'. When the engine is installed at an angle the dipstick has to be re-calibrated to maintain accuracy.

The difference between 'full' and 'low' marks on the dipstick should be large enough to contain the quantity of oil consumed during the service interval. This prevents the operator from being forced to add oil between service intervals. The level of oil in the oil pan at the 'high' mark must not allow the crankshaft to splash the oil.

The level of oil in the oil pan at the 'low' mark must be high enough to ensure that the oil pump inlet is not exposed to air, even when the engine is at the worst possible combination of installation and vehicle angles. Also, there must be a sufficient safety margin to prevent this from happening for a considerable length of time. Typically the dipstick and oil pan are designed so that the engine should be able to operate for an additional oil change interval after the low mark is reached, without exposing the pump inlet to air.

23.15.7 Oil pan

The height of the oil pan needs to be minimized to provide optimum road clearance. The shape of the oil pan is dependent on sump location, which is customer driven in order to avoid drive axles or not interfere with departure angles. To satisfy customer requirements various oil pan configurations may be required. When installed, the oil pan has to provide oil to the oil pump inlet, when the oil is at the 'low' level.

The oil pan material has to be sufficiently resistant to fracture to withstand impacts with road debris and not be a noise attenuator. A vinyl ester is common. The oil pan needs to be isolated from the engine block to reduce noise emissions. A seal between the oil pan and cylinder block along with the use of rubber washer assemblies to fasten the oil pan to the cylinder block should be sufficient. If satisfactory noise attenuation is not achieved by isolating the oil pan in this manner, an oil pan with a permanent enclosure, consisting of an outer barrier to reflect sound waves and a foam material between the outer barrier and the oil pan to dissipate sound wave energy may be installed on the engine.

The engine, oil, when not being circulated in the engine, is stored in the oil pan. The oil pan must have sufficient capacity (retained volume) to carry enough oil to last at least one service interval. The volume also needs to be high enough to minimize oil aeration (the period of time the oil spends in the oil pan while the engine is running allows the air to separate from the oil).

For reference, Mack has announce a 30 000 mile (48 000 km) oil drain interval for their line-haul E7 engines that have V-MAC II electronic controls, Mack CentriMax filtration, and approved oils. Volvo VE D12 engine oil change interval is 25 000 miles (40 000 km) when Volvo drain specification oil and filters are used. Most other manufacturers recommend 15 000 mile (24 000 km) oil change intervals.

23.15.8 Crankcase ventilation

Vapours, formed within the engine, need to be removed from the crankcase, gear train, and valve compartment. This can be accomplished with a slight pressure that is maintained in the engine crankcase by the normal seepage of air and combustion gases past the piston rings. These gases sweep through the engine and exit through a crankcase breather, usually located in the rocker cover.

The crankcase pressure is determined by the rate of combustion gas leakage past the piston rings and by the restriction of the gases as they pass through the breather system. The restriction of the crankcase gases is therefore sized for the worst case: maximum engine power, and a worn engine near the end of the piston ring service life. Over the life of the engine the crankcase pressure can range up to one lb/in^2 (7 kPa), excluding effects of the air compressor, a damaged turbocharger, or a clogged crankcase ventilation filter element.

The vapours passing through the crankcase ventilation system contain some oil from the engine lubrication system. Oil is removed from the vapours by a wire mesh element in the rocker arm breather. Oil collects on the wire mesh and drains back into the valve train.

Closed crankcase ventilation systems, which vent the crankcase vapours back into the air inlet stream of the engine, are becoming increasingly popular for diesel engines. There is no legislation covering breather systems the United States, although this technology has been demonstrated for use in Europe.

23.15.9 Oil quality

The life of the components of the diesel engine is dependent on good lubrication. Customer demand for extended warranty periods and expected service life of diesel engines makes the lubrication system and oil quality even more critical. At the same time, lubrication systems must be able to handle more severe operating conditions, a result of customer demand for more power and more stringent emissions standards.

The demand for increased power results in higher in-cylinder temperatures, requiring oxidative and thermal stability of the lubricant.

As a result of the lower emissions of Nitrous Oxides required by today's emissions standards, diesel engines operate with increasingly retarded injection timing schedules. This results in higher soot content. Some engine manufacturers have tried lowering the engine's oil consumption to conform to lower particulate emission standards. This also results in higher soot content. Other manufacturers have increased fuel injection pressure to lower particulate emissions, which puts higher loads on the injector operating mechanism, causing higher shear rates in the lubricating oil. This breaks down the oil's viscosity improver additive more quickly. The oil must therefore meet the new demands associated with emission reduction strategies.

The use of exhaust aftertreatment, such as catalytic converters or particulate traps, may also force new requirements on the lubricating oil. The increased back pressure associated with some exhaust aftertreatment devices increases in-cylinder temperature.

Synthetic oils may provide the increased high temperature oxidation resistance needed with the increase in engine power ratings. In general new and more stringent requirements for diesel engine lubricating oil are required concurrently with reductions in diesel engine emissions standards. The next reduction in diesel truck engine emissions standards takes place in 1998.

Future developments in the lubricating system may include an oil quality sensor and an oil level sensor. The oil quality sensor could monitor critical oil performance factors such as viscosity, soot content, water content, and iron content. The oil level sensor will warn the operator when the oil level becomes too low, and provide an indication of current oil levels. Engine operators will save service costs by changing oil and filters at intervals required by their particular application.

Less severe applications may be able to implement longer service intervals. More severe applications will increase the engine service life and decrease warranty costs by decreasing the service interval as required by the severity of the application.

The performance standards and testing requirements for motor oil are currently generated by a tripartite combination of technical committees from ASTM, SAE, and API. More information or their standards and requirements can be found in the following documents: SAE J183, SAE J1146, ASTM D4485, and AIT 1509.

23.16 Coolant system

The primary function of the coolant system is to sufficiently dissipate heat energy from circulating engine coolant to the ambient surroundings, regardless of the engine operating condition. Insufficient cooling capability can adversely impact engine performance, fuel economy, emissions, and durability. It is therefore important to understand the factors affecting the cooling system used in truck and bus applications. Vehicle heat source requirements environmental and operating conditions, structural durability, packaging constraints, corrosion and erosion resistance, and coolant deaeration are each important considerations when designing a cooling system.

Proper replacement of components like transmission coolers, torque converters, and retarders is critical in minimizing water pump cavitation and loss of coolant flow.

Differences in ambient temperature, altitude and grade of the road can impact the cooling system. Therefore, accommodations must be made to account for the extreme variations in environmental and operating conditions. A reserve drawdown capacity should be considered during extremely hot ambient temperatures, high load duty cycles, and at high altitudes, so the maximum coolant temperature is not exceeded.

The most popular radiator used in trucks is a bolted radiator. The use of air-to-air charge coolers, the desire to reduce vehicle weight, and aerodynamic vehicles have fostered the need for smaller radiators, though. Advanced heat exchangers are now being used that have a 30% reduction in area, compared to the traditional bolted design, yet are still able to meet the vibrational, thermal, and pulsating pressure input loading observed in the operation.

23.16.1 Coolant

Water and water/ethylene glycol mixtures act as a suitable heat transfer mediums, however the mineral content in water can produce water pump seal wear and contribute to corrosion within the cooling system thereby reducing the effectiveness. Limits of mineral content in water is therefore critical. The use of inhibitors is required to control corrosion, cavitation, and deposit build-up. Three to six per cent is not uncommon.

Deaeration, the separation of gases entrained in the coolant, is critical to both performance and durability of the engine. Without deaeration the properties of the coolant will be diluted and possibly lead to coolant circulation deficiencies.

23.16.2 Coolant filter and conditioner

Coolant filtration is needed to remove impurities, such as sand and rust particles, which have been suspended in the cooling system. The filter should also act as a water softener to minimize mineral scale deposits, maintain coolant pH, and prevent metal corrosion.

23.16.3 Water pump

The water pump circulates the coolant throughout the cooling system. It is sensitive to inlet restriction, coolant temperature, and aerated coolant. Failure to control these parameters can result in cavitation, caused when partially vaporized coolant cavities collapse near regions of high pressure within the closed loop cooling system.

Water pump wear can be reduced by minimizing inlet pump restriction. All restrictions within the cooling circuit, such as radiators and associated plumbing, as well as customer add-on features (i.e. cab heaters, auxiliary oil coolers, filters), must be considered. Coolant pressure can be well below atmospheric, especially on water pumps that utilize rapid warm-up systems, which can lead to cavitation at coolant temperatures below 100°C (212°F). This condition is even more prevalent at high altitude conditions. Water pump inlet line diameters should be at least the same as that of the pump inlet. Bends within the coolant loop should be avoided. When necessary, they should have a large radius of curvature.

23.16.4 Thermostats

Thermostats are required to automatically regulate the coolant temperature by defining the coolant flow path. At high coolant temperature conditions, the thermostat directs coolant through the radiator. Once the coolant temperature is within the recommended region, the thermostat redirects coolant flow through a bypass circuit, recirculating the coolant through the engine.

If coolant temperature exceeds the fully open temperature of the thermostat, the cooling fan is turned on to increase heat transfer across the radiator. An additional safety measure has been placed on some electronic controlled engines, such as the Series 60, whereby overtemperature conditions on the engine have been significantly reduced by incorporating cooling system diagnostics into the engines' electronic control strategy. Above a predefined coolant temperature, the electronic control system sends a signal to reduce engine load and speed in order to avoid the risk of engine damage.

23.17 Typical engines

While the information in this section will be quickly out-dated, it does show the varied approaches used in this highly competitive area. A summary based on the information made available is provided. The reader should request detailed information from the appropriate source when decisions are to be made. A limited amount of non North American information is provided for comparison.

23.17.1 Product overview

The major heavy-duty diesel truck and bus engine manufacturers within North America are: Caterpillar, Cummins, Detroit Diesel, Navistar, Mack, and Volvo. Of these, Navistar Mack and Volvo also manufacture vehicles. A comparison of products offered follows:

Caterpillar Engines	*Power Range, hp*	*Torque Range, lb.ft*	*Weight, lb*
3126-6L	175–300	420–860	1250
C-10-6L	280–370	975–1350	2050
3176B-6L	275–365	975–1350	1945
3306C-6L	300	1150	1973
G3306-6L (*)	235–250	800–820	1975
C-12-6L	355–410	1350–1550	2070
3406C-6L	350–425	1350–1650	2926
3406E-6L	310–550	1150–1850	2867

(*) 235 hp operates on LPG, 250, hp operates on CNG and LNG

Cummins Engines	*Power Range, hp*	*Torque Range, lb-ft*	*Weight, lb*
C-8.3-6L	210–300	605–860	1330
L-10–6L	260–300	975–1150	2040
L-10G-6L(*)	235–260	750–850	1930-1970
M-11-6L	280–400	1050–1450	2070
N-14-6L	330–525	1350–1850	2805

(*) operates on CNG and LNG

Detroit Diesel Engines	*Power Range, hp*	*Torque Range, lb.ft*	*Weight, lb*
30G-8v (*)	210	450–485	980
40E-6L	175–300	520–1050	1275
50–4	250–315	780–1150	2250
50G-4L (*)	250–275	780–890	2250
55-6L	330–400	1250–1450	2080
60-6L	300–500	1150–1550	2580 - 2610
92-6.8v	300–500	975–1470	2020–2,040

(*) operates on CNG and LNG

Mack Engines	*Power Range, hp*	*Torque Range, lb.ft*	*Weight, lb*
E7-6L	250–350	975–1425	2150
E7 V-MAC-6L	250–454	1160–1560	2190
E9-8v	500	1660	2907

Volvo Engines	*Power Range, hp*	*Torque Range, lb.ft*	*Weight, lb*
VE D7-6L	230–280	660–850	1540
VE D12-6L	310–425	1250–1450	2337

Mercedes-Benz Engines	*Power Range, hp*	*Torque Range, lb.ft*	*Weight, lb*
OM 366-6L	211–240	545–620	979
OM 447-6L	213–295	700–922	1848
OM 441-6v	272–340	922–1180	1606
OM 442–8v	381–530	1364–1696	1947

Navistar Engines	*Power Range, hp*	*Torque Range, lb.ft*	*Weight, lb*
466E-6L	175–250	520–660	1275
530E-6L	250–300	800–1050	1275

VarityPerkins Engine	*Power Range, hp*	*Torque Range, lb.ft*	*Weight, lb*
Peregrine-6L	175–300	520–1050	1275
Eagle-6L	300–410	999–1418	2332

Mitsubishi Engines	*Power Range, hp*	*Torque Range, lb.ft*	*Weight, lb*
6D22-6L	288–446	550–1035	2050–2,180
8DC9-8v	415–509	911–1473	2575–2950

23.17.2 Caterpillar engines

Cat has several heavy-duty truck and bus engines; the 3126, the C-10, the 3176B, the 3306C, the G3306, the C-12, the 3406C, and the 3406E covering the 175 to 550 hp (130 to 410 kW) power range with between 23 and 52% torque rise. All are turbocharged in-line six-cylinder engines and have air-to-air charge cooling. The 3126, the C-10, the 3176B, the G3306, the C-12, and the 3406E are electronically controlled with flash memory technology; the others are mechanically governed.

The B-50 life of Cat engines continues to improve. As shown in *Table 23.2*, 50 per cent of the 3406E engines will be operational, without requiring an in-frame overhaul, at 1 000 000 miles (1 600 000 km) and the new C-10 and C-12 engines are expected to have B-50 lives of 900 000 miles (1 440 000 km).

Table 23.2 B-50 life of cat engines

Engine	*B-50 Life, miles (km)*
3176B	800 000 (1 280 000)
C-10	900 000 (1 440 000)*
C-12	900 000 (1 440 000)*
3406E	1 000 000 (1 600 000)

(*) denotes expected life

The 3126 truck and bus engines cover the 175 to 300 hp (131 to 224 kW) power range with between 19 and 52% torque rise. These engines have two-piece articulated pistons, with forged steel domes and cast aluminum skirts and three piston rings. Electronically controlled high pressure unit injectors are hydraulically activated. The one-piece cylinder head incorporates a quiescent air intake design. These engines have an altitude capability of 10 000 ft (3050 m).

The C-10 (formerly the 3176C) truck engines cover the 280 to 370 hp (209 to 276 kW) power range with between 25 and 53% torque rise. These engines have a one-piece cylinder head and have two-piece articulated pistons, with forged steel domes and cast aluminum skirts. Engine control is electronically controlled by dual microprocessors. They use high pressure unit injectors. One multi-torque rating is available on this model. These engines have an altitude capability of 7500 ft (2288 m).

Iveco Engines	*Power Range, hp*	*Torque Range, lb.ft*	*Weight, lb*
8469-6L (*)	219	643	1892
8360-6L	238–266	702	1012–1056
8460-6L	260–375	708–1253	1859–1896
8210-6L	260–470	712–1534	2310–2398
8280-8v	514	1622	2970

(*) operates on CNG

KHD Duetz Engines	*Power Range, hp*	*Torque Range, lb.ft*	*Weight, lb*
BFM1013-6L	191–258	1294–1401	1254–2860
BFM1015-6v	316–395	1086–1356	1826
BFM1015-8v	379–526	1312–1878	2332
FL513-8v	236–255	656–681	1 837–1936
FL513-10v	319	820	2189
FL513-12v	383	984	2486
BFL513-8v	330–360	862–1106	2024–2134
BFL513-10v	399	1076	2508
BFL513-12v	480–525	1294–1401	2750–2860

The 3176B truck engines are dual-fuel systems covering the 275 to 365 hp (205 to 272 kW) power range with between 11 and 53% torque rise. These engines have two-piece articulated pistons, with forged steel domes and cast aluminum skirts. The one-piece cylinder head has siamese plenum intake ports and series flow exhaust port design into a pulse exhaust manifold system. These engines operate on a combination of natural gas and diesel fuel simultaneously and have two independently controlled fuel systems. Both systems are controlled by ADEM. A natural gas electronic control unit (ECU) and the diesel electronic control module (ECM) provide the processing power and the data storage/retrieval capability for the dual-fuel system. The majority of the combined fuel is gas with the diesel portion acting as the pilot for combustion. The combination allows retention of the diesel compression ratio and substitution of diesel fuel with less expensive, cleaner burning, natural gas. Average substitution of diesel fuel is approximately 80%.

The C-12 truck engines cover the 355 to 410 hp (265 to 306 kW) power range with between 21 and 51% torque rise. These engines have a one-piece cylinder head, two-piece articulated pistons with forged steel domes and cast aluminum skirts, and high pressure unit injectors. Engine control is electronically controlled by dual-microprocessors. Three multi-torque ratings are available on this model. These engines have an altitude capability of 7500 ft (2288 m).

The 3306C engine operates at 300 hp (224 kW), uses pump-line nozzle fuel injection technology and has aluminum pistons with a Ni-Resist insert. The direct injection G3306 gas truck engine operates at 235 hp (175 kW) on LPG with a 31% torque rise and at 250 hp (186 kW) on CNG and LNG with a 36% torque rise. It uses an air–fuel mixer and requires a catalytic converter. It has an altitude capability of 5000 ft (1524 m).

The 3406E overhead cam truck engines cover the 310 to 550 hp (231 to 410 kW) power range with between 21 and 52% torque rise. These engines have a one-piece cylinder head/intake manifold casting with cross-flow ports; two-piece articulated pistons, with forged steel domes and cast aluminum skirts with three piston rings; high pressure, electronically controlled unit injectors; and boreless compressor wheels. Engine control is electronically controlled by dual-microprocessors. These engines also have a camshaft drive gear with an integral pendulum absorber to reduce shock loading caused by the unit injectors. Engines producing 410 hp (306 kW) or more have wastegate turbochargers.

The 3406C truck engines are available with 350 and 425 hp (261 to 317 kW) with 31 and 40% torque rise. These engines have pump-line-nozzle fuel injection technology and have limited certification.

Caterpillar is expected to announce the production of a six-cylinder, lean burn 14L G3406 engine with a catalytic converter having a 350 hp (261 kW) rating at 2000 r/min.

23.17.3 Cummins engines

Cummins has five heavy-duty truck and bus engines, the C-8.3, the L-10, the L-10G, the M-11, and the N-14. The mechanically governed engine ratings cover the 210 to 300 hp (156 to 224 kW) power range. All are turbocharged, air-to-air charge air cooled, in-line six-cylinder engines. The L-10, the M-11 and the N-14 are electronically controlled with CELECT or CELECT PLUS flash memory covering the 260 to 525 hp (194 to 392 kW) power range.

The in-line six-cylinder C-8.3 engine requires a catalytic converter, has a wastegate turbocharger, and covers the 210 to 300 hp (156 to 224 kW) power range. The one-piece aluminum pistons have cut-outs for the valves and three piston rings with dual Ni-Resist inserts. The intake manifold and thermostat housing are integrally cast into the cylinder head. The one-piece cross-flow cylinder head incorporates a swirl port design.

The in-line six-cylinder L-10 engine covers the 260 to 300 hp (194 to 224 kW) power range. This engine has one-piece aluminum pistons with Ni-Resist inserts and is offered with mechanical or electronic unit injectors. The spark ignited L-10G engine covers the 235 to 300 hp (175 to 224 kW) power range. It has an electronically controlled air–fuel mixer and wastegated turbochargers. The 300 hp (224 kW) L-10G engines meet emission standards without a catalytic converter.

The in-line six-cylinder M-11 engine has electronically controlled high pressure unit fuel injectors and covers the 280 to 400 hp (209 to 298 kW) power range. The two-piece articulated piston design has a cast aluminum skirt and a forged steel crown with cut-outs for the valves. The rocker cover isolated from the cylinder head. The one-piece cross-flow cylinder head incorporates a swirl port design. The B-50 life of M-11 engines is 800 000 miles (1 280 000 km).

The in-line six-cylinder N-14 engine has electronically controlled high pressure unit fuel injectors and covers the 330 to 525 hp (246 to 392 kW) power range. The two-piece articulated piston design has a cast aluminum skirt and a forged steel crown with cut-outs for the valves. The one-piece cross-flow cylinder head incorporates a swirl port design. The B-50 life of N-14 engines is 1 000 000 miles (1 600 000 km).

23.17.4 Detroit diesel engines

Detroit Diesel has several heavy-duty truck and bus engines, the Series 30G™, the Series 40™, the Series 50®, the Series 50G™, the Series 55™, the Series 60®, and the Series 92™. These engine ratings cover the 175 to 500 hp (130 to 373 kW) power range. All are turbocharged and are electronically controlled by DDEC®III, except the Series 40 that is mechanically governed. The Series 50 and the Series 60 engines have electronic unit injectors; the Series 40 engines have pump-line nozzle fuel injection; and the Series 55 engines have electronically controlled unit pumps. The Series 30G and Series 50G engines have electronically controlled air–fuel mixers. All have air-to-air charge air cooling, except the Series 92 that has jacket water after cooling.

The spark ignited vee-eight Series 30G truck and bus engine has a wastegate turbocharger and is available at 210 hp (156 kW) with 6 and 14% torque rise. It operates on CNG and LNG. DDEC is used to control spark timing, air-to-fuel ratio, throttle position, and speed governing. Features of the Series 30G include: gravity cast aluminum alloy pistons incorporating a cast-in Ni-Resist insert; stellite valves and inserts; hydraulic valve lifters and rotators; forged and hardened gear train and crankshaft; and air-to-air charge cooling. This engine has an altitude capability of 5000 ft (1575 m).

The in-line six-cylinder Series 40E specialty truck and bus engine is available in two displacements covering the 175 to 300 hp (130 to 224 kW) power range. This engine has a one-piece swirl designed cylinder head; aluminum alloy pistons with a Ni-Resist insert; hydraulically actuated, electronic unit injectors; and air-to-air charge cooling. Engines producing 270 hp (202 kW) or more have articulated iron pistons with three piston rings. All are electronically controlled and have an altitude capability of 10 000 ft (3050 m).

The in-line four-cylinder Series 50 truck and bus engine is a derivative of the popular Series 60 engine covering the 250 to 315 hp (186 to 235 kW) power range. It has an overhead camshaft, a one-piece cylinder head, and two-piece articulated malleable iron pistons with three piston rings, top liner cooling, an isolated rocker cover, and an isolated oil pan. The exhaust manifold has pulse recovery. DDEC offers engine control and protection. The engine has a 12 000 ft (3660 m) altitude capability.

The spark ignited Series 50G bus engine covers the 250 to 275 hp (186 to 205 kW) power range, operates on LNG and CNG, has wastegate turbocharging, and does not require a catalytic converter. DDEC is used to control spark timing, air-to-fuel ratio, throttle position, and speed governing. Series 50G engines have: an overhead camshaft, a one-piece cylinder head, two-piece articulated malleable iron pistons with three piston rings, top liner cooling, a pulse recovery exhaust manifold, an isolated rocker cover, and an isolated oil pan. This engine has an altitude capability of 3000 ft (915 m).

The in-line six-cylinder Series 55 truck engine covers the 330 to 400 hp (246 to 298 kW) power range with a constant peak torque range from 1100 to 1500 r/min giving it outstanding driveability. This engine has a unique engine brake that includes a fifth valve integrated into the cylinder head for braking. This engine has one-piece aluminum trunk style pistons with three piston rings. DDEC offers engine control and protection. The engine has a 12 000 ft (3660 m) altitude capability.

The in-line six-cylinder Series 60 truck and bus engine has electronically controlled high pressure unit fuel injectors and covers the 300 to 500 hp (224 to 373 kW) power range with two displacements. The 11.1 l version covers the 300 to 365 hp (224 to 272 kW) power range with a 19 to 64% torque rise. This engine has an altitude capability of 10 800 ft (3290 m). The 12.7 l version covers the 370 to 500 hp (276 to 373 kW) power range with a 6 to 44% torque rise. All Series 60 engines have: an overhead camshaft, a one-piece cylinder head, two-piece articulated malleable iron pistons with three piston rings, top liner cooling, a pulse recovery exhaust manifold, an isolated rocker cover, and an isolated oil pan. DDEC offers engine control and protection. This engine has an altitude capability of 12 000 ft (3660 m).

The Series 92 truck and bus engine is available in six-and eight-vee cylinder configurations covering the 300 to 500 hp (224 to 373 kW) power range. It is a two-stroke cycle engine. All Series 92 engines have: in-block camshafts, two-piece articulated malleable iron pistons with five piston rings, and bypass blowers. DDEC offers engine control and protection. This engine has an altitude capability of 12 000 ft (3660 m).

23.17.5 Mack engines

Mack has two heavy-duty truck engines, the E7 and the E9. The E7 engine has 13 electronically controlled V-MAC II engine ratings and six mechanically governed ratings covering the 250 to 454 hp (185 to 338 kW) power range. The E9 engine expands Mack's power range to 500 hp (373 kW).

Mack has announced plans to market a Mack/Renault mechanically injected 10 l diesel engine, the E5 engine, covering the 250 to 290 hp (185 to 216 kW) power range. The use of electronically controlled Bosch pump/injectors for the E5 engine in 1998 has also been announced.

The in-line six-cylinder E7 engine has a new swirl injection system, twin inlet ports, and a unique piston bowl design. The pistons are a two-piece articulated design having a forged steel crown and an aluminum skirt with three piston rings. The engine's tightness is further aided by a new manufacturing process called Torque Plate Honing, which simulates cylinder head clamping forces during the block machining process.

The E7 power ratings are categorized into three classes: Maxidyne engine technology delivers 58% torque rise and facilitates operation at lower speeds resulting in fuel efficiency; Maxicruise engine technology delivers 53% torque rise and is designed to match the needs of highway hauling, and Econodyne engine technology delivers conventional torque rise (18 to 33%).

The vee-eight E9 engine has one rating, a 500 hp (373 kW) power rating with a 20% torque rise. It is categorized as a Econodyne rating. The pistons are an aluminum alloy and have three piston rings. The four cylinder heads (two per bank) are alloyed cast iron.

23.17.6 Mercedes-Benz engines

Mercedes-Benz has two heavy-duty truck and bus engine families, the Series 300 and the Series 400. Both meet EURO II standards.

The Series 300 family consists of the OM 336 engine, an in-line six-cylinder engine with wastegate turbocharging. It covers the 211 to 240 hp (155 to 177 kW) power range. This engine is water cooled, has a mechanical governor, and uses high pressure pump-line-nozzle fuel injection technology.

The Series 400 family consists of the OM 441 engine, a vee-six; the OM 442 engine, a vee-eight; and the OM 447 engine, a horizontally configured in-line six-cylinder engine, covering the 213 to 530 hp (157 to 390 kW) power range. These engines are water cooled and utilize electronically controlled high pressure pump-line-nozzle fuel injection technology.

23.17.7 Navistar engines

Navistar has two heavy-duty truck engines, the 466E and the 530E. The 7.6 l 466E engine covers the 195 to 250 hp (145 to 186 kW) power range. The 8.7 l 530E engine covers the 250 to 300 hp (186 to 224 kW) power range. Both are electronically controlled and use hydraulically actuated, electronically controlled unit injectors. These engines have one-piece swirl designed cylinder head; aluminum alloy pistons with a Ni-Resist insert; and air-to-air charge cooling. Engines producing 270 hp (306 kW) or more have articulated iron pistons with three piston rings.

23.17.8 VarityPerkins engines

VarityPerkins has two heavy-duty truck engines, the Eagle and the Peregrine. Both meet EURO II standards. The Eagle engine covers the 300 to 410 hp (224 to 306 kW) power range. Both are mechanically governed and use in-line Bosch injection pumps. The Peregrine engine is available in two displacements covering the 175 to 300 hp (130 to 224 kW) power range. This engine has a one-piece swirl designed cylinder head; aluminum alloy pistons with a Ni-Resist insert; and air-to-air charge cooling. Engines producing 270 hp (201 kW) or more have articulated iron pistons with three piston rings.

23.17.9 Volvo engines

Volvo has two heavy duty truck engines, the VE D7 and the VE D12 covering the 230 to 425 hp (171 to 317 kW) power range. Both of these engines are turbocharged. The VE D7 is mechanically controlled and supports the 230 to 280 hp (171 to 209 kW) power range. The VE D12 has an overhead camshaft with electronically controlled high pressure unit injectors via VECTRO and supports the 310 to 425 hp (231 to 317 kW) power range. It has two-piece forged steel pistons, has air-to-air charge cooling, a one-piece steel head gasket, a new two-piece oil pan, and a new patented engine brake that uses a third cam lobe with a special control mechanism to allow back flow of gases generating 350 bhp (261 kW) at 2300 r/min.

Table 23.3 Cat engine configuration summary

Description	*3126*	*C-10*	*3176B*	*3306C*
Configuration	In-line six-cylinder			
Displacement	7.2L	10.3 L	10.3 L	10.5 L
Camshaft location	In-block			
Air system	Turbocharged, Air-to-air charge cooling			
Controls	Electronic fuel injection and engine control			Mech. governed
Injection system	HEUI	EUI	EUI	Pump-line-nozzle
Valves	2 per cylinder	4 per cylinder	4 per cylinder	2 per cylinder
Weight	1250 lb	2050 lb	1945 lb	1973 lb

Table 23.3 (Continued) Cat engine configuration summary

Description	*G3306*	*C-12*	*3406C*	*3406E*
Configuration	In-line six-cylinder			
Displacement	10.5 L	11.9 L	14.6 L	14.6 L
Camshaft location	In-block			Overhead cam
Air system	Turbocharged, water cooled	Turbocharged, air-to-air charge cooling		
Controls	Electronically controlled fuel system	Electronic fuel injection and engine control	Mech. governed	Electronic fuel injection and engine control
Injection system	Air/fuel mixer	EUI	Pump-line-nozzle	EUI
Valves	2 per cylinder	4 per cylinder	4 per cylinder	4 per cylinder
Weight	1975 lb	2070 lb	2926 lb	2867 lb

Table 23.4 Cummins engine configuration summary

Description	*C-8.3*	*L-10*	*L-10G*	*M-11*	*N-14*
Configuration	In-line six-cylinder				
Displacement	8.3 L	10.0 L	10.0 L	10.8 L	14.0 L
Camshaft location	In-block				
Air system	Turbocharged, Air-to-air charge cooling				
Controls	Mech. governed, the L10 also has an electronically controlled version		Electronically controlled fuel system	Electronic fuel injection and engine control	
Injection system	Pump-line-nozzle	MUI, EUI optional	Air/fuel mixer	EUI	EUI
Valves	2 per cylinder	4 per cylinder	4 per cylinder	4 per cylinder	4 per cylinder
Weight	1330 lb	2040 lb	1930–1970 lb	2070 lb	2805 lb

Table 23.5 Detroit Diesel engine configuration summary

Description	*30G*	*40E*	*50*	*50G*
Configuration	Vee-eight	In-line six cylinder	In-line four cylinder	
Displacement	7.3 L	7.6 and 8.7 L	8.5 L	8.5 L
Camshaft location	In-block		Overhead cam	
Air system	Turbocharged, Air-to-air charge cooling			
Controls	Electronically controlled fuel system	Electronic fuel injection and engine control	Electronic fuel injection and engine control	Electronically controlled
Injection system	Air/fuel mixer	HEUI	EUI	Air/fuel mixer
Valves	2 per cylinder with lifters and rotators	2 per cylinder with rotators	4 per cylinder with rotators	4 per cylinder with rotators
Weight	980 lb	1275 lb	2250 lb	2250 lb

Table 23.5 (Continued) Detroit Diesel engine configuration summary

Description	*55*	*60*	*6v-92*	*8v-92*
Configuration	In-line six-cylinder		Vee-six	Vee-eight
Displacement	12.0 L	11.1 and 12.7 L	9.0 L	12.1 L
Camshaft location	In-block	Overhead cam	In-block	
Air system	Turbocharged, Air-to-air charge cooling		Turbocharged, water cooled	
Controls	Electronic fuel injection and engine control			
Injection system	EUP	EUI	EUI	EUI
Valves	4 per cylinder	4 per cylinder with rotators	4 per cylinder	4 per cylinder
Weight	2080 lb	2580–2610 lb	2020 lb	2420 lb

Table 23.6 Mack engine configuration summary

Description	*E7*	*E7 V-MAC II*	*E9*
Configuration	In-line six-cylinder		Vee-eight
Displacement	12.0 L		16.4 L
Camshaft location	In-block		
Air System	Turbocharged, remote-mounted air-to-air charge cooling		
Controls	Mech. governed	Electronic fuel injection and engine control	Mech. governed
Injection system	Pump-line-nozzle	EUP	Pump-line-nozzle
Valves	4 per cylinder with rotators	4 per cylinder with rotators	4 per cylinder with rotators
Weight	2150 lb	2190 lb	2907 lb

Table 23.7 Mercedes-Benz engine configuration summary

Description	*OM 366*	*OM 447*	*OM 441*	*OM 442*
Configuration	In-line six cylinder	Horizontal in-line six cylinder	Vee-six	Vee-eight
Displacement	6.0 L	12.0 L	11.0 L	14.6 L
Camshaft location	In-block			
Air system	Turbocharged, water cooled			
Controls	Mech. governed	Electronic control	Mech. governed	
Injection system	Pump-line-nozzle			
Valves	2 per cylinder	2 per cylinder	2 per cylinder	2 per cylinder
Weight	979 lb	1848 lb	1606 lb	1947 lb

Table 23.8 Navistar engine configuration summary

Description	*466E*	*530E*
Configuration	In-line six-cylinder	
Displacement	7.6 L	8.7 L
Camshaft location	In-block	
Air system	Turbocharged, Air-to-air charge cooling	
Controls	Electronic fuel injection and engine control	
Injection system	HEUI	
Valves	2 per cylinder with rotators	
Weight	1275 lb	1275 lb

Table 23.9 VarityPerkins engine configuration summary

Description	*Peregrine*	*Eagle*
Configuration	In-line six-cylinder	
Displacement	7.6 and 8.7 L	12.1 L
Camshaft location	In-block	
Air system	Turbocharged, Air-to-air charge cooling	
Controls	Mech. governed	
Injection system	Pump-line-nozzle	
Valves	2 per cylinder with rotators	2 per cylinder
Weight	1275 lb	2332 lb

Table 23.10 Volvo engine configuration summary

Description	*VE D7*	*VE D12*
Configuration	In-line six-cylinder	
Displacement	6.7 L	12.1 L
Camshaft location	In-block	Overhead cam
Air system	Turbocharged, Air-to-air charge cooling	
Controls	Mech. governed	Electronic fuel injection and engine control
Injection system	Pump-line-nozzle	EUI
Valves	2 per cylinder	4 per cylinder
Weight	1540 lb	2337 lb

Bibliography

HEYWOOD, J. B., *Internal Combustion Engine Fundamentals*, McGraw-Hill New York, (1988)
MERRION, D. F., 'Diesel Engine Design for the 1990s, The Fortieth L. Ray Buckendale Lecture', Society of Automotive Engineers, Warrendale, PA, SP-1011 (940130) (1994)
OBERT, E. F., *Internal Combustion Engines*, third edition, International Textbook Company, Scranton, PA (1968)
TAYLOR, C. F., *The Internal Combustion Engine in Theory and Practice, Volume 1: Thermodynamics, Fluid Flow, Performance*, second edition, The MIT Press, Cambridge, MA (1985)
TAYLOR, C. F., *The Internal Combustion Engine in Theory and Practice, Volume 2: Combustion, Fuels, Materials, Design*, revised edition, The MIT Press, Cambridge, MA (1987)
WILSON, K., *Practical Solutions of Torsional Vibration Problems, Volume 2*, John Wiley & Sons New York, (1963)

24

Locomotives

Contents

24.1 Introduction

The choice facing any Railway Administration as to whether their locomotive should be powered by diesel engines or by electricity drawn from an external supply is a complex one, with many factors to be considered. In general, most administrations would like to electrify the majority of their networks, but the very high capital cost of doing this means that only those routes carrying the most traffic can be considered in order to generate sufficient revenue to cover, at least in part, the capital cost of the installation of transformers, switchgear, and lineside equipment.

Electric traction reduces the dependence on petroleum-based fuels. A higher installed horsepower per locomotive is usually available, and the maintenance requirement when compared to that of a diesel powered locomotive is considerably less. As railway administrations are tending to concentrate their networks into arterial links between their main cities and ports to improve their utilization factors, the use of electric traction is becoming more widespread and will continue to do so in the foreseeable future.

The majority of European countries have expanded their network of electrification. In France, for instance, it is claimed that one third of their network is electrified and this carries 80% of the traffic. In countries outside Europe, electrification is often used for urban transport systems; but because of the immense distance involved in countries such as the USA, Canada, South America, China, India, Australia, the African Continent and Russia, the capital cost of installing main line electrification would be extremely expensive. In some of the countries mentioned, a small proportion of the most heavily used parts of their networks are already electrified, and plans are in hand for further electrification but the diesel powered locomotive will be the mainstay of most railway networks for the foreseeable future.

The locomotive market is a specialist one and is well served by the major manufacturers of diesel engines throughout the world. To give some idea of the size of the market, there are approximately 25 000 diesel locomotives in the USA alone.

The majority of main line locomotives are powered by medium speed diesel engines running at 900–1050 rev/min.These engines are of robust construction in order to meet both the durability and reliability requirements. For instance, in the USA it is not uncommon for main line locomotives to run 800 000 to 1 000 000 km between overhauls and the aim is to be able to run for 1 500 000, to 1 600 000 km between overhauls in the not too distant future. This significant increase in overhaul life is brought about by continuous improvements in design, materials, lubricating oil quality and condition monitoring.

High speed engines, namely those running at speeds up to 1800 rev/min continue to find applications where light weight, compactness and first cost are important. Light weight engines find applications in high speed passenger service where low axle loads are necessary to control the amount of track maintenance required to support a high speed service. High speed engines also find application in narrow gauge and weight restricted lines where their low weight and small size enable a smaller, lighter locomotive to be built compared to one using a medium speed engine of the same power output.

The majority of diesel engined locomotives use electric transmissions. Very few locomotives with hydraulic or mechanical transmissions are produced today. However, hydraulic and mechanical transmissions are still widely used in railcars where the engine power rarely exceeds 750 kW per unit. During the last 20 years the standard form of electric transmission for locomotives consisted of an alternator coupled directly to the engine crankshaft. The AC output was then rectified and fed to individual DC traction motors on each axle of the locomotive. Over the years this system has been developed to give a coefficient of adhesion of up to 28% under optimum conditions. Coefficient of adhesion is defined as the ratio of tractive effort exerted by the locomotive to gross locomotive weight. However under conditions of high weight transfer between the locomotive bogies and/or low wheel to rail coefficient of friction due to rain, ice, dirt or wet vegetation the system performance in terms of coefficient of adhesion is significantly degraded. It is worth noting as well that DC traction motors are bulky and heavy and have brush gear and commutators which require maintenance.

Developments in compact AC/AC frequency converters have enabled the DC traction motors to be replaced by AC motors which are lighter, more compact and do not use brush gear. Improvements in control technology now make it possible to monitor individual locomotive wheel rim speeds relative to rail speed so that each wheel can be powered up to the point where the wheel rim speed exceeds the rail speed by between 5 and 8%. At this degree of wheel to rail slip the tractive effort is maximized. The use of such a system enables each axle on a locomotive to be powered up to its maximum potential, thus maximizing the tractive effort of the locomotive. Using AC/AC traction control, typical coefficients of adhesion are up to 36–38%. The use of individual axle load control enables weight transfer, wheel sets of different diameter due to rim wear and wet rails to be taken into account to optimize the tractive effort of the locomotive under all conditions.

The significant improvement in coefficient of adhesion brought about by the adaption of AC/AC technology has made increases in locomotive power possible. It is of interest that for 15 years or so, diesel locomotives of 2750 to 3300 kW were the most powerful available. However, during 1998 the American locomotive builders GM-EMD and General Electric will put into production locomotives using AC/AC traction technology and a 4700 kW diesel engine. See *Figures 24.1* and *24.2*. These locomotives will have a significant impact on the economics of motive power as for instance two 4700 kW units will replace three 3100 kW units with significant savings in maintenance costs.

Railcar engines are normally of the horizontal type and are mounted under the carriage floor between the axles. This avoids any intrusion into the passenger carrying space. They normally have outputs ranging from 160 to 750 kW. At the lower end of the power range (160–350 kW) the engines are derived form heavy duty truck designs, and the upper end of the power scale is catered for by quick running engines having a bore in the range 150–185 mm.

24.2 Development trends

Over the last ten years much effort has been expended by the engine manufacturers serving the locomotive market. To increase the power of their engines; reduce the fuel consumption, reduce the emissions of particulates (black smoke) and improve their reliability and durability.

For the future these issues will still be vigorously pursued, but the introduction of both gaseous and particulate emissions will have a major impact on design and development efforts. The introduction of a number of new designs in the bore size range 250 to 280 mm aimed at the 4700 kW requirement will require significant effort and investment. The need to keep the weight of these new 4700 kW engines to around 25 000–27 000 kg will prove a significant challenge while at the same time improving the durability and reliability.

24.2.1 Emissions

The Union International de Chemins de Fer (UIC) have for

Figure 24.1 GM-EMD SD90MAC locomotive

Figure 24.2 General Electric AC 6000^{TM} locomotive

many years specified emissions standards for diesel powered locomotives and rail vehicles having an engine power greater than 100 kW. The limits are shown on *Table 24.1*. These limits have been progressively tightened with time and the latest values were published in 1997. In this instance UIC co-operated with the European Railway Research Institute (ERRI) in setting the limits and defining the test procedure for measuring the results. It has been agreed that cycle F of ISO 8178 should be used.

In the USA the Environmental Protection Agency (EPA) have been reviewing the subject of locomotive emissions legislation for several years and in 1997 published draft legislation which became law in the USA in March 1998. The limits are shown in *Table 24.2*.

Tier 0 and 1 are fully effective from 1/1/2002. However, between 1/1/2000 and 31/12/2001, manufacturers have essentially three options.

(1) For the 'primary engine family' produced between 1994 and 1997, provide a retrofit kit of parts to enable the remanufactured engine to meet Tier 0 Standards, and meet Tier 0 Standards for all new production of this engine family after 1/1/2000.
(2) On a fleet average basis, meet Tier 0 Standards for all new production and remanufacture locomotives after 1/1/2001.
(3) 'Formulate an alternative plan that would have equal or greater reduction in emissions than either Tier 0 or 1 and submit this course of action to the EPA for approval.

A manufacturer that follows option 1 does not have to make an engine that is introduced onto the market after 1/1/98 comply with Tier 1 until 1/1/2002. After 1/1/2000, if a retrofit kit to comply with Tier 0 Standards is available for any engine family, a railroad overhauling a member of that family must utilise it. Compliance with both Tier 0 and Tier 1 will require a higher compression ratio, retarded injection timing and rematched combustion chamber and fuel injection equipment. Some increase in fuel consumption is likely to occur due to the use of retarded injection timings to comply with the NOx requirements.

Tier 2 is effective for all new locomotives from 1/1/2005. To achieve compliance with the proposed Tier 2 levels will necessitate the use high compression ratios (> 16 : 1), high injection pressures (> 1700 bar), electronic control of injection timing and probably exhaust gas recirculation (EGR). The task facing the engine manufacturers in developing their engines to meet the Tier 2 limits should not be underestimated, particularly as the limits are calculated from results obtained from operation over all the control settings of the engine.

The stringent particulate requirements to Tier 2 will also necessitate significant reduction in lubricating oil consumption as the combustion of lubricating oil contributes to particulate emissions. The treatment of crankcase blowby gases will also pose problems as their discharge directly into the atmosphere will certainly be regulated by legislation to reduce levels of hydrocarbons emitted into the atmosphere.

24.2.2 Engine weight

The challenge of building a 4700 kW engine with a weight of less than 25 000–27 000 kg is significant. Close control of component weight throughout the design process is vital and all the major components must be analysed using Finite Element Analysis (FEA) to ensure that both the stiffness and strength criteria are met with the minimum use of material. To that end, the use of nodular (ductile) cast iron for engine and other stressed components will become common practice. Aluminium will continue to be used for non-structural components in the interests of weight control.

As it is likely that all new designs will have a maximum cylinder pressure capability of at least 180 bar, the strength versus weight trade-off will provide demanding challenges to the engine designers.

24.2.3 Reliability and durability

Reliability and durability are only achieved by rigorous analysis of every engine component during the design phase together with the selection of materials and surface treatment and appropriate to each component.

Once the prototype engine has been built it must be tested in such a manner as to impose realistic operational parameters imposed by locomotive duty which is in fact, the most arduous

Table 24.1 Exhaust emissions limits according to UIC 623 code

Constituent	*Year*			
	pre-1982	*1982*	*1993*	*1997*
CO [g/kW.h]	12.0	8.0	4.0	3.0
HC [g/kW.h]	4.0	2.4	1.6	0.8
NOx [g/kW.h]	24.0	20.0	16.0	12.0
Smoke [Bosch index]	—	—	—	1.6–2.5*

*according to air intake flow at nominal power

Table 24.2 EPA locomotive emissions standards

Constituent	*Remanufactured locomotive*		*New locomotive—Tier 1*		*New locomotive—Tier 2*	
	Freight	*Switcher*	*Freight*	*Switcher*	*Freight*	*Switcher*
NOx	9.5	14.1	7.4	8.5	5.5	6.0
Pm	0.6	1.0	0.5	0.6	0.2	0.2
HC	1.0	2.0	0.6	0.8	0.3	0.3
CO	5.0	6.0	2.2	2.5	1.5	1.5
Implementation date		01/01/2002*		01/01/2002		01/01/2005

Emissions are quoted in g/bhp.h on EPA duty cycles

duty that any diesel can be called upon to operate in. It should be remembered that locomotives operate in ambient conditions that vary from – 40°C to 50°C and at altitudes up to 3500 m, with at times, high levels of dust in the air.

The cyclic thermal loading on a locomotive engine imposes severe thermal stresses on the hot components of the engine, namely the piston, liner, cylinder head and valves, exhaust system and turbocharger. Some form of cyclic thermal testing is necessary to ensure that these components are fit for purpose. The cycle testing procedure developed by the then British Railways has combined the requirements of high cycle fatigue as well as low cycle thermal fatigue in a single test procedure that takes less than 1000 engine operating hours to complete. See Reference 1.

Manufacturing quality control is vital to the achievement of reliability and durability and this calls for the use of high quality machine tools as well as first class measurement equipment to ensure that all the components conform to the sizes and tolerance specified on the drawings. Consistent quality of raw materials, namely castings, forgings; bar stock and bought out components is vital.

24.3 Engine descriptions

A number of designs of diesel engines used in locomotives around the world are described in this section. *Table 24.3* shows the bore, stroke, cylinder configuration, power output, speed and other related data for the engines described in this section.

24.3.1 Caterpillar 3500 *(Figure 24.3. References 2, 3 and 4)*

The engine housing is a single piece iron casting comprising the cylinder blocks and crankcase. The bolts fastening each main bearing cap to the housing are angled to absorb both vertical and horizontal forces. The crankshaft, which is underslung, has induction hardened pins and journals. The pistons are aluminium alloy castings with an alloy cast iron insert for the top two rings. The piston also incorporates a cast in cooling gallery which is supplied with oil from a fixed jet mounted at the bottom of each liner.

The cast iron liners are held by a flange at the top of the liner, the liner flange seats directly on the top of the block with no recess. The connecting rods have horizontally split big ends and the caps are retained by four angled bolts. The rods are arranged side-by-side on a common crankpin.

The camshafts are located on the outside of the engine and are driven by a train of gears from the drive end of the engine. Each camshaft operates the push rod operated valve gear through roller followers. Individual unit fuel injectors, operated by a push rod from the camshaft are fitted in each cylinder head. The intake and exhaust manifolds and the turbochargers are mounted in the centre of the vee. The space between the cylinder banks serves as an air manifold and the aftercooler is mounted lengthways along the centreline of the manifold. The oil and water pumps are mounted at the front of the engine and are driven by a train of gears from the crankshaft.

24.3.2 Caterpillar 3600 *(Figure 24.4. References 5, 6 and 7)*

The Caterpillar 3600 engine has a single piece grey cast iron housing comprising cylinder blocks and crankcase. The crankshaft which is underslung has induction hardened pins and journals. The piston has a forged aluminium body with a steel crown attached to it by four bolts. The pistons are cooled by a fixed jet mounted at the bottom of each liner. The cast iron liners are located by a flange that sits directly on the top face of the cylinder block. There is no recess for the flange in the cylinder block. The connecting rods are machined from steel forgings and have a horizontal split big end which is located by four bolts. On the Vee engines the connecting rods are arranged 'side-by-side' on a common crankpin.

The camshafts are located on the outside of the engine and are driven by a train of gears from the flywheel end of the engine. Each camshaft operates the push rod operated valve gear through roller followers as are the push rod operated unit injectors. Air and exhaust ports are located in the centre of the vee and the space between the cylinder banks serves as an air manifold. The aftercooler is mounted longitudinally above the air manifold. On the 12 and 16 cylinder engines two turbochargers are used, mounted at the flywheel end of the engine. The oil cooler and oil filter are mounted horizontally above the front end cover.

The oil and water pumps are mounted at the front of the engine and are driven by a train of gears from the crankshaft.

24.3.3 Dalian 240 ZD *(Figure 24.5. Reference 8)*

The Dalian 240ZD engine has a single piece welded steel housing comprising the cylinder blocks and crankcase. The crankshaft, which is underslung, is cast in nodular iron and is fillet rolled and nitrided. The pistons are two piece with a steel crown and an aluminium body. Cooling oil is supplied to the piston via drillings in the piston pin which are in turn fed by a drilling up the shank of the connecting rod. The connecting rods are machined all over from steel forgings and have an angled split, serrated

Table 24.3 Locomotive engine specifications

Make and Type	Bore × Stroke [mm]	Stroke/Bore ratio	Swept volume per cylinder [l]	Speed [rev/min]	Mean Piston speed [m/s]	BMEP [bar]	Power/ Cylinder [kW]	Cylinder Configuration	Power range [kW]	Weight of 16 cylinder [kg]	Specific weight of 16 cylinder [kg/kW]
CAT 3500	170 × 190	1.12	4.31	1800	11.40	11.54	75	8V, 12V, 16V	597–1195	7900	6.6
CAT 3600	280 × 300	1.07	18.48	1000	10.00	19.97	308	6L, 8L, 12V, 16V	1850–4920	29000	5.9
Dalian 240ZD	240 × 275	1.15	12.45	1000	9.17	17.73	184	6L, 8L, 12V, 16V	1471–2942	27000	9.2
GE 7FDL	229 × 267	1.17	10.95	1050	9.33	21.89	210	8V, 12V, 16V	1678–3356	20000	6.0
GE 7HDL	250 × 320	1.28	15.71	1050	11.20	21.35	294	6L, 8L, 9L, 12V, 16V	1762–4698	22955	4.9
GM-EMD 645	230.2 × 254	1.10	10.57	900	7.62	10.53	153	8V, 12V, 16V, 20V	1230–2908	16522	5.7
GM-EMD 710	230 × 279	1.12	11.64	900	8.38	10.71	179	8V, 12V, 16V, 20V	1567–3578	17962	5.7
GM-EMDH	265 × 300	1.13	16.55	1000	10.00	21.29	294	12V, 16V	3524–4698	24090	5.1
Kolomna D49	260 × 260	1.00	13.81	1000	8.67	20.72	220	8V, 12V, 16V, 20V	1620–4770	17460	5.0
MTU/DDC 4000	165 × 190	1.15	4.06	1800	11.40	20.50	125	8V, 12V, 16V	1000–2000	7400	3.7
Paxman VP 185	185 × 196	1.06	5.27	1800	11.76	21.71	172	12V, 18V	2060–3090	11400 (18V)	3.6
Pielstick PA4-200	200 × 210	1.05	6.60	1500	10.50	17.10	141	8V, 12V, 16V, 18V	1130–2540	7800	3.5
Pielstick PA6B	280 × 330	1.18	20.32	1050	11.55	25.30	405	12V, 16V, 20V	4860–8100	34000	5.2
Ruston RK215	215 × 275	1.28	9.99	1000	9.17	23.73	198	6L, 8V, 12V, 16V	1185–3160	14600	4.6

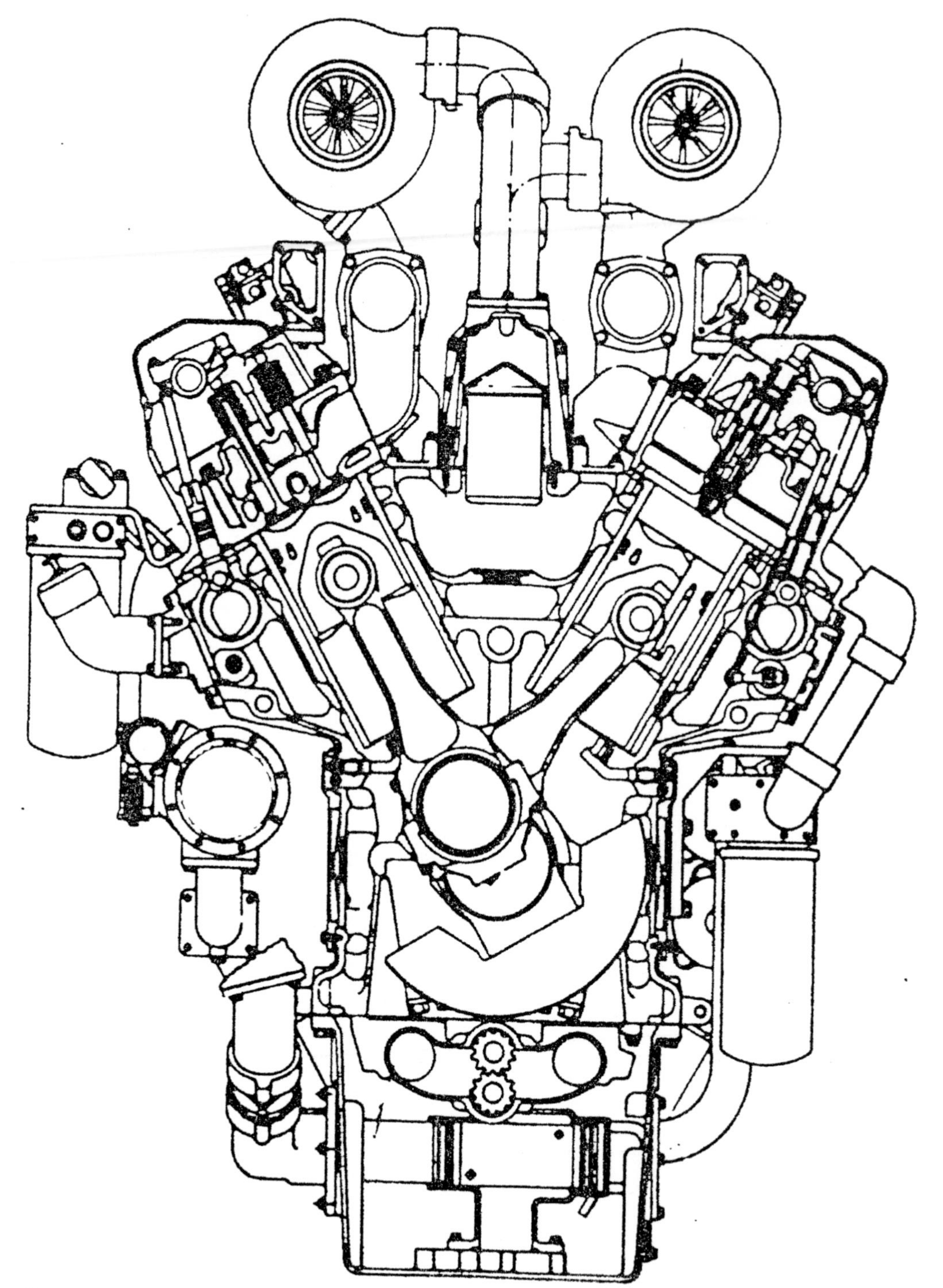

Figure 24.3 Caterpillar 3500 engine

joint. On the vee engines the rods are arranged 'side-by-side' on a common crankpin.

The camshafts are located on the outside of the engine and are driven by a train of gears from the flywheel end of the engine. Each camshaft operates the push rod operated valve gear through roller followers, as are the individual fuel injection pumps. The air and exhaust ports are located in the centre of the vee. The space between the cylinder banks serves as an inlet air manifold and a single constant pressure exhaust manifold is mounted on the centreline of the engine. On the 16-cylinder engine two turbochargers and aftercoolers are used; one of each being positioned at either end of the engine. The oil and water pumps are mounted at the front end of the engine and are driven by a train of gears from the crankshaft.

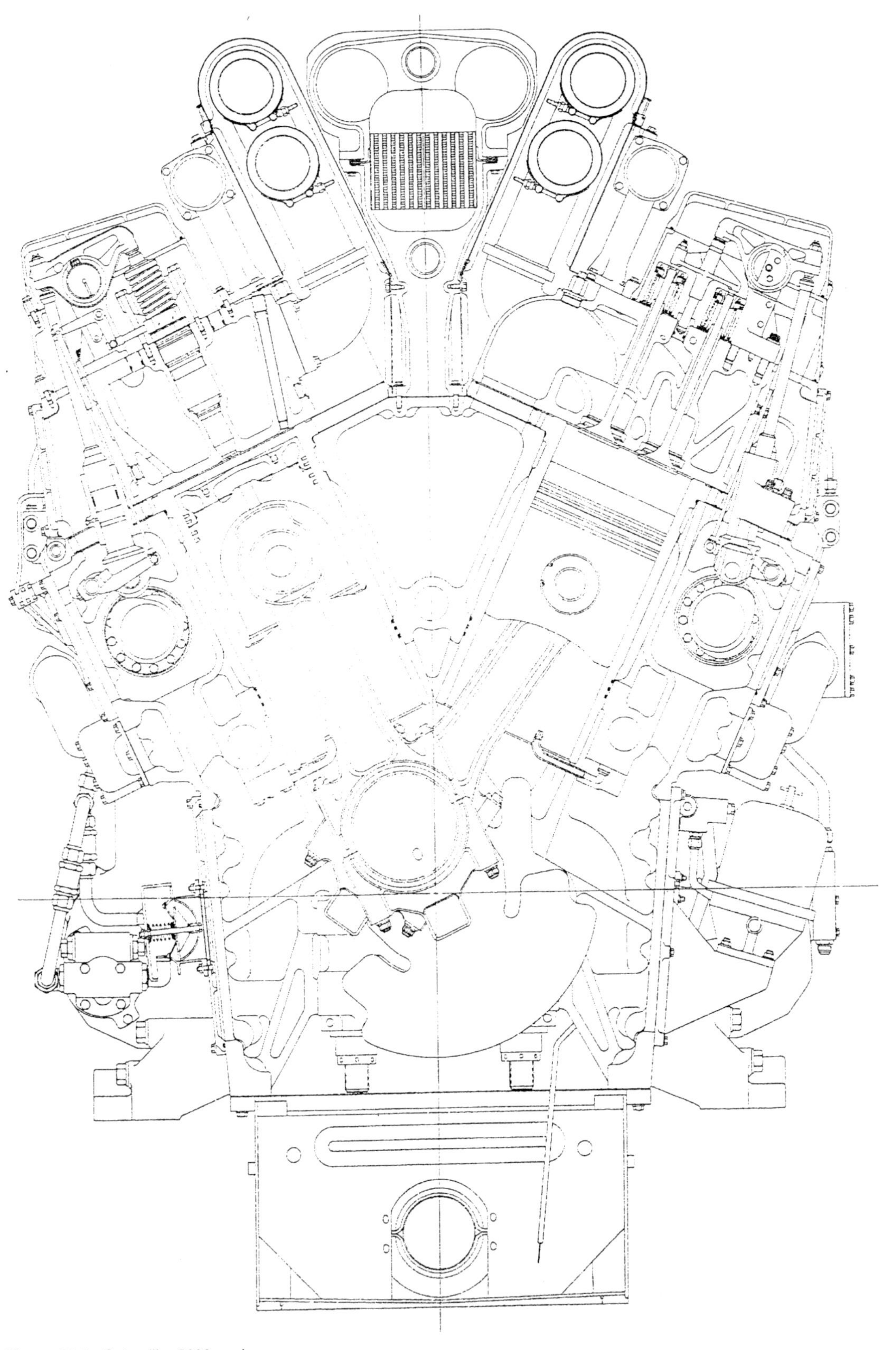

Figure 24.4 Caterpillar 3600 engine

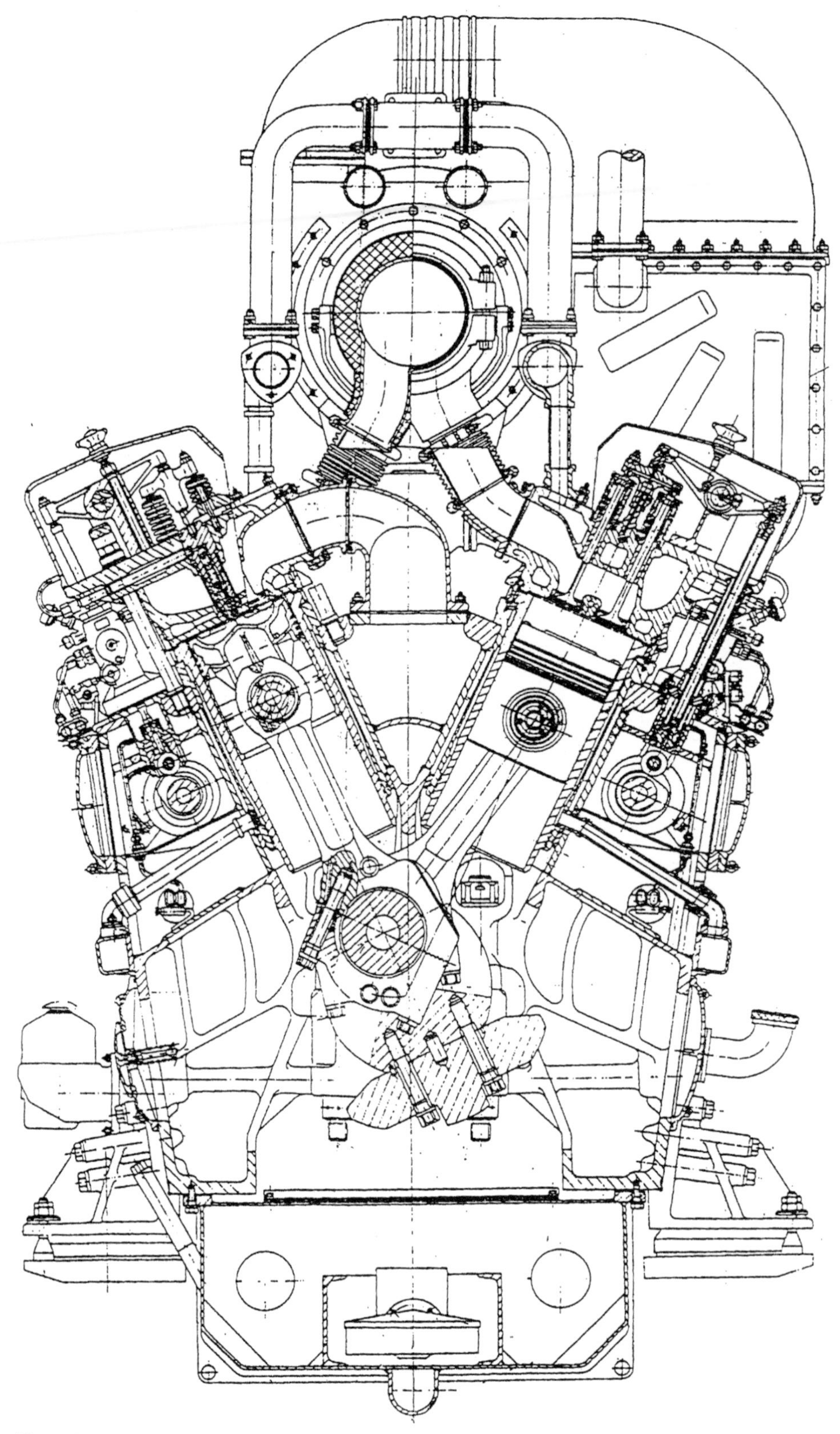

Figure 24.5 Dalian 240ZD engine

24.3.4 General Electric 7FDL™ *(Figure 24.6. References 9, 10 and 11)*

The General Electric 7FDL engine has a substantial cast iron crankcase. The crankshaft, which is underslung, is nitrided. It is unusual in that it is the only four-cycle engine in which each cylinder assembly is self-contained and is attached to the crankcase by four bolts.

The cylinder assembly comprises a single-piece iron casting encompassing the water jacket, air and exhaust porting, and the rocker gear supports. The pot type cylinder head is made up from a number of steel castings, welded together. The liner is

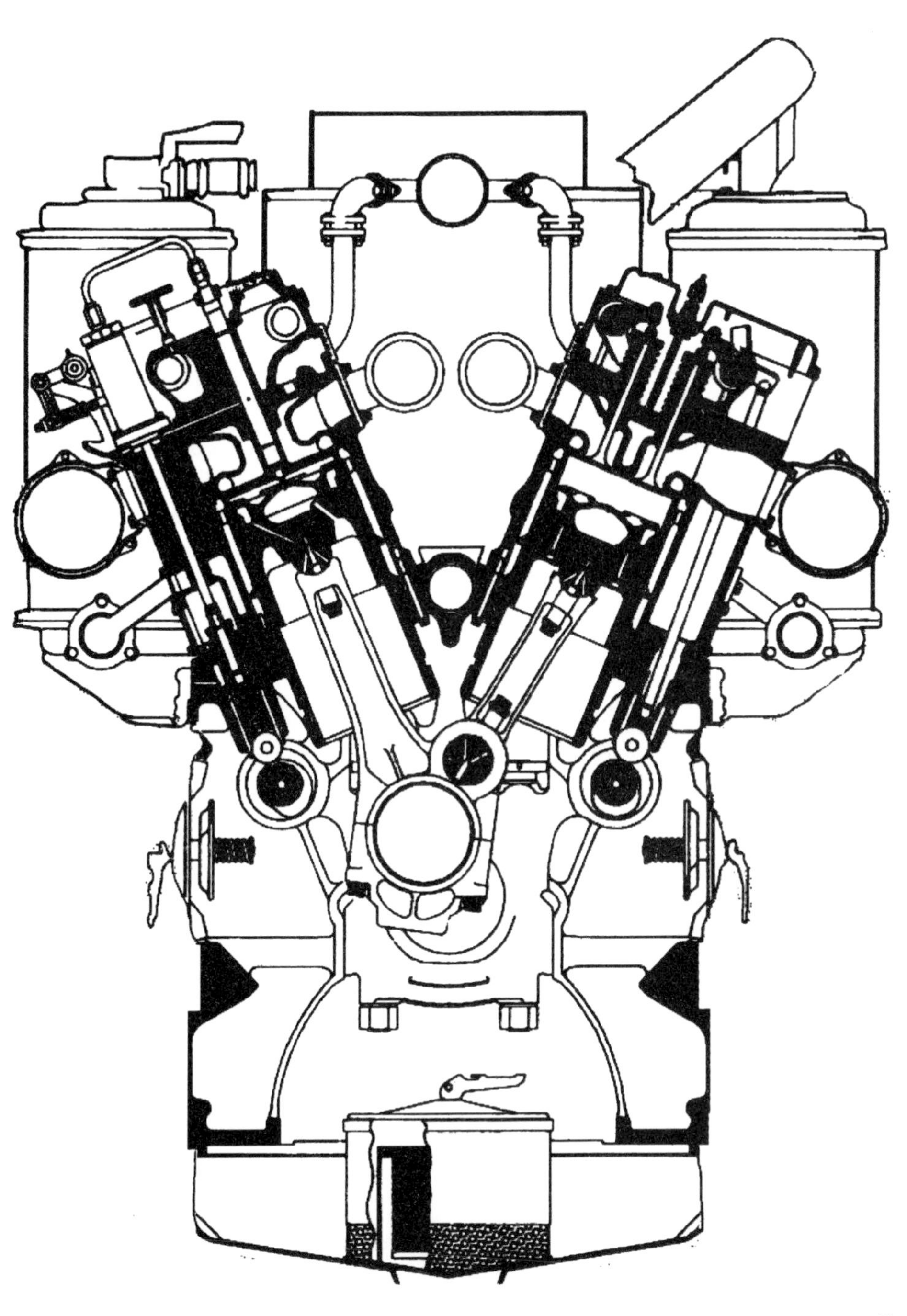

Figure 24.6 General Electric 7FDL™ engine

made from drawn steel tube and is nitride hardened after machining, it is then welded onto the steel cylinder head assembly, thus eliminating the gas joint between cylinder head and liner. To put together a power assembly, the cylinder jacket is heated in an oven and then the head/liner assembly is inserted into it and retained by a clamp ring around the bottom of the liner. The piston and connecting rod are inserted into the cylinder liner and held in place by a special clamp. After the complete cylinder assembly is bolted onto the crankcase, the connecting rod clamp can then be released and the connecting rod bolted onto its respective crankpin. The connecting rods are of the articulated type with a master and slave rod (there is no bank-to-bank offset with the design). The piston pin is bolted to the connecting rods, and this allows a full width piston pin bearing to be used.

The camshafts are located on the outside of the engine and run in bearings in the crankcase. Each camshaft is driven by a gear from the flywheel end of the engine and operates both the valves and the fuel injection pumps through roller followers and pushrods. The individual fuel injection pumps are mounted high up at cylinder head level and are controlled mechanically or electronically. The electronic controls enable the injection timing to be varied to optimize engine performance and emissions. The cylinder heads have four valves and a centrally mounted injector. The air manifolding is located on the outside of the engine, the exhaust manifold is located in the centre of the vee and the turbocharger and aftercoolers are mounted over the front end of the engine. The 8-cylinder engine uses a pulse system and the 12- and 16-cylinder engines use a twin pipe MPC system. Only one turbocharger is used per engine, whether it has 8, 12, or 16 cylinders.

24.3.5 General Electric 7HDL™ *(Figures 24.7 and 24.8. References 12)*

The General Electric 7HDL engine has a single piece nodular cast iron housing comprising the cylinder blocks and crankcase. The crankshaft is underslung and is unhardened. The pistons have a forged aluminium body and the steel crown is attached to the body by four bolts. The piston cooling oil is supplied via the piston pin from an oil drilling up the shank of the connecting rod. The cast iron liners are strategically cooled at the top only. The liner is both supported by a close fitting water jacket that seats on the top of the cylinder block. This water jacket also incorporates location bores for the fuel injection pump and the roller followers for the valve gear. The connecting rod is machined from a steel forging and has an angle split joint with serrations and two big end bolts. The connecting rods are arranged 'side-by-side' on a common crankpin.

The camshafts are mounted on the outside of the engine and are driven by a train of gears from flywheel end of the engine. Each camshaft operated push rod operates valve gear through roller follower tappets. The individual fuel injection pumps are electronically controlled.

The inlet and exhaust ports are located in the centre of the vee and a circular air manifold made of cast aluminium, feeds air to all cylinders. A two-pipe MPC exhaust manifold conveys the exhaust gases to the two turbochargers which are mounted at the front of the engine, together with the aftercoolers on an integrated front end cover which incorporates the turbocharger and aftercooler supports as well as drillings and cast in pipes for fluid transfer into a single casting. The water and oil pumps are mounted at the front end of the engine and are driven by a train of gears from the crankshaft.

24.3.6 General Motors EMD 645 and 710 *(Figure 24.9. References 13, 14 and 15)*

The General Motors EMD 645 has been in production sincc 1965. The 710 version was released onto the market in 1985. The 710 follows the concept of the 645, the main difference being an increase in stroke of 25.4 mm (1″). Both variants are available for locomotive duty.

The 645 and 710 engines are the only two-stroke locomotive engines in large-scale production at the present time. It is a uniflow design with a ported liner and four exhaust valves in the cylinder head.

The engine housing is made up from steel forgings and plate welded into a single assembly. The crankshaft is underslung and has induction hardened pins and journals. On the 8- and 12-cylinder engines a single-piece crankshaft is used, while on the 16- and 20-cylinder engines a two-piece crankshaft is used, the two halves being flanged and bolted together in the centre of the engine. The alloy cast iron liner has two steel sleeves brazed onto the outside to form the water jacket. The eighteen inlet ports are positioned at about the mid-point of the liner and are directed at an angle of 15° with respect to radial entry. The upper half of the liner as well as the relieved area in the belt adjacent to the ports is laser beam hardened.

The composite piston has a cast iron body, with a number of radial ribs under the crown for support. The crown thickness has been kept low in the interest of good heat transfer and low crown temperatures. The piston pin is supported by a full width forged steel carrier which takes the piston crown thrust loads through a large diameter thrust washer. In the interests of even temperature distribution, the piston body is able to rotate relative to the pin carrier. The connecting rod is bolted to the piston pin with high tensile bolts. The piston pin incorporates a 'rocking' design of pin and bearing whereby the bearing surfaces are alternately loaded and unloaded during one revolution of the crankshaft in order to improve the bearing lubrication and hence its load-carrying capacity. The connecting rods are of the fork and blade type (no bank-to-bank offset required).

The cylinder head is made of cast iron and carries the four exhaust valves and the central unit injector. The outside of the cylinder head is round and fits into a recess in the top of the housing. It is bolted to the liner via eight bolts, and the complete cylinder assembly is held into the housing by four large studs per head. For servicing, a power assembly, consisting of head and liner complete with piston and rod in the liner can be removed very easily.

Two camshafts (one per cylinder bank) are located at the top of the housing and actuate the valves and unit injectors through rocker levers with roller followers. The camshafts are gear driven from the flywheel end. The engine is supplied with scavenge air from a turbocharger which has a gear drive to it. This gear drive incorporates an over-running clutch which uncouples the drive when there is sufficient energy in the exhaust gas to the turbine to make it self-sustaining. A single turbocharger is used with two aftercoolers, one feeding each cylinder bank. The oil and water pumps are mounted at the front end of the engine and are gear driven by a train of gears from the crankshaft.

24.3.7 General Motors EMD H engine *(Figures 24.10 and 24.11)*

The General Motors 16 Cylinder H engine is a new design of 4 stroke engine designed to deliver 4700 kW from the same envelope size of the 20 cylinder EMD 710 engine which is rated at 3750 kW.

The engine housing is a single piece nodular iron casting incorporating an integral turbocharger support at the flywheel end of the engine. The crankshaft is underslung and has induction hardened pins and journals. The liner is strategically cooled at the top only and is bolted to the cylinder head directly. The piston is a two-piece design with a steel crown and an aluminium body. The connecting rods are machined from steel forgings

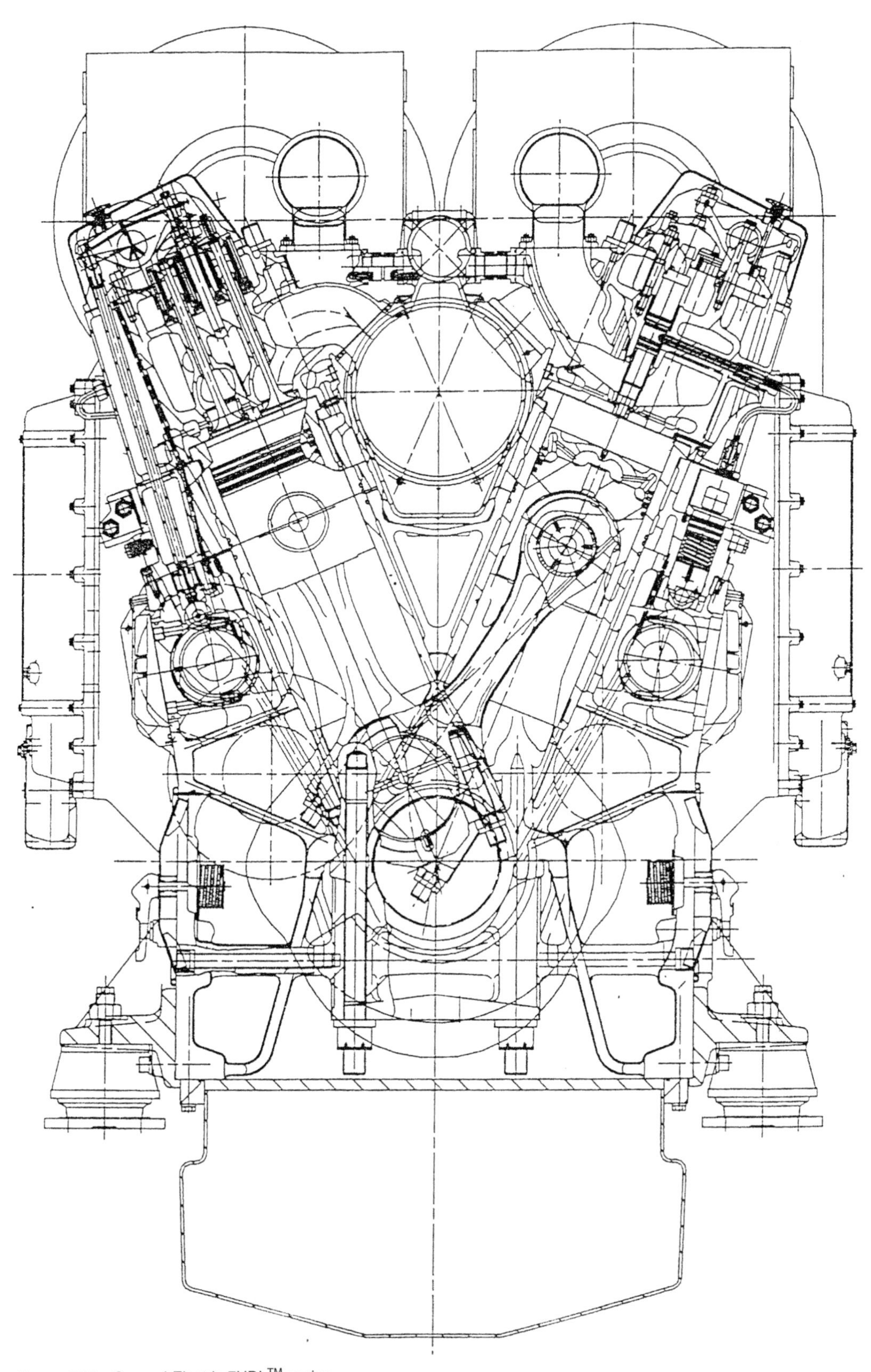

Figure 24.7 General Electric 7HDL™ engine

Figure 24.8 View of General Electric 7HDL™ engine

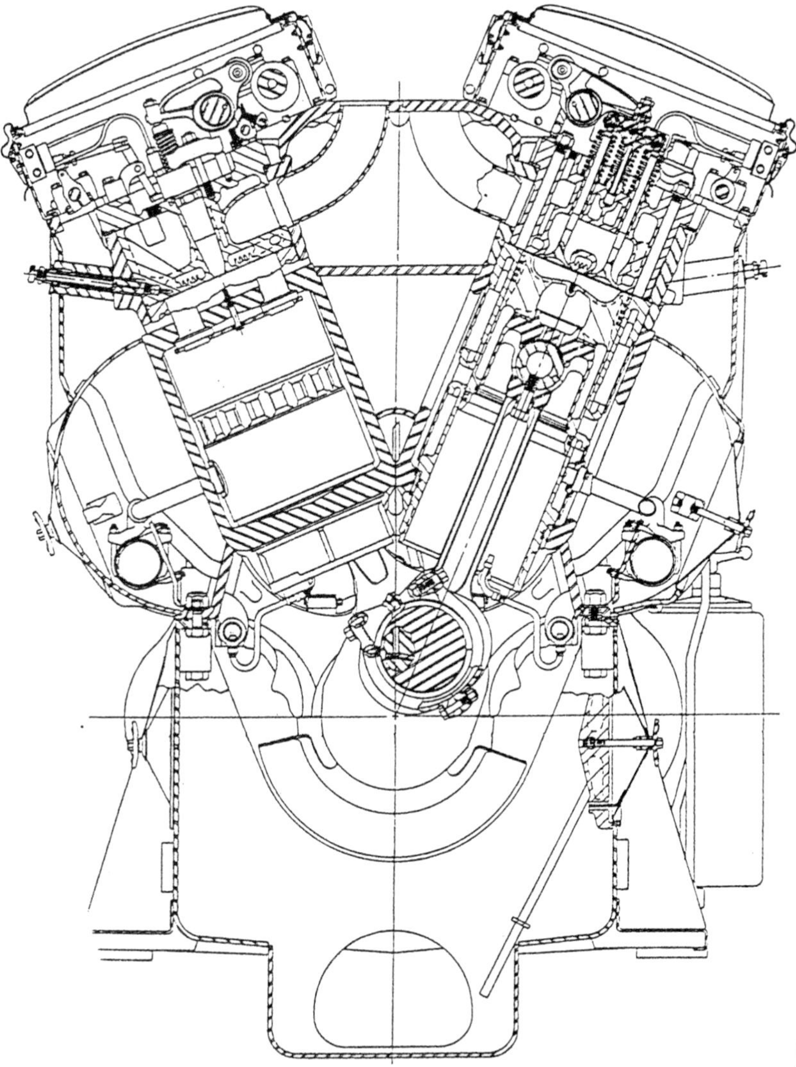

Figure 24.9 General Motors EMD 645 and 710 engine

and have a horizontal split big end and each cap is retained by four bolts. The rods are arranged side-by-side on a common crankpin. The cylinder head, liner and piston can be removed as a single assembly from the engine for servicing purposes.

The single central camshaft is driven by a train of gears from the flywheel end of the engine. The camshaft operates the push rod operated valve gear and the individual electronically controlled fuel injection pumps.

The twin pipe MPC exhaust system is situated in the centre of the vee and the two turbochargers are mounted over the flywheel end of the engine. The air manifold/aftercoolers are mounted on the outside of each cylinder bank of the engine.

The oil and water pumps are mounted at the front end of the engine and are driven by a train of gears from the crankshaft.

24.3.8 Kolomna D 49 *(Figure 24.12. References 16 and 17)*

The Kolomna D49 engine has a single piece fabricated steel housing comprising the cylinder blocks and crankcase. The crankshaft, which is underslung, is nitrided. The pistons have a forged steel crown attached to an aluminium body and the piston is cooled by oil fed through drillings in the piston pin and connecting rod. The cast iron liners are located by deep flange at the top of the liner and a steel water jacket is pressed onto the liner.

The connecting rods are of the articulated type with a master and slave rod. The big end of the master rod has an angled split big end with a serrated joint.

A single camshaft is located in a housing at the top of the engine above the cylinder heads. It is driven by a train of gears from the flywheel end of the engine. The camshaft operates the push rod operated valve gear through lever followers; these pushrods operate in a nearly horizontal plane and activate the valve gear through cranked rockers. The central camshaft also drives the individual fuel injection pumps which are also attached to the camshaft housing.

The exhaust manifolds are located on the outside of the engine and the intake manifold in the centre of the engine between the cylinder banks. A constant pressure exhaust system is used and the turbocharger and aftercooler are mounted at the front of the engine. The oil and water pumps are mounted at the front of the engine and are driven by a train of gears from the camshaft.

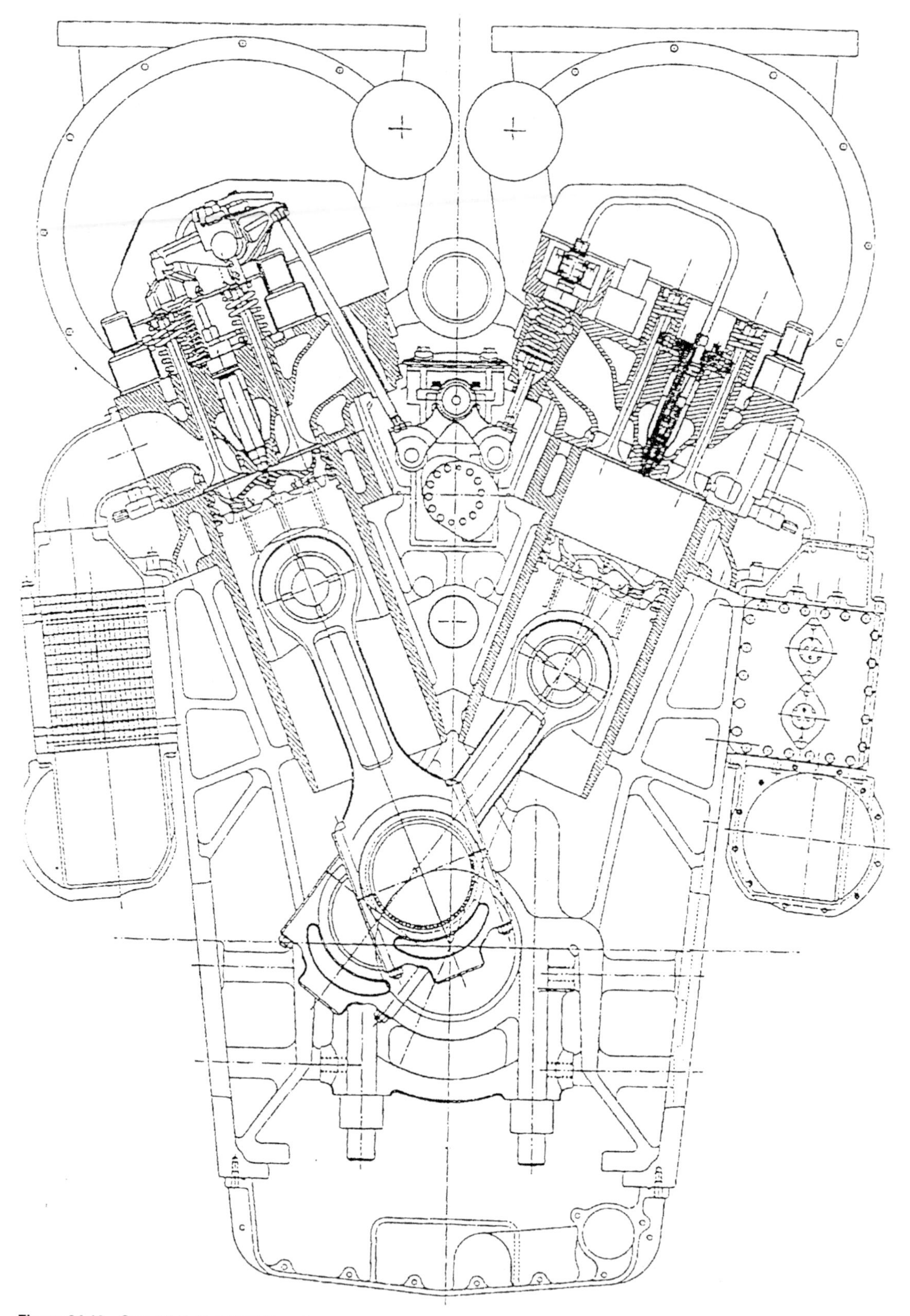

Figure 24.10 General Motors EMD H engine

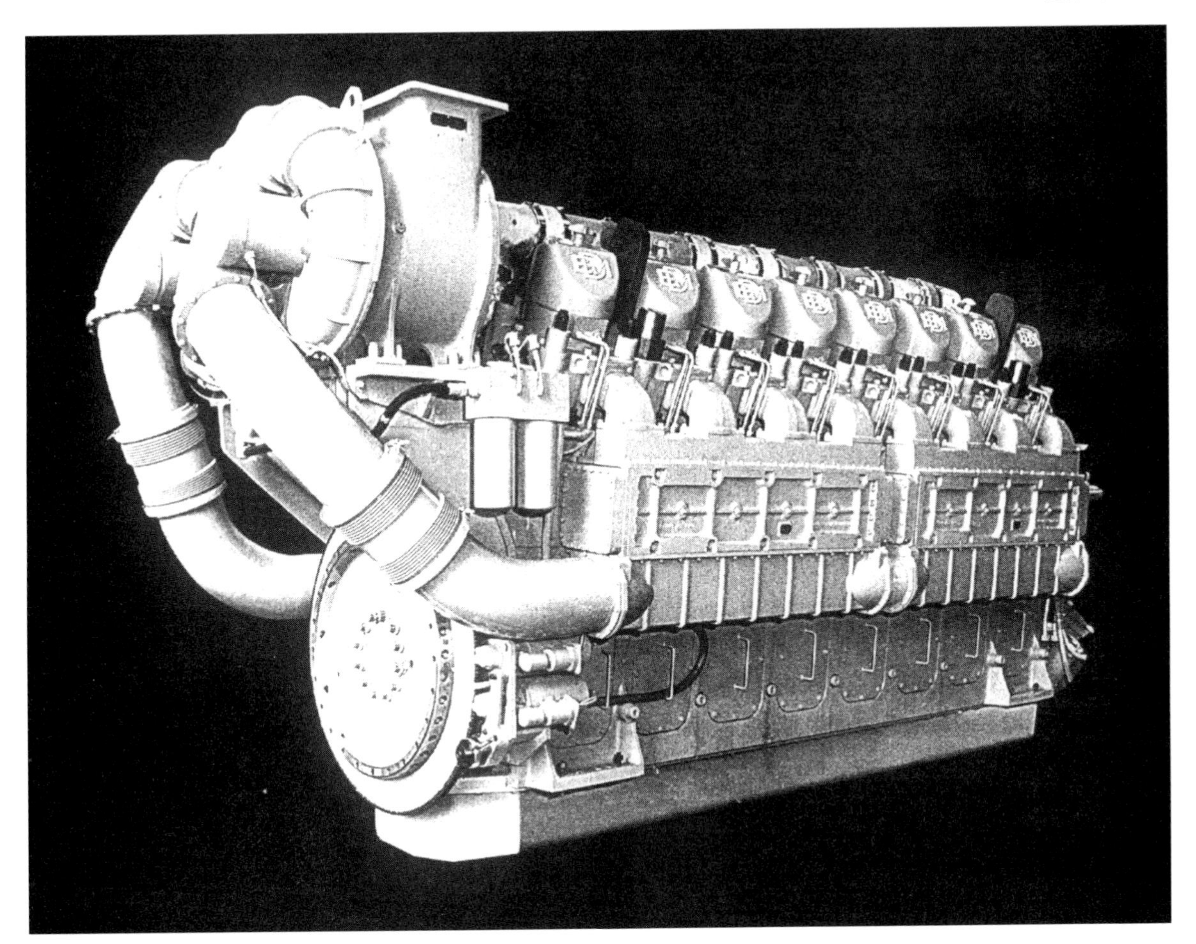

Figure 24.11 View of General Motors EMD H engine

24.3.9 MTU/DDC 4000 series *(Figure 24.13. Reference 18)*

The MTU/DDC 4000 series has a single piece grey iron casting comprising cylinder blocks and crankcase. The crankshaft, which is underslung has hardened pins and journals. The piston has a forged aluminium body with a steel crown attached to it by four bolts. The pistons are cooled by a fixed jet mounted at the bottom of each cylinder liner. The cast iron liners are located by a flange that sits directly on the top face of the cylinder block. The connecting rods are machined from steel forgings and have an angled split big end which is retained by two bolts. The rods are arranged 'side-by-side' on a common crankpin.

A single camshaft is located in the centre of the vee and is driven by a train of gears from the front end of the engine. The valve gear is operated by pushrods and lever followers. The individual cylinder heads are to a cross-flow design with the air manifolds on the outside of the cylinder banks and the exhaust manifolds (and the turbochargers) in the centre of the vee. The aftercooler is mounted at one end of the engine.

A common rail fuel injection system manufactured by L'orange is used. The high pressure fuel pump is mounted at the from of the engine and is gear driven. One common rail per bank is used and it is located underneath the air manifold and individual, short high pressure pipes are used to connect the common rail to the electronically controlled injectors.

The oil and water pumps are mounted at the front end of the engine and are driven by a train of gears from the crankshaft.

24.3.10 Paxman VP 185 *(Figure 24.14. Reference 19)*

The Paxman VP 185 has a single piece nodular cast iron housing comprising the cylinder blocks and crankcase. The crankshaft is underslung and is machined all over and then full nitrided. The piston is a single piece nodular iron casting with integral cooling gallery cast in. The piston cooling oil is supplied from a fixed jet located at the bottom of the liner. The cast iron liners are located in the housing by a flange at the top of the liner. The connecting rods are fully machined from steel forgings and are then nitro carburized; they have an angled split big end with a serrated joint. The connecting rods are arranged side-by-side on a common crankpin.

A single camshaft is located in the centre of the vee and is driven by a train of gears from the front end of the crankshaft. The camshaft operates lever followers to operate both the valve gear and the individual unit injectors.

The individual cylinder heads are retained by eight studs, four large and four small, the large studs are 'shared' in that they clamp adjacent cylinder heads through a bridge piece with a spherical washer arrangement under the nut to allow the stud load on adjacent heads to be equally shared. The air manifolds are located on the outside of the vee and the exhaust system inside the vee. A two-stage turbocharging system is used and it is enclosed in a gas-tight enclosure. The walls of this enclosure are water cooled. On the 12-cylinder version six turbochargers are used. Two high pressure turbochargers and four low pressure turbochargers are used. They are of the radial type and all six

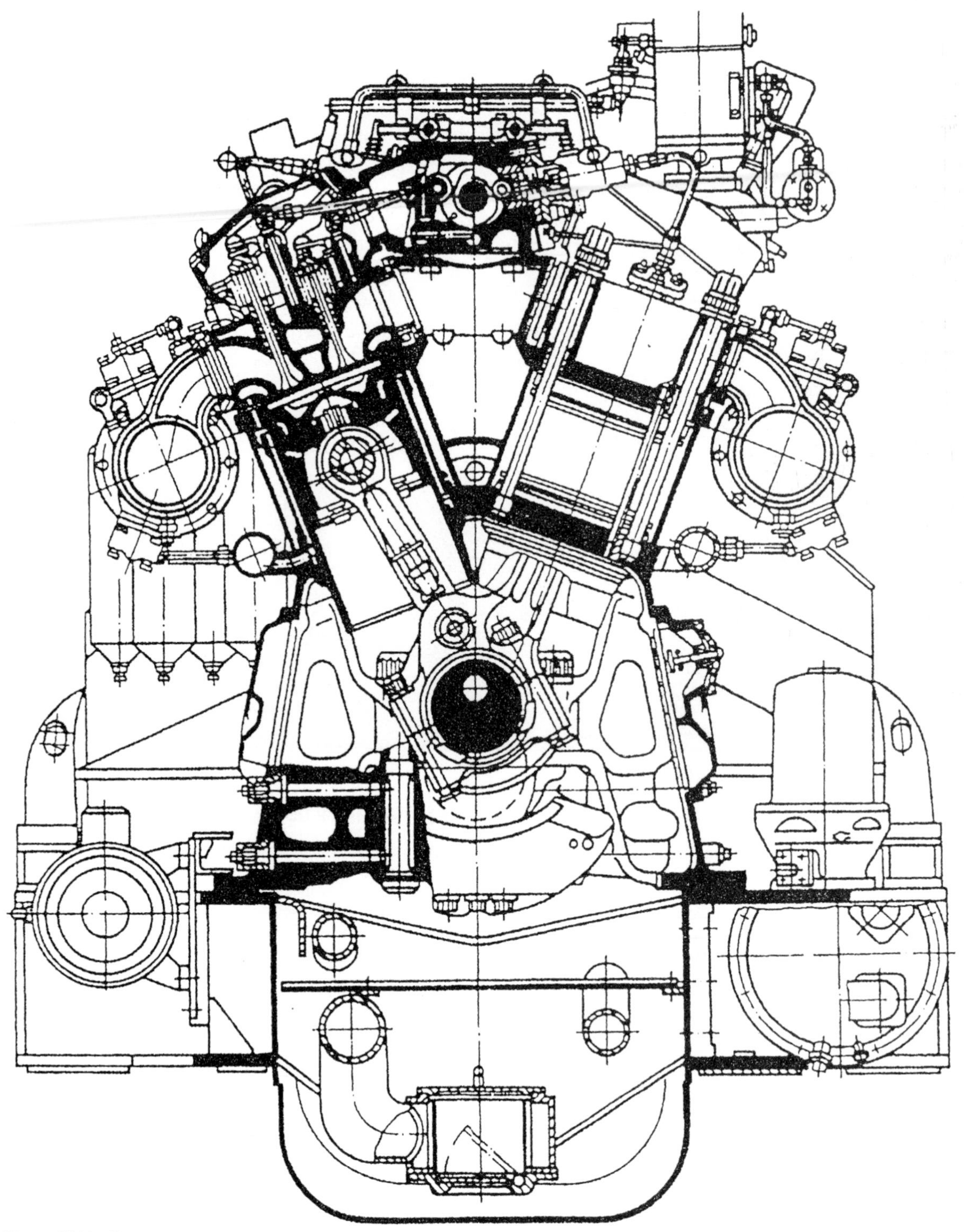

Figure 24.12 Kolomna D49 engine

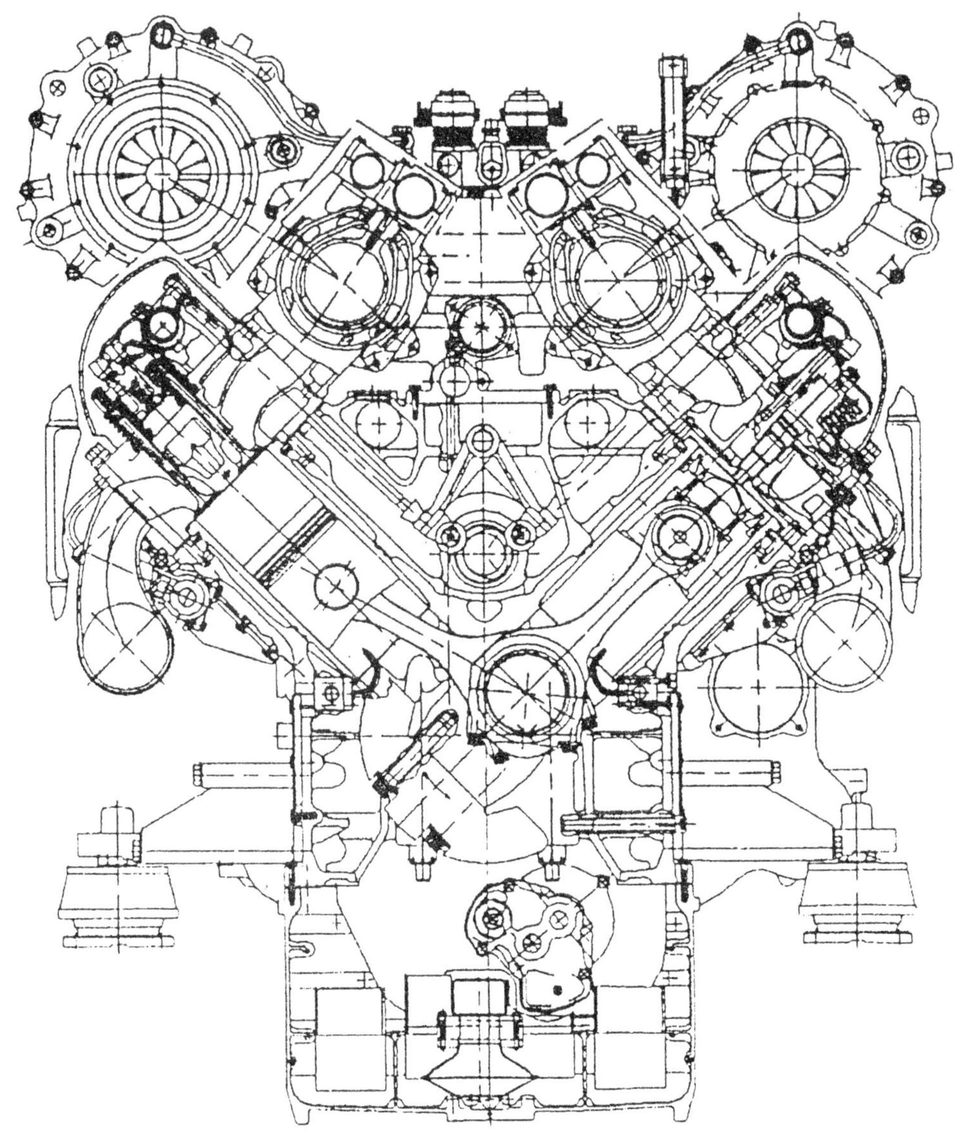

Figure 24.13 MTU/DDC 4000 series engine

are identical. Both intercoolers and aftercoolers are used to control the charge air temperatures. The water and oil pumps are mounted at the front end of the engine and are driven by a train of gears from the crankshaft.

24.3.11 Pielstick PA4 200 VG *(Figure 24.15. References 20, 21, 22 and 23)*

The Pielstick PA4 200 VG has a single-piece nodular iron housing, comprising the cylinder blocks and crankcase. The crankshaft, which has induction hardened pins and journals, is inserted into the tunnel crankcase from one end, and the main bearing shells are housed in split circular bearing blocks which are clamped by circular wedges in the crankcase bores.

The pistons are cast in aluminium alloy and have cooling galleries fed with oil through drillings in the piston pin. An obturater plug, made in heat resisting steel, is mounted in the centre of the piston crown. This plug fits into the throat of the prechamber when the piston is at the top of the stroke. The cast iron liners have a flange at the top for location, and each liner is surrounded by its individual cast iron water jacket. The liner/jacket assembly is then fitted into location diameters in the engine housing. The connecting rods are machined from steel forgings and have an angled split joint, with serrations and four big end bolts. On the Vee engines the connecting rods are arranged 'side-by-side' on a common crankpin.

The camshaft is located in the centre of the Vee and is driven by a train of gears from the front of the engine. The block type fuel injection pump is located in the Vee above the camshaft and is driven by a tubular shaft, gear-driven from the camshaft gear. The camshaft operates the pushrod-operated valve gear through roller follower trappets.

The individual cylinder heads have four valves and a centrally located prechamber, the injector is mounted in the top of the prechamber. The intake manifolds are located on the outside of the engine, and the exhaust manifolds and the coolers are mounted at the front of the engine.

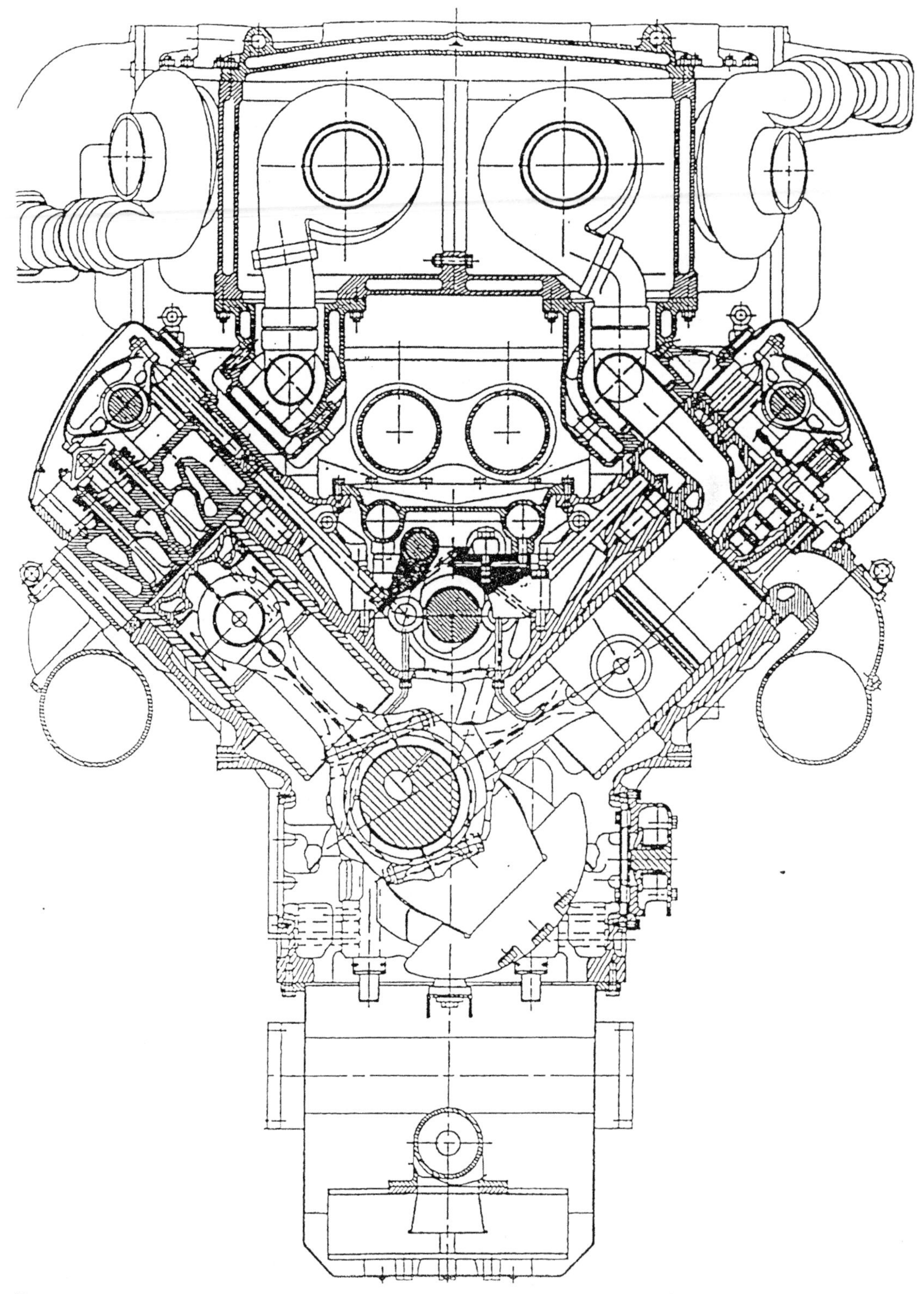

Figure 24.14 Paxman VP 185 engine

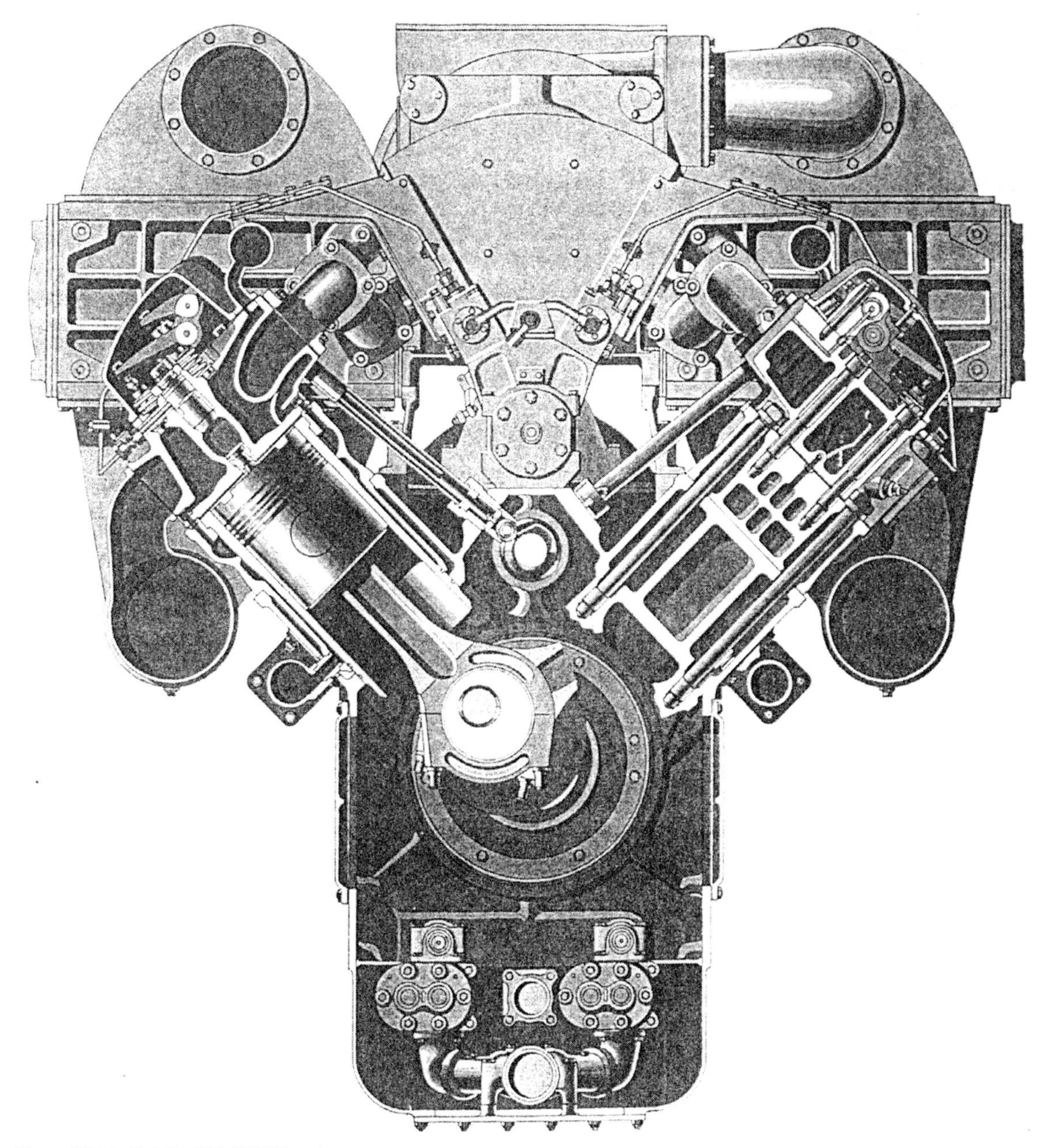

Figure 24.15 Pielstick PA4 200 VG engine

The oil and water pumps are mounted at the front of the engine and are driven by a train of gears from the crankshaft.

24.3.12 Pielstick PA6B *(Figure 24.16. References 24 and 25)*

The Pielstick PA6B engine has a single-piece nodular iron housing, comprising the cylinder blocks and crankcase. The crankshaft, which is underslung, has induction hardened pins and journals. The two-piece pistons have steel crowns and aluminium skirts. The piston cooling oil is supplied through drillings in the piston pin. The cast iron liners are held by a flange at the top of the liner. The connecting rods are machined from steel forgings; they have a horizontal split big end and each cap is retained by two studs. On the vee engines the connecting rods are arranged 'side-by-side' on a common crankpin. The piston, liner and connecting rod have to be removed from the engine as a single assembly for servicing.

The camshafts are located on the outside of the engine and are gear driven from the front of the engine. Each camshaft operates the pushrod-operated valve gear through roller follower tappets. The individual fuel injection pumps have integral roller tappets.

The cylinder heads have four valves and a centrally mounted injector. Both the intake and exhaust ducting is located in the centre of the Vee, as are the turbochargers.

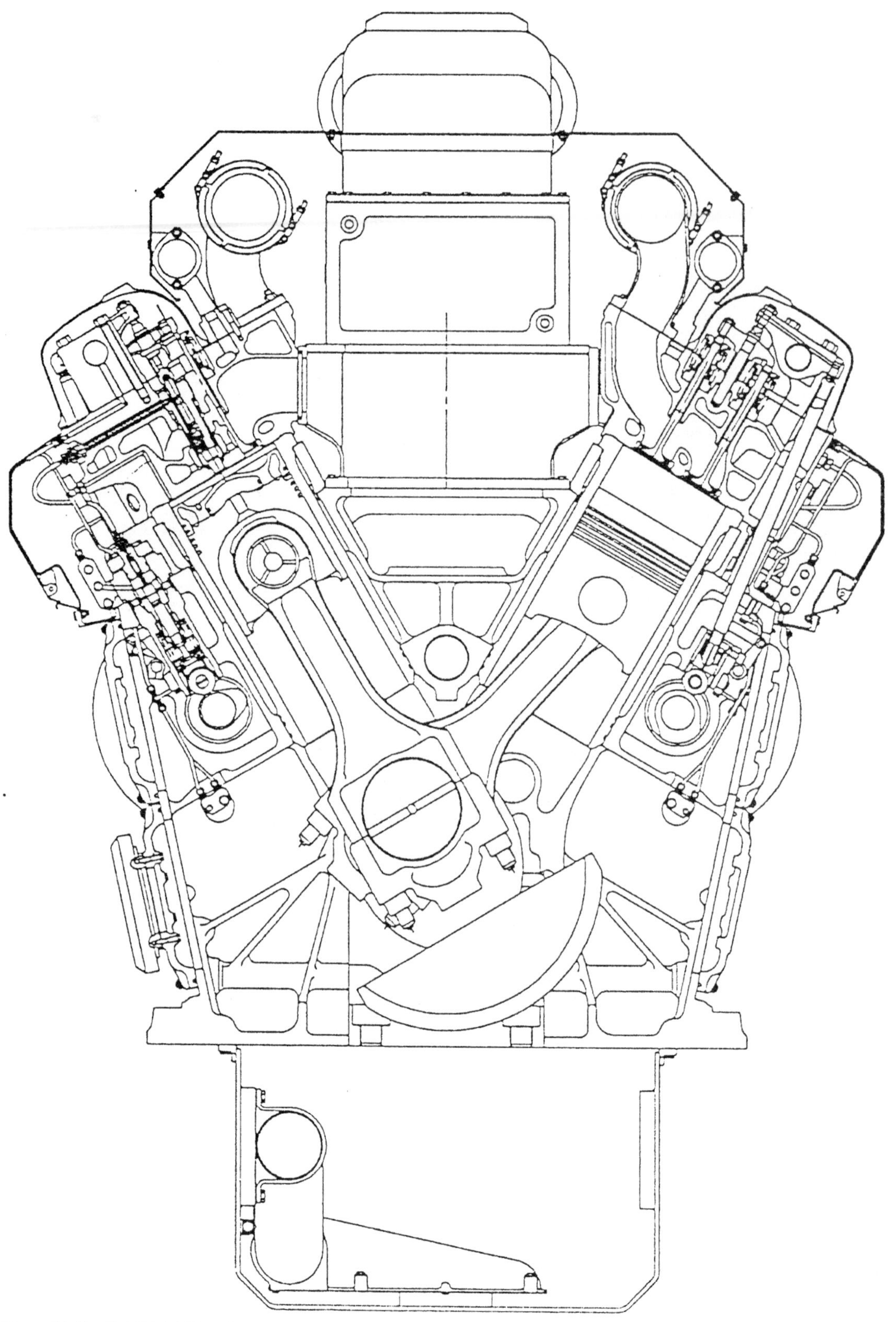

Figure 24.16 Pielstick PA6B engine

The water and oil pumps are mounted at the front end of the engine and are driven by a train of gears from the crankshaft.

24.3.13 Ruston RK215 *(Figure 24.17. Reference 26)*

The Ruston RK215 engine has a single piece nodular iron casting comprising cylinder blocks and crankcase. The crankshaft is underslung and is unhardened. The piston has an aluminium body and a steel crown attached to it with four bolts. Cooling oil is supplied to the piston via drillings in the piston which are in turn fed by a drilling up the shank of the connecting rod. The connecting rods are machined from steel forgings and have an angled split, serrated joint. On the vee engines the rods are arranged 'side-by-side' on a common crankpin.

The crankshafts are located on the outside of the engine and are driven by a train of gears from the front end of the crankshaft. Each camshaft operates the push rod operated valve gear through lever followers as are the push rod operated unit injectors. The air and exhaust ports are located in the centre of the vee and the space between the cylinder banks serves as an air manifold.

A single turbocharger is fitted on the 8, 12 and 16-cylinder engines at the front end of the engine and pulse type exhaust manifolds are used. The aftercooler is mounted at the front end of the engine below the turbocharger.

The oil and water pumps are fitted at the front of the engine and are driven by a train of gears from the crankshaft.

24.4 Summary of engine design features and future trends

Locomotive engines have to meet constraints on engine width, engine weight and to a lesser extent height. These constraints result in a design that has a high power-to-weight ratio (i.e. typically should be less than 6 kg per kW output) and the width less than 1725 mm.

The operating conditions can vary widely from –40°C to + 50°C and altitudes of up to 3500 m have to be taken into account. The limitations on space within the locomotive make it difficult to fit a radiator system that can give air manifold temperatures of less than 60–65°C. This together with the need to allow for running at altitude have limited brake mean effective pressures (bmep) to around 21.5–22.0 bar. This is 2–3 bar less than would be typical for a modern design aimed at the industrial or marine markets.

The largest cylinder bore that can be accommodated in a locomotive is 280 mm and this in turn limits the engine speed to 1000–1050 rev/min. For small bore engines, namely those having a cylinder bore in the range 170 to 190 mm, engines speed of up to 1800 rev/min are common. Piston speeds of between 10.5 and 11.5 m/s have been used for some time. Vee engines predominate in locomotive service, whether they are medium or high speed types.

Examination of the engines described in the previous section shows that there are a number of design features which are common particularly on the newer engines.

- The power assembly concept, where the cylinder head, liner, piston and connecting rod are removed from the engine as a single unit, for repair or refurbishment, without dismantling adjacent components, so the engine can be returned to service with the minimum of delay. The worn power assemblies can then be refurbished at will and stored for future use.
- Nodular (which is also called spheroidal or ductile) cast iron has largely replaced flake or grey cast iron as the material for engine housings. The better material properties of nodular iron compared to grey iron enable a lighter casting of the engine housing to be made if the loads are not increased or alternatively if the loads are increased then a stronger casting can be made while not exceeding the weight of a grey iron casting. As most of the new designs are using a maximum cylinder pressure of 180–200 bar, the use of nodular iron for the housing casting is the usual practice.
- More engine designs are incorporating the 'dry block' principle, i.e. there is no cooling water in the cylinder block. This has been brought about by the use of strategically cooled liners where only up to 20–30% of the axial length of the liner is cooled. With such a design it is common to use a well directed flow of high velocity water to cool the liner. Significant efforts are being made on new designs to incorporate as many fluid (oil, water, fuel) transfer functions, either by drilling or cast in ducts to eliminate as far as possible external pipe work. This is being done to increase reliability and to reduce costs as pipework is very often a high maintenance item due to leaks and vibration problems.
- The use of a common design of connecting rods for both banks of the engine is standard practice for new designs; articulated and fork and blade designs are not specified for new designs.
- Fuel injection systems either consist of the traditional pump-pipe-nozzle system or unit injectors. Maximum injection pressures used currently are around 1500–1700 bar, but these will rise to 2000 bar within the next 5–8 years. Higher pressures are necessary to enable larger volumes of fuel to be injected within the same overall injection period (30–35°C crank) while keeping the fuel droplet size small in order to aid clean combustion to minimize smoke levels. Both these designs of injection equipment are on offer today with electronic control. This enables the timing and quantity of fuel injected to be accurately controlled to enhance economy and minimize smoke levels. As emissions levels are legislated for, this ability to vary the timing and possibly other injection parameters as well, will help in achieving compliance with these standards.
- Single stage turbocharging is still the preferred choice, usually coupled to a modular pulse converter (MPC) design of exhaust manifold. The MPC system gives the best compromise between cost, performance, reliability and ease of maintenance. However, the increasing throughput of air and pressure ratio in response to increases in bmep has resulted in turbochargers and after-coolers becoming a larger proportion of the total engine package both in terms of weight and volume. As these assemblies are mounted high up on the engine, they are often subjected to high levels of vibration if they are not supported on a rigid structure that is bolted to the engine housing so that the imposed loads are fed into the housing over a wide area. Modern designs of turbocharger and aftercooler supports are tending to move towards an integrated approach where one casting serves both purposes, saving in parts count, enhancing stiffness and reducing levels of vibration.

24.5 Railcar engines

Railcars are self powered and the engines are usually mounted under the floor of the vehicle between the wheel sets. In order to avoid intrusion into the passenger space, it is usual to specify an engine with its cylinders mounted horizontally in order to minimize the overall height of the engine. As access through the floor of the vehicle is usually impracticable, it is normal to mount all the serviceable items such as oil, air and fuel filters, dipstick and oil filler cap at the side of the engine for easy access.

Railcars have engines ranging from 160 to 750 kW. The engines used are usually derived from heavy duty truck designs and

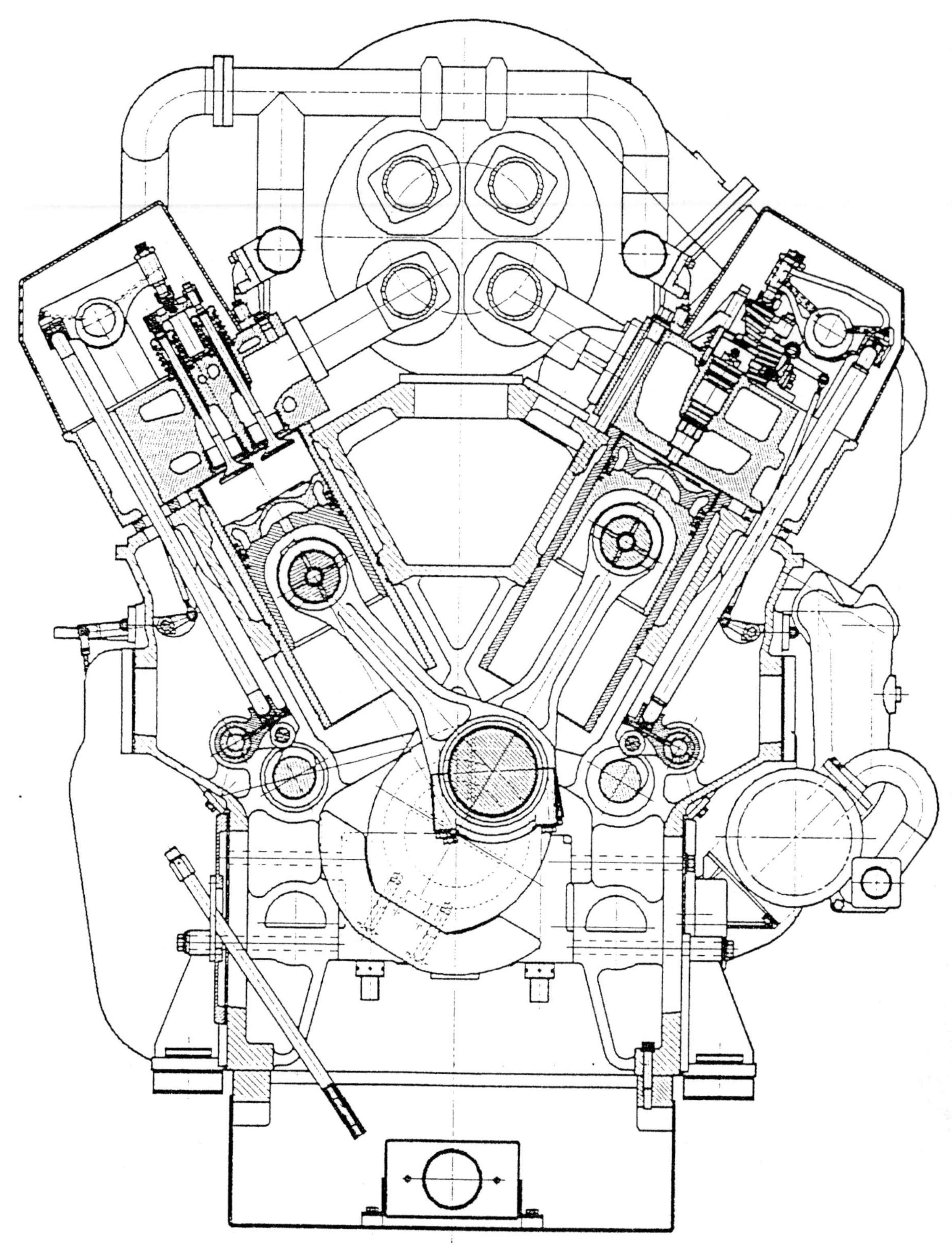

Figure 24.17 Ruston RK215 engine

have 6 or 8 cylinders of 130–160 mm diameter, mounted horizontally, or 12 cylinders in vee form.

Brief descriptions of some of the engines available for the railcar market are set out in the following paragraphs.

24.5.1 Cummins

Cummins offer three engine types for underfloor applications. All have six cylinders mounted horizontally. All three engines have 4-valve cylinder heads and unit injectors. Two types of control systems are offered, the first is CELECT a full authority electronic engine management system. The second is CENTRY, an electronic system offering many of the features of CELECT without the diagnostic or prognostic capabilities of CELECT.

- The M11 has a bore of 125 mm and a stroke of 147 mm giving a displacement of 10.8 litres, and is rated between 185 and 274 kW at 2100 rev/mm.
- The N14 has a bore of 140 mm and a stroke of 152 mm giving a displacement of 14 litres, it is rated between 259 and 384 kW at 2.00 rev/mm.
- The QSK19 has a bore and stroke of 159 mm giving a displacement of 19 litres; it is rated between 445 and 556 kW at 2100 rev/mm.

24.5.2 MAN

MAN offer their D2866 and D2842 series for underfloor installation in railcars. Both series have a bore of 128 mm, the D2866 is made in 6-cylinder horizontal form with a stroke of 155 mm, giving a displacement of 12 litres; the D2842 is made in 12-cylinder vee form and has a displacement of 22 litres. Individual cylinder heads with two valves are fitted. Both series are turbocharged and aftercooled.

The 6-cylinder D2866 series are rated between 210 and 300 kW at 2100 rev/min and the 12-cylinder D2842 series are rated between 302 and 588 kW at 2100 rev/min.

24.5.3 MTU

MTU offer their 183 series engine for underfloor installations in railcars, either in six cylinders in horizontal form or in 12-cylinder vee form. The 183 series is the industrial version of the 400 series truck engine having a bore of 128 mm and a stroke of 142 mm. The vee versions have a bank angle of 90°. Individual cylinder heads with two valves are used. The engines are turbocharged and after cooled. The ratings of the engines are as follows: 6-cylinder, 315 kW at 1900 rev/min and for the 12-cylinder versions, 485 to 550 kW at 2100 rev/min.

24.5.4 Niigata

Niigata have co-operated with the Japan Railway Company (JR) to develop engines for underfloor installation in railcards. Two models are offered, both are 6-cylinder horizontal layouts. The DMF 13HS version has a bore of 130 mm a stroke of 160 mm and a swept volume of 12.7 litres. Its output is 183 to 242 kW at 2000 rev/min. The DMF 13HZA has a bore of 132.9 mm and a stroke of 160 mm giving a displacement of 13.3 l. Its output ranges from 330 to 368 kW at 2000 rev/min.

Acknowledgements

The author would like to warmly thank all those companies who have provided information and assistance in the preparation of this chapter.

References

1 BENYON, J., RAZDAN, D. and DRAPER, E., 'The development of a diesel engine acceptance test for British Railways', *IMechE C70/87* (1987)

Caterpillar 3500

2 McCLUNG, C. L. and GRURICH, W., 'Continued development of the Caterpillar 3500 diesel engine family', *SAE 860879* (1989)

3 WELLER, B. and SELBY, W., 'The Caterpillar 3500 series B electronic engine system', *ASME Conference on New Technology Challenges for the Future*, Milwaukee (1995)

4 KNOLHOFF, D., McCLUNG, C. and WELLER, B., 'The new Caterpillar 3500 Series B diesel engine', *ASME Conference on New Technology Challenges for the Future*, Milwaukee (1995)

Caterpillar 3600

5 KIESER, R., 'Development of the Caterpillar 3600 series engine', *CIMAC D87* (1985)

6 BURNS, S. and EVANS, R., 'Caterpillar 3600 PEEC (Programmable Electronic Engine Control)', *ASME Engine Components Conference*, Illinois (1986)

7 AMDALL, J., 'Initial operating results from Caterpillar 3600 diesel engine', *ASME 87-ICE-30* (1987)

Dalian 16V240ZJD

8 YUKAN, T. and THOMAS, J. R., 'Development of a locomotive for Chinese railways', *CIMAC D114* (1989)

GE-7FDL

9 DEAN, J. and JOHNSON, B., 'Reliability and high specific output—an American manufacturers viewpoint', *IMechE C90/82* (1982)

10 FRITZ, S. and CATALDI, G., 'Gaseous and particulate emissions from diesel locomotive engines', *ASME Conference on Large Bore Engines*, Rockford (1990)

11 FRITZ, S., 'Exhaust emissions from two intercity passenger locomotives', *ASME Conference on Alternative Fuels*, Morgantown (1993)

GE-7HDL

12 HAPEMAN, M. and HITZINGER, H., 'The General Electric AC6000 diesel electric locomotive', *Dalian International Symposium on Locomotives and Motor Cars*, DL-21 (1994)

GM-EMD 645

13 KOTLIN, J., WILLIAMS, H. and DUNTEMAN, N., 'Higher Fuel efficiency for EMD diesel locomotives', *ASME 78-RT4* (1978)

GM-EMD 710

14 UZKAN, T., 'An analysis of the engine blowdown process using multidimensional computations', *ASME 87-ICE-17* (1987)

15 KOTLIN, J., DUNTEMAN, N., CHEN, J. and HEILENBACH, J., 'The General Motors EMD model 710G series turbocharged 2 stroke cycle diesel engine', *ASME, 85-DGP-24* (1985)

Kolomna D49

16 NIKITIN, E., IVANTCHENKO, N., SOKOLOV, S. and NIKOLOSKY, N., 'Development of a four-stroke locomotive diesel engine type CHH26/26', *CIMAC* (1975)

17 NIKITIN, E., ULANOVSKY, E., BALAKIN, V. and IVANTCHENKO, N., 'Test results of CHH26/26 diesel engine 6000hp with two-stage turbocharging system', *CIMAC D18* (1979)

MTU 4000

18 'The new diesel engine range 2000 and 4000 from MTU and DDC', *MTZ special edition* (1997)
containing:
Production concepts for new engine series
Development of the 2000 series engines
Development of the 4000 series engines
Simultaneous engineering
Combustion development with common rail injection system
Development of the common rail injection system for the new series 4000 engines
Development of a new maintenance concept based on operational experience gained with series 396 engines
New anti-friction bearing technology for increased service life
Piston development for series 2000 and 4000 engines
New engines—New electronics. The ECU—the "brain" of the electronic fuel injection system.

Paxman VP185

19 RAMSDEN, J., 'The design and development of the 12VP185', *Journal of the IDGTE*, April (1997)

Pielstick PA4-200

20 BRISSON, B., ECOMARD, A. and EYZAT, P., 'A new diesel combustion chamber—the variable-throat chamber', *SAE 730167* (1973)

21 GALLOIS, J., 'Moteurs diesel a pression moyenne effective de 256KG/sq cm', *CIMAC A4* (1977)

22 'More punch for the PA4', *Motor Ship*, July (1978)

23 'Performance of the PA4 with two-stage turbocharging and variable geometry', *IPG*, April/May (1978)

Pielstick PA6B

24 GALLOIS, J., 'Pressure charging systems for highly rated traction engines', *IMechE C89/82* (1982)

25 HERRMANN, R., 'Sequential turbocharging for PA6 engines', *IMechE Seminar on Sprint Rated Diesel Engines*, London (1989)

Ruston RK215

26 JACKSON, P., PUGH, W. and POULSON, R., 'The Ruston RK215 Series engines', *CIMAC D21* (1991)

Further reading

BROWN, W., SCOTT, M. and WARRINER, J., 'Development of bi-metal bearing alloys for diesel engines' *CIMAC D74* (1995)

NIVEN, H., 'The control of emissions from diesel locomotives', *IMechE C478/123/94* (1994)

25 Dual fuel engines

Contents

25.1 What is a dual fuel engine?

In a Paper published in 1949[1], when the development of dual fuel engines was in its infancy, Ralph Boyer of Cooper-Bessemer wrote: 'Over a period of years there have been a number of engine developments referred to as Dual Fuel. Unfortunately this terminology has been used in many cases where more than one fuel is involved, but where the fuels could be gasoline versus fuel oil, gasoline versus kerosene, and many other combinations. The term Dual Fuel is therefore perhaps in need of some explanation.' He went on to describe the further confusion arising from the use of the term 'Gas Diesel'.

The situation today is even more confusing. Engines capable of using more than one fuel, and therefore entitled to be referred to as dual fuel engines, have been produced in all shapes and sizes, from the smallest industrial engine to the largest marine engine. Since this is a diesel engine reference book, we need not concern ourselves with spark ignited Otto cycle engines. However, there is a certain irony in the fact that the most common type of dual fuel engine, and the type with which we are primarily concerned here, actually operates predominantly on the Otto cycle when running on gas.

This "typical" dual fuel engine is a turbocharged medium speed diesel engine or between 200 and 500 mm bore, equipped with an additional natural gas fuel supply system. The gas is mixed with the combustion air by timed injection into the inlet ports, and the mixture is ignited by the injection of a small quantity of diesel fuel. Compression ignition of this "pilot" fuel then provides the ignition energy for combustion of the gas-air mixture. As we shall see, there are many variations on this basic system, and many different fuel gases which may be burned.

In recent times there has been increasing interest in a different type of dual fuel engine, known as the "Gas Diesel". This operates on the true Diesel cycle whether running on gas or fuel oil. The gas is injected directly into the combustion chamber and burns as it is injected, in exactly the same way as the liquid diesel fuel. A pilot injection of diesel fuel is still required to provide ignition. Although this approach overcomes a number of drawbacks associated with the conventional dual fuel engine, it requires the gas to be injected at high pressure (up to 350 bar), which brings problems of its own.

25.2 Combustion in dual fuel engines

In the conventional dual fuel engine, combustion is initiated by the compression ignition (auto-ignition) of the liquid fuel pilot. This then serves as the ignition source for the pre-mixed charge of air and gas.

Efficient pre-mixed combustion depends on turbulent flame propagation which in turn is dependent on the equivalence ratio of the gas air mixture. If the mixture is too lean flame propagation will be erratic, leading to misfire. Fuel unburned in the cylinder may then burn in the exhaust manifold. This provides more energy to the turbocharger turbine, increasing the air boost pressure and making the mixture leaner still. In these circumstances the engine can rapidly go out of control.

Conversely, if the mixture is too rich, detonation (or "knock") will occur. It has been shown[2] that, in dual fuel engines, detonation is due to the auto ignition of the premixed air and fuel in the neighbourhood of the pilot fuel sprays. During compression, pre-ignition reactions take place in the mixture. Then, once the pilot fuel ignites both pilot fuel and gas burn very rapidly and simultaneously. This causes excessive rates of pressure rise, high temperature and increased heat transfer often leading to engine damage.

The misfire and detonation limits are strongly affected by the properties of the gas, engine inlet manifold temperature (illustrated in *Figure 25.1*), and compression ratio. The relative susceptibility to detonation of fuel gases is often measured by methane number (see Section 25.3).

Engine control systems need to protect against misfire and detonation and ensure operation at maximum efficiency. In recent times the requirement to do this without exceeding NOx limits has led to increasingly complex and sophisticated engine management systems, such as that shown in *Figure 25.2*.

The requirement to operate within a limited range of equivalence ratio means that, as with spark ignited engines, some means of controlling air flow into the engine is needed. Methods of achieving this are discussed in Section 25.5. To minimize the risk of misfire the pilot quantity as a percentage of the total fuel may be increased. In practice this usually means that the pilot fuel quantity is kept constant throughout the load range, and light load running (below 25% of full load) is done on diesel fuel.

As already noted, the Gas Diesel engine operates on an entirely different principle. The gas is not premixed with the air so control of equivalence ratio is not necessary. Furthermore, it has been shown[3] that the combustion of the gas takes place by the diffusion process, and that the thermal efficiency, cylinder pressure and temperatures are all similar to normal diesel operation.

The fundamentals of combustion processes in dual fuel engines have been investigated in great depth by Karim and his colleagues at the University of Calgary[2,4,5,6].

25.3 Gas properties and their effects

A large variety of fuels are currently being used with medium speed gas engines and these are very dependent on availability and the location of the installation. Details of the major gas types are listed in *Table 25.1* and specific properties and characteristics are shown in *Table 25.2*.

The most important parameters which affect engine operation are as follows:

25.3.1 Heat value of a stoichiometric mixture volume

This will determine the amount of mixture required to achieve a given output. This is of particular importance for naturally aspirated engines as it will directly affect full load output. On a turbocharged engine this is not so critical and will only affect the turbocharger matching. This level remains similar for all hydrocarbon fuels but reduces if significant amounts of inert gas are present, as in digester gas or gasifier gas.

25.3.2 Net heating value (kJ/m^3)

This parameter indicates the volume of gas required for a given engine output. This will affect the design of the gas admission system. It is a major consideration when the engine is to be fuelled by a variety of gases or when the quality of the in-service gas supply varies significantly.

25.3.3 Anti-detonation properties

This is a measure of the auto-ignition properties of the gas ahead of the flame. It is dependent on the chemical constitution of the end gas and the local temperatures and pressures.

The chemical constitution of the end gas will be a function of the air–gas ratio and the gas constitution. As a general rule long-chain gases tend to have poor auto-ignition characteristics. This is reflected by the heavy hydrocarbon gases having poor detonation properties, while CO and CH_4 tend to have good anti-detonation properties. Various methods exist for assessing the

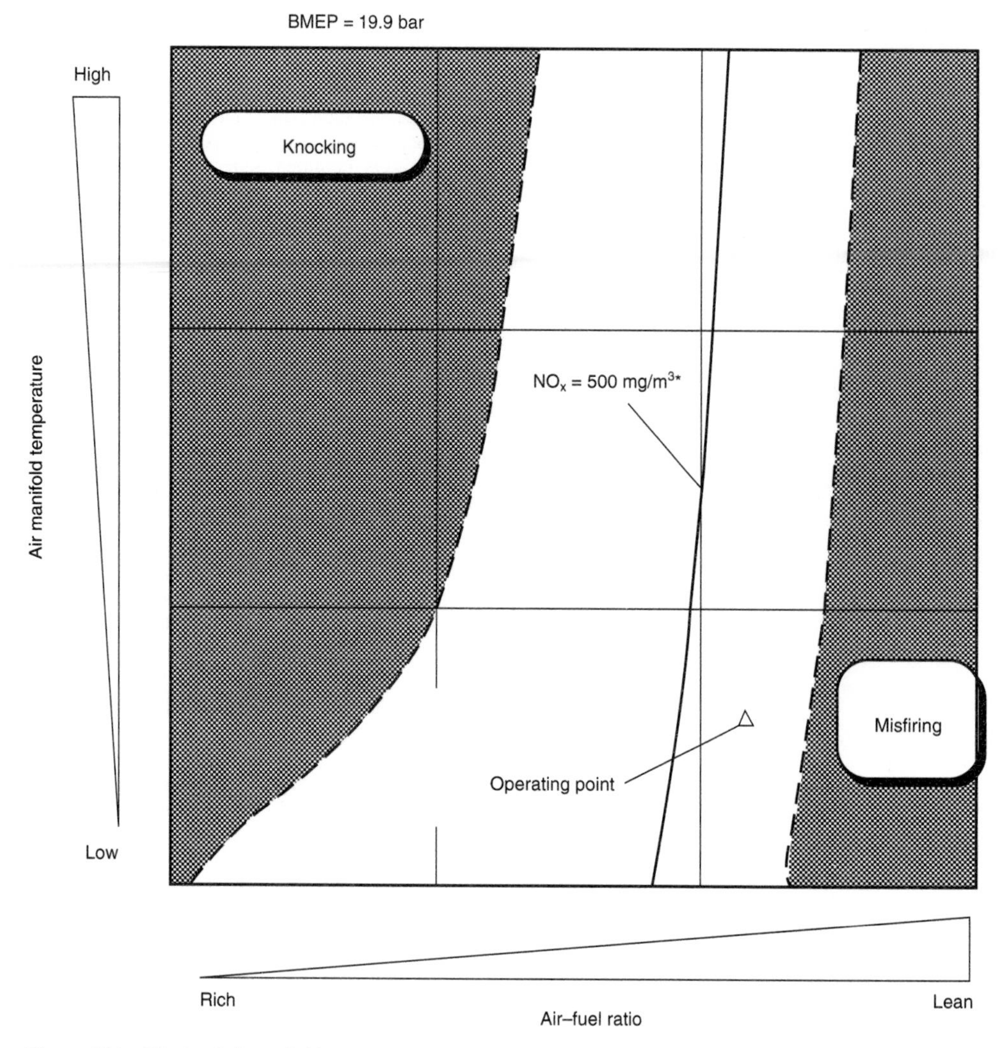

Figure 25.1 Effects of air manifold temperature and air–fuel ratio on the operating range between misfire and knock (MAN B&W)

*Referenced at 5% O_2

detonation tendency of different gases, the most common being the Methane Number rating system, which has some similarity to the octane rating system used for gasoline fuels. Higher numbers indicate greater resistance to detonation, referenced to pure methane which is given the value 100. An extensive test programme to investigate the effects of gas composition (on spark ignited gas engines) was jointly undertaken by Ricardo Consulting Engineers and Tokyo Gas Company in 1994[7]. This confirmed the validity of Methane Number as a general measure of the knock resistance of a gas fuel, but also showed that there can be significant differences in the actual BMEP achievable with gas mixes of similar Methane Number.

The local temperatures and pressures in the combustion chamber are dependent on a complex relationship between engine variables, such as engine speed, bore size, compression ratio, combustion chamber shape, etc. and combustion variables such as air temperature, start of combustion, flame speed, fuel heat input, etc.

25.3.4 Pre-ignition tendency

This is the tendency of the gas to ignite spontaneously owing to local high temperatures of 'hot spots'. These normally occur in the vicinity of the exhaust valve and can be minimized with correct engine operation and adequate combustion chamber cooling. As a general rule, gases with low resistance to detonation will also be susceptible to pre-ignition. Hydrogen in particular has very poor pre-ignition characteristics. Because of the likelihood of local hot spots caused by combustion chamber deposits, low ash content lubricants are normally recommended.

25.3.5 Flame speed

Gases which burn with very high flame speeds, such as hydrogen, promote very rapid rates of burning within the chamber. The laminar flame speed of hydrogen is typically seven times that of methane or natural gas[8], while digester gas burns at half the speed of natural gas. If any hot spots occur within the combustion chamber, backfiring through the intake valve can occur during the induction stroke. Gases with very slow burning speeds, such as those with a high level of inert gases present, can delay buming to such a degree that the combustion is incomplete when the exhaust valve opens. This can cause extensive heating of the exhaust valve which may promote backfiring in the exhaust, or pre-ignition. Combustion flame speed is also very dependent

Table 25.1 Details of gaseous fuels used in reciprocating engines

Gas type	*Chemical composition*	*Gas net heating value kJ/m³ (15°C, 1 atm.)*	*Comments*
Methane	CH_4	33 900	Pure methane is often used as a reference gas for engine operation. Good detonation characteristics.
Propane	C_3H_8	86 400	Poor detonation characteristics compared to methane.
HD-5	≈ 90% C_3H_8 + other paraffins	≈92 000	Stored as a liquid (under pressure)
Commercial propane (LPG)	(70–90%) C_3H_8 + C_3H_6	≈86 000	Can contain C_3H_6 and C_4H_8 if produced from chemical plant. Stored as a liquid.
Butane	C_4H_{10}	112 400	Poor detonation characteristics.
Commercial butane (LPG)	(70–90%) C_4H_{10} + C_4H_8 + C_3H_8 + C_3H_6	≈110 000	Stored as a liquid
Natural gas (Pipeline/CNG)	CH_4 (90%) + heavier hydrocarbons + CO + N + CO_2	≈34 000–42 000	Composition can vary depending on location but predominantly methane. Pipeline gas is generally cheaper than refined liquid fuels.
Coal gas/town gas	H_2(30–55%) + CH_4(20–50%) + CO(5–15%) + CO_2(4%)	≈16 000–19 000 (30 000 if CH_4 content is high)	Traditional method of producing gas from coal: low conversion efficiency (≈25%) and coal with a high volatile content is required.
Wellhead gas/field gas	CH_4(60–95%) + heavier hydrocarbons + CO + N + CO_2 + H_2S	≈ 30 000–45 000	Can contain small amounts of H_2S (≈1%) These must be removed.
Digester gas (sludge/landfill/ sewage/bio gas)	CH_4(≈60–70%) CO_2(30–40%) H_2S (< 1%)	≈20 000–23 000	Anaerobic digestion of sewage waste. Conversion efficiencies typically 50%. Good detonation tendency due to CO_2. Low heating value requires an increase in gas supply to maintain output relative to methane. The heating value of a stoichiometric mixture is lower than for natural gas which will limit naturally aspirated output to typically 85%. Gas available at pressures just above atmospheric and so is generally used with naturally aspirated engines.
Gasifier gas	N_2 (50–60%) + CO (10–25%) H_2 (5–15%) + CH_4 + CO_2	≈3000–6000 (varies depending on feedstock)	Produced from carbonaceous (wood. coal) or vegetation waste. Gasification method is generally used. Good detonation characteristics due to large amounts of inert gases present. The stoichiometric mixture heating value is typically 75–85% compared to methane which causes a subsequent reduction in achievable naturally aspirated full load output. Gases available at atmospheric pressure or just above. Gas conversion efficiencies are typically 50–60%.
Blast furnace gas (see Gasifier gas)	N_2(50–60%) + CO(20–30%) + CO_2(8–15%) + H_2(2–4%)	≈3000–4000	Produced by pyrolysis of ligneous materials.
Producer gas (see Gasifier gas)	N_2(50–55%) + CO (30%) + H_2(12%) + CO_2(4%)	≈ 3200–6 500	Air is drawn past incandescent fuels such as coke charcoal, wood, etc. High conversion efficiencies achievable (90% +).
Refinery gas	C_3H_8(18–34%) + C_4H_{10}(45%) +CH_4(4–12%) + C_2H_6(10–20%) +H_2 (<60%) + CO_2	≈32 000–56 000	Gas type very dependent on location.

on the air–fuel ratio (lean mixtures reduce flame speed) and on turbulence.

As can be seen from *Table 25.1* there are many different gaseous fuels available. These can be split into two major types. The first contains gases which are either naturally occurring or are readily available. These tend to be predominantly hydrocarbon gases with high calorific values. Natural gas is the most commonly used fuel gas, and has methane as the major constituent. Owing to the presence of certain heavier hydrocarbon constituents, the detonation resistance of some gases can be low. Even natural gas can be affected in this way. It normally has a methane number of between 80 and 100, but it can be as low as 65. The effect of gas composition on (detonation limited) power output has been investigated by Gillispie and Jensen[9].

Gases of the second type are produced from various carbonaceous/biomass materials[10]. The various methods of production include thermochemical methods (dry) such as direct combustion, gasification and pyrolysis. and biochemical methods (wet) such as anaerobic digestion of bacterial fermentation. All these methods require specialized plant to produce the gas, so the cost effectiveness of such plant must be considered against the conversion efficiency and the plant size. The type of fuel to be used also has a significant effect on prime mover selection. These gases contain varying proportions of inert constituents (CO_2 and N_2) which tend to reduce the gas heating value, improve the detonation tendency, reduce the combustion flame speed and reduce the stoichiometric mixture heating value.

25.4 Combustion systems

There are three basic types of combustion system used in dual

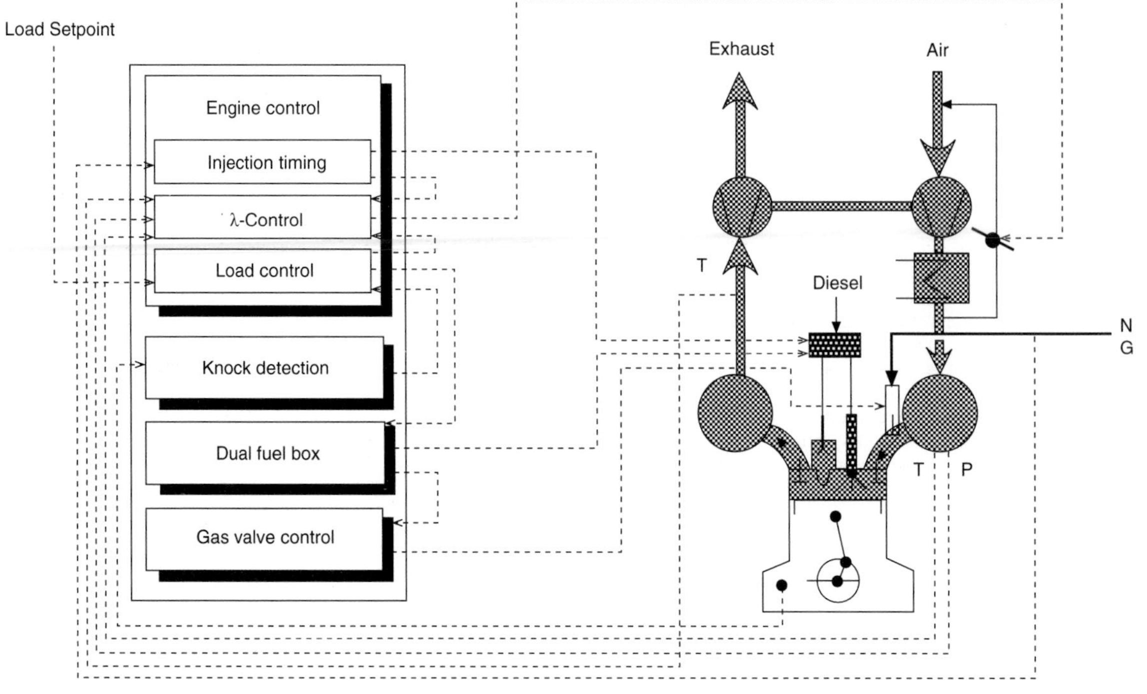

Figure 25.2 Engine management system (MAN B&W)

Table 25.2 Gaseous fuels—properties and characteristics

Gas type	*Chemical composition*	*Relative molecular mass*	*P** *(kg/m³)*	*LCV (kJ/kg)*	*LHV** *(kJ/m³)*	*Stoichiometric air–fuel (kg/kg)*	*Stoichiometric air–fuel (m³/m³)*	*Heat content of stoichio-metric mixture** *(MJ/m³)*	*Heat energy required to burn 1kg ideal air (kJ)*	*Methane Number (MN)*	*Minimum auto-ignition temp. in air (Ref. 2) (k)*
Methane	CH_4	16.04	0.6785	50 040	33 950	17.25	9.52	3.227	2900	100	813
Propane	C_3H_8	44.1	1.865	46 338	86 420	15.65	23.81	3.483	2960	35	723
Butane	C_4H_{10}	58.12	2.458	45 732	112 410	15.43	30.95	3.518	2964	10.5	678
Propene	C_3H_6	42.08	1.780	45 758	81 450	14.82	21.43	3.631	3087	20	733
Hydrogen	H_2	2.0	0.0853	119 810	10 220	34.18	2.38	3.023	3505	0	–
Carbon monoxide	CO	28.0	1.185	10 101	11 970	2.46	2.38	3.54	4106	73	–
Nat. gas	≈90% CH_4 + others	= 17.5	≈0.74	≈46 000	≈ 34 000	≈17.0	≈10.3	≈3.01	≈2705	≈90	–
Digester gas	≈65% CH_4 ≈35% CO_2	= 24.8	≈1.05	≈20 000	≈21 100	≈7.7	≈ 6.6	≈ 2.78	≈ 2600	≈ 130	–
Gasifier gas	≈55% N_2 ≈25% CO ≈10% H_2 + others	≈26.0	≈1.10	≈ 4700	≈ 5200	≈ 1.23	≈ 1.1	≈ 2.47	≈ 3821	≈ 130	–
Air	≈79% N_2 ≈21% O_2	28.96	1.225	–	–	–	–	–	–	–	–

*1 atmosphere 15°C.

fuel engines, each of which has advantages and disadvantages compared to the others. The performance of the three types in terms of BMEP and NOx emissions is compared with equivalent diesel and spark ignited (SI) gas engines in *Figures 25.3* and *25.4*. A description of the main characteristics of each type of combustion system, and some of their variants, is given in the following paragraphs.

25.4.1 The 'conventional' dual fuel engine

Early dual fuel engines were naturally aspirated and gas was mixed with the intake air by means of a carburettor. Today's turbocharged dual fuel engine, produced by Ruston, Allen and Niigata among others, uses timed gas admission into the inlet ports. Timed admission is used to prevent gas being lost in the scavenge air. The pressure of the gas supply to the injection valves needs to be a little above the boost air pressure, hence typical gas supply pressure is between 3 and 5 bar. The engine retains its standard diesel injection system, with perhaps some modification to the injection pump and nozzle to improve their performance at low fuelling rates. To prevent overheating of the nozzle due to prolonged operation with small quantities of fuel, cooled nozzles may be fitted. Depending on the quality of the gas fuel to be used, compression ratio may be lowered to avoid detonation, but must still be high enough to ensure reliable auto-ignition of the diesel pilot: 11 or 12 to 1 is normally considered to be the minimum. Maximum power output is typically 10-20% lower than the equivalent diesel engine. This is detonation limited and depends on the gas methane number.

For light load operation and start up, the engine is generally operated in the diesel mode only and, when at operating temperature, the diesel fuel is switched to the pilot fuel quantity (equal to or less than the idle fuelling level), the balance of the fuel requirement being provided by the gas. Any reduction in load at rated speed is then achieved by reducing the gas supply. A control system sometimes based on exhaust temperature is then used to regulate the air supply to maintain the required equivalence ratio.

The gas and air mixture is ignited by a finely atomized diesel fuel spray which is typically between 5% and 8% of the full load fuelling. By careful attention to the governing control system, the engine can, therefore, be operated at full load either as a conventional diesel engine or as a dual fuel engine. Furthermore the engine can automatically revert to full diesel operation if the gas supply is lost. This is beneficial if the full supply of gas to the engine cannot be guaranteed, as is often the case in dual fuel installations.

As these engines are normally turbocharged, equivalence ratios of 0.6 can be maintained to avoid excessive loss in output due to detonation. The high ignition energy from the pilot injection minimizes cycle-to-cycle variation in combustion, and the effects of changes in gas quality. However, any large change in air–fuel ratio may promote either detonation (richer) or misfire (leaner). Increasing the pilot fuel quantity may enable some extension to the lean misfire limit to be achieved owing to the higher ignition energy available and the reduced combustion period, but these effects are attenuated with increases in pilot fuel quantities above 15%. As pilot fuel quantities increase, NOx emissions will also increase owing to the greater amount of high temperature ignition sources present.

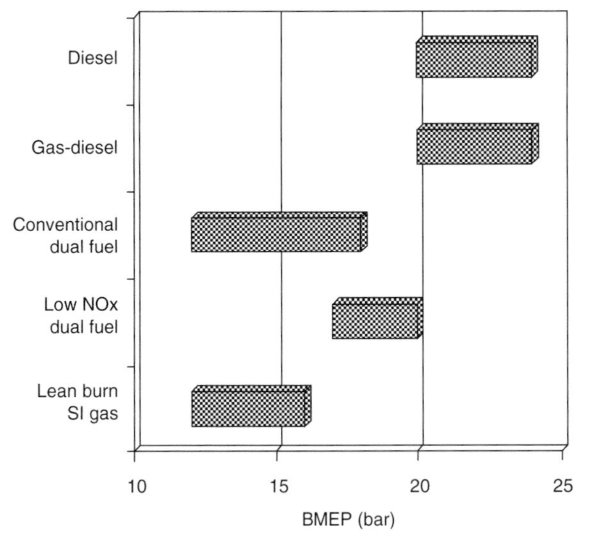

Figure 25.3 Typical BMEP range for various engine types

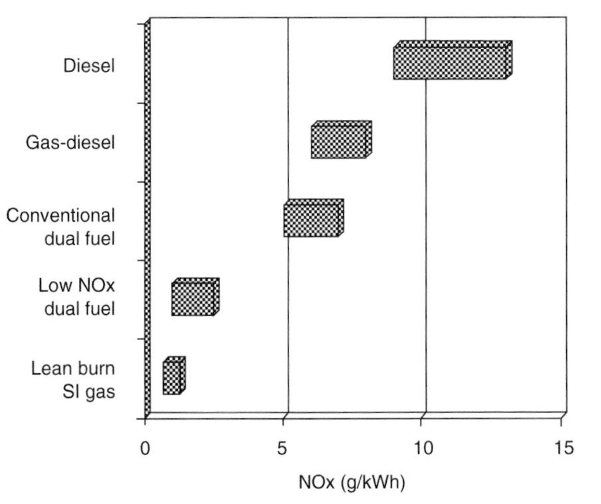

Figure 25.4 Typical NOx range for various engine types

25.4.2 The 'low NOx' dual fuel engine

Lean burn gas engines have inherently lower NOx emissions than diesel engines, by virtue of lower peak combustion temperatures. However, in the case of the dual fuel engine, this is compromised to some extent by the pilot injection requirement. As a result, the Nox emissions of dual fuel engines lie somewhere between those of diesel engines and lean burn spark ignited gas engines (see *Figure 25.4*). Furthermore, very low values of equivalence ratio (of the order of 0.4) are required to achieve very low NOx levels. Lower equivalence ratio requires higher ignition energy, but in the dual fuel engine, this means more pilot fuel which itself is a significant contributor to overall NOx emissions. Several manufacturers, notably Cooper Energy Services[11], Coltec/Fairbanks Morse[12] and MAN B&W[13], have developed combustion systems which reduce the quantity of pilot fuel required, and hence the overall NOx emissions.

As described earlier, engines which use the normal fuel injection system to deliver the pilot fuel for gas ignition require at least 5% of the full load fuelling to achieve reliable combustion. This can be reduced to about 2% if a separate injection pump and nozzle, optimized for the pilot fuel quantity, are used. Higher injection pressures and smaller nozzle holes give better penetration and atomization, making more effective use of the fuel to provide a concentrated high energy ignition source.

Even better use of the small quantity of pilot fuel can be made by injecting the fuel into a prechamber. This amplifies and intensifies the available ignition energy, allowing pilot quantity to be reduced to less than 1%. Moreover, this can be achieved together with equivalence ratios of 0.4 or less. However, a sophisticated engine management system is required to tread the fine line between detonation and misfire, while maintaining low NOx, as illustrated in *Figure 25.1*.

In the 1980s in the United States, increasing concern over engine emissions led gas engine manufacturers such as Cooper Bessemer and Fairbanks Morse to seek ways to overcome the relatively high NOx emissions from dual fuel engines. In the early 1990s, both companies published work on prechamber engines[12,14]. Subsequently, Fairbanks Morse have successfully applied this approach to their opposed piston, uniflow scavenged two-stroke engine. The general arrangement of this engine and its intake and exhaust systems is shown in *Figure 25.5*. Gas admission to a two-stroke engine cannot be made with the intake air, because a significant amount of this air escapes from the cylinder during exhaust scavenging. Gas must therefore be injected into the cylinder after the exhaust valves have closed.

The opposed piston arrangement means that the fuel injectors are located at the periphery of the combustion chamber. In the standard dual fuel engine, there are two liquid fuel injection nozzles and one gas injector. These are retained in the prechamber version, and a separate prechamber with pilot injector is added, as shown in *Figure 25.6*. Development of the prechamber system is described by Blythe[15]. The energy efficiency of the standard dual fuel engine is retained, with NOx levels comparable to the lean burn spark ignited engine (see *Figure 25.7*).

In co-operation with Fairbanks-Morse, MAN B&W have developed a prechamber dual fuel version of their 320 mm bore 32/40 engine. Development work focused on meeting the stringent German TA Luft limits for NOx emissions from gas engines (0.5 g/m^3 @ 5% O_2, equivalent to about 1.5 g/kWh) without exhaust gas after treatment. The main injector for diesel fuel is retained to permit operation on diesel fuel if required. A separate prechamber and pilot injector is fitted into the cylinder head near the outside of the combustion chamber, as shown in *Figure 25.8*. Fuel to the pilot injector is provided by a separate injection pump mounted alongside the main pump and operated by the same camshaft (*Figure 25.9*). An electronically controlled, hydraulically actuated gas valve is fitted to each inlet port as shown in Figure *25.10*.

The success of the prechamber approach is crucially dependent on getting the prechamber design right. The geometry of the prechamber profoundly affects the combustion process. The jet, or jets, of hot gas must have high energy and be spread over

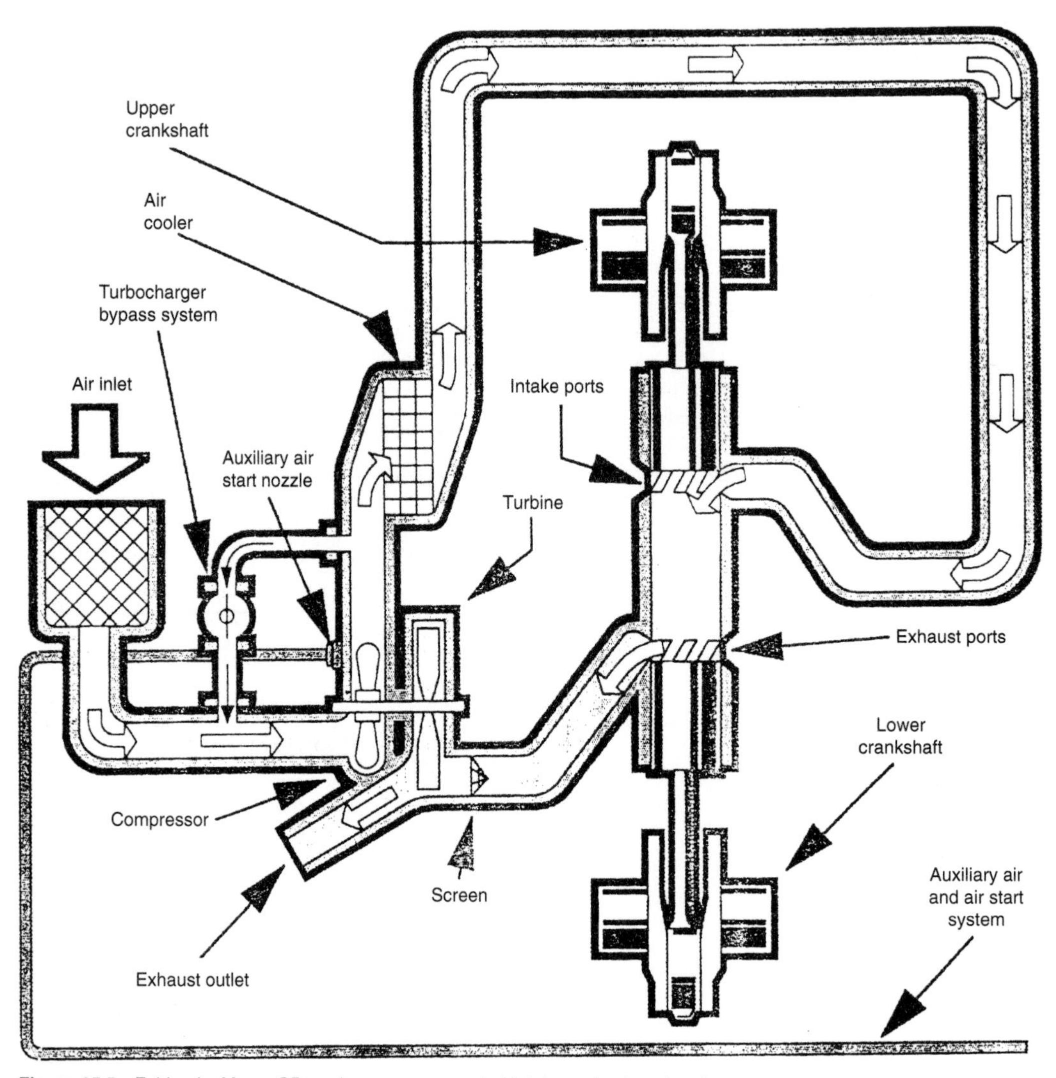

Figure 25.5 Fairbanks Morse OP engine—arrangement of intake and exhaust systems

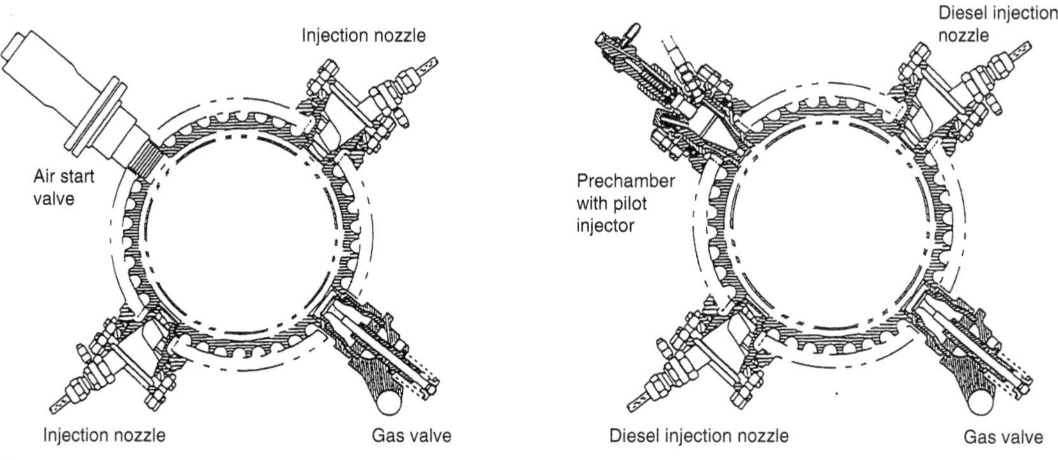

Figure 25.6 Fairbanks Morse OP engine—arrangement of fuel injectors

the combustion chamber space to ensure reliable and consistent ignition of a lean air/gas mixture of low flammability. Also the prechamber is subject to very high thermal loads, requiring careful optimization of geometry, materials and cooling to ensure adequate life.

25.4.3 The 'gas diesel' engine

A fundamentally different approach to burning gas in a diesel engine is taken by the 'gas diesel' engine. The concept is, quite simply, to replace the direct injection of liquid diesel fuel into the combustion chamber with the direct injection of gas fuel. In practice, of course, things are a little more complex than this.

Pilot injection of diesel fuel is retained to provide an ignition source. The required pilot quantity is 3 to 5% of full load fuelling, and NOx emissions are similar to the conventional dual fuel engine. There is no equivalent to the 'low NOx' engine. Because combustion is diffusive rather than pre-mixed, the problems of detonation and part load air–fuel ratio control are largely eliminated. The part load efficiency and full load BMEP of the diesel engine are retained and operation is largely unaffected by variations in gas quality.

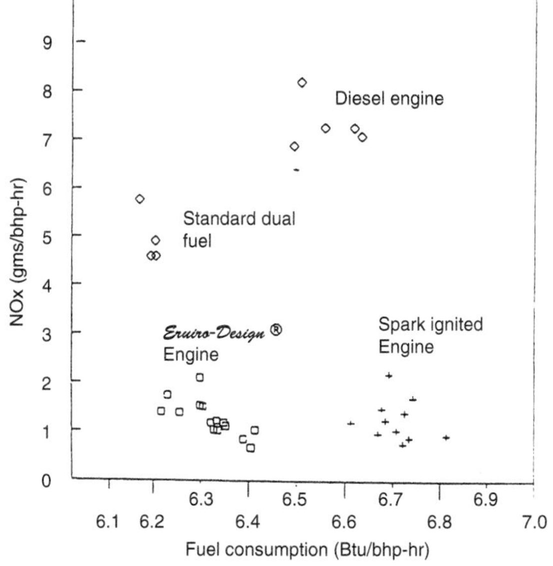

Figure 25.7 Fairbanks Morse OP engine—NOx emissions for different versions

However, there is a price to pay for these advantages. Direct injection of gas requires a high pressure gas supply (250 to 350 bar). The capital cost of the necessary gas compressor is high, and the power required to drive it can approach 5% of the engine output. Safety systems associated with the high pressure gas supply also add to the cost. The higher output and efficiency of the engine offset these costs to some extent, but the result is that the gas diesel is best suited to applications where high pressure gas is cheaply available, and diesel fuel is costly.

The main exponent of gas diesel technology for medium speed engines in recent times has been Wartsila Diesel of Finland. Development of a gas diesel variant of their 320 mm bore engine began in 1987. A paper presented in 1993[16] described the development work and field trials at three installations. The motivation for this development was a concept for power generation on floating oil production vessels called 'fuel off the well', the idea being that these vessels could produce their own energy from whatever gas or crude oil was available from the well.

A common central injector is used with twin nozzles (*Figure 25.11*). The gas injection is controlled electronically with hydraulic oil providing the actuating force for the gas needle. To prevent gas leakage through the needle clearance, sealing oil with higher pressure than the gas is applied in the clearance between the needle and the nozzle body.

The diesel fuel nozzle is designed to provide both good atomization with small pilot fuel quantities, and the same characteristics in full load operation on diesel fuel as a standard injector nozzle.

An important safety feature is an automatic quick closing shut-off valve in the high pressure gas line to each cylinder. If the gas injector needle does not close properly at the end of the injection period, the resulting pressure drop will immediately close the shut-off valve. The high pressure gas pipes are double walled. The space between the walls is continuously vented and equipped with gas sensors to detect any leakage. These and other safety systems are integrated into a comprehensive electronic control system. This automatically compensates for changes in fuel quality, and provides instantaneous switching from gas to diesel fuel if the gas supply is interrupted. Engines of 320 mm and 460 mm bore cover the power range 2 to 16 MW.

The gas diesel concept has also been applied to low speed two-stroke engines, notably by MAN B&W and Mitsui. Following the joint development by these two companies of a system for medium speed four stroke engines, the technology has been adapted to the MAN B&W MC range of engines[17,18]. These

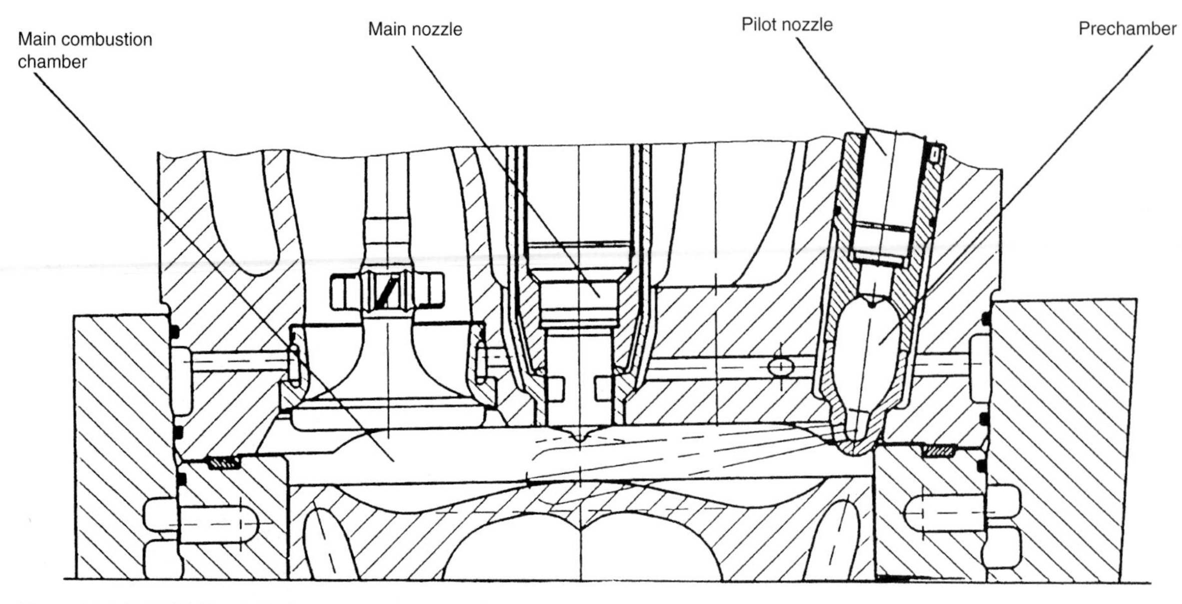

Figure 25.8 MAN B&W 32/40 DG combustion chamber (MAN B&W)

operate at speeds from 75 to 250 rev/min, with power output ranging from less than 10 MW to nearly 70 MW. A cross-section of one of these engines is shown in *Figure 25.12*, indicating the additional or modified parts for gas diesel operation. Separate injectors are used for diesel fuel and gas. The diesel fuel pump and injector may be either the standard items, where full load operation on diesel fuel alone is required, or smaller versions optimized for pilot fuel delivery, With the latter, the quantity of pilot fuel required to achieve stable gas combustion (and hence NOx emission) is reduced to about 8%. The pilot fuel may be either distillate diesel fuel or heavy fuel oil (HFO).

The ability of the gas diesel to burn gas that has poor detonation characteristics has been exploited for shuttle tanker propulsion. Shuttle tankers are widely used to serve offshore oilfields from which pipeline connections to the shore are not feasible. The shuttle tankers load their crude oil cargo either from storage facilities at the oilfield or directly from the production platforms and loading buoys at the oilfield. The crude oil cargo is then transported to oil refineries or terminals ashore close to the oilfield. During the handling of the oil, and in particular during loading, large quantities of the light components of the oil evaporate. These oil vapour are known as Volatile Organic Compounds (VOC). The VOC emission during the voyage is not normally significant due to the short sailing distance.

The high content of propane and the higher hydrocarbons in such gases lead to methane numbers which may be very low (close to zero) and which vary from one oilfield to another, making it unrealistic to utilize this in an engine employing the pre-mixed combustion system. Thus, the only technology available for using VOC as fuel is high pressure injection of the VOC directly into the cylinders.

In a joint venture with Statoil, MAN B&W have developed and adapted the MC gas injection systems for VOC use[19]. The normal (HFO) fuel system is retained for use when the VOC gas is not available.

The VOC utilization system is shown schematically in *Figure 25.13*. Because the tendency to release VOC is greatest during the loading of the crude oil, the VOC has to be converted to a form that can easily be collected and stored until the engine can use it.

The VOC and inert gases emitted from the crude oil tank are therefore conducted through gas pipes from the crude oil tanks to the VOC treatment and collection facility consisting of a cleaning system and a gas condensation system. The condensed hydrocarbons are separated from the lighter hydrocarbons and inert gas, and transported to a storage tank. Liquid VOC is taken from the tank, pressurized and supplied to the engine.

25.4.4 Other combustion systems

The General Electric 'H-Process'[20] is a hybrid of the dual fuel and gas diesel approaches, in an attempt to combine the benefits of both. The combustion chamber layout is shown in *Figure 25.14*. At light loads it operates as a gas diesel, with high pressure direct injection and diffusion burning, thereby avoiding the poor efficiency and other problems associated with operating dual fuel engines at part load. At high loads, however, all the gas is injected before TDC so that homogeneous charge combustion is obtained with associated low levels of NOx emissions. Because the gas is injected late in the cycle, pre-mixing time is limited and detonation problems can be avoided without the need to reduce compression ratio. Hence the same full load power and efficiency of the diesel engine can be achieved on gas. Full load NOx is about half that of the equivalent diesel engine.

Furthermore, because the gas is injected when cylinder pressure is lower than is the case for the gas diesel, the pressure of the gas supply can be correspondingly lower, reducing the amount of energy required to compress the gas and hence improving the overall system efficiency.

The EMD two-stroke diesel engine is widely used as a locomotive power unit in North America and elsewhere. In the interests of both utilizing alternative fuels and reducing NOx emissions, a number of dual fuel combustion systems have been investigated for this engine.

As described earlier, gas admission to a two-stroke engine must be made into the cylinder after the exhaust valves have closed. If the gas is injected as early as possible, this can be done at low pressure (about 7 bar) and conventional, homogeneous charge combustion results. Alternatively, the gas diesel approach of high pressure injection and diffusion combustion can be used. A number of variations on these themes were investigated in a research programme called "GasRail USA" in 1993[21].

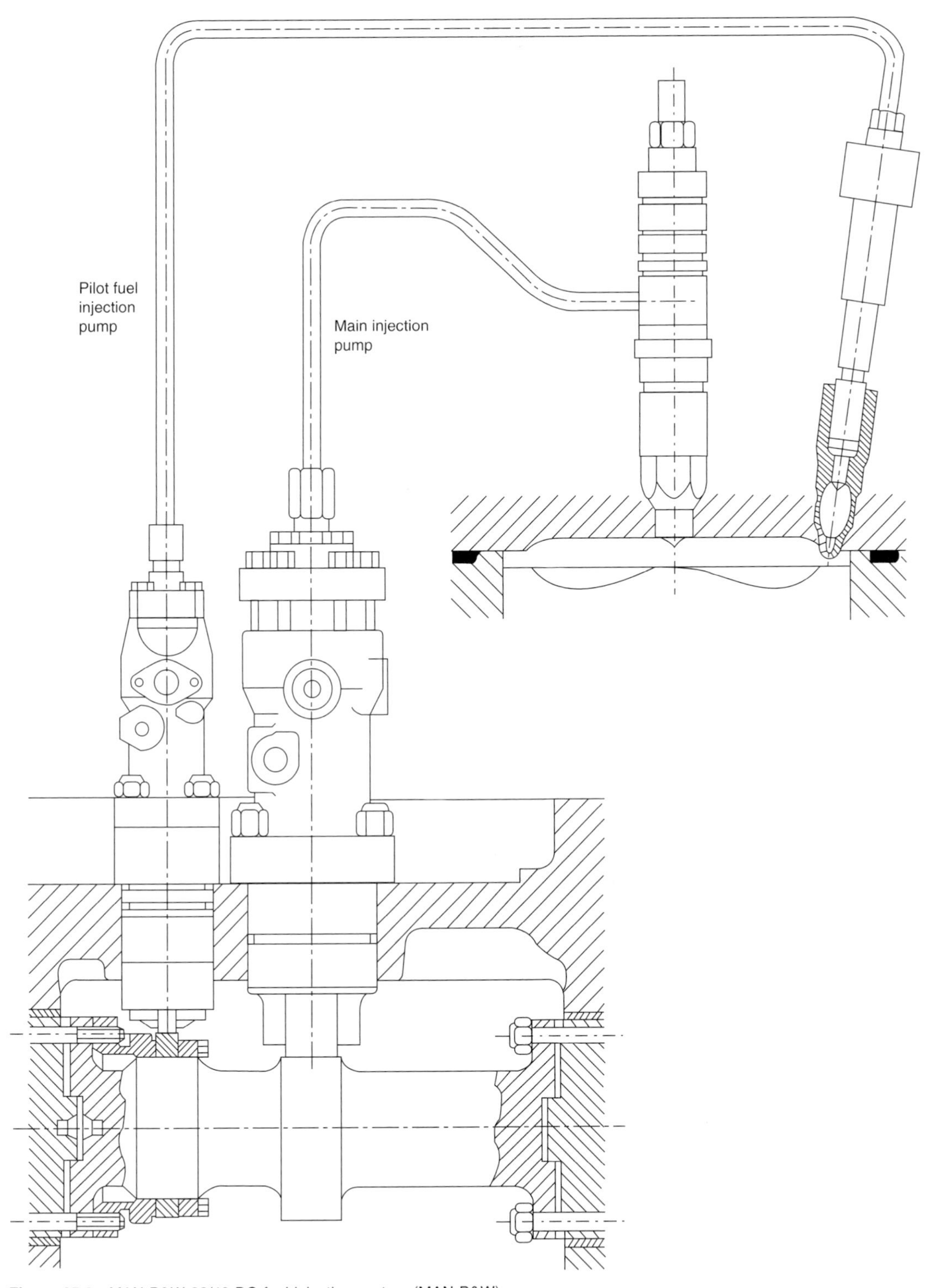

Figure 25.9 MAN B&W 32/40 DG fuel injection system (MAN B&W)

Subsequently, locomotive trials have been conducted using both the dual fuel and the gas diesel approach. The gas diesel was chosen for a trial with four locomotives[22] using liquefied natural gas (LNG). NOx emissions of below 5 g/kWh were reported. A beneficial side effect of using LNG is that the gas can be pumped to the required pressure whilst in the liquid state, then heated to turn it into a gas.

EMD two stroke engines used in stationary applications have also been converted to dual fuel operation using low pressure gas[23,24].

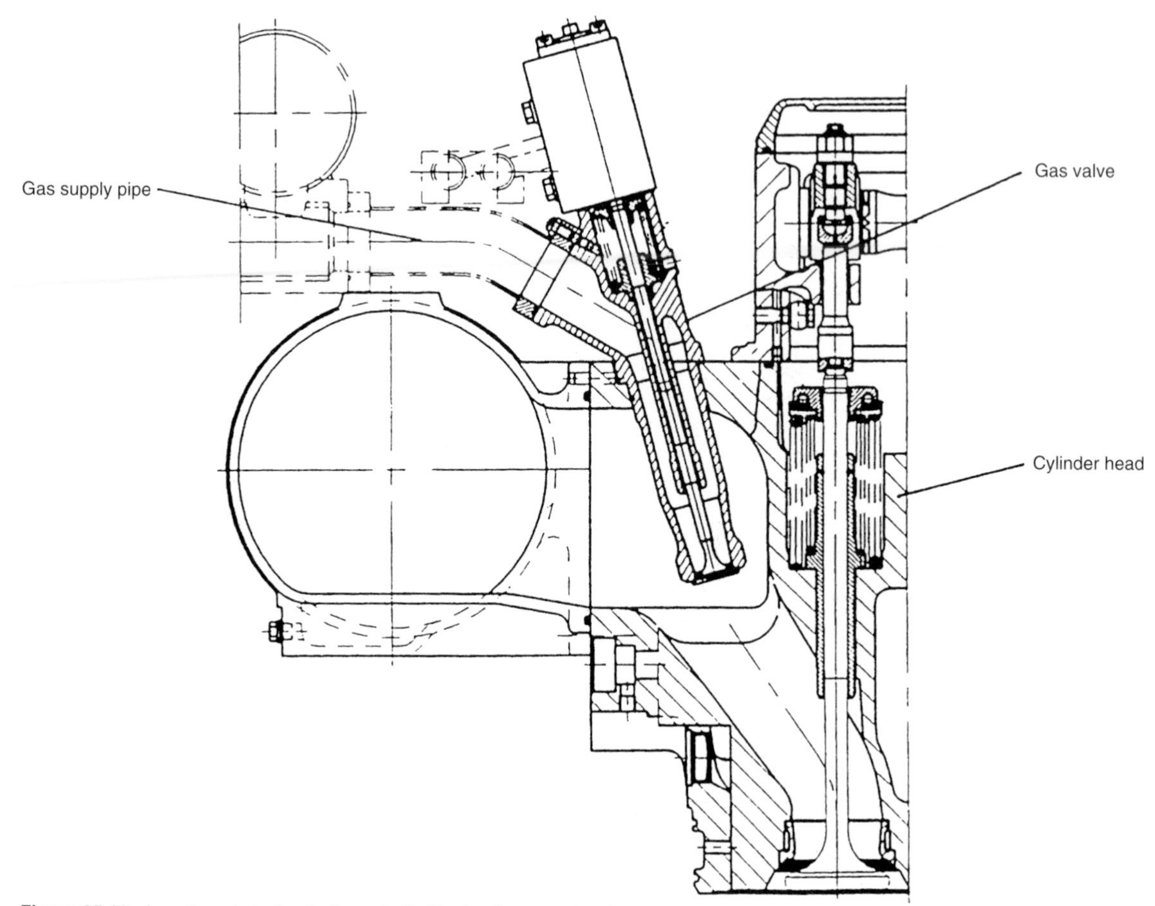

Figure 25.10 Location of electronically controlled hydraulic gas valve (MAN B&W)

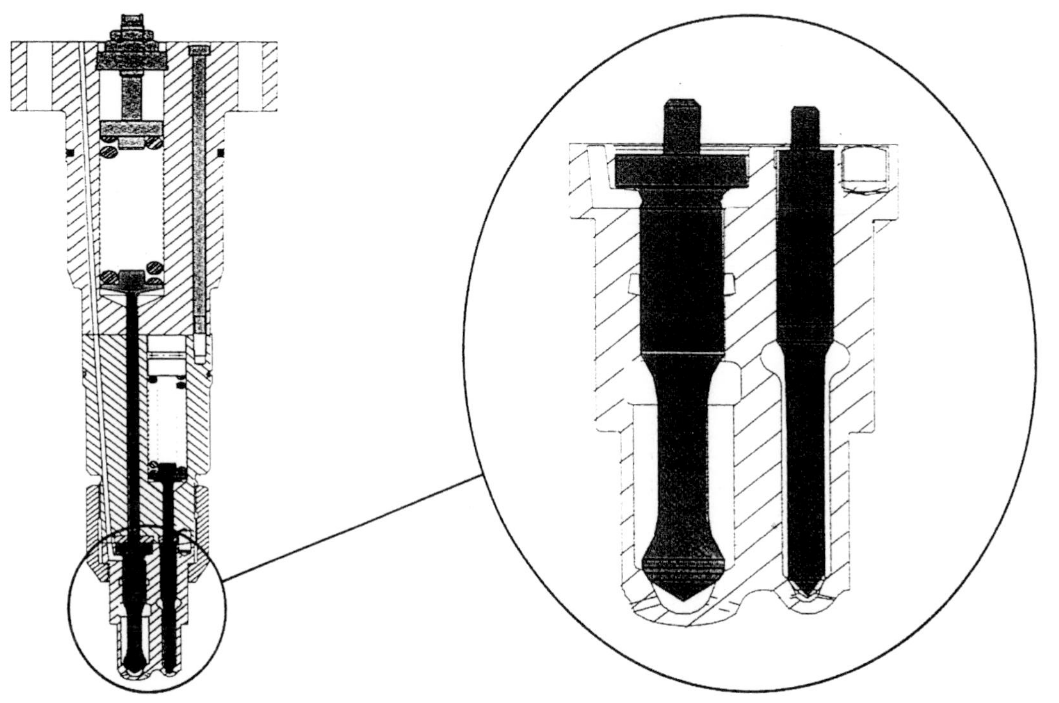

Figure 25.11 Twin needle combined gas and liquid fuel injector (Wartsila)

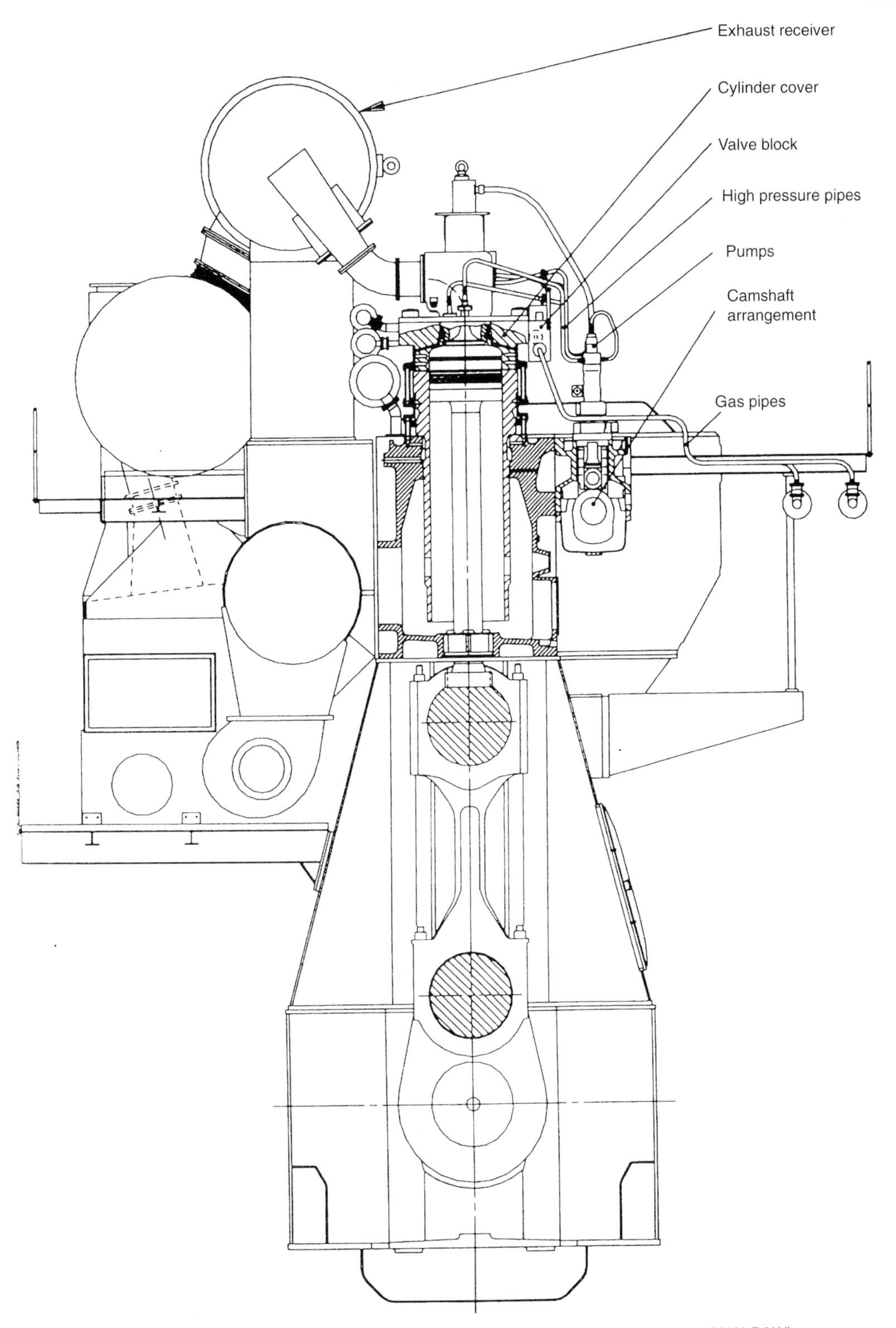

Figure 25.12 MAN B&W MC-GI dual fuel engine, showing additional or modified components (MAN B&W)

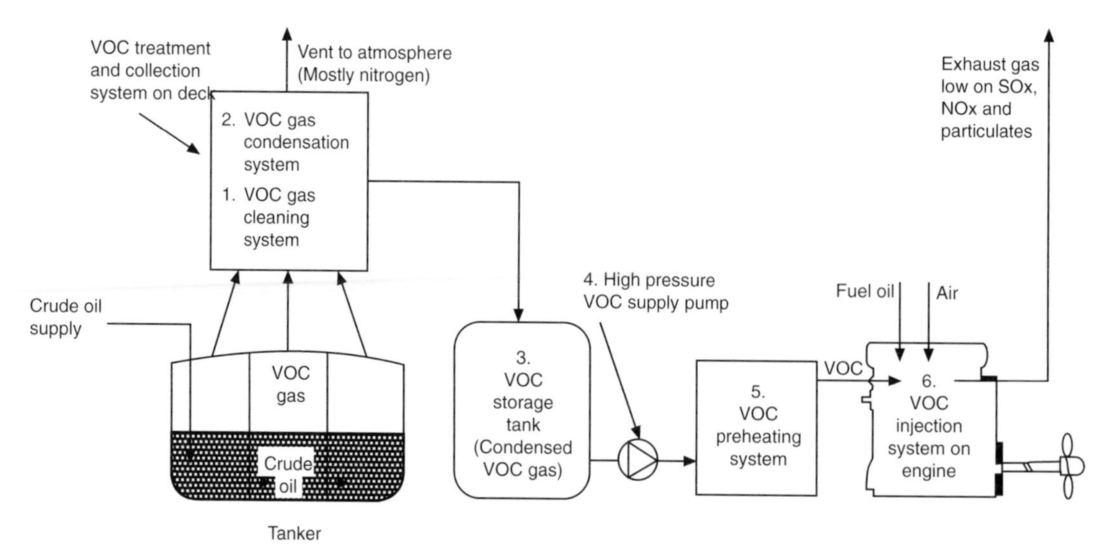

Figure 25.13 Schematic diagram of VOC utilization system (MAN B&W)

25.5 Air–fuel ratio control systems

With the exception of the gas diesel, dual fuel engines utilize pre-mixed combustion of the gas and hence require some form of air flow control to maintain air–fuel ratio below the misfire limit. There are several methods of achieving this.

25.5.1 Intake throttle

Such a system is necessary for engine operation over a wide range of constant air fuel ratios. As the throttle will act to increase the compressor discharge pressure and reduce the air flow, it will move the operating point towards surge. It is therefore important to match the compressor to give surge-free operation across the engine range while maintaining good turbocharger efficiencies at rated conditions.

25.5.2 Exhaust by-pass

This method is most suited to constant load control because a boost pressure signal can wastegate the turbocharger to maintain constant load. This system is also used by one American manufacturer to regulate air manifold pressure as a function of gas pressure (load) and engine speed.

25.5.3 Compressor by-pass

This can enable control to be applied without affecting the compressor operating line. It is widely used on dual fuel engines where the control of air supply at very low load is not required, because most dual fuel engines revert to diesel operation under these conditions.

25.6 Safety systems

The following paragraphs are derived from an MAN B&W brochure[25] describing the gas diesel variant of their low speed two-stroke MC engines.

All the normal safety systems incorporated in the diesel engine are retained for dual fuel operation; However, additional safety devices and systems are incorporated to avoid situations which might otherwise lead to failures.

There are numerous systems around the engine which have to be incorporated in an overall control and safety system and which could trigger off a shutdown of the gas operation or even a complete engine shutdown. The most complicated of these is the gas compressor system, while the sealing oil system, the ventilation system for the double wall piping, and the inert gas system are relatively simple. The engine systems also have to be integrated with, or at least connected to, the plant's overall control and safety system, so it is important that the interface between the systems is clearly defined in each case.

In the engine room of a ship or the engine hall of a power station, the presence of gas will be detected by a hydrocarbon (HC) analyser. The ventilation system of the intervening space of the double wall piping also incorporates an HC analyser, where an alarm is set off at a gas concentration of 30% of the lower explosion limit and shutdown occurs at 60%.

During operation in the dual fuel mode, malfunctioning of the pilot fuel injection system or gas injection system may involve a risk of uncontrolled combustion in the engine.

In the event of a damaged gas spindle seat or the sluggish operation of the gas valve, the exhaust gas temperature will increase considerably, and an exhaust temperature sensor will set off an alarm with subsequent shutdown of the engine. In the case of very sluggish operation or even seizure of the gas valve spindle in the open position, there is a risk that large quantities may be injected into the cylinder and when the exhaust valve opens, a hot mixture of combustion products and gas flows out into the exhaust pipe, where it will burn relatively slowly, when mixing with the scavenge air/exhaust gas there. The temperature sensor will then set off an alarm for high exhaust gas temperature for the cylinder in question, with subsequent engine shutdown. In the highly unlikely event of a more intensive combustion in the exhaust gas receiver, the pressure will increase considerably. For stoichiometric reasons, this pressure will not exceed 15 bar. The exhaust gas receiver is designed to maintain its integrity at this pressure.

However, the engine safety system will detect defective gas valves. The gas flow to each cylinder during one cycle is monitored by measuring the pressure drop in an accumulator in each gas supply line. If the pressure drop exceeds the value at full load, the engine will shut down immediately. With this system, any abnormal gas flow, whether due to seized gas injection valves or fractured gas pipes, will be detected immediately, the gas supply will be discontinued, and the gas lines will be purged with inert gas.

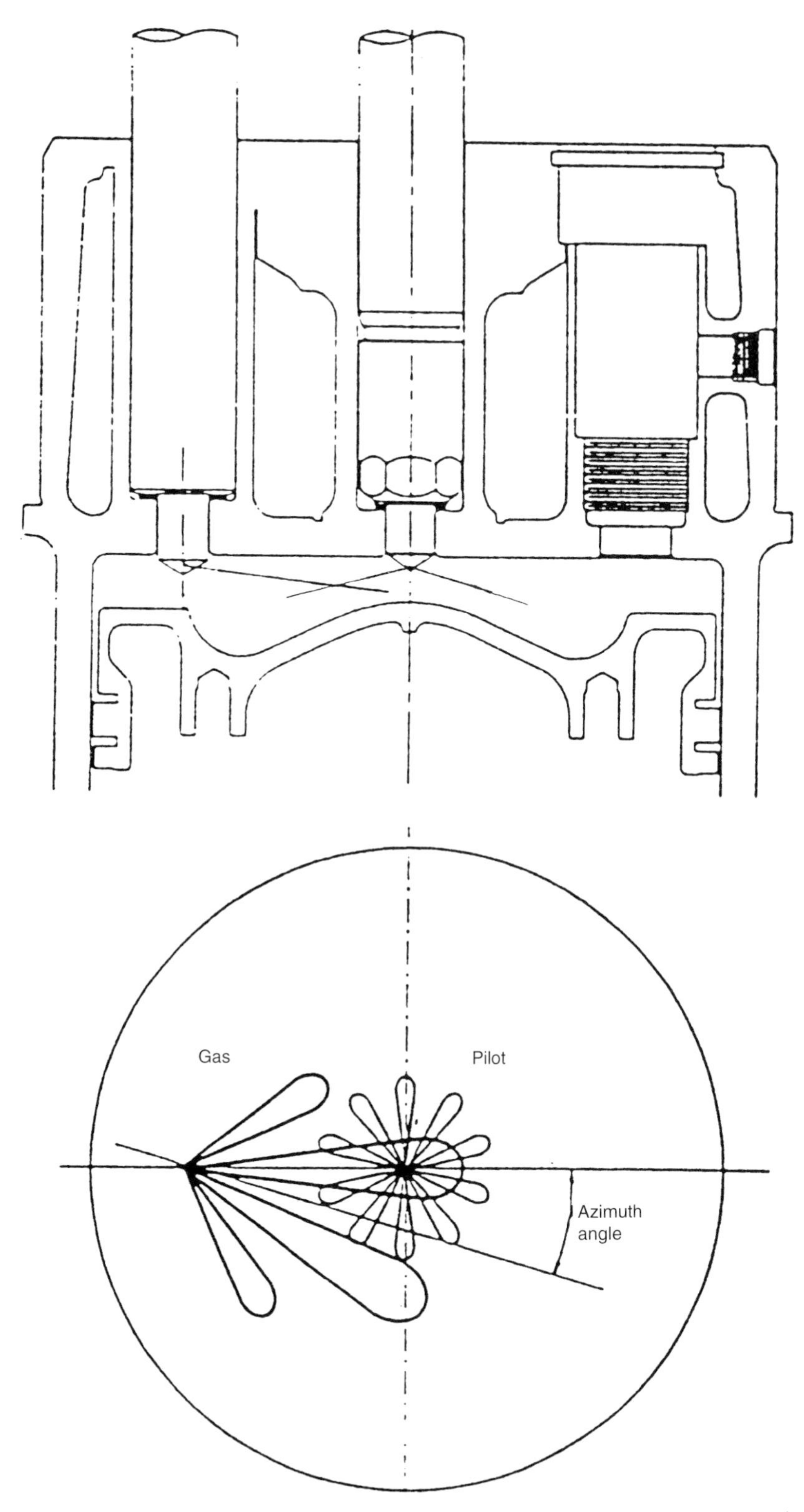

Figure 25.14 The 'H-process' combustion chamber layout (General Electric)

Ignition failure of the injected natural gas can have a number of different causes, most of which, however, are the result of failure to inject pilot oil into a cylinder. Examples are; leaky joints or fractured high-pressure pipes, a seized plunger in the pilot oil pump, a malfunctioning suction valve in the pilot oil pump, a failure of pilot oil supply to the engine, or other faults in the engine forcing the fuel pump to the zero fuelling position. If any of these faults are detected the gas injection is stopped immediately.

A safety valve ensures that gas injection can only take place after pilot oil injection. Between injections, the safety valve is in the open position and to enable gas injection, the safety valve must close. When the pressure on the pilot side rises to the opening pressure of the pilot valve, pilot oil is injected and the oil pressure acting on the safety valve spindle closes it, thus enabling the control oil pressure to build up and open the gas valve.

Despite this system, pilot fuel can be injected without being ignited if there is a sticking or severely burned exhaust valve, provided that this involves so much leakage that there is not enough compression pressure to ensure ignition of the pilot oil. However, the burning of an exhaust valve is a rather slow process extending over a long period, during which time the exhaust gas temperature rises and sets off an alarm well in advance of any risk of misfiring.

25.7 Applications

25.7.1 Automotive

Dual fuel engines have not been widely used in automotive applications. The relatively high cost of two separate fuel systems, and the problems of finding space for adequate quantities of

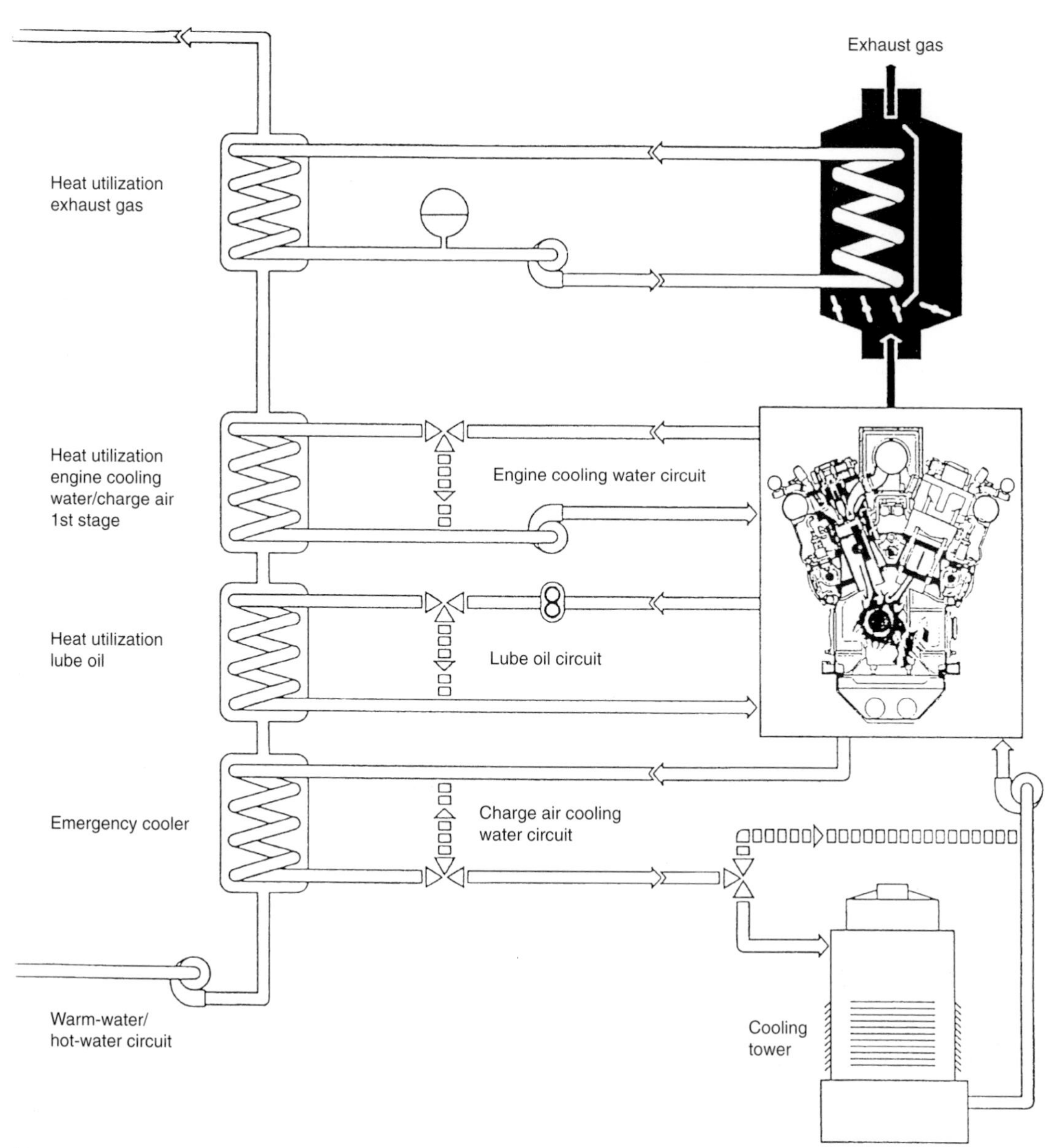

Figure 25.15 Typical waste heat utilization system based on a dual fuel engine (MAN B&W)

each fuel, have restricted their use to special applications such as city buses[26].

Conversely, where there are strong incentives for gas fuelled vehicles, such as emissions reduction and alternative fuel programmes, having a dual fuel engine does mean that the vehicle can continue to operate if its gas supply runs out. Target vehicles for alternative fuels programmes are normally those which return to a depot each day where refuelling facilities can be made available. Dual fuel vehicles give more flexibility.

Other advantages over spark ignited gas engines include higher thermal efficiency and lower conversion costs where a diesel engine is used as the starting point. However, the problems of part load operation have been a barrier; automotive engines spend much of their time operating at light load. Much research and development work has been done to try to resolve these problems, notably by BKM Inc.[27,28].

A significant step forward on the automotive front took place in 1996, when Caterpillar announced a new series of dual fuel engines in the power range 140 to 320 kW[29]. These engines are variants of existing electronically controlled diesel engines, with a gas fuel injection system and additional electronic control module. These engines are conventional dual fuel engines (without prechamber) and pilot fuel quantity is about 15% in most applications. Prechamber versions are being developed.

25.7.2 Locomotive

Many of the arguments stated above for automotive applications also apply to railway locomotives. However, there has been a strong focus on locomotive emissions in California, and this has led to a number of development programs aimed at utilizing CNG and LNG in locomotives[20,22,30,31]. A significant part of this effort has been devoted to the widely used General Motors EMD two stroke engines, as described earlier.

25.7.3 Stationary (power generation and mechanical drive)

The market for dual fuel engines for power generation purposes has increased as a result of more widespread introduction of emissions regulations for stationary plant, and the increasing availability of pipeline natural gas. This is the main area of application for the prechamber low NOx engine.

The dual fuel engine also finds favour in situations where gas is cheaper than diesel fuel, but the supply is unreliable, such as oil and gas pipelines and wellheads[24], offshore platforms[32], and biogas sources (landfill sites and sewage works)[10]. Also falling into this category are the increasing number of situations where gas is available at relatively low cost on an "interruptible supply tariff", whereby the gas supply may be cut off from time to time.

Dual fuel engines are well suited to cogeneration applications, providing both heat and electrical power from a low emissions prime mover. A typical waste heat utilization system to provide hot water for process or district heating is shown in *Figure 25.15*. One application where there is a true synergy between engine and fuel source is in sewage treatment works, where recovered heat from engine exhaust and jacket water can be used to maintain digester tanks at the required temperature for optimum gas production.

In addition to power generation, dual fuel engines are also used for driving pumps and compressors. Gas pipelines require compressor stations at intervals along their length. Because the gas is available at a relatively high pressure (typically 70 bar), this is an area where the gas diesel is sometimes preferred[33].

25.7.4 Marine and offshore

Marine applications of dual fuel engines are limited to those where gas is available from the cargo, such as LNG carriers. Both conventional dual fuel[34] and gas diesel[25] variants of low speed two-stroke propulsion engines have been developed to compete with the traditional steam turbine propulsion systems for this application.

The system developed by Statoil and MAN B&W for recovering and using the hydrocarbon gases (VOC's) given off by the cargo of crude oil shuttle tankers[19] has already been described.

Offshore oil platforms provide an ideal application for dual fuel engines. Gas is available at little or no cost, but the supply cannot be relied on. Likewise, crude oil may be available from the well or from storage tanks, and a certain amount of (expensive) distillate or heavy fuel is available for use when no other fuel is available. This is the application for which Wartsila developed their gas diesel engine. A modern alternative to the fixed oil platform is the floating oil production vessel, with multiple production, storage and processing functions[32], where a multi-fuel engine is even more beneficial. The ability of the gas diesel to deliver consistent performance on a range of different fuels, and to switch quickly from one to another, makes it very well suited to this application.

References

1 BOYER, R. L., 'Status of Dual Fuel Engine Development', *SAE 490018* (1949)
2 LIU, Z. and KARIM, G. A., 'Knock in Dual Fuel Engines', *COMODIA 94* (1994)
3 MTUI, P. L. and HILL, P. G., 'Ignition Delay and Combustion Duration with Natural Gas Fueling of Diesel Engines', *SAE 961933* (1996)
4 KARIM, G. A., LIU, Z. and JONES, W., 'Exhaust Emissions from Dual Fuel Engines at Light Load', *SAE 932822* (1993)
5 LIU, Z. and KARIM, G. A. 'The Knock and Autoignition Characteristics of Dual Fuel Engines', *ISATA 94EN054* (1994)
6 LIU, Z. and KARIM, G. A. 'An Examination of the Exhaust Emissions of Gas Fuelled Diesel Engines', *ASME ICE Vol. 27–3* (1996)
7 THOMAS, J. R., SAKONJI T., *et al.* 'A Test Program to Evaluate the Influences of Natural Gas Composition on SI Engine Combustion and Knock Intensity', *ASME ICE Vol. 22* (1994)
8 ROSE and COOPER Technical Data on Fuel
9 GILLISPIE, M. J. and JENSEN, M. A., 'Effects of Fuel Gas Mixtures on Power Limits in a Dual-Fuel Engine', *ASME ICE Vol. 21* (1994)
10 BLOWES, J. H., 'Power Generation from Biogas', IDGTE Publication 496, Jan/Feb 1997 (1997)
11 CHRISMAN, B. R., CALLAHAN, T. J. and CHIU, J. P., 'Investigation of Micro Pilot Combustion in a Stationary Gas Engine', *ASME 98 ICE 106 Vol. 30–3* (1998)
12 DANYLUK, P. R., 'Development of a High Output Dual Fuel Engine', *ASME 93-ICE-20* (1993)
13 SCHIFFGENS, H.-J., *et al.* 'Development of the New MAN B&W 32/40 Dual Fuel Engine', *ASME ICE Vol. 27–3*
14 BLIZZARD, D. T., *et al.* 'Development of the Cooper Bessemer Cleanburn Gas-Diesel (Dual Fuel) Engine', *ASME ICE Vol. 15*
15 BLYTHE, N. X., 'Development of the Fairbanks Morse Enviro-Design Opposed Piston Dual Fuel Engine', *ASME ICE Vol. 22* (1994)
16 NYLUND, I. and ROSGREN, C.-E. 'The Latest Achievements in Gas Diesel Technology', IDGTE Publication 476 October 1993 (1993)
17 FUKUDA, T., PEDERSEN, P. S. *et al.* 'Development of the World's First Large-Bore Gas-Injection Engine', *CIMAC 95 Paper D51* (1995)
18 BEPPU O. *et al.* 'Service Experience of Mitsui Gas Injection Diesel Engines', *CIMAC Congress 1998* (1998)
19 PEDERSEN, P. S. and RUCH, O. 'Environmentally Friendly Propulsion System for Shuttle Tankers', *CIMAC Congress 1998* (1998)
20 HSU, B. D., CONFER, G. L. and McDOWELL, R. E., 'The 'H-Process' Dual Fuel Diesel Engine' *ASME ICE-Vol. 24 'Natural gas and alternative fuels for engines', (1994)*
21 MEYERS, D. P., BOURN, G. D., HEDRICK, J. C. and KUBESH, J. T. 'Evaluation of Six Natural Gas Combustion Systems for LNG Locomotive Applications', *SAE 972967* (1997)
22 'Union Pacific's New LNG Road Locomotives Feature High-Pressure Design for More HP', *LNG Express*, February 1995 (1995)

23 'Dual Fuel Engine Conversions Evaluated by U.S. Navy *D>WW Oct. 1996 pp. 38–39* (1996)
24 OLSON, L. E. and JENSEN, S. P., 'Application and Maintenance of a Low Pressure Dual Fuel System for Offshore Drilling Rig Power Generation', *ASME Paper no. 97–ICE-14* (1997)
25 'Large Diesel Engines Using High Pressure Gas Injection Technology', MAN B&W Diesel A/S P206–96.02 (1996)
26 WELCH, A. B., 'Development of a Heavy-Duty Two-Stroke Dual Fuel Natural Gas/Diesel Engine', *SAE 912475* (1991)
27 GEBERT, K, BECK, N. J., BARKHIMER, R. L. and WONG, H.-C. 'Strategies to Improve Combustion and Emission Characteristics of Dual-Fuel Pilot Ignited Natural Gas Engines', *SAE 971712* (1997)
28 GEBERT K, BECK, N. J. BARKHIMER, R. L., WONG, H.-C and WELLS, A. D., 'Development of Pilot Fuel Injection System for CNG engine', *SAE 961100* (1996)
29 OSENGA, M., 'CAT Introduces Dual Fuel Truck Engines', *Diesel Progress Engines and Drives,* August 114–116 (1996)
30 JANSEN, S. P., 'A Retrofit System to Convert a Locomotive to Natural Gas Operation', *ASME ICE Vol. 21* (1994)
31 WEAVER, C. S., 'Controlling Locomotive Emissions in California', *CARB* 29 March (1995)
32 MIEMOIS, M., 'Gas Diesels in Offshore Applications', *Wartsila Diesel Marine News,* 1–96 (1996)
33 KUNBERGER, K., 'Gas-Diesel Engines for Compressor Station Applications' *D>WW*, July-August, 50–51 (1994)
34 GROSSHANS, G., 'Development of a 1200 kW/cyl Low Pressure Dual Fuel Engine for LNG Carriers' *CIMAC Congress* (1998)

Bibliography

'Advances on Low Speed Application', by MAN B&W *International Power Generation,* March, 67–68 (1993)

BROGAN R. T. *et al.* 'Operation of a Large Bore Medium Speed Dual Fuel Engine on Low BTU Wood Gas' *ASME Conference on Alternative Fuels, Engine Performance and Emissions* (1993)

ALY H. and SIEMER G. 'Experimental Investigation of Gaseous Hydrogen Utilization in a Dual Fuel Engine for Stationary Power Plants' *ASME Conference on Alternative Fuels, Engine Performance and Emissions* (1993)

KAWAKAMI M., TANABE H., MATSUDA S., SATO K. 'Environmental Control of Advanced Medium Speed Engines' *ASME 94–ICE–14* (1994)

'Gas Fuelled Engines' (Brochure) GEC Alsthom Ruston Diesels Publication RD 144 (1994)

LIU Z. and KARIM G. A. 'A predictive Model for the Combustion Process in Dual Fuel Engines' *SAE 952435* (1995)

DAISHO Y. *et al.* 'Controlling Combustion and Exhaust Emissions in a Direct Injection Diesel Engine Dual Fuelled with Natural Gas' *SAE 952436* (1995)

VESTERGREN R. 'The Merits of the Gas-Diesel Engine' *ASME ICE Vol. 25–3* (1995)

LIU Z. and KARIM G.A. 'An Examination of the Exhaust Emissions of Gas Fuelled Diesel Engines' *ASME Conference on Advancements in Engine Design and Application* (1996)

LIU Z. and KARIM G.A. 'Simulation of Combustion Processes in Gas Fuelled Diesel Engines' *Journal of Power and Energy,* 159–169 (1997)

DAISHO Y. *et al* 'Combustion and Exhaust Emissions in a Direct-Injection Diesel Engine Dual-Fuelled with Natural Gas' *SAE 950465* (1995)

SCHIFFGENS H.–J. *et al* 'Low NOx Gas Engines from MAN B&W' *CIMAC Congress* (1998)

HILL P.G. and DOUVILLE B. 'Analysis of Combustion in Diesel Engines Fuelled by Directly Injected Natural Gas', *ASME ICE Vol. 30–3* (1998)

STAN C. and HILLIGER E. 'Pilot Injection System for Gas Engines Using Electronically Controlled Ram Tuned Diesel Injection' *CIMAC Congress* (1998)

26 Marine engine applications

Contents

26.1 High speed engines

High speed four-stroke trunk piston engines are widely specified for propelling small, generally specialized, commercial vessels and as main and emergency genset drives on all types of tonnage. The crossover point between high and medium speed diesel designs is not sharply defined but for the purposes of this chapter engines running at 1000 rev/min and over are reviewed.

Marine high speed engines traditionally tended to fall into one of two design categories: high performance or heavy duty types. High performance models were initially aimed at the military sector, and their often complex designs negatively affected manufacturing and maintenance costs. Applications in the commercial arena have sometimes disappointed operators, the engines dictating frequent overhauls and key component replacement.

Heavy duty high speed engines in many cases were originally designed for off-road vehicles and machines but have also found niches in stationary power generation and locomotive traction fields. A more simple and robust design with modest mean effective pressure ratings compared with the high performance contenders yield a comparatively high weight/power ratio. But the necessary time-between-overhauls and component lifetimes are more acceptable to civilian operators.

In developing new models, high speed engine designers have pursued essentially the same goals as their counterparts in the low and medium speed sectors: reliability and durability, underwriting extended overhaul intervals and component longevity and hence low maintenance costs; easier installation and servicing; compactness and lower weight; and enhanced performance across the power range with higher fuel economy and reduced noxious emissions.

Performance development progress over the decades is highlighted by considering the cylinder dimension and speed of an engine required to deliver 200 kW/cylinder (*Figure 26.1*). In 1945 a bore of 400 mm-plus and a speed of around 400 rev/min were necessary; in 1970 typical medium speed engine parameters resulted in a bore of 300 mm and a speed of 600 rev/min, while typical high speed engine parameters were 250 mm and 1000 rev/min to yield 200 kW/cylinder. Today, that specific output can be achieved by a 200 mm bore high speed design running at 1500 rev/min.

Flexible manufacturing systems (FMS) have allowed a different approach to engine design. The reduced cost of machining has made possible integrated structural configurations, with more functions assigned to the same piece of metal. The overall number of parts can thus be reduced significantly over earlier engines (by up to 40 per cent in some designs), fostering improved reliability, lower weight and increased compactness without compromising on ease of maintenance. FMS also facilitates the offering of market-adapted solutions without raising cost: individual engines can be optimized at the factory for the proposed application.

A widening market potential for small high speed engines in propulsion and auxiliary roles encouraged the development in the 1990s of advanced new designs for volume production. The *circa*-170 mm bore sector proved a particularly attractive target for leading European and US groups which formed alliances to share R&D, manufacture and marketing–notably Cummins with Wärtsilä Diesel, and MTU with Detroit Diesel Corporation.

High speed engine designs have benefited from such innovations as modular assembly, electronically controlled fuel injection systems, common rail fuel systems and sophisticated electronic control/monitoring systems. Some of the latest small bore designs are even released for genset duty burning the same low grade fuel (up to 700 cSt viscosity) as low speed crosshead main engines.

26.1.1 Caterpillar

A wide programme of high speed engines from the US designer Caterpillar embraces models with bore sizes ranging from 105 mm to 170 mm. The largest and most relevant to this review is the 170 mm bore/190 mm stroke 3500 series which is produced in V8-, V12- and V16-cylinder versions with standard and higher B-ratings to offer outputs up to around 2200 kW. The engines, with minimum/maximum running speeds of 1200/1925 rev/min, are suitable for propelling workboats, fishing vessels, fast commercial craft and patrol boats. Genset applications can be covered with ratings from 1000 kVA to 2281 kVA.

The series B engines (*Figure 26.2*) benefited from a number of mechanical refinements introduced to take full advantage of the combustion efficiency improvement delivered by an electronic control system. Electronically controlled unit fuel injectors combine high injection pressures with an advanced injector design to improve atomization and timing. Outputs were raised by 17 to 30 per cent above earlier 3500 series models.

A special high performance variant of the V16-cylinder 3500 series model was introduced to target niche markets, the refinements seeking increased power, enhanced reliability and lower fuel and lubricating oil consumptions without undermining durability. This Phase II high performance version of the 3516 has an upper rating of 2237 kW at 1925 rev/min. It was released for fast passenger vessels with low-load factors with a standard maximum continuous rating of 1939 kW at 1835 rev/min and

Bore	(mm)	420	300	(250)	200
Stroke	(mm)	500	360	(300)	240
Speed	(rpm)	428	600	(1000)	1500
Year		1945	1970		1995

Figure 26.1 Cylinder dimensions and speeds for medium and high speed engines delivering 200 kW/cylinder (1945, 1970 and 1995). Reference Wärtsilä Diesel

Figure 26.2 Caterpillar 3512B engine with electronic control system

a 'two hours out of 12' rating of 2088 kW at 1880 rev/min. Optional higher ratings up to 2205 kW at 1915 rev/min can be specified for cooler climate deployment, with revised turbocharger, fuel injector and timing specifications.

Key contributions to higher performance came from high efficiency ABB turbochargers, a seawater aftercooler to supply colder air to the combustion chambers, larger and more aggressive camshafts, and a new deep crater piston design. The fuel is delivered through strengthened unit injectors designed and manufactured by Caterpillar to secure injection pressures of 1380 bar.

An optimum air–fuel mixture which can be burned extremely efficiently is fostered by the combination of a denser air intake and the high injection pressure. The reported result is a specific fuel consumption range at full load of 198–206 g/kWh with all fuel, oil and water pumps driven by the engine. Modifications to the steel crown/aluminium skirt pistons and rings lowered lubricating oil consumption to 0.55 g/kWh.

A particularly desirable feature for fast ferry propulsion is underwritten by the high efficiency combustion and low crevice volume pistons which help to eliminate visible exhaust smoke at all steady points along the propeller demand curve. The rear gears were widened and hardened to serve the higher pressure unit injectors. New gas-tight exhaust manifolds with bellow expansion joints and stainless steel O-rings improved engineroom air quality by eliminating exhaust gas leaks.

All Caterpillar 3500 series-B engines are controlled by a microprocessor-based electronic control module (ECM). Information is collected from engine sensors by the ECM which then analyses the data and adjusts injection timing and duration to optimize fuel efficiency and reduce noxious exhaust emissions. Electronic control also supports onboard and remote monitoring capabilities, the ECM reporting all information through a two-wire Cat Data Link to the instrument panel. The panel records and displays faults as well as operating conditions. An optional Customer Communications Module translates engine data to standard ASCII code for transmission to a PC or via satellite to remote locations.

Caterpillar's Engine Vision System (EVS) is compatible with the high performance 3500 series-B engines and the company's other electronically controlled engines. The EVS displays engine and transmission data, vessel speed, trip data, historical data, maintenance intervals, diagnostics and trouble-shooting information. Up to three engines can be monitored simultaneously, the system transferring between the vision display and individual ECMs via the two-wire data link.

26.1.2 Cummins

The most powerful own-design engine in Cummins' high speed programme, the KTA50-M2 model, became available from early 1996 (*Figure 26.3*). The 159 mm bore/159 mm stroke design is produced by the US group's Daventry factory in the UK in V16-cylinder form with ratings of 1250 kW and 1340 kW for medium continuous duty and 1030 kW and 1180 kW for continuous duty applications. The running speeds range from 1600 rev/min to 1900 rev/min, depending on the duty, typical applications including fishing vessels, tugs, crewboats and small ferries.

The KTA50-M2 engine benefited from a new Holset turbocharger, low temperature after-cooling and gallery-cooled pistons. Cummins's Centry electronics system contributes to enhanced overall performance and fuel economy, providing adjustable all-speed governing, intermediate speed controls, dual power curves, a built-in hour meter and improved transient response. Diagnostic capabilities are also incorporated.

Cummins offers a more powerful engine from Daventry in the shape of the QSW family, released in 1997 as the result of a joint venture with Wärtsilä Diesel which markets the design as the W170 series. The 170 mm bore/200 mm stroke engine was designed for continuous heavy duty running and features full-authority electronic controls and advanced fuel system technology. The range embraces six and eight in-line and V12-16- and 18-cylinder models covering a power band from 690 kW to 2340 kW at 1800 rev/min.

26.1.3 Deutz MWM

A long tradition in high speed engine design is maintained by Deutz MWM of Germany whose current programme is focused on the 616, 620 and 628 series with an upper output limit of 3600 kW for propulsion and genset drives. Higher power demands (up to 7400 kW) can be met by the 632 series, a 250 mm bore high/medium speed design.

The 616 series is a 132 mm bore/160 mm stroke design with a maximum output per cylinder of 85 kW at 2300 rev/min; propulsion plant applications from 480 kW to 1360 kW are covered by V8-, 12- and 16-cylinder models (*Figure 26.4*).

The 170 mm bore/195 mm stroke 620 series design offers a maximum rating of 140 kW/cylinder at 1800 rev/min, V8-, 12- and 16-cylinder versions covering a power band from 1016 kW to 2032 kW for propulsion duty; special drives are addressed by a higher upper limit of 2240 kW at 1860 rev/min for the V16 model (*Figure 26.5*).

Figure 26.3 Cummins' largest own-design engine, the KTA50-M2 model

Figure 26.4 V12 cylinder block of Deutz MWM 616 series engine

Outputs from 1350 kW to 3600 kW at 1000 rev/min are delivered by six, eight and nine in-line and V12- and V16-cylinder versions of the 240 mm bore/280 mm stroke 628 series design.

Other high speed engines produced under the Deutz banner cover a power range from 30 kW to 400 kW.

26.1.4 GMT

Grandi Motori Trieste (GMT) of Italy has focused its high speed engine developments on a 230 mm bore design which is offered in several versions for commercial and naval propulsion applications as well as genset drive duties (*Figure 26.6*). A

Figure 26.5 V16-cylinder version of Deutz MWM 620 series engine

Figure 26.6 GMT BL230 engines in a genset installation

heavy fuel-burning model is available for unifuel machinery installations while a non-magnetic version can be specified for minehunters.

The series is produced in standard (B230) form with a stroke of 270 mm and in BL230 long stroke (310 mm) form, later versions benefiting from improved cooling and turbocharging arrangements for higher outputs at speeds up to 1200 rev/min. The B230 has a rating of 210 kW/cylinder and the BL230 a rating of 225.5 kW/cylinder. Both types are produced in 4L- to V20-cylinder configurations. A two-stage turbocharged version with variable compression ratio, the BL 230 DVM, develops 283 kW/cylinder at 1050 rev/min on a mean effective pressure of 25.1 bar to provide a compact plant for corvettes and frigates. It is produced in V20-cylinder form to deliver 5660 kW.

A BL230P version was developed for operation on heavy fuel up to IFO 500 with a maximum speed of 1000 rev/min and a power rating—190 kW/cylinder—lower than the gas oil-fuelled models. Its special features include: composite pistons with forged steel crowns and plasma-coated rings in the first groove for wear resistance; Nimonic A material exhaust valves; valve rotators; cooled fuel injection valves; and oversized fuel injection pumps for enhanced atomisation and combustion. Other variants in the engine programme target natural gas and LPG burning installations.

A highly rigid cast iron main structure for the 230 mm bore engine incorporates the water, oil and air manifolds, virtually eliminating external pipes and fostering compactness. Camshafts, oil cooler, oil, water and fuel pumps, and filters are arranged for accessibility; and pistons and connecting rods are withdrawable from the top.

GMT's high speed portfolio also includes the A210 series, the 210 mm bore/230 mm stroke design developing 170 kW/cylinder at 1500 rev/min on a mean effective pressure of 16.2 bar. A programme embracing V6- to V20-cylinder models covers output demands up to 3400 kW. A special version designed for submarine propulsion—the compact and lightweight A210SM–benefits from anti-noise, shock and vibration characteristics. An ability to operate in severely inclined positions was also addressed, along with ease of access for maintenance.

26.1.5 Isotta Fraschini

A family of high speed engines from Isotta Fraschini, part of Italy's Fincantieri group, includes the 170 mm bore/170 mm stroke ID 36 series which is available in different versions for light and heavy propulsion duties as well as for genset drives. A non-magnetic variant was developed for mine warfare vessels.

The ID 36 series embraces V6- to V16-cylinder models, all with a 90-degree configuration arranged on a high tensile alloy iron crankcase and featuring a direct injection system and four valves per cylinder. The power band extends to 2350 kW at 2100 rev/min.

26.1.6 MAN B&W Holeby

In 1995 MAN B&W Diesel's genset engine specialist Holeby Diesel of Denmark supplemented its popular L23/30H and L28/32H medium speed auxiliary prime movers with the innovative high speed L16/24 series. The 160 mm bore/240 mm stroke design (*Figure 26.7*) was conceived as a new generation 1000/1200 rev/min engine dedicated to genset drives and capable of operating on an unrestricted load profile on heavy fuel up to 700 cSt/50°C viscosity. The programme embraces five-, six-, seven-, eight- and nine-cylinder models covering a power range from 450 kW to 900 kW.

The main problem in burning heavy fuel oil in small high speed engines is ignition delay. Smaller quantities of the volatile, easily combustible hydrocarbon fractions are present in such fuels than in lighter diesel fuels. The L16/24 designers addressed the problem by adopting a number of measures to achieve excellent heavy fuel combustion even at part- and low-load operation: a higher injection pressure (1500 bar), a higher opening pressure for the fuel valve, smaller nozzle hole diameters in the valve, and a higher compression ratio (15.5:1). The maximum combustion pressure is 180 bar. Another contribution came from a cylinder head design refined to improve the swirl of fuel in the combustion chamber.

Support-function components were traditionally distributed around the engine block and connected with externally mounted supply pipes. The practice was reversed for the L16/24 engine: all support elements—oil and water pumps, coolers, filters, and safety and regulator valves—are arranged in a single front-end box for ease of accessibility and maintenance (*Figure 26.8*); and the supply channels are cast into the block for maintenance-free operation. Virtually all the engine's internal supply lines are channelled through the cooling water jacket and cylinder head. The arrangement considerably simplifies the overall design, reducing the overall number of components by some 40 per cent compared with earlier engines. The front-end box system components can be exchanged using clip-on/clip-off couplings without removing any pipes.

High rigidity was sought from a monobloc cast iron engine frame whose elements are all held under compressive stress. The frame is designed to accept an ideal flow of forces from the cylinder head down to the crankshaft and to yield low surface vibrations from the outer shell. Two camshafts are located in the frame: the camshaft for the inlet/exhaust valves is arranged on the exhaust side in a very high position; and the fuel injection camshaft is on the service side of the engine.

Covers in the frame sides offer access to the camshafts and crankcase; some of the covers are arranged to act as relief valves. There is no cooling water in the frame. The framebox is designed to minimize noise emission, the inner frame absorbing all the engine forces and the outer frame forming a stiff shell with minimal vibration (*Figure 26.9*). The main bearings for the underslung crankshaft are carried in heavy supports by tie-rods from the intermediate frame floor and secured by bearing caps. The caps are provided with side guides and held in place by studs with hydraulically tightened nuts. The main bearing features replaceable shells which are fitted without scraping.

Both engine and alternator are mounted on a rigid baseframe that acts as a lubricating oil reservoir. Specially designed engine mounts reduce noise and vibration.

The centrifugal cast iron cylinder liner, housed in the bore of the engine frame, is clamped by the cylinder head and rests on its flange on the water jacket; it can thus expand freely downwards when heated during engine operation. The liner is of the high flange type, the height of the flange matching the water-cooled area to give a uniform temperature pattern over the entire liner surface. The liner's lower part is uncooled to secure a sufficient margin for cold corrosion at the bottom end. There is no water in the crankcase area. Gas sealing between liner and cylinder head is effected by an iron ring. The liner is fitted with a slip-fit-type fire ring at its top to reduce lubricating oil consumption and bore polishing.

The cast iron cylinder head, with integrated charge air receiver, is made in one piece and incorporates a bore-cooled thick-walled bottom. It has a central bore for the fuel injection valve and a four-valve cross-flow configuration with a high flow coefficient. The valve pattern is turned about 20 degrees to the axis to achieve an intake swirl promoting optimised combustion. The head is hydraulically tightened by four nuts acting on studs screwed into the engine frame. A screwed-on top cover for the head has two main functions: oil sealing the rocker chamber and covering the complete top face of the head.

Air inlet and exhaust valve spindles are of heat-resistant

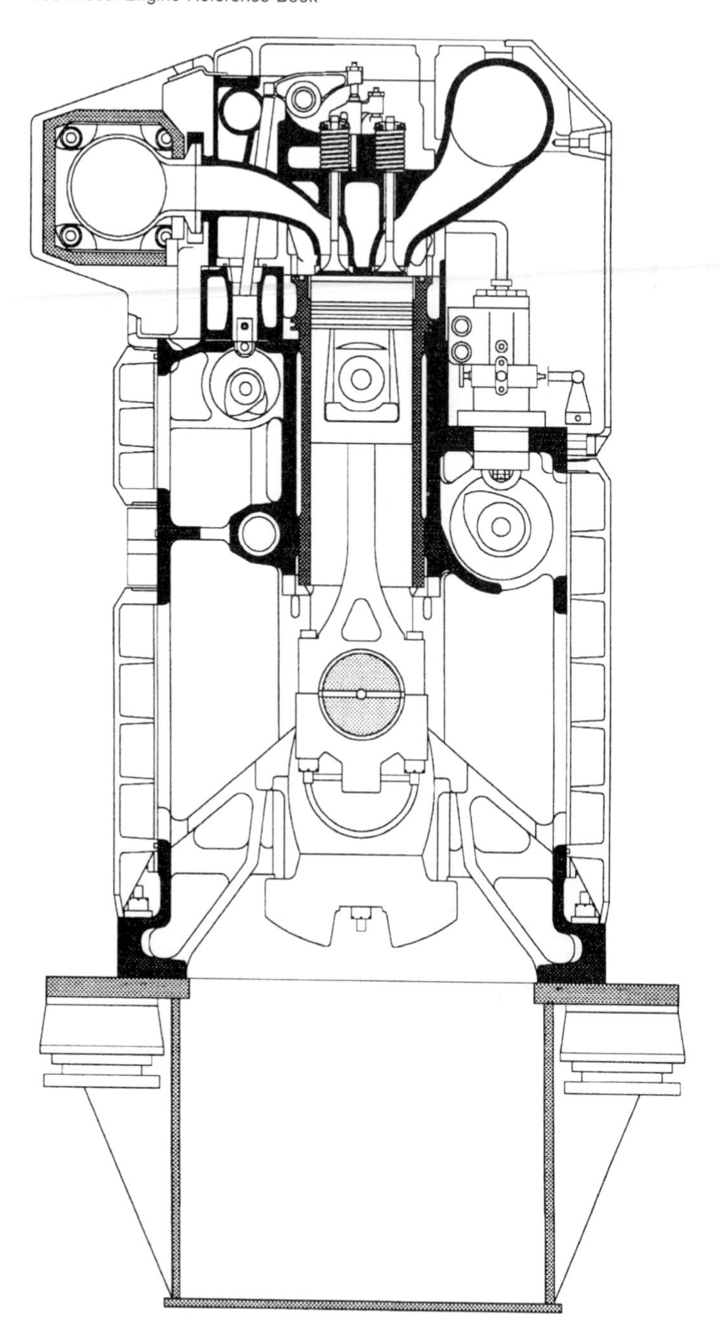

Figure 26.7 Cross-section of MAN B&W Holeby L16/24 genset engine. Note the separate camshafts for the gas exchange valves (left) and fuel injection (right)

material and their seats armoured with welded-on hard metal. All the spindles are fitted with valve rotators to ensure an even temperature on the valve discs and to prevent deposits forming on the seatings. The cylinder head is provided with replaceable valve seat rings of heat-resistant steel, and the exhaust valve seat rings are water cooled. The seating surfaces are hardened to minimise wear and prevent dent marks.

The valve rocker arms are actuated through rollers, roller guides and pushrods. The roller guides for the inlet and exhaust valves are mounted in the water jacket part. Access for dismantling is provided by a side cover on the pushrod chamber. Each rocker arm activates two spindles via a spring-loaded valve bridge with thrust screws and adjusting screws for valve clearance. The valve actuating gear is pressure feed-lubricated from the centralized lubricating system, through the water chamber part and from there into the rocker arm shaft to the rocker bearing.

The oil-cooled pistons comprise a nodular cast iron body and forged steel crown with two compression rings and one scraper ring fitted in hardened grooves. The different barrel-shaped profiles of the compression rings and their chromium-plated running surfaces aim to maximize sealing and minimize wear. The piston has a cooling space close to the crown and the ring zone which is supplied with oil from the engine lubricating system. Heat transfer, and thus cooling, is promoted by the shaker effect stimulated by the piston movement.

Oil is supplied to the cooling space through channels from the oil grooves in the piston pin bosses. Oil is drained from the space through ducts located diametrically to the inlet channels.

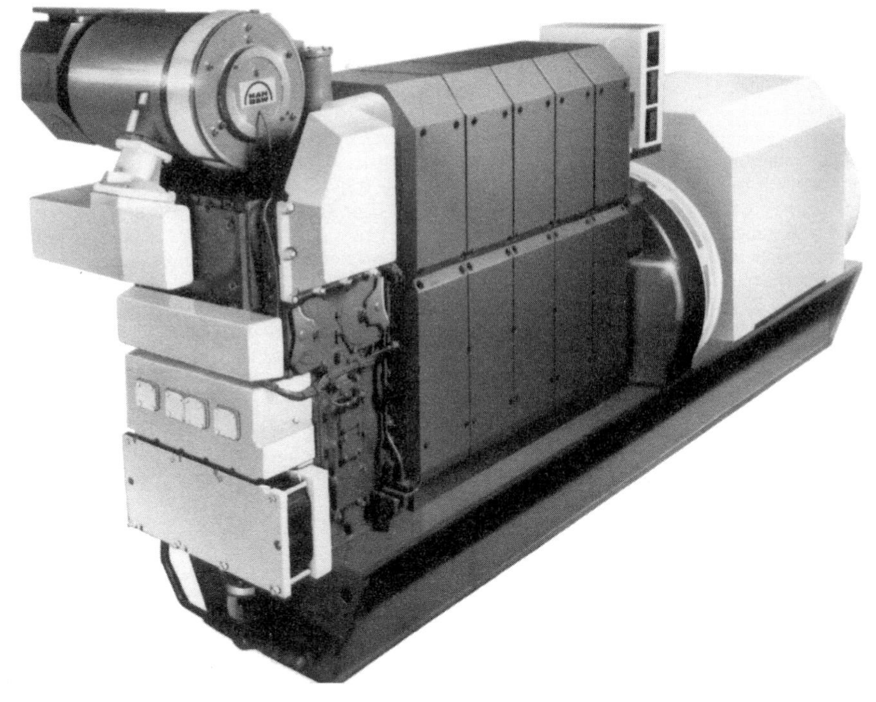

Figure 26.8 MAN B&W Holeby L16/24 engine. All main support ancillaries are grouped in a single front-end box for ease of access

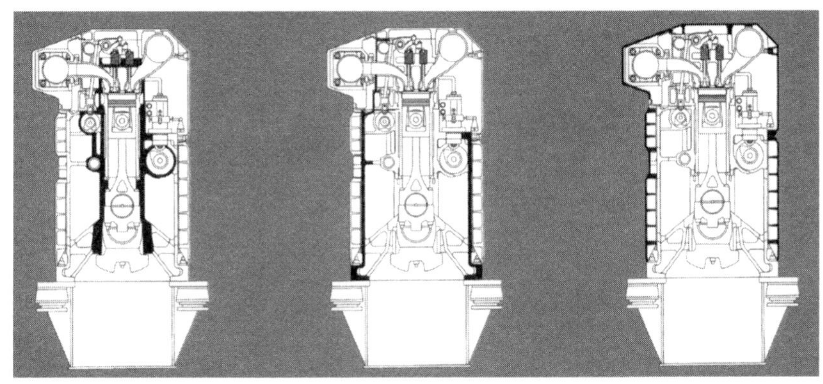

Figure 26.9 Noise emission and vibration from MAN B&W Holeby L16/24 engine are minimized by the inner frame, outer shell and cover arrangement

The piston pin is fully floating and kept in position in the axial direction by two circlips.

The die-forged connecting rod has a big end with a horizontal split and bored channels to transfer oil from the big end to the small end. The big end bearing is of the tri-metal type coated with a running layer. The bearing shells are of the precision type and can therefore be fitted without scraping or any other adaptation. The tri-metal small end bearing is pressed into the connecting rod. The bush is provided with an inner circumferential groove and a pocket for distributing oil in the bush itself and for supplying oil to the pin bosses.

A one-piece forged crankshaft with hardened bearing surfaces is suspended in underslung, tri-metal main bearings coated with a running layer. To attain a suitable bearing pressure and vibration level the crankshaft is equipped with counterweights which are attached to the shaft by two hydraulic screws. At its flywheel end the crankshaft is fitted with a gearwheel which, through two intermediate wheels, drives the twin camshafts. Also mounted here is a coupling flange for the alternator. At the opposite (front) end is a gearwheel connection for the lubricating oil and water pumps.

Lubricating oil for the main bearings is supplied through holes drilled in the engine frame. From the main bearings the oil passes through bores in the crankshaft to the big end bearings and thence through channels in the connecting rods to lubricate the piston pins and cool the pistons.

Separate camshafts for the inlet/exhaust valves and the fuel pump facilitate adjustment of the gas exchange settings without disturbing the fuel injection timing (*Figure 26.7*). Likewise, it is possible to adjust fuel injection without disturbing gas exchange parameters. The resulting flexibility allows engine operation to be adjusted and optimized for fuel economy or low NOx emissions. The camshafts are mounted in bearing bushes fitted in bores in the engine frame. The valve camshaft is arranged in a very high position on the engine exhaust side to secure a short and stiff valve train and reduce moving masses. The fuel injection camshaft is arranged on the service side of the engine.

Both camshafts are structured in single-cylinder sections and bearing sections in such a way that disassembly of individual cylinder sections is possible through the side openings in the crankcase. The camshafts and governor are driven by the main gear train at the flywheel end of the engine, rotating at a speed half that of the crankshaft. The lubricating oil pipes for the gearwheels are equipped with nozzles adjusted to apply lubricant at the points where the gearwheels mesh.

All fuel injection equipment is enclosed securely behind removable covers. Each cylinder is individually served by a fuel injection pump, high pressure pipe and injection valve with uncooled nozzle. The injection pump unit, mounted on the engine frame, comprises a pump housing embracing a roller guide, a central barrel and a plunger. The pump is activated by the fuel cam and the volume injected controlled by turning the plunger. The fuel injection valve is located in a valve sleeve in the centre of the cylinder head, its opening controlled by the fuel oil pressure and closure effected by a spring.

The high pressure fuel pipe is led through a bore in the cylinder head surrounded by a shielding tube. The tube also acts as a drain channel to ensure that any leakage from the fuel valve and the high pressure pipe is drained off.

A lambda controller ensures that all injected fuel is burnt, countering the internal engine pollution and increased wear that might otherwise result from genset step loading.

A constant pressure turbocharging system embraces an MAN B&W NR/S turbocharger purpose-designed for the L16/24 engine, charge air cooler, charge air receiver and exhaust gas receiver. The charge air cooler is a compact two-stage tube unit deploying a large cooling surface.

A patented 'intelligent' cooling water system was designed to secure an optimized temperature across the engine operating band from idling to full load. The system, which accepts fresh water within the 10–40°C temperature range, has one inlet and one outlet connection. Its two pumps, in combination with thermostatic valves, continuously regulate cooling water temperature to achieve the optimized operating condition. Since charge air from the turbocharger never falls below the dew point there is no danger of water condensation in the cylinders.

The cooling water system comprises a low temperature (LT) system and a high temperature (HT) system, each cooled by fresh water. The LT circuit is used to cool charge air and lubricating oil. The HT circuit cools the cylinder liners and heads, fostering optimized combustion conditions, limiting thermal load under high load conditions, and preventing hot corrosion in the combustion area. Under low load, the system is designed to ensure that the temperature is high enough for efficient combustion and that cold corrosion is avoided.

Water in the LT system passes through the low temperature circulating pump which drives the water through the second stage of the charge air cooler and then through the lubricating oil cooler before the water leaves the engine together with the high temperature water. The amount of water passing through the second stage of the charge air cooler is controlled by a three-way valve dependent on the charge air pressure. If the engine is operating at low-load condition the temperature regulation valve cuts off the LT water flow, thus securing preheating of the combustion air by the HT water circuit in the first stage.

The HT cooling water passes through the high temperature circulating pump and then through the first stage of the charge air cooler before entering the cooling water jacket and the cylinder head. It then leaves the engine with the low temperature water. Both LT and HT water leaves the engine via separate three-way thermostatic valves that control the water temperature.

All moving parts of the engine are lubricated with oil circulating under pressure, the system served by a lubricating oil pump of the helical gear type. A pressure control valve built into the system reduces the pressure before the filter with a signal taken after the filter to ensure constant oil pressure with dirty filters. The pump draws oil from the sump in the baseframe, the pressurized oil then passing through the lubricating oil cooler and the filter. The oil pump, cooler and filter are all located in the front box. The system can also be provided with a centrifugal filter. Lubricating oil cooling is carried out by the low temperature cooling water system, with temperature regulation effected by a thermostatic three-way valve on the oil side (see above). The engine is equipped as standard with an electrically driven pre-lubricating pump.

The L16/24 engine is prepared for MAN B&W Diesel's CoCoS computerized surveillance system, a Microsoft Windows-based program undertaking fully integrated monitoring of engine operation, maintenance planning, and the control and ordering of spares. The four CoCoS software modules cover engine diagnosis, maintenance planning, spare parts catalogue, and stock and ordering.

Each cylinder assembly (head, piston, liner and connecting rod) can be removed as a complete unit for repair, overhaul or replacement by a renovated unit onboard or ashore. Replacing a cylinder unit (*Figures 26.10, 26.11* and *26.12*) is accomplished by removing the covers and high pressure fuel injection pipe, and disconnecting a snap-on coupling to the exhaust gas pipe. The only cooling water connections are to the cylinder unit as there is no cooling water in the baseframe. Inlet and outlet cooling water passes between cylinder units via bushes which are pushed aside in disassembling a unit. The charge air connections are dismounted in the same way. The four hydraulically fastened cylinder head nuts and the two connecting rod nuts (all six are of the same size) are then removed, allowing the 200 kg unit to be withdrawn from the engine.

The design principles of the L16/24 engine were later applied in the creation of MAN B&W Holeby's L27/38 medium speed engine, announced in 1997.

26.1.7 Mitsubishi

Fast ferry propulsion business potential stimulated the development in the early 1990s of the Mitsubishi S16R-S engine by the Japanese group's Sagamihara Machinery Works. The higher performance V16-cylinder model was evolved from the established S16R design which had been in production since 1989 as a general purpose marine engine. A constant pressure turbocharging system based on a newly developed turbocharger and a revised fuel injection system contributed to a 20 per cent rise in the power output. The 170 mm bore/180 mm stroke design has a maximum continuous rating of 2100 kW at 2000 rev/min with overload ratings up to 2300 kW.

The weight was reduced to 89 per cent of the original engine, primarily through the adoption of aluminium alloy components optimized in size for the duty. An overall weight of 5500 kg underwrites a power-to-weight ratio of 2.62 kg/kW.

26.1.8 MTU

A portfolio of high performance high speed engines with an upper output limit of 7400 kW is offered by MTU (Motoren-und Turbinen-Union) Friedrichshafen. MTU was created in 1969 when Daimler-Benz and MAN consolidated the development and production of relevant engines from MAN and Maybach Mercedes-Benz. MTU Friedrichshafen became part of the Deutsche Aerospace group in 1989.

Series 396

Addressing lower power demands, MTU's Series 396 engine is long established as a propulsion and genset drive, the 165 mm bore/185 mm stroke design delivering up to 2560 kW at 2100 rev/min from V90-degree configuration 8-, 12- and 16-cylinder models (*Figure 26.13*). Engines are available in three different versions:

- TB04, with external charge air cooling (intercooler in raw water circuit).

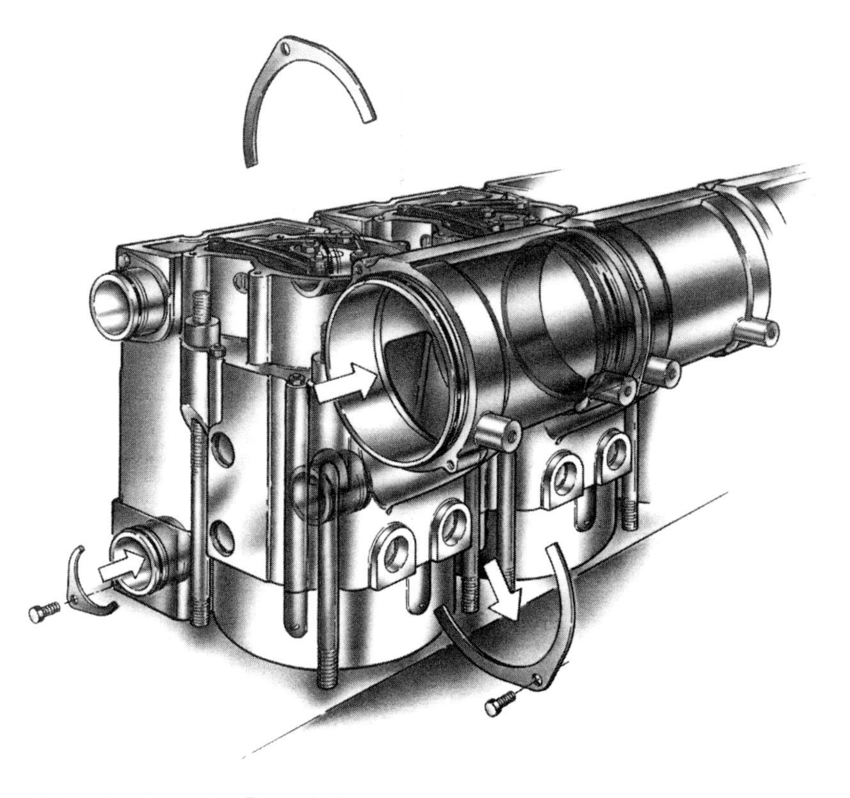

Figure 26.10 MAN B&W Holeby L16/24 engine: preparing for cylinder unit removal

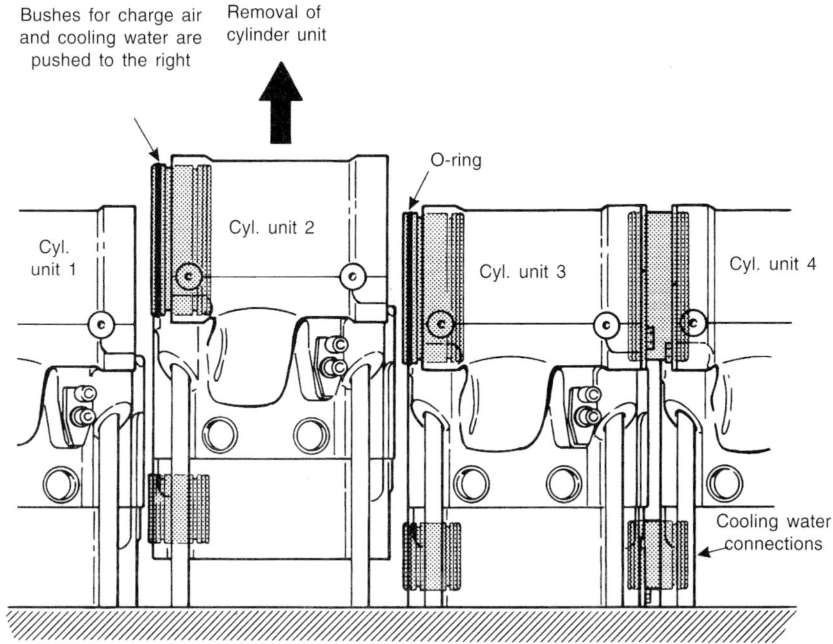

Figure 26.11 MAN B&W Holeby L16/24 engine: preparing for cylinder unit removal

- TC04, with internal charge air cooling (intercooler in engine coolant circuit).
- TE04, with internal charge air cooling (intercooler in engine coolant circuit; split-circuit coolant system).

TE split-circuit cooling system: the Series 396 engine–and other MTU designs–can be supplied with the TE split-circuit coolant system with a power-dependent subcircuit to cool the combustion air. Optimum performance is fostered throughout the engine's power range: at idling speed the air supply is heated to achieve complete fuel combustion; in the medium power and full load range conditions are optimized for high output while keeping thermal stress on engine components at a low level.

Coolant flow from the engine is split in two. Approximately two-thirds of the flow passes through a high temperature (HT) circuit and returns directly to the engine inlet, while the remainder

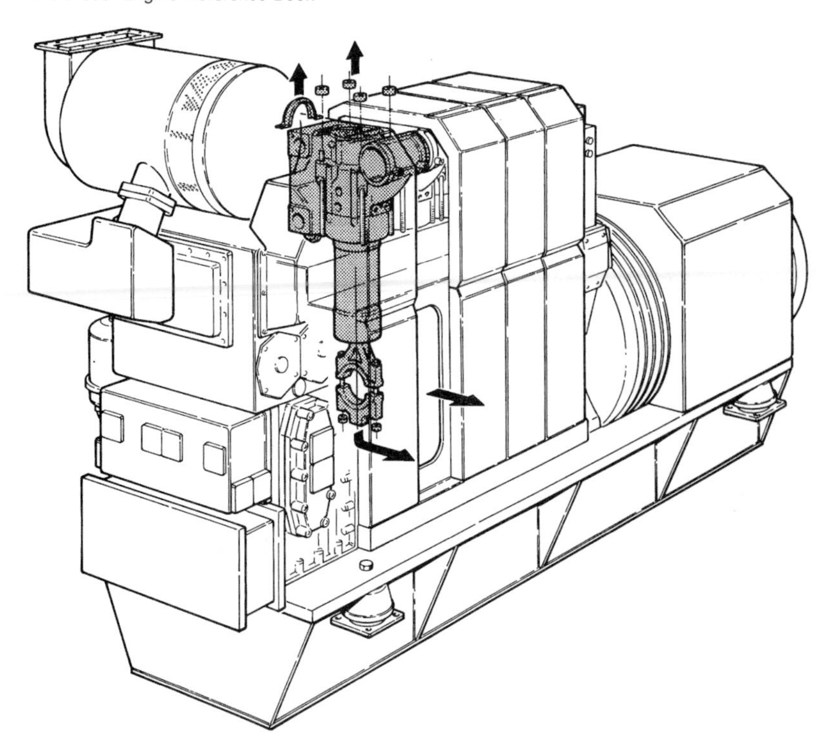

Figure 26.12 MAN B&W Holeby L16/24 engine: removing a complete cylinder unit

Figure 26.13 MTU 16V 396 TE94 engine with split-circuit cooling system

is fed into a thermostatically controlled low temperature (LT) circuit. During engine idling or low-load operation the thermostat allows heated coolant to bypass the recooler on its way to the intercooler in order to warm up the combustion air and prevent white smoke in the exhaust.

An annular slide valve in the thermostat remains in its initial position until increasing power raises the coolant temperature, causing the wax pellet in the thermostat to expand. Gradual closing of the bypass line now directs the coolant stream through the recooler. As a result, coolant entering the intercooler is at a low temperature which, in turn, underwrites a high combustion air volume and, consequently, maximum engine power. After flowing through the intercooler and the oil heat exchanger, 'cold' coolant rejoins the uncooled HT circuit, thereby cooling the total volume flow before it re-enters the engine.

Sequential turbocharging is exploited for propulsion engines required to deliver high power in the lower and medium speed ranges. The system incorporates two or three turbochargers with automatic on/off control as a function of engine speed, power demand and turbocharger maximum efficiency. In addition to increased torque the system yields reduced fuel consumption and lower exhaust temperatures.

Cylinder cut-out (no fuel injection into selected cylinders) is adopted for engines operating under varying speed conditions (low idling speed). The system enhances combustion at idle, thus shortening the warm-up phase and avoiding white smoke from unburnt fuel, and generally contributing significantly to environmental compatibility.

Series 595

A power range from 2000 kW to 4320 kW at 1500–1800 rev/min is covered by the 190 mm bore/210 mm stroke Series 595 engine, introduced in 1990 in V12- and V16-cylinder versions for propulsion and genset applications (*Figure 26.14*).

Contributing to a compact design yielding power-to-weight ratios of under 3 kg/kW and a power-to-space ratio of 250 kW/m^3 (including all ancillaries) are the following features: a V72-degree cylinder configuration with all ancillaries arranged for space saving within the engine contour; nodular cast iron crankcase extending below the crankshaft centreline to maximize rigidity; crankshaft with hardened main and connecting rod bearing radii, designed to withstand firing pressures of up to 180 bar; individual fuel injection pumps designed for injection pressures of up to 1500 bar; plate-type coolant heat exchanger integrated with the engine; split-circuit coolant system; and two-stage sequential turbocharging system with charge air cooling for boost pressures of up to 4.8 bar. The turbocharging sequencing is electronically controlled.

Components requiring regular servicing are located to provide good access at the engine's auxiliary power take-off end. Maintenance is also smoothed by the modular arrangement of functionally interlinked components, plug-in connections and the omission of complex pipework, hose couplings and cabling. A new electronic control system (ECS) was purpose-developed by MTU for the Series 595 engine to provide information for the operator, promote easier operation and enhance safety, reliability and economy. The safety functions embrace engine and plant monitoring, overload protection, diagnostics, automatic start-up and load control.

Series 1163

The highest power outputs from the MTU programme are offered by the Series 1163 engines (*Figure 26.15*). The 230 mm bore/280 mm stroke design is produced in V60-degree configuration 12-, 16- and 20-cylinder versions delivering up to 7400 kW at 1300 rev/min for commercial and naval propulsion installations. The key elements of the design are summarized as:

Crankcase: nodular cast iron structure with access ports on both sides and a flange-mounting facility for alternators or other driven machinery; a welded sheet steel oil pan is provided.

Cylinder liner: replaceable, wet-type centrifugally cast.

Cylinder head: cast iron component arranged with two inlet and two exhaust valves, all equipped with Rotocap rotators, a centrally located fuel injector separated from the rocker area, and a decompression valve.

Valve actuation: by two camshafts, roller tappets, pushrods and rocker arms. Valve clearance adjustment is performed via two screws located in the rocker arms.

Crankshaft: a single-structure forged component finished all over and featuring bolted-on counterweights; axial crankshaft alignment is effected by a deep-groove ball bearing.

Bearings: thin-walled, two-piece, replaceable steel-backed tri-metal sleeve bearings for the crankshaft and connecting rod, with cross-bolted bearing caps.

Connecting rods: forged and finished all over and grouped in pairs to serve two opposite cylinders.

Piston: composite-type with light alloy skirt and bolted-on steel crown, cooled by oil spray nozzles; three compression rings are fitted in the crown, with one oil control ring between crown and skirt; all rings replaceable after piston crown removal.

Fuel injection: direct injection by individual pumps and via short high pressure lines; pump replacement requires no readjustment; gear-type fuel delivery pump; two duplex fuel

Figure 26.14 V16-cylinder MTU Series 595 engine

Figure 26.15 MTU Series 1163 engine in V20-cylinder form

filters; cylinder cut-out facility available (see under Series 396 engine section).

Governor: hydraulic MTU unit mounted on the gearcase, with the linkage between governor and fuel injection pumps accommodated in the gearcase and camshaft space. Engine shutdown solenoids act on the fuel rack; in addition, independent emergency air shut-off flaps are arranged to block the engine's air supply.

Lubricating oil system: self-contained forced feed system, with oil flow progressing through gear-type pumps, heat exchanger and filters to the engine lubrication points. Oil for piston cooling is tapped off after the heat exchanger and filters; the engine-mounted oil heat exchanger is integrated in the engine coolant circuit (Series 1163-02 model) or raw water circuit (Series 1163-03 model).

Cooling system: two-circuit system; closed engine coolant circuit using two centrifugal pumps; thermostatic control; re-cooling provided by raw water heat exchanger or fan cooler.

Turbocharging: the Series 1163-02 model exploits pulse charging based on two turbochargers (one turbocharger for the 12-cylinder engine) with intercoolers incorporated in the raw water circuit. The Series 1163-03 model exploits constant pressure turbocharging with two-stage air compression and interstage cooling; four or five charger groups (depending on the application) are under sequential turbocharging control; the constant pressure exhaust manifold is located in the vee-bank, and charge air pipework mounted externally; high pressure (HP) and low pressure (LP) intercoolers are incorporated in the raw water circuit. For start-up and low-load operation the Series 1163-03 model features a charge air preheater, fed with engine coolant, in each HP intercooler outlet. Coolant jackets for the exhaust manifolds and turbocharger turbines of this model secure reduced heat rejection.

Like the Series 595 engine, the Series 1163 design features inboard high pressure fuel injection lines and the enclosure of all hot exhaust components in water-cooled and gas-tight casings: valued for unmanned enginerooms by helping to reduce the possibility of fire in the event of fuel or lubricating oil leakages. A triple-walled insulation design also maintains surface temperatures well within the limit of 220°C dictated by classification societies, as well as considerably reducing radiant heat in the engineroom.

Sequential turbocharging (output-dependent control of the number of turbochargers deployed) fosters high torque at low rev/min (wide performance band) and hence good acceleration. Optimization during sea trials, with assistance from an electronic control system, can eliminate black smoke emissions.

An uprated Series 1163 engine for fast ferry propulsion benefits from:

- An improved fuel injection system to optimize consumption and exhaust emissions reduction.
- A higher cylinder head fatigue strength, thanks partly to bore cooling.
- Improved power and torque characteristics achieved by modifying the two-stage sequential turbocharging system.
- Adaptation and optimization of the turbochargers to match the increased cylinder output (325 kW).
- Intelligent electronic engine management.

26.1.9 MTU/DDC designs

An alliance formed by MTU in 1994 with Detroit Diesel Corporation (DDC) resulted in the creation of two advanced new high speed designs for joint marketing by the German/US partners: the Series 2000 and Series 4000 engines which were launched in 1996. MTU was responsible for the basic design development of both engines and the variants for propulsion applications.

Series 2000

The Series 2000 design, with a 130 mm bore/150 mm stroke, is based on the Mercedes-Benz 500 commercial vehicle engine (*Figure 26.16*). Models with V90-degree 8, 12 and 16 cylinders cover a propulsion band from 400 kW to 1343 kW, while genset drive applications start from 270 kW. The most powerful version, intended for yachts, has a maximum speed of 2300 rev/min. A single-stage turbocharging system is specified as standard but, when special requirements regarding power band width and acceleration characteristics are dictated, Series 2000 engines can benefit from MTU's sequential turbocharging system. Fuel injection, adapted from the commercial vehicle engine, is a solenoid valve-controlled pump/fuel line/injector system, each cylinder having its own separate injection pump.

Figure 26.16 MTU Series 16V 2000 engine

Marine versions are fitted with MTU's TE twin-circuit cooling system (detailed above in the Series 396 engine section) which fosters an optimum temperature for all operating conditions: intake air can be heated at idling speeds or at partial throttle and cooled at full throttle.

Series 4000

A number of notable design and performance features were highlighted by MTU for the 165 mm bore/190 mm stroke Series 4000 engine whose designers focused on high reliability and ease of maintenance without compromising compactness (*Figures 26.17* and *26.18*). A propulsion power band from around 840 kW to 2720 kW at 2100 rev/min is covered by the V90-degree 8-, 12- and 16-cylinder versions.

Testing confirmed fuel economy figures that reportedly set a new standard for compact engines in the design's performance class and speed range. A specific fuel consumption of 194 g/kWh was considered exceptional for an engine with a power-to-weight ratio of between 2.7 and 3.5 kg/kW.

A key contribution, along with a high peak firing pressure potential, is made by the common rail fuel injection system–an innovation in this engine category–whose development was carried out in conjunction with MTU's specialist subsidiary company L'Orange (*Figure 26.19*). The system, which allows infinite adjustment of fuel injection timing, volume and pressure, embraces a high pressure pump, a pressure accumulator, injectors and an electronic control unit. The tasks of pressure generation and fuel proportioning are assigned to different components: pressure generation is the job of the high pressure pump while metering of the fuel relative to time is the job of the injectors.

The flexibility of the common rail system across the engine power band enables it to deliver the same injection pressure (around 1200 bar) at all engine speeds from full rated speed down to idling. The high pressure fuel pump and electronically controlled injectors are fully integrated in the electronic control

Figure 26.17 MTU Series 4000 engine in V16-cylinder form. The design features a common rail fuel injection system, an innovation in its class

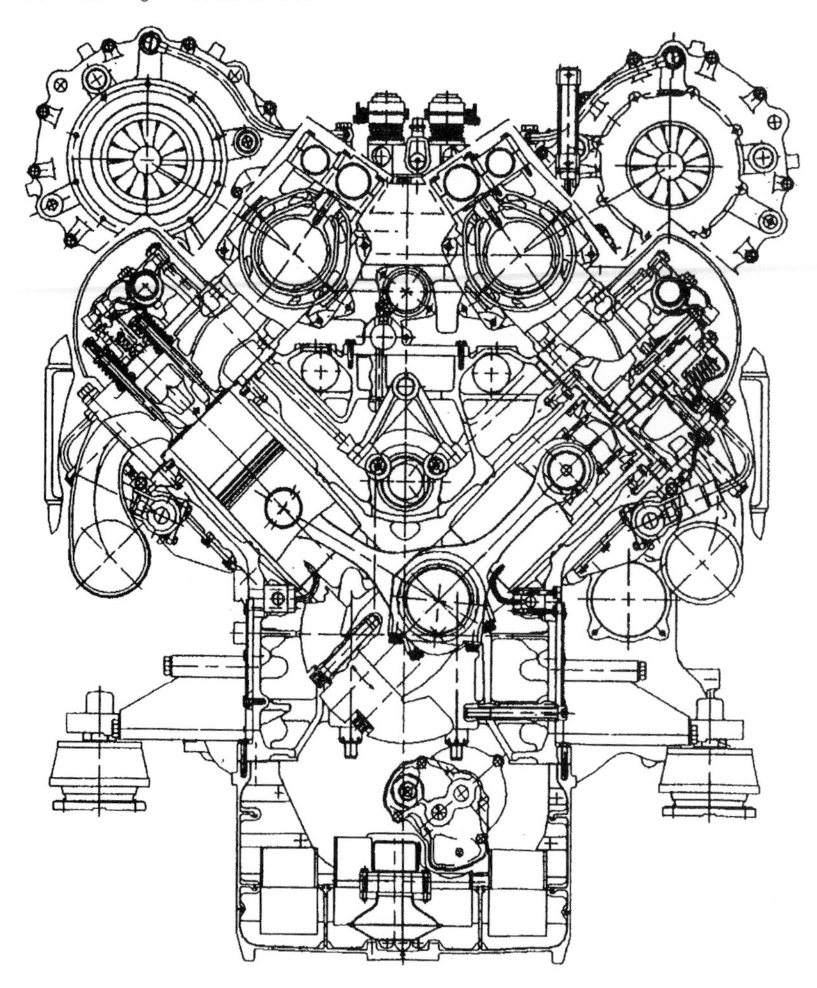

Figure 26.18 Cross-section of MTU Series 4000 engine

system. Peak pressures with the common rail system are around 20 per cent lower than with conventional systems, reducing stress on the high pressure components. The pump is also simpler and its plunger has no helices. Since the pressure-relief requirement of conventional systems is eliminated the mechanical effort is reduced, further enhancing fuel economy.

One fuel rail is provided for the injectors and their respective solenoid valves on each bank of cylinders. The fuel pressure is generated by the high pressure pump driven via a gear train at the end of the crankcase. The electronic system controls the amount of fuel delivered to the injectors by means of the solenoid valves, while the injection pressure is optimized according to the engine power demand. Separate fuel injection pumps for each cylinder are thus eliminated, and hence the need for a complicated drive system for traditional pumps running off the camshafts. A reduced load on the camshaft and gearing system is therefore realized, and the removal of the mounting holes for separate injection pumps enhances the rigidity of the crankcase.

Maximum strength and optimized rigidity were sought from the crankcase, with integration of the main lubrication and coolant circulation channels reducing the number of separate components (*Figure 26.20*). The crankshaft, machined all over and with bolted-on counter-weights, is designed to withstand maximum combustion pressures of up to 200 bar. Wear-resistant sleeve bearings contribute to longer service intervals. The cylinder head design, with integrated coolant channels and stiffened base, also addresses a 200 bar pressure. Provision is made for two inlet and two exhaust valves, and a fuel injector located in the centre of the head (*Figure 26.21*).

A piston comprising an aluminium skirt and a bolted-on steel crown is provided with chrome-ceramic coated rings which, in combination with the plateau-honed cylinder liner, foster low lubricating oil consumption and an extended service life. An optimized combustion chamber shape and the bowl of the composite piston promote low fuel consumption and low emission levels.

Exhaust and turbocharging systems on the Series 4000 engine benefit from established MTU practice. A triple-walled exhaust system, with an outer water-cooled aluminium casing, ensures that the surface temperatures do not exceed permissible levels at any point while also securing gas tightness. Heat dissipation via the cooling system is reduced. The exhaust pipework is positioned centrally between the V-cylinder banks. The turbochargers are mounted in a water-cooled housing. A sequential turbocharging system is exploited to deliver high torque at low engine speeds. A wide power band at high fuel economy is also claimed, along with excellent acceleration capabilities without black smoke emission.

A split-circuit cooling system for the engine and intake air complements the sequential turbocharging system. It acts as an 'intelligent' cooling system to maintain engine coolant, lubricating oil and intake air at an optimum temperature for all operating conditions. When the engine is idling or running at low load the temperature of the intake air is raised in order to ensure smooth

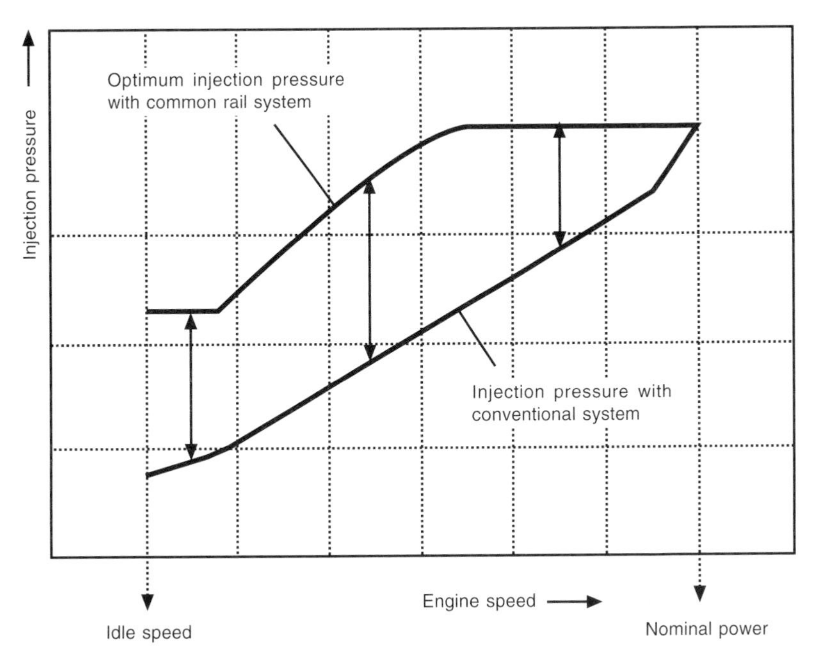

Figure 26.19 Performance of common rail fuel injection system of MTU's Series 4000 engine

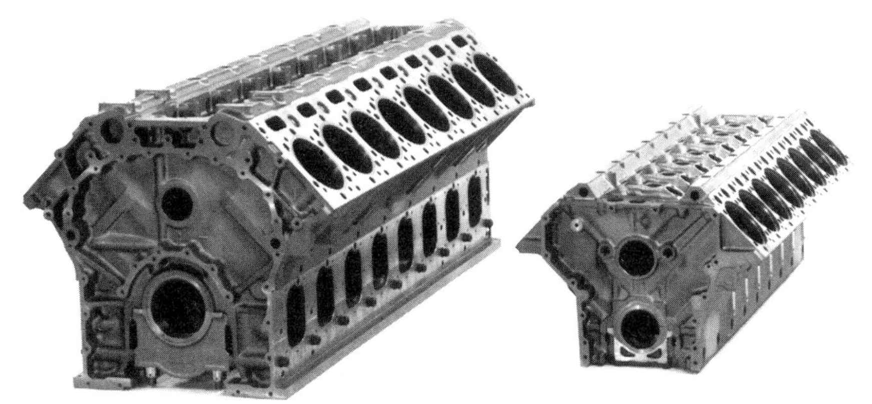

Figure 26.20 The crankcases of MTU's Series 2000 (right) and Series 4000 engines

and complete combustion without generating white smoke emissions.

On the front end of the engine is an integral service block providing ease of inspection and maintenance of the seawater cooler, oil cooler, oil filter, oil centrifuge and fuel filter. The turbochargers and charge air coolers are located at the flywheel end.

Access openings in the crankcase are said to be large enough to allow all running gear servicing to be performed without removing the engine, even under restricted machinery space conditions. The application-tailored service block integrated in the auxiliary power take-off end also simplifies routine maintenance tasks. A variety of ancillary mounting options enable the engine to be matched to specific customer requirements.

Early indication of necessary maintenance, based on the actual duty profile, is provided by an electronic engine management system designed to foster reduced downtime periods and lower servicing costs. A time-between-major-overhauls of 18 000 hours was anticipated by MTU.

26.1.10 Niigata

The FX series from Niigata Engineering was completed in 1996 with a 205 mm bore model to complement the established 165 mm and 260 mm high speed engines in a programme developed for fast commercial and military vessel propulsion. The resulting 16FX, 20FX and 26FX models offer maximum continuous outputs from 1000 kW to 6200 kW in commercial service, with slightly higher ratings available for naval propulsion. The 16FX engine is produced in eight in-line, V12- and V16-cylinder versions running at 1950 rev/min, the V20FX in V12 and V16 versions running at 1650 rev/min, and the V26FX (*Figure 26.22*) in V12-, 16- and 18-cylinder versions running at 1250 rev/min.

Compact and highly rigid engines were sought by the Japanese designer from a monobloc structure fabricated from nodular cast iron, with the extensive use of light alloy elements contributing to a modest overall weight. Engine width was minimized by adopting a V60-degree cylinder bank and overall length reduced by minimizing the distance between cylinders while maintaining the required bearing width.

26.1.11 Paxman

Over 60 years experience in high speed diesel design was exploited by Paxman Diesels in creating the VP185 engine which was launched in 1993 to join the British company's established 160 mm bore Vega and 197 mm bore Valenta series (*Figure 26.23*), which, respectively, offer specific outputs of 107 kW/cylinder

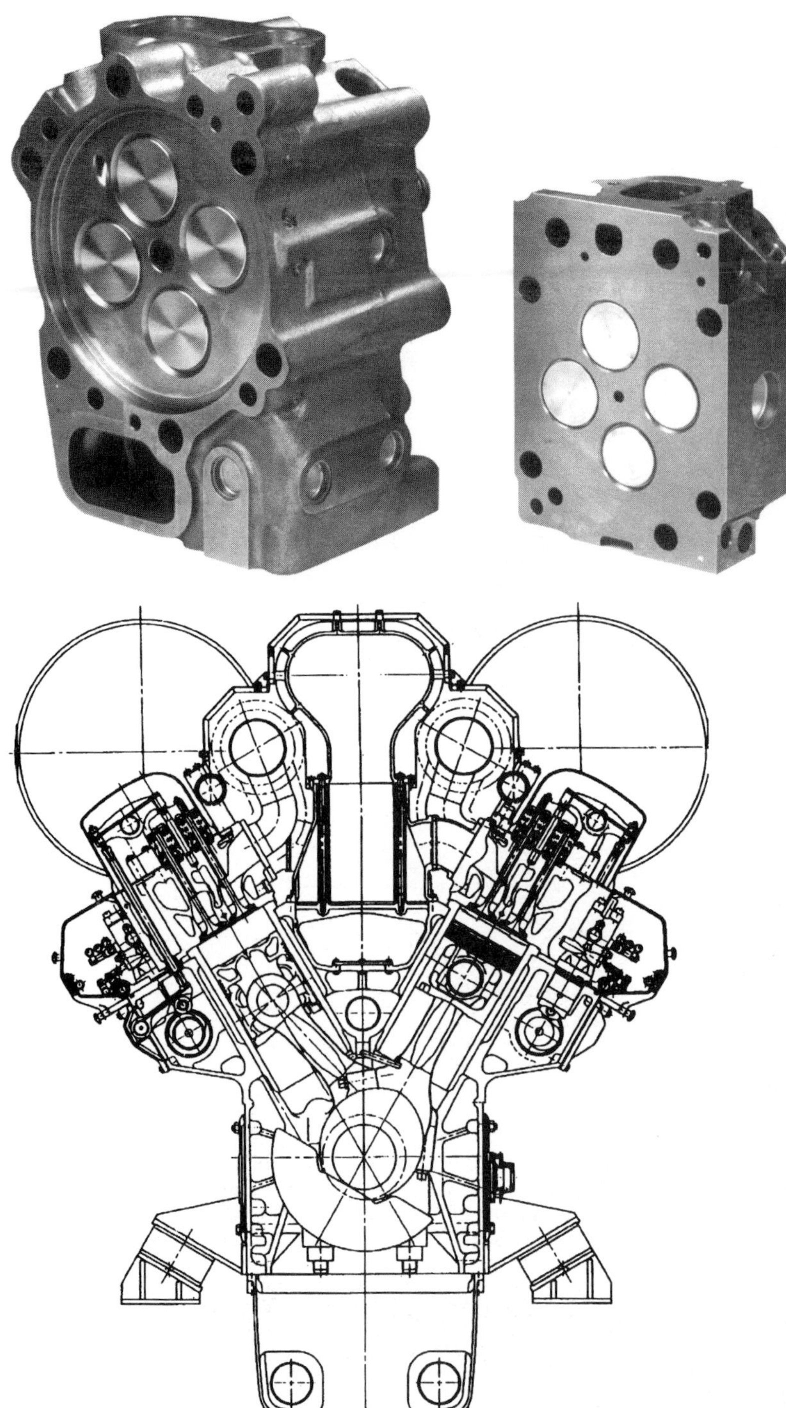

Figure 26.21 The four-valve cylinder heads of MTU's Series 2000 engine (right) and Series 4000 engine (left) with central fuel injector

Figure 26.22 Cross-section of Niigata V26FX engine

and 206 kW/cylinder. The new 185 mm bore design was introduced initially in V12-cylinder form with outputs up to 2610 kW and complemented in 1998 by a V18-cylinder version extending the power limit of the series to just over 3900 kW at a maximum speed of 1950 rev/min.

In designing the 12VP185 engine (*Figure 26.24*), Paxman sought improvements in compactness and lightness over earlier models, goals dictating a smaller swept volume, slightly higher ratings and a review of piston speeds. The first parameter to be fixed was the stroke which was set at 196 mm; this, linked to a crankshaft speed of 1800 rev/min, gave a mean piston speed of 11.8 m/s for continuous duty. An improved speed platform for marine applications was derived, the edge of which is bounded by a maximum speed of 1950 rev/min with a mean piston speed of 12.8 m/s.

The cylinder bore was fixed at 185 mm which, coupled with

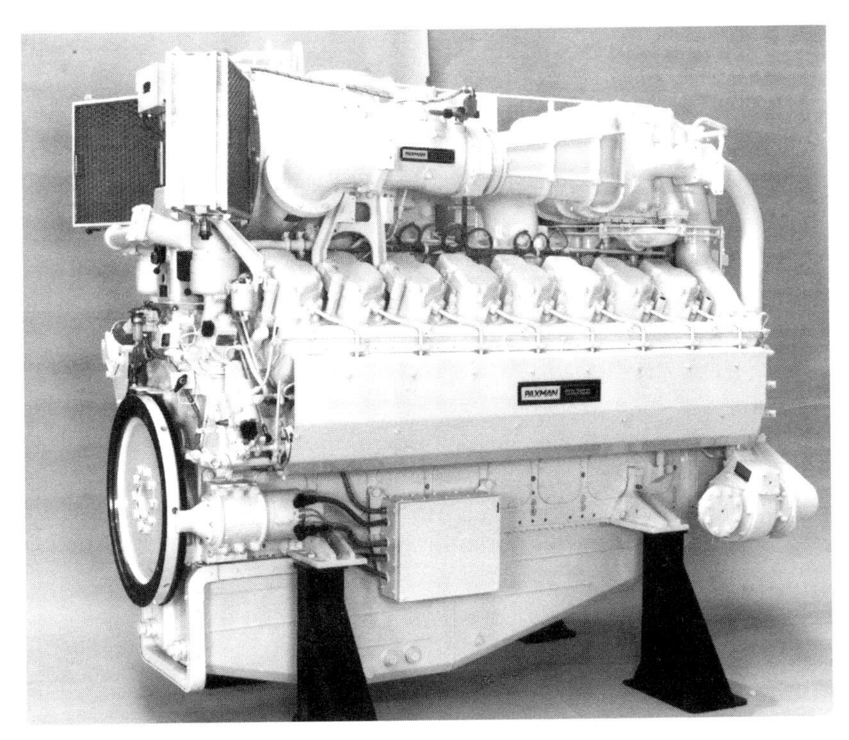

Figure 26.23 Paxman V16-cylinder Valenta engine

Figure 26.24 Paxman 12VP185 engine. Note the compact turbocharging arrangement along the top

a design maximum mean effective pressure of 25.3 bar, delivered a maximum power of 2610 kW at 1950 rev/min from the V12-cylinder configuration: a rating primarily targeting high speed military and commercial marine markets, as well as megayacht propulsion. Lower down the power scale, 2180 kW at 1800 rev/min is available (at 45°C air and 32°C sea water conditions) for continuous marine duties (for example, fast ferry propulsion) where the engine's high torque capability over a broad speed range is considered especially attractive for hydrofoil service. The 18VP185 model offers up to 3916 kW at 1950 rev/min, with an output of 3265 kW at 1835 rev/min quoted for fast ferry propulsion.

A smaller bore and stroke than the Valenta design, combined with a switch from a 60-degree cylinder bank angle to a 90-degree configuration, achieved a layout which was short and with a width under 1.5 m while providing an adequate platform for the charge air system mounted above the engine. The 90-degree bank angle itself also lowered the profile of the charge air system to foster space efficiency.

A higher mean effective pressure normally leads to a lower compression ratio which in turn promotes poor starting characteristics, reduced thermal efficiency and possible exposure to ignition delay damage. Some of these problems can be addressed by the use of inter-cylinder charge air transfer arrangements but Paxman decided to adopt a compression ratio in excess of 13:1. The resulting VP185 engine was claimed to be simple and easy to start under cold conditions, with good thermal efficiency and the potential to satisfy low NOx emission limits. The combination of high boost levels and compression ratios demands a robust construction but Paxman sought to avoid a heavy and cumbersome engine by securing robustness within a small and hence a stiff envelope.

The reciprocating assembly of the engine was designed for high strength and rigidity in handling a high maximum cylinder pressure and securing high reliability with a long lifetime. The backbone is a compact crankcase cast in high strength spheroidal graphite iron to provide a stiff and solid support for the underslung crankshaft. Crankcase doors along each side of the engine give access to the connecting rod large ends for *in situ* servicing and piston removal.

The crankshaft is a fully machined steel forging, fully nitrided to yield strength and durability. It is secured by main bearing caps which are drawn internally against the deep-fitting side faces of the crankcase by high tensile set screws and hydraulically tensioned main bearing studs. Such a configuration, the designer asserts, delivers good strength and stiffness characteristics to the bottom end of the crankcase and provides solid support for the crankshaft. A generous overlap between the crankpin and main bearing journals adds to stiffness and strength. Both main and big end bearings are of steel-backed aluminium–tin type with thrust washers controlling axial location. The crankshaft is provided with a viscous torsional damper totally enclosed within the gearcase.

Connecting rods of side-by-side design are forged in high tensile steel, fully machined and ferritic nitro-carburized. The large end is obliquely split to allow the rod to pass through the cylinder liner and the joint faces are serrated. The one-piece nodular cast iron pistons incorporate a large oil cooling galley above the ring belt, oil being supplied by accurately aligned standing jets mounted in the crankcase. The pistons run in cylinder liners made of centrifugally cast high grade iron.

Individual cast iron cylinder heads are designed to withstand the high firing pressures and incorporate two exhaust valves, two inlet valves and a centrally mounted unit fuel injector (combined pump and injector) which is fixed with a single screw. The valves and injectors are actuated through a pushrod arrangement from a single large diameter camshaft mounted centrally in the vee of the engine. The stiffness of the actuating system reportedly achieves valve control similar to that of an overhead camshaft while the specially designed unit injectors yield very high rates of fuel injection with clean cut-off, contributing to low NOx generation and good fuel efficiency. The fuel cam design was based on an optimized width for long life and a large base circle, coupled with high rates of lift to secure the key injection characteristics.

The unit injector is rack controlled, the rack itself being controlled by a linkage operated from twin shafts in the centre vee of the engine. The shafts in turn are coupled to the governor via an overspeed protection assembly which closes the fuel racks independently of the governor in the event of the engine overspeeding.

Low pressure fuel oil is supplied from a gear-type lift pump mounted on the main gearcase. The fuel is fed to a fuel service sub-assembly comprizing cooler, filter (duplex optional), reservoir and solenoid-operated shut-off valve. The sub-assembly is mounted high at the free end of the engine for convenient access.

A single camshaft reflects the policy adopted by the designers of minimizing the number of components. This, allied to the 90-degree cylinder bank angle, allowed the camshaft gear to mesh directly with the crankshaft gear and so eliminate the need for idlers in a critical area and reduce the component count even further.

Charge air is delivered by a 'valve-less' two-stage turbocharging system designed for simplicity and ease of maintenance. Preliminary explorations of single-stage high pressure ratio turbocharging showed poor low speed torque characteristics due to surge limitation, along with poor acceleration and load acceptance. A two-stage turbocharging arrangement with intercooling and aftercooling was finally adopted, based on six large automotive-type turbochargers with broad and stable operating characteristics: four turbochargers provide low pressure charge air and two provide the high pressure air.

Inlet air compressed in the low pressure turbochargers is fed through a raw-water intercooler to the compressors of the high pressure turbochargers; high pressure air is then passed through the jacket water aftercooler into the air manifolds on either side of the engine. The high pressure stage exploits a pulsed exhaust system to give good low-end performance without recourse to complex valve systems associated with sequentially turbocharged configurations. The turbocharging system's high air–fuel ratios, coupled with high pressure/high rate fuel injection, address low NOx emission requirements. A highly responsive performance with good engine torque characteristics is reported throughout the speed range.

All six turbochargers are mounted in the walls of a water-cooled gas-tight casing, the turbine sides arranged on the inside and the compressors on the outside. The rotating assembly and compressor casing of each turbocharger form a cartridge which can be replaced quickly without the need for lifting gear and without disturbing the rest of the engine. The exhaust manifolds feature sliding joints and, like the turbochargers, are housed in water-cooled gas-tight aluminium casings; together with the turbocharger enclosure, they form a single, cool gas-tight shell around the hot parts of the engine. The casings foster low radiant and convective heat losses from the engine.

The gear train is mounted at the engine's free end and the camshaft gearwheel meshes directly with the crankshaft gearwheel. Auxiliary drives are provided for the externally mounted lubricating oil and jacket water pumps, the governor, overspeed governor and fuel lift pump; further auxiliary power take-off capability is incorporated for different applications. The PTO at the free end will accept either gear-driven or belt-driven pumps and alternators. The VP185 is supplied as a complete assembly with engine-mounted jacket water and lubricating oil heat exchangers to simplify installation procedures. A choice of air, electric or hydraulic starting systems is available. Electronic or hydro-mechanical governing is based on governors from Regulateurs Europa, a fellow member with Paxman Diesels of the Anglo-French GEC Alsthom group.

26.1.12 SEMT-Pielstick

The Paris-based designer SEMT-Pielstick fields three long-established high speed programmes, the PA4, PA5 and PA6 series, now under the umbrella of the German group MTU. The PA4 is produced in 185 mm and 200 mm bore versions (both having a stroke of 210 mm) and features a variable geometry (VG) pre-combustion chamber; and the PA5 is a 255 mm bore/270 mm stroke design.

Evolution has benefited the 280 mm bore PA6 design, introduced in 1971 with a stroke of 290 mm and a maximum continuous rating of 258 kW/cylinder at 1000 rev/min. The output was raised to 295 kW/cylinder in 1974 and to 315 kW/cylinder in 1980. Extensive service experience in naval vessels moulded progressive refinements over the years, including the creation of a BTC version in 1980 which yielded 405 kW/cylinder at 1050 rev/min through the adoption of a reduced compression ratio together with two-stage turbocharging. This PA6 BTC model was released in 1985 with a higher output of 445 kW/cylinder. The series was extended in 1983 by a longer stroke (350 mm), slower speed PA6 CL variant with a rating of 295 kW/cylinder at 750 rev/min.

A sequential turbocharging system (STC) was introduced to the 290 mm stroke engine in 1989, this PA6 STC model developing 324 kW/cylinder at 1050 rev/min. Performance was enhanced in a 330 mm stroke B-version from 1994, an STC system and a cylinder head with improved air and gas flow rates contributing to a nominal maximum continuous rating of 405 kW/cylinder at 1050 rev/min. A maximum sprint rating of 445 kW/cylinder at 1084 rev/min equates to an output of 8910 kW from the V20-cylinder PA6B STC engine which targets high performance vessels.

At the nominal mcr level the V12-, 16- and 20-cylinder models, respectively, offer 4860 kW, 6480 kW and 8100 kW for fast ferry propulsion, and are released for sustained operation at these ratings. The power-to-weight ratio of the 20PA6B STC model at 8100 kW is 5.2 kg/kW (*Figure 26.25*).

PA6B STC design

High rigidity without compromising overall engine weight is achieved by a stiff one-piece crankcase (*Figure 26.26*) of nodular cast iron specially treated for shock resistance, with transverse bolt connections between both crankcase sides through the underslung-type bearing caps. Integrated longitudinal steel piping supplies lubricating oil to the main bearings. Large-dimension main journals yield a particularly large bearing surface and conservative pressures which underwrite prolonged bearing life, as do the large surface area connecting rod bearing shells. The alloy steel one-piece forged crankshaft (*Figure 26.27*) has high frequency hardened crankpins and journals, and is bored to feed lubricating oil to the connecting rods.

A nodular cast iron cylinder head was configured to achieve improved air and gas flow rates for the PA6B STC version. The composite piston (steel crown and aluminium skirt) is fitted with five rings. The forged steel connecting rod is fitted with large surface bearing shells to yield high durability under high firing pressures. Special low wear characteristics were sought from the centrifuged cast iron material specified for the cylinder liner (*Figure 26.28*). The camshaft is formed from several sections for ease of dismantling.

Operational flexibility and power output are fostered by the single-stage sequential turbocharging system which is based on two turbochargers and delivers a large combustion air excess at partial loads. Supercharging is effected by one turbocharger for engine loads up to 50 per cent of the nominal power, its effort boosted by the second identical turbocharger at higher loads. Switching from single to twin turbocharger mode is performed

Figure 26.25 SEMT-Pielstick V20-cylinder PA6B STC engine with sequential turbocharging system. A rating of 8100 kW is offered for fast ferry propulsion

Figure 26.26 SEMT-Pielstick PA6B STC crankcase

Figure 26.27 SEMT-Pielstick PA6B STC crankshaft

Figure 26.28 SEMT-Pielstick PA6B STC cylinder liner

automatically by opening two flap valves. Engine performance at prolonged low load is improved with respect to fuel consumption, smoke emission, fouling resistance and transient performance. Additionally, engine utilization is expanded towards the high torque/low rev/min area.

26.1.13 Wärtsilä Diesel

The medium speed specialist Wärtsilä Diesel added high speed designs to its four-stroke engine portfolio with the acquisition in 1989 of SACM Diesel of France whose 142 mm bore UD23 and 150 mm bore UD25 models remain in the programme. The Finnish parent group saw the market potential for a new generation engine blending the best features of high and medium speed designs for continuous duty applications, resulting in the 1994 launch of the Wärtsilä 200 series (*Figures 26.29* and *26.30*).

The 200 mm bore/240 mm stroke engine is produced by Wärtsilä France for diverse propulsion and genset drive applications with an output band from 2100 to 3600 kW at 1200 or 1500 rev/min covered by V12-, 16- and 18-cylinder models. Wärtsilä Diesel markets the engine as the W200 series while, under a joint venture partnership, it is fielded in the US-based Cummins' portfolio as the QSZ family.

High reliability in continuous duty (defined as 24 hours a day operation with an annual running period of over 6000 hours) was sought from an engine structure and main components designed for a maximum cylinder pressure of 200 bar. Wärtsilä Diesel's medium speed engine technology was exploited to achieve a high power density, low emissions and fuel consumption, and ease of maintenance.

The nodular cast iron engine block (*Figure 26.31*) is designed for maximum stiffness with a V60-degree configuration optimizing balancing and hence limiting vibration. Bolted supports dictate only four to six fixing points, and provision is made for elastic mounting. The locations of the camshaft drive (integrated in the flywheel end of the block) and the air receiver channel (in the middle of the vee-bank and also integral with the block) reflect solutions adopted in Wärtsilä Diesel medium speed engines. Arranging the oil lubrication distribution through a channel in the middle of the vee, however, was a new idea. The areas of the block that support the camshafts and fuel injection pumps are designed for high strength to accept the forces created by the high injection pressures. Large openings on both sides of the block facilitate access for inspection and maintenance.

The cylinder unit also benefited from solutions validated in other Wärtsilä Diesel designs, notably: a plateau-honed cast iron cylinder liner incorporating an anti-polishing ring at the top to eliminate bore polishing; a composite piston (steel crown and aluminium alloy skirt) with a three-ring pack and optimized cooling; and Wärtsilä Diesel's patented piston skirt lubrication system.

The connecting rods, placed side-by-side and axially guided by the piston, are designed to facilitate oil lubrication to the small-end bearings and pistons. A stepped split allows their removal up through the cylinder liners.

The strong 220 mm diameter forged crankshaft (*Figure 26.32*) has gas-nitrided surfaces for added safety and a high degree of balancing from two bolted counterweights per crank throw. The generous diameters of both crankpin and journal achieve a large bearing surface while allowing a reduced cylinder spacing ratio (1.5 times the bore) to minimize engine length and weight.

Supporting the loads generated by the high injection pressure, the well-dimensioned camshafts are formed from segments bolted together, each segment serving three cylinders. The camshafts, located on both sides of the engine block, are driven by case-hardened gearwheels arranged inside the engine block at the flywheel end.

Among the measures designed to ease inspection and maintenance are: cylinder heads hydraulically tightened with four studs on the cylinder block; and connecting rod big ends and main bearing caps fastened by two hydraulically tightened studs.

Fuel injection starts slightly before TDC so that combustion takes place during the beginning of the expansion phase in lower temperatures. The combination of a high compression ratio (16:1) and late fuel injection fosters low NOx generation without raising fuel consumption since the engine's injection period at 1500 rev/min is sufficiently short. Applying this low-NOx principle requires the engine to be able to inject fuel late in the cycle and over a short duration without undermining performance: high pressure capacity injection equipment is therefore dictated. The engine is released for operation on marine diesel oil (ISO 8217, F-DMX to F-DMB).

Individual fuel injection pumps are integrated in the same cast multihousing (*Figure 26.33*) as the inlet and exhaust valve tappets and the inlet and outlet fuel connections, and can be

Figure 26.29 Wärtsilä W200 engine

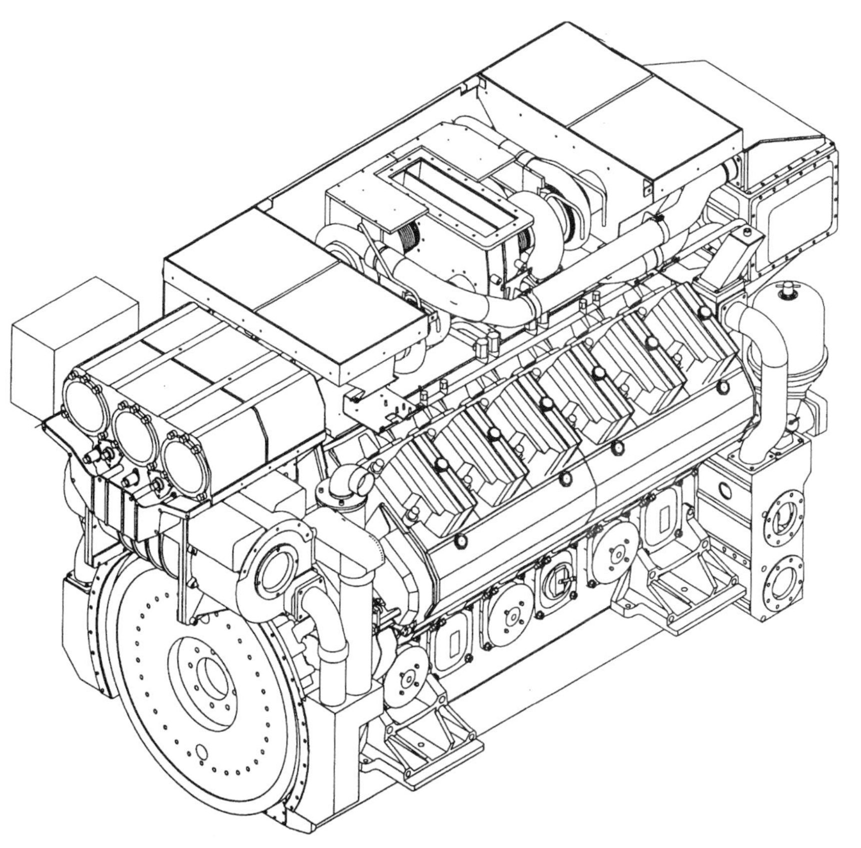

Figure 26.30 Wärtsilä W200 engine. The original automotive-type turbochargers shown here were replaced by ABB or Holset models

Figure 26.31 The nodular cast iron block of the Wärtsilä W200 engine incorporates fluid channels

Figure 26.32 The Wärtsilä W200 engine crankshaft has gas-nitrided surfaces

removed directly without touching the housing or any piping. The coated pump plungers are designed for pressures up to 2000 bar and for a high flow capacity. Fuel is injected into the cylinder through a nitrided eight-hole nozzle designed to yield optimal performance in combination with the combus-tion bowl and swirl level, and to ensure good resistance to wear and thermal loads. The feeding parts are optimized to eliminate the risk of cavitation. Injection timing is adjusted from the camshaft drive.

An important contribution to engine performance is made by the cooling water system which is divided into low temperature (LT) and high temperature (HT) systems. The LT system cools both the charge air and the lubricating oil; the charge air and oil coolers are placed in parallel to achieve maximum cooling efficiency. Increased engine component reliability is reported due to the low lubricating oil temperature (70°C after cooler) and reduced NOx emissions due to the cold charge air temperature (60°C after cooler). The HT system cools the engine block and also the first stage of the charge air when optional maximized waste heat recovery on the HT line is specified. Both charge air cooler and lubricating oil cooler are tube-type units with horizontal tubes; only the end covers have to be removed to clean the water side.

Integration of lubricating oil, fuel oil and cooling equipment in modules mounted on the engine achieves a significant reduction in external connections: only two for water, two for fuel oil and one for starting air. The lubricating oil filters can be changed one by one, even while the engine is running, using a three-way valve to interrupt the oil flow to the filter chamber. The chamber is drained externally by a tap and cleaned without risk of contaminating the lubricating oil system. The system is equipped with a bypass centrifuge filter, located to facilitate ease of maintenance while the engine is running. The washable cell-

Figure 26.33 The fuel injection pump of Wärtsilä's W200 engine is integrated in the multihousing

type air filters are mounted on top of the engine and fixed with a quick clamping arrangement.

The lubricating oil pump, electrically driven pre-lubricating pump, cooling water pump, fuel pump and optional priming fuel pump, and seawater pump are all mounted externally at the free end of the engine. The fuel pump is placed on the same shaft as the lubricating oil pump. The twin water pump for both LT and HT cooling water circuits has only one shaft for both circuits; this means fewer gearwheels need checking and less changing of bearings and sealings. The pre-lubricating pump can also be used for draining the oil sump.

An engine flywheel housing of multi-functional design supports starters and turning gear, and conveys fluids between the different modules. The starting motor is of the air turbine type, operating effectively even at low pressure.

Marine versions of the W200 were originally turbocharged with four automotive-derived turbochargers but these were replaced by ABB or Holset marine models.

26.1.14 Automotive-derived engines

Marinized automotive-derived high speed engines are popular for small craft propulsion and genset drives. VOLVO PENTA's currently most powerful design, the TAMD163P, is rated for 566 kW at 2100 rev/min in special light duty mode (*Figure 26.34*). The in-line six-cylinder 16-litre engine exploits a four-valve head and advanced combustion technology. Another Swedish truck engine specialist, SCANIA, offers its marinized DSI14 model with a rating of 551 kW. The V8-cylinder 14-litre model features twin water-cooled turbochargers and a triple charge cooling system: a freshwater intercooler integrated into each of the air inlet manifolds and a seawater intercooler at the aft of the engine for primary cooling of the charge air.

26.2 Low speed engines

26.2.1 Introduction

Low speed two-stroke engine designers have invested heavily to maintain their dominance of the mainstream deepsea propulsion sector formed by tankers, bulk carriers and containerships. The long-established supremacy reflects the perceived overall operational economy, simplicity and reliability of single, direct-coupled crosshead engine plants. Other factors are the continual evolution of engine programmes by the designer/licensors in response to or anticipation of changing market requirements, and the extensive network of enginebuilding licensees in key shipbuilding regions. Many of the standard ship designs of the leading yards are based on low speed engines.

The necessary investment in R&D, production and overseas infrastructure dictated to stay competitive, however, took its toll over the decades. Only three low speed engine designer/licensors–MAN B&W Diesel, Mitsubishi and New Sulzer Diesel–survived into the 1990s to contest the international arena.

The roll call of past contenders include names either long forgotten or living on only in other engineering sectors: AEG-Hesselman, Deutsche Werft, Fullagar, Krupp, McIntosh and Seymour, Neptune, Nobel, North British, Polar, Richardsons Westgarth, Still, Tosi, Vickers, Werkspoor and Worthington. The last casualties were Doxford, Götaverken and Stork whose distinctive engines remain at sea in diminishing numbers. The pioneering designs displayed individual flair within generic classifications which offered two- or four-stroke, single- or double-acting, and single- or opposed-piston configurations. The Still concept even combined the diesel principle with a

Figure 26.34 Volvo Penta's automotive-derived TAMD 163P marine engine

steam engine: heat in the exhaust gases and cooling water was used to raise steam which was then supplied to the underside of the working piston.

Evolution decreed that the surviving trio of low speed crosshead engine designers should pursue a common basic configuration: two-stroke engines with constant pressure turbocharging and uniflow-scavenging via a single hydraulically operated exhaust valve in the cylinder head. Current programmes embrace mini-to-large bore models with short, long and ultra-long stroke variations to match the propulsive power demands and characteristics of most deepsea (and even some coastal/shortsea) cargo tonnage. Installations can be near-optimized for a given duty from a permutation involving the engine bore size, number of cylinders, selected output rating and running speed. Bore sizes range from 260 mm to 980 mm, stroke/bore ratios up to 4.2:1, in-line cylinder numbers from four to 12, and rated speeds from around 55 to 250 rev/min. Specific fuel consumptions as low as 154 g/kW h are quoted for the larger bore models whose economy can be enhanced by optional Turbo Compound Systems in which power gas turbines exploit exhaust energy surplus to the requirements of modern high efficiency turbochargers. A trend in recent years has seen the addition of intermediate bore sizes to enhance coverage of the power/speed spectrum and further optimize engine selection.

Both MAN B&W Diesel and New Sulzer Diesel also extended their upper power limits in the mid-1990s with the introduction of super-large bore models–respectively of 980 mm and 960 mm bore sizes–dedicated to the propulsion of new generations of 6000 TEU-plus containerships with service speeds of 25 knots or more. The 12-cylinder version of MAN B&W's current K98MC-C design, delivering 68 520 kW, highlights the advance in specific output achieved since the 1970s when the equivalent 12-cylinder B&W K98GF model yielded just under 36 800 kW. Large bore models tailored to the demands of new generation VLCC propulsion have also been introduced.

Parallel development by the designer/licensors seeks to refine existing models and lay the groundwork for the creation of new generations of low speed engine. Emphases in the past have been on optimizing fuel economy and raising specific outputs but reliability, durability and overall economy are now priorities in R&D programmes, operators valuing longer component lifetimes, extended periods between overhauls and easier servicing. Lower production costs and easier installation procedures are also targeted, reflecting the concerns of enginebuilder/licensees and shipyards. More compact and lighter weight engines are appreciated by naval architects seeking to maximize cargo space and deadweight capacity within given overall ship dimensions. In addition, new regulatory challenges–such as noxious exhaust emission controls–must be anticipated and niche market trends addressed if the low speed engine is to retain its traditional territory (for example, the propulsion demands of increasingly larger and faster containerships which might otherwise have to be met by multiple medium speed engines or gas turbines).

Computer software has smoothed the design, development and testing of engine refinements and new concepts but the low speed engine groups also exploit full-scale advanced hardware to evaluate innovations in components and systems. New Sulzer Diesel began operating its first Technology Demonstrator in 1990, an advanced two-stroke development and test engine designated the 4RTX54 whose operating parameters well exceeded those of any production engine (*Figure 26.35*). Until then, the group had used computer-based predictions to try to calculate the next development stage. Extrapolations were applied, sometimes with less than desirable results. The 4RTX54 engine, installed at the Swiss designer's Winterthur headquarters, allowed practical tests with new parameters, components and systems to be carried out instead of just theory and calculations. Operating data gathered in the field could be assessed alongside results derived from the test engine.

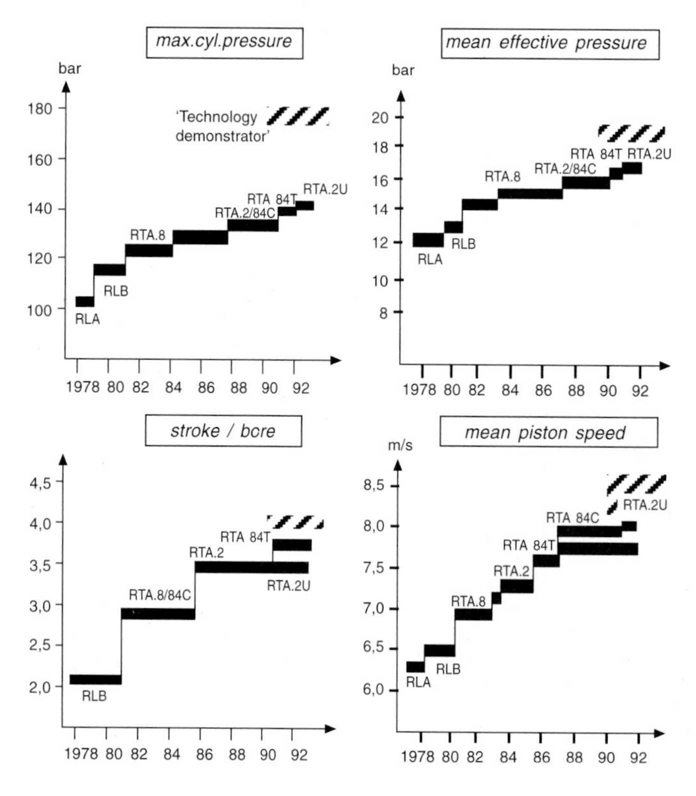

Figure 26.35 Evolution of New Sulzer Diesel low speed engine parameters from the late 1970s in comparison with those for its RTX 54 Two-stroke Technology Demonstrator engine

The four-cylinder 540 mm bore/2150 mm stroke engine had a stroke/bore ratio approaching 4:1 and could operate with mean effective pressures of up to 20 bar, maximum cylinder pressures up to 180 bar and mean piston speeds up to 8.5 m/s. Operating without a camshaft—reportedly the first large two-stroke engine to do so—the RTX54 was equipped with combined mechanical, hydraulic, electronic (mechatronics) systems for fuel injection, exhaust valve lift, cylinder lubrication and starting, as well as controllable cooling water flow. The systems underwrote full flexibility in engine settings during test runs.

New Sulzer Diesel's main objectives from the Technology Demonstrator engine were to explore the potential of thermal efficiency and power concentration; to increase the lifetime and improve the reliability of components; to investigate the merits of microprocessor technology; and to explore improvements in propulsion efficiency. A number of concepts first tested and confirmed on the 4RTX54 engine were subsequently applied to production designs. The upgraded RTA-2U series and RTA84T, RTA84C and RTA96C engines, for example, benefit from a triple-fuel injection valve system in place of two valves. This configuration fosters a more uniform temperature distribution around the main combustion chamber components and lower overall temperatures despite higher loads. Significantly lower exhaust valve and valve seat temperatures are also yielded. An enhanced piston ring package for the RTA-2U series was also proven under severe running conditions on the 4RTX54 engine. Four rings are now used instead of five, the plasma-coated top ring being thicker than the others and featuring a pre-profiled running face. Excellent wear results are reported. The merits of variable exhaust valve closing (VEC) were also investigated on the research engine whose fully electronic systems offered complete flexibility. Significant fuel savings in the part-load range were realized from the RTA84T 'Tanker' engine which further exploits load-dependent cylinder liner cooling and cylinder lubrication systems refined on the 4RTX54. The 4RTX54 engine was replaced as a research and testing tool in 1995 by the prototype 4RTA58T engine adapted to serve as Sulzer's latest Two-stroke Technology Demonstrator (*Figure 26.36*).

The widest flexibility in operating modes and the highest degree of reliability are cited by Copenhagen-based MAN B&W Diesel as prime R&D goals underwriting future engine generations, along with:

- Ease of maintenance.
- Production cost reductions.
- Low specific fuel consumption and high plant efficiency over a wide load spectrum.
- High tolerance towards varied heavy fuel qualities.
- Easy installation.
- Continual adjustments to the engine programme in line with the evolving power and speed requirements of the market.
- Compliance with emission controls.
- Integrated intelligent electronic systems.

Continuing refinement of MAN B&W Diesel's MC low speed engine programme and the development of intelligent engines (see below) are supported by an R&D centre adjacent to the group's Teglholmen factory in Copenhagen. At the heart of the centre is the 4T50MX research engine, an advanced testing facility which exploits an unprecedented 4.4:1 stroke/bore ratio. Although based on the current MC series, the four-cylinder 500 mm bore/2200 mm stroke engine is designed to operate at substantially higher ratings and firing pressures than any production two-stroke engine available today. An output of 7500 kW at 123 rev/min was selected as an initial reference level for carrying out extensive measurements of performance, component temperatures and stresses, combustion and exhaust emission characteristics, and noise and vibration. The key operating parameters at this output equate to 180 bar firing pressure, 21 bar mean effective pressure and 9 m/s mean piston speed. Considerable potential was reserved for higher ratings in later test running programmes.

A conventional camshaft system was used during the initial testing period of the 4T50MX engine. After reference test-running, however, this was replaced by electronically controlled fuel injection pumps and exhaust valve actuators driven by a hydraulic servo-system (*Figure 26.37*). The engine is prepared to facilitate extensive tests on primary methods of exhaust emission reduction, anticipating increasingly tougher regional and international controls in the future. Space was allocated in the R&D centre for the installation of a large NOx-reducing selective catalytic reduction (SCR) facility for assessing the dynamics of SCR-equipped engines and catalyst investigations.

The research engine, with its electronically controlled exhaust valve and injection system, has fully lived up to expectations as a development tool for components and systems, MAN B&W Diesel reports. A vast number of possible combinations of injection pattern, valve opening characteristics and other parameters can be permutated. The results from testing intelligent engine concepts are being tapped for adoption as single mechanical units as well as stand-alone systems for application on current engine types. To verify the layout of the present standard mechanical camshaft system, the 4T50MX engine was rebuilt with a conventional mechanical camshaft unit on one cylinder. The results showed that the continuous development of the conventional system seems to have brought it close to the optimum, and the comparison gave no reason for modifying the basic design.

An example of the degrees of freedom available is shown by a comparison between the general engine performance with the firing pressure kept constant in the upper load range by means of variable injection timing (VIT) and by variable compression ratio (VCR). The latter is obtained by varying the exhaust valve closing time. This functional principle has been transferred to the present exhaust valve operation with the patented system illustrated in the diagram (*Figure 26.38*). The uppermost figure shows the design of the hydraulic part of the exhaust valve; below is the valve opening diagram. The fully drawn line represents control by the cam while the dotted line shows the delay in closing, thus reducing the compression ratio at high loads so as to maintain a constant compression pressure in the upper load range. The delay is simply obtained by the oil being trapped in the lower chamber; and the valve closing is determined by the opening of the throttle valve which is controlled by the engine load.

Traditionally, the liner cooling system has been arranged to match the maximum continuous rating load. Today, however, it seems advantageous to control the inside liner surface temperature in relation to the load. Various possibilities for securing load-dependent cylinder liner cooling have therefore been investigated. One system exploits different sets of cooling ducts in the bore-cooled liner, the water supplied to the different sets depending on the engine load. Tests with the system have shown that the optimum liner temperature can be maintained over a very wide load range. The system is considered perfectly feasible but the added complexity has to be carefully weighed against the service advantages.

The fuel valve used on MC engines operates without any external control of its function. The design has worked well for many years but could be challenged by the desire for maintaining an effective performance at very low loads. MAN B&W Diesel has therefore investigated a number of new designs with the basic aim of retaining a simple and reliable fuel valve without external controls. Various solutions have been tested on the 4T50MX engine, among them a design whose opening pressure is controlled by the fuel oil injection pressure level (which is a

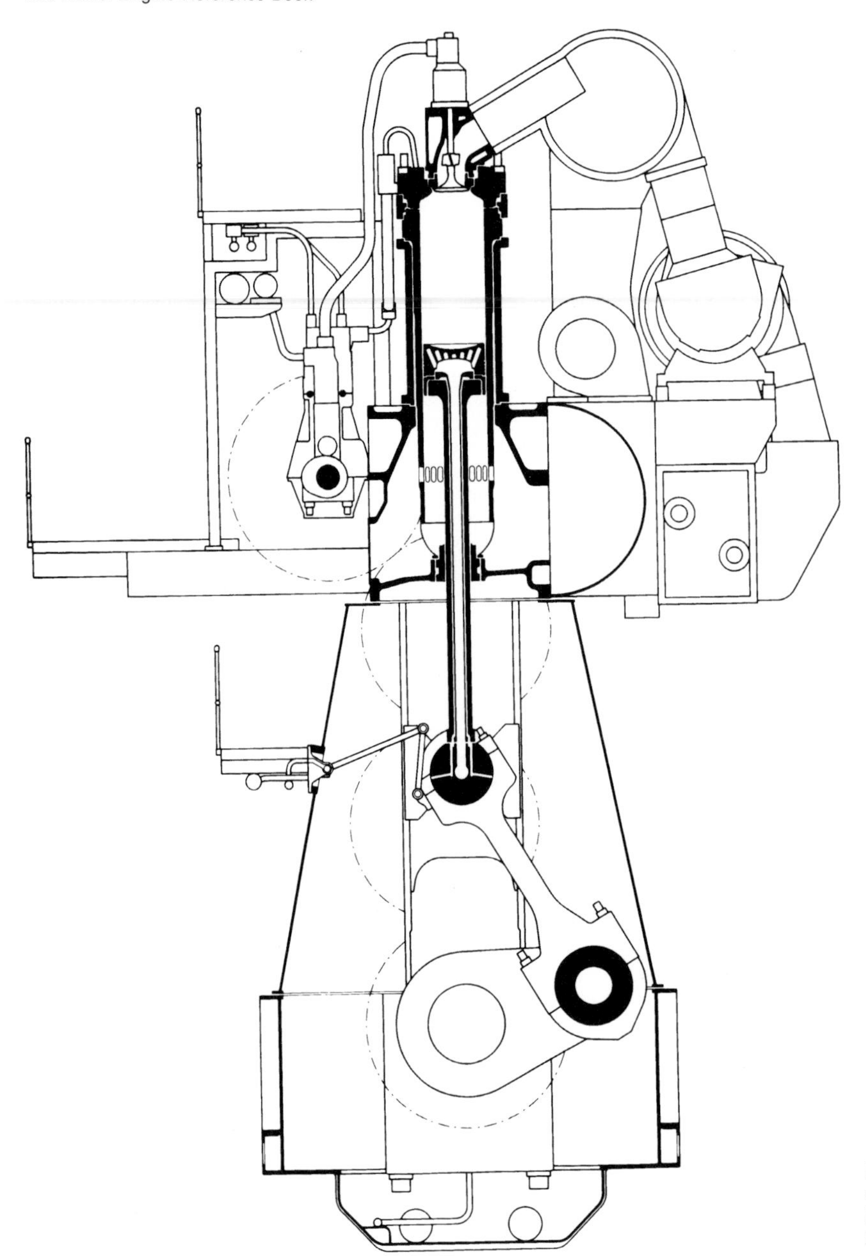

Figure 26.36 New Sulzer Diesel's current Two-stroke Technology Demonstrator research engine is based on an RTA58T model

function of the engine load). At low load the opening pressure is controlled by the spring alone. When the injection pressure increases at higher load, this higher pressure adds to the spring force and the opening pressure increases.

Another example of fuel valve development is aimed at reduced emissions. This type incorporates a conventional conical spindle seat as well as a slide valve inside the fuel nozzle, minimising the sack volume and thus the risk of after-dripping. Significantly lower NOx emissions are reported, as well as reduced smoke and even carbon monoxide, but at the expense of a slightly higher fuel consumption. This type of fuel valve is now included in the options for special low NOx applications of MC engines. The 4T50MX engine was used to test a three-fuel-valve-per-cylinder configuration, the measurements mirroring New Sulzer Diesel's results in yielding reduced temperature levels and a more even temperature distribution than with a two-valve arrangement. The K80MC-C, K90MC/MC-C, S90MC-T and K98MC-C engines are now specified with triple fuel valves to enhance reliability.

26.2.2 Intelligent engines

Both MAN B&W Diesel and New Sulzer Diesel have demonstrated 'camshaftless' operation with their research engines, applying electronically controlled fuel injection and exhaust valve actuation systems. Current R&D will pave the way for a future generation of highly reliable 'intelligent engines': those which monitor their own condition and adjust parameters for optimum performance in all operating regimes, including fuel-optimized and emissions-optimized modes. An 'intelligent engine-management system' will effectively close the feedback loop by built-in expert knowledge. Engine performance data will be

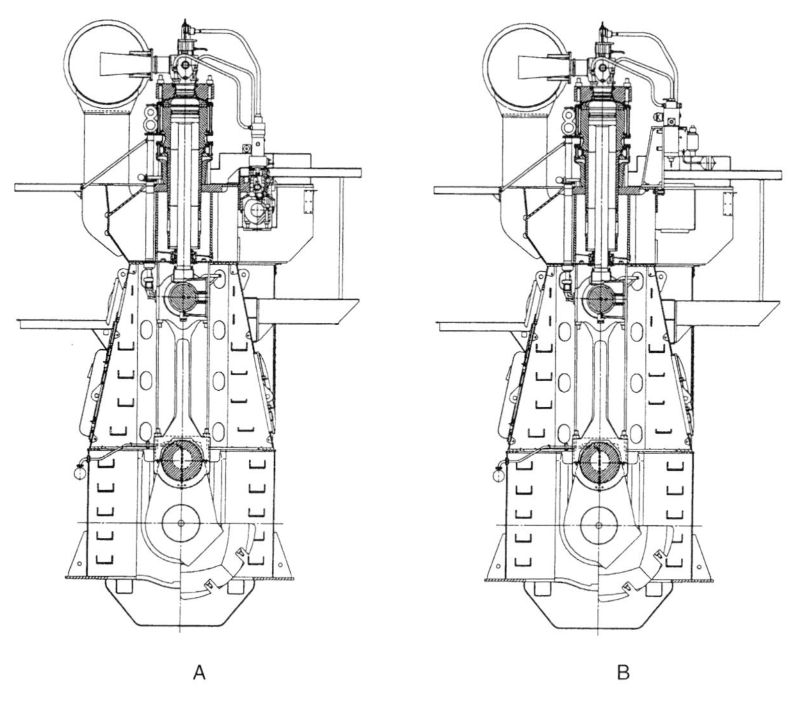

Figure 26.37 MAN B&W Diesel's 4T50 MX low speed research engine arranged with a conventional camshaft (A) and with electronically controlled fuel injection pumps and exhaust valve actuating pumps (B)

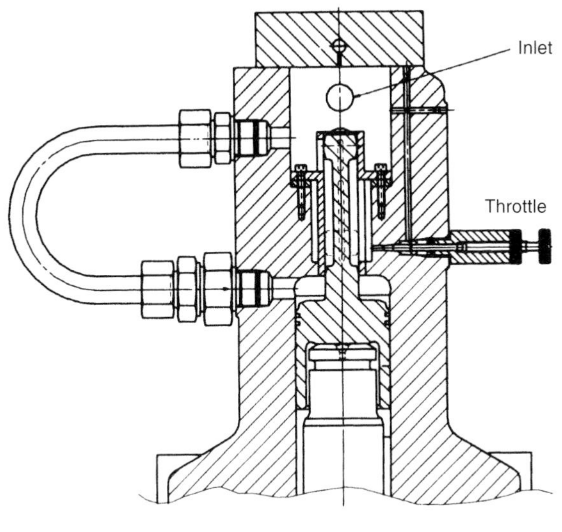

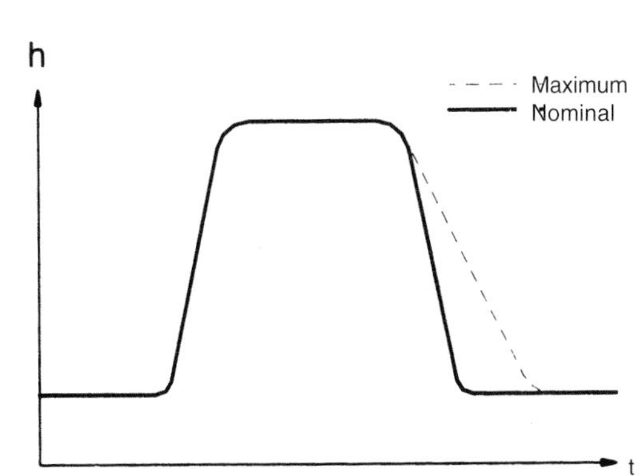

Figure 26.38 Mechanical/hydraulic variable compression ratio (MAN B&W Diesel)

constantly monitored and compared with defined values in the expert system; if deviations are detected corrective action is automatically taken to restore the situation to normal. A further step would incorporate not only engine optimizing functions but management responsibilities, such as maintenance planning and spare parts control.

To meet the operational flexibility target, MAN B&W Diesel explains, it is necessary to be able to change the timing of the fuel injection and exhaust valve systems while the engine is running. To achieve this objective with cam-driven units would involve a substantial mechanical complexity which would undermine engine reliability. An engine without a traditional camshaft is therefore dictated.

The concept is illustrated in *Figure 26.39* whose upper part shows the operating modes which may be selected from the bridge control system or by the intelligent engine's own control system. The centre part shows the brain of the system: the electronic control system which analyses the general engine condition and controls the operation of the engine systems shown in the lower part of the diagram (the fuel injection, exhaust valve, cylinder lube oil and turbocharging systems).

To meet the reliability target it is necessary to have a system which can actively protect the engine from damage due to overload, lack of maintenance and maladjustments. A condition monitoring system must be used to evaluate the general condition of the engine, thus maintaining its performance and keeping its operating parameters within prescribed limits. The condition monitoring and evaluation system is an on-line system with automatic sampling of all 'normal' engine performance data, supplemented by cylinder pressure measurements. The system will report and actively intervene when performance parameters show unsatisfactory deviations. The cylinder pressure data delivered by the measuring system are used for various calculations:

- The mean indicated pressure is determined as a check on cylinder load distribution as well as total engine output.
- The compression pressure is determined as an indicator of excessive leakage caused by, for example, a burnt exhaust valve or collapsed piston rings (the former condition is usually

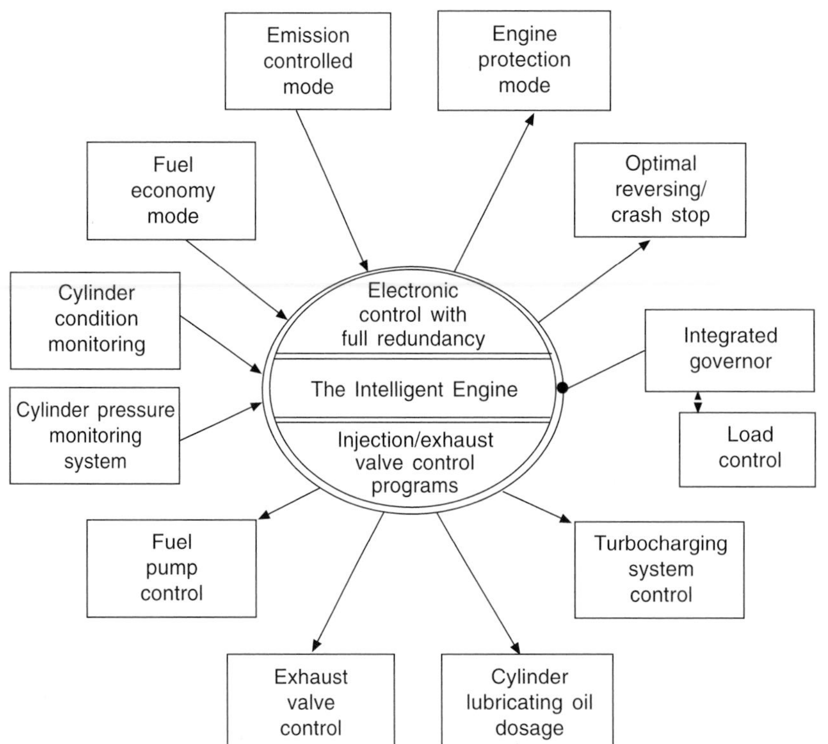

Figure 26.39 Schematic of the Intelligent Engine (MAN B&W Diesel)

accompanied by an increased exhaust gas temperature in the cylinder in question).

- The cylinder wall temperature is monitored as an additional indicator of the piston ring condition.
- The firing pressure is determined for injection timing control and for control of mechanical loads.
- The rate of pressure rise (dp/dt) and rate of heat release are determined for combustion quality evaluation as a warning in the event of 'bad fuels' and to indicate any risk of piston ring problems in the event of high dp/dt values.

The cylinder condition monitoring system is intended to detect faults such as blow-by past the piston rings, cylinder liner scuffing and abnormal combustion. The detection of severe anomalies by the integrated systems triggers a changeover to a special operating mode for the engine, the 'engine protection mode'. The control system will contain data for optimum operation in a number of different modes, such as 'fuel economy mode', 'emission controlled mode', 'reversing/crash stop mode' and various engine protection modes. The load limiter system (load diagram compliance system) aims to prevent any overloading of the engine in conditions such as heavy weather, fouled hull, shallow water, too heavy propeller layout or excessive shaft alternator output. This function will appear as a natural part of future governor specifications.

The fuel injection system is operated without a conventional camshaft, using high pressure hydraulic oil from an engine-driven pump as a power source and an electronically controlled servo system to drive the injection pump plunger. The general concept of the InFI (intelligent fuel injection) system and the InVA (intelligent valve actuation) system for operating the exhaust valves is shown in *Figure 26.40*. Both systems, when operated in the electronic mode, receive the electronic signals to the control units. In the event of failure of the electronic control system the engine is controlled by a mechanical input supplied by a diminutive camshaft giving full redundancy.

Unlike a conventional, cam-driven pump the InFI pump has a variable stroke and will only pressurize the amount of fuel to be injected at the relevant load. In the electronic mode (that is, operating without a camshaft) the system can perform as a single injection system as well as a pre-injection system with a high degree of freedom to modulate the process in terms of injection rate, timing, duration, pressure, single/double injection, cam profile and so on. Several optimized injection patterns can be stored in the computer and chosen by the control system in order to operate the engine with optimum injection characteristics at several loads: from dead slow to overload as well as for starting, astern running and crash stop. Changeover from one to another of the stored injection characteristics is effected from one injection to the next. The system is able to adjust the injection amount and injection timing for each cylinder individually in order to achieve the same load (mean indicated pressure) and the same firing pressure (P_{max}) in all cylinders; or, in protection mode, to reduce the load and P_{max} on a given single cylinder if the need arises.

The exhaust valve system (InVA) is driven on the same principles as the fuel injection system, exploiting the same high pressure hydraulic oil supply and a similar facility for mechanical redundancy. The need for controlling exhaust valve operation is basically limited to timing the opening and closing of the valve. The control system is thus simpler than that for fuel injection.

Cylinder lubrication is controllable from the condition evaluation system so that the lubricating oil amount can be adjusted to match the engine load. Dosage is increased in line with load changes and if the need is indicated by the cylinder condition monitoring system (in the event of liner scuffing and ring blow-by, for example). Such systems are already available for existing engines.

The turbocharging system control will incorporate control of the scavenge air pressure if a turbocharger with variable turbine nozzle geometry is used, and control of bypass valves, turbocompound system valves and turbocharger cut-off valves

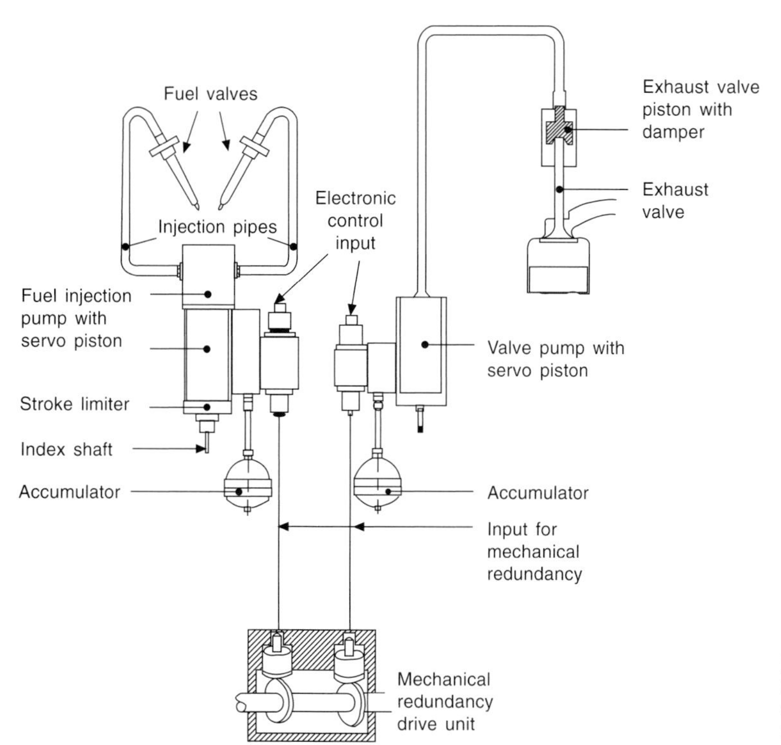

Figure 26.40 Electronically controlled hydraulic systems for fuel injection and exhaust valve operation on MAN B&W Diesel's 4T50MX research engine

if such valves are incorporated in the system. Valves for any selective catalytic reduction (SCR) exhaust gas cleaning system installed will also be controlled.

Operating modes may be selected from the bridge control system or by the system's own control system. The former case applies to the fuel economy modes and the emission-controlled modes (some of which may incorporate the use of an SCR system). The optimum reversing/crash stop modes are selected by the system itself when the bridge control system requests the engine to carry out the corresponding operation. Engine protection mode, in contrast, will be selected by the condition monitoring and evaluation system independently of actual operating modes (when this is not considered to threaten ship safety).

Research and development by Mitsubishi, the third force in low speed engines, has successfully sought weight reduction and enhanced compactness while retaining the performance and reliability demanded by the market. The Japanese designer's current UEC-LS type engines yield a specific power output of around three times that of the original UE series of the mid-1950s. The specific engine weight has been reduced by around 30 per cent over that period and the engine length in relation to power output has been shortened by one-third.

Part 6

Operation

27 Condition monitoring

Contents

27.1 Introduction

Engine health monitoring, more usually described as performance or condition monitoring, has been totally transformed by the development of the microprocessor and associated software programs. Condition monitoring has perhaps always had higher priority on ships at sea than in any other application; for obvious reasons since here an engine failure can (and still sometimes does) result in the loss of the vessel. The engine room log, in which the watchkeeper records loads, speeds, key pressures and temperatures and the results of occasional manual checks on such measurements as maximum cylinder pressures, is a primitive example of condition monitoring.

Such monitoring however has limitations:

- the number of quantities that it is practicable to observe is strictly limited;
- more complex and derived quantities, such as fuel consumption rate, turbocharger performance, exhaust analysis and vibration, are not accessible to such observation;
- remedial action and the issue of warnings, depends on the alertness of the watchkeeper;
- most important: long term trends may go unnoticed.

All these limitations may now be transcended as a result of two main developments: the vast range of robust and reasonably cheap transducers that are now available, and the ready availability of computer systems capable of storing a vast amount of data, with the ability to recognize complex patterns and trends of behaviour that may represent deterioration or actual danger, and with the capability of issuing appropriate warnings.

Nevertheless, when making the change from a manned to an unmanned or partially mannned system it is well to be aware of what may be lost, as well as what it is hoped to gain. In a manned installation, the 'expert system', the engineer, is capable of a range of diagnostic skills and responses that cannot be fully equalled artificially; he will be, for example, sensitive to small changes in engine sound, to small increases in vibration, even to the smell of overheating machinery, signals not yet accessible to a transducer.

It will be evident that the 'intelligence' of an automated or so-called 'expert' system will be only as good as the model embodied in the computer software; there are many possible models just as there are many possible engine installations. An engineer responsible for choosing such a system for a given application should check that it is capable of recognizing all the types of malfunction and deterioration that he can foresee, also that it has sufficient built-in flexibility to allow development as experience with the particular installation is accumulated.

27.2 A typical condition monitoring system

A typical condition monitoring system for a medium or large marine engine will monitor continuously all or most of the following functions:

- fuel injection pressure profile and quantity;
- cylinder pressure profile with calculation of i.m.e.p., rate of heat release and specific consumption;
- turbocharger performance, with measurements of air flow rate, pressures and temperatures and calculation of turbine and compressor efficiencies;
- evaluation of air filter, air cooler and heat exchanger performance by appropriate measurements of pressures, temperatures and flow rates;
- cooling water, primary and secondary flow and temperatures;
- fuel supply and lubricating oil system performance;
- vibration, at certain selected locations;
- temperature at certain selected locations.

In addition the following long-term trends will be monitored:

- lubricant condition, a number of different tests;
- crankcase blow-by.

This flow of information is fed to a central data acquisition system, where it is compared with predefined parameters. If any departure from these acceptable limits is detected its significance is assessed and the result is presented, in one typical modern system[1], on a series of displays. The 'Summary' screen indicates concisely the nature and severity of the malfunction, with approved actions required to contain or correct it. A second screen entitled 'Supporting Evidence' lists the operating data that gave rise to the signalling of a malfunction, thus giving the engineer in charge the opportunity to judge for himself the seriousness of the problem. A third screen, 'Details', gives background information and may be used for training purposes. A final screen, entitled 'Notes', is for the use of the engineer and may contain the kind of information typically recorded in the traditional log.

The logic of this system, with its encouragement of an effective dialogue between engineer and computer, will be apparent. However, the complexity of the diagnostic process must not be underestimated. Reference 2 describes the evolution of such a system for a medium-speed four-stroke turbocharged marine diesel engine, a type of engine which, because of its relatively high speed, is very sensitive to adjustments and the condition of sub-systems, such as the fuel injection system and turbocharger. The engine in this case was an 18-cylinder four-stroke, bore 400 mm, stroke 460 mm, running at 500 rev/min. *Figure. 27.1* shows indicator diagrams, averaged over 20 cycles, *Figure 27.2* shows corresponding maximum pressures, and *Figure 27.3* corresponding cylinder powers. It will be apparent that there was a wide variation in performance.

The diagnosis of the reasons for this variation, and also for the failure of the engine to deliver rated power, was by no means simple. The main reasons were:

- variation in compression ratio due to variations in (individual) cylinder head gasket thickness;
- variation in fuelling rate between cylinders due to variable pump rack settings;
- variations in injection timing (this has a major effect on peak pressure, exhaust gas temperature and power output).

It was clear that a simple measurement of exhaust gas temperature, *Figure 27.4*, the easiest observation to make, would have been of no help in diagnosing the problem, since it is affected by all the other parameters. This comparatively simple example serves to illustrate some of the pitfalls in designing a condition or performance monitoring system.

27.3 Instrumentation for condition monitoring

The measurements to be made may be broadly classified as either direct, concerned with the actual performance of the engine, or indirect, concerned with the state of the lubricant and cooling water. Fuel properties may be regarded as coming into the latter category:

- direct: pressures, temperatures, torques, speeds, flows, vibrations, diamensional changes, exhaust condition;
- indirect: a wide range of observations on the condition of the lubricating oil (most important), fuel (vital when, for example, taking aboard fuel from an unfamiliar source) and cooling water.

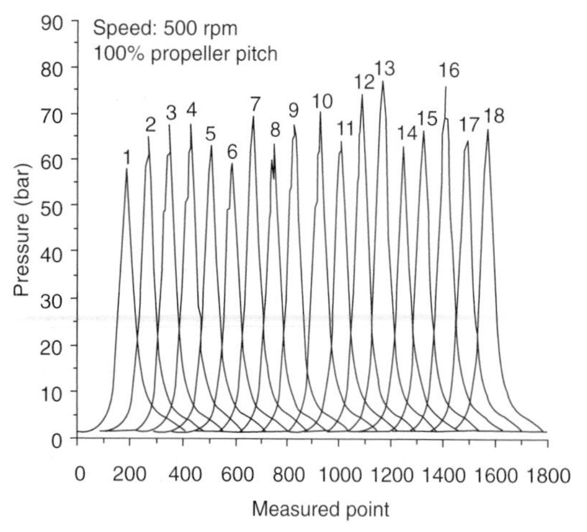

Figure 27.1 Pressure diagrams for all cylinders at 500 rpm. Reproduced from TriboTest journal by kind permission of Leaf Coppin Publishing Ltd

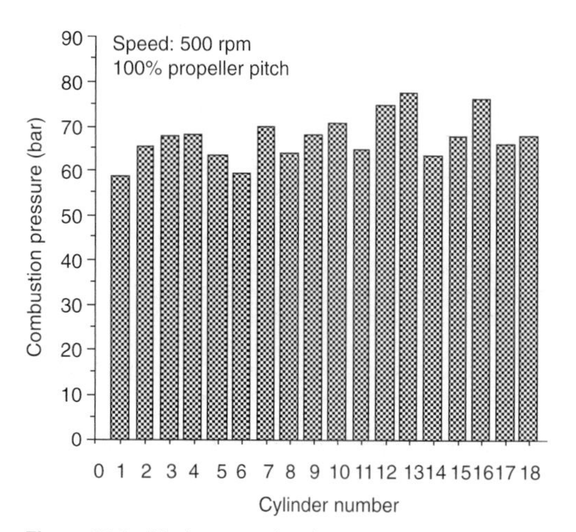

Figure 27.2 Maximum combustion pressures for all cylinders at 500 rpm. Reproduced from TriboTest journal by kind permission of Leaf Coppin Publishing Ltd

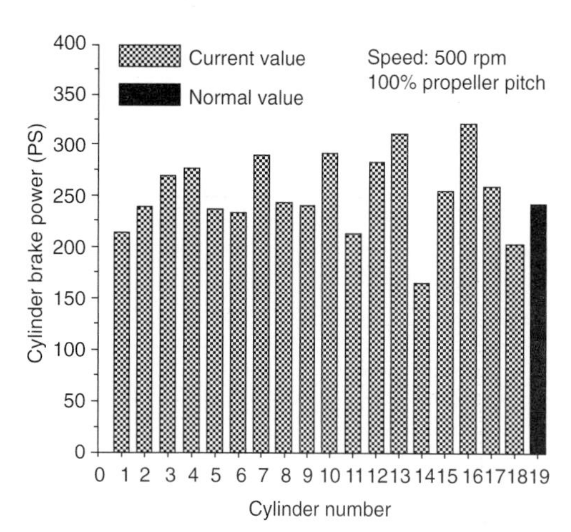

Figure 27.3 Brake power output for all engine cylinders at 500 rpm. Reproduced from TriboTest journal by kind permission of Leaf Coppin Publishing Ltd

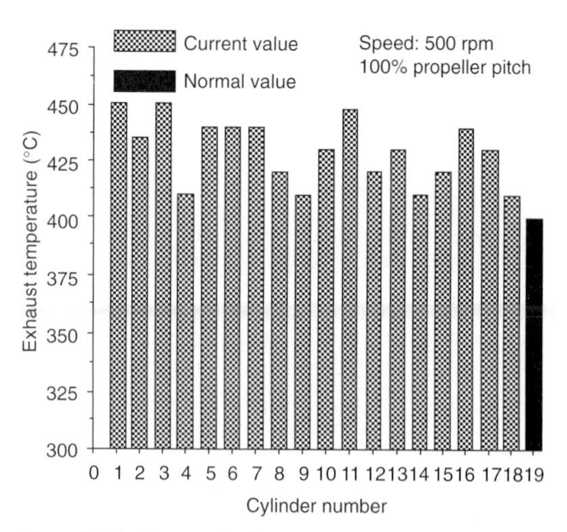

Figure 27.4 Measured exhaust gas temperatures for all cylinders at 500 rpm. Reproduced from TriboTest journal by kind permission of Leaf Coppin Publishing Ltd

While most of the instrumentation and transducers employed to capture direct monitoring data are conventional in form, the processing of vibration measurements requires more specialized devices, while certain temperature measurements call for comment.

27.3.1 Vibration monitoring[3,4,5]

An increase in the level of vibration normally associated with any particular engine and its associated driven machinery may arise from a number of different causes:

- abnormal combustion conditions, cylinder wear;
- wear or partial seizure of plain bearings;
- wear and pitting of rolling bearings;
- damage to gear teeth;
- damage or fouling of turbocharger elements;
- defects in engine mountings or drive line.

Essentially, vibration pick-ups (accelerometers) may be located at strategic points on the engine structure. One of the most informative is at about mid-cylinder height, on the thrust side: this will sense abnormal combustion conditions and piston wear. However, the interpretation of accelerometer signals is by no means a simple matter, particularly with machines such as diesel engines having a high level of background vibration. In some cases increasing wear is indicated by a general increase in vibration signal level, while in others defects give rise to a new, commonly spiky, pattern of wear, which may be overlooked in the absence of specialized analysis.

Rolling bearings present a particular problem as the presence of pitting and wear gives rise to complex patterns of vibration[3], which call for particular methods of analysis; *Figure 27.5* (from Reference 3) shows an example of the response to an artificial pit in a deep-groove ball race of 50 mm bore. The processing of signals from a damaged bearing, by a process known as 'envelope analysis' is a somewhat complex matter, and is probably only justified where exceptionally critical bearings are concerned.

27.3.2 Temperature measurements

The accurate measurement of exhaust gas temperature is by no means a simple matter: there are fundamental difficulties in even the definition of the temperature of a gas stream in which reactions may be continuing. See Reference 10 for a discussion

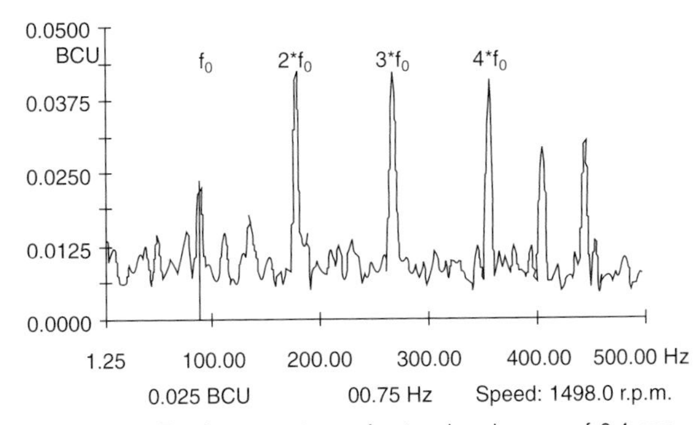

Figure 27.5 Envelope spectrum of outer ring damage of 0.4 mm diameter at a load of 0.5 kN and a speed of 1500 rpm. Reproduced from TriboTest journal by kind permission of Leaf Coppin Publishing Ltd

of this problem. The location and shielding of the sensor need careful consideration.

Thermocouples mounted flush with the cylinder wall and near the top of ring travel will give a good indication (by sensing increased temperature) of increased blow-by and deterioration of cylinder lubrication. Any sign of distress in critical plain bearings may be picked up very rapidly by a thermocouple installed in contact with the outside of the bearing shell. Similarly, where critical rolling bearings are concerned, a thermocouple in contact with the outer race gives an indication of any deterioration.

27.4 Instrumentation for condition monitoring: indirect methods

Indirect methods of monitoring engine condition, and in particular the monitoring of the condition of the lubricant, are the most widely used, in fact in many applications involving diesel engines, for example in rail traction, they are employed almost exclusively.

An engine lubricant in service is subject to the following adverse influences:

- dilution by fuel, coolant, sea water;
- changes in viscosity;
- oxidation, depletion of anti-oxidants and other additives;
- contamination by suspended particles: combustion products and iron and other particles resulting from wear.

Periodical sampling with examination of as many as possible of these factors can provide many indications and warnings of deteriorating engine condition, while of course also showing the state of the lubricant and indicating when oil change is necessary. Some of the test methods are complex and only suitable for laboratory use, but valuable information can be obtained from comparatively simple tests. For example Reference 4 describes test equipment devised for use at sea by the U S Navy and which makes the measurements defined in *Table 27.1*. These tests monitor the principal significant changes in the chemistry of the lubricant. Viscosity increases with ageing and oxidation of the oil and with the presence of contaminants, notably of carbon particles arising during combustion. Contamination by fuel can seriously affect the load-carrying capability of the lubricant. A dilution of 4% by volume of fuel is usually regarded as the acceptable maximum.

The Total Base Number (TBN) falls as the anti-oxidant additives, essential in engine lubrication, are depleted. The alkalinity of the oil is measured by titration through an acid and expressed in mg KoH/gm. Comparison with the TBN of fresh oil determines whether the oil is still capable of neutralizing the acid residues produced by combustion (particularly of sulphur) and oxidation.

A test for the water content of the oil, usually performed with a Karl Fisher apparatus, is also desirable. Besides leakage from the cylinder head gasket, causes include running at too low a temperature, defective crankcase ventilation and obstruction of the exhaust system.

These tests, while giving a good indication of the state of the lubricant, do not deal with the state of wear in the engine. Valuable indications in this area may be given by study of the suspended materials in the lubricant and a wide range of techniques is available:

- Magnetic plugs. Fitted universally in the lubricant systems of large engines, periodical examination of the amount of ferrous material captured gives a useful indication of the state of wear, particularly of such components as cylinder liners.
- Spectrometric analysis. Essentially a laboratory procedure. Various techniques are available and involve burning of an oil sample and subsequent observation of the emission spectrum. Most common engineering metals are detected but particles of diameter greater than about 10 micron are not recorded.
- Ferrography. A comparatively simple and very valuable technique. A diluted sample of the oil flows across an inclined glass plate above a powerful magnet; ferrous particles are deposited with a distribution related to their size.
- Visual examination of wear particles under the microscope. This calls for experience. Useful guidance is given in Reference 7, with illustrations of many types of wear debris.

Table 27.1 Lubricant quality test equipment

Measurement	*Sensor*	*Range*	*Alarms*
Kinematic viscosity	Magnetic cylinder	100-225 cSt at 104°F (40°C)	Less than 100, more than 225 cSt
Fuel dilution	Scanning Acoustic Wave device (SAW detector)	0.0–5.0% fuel JP-5 OR F-76 OR	0.0–2.4% Accept oil 2.5–4.9% Marginally accept oil 5.0% + Reject oil
Total base number	Pressure transducer	0.0–14.0 TBN	14.0-8.0 TBN Accept oil 7.9–2.1 TBN Marginally accept oil 2.0 or lower TBN Reject oil

Reproduced from TriboTest journal by kind permission of Leaf Coppin Publishing Ltd.

27.5 Fuel monitoring[8]

This is a subject of particular importance in the marine world, where ships may be required to take aboard fuel from questionable sources. This rapid analytical service provided by Lloyd's Register is widely used. The properties of particular significance where engine performance and reliability is concerned are:

- cetane number
- sulphur content
- viscosity and pour point
- water and sediment content.

It is clearly essential, for the 'health' of a marine engine, both that the fuel it is required to burn does not lie outside the limits specified as acceptable by the engine builder, also that the operator should be in a position to optimize such parameters as fuel temperature, injection timing and pressure, to achieve optimum results with the particular fuel that is in use.

27.6 Exhaust emissions[9]

The specialized aspect of performance monitoring, principally with reference to road vehicles, has become of very great importance in recent years, largely driven by an obsessive public concern with the possible effects of emissions, particularly from diesel engines, on human health. It is not possible, within the limits of this book, to give more than a very brief summary of the current situation. A more comprehensive account is given in Reference 9, which summarizes International Standards dealing with the subject originating from the USA, the UN Economic Commission for Europe (ECE) and the European Union (EU).

The majority of this legislation is concerned with specifying limits for the main pollutants: in the case of the diesel engine oxides of nitrogen (NOx) and particulates. The latter are very sensitive to the level of maintenance of the engine, particularly of the fuel injection system, and in the USA annual tests are becoming mandatory. The techniques involved are complex (see Reference 10 for a summary) and call for a degree of skill in their operation. The vehicle is mounted on a rolling road or chassis dynamometer and is put through a standard driving cycle, lasting typically some four minutes, in the course of which the exhaust gas is subjected to various tests. There are essentially three methods in use for measuring particulate emissions, and these methods cannot readily be related one to the other:

- the 'traditional' method by use of a smoke-meter which measures the opacity of the undiluted exhaust by the degree of obscuration of a light beam;
- measurement of the particulate content of an undiluted sample of exhaust gas by drawing it through a filter paper of specified properties and estimating the consequent blackening of the paper against an agreed scale;
- measurement of the actual mass of particulates trapped by a filter paper during the passage of a specified volume of diluted exhaust gas.

It seems inevitable that the incidence of health monitoring of this kind, having legislative backing, is bound to increase.

27.7 Conclusion

The advantages of performance or health monitoring when applied to prime movers such as the diesel engine have long been appreciated: reduced unscheduled down-time and increased availability[7]; reduced maintenance costs and spares consumption and, in critical cases, avoidance of serious danger. It is, however, only with the arrival of low-cost computer systems and transducers that its use has become widespread, to such an extent that all road vehicles now incorporate some aspects of health monitoring into the engine management system.

References

1 Advances in machinery management', *Diesel and Gas Turbine Worldwide*, May (1998)
2 KOUREMENOS, D. A. and HOUNTALAS, D. T., 'Diagnosis and condition monitoring of medium-speed marine diesel engines', *TriboTest*', **4**, No. 1, September (1997)
3 TOERSEN, H., 'Application of an envelope technique in the detection of ball bearing defects in a laboratory experiment', *TriboTest*, **4**, No. 3, March (1998)
4 McFADDEN, P. D., 'Condition monitoring of rolling element bearings by vibration analysis', *Machine Condition Monitoring*, January, I. Mech. E. London (1990)
5 RATCLIFFE, G. A., 'Condition monitoring of rolling element bearings using the enveloping technique', *Machine Condition Monitoring*, January, I. Mech. E. London (1990).
6 GORIN, N. and SHAY, G., 'Diesel lubricant monitoring with new-concept shipboard test equipment', *TriboTest*, **3**, No. 4, June (1997)
7 NEALE, M. J., *The Tribology Handbook*, Butterworth-Heinemann, Oxford
8 VERLINDEN, A., 'Laboratories at work: used oil analysis at Wear Check, Belgium', *TriboTest*, **1**, No. 4, June (1995).
9 *SAE Surface Vehicle Emissions Standards Manual*, Society of Automotive Engineers Inc.
10 PLINT, M.A. and MARTYR, A. J. *Engine Testing: Theory and Practice, 2nd edn* Butterworth-Heinemann (1999).

Units and conversion tables

A compilation of assorted engineering units and conversions

Convert (Unit)	*To*	*Multiply by*
atmosphere	inches water (at 0°C)	29.92
atmosphere	kg/m^2	10 330
atmosphere	N/m^2	101 320
atmosphere	pounds/sq. in.	14.7
BTU	joules	1054.8
BTU	kcal	0.252
centigrade	Fahrenheit	(C × 1.8) + 32°
centimetres	feet	0.03281
centimetres	inches	0.3937
cubic cm	cubic inches	0.06102
cubic feet	cubic cm	28 320
cubic feet	litres	28.32
cubic feet/min	litres/s	0.4717
cubic inches	cubic cm	16.387
cubic metres	cubic feet	35.31
cubic yards	cubic metres	0.7646
degrees	radians	0.01745
Fahrenheit	centigrade	(F – 32) × 0.555
fathoms	feet	6
feet	cm	30.48
feet of water (at 4°C)	inches of mercury (0°C)	0.8826
feet of water (at 4°C)	kg/m^2	304.8
feet of water (at 4°C)	lbs/in^2	62.43
feet/min	m/s	0.00508
foot-pounds	kilogram-metres	0.1383
g/kW-hr	lb/HP-hr	0.00164
gallons (Imperial)	cubic metres	0.003785
gallons (Imperial)	gallons (USA)	1.201
gallons (Imperial)	litres	4.545
gauss	webers/sq.m.	0.0001
grams of air	litres (NTP)	0.7734
horsepower	foot-pounds/min	33 000

Convert (Unit)	*To*	*Multiply by*
horsepower	kilowatts	0.7457
inches	centimetre	2.54
inches of water (at 4°C)	pounds/square foot	5.202
J/g	BTU/lb	0.4299
joules	foot-pounds	0.7376
kcal	BTU	3.9683
kg/cubic metre	pounds/cubic foot	0.06243
kg/m^2	lbs/in^2	0.001422
kg-calories	kilogram-metres	426.9
kilograms	ponds	2.2046
kilograms	pounds	2.2046
kilometres	miles	0.6214
kJ	BTU	0.9478
knots	kph	1.8532
kW	HP	1.341
kWhr	MJ	3.6
kW-hrs	joules	3.6 + 1000
kW-hrs	kg-metres	367 100
lbf	N	4.4482
litres	cubic feet	0.03532
litres	cubic inches	61.03
litres	gallons (Imperial)	0.2199
litres	pints	1.759
m/s	fpm	196.85
metre/s	feet/min	196.65
metres	inches	39.37
metres	feet	3.281
metres	yards	1.0936
miles	kilometres	1.6093
miles (nautical)	feet	6080
miles/hour	feet/min	88
miles/hour	m/s	0.44704
millimetres	inches	0.03937
mm of water gauge (at 4°C)	N/m^2 (Pascals)	9.807

Convert (Unit)	*To*	*Multiply by*
N	lbf	0.2248
newtons/m^2	pounds/in^2 (psi)	0.000145
pounds	kilograms	0.4536
pounds (weight)	grams	453.6
pounds of air	cubic feet (0°C, 1 atm.)	12.4
pounds of water	cubic feet	0.01602
pounds/in^2 (psi)	newtons/m^2 (Pascals)	6894
radians	degrees	57.296
square miles	acres	640
temperature Centigrade	kelvin	K = °C + 273.15
tons (2240 lbs)	tonnes (1000 kg)	1.016
W/m^2	BTU/ft^2 hr	0.3170
water gauge (inches)	N/m^2	249
watts	ft-lbs/min	44.26
yards	metres	0.9144

Abbreviations

C = Celsius	kg = kilogram
cP = centipoise	K = kelvin
cSt = centistokes	m = metre
d = day	ml = millilitre
g = gram	min = minute
ha = hectare	N = newton
h = hour	rev = revolution
Hz = hertz	s = second
J = joule	

SI factors

T	=	tera	=	10^{12}	c	=	centi	=	10^{-2}
G	=	giga	=	10^{9}	m	=	milli	=	10^{-3}
M	=	mega	=	10^{6}	μ	=	micro	=	10^{-6}
k	=	kilo	=	10^{3}	n	=	nano	=	10^{-9}
h	=	hecto	=	10^{2}	p	=	pico	=	10^{-12}
da	=	deca	=	10	f	=	femto	=	10^{-15}
d	=	deci	=	10^{-1}	a	=	atto	=	10^{-18}

Index